and its enforcement aborted. (*Suara Pembaharuan*: PTUN instru[...] restored, 3 May 1995).

On Tuesday, the 21st November 1995, the State Administrative High Court (PTTUN) Jakarta upheld the verdict in the case of revocation of the license of Tempo. In its decision, PTTUN Jakarta declared that the decree of the Minister of Information on the revocation of the license was void and ordered the defendant (the Minister of Information) to revoke the decree. A press observer, Ashadi Siregar, said whatever court decision and whatever the attitude of the Ministry of Information in response with the decision of PTTUN, Jakarta should reinforce the importance of the courts in cases related to the press. Furthermore, he said that the journalist and the chief editor of a publication as a citizen, of course, could be guilty. However, it is important to remember that, as the law states, every decision can only be made after a court process. (Suara Pembaharuan: PPTUN reinforces administrative court ruling regarding the license of "Tempo", 22 November 1995).

After two consecutive wins of his case in court, on Thursday, 13 June 1996, the Panel of Judges of the Supreme Court (MA) in its verdict on cassation rejected the lawsuit against the revocation of the permit of Goenawan's *Tempo*. Instead, the justices granted the appeal filed by the Indonesian Information Minister on this permit revocation. According to the Supreme Court, Goenawan had no authority and credibility to sue the Minister of Information because the constitution of PT Grafiti Pers, the publisher of *Tempo*, stated that Goenawan only served as the Editor-in-Chief, while the President Director of PT Grafiti Pers was Eric Samola.

Goenawan's persistence against the abuse of power by the regime did not stop. He took other ways to uphold freedom of information.

Stepping outside the courtroom, Goenawan Mohamad was immediately thronged by journalists. Wearing a white shirt, black armband, and *peci* (a fez-like hat), Goenawan Mohamad raised his arms and declared, "For me, the struggle for freedom of the press by legal means ends here. Now the struggle must take another form." (Janet Steele. *Wars Within: The Story of Tempo, an Independent Magazine in Soeharto's Indonesia*, 2005, p. 258).

In December 1994, Goenawan called a meeting at Koi restaurant in the Blok M area of South Jakarta. He invited journalist and intellectuals, such as Fikri Jufri, Aristides Katoppo, Mochtar Pabotinggi, M. Dahana, and Andreas Harsono, and explained that he wanted to set up a foundation. The legal document creating ISAI was signed in January 1995. The biggest funder of ISAI was the U.S. Agency for International Development (USAID). (*The Story of Tempo, an Independent Magazine in Soeharto's Indonesia*, 2005, p. 260).

1.4 *Resistance through the internet*

Since 1995, the Internet has been playing an important role in the dissemination of information among activists and Internet users in Indonesia. Internet fever in Indonesia was sparked by the presence of *Apa Kabar*, a mailing list managed by John McDougall of America. *Apa Kabar* disseminated a variety of opinions, from the most radical to the purists and from pro-democracy activists to the military-intelligence apparatus. Besides various polemical opinions and views, *Apa Kabar* disseminated information from the mass media, at home and abroad, with regard to the latest situation in Indonesia.

Apa Kabar's success was then followed by the emergence of a variety of Internet sites and mailing lists that were managed by activists in Indonesia. Ex-*Tempo* journalists managed *Tempo Interactive*, followed by a number of mailing lists, such as *SiaR*, *KDPnet*, *AJInews*, *X-post*, *Demidemokrasi*, and *Indo-News.com*. The information delivered via the Internet could satisfy the people's appetite for information. The material content from the Internet was often downloaded and photocopied so that they could be read by those who had no access to the Internet. In addition, the regime's censoring system that was operated by blocking out the pages of foreign newspapers or magazines reporting the latest situation in Indonesia cannot be applied to the Internet.

The New Order government was just starting to focus the development of the Internet in the national agenda in 1996 through *N21 Vision* or *Nusantara Vision* program pioneered by Jonathan Parapak, the director of the state enterprise Indosat. The program was expected to coordinate the activities of various actors of the development of the Internet network and to obtain a loan from the World Bank. However, the Internet-based network development agenda of N21Vision faltered, and the loan from the World Bank could not be obtained because of the economic crisis in 1997.

In the midst of the government's failure to provide Internet services due to the lack of facilities and availability of personal computers, the emergence of *Warnet* (Internet Café) in 1996 was able to offer Internet services at a relatively low cost to the middle and lower segments of the society. It had emerged into a democratic public sphere for the majority of people where they could access information without any restriction from the government. The *Warnet* had become a medium for freedom of expression and opinion spread across the world. Meanwhile, various forms of information and messages whose content discredited the regime could easily be accessed.

In January 1998, Soeharto accused press reports of triggering public panic. As a result, a large number of people stormed stores and supermarkets to buy essential supplies. In February, after signing a memorandum of understanding with the International Monetary Fund (IMF), Soeharto accused the Indonesian press of having made the situation in Indonesia worse with regard to the financial crisis that was going on and caused rupiah to decline sharply. The pressure on the press did not lead to any banning because Soeharto's regime started to get serious and massive resistance and began to undermine its legitimacy.

2 CONCLUSION

The New Order regime exercised its strategic values through tight control of information media. They achieved this goal by making uncompromising regulations to restrict information, mainly concerning national security. They were also manipulative in their selection of information to be published and made available to the public and the press. The regime exercised freedom of expression "irresponsibly", which was considered as destabilizing the national development or, worse still, classified as a subversive action. There were links between the press control policy and the ability of the regime to sustain power for more than 32 years.

For 32 years, the Indonesian press had been used by the regime to cover up scandals and chaos in the state administration, to conceal true facts and to ignore the harsh reality or concerns. News coverage was based solely on what the government desired to convey. This led to a condition where information media contained news and propaganda from the standpoint of the New Order regime. Consequently, people could not obtain the true picture of what was actually going on around them. They could not distinguish facts from fiction, truth from lies, and national interests from the interests of a few elites.

With the emergence of the Internet, opposition groups could successfully eliminate the New Order regime's tight control of information media. Information about the dark sides of the regime was revealed, resulting in a widespread fight against it throughout the country and the world. The Internet boom in Indonesia during 1995–1998 significantly contributed to the fall of Soeharto's New Order.

REFERENCES

Abdullah, T. (2009) *Indonesia towards Democracy*. Singapore, ISEAS.
Bujono, B. (1995) *Buku Putih Tempo, Pembredelan ITU* (Tempo White Book, ITU Banning). Jakarta, Yayasan Alumni Tempo.

Bujono, B. (1995) *Mengapa Kami Menggugat* (Why We Pressed Charges). Jakarta, Yayasan Alumni Tempo.

Casci, R.O. (2006) *Organising and Sustaining Hegemony: A Gramscian perspective on Suharto's New Order Indonesia*. Research Commons at the University of Waikato.

Hill, D.T. (2005) *The internet in Indonesia's new democracy*. New York, Routledge.

Hill, D.T. (2011) *Pers di Masa Orde Baru (*The Press in New Order in Indonesia). David T. Hill; translator, Gita Widya Laksmini Soerjoatmodjol. Jakarta, Yayasan Obor Indonesia.

Hisyam, M. (2003) *Krisis Masa Kini dan Orde Baru* (The Crisis Today and the New Order). Jakarta, Yayasan Obor.

Kasenda, P. (2013) *Soeharto: Bagaimana Ia Bisa Melangengkan Kekuasaan Selama 32 Tahun*? (Soeharto: How Could He Maintain Power for 32 Years?) Jakarta, Penerbit Buku Kompas.

Klin, A. & Volkmar, F.R. (2000) Treatment and Intervention Guidelines for Individuals with Asperger Syndrome. In: Klin, A., & Volkmar, F.R., Sparrow, S.S., eds. *Asperger Syndrome*. New York, Guilford Press, pp. 340–66.

Lehiste, I. (1970) *Suprasegmentals*. Oxford, England, MIT Press.

Lim, M. (2006) Lost in Transition? The Internet and Reformasi in Indonesia. In: Dean, J. & Anderson, J.W. (eds.). *Reformatting Politics: Networked Communications and Global Civil Society*, 85–106. London, UK, Routledge.

Mallarangeng, R. (2010) *Pers Orde Baru* (New Order Press). Jakarta, Gramedia.

Robinson, R. & Hadiz, V.R. (2004) *Reorganising Power in Indonesia: The Politics of Oligarchy in an Age of Markets*. London, Routledge Curzon.

Schwarz, A. (2000) *A Nation in Waiting: Indonesia's Search for Stability*. USA, Westview Press.

Sen, K. & Hill, D.T. (2007) *Media, Culture and Politics.* Jakarta, Equinox Publishing Indonesia.

Soeharto. (1989) *Soeharto: Pikiran, Ucapan dan Tindakan Saya/Soeharto* (Soeharto: My Thought, Speech and Action/Soeharto), as presented to G. Dwipayana dan Ramadhan K.H. Jakarta, Citra Lamtoro Gung Persada.

Steele, J. (2005) *Wars Within: The Story of TEMPO, an Independent Magazine in Soeharto's Indonesia.* Jakarta, Equinox Publishing Indonesia.

Cultural Dynamics in a Globalized World – Budianta et al. (Eds)
© 2018 Taylor & Francis Group, London, ISBN 978-1-138-62664-5

Public reason and the rise of populism in post-reformation Indonesia

G.C. Acikgenc & H.S. Pratama
Department of Philosophy, Faculty of Humanities, Universitas Indonesia, Depok, Indonesia

ABSTRACT: One of the significant characteristics of the post-reformation era in Indonesia is the religion-based social morality that has been a dominant discourse in the public sphere. The meaning of being moral is often perceived as an association between morality and religiosity. A moral citizen is one who practices his/her religious teachings in their private space and in the public domain. Using the conceptual framework of Nadia Urbinati on populism, this paper aims at analyzing how the current discourse of social morality shapes the public sphere in the post-reformation era in Indonesia. This paper will contribute to the contemporary theoretical discussion about the public reason and populism in general.

1 INTRODUCTION

The discussion about morality issues mainly focuses on the nature and the source of individual moral beliefs. One of the main debates is an assumption that there is an individual autonomy in making a moral verdict that is confronted by communitarian critique arguing that a morality source clings to the socials. "The socials" locus triggers arguments on how every free individual with his/her own definitions of righteousness and kindness can live side-by-side with other individuals with different points of view. The concept of social morality exists to respond to the deliberation by providing a foundation of life in unison (Gaus, 2011, p. 56). In its development, this social morality concept will always form the basis for contestation of ideas about kindness and righteousness in the public sphere.

In the context of post-reformation Indonesia, this contestation is currently won by a point of view according to which the religious teachings espoused by the majority are a valid basis for determining normative criteria of what is moral and what is not. This view is reflected in public policies at the regional level, verdicts of judicial reviews in the Constitutional Court, and can also be found in the amendments of the national education curriculum design. This victory comes from civil liberties gained after the post-reformation era in 1998. In the course of 32 years under the authoritarian New Order government, morality ideas leaned on the more nationalist principle of Pancasila. The individuals' and groups' space to express opinions that were related closely with a particular religious identity and symbols tend to be limited.

After Suharto was ousted, the organization of the government was no longer centralistic. Constitution regarding regional autonomy gave the head of city or district-level executive agencies a more flexible authority to issue regulations according to the society's needs and aspirations in each district. The values that were carried by regional politicians during the regional head elections also accentuate the religious side for the sake of gaining the majority vote (Aspinall, 2011). The recent dominance of religion consequently blurs the barrier between religion (private domain) and country (public domain) in the society's socio-political life.

Meanwhile, a discourse on non-religious morality foundation has not been operating optimally when compared with the discourse of religion-based public's morality views. The failure in conducting a rational open debate on this matter is inseparable from the repression of the New Order government that limited the civilians' freedom of speech. This repression shut down the principle of the importance of the rationality test in every public policy that controls the life of the civilians. Thus, what happened in the post-reformation era in Indonesia was a democratization of the new public policy maker election instrument reaching the institutional level. Culturally, Indonesia still holds on to the social life from the New Order era, which is a principle that is not democratic; this tends to be close-minded on differences in perspectives for the sake of maintaining "harmony" in their socio-political life. The transition that has not occurred during the post-reformation era in Indonesia is the application of principles of the community that emphasizes the importance of commitment toward the public use of reason/public reason in order to protect plurality values and facts in the process of democracy.

In this study, we will discuss theoretical implications regarding the lack of commitment toward a public reason that indirectly affects the continuity of populism practice in the post-reformation era in Indonesia. First, technical terms such as populism and public reason that form the theoretical framework of this study will be explained. Second, the importance of a strict segregation between religion (private domain) and state (public domain) will be elaborated and the notion that the morality definition in the public sphere is a theme of conversation among citizens that needs to be sustained will be discussed. Then, the relationship between religious dominance in the basic discourse of morality ideas in public policy-making and the potential of majoritarianism that might be caused by the practice of populism will be analyzed using the conceptual framework of populism.

1.1 Populism: Polarization of public opinion

The term populism used in this article does not refer to the grassroots movement that protests the elitist political practice in running a democratic state. Based on an illustration from Nadia Urbinati in her book *Democracy Disfigured: Opinion, Truth, and the People* (2014), there are two distinctive differences between the grassroots movement and populism practice that can be explained through political phenomena in the United States, namely the Occupy Wall Street movement and the Tea Party movement. First, the Occupy Wall Street movement that mobilized the mass was not organized centrally to conquer political powers at the government level, whereas the Tea Party movement was organized to look for leadership that could win the majority and occupy a strategic position at institutional level in order to transform the United States, making it consentaneous with the Republican Party's ideology. Second, the popular grassroots movement that is similar to the Occupy Wall Street movement is still consistent with a representative democracy. This movement reflects antagonistic forms in a democratic society as a critique to supervise the accountability of the chosen state administrators and protect constituent independence from elitist political interests. On the contrary, the practice of populism considers representative democracy as an obstacle. Thus, in order to win the majority support, populism practice polarizes public opinions and claims that it represents a majority opinion as a materialization of the representation of people's will.

Deviating from the moderation and compromise process in the representative democracy system that can accommodate the minority's opinion, Urbinati (2014) explained that populism practice runs the formation of consensus politics vertically through charismatic (Caesarist) leaders using a political rhetoric strategy that operates through the discourse of "we/them." Polarization becomes a means of formatting identity politics by creating two sides that seem to contradict each other by raising an issue that will gain the majority's sympathy and support. This polarization reduces the role of the opposition party—that is an important pillar in the democratic state system—to a meaningless entity that loses its role in political dynamics in governing a democratic state. Polarization excludes opinions coming from outside the majority by positioning it as partisan-enemies. Consequently, symbolical erosion happens to the institution of the representative democracy system that necessarily works as a liaison medium as well as an arbiter between social needs (constituent) and country's interests due to populism that actually makes the power of the country an extension of the major constituent interests. This has an impact on the civilians' political freedom as a result of the limitation of communication in the public sphere that is supposed to be open for every kind of group existing in the society, including aspirations that do not intersect with the majority. Besides posing a risk to the sustainability of the plurality of opinions in the socio-political domain, the ambition of populism practice rules out the principle of the majority rule in the democratic system into the rule of the majority that is, finally, prone to falling to majoritarianism (Urbinati, 2014, p. 139). At an extreme point, populism practice that is allowed can risk transforming political communities into entities that are against ideological differences for the sake of fulfilling the myth of a national totality and comprehensive society.

In order to contextualize the concept of populism in the socio-political situation in the post-reformation era in Indonesia, the distinction concept of the public–private sphere and rationality that works from within is inseparable from the main topic of this article. Differentiating the two aims to help us see how religion holds an important role in public discourse and how it is related to the practice of populism in the post-reformation era in Indonesia. The use of the term "public sphere" here refers to the location where those argumentative interactions consisting of explicit political intentions, such as modifying the rule of law, take place. These interactions that happen in the public sphere can be in the form of making petitions, composing political speeches, protesting through peaceful rallies, and so on (Barthold, 2015). Referring to the distinction used by Richard Rorty (2003), the private sphere is an individual's sphere that does not need rational argumentation to justify his/her belief or action—or it can also be called the irrational domain. The belief includes the individual's religious belief. The distinction of this public–private sphere is based on the belief that first, as a doctrine, religion has a tendency to refuse argumentative justifications so that it has the potential to close an open discussion in the public sphere (in other words, it is a conversation-stopper); second, the personal individual agenda does not possess any relevance to the actions taken in the public sphere; third, religious practices in the public sphere tend to be addressed to establish political relations based on an essential and metaphysical humanity conception; fourth, the religion form is organized hierarchically so that the process of argumentation consists of *appeal to authority* logical fallacy and actually contradicts the equality principles that are supposed to be upheld in the public sphere. The distinction between the public and private spheres is a gateway for public reasons or the public use of reason that becomes a meeting point between ideas about rationality and morality. Gerald Gaus in his book *The Order of Public Reason: a theory of freedom and morality in a diverse and bounded world* (2011) explained that the idea about morality in the public sphere or social morality is based on the influence of the modern contractarian theory tradition pioneered by the philosopher Thomas Hobbes. As an attempt to achieve a peaceable, sociable, and comfortable living that is the best alternative to the nasty, poor, short, and brutish situations of human nature, a set of moral norms becomes a necessity in order to achieve conditions of coexistence. Therefore, the concept regarding social morality involves particular interpersonal relations, wherein our requests to make other people agree with our own moral convictions can also apply when other people ask us to agree with the values that they believe in. Gaus noted that in order to materialize this relation without any coercion, the party, whose willingness to agree with our moral convictions is requested, must have a sufficient and plausible reason. However, because everyone is in an epistemic position with a set of different individual beliefs, which is where public reason is needed in the dynamics of the aforementioned interpersonal relation, "the moral demands we make on others must be justifiable to those others by appeal to reasons they have, and not simply by appeal of truth as we see it" (Gaus, 2011).

1.2 *Public reason in the dynamics of social morality discourse*

The issue that will be elaborated in this part highlights the dynamics of social morality discourse in the post-reformation era in Indonesia. Ever since Suharto stepped down from his autocratic presidency, there has been a significant change in the life of the civilians covering the field of freedom of speech, assembly, religion, and the media. In addition, influential friction also occurred in the government system itself, which is the decentralization of the public space governance. This is reflected in the figure of the regional head at a district/province level who obtains greater authority to issue regional regulations according to the needs of each region. A problem occurs when the freedom experienced by the civilians is considered absolute and does not go hand-in-hand with the constitutional mandate that is the foundation of the establishment of a country. The Constitution clearly states that Indonesia is not a religious state (Assyaukanie, 2011). Proponents of this view blur a clear boundary between religion and state by justifying that religious teachings are valid to be the cornerstone of public policy-making. An example of this can be observed in a regional regulation in Serang, Banten Province, No. 2/2010 concerning prevention, eradication, and alleviation of social ill-nesses that manages and gives state actors the legitimacy to limit food stalls' business hours during the Ramadhan, the holy month of the Moslems. This regulation is supported by the majority and later turns to be an indicator of the Islamic government as a benchmark of qualified governance (Tanuwidjaja, 2010). In other words, there is a misconception in com-prehending the intention of the majority rule in the process of the public sphere governance in the democratic state. The majority's normative interpretation is used as a foundation of legal regulations without considering the possibility of majoritarianism. Another misconcep-tion is the understanding of this segregation of state and religion as a manner of anti-religion. Deliberation regarding this segregation actually does not lead to an understanding that the segregation between the two is intended to avoid the country from being used as an exten-sion for the interests of fundamentalist religious teaching groups. This segregation functions to protect the principles of liberalism in order for them to continue to be present in the life of the democratic state's civilians. The principles of liberalism such as equality, freedom of speech, and assembly are also what enables a proponent's view that states that religious teach-ings can become the base of the formulation of policies for the voiceless, the right to always be assured. In fact, this proponent's voice gains sympathy and support from the majority of people who do not yet understand the basic Constitution and urgency of segregating the national and religious domains. The support is collected in various forms, such as political parties, community organizations, social fund channeling institutions, educational institu-tions, and study groups (Mauleman, 2011). In general, this institutionalization uses religious teaching principles that cannot be argued rationally as a standard to define morality—this is why Rorty (2003) stated that religion is one of the conversation-stoppers. Consequently, the temporary discussion on morality in the public sphere becomes final because the ideas are based on divine kindness and righteousness criteria. As a result, the concept of social moral-ity that is supposedly sourced from the consensus based on public reason is not materialized.

The institutionalization of ideas about social morality that is no longer flexible becomes an effective means of changing its status from a norm into a rule of law. This can be illus-trated through examples of regulations that are, in fact, discriminative and contradictive to the principle of individual freedom, especially those implemented in the Aceh Province. This phenomenon has flourished in the post-reformation era in Indonesia through the practice of populism that is sustained through the use of the rhetoric of morality. Critiques regard-ing the proponent's perspective that justifies religious norms to become the rule of law are positioned as critiques that do not represent the majority. Populism practices work by polar-izing the diversity of opinions regarding the definition of morality in the public sphere and depoliticizing the constructive function of critiques and oppositional critiques. The plurality of society's opinions that may disagree with a proponent's views should be freely expressed without being positioned as an opponent of the will of the majority. However, the practice of populism, in fact, desires the opposite because it has the ambition to occupy the strategic position at a government level by achieving the support of the majority. This ambition is basically facilitated by procedures in the representative democracy that opens up contesta-tion of every kind of ideas. However, this process does not promise an outcome that can be achieved instantly. Polarization is prone to resulting in a stigmatization of minority opinions that should have a position in the public sphere. Polarization also has a contra-productive potential to deliberative decision-making as expected through the process of rational argu-mentation. The practice of populism in a representative democracy system operates with an approach parallel to Aristotle's description of demagoguery in a direct democracy (Urbinati, 2014, p.138). He stated that demagoguery is an example of bad democracy because it replaces the consent-seeking process in an assembly by making the people's minds conform to the political intentions of an orator through rhetoric that is manipulated in accordance with his/her interests. A sample case in Indonesia that can be used as an illustration is how an institution of a community organization conducts political campaigns in the public sphere consisting of racist contents toward the governor of DKI Jakarta Basuki Tjahaja Purnama who is also a candidate in the next governor election in 2017. In this case, the organization acts on behalf of Moslems as the majority that based on their group's belief they cannot be led by one who does not have the same religion as them. There is no process of reasoning proposed in this type of campaign model. What is done is homogenization of the political position of Moslems and justification of their claim on behalf of the majority. At a more serious level, the practice of populism works through rhetoric morality, which reaches an official institution, i.e. Constitutional Court. The first case was a judicial review that attempted to increase the minimum age of a girl in the Marriage Act that was declined by the judge of the Constitutional Court on the grounds that "it is better for a girl to get married rather than to open a chance to commit adultery that is clearly forbidden by religious teachings." The second case was an attempt to make a constitutional amendment on decency that was recently proposed by *Aliansi Cinta Keluarga Indonesia*, which encouraged the state to have a legal force in banning sexual orientations other than heterosexual. The panel of the judges responded to this judicial review by giving the same argumentation as in the previous case, which is by bringing in the basic argument of morality that leads to God's teachings. In the trial, the Chief Judge of the Constitutional Court himself stated that "the legal system in Indonesia has to be 'illuminated' with values of divinity."

The lack of commitment to public reason that is supposed to be the base of the foundation of social morality influences the process of deliberation up to the realm of law. This practice does not only use the majority as a single measure of the pseudo-will of the society, but also turns a liberal democracy into an illiberal and unconstitutional one (Urbinati, 2014, p.138). The application of democratic principles that respect individual freedom and guarantee plurality in Indonesia is obscured by the practice of populism that homogenizes the society's opinion by claiming to be a representation of the majority. In the short term, this practice draws the public energy that is supposedly reserved to oversee more crucial issues such as ensuring the government's accountability, urging the transparency in the absorptions of state budget (APBN), and especially pursuing social justice and realizing a democracy that respects human rights. Meanwhile, in the long term, the worst-case scenario of populism practice if it is allowed to continue is that the country will be used as an instrument of fascist politicians who want to establish a living in a unison system that does not welcome differences in ways of life—in other words, this will open up a loophole for Indonesia to return to be a country that is led by a totalitarian regime.

2 CONCLUSION

The spirit and efforts of the reformation desiring the civilians' openness and freedom of speech as the governing principle of a democratic country are facing the worst-case scenario caused by the practice of populism. Today, the boundary between country and religion in Indonesia is obscured by the practice of populism that brings out religious sentiments in morality discourse concealing the urgency of the public use of reason as a basis for contestation of ideas about social morality. The polarization of public opinion translates opinions as the majority's will without considering the representation of the minority's voice. As a result, manifestation of the freedom of individual politics outside the voting booth that supposedly delivers plurality of opinions is violated by the practice of populism. The freedom of individual politics is an important component in the representative democratic system because it functions to supervise the accountability of state administrators and creates a gap between individuals and the state. However, this component is missing when opinion formation is taken over by the practice of populism that does not give individuals an equal space to speak. In addition, the power of ambition hiding behind the practice of populism that uses the rhetoric of morality has a potential to conceal urgent issues such as alleviation of poverty, alleviation of corruption, equity of educational access, and improvement of the quality of public facilities that should become priorities in the society.

The rampant practice of populism in the post-reformation era in Indonesia is related to the lack of commitment to the public use of reason that is critical in responding to religious sentiments in the discourse of social morality. The key role in this practice is played by politicians to gain support from the majority through polarization that undermines the process of opinion formation. The benchmark for equal freedom of individual politics is stopped at the maxim, "one man, one vote." Outside the voting booth, individual opinions that do not intersect with the majority's voice are depoliticized and considered as the antithesis of the citizens' will. The individual political opinion outside the voting booth is also a means of social integration. It is a manifestation of a representative healthy opinion formation. Political opinions are also a means of exchanging information, values, and knowledge that enable citizens to have a conversation in the public sphere in order to stimulate public reason (Urbinati, 2014, p. 36). The practice of populism using the rhetoric of morality through religious sentiments is political personalization. This practice justifies the private agenda (faith) to penetrate the public sphere, therefore preventing the proper function of the democratic state system. This practice is prone to majoritarianism when public policies are not formulated for social engineering and to seek justice, but to gain support of the majority and use it as an electoral machine especially in the conditions of economic inequality (Pepinsky, et al., 2012). Further research is necessary to conduct a more thorough study on the strategic moves that can be taken to put an end to the practice of populism in the democratic political system.

REFERENCES

Adimaja, M. (2016) Massa Hizbut Tahrir dan Barisan RT-RW Demo Tolak Ahok. *Tempo. (Hizbut Tahrir and Neighbourhood Associations Demonstrating against Ahok)* [Online] Available from: https://m.tempo.co/read/news/2016/09/04/214801524/massa-hizbut-tahrir-dan-barisan-rt-rw-demo-tolak-ahok.

Aspinall, E., Dettman, S., Warburton, E. (2011) When Religion Trumps Ethnicity: A Regional Election Case Study from Indonesia. *South East Asia Research*, 19 (1), 27–58.

Assyaukanie, L. (2011) *Ideologi Islam dan Utopia: Tiga Model Negara Demokrasi di Indonesia (Islamic Ideology and Utopia: Three Models of Democratic States in Indonesia)*. Jakarta, Freedom Institute.

Barthold, L. (2012) Rorty, Religion and the Public–private Distinction. *Philosophy and Social Criticism*, 38(8), 861–878.

Florene, U. (2016) *Lanjutan sidang uji materi KUHP di MK. (Continued Session on the Judicial Review of the Indonesian Penal Code in Constitutional Court) RAPPLER.* [Online] Available from: http://www.rappler.com/indonesia/145575-live-blog-uji-materi-kuhp-mk-aila.

Gaus, G. (2011) *The Order of Public Reason: A Theory of Freedom and Morality in a Diverse and Bounded World.* Cambridge, Cambridge University Press.

Meuleman, J. (2011) Dakwah, Competition for Authority, and Development. *Bijdragen tot de Taal-, Land—en Volkenkunde*, 167 (2/3), 236–269.

Pepinsky, T., Liddle, W., Mujani, S. (2012) Testing Islam's Political Advantage: Evidence from Indonesia. *American Journal of Political Science*, 56 (3), 584–600.

Policing Morality Abuses in the Application of Sharia in Aceh, Indonesia (2010) [Online] Available from: http://www.hrw.org.

Rorty, R. (2003) Religion in the Public Square: A Reconsideration. *Journal of Religious Ethics*, 31 (1), 141–149.

Sahbani, A. (2016) Ketua MK: Sistem Hukum Indonesia Mesti 'Disinari' Nilai Ketuhanan. (Chief of Constitutional Court: Indonesian Legal System Should be "Illuminated" by Divine Values) *Hukum Online.* [Online] Available from: http://www.hukumonline.com/berita/baca/lt57e75880d9bc6/ketua-mk--sistem-hukum-indonesia-mesti-disinari-nilai-ketuhanan.

Tanuwidjaja, S. (2010) Political Islam and Islamic Parties in Indonesia: Critically Assessing the Evidence of Islam's Political Decline. *Contemporary Southeast Asia*, 32 (1), 29–49.

Urbinati, N. (2014) *Democracy Disfigured: Opinion, Truth and the People.* Cambridge, Harvard University Press.

Cultural Dynamics in a Globalized World – Budianta et al. (Eds)
© *2018 Taylor & Francis Group, London, ISBN 978-1-138-62664-5*

Construction of grief in obituary texts in German (2012–2015)

S.P. Suganda
Department of Linguistics, Faculty of Humanities, Universitas Indonesia, Depok, Indonesia

ABSTRACT: The purpose of this study is to examine 'death' and other aspects evolving around this theme. The objective is acquired through a text analysis. Death itself is a natural phenomenon related to a certain emotion, namely grief. While grief is a state of mind, mourning is constructed by culture. Recent studies on German obituaries show that on textual level, there have been attempts from the survivor's side to reflect their perception on death, but the aspect of emotion on texts is yet to explore. My preliminary observation shows that in German obituaries the aspect of mourning is still more dominant than grief. Using the framework of critical discourse analysis and psychological notions on emotion, a final conclusion will be drawn. Valuable information regarding contemporary socio-cultural context in Germany will also be elaborated and highlighted.

1   INTRODUCTION

The concept of life cycle is widely recognized in the society. Normally any society either traditional or modern recognizes concepts of the phases of life undergone by mankind in their worldly life, i.e. birth, marriage, and death. Crapo (2002, p. 298) and Miller (2004, p. 304) suggest that several phases taking place in or undergone by humans are often marked by rituals or ceremonies. These two cultural anthropologists base their ideas on the life cycle concept initially introduced by Arnold van Gennep, a Belgian anthropologist in 1909 (van Gennep 1960, pp. 2–3).

In this paper I will narrow down my study in death as part of a human life cycle. Data used in this study comprise obituary texts published in various newspapers circulated in the Federal Republic of Germany within the span of the last three years (2012–2015). The collected data are divided into three perspectives, namely family, friend/colleague, and personal (self) perspectives.

In the context of the current German society, death is no longer connected with the mandatory traditional or ritual practices as previously performed in the past (Sörries 2012, p. 26). Formerly, when a member of a family passed away, the other members had to don mourning clothes (Trauer kleidung), normally traditional black-coloured clothes worn by the family as a symbol of sorrow for the loss of their family member. In European culture, black is a colour that is identical with sorrow or death. This tradition is normally performed by the women (widows), and beginning from the 20th century it has been forsaken as a result of women emancipation movement (Sörries 2012, pp. 36–42). These days, sorrow due to the death of a person is no longer symbolized by plain clothing or ritual performance.

One of the reasons of the fading away of this used-to-be mandatory tradition is demographic consideration (Sörries 2012, p. 26). At the moment, German population is dominated by senior citizens. They are viewed as people who have undergone a great number of phases in their lifetime. Some of these phases may be seen as a "farewell" (sich verabschieden) to the ongoing life. Such phases include the point when one enters his or her retirement, the 60th or 65th birthday, or the transfer of senior citizens to an old people's home. With these phases of life, when a person passes away, the bereaved are not required to mourn his or her death since the deceased has actually "bid farewell" to those who survive him or her and still continue living their lives (Sörries 2012, p. 26). Judging by the anthropological point of view, the phenomenon of "death" in the society as described above constitutes an example of "social death" 'sozialer Tod', implying that the individual is biologically alive but considered "dead" as he or she is no longer included in the category of productive people in the social order.

In the current German society, mourning has a different meaning. In the past the mourning process used to be closely connected with a certain span of time or phase (for instance 7 days, 40 days, or one year). Nowadays it is no longer marked by any particular ritual in any particular span of time. The reason for this is because the duration of the grief or sorrow caused by death can no longer be fixed, as it depends a lot on the person's psychological state. As such, it is possible for grieving to reappear after the end of a mourning period.

It indicates that at present people distinguish sorrow as an attitude (mourning) and sorrow as a psychological state (grief). 'Mourning' takes a more social nature, whereas 'grief' is more personal as it is related to the psychological state of a person. This psychological state is closely related to human characters, and thus it allows the possibility for 'grief' to go beyond 'mourning'. With a commonly accepted view of 'mourning' lasting for 40 days, there is no certainty that 'grief' discontinues on the 40th day. It all depends on the characters of the bereaved, the person's psychological and emotional states in general, and the relation between the bereaved and the deceased in particular (Sörries, 2012, p. 28).

The fact that no ritual as formerly known in the society is performed by a bereaved by no means suggest that he or she does not grieve or mourn. Basically every individual is entitled to mourn. However, death rituals are beginning to lose their significance. How ritual practices are performed nowadays depends a lot on the family, social, and work environments. On this basis, various mourning behaviours that are emphasized more on the psychological state of the bereaved rather than constituting mere rituals can be analysed (Sörries 2012, pp. 31–33).

Rituals recognized by the modern German society comprise, among others, the planting of trees to commemorate a deceased (*Erinnerungsbaum*), the composing of obituary texts and the publication in mass media, and the writing of letters (*Trauerbrief*) or the preparation of a certain website containing the impressions or memories of the deceased and of what he or she did during his or her life time, and the expressions of sorrow (*Gedenkseite*) in the internet (http://www.bestatter-in-deutschland.de/informationen/trauersitten. Accessed on Monday, 28th September 2015 at 15:03). Visitors to the website can write similar expressions and light the virtual candle as an expression of condolences.

The formats of these examples indicate that death rituals performed by the Germans nowadays are simpler and more pragmatic. Many people make use of the technological progress (internet) to express their sorrow. Those who intend to participate in expressing grief yet fail or lack the opportunity to be present at the funeral should no longer worry about "breaching" the prevailing custom that is practiced in death rituals. People who are prevented from attending the funeral can make use of the new media, for instance by writing messages or lighting virtual candles in internet website (in the case of the surviving family announcing the decease of their family member through the internet).

2 CONSTRUCTION OF GRIEF IN OBITUARY TEXTS

As mentioned before, an obituary text is also seen as an effort made by people to overcome their sorrow. In psychology, sorrow is categorized as an emotion caused by, among others, death. Paul Ekman is a psychologist that undertook numerous studies on non-verbal behaviours, facial expressions, and gestures. Ekman suggests that sorrow contains two emotions, namely sadness and agony. He states that at the moment of death a deeply felt pain arises resulting in a protest within a person. While agony is an active attempt made to overcome human sadness by understanding what is actually happening (the source of loss), sadness is more passive in nature (Ekman 2003, p. 85).

In terms of language, there are numerous ways of expressing grief. Examples provided by Ekman in English comprise: distraught, disappointed, dejected, blue, depressed, discouraged, despairing, grieved, helpless, miserable, and sorrowful (Ekman 2003, p. 85). However, Ekman is also of the opinion that sometimes words that are familiar to us in expressing sadness turn out to be insufficient in representing the sadness felt by the bereaved. Psychological explanation states that the grieving person may also feel a sort of anger. This anger may be addressed to life, to God, to people, or things that are seen as the cause of the death, or even the deceased. The latter mentioned anger is particularly caused by the death of a person who purposefully placed himself or herself in a position susceptible to danger (Ekman 2003, p. 86).

This study proposes the manner of sadness as constructed in obituary texts. The term used for such texts in German is "*Traueranzeige*", which literally means 'advertisement of sadness'. This term contains the word "sadness", and as such, the language users expect to find substantially numerous elements in such texts that express sadness as an emotion (grief). To answer questions in this study, I use a critical discourse analytical approach, i.e. by means of the tridimentional analytical frame by Fairclough (1995) that is supported by a theory of emotion by Ekman (2003).

This study uses Fairclough approach (1995) as its main analysing frame. To analyse the texts as discourses, three phases are proposed by Fairclough, as follows:

1. Description. This phase is related to the analysis of the formal characteristics of a text.
2. Interpretation. This phase describes the relation between a text and the social interaction by viewing the text as a result of a production process, and the resources used in the interpretation process.
3. Explanation. This phase describes the relation between the interaction (text) and the social context that affects it, namely the political, social, cultural, religious, and economic contexts, etc.

3   SAMPLES OF OBITUARY TEXTS IN GERMAN

3.1   *Text type A*

Figure 1.   *Rheinische Post,* 14th July 2012.

3.1.1 *Descriptive analysis of text A*

Death in this text is represented by the word gehen that lexically means go or move. The text starts with a farewell greeting (*Abschiedsgruß*), "Tschüß". This expression is a daily form of language (*umgangssprachlich*) that has a meaning similar to "until we meet again" '*auf Wiedersehen*' (Wahrig 1986, p. 1302). The expression tschüß starts with the conjunction *und*. In German, *und* as a conjunction is an equal conjunction (*nebenordende Konjunktionen*). The function of this conjunction is to integrate or combine words or sets of words (Duden Band 4 1997, p. 373).

The conjunction *und* in this obituary is used as if to combine one statement with another. In fact, the text shows that the conjunction *und* is not preceded by any statement, and as such appears to be independent. In its relation to the above samples of forms in regard to how the conjunction *und* is used, it may be concluded that the use of the conjunction *und* as a copulative conjunction relates more to the stylistic aspect.

The conjunction *und* is also used in the following sentence "...*und dann ist es alles ganz schnell gegangen*" '...and then all too soon it all goes away'. The function *und* in this sentence is as a copulative conjunction, and the purpose is to provide an emphasis. This function also relates to the stylistic aspect of the text. *Und* appears to integrate two or more statements. However, these statements are not there to be found existing in the text.

In this text, the name of the deceased is positioned at the centre of the text in a bigger font compared to the others and is written in bold. The names of the surviving family members are placed under the name of the deceased and are written in bold letters as well.

This text also contains an expression of grief as stated by the phrase "*in stiller Trauer*" 'in reverent grief', and is directly followed by the names of those who wrote the obituary text. This text ends with a part that contains the information of the farewell ritual (*Trauerfreier*). Still lexically means 'silent' or 'calm' (Wahrig 1986, p. 1230). Furthermore, *still* may also mean 'secretly' or 'hidden'. The referential meaning of *in stiller Trauer* is 'in reverent grief'. The person who wrote this obituary text implies that the feeling of sadness that he or she feels is of personal nature that he or she does not wish to announce it to the public.

3.1.2 *Interpretative analysis of text A*

The text about Günther Vieth contains an expression that uses a phatic function, namely "*und tschüß!*" By placing it in the beginning of the text, this part quickly draws the readers' attention. This part is a form of interaction between the writer of the text and the deceased. The text writer seems to talk directly with the deceased. It indicates that there is sometimes a two-way communication in an obituary text, between the text writer and the deceased, as well as between the text writer and the text readers.

The information addressed to the text readers is matters related to the information regarding the deceased, for instance the name, the date of birth, and the date of death. Otherwise, information related to the death rituals is also included in the communication between the text writer and the text readers. Information regarding the death ritual is placed on the lower part of the text and is stated in past tense (*hat...stattgefunden*), and as such functions are more as an announcement rather than an invitation.

3.1.3 *Explanatory analysis of text A*

Based on this text, death for the surviving family is a private moment that does not have to be part of the public domain. It is reflected in the expression of grief phenomenon, "*in stiller Trauer*" 'in reverent grief' and in the manner of how the death ritual is performed "*hat im engsten Familien-und Freundekreis stattgefunden*", or 'conducted in the midst of family and friends'. Contextually, "*still*" or 'reverent' does not only imply a situation that is far from any outburst, anxiety, tumult, etc.; "*still*" also refers to the fact that the moment was spent only by those who were really close to the deceased as they needed time and privacy to mourn.

Wir hätten ihr mehr Zeit gewünscht.

Astrid Hey
geb. Spickschen

* 8. Juni 1943   † 18. Februar 2014

Wir trauern um sie
Ihre
3 Töchter, Schwiegersöhne, Enkelkinder,
Geschwister, Nichten und Neffen

Die Trauerfeier findet statt am 7. März 2014, um 12.30 Uhr in der Kapelle auf dem Friedhof Rahlstedt, Am Friedhof 11, 22149 Hamburg. Anstatt Blumen bitten wir um eine Spende an Plan International, Bank für Sozialwirtschaft, IBAN DE92 251205100009444933, BIC BFSWDE33HAN, Stichwort „Beisetzung/Astrid Hey".

Translation

If only she had more time

Astrid Hey

born: Spickschen

*8th June, 1943 †18th February, 2014

We, the bereaved:

Her three daughters, sons-in-law, and grandchildren,

Her siblings, nieces, and nephews.

The funeral will take place on 7th March, 2014 at 12.30 at the cemetery chapel of Rahlstedt, am Friedhof 11, 22149 Hamburg.

Instead of flowers, we would appreciate your donation to be transferred via bank: Plan International, Bank für Sozialwirtschaft IBAN, DE 92 251205100009444933 BIC BFSWD33HAN, password "Funeral/Astrid Hey"

Figure 2.   *Hamburger Abendblatt*, 22nd–23rd February 2014.

3.2.1   *Descriptive analyis of text B*

This obituary text starts with a sentence that reads *"wir hätten ihr mehr Zeit gewünscht"* 'if only she had more time'. This sentence uses the second form conjunctive mode that functions, among others, to express some kind of hope. However, it is an unreal one (*irrealer Wunschsatz*). In addition to the use of the second form conjunctive mode, this sentence also uses past tense. This sentence is placed on the uppermost part of the text that has a phatic function, i.e. to draw the attention of the readers.

This text also contains cultural rituals related to condolences. Normally condolences can be expressed directly face to face, or in case of a situation that does not allow one to be present at the funeral, it may be represented by funeral flower arrangements, and letters/cards would be delivered to the surviving family. Through this text, the surviving family would prefer that those who would express their sympathy could send their donation in the form of money instead of flowers.

3.2.2   *Interpretative analysis of text B*

The sentence "if only she had more time" implies an unreal hope. It indicates that obituary texts in German do not only express death but also hope. Usually the said hope relates to a future after death. However, this text also indicates that such hope has an implication of the past.

This text contains an expression of grief as seen in the use of the sentence "wir trauern um sie" 'we, the bereaved'. Normally this part is followed directly by the names of the persons who wrote the text. In this particular text, the names of the surviving family members are not mentioned, and only their social status is mentioned in relation to the deceased, namely her children, siblings, grandchildren, and so on.

3.2.3 *Explanative analysis of text B*

This text includes a part that reads "instead of flower arrangements, we would appreciate your donation to be transferred through the bank, Plan International, Bank für Sozialwirtschaft IBAN, DE 92 251205100009444933 BIC BFSWD33HAN, password "Funeral/Astrid Hey". Appeal for donation in the form of money is often seen in obituary texts in German. This text also contains such appeal. Unlike similar appeals that mention the donation being designated to be distributed to charity foundations or community organizations, the appeal in this text mentions that the fund is meant for a specific purpose, namely to cover the funeral costs.

It is mentioned in the Introduction that sadness nowadays has two facades, namely mourning and grief, as suggested by Sörries (2012, p. 187). Mourning is a process that is socially arranged (*sozial verordnet*) and is conducted collectively, and hence it can be visible for others. On the other hand, grief is an emotion (*emotionale Befindlichkeit*) that is felt privately by a person; hence, it is not visible to others. With the possibility being open for participating in a donation for the funeral of the deceased, there is an indication that the participants (the readers) can also take part in mourning together with the surviving family (*die Hinterbliebenen*).

In addition to the abovementioned description, another explanation can be made of this phenomenon by taking a closer look at the German social and cultural context of today. Gronemeyer (2010, p. 303) states that at the moment the theme of "death" in Germany is frequently related to institutional issues (*Institutionalisierung*), medical issues (*Medikalisierung*), and economic issues (*Ökonomisierung*). The phenomenon that is particularly related to text type B is an economic one.

According to Gronemeyer (2010, p. 305), death requires substantial costs (*sterben ist teuer*). Based on an internet website (www.bestattungen.de/ratgeber/bestattungskosten. Accessed on Monday, 20th October 2014 at 2:20 p.m), the cost for funeral varies, depending on the type, execution, and venue of the burial. The largest cost to spend is for the funeral (*Friedhof*). Other costs involved comprise costs for the service of the funeral parlour, funeral flowers, the preparation and publication of obituary at the mass media, and the reception book. Based on this website, we can have information that the cost of burial in Germany at the moment would be at least 4,000 Euro, or around 60 million IDR using the current Euro-IDR exchange rate of October 2014.

4   CONCLUSION

Obituary texts published in Germany contain emotional aspects. However, they are not overly emphasized. The term *Traueranzeige*, which in the present standard German is more dominantly used than the term *Todesanzeige* (obituary texts), led me to the assumption that obituary texts written in Germany would highly emphasize emotions of sadness. My analysis of this phenomenon has proven otherwise.

Expressions of sadness that are often found in these texts constitute a phrase that is followed by the information about the writer of the obituary text ("*in tiefer Trauer*" 'in deep grief', "*in stiller Trauer*", 'in reverent grief'). This phrase acts as an introduction that leads to the part that describes the writer of the text. As such, the phrase that contains the noun "*Trauer*" 'sadness' actually puts more emphasis on "the person who feels" rather than "what is felt".

In the analysed obituary texts in German, it is obvious that the aspect of 'mourning' is more dominating than 'grief'. The findings of the analysis of the three perspectives of the whole obituary texts lead us to the conclusion that the mourning process that is constructed in the texts can be divided into four categories, namely matters related to the execution of the funeral rituals, those related to the expression of condolences, those related to the mourning clothes, and those related to the donation.

The present trend shows that the surviving families opt for donation rather than flowers or flower arrangements as tokens of condolences, which would be used to support the costs for the funeral procession or to be distributed to particular charity foundations. It indicates the increasingly pragmatical attitude of the German people that leads to the appeal for forms of condolences that may have a direct use in terms of the living ones (*die Hinterbliebenen*) instead of the deceased (*die Verstorbenen*).

Privacy is also something of high significance to the Germans. In many obituary texts we can see that the surviving family members explicitly state that the funeral ritual is conducted only within a circle of closest family. This expression is implicitly understandable as death is taken as an intimate moment that is expected to be experienced by the closest family and friends. Mourning is not viewed as a public domain, and as such the surviving family does not wish to involve other people in the mourning process.

REFERENCES

Crapo, R.H. (2002) Understanding ourselves and others. *Cultural Anthropology*. Boston, McGraw Hill.

Duden, B. (1997) *Grammatik*. Duden, Mannheim.

Ekman, P. (2003) Understanding faces and feelings. *Emotions Revealed*. London, Phoenix.

Fairclough, N. (1995) The critical study of language. *Critical Discourse Analysis*. London, Longman.

Gennep, A.V. (1960) *The rites of passage*. Chicago, the University of Chicago.

Gronemeyer, R. (2010) Hospiz, Hospizbewegung und Paliative Care in Europa. In: Elsas, Christoph. (2010). *Sterben, Tod und Trauer in den Religionen und Kulturen der Welt*. Berlin, EB-Verlag, pp. 199–308.

Miller, B.D. (2004) *Cultural anthropology*. 3rd. Edition. Boston, George Washington University.

Sörries, R. (2012) *Herzliches Beileid. Eine Kulturgeschichte der Trauer.* Darmstadt, Primus.

Wahrig Deutsches Wörterbuch. (1986) Gütersloh, Bertelsmann.

http://www.bestatter-in-deutschland.de/informationen/trauersitten [Accessed on Monday, 28th September 2015 at 15:03].

Cultural Dynamics in a Globalized World – Budianta et al. (Eds)
© 2018 Taylor & Francis Group, London, ISBN 978-1-138-62664-5

The political as the ontological primacy: On Ernesto Laclau's thoughts

D. Hutagalung & A.Y. Lubis
Department of Philosophy, Faculty of Humanities, Universitas Indonesia, Depok, Indonesia

ABSTRACT: Ernesto Laclau has opened up a new horizon in understanding the concept of *the political* in a totalitarian system that constantly depoliticises the whole aspect of life and treats all problems merely as administrative, bureaucratic and technocratic issues. Totalitarianism reduces the term 'politics' merely to 'political interest', which means that achieving these interests is different and determined in advance and separated from its possible articulation among competing discourses. Therefore, according to this reasoning, the specific characteristics of the political arena, namely conflicts, antagonisms, power relations, forms of subordination and repression,, disappear from the equation. According to Laclau, *the political* can be understood only through the logic of populism. Laclau viewed populism as the best way to understand the ontological formation of *the political*. Laclau's thought originated from his dissatisfaction with the sociological perspective, which considers a group only as a base unit of its social existence before its political construction or incorporates the political subject and political construction into the functionalist or structuralist paradigm in *the social*. He viewed populism from three perspectives: psychoanalysis, linguistic (rhetoric) and politics (hegemony). Laclau concluded that popular subjectivity can only be formed based on the creation of an empty signifier in a discursive way.

1   INTRODUCTION

In his article *Glimpsing the Future*, Ernesto Laclau concluded with the statement, 'Rhetoric, psychoanalysis, and politics (conceived as hegemony): in this triad I see the future of social and political thought' (Laclau, 2004, p. 36). This statement is the conclusion of Laclau's quest and thought that are heavily influenced by post-structuralism philosophy, Lacanian psychoanalytic, Gramsci's thought and critics on classical Marxism.

This paper analyses Ernesto Laclau's thought on *the political* as the ontological primacy, by tracking through the concept of rhetoric, psychoanalysis and politics, as well as populism, which Laclau believed as the royal way to understand *the political*. By reading some of Laclau's major works, we try to re-read the conceptualisation of *the political* in the political arena.

1.1   *Deconstruction, hegemony and discourse*

In his theory, Laclau set the importance of deconstruction in analysing politics. Deconstruction, which is the highlight in the theory of rhetoric, according to Laclau, is fundamental in analysing politics. He emphasised this in his argument:

> Deconstruction is a primarily political logic in the sense that, by showing the structural undecidability of increasingly large areas of the social, it also expands the area of operation of the various moments of political institution [...] The central theme of deconstruction is the politico-discursive production of society (Laclau, 1996a, p. 52).

The question is: how can deconstruction work in analysing politics concretely? Is deconstruction an approach relevant to political analysis? To discuss this question, it is interesting to see Laclau's analysis on deconstruction in political analysis. Laclau believed that deconstruction approach is relevant to explain a political phenomenon. Laclau suggested two main reasons to support his argument. The first is the idea of *the political* as an institutionalised moment in a society. His argument is based on the situation encountered back then, which tended to immerse the social life and then reactivate it by putting it back to the initial political condition. Such a situation does not foster understanding to the society as a whole, which is based on the original logic. This situation also adds the contingency character to many actions of political institutions. According to Laclau, the unfinished formation of social class is essential in understanding how the logic of hegemony works (Laclau, 1996a, p. 47). Second, all political institutions are not developed completely. From this point of view, public 'polarisation' seems to experience double transitions; on the one hand, it is viewed as an expansion of *the political* against social life; on the other hand, politicisation also comprises contingency of social networking formation as a decentring of society (Laclau, 1996a, p. 48).

From Laclau's point of view, the democratic politics is unable to do anything without philosophical reflection. It is difficult for the democratic politics to understand the dynamics within; therefore, it is critical to describe all consequences considering the fact that power and antagonism are ineradicable things in society. This is the importance of deconstructive approach in understanding the objectives of the consensus comprehensively.

Hegemony is one of the key factors in Laclau's philosophy. This concept refers to Gramsci's thought, with which Laclau has made the epicentre of all his theoretical analyses. The concept of politics is inseparable from the concept of hegemony and discourse because all Laclau's political philosophies and theories refer to those two concepts in formulating *the political* as the key to ontology.

From Lacan's and Gramsci's psychoanalyses, Laclau formulated his idea on hegemony. Laclau borrowed the conception of *jouissance* from Lacan to explain how hegemonic discourses form in an empty signifier, which integrates various discourses into one hegemonic discourse. *Jouissance* means excessive pleasures which can bring a very exciting feeling or sickening, but simultaneously become the sources of charms (Fink, 1995, p. xii). According to Laclau, the relationship between the lack of *jouissance* and fantasy in embracing a certain image or picture that universalises both tends to transform them into an empty signifier, which is the relationship between *jouissance* and repression in the notion of 'social symptom' (Laclau, 2004, p. 300).

According to Laclau, hegemonic relationship requires something asymmetrical between universality and particularity. It means, every group is particular in *the social* and is created around certain interests. Hegemony can only be created if the group takes some degree of representation of universality from all of the community as a whole. The main argument Laclau suggested is that the asymmetry between the universality of social agent's task and the particularity of social agent's ability to accept tasks is the main condition to politics. Whenever certain particularity is able to lead the struggle against a regime, it can be considered a 'common' or 'harsh' crime because only the regime can have the power to do that. Laclau formulated hegemonic relations as: 'unevenness of power is constitutive of it' (Laclau, 2000, p. 141).

Laclau took democracy as an example by referring to Claude Lefort where democracy assumed power as an empty place and is not predetermined by the structure of enforcer of power that tries to occupy that empty place. In order to achieve democracy, the particular power that occupies the empty room is needed, but not to be identified by it, which means: democracy can only exist if the gap between universality and particularity is never bridged, but has been reproduced. Laclau then emphasised that democracy can only be possible in a hegemonic field.

In Laclau's opinion, a hegemonic operation can only be possible as long as it never achieves what it tries to achieve, that is, a total unification of universality from communitarian space and the power that shapes universal moments. If total adhesion is made possible, then the universality will find its absolute body, and there will be no possible hegemonic variation anymore. However, according to Laclau, it also means that democracy is the only true political society because it is the only room where the gap between universality of power and substantive power, which are contingent in nature, can occupy. Laclau defined hegemonic relations as:

> [...] there is only hegemony if the dichotomy universality/particularity is constantly renegotiated: universality only exists incarnating—and subverting—particularity but, conversely, no particularity can become political without being the locus of universalizing effects. Democracy, as a result, as the institutionalisation of this space of renegotiation, is the only truly political regime (Laclau, 2000, p. 142).

In his early writings, Laclau developed and focused his attention strongly on the idea of *the political*. This can be seen in the introduction of the second edition of *Hegemony and Socialist Strategy,* which he wrote together with Chantal Mouffe, where he clearly stated that the theoretical perspective he developed in that book could contribute to restore *the political* as the centre point or centrality (Laclau and Mouffe, 2001). A number of Laclau's earlier works about *the political* are also related to this theme, especially his writing about a debate between Nicos Poulantzas and Ralph Miliband, as well as his first important book on politics and ideology (Laclau, 1975 and Laclau, 1987). In *Hegemony and Socialist Strategy*, Laclau emphasised that the political identity is not something given, instead it is something created and recreated through debates in public domain. *The political* does not merely draw conclusions from the existing interests, but plays an important role in creating political subjectivity. At this point, there is a significant divergence between Laclau and Habermas and the others, where Laclau stated, 'the central role that the notion of antagonism plays in our work forecloses any possibility of a final reconciliation, of any kind of rational consensus. Conflicts and division, in our view, are neither disturbances that unfortunately cannot be eliminated, nor empirical impediments that render impossible the full realization of a harmony' (Laclau and Mouffe 2001, p. xvii). According to Laclau, we cannot achieve such harmony because 'we will never be able to leave our particularities completely aside in order to act in accordance with our rational self'. Laclau then suggested that 'without conflict and division, a pluralist democratic politics would be impossible' (Laclau and Mouffe 2001, p. xvii). In the context of 'divergence', it is important to add that in the 'wars of interpretation' it is needed to include the relation with struggle against racial hierarchy and patriarchy, and therefore, feminist and post-colonial perspectives become very important.

In this context, Laclau thought that one of the main principles in his work is to underline the need to create the chains of equivalences among various democratic movements in fighting against different types of subordination. For example, the struggle against sexism, racism and environmental movement must be articulated together with labour movement in every hegemonic project of the left-wing movement. Laclau drew a conclusion that the need to create a chain of equivalences among the various democratic struggles against different forms of subordination is to create a frontier and to define adversaries. The term 'adversary' is a deeper idea elaborated and developed by Chantal Mouffe to describe the idea of radical plural democracy she coined together with Laclau. According to Mouffe, adversary is understood as the similarity and loyalty to the principles and values of democracy: 'freedom and equality for all'. However, a different opinion in the interpretation of those principles and values exists. The adversaries fight with each other because they want their interpretations on those principles and values to be hegemonic, but they do not criticise the legitimacy of their adversaries on their rights to fight and win their position (Mouffe, 2013, p. 7). On another part, Mouffe also stated that '... an "adversary", which refers to somebody whose ideas we combat but whose right to defend those ideas we do not put into question' (Mouffe, 2000, pp. 101–102). However, it is important to consider the form of society they want to build, and in his analysis, Laclau emphasised that 'this requires from the Left an adequate grasp of the nature of power relations, and the dynamics of politics' (Laclau and Mouffe, 2001, p. xix).

1.2 Politics as the ontological primacy

In his theoretical approach, Laclau attempted to restore the subject omitted by the post-structuralism in his analysis and political theory. Laclau's conception of subject is inseparable from politics, as Laclau stated:

> 'Politics' is an ontological category: there is politics because there is subversion and dislocation of the social. This means that any subject is, by definition, political. Apart from the subject, in this radical sense, there are only subject positions in the general field of objectivity. But the subject, as understood in this text, cannot be objective: it is only constituted on the structure's uneven edges. Thus, to explore the field of the subject's emergence in contemporary societies is to examine the marks that contingency has inscribed on the apparently objective structures of the societies we live in (Laclau 1990, p. 61).

The subject in Laclau's view arises from a movement that subverts symbolic order. Thus, the subject is a distance between the undecidability from structure and decisions. Laclau defined *undecidability* as a destructibility of social area where the political logic was previously questioned by an occurrence of dislocation. Therefore, according to Laclau, *undecidability*: 'should be literally taken as that condition from which no course of action necessarily follows. This means that we should not make it the necessary source of *any* concrete decision in ethical or the political sphere' (Laclau, 199b, p. 78).

The subject can arise from a dislocated structure, so this dislocated structure forms the subject outside itself. In his conception of subject, Laclau observed that in hegemonic practice, the rise of political subjects is inevitable, whose task is to create structures in various new forms. Laclau refused the essentialist approach on subjectivity that sees individual merely pursues interests maximally, or the agents' role is reduced to reproducing structures in the making. According to Laclau, whenever human is formed as subjects in different discursive structures, those structures inherently become contingent and pliable. Thus, when undecidability from a subject can be seen in dislocated situations, where structure cannot function anymore to give identity, the subject becomes the political agent in a stronger term, as they identify themselves with new discursive objects and act to reorganise various structures.

The concept of *the political* in Laclau is related to the ideas on the meaning and differences in popular and democratic subjects' positions. Through his observations, Laclau gave an example about the important thing in terms of 'differential characteristic may be established between advanced industrial society and the periphery of the capitalist world', and he illustrated that in the developed industrial society the number of antagonistic points enable the formation of several democratic movements. However, such diverse movements tend not to constitute the 'people', meaning that those movements tend 'not to enter into equivalences with one another and to divide the political space into two antagonistic fields'. The difference between the Third World society, where 'imperialist exploitation and the predominance of brutal and centralised forms of domination tend to form the beginning to endow the popular struggle with a centre, with a single and clearly defined enemy', and here is that 'the division of the political space into two fields is present from the outset, but the diversity of democratic struggle is more reduced'. The term 'popular subject position' is used 'to refer to the position that is constituted through dividing the political space into two antagonistic camps', whereas the term 'democratic subject position' refers to 'the locus of a clearly delimited antagonism limit, which does not divide the society in that way' (Laclau and Mouffe, 2001, p. 131).

Laclau further disagreed with the opinion saying that democracy is an attempt to manage the political space around the society's communality by bringing unity to the people. On the contrary, it can also be seen that democracy is understood as an expansion from the logic of equivalence to a wider domain in social relations, such as socio-economic equality, racial equality, gender equality, special equality and so on, which in this regard democracy also means respect toward differences. Therefore, as Laclau once stated, 'ambiguity of democracy' can be formulated as 'the need of unity but only when it is thought by differences' (Laclau, 2001, p. 4).

The concept of *the political* in Laclau's idea is possibly influenced by Carl Schmitt's notion, which is written in his masterpiece *The Concept of the Political* published in 1932. However, Laclau rarely mentioned or referred to Carl Schmitt in his works. Only in his article *'On 'Real' and 'Absolute' Enemies'*, Laclau reviewed Schmitt's work *Theory of Partisan*. Even in the book *Hegemony and Socialist Strategy*, Schmitt's name is never mentioned or referred to. It was Lucio Colleti's social antagonism that Laclau referred to. Schmitt influenced Chantal Mouffe's thought more, who set down the notion of plural and radical democracy in antagonism/agonism relation or friend–enemy/friend–adversary on ideas developed by Carl Schmitt. According to Schmitt, *the political* should be put in special criteria, which ensure the autonomy of *the political* against other social domains, where Schmitt put the uniqueness of *the political* in a distinction between friends and enemies. It means that *the political* should always be about antagonism, the creation of the friend–enemy relation or the association–disassociation relation. Schmitt described that: 'The distinction of friend and enemy denotes the utmost degree of intensity of a union or separation, of an association or disassociation' (Schmitt, 1996, p. 26). As stated by Schmitt:

> *The political* is the most intense and extreme antagonism, and every concrete antagonism becomes that much more political the closer it approaches the most extreme point, that of the friend-enemy grouping (Schmitt, 1996, p. 29).

From the Schmitt's concept, Laclau (also Chantal Mouffe) started the idea of *the political* in his thinking. Schmitt differentiated 'the political' from 'politics' exactly on the point of antagonism, on the formation of friend–enemy relation. In Schmitt's opinion, if friend–enemy relations cannot be built anymore, then precisely at that point, the politics in a radical meaning disappears automatically, and what is left is how to keep things in order and free from rivalry, conspiracy or insurgency (Marchart, 2007, p. 43).

In fact, the distinction between 'the political' and 'politics' is never explicitly explained by Laclau. The explanation on this appears mostly in the works of Mouffe, who was Laclau's intellectual partner almost for all his life. Through Mouffe, the explanation about the difference of both concepts becomes clear. Mouffe explained that:

> By the 'political', I refer to the dimension of antagonism that is inherent in human relations, antagonism that can take many forms and emerge in different types of social relations. 'Politics', on the other side, indicates the ensemble of practices, discourses and institutions which seek to establish a certain order and organize human coexistence in conditions that are always potentially conflictual because they are affected by the dimension of 'the political'. I consider that it is only when we acknowledge the dimension of 'the political' and understand that 'politics' consists in domesticating hostility and in trying to defuse the potential antagonism that exists in human relations, that we can pose what I take to be the central question for democratic politics (Mouffe, 2000, p. 101).

In another part, Mouffe added that, in order to differentiate 'politics' from 'the political', as far as she is concerned, she further indicated the difference between the political science approach, which limits itself within empirical domain, and the political theory approach, 'which is the domain of philosophers, who enquire not about facts of "politics" but about the essence of "the political", or what Mouffe borrows from Heidegger's vocabulary, which says that politics refers to the "ontic" level while "the political" is to do with the "ontological"' (Mouffe, 2005, p. 8).

With such understanding about it, *the political* will always relate to the dimension of antagonism and conflictual. Dimension of antagonism in Laclau is part of the emancipation effort through democratic revolution. According to Laclau, in the classical idea about emancipation, humans consider emancipation and power as two antagonistic conceptions. This idea deals with the following paradox: the limit of the word freedom—or power—is also a thing that makes freedom possible. Similar to the other two previous concepts where one condition which enables something to happen is also the impossibility on that condition, in explaining an undecidable domain, Laclau stated, 'I am exercising a power which is, however, the very condition of my freedom' (Laclau, 1996a, p. 52). Furthermore, the relation between power and freedom is a permanent renegotiation and a shift from a contradictory situation.

According to Laclau, *the political* can be understood only through the logic of populism. Laclau viewed populism as the most appropriate way to understand the ontological formation of *the political* (Laclau, 2005, p. 67). Laclau's view originated from his dissatisfaction with the sociological perspective, which considers a group only as a base unit of its social existence before its political construction or incorporates the political subject and political construction into the functionalist or structuralist paradigm in *the social*. Laclau understood populism as part of democracy, which is an attempt to manage the political space around the people's communality to bring unity to 'the people'.

Laclau built a populism theory, which focuses on the chains of equivalent creation between various political and social demands that scatter, where political demand requires the result of certain meanings of representation (i.e. floating signifier and empty signifier), which can be presented as subjective identification points. In such a conception, populism is not a specific ideology, but rather it materialises in a set of rhetorical demands, and according to Laclau, rhetoric is constitutive in all political practices. Discourses and populist practices are directly aimed at the political dimension of social relations (Howarth, 2015, p. 13). Therefore, when *the political* refers to contestation and institution from different social relations, the logic of populism catches the practices through the society, which is divided into opposing groups in an endless struggle to reclaim the hegemony. Populist politics involve the formation of collective agency—the people—by constituting a political frontier between 'we' and 'they' in a social formation.

2 CONCLUSION

According to Ernesto Laclau's view, the ontological primacy of *the political* against *the social* is based on the claim that all social relations throughout history are basically antagonistic in nature and part of the political history. Laclau articulated the idea of *the political* in the totalitarian system, with its mask of liberal democracy, which constantly depoliticises all aspects of life and treats all matters as only administrative, bureaucratic and technocratic issues. Totalitarianism reduces 'politics' into merely 'political interests', which means that achieving these interests is different and determined in advance and separated from its possible articulation among competing discourses,. According to this reasoning,, the characteristics of the political domain, namely conflict, antagonism, power relations, forms of subordinations and repression, disappear from the equation. According to Laclau, *the political* can be understood only through the logic of populism. He viewed populism as the best way to understand the ontological formation of *the political*. Laclau's view originated from the dissatisfaction with the sociological perspective, which considers a group as the base unit of its social existence before its political construction or incorporates the political subject and political construction into the functionalist or structuralist paradigm in *the social*. Laclau understood populism as part of democracy which is an attempt to manage political space around the society's communality to bring unity to 'the people'. He observed *the political* through three perspectives: psychoanalysis, rhetoric and politics (hegemony). Laclau concluded that the formation of popular subjectivity can only be made possible based on the creation of an empty signifier in a discursive way.

REFERENCES

Fink, B. (1995) *The Lacanian subject: between language and Jouissance*. New Jersey, Princeton University Press.
Howarth, D. (2015) Discourse, hegemony and populism: Ernesto Laclau's political theory. In David Howarth (ed.) *Ernesto Laclau: Post-Marxism, Populism and Critique*. London, Routledge.

Laclau, E. & Chantal, M. (2001) *Hegemony and socialist strategy: Towards a radical democratic politics.* London, Verso.

Laclau, E. (1975) The specificity of the political: The Poulantzas-Miliband debate. *Economy and Society,* 4 (1) (February).

Laclau, E. (1987) *Politics and ideology in Marxist theory: Capitalism-fascism-populism.* London, Verso.

Laclau, E. (1990) *New reflections on the revolution of our time.* London, Verso.

Laclau, E. (1996a) Deconstruction, pragmatism, hegemony. In Chantal Mouffe (ed.) *Deconstruction and Pragmatism.* London, Routledge.

Laclau, E. (1996b) *Emancipation(s).* London, Verso.

Laclau, E. (2000) Power and social communication. *Ethical Perspective,* 7 (2–3).

Laclau, E. (2001) Democracy and the question of power. *Constellation,* 8 (1).

Laclau, E. (2004) Glimpsing the future. in Oliver Marchart and Simon Critchley (eds.) *Laclau: A Critical Reader.* London, Routledge.

Laclau, E. (2005a) *On populist reason.* London, Verso.

Laclau, E. (2005b) On 'real' and 'absolute' enemies. *The New Centennial Review* 5 (1).

Marchart, O. (2007) *Post-foundational political thought: Political difference in Nancy, Lefort, Badiou and Laclau.* London: Routledge.

Mouffe, C. (2000) *The democratic paradox.* London, Verso.

Mouffe, C. (2005) *On the political.* London, Routledge.

Mouffe, C. (2013) *Agonistics: Thinking the world politically.* London, Verso.

Schmitt, C. (1996) *The concept of the political.* Chicago, The University of Chicago Press.

Cultural Dynamics in a Globalized World – Budianta et al. (Eds)
© 2018 Taylor & Francis Group, London, ISBN 978-1-138-62664-5

Roles and contributions of participants in broadcast talk

M.K. Wardhani & B. Kushartanti
Department of Linguistics, Faculty of Humanities, Universitas Indonesia, Depok, Indonesia

ABSTRACT: As a public discourse, an interview in a talk show cannot be simply perceived as an interaction between the host and the guest. Both parties contribute towards the conversation; however, the audience also has a role in it. The televised interview is purposefully designed to appeal to the target audience. Thus, all utterances produced by both the interviewer and the interviewee are designed not only for each other, but also for the audience. It means that the audience is also considered as one of the participants making a certain contribution to the conversation. This article aims at highlighting roles and contributions of participants in an American talk show, *Late Show with David Letterman*. The researchers focus on the conversation between David Letterman as the host of the show and John Oliver as the guest. This study employs Conversation Analysis (CA) approach, applying Clark's (1994/1996) theory on participation roles. The conversation is transcribed using a certain transcription convention. The findings of this study show the designs of participants in a broadcast talk that can lead to further development of research on conversation in talk shows.

1   INTRODUCTION

Talks that occur in a talk show may bear a remarkable similarity to everyday conversation in terms of the involvement of the speaker and the addressee. However, the talks in TV talk shows cannot be merely seen as an exchange of information between the two parties. The audience in the studio is the third party that also has a role in the talk since the conversation is a public discourse.

According to Hutchby (2005), as a public discourse, talk shows involve the audience along with the host and the guest. It means that the audience is acknowledged as one of the participants contributing to the conversation. It is also stated that talk shows are public discourse because the fact that they are broadcast makes them accessible to the audience outside of the studio (Hutchby, 2005). This shows that the conversation in a talk show is purposefully designed to appeal to "a wider audience which is not co-present, invisible, and (usually) unheard" (Tolson, 2001). In other words, the conversation in a talk show is developed by the host and the guest by taking the 'existence' of the audience into account.

Since the 'existence' of the audience counts quite much in talk shows, the audience can be considered as one of the personnel in the conversation. Clark (1994, 1996) states that the personnel of a conversation have participation roles. There are some roles which can be categorized as participants and non-participants or over-hearers. Participants consist of speakers, addressees, and side participants. Meanwhile, non-participants or over-hearers include bystanders and eavesdroppers. In a conversation, the participants involved make some contribution. They give signals to ensure that they have "the mutual belief that they have understood well enough for current purposes" (Clark, 1996).

In this article, we aim to elucidate the roles and the contributions of those involved in the conversation in a talk show. The talk show studied here is *Late Show with David Letterman*, an American comedy talk show broadcast on 29th January 2015. We focus on the conversation between David Letterman as the host of the show and John Oliver, who was invited as the guest to be interviewed. Here, we address two questions: (1) what role each participant in the conversation has and (2) how their contribution to the conversation shows the design of the relationship between participants in the conversation. The content of this article is partly based on the Master's thesis of the first author (Wardhani, 2016) entitled *Triggers of Laughter in Late Show with David Letterman: an Analysis of Humour in Conversation*.

1.1 Short profiles of David Letterman and John Oliver

David Letterman is known as a senior comedian and talk show host in the USA. He graduated in radio and television from Ball State University in Muncie, Indiana ("Bio"). He began hosting *Late Show with David Letterman* in 1993, and he retired in 2015. The last episode of the show he hosted was aired on 20th May 2015. During his time with this show, *Late Night Show with David Letterman* won six Emmy Awards for *Outstanding Variety, Music, or Comedy Program* ("Late Show").

Meanwhile, John Oliver is a British stand-up comedian who is popular not only in Britain, but also in America. He is known as a comedian who likes to bring up political issues in his performances. He began hosting *Last Week Tonight*, a talk show "presenting a satirical look at the week in news, politics and current events" ("Biog") in 2014. His appearance in this show has been considered to make a real-life impact on the United States (Luckerson, 2015). John Oliver has won several awards including Writers Guild Award and Emmy Award.

2 METHODS

This is a qualitative study in which we employ the Conversation Analysis (CA) approach and apply Clark's theory (1994, 1996) on participation roles and contributions. As mentioned earlier, the data of this research is the conversation between David Letterman and John Oliver in *Late Show with David Letterman* that was broadcast on 29th January 2015. We obtained the video of John Oliver's interview with David Letterman from YouTube since the one on the official site (http://www.cbs.com/shows/late_show/) was not available for some regions including Indonesia. In order to make sure that the video contained the full interview, we compared the content of the video to the summary of the interview on http://www.cbs.com/shows/late_show/wahoo_gazette/.

After obtaining the video, we transcribed the interview using Discourse Transcription, a transcription convention from Du Bois (2006). It is the revised version of Du Bois (1991). We then divided the conversation into five segments: opening, topic of tasing (TS), topic of childhood (CH), topic of Super Bowl (SB), and closing. Then, the sentences in each topic were divided into some series of exchanges. The segment of TS consisted of 13 exchanges, CH contained six exchanges, and SB had two exchanges.

After grouping the utterances, we did the analysis using Clark's theory (1994, 1996). We started by identifying the roles of those involved in the conversation. The purpose was to find out how each partaker in the conversation in the talk show was positioned. Here, we also tried to explain the relationship between partakers. It was done by analyzing their contributions to the conversation. This analysis would show how the speakers designed their utterances to reach a certain purpose and what kind of interaction participants create in the conversation.

3 ANALYSIS

As for the data, there are four partakers in the conversation that can be identified, namely David Letterman (DL), John Oliver (JO), studio audience (SA), and television audience (TA). They have different roles in the conversation. Based on Clark (1994), a conversation includes at least two participants whose participation roles can "change from one action to the next". These participants are the speaker and the addressee. In the data, DL and JO become the speaker and the addressee interchangeably. Take the following example:

CH3
[...]
DL; And then you moved to a: home of the Beatles,
Liverpool?
JO; I never—I never lived there.
DL; Never lived there?
JO; Yeah.
[...]

When DL asks *and then you moved to a home of the Beatles, Liverpool* in (312)–(313), DL is the speaker and JO is the addressee. However, when JO says *I never lived there* in (314), JO becomes the speaker and DL is the addressee.

As the speaker and the addressee, both DL and JO are responsible for making each other understand what they utter, and they have the rights to take part in the conversation and get adequate signals, "any actions by which one person means something for another person" (Clark, 1996), from their interlocutor. When DL and JO become speakers, they do not just speak. They design their utterances in a way that can be understood by the addressee. When DL and JO become the addressees, they not only listen to the speaker's utterances, but also take up the speaker's signals, try to comprehend them, and prepare themselves to give evidence that they understand what the speaker means. In other words, both participants make some efforts to contribute to the conversation; they work together to make sure that both of them believe that the signals they present have been understood well enough for current purposes.

Contributions to the conversation include presentation and acceptance phases that function in this way (Clark, 1996):

Presentation phase. A presents a signal *s* for B to understand. He assumes that, if B gives evidence *e* or stronger, he can believe that B understands what he means by it.

Acceptance phase. B accepts A's signal *s* by giving evidence *e* that she believes she understands what A means by it. She assumes that, once A registers *e,* he too will believe she understands.

In the earlier example, DL and JO can reach the mutual belief that they understand each other through the two phases. The utterances of DL and JO can be classified under those two phases:

Presentation phase
DL; And then you moved to a: home of the Beatles,
Liverpool?
Acceptance phase
JO; I never—I never lived there.
DL; Never lived there?
JO; Yeah.

In the presentation phase, when DL asks JO his question in (312)–(313), DL designs his utterance in a way that he believes JO can comprehend. He takes their common ground, the mutual knowledge, into account. Since DL knows that both DL and JO are native English speakers, and he knows that JO is from Britain, he assumes that JO must know who the Beatles is and what Liverpool is. Thus, DL thinks that he has presented all JO needs for recognizing what he means, and he expects JO to understand it.

JO gives evidence of understanding by uttering *I never, I never lived there* in (314). This utterance shows that JO really understands what DL means in (312)–(313). According to Clark (1996), the reasons can be explained as the following:

1. JO passes up the opportunity to ask for clarification. He thereby implies he believes he understands what DL means.
2. JO initiates an answer as the next contribution. He thereby displays that he has construed DL as having asked a question.
3. JO provides an appropriate answer. He thereby displays his construal of the content of DL's question.

Although JO has presented the evidence through (314), DL does not immediately accept it. He may not be sure that JO never lived in Liverpool; thus, he confirms JO's answer by asking *never lived there* in (315). He gets the evidence of JO's understanding of his utterance when JO answers *yeah* in (316). Clark (1996), using Jefferson (1972)'s term, calls this question and answer a side sequence, the device used by the two participants for resolving problems that may appear in the acceptance phase.

The example above shows how two participants contribute to the conversation—how DL and JO complete their signals and construal in the conversation. When there is a problem hindering their understanding, they can take an action to clear it up. This indicates that the speaker and the addressee build direct interaction in the conversation. Figure 1 below provides the illustration.

Figure 1 shows the design of the relationship between the speaker and the addressee in the conversation in *Late Show with David Letterman*. When DL becomes the speaker, he directs his utterance toward JO, presents some signals to make JO as the addressee understand what he means, and expects JO to display his construal. When JO becomes the speaker, he also does the same thing. It means that the two participants here can express things and respond to each other immediately. In other words, they interact directly in the conversation.

Meanwhile, studio audience (SA) and television audience (TA) have different roles and designs. The role of SA in the interview between DL and JO in *Late Show with David Letterman* is as a bystander. Bystanders can access the speakers' utterances (Clark, 1994), and they "are openly present but not part of the conversation" (Clark, 1996). Here SA can access what DL and JO utter since they are in the same studio as DL and JO. They are openly present—their presence is fully recognized by the speakers, but they are considered non-participant or over-hearer because they are not given an opportunity to speak in the conversation. Moreover, since strangers have a higher possibility to be bystanders (Clark & Schaefer, 1992), SA can be categorized as bystanders.

We also view TA as bystanders because of two reasons. First, they have access to what the speakers say through television. Second, although they are not present in the same place as the speakers, the speakers are fully aware of their presence too. It happens because both DL and JO know well that their conversation is purposefully broadcast; there will be some audience who watch and listen to what they say. We do not categorize TA as eavesdroppers, for the characteristic of eavesdroppers does not match TA. Clark (1996) mentions that "eavesdroppers are those who listen in without the speaker's awareness". Here the speakers are completely aware that TA watch and listen to the conversation.

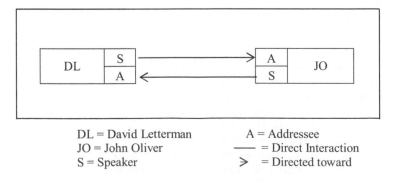

DL = David Letterman   A = Addressee
JO = John Oliver   —— = Direct Interaction
S = Speaker   > = Directed toward

Figure 1.   The design of the relationship between speaker and addressee.

DL and JO's awareness of the existence of SA and TA can be viewed from the example below:

TS1

[...]
(55) DL; I was told [that:]
(56) JO;   [Ya.]
(57) DL; recently I believe you're performing for troops at the USO function...
(58) JO; Ya.
(59) DL; that you: took a taser...
(60) JO; Ya.
(61) DL; to yourself.
(62) JO; I did.
[...]

TS3

(80) DL; Is this recent?
(81) JO; This is—this is like a year ago.
(82) I was in Afghanistan,
(83) and the—so I was doing gigs for the troops,
(84) and all you wanna do...
(85) all you wanna do is make'em laugh.
(86) DL; Mhm.
(87) JO; And whatever..whatever it takes you wanna make'em laugh.
(88) So I finished one gig,
(89) and I walked off stage and this guy said,
(90) <VOX > I really enjoyed that,
(91) do you wanna tase yourself?</VOX>
(92) SA; @@@@@
[...]

When DL produces his utterances in (55), (57), (59), and (61), he directs them not only toward JO as the addressee, but also toward SA and TA as bystanders at the same time because of two different purposes. First, at the mention of *you took a taser to yourself* in (59) and (61), DL shows his intention of asking JO about the tasing occurrence. He gives brief information on it by saying *you're performing for troops at the USO function* in (57) and *you took a taser to yourself* in (59) and (61) to lead JO to tell his story about the tasing occurrence in detail. Second, DL wants to present the initial background information of what he wants to talk about with JO to SA and TA. He expects SA and TA to be able to follow and understand his conversation with JO. Here, DL chooses an attitude of disclosure toward SA and TA. When the speaker discloses his utterances, he designs his utterances in a way that over-hearers can comprehend (Clark & Schaefer, 1992).

JO gives evidence that he understands what DL means through his two types of contribution. He makes continuing contribution, displaying his understanding by giving background acknowledgement (Clark, 1996), in the first acceptance phase which can be seen in (56), (58), (60), and (62). When DL is talking, JO is giving acknowledgements like *ya* in (56), (58), and (60) and *I did* in (62). With those acknowledgements, JO asserts that he has understood DL's utterances in (55), (57), (59), and (61). Next, he provides concluding contribution, displaying his understanding by making his contribution "at the same level" as the interlocutor's contribution (Clark, 1996). The acceptance phase of DL's presentation is in (55), (57), (59), and (61) includes utterances (82)–(85) and (87)–(91). What JO utters in (82)–(85) and (87)–(91) is a story about the tasing occurrence since it contains the background of place (I was in Afghanistan in (82)) and situation (I was doing gigs for the troops in (83)) and also the sequence of events (I finished one gig in (88), I walked off stage in (89)). JO's utterances show that he has understood that DL wants him to tell his story about the tasing occurrence.

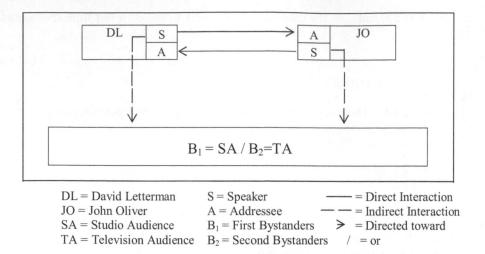

DL = David Letterman	S = Speaker	—— = Direct Interaction
JO = John Oliver	A = Addressee	— — = Indirect Interaction
SA = Studio Audience	B_1 = First Bystanders	➤ = Directed toward
TA = Television Audience	B_2 = Second Bystanders	/ = or

Figure 2. The design of the relationship between speaker, addressee, and bystanders.

JO's utterances can also show that he is aware of the presence of SA and TA, he directs his utterances toward SA and TA, and he anticipates their response. He purposefully designs his utterances to make SA and TA laugh. When he mentions *all you wanna do is make'em laugh* in (85) and *whatever it takes you wanna make'em laugh* in (87), he actually indicates that his story will be funny, for he mentions the word *laugh* two times. Then, when JO utters (89)–(91), he expects SA and TA to grasp the context in (89)–(91) based on their understanding of the previous utterances and their background knowledge and understand the soldier's request that JO tases himself is for making JO feels the pain he felt and making him entertained due to JO's action. SA's laughter in (92) becomes the evidence that they understand what JO means, for SA present the response that JO anticipates.

JO can get the evidence that SA comprehend his utterances since he can see and hear SA's response immediately. However, he cannot see and hear TA's reaction to (89)–(91). He cannot check whether TA gives the response he anticipates, for TA is not present in the same place as he is. This makes SA and TA different, although they are all bystanders. Since SA has the capacity that TA does not have as bystanders, we can call SA 'first bystanders' and TA 'second bystanders'. We coin these two terms as they are not available in Clark's theory on participation roles.

Although SA and TA have different capacity, they have the same type of interaction with the speakers. The speakers interact indirectly with SA and TA. Their relationship can be illustrated in Figure 2 below:

While speakers directly interact with their addressees, they have indirect interaction with first bystanders and second bystanders. It means that when speakers produce their utterances and direct them toward their addressees, they also direct their utterances toward both first bystanders and second bystanders. Through their utterances, they have certain goals that they want to reach, such as giving adequate information to allow bystanders to follow the conversation and making them produce some responses that they expect, like laughter. Speakers can get immediate response only from the first bystanders; however, they can gauge whether they succeed in making the second bystanders give the response they want by checking the response of the first bystanders.

4 CONCLUSION

This research illustrates how each participant in a conversation in a talk show is positioned. We view audience, both studio audience and television audience, as bystanders with different capacity in contributing to the conversation. Television audience should be seen as bystanders since a talk show is designed with the purpose of making people watch so that speakers are fully aware of the presence of the audience, although they cannot directly see them. We feel that it is necessary to conduct follow-up research on the roles of participants in a talk show. Carrying out a study of other talk shows with different formats will provide varied results of the roles of participants in conversations.

REFERENCES

Bio. http://www.biography.com/people/david-letterman-9380239. [Accessed on 28th February 2016].

Biog. http://www.iamjohnoliver.com/*Late Show.* [Accessed on 28th February 2016]. http://www.cbs.com/shows/late_show/. [Accessed on 28th February 2016].

Clark, H.H. & Schaefer, E.F. (1992) Dealing with overhearers. In H.H. Carlson (Ed.), *Arenas of Language Use,* 248–274. Chicago, The University of Chicago Press.

Clark, H.H. (1994) Discourse in production. In M.A. Gernsbacher (Ed.), *Handbook of Psycholinguistics,* 985–1021. California, Academic Press.

Clark, H.H. (1996) *Using language.* Cambridge, Cambridge University Press.

Du Bois, J.W. (2006) *Transcription in action: Representing discourse.* [Online]. Available from http://www.linguistics.ucsb.edu/projects/transcription/representing. [Accessed on 19th January 2016].

Hutchby, I. (2005) *Media talk: Conversation analysis and the study of broadcasting.* Berkshire, Open University Press.

Luckerson, V. (2015) *How the 'John Oliver Effect' is having a real-life impact.* Available from http://time.com/3674807/john-oliver-net-neutrality-civil-forfeiture-miss-america/. [Accessed on 20th January 2016]

Tolson, A. (2001) Talking about talk: The academic debates. In: A. Tolson (Ed.), *Television Talk Shows: Discourse, Performance, Spectacle,* 7–30. New Jersey, Lawrence Erlbaum Associates.

Wardhani, M.K. (2016) *Pemicu Tawa dalam Gelar Wicara Late Show with David Letterman: Sebuah Analisis Percakapan Humor.* [Thesis]. Depok, FIB, Universitas Indonesia.

Representing nation/homeland, gender, and religion

Cultural Dynamics in a Globalized World – Budianta et al. (Eds)
© 2018 Taylor & Francis Group, London, ISBN 978-1-138-62664-5

Foreword by Melani Budianta

Nation-states towards the second decade of the 21st century face a world that is growingly globalized – with the capitalist system entrenched – and also more interconnected, with the internet, social media, waves of migration and the random spread of terrorism. Scholars have discussed the contradictions of this global complexity which can engender an inward looking, tightening of national security and surveillance of borders (Sabanadze, 2010; Kuortti & Dwivedi, eds., 2012).

This section addresses this issue of nation in flux, examining citizen's sense of belonging in a permanent transit space, and discussing how identity politics, religion and gender are implicated in this process. The papers, coming from the field of social and humanities, are diverse not only in their methods and approaches, but also in the kinds of data they are looking at. Discourse analysis, speech act, critical theories of hegemony and ideology, corpus linguistics, philology, ethnography, historical analysis, are used to analyze manuscripts, epics, rituals films, novels, magazine, speeches, blog writings, social media and also diaspora and online communities. Different in their methodologies and disciplines, these papers share a similar perspective that the texts, expressions, policies, teachings and praxis are all historically and socially constructed. Another common thread that binds these papers together is the focus on representation of national identities or gendered and religious subjectivities in a transnational context.

Nation in Transnational Connection

The first 13 papers in this section examine the representation of identities in relation to a nation/homeland in transnational, colonial, pre-national or transnational contexts. Khusna's paper shows how the colonial heritage, that serves as a central focus of the National Museum of Indonesia in Jakarta, is at present put in negotiation by the two flanking wings with a more nationalistic orientation. This paper underlines the presence of the looming colonial past in the official representation of the cultural identity of a Nation-state. Sunarti's paper on "Javanese Malay identities in Malay Peninsula takes us back to the pre-national era of the Malay peninsula in the 18th and 19th century to see the porous cultural borders between the Malay and the Javanese world. Her paper illustrates how the Javanese-Malay diaspora culture contributes to the construction of the heritage of the Malay Peninsula – that is now the territory of a nation-state called Malaysia. Sunarti's paper serves as a contrasting historical backdrop to two other papers. Limanta and Kurnia's, and Kurnia's, which examines the filmic representation and condition of the Malaysia-Indonesia borderlands in the 21st century. Both papers address the social gap between the developed Malaysian side of the border and the less governed side of the Indonesian side. While the film discussed by the first paper show the multiple identities and loyalties of the people inhabiting the Indonesian border zone, the second paper discusses management problems of a community radio in facilitating the need of the people isolated in the periphery of the Indonesian nation state.

The way people in the border zone negotiate their national loyalty is comparable to the way diaspora communities juggle their attachment to their old/new or in between homeland. Widhiasti and Budianta's discussion of the works of second and third Turkish German film makers, and Setyaningrum and Hapsarani's paper on the first-generation Indian-American film maker adaptation of the second-generation Indian American novelist discuss the way cultural homeland is constructed. While these papers discuss the perspective of the immigrants, Baga and Budianta's paper on the contested representation of the Dutch American in the history of New York – like Sunarti's on the Javanese in Malaysia – reminds us of the forgotten, marginalized or misrepresented Other in the mainstream historical writing. The interaction between the Others within a nation state, or the cross-cultural mingling and encounters of diverse ethnic groups – not free from power relations – is the subject of Triadnyani's paper.

The papers above clearly show that the nation cannot be isolated from its transnational network that is colored by power relations shown in capital competition, political influence and cultural politics. Three papers highlight the connection of the transnational influence on identity: Ihwany and Budiman's paper on funding politics of European film festivals, Hadi and Budiman's analysis on China's cultural diplomacy through the China Radio International, and Qian and M. Budiman's discussion on the way Chinese Indonesian position themselves through Mandarin education. The 21st century is also marked by the heyday of the East Asian transnational popular culture – a realm where popular culture industries and youth culture reign – almost independently within the nation-state. Two papers, Tambunan's and Arisun, and M. Budiman's, illustrate how Indonesians insert their cultural perspectives in appropriating East Asian popular culture.

Gendered Dimension

What occurs in the transnational flow of popular culture and social media is not merely locally negotiated appropriation of the global, but also the reconstruction of gendered subjectivities. Three papers in this section, Wuri and Tambunan's positive body image activism in Instagram, Azizi and Tambunan's female football fans' social media use, and Angajaya and Tambunan's analysis of video games, show how consumers creatively and critically deconstruct dominant gender ideology – albeit not free from ambivalence.

Some papers in this section problematize the way the nation state, often acts as the embodiment of Patriarchy through militaristic oppression and violence. From the department of criminology, Sari uses Nussbaum's standard of central capabilities to measure the data of violence against women in Indonesia, while Agustina and Suprihatin reflect on Afghan's women's repression through a novel. Sukham and Hapsarani's work on Catherine Liem's novel argues that memory of oppression offers a double-bind function of both liberating and constraining the female subjects. Albeit women's susceptibility to be victimized by Patriarchy, Akun and Budiman point out that women could also participate in the corrupt system around them. Further problematizing the dichotomy is Hapsarani and Budianta's paper on Hillary Clinton's memoir, with its representation strategies to negotiate with the dominant gender ideology of the U.S. politics.

The papers on gender also provides a diachronic, historical dimension to show that gender construction is not static, but develops through time, and in terms of space, contextually situated. Five papers particularly deal with the development and local construction of gender through magazines, manuscripts and ritual in local languages (Sundanese magazine, Javanese manuscript, Butonese ritual). The methodology of corpus linguistics serves as a handy tool to provide statistical and collocation mapping of various words denoting women – with different connotations (Yuliawati and Hidayat, Alkautsar and Suhardijanto, Bagasworo and Suhardijanto). Philological method is employed in showing the different gendered versions of one Javanese manuscript (Sudarmadji and Prasetiyo), while ethnographic account gives a vivid rendering of the gendered dimension of a Butonese ritual (Ibrahim and Budiman).

These papers, compared to the previous discussion of patriarchy, offers a hopeful view, that linguistic, philological and ethnographical evidences show a diachronic progress towards a more emancipative concept of women. Papers by Agus and Budianta and Hodijah and Suprihatin show how women – within the religious space – and in transnational journeys, can creatively and critically engage with their existing cultural resources and the cross-cultural knowledge gained from their travels to empower themselves.

The final part of this section deals with the issue of religion as a transnational force that has played its role even before the nation-state came into being, and that still exerts significant influence upon the lives of the people in the 21st century. Ten papers in this section pay attention specifically to the way diverse Islamic ideologies are socialized through popular religious self-help book (Altiria and Muta'ali), film adaptation of a novel to film (Jinanto and Tjahjani), German on line mass media (Liyanti), and education (Wibowo and Naupal). Using speech act theory and discourse analysis, some papers read speeches to show main ideas in asserting peace and justice (Lu'lu and Wastono), or deconstruct the Wahabi moral teaching to critique its tendency towards fundamentalism and radicalism (Mardliyati and Naupal, Naupal).

Philological reading of an Islamic classical text from 18th century copied and kept in Ambon shows a local interception in "de-Shiazation" and influence of Ambonese language. At the same time, the text functions as a bond for heterogenous Muslim population in the area (Karmadibrata). Liyanti examines articles written by three German Muslims. Using Van Dijk's ideological square, she finds shifting positions of the writer in the Us-Them relations between German and Muslim migrants. Syarif and Machmudi's paper examine the biography of an influential Sufi kiyai (religious leader) in Banten, who was detained by the New Order regime in 1977 for standing up against the coercion for Islamic leaders to vote for Golkar, then the state sanctioned party.

The above papers construct, first of all, not a monolithic ideology of Islamism, but a plural Islamic expressions and ideologies. Secondly, Islam as a transnational religion complicates the internal dynamics of the nation state, between different Islamic *tarekats* or *mazhabs*, between the liberal and the more conservative Moslem, between the host society and the Islamic migrants in Europe, between the government and religious leaders and groups, between the locally grown Moslem groups and the transnational fundamentalism. Third, these positions cannot be reduced in a dichotomy. Instead, there are shifting positions, border or mediating bridges between different orientations, or third spaces that mix or go beyond the differences. With all their different topics and themes, these papers agree on one thing: the importance of socializing a more tolerant religious attitude that embraces difference and dialogue.

Compiled together, the papers in this section engage with the dynamics of nation states in transnational era, when capital, information, popular culture, social media messages, migrating people, religious ideologies, activism, flow seamlessly across borders. During the time where nation states reacted to global terrorism by looking inward and building walls against the unwanted Others, these papers pose their humanistic critique. Using various methodologies and approaches, they deconstruct dichotomies, unfold complexities, unpack ideologies, show the way texts, citizens, praxis discursively negotiate and continuously transform identities and positions. Coming from the disciplines of social and humanities, the papers all point at the central role of the media – in various forms and genres – to construct and shape realities. At the same time, the papers also show the contingency of such representation to multidimensional forces of capital and politics. As conference papers, the questions that the papers pose are still open to further inquiry. This proceeding, therefore, serves as an invitation to future scholarship.

Melani Budianta
Universitas Indonesia

References

Kuorrti, J. & Dwivedi, O.P. (2012) *Changing World Changing Nations: The Concept of Nation in Transnational Era*. Jaipur, Rawat Publications.
Sabanadze, N (2010) *Globalization and Nationalism, The Cases of Georgia and Basque Countries*. Budapest, Central European University Press.

Cultural Dynamics in a Globalized World – Budianta et al. (Eds)
© *2018 Taylor & Francis Group, London, ISBN 978-1-138-62664-5*

Beyond the collections: Identity construction at the National Museum of Indonesia

M.K. Rizqika
Faculty of Humanities, Universitas Indonesia, Depok, Indonesia

ABSTRACT: The National Museum of Indonesia is one of the products from the long history of the Dutch colonial ruling in the Indonesian archipelago. This is the first and biggest museum established in 1778 by the Dutch East Indies government in Indonesia. Based on the collections, the museum was initially set out to educate the local people in the context of colonial history. The colonial government's practice of acquiring objects reflects the colonization of the local people. Following the independence of Indonesia in 1945, the museum changed its discourse. The postcolonial reinterpretation of imperial history affected and changed curatorial practices in the museum. Through museums, Indonesia can construct a new identity that is free from the shadow of the colonial government. Today, the museum is changing and focusing on how to represent the cultural diversity of Indonesia, which is an enriching and the most important element in showing the new identity of Indonesia. Museums can connect the past and the present for a better future. This article aims to address the underlying and interrelated aspects that bring all of these elements to reveal the complexity of today's identity.

1 INTRODUCTION

Postcolonial studies challenge the outdated view about colonies (Eastern world), which is full of stigma, biased, and narrow. In Indonesia, a nation which was formerly a Dutch colony, traces of colonialism can still be found. One of them is the National Museum of Indonesia. This article is aimed at comparing two permanent exhibition rooms in the National Museum of Indonesia, namely the Room of Historical Relics in Building A (old building) and the Room of Gold Ethnography Repertoire in Building B (new building), from the viewpoint of the postcolonial theory. The exhibition's arrangements of both rooms are compared to show how the Indonesian Government strives to liberate itself from the identity as a former colony and how it has started to build its own identity. A qualitative approach is used in writing this article. In order to obtain data for this article, the author conducted direct observation in the Room of Historical Relics in Building A and the Room of Gold Ethnography Repertoire in Building B of the National Museum. The author also conducted bibliographical studies and unstructured interviews with several museum staff. The data were enriched by the author's own experience as a curator in the National Museum of Indonesia.

1.1 *Museum and construction*

Etymologically, the word "colony" originates from the Latin word *colonia*, meaning a land for cultivation or habitat (Sutrisno and Putranto, 2008). Loomba (2003) described that originally the term *colony* referred to the Romans who lived outside their country, but were still registered as citizens of Rome and related to their homeland (Lubis, 2015, p. 127). Afterward, the term developed, and it was associated to the concept of power and conquest, that is, the settlers who wished to conquer and dominate the colonies. This incited problems, such as oppression and control of human and natural resources. Said (1993) indicated that this view motivated the Western people to dominate distant Eastern countries. They considered the natives of the colonies as barbaric and primitive; thus, they had to be civilized according to the Western culture. This view was legitimized in several ways by Western academics, for example, by producing numerous writings about Eastern world and by establishing museums to fortify the Western power over the colony.

In this regard, Bouquet (2012) described that several ethnographic museums started to be founded in the 19th century in Europe. One of them was the Royal Danish Ethnographic Museum, established in 1825. Collecting unique objects from distant colonized countries was a prestigious matter among European elites. The collections were exhibited and interpreted in the national context of European countries. Most of them were daily appliances from Asia, Africa, Australia, and America. They represented the superior values of the Western culture over the colony's culture.

Afterward, the natives actively fought against the colonists. Said (1993) wrote that physical confrontations broke out in the 19th century in countries such as Indonesia and Algeria, which resulted in their independence. Much effort was also made in defending the national culture and reinforcing the national identity. Such situations were the focus of postcolonial academicians. Lubis (2015) divided the postcolonial concept into two main aspects. The first aspect regards the postcolonial as a period after the colonial era (colonialization). One of the important accounts during this period was the introduction of cultural elements or the system by the colonists to be imitated or applied as new elements by the natives. The postcolonial is a study about the effects of colonialism after the colonial period. In addition, new forms of colonialism practices were studied, for example, the colonization of minority groups by dominant groups. Academicians of this theory include Frantz Fanon and Gayatri C. Spivak.

The second aspect of postcolonial theory is considered as the continuation of the discussion on the previous aspect. Postcolonialism is a form of confrontation against the domination of colonists' thought by the Western world over colony culture. This study criticizes the inequality of Western-dominated knowledge, social, and cultural matters—Western hegemony—resulting in a bias in viewing the Eastern or colony culture, particularly in early modern centuries. Edward Said is one of the figures who avidly expressed his view about the conspiracy of colonists' or orientalists' power and knowledge over the Eastern world.

With regard to identity in the present time, Maunati (2004) highlighted that the concept of identity and identity itself are viewed as the result of dynamic interaction between context (history) and construction. Identity might change in different time and space. Identity is continuously formed and transformed, which is related to how it is represented or highlighted in the surrounding cultural system. With regard to nationality, Woolf (1996) wrote that national identity is an abstract concept, which combines individual sense with a socio-political unit (Graham, Ashworth, Tunbridge, 2005).

It is very interesting to relate the arrangement of the exhibition in the National Museum of Indonesia to Said's thoughts about orientalism, as indicated by King (2001), which are related to three phenomena (Lubis, 2015, p. 139). First, an orientalist taught, wrote, and examined the Eastern world. An orientalist can be an anthropologist, archaeologist, historian, philologist, or anyone from other disciplines. In its early period, members of the *Bataviaasch Genootschap van Kunsten en Wetenschappen* (the pioneer of the National Museum of Indonesia) were the elites from diverse professions, such as bureaucrats, anthropologists, historians, and missionaries, who were interested in studying cultures in various parts of the Dutch East Indies.

Second, orientalism refers to the Western and Eastern thoughts, which are based on different ontology and epistemology. Therefore, articles on research results or expedition reports written by members of the *Bataviaasch Genootschap van Kunsten en Wetenschappen* (hereinafter referred to as BG) were said to be bipolar in nature and were divided into two perspectives, namely Eastern and Western. Third, orientalism can be seen as a legal institution, which was established to conquer the East; to have interest, power, and authority to make representations about the East; to justify the view about the East; and to describe, teach, publicize, position, and master it. BG was the institution to facilitate the various interests of the orientalists in the past.

Pearce (1994) indicated that the collections in a museum are the results of the process of collecting. There is always a relationship between collections and the ideas that museum administrators want to convey, as Impey and Macgregor (1985) pointed out "... *the nature of the collections themselves and the reasons and the more obscure psychological or social reasons*" (Pearce, 1994, p. 194). The collections are bound with various elements in the history of their former owners and in how they are acquired by the museum. Pearce (1994) wrote that a curator must comprehend aspects of history and the nature of the collections, as well as their background information, in order to have better assumptions about the value and knowledge inherent in them. The collections have their own initial contexts, and the museum can provide new contexts to them. Mullen (1994) explained that museums play a critical role in constructing cultural knowledge. He wrote:

> "*That role needs not be confined to preserving and disseminating established, legitimated cultural knowledge; it can also be one of the facilitation for the social construction of a broad range of cultural knowledge, and so contribute to the cultural empowerment of a broad range of people.*"

1.2   National Museum of Indonesia from time to time

Since its establishment on 24 April 1778, *Bataviaasch Genootschap van Kunsten en Wetenschappen* or the Batavia Society for Arts and Sciences started to gather collections, which were mostly donations from its members. The collections are full of historical values. They were also gathered from scientific and military expeditions, religious missions, and purchases (Soebadio, 1985; Hardiati, 2005). The members of the Society were categorized into four groups, namely administrative officers (*pamong praja*), church clergymen (Bible missionaries), military officers, and academicians (Jonge, 2005).

Initially, the Society occupied a building owned by J.C.M. Radermacher in Kalibesar Street, Oud Batavia. Later, during the British administration in Java (1811–1868), Sir Thomas Stamford Raffles provided a building (*Societet de Harmonie*) for this institution, located on Majapahit 3 Street (currently the State Secretariat/*Sekretariat Negara* building). In 1862, the construction of the museum building in *Koningsplain West* Street, currently Medan Merdeka Barat Street, began. The museum was opened for public in 1968 (Rufaedah et al., 2006).

A change in the museum administration also brought a change in the name of the institution. In 1933, the Society was awarded the title "*koninklijk*" (royal); thus, its name was changed to *Koninklijk Bataviaasch Genootschap van Kunsten en Wetenschappen*. Then, on 26 January 1950, the name of the Society was changed to *Lembaga Kebudayaan Indonesia* (the Cultural Institute of Indonesia). On 17 September 1962, *Lembaga Kebudayaan Indonesia* consigned the museum to the Indonesian Government and thus its name was changed to *Museum Pusat* (the Central Museum). The last change of the name to *Museum Nasional* (the National Museum) occurred on 28 May 1979 (Rufaedah et al., 2006).

The National Museum consists of two main exhibition buildings: Building A and Building B. The construction of Building A began in 1862. Documentation photos of the building show that its exhibition pattern did not undergo crucial changes, for example, in the ethnography room and the rotunda. The collections were arranged by applying quite an unclear approach, which is based on regional classification (ethnographic), chronological classification (prehistoric, archeologic, historic), material classification (stone, bronze, gold, terracotta, ceramics), and scientific discipline (numismatics, geography). The collections were placed systematically similar to what has been applied since the colonial era. Substantially, they did not experience much change although the textile and ceramics rooms were renovated several times. Meanwhile, in Building B, the collections are exhibited thematically based on the concept of universal elements of culture. The purpose of this type of arrangement is to portray the history and diversity of the nation's culture (Museum Nasional Indonesia, 2013). This arrangement will also be applied in Building C, which, at the time of the writing, is under construction.

1.3 *Western domination in the room of historical relics*

The Room of Historical Relics is located in Building A, which was constructed in 1930. The architecture of the building is strongly dominated by the European Neoclassicism style with a mixed touch of Dutch and Greek. In the National Museum building, the style was materialized in the cylindrical pillars (in the front part of the building), cascading roofs, large windows, and white-painted walls to portray grandiosity (Fahnoor & Furihito, 2007). The position of the Room of Historical Relics is very strategic, and its interior design appears to be the most exceptional one compared to other rooms. An atmosphere of colonialism is evident in the front part of the room. Its door together with the frame has an interesting history regarding colonialism because parts of this interior were formerly parts of *Toko Merah* (Red Store). Entering the room, elements of Western domination are more noticeable. The room's floors have a chessboard-like pattern, but the colors are black and dark red.

Toko Merah was present during the heyday of old Batavia on the banks of the Muara Ciliwung River, which was precisely located at Kali Besar Street No. 11, West Jakarta. The building was once used as the residence of the VOC Governor General, Baron van Imhoff (1705–1751). It also underwent several transfers of ownerships, ranging from Phillippine Theodore Mossel (the son of Governor General Mossel), Governor General Petrus Albertus van der Parra, and Renier de Klerk to Nicolaas Hartingh and many others. During 1813–1851, there had been some changes of the building ownerships before it was later owned by Oey Liauw Kong who turned it into a *taka* (*toko*/store); thus, there came the popular term "*Taka Merah*" (*Toko Merah*, 2015).

In 1898, a retired member of KNIL who was also a member of BG called H. D. H. Bosboon initiated the relocation of several interior components of Toko Merah to the museum. It was an effort to protect and preserve the building with the characteristic of the era. It also served as an example of the grandeur residential style of the elites. The components included doors, windowsills, ceilings, window panes, the balcony, and the stairway (Hardiarti et al., 2014). In the museum, the interior components were placed in a *compagnie kammer* (*Toko Merah*, 2015). Currently, the room functions as the Room of Historical Relics, at the Auditorium of Building A, which was formerly the Numismatic Room.

During the BG era, the room was used as the director's room. Thus, it is not surprising that its interior architecture was made meticulously, by paying special attention to the artistic aspect and displaying the domination of the Dutch East Indies Government in the past. It can also be perceived that they had the legitimacy to present symbols of Dutch glory in its colonies. The interior architecture and furniture in this room displayed the domination of the colonist government. This is the grandiosity of the Western-styled building displayed without any regard to elements of Indonesia and its cultures.

All collections exhibited in the room were material cultures made by the colonists. This is allegedly to confirm that the specially designed room was only for goods produced by the Western culture, which were considered special. The collections were related to the arrival of other Western countries, such as Portugal and England. For example, there is a half-body bust of Raffles on the right side of the entrance. England dominated Java in 1811–1816, when Sir Stamford Raffles took over the leadership of the society. Other collections are the two *padrao* (inscription related to the arrival of the Portuguese), exhibited on the left side of the entrance, a cannon, a display cabinet, and a banquet table.

The arrangement of the exhibition in this room is collection-oriented, without displaying the context of the collection. For years, the exhibition's arrangement in Building A, including the one in the Room of Historical Relics, has not experienced any change (Hardiati et al., 2014). One of the examples is a text on collection label of Padrao Lonthoir, which reads as follows:

> "This *padrao* was the legacy of the Portuguese as seen from the writings which can still be vaguely read, although they have faded. In 1705, someone named Jan Van de Broke inaugurated the establishment. Jan Van de Broke was appointed by the Dutch government to be in charge of the garden."

The above narration highlights the existence of the Portuguese and neglects the present context of Indonesia. The exhibition's arrangement of the room shows that the present museum administrators cannot fully liberate themselves from the domination of the Dutch who administered the museum in the earlier period.

1.4 *From Nusantara to Indonesia*

The second permanent exhibition room analyzed in this study is the Room of Gold Ethnography Repertoire located on the fourth floor of Building B, which was constructed in 2006. Physically, the appearance of the building is different from that of Building A. The exhibition room can only be accessed by visitors through two available elevators. This is different from the rooms on the second and third floors, which can be accessed using both the elevator and the escalator. This is due to stricter security procedures applied to the Room of Gold Ethnography Repertoire. The interior design of this room looks more modern. Large vitrines are attached on every wall of the room. Some unattached vitrines are placed in several corners. A roomy and spacious impression is present when entering the room. The various shining exquisite gold objects can instantly amaze the visitors.

The gold collections exhibited in the room are classified into repertoire groups because they are highly valued, particularly in terms of material, shape, and function. The collections of gold ethnography repertoire originated from the 17th century to the 20th century. Most of them are objects originating from monarchs or sultanates, and some others are ethnographic objects related to the identity of an ethnic group. Royalty regalia objects are symbols of grandeur and social status of the king or ruler, while also possessing outstanding artistic values. Some distinguished collections in the room include the *Mahkota Banten, Keris Si Gajah Dompak* from North Sumatra, various pieces of jewelry from Cakranegara Castle Lombok, spears from Kalimantan, treaty scriptures from Sulawesi, and Puputan *keris* (traditional dagger with a wavy pattern) from Bali. In addition to those are the *talam* (inscribed slabs) from Madura and jewelry from Aceh. All these objects possess rich historical values, particularly regarding the past relation of Indonesia and the Dutch colonies.

These objects are presented in the National Museum as a result of the distressing incidents in the past. Some of the objects were loots from the war between the Dutch and various local monarchs or rulers in Nusantara. In the annual minute books (Dutch: *jaarboek*) and the collection inventories, it was mentioned that the Dutch conducted military exhibition to expand their colonies and to overthrow the local rulers. One of the examples is a collection from Aceh, a *sunting* (crown worn on women's head embellished by gold) (E.151/11750), which was obtained from a raid carried out by the Dutch colonists on the house of Teuku Umar's wife before 1905. This object was then accorded and became the collection of BG.

In the context of the exhibition's arrangement of this room, many lessons can be learned from the past. What needs to be encouraged and passed on to the visitors is the spirit of being productive and proud of the country. For example, one of the values worth to be maintained is the one inherent in the *sunting*. The *sunting* functions as a hair adornment for women placed above the ears. Its shape is inspired from the *cempaka* flower (*Magnolia champaca/* Aceh: *bungong jeumpa*). For the Acehnese, the meaning of *jeumpa* goes beyond simply a flower. Philosophically, this flower inspires Aceh people to have good morals and personalities; thus, it is worth being remembered. This positive value can be universally understood within the framework of the current Indonesia.

The enthusiasm to bring out and assert national identity is also seen in the text on group labels and collection labels. In the group label of *"Ekspedisi Militer Lombok"* (Lombok Military Expedition) vitrine, for example, it is assertively narrated that the Dutch was the colonist and that the Indonesian government currently has equal power with them. It is precisely stated as follows:

"During the aggression, the Dutch took away royalty heritage which was gold jewelleries and other prized possessions. The properties looted by the Dutch army were brought to Batavia and divided into two groups. The first group was stored in Bataviaasch Genotschap

Museum, and the second one was brought to Rijkmuseum voor Volkenkunde, Leiden. In 1977, the Indonesian government and the Dutch government entered into an agreement to repatriate some of the Cakranegara Castle heritage which was taken away by the Dutch to be returned to the Indonesian government. In 1978, the agreement was successfully fulfilled by the return of the Lombok repertoire which consists of 243 gold and silver collections to be surrendered to the National Museum in Jakarta."

In terms of the exhibition's arrangement, the regalia objects in this room are displayed by emphasizing the majestic side of the objects. The light beams on the objects accentuate the glittering gold and the various precious stones as embellishments from the past. These show the artists' expertise in creating numerous masterpieces during the Nusantara period. In addition to their highly valuable material, i.e. gold, the objects look even more exceptional due to their marvelous designs and the highly complicated craftsmanship. The arrangement of the exhibition in this room is expected to promote the pride of Indonesians visiting the museum and understanding that Indonesia has always been culturally rich since the past.

2   CONCLUSION

From a historical point of view, the National Museum is indeed a legacy from the Dutch colonists. It was established with the aim of displaying the Dutch's power, glory, and greatness over their colonies. This aim is still evident in the Room of Historical Relics in Building A. The administrators still maintain the arrangements and collections of the exhibition that date back to several years ago. However, as the museum's development took place with the construction of Building B beginning in 2006, the administrators have more opportunities to make new arrangements of the exhibition that are tailored to the current circumstances to express the pride of being an Indonesian. The exhibition's arrangement endued with a new atmosphere and spirit can now be seen in the Room of Gold Ethnography Repertoire in Building B.

To strengthen the national identity as Indonesians, a new exhibition room (Building C) is under construction. This exhibition building, which is expected to be completed by 2018, will present a new storyline concerning Indonesia's culturally rich national identity, including various themes such as *Pusaka Indonesia* (Indonesian Heritage), *Menjadi Indonesia* (Becoming Indonesia), and *Lestari Indonesia* (Sustainable Indonesia) (Museum Nasional Indonesia, 2016). These themes are the results of several focus group discussions conducted in 2013 and 2014 between the museum staff and several interviewees from various backgrounds such as academicians, anthropologists, archaeologists, historians, and communities.

The cultural diversity of Indonesia is a source of national identity that should be strengthened in the current era of globalization. This can be done through the narration of museum collections used as the medium of communication. The narration of museum collections highlights the values of the people's dignity and equality in a community. Therefore, it is important that the narration of the negative past inherent in the collection be revised to reflect a sense of pride in this culturally rich country. This conforms to the postcolonial ideas that ex-colony countries can free themselves from the shadow of their past as colonies and that they have the right to be equal with other countries. It is expected that this exhibition in the museum will act as a channel for visitors to understand the values of pride of Indonesia, make self-interpretation of museum collections, and be inspired.

REFERENCES

Bouquet, M. (2012) *Museum: a visual anthropology.* London, Berg.
Fahnoor, F. and Furihito, X. (2007) *Penerapan arsitektur neo-klasik Yunani pada fasade bangunan Museum Nasional Jalan Merdeka Barat No. 12 - Jakarta.* (The Application of Greek neo-classical architecture on the building facade of Museum Nasional Jalan Merdeka Barat No. 12 – Jakarta).

Universitas Gunadarma. [Online] Available from: http://www.gunadarma.ac.id [Accessed 13th November 2015].

Graham, B., Ashworth, G.J., & Tunbridge, J.E. (2005) The uses and abuses of heritage. In: Corsane Gerard (Ed.). *Heritage, museums and galleries, an introductory reader*. New York, Routledge. 26.

Hardiati, E.S. (2005) Dari Bataviaasch Genootschap sampai Museum Nasional Indonesia. (From Bataviaasch Genootschap to Museum Nasional Indonesia). In: Hardiati, Endang Sri, & Keurs, Pieter (Ed.). *Warisan budaya bersama*. Amsterdam, KIT Publishers. 11–13.

Hardiati, E.S., & Supardi, N. (2014) *Potret Museum Nasional Indonesia, dulu, kini, & akan datang.* (Portrait of Museum Nasional Indonesian, past, present, & future) Jakarta, Museum Nasional.

Jonge, N. (2005) Para kolektor di kepulauan yang jauh. (Collectors from far islands). In: Hardiati, Endang Sri, & Keurs, Pieter (Ed.). *Warisan budaya bersama*. Amsterdam, KIT Publishers.

Lubis, A.Y. (2015) *Pemikiran kritis kontemporer, dari teori kritis, culture studies, feminisme, postkolonial hingga multikulturalisme.* (Contemporary critical thinking, from critical theory, culture studies, feminism, postcolonialism, to multiculturalism). Jakarta, Raja Grafindo Persada.

Maunati, Y. (2004) *Identitas Dayak, komodifikasi & politik kebudayaan.* (The Dayak identity, commodification & cultural politics).Yogyakarta, LKiS.

Mullen, C. (1994) The people's show. In: S. Pearce (Ed.). *Interpreting objects and collections.* London, Routledge.

Museum Nasional Indonesia. (2013) *Pengembangan Museum Nasional.* (The development of Museum Nasional). Unpublished.

Museum Nasional Indonesia. (2016) *Daftar awal rencana koleksi alur kisah pameran tetap MNI. (The* preliminary list of MNI's permanent show exhibition collection plots). Unpublished.

Pearce, S. (1994) Collecting reconsidered. In: S. Pearce (Ed.). *Interpreting objects and collections.* London, Routledge.

Said, E.W. (1993) *Culture and imperialism.* London, Vintage.

Soebadio, H. (1985) *Studies and documents on cultural policies, cultural policy in Indonesia.* France, UNESCO.

Sutaarga, A. (1990) *Studi museologia.* (The study of museology). Proyek Pembinaan Permuseuman Jakarta, Direktorat Jenderal Kebudayaan, Jakarta, Departemen Pendidikan dan Kebudayaan.

Sutrisno, M., & Putranto, H. (2008) *Hermeneutika pascakolonial: soal identitas.* (Postcolonial hermeneutics: a matter of identity). Yogyakarta, Kanisius.

Toko Merah. (The red store). (n.d.). [Online] Available from: http://www.jakarta.go.id/web/encyclopedia/detail/3429/Toko-Merah [Accessed 13th November 2015].

Cultural Dynamics in a Globalized World – Budianta et al. (Eds)
© *2018 Taylor & Francis Group, London, ISBN 978-1-138-62664-5*

Formation of Javanese Malay identities in Malay Peninsula between the 19th and 20th centuries

L. Sunarti
Department of History, Faculty of Humanities, Universitas Indonesia, Depok, Indonesia

ABSTRACT: This paper aims to explain the cultural ties between Indonesia and Malaysia by examining the diaspora or migration of Indonesian people across the archipelago, especially the Javanese to the Tanah Melayu peninsula (Malaysia), from the early 19th century to the early 20th century. It focuses on searching the traces of Javanese diaspora in the states of Selangor and Johor (two areas in Malaysia with the largest population of Javanese descendants) and the culture they brought and developed in their new places in that period. This paper also seeks to discuss how they adapted, assimilated and formed their identities in the new environment. This study uses a historical approach involving both qualitative and quantitative data as well as a literature review. Qualitative data were obtained by interviewing cultural actors in both Indonesia and Malaysia. Malaysians who were interviewed were Indonesian descendants, academics and so on. Quantitative data were obtained through questionnaires completed by a group of young people in Indonesia and Malaysia. A literature review was conducted by tracing written sources, especially archives, documents, newspapers and books in both countries. These data were analysed using a historical method which includes several steps, namely heuristics, criticism, interpretation and historiography.

1 INTRODUCTION

1.1 *The arrival of the Javanese in Malaya*

The Javanese are an ethnic group that has high mobility and has spread widely across the globe. According to Lockard (1971, p. 43), the Javanese diaspora had begun before the 19th century, but they migrated on a large scale between 1875 and 1940. The large migration of the Javanese was caused by various factors. In the land of the Malays, the arrival of the Javanese has a long history that makes it an interesting phenomenon. Malaya region, in this case, covers the entire territory of the Federation of Malaya, including Singapore, as depicted in Figure 1.

Although Singapore and North Borneo were also included in the Federation of Malaya, the researchers limit the scope of the discussion to the country of Malaysia. The number of the Javanese who immigrated to the Malay Peninsula reached approximately 190,000 in the 1960s. It shows that the Javanese diaspora in Malaya was larger than that in Suriname and New Caledonia, where there were 43,000 and 3,600 Javanese migrants, respectively (Lockard, 1971, p. 42).

Although it is difficult to know precisely when the first Javanese migrated to the Malay Peninsula, with the establishment of the strait settlements by the British colonial government, the arrival of the Javanese in Malaya was written in the civil records. In Singapore, for example, Kampong Jawa was established in 1836 in the western part of the Rochore River. The growth of Javanese population in the region continued to increase slowly but significantly from 38 in 1825 to 5,885 people in 1881. In Malaya, Penang and Malacca regions, it was recorded that the number of 'Indonesian' residents, mostly from Java, reached 4,683 in 1871 (Bahrin, 1967a, p. 269). Most of the Javanese migrants were workers and traders working in various sectors ranging from the plantation to services. The above situation illustrates that the Javanese had migrated before the introduction of the contract labour system.

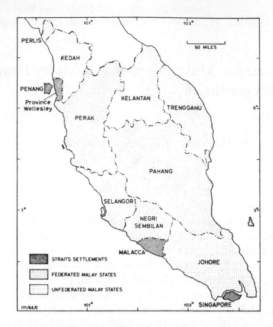

Figure 1. Malaya Federation under British colonial administration (Bahrin, 1967a, p. 268).

Javanese people were familiar with emigration. As previously mentioned, the mobility of the Javanese began to increase rapidly in the late 18th century. In that period, there were several factors that encouraged the Javanese to migrate, one of which was the condition in rural Java where access to various basic amenities was difficult. Therefore, many of the inhabitants of Java were driven to migrate to various areas in search of livelihood that would allow them to be prosperous, either temporarily or permanently. Clifford Geertz (1966) described the situation in Java in his book *Agricultural Involution* that there was a scarcity of land, an exploding population and poverty that hit most areas of Java. This condition encouraged people to migrate to other areas. Hardjosudarmo stated that in 1930 there were 1,200,000 or approximately 2.9% of people living outside Java (Lockard, 1971, p. 44).

The migration of the Javanese was not always done voluntarily. Some people migrated because of the demands of work as labourers in various business sectors. The growth of the plantations in Java and Sumatra, which was driven by the so-called 'liberal policy' in the second half of the 19th century, led the Javanese to search for work as labourers in plantations operated by the Europeans (Lockard, 1971, p. 45). The opening of the plantation in the eastern part of Sumatra in the early 1860s also required a lot of labourers from Java, but they were still less competitive than the plantation workers coming from China.

Most Javanese were under pressure to sign a contract and work in different areas. The workers, who came from Java, were sent to different areas of the European colonies, belonging mostly to the British. The Javanese were then sent to Suriname, New Caledonia and the Federation of Malaya, including Sabah and Sarawak. The distribution of Javanese labourers in Malaya can be seen in Figure 2.

Malaya was the first region to receive the Javanese as workers, as well as the biggest destination for the Javanese migrant workers. Previously, there were thousands of Chinese brought to Malaya to work in tin mining. Then, the expansion of the sugar and coffee industries in the late 19th century accompanied by the rising demand for labourers in the rubber industry caused the Europeans to bring many Javanese to work in Malaya.

In the late 19th century, the Europeans needed Javanese workers. In general, the labourers in Malaya originated from India and China. However, the Europeans feared that they would be too dependent on Chinese labourers. Moreover, the Chinese workers were considered to work optimally only under the supervision of the Chinese people themselves. In addition,

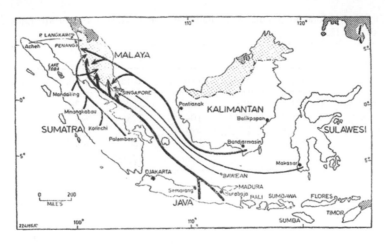

Figure 2. Distribution of Javanese labourers in Malaya (Bahrin, 1967b, p. 235).

the competition for Chinese labourers was quite tough, and so was the competition for the Indian workers. The Europeans preferred Javanese workers because they could easily assimilate with the locals (Lockard, 1971, p. 46). It can also be said that since the living environment in Malaysia is similar to the environment in Java, the Europeans found it easy to open new plantations by employing Javanese workers.

In the migration process of those workers, recruitment procedures were carried out by the Europeans by using agents to collect workers in Java to be sent to various regions. The recruitment agents initially had difficulties in collecting workers from Java because the Dutch East Indies Government at that time did not allow the Javanese to migrate from the Dutch East Indies as enacted in the regulations in 1887 that prohibited the Javanese to work outside the Dutch East Indies (Lockard, 1971, p. 48).

However, after 1900, there was a rising argument from the colonial administration that *outer Indonesia* (Geertz, 1966) (Sumatra, Borneo and others) had a population deficit. Therefore, those regions required labourers from Java. In practice, the workers who came from Java were sent not only to other regions in Indonesia, but also to Suriname, New Caledonia and Malaya.

The recruitment centre of the Javanese workers was located in Central Java. Although many people were sent via Batavia, many companies and recruitment agencies were located in Central Java, especially Semarang (Lockard, 1971, pp. 48–9). There were private recruitment firms operated by the Europeans, and they received high commission for their recruitment services. The recruitment firms had to obtain permission from the government to recruit workers. According to Yusuf Ismael, in the works of Lockard, recruitment firms promised a prosperous life abroad. Sometimes, the agents were forced to cheat and conspire with the village heads. They would do anything to collect a commission worth 80–100 rupiah for each recruit (Lockard, 1971, p. 49).

In the case of workers in Malaysia, the Javanese were generally recruited by the Chinese agents in Singapore. Once registered as workers, they were sent to Singapore and then worked in plantations in Malaysia. After 1902, the planters in Malaya were allowed to bring a large number of the Javanese to work there. The number of Indonesian workers in the early 19th century is given in Table 1.

Before 1902, the Javanese had dominated the total population of Indonesians in Malaya. Based on the data from the civil registry between 1891 and 1901, the Javanese accounted for more than 70% of the total population of Indonesians, and most of them were workers. Then, after 1902, the population of the Javanese was reported to have increased drastically. The population census conducted in 1937shows that there were more than 100,000 people in Malaya Java, especially in the cities mostly settled by Indonesians, such as Johor

227

Table 1.   Indonesian population in the Federation of Malaya, 1891–1901 (Bahrin, 1967a, p. 272).

Community	1891			1901		
	Total	Male	Female	Total	Male	Female
Achinese	621	456	165	1,485	1,301	181
Batak	228	124	104	228	–	–
Boyanese	3,161	2,223	938	3,509	2,307	1,202
Bugis	2,168	–	–	1,133	600	533
Javanese	14,239	–	–	17,578	12,557	5,025

Table 2.   Indonesian immigrants in Malaya 1947: by year of first arrival (Bahrin, 1967a, p. 273).

Year	Number	Percent
1900 and earlier	3,369	3.3
1901–1910	9,931	9.7
1911–1920	26,247	25.6
1921–1930	27,472	26.8
1931–1935	8,515	8.4
1936–1940	13,211	12.9
1941–1947	10,238	9.9
Not stated	3,371	3.3
Total	102,454	99.9

and Selangor. Then, in the following years, the number of people in Malaya Java continued to grow, as outlined in Table 2.

1.2   *Javanese Malay: Adaptation and assimilation*

Like most immigrants from Indonesia, the Javanese in Malaya settled in various coastal areas. Banjarnese and Bugis people were among the many ethnic groups who settled in Malaya, apart from the Javanese. Most of them lived in coastal areas (Ramsay, 1956, p. 124). The distribution map of the Javanese in Malaya until 1947 is shown in Figure 3.

There were three areas with the largest Indonesian population. Among the cities with the highest number of immigrants from Indonesia, Johor was the most important because there were more than 80% of the Indonesian population, which increased every year since 1911, followed by the cities of Selangor and Perak (Bahrin, 1967a, p. 276). Immigrants from Java dominated the number of Indonesian immigrants, and they had a tendency to settle permanently in Malaya. The Javanese in Malaya formed a new identity, which is known as Javanese-Malays.

The process of adaptation and assimilation carried out by the Javanese in Malaya later gave birth to a new group of Malay society called the Javanese-Malays. The Javanese-Malay groups can be defined as the residents of Malaya who have a legal status but still have a close relationship with the cultural roots of origin in Java (Miyazaki, 2000, pp. 76–7). In this case, the Javanese differentiated themselves from other immigrants who came from the other parts of Indonesia. Despite their large number, the Javanese-Malays were included under the category of ethnic Malays, which means that they were not a separate ethnic group among the three major ethnic groups in Malaysia: Malay, Indian and Chinese.

A study conducted by Koji Miyazaki in Batu Pahat, Johor, in 1991 explains that migrants from Java have much in common with the local residents of Malay. Unlike the Chinese and Indians or 'migrants of difference' who had a different cultural background from that of the Malays, the Javanese were considered 'migrants of similarity'. One of the similarities among others was the religious background of Islam, which is the religion adhered to by both ethnic groups. In

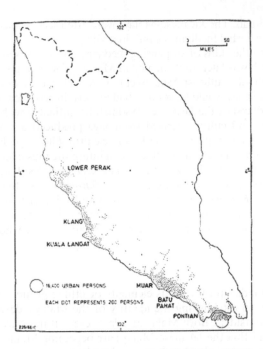

Figure 3.   Distribution of the Javanese in Malaya until 1947 (Bahrin, 1967b, p. 245).

addition, the Javanese and Malays also had language and cultural proximity. It was easier for the Javanese to be accepted in Malaysia compared to immigrants from China and India.

One area where the Javanese-Malays settled in large numbers is Johor. Most Javanese people, who had arrived in the early 20th century, came from Ponorogo, East Java. Johor is the region where most Indonesian descendants inhabit, especially in Batu Pahat, which was occupied by the majority of the Javanese and Bugis. In this region, the Bugis and the Javanese lived in coastal areas and worked in palm plantations. However, there were also some Javanese who lived in hinterlands working in rubber plantations. This was parallel with a moderate rise in the rubber industry from 1911 to 1930, so there were many Javanese workers who were brought by the British colonial government to become workers in rubber plantations.

The term 'Javanese' refers to the Javanese living in Malaya who are considered different from the Javanese people of Indonesia. The Javanese immigrants who came to Malaysia before the Indonesian independence in 1945 refer to themselves as Javanese, not Indonesians.

Javanese-Malays can be identified by their language for they use the Javanese language as their everyday language. In addition, the Javanese-Malays can be identified through their names. Although the second and third generations of the Javanese-Malays and Malay-Arabs have names similar to most Malays in general, their parents still have names that possess the distinctive character of Javanese names of having the suffixes such as -*man* or -*min* in the last syllable of their names. The identification is more visible through the use of the parent's surname in a Javanese-Malay's name.

Although the Javanese-Malays spoke Javanese in their everyday life as a marker of identity, they were capable of speaking Malay fluently. This was because Malay was the lingua franca among people in the Archipelago. The similar structure between both Javanese and Malay also made it relatively easy for the Javanese to learn Malay. However, the majority of the new generation of Javanese-Malays have difficulties in understanding Javanese and only speak Malay.

Furthermore, in general, the Javanese were associated with various mystical beliefs. Among the Malays, the Javanese were believed to have magical 'powers', which made them capable of curing diseases or solving health problems. This stereotype was in line with the fact that the Javanese-Malays considered themselves as having the 'capability' in dispensing medicine and performing spiritual rituals. The Javanese-Malays were often seen as medicine men (*dukun*) as they were able to make herbal medicine (*jamu*) in a village.

In addition to having the skills of dispensing medicine, the Javanese-Malays introduced the Javanese calendar, which was later implemented in Malaya. The dating system does not exist in any other areas throughout Southeast Asia, except in Java. They use a cycle of 5 and 7 days and they can determine which day would be good or bad for certain activities. The dating system and herbal medicines introduced by the Javanese are sometimes affiliated with the Islamic science.

A lot of knowledge and culture derived from Java produced a new culture in Malaya. Besides the language, they kept parts of the Javanese cultural heritage. For example, the Java-nese-Malays in Johor have made a great effort to maintain the music and dance they brought from Java. Several groups of the horse dance (*Kuda Kepang*) are still active in the district of Batu Pahat. Another Javanese dance, Barong dance (*Barongan*), is often performed in a local festival. Notably, *Kuda Kepang* played by the Javanese-Malay girls is now included in the 'traditional arts' in the state of Johor, despite its Javanese origin. The dance groups are often invited to perform in various occasions as the representative of the whole of Johor.

2 CONCLUSION

Malaysian culture is made up of various elements of ethnic groups. The Javanese-Malays play an important role in establishing the identity of the Malays, despite being considered by the Malays as the 'outsider'. Both Malay and Javanese cultures have become increasingly intertwined. This means that the Javanese-Malays are becoming Malays not only by discard-ing their cultural heritage, but also by incorporating their cultural elements into the Malay tradition which is continuously shaped and reshaped.

In addition to adapting and assimilating their own culture with the local community, the Javanese in Malaysia are known as a strong community in maintaining their customs. This cul-tural maintenance is realised through the use of the language, food (*nasi ambeng*), clothing, the way of life (tradition of *'gotong royong'* like *Rewang*, *Pakatan* and *Kondangan*) and art.

REFERENCES

Bahrin, T.S. (1967) *The Growth and Distribution of the Indonesian Population in Malaya.* Bijdragen tot de Taal-, Land—en Volkenkunde (Brill), Deel 123, 2de Afl, (1967), 267–86.

Bahrin, T.S. (1967) The Pattern of Indonesian migrations and settlement in the Malayan. *Asian Studies Journal of Critical Perspectives on Asia,* 5 (2), 233–257.

Blust, R. (1981) *The Reconstruction of Proto-Malay-Javanic: An Appreciation.* Bijdragen tot de Taal-, Land—en Volkenkunde (Brill), Deel 137, 4DE Afl, (1981), 456–69.

Ferawati, L.E., Syed, Md Md A. & Hamzah, A. (2016) *Television Consumption and the Construction of Hybrid Identity among the Female Javanese Descendants in Malaysia.* Retrieved from https:// umexpert.um.edu.my/file/publication/00001218_121472.pdf.

Geertz, C. (1966) *Agricultural Involution.* Berkeley, University of California.

Lockard, C.A. (1971) The Javanese as emigrant: Observations on the development of settlements Javanese overseas. *Indonesia,* 11, 41–62.

Miyazaki, K. (2000) Javanese-Malay: Between adaptation and alienation. *Journal of Social Issues in Southeast Asia,* 15 (1), 76–99.

Ramele, R.B. & Yamazaki, J. (2013) A study on traditional Javanese-Malay Kampung structure, culture and community activities in Kampung Sungai Haji Dorani, Selangor, Malaysia. *Official Conference Proceedings in the Asian Conference on Asian Studies, (2013),* Kobe University, Japan, 65–73.

Ramsay, A.B. (1955) Indonesians in Malaya. *Journal of the Malayan Branch of the Royal Asiatic Society,* 29 (1), 119–124.

Sekimoto, T. (1994) Pioneer settlers and state control: A Javanese migrant community in Selangor, Malaysia. *Southeast Asian Studies,* 32 (2), 173–96.

Sekimoto, T. (1994) Preliminary Report on the Javanese in Selangor Malaysia. *Southeast Asian Studies,* 26 (2), 175–90.

Tirtosudarmo, R. (2005) *The Malays and Javanese in the 'Lands below the Winds.* Centre for Research and Inequality, Human Security and Ethnicity, Paper 14.

Cultural Dynamics in a Globalized World – Budianta et al. (Eds)
© 2018 Taylor & Francis Group, London, ISBN 978-1-138-62664-5

The representation of national identity in the film *Tanah Surga Katanya*

L.S. Limanta & L. Kurnia
Department of Literature, Faculty of Humanities, Universitas Indonesia, Depok, Indonesia

ABSTRACT: Identity is central in steering people's life. However, it is not something that is fixed and solid. As Hall states, identity is more of a process than a product or a fixed entity. National identity is a more sophisticated construction that involves a positioning. It becomes more complicated if it concerns those who live in the border areas. Drawing on the concepts from Anderson, Hall, Kolakowski, Smith, and Edensor on identity, national identity, and representation, this paper analyses how national identity is represented in the film *Tanah Surga Katanya...* (Heaven on Earth, They Say...). From the analysis it can be concluded that the representation of national identity in the film *Tanah Surga Katanya...*, which is carried out through the usage of symbols (flag and currency), memory, and rituals, has been successful in showing that people who live on the border areas may have multiple and competing national identities, and that national identity is a construction that cannot be solidified into a single meaning.

1 INTRODUCTION

Everybody needs an identity. It gives a sense of security for someone to play out his/her role in society. A lack of identity can lead to confusion. However, identity is not something that is fixed and easy to determine. As Hall (1990) argues, identity is a production, not a product, and it is "never complete, always in process, and is always constituted within, not outside, representation" (as cited in Rutherford, p. 222). The key point of this identity construction within representation is *positioning*. Hence, according to Hall, there is always "a politic of identity, a politic of position" (p. 226). Identity is very important since, as Castells (2010) defines it, it is "the people's source of meaning and experience" (p. 6). By this definition, identity gives a sense of direction to people's behaviour and an orientation of their values. Furthermore, in line with Hall's idea, Castells sees identity as "the process of construction of meaning" (*ibid.*).

Thus, identity is always constructed in a certain social context. This paper analyses how national identity is represented by the characters in the film *Tanah Surga Katanya...* (2012), their struggles, experience, and the positioning of themselves as Indonesians.

1.1 The context of the film

This film was produced in 2012 and won five awards in the 2012 Indonesian Film Festival, namely the Best Film, the Best Director, the Best Original Screenplay, the Best Supporting Actor, and the Best Art Direction (Siregar, 2012). It tells a story of an ex-volunteer-fighter of the 1965 Indonesia-Malaysia confrontation, Hasyim, who lives in the border area between West Kalimantan (Indonesia) and Serawak (Malaysia) with his two grandchildren, Salman and Salina. Meanwhile, Haris, Hasyim's son, prefers to live in Serawak as it promises a brighter future. Haris is already married to a Malaysian woman and is quite established economically. One day, Haris comes home to take his father and two children to go to

Serawak with him. However, Hasyim fervently declines Haris' invitation in a spirit of heroism, as he fought against Malaysia in the 1965 confrontation, and nationalism.

From a superficial view, the conflict between Hasyim and Haris can be interpreted as nostalgic and nationalistic attitudes against pragmatic and realistic ones. However, this film presents a more profound, subtle, and complicated positioning of each side. The setting in the film is the border area between West Kalimantan (Indonesia) and Serawak (Malaysia). Historically speaking, Indonesia and Malaysia have had a long history of tensions beginning with the Confrontation from 1961 to 1966 (Omar, 2010). Although the relationship was restored under the late President Suharto, the conflict between the two countries continued over territorial disputes, particularly Ligitan and Sipadan Islands, and the Ambalat sea block. While this short historical survey does not emphasize the practical political tensions between the two countries over territorial disputes, it provides a background to understand Hasyim's attitude and self-positioning towards Malaysia.

Another aspect relevant to the discussion of the construction of national identity and its representation is the borderland itself. As the characters move in and out of the border between Indonesia and Malaysia, the borderland plays an important role in how they position themselves. The border area is actually not natural and difficult to determine. As Diener and Hagen (2010) state, "Professional geographers soon conclude that all borders are arbitrary, subjective, and the result of human decisions, not forces of nature" (p. 8). Furthermore, Diener and Hagen emphasize the cultural perspective of border and conclude that recent studies view border areas "as sites of cultural interaction, exchange, and possibly hybridity" (p. 10). In this connection, Puryanti and Husain's study (2011) on the negotiation between people and the state in the borderland of Indonesia-Malaysia in Sebatik Island is important for the discussion:

> "The relationship with Malaysia, for example, cannot only be viewed in terms of 'us' and 'them', as is suggested when people talk about national identity. ... It would be illusive to measure their nationalism only from their use of the Ringgit in their daily life. ... A focus on borders can show how citizens relate to their nation-state in which loyalties are competing and multiple identities are managed on a daily basis, especially when there are economic disparities between two neighbouring countries like Indonesia and Malaysia in their border areas." (p. 117)

From this quotation, it is clear that someone's identity cannot be determined easily on the surface from what one does. However, as in the case of Haris in the film, he can position himself in such a way as to form multiple identities or to contain the contestation of identities within him.

1.2 *Nation and national identity*

Before analysing the representation of national identity in the film *Tanah Surga Katanya* ...., it is necessary to clarify the idea of nation and national identity. Anderson (2006) defines a nation as "an imagined political community" (p. 6). Furthermore, he explains that it is imagined because each member of the community has the idea that s/he belongs to a community in which most of its members very likely never meet each other. Thus, her/his membership to the community practically exists in her/his mind. The effect of this idea is that people might sometimes adhere to the idealization of the imagined community (its values, norms, rules), but at some other time they might contest it and prefer to live her/his concrete and practical daily life without necessarily complying with the idealization. This will be elucidated further in the analysis of the film, particularly in the case of Haris.

The next clarification needed is the idea of national identity which is more difficult to define. Some theorists have tried to identify the elements that construct national identity. For Stuart Hall, national identity is "an example and, at the same time, a form of national cultural identity" (as cited in Wodak, Celia, Reisigl, & Liebhart, 2009, p. 23). Furthermore, Hall defines national culture as "a discourse—a way of constructing meanings which influence and organize both our actions and our conceptions of ourselves ... National cultures construct identities by producing meanings about 'the nation' with which we can *identify*; these are contained in the stories which we are told about it, memories which connect its present with its past, and imaginings which are constructed of it." (*ibid.*)

Thus, it can be said that national identity enables people to identify themselves with the cultures of a nation through common stories, memories, and imaginings. However, the identification process might be experienced differently by members of the community since national culture is a discourse that "represents differences between social classes, between ethnic groups, or between the sexes" (*ibid.*).

Moreover, Kolakowski identifies five elements that characterize national identity, namely the national spirit or '*Volksgeist*,' as expressed in the collective manners of behaviour, historical or collective memory, anticipation and future orientation, and the national body concerning the national territories, landscapes, and nature, and a nameable beginning, referring to the story of the 'founding fathers' or the beginning of time (*ibid.*, p. 26). The relevant elements for the discussion in this paper are the first, second, and the third elements of *Volksgeist*, especially in Salman's attempt to project the image of Indonesia based on his life experiences and in the conflict experienced by Hasyim and Haris. Furthermore, Anthony D. Smith (1991) explains that members of a community or nation identify themselves as a part of it since they are united in a bond through "the use of symbols—flags, coinage, anthems, uniforms, monuments and ceremonies", and thus they are "reminded of their common heritage and cultural kinship and feel strengthened and exalted by their sense of common identity and belonging" (p. 16–17).

While it is true that national identity can be clearly represented in the use of symbols and in ceremonies, according to Edensor (2002), it is also "grounded in the everyday, in the mundane, details of social interaction, habits, routines, and practical knowledge" (p. 17). What happens in daily social life does not always go smoothly in parallel with the ideals of national culture. As Edensor further states, what is called "national" can even be "constituted and reproduced, contested and reaffirmed in everyday life" (p. 20). The analysis of the characters in the film especially Haris elucidates this point clearly.

1.3   *The representation of national identity in Tanah Surga Katanya...*

The film *Tanah Surga Katanya...* (Heaven on Earth, They Say...) represents the national identity in the quotidian life experiences of Indonesians who live in the borderland between Indonesia and Malaysia. The representation is done through the symbols (usage of national flag and currency), ceremonies (singing national songs, welcoming of the state officials), and the characters' actions (the position and choices they take), and memory (shared memories). The analysis in this paper shows that the representation contains contestation of meanings produced by the film characters.

First of all, the national identity is represented through the use of the Indonesian national flag or the Red-and-White flag. The flag is a common signifier for the Indonesian national identity, but in this film the use of the flag is applied to represent contested meanings. On the one hand, the Red-and-White flag represents a spirit of pride and heroism. This is shown through the characters Hasyim and Salman. The emotional attachment to the national flag is clearly shown first by Hasyim, and he passes it down to Salman later on. Hasyim is the only person in the village who still owns and keeps the national flag. For Hasyim, a former fighter in the conflict between Indonesia and Malaysia, the flag signifies the dignity and greatness of Indonesia. Therefore, he keeps the flag in a special box as a symbol of such dignity and greatness, almost like a sacred object (Figure 1). When the local officials come to the village and for the first time the Red-and-White flag is hoisted again after the Dwikora operation, i.e. the confrontation era, Hasyim attends the ceremony and stands in reverence while looking up at the flag from a distance (Figure 2).

Figure 1.   The Red-and-White flag.

Figure 2.   Showing reverence to the red-white flag.

Hasyim's attitude towards the flag and his memory of fighting against Malaysia is passed down to his grandson Salman who reveres it almost as sacred too. Thus, Salman strongly protests to an Indonesian salesperson (the second owner of the Red-and-White flag in the film) when he goes to the market in the Malaysian territory and sees the salesperson using the Red-and-White flag casually as a mat for his goods (Figure 3). Later on, when Salman has succeeded to collect some money and buys two pieces of sarong (intended formerly for his grandfather, Hasyim) and sees the same salesperson, he offers one piece of sarong to the salesperson in exchange for the Red-and-White flag. Salman's attitude clearly shows how he reveres the Red-and-White flag as a representation of dignity and greatness of the Indonesian national identity.

Ironically, the state elementary school, the representation of official educational institution in the village, and the village head, the extension of the government, do not own a national flag. The state's failure to provide even a single flag in the village, which is a symbol of national sovereignty and dignity, and the salesperson's ignorance and indifference towards the flag represent a contrasting meaning attached to the flag. In this case, the same object that symbolizes a national identity is presented as an unimportant or irrelevant object in people's daily life. The teacher, students, and people in the village are not aware of the need or the existence of the Red-and-White flag until the local government officials visit the school. Hence, although the Red-and-White flag is used in the film as the representation of a national identity imbued with pride and heroism, it is presented as an insignificant object without the weight of those meanings.

The next representation of the national identity in this film is done through the usage of currency. The Malaysian ringgit is depicted to be used within the Indonesian side of the borderland, and Dr. Anwar, the volunteer doctor in the village, finds this outrageous.

Figure 3.   Contrasting meaning construction of the national flag.

He attaches the Indonesian national identity to the usage of currency and considers the use of the Malaysian ringgit in Indonesian land by Indonesians intolerable and dangerous because it can threaten the Indonesians' national identity. In this case, Dr. Anwar represents the superficial and ideal identification between the use of the Indonesian rupiah and national identity. While for Lized, an Indonesian boy who helps him bring his luggage and is tipped with rupiah, and for other people who live on the Indonesian borderland, the ringgit is the common currency they use for daily trading practices because they rely on doing business with their Malaysian neighbours.

As cited by Puryanti and Husain previously, it is "illusive to measure their nationalism only from their use the ringgit in daily life". This film thus represents competing ideas of what it means to be Indonesians. On the one hand, the usage of rupiah may signify the pride of being a part of Indonesia, but on the other hand, for the people who live in the borderland and rely on doing business with their Malaysian neighbours, the use of the ringgit is the only available meaningful economic practice since they depend on their neighbour's more bustling area. Meanwhile, competing and multiple identities are formed among the students who have never seen the rupiah and just learn that they are Indonesians from their teacher, Astuti, but still rely on their Malaysian neighbours. They may realize that they are Indonesians, but they use the ringgit to be part of Malaysia on a daily practical basis. In this case, the film shows that there are contested meanings of being a part of Indonesia through the national identity represented in the use of currency.

A further contestation of meaning in terms of national identity is represented through the ceremony of the singing of the national song. The Indonesian anthem, *Indonesia Raya*, is usually sung in official occasions like during a flag ceremony or welcoming government officials. In this film, the students in the elementary school only know one song, which is *Kolam Susu* (Pool of Milk), which tells the wealth and prosperity of Indonesia that is so abundant that anyone could 'turn sticks and rocks into plants'. While the song captures what is believed to be an image of the reality of the wealth of Indonesia's nature which has become a collective memory, the prosperity, in reality, is not enjoyed evenly by most of the Indonesians, especially those who live in the border area as presented in this film. Thus, this song is mainly used as an irony to criticize the government for not paying enough attention to the borderland.

The image of reality presented in this film is that the village suffers from severe inadequate facilities. The students only know the song *Kolam Susu* instead of the national anthem because they have forgotten it since the last teacher who taught the song has been replaced for a year, and the substitute teacher to teach the anthem has not been provided. The awareness of the importance of singing the national anthem emerges when the government officials visit the village. When welcoming officials, a flag ceremony is usually held, and the local students also perform in a talent show. In this occasion, Salman reads a poem which presents a counter-meaning to the construction of the belief and ideology built in the song *Kolam Susu*. In his poem, Salman tries to provide an image of reality he and his grandfather have experienced in their daily life. Table 1 shows the comparison between the song and Salman's poem to clarify this point.

Salman illustrates the big gap between the image of his grandfather's nationalism and sacrifice in defending his country and the image of reality that he and his grandfather experience. Hasyim has a serious heart problem, and it is very difficult to get proper health treatment for him due to many constraints, namely lack of financial capability, the long distance to the nearest hospital, and inadequate transportation. For Salman, the imagining of Indonesia in the song *Kolam Susu* is not translated in his daily life. Salman has to work hard after school and sometimes has to skip lessons to collect 400 ringgit to take his grandfather to the hospital. The only transportation available is a motorboat passing through the forest. This everyday fact of life that Salman experiences forces him to challenge the common myth about Indonesia as represented in the song *Kolam Susu* and construct a different meaning as represented in his poem. Salman tries to project a new constructed meaning of Indonesia. In other words, the imagined common myth about Indonesia as represented in the song is challenged with a different projected meaning based on the everyday reality.

The next representation of the national identity in this film is carried out through what the characters do and remember. In this case, there is a contrasting representation as shown in the choices and position that Hasyim and Haris take. As Kolakowski suggests, one element of national identity is the existence of the national spirit or *volkgeist* which is expressed in collective behaviour especially in the moment of crisis (as cited in Wodak, Celia, Reisigl, & Liebhart, 2009, p. 25). Hasyim represents this element. Despite his poor health and financial conditions, and also his unfulfilled expectation of the common myth of the prosperity of Indonesia, he still chooses to stay and live in the borderland of Indonesia. Even towards his death, Hasyim is still proud of being an Indonesian and urges Salman to be proud as well.

Hasyim still believes in the common myth of the Indonesian prosperity as represented in the song *Kolam Susu*. This is expressed when Haris presents a different construction of meaning regarding prosperity. Haris says that Malaysia is more prosperous proven by the real condition of which he himself attests. The portrayal of the physical condition between the border area in Indonesia and that in Malaysia helps to build Haris' construction of meaning. Even the differences of the border road condition between the two countries shown in the film support this construction of meaning. The road on the Malaysian side is well-paved, whereas on the Indonesian side, it is rocky (Figure 4). Nevertheless, although Haris insists on choosing to stay in Malaysia and becoming a Malaysian citizen by marrying a

Table 1.   Comparison between the song Kolam Susu (Pool of Milk) and Salman's Poem.

The Song: Pool of Milk	Salman's Poem: Heaven on Earth They Say...
No seas but pools of milk	No seas but pools of milk ... *they say*
	But my grandfather says, only rich people can drink milk
Fishing rod and net provide you with means of living adequately, no storms, no typhoons will you meet	Fishing rod and net provide you with means of living adequately, no storms, no typhoons will you meet ... *they say*
	But my grandfather says, the fish are taken by foreign fishermen
Fish and shrimps will come to you	Fish and shrimps will come to you... *they say*
	But my grandfather says, hush... there is an axe to grind
People say our land is heaven land	People say our land is heaven on Earth ... so they say
	But dr. Intel says only the officials have it
Sticks and rocks turn to plants	Sticks and rocks turn to plants ... *they say*
	But dr. Intel says our logs are sold to neighbouring countries
They say our land is heaven land, Sticks and rocks turn to plants	People say our land is heaven on earth, Sticks and rocks turn to plants ... *they say*
	But my grandfather says, not all people have prosperous life, many officials sell logs and rocks to build their own heaven

Figure 4.   Contrasting infrastructure.

Malaysian woman, he is still emotionally attached to his Indonesian family (i.e. Hasyim and Salman). This can be seen from his expression when he learns about the death of his father. He becomes emotionally disturbed and experiences a moment of loss. Thus, even though Haris positions himself politically and practically as a Malaysian and chooses to stay in Malaysia, he cannot completely sever his cultural identity as an Indonesian. In this sense, he forms multiple identities and undergoes a contestation of identities.

2   CONCLUSION

National identity is a social construction of meaning related to the nation which people can identify with. This meaning construction is not fixed once for all, but it is a process that is ongoing in people's life. Furthermore, Edensor (2002) contends that what is called "national" is "constituted and reproduced, contested, and reaffirmed in everyday life" (p. 20). Moreover, Puryanti and Husain's study (2011) highlights the negotiation that the people living in the borderlands can do in relation to "their nation-state in which loyalties are competing and multiple identities are managed on a daily basis" (p. 117). Since people who live in the borderland go in and out of the border, they may form multiple identities. This is what happens to Haris who becomes a Malaysian citizen in order to gain a better living, but who is still culturally attached to his Indonesian family and heritage. Moreover, from the analysis of the film, it can be concluded that the representation of national identity does not only provide a single construction of meaning.

In conclusion, the film *Tanah Surga Katanya* ... has been successful in representing multiple and competing national identities and raising the awareness of the viewers that although Indonesia has been represented through a myth as a prosperous country in the song *Kolam Susu*, the reality of life in the borderland presents a contrasting picture. This film raises the awareness that the state needs to pay more attention to the people in the borderland and that national identity is a construction which cannot be encapsulated into a single meaning, but it involves anticipation and future orientations which may contest the present one.

REFERENCES

Anderson, B. (2006) *Imagined Communities: Reflections on the Origin and Spread of Nationalism.* London, Verso.
Castells, M. (2010) *The Power of Identity*. Singapore, Wiley-Blackwell.
Diener, A.C. & Hagen, J. (2010) *Borderlines and Borderlands: Political Oddities at the Edge of the Nation-State.* Lanham, Rowman & Littlefield Publishers.
Edendor, T. (2002) *National Identity, Popular Culture and Everyday Life*. Oxford, Berg.

Hall, S. (1990) Cultural identity and diaspora. In: Rutherford, J. (eds.) *Identity: Community, culture, difference*. London, Lawrence & Wishart. pp. 222–237.

Novianto, H. (Director), & Mizwar, D., Brajamusti, G., & Nawawi, B. (Producers). (2012) Tanah Surga Katanya ... (Heaven on Earth, They Say...) [DVD]. Indonesia: Demi Gisela Citra Sinema.

Omar, S. (2010) *The Indonesian-Malaysian Confrontation*. [Online] Available from: http://eresources.nlb.gov.sg/infopedia/articles/SIP_1072_2010-03-25.html.

Puryanti, L., & Husain, S.B. (2011) A people-state negotiation in a borderland: A case study of the Indonesia-Malaysia frontier in Sebatik island. *Wacana: Jurnal Ilmu Pengetahuan Budaya*, 13, 105–120.

Siregar, A.S. (2012) *Daftar Pemenang Festival Film Indonesia 2012* (The List of the 2012 Indonesian Film Festival). [Online] Available from: http://www.flickmagazine.net/feature/1511-daftar-pemenang-festival-film-indonesia-2012.html [Accessed December 2012].

Smith, A.D. (1991) *National identity*. London, Penguin Books.

Wodak, R., Cillia, R. de, Reisigl, M., & Liebhart, K. (2009) *The discursive construction of national identity* (A. Hirsch, R. Mitten, & J. W. Unger, Trans.). Edinburgh, Edinburgh University Press.

Cultural Dynamics in a Globalized World – Budianta et al. (Eds)
© *2018 Taylor & Francis Group, London, ISBN 978-1-138-62664-5*

Building national identity in the border areas: The critical success factors analysis in the management of community radio in Sintang, West Kalimantan

L.H. Kurnia
Department of Indonesian Studies, Faculty of Humanities, Universitas Indonesia, Depok, Indonesia

ABSTRACT: National broadcasting services in the border areas are the primary means of information dissemination that emphasize Indonesian values and develop a sense of national identity in the border areas. Despite all the efforts, citizens living in the vicinity of the national border areas continue to face challenges in accessing the national radio broadcasts. This situation indicates that there are still some issues in the management of broadcasting and information distribution processes in the border areas. Given the situation, the Ministry of Communication and Information (Kemenkominfo) initiated the Community Radio Program (*Radio Komunitas* or *Rakom*) as an effort to overcome those issues. However, this program could not be implemented properly due to some obstacles. Therefore, this study was conducted to identify the problems and to examine and evaluate the existing policy implementation of Rakom in the border areas. This study uses qualitative research methods. The data were obtained from field observation and a study of existing literature on this subject. The study demonstrates that the main problems in the management of the Community Radio Program in Sintang are: (1) the lack of coordination between institutions; (2) the lack of planning and supervision of the program's implementation; and (3) the lack of qualified personnel to manage Rakom. These issues are reviewed by using the Critical Success Factors Analysis.

1 INTRODUCTION

In 2012, the Indonesian Broadcasting Commission (*Komisi Penyiaran Indonesia* or KPI) informed the media about the difficulty of accessing Indonesia's national broadcasting and the domination of the Malaysian broadcasting in Indonesia's border areas in Kalimantan, Riau, and Natuna Islands. There is an assumption that this condition would disrupt national order, including the identity and nationalism, of Indonesian people in the border areas. Dadang Rahmat Hidayat, as the KPI commissioner at that time, highlighted the importance of content information delivered through the national broadcasting system. The information received by the people in the border areas should be correct, decent, and able to maintain national integration and pluralism of the Indonesian society. If the reality does not meet these needs, and the Indonesian people living in the borders still receive improper information, their nationalistic spirit could be undermined. The socio-cultural integrity of the communities living in the border areas could be impacted by the exposure to incorrect information. The nation's cultural values are contained within the socio-cultural aspects of the community, and these values could evoke the spirit of nationalism. This is important so that the nation could remain resilient in the face of foreign influences (Lemhannas RI, 2013).

KPI's press release in 2012 stated that it had found eighteen radio stations and three television stations from Malaysia broadcasting daily in Indonesia's border areas in Kalimantan. The Malaysian shows even became the favourite shows for Indonesian people in those areas (Tempo. co, 2012). KPI also stipulated in the "Profile and Dynamics of Broadcasting in Border Region" in 2012, that KPID (the Indonesian Regional Broadcasting Commission), which served in the border regions, complained about: (i) why TVRI/RRI and LPS had no broadcasts in the border areas; (ii) why foreign broadcasts had higher quality broadcasts compared to the local ones; (iii) why there was a lack of foreign investment; (iv) why the broadcasting commission was located so far away; (v) why there was a lack of support from the local government; (vi) why the parties concerned with broadcasting in the border regions had not been able to work together; (vii) why there were blank spot areas; and (viii) why foreign broadcasts became the favourite shows for the people in the border areas. As an attempt to overcome those issues, the Ministry of Communication and Information launched the Information Village Program in the Medium-Term Development Plan (RPJM) of 2010–2014 and the ministry's five-year program, which is reinforced by the Minister's Instruction No.1 Year 2011 regarding the implementation of the Information Village Program in the border areas. Some of these programs are the empowerment of the Community Information Groups (*Kelompok Informasi Masyarakat*), allowing citizens owning TV sets to be able to receive subscription broadcasts from the Community Broadcasting Institution (*Lembaga Penyiaran Komunitas*).

This study focuses on assessing the implementation of the Community Radio Program as an effort to meet the needs for access to information and to strengthen national identity for people living in the border areas in Sintang, West Kalimantan. This study was conducted to evaluate the existing policy implementation, which was reviewed by the program's management, with the purpose that the policies' and programs' objectives can be achieved successfully.

2 LITERATURE REVIEW AND GENERAL DESCRIPTION OF SINTANG

2.1 Literature review

Critical Success Factors. The use of Critical Success Factors (CSFs) is needed to define the things that must be done for a system, program, or project of an organization to achieve its objectives. CSFs can be used in all aspects of the organization or management, ranging from the highest management to the most technical position. By using the CSFs analysis, organizations can make changes internally and prepare for external changes that will affect how the main objectives of the organization are achieved.

Since 1961, organizations have relied on critical-to-success factors (CSFs) to define the things that must go right—either a system, program, project, process, or job—if their organizations are to achieve their purpose, mission, or objective (Howell, 2010: 1).

In the analysis of CSFs, the data are grouped according to the major themes that are often stated in the source statement. These themes are known as Key Result Areas or KRAs. Based on KRAs, we can determine which CSFs affect the achievement of the objectives. Not all KRAs are CSFs, but only those that play a critical role in the achievement of goals are crucial. KRAs play an important role in the formulation of strategic planning. KRAs statements should be made as simple and short as possible, without the use of verbs. The statements should reflect the results (output or outcome), not the actions.

Prior Research. Research about community radio has been published in the proceedings of the National Seminar "Branding Based on Local Wisdom" held by Universitas Jenderal Soedirman in 2012, which was written by Dede Lilis Ch and Nova Yuliati in their article titled "Carrying the Community Radio as a Base for Local Wisdom," as well as by Mochamad Rochim in his article "The Role of Community Radio in Community Empowerment." These writings focus on the Community Radio's role in the community in general, but do not discuss the role of Community Radio in border areas and its organization management.

2.2 Overview of the region

Sintang. West Kalimantan has five districts that share borders with Malaysia. These are Sambas, Bengkayang, Sanggau, Sintang, and Kapuas Hulu. Sintang is the second largest border district after Kapuas Hulu with a total area of 21,635 km², or 14.47% of the West Kalimantan area. Sintang is located at 1005 'N – 0046' LS and 110 050 'BT – 113 020' BT, and in the north it is bordered by Sarawak, Malaysia (BPS Sintang, 2015, p. 5). Sintang's regions which share borders with the bordering country are Ketungau Hulu and Central Ketungau Subdistricts (Sintang BPP, 2014).

The discussion on broadcasting in the border areas cannot be separated from the supporting infrastructure, i.e. the availability of electricity in border subdistricts in Sintang. Based on the data from the Sintang Border Management Agency in 2016, there are 29 villages in the Ketungau Hulu Subdistrict, six of which are border villages, and only two border villages have access to electricity, namely Jasa Village and Riam Sejawak Village. As for the Central Ketungau Subdistrict, it has 29 villages, three of which are border villages, and they have no access to electricity.

Community Radio and Kimtas. Community Radio or Rakom is one of KPU/USO programs initiated by the Ministry of Communication and Information to support infrastructure development and dissemination of information in the border areas. Under the Law No. 32 Year 2002 on Broadcasting, on the sixth part of Article 21, it is stated that the Community Broadcasting Institution is the official broadcaster, which is a legal entity in Indonesia. The Community Radio program is established by a particular community, independent, not commercial, has low transmitting power and limited coverage areas, and is established to serve the interests of the community. The purpose of the establishment of a community radio is to educate and improve the community's well-being by implementing programs that include culture, education, and information that define the identity of the nation. The establishment and management of Community Radio in each region in Indonesia have different backgrounds and purposes. For example, the Community Radios "Angkringan" in Yogyakarta and "Cibangkong" in Bandung were established to meet the needs of socialization and problem solving for their communities and to monitor the local governments' policies. Meanwhile, the Community Radios "PASS" in Bandung and "Panagati" in Yogyakarta were established because of the support of the local government (Masduki, 2007).

Community Radio is also supported by the instruction of the Minister of Communication and Information Number 01/INST/M.KOMINFO/03/2011 on the implementation of the Information Village in the border areas. Community Radio is managed by the Society Information Group (KIM) and consists of local communities. The Community Radio Sintang and Border Society Information Group (Kimtas) are located in Senaning Village, Ketungau Hulu Subdistrict, with a radio transmitter device of 100–500 watts.

3 DISCUSSION

3.1 *Stakeholders and management of community radio in Sintang*

Central Government. The Ministry of Communication and Information as a central government's body is the leading actor in infrastructure development and provision of information and communication facilities in the border regions through the program Universal Service Obligation or KPU/USO. In 2015, there was an evaluation which resulted in a new policy, which was to redesign the concept of KPU/USO. The main essence of the redesigning of KPU/USO is to continue the existing programs by changing the mechanism, which was adjusted from a top-down one to a mechanism that served local needs and was based on the needs of the ministry, institutions, and communities. In addition, the KPU/USO program in the future is expected to focus not only on infrastructure development, but also on the development of programs such as community empowerment, and the development of content and application. In monitoring and evaluation, the redesigning of KPU/USO involves participation of all stakeholders in planning, implementation, and supervision (Kominfo.go.id, 2015).

Local Government. The government of the West Kalimantan Province, represented by the Governor and Dishubkominfo of West Kalimantan, is one of the stakeholders in the management of broadcasting at the border regions. Agustinus Edi Sukarno (the Head of

Division of Communications and Information in West Kalimantan) said that until 2015, for the development of infrastructure of information and communication, the Dishubkominfo of West Kalimantan had only enough budget for human resources development in communication and broadcasting, including in maintaining Kimtas.

The Local Government of Sintang just combined the Department of Communication and Information with the Department of Transportation in 2015. The Sintang's Department of Communication and Information is still in the stage of studying and coordinating with other districts, so it has not yet run any programs related to Rakom and broadcasting in the border areas of Sintang. The Department of Communications and Information does not even have information with regard to assistance and central government programs in the previous years that were carried out in the district of Sintang.

Community and Kimtas. Basically, people in the border areas have a high curiosity about the information regarding the central government in Jakarta. It can be seen from their enthusiasm to have parabolic antennas and cable TV as well as to listen to RRI (the Radio of the Republic of Indonesia) as a means to obtain information and a medium of communication in the villages. People's role as actors in information dissemination is facilitated by Kimtas. Kimtas functions together with Community Radio as part of the Information Village Program KPU/USO run by the Ministry of Communication and Information, which was inaugurated by the minister for the entire border districts in West Kalimantan on 11th December 2010, in Jagoi Babang, Bengkayang (Press Release Kominfo 37, 2010).

Problems in the Community Radio and Kimtas Management in Sintang. Based on the results of interviews with various institutions at the provincial, district, and sub-district levels, all of them stated that assistance from the central government for the Community Radio program through third parties is received directly by the public without coordinating with the local government. The Rakom program is considered unsustainable because it did not consider the equipment needed for the program's operation, such as generators. It also did not consider the difficulty of finding diesel fuel for local people. Operators received devices to broadcast from subdistrict offices or other appointed places, but there was no further monitoring from the central government. The operators did not receive any further training and guidance if the devices were broken. There was only one training in the form of technical assistance by the ministry, which was given at the start of the program. After that, both the ministry and third parties had never conducted further monitoring or coordination of the program. This led the operators not to have a sense of belonging with the program or the call to assist if there were problems. As a result, the program could not be continued.

The operators also did not know how to process the license, so the legal aspect of their Community Radio was still abandoned. Operators refused to continue to take care of the assets because there was no budget to pay for electricity and diesel fuel. The district government was not able to provide any assistance to it because those were not the assets of the district government, and the ministry went directly to communities without consulting with the district government. Problems like these were found in Kimtas management in all border districts in West Kalimantan, including Sintang. Six Kimtas which had previously been in six border districts now were no longer in operation, including the ones located in Senaning, Ketungau Hulu Subdistrict, Sintang.

According to the Head of the Department of Communication and Information, Edi Sukarno, one of the causes that make Kimtas in the border areas unsustainable is that its operators do not get financial compensation or salary. The Community Radio operators are social workers who do not receive rewards. The Provincial Government of West Kalimantan cannot provide such compensation because they think this is the central government's program and not their responsibility, so they cannot use the provincial budget to help.

Besides, the implementation of Rakom and Kimtas also faced challenges in the social and cultural aspects. Public awareness of the importance of Kimtas is still low, and Rakom operators do not have broadcasting capability. People in the border regions do not think that Kimtas is important because they still prioritize meeting their daily basic needs. They are not aware that with Kimtas they could receive and distribute information that is useful to support the economy and needs of the community. The challenge of implementing the program is to be able to educate people in the border regions about the importance of broadcasting as a medium of information dissemination that can be useful for their everyday life. The issuance of Presidential Decree No. 96 Year 2014 on Indonesian Broadband Plan 2014–2019 made Kemenkominfo currently focus more on building internet networks in the border regions. The Information Village Program initially had no supervision, but with the publication of the new presidential regulation, it is now becoming abandoned even more. This is a challenge for the local government, operators, and program managers from the ministry to be able to remain consistent in developing the Rakom and Kimtas programs which are part of the Information Village Program.

According to Gambang (the Head of Ketungau Hulu Subdistrict), the Community Radio in the Senaning Village, Ketungau Hulu, ran for one year and people received benefit from it. In the second year, the Community Radio devices started to malfunction, and subdistricts did not have both the budget and capability to fix these tools. In addition to the problem of devices' durability, the lack of human resources also became an obstacle in managing Community Radio. Operators did not have the competence in broadcasting, so Rakom ran without ordinances and broadcasting agenda. Gambang also said that the ministry did not make it clear or make an agreement at the beginning of the program about the local government's responsibility and the standard operating procedures (SOP) of the management of Community Radio.

Community Radio should be an asset that can support broadcasting in the border regions. The Rakom program that has run for one year in Sintang, and in other border regions, reflects the desire of the community to manage and develop community radio. Unfortunately, the government that initiated Rakom did not use this program to its maximum potential.

Critical Success Factors Analysis in the Management of Community Radio in Sintang. The main determinant of success, or Critical Success Factors, is a concept in management to determine what things should be done so that organizations can successfully reach their goals. As mandated by the law, the objectives and duties of the state are to provide access to information, which is the right of all citizens, including citizens in the border regions. This section discusses the internal institutional factors that affect the realization of the objectives of broadcasting in the border regions. Based on the results of the field observation and in-depth interviews with stakeholders, we obtained five Key Result Areas in the management of Community Radio in Sintang:

Planning. Planning is an important aspect in strategic management and implementation of the program. Without careful planning, programs and plans cannot run well or can even fail. Although the Ministry of Communication and Information has redesigned the KPU/USO program in order to be more bottom-up, comprehensive, piloting, and synergistic, in practice the concept of the redesign does not work properly. In the presentation of the concept of redesigning KPU/USO, there is no explanation on how to improve the coordination system. Therefore, it is necessary to have a specialized management system design that integrates all stakeholders in order to carry out the functions of coordination and supervision properly. Planning is also needed in the management of broadcasting assets in Sintang, such as Community Radio. Long-term planning is not only about the execution of the program, but it also considers maintenance, monitoring, improvement, and development of assets. Planning that is not long-term can lead to failure of the program and a waste of time, devices, cost, and labour force.

Coordination. The core issues learned from the in-depth interviews is a lack of coordination between the central and local governments that resulted in the failure of the Rakom program. Therefore, the action that needs to be taken is building the coordination between the Ministry of Communication and Information and the local government, as well as coordination with other related ministries or institutions that are in charge of managing infrastructure, such as electricity-supplying facilities. In the implementation of the program, the local government as the land owner also should play an active role, and not simply wait for the initiative of the central government. The central government should state clearly the division of tasks and responsibilities of the Community Radio asset management, so that the local government is able to carry out their responsibilities and duties properly and not scapegoat each other anymore. The weakness of the KPU/USO program that was launched by the Ministry of Communication and Information is that asset management was handed over entirely to a third-party service provider.

The central government also needs to coordinate with third parties during the implementation of the program because discontinuation and poor coordination with the central government are the result of the third parties' action of going straight to the communities without any supervision from the ministry. Thus, the program seems like short-term programs without the possibility of sustainability and long-term benefits. In fact, the management of broadcasting, the construction of identity, and the development of border areas are long-term programs that cannot be realized only within one or two years, and then abandoned.

Implementation and Program Monitoring. The process of granting assets from the central to the local governments is also one of the main critical factors that must be dealt with immediately. Assistance that is directly provided by the central government through a third party to the communities has caused problems in the management of broadcasting assets, which in this case is Community Radio. A lack of budget is a major constraint on the sustainability of the program in the border regions, not only in the Rakom program case, but also in almost every program organized by the central government. Rakom devices provided by the central government are still owned by the central government, so the Sintang and West Kalimantan local governments cannot manage and follow up on the Rakom program. If the West Kalimantan or Sintang local government takes over the management of assets without the grants from the central government, it would be something illegal and could be investigated by BPK (the Audit Board of the Republic of Indonesia) (Asmidi, the Head of Department of Communications and Information, Direct Interview, 4th February 2016).

Maximizing Broadcasting Agency. The existence of the Rakom program is a capital that should be used for the management of broadcasting in the border regions. One of the weaknesses of this program is the lack of qualified personnel in broadcasting. Kemenkominfo as the leading actor should cooperate with existing public broadcasters which have experience in the management of broadcasting, namely RRI Sintang. The realization of cooperation between Rakom and RRI Sintang could bring about a reciprocal positive relationship and synergy to meet the needs of access to information for people in the border regions.

The Rakom operators could increase their capacity in broadcasting by getting guidance from RRI Sintang, both in broadcasting management and in device maintenance. RRI Sintang could put a relay transmitter in Rakom to reach areas that are not covered by RRI. In addition, if Rakom operators are not able to broadcast, operators could broadcast relays from RRI, and it will keep Rakom running even without operators. Another benefit gained from this cooperation is that RRI Sintang could overcome the problem of lack of journalists in border regions. The Rakom operators in the border regions could become citizen journalists for RRI Sintang. If RRI Sintang and Rakom could work synergistically, it can help solve the problems faced by RRI Sintang, namely limited human resources and limited capability of operators in broadcasting management and maintenance of broadcasting devices.

Regulations. Kemenkominfo's consistency in regulating broadcasting in the border regions is necessary. Despite the new directives from the president, it does not mean that the old programs should stop right away, and the new program becomes the only priority. Based on the interview with the Head of Ketungau Hulu Subdistrict, the main problems of the Community Radio are malfunctioned devices, the lack of capability of operators in broadcasting, and the discontinuation of the program. The Community Radio licensing issues can be resolved over time, as long as there is a continuity of the Community Radio broadcasting.

4 CONCLUSION

Planning, coordination, implementation, and monitoring of programs, maximizing of broadcaster agencies, and regulations are the five Key Result Areas in the management of the Community Radio in Sintang. Communication and coordination are the key determinants of success in the implementation of broadcasting programs from Kemenkominfo. Poor communication and coordination in the implementation of the program among the central government, local governments, and service providers are an internal problem in organization that needs to be fixed, focusing specifically on the coordination line and division of responsibilities. As one of the strategies to support the implementation of management and broadcasting in the border regions, it is necessary to set up a special team consisting of the stakeholders.

REFERENCES

Badan Pusat Statistik Kabupaten Sintang (2015) *Kabupaten Sintang dalam Angka 2015* (Sintang District in Numbers 2015). Sintang, Author.

Hidayat, D.R. (2012) Hak Informasi Bagi Masyarakat Perbatasan. Komisi Penyiaran Indonesia. *Profil dan Dinamika Penyiaran di Daerah Perbatasan Negara Kesatuan Republik Indonesia.* (Rights to Information for Border Societies. *Broadcasting Profile and Dynamics of the Border Areas of the Republic of Indonesia*) Jakarta, Komisi Penyiaran Indonesia.

Howell, M.T. (2010) *Critical Success Factors Simplified, Implementing the Powerful Drivers of Dramatic Business Improvement.* New York, Taylor and Francis Group, LLC.

Instruksi Menkominfo No.1 Tahun 2011, on Pelaksanaan Desa Informasi di Daerah Perbatasan (The Decree of the Minister of Communication and Information Technology No. 1 Year 2011 on the Implementation of Information Village in Border Areas).

Kominfo. (2014) *Redesign Program USO Didukung Dinas Kominfo Seluruh Indonesia (USO Program Redesign Supported by Communication and Information Technology Services in Indonesia).* [Online] Available from: https://kominfo.go.id/index.php/content/detail/4204/Redesign+Program+USO+Did ukung+Dinas+Kominfo+Seluruh+Indonesia/0/berita_satker [Accessed on 17th October 2014].

Komisi Penyiaran Indonesia. (2012) *Profil dan Dinamika Penyiaran di Daerah Perbatasan Negara Kesatuan Republik Indonesia (Broadcasting Profile and Dynamics of the Border Areas of the Republic of Indonesia).* Jakarta, Author.

Lemhannas. (2013) Meningkatkan Bela Negara Masyarakat Perbatasan Guna Mendukung Pembangunan Nasional dalam Rangka Menjaga Keutuhan NKRI (Strengthening the Nationalism of the Border Societies to Support National Development and Maintain the Unity of the Republic of Indonesia). *Jurnal Kajian Lemhan nas RI,* Edisi 15, 88–104.

Lilis Ch, D. & Yuliati, N. (2012) Mengusung Radio Komunitas sebagai Basis Kearifan Lokal: *Prosiding Seminar Nasional Menggagas Pencitraan Berbasis Kearifan Lokal* (Supporting Community Radio as the Basis of Local Wisdom: *Proceeding of National Seminar on Formulating Imageries based on Local Wisdom*), Purwokerto, 26 September 2012, 197–231.

Masduki. (2007) *Radio Komunitas: Belajar dari Lapangan* (Community Radio: Learning from the Field). Jakarta, The World Bank.

No.11/PIH/KOMINFO/3/2015 tentang Suspensi (Penghentian Sementara) Layanan Kewajiban Pelayanan Universal/Universal Service Obligation (KPU/USO) (on Suspension (Temporary Suspension) of Universal Service Obligation). Kementerian Komunikasi dan Informatika.

Rencana Pembangunan Jangka Menengah (RPJM) 2010–2014 and Program 5 Tahun Menkominfo. (Medium-Term Development Plan 2010–2014) and (Ministry of Communication and Information Technology 5-Year Program).

Rochim, M. (2012) Kiprah Radio Komunitas dalam Pemberdayaan Masyarakat: *Prosiding Seminar Nasional Menggagas Pencitraan Berbasis Kearifan Lokal,* (The Achievements of Community Radio in Community Empowerment: *Proceeding of National Seminar on Formulating Imageries based on Local Wisdom*) Purwokerto, 26 September 2012, 565–574.

Tempo, (2012) *Siaran Radio TV Malaysia Favorit Warga Perbatasan (Malaysian TV Radio Broadcasting Popular for Border Societies).* [Online] Available from: http://nasional.tempo.co/read/news/2012/06/22/058412297/siaran-radio-tv-malaysia-favorit-warga-perbatasan [Accessed on 22nd June 2012].

Undang-undang Nomor 32 Tahun 2002 on *Penyiaran* (Broadcasting). (Law No. 32 Year 2002 on Broadcasting).

Cultural Dynamics in a Globalized World – Budianta et al. (Eds)
© *2018 Taylor & Francis Group, London, ISBN 978-1-138-62664-5*

Heimat for three generations of immigrants in Germany: An analysis of Turkish–German films

M.R. Widhiasti & M. Budianta
Department of German Studies, Faculty of Humanities, Universitas Indonesia, Depok, Indonesia

ABSTRACT: Turkish immigrants have been residing in Germany for three generations since the 1960s. The immigrants' different social, political and economic backgrounds have shaped the way the German society treats them. Their relationships with the host society have influenced the way they construct their idealised imagination of their home (*Heimat*), which could be a Turkish, a German or an in-between space of Turkish–German imaginary. This study aims at examining how each generation of Turkish immigrants conceptualises their *Heimat* and the contexts for their different *Heimat*. The objects of this study are three films directed by Turkish–German immigrants in the early to mid-21st century. They are *Gegen die Wand* and *Auf der anderen Seite* directed by Fatih Akin and *Almanya—Willkommen in Deutschland* directed by Yasemin Samdereli. At that time, the issue of integration became highly prominent in the public space and the media in Germany.

1   INTRODUCTION

The concept of borderless Europe championed by the EU is facing its biggest challenge in the 21st century. In two decades, the number of immigrants in Europe has increased by 21 million people. According to the data compiled by the United Nations, DESA, Population Division (2011), there were approximately 28 million immigrants in Europe by 1990, and the number rose by 8 million to 36 million immigrants in 2000. The next 10 years saw a spike of 13 million additional immigrants moving to Europe, totalling to 49 million immigrants living in the continent by 2010. In 2015, more than 1 million people entered Europe, or increased four times than a year earlier (280,000 immigrants). The number continues to increase, with more than 135,000 people coming to Europe only within the first two months of 2016. Germany is one of the countries in Europe that has become one of the main destinations of immigrants from across the world. In 2015, 1.1 million asylum seekers immigrated to Germany, the highest among the EU countries. As a country with a large number of immigrants, Germany (together with other European countries) is facing challenges in integrating these newcomers into the labour market and the society because most of them have minimum working and language skills. Immigrants are in fact an important part of German history. Before the wave of immigrants had hit Germany and Europe in general in the past few years, Germany has already had a long immigration history. One of the big waves started in the 1960s, when the country signed an agreement with several countries, including Yugoslavia, Italy and Turkey, to recruit manpower to meet the demand for labour in the informal sector. Among the million workers who came to Germany, those from Turkey dominated (Gökturk, 2007).

The migrant workers who came to Germany based on the bilateral agreements are known as *Gastarbeiter* or guest workers. After more than three decades since the arrival of the *Gastarbeiter* in Germany, the Turkish immigrants have become the largest migrant group in Germany. According to a census in 2014, 3 million immigrants from Turkey lived in the country. After more than 50 years of living in Germany, the Turkish immigrants have reached their third generation. These three generations have been facing different problems with their identity issues as immigrants. One of the most significant issues of their identity construction is the concept of home (*Heimat*). This issue is closely related to a sense of belonging: these immigrants no longer live in their place of origin, but do not feel completely at home in the country where they have immigrated.

For several years, Turkish immigrants have become part of the German population, and the experiences of the three generations are represented in several media, such as literature and film (Adelson, 2005). Through the portrayals in films, we can see the complexity of the Turkish immigrants' situations in Germany, being a Turk and a German at the same time. Although Turkish film directors have been making films since the 1980s, it is the (second generation of) Turkish immigrants who play an important role and mark the re-awakening of the German cinema in the 2000s. The filmmakers of this generation like Fatih Akın, Kutluğ Ataman, Thomas Arslan, Ayşe Polat and others reflect their own cultural background in a far more relaxed way than the generation before them did. This certainly has to do with the fact that they grew up in the country where they started their professional filmmaking career (Yeşilada, 2008, p. 74).

Among the films produced by the second generation of Turkish immigrants in Germany, there are three films that show the struggle of the Turkish immigrants in interpreting their presence in Germany: *Gegen die Wand* (2004) and *Auf der anderen Seite* (2007) by Fatih Akın, and *Almanya—Willkommen in Deutschland* (2011) directed by Yasemin Şamdereli. Fatih Akın and Yasemin Şamdereli are Turkish–German directors. They were born in Germany in 1973 and started directing German films since the mid-1990s. These three films represent issues on identities faced by the three generations of Turkish immigrants in Germany and their imagination of *Heimat*. In its simplest sense, *Heimat* means home or homeland. However, the term *Heimat* carries a burden of references and implication that is not adequately conveyed by the translation of homeland or hometown. For almost two centuries, *Heimat* has been at the centre of a German moral and, by extension, political discourses about place, belonging and identity (Applegate, 1990).

Berghahn (2006) examined three films by Fatih Akin (*Kurz und Schmerzlos, Solino* and *Gegen die Wand*) to see the significance of the protagonists' homecoming journey. She concluded that the homecoming journey reflects the longing for home and the feeling of displacement. By focusing on these films and connecting them to Akin's migrant experience, Berghahn limited her study to the second generation of Turkish immigrants in Germany, whereas this study aims to examine how the three generations of Turkish immigrants in Germany articulate their closeness or connection, both with Turkey and Germany. It also aims at examining how these films produce *Heimat* visually by examining the language used in the films as well as the daily practices and mobility.

In the three films, the *Heimat* concept of each generation is shown in their daily social practices such as language use, food, clothing and mobility. The analysis of the films—which become the data corpus—shows that the *Heimat* concept in the films is different for each generation because each generation of Turkish immigrants in Germany faces different situations over three decades. For the first generation, the question regarding their sense of belonging is an accumulation of homesickness. For the second generation, it is closely related to dualism in their lives. They still carry the influence of Turkish cultural practices from their upbringing and, at the same time, experience German cultural practices in their daily lives. Meanwhile, born and raised in Germany, most of the third-generation immigrants no longer have any connection with Turkey. Although they never visit the country nor speak the language, they are still considered Turkish.

1.1 Heimat *in language*

Language is one of the ways people articulate their identity or connection to a culture. In the films, we can see different language behaviours of the three generations. The first-generation immigrants almost always use the Turkish language because they have not acquired the German language. The second-generation immigrants have equally good skills in both the Turkish and German languages, so the dominant use of one language is decided by their choice and not by force. Meanwhile, the third-generation immigrants are depicted to have better skills of the German language instead of the Turkish language, and there is a tendency of declining use of the latter among them.

Ali Aksu (in *Auf der anderen Seite*) and married couple Hüseyin and Fatma Yilmaz in *Almanya—Willkommen in Deutschland* use only Turkish in their conversations. The three characters have been living in Germany for years, but grammatically it is apparent that they are not German natives. Fatma is unable to use German in the entirety of the film, and there are two scenes that show Fatma's inability to communicate with German people when she just recently arrived from Turkey. Their lack of ability to communicate in the German language is also underlined by the portrayal of the domestic setting. Meanwhile, in the public space, the characters continue to use Turkish (because the other characters are also Turks) or are altogether silent.

The second-generation immigrants are portrayed to be equally fluent in Turkish and German. They switch between the languages, and we can see their preference in using German and their conscious choice to stop speaking Turkish. There are two characters that represent the language behaviour of the second generation: Cahit Tomruk (in *Gegen die Wand*) and Nejat Aksu (in *Auf der anderen Seite*). In both films, they use German as their main language in their daily lives in Germany. However, there is a difference between the two characters. Cahit deliberately chooses not to speak Turkish. This decision is apparent when Cahit visits Sibel's parents. In that scene, Sibel's brother questions Cahit's poor Turkish skills, and Cahit openly explains that he no longer uses Turkish in his life. This choice shows that he feels uncomfortable with his identity as a Turk and thus consciously decides to neglect all cultural markers of Turkish identity.

The second character that represents the second generation of Turkish immigrants is Nejat Aksu. Unlike Cahit Tomruk, Nejat does not oppose the use of Turkish, but he chooses to always use German, even when he meets Turkish people who speak Turkish. One of the scenes that clearly shows Nejat's attitude is when he is having dinner with his father (Ali Aksu) and Yeter Özturk, his father's female partner. During that meal, Ali and Yeter converse in Turkish, while Nejat always replies in German. Nejat's choice correlates with his occupation as a professor in a German university, which puts him in an equal position with other Germans. By speaking German in a Turkish environment, Nejat marks his difference from other immigrants.

Unlike the first-generation immigrants who are more fluent in Turkish, the third-generation immigrants are more fluent in German. Although their Turkish language skills are portrayed to be not completely erased, immigrants of this generation are more fluent in German than in Turkish. This domination of the German language over Turkish poses a question for the third-generation immigrants: whether they can still be called Turks when they can no longer speak Turkish. Such question arises in a dining scene in *Almanya—Willkommen in Deutschland*, when Cenk Yilmaz asks all of his family members, '*was sind wir den jetzt, Türken oder Deutsche?*' (So, what are we? Turks or Germans?). The question is answered simultaneously by Cenk's parents. Cenk's father, who is a Turkish descendant, answers 'Turks', while his mother—a German woman—replies 'Germans'. The two answers prompt Cenk to fire more questions: '*wenn Oma und Opa Türkensind, warum sind sie den hier?*' (If Grandma and Grandpa are Turks, why are they here?) and '*warum kann ich eigentlich kein Türkisch?*' (Why can't I actually speak Turkish?). In Cenk's understanding, they are not Turks because they are in Germany. Meanwhile, if Cenk is part of a Turkish family, he questions why he cannot speak Turkish. For the third-generation immigrants, who were born and raised in Germany, Turkish culture is no longer part of their daily lives. However, as Turkish descendants who live in Germany, the third generation continuously face questions about their identities. While they no longer feel the connection with the Turkish culture, the identity as Turk is still embedded within them.

By showing differences in using languages, *Gegen die Wand, Auf der anderen Seite* and *Almanya—Willkommen in Deutschland* show that each generation has different attachment towards Germany and Turkey, which also reflects their attachment towards the German and Turkish cultures.

1.2 Heimat *in daily practices: Clothes and food*

In daily lives, that 'at home' feeling is represented by parts of daily routines, such as clothing and variety of food. Choices of clothing represent the comfort of being part of a certain culture. Fatma Yilmaz, the main character in *Almanya—Willkommen in Deutschland*, who came to Germany in her 20 s in 1963, always wears clothes with similar models and colour worn by her fellow female villagers, a long-sleeved blouse with a small floral pattern, in Germany in the present as well as in scenes showing her past in Turkey. Fatma's clothing (that she wears until her old age) shows the dominant Turkish culture in her life. In this case, clothing is a signifier for Fatma's lasting connection with her hometown. Fatma's clothes, which function as identity markers, are also used to show identity change in a hyperbolic dream scene. Fatma and her husband, Hüseyin, are applying for German citizenship, which has caused Hüseyin's anxiety. In a dream, he saw Fatma wearing *dirndl*, German's traditional clothes, exposing her neck and breast. Similarly, Hüseyin's moustache style changes into that of Hitler's. In *Almanya—Willkommen in Deutschland*, moustache is the characteristics of Turkish men. In a scene when the boys of the Yilmaz family arrive in Germany, they are confused to see that German men do not have a moustache. This signifier is then used in the film to heighten Hüseyin's fear of identity change as they become German citizens.

Hüseyin's uneasiness regarding their citizenship change is related to Fatma's role as an immigrant woman who preserves Turkish culture by passing the values and tradition to her children (Levitt, Lucken & Barnett, 2015). Hüseyin's fear that stems from his wife's wish to have a German passport represents the fear and anxiety of immigrants about the loss of their cultural roots amongst the future generation.

In addition to clothes, food represents the Turkish culture. Having meal together at home is one central scene in the films, and the meals always consist of Turkish delicacies, such as *börek or lahmacun*. Symbolically, these films use food as a factor that brings together all the members of the immigrant families. As David Morgan stated, the food served in a family represents strong connection to the past (Bell & Valentine, 1997).

In this case, through food, immigrants 'preserve' their *Heimat* in a foreign place. However, there is also a portrayal of rejection of Turkish food. Ali in *Almanya—Willkommen in Deutschland* is the youngest son of the Yilmaz family. Born and raised in Germany, he develops an allergy to spicy Turkish dishes. There is a scene that shows Ali bringing a bag full of medicine before leaving for Turkey, to anticipate possible food allergy he may suffer there. Compared to the first-generation immigrants who always eat Turkish dishes (even though they have lived in Germany for many years), the portrayal of Ali as a Turkish descendant who is allergic to Turkish food shows a detachment to his parents' cultural roots.

Through the visualisation of Hüseyin's fear and the depiction of Ali's allergy to Turkish food, we can conclude that the films show differences among generations of immigrants in their daily lives. In terms of food and clothing, for example, the first generation of immigrants still cling to their homeland. They continue to practice the habits from Turkey in their lives in Germany. Meanwhile, the second generation (who were born in Turkey and moved to Germany in their childhood) receive German cultural influences from the environment outside their domestic lives. German influence is reflected from their clothing. If we connect this with comfort and sense of belonging, it can be seen that the first generation continues to feel comfortable living their lives in Germany, a place they do not consider their *Heimat*, in the same way as they lived in Turkey. The second generation, on the other hand, have started to feel some distance with the Turkish culture, but at the same time are not entirely at home with the German culture.

1.3 Heimat *in mobility*

The concept of *Heimat* in the three films can also be seen from the mobility of the characters from Turkey to Germany and then to Turkey again. The mobility can be translated as a search for a place that gives them a sense of belonging. Almost all of the main characters in the films were born in Turkey and moved to Germany at a certain age. The first-generation immigrants moved to Germany in their 20 s, whereas the second generation moved when they were children. Several decades later, when the first generation entered their pension age (or in their 60 s) and the second generation were in their productive age (in their 30–40 s), they moved back to Turkey. The moves by the two generations were for different reasons and thus resulting in different situations.

The mobility among the first generation of Turkish immigrants can be separated into two categories. The first category, which shows that the immigrants return to Turkey on their own choice, is represented by Hüseyin Yilmaz (of *Almanya—Willkommen in Deutschland*). The second category is represented by Ali Aksu (from *Auf der anderen Seite*), who has to return to Turkey by force. In both cases, they do not find their *Heimat* in Turkey.

Hüseyin's and Ali's return is prompted by nostalgia in their retirement age. Coming to Germany as guest workers, they no longer have a meaningful activity when they retire, and they start to reminiscent their home villages. Hüseyin realises his desire to return to his village by taking all of his family members to visit. However, he dies on the return trip from his village. Hüseyin's death depicts how he never reaches his destination. Hüseyin is a representation of the first-generation immigrants who do not have home (*Heimat*) in both Germany and Turkey and occupy an in-between space.

As a retiree with much free time on his hand, Ali Aksu spends most of his time betting at horse race tracks or visiting brothels. At one time, Ali causes the death of someone and has to be deported back to Turkey. After he returns to Turkey, we can see notable differences in his character through close-up shots. Ali's character is never shown in close-up shots during his time in Germany. It suggests that in Germany his feelings are unimportant because he is only part of that place. When Ali is in Turkey, the film focuses on his emotional state. However, the close-up depicts Ali's sad expression, looking morose with glassy eyes, his long, greying moustache suggesting a much older-looking person.

Ali's sadness is also shown symbolically in the scene where he looks outside the small window of his dark prison cell. The small window lets rays of sunshine in. The rays of light that shine through the window symbolise a better condition outside. Ironically, after he is released from prison and returns home to Turkey, Ali remains as an unhappy person. The final scene of *Auf der anderen Seite* shows Ali going fishing. Fishing is a tranquil activity usually done in places far from noisy crowd. The ending suggests loneliness and a sense of estrangement that Ali feels since his return to Turkey.

The portrayal of Ali and Hüseyin who do not find their *Heimat* in Turkey contradicts with what happens to Nejat Aksu (in *Auf der anderen Seite*) and Cahit and Sibel (from *Gegen die Wand*). These characters, who represent the second-generation immigrants, find their *Heimat* in Turkey. Nejat moves back to Turkey in an aim to 'repent' against what his father has done. Nejate leaves a very comfortable life in Germany, and his journey to Turkey is shown in a lengthy manner. This can be interpreted as a symbol of the difficulties of Nejat's search for his *Heimat*. The clearest change that happens after he moves to Turkey is Nejat's face that turns brighter. Nejat is shown to be happier after he lives in Turkey. In spite of this, there is something that does not change, which is Nejat's love for the German literature. During his stay in Germany, Nehat is a professor in the German literature. After he lives in Turkey, he becomes an owner of a German bookstore in Istanbul. This shift shows Nejat's efforts to continue his link with Germany within him. Although he moves to Turkey, he still brings the German side in him and bases his new life on that. With his move, Nejat finds his *Heimat* in Turkey with a strong German element in it.

Other representations of the second-generation immigrants who find their *Heimat* in Turkey are Cahit and Sibel. Similar to Nejat, Cahit and Sibel start new lives in Turkey. However, unlike Nejat who lives a better life in Germany, Cahit and Sibel say goodbye to their awful lives there. During their stay in Germany, Cahit and Sibel's lives are portrayed to be full of troubles. They are hospitalised because of suicidal tendencies. Cahit is also once imprisoned for murder. Thus, their move can be seen as their efforts to secure better lives, and they find that in Turkey. By moving to Turkey, Cahit and Sibel are portrayed as two persons who are calmer and live better lives. The two characters are depicted as people who want to detach from the Turkish culture during their stay in Germany, but then move back to Turkey for better lives.

2 CONCLUSION

This study examined the portrayal of the lives and problems of Turkey immigrants for three generations in Germany. It shows that each generation has a different idea about *Heimat*. The differences, for example, in language, clothing and food preferences are shown through the portrayal of the immigrants' daily lives. The films show the immigrants' relationships with *Heimat* using the mobility of characters. The first generation lose their *Heimat* in Turkey and do not see Germany as their home, while the second generation are influenced by the cultural practices of both Turkey and Germany. However, they are facing challenges because they do not have the closeness to the Turkish culture, unlike the previous generation, nor do they feel completely at home in Germany. The third generation is no longer associated with Turkish culture but cannot free themselves from being identified as Turks. This in-between space represents the symbolic *Heimat* of the three generations of Turkish immigrants in Germany.

REFERENCES

Adelson, L.A. (2005) *The Turkish Turn in Contemporary German Literature. Toward a New Critical Grammar of Migration*. New York, Palgrave Macmillan.
Applegate, C. (1990) *A Nation of Provincials. The German Idea of Heimat*. Berkeley & Los Angeles, University of California Press.
Bell, D. & Valentine, G. (1997) *Consuming Geographies: We Are where We Eat*. Psychology Press.
Berghahn, D. (2006) No Place Like Home? Or Impossible Homecomings in the Films of Fatih Akin. *Journal of Contemporary Film*, 4 (3), 141–157. New Cinemas.
Gökturk, D., Gramling, D. & Kaes, A. (eds.) (2007) *Germany in Transit: Nation and Migration 1955–2005*. Berkeley, Los Angeles & London, University of California Press.
Levitt, P., Lucken, K. & Barnett, M. (2015) Beyond home and return: Negotiating religious identity across time and space through the prism of the American experience. In: A.C. Russell King, *Links to the Diasporic Homeland: Second Generation and Ancestral 'Return' Mobilities*, 17–32. Routledge.
Yeşilada, K.E. (2008) Turkish-German Screen Power—The Impact of Young Turkish Immigrants on German TV and Film. *German as a Foreign Language Journal*, 1, 72–99. Available from: http://ec.europa.eu/eurostat/statisticsexplained/index.php/Migration_and_migrant_population_statistics.

Cultural Dynamics in a Globalized World – Budianta et al. (Eds)
© *2018 Taylor & Francis Group, London, ISBN 978-1-138-62664-5*

Representation of India as the root of identity in the film adaptation of *The Namesake*

R.A. Setyaningrum & D. Hapsarani
Department of Literature, Faculty of Humanities, University of Indonesia, Depok, Indonesia

ABSTRACT: This paper investigates the different points of view in seeing India as the root of identity portrayed in Jhumpa Lahiri's novel *The Namesake* (2003) and its film adaptation that goes by the same name directed by Mira Nair (2007). Both the author and director of the two works are of Indian descents, but they come from different generations. Lahiri belongs to the second generation immigrants, whereas Nair comes from the first generation. By using essentialist and non-essentialist perspectives on the concept of identity, the analysis denotes that while the novel stresses identity as a fluid notion that always undergoes changes and transformations which suggest the non-essentialist point of view, its film adaptation accentuates India as the essence of identity that is unchanged, timeless, and stable from time to time, reflecting the essentialist point of view. By comparing on how the author and the director make use of two different media in portraying the same story, this paper highlights the difference of loyalty toward India as the root of identity for each diasporic generation.

1 INTRODUCTION

In 2015, a Bengali-American writer, Jhumpa Lahiri, published her memoir written in the Italian language titled *In Altre Parole* (In Other Words). Lahiri herself considered the book as her efforts to detach herself from her given languages, Bengali as her mother tongue and English which she acquired from her upbringing. The book is about her linguistic journey in the pursuit of a language that she feels she can identify herself with, a language that she chooses instead of choosing her. Lahiri, known to be eloquent in deciphering the immigrants' experiences, illuminates themes, such as estrangement, belonging, uprooting, and identity through this memoir as well as in her other successful fiction books and short story anthologies.

The urge to find roots, an origin point where someone feels a sense of belonging, is an unending journey for second generation immigrants like Lahiri. The desire of finding roots arises because second generation migrants were born in two—or more—different cultures; the culture of their parents and the dominant culture they grew up with. Being born in such a condition brings at least two cultural internalisations. First, they obtain the culture of their ancestors mediated by the parents. Second, along with the first internalisation, they also absorb the dominant culture of their everyday lives. The possible result is that their loyalty is more pronounced to the culture of their birth country and distanced them from their ancestral culture (Grewal, 1996). Nevertheless, it does not make them a complete person regarding identifying themselves to one of the cultures. In Lahiri's case, she can neither claim as a Bengali as much as her parents nor assert herself as fully an American. She is eager to find something that can represent herself as a whole, and she chooses a language as her first experiment in a journey to find her true self, as said in one of her interview "*I waited a very long time to go away from the world I knew,*" she says. "*Rome has given me a sense of belonging*" (Pierce, 2015). Her attempts to master Italian is her way to liberate herself from the other two languages she has been familiar with, to give her a chance to feel that she belongs to something completely new to her.

The idea of liberating one's identity from two different cultures is depicted in one of Lahiri's renowned novel, *The Namesake* (2003), which tells a story of a Bengali immigrant family in America, the Gangulis. The novel problematizes the conflict of cultural identity experienced by the characters who represent different generations. Gogol Ganguli, the main character of the story, represents the portrayal of the second generation migrant, while Ashima Ganguli, the mother, stands as the representation of the first generation migrant. Each generation deals with the different conflicts caused by cultural differences which require different ways in the effort to solve the problems.

Cultural identity, according to Hall (1990), can always undergo some changes depending on the space and time inhabited by the individual. Hall criticizes the essential point of view that assumes the identity as a fixed entity based on a set of characteristics that is unchanged throughout time. According to his non-essential point of view, identity is seen as something that is unstable and as an ongoing transformation, *"cultural identity is not a fixed essence at all, lying unchanged outside history and culture. It is not once-and-for-all"* (Hall in Woodward, 1997, p. 53). *The Namesake* is regarded as a novel that depicts the spirit to celebrate the transformation of cultural identity shown through the development of Gogol and Ashima who in the end adopt a flexible sense of belonging to more than one culture (Friedman, 2008).

In 2007, *The Namesake*'s film adaptation directed by Mira Nair hit the box office. Some of the previous studies that examined this film adaptation focused on the intertextuality between the two works (Mani, 2012; Nagajothi, 2013). Another study (Das, 2013) examined this film adaptation through the lens of Roman Jacobson's inter-semiotic translation to show that Bengali culture is used by the film as the strategy to translate the novel into moving images. This study aims to fill the gap in the discussion about *The Namesake* film adaptation by focusing on the generational differences between the two makers, the author and the director. Lahiri belongs to the second generation immigrants, whereas Nair comes from the first generation.

Because Nair directly internalized India as part of her cultural identity in most of her adult life before she moved to America, her loyalty to India presumably will be different from Lahiri's. This will lead us to the question, whether the adaptation shows a different portrayal of India as the root of identity and changes the idea of the novel regarding its view on cultural identity. The changes in film adaptation are likely influenced by the generation gap spans between the author and the director. To support the hypothesis, the analysis will focus on the visual and audio use, as the specificity of a film as a medium that converts words into moving images, in order to show how different generations have different points of view in seeing India as the root of identity by comparing the novel and its film adaptation.

1.1 *Visual specificity: Indian cultural icons*

The film makes use of several Indian cultural icons to stress on India as the root of identity. The first visual cultural icon that appears in the film that is unwritten in the novel is the statue of the goddess Saraswati. The statue appears in scenes that depict Ashima's important stages of life. The statue shows up three times in the film, and I assume that since it appears as a recurring image, it must have an important meaning regarding the film's portrayal of India as the essence of Ashima's identity. The first one appears when the film introduces Ashima as an Indian classical music singer in which the novel provides no such information right before the scene that shows her arranged marriage with her then husband, Ashoke. By portraying Ashima as a traditional singer, the film created an allusion between Saraswati and Ashima through arts, specifically through music. The allusion is obvious because Saraswati is widely known as the goddess of knowledge, arts, and music. The film also constructs the allusion through their similar appearances through the use of the same icon, which is *Veena*. *Veena* is one of Indian's traditional music instruments, the same instrument that is attached to the popular depiction of Saraswati which symbolizes her as the goddess of arts and music.

The second appearance of the statue occurs in the scene that shows Ashima who is about to deliver a baby (Gogol). The statue itself is shown in the film as being moved on a cart, crossing the road. The movement of the statue functions as a metaphor of Ashima's current role from being a wife to a whole new role of being a mother and entering motherhood. The third appearance is the most important of all because it concludes the whole idea of the film in presenting the goddess statue. The statue emerges in the last scene of Ashima who is back in India and once again resumes her first position as an Indian classical music singer as shown at the beginning of the film. This last scene is rather different from the description given in the novel which shows Ashima's identity transformation, "*she is not the same Ashima who had lived in Calcutta*" (Lahiri, 2003, p. 276). By putting Ashima back to her first appearance, the film asserts its goal in showing Ashima's loyalty toward her ancestral culture. The appearance of the goddess Saraswati statue in the film itself can be concluded to represent several things. Firstly, it represents India's dominant spiritual identity that is Hinduism. Secondly, it functions as an allusion between Ashima and the goddess. Thirdly, it represents Ashima's loyalty to the Indian's culture as the essence of her identity because the statue remains the same, neither changed nor shattered until the end of the film. The meaning of the goddess Saraswati statue in the film also hints the desire of the first generation migrant to return to its cultural root.

Taj Mahal is the second Indian cultural icon used by the film to replace the influence of Yale buildings on Gogol, who decides to pursue his career as an architect. The novel elaborates the buildings, which were built in the style of European Gothic architecture, as the source of inspiration for Gogol and had affected his decision to be an architect,

> *For his drawing class, in which he is required to make half a dozen sketches every week, he is inspired by the details of the buildings: flying buttresses pointed archways filled with flowing trace, thick rounded doorways, squat columns of pale pink stone. In the spring semester, he takes an introductory class in architecture. (Lahiri, 2003, p. 108)*

In the novel's depiction, it is clear that Gogol's attachment to the buildings brings him a sense of belonging to a particular place. According to May, a sense of belonging can be constructed through someone's attachment to a place which eventually makes him identify himself with such a place, "*we build sense of belonging in the world based on the meanings we give our environment by moving through and engaging with it*" (2011, p. 371). Gogol's comfortable feeling, familiarity, and a sense of root that he feels toward America—or in a broader sense, Western culture—can be seen from this following description, "*but now it is in his room at Yale where Gogol feels most comfortable...He has fallen in love with the Gothic architecture of the campus...that roots him to the environment in a way he had never felt growing up on Pemberton Road*" (Lahiri, 2003, p. 108).

The film replaced the function of place related to Gogol's sense of belonging by changing the building with one of the most famous Indian's cultural icons, Taj Mahal. Although the film could not escape from showing Gogol's tourist-gaze when he stares dazzlingly at the building, the heart of the matter is that the film replaced the Gothic Western architecture with the Indian's ancient Mughal architecture which in the film is depicted as a strong influence on Gogol's further education. The film shows more that Gogol, who is wearing his American football shirt and a Jansport backpack, is pulled back to his ancestor's civilization rather than shows the superiority of Western culture that affects Gogol as portrayed in the novel.

The positive effect of Taj Mahal toward Gogol in the film also reversed the novel's description to illustrate Gogol's distance feeling against the building,

> *Their second day at the Taj he attempts to sketch the dome and a portion of the facade, but the building's grace eludes him, and he throws the attempt away. Instead, he immerses himself in the guidebook, studying the history of Mughal architecture, learning the succession of emperors' names: Babur, Humayun, Akbar, Jahangir, Shah Jahan, Aurangzeb. (Lahiri, 2003, p. 85)*

The above quotation shows how Gogol reacts while he is surrounded by the building. Indeed Taj Mahal inspires him to draw some sketches, but the word 'eludes' implicates something that Gogol fails in achieving. This may be interpreted as the lost connection between Gogol and his ancestral cultures in which Gogol fails to identify himself with. Moreover, the distance is strengthened by the novel as it describes Gogol's position as a tourist, by making him drawn to the guidebook. The novel also shows a contrasting attitude in the description of each building. For Gothic architecture, the novel describes it meticulously, such as 'flying buttresses, pointed archways filled with flowing trace, thick rounded doorways, squat columns of pale pink stone', while in describing the Mughal architecture, the novel does not touch the details of the building. Instead, it only mentions prominent figures of Mughal emperors, such as Babur, Humayun, Akbar, Jahangir, Shah Jahan, and Aungrazeb.

Taj Mahal does not only affect Gogol's future career but also influences him to express his personal feeling toward his wife, Moushumi. Taj Mahal itself is known as one of the symbols of eternal love. The symbolism inspires Gogol to name his dream house design after the building, 'Moushumi Mahal', as a wedding anniversary gift. On the other hand, the novel clearly states that the gift is "a lilac pashmina shawl" (Lahiri, 2002, p. 247). A house and a shawl might implicate the same meaning of protection and warmth. However, as a material object, pashmina shawl represents one's social status. Meanwhile, the film replaces the shawl with 'Moushumi Mahal', a direct reference to the symbol of love, Indian national symbol, and the pride of the nation.

Another cultural icon used by the film is the tradition of tonsure, a ritual of head shaving after the death of someone's father. The film shows Gogol practicing this tradition, an act that is not mentioned in the novel. The novel only stresses that "Gogol had learned the significance, that it was a Bengali son's duty to shave his head in the way of parent's death" (Lahiri, 2003, p. 179). Even so, the novel gives no further information whether Gogol shaves his head or not. It is only implied that Gogol recognizes the ritual but does not mention that he does it. Meanwhile, in the film, the ritual is depicted elaborately soon after Gogol learns about his father's death. The ritual is chosen by the film to emphasize Gogol's spiritual identity. Other than that, the ritual also enables the film to visualize Gogol's pivotal moment that marks his homecoming to his Bengali identity.

In the scene where the film shows the process of the tonsure, it is interesting to notice how the film uses a mirror to symbolize the transformation of Gogol's cultural identity. In one of the scenes at the beginning of the film, little Gogol is shown looking through a window glass. The textured glass of the window makes Gogol's reflection seen as scattered pieces as if indicating Gogol's fragmented cultural identity as it is divided between India and America. Meanwhile, when the adult Gogol shaves his head, the mirror reflects Gogol as a whole indicating his new found complete self-identity.

Each picture also uses different points of view. The scene with Gogol's scattered reflections is shown from the audience's point of view. It means the others see Gogol as having a fragmented identity. On the contrary, the scene with Gogol's reflection on the mirror uses Gogol's point of view. It can be interpreted that Gogol sees himself having a whole identity. The mirror also functions as a symbol of Gogol's awareness of himself as a Bengali descent. It also reflects the way he wants other people to see him.

Based on the differences between the novel and the film adaptation, it can be concluded that the film uses its visual specificity to stress its goal on displaying India as the essence of Ashima and Gogol's identity. The film uses Indian cultural icons as the strategy to illuminate the meaning of India as the root of identity to each aforementioned character. The goddess of Saraswati and Taj Mahal are two powerful cultural icons that bind Ashima and Gogol to their ancestors' culture. The ritual of tonsure, which is elaborately depicted in the film, serves as the film's strategy to show Gogol's spiritual identity and his acceptance of his Bengali identity.

1.2   *Audio specificity: Language, Indian classical music, and Bengali ethnic music*

The audio department also takes an important role in the film's accentuation of India as the essence of each character's identity. The aforementioned Indian classical music is one of the examples of how the film utilizes audio as its strategy in portraying Indian culture that binds Ashima with her ancestral culture. Another type of music that the film used is Bengali ethnic music. Ashima sings a Bengali lullaby to soothe Ashoke's feeling in a scene when Ashoke is hysterical because of a nightmare caused by the traumatic train incident he experienced when he was young. This scene is strongly intertwined with Indian's culture that posits Indian women as the source of love, care, sacrifice, creation and life-giver in the family and elevates them to the level of a goddess-like, "*in the Aryan home, a woman stands supreme. As the wife in the West—lady and queen of her husband—as a mother in the east, —a goddess throned in her son's worship—she is the bringer of sanctity and peace*" (Nivedita in Bagchi, 1990, p. 68).

The scene discussed above is in contrast to the description given in the novel. Ashima's reaction toward Ashoke's trauma about the accident described in the novel as "*polite new-lywed sympathy*" (Lahiri, 2003, p. 29). The quoted description creates a sense of obedience from a wife to a husband. Meanwhile, in the film, the lullaby that Ashima sings has created an intimacy between her and her husband. Moreover, Ashoke's reaction in the film appears to be powerless and in need of help from his wife, while the novel describes him to be "*suddenly pensive, aloof*" (Lahiri, 2003, p. 29). Compared to the description given in the novel, it can be concluded that the film attempts to show the ability of Ashima in bringing peace and in purifying her husband from his trauma. Through the use of Bengali lullaby, this scene emphasizes Ashima's portrayal as goddess-like, a social construction of Indian women being placed at the same level as a goddess, the bringer of sanctity, love, and peace in the family.

The film also makes use of the audio when it comes to a scene that shows Gogol's acceptance with his Bengali identity. After having his head shaved, Gogol comes up to his mother's surprise seeing the bald son. In this scene, Ashima speaks in Bengali saying "*you didn't have to do this*" followed by Gogol's answer which also speaks in Bengali "*I wanted to*". The dialog marked the first time Gogol speaks in Bengali, in which the novel does not give such illustration. At this point, it can be seen that the film makes use of the audio to point out that Gogol starts to openly embrace his Indianness, either visually (with the bald head) and aurally (by speaking Bengali). Through this scene, the film accentuates that the trauma of loss caused by the parent's death becomes a bigger sentimental force for the second generation immigrants to start respecting and embracing their ancestral culture.

Based on the comparison above, the film uses its audio specificity to fill the gap that cannot be obtained through words written in the novel. The addition of music, especially Indian classical music, to Ashima's portrayal in the film interweaves her with Indian's culture. It can be seen through her position in the beginning and in the ending of the film which emphatically shows her as a traditional Indian singer. Bengali ethnic music is another kind of music used by the film to depict Ashima as the manifestation of a goddess. The film seems to use India's social construction of women that posits them as the mother goddess to highlight Ashima's essential identity. Bengali native language also serves to stress Gogol's willingness to accept his Bengali identity.

2 CONCLUSION

In conclusion, *The Namesake*'s film adaptation shows that the portrayal of India as the root of identity is shown more boldly. It shifts the allegiance of the film, so to speak, as it is inclined to embrace India as the origin of identity for Ashima and Gogol. For Ashima's case, the film stresses that she is the manifestation of India's ideal woman based on the social construction of the mother goddess permeated in India. The film also shows her as the bearer of Indian culture by modifying her role as a traditional singer. On the other hand, Gogol, who represents the second generation immigrants, is portrayed to be closer and more comfortable with his ancestral culture compared to the description given in the novel. The analysis discussed above shows that the novel and its adaptation have different points of view in translating India as part of the diasporic identity. It can be said that the differences reflect the views of the author and the director who come from two different generations. The director of the film adaptation shows her subjectivity as an immigrant of the first generation who tends to see India as her origin culture. Meanwhile, the novel, which is written by an immigrant of the second-generation, perceives cultural identity as fluid and more flexible.

REFERENCES

Bagchi, J. (1990) Representing Nationalism: Ideology of motherhood in colonial bengal. *Economic and Political Weekly*, 25 (42), 65–71.

Das, S.K. (2013) Bengali diasporic culture: A study of the film adaptation of jhumpa lahiri's *the namesake* (2003). [Online] Available from: http://www.academia.edu [Accessed 17th August 2017].

Friedman, N. (2008) From hybrids to tourists: Children of immigrants in Jhumpa Lahiri's the namesake". *Critique-Studies in Contemporary Fiction*, 50 (1), 111–126.

Grewal, G. (1996) Indian-American literature. In Knippling, A.S. (ed.) *New immigrant literatures in the united states: Sourcebook to our multicultural literary heritage*. Westport, Greenwood Publishing Group. pp. 91–108.

Lahiri, J. (2003) *The Namesake*. Boston, Houghton Mifflin.

Mani, B. (2012) Novel/cinema/photo: Intertextual reading of the namesake. In: Dhingra, L., & Cheung, F. (ed.). *Naming jhumpa lahiri: Canons and controversies*. United Kingdom, Lexington Books. pp 75–96.

May, V. (2011) Self, belonging and social change. *Sociology,* 45 (3), 363–378.

Nagajothi, N. (2013) Transposal from fiction to motion picture: Creating jhumpa lahiri's *the namesake* on celluloid. *Language in India*. [Online] 13 (3). Available from: www.languageinindia.com/march2013/nagathinamesake.pdf [Accessed 17th August].

Nair, M. (2007). *The Namesake*. New York, Mirabai Films and Cine Mosaic.

Pierce, S. (2015) Why Pulitzer Prize-winner Jhumpa Lahiri quit the US for Italy. [Online] Available from: https://www.ft.com/content/3b188aec -f8bf-11e4-be00-00144feab7de [Accessed 23rd August 2016].

Woodward, K. (1997) Concepts of identity and difference. In Kathryn Woodward (ed.). *Identity and difference*. London, SAGE Publications. pp. 8–50.

Cultural Dynamics in a Globalized World – Budianta et al. (Eds)
© *2018 Taylor & Francis Group, London, ISBN 978-1-138-62664-5*

Contested representation of Dutch Americans in Washington Irving's *A History of New York*

M. Baga & M. Budianta
Department of Literature Studies, Faculty of Humanities, Universitas Indonesia, Depok, Indonesia

ABSTRACT: This study aims at analyzing the representation of Dutch Americans in New York in Washington Irving's (1809) historical text entitled *A History of New York*. It argues that Irving's text engages in the spatial and cultural dispute between the Dutch and English colonies in America in the 17th and 19th centuries. The study shows how the historical depiction of the early Dutch colony in the 17th-century America in Irving's text serves as a political critique of the 19th-century America. It also shows the reassertion of Irving's text about the contribution of Dutch culture to American history, which was not acknowledged in any of the historical writings of his time. Furthermore, this study shows that Irving's multilayered strategy of using Dutch colonists both for portraying a more inclusive and pluralist America and for a political parody complicates his representation of the Dutch ethnic community and thus results in their stereotypical caricatures. The study uses the concept of representation and the New Historicism method to reveal the connection between the depiction of the 17th century and the context in the 19th century and to unmask the underlying ideological biases in the text.

1 INTRODUCTION

The American society experienced what was known as *Holland Mania* in the late 19th century and the early 20th century (Goodfriend, 2008: 17). The term Holland Mania is labeled by Annette Stott in her book as the time when a craze about Dutch things colored New York life. Literary works, paintings, and advertisements reflected the popularity of the Dutch and their antique cultural characteristics (Stott, 2005: 14–15; Bradley, 2009: 41–46). This craze was attributed to the publication of *A History of New York* (AHONY) in 1809 by Washington Irving. Ironically, the same book ignited anger of the Dutch community in New York because of its negative stereotypical descriptions of the Dutch (Bowden, 1975: 159).

In the opening paragraph of chapter III, Irving, through his narrator named Knickerbocker, criticized the *literati* in the New York Historical Society (NYHS) who only highlighted warfare while failing to notice the role of the Dutch community in writing the history of New York (AHONY, 475). Through AHONY, Irving reminded the English people in his contemporary works in the 19th century that New York was once a Dutch colony in the past. In the 21st century, the United States has transformed into a large and established country dominated by white people. The arrival of the European immigrants to the land around 400 years ago recast the Indians as Native Americans through a struggle for space. In practice, the spatial struggle took place not only between the European immigrants and the Indian tribes, but also among the European immigrants. However, the history of territorial dispute among the Europeans is rarely discussed in American history. The history of the United States begins invariably from a territorial dispute among white people in the New England colonies despite the fact that there were other European colonies that existed before the New England colonies.

The problem that arises in this study is that AHONY attempts to reconstruct that very critical part of colonial history—the spatial competition between the Dutch and the English—that has been forgotten. However, Irving used a parody to deliver the historical reconstruction in AHONY. In many cases, the use of parody is associated with sheer wisecracking; thus, it obfuscates the purpose of reconstructing the history of the Dutch in AHONY. Therefore, on the one hand, Irving offered "advocacy" for the Dutch society, being one of the forgotten communities in the American colonies in the writing of history after the nation's declaration of independence; on the other hand, he staged the problem in a wisecracking manner by making the Dutch ethnic group in New York as the object of ridicule. Irving's wisecracking refers to stereotypes of the Dutch ethnic group that are shared in the mind of the dominant Anglo-American society. At the same time, he used caricature and jokes about the Dutch society in his characters to criticize the New American government in the 19th century, particularly the presidency of Thomas Jefferson. This study seeks to address the ambivalence of AHONY by pointing out how the comical depiction of the Dutch historical figures from the 17th century was used as a means to criticize the 19th-century America in a complex and multilayered narrative strategy.

Stuart Hall's concept of representation (1997) is used as a basis in this study to see how Irving brought back the past of New York throughout the 17th-century Dutch society, to generate a certain image representation of this ethnic group. The method used in this research is New Historicism, whereby the AHONY is read against the historical background of the 19th century, to see how Irving's contemporary context feeds into his writings. In the framework of New Historicism, authors constructed reality through their works as a means to respond to issues of their time (Gallagher and Greenblatt, 2000).

2 THE STRUGGLE FOR SPACE

AHONY tells the story of the territorial dispute between the English and the Dutch. In AHONY (498–499), the English, who came from the New England colonies and were nicknamed the "Yankees", performed systematic expansion toward the Dutch colonies (New Netherland). They cut down trees to open up new areas and settled there for a while before moving westward by selling their land. This arrangement was continuously repeated. From another historical source (Michael Kammen, 1975: 43–44; Van Rensselaer, 1909: 128–129), we obtain a detailed description about how the Yankees gradually infiltrated the New Netherland colonies. They took over the unoccupied areas near Fort Goede Hoop, which belonged to the Dutch but were not turned into settlement areas. Fort Goede Hoop, which was both a fort and a trade post, was located on the banks of the Connecticut River (*Fresh River*) and bordered by the New England colonies. Today, the area is named Hartford.

AHONY described how the vast New Netherland populated by a small number of Dutch inhabitants was brought to bay by the more populous English who came from the East, who were constantly moving toward the Dutch colonies (see the left map in Figure 1). The Dutch colonies at that time owned several unoccupied areas in North America. The three governors who presided over New Netherland were described to be helpless in preventing the flow of immigrants from New England into their territory. For the English, the area of New Netherland was considered to belong to England. Territorial dispute began when Governor Wouter van Twiller was in office and continued until the administration of Governor Willem Kieft. During the administration of the second governor, Fort Goede Hoop fell into the hands of New England and the area became known as Hartford. The Dutch lost many more of their territories during the administration of the last governor, Governor Peter Stuyvesant. Fort Goede Hoop was not the only landmark that was captured by the English. Other territories that were rich in *wampum* (traditional shell beads that were the currency used by the Indian tribes) in Oyster Bay, Long Island, were captured by the English (see the right map in Figure 1).

In historical records about New Netherland, the treaty signed by Governor Stuyvesant with the New England confederation addressed the East border that had been a problem since the administration of Governor Wouter van Twiller, or known as the Hartford or Fort Goede Hoop matter. Governor Stuyvesant wanted to resolve the border dispute; thus, the Hartford Treaty was signed in 1650 (see the map for comparison). This treaty made New Netherland lose a large part of its territory in the east (Van der Zee, 1978: 219–221). From the English's side, the entire east coast of America belonged to the English. The presence of the Dutch

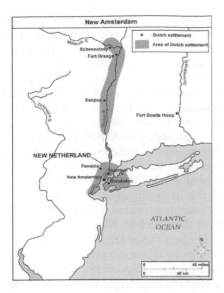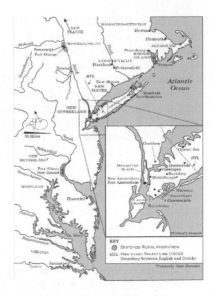

Figure 1. **Left map:** The New Netherland map. As seen on the map, the Hudson River stretches out from the north to the south and flows into New Amsterdam, the capital city of New Netherland. In the north, we can see Fort Orange, and in the east, we can see Fort Goede Hoop that is located near the Fresh River or the Connecticut River (source: Tim McNeese. *New Amsterdam*, 2007), 2. **Right map:** The New Netherland map after the Treaty of Hartford in 1650. As seen on the map, the borders are established based on the Treaty of Hartford (Hartford Treaty Line) and the former territory of New Sweden (source: Evan Haefeli 2012, 97).

there was disregarded and as a territory it was not acknowledged by the English. In AHONY, Knickerbocker mentioned the treaty, although he did not refer directly to its name.

In the 17th century, the dispute between the Dutch and the English could be traced beyond territorial issues. From the claiming of Fort Goede Hoop, one can infer that there were economic motives behind the struggle for space. The English took over the Dutch's trade route, because Fort Goede Hoop was the Dutch's trade post in the east of mainland North America. Economically, a large population requires a strong economy as well. New England colonies were required to expand their territories because of the population growth, as well as to increase their economic capacities. At the same time, their neighbor owned unoccupied areas with a small population.

In AHONY, Irving showed that the dispute which was originally about territories in the 17th century became a struggle for space for cultural representation when territories were no longer the issue because the ethnic groups became one new nation, the American nation. However, the competition did not end there. The struggle for space for cultural representation in the 19th century of New America is illustrated in AHONY, but the complicated structure of the text made the spatial struggle for representation not easily picked up by the readers.

The 19th-century background in AHONY illustrates the situation in the United States that was about to hold an election marked by full-blown competition from two parties (Republican and Federalist), as narrated through the stories told by Knickerbocker. This character is the one who later tells the story of New Netherland in the 17th century. However, in the story, this character disappeared and was reported to have died. Stories about New Netherland were only found in his notes. Thus, there is a skip in the story of AHONY, which initially attempted to tell the story of the early 19th-century America in the narrative frame to the story of New Netherland in the 17th century.

In AHONY, Knickerbocker is described as a member of the American congress and served as the Dutch New York representation after the declaration of independence. His opinions were never considered and he always lost debates in the US congress. By mentioning this, AHONY shows how marginalized the Dutch were in American politics. Not recognized by the congress, Knickerbocker was determined to rewrite the history of the Dutch power in North America in the 17th century so that their history would not be forgotten. Knickerbocker sought to enter the space of representation that was controlled by the dominant society to demonstrate that a small community within the American nation once had a role in the formation of the United States. AHONY aims to remind its readers that the presence of the Dutch minority community was prior to any other communities in mainland America, and they also played a part in the American revolutionary war. Because of their role in the formation of the nation-state, they should be given space in American history.

Knickerbocker's writing about Dutch Americans—according to AHONY—was found by the character Seth Handaside, the owner of the hotel where Knickerbocker stayed in, after Knickerbocker disappeared and was reported to have died. In telling this imagined character and his mysterious death, Irving symbolizes the end of an era. The death of Knickerbocker is a symbol of the end of the Dutch ethnic group's role in the life of the New America.

The publication of AHONY in 1809 is a form of resistance toward the forsaking of Dutch history by bringing a different representation about the Dutch in New York. Judith Richardson (2008) in her article "The Ghosting of the Hudson Valley Dutch" proposed that the ghost stories are always related to Hudson Valley where the Dutch community lived. According to Richardson, the ghost stories circulated in the Dutch community in the United States were probably originally from the mother country. Richardson interpreted that on a different level it was the Dutch society themselves (which according to Richardson were set aside as "the second native people" after the Indians") who haunted the Anglo-Americans.

Richardson's interpretation shows that the competition and dispute between the English and the Dutch were no longer about territorial matters. Even after becoming part of the American nation, the Dutch are still considered a threat by the dominant Anglo-American society. In reality, the territory that belongs to the Dutch in the United States does not exist any longer, and it has already been integrated into the territory of the United States, but the assumption that the Dutch are a threat remains.

3 IRVING'S AMBIVALENCE

The use of parody by Irving in writing AHONY is a strategy in the writing of a literary work. Irving had intentions behind his choice of strategy. Irving found the moment to use his strategy when a book that was published by NYHS did not include a detailed history of the Dutch who once occupied New York. Thus, he chose to parody this book.

According to Bradley (2009, 24; 2008, xviii), Irving parodied the book by Dr. Samuel Latham Mitchill because the book "removed fifty years off the Dutch's power over New York (New Netherland)". Mitchill only mentioned the warfare that led to the English's victory over New Amsterdam (the capital city of New Netherland), which was under the administration of the Dutch Governor Stuyvesant. After a transition of power, New Netherland was captured by the English who was represented by Richard Nicholls in 1664. The period when New Netherland became a Dutch colony was not described at all, particularly the part related to governance, society, and culture. For this reason, Irving published AHONY in December 1809.

Bradley (2008, xvii) also argued that AHONY was a direct attack toward NYHS. The organization was founded in 1804 to gather every bit of information related to the history of America. However, information about the settlements in New Amsterdam was not brought up. Irving sought to reveal New York's past that has been forgotten. The publication of AHONY is one of Irving's reactions toward NYHS for what the organization did. NYHS, whether they realize it or not, indirectly approved the action of one of their members who annulled the history of New Netherland from the collection of notes. This is the same as covering up America's history and annulling the foundation behind the rise of America as a new nation.

The contested representation that Irving sought to show could be considered quite successful. He presented a piece of work that contradicted from the dominant perspective in a history that remained very much influenced by the concept of WASP (White Anglo-Saxon Protestant), or in other words the history of the Anglo-Saxons and their descendants. However, at the same time,

Irving had another purpose, that is, to criticize the arena of the 19th-century American politics. To achieve this, he used characters from the history of the Dutch colonies in the 17th century as a vehicle to quip the leaders and criticize the American government's policy in the 19th century. Characterization of the Dutch in a caricature manner in Irving's work reaffirms negative stereotypes about the Dutch that circulated in that period. The three New Netherland governors and the important figures surrounding the governors are portrayed with physical stereotypes and dressed in funny attire. They are also depicted to lack intelligence as leaders, because the decisions that they made always end in mistakes and failures. These New Netherland characters are a mockery of the New America government, particularly the presidency of Thomas Jefferson, who Irving considered, like the New Netherland government, incompetent and a failure.

One cannot deny that the parody used by Irving implicates the representation of the Dutch ethnic community in his work. Instead of showing appreciation to the Dutch colonies as the pioneers of American history, he resonated stereotypes of the Dutch ethnic community within the American society and turned them into a joke. The Dutch New York representation in AHONY has become ambiguous because, on the one hand, their existence and role in the formation of the New America are acknowledged, but, on the other hand, they are mocked and ridiculed.

In "The Author's Apology" written by Irving in 1848 at the same time when the revised edition of AHONY was published by the publishing house G.P. Putnam and Sons around 40 years after the first edition in 1809 (Bowden, 1975: 160; Bradley, 2008: xxii), Irving told why he chose the Dutch as the subject of his book. It turns out that Irving himself lived in a neighborhood inhabited by descendants of the Dutch. Irving also had a brother-in-law who was of Dutch descent. Thus, he had been in contact with Dutch culture since a very young age.

The struggle for spatial representation had been apparent since the publication of AHONY when Irving tried to attract attention through advertisements in newspapers. Irving used the clever strategy to draw the attention of not only the group of American historians that are the subject of his criticism but also the American people. Irving succeeded in doing so as AHONY became popular in the 19th century, and the Dutch community began to receive some attention.

The announcement of the missing person, who was Knickerbocker the character and the narrator in AHONY, was in fact a hoax. The purpose was to promote the publication of AHONY. Not only did Irving promote fabricated news about a missing person in *Evening Post*, but he also wrote a fictional response about the missing person in *Traveller*. He also calculated the time of publication of these advertisements that was a month after the commemoration date of the discovery of Manhattan by Henry Hudson, which was marked by the annual NYHS meeting on 4 September 1809. In addition, he also chose 6 December 1809 as the publication date of AHONY. This date is also the anniversary of the NYHS meeting (Bowden, 1975: 161).

The American public's attention toward this minority group rose through the writing of AHONY. This was acknowledged by Irving in "The Author's Apology" on how since the publication of his book the attention of the American people toward the Dutch ethnic group increased. Therefore, AHONY shows the struggle for not only territorial space in the history of America but also representation space. This struggle is in fact apparent in the AHONY text, which reflects the ethnic bias of the writer. One the one hand, Irving believed that it was necessary to fight for acknowledgment of the Dutch ethnic group's presence in New York within American history; on the other hand, he was entrapped in the negative stereotypes constructed by the dominant American society toward the Dutch so that his noble purpose to attain equality between the Dutch and the dominant American society at that period of time became buried by his own ethnic bias.

4 CONCLUSION

History has always been written from the perspective of the conqueror. Apparently the 19th-century American history was always written from the perspective of Anglo-Americans. As a young American in the 19th century, Irving appeared to be anti-mainstream. He strived to remind others that the American country and nation were created by various ethnics and cultural communities. As a creative youth, Irving used unconventional methods to deliver his point.

In the 19th century, books or reading materials were the fastest way of influencing people; therefore, Irving adopted a different way to reintroduce the contribution of the Dutch community to American history. He used literary works to provide information about what took place in the past and placed it side-by-side with the contemporary situation of his time.

Narrations about the early history of America always began from Puritan England. However, Irving shattered this paradigm by demonstrating that there were other European nations in addition to the Native Americans. Irving wanted to show that the Americans were not the only English-speaking people. He asserted that before the declaration of America's independence, the society in America was already multiethnic. Irving showed in AHONY that the Dutch community in New Netherland did not move anywhere since the area was taken over and turned into an English colony. They remained in the former area of New Netherland and changed the name into New York, even though the nation remained a defeated one.

The contested representation demonstrated by Irving can be considered "successful" because the Dutch ethnic community in New York, which had almost been forgotten, became the center of attention after the publication of AHONY. Ironically, at the same time, Irving also "succeeded" in othering the Dutch ethnic community through his parodical style and caricatured characterization. Although the Dutch ethnic community was brought into memory, it was remembered as failed and incompetent.

The contribution of the Dutch ethnic community to American history is one of the aspects revealed by Irving about minority communities. However, there are also other ethnic communities in addition to the Europeans that came and became part of the American nation, including Chinese, Korean, Indian, and Egyptian ethnic communities, that settled in the United States earlier than predicted. The Islamic culture that is today propagandized as a foreign entity in the United States could have existed in the land hundreds of years ago. Their history could serve as the very site for the struggle for representation. This study contributes to the existing scholarship of American and Dutch immigration history by highlighting the complexity of a historical narrative. Similar studies can be carried out to reveal the strategies for representing other minorities in the US literature and history books.

REFERENCES

Bowden, M.W. (1975) Knickerbocker's History and the 'Enlightened' Men of New York City. *American Literature*, 47 (2), 159–172.

Bradley, E.L. (2008) *Introduction and Notes. History of New York by Washington Irving*. New York, Penguin Books.

Bradley, E.L. (2009) *Knickerbocker: The Myth behind New York*. New Jersey, Rutgers University Press.

Goodfriend, J.D., Schmidt, B., & Stott, A. (Eds.). (2012) *Going Dutch: The Dutch presence in America 1609–2009*. Leiden Boston, Brill.

Haefeli, E. (2012) *New Netherland and the Dutch Origins of American Religious Liberty*. Philadelphia, University of Pennsylvania Press.

Hall, S. (1997) *Representation: Cultural Representations and Signifying Practices*. London, Sage Publication Ltd.

Irving, W. (1983) A History of New York: From the Beginning of the World to the End of the Dutch Dynasty. In: *Washington Irving: History, Tales and Sketches*. New York: Literary Classics of the United States, Inc.

Irving, W. (2004) The Author's Apology. *The Project Gutenberg eBook of Knickerbocker's History of New York*. Retrieved from: http://www.gutenberg.org/files/13042/13042-h/13042-h.htm.

Kammen, M. (1975) *Colonial New York: A History*. New York, Charles Schribner's Sons.

Mc Neese, T. (2007) *New Amsterdam*. New York, Infobase Publishing.

Richardson, J. (2008) The Ghosting of the Hudson Valley Dutch. In: Joyce D Goodfriend, Benjamin Schmidt & Annette Stott (eds.), *Going Dutch: The Dutch Presence in America, 1609–2009*, pp. 87–107. Leiden-Boston, Brill.

Stott, A. (2005) Inventing Memory: Picturing New Netherland in the Nineteenth Century. In: Joyce D. Goodfriend (ed.), *Revisiting New Netherland: Perspectives on New Netherland*. Leiden Boston, Brill. pp. 13–39.

Van der Zee, H. & Barbara (1978) *A Sweet and Alien Land: The Early History of New York*. New York, The Viking Press.

Van Rensselaer, S. (1909) *History of the City of New York in the Seventeenth Century Vol. 1*. New York, MacMillan Company.

Cultural Dynamics in a Globalized World – Budianta et al. (Eds)
© *2018 Taylor & Francis Group, London, ISBN 978-1-138-62664-5*

Indonesian novels' phenomenon of mingling and competing with cultural exposure

I.G.A.A.M. Triadnyani
Department of Indonesian Studies, Faculty of Humanities, Udayana University, Bali, Indonesia

ABSTRACT: Although the interaction among literary works has in fact been one of the significant areas to study Indonesian novels, it is still little known due to the way a phenomenon of mingling and competing is developed to expose cultural differences. This study discusses this phenomenon by using an interaction analysis to investigate the potential or actual problems of the communication experienced by characters in three novels, namely *Api Awan Asap* (from Dayak's cultural view), *Lampuki* (from Aceh's cultural view) and *Puya ke Puya* (from Toraja's cultural view). Furthermore, the methodology of the interaction analysis, which was developed by Bateson (1958, 1970), Foster (1979) and Stephen (2001), provides an empirical basis for any claims on the problems or conflicts of communication. By conducting a corpus-based analysis of the interaction among the characters' different cultural backgrounds in those novels, this study will show the types of conflicts displayed through their communications. Interestingly, conflicts do not take place among characters from different social backgrounds, such as profession, age, sex and education, but among those from the same background. This study also applies Deutsch theory (2000) to conduct competition and establish cooperation, and Meyer's cultural scale (2015) to help analyse the cultural differences among those characters.

1   INTRODUCTION

Indonesia is a big country with a large population of almost 260 million people. It is also endowed with natural and cultural diversities together with complexity. The existence of various ethnicities, such as Aceh, Toraja, Dayak or Sunda, which are different from one another, proves the condition. Each ethnicity is unique because of its own characteristics that often generate cultural pride and dignity. Differences among the ethnicities can be viewed from many angles, starting from the nature and local community to the characteristics of the individuals. Indonesian natural landscapes, for example, also vary and are often characterised by mountains, valleys, seas and forests. These natural differences certainly shape the way of living of those ethnic groups. Because of different environmental conditions, people who live in mountain areas will show different characters from those living near the beach, where there is an easy access to crowds. In other words, nature moulds cultural characteristics of the people.

When two groups of people coming from different natural landscapes and cultures mingle to do things together, it is likely that there would be conflicts caused by miscommunications and controversies. The condition can grow worse when the conflicts are rooted in the different personalities of individuals. Indeed, individual personalities could actually be the trigger of the confusing and complicated pattern of communications. However, Meyer (2015, p. 11) stated that, in fact, difference, controversy and misunderstanding occur because of cultural differences. Specific cultural patterns have great effects on an individual's perception of what s/he sees, the cognition of what s/he thinks and the action of what s/he does (Meyer, 2015, p. 12). For example, the Balinese believe in the law of Karma that brings consequences to human attitudes. An event may occur; be dealt with individual differences, about where he/she was born, grew up, went to school and works; by individual interactions leading to a particular situation, either tensed or relaxed and by cultural contexts that would help to shape situation. The Javanese appear shy and softly speaking compared with the Batak people. In general, children reconstruct their cultural identity in the community in which they live, such as family and friends. This may include religion, language, customs and traditions. Similarities in blood, language, homeland and beliefs affect emotional coerciveness (Kemmel, 2000, p. 456). All of those factors mould an individual with cultural characteristics and uniqueness.

The world of literature is different from the world of reality, because each has its own notion of truth. Literary work has figures and events created by the author to represent thoughts, experiences and knowledge about his/her struggles in life. Therefore, actions experienced by the characters of the story form specific patterns that can be investigated and observed. This study aims to identify the pattern of interactions among the characters in the novels *Lampuki*, *Puya ke Puya* and *Api Awan Asap*, especially those leading to the types of cooperation and competition. Through this study, the way traditional culture contributes to the global culture can also be observed by exposing the aspect of competition, which is better than the aspect of cooperation. Those three novels are selected because they represent the local wisdoms of the Indonesian culture. The novel *Lampuki* is locally characterised by traditions of Aceh in Sumatra, the novel *Puya ke Puya* (Puya to Puya) depicts the community of Toraja in Sulawesi and the novel *Api Awan Asap* (Fire Cloud Smoke) raises the issue of the ethnicity of Dayak in Kalimantan.

2 THEORY

This study applies a structural approach by conducting an interaction analysis of the characters in each novel and then identifying the types of competition and cooperation in accordance with Deutsch theory. The analysis of social interaction is conducted as it commonly occurs in communication sciences. According to Anderson (1972), power is a formulation of patterns of social interaction that can be investigated. The existence of power is real in a situation in which a person or people obey another person or people, either consciously or unconsciously and willingly or unwillingly. Power is shown in the form of a causal relationship between the command and implementation. By employing the concept of duality that emphasises interaction, we can see pairs of characters together with their actions.

The character interaction model contains two relationships: complementary and symmetrical (Bateson, 1958, p. 176). The complementary relationship implies a hierarchical relationship, that is, individuals or groups with one individual or group dominating the other. The relationship between teacher and student or between husband and wife exemplifies this kind of relationship. Meanwhile, symmetrical relationships occur between individuals or groups that are equal, like among peers or fellows. In dealing with the element of complementary relationship, Deutsch called it asymmetry (2000, p. 23). Asymmetry exists with regard to interdependence, which means someone has a greater power and influence than the others.

2.1 *Model of interaction*

Allu (son) >< Rante Ralla (father) ----> Complementary Relationship

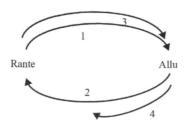

Notes:
1. Rante is advising Allu
2. Allu agrees but on condition
3. Rante explains
4. Allu is silent
   (page 5)

The above interaction scheme shows that Rante explains the custom that must be maintained. Allu agrees with the custom but doubts that it would be a burden on the family. After more explanation, he becomes silent. Thus, there is no increasing conflict in the interaction and the conflict is terminated.

Mary Foster (1979) applied the principle of unity and conflict to analyse the events in Balinese rituals based on Bateson's (1970) argument that Balinese people do not have the characteristic of *schismogenesis*. Foster conducted anthropological studies on the rituals of the Balinese traditions (such as *tajen, pawiwahan* and *med-medan*) by using the concept of Bateson's *schismogenesis*. The conclusion shows that the third tradition is the climax. Meanwhile, Stephen (2001) revealed that one of the characteristics of Balinese people is to live in peace. Foster (1979) and Stephen (2001) seemed to have opposite opinions on the nature of the Balinese. In my dissertation, I suggest that this difference occurs because the climax in Foster's research might not be supplied by the idea of competition but merely by the idea of harmony (Triadnyani, 2014, p. 18).

Theories about cooperation and competition built by David W. Johnson (Deutsch, 2000, p. 22) are associated with two things: the type of dependence of people's interest involved in a given situation and the type of actions carried out by them. Deutsch then combines the two into three social psychological processes when someone does cooperation and competition, which are substitutability, attitude and inducibility. Deutsch (2000, p. 23) defined substitutability as 'how a person's actions can satisfy another person's intensions is central to the functioning of all social institutions (family, school, industry), the division of labour, and to the role of specialization.' Substitutability allows someone to accept the actions of others in fulfilling his/her needs. Furthermore, 'attitudes refer to predisposition to respond evaluatively, favourably or unfavourably, to the aspects of one's environment or self' (p. 23). Humans have the capacity to respond positively to stimuli that are beneficial and negatively to those that are not good. Those responses appear through the natural selection, which is the convincing evolutionary process. Inborn tendency of this (about the actions of both positive and negative) is the basis for human potential to cooperate and compete. Inducibility refers to our attitude to accept the influence of others. Negative inducibility is an attitude of rejection towards one's desire.

In order to examine the interpretation of cultural interest in the interaction, the parameters built by Erin Meyer can be applied. Meyer (2015, p. 14) described eight cultural scales to see how people from different cultural contexts interact, that is, communicating, evaluating, persuading, leading, deciding, opposing, trusting and scheduling. In this study, the scale of the culture is elaborated and adapted with an interaction interest, which leads to the competition and cooperation. Thus, the parameters that will be used are communicating (explicit–implicit), criticising (direct–indirect), being a leader (egalitarian–hierarchical), deciding (negotiation–individual), confronting (direct–dodge), competing (principle–envy) and collaborating (because of task or kinship).

3   INTERACTION ANALYSIS

3.1   *The novel* Puya ke Puya

The novel *Puya ke Puya* depicts the character of Allu Ralla who has a dilemma of whether or not to perform a ceremony for his father's funeral. He is described as a young man who is obedient, brave and does not like to be indebted. He is a student at a university in the city of Makassar. Unfortunately, he easily gets carried away when he fails to marry Malena. He has stolen a dead baby and is forced to sell it to the owner of foreign companies in order to hold the funeral ceremony 'Rambu Solo' that requires a huge cost.

His father is Rante Ralla, a nobleman from Toraja who has a large inheritance, such as land, buffalos and paddy field. Rante is described as being very strict in upholding customs and traditions, although he also likes gambling and drinking *tuak* (a kind of local alcoholic drink). He is eventually poisoned by the village head because he refuses to sell his land. Allu's mother, Tina Ralla, is a quiet and compassionate woman. She witnesses her husband's death and tries hard to defend their land. Marthen Ralla is Allu's uncle. From the beginning, he has been hostile to Rante's family members who do not want to join their group. Marthen is described as rough and willing to maintain prestige.

Maria is Allu's sister, who dies when she was just 5 months old. She is buried at the top of a big tree, called Tarra, which is a *passilirian* tomb. Malena is Allu's former lover who returns from Java after finishing her Master's degree program. Malena is the village head's daughter. She pretends to marry Allu upon his father's persuasion. At the end of the story, Allu and the villagers are against the presence of a company.

Based on the interaction analysis of figures, all show some types of interaction that lead to forms of cooperation and competition. Allu, as the son, and Rante Ralla, as the father, do not have any conflicts because they agree to continue the tradition. However, Allu imposes a require-ment that they would not burden the family. Meanwhile, there is a conflict between Allu and Marthen, the uncle, because of the strong principle each of them holds. However, when Marthen finds out that his brother is poisoned by the village head, he changes and joins Allu's group. Allu and his mother do not have any conflicts, but their interaction intensifies when Allu insists on marrying Malena. Because of this problem, Allu's mental attitude changes. He dares to steal dead bodies to be sold to cover his wedding costs. There is no conflict between Allu and Malena because Malena only pretends to marry Allu. Interactions between Allu and a group of foreign companies worsen as both are uncooperative. Interactions between Allu and his relatives also end with a conflict because both parties hold strongly to their individual principles. Meanwhile, between Rante and the village head, there is an intense conflict, which causes the death of Rante. This event shows what Bateson states about the phenomenon of *schismogenesis*.

3.2 *The novel* Api Awan Asap

The novel *Api Awan Asap* illustrates the struggle of the Dayak Benuaq people who live in the downstream of Nyawatan River. They fight to defeat the invasion of capitalism. Logging operations on a large scale, ignorance of the law and ownership, as well as the exploitation of natural resources and agricultural products are the issues raised in this environmental novel. Environmental problems that occur in the Dayak community are sharply disputed without any disturbance against the idea of beauty in describing the characters and their actions. Figures like Jepi and Nori are examples of leaders who articulate the idea of preserving the environment. The novel also highlights the strong bond that exists in the community as a form of expression of solidarity that characterises traditional societies.

Based on the interaction analysis of the characters in the novel, it can be found that there are some attitudes that lead to compliance and the tendency to cooperate. The interaction between Nori and her father (Petinggi Jepi) is described without any conflicts. Both could work together, especially when Nori expresses her desire to clear the land. Although Nori is a woman, she is given the power to execute her plan. The interaction between Nori and her mother shows no conflicts because her position as a child causes her to be obedient. Meanwhile, there is a symmetrical rela-tion between Nori and Sakatn because both are equally stubborn. In the next interaction, Nori dominates and Sakatn seems submissive. Sakatn is always ready to help Nory because he wants to marry her. The interaction between Sakatn and his parents shows no conflict. His parents under-stand Sakatn's intention to marry Nori. Jepi and the community could work together. Jepi as a person who has a position at the village is able to calm the people and provide solutions. The resi-dents are obedient. Between Sakatn and Jue an indirect conflict occurs. Sakatn holds a grudge for a long time against Jue who successfully steals the heart of Nori. Sakatn finally takes an action to kill Jue.

3.3 *The novel* Lampuki

The interaction of the characters in the novel *Lampuki* begins with the existence of Aku. He is concerned about the situation in the village where there is fear of war between the govern-ment and a rebellious group. He plans to open a religious school in the village while Ahmadi is fond of inciting people to revolt. Although Ahmadi is very stubborn and always creates problems, Aku could control himself. There is no negative response to Ahmadi's actions. Musa is one of the students who has come to the village to learn the Holy Scripture. Musa

Table 1. Cultural identification.

Cultural scale	*Puya ke Puya* (Toraja)	*Api Awan Asap* (Dayak)	*Lampuki* (Aceh)
1. Communicating	to parents: implicit to uncle: explicit to relative: explicit to friend: explicit	to parents: explicit to couples: explicit to relative: explicit to friend: explicit	to parents: -- to couples: explicit to relative: explicit to friend: implicit
2. Confronting	to parents: dodge to uncle: direct to relative: direct to friend: direct	to parents: -- to couples: direct to relative: direct to friend: direct	to parents: -- to couples: direct to relative: direct to friend: dodge
3. Criticising	to parents: direct to uncle: direct to relative: direct to friend: direct	to parents: direct to couples: direct to relative: direct to friend: direct	to parents: -- to couples: direct to relative: direct to friend: indirect
4. Collaborating	to parents: kinship to uncle: task/kinship to relative: kinship to friend: task	to parents: kinship to couples: kinship to relative: kinship to friend: task	to parents: -- to couples: kinship to relative: kinship to friend: task
5. Competing	to parents: -- to uncle: principle to relative: principle to friend: envy	to parents: -- to couples: envy to relative: -- to friend: envy	to parents: -- to couples: principle to relative: -- to friend: envy
6. Deciding	to parents: negotiation to uncle: individual to relative: negotiation to friend: negotiation	to parents: negotiation to couples: negotiation to relative: negotiation to friend: negotiation	to parents: -- to couples: individual to relative: individual to friend: individual
7. Being a leader	to parents: hierarchy to uncle: hierarchy to relative: hierarchy to friend: egalitarian	to parents: egalitarian to couples: egalitarian to relative: hierarchy to friend: egalitarian	to parents: -- to couples: hierarchy to relative: hierarchy to friend: hierarchy

finally decides to join Ahmadi's followers because he feels lost after his father was killed by the government soldier. Karim and Aku have a close relationship, although the former is a marijuana dealer. The interaction between the two characters is like an interaction between friends. According to the common view, a teacher should not be friend with a vendor of illicit goods. Sulaiman is a friend of Aku who was once a foreman. The relationship is based on a shared bad luck. The characters Aku and Waluyo have a complementary relationship. Waluyo is the commander of government troops sent to safeguard Lampuki village. Aku has to ask permission to Waluyo in dealing with reopening the recital at the hall. Ahmadi does not like this action and feels jealous.

The interaction between the people of Lampuki and Ahmadi is complementary. This can be seen when people easily follow Ahmadi's instruction to return home. The interaction between Ahmadi and the army is caused by the constant uprising carried out by Ahmadi's group. Ahmadi thinks that the soldiers are occupiers who have to be destroyed. On the contrary, the soldiers assume that, because Ahmadi and his followers are rebels, it is reasonable for them to be obliterated. Ahmadi's hatred to the soldiers can be seen in the following citation:

'Colonizer! Colonizing dog! Damn you all, the evil colonizer!' (p. 72)
'We just need to wait a moment, then all will be freed, and we are independent from those uncivilized, greedy, and closed-hearted people,' Ahmadi said loudly (p. 405).

Competition occurs between Ahmadi and Waluyo. Ahmadi who hates army personnel gets irritated because Aku prefers to ask for permission to Waluyo to himself. The interaction leading to the competition is also shown by Jibral and Aku. Aku feels jealous because the students seem to like Jibral more and be happier to be taught by him.

4   CONCLUSION

Communication is one of the ways through which one can interact with others. The success of communication depends on some factors, including an individual's behaviour towards his/her acts of communication. An individual's personality is generally molded by a certain cultural system where s/he lives. Therefore, it is important for everyone to understand better the cultural aspects of a place in order to minimise the possibility of miscommunication.

Indonesia is home to many local cultures. This cultural diversity is Indonesia's strength, as established by many Indonesian literary works. From the discussions on the three novels, *Puya ke Puya*, *Api Awan Asap* and *Lampuki*, an interesting and important conclusion can be drawn, which will allow us to better understand the interaction among those local cultures.

It can be concluded that the communication experienced by the characters from the cultures of Toraja, Dayak and Aceh is explicit. It means those characters openly speak face to face without hiding the true intention of speaking. Nevertheless, in *Puya ke Puya*, the communication between Allu and his parents is implicit because Allu does not want his parents to know his plan. Regarding confrontation, generally, the characters try to avoid conflicts except when confronting a strong principle, such as Allu's case against his uncle. Uplifting criticism is usually direct and straight. Negative feedback, however, is delivered with explicit words, such as 'evil'.

The interaction of working together is based more on a duty or purpose than on kinship. For example, the village head would work together for the sake of money or for the inhabitants who are obedient to the government. Meanwhile, the competition among characters is shown to occur due to jealousy rather than a strong desire to maintain one's principle. Competition is stirred by envy and revenge. Concerning decision-making, the novel *Puya ke Puya* shows that it is based on consensus. However, in *Lampuki*, decision-making is influenced by the leader (the top). Regarding the act of leading, the Toraja and Aceh cultures still hold the tradition that puts the leader at the top of the hierarchy. However, the Dayak culture is depicted to be more egalitarian, where the leader and the people have the same responsibility to participate in decision-making. This could have been the author's idea to impose the role on Dayak women.

REFERENCES

Anderson, B.R.O'G. (1972) The idea of power in Javanese culture. In: Holt C. et al. (eds.) *Culture and Politics in Indonesia*. Ithaca, New York, Cornell University Press.
Bateson, G. (1958) *Naven: A Survey of the Problems Suggested by a Composite Picture of the Culture of a New Guinea Tribe Drawn from Three Points of View*. California, Stanford University Press.
Bateson, G. (1970) Bali: the value system of a steady state. In: Jane Belo (eds.) *Traditional Balinese culture*. New York, Columbia University Press.
Deutsch, M. (2000) Competition and cooperation. In: Morton, D. & Peter, T.C. (eds.) *The handbook of conflict resolution: theory and practice*. San Francisco, Jossey-Bass. pp. 21–39.
Foster, M. (1979) Synthesis and antithesis in Balinese ritual. In: Becker, A.L. & Aram, A.Y. (eds.) *The imagination of reality: essays in Southeast Asian coherence systems*. New Jersey, Ablex Publishing Corporation.
Kimmel, P.R. (2000) Culture and conflict. In: Morton, D. & Peter T.C. (eds.) *The handbook of conflict resolution: theory and practice*. San Francisco, Jossey-Bass. pp. 453–473.
Meyer, E. (2015) The Culture Map. (S.N. Verawaty, Trans). Jakarta, KPG (Kepustakaan Populer Gramedia).
Nur, A. (2011) *Lampuki*. Jakarta, PT Serambi Ilmu Semesta.
Odang, F. (2015) *Puya ke Puya* (Puya to Puya). Jakarta, KPG (Kepustakaan Populer Gramedia).
Rampan, K.L. (2015) *Api Awan Asap* (Fire Cloud Smoke). Jakarta, Grasindo.
Stephen, M. (2001) Barong and Rangda in the context of Balinese origin. *Review of Indonesian and Malaysian Affairs*, 35, 137–194.
Triadnyani, I G.A.A.M. (2014) *Fenomena Rangda dalam Teks Calon Arang dan Janda dari Jirah: Kajian Hermeneutik Ricoeur* (The Phenomenon of Rangda in Calon Arang and Janda dari Jirah: A Ricoeur Hermeneutics Study). [Dissertation]. Indonesia, Universitas Indonesia.

Cultural Dynamics in a Globalized World – Budianta et al. (Eds)
© *2018 Taylor & Francis Group, London, ISBN 978-1-138-62664-5*

Funding politics in European film festivals and its impacts on the development of Indonesian cinema

R. Ihwanny & M. Budiman
Department of Literature, Faculty of Humanities, Universitas Indonesia, Depok, Indonesia

ABSTRACT: Nowadays film festival is no longer a mere showcase for films; it also engages in production and distribution. It all began with the Rotterdam film festival with its Hubert Bals Fund, where Indonesian filmmakers began to receive funding from European film festivals. This raised a number of issues, which become the focus of this paper. Two of such issues are the suspicion about hegemony practices which operate between the funding providers and the receivers, and whether the funded films contribute to the development of Indonesian cinema, and, if so, what the contributions are. This study reveals that a funding program is like a double-edged sword. On the one hand, the hegemony practices have brought some impacts on the representation of Indonesia. Funded films tend to perpetuate the stigma of Indonesia as a third world country. On the other hand, the funding also brings positive impact to the development of Indonesian cinema. The funding becomes a training ground to sharpen the skills of filmmakers, as well as an arena to improve, develop, and promote themselves. As a result, the Indonesian cinema has been enriched by a significant number of experienced filmmakers, a condition which will ultimately make a positive impact to the development of the cinema itself.

1 INTRODUCTION

Film festival nowadays is not only an exhibition site; it also engages in producing and distributing films (Iordanova, 2015). These two aspects are apparent in programs which are intended to fund the production and distribution of films. Such programs began in 1988 with the Rotterdam film festival under the Hubert Bals Fund. Indonesia is one of the countries whose many of its filmmakers have benefitted from the funding. They include names such as Riri Riza, Ismail Basbeth, Yosep Anggi Noen, Mouly Surya, Ravi Bharwani, Kamila Andini, and Edwin, to name a few. These names are usually associated with independent cinema. The common issue around independent cinema is the fact that these films are successful in international film festivals, yet they are not shown in their own country. Mainstream Indonesian films become the centre of public attention and have established their position in the domestic market, while independent films are pushed aside into the periphery and become known only on the international stage (Aartsen, 2011).

The study will focus on the following European film festivals: Cannes (France), Berlin (Germany), and Rotterdam (the Netherlands). Cannes and Berlin film festival have been chosen because they are the biggest film festivals in Europe. Meanwhile, the Rotterdam film festival has been chosen because it is the first film festival which grants funding to independent filmmakers. Funding schemes offered by those film festivals are; the Berlinale Residency and World Cinema Fund (Berlin), the Residence (Cannes), and Hubert Bals Fund (Rotterdam).

The Berlinale Residency and the Residence are fellowship programs. They are the forums where new talents in cinema can train and meet professionals in the field. The Residence was established in 2000; it is a program for twelve filmmakers from all over the world who are selected to develop their first or second film scripts. This program is intended for young filmmakers who have directed one or a few short films or one feature film and are working on their new feature film (Cinéfondation, 2016b). They are given the opportunity to stay in Paris for four and a half months. During that time, these filmmakers will be assisted in preparing their film scripts. A cultural session will be given at the beginning of the program, where the participants will be shown several cultural objects around Paris. They will get a place to stay in Paris, as well as personal assistance in writing their scripts; they will also be given an allowance to the amount of eight hundred Euros a month, free access to all film theatres in Paris, and a French language class (Cinéfondation, 2016a).

The Berlinale Residency, one of the programs of the Berlin film festival, was initiated in 2012 for the purpose of providing support to filmmakers from all over the world who have gained success in their first film festival to develop their new project. Unlike the Residence, Berlinale Residency only accepts a film project that already has a producer. Three filmmakers are selected annually and will receive a fund to the amount of one thousand five hundred Euros a month. The program provides mentoring sessions with renowned scriptwriters and social gatherings with other filmmakers.

The World Cinema Fund (WCF) and the Hubert Bals Fund (HBF) were initiated to provide financial assistance to films that could not be produced without additional funding. Each year WCF provides funding for filmmakers from Latin America, Central America, the Caribbean, Africa, the Middle East, Central Asia, South East Asia, Bangladesh, Nepal, Sri Lanka, and countries from the Caucasus region. Meanwhile, HBF provides funding for filmmakers from Africa, Asia, Latin America, the Middle East, and a few East European countries. The countries receiving this funding are also referred to as the 'third world' countries.

WCF was established in 2004. WCF is divided into production and distribution funding. The maximum amount of production funding for one film is eighty thousand Euros, while ten thousand Euros is allocated for distribution (Internationale Filmfestspiele Berlin, 2016). The production funding must be used in the recipient's country. As for distribution, only a German distributor may apply. WCF launched an additional program in 2015 called WCF Europe. There is no significant difference between the two programs, except for the partners. If WCF requires a German partner, WCF Europe allows for cooperation with other European countries outside Germany.

The Rotterdam film festival was the first festival that offers funding for film. HBF, which was established in 1988, provides two types of funding: (1) script and project development support and (2) the fund for post-production works (IFFR, 2016). The fund allocated for script and project development may be used for developing film scripts, including for conducting the necessary research, script writing, and translation, or for hiring a script consultant. The fund provided for this purpose shall not exceed ten thousand Euros. The post-production funding can be used for covering expenses during post-production stages, such as editing or dubbing. The maximum amount is twenty thousand Euros.

HBF also has another program called NFF+HBF, which was established in 2006. This program is a collaboration between HBF and the Netherlands Film Fund. The program is intended to promote cooperation between Dutch producers and film projects from the countries funded by HBF. Only a Dutch producer is allowed to apply for funding and select a film for the collaboration project, and also will receive the funding not exceeding fifty thousand Euros. The selected film must be a project that is already receiving script and project development funding.

2 METHOD

The funding is one of the key factors in the continuity of independent cinema. Nonetheless, funding is not the last solution, because the conferring of the funding is followed by a suspicion regarding hegemony practices operating between the fund providers and the receivers, which will in turn have an impact on the representation of Indonesia. Another issue that concerns films which are funded by this funding and their acceptance in their own country is, do they make significant contribution to the development of Indonesian cinema? And if they do, what is their contribution? This paper seeks to investigate these two main issues: (1) the hegemony practices that operate within the funding and (2) the impact of the funding on the development of Indonesian cinema. This paper will answer those questions using Gramsci's theory of hegemony.

Aeron Davis (2008) mentions three approaches that can be used to examine the production of culture, namely political economy, textual analysis, and sociological/ethnographic. Political economy is used to analyze the economic, political, and industry influences to the production of a cultural product, while the textual approach analyzes the code, ideology, discourse, and individuals that affect the production of cultural products. The sociological/ ethnographic approach focuses on observing and documenting the production process and the people involved in a cultural product (Davis, 2008). This paper uses the textual approach because it focuses on analyzing the discourses and individuals that affect the funded film production.

In this paper, funding refers to the fund provided by European film festivals. Europe is picked for a number of reasons: the continent is the birthplace of film festival and has now become the dominant venue in festival film circuit. Europe is also picked for its long history of colonialism and orientalism. Scholars have long expressed their concern with regard to the funding provided by European film festivals. Shaw (2015) suspects that funding from Europe is a form of post-colonialist intervention and an effort to perpetuate a Eurocentric worldview. Halle (2010) suspects that it is a new form of orientalism.

3 RESULTS AND ANALYSIS

Antonio Gramsci, the scholar who popularized the theory of hegemony, stated that popular culture is the arena where the battle for hegemony occurs (Procter, 2004). Hegemony is described as a process for creating an effective domination by using culture instead of physical force—a domination which is achieved through voluntary processes involving negotiation as opposed to oppression (Procter, 2004). A film festival is part of popular culture that serves as the arena for hegemony between the provider and receiver of the funding. A funded film becomes the product of hegemony, which is the end result of a process of negotiation and intervention.

At a glance, there is nothing dangerous about the funding provided by these festivals. Terms of agreement consist of nothing more than technical requirements of a film or a script to be funded. However, some of the conditions could potentially be seen as hegemony practices. The fund provider requires two things as the conditions for submitting a proposal: the script or the rough cut of the film. Because script and rough cut are an unfinished form of film, there is an opening for negotiations or even interventions by the fund provider to make some fundamental changes to the project, specifically through the script writing assistance program and an explicit request that the content, topic, and location of the project must be kept open.

A number of filmmakers and scholars have spoken about this intervention. Gaston Kabore, a film director from Burkina Faso, said that the danger from accepting the funding is the attached condition that African filmmakers must depict Africa only to the extent the Europeans are willing to accept. He said, "When you write a script to please European producers, you take their expectations into consideration. Our films can become unbalanced; we are so weak that we are turning like this and like that. The danger is forgetting your own people, your own fundamental vision, and presenting Africa only as Europe is prepared to receive it. The danger is we will lose our souls" (Turan, 2002). Falicov (2010) talks about a compromise made during the script and idea development to fulfil the wishes of the fund provider, "In some cases, global south filmmakers are asked to change their scripts and ideas to curry favour with international funders." De Valck made an important note about the compromise: "Many festival films nowadays demand a say in which films are artistically interesting, before they are made, and with these funds the festivals, in fact, influence which films will be realized" (Falicov, 2010). These interventions can be seen in the themes of the funded films. Some of them picked war, terror, or occupation as their themes, or any other themes that depict the horrible living condition in a country without a good democracy.

Darmawan (2015) is infuriated because more than a few Asian films are portraying a lonely individual wandering all over the world without purposes and often the films end with the death of the character. She went on as far as questioning this phenomenon in a forum, "Is this how the West perceives the East?" Chalida Uabumrungjit, Thailand filmmaker, answered the question, "Blame the festival's programmer, especially the Western, because they always choose films of that sort. I always submitted films with happy themes with a lot of dialogues and they hardly ever noticed them" (Darmawan, 2015). Such an intervention raises the concern on the rise of new orientalism, "The dynamic of Orientalism at work here supports the production of stories about other peoples and places that the funding source wants to hear." (Halle 2010). The new orientalism is different from the one in the past. The "old" orientalism provides an image of the East as depicted by Western scholars, but now the depiction of the East is provided the filmmakers from the East through their films, which are funded by the West. Hence, the new orientalism may be defined as a depiction of the East as presented by Easterners, with Western intervention and involvement in it.

Within the context of hegemony, the subordinate group will have a strong bond and association with the values and ideals proposed by the dominant (Jones, 2006). If we were to apply this concept on the dynamics of funding, these filmmakers are the subordinate, and they share the values and ideals of the fund provider. The simplest way to instil a value is by involving those filmmakers in a cultural program. Those filmmakers receiving the fellowship will be enrolled in French language classes, participate in field trips to cultural heritage sites in Paris, and be given free access to watch films at film theatres in Paris.

Baumgärtel (2011) sees these 'third world' filmmakers as being in an 'imagined world'. They are entering a space where they are not bound by the regulations that restrict them in their own countries (Baumgärtel, 2011). One of the spectres that haunt these filmmakers is the censorship. For instance, in Indonesia, LSF (*Lembaga Sensor Film*) or the Film Censorship Board is the state's representative that controls what the audience can or cannot see in a film. The Indonesian filmmakers also have to deal with *UU Pornografi* (Pornography Law), which is considered as a limitation of their movement and freedom of expression. Meanwhile, the European film festivals are offering a forum where these filmmakers can freely express their opinions. These are the conditions that ultimately provide the justification for the 'imagined world' and the similarity of values and ideas between independent Indonesian filmmakers and their European funders.

Throughout its twelve years of existence, only two Indonesians filmmakers have received funding from WCF; they are Edwin and Ravi Bharwani. Kamila Andini is the only Indonesian filmmaker receiving fellowship from the Residence. These three filmmakers also received funding from HBF. Five other Indonesian filmmakers have also received funding from HBF in the last five years. Below is the list of them and their works.

This paper specifically focuses on Jermal. *Jermal* means a fishing platform in the open sea. The film tells a story about Jaya, a boy who, after the death of his mother, went to look for his father working as a foreman on a *jermal*. In his quest, Jaya had to confront the harsh life on a jermal. Not only did his father refuse to acknowledge him as his son, Jaya also had to deal with the abusive treatments by *jermal* child workers. This film clearly highlights the issue on child labour, mixed with an exotic background of Indonesian seascape. Those poor children have to work on a *jermal* without any opportunity to attend school; they are also unable to read or write.

Issues on exploitation are presented with a depiction of the miserable living conditions on a jermal. Drinking water is rationed, while the child workers are wearing tattered clothes and are often bare-chested. They live in a wooden hut without any partition or bathroom, and they sleep on a thin sheet. Smoking and alcohol consumption are common among them. The film also deals with violence. Jaya had to face physical and verbal abuse not only from his father but also from the child workers. The harsh life on a *jermal* ultimately turns these child workers into violent and aggressive adults.

Films funded by these festivals raised the darker side of human life—a child that has to deal with loneliness, abandonment, father and child conflict, poverty, child labour, women

Table 1. Funding recipients.

Title	Theme/Issue/Story	Director	Year	Funding
The Seen and Unseen	Loss, loneliness	Kamila Andini	2011 2012–2013	HBF The Residence
Jermal	A father who does not acknowledge his own child; child labour and poverty	Ravi Bharwani	2005 2007 2008	HBF NFF + HBF WCF
Postcards from the Zoo	Child abandonment; women trafficking	Edwin	2011 2012	HBF WCF
Peculiar Vacation and Other Illnesses	A lower class female worker supporting her husband	Yosep Anggi Noen	2012	HBF
Atambua 39 Derajat Celsius	Day to day struggle of the people in Atambua	Riri Riza	2012	HBF
What They Don't Talk about When They Talk about Love	A love story between disabled teens	Mouly Surya	2012	HBF
On Mother's Head	A story about three women confronting a hard life in Bali	Putu Kusuma	2012	HBF
Another Trip to the Moon	A story about two women who live in the forest	Ismail Basbeth	2014	HBF

trafficking, exploitation, adultery, frontier life, and alienation. Locations selected for the scenes are situated far from modern city environment: in Bali, on the open sea, at a zoo, in Yogyakarta, in Atambua, in a School for Children with Special Needs, and in the forest. Halle (2010) states, "The coproduced films must tell stories that offer to European audiences the tales they want to hear." The other thing the European moviegoers want to see is a confirmation that Indonesia is a third world country with an exotic and beautiful landscape and a widespread poverty, ignorance, and weak enforcement of human rights. It is an obvious sign of orientalism-romanticism that is still deeply rooted in European minds.

According to Lent (2012), a film is classified as independent when it is not bound by government regulations or censorship, or not produced by a big production house, or not adhering to the style or method of conventional filmmaking. The films funded by the film festivals can be classified as independent because they reject government's censorship, are funded by several sources, and are made unlike commercial films. Aartsen (2011) states that Indonesian films can be classified into two groups, i.e. national commercial films and independent films. Commercial films become the centre of public attention and have established their position in the domestic market, while independent films are pushed aside into the periphery and become known on the international stage.

The funding program became one of the saviours for an independent filmmaker who has a financial problem. For example, the film entitled The Seen and Unseen by Kamila Andini. Andini received funding to the amount of ten thousand Euros from HBF, a fellowship program from the Residence, and funding from APSA Children Film Fund. However, this has also brought in new problems, due to suspicions about the presence of hegemony practices and the way Indonesia is represented in the film. A film can carry a strong symbolic message with a powerful impact, especially when it comes to an image of a country (Herold 2004). Considering the fact that an independent film is enjoyed by international moviegoers, the issue of representation becomes a crucial one. The theme and story of the funded film highlighted a darker side of Indonesia with its widespread poverty and daily hardships. Ultimately, these films are perpetuating the stigma of Indonesia as a 'third world' country. A film funded by such programs becomes a double-edged sword: on the one hand, it promotes Indonesia to the international world; on the other, it may be considered as creating a problematic representation.

Some of independent filmmakers later went on making commercial films. Ismail Basbeth directed a commercial film entitled *Mencari Hilal* (2015). The film *Ada Apa dengan Cinta 2* (2016), which was seen by more than three million moviegoers and distributed to Malaysia and Brunei Darussalam, was directed by Riri Riza. These achievements show that the funding programs provided by European film festivals have actually made positive contributions to the development of Indonesian filmmakers, as well as the Indonesian cinema. The funding from European film festivals becomes the training ground for Indonesian filmmakers to sharpen their talents and skills. The festivals also serve as an arena to improve, to develop, and to promote themselves on the international stage. The Indonesian cinema has been enriched by a significant number of experienced filmmakers, a condition which will ultimately make a positive impact to the development of the cinema itself.

4 CONCLUSION

The provision of the funding from European film festivals has become a problematic issue. No one would argue that such funding programs can boost the skills and establish the career of the filmmakers who are receiving the fund and promote the Indonesian cinema on the world stage. However, the hegemony practices related to such programs are also perpetuating the image of Indonesia as a land burdened with problems. In other words, the funded films tend to propagate the stigma of Indonesia as a third world country. De Valck voiced her suspicion by stating, "It may not be a representation; maybe it is a repression" (Mubarak & Ageza, 2015). Such problematic representation also raises a new concept of orientalism or, "self-exoticization" (Taymuree, 2014).

Mubarak and Ageza (2015) said that there are only few studies that examine the relationship between Indonesian films and the taste of foreign film festivals. They asked whether the Indonesian films, which are often shown and admired by the programmers at European film festivals, belong to the "self-exoticization" group. They were also concerned that there could be an attempt to erase the diversity of world's independent films, one of which was through the politics of providing funding that would steer the direction of the films receiving those funds. This study seeks to fill the gap in the study of Indonesian films and the taste of European film festivals by trying to answer those above questions using Gramsci's theory of hegemony. This study can be expanded even further by, among others, using the ethnographic and political economy approach, namely by interviewing the recipients of the fund, film critics, observers, and other players in the film industry and festivals.

REFERENCES

Aartsen, J. (2011) *Film World Indonesia the Rise after the Fall*. [Online] Available from: http://dspace.library.uu.nl/bitstream/handle/1874/205138/thesis_filmworldindonesia_jv_aartsen.pdf?sequence=2.

Baumgärtel, T. (2011) Imagined communities, imagined worlds: Independent film from South East Asia in the global mediascape. *Transnational Cinemas*. 2, 57–71.

Cinéfondation. (2016a) *Presentation*. [Online] Available from: http://www.cinefondation.com/en/generalinformation.

Cinéfondation. (2016b) *Rules and Regulations*. [Online] Available from: http://www.cinefondation.com/en/rrules

Darmawan, A. (2015) Festival film, pendanaan film, dan ideologi ber-film (Film festivals, film funding, and the ideology of filmmaking). *Cinema Poetica*, 15th December. [Online] Available from: https://cinemapoetica.com/festival-film-pendanaan-film-dan-ideologi-ber-film/.

Davis, A. (2008) Investigating cultural producers. In: Pickering, M. (eds.) *Research methods for cultural studies,* Edinburgh, Edinburgh University Press Ltd. pp. 53–67.

Falicov, T.L. (2010) Migrating from South to North: The role of film festivals in funding & shaping global South film & video. In: Elmer, G., Davis, C.H., Marchessault, J. & McCullough, J. (eds.) *Locating Migrating Media*, Lanham, Lexington Books. pp. 3–22.

Halle, R. (2010) Offering Tales They Want to Hear: Transnational European Film Funding as Neo-Orientalism. In Galt, R. & Schoonover, K. *Global Art Cinema: New Theories and Histories*. New York, Oxford University Press. pp. 303–319.

Herold, A. (2004) EU Film Policy: Between Art and Commerce. *EDAP: European Diversity and Autonomy Paper* 3, 1–21. [Online] Available from: http://aei.pitt.edu/6160/1/2004_edap03.pdf.

IFFR [International Film Festival Rotterdam]. (2016) About HBF. [Online] Available from: https://iffr.com/en/professionals/iffr-industry/hubert-bals-fund/about-hbf.

Internationale Filmfestspiele Berlin. (2016) *World Cinema Fund.* [Online] Available from: https://www.berlinale.de/en/branche/world_cinema_fund/wcf_profil/index.html.

Iordanova, D. (2015) The Film Festival as an Industry Node. *Media Industries Journal* [Online] 1 (3), 7–11. Available from: http://www.mediaindustriesjournal.org/index.php/mij/article/view/98/123.

Jones, S. (2006) *Antonio Gramsci*. Oxon, Routledge.

Lent, J.A. (2012) "Southeast Asian Independent Cinema: Independent of What?" In: Baumgärtel, T. *Southeast Asian Independent Cinema,* 13–19. Hong Kong, Hong Kong University Press.

Mubarak, M. & Ageza, G. (2015) Agenda politik selera dalam festival film dunia (The political agenda of tastes in the world's film festivals). *Cinema Poetica* [Online] Available from: http://cinemapoetica.com/agenda-politik-selera-dalam-festival-film-dunia/. [Accessed on 30th September].

Procter, J. (2004) *Stuart Hall*. London, Routledge.

Shaw, D. (2015) European Support for Latin America Cinema. *Mediating Cultural Encounters through European Screens*. [Online] Available from: http://mecetes.co.uk/european-support-latin-american-cinema/. [Accessed on 6th February].

Taymuree, Z. (2014). Self-exoticization for the film festival. *Avicenna: The Stanford Journal on Muslim Affairs* [Online] 4 (1), 26–29. [Online] Available from: http://stanford.edu/group/avicenna/cgi-bin/wordpress/wp-content/uploads/2014/04/4.1-Self-Exoticization.pdf.

Turan, K. (2002) *Sundance to Sarajevo: Film Festivals and the World They Made*. Berkeley, Los Angeles & London, University of California Press.

The Lunar New Year and *Guanggunjie* 光 棍 节 tradition: The representation of China in the *Lentera* broadcast by China Radio International

R.P.S. Hadi & M. Budiman
Department of Literature, Faculty of Humanities, Universitas Indonesia, Depok, Indonesia

ABSTRACT: The cooperation between China Radio International (CRI) as the official radio of the PRC with the Indonesian national private radio began in 2010. One of the broadcast programs, known as "Lentera," is an on-air talk show that discusses various aspects of social life, politics, economy, history, and culture. This research specifically examines two broadcast topics in the cultural section, titled "the Lunar New Year celebration" and the "Singles Day/*Guanggunjie*". Through these two topics, the representations of China will be analysed, especially for the listeners in Indonesia. The Lunar New Year is also celebrated in Indonesia as a cultural festival and a tradition, particularly by the Chinese ethnic group. This research examines the correlation between the Lunar New Year celebration in China and the same celebration in Indonesia, in the context of the Indonesia-China relationship. This study also examines the cultural celebration emerging in this era of globalisation, which is the celebration of the "*Jomblo* (Singles') Day," or *Guanggunjie*. This festivity is held specifically for the youths who are yet to find a life partner. Through the celebration of traditional and modern festivities, the representation of China presented in the radio program in "Lentera" by CRI for the Indonesian listeners will be observed.

1 INTRODUCTION

Globalisation has enabled countries around the world to communicate and interact regardless of time and distance. One important tool to communicate a message from one country to another is the media. Through the media, messages can be delivered easily and directly from one place to another. Media may take different forms including the classic forms of media, such as the radio and television, as well as the media of mobile phones, computers, and the Internet.

As explained by Medeiros (2003), the People's Republic of China (PRC) began to undergo a transformation process by opening up to various countries in the world in the late 1970s, as a realisation of the "Reform and Opening up" Policy (*Gaige Kaifang* 改革开放) introduced by the leader of PRC at that time, Deng Xiaoping 邓小平. This transformation was followed by China's policies to open more bilateral relations aggressively in the 1990s. In this era of globalisation, China, as the most populous country in the world, also makes use of the advances in technology through the medium of CCTV4 and CCTV9 TV stations to broadcast news on China to the world community. Besides television, China also utilises radio programs to disseminate information through the China Radio International (CRI) or *Zhōngguó Guójì Guǎngbō Diàntái* 中国国际广播电台 network, the official radio station of the government under the State Administration of Radio, Film, and Television or *Guojia Guangbo Dianying Dianshi Zongju* 国家广播电影电视总局. This radio station is dedicated to establish cooperation with foreign radio stations and is used as a *soft-power* tool for the PRC to promote the country's economic, political, social, cultural, and defence aspects to the world. The CRI broadcasts are designed using the language of the destination country and present programs related to PRC's particular interests in a country. China affirms its presence in each country by establishing cooperation with local radio stations. Through this medium, China disseminates information on the situation, condition, and transformation in China from the past to the present, the economic growth, infrastructure development, lifestyle, international relations, and even the political situation of PRC. This is part of China's development process, emerging from a regional force to become a global force.

By observing the topics of the programs broadcasted in 2011–2013, the author analyses a variety of topics covering social, cultural, economic, and political themes which were aired in the 30-minute talk shows or interviews. These themes were broadcasted in fluent Bahasa Indonesia. Besides that, the CRI also hosts a website that is run in parallel with the radio broadcast to provide virtual information, supplementary articles and visuals such as photographs and pictures. Through this media, China enters its target countries by exposing various aspects of China to be more widely known and understood by the world community.

As mentioned by Storey (1996), the media provides information that can be interpreted ideologically because the media serves as an arena for various symbols dominating other symbols to represent a certain ideology. Since the media is a powerful tool to deliver messages, the cooperation between PRC radio stations and the Indonesian private radio stations can be interpreted as a means to convey certain messages related to the interests of China toward the Indonesian public. This study, therefore, is aimed to further investigate the representation of China in the selected topics broadcasted in CRI's "Lentera" program.

2 METHOD

The broadcast programs are selected by sorting out topics into the categories of celebrations/holidays/festivals throughout 2011–2013. The selected topics as the study focus are the Chinese New Year festivity and the newly celebrated "*Jomblo* (Singles') Day" or *Guanggunjie* 光棍节. The reason behind the selection of these two celebrations is to observe the messages delivered through CRI's "Lentera" program in relation to both the traditional and modern celebrations. However, this study is not intended to examine the link between the two celebrations, but to observe the ideology conveyed by each broadcast program, representing the traditional culture and the modern development in China. In this research, the dynamics behind the delivered message are observed and then followed up by a study on the representation of the two broadcast topics.

3 RESULTS AND ANALYSIS

The "Lentera" radio program of the Indonesian CRI generally presents topics related to culture. For instance, the topics of the Spring Celebration and the *Cap Go Meh* festivals are among the topics of discussion featured in the radio programs. These celebrations are known as typical Chinese celebrations that have long been known since the days of the dynasties. Nevertheless, aside from the discussion on traditional celebrations, several broadcasts of "Lentera" also presented other interesting celebrations that are considered a novelty, including celebrations adopted from outside of China and celebrations that are created in this globalisation era. Considering the popularity of the traditional Chinese celebrations in Indonesia, this research focuses on the topic of the Spring/New Year/Lunar celebration, which represents the traditional celebrations, and the *Guanggunjie* festivity that represents one of the new celebrations in this era of globalisation.

Broadcast programs featuring the theme of Chinese traditional celebrations generally focus on the Chinese traditional values as a cultural heritage that has long been known since the days of the dynasties. Until this day, these traditions are still upheld as part of the Chinese culture in Indonesia. The celebration to welcome the spring and herald the arrival of a New Year is known as the Chinese Lunar New Year which is specifically celebrated by the Chinese ethnic group, while the Solar Calendar New Year that falls on 1st January is also celebrated in Indonesia. However, the celebration of the Lunar New Year in Indonesia was just recently recognised openly, after the revocation of the *Inpres* (Presidential Decree) No. 14/67 of 2000 that previously restrained the activities related to the Chinese Culture. This policy annulment can be traced back to the history of the diplomatic relationship between Indonesia and China that has experienced ups and downs and has affected the freedom to openly express the Chinese culture in Indonesia. The impact from the annulment of *Inpres* No. 14/67 is indicated by the resumption of the public celebrations of Chinese culture in Indonesia. The Chinese ethnic group in Indonesia welcomed this freedom by exhibiting the performance of the *Barongsai* (Lion Dance) and openly displaying red-colored ornaments to celebrate the Lunar New Year day. These festivities gained popular attention from the Indonesian people in general, especially after the government officially designated the Lunar New Year day as a national holiday in Indonesia. As an intrinsic but exotic culture from outside of Indonesia, the celebration of the Lunar New Year has earned significant recognition involving the general public of Indonesians in expressing their respect toward the celebration of the Lunar New Year. This development has obviously made it easier for "Lentera" to present this topic to the radio listeners in Indonesia. The broadcast on the topic of the Lunar New Year is selected based on the similar cultural codes of China and Indonesia to identify the comprehension patterns and the renewed relationship between China and Indonesia. In binding the cultural pattern between China and Indonesia, this broadcast aims to revive the native culture of the Lunar New Year celebration, associated with the Chinese culture in Indonesia. The tracking of the Chinese cultural traces in other countries would also make it easier for China to "recall" the originality and the origin of their culture. By applying this method, the once disrupted relationship between Indonesia and PRC could be gradually restored through the power of culture. The influence of the Chinese culture is evidence of China's success in employing its *soft power* introduced to Indonesia. The representation of China through these radio programs carries the meaning of *re-presentation* or to reintroduce China to Indonesia. Thus, the dominant ideology disseminated through these radio programs is aimed to restore China's cultural power in Indonesia.

Unlike the traditional celebrations, the *Guanggunjie* festivity, which is a relatively new celebration, provides a different cultural angle. *Guanggunjie* is a special celebration for people who do not have a partner or, in other words, who are single. As elaborated in the "Lentera" program, this celebration is a fairly new festivity initiated by a group of university students. *Guanggunjie* is translated as "*Hari Jomblo*" – a slang word in Indonesian language which is used as a lingua franca among Indonesian youths—by the radio announcer of "Lentera". The use of an informal term shows that "Lentera" not only serves as the source for official news, but also presents trivial and current trends. This topic has gained wide attention, as it shows China's new paradigm of becoming more open to new and trivial issues and current trends, which are quite contrary to the conventional, closed and traditionally strict attitude, which is closely associated to the past era of the Chinese. This topic provides a contrasting view toward China that is commonly perceived as a closed culture. In regards to this phenomenon, the broadcaster stated:

> "...Khususnya di kalangan muda, banyak sekali festival-festival. Ada Halloween, Thanksgiving, Valentine. Pertama kali itu trend-nya dari Amerika.... dan banyak makanan. Mengajari kita untuk bersyukur. Kalangan muda di China membuat trend sendiri, ada suatu perayaan yang menetapkan tanggal 11 bulan November sebagai hariJomblo. *Guānggùn Jié* 光棍节 dimulai dari kalangan mahasiswa. Kalau di Jakarta namanya jomblo, artinya belum punya pasangan. Kalau Jumat Sabtu di rumah, pasti jomblo. Banyak sekali kota besar di dunia, banyak orang tidak berhasil menemukan pasangan..."

> ("..Particularly among the young people, there are a lot of festivals. There are the Halloween Day, Thanksgiving Day, and Valentine's Day. Initially, this trend comes from America.... and lots of food is involved. They teach us to be grateful. The young people in China created their own trend. There is a celebration that is set on 11 November as the day for single people. The *Guānggùn Jie* 光棍节 was initiated by university students. In Jakarta, the singles are known as *jomblo*, meaning not having a partner. On Fridays or Saturdays, if young people stay at home, they are

definitely single. In many major cities in the world, many people are unable to find a partner...").

This statement illustrates how the Chinese younger generation of today is influenced by globalisation, where the celebrations of the western world have penetrated the young people in PRC. These celebrations have inspired the young Chinese people to create their own unique celebration, the *Guanggunjie* celebration every 11th November. The following statement explains why the broadcast program on the 11th day of the 11th month has been selected:

"... Dulu festival ini diciptakan oleh mahasiswa Universitas. Di Universitas banyak yang belum pacaran. Satu satu itu kan sebelas. 11 itu kan satu satu. Tanggal 11–11 mewakili satu satu satu satu. jadi 4 – 1. Perayaan paling meriah itu tahun 2011. Jadi ada 6 angka 1...."

("...This festival was initiated by university students. At the time, the university students were not dating anyone. The number eleven is formed by two number "1"s. The date 11–11 represents one, one, one, and one. So, there are four digits representing "1"s. The most festive celebration was in 2011 because the date was formed by six digits of "1"s.")

As explained in the broadcasted program, the date of 11 November was selected based on the Roman numerical characters of "1111" that represents 11 November, and according to the young generation in China, these numbers symbolise four people standing in a row. The four people standing in a row are associated to the idea of four people standing with no partner. The broadcaster also explained that the most celebrated occasion was in 2011, at which the date was 11th November 2011 represented by "111111". Therefore, in their opinion, the more numbers that are lined up in a row (there are six "1"s), the "more single people" there are.

In the following segment, the broadcast program discussed the activities of some of the young people during the *Guanggunjie* festival which included online shopping as a spare time activity for those that do not have a partner. The discussion about online shopping used two out of the total three broadcast segments. The shift of topics from the celebration to the lifestyle of shopping by the sender or the encoder shows that the delivery of messages is completely controlled by the encoder. What is produced through the media is a construct of the message sender (encoder) to be delivered to the message receiver (decoder). The following is the narration from the broadcaster:

"...Semua orang bisa masuk internet dan ikut promosi ini, dimulai jam 24.00. Promosi ini sangat sukses. Karena 1 bulan sebelumnya sudah ada iklan. Mereka kasih diskon 50%, jadi kasih separuh..... Saya sudah lama memperhatikan satu brand, dia hari itu bukan saja kasih diskon, tapi juga kasih kupon. Hampir 2000 yuan. Jadi, hematnya 2000 juga, ya. Ada banyak kupon dan memilih item-item yang kamu suka sebelumnya. Mulai jam 12 bisa beli, bisa bayar. Saya, semuanya, sudah mempersiapkan. Saya udah rebut kupon, sudah pesan, jam 12 tinggal klik di online..."

"...Everyone can get on the Internet and join this campaign, starting at midnight. This promotion was very successful because it has been advertised since the previous month. They are giving 50% discounts, so we are given half the price. I have long observed a brand which, on that day, did not only give a discounted price, but also offered coupons nearly worth 2000 Yuan. So, the customer can save 2000 Yuan. There are plenty of coupons and you could select the items you like. You can just buy and pay for the items, starting at midnight. I have prepared everything. I"ve managed to get a coupon and made an order, and at midnight I just need to click..."

From this statement, it is apparent that there is an opportunity to promote bargain shopping via the internet during the celebration of *Guanggunjie*. The vendors took advantage of this opportunity to reap profitable results and they gained a huge amount of profit from the *Guanggunjie* special offer, as reflected in the following statement:

"... Hasil promosi kali ini menurut statistik 19.1 milyar yuan. Bisa dibandingkan dengan toko-toko lain. Hampir sama dengan 1400 toko McDonald. Keuntungannya berapa kali lipat dari McDonald. Hampir sama dengan 130 perusahaan bisnis di Tiongkok. Dimulai pada tahun lalu. Gaobao ini sudah belajar banyak dari tahun lalu. Internet macet sekali. Dua sifat yang utama, harga benar-benar murah, item-nya bener-bener buat orang biasa. Bukan yang mewah-mewah. Tetapi barang-barang bermerek juga ada. Ada, tapi jumlahnya kecil. Yang rebutan itu Ferragamo. Cuma ada 1 tas. Harganya 10.800 yuan. Tapi harga nya saat itu cuma 1111..."

"... The profit from the promotion this time, based on the statistics, was 19.1 billion Yuan. Comparing it with other businesses, this is almost equal to the amount from 1,400 McDonald outlets. The profit was several times higher than that of McDonald's. It is almost equal to 130 business enterprises in China. Since last year, Gaobao has learned a lot. The website became jam packed or overcrowded for two main reasons: either the price was really cheap, or the items were really popular for the common people. Not too fancy. However, the designer goods were also available although in limited quantity. The most wanted item is the Ferragamo goods. There was one bag which is usually priced at 10,800 Yuan, but the special price was only 1111 Yuan..."

The above statement of the CRI announcer shows that online shopping via the internet has become a profitable business in China. Besides that, it reflects how the *Guanggunjie* event is considered a worthy opportunity to promote online businesses because it is a good opportunity to double the profit. "Lentera" even compared it to the sales of 1,400 outlets of a well-known American franchise restaurant. Apparently, in addition to featuring the *Guanggunjie* festivity every 11 November, "Lentera" also shows another aspect of the celebration, namely the business aspect as indicated by the increase in the sales turnover and the profit reaped by the online entrepreneurs. Hence, the *Guanggunjie* celebration discussed in the "Lentera" radio program was described as a stimulating celebration. Nevertheless, the reason for this celebration is actually linked to the historical background of the cultural phenomenon, which is signified by the surplus number of single people in China in the 1990s. This phenomenon only appeared after the 1990s at which time the Chinese government enacted the "One Child Policy" to limit the population growth in China which is one of the highest in the world.

The *Guanggunjie* celebration that emerged in the 1990s in PRC was an after-effect of the growing number of single people which was prompted by the rapid economic growth, creating a fully busy work force of today's Chinese society. The cultural background of this issue started in 1979 when the PRC government implemented an economic reform. This economic reform was a major progress for China after experiencing stagnancy due to the Cultural Revolution (1966–1976). Hesketh *et al.* (2005) state that:

"...At the time, China was home to a quarter of the world's people, who were occupying just 7 percent of world's arable land. Two thirds of the population was under the age of 30 years, and the baby boomers of the 1950s and 1960s were entering their reproductive years. The government saw strict population containment as essential to economic reform and to an improvement in living standards. So the one-child family policy was introduced..."

However, the success in controlling the population growth was not matched by overcoming the social inequality. This inequality particularly refers to the psychological problems arising from this policy. When the children that became the only child of the family had grown up, the population of men and women became unbalanced. This ratio imbalance has resulted in a greater number of males compared to the female population. Thus, the broadcast of *Guanggunjie* basically represents the impact of the government's One Child Policy.

4 CONCLUSION

The CRI represents China in its every broadcast and reflects China's interest for the Indonesian listeners. The traditional Lunar New Year celebration, which was illustrated through the broadcast of "Lentera", refers to an important celebration in the traditional Chinese society to welcome the Lunar New Year. The hope for happiness and prosperity as part of this tradition has also become part of the Lunar New Year celebration in Indonesia. The broadcast of "Lentera" transmitted the cultural codes through the features of the Lunar New Year celebration that can be directly and clearly adopted by the Indonesian listeners. This research has revealed the significance of this broadcast for the Indonesian listeners, which is to reconnect the traces of the once lost Chinese culture in Indonesia and to recall the originality of this Chinese celebration.

Meanwhile, the broadcast topic of the *Guanggunjie* celebration illustrates the openness toward the Chinese non-traditional and contemporary celebration that exhibits a festive mood and describes how exciting the situation is, which is associated with the increasing number of people "that do not have partners." The festivity is depicted by those single people going out to the cinema, with no partner, and doing online shopping via the internet as featured in the general overview in this broadcast. The fact that there is an increasing number of young people in China who are single has roused the curiosity to understand the background of this event. This research found that the significant increase of the young people that do not have partners is closely related to the "One Child Policy" implemented since 1979. The impact of this policy has led to the emergence of social problems, such as the unbalanced ratio between women and men, the rising number of abortion cases, the raising age of people getting marriage, and other social issues. The problems of this generation of singles began in the 1990s, when the children born in the 1980s became teenagers or were at the "marrying" age.

Thus, this research shows that the representation of China in the broadcast programs reflects a modern China that is respectful of both the traditions and the new trends. The relations between countries such as China and Indonesia can be improved through the exposure of the same cultural codes in celebrating the Chinese tradition in Indonesia, such as the Lunar New Year celebrations. Therefore, the radio can be an effective instrument to convey the Chinese representation to the Indonesian listeners. Nonetheless, China's representation in the broadcast of "Lentera" cannot be separated from the events behind it. The relationship between China-Indonesia that once was deranged should be restored by sharing a common understanding of traditional celebrations, such as the Chinese New Year festivity, and the excitement of modern celebrations such as the *Guanggunjie*, which actually represents a social issue in China"s progress in this era of globalisation.

REFERENCES

China Radio International Indonesia. (2011) *Profil Lentera (The profile of Lentera)*. [Online] Available from: http://indonesian.cri.cn/481/2011/05/10/1s118255.htm.
China Radio International Indonesia. (2012) *Hari Jomblo (Singles' Day)*. [Online] Available from: http://indonesian.cri.cn/201/2012/12/10/1s133678.htm.
Hall, S. (1997) *Representation: Cultural Representations and Signifying Practices*. London, Sage Publications Ltd.
Hesketh, Li Lu, M.D. & Zhu Wei Xing. (2005) *The Effect of China's One-Child Family Policy after 25 Years*. [Online] Available from: http://www.nejm.org/doi/full/10.1056/NEJMhpr051833.
Medeiros, E.S. & Fravel, M.T. (2003) China's New Diplomacy. *Foreign Affairs*, 82(6), 22–35.
Storey. (1996) *Cultural Studies and the Study of Popular Culture: Theories and Methods*. Edinburgh, Edinburgh University Press.
Zlatar, A. (2003) *The role of media as an instrument of cultural policy: An inter-level facilitator and image promoter*. Bucharest: Policies for Culture—European Cultural Foundation, Amsterdam & ECUMEST Association.

Cultural Dynamics in a Globalized World – Budianta et al. (Eds)
© 2018 Taylor & Francis Group, London, ISBN 978-1-138-62664-5

Mandarin education and contemporary Chinese-Indonesian identity repositioning: Between *recinicization* and cosmopolitanism

Q.Q. Luli & M. Budiman
Department of Literature, Faculty of Humanities, Universitas Indonesia, Depok, Indonesia

ABSTRACT: Over the last few years, Indonesia has witnessed the booming of Mandarin schools in its major urban settings. This emerging phenomenon is especially obvious among Chinese-Indonesian parents who were born after the 1966 ban on communism in Indonesia. Children aged 6–13 years are sent to modern Mandarin-based international schools that use not only the Indonesian language as the instructional language, but, more importantly, Mandarin and English as well. This cultural trend in contemporary Indonesia has raised a question: are we seeing the reassertion of 'Chineseness' among Chinese-Indonesians, which is oriented towards the 'ancestral homeland' (mainland China), or is it based more on a cosmopolitan urban outlook, which does not consider cultural boundaries as an essential part of their identity markers? This paper seeks to address this issue based on field research conducted in two of the largest cities on Java Island, the primary island of Indonesia. The concepts of Chineseness and cosmopolitanism developed by Ien Ang and other Chinese diaspora scholars will be used as the main framework to discuss the issue. This paper argues that the kind of 'Chineseness' constructed by contemporary Chinese-Indonesian identity tends to be less closely related to the essentialist concept of 'recinicization' but more closely related to a fluid idea of cosmopolitanism, which, in turn, helps redefine nationalism in the specific Chinese-Indonesian context.

1 INTRODUCTION

This study is based on field research conducted in the cities of Jakarta and Surabaya. The criteria of target schools in the chosen cities are those with primary and secondary schools that use international curricula, in which the teachings are carried out in three languages, namely English, Mandarin and Indonesian language, with a significant number of students.

This type of schools began to emerge not long after the collapse of the New Order regime in 1998, which had ruled the country for 32 years and strictly prohibited the teaching and dissemination of the Chinese language and culture. The end of the authoritarian rule in Indonesia was quickly followed by a brief political turmoil, which included the lootings and burnings of Chinese-Indonesian properties, as well as the infamous and controversial cases of mass rape of Chinese descent women.

The tragedy left a scar in the collective memory of the nation, particularly among the Chinese-Indonesian communities across the archipelago; however, as Melani Budianta suggested, it also serves as a sort of 'blessing in disguise' (2003) because, for the first time after 32 years of discrimination and restriction, the Chinese-Indonesians have found a determination to voice their political views and urged the government to lift the ban on the Chinese language and culture.

The ban was eventually lifted by former President Abdurrahman Wahid in 1999, and there has been a booming interest among the Chinese-Indonesian communities in learning the Mandarin language and revitalising their cultural traditions ever since. Wahid took a step further by officially recognising Confucianism as one of the major religions in Indonesia. The reintegration process of Chinese cultural tradition into the Indonesian socio-cultural system culminated with the release of a presidential decree in 2003 by former President Megawati Sukarnoputri, which regulated that the Chinese New Year (*Imlek*) is celebrated as a national holiday.

One of the consequences of the lifting of the ban on the Chinese language and culture is the proliferation of kindergarten, primary and secondary schools that offer classes using Mandarin—along with other languages. This is viewed as the revival of the enthusiasm and curiosity to understand more about many aspects related to the Chinese culture both among Chinese-Indonesians and the indigenous population of Indonesia. In addition to formal schools, informal private institutions also offer courses of Mandarin that caters to different groups and needs, ranging from young children to housewives to professionals. Furthermore, Chinese-language newspapers and magazines also began to be published widely (Hoon, quoted in Tsuda, 2012, p. 192). At the international level, as Kitamura (2012) suggested, the 'rising China' factor, which refers to the rise of China as a new superpower of the world, also plays a part in the promotion of Mandarin in Indonesia.

The significant surge of interest in the Chinese language and culture since the Reform period raises a few questions: What is actually happening in the socio-cultural context of Chinese-Indonesians regarding their positioning in terms of their status as 'Chinese diaspora' in Indonesia? Does the renewed interest in the culture of their ancestors signify an emerging phenomenon of 'recinicization' (the formation of 'Chineseness' that is primarily oriented towards the mainland China) or is it part and parcel of a larger, global phenomenon of cosmopolitanism as a result of globalisation (individuals no longer see themselves as defined by national boundaries and identities but transcending those traditional borders and claiming a global citizenship)? Furthermore, if a search for 'Chineseness' indeed takes place, this study tries to investigate what kind of 'Chineseness' that is evolving in the Indonesian context in the post-New Order era.

In a research published in 2012, Koji Tsuda asked what motivated Chinese-Indonesian students in the town of Rembang, Central Java, to attend Chinese-language classes and if 'Chineseness' had anything to do with it and whether it was part of the attempt to acquire more 'authenticity' in terms of their Chinese identity. Tsuda's research results were ambiguous. The results in the sense point towards a possibility of 'recinicization,' but nothing was conclusive because learning the language at that point was simply part of the students' effort to gain experience.

2 METHOD

On the basis of the field research done in trilingual schools in Surabaya and Jakarta through interviews with the students' parents, almost two-thirds of whom were born after 1967; when communism and anything related to the Chinese culture were banned, this paper develops the results of Tsuda's research further by connecting the emergence of international schools offering Mandarin courses not only in the efforts to rediscover 'Chineseness', but also with the aim of becoming global citizens of the world, in which 'Chineseness' is just one out of many building blocks.

Similar research has been done by Lee and Chen (2007), who interviewed 29 parents and students of elementary and secondary schools to find out whether Chinese-Indonesian parents preferred sending their children to 'internationally oriented' schools or 'Chinese-oriented' schools. The results show that only two respondents chose 'Chinese-oriented' schools because they wanted to better prepare their children for further studies in Western countries and, if possible, to settle down there afterwards. Our research aims to explore more in-depth and reveal the complexity of the process to avoid any simplistic generalisation in our understanding of the situation.

3 RESULTS AND ANALYSIS

This paper sets out to find if the growing interest among Chinese-Indonesian parents to send their children to trilingual schools with international curricula is related to the rediscovery of 'Chineseness' ('recinicization') oriented towards mainland China or whether it is an indication of a cosmopolitan lifestyle as a result of globalisation. The two surveys conducted in Surabaya and Jakarta provide some interesting insights into how Chinese-Indonesians growing up in the New Order era envision themselves in three different worlds: Indonesia, China and the global world. Some important observations regarding these parents that can be explored and discussed further are as follows:

1. They send their children to quality schools with international standards not only to master Mandarin but also to learn English;
2. They are proud of their 'Chineseness' and view the ability to speak Mandarin as an added value to it;
3. Many of them have travelled to China and some other countries;
4. Most of them consider Indonesia as their permanent homeland;
5. Many of them could no longer speak or write in Mandarin, but they want their children to be able to do so;
6. Respondents in Jakarta come from lower middle-class background and are more proud of being Chinese than being Indonesian, whereas those in Surabaya come from middle to upper-middle-class background and are more proud of being Indonesian than being Chinese;
7. Two respondents in Jakarta accuse the contents of the questionnaire to be too racist and prejudiced against the Chinese, but none in Surabaya.

There are similar patterns that create a coherent picture of urban middle-class Chinese-Indonesians living in the two major cities in Indonesia. As such, how they perceive themselves in a cosmopolitan setting is also shaped by the rise of China in Asia and the world, as well as by globalisation. At the same time, there is also a contradiction and perhaps, some ambivalence, with regard to their positioning within the Indonesian context. The importance of mastering the Mandarin language is seen as an absolute necessity, and while not all of them are able to speak Mandarin, they want to ensure that their children are able to acquire the language. Not only do they think that Mandarin has become one of the global languages, but also they believe that it is part of their cultural identity, an identity that has temporarily faded, but can be regained by the next generation of Chinese-Indonesians who are now attending trilingual schools. The mastery of Mandarin then serves a dual purpose: on the one hand, it serves as one of the success factors and an important skill in the future global setting, while functioning as a marker of their cultural identity on the other.

Fransisca Handoko even drew a more local and pragmatic conclusion in her study on why there has been a boom of Mandarin informal schools after the end of the New Order era. She argued that the mastery of the Mandarin language is considered strategically important by not only the Chinese-Indonesians but also Indonesians with non-Chinese background of all age groups. Their primary motivation, according to Handoko, is basically economic-driven, in the sense, that the ability to speak Mandarin would enhance their opportunity in the job market rather than having anything to do with, say, a yearning for the fatherland or rediscovery of the ancestral traditions (2010, pp. 368–369). Such a pragmatic motivation is naturally triggered also by the rising economy of China, and as a result, the mastery of Mandarin language becomes one of the requirements in many vacancies in Indonesia. The study conducted by Handoko indicates a fundamentally different type of conclusion from what this research discovers.

Travelling seems to have opened up the parents' horizon of the world. They see that America, Europe and Australia still offer good living standards; therefore, many of these parents prefer living in those places to living in China, although a significant number of the parents still seriously consider Indonesia as their permanent home country. The parents certainly have an expectation that their children would later be able to survive and succeed living overseas in a very competitive atmosphere alongside higher standards of living. This could be the reason why, in addition to Mandarin, English appears to be highly important. They strongly believe that it is essential for their children to master these two international languages in order to make a good living, whereas non-Chinese-Indonesian parents seem to have taken that issue for granted. It happened because, after all, their home base would still be in

Indonesia no matter where their children would decide to live in the future. China is not the favourite option as a permanent home country for most of the Chinese-Indonesian parents. Even though the rise of China has brought awe and praise and helped boost their pride in having Chinese ancestral background, they seem to be fascinated more by the achievements of the Western world.

The fact is that many of the parents do not speak or write Mandarin (except, perhaps, their own Chinese names), but they still feel proud of their Chinese origin, especially those living in Surabaya, and this speaks volume of the Chinese-Indonesians' unique identity construction. This is certainly different from the kind of experience with the 'Chineseness' that Ien Ang discussed in her celebrated book, *On Not Speaking Chinese* (2001), in which she recounts her feeling of being categorised as an 'overseas Chinese' who looks Chinese, but was unable to speak Chinese during her visit to China as a tourist. Ang wrote that she considers the experience as an 'embarrassment'. Ang is the daughter of a *peranakan* (of Chinese descent) family who was born in Indonesia and whose family has embraced fully the local traditions of their new homeland and has practically cut off any connection they had with their land of origin. Thus, the loss of language and cultural knowledge for Ang's family and perhaps for many other *peranakan* families is not the result of the repressive policy of the New Order towards Chinese-Indonesians. Such disconnection, in fact, was largely spurred by an imperial policy in China in the 18th century, which prohibited Chinese from leaving or re-entering China, and this strict regulation was even reinforced with a death-penalty sanction (2001, p. 26). This is probably why, after they have left their homeland, the Chinese people who have migrated to other parts of Asia typically no longer dream of being able to come back or hope to still belong to China.

Much of the literature about the loss of ancestral knowledge and traditions experienced by Chinese-Indonesians puts the blame on the New Order's 32 years of rule (among others, see Thung, 2010 and Dawis, 2010). The most frequently referred to is the infamous Presidential Instruction issued in 1967 on the 'Chinese Problem', which stipulated that Chinese-Indonesians had to assimilate fully into the Indonesian culture. This means they had to practically abandon or hide the language and traditions of their ancestors. This policy may have had a much greater impact on the *totoks*, the Chinese who had only left China in the late 19th century, more than on the *peranakan*. The Chinese-*peranakan*, judging from their history of migration that Ien Ang retells, might have already lost much of their collective memory regarding China even long before the New Order came into power. For many of them, that Indonesia is now their homeland is accepted as a fact. This probably explains why most of the respondents of this research, particularly those living in Surabaya, claim that they felt more proud of being Indonesian than being Chinese. However, by contrast, why do the respondents in Jakarta claim that they are more proud of their ancestral land?

The May 1998 riot, followed by the alleged mass rape of Chinese-Indonesian women, was unequalled in Jakarta, where hostility towards Chinese-Indonesian communities during the transition period was considerably strong, compared with the other major cities. In Jakarta, more than 1,000 people died during the lootings and burnings of mostly Chinese properties, and the joint-fact finding team (TPGF) formed by the government to investigate the sexual assaults on Chinese-Indonesian women found at least 100 confirmed victims of mass rapes in Jakarta area. The Chinese-Indonesian communities' experience of the riot is extremely traumatic, and the 'collective scar' among them, as Aimee Dawis referred to in her study (2010, p. 35), is significantly deeper. This is especially true among the lower middle-class Chinese-Indonesians who were vulnerable and weak because they received very little protection from the security apparatuses compared with those wealthy Chinese-Indonesian families living in gated communities that were heavily guarded during the riot. The 1998 riot would serve as a reminder, particularly for those living in North or West Jakarta, that—as Chinese-Indonesians—they would always be viewed as outsiders, estranged from their own homeland far away in China. This probably also explains why the questionnaire received some negative responses and being labelled as 'racist' as the ethnicity issues seem to be highly sensitive to them, even more so than to their fellow respondents in Surabaya.

We believe that the trauma caused by the riot plays an instrumental role in the perception among lower middle-class parents in our survey, who mostly come from the surrounding riot-hit areas in Jakarta. Although they feel prouder of their ancestral land than Indonesia, they share the same sense of attachment to Indonesia as those in Surabaya, and have motivation to send their children to trilingual schools that teach that Mandarin has little to do with the nostalgic sentiment towards China. In fact, more than 50% of the parents stated that they want their children to continue living in Indonesia after they finish school, and only 11% would identify themselves as Chinese (instead of Indonesian or Chinese-Indonesian when travelling to other countries). Moreover, even though many of them come from the lower middle-class background, they have developed a cosmopolitan tendency towards envisioning the future of their children. More than 50% of the respondents have travelled abroad despite their claim of low income in the questionnaire. This shows that either they might not have been completely candid when answering the questionnaire or they may not have fully regained their trust in others outside their in-group and are still concerned with their safety.

The most important factor is the lack of the essentialists' attitude about the Chinese cultural identity among the respondents. Although they openly claim to be proud of their 'Chineseness', they definitely do not see such 'Chineseness' in a deterministic sense. It is the kind of 'Chineseness' that will only begin to make sense when understood as rooted in their adopted homeland of Indonesia. The data obtained from both sites of our research show the focus of attention as 'selective Chineseness' that takes into account the relationship of hybridity with their 'Indonesianess' that is also distinct from what would be taken for granted by non-Chinese-Indonesians. This is the sort of 'Indonesianess' that has been seeped historically, politically and economically into the psyche of our Chinese-Indonesian respondents. Ien Ang wrote, 'If I am inescapably Chinese by descent, I am only sometimes Chinese by consent. When and how is a matter of politics' (2001, p. 51). The Chinese-Indonesian respondents in our research would have no problem in understanding what Ang was talking about. Their answers to the points in the questionnaire clearly indicate that they take full advantage of the flexibility provided to them by their hybrid identity. For certain issues, they would without a doubt identify themselves as Indonesians; however, for some other issues, they would quickly 'shift' to their Chinese identity. Their cultural identity, in short, is extremely fluid and versatile.

Gergen (quoted in Dawis, 2010, p. 36) referred to such an ability to switch from one aspect of cultural identity to another, which appears to be incompatible to each other, as 'pastiche personality'—a chameleon-like feature that enables one to draw resources from various identities to take as much advantage as possible from the situation. However, we think that 'pastiche' and 'chameleon' are not the accurate terms to describe our respondents. 'Chineseness' and 'Indonesianess' in their cultural identity are visible and in action simultaneously, instead of showing only one identity, while the other is hiding and vice versa. Their hybridity creates a bridge that connects the two aspects of their identity and allows them to blend as well as synergise with each other. The respondents may have sounded very nationalistic, but at the same time they are willing to transcend any national boundaries to embrace the world. Traumatic experience and years of discriminatory practices against them have not weakened their sense of being Indonesians, and the 'rising China' phenomenon has not blinded them with infatuation with their ancestral land.

The question of language remains a central one in this study. What role does language really play in the whole scheme of things of Chinese-Indonesians' self-positioning? The growing interest among parents in sending children to schools that offer not just any foreign language but particularly Mandarin may give an impression that 'recinicization' is occurring, especially as the mastery of Mandarin is associated with the sense of pride regarding their 'Chineseness'. Nevertheless, on other points in the questionnaire, the responses do not show any indication of a consistent pattern that would otherwise reinforce such initial impression. English is equally deemed important, and the number of parents wanting to send their children to developed countries in the West is significantly higher than those wanting their children to continue their education in China or live in China. Similarly, on whether these parents have totally embraced the cosmopolitan sensibility by not shying away from revealing their wish for their children to be able to go and live in other countries, we find that while a cosmopolitan outlook is visibly present, the majority of the respondents remain convinced in their choice of Indonesia as their permanent homeland. The readiness to cross-national boundaries is there, but they are not ready to leave for good. English becomes instrumental, yet that does not mean they undermine the importance of the Indonesian language. For them, the Indonesian language is the mother tongue, but Mandarin and English are foreign languages that need to be mastered with some extra effort.

4 CONCLUSION

As this is an ongoing research, our qualitative interpretation of the data will need to be further verified through other means such as interviews and closer observations. At this stage, we are likely to prove that the growing phenomenon among Chinese-Indonesians who send their children to trilingual schools, with Mandarin as one of the languages being taught, need not be interpreted as part of a 'recinicization' process together with the rise of China lately. It should also not be regarded as a wholesale cosmopolitanism as a result of encounters with globalisation, for Indonesia remains a pivotal point in the lives of our Chinese-Indonesian respondents. We hope that with the additional data obtained from two other cities, Medan and Pontianak, some coherent patterns may come up, which will help us to draw our conclusions in a reliable manner. If, at this point, identity positioning among the respondents can be detected from an ambivalent standpoint, we do not view it in the negative light because we believe that, in terms of hybridity, such ambivalence serves as a resource rather than a weakness or constraint. In the next stage of our research, we may even have to pay more attention to this ambivalent positioning in order to reveal the complexity of such a positioning process, which may be key to understanding contemporary Chinese-Indonesian cultural identity.

REFERENCES

Ang, I. (2001) *On Not Speaking Chinese: Living Between Asia and the West*. London, Routledge.
Budianta, M. (2003) The blessed tragedy: the making of women's activism during the Reformasi years. In: Heryanto, A. & Mandal, S.K. (eds.) *Challenging Authoritarianism in Southeast Asia: Comparing Indonesia and Malaysia*. London & New York, Routledge Curzon, pp. 145–176.
Dawis, A. (2010) *Orang Indonesia Tionghoa: Mencari Identitas (Chinese Indonesians: Searching for Identity)*. Jakarta, Gramedia Pustaka Utama.
Gergen, K.J. (2000) *The Saturated Self: Dilemmas of identity in Contemporary Life*. New York, Basic Books.
Gudykunst, W. & Ting-Toomey, S. (1990) Ethnic identity, language and communication breakdowns. In: Giles, H. & Robinson, P. (eds.) *Handbook of Language and Social Psychology*. New York, Wiley, pp. 309–327.
Handoko, F. (2010) Perjalanan bahasa Mandarin dalam dunia pendidikan di Indonesia (The journey of Mandarin in the education sector in Indonesia). In: Moriyama, M. & Budiman, M. (eds.) *Geliat bahasa selaras zaman: Perubahan bahasa di Indonesia pasca-orde baru (The twists and turns of languages according to the eras: Language changes in post-new order Indonesia)*. Tokyo: ILCAA, Tokyo University of Foreign Studies, pp. 355–371.
Hoon, C.Y. (2008) *Chinese Identity in Post-Suharto Indonesia: Culture, Politics, and Media*. Eastbourne, Sussex Academic Press.
Kitamura, Y. (2012) Chinese in the linguistic landscape of Jakarta: Language use and signs of change. In: Foulcher, K., Moriyama, M. & Budiman, M. (eds.) *Words in Motion: Language and Discourse in Post-New Order Indonesia*. Singapore, NUS Press, pp. 191–211.
Lee, M.H. & Chen, Y.L. (2007) Identity and educational choices: Middle-Class Chinese Indonesians in the Post-Suharto era. *Journal of Southeast Asian Studies*, 4(2), 25–52.
Thung, J.L. (2010) Orang Cina dalam bahasa politik orde baru (The Chinese in the New Order's Political Language). In: Moriyama, M. & Budiman, M. (eds.) *Geliat bahasa selaras zaman: Perubahan bahasa di Indonesia pasca-orde baru (The twists and turns of languages according to the eras: Language changes in post-new order Indonesia)*. Tokyo: ILCAA, Tokyo University of Foreign Studies, pp. 275–305.
Tsuda, K. (2012) Chinese-Indonesians who study Mandarin: A quest for 'Chineseness'? In: Foulcher, K., Moriyama, M. & Budiman, M. (eds.) *Words in motion: Language and discourse in post-new order Indonesia*. Singapore, NUS Press, pp. 191–211.

Cultural Dynamics in a Globalized World – Budianta et al. (Eds)
© *2018 Taylor & Francis Group, London, ISBN 978-1-138-62664-5*

Appropriating South Korean popular culture: I-pop and K-drama remakes in Indonesia

S.M.G. Tambunan
Department of English, Faculty of Humanities, Universitas Indonesia, Depok, Indonesia

ABSTRACT: The dynamic flow of South Korean popular cultural products has transformed the way "East Asia" is perceived in Indonesia. Consumption of these products greatly influences consumers' imagination, as argued by Appadurai, which becomes an arena of negotiation as well as contestation in the sites of individual and communal agency. On the one hand, consumers in Indonesia perceive these products as a representation of "East Asia," which is considered a well-known entity that is often used as a strategic defense mechanism (i.e., against Western cultural domination). On the other hand, consumers are seeking for more familiar goods in the form of mimicking products, such as I-pop's boy/girl band or Indonesian television *sinetron* copying plots from K-dramas. This paper investigates how cultural borrowing and/or appropriating are strategically used in the meaning-making process of the new global and the modern portrayed by South Korean popular cultural products and their copycat versions. The author argues that, instead of focusing on the authentic and inauthentic cultural products that oversimplify the debate, this phenomenon should be analyzed as a form of pastiche, which reveals the repetitive nature of popular culture flows in Asia as well as of cultural borrowing and/or appropriating.

1   INTRODUCTION

On April 28, 2014, RCTI, one of the biggest television stations in Indonesia, broadcast a local television soap opera, also known as *sinetron*, entitled "Kau yang Berasal dari Bintang," which can be literally translated to "You who Came from the Star." Within the first week of its release, there was a massive outcry accusing the television soap opera as a plagiarized version of a South Korean soap opera, "Man from the Stars" (SBS), which was very popular in 2014. The outcries came from Indonesian fans as well as those from other Asian countries who claimed that the Indonesian version had copied the storyline, the characters, and even the opening scenes. SBS, who had the official intellectual property right of the television show, threatened to take legal actions, which made RCTI decide to withhold the broadcast and finally change the storyline.

Indonesia has been accused, even by Indonesians themselves, as a country that plagiarized a large variety of popular cultural products from many countries, especially East Asian countries. This reflects how the flow of East Asian cultural products in Indonesia has not only ignited the rapid rise of Korean/Japanese wave in Indonesia but also the scandalous production of these plagiarizing products. Besides *sinetron*, the rising popularity of I-pop, which could be seen as the Indonesian version of South Korean idol girl/boy bands, is also an indication of how the popular culture industry in Indonesia constantly reworks and remakes products from South Korea as a ripple effect of K-wave global invasion. Therefore, claims and accusations of Indonesia as a copycat nation have always been the main narration to make meaning out of these recent cultural phenomena.

As an end result of the popular culture traffic in Asia, plagiarizing products reflect an ambiguous meaning-making process, as consumers, while longing for the real products, are actually seeking for more familiar goods, such as I-pop's boy/girl band or Indonesian television *sinetron* copying plots from K-dramas. In this study, the author argues that the copycat texts could be seen as a pastiche revealing the repetitive nature of popular culture flows in Asia. Furthermore, I-pop and K-drama remakes in Indonesia have strategically used the means of cultural borrowing and/or appropriating. As I-pop, for example, replicates the K-pop industry, the mimicking act is no longer a work of plagiarism, as it becomes a way for the Indonesian music industry to rework the already successful formulae. By analyzing these mimicking cross-cultural products, the study reveals that there is a continuous negotiation between the actors involved in the invasion of K-wave and the process of reinscribing these cultural products.

1.1 *Conceptualizing and contextualizing South Korean pop culture in Indonesia*

It is of utmost importance to see how the flow of South Korean products, or other East Asian countries' products, such as those from Japan and Taiwan, actually works in the Indonesian context. Intra-Asian cultural traffic, which has intensified in the last decade, does not imply that the emerging regional connection between Asian countries is happening because of its geo-cultural connection. The regional dynamics occurs because "Asia" is considered a depository of in-betweenness. There has been a capitalization of Asian cultural affinities in order to create a suppositious network of intra-Asian cultural traffic. Chen Kuan-Hsing argued that historical memories within Asia have shaped the redefinition of Asia in each context. "Inside the region itself, anxiety over the meaning of Asia arises from the politics of representation" (2010, p. 215). The term Intra-Asia, in the case of Indonesia, actually emphasizes the construction of a bordered imagined space of "Asia" in order to ensure the flow of unthreatening cultural products. It reflects not only the geographical space of the flow itself but also the setting of the boundaries in the distribution and consumption of East Asian cultural products in Indonesia to separate clearly what is considered "Asia" and not "Asia." During the time when Indonesia's social and political condition is overpowered by escalating politics of morality and anti-West rhetoric, East Asian television series, for example, is used as a strategic defense mechanism, creating a bordered imagined space of "Asia" to ensure the flows of unthreatening cultural products (Tambunan, 2013). Furthermore, East Asian popular culture should not be seen as an essentialized materialization of Japanese/Taiwanese/Korean culture. Iwabuchi argued that:

> "These popular culture are undoubtedly imbricated in U.S. cultural imaginaries, but they dynamically rework the meanings of being modern in Asian contexts at the site of production and consumption. In this sense, they are neither 'Asian' in any essentialist meaning nor the second-rate copies of 'American originals.' They are inescapably 'global' and 'Asian' at the same time, lucidly representing the intertwined composition of global homogenization and heterogenization …" (Iwabuchi, 2002, p. 16)

East Asian cultural products are the embodiments of global and Asian, so they could not be simply categorized as an antithesis of Western cultural products. In other words, these products, including the television dramas or idol bands, are already hybridized products.

To theorize and make sense of these copycat idol bands or television series remakes are a complicated task. Most research mainly deals with

> "… a superficial point-by-point, pluses-and-minuses kind of analysis. Often this kind of discussion employs a common strategy: the critic treats the original and its meaning for its contemporary audience as a fixity, against which the remake is measured and evaluated. And, in one sense, the original is a fixed entity" (Horton & McDougal, 1998, p. 15).

In previous research, the original text is often considered a fixed authentic text and then the researcher would analyze and evaluate the remake or the copycat text based on the original text. Therefore, even though it is significant to refer to the original text, as argued earlier to say that something is original or not is contestable, it would be more fruitful to analyze the remake/copycat text as an entity in itself. In other words, the analysis needs to go beyond the superficial level of what is original and what is not to see how the new text has reworked the formulae or structure of the original text. Therefore, the research will explore the underlying agency in response to the original text's overbearing recognition.

1.2   I-pop: Mimicking K-pop cross-bordering characteristics

The first case study to be explored in this study is I-pop, which is an amalgamation of the word Indonesia and K-pop. I-pop has become a significant part of the Indonesian music industry, although there is a strong resistance from Indonesian consumers. As reported by Arientha Primanita in a Jakarta Globe article (2012), the Tourism Minister, Mari Elka Pangestu, speaking at the Indonesia Creative Products Week (PPKI), pointed out that Indonesia's creative industry would be able to craft I-pop as an innovative commodity.

There are two main developments of I-pop to be highlighted in this discussion, which are the earliest wave in 2010–2011 (SM*SH and Cherrybelle) and the latest ones, which would be categorized as the made-in-Korea version of I-pop (S4 and SOS). The four boy/girl bands mentioned are examples from each stage of development. They are considered the most representative in terms of popularity and recognizability. The first two, SM*SH and Cherrybelle, are the earliest K-pop "look and sound-alike" groups, which receive a lot of resistance from Indonesian consumers. Local media entertainment companies produce these two groups. Meanwhile, S4 and SOS are groups formed, trained, and produced by a local entertainment company in cooperation with a South Korean music agency, emphasizing the made-in-Korea distinctiveness. The discussion will now focus on the characteristics of these two waves of I-pop as they resonate K-pop's global and cross-bordering characteristics.

A palpable characteristic of SM*SH and Cherrybelle, which enunciates a K-pop zest, is the standardized practice of idol fabrication. Soon after they release their debut singles, both groups could be seen in all kinds of media format. They became spokespersons for a number of products. SM*SH members starred in their own television drama (Cinta Cenat Cenut, 2011), and Cherrybelle had a television reality show (Chibi Chibi Burger, 2012). By using all of these media formats as promotion and marketing tools, there is a process of image fabrication. Group members are made into idols through how they look, which highlights the similarities to K-pop idols. The necessity to make them look like K-pop idols is shaped through the aforementioned multiple media formats. Furthermore, the image of a group is kept accentuated as most of these media promotion strategies focus on the group and not on individual members.

One of the most identifiable characteristics of K-pop is the "idol-making system and global marketing strategies" (Shim, 2008), as well as the "star manufacturing system" (Shin, 2009). These strategies integrate the production process, such as in the training of talents, as well as the management system. "The recent success of K-pop acts in Japan exemplifies the typical idol-making strategies of Korean entertainment companies, which is mainly driven by capitalist desires. They do this first by adopting globally popular cultural elements from Japan and the United States, then repackaging and manufacturing culturally hybridized products" (Jung & Hirata, 2012, p. 6). It is through this idol-making system that entertainment companies, driven by capitalist desire, manufacture and repackage K-pop products before circulating them to other parts of the world. In the production process, these entertainment companies train future idols in long training periods, which last from 5 to 7 years. They are not only trained in singing and dancing, but also on how to be multiskilled entertainers. Therefore, an analysis on recent development of I-pop must not only scrutinize the aspect of plagiarism because it would only pore over what is considered as authentic and local in I-pop. Seeing that I-pop is a reaction and reproduction of K-pop, what needs to be examined is the elements that constitute I-pop, which could also be identified in K-pop.

The latest development of I-pop, S4, and SOS could be considered as made-in-Korea I-pop products, taking the mimicking process into a different level. The boy band, S4, was formed from the winners of Galaxy Superstar, a reality show in one of the local television stations in cooperation with a South Korean music agency. The reality "talent" show was produced by the Indonesian YS Media Entertainment and the winners of the show were sent to Seoul to go through an 8-month training camp run by Rainbow Bridge Artist Agency, one of the K-pop training agencies in Seoul, South Korea. S4 became one of the first Indonesian boy bands trained in South Korea, who underwent an excessive training process like K-pop idol groups but within a shorter period of time than K-pop groups. The members also experienced physical transformation. For instance, one of the members lost 20 kg in the training camp. They dyed their hair and changed their fashion style, which makes them look exactly like K-pop idols, even though they are Indonesians.

Besides their appearances, S4 members were trained to sing in Korean and learned basic Korean phrases. As reported by Budi Suwarna in a Kompas article (27 January 2013), the four members would use the Korean greeting style when they met reporters or fans. They would bow down and say "Annyeong Haseyo!" This intensification of Korean-ness in their greetings and how they learned to sing in Korean echo a different form of articulating K-pop elements in this newest I-pop product. Compared with earlier I-pop groups, S4 amalgamates the actuality that they are made in Korea. The idol-making system in which each boy/girl band is fabricated through multiple media activities resonates with the K-pop idol-making system, even though the Indonesian version is done within a shorter time period. It is not only about the music, but, more importantly, about the process of making the members as idols as well. What makes up the complete idol package is their songs and dancing ability, fashion style, and even their manner of greeting people, such as that of S4.

It could be concluded that there are two different yet comparable ways in mimicking K-pop highlighting different aspects of the cross-bordering characteristics. Earlier waves of I-pop have reproduced K-pop products by making carbon copies of the songs, dance movements, and the construction of idols. However, for its later predecessors, it is more significant to enhance the Korean-ness of their products. As entertainment companies fabricate different ways of mimicking K-pop, it also reflects the elements that constitute K-pop as a new global product. Frederic Jameson (cited in Dyer, 2007) proposed that, in the postmodern era, the world is starved of new ideas. The most palpable option is to attempt to recreate and reinvent already existing products, especially those coming from the past, even if this past is not our own, which could explain the repetitive phenomenon of replicating products just like I-pop.

1.3 *Remaking/appropriating Korean trendy drama narrative elements*

Besides I-pop, Indonesian *sinetron* or soap operas, which are remaking K-dramas, are another way to mimic Korean popular cultural products in Indonesia. Indonesian *sinetron* throughout its development since the early 1990s has always been considered a product of plagiarism, even though it has been one of Indonesian television prime time shows. These *sinetrons*, copying either Korean, Japanese, Indian, or, recently, Turkish television soap operas, are very successful and widely accepted, while, at the same time, they are also receiving criticism as plagiarizing products. One of the reasons why these *sinetrons* are copying the storylines from other countries' television dramas is:

> "Only high demanded *sinetron* will be prolonged by the TV station. This situation made producers of production houses seek proven saleable drama series from abroad. The successful TV series from Japan, Taiwan, India, Hong Kong, and nowadays Korea becomes their target. It is not because those countries have a similar cultural background as Asian, but also because those series have proved themselves as a money maker in their own country and even abroad." (Gunawan, 2011, p. 4).

These *sinetrons* produce a lot of money for the production house and the television station. The different ways in which television drama production is done in Korea and Indonesia have greatly influenced the process of remaking/appropriating. For example, the original usually consists of 25 episodes, while the remakes could have more than 50 episodes or could even reach 100–200 episodes if the *sinetron* is successful.

To see how these remakes actually work on the narrative structure and elements from the original television dramas, the author looked at one sample of data, which has already been mentioned at the beginning of this article. The Indonesian *sinetron* entitled *"Kau yang Berasal dari Bintang"* (KYBDB) is a remake of a K-drama, "You who Came from the Stars" (YWCFTS). The remaking elements can be seen even from the opening scenes. The title itself is a direct translation from the Korean version. Watching the first few episodes of KYBDB, one would directly notice that there are a lot of similarities between the Indonesian and Korean versions. For example, in the first episode, almost all of the scenes are similar in both the narrative sequence and the cinematography (e.g., the choice of setting). Every scene looks and sounds similar; however, KYDDB actually made some significant changes.

As mentioned earlier, Indonesian *sinetron* has specific characteristics in the narrative elements that have become a formula even in remaking K-dramas. In KYDDB, the core of the story is the male protagonist's romantic journey to get his "wings" back by finding the one girl who can teach him how to love. He was depicted as an angel who fell from the sky 364 years ago and lost his wings. This is entirely different from the male protagonist's back-story in YWCFTS, who was an alien whose space ship landed on Earth 400 years ago during the Joseon Dynasty. His job is to wait until a meteor hits or arrives on Earth at a particular time in the future. The completely dissimilar way of representing the male protagonists as an alien and an angel could be seen as how the Indonesian *sinetron* industry appropriated the narrative elements in the Korean version to fit its own formulae.

Research findings reveal that these *sinetrons,* as they remake K-dramas, are appropriating their narrative elements in order to fit into the Indonesian *sinetron* structure and, in doing so, reconstructing the K-dramas' formulae. The narrative structure of *sinetron* usually revolves around the representation of black and white characters. The protagonist is usually portrayed as the good character experiencing bad things in his or her life. The antagonist is bad, manipulative, and causes many difficulties for the protagonist. Rarely are there characters outside these two categories. Most of the *sinetron* films focus and evolve around this kind of story … The message mostly explained in such *sinetron* story is that "patience and goodness will get one to happiness" (Udasworo, 2013, p. 159–160). In KYDBD, the female protagonist is represented as a snobbish actress just like the protagonist in YWCFTS. However, in the Indonesian version, there are some scenes portraying her as someone with morals. As she was introduced as a snobbish actress, she was also depicted as a woman with compassion. In one scene, she stepped off her car unexpectedly when she saw an old lady eating leftover food from the trashcan. This scene does not exist in the Korean version. She then gave her money and went back to her car. The story then moves on to other similar scenes that could be found in YWCFTS.

The addition of that short scene to portray a protagonist, who, in spite of being snobbish, actually has a sense of compassion, in fact shows there is a conscious effort to emphasize the *sinetron* formula that the protagonist is a nice person. This actually encourages K-drama enthusiasts to reflect upon this deliberate wandering from the original version. The sharp characterization of the snobbish actress as the protagonist was temporarily disturbed when she did that particular act of kindness. The different core stories of the alien and the angel are also a form of appropriation, as the idea of an angel seems to fit the cultural framework of the Indonesian *sinetron*. The storyline of an angel trying to find his wings back by finding his one true love is parallel with most Indonesian *sinetron* romantic storylines. Meanwhile, the back-story for the protagonist in the Korean version is completely different, even though a romantic storyline is still the main narrative pattern, just like any other Korean trendy dramas.

Looking at the remaking process as a form of cultural borrowing as in borrowing the narrative structure and cinematographic elements as well as appropriating some of these elements to fit the *sinetron* formulae, we can see that there is more to this moment of pastiche. The remakes offer a way of enjoying, understanding, and accessing these foreign texts, K-dramas, while conforming to a specifically *sinetron* language. "…the act of pastiching can also affirm the position of the pasticheur and may consequently form part of a politics of undermining and overthrowing the original" (Dyer, 2006: 157). On the surface, it looks like any other acts of pastiche, while at the same time producing a commercially successful *sinetron* hit while de-authenticating and deconstructing K-dramas in the process.

2 CONCLUSION: "COPYCAT" PRODUCTS AS THE GLOBAL PASTICHE

Copycat products are viewed by some as a symptom of creativity collapse of the Indonesian popular culture industry, especially in the wake of globalization, creating multiple opportunities for cross-cultural amalgamation. I-pop and *sinetron* that remake K-dramas are appropriating Korean cultural products in such a way that this cross-cultural appropriation allows the translation of East Asian cultural flow into the Indonesian context. The remake texts, especially in music and television, alter the original version to better suit the Indonesian consumers. It is clear that the Indonesian remakes offer the consumers a unique pleasure that could not be produced by the original.

REFERENCES

Bhabha, H. (1984) Of mimicry and man: The ambivalence of colonial discourse. *October, 28,* 125–133.

Chen, K.H. (2010) *Asia as Method: Toward Deimperialization.* Durham, Duke UP.

Cho, Y. (2011) Desperately seeking East Asia amidst the popularity of South Korean pop culture in Asia. *Cultural Studies,* 25(3), 383–404.

Chua, B.H. (2010) Korean popular culture. *Malaysian Journal of Media Studies,* 12(1), 15–24.

Dyer, R. (2007) *Pastiche.* London, Routledge.

Gunawan, E. (2011) *Against Imperialism of Bourgeois Image.* [Online] Available from: http://s3.amazonaws.com/academia.edu.documents/37597300/makalah_imperialism_bourgeois.pdf?AWSAccessKeyId=AKIAJ56TQJRTWSMTNPEA&Expires=1467559224&Signature=wZEt2Egs69UsDLRx9ysa4bz9p6A%3D&responsecontentdisposition=inline%3B%20filename%3DAgainst_Imperialism_of_Bourgeois_Image.pdf. [Accessed 2nd July 2016].

Horton, A. & McDougal, S.Y. (1998) *Play It Again, Sam: Retakes on Remakes.* Berkeley, Univ. of California Press.

Iwabuchi, K. (2002) *Recentering Globalization: Popular Culture and Japanese Trans-nationalism.* Durham, Duke UP.

Jackson, J. (2014) *More TV Dramas Accused of Plagiarism. Korea Herald.* [Online] Available from: http://www.koreaherald.com/view.php?ud = 20131224000709. [Accessed 10th July 2016].

Jung, S. & Y. Hirata. (2012) Conflicting desires: K-pop idol girl group flows in Japan in the era of Web 2.0. *Electronic Journal of Contemporary Japanese Studies.* 12: (2).

Morris, M. (2004) On Trans-Asian cultural traffic (participating from a distance). In: Iwabuchi, K., Muecke, S. & Thomas, N. (eds.) *Rogue flows: Trans-Asian cultural traffic.* Hong Kong, Hong Kong University Press.

Primanita, A. (2002) Can 'I-Pop' Emulate the Success of K-Pop? Indonesian Tourism Minister Says Yes. *Jakarta Globe.* [Online] Available from: http://jakartaglobe.beritasatu.com/archive/can-i-pop-emulate-the-success-of-k-pop-indonesian-tourism-minister-says-yes/.[Accessed 28th July 2016].

Sen, K. & Hill, D.T. (2007) *Media, Culture and Politics in Indonesia.* Jakarta, Equinox Publishing.

Shim, D. (2008) The growth of Korean cultural industries and the Korean wave. In: Huat, C.B. & Iwabuchi, K. (eds.) *East Asian Pop culture: Analysing the Korean wave.* Hong Kong, Hong Kong University Press. pp. 15–31.

Shin, H. (2009) Have you ever seen the rain? And who'll stop the rain? The globalizing project of Korean pop (K-pop). *Inter-Asia Cultural Studies,* 10(4), 507–523.

Stam, R. (2000) Beyond fidelity: The dialogics of adaptation. *Film adaptation,* 54–76.

Tambunan, S.M.G. (2013) *Intra-Asia cultural traffic: Transnational flow of East Asian television dramas in Indonesia* [Dissertation] Hong Kong, Lingnan University. Available from: http://dx.doi.org/10.14793/cs_etd.18.

Udasmoro, W. (2013) Symbolic violence in everyday narrations: Gender construction in Indonesian television. *Asian journal of social sciences & humanities,* 2(3), 155–165.

Wright, N.S. (2009) Tom Cruise? Tarantino? ET?... Indian!": Innovation through Imitation in the Cross-Cultural Bollywood Remake. *Cultural borrowings: Appropriation, reworking, transformation,* [Online] 194–210. Available from: https://www.nottingham.ac.uk/scope/documents/2009/culturalborrowingsebook.pdf. [Accessed 15th August 2016].

Cultural Dynamics in a Globalized World – Budianta et al. (Eds)
© 2018 Taylor & Francis Group, London, ISBN 978-1-138-62664-5

Positive body image activism in collective (@effyourbeautystandards) and personal (@yourstruelymelly) Instagram accounts: Challenging American idealized beauty construction

D.M. Wuri & S.M.G. Tambunan
Faculty of Humanities, Universitas Indonesia, Depok, Indonesia

ABSTRACT: Women who have been exposed to the construction of idealized beauty standards have suffered from distorted ideals of body image. This leads to multiple psychological and physical disorders such as anorexia, bulimia, or dangerous cosmetic surgeries and harmful diet regimens. The idealized beauty standard is, at first, disseminated through traditional media such as magazines, newspapers, televisions, movies, videos, and advertisements. The emergence of social media, especially those with a photo-sharing feature, has affected the pattern of dissemination, thus becoming a new area for spreading idealized beauty. Social media provides their users with various features and opens new spaces for interaction and discussion. Instagram, a photo-sharing-based application, has been used to challenge the idealized beauty standard by its users. @effyourbeautystandards, a collective account, and @ yourstruelymelly, a personal account, are the subjects of this research, as both accounts are promoting positive body image through Instagram. Using a textual analysis, this research examines the two accounts to reveal the complexities of positive body image activism through their posts and comment sections. Furthermore, it also analyzes the similarities and differences of the body image issues that have been brought up by these accounts.

1 INTRODUCTION

Since its launch in October 2010, Instagram has grown as one of the most popular social media platforms that specializes in sharing pictures and videos (Salomon, 2013). With its "conversational" features, which mean every feature that enables interactions among Instagram users, such as comment column, direct message, tag people, and mention, Instagram has created a virtual discussion space. One of the much-discussed issues in Instagram is positive body image. Instagram users who are concerned with idealized beauty construction started a positive body image campaign by making accounts and dedicating them to challenge the dominant discourses. Among the excessive number of studies about body image, there has not been much scientific discussion on positive body image accounts in Instagram as a form of social activism. By using two Instagram accounts promoting positive body image, @effyourbeautystandards, a collective account, and @yourstruelymelly, a personal account, this article examines the complexity of the positive body image in the social media campaign, and such accounts challenge the idealized beauty construction.

Researchers have argued that social media users are more vulnerable to the dominant discourse of body image resulting in body dissatisfaction (Andsanger, 2014, Tiggemann & Slater, 2013). Instagram, which offers interactions through pictures and comment sections, has been said to bring a more significant impact on body dissatisfaction among women because the users can easily compare their pictures with others (Ghaznavi & Taylor, 2015, Fardouly & Vartanian, 2016). Although social media is associated with the increasing body dissatisfaction among women, it is also believed that social media could play a critical role in spreading messages of body positivity (Rice & Atkin on Perloff, 2014). Positive body image means positive attitude toward our own body. It could briefly say that positive body image includes body acceptance, body appreciation, confidence, and filtering negative information about our body (Wood-Barcalow, Tylka, & Augustus-Horvath, 2010). Related to the previous statement, some Instagram accounts are found promoting positive body image. Thus, there has been a shift, as Instagram is no longer merely a photo-sharing application, but it has become a tool to spread contesting ideals. For example, by using the comment sections, users are able to discuss the ideas of positive body image and even spread such ideas.

The main data for this study have been taken from the two aforementioned Instagram accounts. Although both collective and personal accounts have similar idealism in promoting a more positive body image, there are some complexities that could be further investigated. This study focuses on how both accounts cover issues about body image and how they deliver their views to the followers or to the public audience, as both accounts can be accessed by non-followers. The comment section on each post is analyzed to see how other users react on the issues. By contrasting those two different accounts, research findings reveal that each account chooses different issues they would like to bring up and how other users or followers also choose the issues they would like to support or oppose. As a collective account, @effyourbeautystandards chooses to cover more issues related to positive body image, whereas @yourstruelymelly as a personal account chooses to focus on the body size issues.

1.1 *Body size and health issues*

Ideal beauty is contextual because it depends on the context of time and place in which the ideals are constructed (Calogero, Borough, & Thompson, 2007). As both @effyourbeauty-standards and @yourstruelymelly originated from the United States, we can argue that it is the American beauty standard being challenged by both accounts. However, their campaign might still be significant for Instagram users outside the United States, because currently the American or Western beauty standard is still dominating the perspective of most women in the world (Evans and McConnell, 2003; Bissell & Chung, 2009). In order to be considered beautiful, women are required to have certain physical characteristics, such as being slender, thin, and having a flat stomach and a narrow waist. These characteristics are generally called as "thinness ideal" (Groesz, Levine, & Murnen, 2002; Harrison, 2003). The pressure of achieving an idealized beauty standard among women often puts them in danger. Some will suffer from anorexia and bulimia, or they will do dangerous diet regimens and even risky plastic surgeries (Tiggemann & Miller, 2010; Tiggemann & Slater, 2013, 2014, Cash & Roy, 1999).

When talking about idealized beauty, one must also talk about body size because it has become one of the most important indicators to determine whether someone is considered beautiful or not. In the United States, most women are struggling to achieve the ideal body according to Western beauty standards (Jackson, Hodge, & Ingram, 1994). The perception of ideal woman's body in Western culture has been changing throughout time. An ideal shape of female body from each era could be seen through the fashion at that particular time, for example, the popularity of corsets for providing the illusion of the hourglass body on females in the 19th century (Kunzle, 2004). Although the ideal shape of female body keeps shifting, being thin is one of the characteristics still used in the society. The pursuit of the thinness ideal also often leads women to various health problems like bulimia or anorexia. Thinness ideal also becomes a major factor for women to suffer from distorted body image. What both @effyourbeautystandards and @yourstruelymelly try to challenge, among others, is thinness ideal. Body size becomes an important issue for both accounts because most women are obsessed with unrealistic goals on body weight (Bordo, 1993).

As with the ideal body standard, US culture has gone through a long history concerning plus-size body. It has been told that plus-size body nowadays is "associated with laziness and low self-discipline" (Puhl & Brownell, 2003) and people with plus-size bodies deserve to be targets of body shaming and hate speech because they have violated the thinness norm (Eller, 2014). Even though plus-size body is considered a sign of wealth in some cultures, there are perceptions about plus-size body in the US society nowadays that plus-size women are "ugly, dangerous, or unhealthy" (Farrell, 2011). It is believed that the US culture is now suffering from fat phobia by how they react toward plus-size people (Wann, 2009). Related to the previous statements, the effort of @effyourbeautystandards on campaigning positive body image through the body size issues is often confronted by health issues. On several posts that show pictures of plus-size-figure women, which are intended as messages to other women not to be ashamed of their body size, some followers who are in a negotiating position criticize @effyourbeautystandards' point of view by bringing up health issues. Although there are many supports from the followers of this account, there are many criticisms among the supportive comments, as quoted below:

> **@carleyito:** "that much extra mass around your middle IS NOT HEALTHY use your brain and stop hiding behind 'body confidence' it's just lazy"; **@isabelginiguez:** "yes I agree fuck your beauty standards but this is overdoing it, it's not even about beauty anymore it's about putting your life at risk"; **@laurie_doyle:** "I also didn't call her names or fat. I said obesity is unhealthy..."; **@brpsd:** "I just hate shit like this, being fat isn't something you should adore. It's unhealthy...".

The above responses show us that other users instantly connect plus-size body pictures with unhealthiness. They also refuse to look at plus-size body as one of the issues that should be acknowledged in the positive body image campaign. It is implied that even though they are aware of positive body image issues, they are still using the dominant perspective to plus-size body issue portrayals with health issues. The media in the US society has constructed that plus-size body is associated with unhealthiness and laziness.

On the contrary, @yourstruelymelly rarely receives rebuttals concerning health issues on her posts that display plus-size woman's photographs. In the comments, she is being praised because of her bravery and confidence on showing her stance toward body size and body image issues through her self-photographs, as quoted by the users:

> **@esther1092:** "there should be more people like you in the world [...] and teach young girls about the importance of positive body image"; **@carriane80:** "beauty and confidence in one amazing package! I love your account"; **@msquigley:** "so incredibly well-written & moving, thank you for sharing".

Although both accounts have the same positive body image issue, which is embracing plus-size body, responses from other users are very different. In other findings, there is one case which clearly shows different responses from Instagram users when @effyourbeautystandards reposted @yourstruelymelly's picture of her standing in front of a full-body mirror wearing a tank top and a tight knee-length skirt. Melissa received supportive comments and no criticisms for that picture. For example, **@biggygross**commented:"wonderful", while **@ohmerrguhr_its_jess**commented:"omg, I love that top! You look great in it!". However, when it was reposted by @effyourbeautystandards, they received criticism by bringing up health issues:

> **@clobo_yo:** "[...] I'm all for loving yourself and curves but being morbidly obese is not something to teach people to 'love yourself' about"; **@devil_neville:** "I'm all about body positivity but this is just as bad as the starving underweight 'Victoria Secret' models. Let's promote HEALTHY individuals. Not somebody that clearly will have health issues in the near future".

The above comments show us about other users' stance in the issues of positive body image. Both users agree with the positive body image campaign, as they said, yet they do not share the same idea about the body size issues. One of them clearly stated that being "morbidly obese" cannot be part of the positive body image campaign, while another said that the picture is "as bad as the starving underweight Victoria Secret models as it is promoting unhealthy body that will have health issues in the near future". In short, they both disagree with the plus-size body on the positive body image campaign as they associate plus-size body with health disturbance.

On the basis of the findings, both accounts are promoting the same idea on issues of positive body image, which are self-acceptance in terms of body size and being confident in having a plus-size figure. Moreover, some pictures posted by them are similar. However, they received different responses from other Instagram users over the same issues and similar pictures. Observing from each comment, the followers of those positive body image accounts could be categorized into two: users who concur with positive body image and those who are still in a negotiating position, which means they are, on the one hand, trying to accept the positive body image campaign while still being trapped in the dominant construction of ideal body image in the society.

1.2   *Personal and collective activism in the positive body image campaign*

The involvement of @effyourbeautystandards and @yourstruelymelly on the positive body image campaign could be seen from almost every aspect of both accounts. It is crucial for activism accounts to clearly show their standpoint to other users as Erving Goffman (1959) stated that ones who enter the presence of others will tend to seek information about them. Furthermore, he also wrote: "information about individual helps to define the situation, enabling others to know in advance what he will expect of them and what they may expect from them" (p. 1). Both @effyourbeautystandards and @yourstruelymelly put brief information about their main purposes and ideas on the front-page so that other Instagram users who visit their page could see the overviews and it will be easier for users who are interested in their campaigns to follow, to raise other's awareness, or even to join their activism.

Even though the idea of movement comes from an individual, to build a collective movement, one should construct a collective identity that is distinct from a personal identity. As argued by Polletta and Jesper about a collective identity, "although it may form part of a personal identity" (2001: 285), it has to be different from one's personal identity. In the @effyourbeautystandards front page, it is stated that this account initially belonged to Tess Holiday from Mississippi, but it is currently managed by eight administrators. Polletta and Jesper also stated that "collective identities are expressed in cultural materials—names, narratives, symbols, verbal styles, rituals, clothing, and so on" (2001, p. 285), which, in this case, @effyourbeautystandards expresses it through its name, profile picture, account bio, captions, and every picture it posts. From the name of the account, @effyourbeautystandards makes a strong statement that it is made to challenge the idea of beauty standards in society. Word "eff" on that name is a refined term from the word "fuck", which refers to their annoyance or anger. The word "eff" that follows by "your beauty standards" indicates that the account would like to express their annoyance and anger to the concept of beauty standards to other Instagram users who visit their page. Moreover, to build their collective identity, @effyourbeautystandards uses a "neutral" profile picture that only contains their account's name. On their account bio, they directly inform that they "will not tolerate hate/body shaming" to warn other Instagram users who visit or follow the account. They also put a fist symbol to emphasize the purpose of challenging beauty standards. @effyourbeautystandards raises their sense of collectiveness by reposting other women's photograph or illustrations, which represent their stance on issues of body image with a general point of view for the captions, instead of a personal point of view. For example, when they reposted one's picture, they often directly quoted what the person said and put it on their caption section, instead of making their own caption. Moreover, they also often put motivational quotes as captions, instead of their personal expressions about such a particular issue.

On the other hand, @yourstruelymelly is a personal account directly maintained by the account's owner, Melissa. There are signs signifying that this is a personal account, such as Melissa's profile picture, and the account's name. Furthermore, this account puts an explicit pronouncement on its bio saying that the owner is a "body positive feminist". The use of word true on @yourstruelymelly is to convince the followers on her authenticity and the fact that she is showing her true self through this account (explained in one of her posts). In order to deliver the campaign about positive body image, she often uses her own photographs or illustrations. Melissa posted a statement on a caption section when she decided to dedicate her personal account to promote positive body image, and how she defines positive body image.

Other than building a collective identity, @effyourbeautystandards as a collective account also covers more body image issues compared to @yourstruelymelly. From the findings, body image issues that appear in @effyourbeautystandards are body size, body diversity, body diversity, and liberation of female body, while @yourstruelymelly only focuses on the issue of body size. The nature of collective movements is collecting participants and supporters. People are naturally more interested to join collective movements because of a sense of solidarity (Taylor and Whittier on Polletta & Jasper, 2001) and as humans, they tend to get emotional satisfaction from a collective identity (Jasper on Polletta & Jasper, 2001). Thus, covering many issues by reposting various women's photographs is a strategic move by @effyourbeautystandards to "mobilize more participants", or in this case, other Instagram users, to join the positive body image campaign. It is feasible because Instagram has several features to connect people, such as the "repost" feature. As someone reposts pictures from other accounts, more notifications will automatically appear on others' news feeds and indirectly promote the account to other users. However, the large number of followers also results in more complexity, as each of them shares different perspectives about positive body image and some of them are in the negotiating position, as explained in the previous section. As a result, the followers who disagree with some issues, such as plus-size figure and liberation of female body (including growing armpit hair, bare-chested, free the nipple), raise debates in the comment sections. On the contrary, as a personal account with selected issues, users who follow @yourstruelymelly are aware of the account's stance and tend to share the same perspective with the account. Furthermore, like any other personal accounts on Instagram, some of the comments came from Melissa's friends or social circle, reducing the possibilities of opposing arguments.

The nature of social media activism, as stated by Merlyna Lim, requires several characteristics for an issue to strive through society successfully. She argues that:

> "... they have to embrace the principles of contemporary culture of consumption: light package, headline appetite, and trailer vision. Beyond that, the activism must neither be associated with high-risk actions nor ideologies that challenge the dominant meta-narratives" (Lim, 2013, p. 18).

Therefore, to successfully convey the activism, the issues must be in a light package and associated with low-risk actions. According to the findings of this research, body size issues could be categorized as a high-risk activism because it challenges the dominant meta-narrative as people believe that plus-size body is a result of laziness and is related to unhealthiness. Thus, the issue of body size in the positive body image campaign received harsh criticism from Instagram users. However, the result of this analysis also found that the issue and the package do not determine the success of the activism. The medium of the activism could affect the success because both personal and collective accounts bring up the same issue but receive different responses.

2 CONCLUSION

Instagram is one of the most popular social media that has played a role in the emergence of the positive body image campaign. The research on two Instagram accounts, @effyourbeautystandards and @yourstruelymelly, that actively promote positive body image reveal the complexities of the positive body image campaign in terms of other users' responses to the issue, which are either supporting or negotiating about the campaign's purpose. The extensive coverage of body image issues on the collective account @effyourbeautystandards apparently raises more debates among Instagram users than the personal account @yourstruelymelly that focuses on only a few body image issues. Some findings from @effyourbeautystandards reveal that even users who seem to take sides in the positive body image campaign are actually still in a negotiating position. It means that although they agree with positive body image, they are still opposing some of its issues that are unfit for their constructed idealism. Although the collective account is more effective in spreading the positive body image campaign as it is seen and followed by more number of users, it has come under several criticisms for the body size issues, unlike @yourstruelymelly that also covers the same issues.

Furthermore, according to other users' responses to the presentation of the positive body image by both accounts, the supporters of the positive body image could be categorized into two types: (1) those who agree with the positive body image campaign, including any issues it conveys, and (2) those who are in the negotiating position. There are some more findings such as sexualizing images, liberation of female body, and body diversity that will be elaborated further in future study.

REFERENCES

Andsager, J. (2014) Research directions in social media and body image. *Sex Roles,* 71(11–12): 407–413.

Bissell, K.L. & Chung, J.Y. (2009) Americanized beauty? Predictors of perceived attractiveness from US and South Korean participants based on media exposure, ethnicity, and socio-cultural attitudes toward ideal beauty. *Asian Journal of Communication,* 19(2), 227–247.

Bordo, S. (1993) Feminism, Foucault and the politics of the body1. Up against Foucault: Explorations of some tensions between Foucault and feminism, 179.

Calogero, R.M., Boroughs, M & Thompson, J.K. (2007) *The impact of Western beauty ideals on the lives of women: A sociocultural perspective*. UK, Palgrave Macmillan. pp. 259–298.

Cash, T.F. & Roy, R.E. (1999) Pounds of flesh: Weight, gender, and body images. Interpreting weight: The social management of fatness and thinness, 209–228.

Eller, G.M. (2014) On fat oppression. *Kennedy Institute of Ethics Journal*, 24(3), 219–245.

Evans, Peggy Chin, & McConnell, A.R. (2003) Do racial minorities respond in the same way to mainstream beauty standards? Social comparison processes in Asian, Black, and White women. *Self and Identity,* 2(2), 153–167.

Fardouly, J. & Vartanian, L.R. (2015) Negative comparisons about one's appearance mediate the relationship between Facebook usage and body image concerns. *Body image,* 12, 82–88.

Fardouly, J. & Vartanian, L.R. (2016) Social Media and Body Image Concerns: Current Research and Future Directions. *Current Opinion in Psychology*, 9, 1–5.

Farrell, A.E. (2011) Fat shame: Stigma and the fat body in American culture. New York, NYU Press.

Ghaznavi, J. & Taylor, L.D. (2015) Bones, body parts, and sex appeal: An analysis of thinspiration images on popular social media. *Body image,* 14, 54–61.

Goffman, E. (1959) *The Presentation of Self in Everyday Life*. New York, Anchor Books.

Groesz, L.M., Levine, M.P. & Murnen, S.K. (2002) The effect of experimental presentation of thin media images on body satisfaction: A meta-analytic review. *International Journal of Eating Disorders,* 31(1), 1–16.

Harrison, K. (2003) Television viewers' ideal body proportions: The case of the curvaceously thin woman. *Sex Roles,* 48(5–6), 255–264.

Jackson, L.A., Hodge, C.N. & Ingram, J.M. (1994) Gender and self-concept: A reexamination of stereotypic differences and the role of gender attitudes. *Sex Roles,* 30(9–10), 615–630.

Kunzle, D. (2004) *Fashion & Fetishism: Corsets, Tight-Lacing and Other Forms of Body-Sculpture* (Thrupp, UK).

Lim, M. (2013) Many clicks but little sticks: Social media activism in Indonesia. *Journal of Contemporary Asia,* 43(4), 636–657.

Perloff, R.M. (2014) Social media effects on young women's body image concerns: Theoretical perspectives and an agenda for research. *Sex Roles,* 71(11–12), 363–377.

Polletta, F. & Jasper, J.M. (2001) Collective identity and social movements. *Annual review of Sociology,* 283–305.

Puhl, R.M. & Brownell, K.D. (2003) Psychosocial origins of obesity stigma: toward changing a powerful and pervasive bias. *Obesity reviews,* 4(4), 213–227.

Salomon, D. (2013) Moving on from Facebook Using Instagram to connect with undergraduates and engage in teaching and learning. *College & Research Libraries News,* 74(8), 408–412.

Tiggemann, M. & Miller, J. (2010) The Internet and adolescent girls' weight satisfaction and drive for thinness. *Sex Roles,* 63(1–2), 79–90.

Tiggemann, M. & Slater, A. (2013) NetGirls: The internet, facebook, and body image concern in adolescent girls. *International Journal of Eating Disorders,* 46(6), 630–633.

Tiggemann, M. & Slater, A. (2014) NetTweens the internet and body image concerns in preteenage girls. *The Journal of Early Adolescence*, 34(5), 606–620.

Wann, M. (2009) The fat studies reader. In: Rothblum, E. & Solovay, S. (eds.). New York, NYU Press.

Wood-Barcalow, N.L., Tylka, T.L. & Augustus-Horvath, C.L. (2010) But I like my body: Positive body image characteristics and a holistic model for young-adult women. *Body Image* 7 (2), 106–116.

Problematizing sexualized female images and Donald Trumps' immigration stance in video games: An analysis of Let's Play videos on YouTube

S. Angjaya & S.M.G. Tambunan
Department of Literature Studies, Faculty of Humanities, Universitas Indonesia, Depok, Indonesia

ABSTRACT: As the use of social media becomes more common, there has been a massive development of creativity in the form of online videos, specifically on YouTube. It provides the opportunity for the fans to express themselves in a variety of ways in regard to their object of fascination such as video games. One video genre that gains a great amount of popularity among YouTube users in recent years is the Let's Play (LP) videos. It is a type of video where people record themselves while playing video games and give their comments during the filming to share their gaming experience with other video gamers. People often assume that video games could bring negative influence and they are considered as an activity that could have harmful effects on the player's social lives, creativity, productivity, and literacy. However, based on the analysis of several LP videos from two YouTube channels namely PewDiePie and Cryaotic, this article examines how, through the appropriation process in the form of LP videos, video games could be utilized by fans to construct and articulate new meanings that are related to various social issues. It could even open up dialogs to discuss critically and problematize particular issues with a wider range of audiences who are not categorized as video game fans.

1   INTRODUCTION

Video game is a product of popular culture that has been a part of human culture for more than a half of a century. The general perception regarding video games is that it is a product intended for children, and it is also closely associated with nerd or geek culture. Furthermore, like most popular culture products, video games and the players are often associated with negative stereotypes. Grossman (2001) compares video games with drugs for their capability to make their players addicted and incapable of doing any other activities. Meanwhile, Kline (2000) and Jessen (1995) argue that playing video games is a solitary activity that pulls the players away from reality as they are becoming more and more immersed in the virtual world. However, Turkle (1984), Ivory (2001), and Dorman (1997) present different findings as they argue that playing video games is not as anti-social as people might think. They consider it as a social activity that often enables the players to interact with other players. Another assumption is that playing video games has psychological effects on the players as it could cause them to show aggressive and violent behaviour. Both assumptions, according to Ivory (2008) and Griffiths (1999), are often inconclusive or even contradictory. There have been an extensive number of studies analysing video games focusing on various aspects which do not focus solely on the effects of playing video games. Dickerman (2008), Leonard (2006), and Dunlop (2007) examine and problematize how gender and race are represented in video games. From the gender perspectives, several studies conclude that video games act as socializing agents that teach both boys and girls about their gender role in society (Dietz, 1998; Espejo, 2003; Provenzo, 1991). However, among those studies, the act of appropriation of video games by fans in the form of video, namely *Let's Play*, in social media platforms like YouTube is largely unexplored because it is a relatively new phenomenon.

Let's Play (hereinafter referred to as LP), also known as video gaming commentary, is a type of video where people (who are called Let's Players) record themselves while playing video games and give their comments during the filming. Generally, this kind of video is uploaded by video game fans on video-sharing or streaming sites such as YouTube, Twitch, Vimeo, or Dailymotion. In the last couple of years, LP has become one of the most popular genres on YouTube, which is shown from the number of LP videos being watched every day—reaching a third of the site's total view numbers everyday (Ault, 2016). The purpose of creating LP is usually to show the audiences their experience, progress, achievements, or difficulties when playing a video game. However, as it progresses, LP has not only become a way for the players to share gaming experience and information about video games with their peers, but it has also turned into a new form of entertainment as well as a commodity.

Through a textual analysis of several LP videos and YouTube comments, this article examines how the appropriation process in these LP videos could be used as a medium to address various social issues, such as political expression and female representation in video games. The goal is to show how the audience becomes an essential part in the meaning-making process as gamers or other audience members make sense of those video contents by actively commenting and participating in the discussion. By analysing these LP videos, the article aims to provide an alternative perspective regarding the common argument that views playing video games as an activity without productivity and creativity. Moreover, it also aims to explore how video game players are not merely passive consumers that are separated from reality, but they actually have active agencies as they produce new meanings when playing video games by connecting it to a wider social context and incorporating it in their creative process.

2 PROBLEMATIZING FEMALE SEXUALIZED REPRESENTATIONS IN VIDEO GAMES

The female representation in video games is one of the most prominent issues that are frequently analysed in video game studies. A study by Heintz-Knowles *et al.* (2001) shows that women are being underrepresented in video games compared to men. Furthermore, when female characters do show up in video games, their role is limited to either the helpless damsel-in-distress or the supporting role for the male characters, and they are often represented in a sexualized way (Dickerman, *et al.* 2008). Many studies have concluded that the female representation showcased in video games is often problematic as they commonly portray female characters in a 'hypersexualized' manner. Hypersexualized is a term coined by Heintz-Knowles *et al.* (2001) that is used as a synonym for the depiction of characters that are overtly sexualized in video games. In many games, female characters are often objectified. Their physical features are exposed in sexually revealing clothing, with big breasts, small waist, voluptuous body, unrealistic body proportion, and even nudity (Downs & Smith, 2009). As most of the video games are aimed towards heterosexual male players, these sexualized female images are used as a marketing strategy (Entertainment Software Association, 2004).

There are several possible effects of sexualized female representations on video game players. Dunlop (2007) argues that such representations could affect an individual's learning process of a gender role and reinforce the patriarchal ideology in which females are positioned inferior to males. For male players, such representations may reinforce the belief that women are sexual objects and may cause unrealistic expectation of the feminine body (Downs & Smith, 2009). Meanwhile, for female players, being exposed to an unrealistic body representation may lead to body dissatisfaction, low self-esteem, or even eating disorder due to the unattainable body standard shown in the games (2009). Most of the studies mentioned above have conclusively argued that sexualized female representations are frequently present in video games, and they have the potential to bring negative effects to the players. However, none of them could actually show the tangible responses from the players. By analysing the content of LP videos, the players' unfiltered and immediate responses to such representations as they are commenting while playing video games could be observed. Furthermore, since the LP video is uploaded to YouTube, the meaning-making process does not stop in the production of LP video by the Let's Players. By analysing the comments section, the way audience members participate in the meaning-making process and how they make sense of these LP videos can also be analysed.

A sample from the data, a video from Cryaotic channel entitled *Cry Plays: Heavy Rain [P5]*, reflects the way LP problematizes female representations in video games. *Heavy Rain* is an award-winner thriller game, which was created by David Cage and developed by Quantic Dream3 (Heavy Rain for Playstation 3 Reviews, 2016). The game revolves around Ethan Mars, a father who tries to save his child from a serial killer called the Origami Killer. In an attempt to reveal the real identity of the killer, the player could control four characters including Ethan Mars, Paige Madison, a journalist, Norman Jayden, an FBI Agent, and Scott Shelby, a private investigator. This video is chosen because there is one particular scene that is related to the issue of sexualized female representations. This LP video basically introduces the female lead character, Paige Madison, as she wakes up from her sleep in her apartment. What is interesting is that the sexualized image is so perceptible even in the way she is introduced for the very first time. She is portrayed as a female with an ideal body proportion wearing a white tank top without any bra and white panties.

Afterwards, there is one particular scene when she is taking a shower which shows her naked body. In the description of the video, Cryaotic wrote the following, "So breasts. How long until you think people flip a shit and freak out about censorship? I give it about tree fiddy hours" (Cryaotic, 2016) because he does not censor the naked scene in the video. As he watched the shower scene, Cryaotic stated (direct comment), "We are born to stay clean, and human form is an important thing to be respect(ed). It's not just some sexualized…It's not some just be like oh man that's so bad. It's natural!" (Cryaotic, 2016).

From his description and commentary on the video, he wants to emphasize that he does not see the scene as showing a sexualized or objectified female body because he perceives it as something natural. This is in accordance with Jenkins' (1992) argument that fans' productivity does not always try to resist or modify the meaning of the original text that they borrow. In this particular case, it is the audience who is more active in the meaning-making process by problematizing and addressing the issue of sexualized female representations displayed through the LP video by giving their commentaries. This is only possible because social media, such as YouTube, helps fans' activities to gain more visibility and become more accessible to the public, while at the same time facilitating the many-to-many communication that enables broader participation in discussions as well as in the exchange of information (Jenkins, 2002).

Based on the audiences' commentaries, it can be seen that their stance on the issue of female sexual representations in video games is not as black and white or predictable because each audience provides logic to back-up their arguments. Research findings reveal that most of the comments do not passively accept the images that are given to them. On the contrary, they are aware that the image of Madison is sexualized and they try to problematize it, and most of these comments actually come from the male audiences:

> "Of course, bodies are just bodies. However, they went way overboard and treated the woman like an object. Unfortunately, breasts are never shown for any sort of "body freedom" message, but instead they are shown in order to make a show out of it for men. Female nudity in media is almost always for the male gaze - especially when made by men. And I wouldn't even have an issue with it if they had treated the man the same way, but they didn't." (Fullmetal Triforce)

The commentary above clearly shows that as a male, instead of enjoying the nudity, he tries to resist such a representation because he perceives it as a form of female objectification. Moreover, he is also aware that the sole purpose of the shower scene in the game is only for the pleasure of the male players and does not contribute to either character development or plot progression in the game; thus, he thinks that it does not qualify as a body freedom message.

Contrary to the above commentary, most female audiences are the ones who support and thank Cry for not censoring the shower scene. The following is one of the commentaries from a female audience regarding the issue,

"Kinda glad that Cry didn't pixel the boobs out (not because I'm oogling them) but mainly because boobs are very natural body parts. He even makes a whole tiny speech about boobs being natural and how they shouldn't be sexualized... I'll just put it simply: for the love of god boobs are not "sexual"... pls realize that 1) Boobs are considered sexual because society has smashed that idea into our heads and 2) Educate your children about boobs. Ain't nothing wrong with them." (Amy Huynh)

The commentary above shows that as a woman, she does not find the sexualized image offensive as she considers it a natural part of the human body. She argues that it is the society that views female breasts as something sexual and thus should be censored, which makes the body image sexualized in the first place.

In the meaning-making process of LP videos, the Let's Player is not the only one who has the capability to construct and articulate meaning from video games. The audiences could also take an active role in meaning-making processes as they are able to problematize the issue that fails to be recognized even by the Let's Player himself through their commentaries. Jenkins (2002, p. 164) considers the various interpretations by the audiences as "a collective event that implies the recipients, transforms interpreters into actors, and enables interpretation to enter the loop with collective action." It also shows that LP videos may not always be used by the Let's Player as a way to address a certain issue, but it still has the potential to open up a dialog with other audiences, who may or may not be video game players, about various social issues exhibited in the video. In this case, the LP video has managed to open-up a critical dialog among the audiences regarding the issue of female representations in video games and deconstructed that notion in the process. Such potential is in accordance with Jenkins' (1992) statement that sees fandom as "a space within which fans may articulate their specific concerns about sexuality, gender, racism, colonialism, militarism, and forced conformity"(p. 289).

3 DONALD TRUMP'S IMMIGRATION STANCE: APPROPRIATING VIDEO GAMES AS A FORM OF POLITICAL EXPRESSION

In the case of LP videos, the video game fans are able to produce new meanings through a video editing process and their personal commentary. A video from the PewDiePie channel will be used as a sample to show how fans appropriate video games and construct a completely new meaning out of it. In his book *Textual Poachers* (1992), Henry Jenkins, a scholar working on a fans study, explains that far from being passive cultural dupes, fans are able to actively construct and circulate new meanings by borrowing and recreating from the textual materials that become their object of fascination in order to form a basis for their own cultural creation and social interaction. By using the concept of "textual poaching" proposed by Michel de Certeau (1984), he posits that fans are "readers who appropriate popular texts and reread them in a fashion that serves different interests" (p. 23). Although Jenkins' statement is created in the context of TV series fans in the early 90's, it is still relevant in addressing how media fans conduct their activities in the digital era.

One of the ways fans appropriate a popular text is by "making intertextual connection across a broad range of media texts" (Jenkins, 1992, p. 37) and that is exactly what PewDiePie did in one of his videos entitled 'Naked Donald Trump Space Simulator.' As the title implies, PewDiePie uses Donald Trump as a reference throughout the video. For the last couple of years, Donald Trump draws the public attention from both the US and many parts of the world. In 2015 he became one of the lead Presidential candidates from the Republican Party for the 2016 US Presidential Election. Furthermore, in July 2016 he was officially nominated as the Presidential Candidate from the Republican Party (Collinson & Kopan, 2016). However, during his presidential campaign, Trump created a lot of controversies from his statements and speeches regarding various issues. One of them is his statement about the immigration policy. He stated that he would build a wall along the Mexican border and regards the Mexican immigrants who came to the US as criminals, drug dealers, and rapists (Donald Trump Speech, debates, and campaign quotes, 2016).

Contrary to what Cryaotic did in his video, PewDiePie edited the video game and used Trump as the main reference. He did it as a form of critique for Trump's controversial statement regarding the immigration issue as well as his participation in the US Presidential Election. In 'Naked Donald Trump Simulator', he used an indie PC game called Cosmochoria in which the players take the role of a naked cosmonaut, whose tasks are to explore the space, claim planets that he found, plant crops to create life, as well as attacking the aliens that try to invade the planets (Cosmochoria by Nate Schmold, 2016). The game itself does not have any connection with Trump, but PewDiePie managed to create that connection through the editing process and his direct commentary which was recorded in the video. He edited the cosmonaut and changed him to become Donald Trump. Every time aliens try to attack the planet, an image of Donald Trump will appear and Trump's head will replace the head of the cosmonaut. At this point, PewDiePie will shout 'Goddamn Immigrants!' Through this video, PewDiePie tries to critically address Trump's oversimplified views regarding the immigration issue. He also shows how Trump sees the Mexican immigrants as nothing but threats to the United States that must be exterminated just like how the aliens are portrayed in the game. By using the features offered by LP videos, PewDiePie managed to articulate his concern in an informal, humorous, and playful manner in order to mock Donald Trump's views. Moreover, in the video, it is shown that as the cosmonaut kills more aliens, he gains more points in the game. PewDiePie uses the game's mechanism as a critique to Donald Trump who made his controversial stance regarding the Mexican immigrants.

Through the appropriation process in the form of video editing and commenting, PewDiePie managed to produce a whole new meaning from the video game. In doing so, he was able to create an inter-textual connection between the game's mechanism and a real-life issue that comes to his attention. Such an appropriation process does not necessarily ruin the original text rather, "the text becomes something more than what it was before, not something less" (Jenkins, 1992, p. 53). The example above also demonstrates that LP videos are not created for the purpose of entertainment or the source of gaming information only but can also be used as an outlet to address and express one's concern about a certain issue in a creative way. PewDiePie uses LP videos as a vehicle to articulate his political concern and a critique against Donald Trump, even though he is not a US citizen; he is a Swedish who immigrated to England. Thus, it can be said that the LP video itself is the reflection of the Let's Player's own identity and value. As Jenkins (1992) states that it is through activities, such as borrowing, reproducing, and modifying popular texts that the fans construct their own cultural and social identity.

4 CONCLUSION

As stated by Burgess & Green (2013, p. 77) YouTube is "big enough, and global enough, to count as a significant mediating mechanism for the cultural public sphere…under the right conditions, the website is an enabler of encounters with cultural differences and the development of political 'listening' across belief systems and identities." In the case of Cryaotic, the Let's Player himself might have not used the LP video to directly address the dubious issue of female representation in video games. However, it managed to open-up a critical and open-ended dialog among the audiences to problematize the issue. This shows that the audiences could also participate actively in making meanings out of the video games, even though they are not video game players. They do this while consuming the LP video because *YouTube* as a social media naturally provides the space for dialogs and gives opportunities to the non-gamer audiences to voice their opinion publically in the commentary section. This shows that in co-creative environment, such as YouTube, "the participants are all at various times and to varying degrees of audiences, producers, editors, distributors, and critics" (Burgess & Green, 2013, p. 82). Such critical dialogs could not occur if the video game content is discussed privately among video game players. The discussion about PewDiePie's critique towards Trump's stance on immigration issues reflects how a Let's Player can be both a fan and a prosumer (a producer and a consumer) who is able to construct and articulate new meanings by appropriating video games.

Through innovation in a creative process, LP videos have been used as a vehicle to address and re-problematize social issues that are meaningful for the Let's Players and in turn reflect their own personal identity and value in the process. In this case, PewDiePie is able to creatively make inter-textual connections between the game's mechanism and Donald Trump's statements regarding immigration to produce a LP video that tries to resist the ideological stance that Donald Trump made in his electoral campaign. PewDiePie exhibited how a LP video reflects the Let's Player's personal identity as well as the value of everyday life, and YouTube allows it to be articulated "to more 'public' debates around social identities, ethics, and cultural politics" (Burgess & Green, 2013, p. 80). Contrary to the common perception that sees playing video games as an unproductive, uncreative, and anti-social activity, the meaning-making process in the case of LP videos shows exactly the opposite of that. Instead of being a fan product that is only used as either a source of information or entertainment, it turns out that LP videos provide a platform for dialogs about various social issues in a public space (YouTube) and become a space to voice one's concern creatively using video games.

REFERENCES

Ault, S. (2016) YouTube's Top Genre. *Variety;* 330, 13; Available from: Pro-Quest Research Library. p. 29. [Accessed 5th January 2016].

Burgess, J., & Green, J. (2013) *YouTube: Online video and participatory culture.* Chicago, John Wiley & Sons.

Collinson, S. & Kopan, T. (2016) *It's official: Trump is Republican nominee.* Retrieved from: http://edition.cnn.com/2016/07/19/politics/donald-trump-republican-nomination-2016-election/. [Accessed July 2016].

Cryaotic (2016) *Cry Plays: Heavy Rain [P5].* Retrieved from https://www.youtu-be.com/watch?v=KX mw4RaEkwo [Accessed 9th March 2016].

DeCerteau, M. (1984) *The Practice of Everyday Life.* Berkeley, University of California Press.

Dickerman, C., Christensen, J., & Kerl-McClain, S.B. (2008) Big breasts and bad guys: Depictions of gender and race in video games. *Journal of Creativity in Mental Health,* 3(1), 20–29.

Dietz, T.L. (1998). An examination of violence and gender role portrayals in video games: implication for gender socialization and aggressive behaviour. *Sex Roles,* 38, 425–442.

Donald Trump speech, debates and campaign quotes. (2016). Retrieved from: http://www.newsday.com/news/nation/donald-trump-speech-debates-and-campaign-quotes-1.11206532 [Accessed August 2016].

Dorman (1997) Video and computer games: effect on children and implications for health education. *Journal of School Health,* 67 (4), 133–138.

Downs, E., & Smith, S.L. (2010) Keeping abreast of hypersexuality: A video game character content analysis. *Sex Roles,* 62(11–12), 721–733.

Dunlop, J.C. (2007) The US video game industry: Analysing representation of gender and race. *International Journal of Technology and Human Interaction,* 3(2), 96–109.

Entertainment Software Association. (2004) *Top ten industry facts.* Retrieved from: http://www.theesa.com/pressroom.html [Accessed 16th August 2016]

Espejo, R. (2003) *Video games.* Greenhaven Press.

Griffiths, M. (1999) Violent video games and aggression: a review of the literature. *Aggression and Violent Behavior,* 4 (2), 203–212.

Grossman, D. (2001) Trained to kill. *Das Journal des Professoren forum,* 2 (2), 3–10.

Heavy Rain for Playstation 3 Reviews (2016) *Heavy Rain Playstation 3.* Retrieved from: http://www.metacritic.com/game/playstation-3/heavy-rain

Heintz-Knowles, K., Henderson, J., Glaubke, C., Miller, P., Parker, M.A. & Espejo, E. (2001) Fair play? Violence, gender and race in video games. *Children Now.*

Ivory, J.D. (2001). *Video Games and the Elusive Search for their Effects on Children: an Assessment of Twenty Years of Research.* National AEJMC Conference, available from: http://www.unc.edu/ ̃jivory/video.html.

Ivory, J. D. (2008) The games, they are a changin': Technological advancements in video games and implications for effects on youth. In: P. Jamieson & D. Romer (Eds.), *The changing portrayal of adolescents in the media since 1950.* New York, Oxford University Press. pp. 347–376.

Jenkins, H. (1992) *Textual poachers: Television fans and participatory culture.* Routledge.

Jenkins, H. (2002) Interactive audiences? The collective intelligence of media fans. *The new media book*, 157–170.

Jessen, C. (1995) *Children's Computer Culture*. Retrieved from: http://www.hum.sdu.dk/center/kultur/buE/articles.html.

Kline, S. (2000) Killing Time: a Canadian meditation on video game culture. *Children in the new media landscape. Games, pornography, perceptions. Göteborg (UNESCO International Clearinghouse on Children and Violence on the Screen*, 35–59.

Leonard, D.J. (2006) Not a Hater, Just Keepin' It Real the Importance of Race- and Gender-Based Game Studies. *Games and Culture*, *1*(1), 83–88.

PewDiePie. (2015) Naked Donald Trump Space Simulator. Retrieved from: https://www.youtube.com/watch?v = mdlfmc1QeCQ [Accessed October 15, 2015].

Provenzo Jr, E. F. (1991) *Video kids: Making sense of Nintendo*. Harvard University Press.

Schmold, Nate. (2016) *Cosmochoria*. Retrieved from: https://www.kickstarter.com/projects/nateschmold/cosmochoria.

Turkle, S. (1984) *The Second Self: Computers and the Human Spirit*. New York, Simon & Schuster.

Cultural Dynamics in a Globalized World – Budianta et al. (Eds)
© *2018 Taylor & Francis Group, London, ISBN 978-1-138-62664-5*

Constructing an imagined community of fandom and articulating gender identity: A case study on Indonesian female football fans' participation in social media

N. Azizi & S.M.G. Tambunan
Department of English Studies, Faculty of Humanities, Universitas Indonesia, Depok, Indonesia

ABSTRACT: Even though nowadays it seems that there is no limitation for women to participate in football, it is still a sport dominated by male players and even male football fans. However, there has been a significant development in the number of female football fans in the last decade. These days, social media has been utilised as a virtual communication medium bringing out a homophily tendency. Groupings of football fan communities are based on similarities in accessing information related to their favourite football club. This study examines two female football fan communities: @JCI_Juvedonna (Twitter) and @babesunitedindo (Instagram). By utilising textual analysis and netnography methods (interviews and observations), the aim is to explore how female football fans construct contesting gender identities in social media. Research findings reveal that there are strategies used in the communities' social media, such as the construction of an imagined community that unites the members with a sense of belonging to the micro-community and the negotiation done to make sense of the dominant existence of the male fan community by using multiple signifying practices in articulating their feminine identity.

1 INTRODUCTION

European football clubs have a great number of fans in Indonesia reflected by a variety of football fan communities established by their loyal fans. Unfortunately, these communities are still dominated by male fans. Compared with female fans, male fans usually overbear the activities in the community making female fans look as if they are flattering mirrors in their own communities (Bourdieu, 1984). These communities, nowadays, rely heavily on social media to ensure constant interactions with other fans (Bruns, Weller, Harrington, 2014). Social media is able to give distinguished features to reflect the individuals or groups' identities, which is called homophily (McPherson, Smith-Lovin, and Cook, 2001). This research will focus on the transpiring Indonesian female football fans in social media, which are Juvedonna Indonesia/@JCI_Juvedonna (Twitter) and Babes United Indonesia/@babesunitedindo (Instagram), as the most active social media-based football fan communities in Indonesia. First of all, @JCI_Juvedonna is a Juventus female fan community established on 7th April 2014, which is part of Juventus Club Indonesia. Meanwhile, @babesunitedindo is a female fan community for Manchester United (MU), which has been very active in their Instagram account since early January 2016. It is part of United Indonesia, the biggest MU football fan community in Indonesia dominated by male fans; however, United Indonesia does not have full control on the activities carried out by @babesunitedindo. The members in both female football fan communities build an association with other members in the same community through the commoditisation of their club identity (Giulianotti, 2002).

Even though these two communities share similar traits, there are some differences, for example in how each community constructs particular gender identities and how they share information on how well they know their favourite football club. For example, members of @JCI_Juvedonna show that they have extensive knowledge about the history of the club, the latest development of the club, community events, and the structure of the club's management. Interestingly, when it comes to talking about legendary players, they usually retweet information from the Juventus Club Indonesia twitter account. On the contrary, @babesunitedindo does not rely on United Indonesia to show how knowledgeable they are about Manchester United. This certainly raises the question: why does @JCI_Juvedonna rely on the male dominated community for this type of information? Is there any knowledge gap between the male and female football fan communities and how does this reflect the communities' construction of gender identity? This will be problematised further in this article by doing netnography as the researchers have done an extensive observation on both communities' social media. Kozinets (2002) suggests that netnography can investigate specific instances in which meaning is produced in the community through computer-mediated communication (CMC) as it is used to manifest culture. Indirectly, netnography will be helpful to uncover the identity of female fans reflected in both communities. Moreover, the activities undertaken by them either in social life or in offline activities as proposed by Poster (1995) often portray disconnections between construction and gender categories in offline and online realm. The research aims to reveal how female fans make meaning out of their fandom activities in the digital era, how nation-branding in an imagined community is used to unify the community in order to negotiate with gender issues and how female fans respond to masculine pride.

1.1 *Female football fans in the digital era*

Being a football fan does not only mean that someone is interested in football as there are a lot of reasons why someone chooses to be a football fan. Wann as cited by Pegoraro (1997) identifies eight common motivations for becoming a sport fan: group affiliation, family, aesthetic, self-esteem, entertainment, escape, economy, and eustress. For female fans, their love to football has a more substantial meaning such as to express their identity (Pope, 2013; James and Ridinger, 2002; Jones, 2008; Regev: 2013). This way, they are able to express or reflect on their affiliation to a particular social class. Therefore, the concept of fandom in the case of female football fans should be problematized in an analytical way (Crawford, 2004). Fandom is a term, which is used to portray a fan's connection to other fans of the sport team. It is different from the fanship that only discusses about a particular group.

Because of the growth of the Internet, the opportunity to interact with others has been made possible and it has become more intense. Both communities, Juvedonna Indonesia and Babes United Indonesia, utilise technological advances to create space and social networks for people who have a common interest, in this case, the same interest on football clubs. The ease of connection facilitates new forms of intimacy and the formation of collective identity to build homogeneity, and this is done through signifying practices, such as wearing the same jerseys (clothes), sharing the same attitudes, or places of memories to distinguish them from other fans (Lennies and Pfister, 2015). In the digital age, female football fans create a new way of understanding fandom in relation to their loyalty to the club. Furthermore, their relationship with the male fans dominating the 'official' community articulates different ways in making sense of football fandom culture in Indonesia.

Moreover, in the digital era, activities done by communities are always closely hand in hand with economic impact (Hellekson, 2009; Scott, 2009). For example, the rule that each member is required to buy a jersey or to give donations to their community. Of course, besides economic value, there is a construction of sharedness, collectiveness, and a sense of belonging by buying the same jersey for every member of the community or by giving donation for the sake of the community's financial stability.

> "The system of relations in moral economies if formed by 'part-societies', small worlds of personal relationships in every society where individuals have lived out their lives ... and the various fan communities that are within a specific fandom form represent this part-society." (Catherine, 2010: 64)

The visible manifestation of economic activities in communities is done in the form of accessing information and technical knowledge, communication with like-minded individuals

(Raacke and Bonds-Raacke, 2008), and finding entertainment and diversion (Ruggiero, 2000). All three factors initiate a culture of consumption that aims to distinguish classes and culture of living. Pierre Bourdieu states that cultural consumption is "predisposed, consciously, and deliberately or not, to fulfil a social function of legitimating social differences" (Storey 1996: 1995–6). Thus, the members of Juvedonna Indonesia and Babes United Indonesia who could not keep up with the activities, such as buying the same jersey, will appear to have a lack of knowledge and will not be recognised in the same level of classes with the community. Pearson (2010) asserts that communities and fans are the two things that should never be separated because "fandom requires a community and participation in that community and possibly self-identification with that community" (93).

Based on the history of Juvedonna Indonesia and United Babes Indonesia, the idea of both small/"micro-communities" has been actually accommodated by male football fans. Ananta Ariaji as the new head of Juventus Club Indonesia (JCI) in 2014 explained that Juvedonna was established with a clear structure of organisation. However, the members have agreed that Juvedonna Indonesia would still be under on the JCI, the 'original' community. Furthermore, Pelangi, as the one of the initiators of Babes United Indonesia, stated that her community rises as a result of the increasing number of female fans in United Indonesia. These members then, organically, formed a micro-community inside the macro-community, United Indonesia. In the development of activities in social media, these micro-communities usually hold a more important role. They cater to very specific groups' shipper or interpretation of the text (Catherine, 2010: 119). The posts on the @babesunitedindo Instagram page are dominated by the micro-community's activities. No activity from the macro-community, United Indonesia, is put on the Instagram page. It means the micro-community has set boundaries, rules, and hierarchical structures that may not necessarily conform to the wider fan community. This also applies in Juvedonna Indonesia, @JCI_Juvedonna, even though they often tweet/retweet the regional community's activities.

The existence of the micro-community (or chapter) domination is caused by the difficulties to examine large-scale activities because of geographical factors, and this phenomenon has similarities with Anderson's imagined community (Rashif in Pandit, 2016). A fan-community is unlikely to know and identify all of the members. They might not have met face to face and possibly never heard about each member. There is a projection of space and time as people make a collective delusion that they are engaging in a single community; even though they do not know each other, they are united by the love for the club. If in the conceptualisation of an imagined community the imagination of collectiveness and a sense of belonging are built through print capitalism, in the two female football fan communities, the main tool is social media: Twitter and Instagram. Furthermore, the impact of the imagined community is a reinstatement of the traditional ways like face to face interaction in the form of social media interaction. The whole process of interaction aims to maintain the solidarity among members, to maximise the cohesiveness, and to demonstrate their existence (Huda, 2012; Sincere, et al., 2013; Wulansari, et al., 2014; Santoso, 2014; Saputra, 2015).

1.1 *Constructing a female football fan identity: A response to masculine pride*

Pride is the one of the central elements of fandom in accordance with the status of the football club and particularly the team's success fielding the leagues. Furthermore, pride is often identified with masculinity in football (King, 1997). Understanding the social and cultural meanings constructed by Juvedonna Indonesia and Babes United Indonesia is a way to see how they want people to recognise their identity. For that, they have built their own identity to negotiate and contest the dominant discourse of female football fans, which mostly depicts the fans as women craving over the handsome male football players. They also need to deconstruct the dominant conceptualisation of football fandom as a masculine cultural practice. The findings of this research reveal that Juvedonna Indonesia and Babes United Indonesia represent their female football fan identities through several signifying practices.

First of all, a logo becomes an important thing in revealing their identity because it becomes a symbol of pride and a distinction between them and other communities (Hatch and Schultz, 1997; Erel-Koselleck, 2004). Juvedonna Indonesia uses an orchid as their logo. According to interviews done with the founders of the community and the information on their twitter account, six petals of the flower represent the initiation of Juvedonna that has one goal, one thought, and one vision. On the other hand, flowers are also often associated with femininity. Does Juvedonna Indonesia try to emphasise on a more feminine identity? Another piece of evidence is the interface on the @JCI_Juvedonna's Twitter page, which is dominated by pink emphasising a feminine identity. When the members of Juvedonna Indonesia celebrate the International Women Day, Valentine's Day, and National Mother Day, they use a pink-dominated page to express their enthusiasm in celebrating these events. However, the macro-community Juventus FC uses white and black as the dominant colour in their social media account. To sum up, Juvedonna Indonesia articulates their gendering references by choosing pink which portrays a construction of feminine identity often done along with other visual components in their Twitter account. A second function of pink is to index sexual identity, which is complementary but also contradictory to the masculine demarcation line. Finally, pink is also used to associate the community with fun attitudes, independence, and confidence.

Babes United Indonesia chooses more or less similar feminine signifiers in their logo; however, through their Instagram posts, this community also tries to present a collective identity of sisterhood. In their logo, they use two slim women silhouettes and high heels portraying their feminine identity differentiating them from the male dominated community, United Indonesia. However, in their Instagram page, almost all of the posts are group photos of the members of the community from several parts of Indonesia. The captions they use represent a sense of collectiveness or more specifically, sisterhood. For example, in one of their earlier posts, there was a picture of the community's members in an event, Gathbabes 2015 (gathering), and the caption was: "The friendship & sisterhood will make your life so colourful." By choosing images depicting togetherness, @babesunitedindo portrays sisterhood as a significant part of their identity. However, there are also some captions that portray a submission to the dominant patriarchal ideology. One of the things they can do in the community is to find 'a soul mate' presumably from the United Indonesia community. One of the captions was (as translated by the researchers): "Don't forget to eat lunch, babes. If you want to find a soul mate, you will need a lot of energy." The feminine identity constructed in the Instagram page of Babes United Indonesia is portrayed through a dualism between wanting to differentiate themselves from the male dominated community and emphasising on the collectiveness of sisterhood and the repeated idealisation of heteronormative relationship goals.

2 CONCLUSION

This paper has provided insights into the various ways in which an imagined community is constructed through the activities of fandom in social media and how a feminine identity is used, first and foremost, to differentiate the female football fan micro-communities from the male dominated macro-community. The male domination in football fandom in Indonesia encourages female football fans to seek recognition by constructing a social and cultural identity though social media activities. The choice of social media by the female football fans (homophily) will affect the way they construct their identity and the way people perceive it. In conclusion, this paper has shown how football fandom for female fans could have a complex meaning-making process. For a complementary study, the researchers recommend exploring the comparison between female football fans' activities and male football fans' activities to figure out how their gender constructions are determined.

REFERENCES

Abitia, R. (2016) *Sepak Bola sebagai Komoditas Global dan Komunitas-Komunitas Terbayang dalam Suporter Sepak Bola Modern* (Football as a Global Commodity and Modern Football Communities among Supporters). Retrieved from http://panditfootball.com/pandit http://panditfootball.com/pandit-sharing/201119sharing/201119.

Ali, R. (2009) Technological neutrality, *Lex Electronica*, 14 (2). Retrieved from http://www.lex-electronica.org/docs/articles_236.pdf.

Anwar, M.H. & Saryono. (2009) *Kontroversi citra perempuan dalam olahraga* (The controversial image of women in sports). Yogyakarta, Lembaga Penelitian Universitas Negeri Yogyakarta.

Archer, J. & McDonald, M. (1990) Gender roles and sports in adolescent girls. *Leisure Studies*, 9, 225–240.

Bourdieu, P. & Wacquant, L.J.D. (1992) *An Invitation to Reflexive Sociology*. Chicago, US, University of Chicago Press.

Bourdieu, P. (2001) *Masculine Domination (translation)*. Chicago, US, Stanford University Press.

Brown, A. (ed.). (1998) *Power, Identity & Fandom in Football*. New York, NY, Routledge.

Bruns, A., Weller, K. & Harrington, S. (2014) *Twitter and Sports: Football Fandom in Emerging and Established Markets. Twitter and Society*. New York, Peter Lang, pp. 263–280.

Budd, C. (2012) *The Growth of an Urban Sporting Culture—Middlesbrough, c. 1870–1914*. [Dissertation], University De Montfort.

Cecamore, S., et al. (2011) *Sport Fandom: What do Women Want? A Multi-Sport Analysis of Female Fans*. London, International Centre for Sport Studies.

Chin, B.C.L.P. (2010) *From Textual Poacher to Textual Gifters; Exploring Fan Community and Celebrity in the Field of Fun Cultural Production*. Wales, Cardiff University.

Chiwese, M. (2014) One of the boys: female fans' responses to the masculine and phallocentric nature of football stadium in Zimbabwe. *Journal Critical African Studies*, 6 (2–3), 211–222.

Crawford, G. & Gosling, V.K. (2004) The myth of the "puck bunny": Female fans and men's ice hockey. Sociology, 38 (3), 477–493.

Dunn, C. (2014) *Female Football Fans: Community, Identity and Sexism*. United Kingdom, UK, Palgrave MacMillan, 5.

Eagleton, T. (2010) *Football: a dear friend to capitalism*. The Guardian.

Erel-Koselleck, E.S. (2004) *The Rule Power of Symbols in the Identity Formation of Community Members*. Thesis. Retrieved from https://etd.lib.metu.edu.tr.

Gauntlett, D. (2002) *Media, Gender and Identity: An Introduction*. London, Routledge.

Giulianotti. R. (2002) Supporters, followers, fans, and flaneurs: A taxonomy of spectator identities in football. *Journal of Sport & Social Issues*, 26 (1), 25–46.

Hall, D. (1994) *Muscular Christianity: Embodying the Victorian Age*. London, Cambridge University Press.

Hatch, M.J. & Schultz, M. (1997) Relations between organizational culture, identity, and image. *European Journal of Marketing*, 31 (5–6), 356–365.

Hellekson, K. (2009) A fannish field of value: online fan gift culture. *Cinema Journal*, 48 (4), 113–118.

James, J. & Ridinger, L.L. (2002) Female and male sport fans: a comparison of sport consumption motives. *Journal of Sport Behaviour*, 25 (3), 260.

Judith, B. (1990) *Gender Trouble: Feminism and the Subversion of Identity*. New York, NY, Routledge.

Katherine, W.J. (2008) Female fandom: identity, sexism, and men's professional football in England. *Sociology of Sport Journal*, 25, 516–537.

King, A. (1997) The postmodernity of football hooliganism. *British Journal of Sociology*, 48 (4), 576–593.

Kozinets, R.V. (2002) The field behind the screen: using netnography for marketing research in online communities. *Journal of Marketing Research*, 39, 61–72.

Lenneis, V. & Pfister, G.U. (2015) Gender constructions and negotiations of female football fans: A case study in Denmark. *European Journal for Sport and Society*, 12 (2), 157–185.

Lorber, J. (1994) *Paradoxes of Gender*. New Haven, Yale University Press.

McLean, R. & Wainwright, D.W. (2003) Social networks, football fans, fantasy, and reality: How corporate and media interests are invading our life world. *Journal of Information, Communication and Ethics in Society*, 7 (1), 55–71.

McPherson, M., Smith-Lovin, L. & Cook, J.M. (2001) Birds of a feather: homophily in social networks. *Annual Review of Sociology*, 27, 415–444.

Neale, L. & Funk, D. (2005) *Fan Motivation and Loyality: Extending the Sport Interest Inventory (SII) to the Australian Football League on ANZMAC 2005*. Retrieved from http://eprints.qut.edu.au/.

O'Reilly, T. & Milstein, S. (2009) *The Twitter Book*. California, O'Rielly Media, Inc., p. 5.

Ozsoy, S. (2011) Use of new media by Turkish fans in sport communication: Facebook and Twitter. *Journal of Human Kinetics*, 28 (11), 165–176.

Pearson, R. (2010) Fandom in the digital era. *Popular Communication Journal*, 8 (1), 84–85.

Pegoraro, A. (1997) *Sport Fandom in the Digital World*. Retrieved from https://www.academia.edu/6644943/Sport_Fandom_in_the_Digital_World.

Pierre, B. (1991) *Sport and Social Class in Rethinking Popular Culture*. Los Angeles, LA, University of California Press.

Poer, F. (2004) *How Soccer Explains the World: An Unlikely Theory of Globalization*.

Pope, S. & Kirk, D. (2014) The Role of Physical Education and Other Formative Experiences of Three Generations of Female Football Fans'. *Sport, Education and Society Journal*. 19 (2), 223–240.

Pope, S. (2013) "The love of my life": The meaning and importance of sport for female fan. *Journal of Sport and Social Issues* XX(X), 1–20.

Raacke, J.D. & Bonds-Raacke, J.M. (2008) MySpace & Facebook: Applying the uses and gratifications theory to exploring friend networking sites. *Cyber Psychology and Behavior*, 11, 169–174.

Rapaport, T. & Regev, D. (2013) *Female Fans Visibility in the Fandom Field: The Case of H'apoel Katamon*. Vienna: FREE Conference University of Vienna.

Ruggerio, T.E. (2000) Uses and gratification theory in the 21st Century. *Mass Communication & Society*, 3 (1), 3–27.

Scott, S. (2009) Repackaging fan culture: The regifting economy of ancillary content models. *Transformative Works and Cultures*, 3. Retrieved from http://dx.doi.org/10.3983/twc.2009.0150. doi:10.3983/twc.2009.0150.

Siregar, A. (1997) *Ecstasy Gaya Hidup: Kebudayaan Pop dalam Masyarakat Komoditas Indonesia* (The Ecstasy of Lifestyle: Pop Culture in the Indonesian Commodity Society). Jakarta, Pustaka Mizan.

Cultural Dynamics in a Globalized World – Budianta et al. (Eds)
© *2018 Taylor & Francis Group, London, ISBN 978-1-138-62664-5*

Martha Nussbaum's central capabilities read violence against women in Indonesia

H.P. Sari
Department of Criminology, Faculty of Social and Political Science, Universitas Indonesia, Depok, Indonesia

ABSTRACT: The ten central capabilities formulated in Martha Nussbaum's unique approach offer us a way of seeing whether a person is living a flourishing and meaningful life and what a person is actually able to be and to do in life to realize his/her potential as a human being. The underlying idea behind these ten capabilities is Nussbaum's conviction of what a human being is and what kind of life is worthy of human dignity. In effect, the central capabilities may be used to determine whether the life aspects lived by a woman indicate that she is a human being and is living a life worthy of human dignity. In this paper, the national data on violence against women in Indonesia (Catahu Komnas Perempuan 2014, 2015, and 2016) were analyzed using the lenses of the central capabilities to know whether the lives of Indonesian women are worthy of human dignity and whether their life is a human life at all. Finally, a discussion on methodological obstacles to translating abstract philosophical ideas into down-to-earth practical applications is presented.

1 INTRODUCTION

The capability approach is a frame of thinking of what constitutes a good life. The approach aims to answer questions of what a person is actually able to do and to be in life. Amartya Sen and Martha Nussbaum initially developed it to take into consideration the lives of women which are often filled with gender discrimination (2011, p. 19). Nussbaum advanced even further by proposing her version of the capability approach that is captured in her ten central human capabilities. This is a list consisting of ten capabilities that every person must have in order to live a life worthy of human dignity. Through an Aristotelian/Marxian way, Nussbaum founded her central capabilities on human dignity and what it requires. In other words, what the prerequisites for living a life that is fully human rather than subhuman, a life worthy of the dignity of the human being, are (Nussbaum, 2004, p. 13).

Nussbaum's capability approach is intended to be a universal normative theory underlined by an account of a human life (Nussbaum, 1995, pp. 72–86). A human life, according to Nussbaum, can be defined in terms of the most important functions and capabilities of the human being. Such a basic idea is reflected in the questions: "What are the characteristic activities of the human being? What does the human being do, characteristically, as such?" In other words, what the forms of activities, of doing and being, are that constitute a human form of life and distinguish it from other actual or imaginable forms of life, such as the lives of animals and plants, or of immortal gods as imagined in myths and legends. The last forms, Nussbaum believed, frequently have the function of delimiting humans.

Nussbaum (1995, pp. 72–73) offered two more concrete questions to approach those questions. The first question is what she calls a question about personal continuity: what changes or transitions are compatible with the continued existence of that being as a member of the human kind, and what are not? She believes that this is also a question about the necessary conditions for continuing as one and the same individual because continued species identity is at least necessary for continued personal identity. The question, in Nussbaum's view, arises from an attempt to make medical definitions of death in a situation in which some functions of life persist, or to decide, for others or for ourselves, whether a certain level of illness or impairment means the end of the life of the being in question. Thus, Nussbaum is convinced that some functions can be absent without threatening our sense that we still have a human being on our hands, while the absence of other functions signals the end of a human life.

The second question is about the kinds of inclusion by Nussbaum (1995, 73–74): what do we believe must be there if we are going to acknowledge that a given life is human? This relates to recognizing other humans as humans across many different times and places. Her example is that we recognize things differently when we see our child playing with a dog or cat, when we see our dog and cat playing, or when we see anthropomorphic creatures, which are not classified as humans because of the nonhuman features of their form of life and function.

The first and second questions, in Nussbaum's view, are inquired by examining a wide variety of self-interpretations of human beings in many times and places, because people in many different societies share a general outline of such conception. The result is a theory that is not the mere projection of local preferences, but is fully international and a basis for cross-cultural attunement (Nussbaum, 1995, p. 74). Thus, it is Nussbaum's conviction that every person is worthy of equal respect and regard because he/she has human dignity.

The facts that women—as human beings—have human dignity and are worthy of equal respect and regard (Nussbaum, 2011, p. 18) and that philosophy has a lot to offer in approaching pervasive gender inequality all over the world (Nussbaum, 1998, pp. 762–769) have spurred Nussbaum (2011, pp. 18–19) to aim her central capabilities at eliminating gender inequality. She asserted that the approach is *concerned with entrenched social injustice and inequality* [sic], especially capability failures that are the results of discrimination or marginalization. Thus, she sees her capability approach as an approach to the question of fundamental entitlements that is especially pertinent to the issues of sexual equality (Nussbaum, 2003, p. 36). She emphasized that "capabilities can help us to construct a normative conception of social justice with a critical potential for gender issues, only if we specify a definite set of capabilities as the most important ones to protect" (Nussbaum, 2003, p. 33).

Nussbaum offered a list of ten capacities (2011, p. 18). The list works as an indicator of a woman's quality of life (her basic decency or justice), that is, whether her life is a human life at all and/or a good human life. A good human life is a life that enables each person to do and to be. This means a life that provides opportunities for every person to fully realize his/her potential to flourish to the optimum level.

Nussbaum's central capabilities (Nussbaum, 2011, pp. 32–34) consist of ten capabilities that each and every person is entitled to at a bare minimum. The capabilities should be secured by a decent political order—the government—for the citizens to be able to pursue a dignified and minimally flourishing life. The capabilities are as follows:

1. *Life.* Being able to live to the end of a human life of normal length; not dying prematurely, or before one's life is so reduced as to be not worth living.
2. *Bodily health.* Being able to have good health, including reproductive health; being able to be adequately nourished; being able to have adequate shelter.
3. *Bodily integrity.* Being able to move freely from place to place; being able to be secure against violent assault, including sexual assault, marital rape, and domestic violence; having opportunities for sexual satisfaction and for choice in matters of reproduction.
4. *Senses, imagination, and thought.* Being able to use the senses, to imagine, to think, and to reason—and to do these things in a "truly human" way, a way informed and cultivated by an adequate education, including, but by no means limited to, literacy and basic mathematical and scientific training; being able to use imagination and thought in connection with experiencing and producing expressive works and events of one's own choice (religious, literary, musical, etc.); being able to use one's minds in ways protected by guarantees of freedom of expression with respect to both political and artistic speech and freedom of religious exercise; being able to have pleasurable experiences and to avoid non-beneficial pain.

5. *Emotion.* Being able to have attachments to things and persons outside ourselves; being able to love those who love and care for us; being able to grieve at their absence; in general, being able to love, to grieve, to experience longing, gratitude, and justified anger; not having one's emotional development blighted by fear or anxiety. (Supporting this capability means supporting forms of human association that can be shown to be crucial in their development).
6. *Practical reason.* Being able to form a conception of the good and to engage in critical reflection about the planning of one's own life. (This entails protection for the liberty of conscience.)
7. *Affiliation.* (a) Being able to live for and in relation to others, to recognize and show concern for other human beings, to engage in various forms of social interaction; being able to imagine the situation of another and to have compassion for that situation; having the capability for both justice and friendship. (Protecting this capability means, once again, protecting institutions that constitute such forms of affiliation, and also protecting the freedom of assembly and political speech). (b) Having the social bases of self-respect and non-humiliation; being able to be treated as a dignified being whose worth is equal to that of others. (This entails provisions of nondiscrimination.)
8. *Other species.* Being able to live with concern for and in relation to animals, plants, and the world of nature.
9. *Play.* Being able to laugh, to play, and to enjoy recreational activities.
10. *Control over one's environment.* (a) *Political*: being able to participate effectively in the political choices that govern one's life; having the rights of political participation, free speech, and freedom of association (b) *Material*: being able to hold property (both land and movable goods); having the right to seek employment on an equal basis with others; having the freedom from unwarranted search and seizure. In work, being able to work as a human being, exercising practical reason and entering into meaningful relationships of mutual recognition with other workers.

The research question in this study is: how do Nussbaum's ten central capabilities see violence against women in Indonesia from 2013 to 2015?

2 APPROACH/METHODOLOGY

All data used in this study were adapted from annual reports (henceforth Catahu Komnas Perempuan) published in 2014, 2015, and 2016 by the National Commission on Anti-Violence against Women (henceforth Komnas Perempuan). Komnas Perempuan has divided the data of violence against women in Indonesia into three categories: violence against women in the personal sphere, in the community sphere, and in the state (authority) sphere. Each category then is cross-checked with the ten central capabilities to see the existence or nonexistence of each capability. The tables presented in the paper are for highlighting the capabilities and not for statistic measures in whatsoever terms. In each sphere, the data from 2013, 2014, and 2015 are compared to find the forms of violence against women occurring within the three-year period. Then, the forms of violence are cast under each capability to find their existence or nonexistence. Because the data of the state (authority) sphere are more qualitative in nature, they are first summarized and then presented in the table.

3 RESULTS AND DISCUSSION

Utami's (2012) research showed that Indonesian women in general suffer from gender discrimination, including violence, in different levels and forms. Concurring to that, the national data from 2013, 2014, and 2015 (Komnas Perempuan, 2014, 2015, and 2016) indicate no tendency of the decrease of violence against women. Table 1 presents forms of violence against women occurring in the personal sphere in 2013, 2014, and 2015.

As can be seen in Table 1, in the personal sphere, the number of violence cases decreased slightly in 2014 but increased in 2015 to an almost similar number to 2013.

Some research using feminist approach has highlighted that female victims of sexual, psychological, economical, and physical violence experience little to none of each of the ten capabilities (Nussbaum and Glover, 1995). Table 2 shows that in the lives of the victims, capabilities of life, bodily health, bodily integrity, senses, imagination, and thought, emotions, practical reason, and affiliation do exist, but they are hampered at best or non-existing at worst. Table 2 shows whether each capability exists in each form of violence against women.

In the community sphere, there is a similar trend of violence: a slight drop in the number in 2014 but rising again in 2015. A new form of violence, "others", is added here, and it consists of violence experienced by migrant workers and victims of trafficking.

Table 4 shows that all the ten central capabilities do exist, yet they are also hampered by the mere fact that the violence has occurred at a communal level. The female victims then have less than full central capabilities.

In the state (authority) sphere in 2013, there were four cases of criminalizing women who were involved in conflicts of claims over natural resources in West Sumatera and one case of a woman victim of violence in the community sphere who was facing obstacles in reporting it to the justice system (Catahu Komnas Perempuan, 2014, p. 14). Meanwhile, violence against women in the state (authority) sphere in 2014 consists of one case of the rights of a woman to adopt a child in Aceh, four cases of virginity tests in West Java (Bandung district), two

Table 1. Violence against women in the personal sphere in 2013, 2014, and 2015.

Forms of violence	2013 N = 11,719	2014 N = 8,626	2015 N = 11,207
Sexual	2,995	2,274	3,325
Psychological	3,344	2,444	2,607
Economical	749	498	971
Physical	4,631	3,410	4,304

Source: Adapted from Catahu Komnas Perempuan 2014, 2015, and 2016.

Table 2. Existence of central capabilities in the occurrence of violence against women in the personal sphere in 2013, 2014, and 2015.

Capabilities	Sexual			Psychological			Economical			Physical		
	√	×	NA	√	×	NA	√	×	NA	√	×	NA
Life		×		√	×		√	×			×	
Bodily health		×		√	×		√	×			×	
Bodily integrity		×		√	×		√	×			×	
Senses, imagination, and thought		×		√	×		√	×		√	×	
Emotions		×		√	×		√	×		√	×	
Practical reason	√	×		√	×		√	×		√	×	
Affiliation	√	×		√	×		√	×		√	×	
Other species			NA			NA			NA			NA
Play			NA			NA			NA			NA
Control over one's environment			NA			NA		×				NA

Source: Adapted from Nussbaum (2011, 33–34) and Catahu Komnas Perempuan 2014, 2015, and 2016.

Table 3. Violence against women in the community sphere in 2013, 2014, and 2015.

Forms of violence	2013 N = 4,679	2014 N = 3,860	2015 N = 5,002
Sexual	2,634	2,183	3,174
Psychological	248	41	169
Economical	25	7	64
Physical	897	1,086	1,124
Others	875	543	471

Source: Catahu Komnas Perempuan 2014, 2015, and 2016.

Table 4. Existence of central capabilities in the occurrence of violence against women in the community sphere in 2013, 2014, and 2015.

Capabilities	Sexual			Psychological			Economical			Physical			Others		
	√	×	NA	√	×	NA	√	×	NA	√	×	NA	√	×	NA
Life	√	×		√	×		√	×		√	×		√	×	
Bodily health	√	×		√	×		√	×		√	×		√	×	
Bodily integrity	√	×		√	×		√	×		√	×		√	×	
Senses, imagination, and thought	√	×		√	×		√	×		√	×		√	×	
Emotions	√	×		√	×		√	×		√	×		√	×	
Practical reason	√	×		√	×		√	×		√	×		√	×	
Affiliation	√	×		√	×		√	×		√	×		√	×	
Other species	√	×		√	×		√	×		√	×		√	×	
Play	√	×		√	×		√	×		√	×		√	×	
Control over one's environment	√	×		√	×		√	×		√	×		√	×	

Source: Adapted from Nussbaum (2011, 33–34) and Catahu Komnas Perempuan 2014, 2015, and 2016.

cases of criminalizing victims of domestic violence in Jakarta, and 17 cases of state omission of migrant workers in Jakarta (Catahu Komnas Perempuan, 2015, p. 10).

In the state (authority) sphere in 2015, there were two cases of forfeit of marriage certificates in West Java; six cases of similar forfeit in East Nusa Tenggara; two cases of unrecorded marriages, which caused obstacles for the wives who experienced spousal violence to report the crimes as domestic violence to the justice system; one case of trafficking facing obstacles with the police; and three cases of physical violence by a police officer (Catahu Komnas Perempuan, 2016, p. 21).

Table 5 shows whether there are central capabilities in violence against women in the state (authority) sphere in 2013, 2014, and 2015.

In the sphere of the state (authority), due to its huge and pervasive power over the citizens, all central capabilities for the victims are hampered.

The victims of violence against women in all the three spheres have not yet had full central capabilities, as shown in Tables 2, 4, and 5. These women have not lived a life worthy of human dignity; they have lived less than human lives. Thus, they have not been full human beings yet. The data of Catahu Komnas Perempuan published in 2014, 2015, and 2016 are only the tip of the iceberg, as a larger number of incidents against Indonesian women go unreported. In fact, all Indonesian women might have experienced violence at one point or another in their lives. As a matter of fact, Nussbaum and other feminists state that all women in the world might have such experience (Nussbaum, 2011 and 2000). Hence, Indonesian women have not yet lived flourishing life up to their potential.

Table 5. Existence of central capabilities in the occurrence of violence against women in the state (authority) sphere in 2013, 2014, and 2015.

Capabilities	2013			2014			2015		
	√	×	NA	√	×	NA	√	×	NA
Life	√	×		√	×		√	×	
Bodily health	√	×		√	×		√	×	
Bodily integrity	√	×		√	×		√	×	
Senses, imagination, and thought	√	×		√	×		√	×	
Emotions	√	×		√	×		√	×	
Practical reason	√	×		√	×		√	×	
Affiliation	√	×		√	×		√	×	
Other species	√	×		√	×		√	×	
Play	√	×		√	×		√	×	
Control over one's environment	√	×		√	×		√	×	

Source: Adapted from Nussbaum (2011, 33–34) and Catahu Komnas Perempuan 2014, 2015, and 2016.

What should Indonesia do then? In answering the question, we can refer to Nussbaum's opinion (2011, pp. 63–64) that the duties to distribute to all citizens an adequate threshold of the ten central capabilities belong to the nation's basic political structure. She sees that the job of the government is, at a minimum, to make it possible for people to live such a life. In line with that, it is the duties of the Indonesian government to be accountable for the presence of the ten capabilities on Nussbaum's list if Indonesia wants to be even at least a just nation.

Nussbaum (2011, pp. 63–64) acknowledged, however, that poor nations may not meet all their obligations without the aids from richer nations; thus, richer nations have the duties of providing aids. Concurring with that, Indonesia is entitled to be assisted by richer nations to meet its obligation to deliver the capabilities. Meanwhile, other duties to promote human capabilities in Indonesia can be assigned to corporations, international agencies and agreements, and finally the Indonesian individuals (Nussbaum, 2011, pp. 63–64).

In discussing the implementation of the capabilities approach, Nussbaum (2011, p. 97) asserted that the ten capabilities are all important and that subordinating one to another will not be a recipe for achieving full justice. One of the major avenues of the implementation of the central capabilities, in her view, is a nation's system of constitutional adjudication involving fundamental rights. Following that, the *Mahkamah Konstitusi* (Constitutional Court) in Indonesia might be used to implement the central capabilities.

Nussbaum (2011, p. 97) reminded us that the goal of policy makers should always be to present people with choices in the areas the list identified as central, rather than to dragoon them into a specific mode of functioning. She believed that such an emphasis on choices certainly shapes the strategies of implementation that policy makers should consider. In addition, some extent of further recommendation for implementation should be context-specific. To achieve such ideal thoughts, Indonesians will still have to wait.

4 LIMITATIONS AND CONCLUSION

The data of Catahu Komnas Perempuan published in 2013, 2014, and 2015 are processed data that do not provide detailed case-by-case information (for the methods and measurements used by Komnas Perempuan, see the annual reports). Cross-checking them with the ten central capabilities is rather difficult and results in a very crude and superficial analysis. The fact that they are the only available national data on violence against women over a certain period of time has put some limitations on the following issue of Nussbaum's central capabilities.

Nussbaum (2011, pp. 59–62) admitted that there are some challenges faced by the proponents of the ten central capabilities to show whether and how this new standard of values for public action is to be measured. Although the ten central capabilities are plural, she believed that each of them can be measured individually. However, she admitted that the measurement is indeed not an easy task because the notion of capability combines internal preparedness with an external opportunity in a complicated way. This issue has resulted in a still developing literature on the measurement of capabilities. Citing from the work of Hackett (2000, p. 1), Utami (2012, p. 21) pointed out that this methodological issue is an obstacle to translating abstract philosophical theory into practical life-transforming policy applications.

REFERENCES

Komisi Nasional Anti Kekerasan Terhadap Perempuan. (2014) Kegentingan Kekerasan Seksual: Lemahnya Upaya Penanganan Negara (The seriousness of sexual violence: The weakness of state's involvement) *Catatan Tahunan tentang Kekerasan terhadap perempuan.* [Online] Available from: http://www.komnasperempuan.go.id/category/publikasi/catatan-tahunan/.

Komisi Nasional Anti Kekerasan Terhadap Perempuan. (2015) Kekerasan terhadap Perempuan: Negara Segera Putus Impunitas Pelaku (Violence against women: The state must immediately stop impunity for perpetrators). *Catatan Tahunan tentang Kekerasan terhadap perempuan.* [Online] Available from: http://www.komnasperempuan.go.id/category/publikasi/catatan-tahunan/.

Komisi Nasional Anti Kekerasan Terhadap Perempuan. (2016) Kekerasan terhadap Perempuan telah Meluas: Negara Urgen Hadir Hentikan Kekerasan terhadap Perempuan di Ranah Domestik, Komunitas dan Negara (Violence Against Women has Expanded: The state must stop domestic, community and state violence against women). *Catatan Tahunan tentang Kekerasan terhadap perempuan.* [Online] Available from: http://www.komnasperempuan.go.id/category/publikasi/catatan-tahunan/.

Nussbaum, Martha C. (1995) Human Capabilities, Female Human Beings. In: Martha C. Nussbaum and Jonathan Glover (Eds.). *Women, Culture, and Development: A Study of Human Capabilities.* Oxford, Clarendon Press, 62–104.

Nussbaum, Martha C. (1998) Public Philosophy and Feminism. *Ethics,* 108, 762–796.

Nussbaum, Martha C. (2000) *Women and Human Development the Capabilities Approach.* New York, Cambridge University Press.

Nussbaum, Martha C. (2003) Capabilities as Fundamental Entitlements: Sen and Social Justice. *Feminist Economics* 9, 33–59.

Nussbaum, Martha C. (2004) Beyond the Social Contract: Capabilities and Global Justice. *Oxford Development Studies,* 32, 3–18.

Nussbaum, Martha C. (2011) *Creating Capabilities the Human Development Approach.* Cambridge, Massachusetts, and London, England, The Belknap Press of Harvard University Press.

Nussbaum, Martha C. and Jonathan Glover. (1995) *Women, Culture, and Development: A Study of Human Capabilities.* Oxford Clarendon Press.

Utami, Rona. (2012) *The Application of Capabilities Approach,* Master 2 en Philosophie, Sciences et Arts Facultè de Lettre et Sciences Humaines, Universitè de Lorreaine, unpublished.

Cultural Dynamics in a Globalized World – Budianta et al. (Eds)
© *2018 Taylor & Francis Group, London, ISBN 978-1-138-62664-5*

Afghan women's repression by patriarchy in *My Forbidden Face* by Latifa

H.N. Agustina & C.T. Suprihatin
Department of Literature, Faculty of Humanities, Universitas Indonesia, Depok, Indonesia

ABSTRACT: This to show the repression experienced by Afghan women in the novel *My Forbidden Face* written by a young Afghan woman writer. The discuss is the problems that when women can no longer speak of their own freedom and independence. During those dark periods the invasion of the Taliban, women were repressed by the confines of the social system and traditions. *My Forbidden Face* features the repression Afghan women in the name of Islam and under a patriarchy system. Women were restricted only the domestic realm and were not allowed to work, attend school or engage in any other activities outside their home. Only men were allowed in the public domain and could freely travel outside the home, to work and to do other activities. This study a qualitative method, applying the concepts of gender and feminism as tools. The results of this study reveal the repression experienced by many Afghan women the patriarchal culture and gender construction prevailing in Afghanistan women were deprived of their rights part in the public domain.

1 INTRODUCTION

Afghan women's struggle to liberate themselves from the shackles of oppression, gender repression and patriarchy under the Taliban era in 1996–2000 was voiced through several writings of many Afghan diasporas. The story of the women's struggle is interesting to be discussed because they often become the main victim of repression. *My Forbidden Face* tells one of the stories on the suffering of Afghan women. *Growing Up under the Taliban: A Young Woman's Story* by Latifa (2001) articulated the various forms of repression that Latifa and other Afghan women experienced. The efforts and struggle of Afghan women for their rights and freedom to participate actively in the public domain have opened the eyes of women around the world. This article discusses the story of a young Afghan woman's struggle to defend women's rights and represent Afghan women in portraying the many forms of repression in a memoir.

2 METHOD

Gender and feminism are the two issues that are interrelated when discussing women's issues in the Middle East. Different and diverse forms of gender inequality exist in these Middle Eastern countries. This often raises the issues regarding the relationship between women and men. *My Forbidden Face* is written from the perspective of a young woman explaining the forms of repression against Afghan women as a result of the patriarchal cultural construction, which is represented by the presence of the Taliban regime. This novel has a wealth of analysis documenting the situation of Afghan women and placing it in a cultural, historical and political context.

The novel telling the story of repression against Afghan women is a product of several years of research on women and politics in the country. As a researcher, the leading factor that has inspired me to conduct research on this subject is the repression against women in Afghanistan, which has become the most frequently discussed topic. To render justice to the women in Afghanistan, I dedicate my research to make this ignored issue gain international attention. By analysing the various gender issues in Afghanistan in my research, I aim to learn more in depth and also acquire a clearer understanding on the factual effects from the Taliban's ferocity upon Afghan women (Agustina, 2015a-c; 2016). This research is important to show the repression suffered by Afghan women as a result of the patriarchal culture constructed by the Taliban regime and the misinterpretation of the Hadith and Al-Qur'an. The foremost aspect of this study is to present all the determinant factors of the main problems related to the repression against Afghan women.

3 RESULTS AND ANALYSIS

3.1 *Overview of Afghanistan*

Afghanistan is a multi-ethnic country in the Middle East. People in Afghanistan are divided into four major ethnic groups: the Pashtuns (38%), Tajiks (25%), Hazara (19%) and Uzbeks (6%) (Riphenburg, 2005, p. 37). This demography is based only from estimated statistics because Afghanistan has never held a census; however, other data sources presented a ratio of the four ethnic groups of approximately 42%, 25%, 30%, and 10%, respectively (Saikal, 2012, p. 80). The Pashtuns dominate the political and military leadership chair, while the Tajiks mostly are the intellectuals and hold a clerical position. Meanwhile, the Hazara ethnic occupies a lower social class, such as slaves, servants or low-skilled labourers. The complex relationship between the upper- and lower-class ethnics also often triggers conflicts among them (Agustina, 2007, 2015a, 2015b).

Early attempts of reform by the Pashtun rulers of the nascent Afghan State had to face fierce resistance from the rural periphery. In the 1920s, King Amanullah (1919–1929) opened girls' schools in Kabul, advocated against the veil and gender segregation and ordered all Afghans to wear Western dresses and hats (Emadi, 2002, pp. 63–65). He also introduced the most progressive Muslim family legislation seen at that time, banning child marriage and forced marriage and restricting polygamy—which then upset the political economy of tribal kinship relations and men's patriarchal authority (Moghadam, 2003, pp. 218–219).

The government under Prime Minister Mohammed Daoud attempted modest reforms in the 1950s, advocating against compulsory veiling in 1959. Women also were given the right to vote in 1964. These reforms benefitted the Dari-speaking urban elites, but had little impact on women in rural areas. The Communist People's Democratic Party of Afghanistan also attempted wider national reforms when it took power in 1978. Family legislation was overhauled, and compulsory literacy programs were introduced for all Afghan men and women, including those in remote villages.

In the 1970s and the 1980s, a new kind of rebels emerged in Afghanistan. They called themselves the *Mujahideen*, a word originally used to represent Afghan fighters who opposed the British Raj's push into Afghanistan in the 19th century. In the context of Afghanistan during the late 20th century, the *Mujahideen* were Islamic warriors defending their country from the Soviet Union, which invaded the country in 1979 and fought a bloody and pointless war there for a decade. Afghanistan's *Mujahideen* were an exceptionally diverse lot that included the Pashtuns, Uzbeks and Tajiks ethnics and others. Some were from the Shi'a sect, sponsored by Iran, while most factions were made up of Sunni Muslims. In addition to the Afghan fighters, Muslims from other countries volunteered to join the *Mujahideen* ranks. Much fewer Arabs (like Osama bin Laden), fighters from Chechnya and others rushed to aid Afghanistan (Szcepansky, 2016).

The Soviet occupation (1979–1989) created opportunities for women in urban areas, while creating hardships for women elsewhere. Throughout this period, women in the capital city of Kabul were active in public sphere as doctors, journalists and even police officers and soldiers (2003, p. 233). The state's Democratic Organisation of Afghan Women worked to actively promote women's rights. However, the women's gains were tempered by the brutality of the state's secret service, KHAD, which silenced all political dissent (Rubin in Wall, 2012, p. 1).

The state's failure paved the way for the emergence of the fundamentalist Pashtun-based Taliban. Driven to cleanse the Afghan society from the ravages of the civil war, Taliban leaders came from the most rural and conservative Pashtun provinces in southern Afghanistan.

They shared a vision of creating an ideal Islamic society akin to that created by the Prophet Mohammed 1,400 years ago (Rashid, 2000, p. 23). With the support of Pakistan, the Taliban has succeeded in conquering most of the country by 1996. The Taliban's abuses have been well documented. Upon taking power, the Taliban enforced a strict interpretation of the Shari'a law and Pashtun customary law, or *Pashtunwali*, which was an anathema to many other ethnic groups. Men were forced to grow beards and women to wear the *burkha*. Girls' schools were closed and women were banned from working outside the home or leaving their homes without a *mahram*, a male family member (2000, p. 70).

Those descriptions explicate some factors which caused the destructive system of the government, the gender inequality and women's repression in Afghanistan. Wars, ethnic conflicts, the various sects of religion and the patriarchal culture gave rise to the huge impact on social disharmony and breakdown in Afghanistan.

3.2   *Afghan women's repression by patriarchy*

My Forbidden Face carries the voice of a young girl coming from the period of the terror of the Taliban and reveals the complicated touch points between our world and hers. Throughout the book, Latifa also reveals the complex American influence in Kabul, from her crush on Leonardo DiCaprio to her belief and possibility to go to the United States. Latifa's community had been quite exposed to the Western popular culture and political ideals to be left with a stinging feeling of hypocritical abandonment when the world turned a blind eye to the Taliban oppression (Schiff, 2001, pp. ix–xii).

Walby defined patriarchy as a system of social structures and practices where men dominate, oppress and exploit women. The patriarchal relations in the family are common when women are confined to domestic work, as restricted by their husbands or by the men who live with them. The second form of patriarchy exists in the economy sector, namely the relation of a patriarchal system in employment and payment of wages. The state also becomes a patriarch as a capitalist and racist. As the arena of struggle and not as a monolithic entity, the state shows a systematic bias towards the interests of patriarchy (Walby, 2014, pp. 28–30). This form of patriarchy is also apparent in Afghanistan, because the male-oriented policies and rules applied by the Taliban regime became the most powerful restrictions forcing everything under their control.

In Afghanistan, the government was present in the form of the Taliban regime holding power and authority to impose policies for all citizens in the era of 1996–2000. In practice, the regulations and policies adopted by the Taliban were in the name of Islam. In fact, the rules of Islam are very egalitarian and actually embrace the values of equality between men and women. However, the patriarchy system has made a tremendous impact on the lives of Afghan women due to the violence they had to endure. Various studies conducted by researchers show a very significant impact from the patriarchy construction in Afghanistan (Pourzand, 1999, and Povey, 2003).

Taliban's position on this issue seems to be a reflection of their overall ultra-conservative interpretation of Islam combined with a very traditional understanding of the Pashtun tribal code of conduct (*Pashtun Wali*). Added to this is their reactionary notion of what it means to be an Afghan man or woman and their power struggles with other groups in Afghanistan and regional or international stakeholders in the country. They want to demonstrate that they are the 'purest' and 'most authentic' Moslems and Afghans amongst all groups (Pourzand, 1999). As usual, women play an important symbolic role in these assertions. This has very tragic consequences, as now widely known and denounced internationally. It is, no doubt, also an issue of controlling men through controlling women, and controlling various ethnic and linguistic groups and other factors. The human rights of almost all Afghans have been seriously violated.

Most Americans would probably agree that the dominant visual image of the new millennium is the flaming collapse of the twin towers of the World Trade Center in New York City on 11 September 2001. Nevertheless, there is another powerful one: Muslim women wearing the *burkha*, covering their body so completely that we can barely even see their eyes. These women can hardly see where they are going, and it must certainly be difficult for them to

speak clearly. They appear to be trapped physically, economically, emotionally and spiritually (Spector, 2006). The donning of *burkha* by all Afghan women whenever they go out of their homes and accompanied by a *mahram* is a symbol of men's domination and the control over women's rights. These extraordinary changes come with substantial and harsh punishments if the women do not obey the rules, and this has uprooted Afghan women's lives.

Barlas (2002, pp. 2–3) revealed that the status and the role of women in a Muslim society that adopts a patriarchal and gender bias social structure have multiple-factored functions that are almost entirely not religion-oriented. Thus, when the patriarchal and gender bias constructions are openly declared in the name of religion, it is actually contradictory and very opposite of what is written in the Al-Qur'an. The methods and principles, which were used by Asma Barlas in re-reading the Al-Qur'an and its application in the verses on gender, are to establish the principles of egalitarianism and anti-patriarchal as interpreted from the Al-Qur'an closely related to women's liberation.

3.3   *Latifa and her struggle against the repression*

Gender inequality, patriarchal culture and religious values, which are interpreted differently, are some of the issues leading to the repression against Afghan women. The period of darkness in Afghanistan was marked by the arrival of the Taliban regime, which then controlled all forms of rules and policies for the entire population. Prohibition, intimidation and discrimination were often present as a symbol of the Taliban rule to regulate the rights of the civil society. The rights of women were revoked and removed on the pretext of Islamic rules to enforce the Shari'a principles.

The rules imposed by the Taliban have entrapped both men and women in toils and terror. Some of those tormenting rules are mentioned in the following text: 'From now on, the country is ruled by a completely Islamic system. All foreign ambassadors are relieved of their duties. The new decrees in accordance with the Shari'a principles are as follows: 1) Anyone in possession of a weapon must surrender it to the nearest mosque or military checkpoint; 2) Women and girls are not permitted to work outside of home; 3) All women who are obliged to leave their homes must be accompanied by a *mahram*: their father, brother or husband; 4) Public transportation department will provide buses reserved for men and buses reserved for women; 5) Men must let their beards grow and trim their moustaches according to the Shari'a principles; 6) Men must wear a white cap or turban on their heads; 7) The wearing of suit and tie is forbidden. The wearing of traditional Afghan clothing is compulsory; 8) Women and girls are required to wear the *chadri*; 9) Women and girls are forbidden to wear brightly coloured clothes beneath the *chadri*; 10) It is forbidden to wear nail polish or lipstick or make up; 11) All Muslims must offer ritual prayers at the appointed times wherever they may be; 12) It is forbidden to display photographs of animals and human beings; 13) A woman is not allowed to take a taxi unless accompanied by a *mahram*; 14) No male physician is allowed to touch the body of a woman under the pretext of a medical examination; 15) A woman is not allowed to go to a male tailor; 16) A young girl is not allowed to converse with a young man. Infraction of this law leads to the immediate marriage of the offenders; 17) Muslim families are not allowed to listen to music, even during a wedding; 18) Families are not allowed to photograph or videotape anything, even during a wedding; 19) Women engaged to be married may not go to beauty salons, even in preparation for their weddings; 20) Muslim families are not allowed to give non-Islamic names for their children; 21) All non-Muslims, Hindus and Jews must wear yellow clothing or a piece of yellow cloth. They must mark their homes with a yellow flag so that they are be recognisable; 22) All merchants are forbidden to sell alcoholic beverages; 23) Merchants are forbidden to sell female undergarments; 24) When the police punish an offender, no one is allowed to ask a question or complain; 25) All those who break the laws of Shari'a are to be punished in the public square' (Latifa, 2001, pp. 47–49). By the time the bans were announced by the Taliban, they have killed a lot of girls and women stealthily, in silence.

The worst restriction imposed throughout the great majority of the country is the annihilation of the women's rights by locking them out of the society. All women, young and old, are affected. Women may no longer work, which means a collapse of medical services and government administration. There is no more school for girls, no more health care for women and no more fresh air for women. The women went home or disappeared under the *chadri*, out of the sight of men. It is an absolute denial of individual liberty, a real sexual racism (2001, p. 50). Another rule from the Taliban that obliges women to be accompanied by a *mahram* when going out of their homes is an attempt to eliminate the rights of women to participate in the public sphere. They lost their rights to education, to socialise with others in the workplace environment and also to gain access to information. All those rules and strict bans have really 'killed' Afghan women's freedom to participate in the public sphere. They have no more rights to be what they want to be, and they are imprisoned in their own homeland.

The diverse linguistic, cultural and ethnic identities in Afghanistan have been formed and reconstructed as a consequence of a broader historical process involving local and regional wars and colonial and imperialist intervention (Povey in Afshar, 2012, p. 149). In Afghanistan, as in all Muslim majority societies, the interaction of Islamic culture and religion with secularism, nationalism, ethnicity and other important historical, social and economic mechanisms structures the lives of the women and men of the country. Therefore, too often, Islamic culture and religion are considered to be the primary agent in determining the identities of women in Muslim majority societies, and this perception is used to justify war, occupation and invasion of a country. Of course, patriarchal attitudes and structures remain extremely strong in Afghanistan. The public domains for women are strictly forbidden, such as schools, markets and offices.

As written by Latifa in her novel, 'School, college, Sundays at the swimming pool, expeditions with my girlfriends in search of music tapes, film videos, novels to read avidly in bed in the evening ... How I hope the resistance forces haven't abandoned us to our fate' (2001, p. 14). *My Forbidden Face* is an expression of testimony on a variety of issues of violence experienced by the women in Afghanistan. It is a story that shows the struggle of a teenager from Kabul who was educated under the Soviet power and living under the Communist rule, as well as factional disputes over the 4-year civil war and the arrival of the Taliban regime in Afghanistan. Latifa finds herself imprisoned by a frightening power and sees her life taken away at the age of 16 years. The issues of repression, bans and verbal and non-verbal violence that are often experienced by Afghan women lead to the uprooting and loss of the women's rights. The Taliban is the embodiment of the present patriarchal structure in Afghanistan.

As a witness of repression and sexual abuse, Latifa spoke out in front of the international public about the great suffering experienced by many Afghan women. *My Forbidden Face*, a memoir of a young Afghan woman named Latifa, provides an illustration about the many Afghan women whose lives were controlled by a patriarchal structure. This is a common situation in Middle Eastern countries that embrace Islam. Latifa, the writer of *My Forbidden Face*, has spoken for the nation. Coming from a progressive and relatively well-to-do family of seven, she was a 16-year-old student preparing for university life, majoring in journalism, when the Taliban invaded Kabul in September 1996. In May 2001, she and her mother were given the opportunity to travel to Paris to initiate a campaign on women's liberation sponsored by an advocacy organisation, Afghanistan Libre, and *Elle* magazine. While they were in Paris, the Taliban issued a *fatwa* against them for denouncing the regime and stripped their Kabul apartment. Latifa and her parents remained in exile in Paris, where Latifa was given the opportunity to write her book, as an attempt to 'explain how a girl from Kabul, educated first during the Soviet occupation, then under the Communist regimes throughout four years of civil war, was finally locked away by a monstrous power, confiscating her life when she was only sixteen'.

4 CONCLUSION

A noteworthy feature is that the Taliban enforced a strong patriarchal structure that limited the role of women in the public sphere. As a result, women's voices have been silenced. They have no role to play in any decision-making process. Latifa's endeavour to be the voice of

Afghan women through her writing is part of her great struggle to make the people around the world aware of and understand the complex problems prevailing in Afghanistan. The invasion of the Soviet Union, the patriarchal culture and the conflicts among the ethnics, which contributed to the harsh attitude towards Afghan women, worsened the situation. In some cases, the problems arose from the misconception about interpreting the Hadith and Al-Qur'an. Gender inequality between women and men due to the patriarchal structure has inflicted even worse conflicts. The voice of Latifa could promote her nation, Afghanistan, to win world sympathy through her writing to help Afghan women to have faith and hope for the future.

Afghan women should have equal rights to men based on the concept of Al-Qur'an and Hadith. Finally, the book *My Forbidden Face* is a concise memoir that is riveting, insightful and fascinating: a testament to the human capacity to grow beauty on the rockiest soil.

REFERENCES

Afshar, H. (2012) *Women and fluid identities. Strategic and practical pathways selected by women*. United Kingdom, Palgrave MacMillan.

Agustina, H.N. (2007) *Kompleksitas penyajian cerita dan kompleksitas konflik in The Kite Runner*. (The complexity of narration and conflict in The Kite Runner) [Thesis] Universitas Indonesia, Indonesia.

Agustina, H.N. (2015) The kite runner: My passion of literature. *International conference on Social Sciences and Humanities (ICSSH'15), 5–6 May 2015, Bali, Indonesia*. pp. 43–46.

Agustina, H.N. (2015) Dominasi konflik antar-etnis in The Kite Runner karya Khaled Hosseini. (Inter-ethnic conflicts in The Kite Runner by Khaled Hosseini) *Seminar Nasional Bahasa dan Sastra (Senabastra) VII, 10th June 2015, Universitas Trunojoyo Madura, Indonesia*.

Agustina, H.N. (2015) Inspiring woman against the terror in A Dressmaker from Khair Khana. *4th international Multi-Conference on Humanities, Law, Literature and Management Sciences (IJHMS) Pattaya, Thailand* 3(4), pp. 182–187.

Agustina, H.N. (2016) When desperate and guilty feeling destroy humans' life in *And the Mountains Echoed* by Khaled Hosseini. *2nd International Conference on Culture, Languages and Literature (ICCLL), 22–24th June 2016, Singapore*.

Barlas, A. (2002) *Believing women in Islam: Unreading patriarchal interpretations of the Qur'an*. USA, University of Texas Press.

Emadi, H. (2002) *Repression, resistance, and women in Afghanistan*. USA Praeger.

Latifa. (2001) *My Forbidden Face. Growing up under the Taliban: A young woman's story*. USA, Hyperion.

Moghadam, V.M. (2003) *Modernizing women: Gender and social change in the Middle East*. Boulder, CO, Lynne Rienner Publishers.

Pourzand, N. (1999) The Problematic of Female Education, Ethnicity and National Identity in Afghanistan (1920–1999), *Social Analysis: The International Journal of Social and Cultural Practice*, 43 (1), 73–82.

Povey, E.R. (2003) Women in Afghanistan: Passive victims of the Borga or Active Social Participants? *Development in Practice*. [Online] 13 (2/3), pp. 266–277. Taylor & Francis, Ltd. Available from: http://www.jstor.org/stable/4029597. [Accessed 14th March 2015].

Rashid, A. (2000) *Taliban: Militant Islam, oil, and fundamentalism*. New Haven, CT, Yale University Press, 23.

Riphenburg, C.J. (2005) Ethnicity and civil society in contemporary Afghanistan. *Middle East Journal*, 59 (1), 31–51.

Rubin, B. (2002) *The Fragmentation of Afghanistan: State formation and collapse in the International System*. New Haven, CT, Yale University Press.

Saikal, A. (2012) Afghanistan: The status of the Shi'ite Hazara minority. *Journal of Muslim Minority Affairs*, 32: 80, No. 1.

Spector, B. (2006) Dionysus, revenge, and the woman in the Burkha. *San Francisco Jung Institute Library Journal*, 25 (1), 33–41.

Szczepanski, K. (2016) *Asian History Expert: The Mujahideen of Afghanistan*. [Online] Available from: http://asianhistory.about.com/od/glossaryko/g/Who-Were-the-Mujahideen-of-Afghanistan.htm [Accessed on 25th June 2016].

Walby, S. (1990) *Theorizing Patriarchy*. Oxford UK, Blackwell. pp. 28–30.

Wall, K. & David C. (2012) *Afghan women speak. Enhancing security and human rights in Afghanistan*. USA, University of Notre Dame.

Cultural Dynamics in a Globalized World – Budianta et al. (Eds)
© *2018 Taylor & Francis Group, London, ISBN 978-1-138-62664-5*

Memory of the female protagonist in relation to gender oppression in Catherine Lim's *The Teardrop Story Woman*

J.D. Sukham & D. Hapsarani
Department of Literature, Faculty of Humanities, Universitas Indonesia, Depok, Indonesia

ABSTRACT: The aim of this study is to find out the connection between memory and gender oppression as experienced by the female protagonist in *The Teardrop Story Woman* (1998), a novel written by a Singaporean author, Catherine Lim. Using Kate Millett's concept of sex/gender and Daniel Schacter's memory theory, the analysis shows that memory serves as a means of internalising gender-oppressive social constructions that eventually prevent the female protagonist from rebelling against gender oppression. However, at the same time, memory—through a selection process—also serves as a defence strategy that helps the female protagonist to cope with gender oppression in a situation where real actions of resistance are not possible. In relation to this, memory can also be seen as a form of female subjectivity. By consciously using her memory as a defence strategy, the female protagonist positions herself as a subject that has the right and ability to choose—thus performing her subjectivity as a woman despite living under the constant pressure from gender oppression.

1 INTRODUCTION

Singaporean literature basically refers to a group of literary works written by Singaporean authors in any of Singapore's four official languages: English, Malay, Mandarin, and Tamil ("Literature of Singapore", 2016). However, a long period of British colonialism causes Singaporean literature to be dominated by literary works written in English. Singaporean literature in English itself can further be divided into two periods—colonial and postcolonial—each with its own characteristics. In the colonial period, Singaporean literature in English was produced in an academic environment, written by a few people who had access to the Western education. The literature was dominated by poetry and shorter prose (short stories), and was imitative of the Western writing styles and themes (Patke & Holden, 2010). Meanwhile, in the postcolonial period, Singaporean literature in English was produced in a broader environment, written by people who use English as their *lingua franca*. It was dominated by the longer type of prose (novels) and adjusted to the writing styles and themes relevant to the social issues the authors wanted to discuss (Patke & Holden, 2010).

Catherine Lim is one of the most prolific and significant Singaporean authors of Chinese descent whose works show the characteristics of Singaporean literature in English produced in the postcolonial period. Her works often depict the social issues that happen in everyday Singapore and offer new points of view in response to the social phenomenon. This "socially-engaged" characteristic is then reflected as well in *The Teardrop Story Woman* (1998), one of Lim's most popular novels. First published by Orion in London, the novel is set in Luping—a small town in Malaya—circa 1940. The novel tells a story of a young Chinese woman named Mei Kwei, who experiences gender oppression all her life. Unable to rebel against the oppression (as she lives in a Chinese community well-known for its patriarchal system), Mei Kwei selects and relives certain memories that can help her cope with the gender oppression.

The above explanation shows the possibility of a connection between memory and gender oppression—the two main elements in the novel—and this study aims to analyse that particular connection. The study itself is conducted using a textual analysis and supported primarily by Kate Millett's concept of sex and gender and Daniel Schacter's memory theory.

In *Sexual Politics* (1970), Millett, using Robert Stoller's statement, differentiates the term "sex" from "gender". She explains that sex is biologically determined, while gender is socially constructed. However, a patriarchal society uses the difference between sex and gender to create an environment where it is possible for men to dominate women based on sex. This leads to the thought that sex is actually political—in terms of a power relation where one group (male) subjugates the other (female).

Meanwhile, when discussing memory, Schacter (1996), in his book *Searching for Memory: The Brain, the Mind, and the Past*, states that memory is a "fragile power". Memory is considered to be powerful because it is able to influence people greatly. He explains that "the past shapes the present" (p. 7) in terms that our memory—our accumulated experiences in the past—is able to influence our present and future. Schacter then states that the elderly members of the society use this influential quality of memory to internalise certain knowledge—including the gender-oppressive social construction—into the minds of its younger members through storytelling: "In many societies, the primary function of elderly adults is to pass on significant personal and cultural lore to younger members of the group—to tell stories about their own experiences and about the traditions and momentous events of the society" (1996: 300). This, consequently, leads to the regeneration and reinforcement of gender-oppressive social constructions. Memory facilitates the gender-oppressive social constructions to operate not only in the public sphere (in society) but also in the private domain—in the mind of each member of the society.

However, Schacter explains that memory is also "fragile" (1996: 7)—in terms that it can be modified easily. Memory is built through a construction process, and any deviations in the construction may result in the change of memory. Anderson and Levy (2009) explain that people usually see this vulnerability of memory as "a human frailty to be overcome" (p. 189). Nevertheless, they also state that this vulnerability may instead be useful to us. They explain that forgetting, as an example of memory's vulnerability, may actually help us to cope with our negative experiences. By forgetting something, we sometimes prevent ourselves from being exposed to the disadvantageous side effects of a certain negative experience. Schudson (1995) states that this is possible to happen because our memory as a human is "selective". Therefore, we naturally select memories that are needed for our survival and repress—or even omit—the ones that are dangerous for our mental health. Schudson (1995) also describes that these two processes happen at the same time: "Whenever we selectively retrieve some memories in response to a particular cue, but not others, inhibition of the non-retrieved information occurs" (p. 360). The increased remembrance of certain memories will reduce the appearance of other unwanted memories. Accordingly, this process may help us balance and protect our mental condition. As a whole, what we previously think as the vulnerability of memory can thus be seen as an advantageous flexibility. This study will then make use of the dualism of memory as a starting point to analyse the memory of the female protagonist.

1.1 *Memory as a means of internalising gender-oppressive social constructions*

It has been mentioned in the previous part of this study that Mei Kwei, the female protagonist, falls victim to the oppressive male characters in the novel. Nevertheless, Mei Kwei is unable to rebel against the oppression as she is confined by her personal memories, which are influenced by the concept of male domination. These memories work implicitly and cause her to submit to the male characters' oppression. These memories can be divided into two types: the stories told by the other female characters (Ah Oon Soh/Mei Kwei's mother, Second Grandmother/Mei Kwei's grandmother, Sister St. Elizabeth/Mei Kwei's teacher at school) and her personal life experiences. The concept of male domination itself can be seen more clearly when it is manifested in the rules correlated with women's status and body. The following table lists the memories that internalise the gender-oppressive social constructions.

Mei Kwei's memories prevent her from rebelling against the gender oppression. The role of these memories can be seen through Mei Kwei's interactions with the male characters when

Table 1. Memories related to women's status and body.

Memories related to women's status	Memories related to women's body
The story of Mei Kwei's older sister.	The story of the Second Grandmother's lotus feet.
The story of Mei Kwei's adoption.	The story of an impure bride caged in a pigsty.
Mei Kwei's experience of being discriminated by her father.	The story of St. Agatha.
Mei Kwei's experience of being expelled forcefully from school by her father.	Mei Kwei's interaction with Uncle Big Gun.
Mei Kwei's experience of being slapped by her father for opposing Big Older Brother.	Mei Kwei's interaction with Big Older Brother.

Table 2. The male character's gender oppression.

Male characters	Manifestations of the gender oppression
Big Older Brother	Stalking Mei Kwei outside the bathroom.
	Sexually abusing Mei Kwei when they were sleeping together with their mother on the same bed.
Old Yoong	Treating Mei Kwei as a sexual object.
	Using Mei Kwei as a property to be flaunted to his rival Old Tang.
	Sexually abusing Mei Kwei.
	Trying to mark Mei Kwei symbolically as his property by giving her a jade bracelet.
Austin Tong	Treating Mei Kwei as a sexual object.

she has grown up: Big Older Brother (Mei Kwei's older brother), Old Yoong (Mei Kwei's ex-fiancé to whom Mei Kwei becomes a mistress at the end of the story), and Austin Tong (Mei Kwei's husband). The male characters put Mei Kwei under their dominance, but Mei Kwei is unable to oppose their subjugation. The various manifestations of the gender oppression are listed in the following table.

The story of Second Grandmother's lotus feet and its correlation with Old Yoong's acquisition is chosen to show how one's personal memory can serve as a means of internalising gender-oppressive social constructions.

In the novel, the Second Grandmother is portrayed as having tiny lotus feet that are claimed to be able to make her husband happy and satisfy him sexually. The lotus feet also enable the Second Grandmother to become the favourite among her husband's four wives. In this context, the lotus feet can be seen as symbolising women's position in the society. First, just like tiny lotus feet, women are seen as small, insignificant, and inferior. Second, the lotus flower symbolises purity in the Chinese culture (Qu, 2010), as women are expected to be virgin even if they are considered as men's sexual objects. Women who possess lotus feet are thus seen as having the necessary qualities to please men. This leads to the thought that women are positioned as inferior beings whose only function is to serve men.

The novel then indicates that Mei Kwei knows about the Second Grandmother's story, even though it does not identify the person who tells this story to Mei Kwei. This story later influences Mei Kwei when she must face the gender oppression from Old Yoong—an old, rich man from Penang who becomes Mei Kwei's fiancé. In one part of the novel, Old Yoong gave Mei Kwei a very expensive jade bangle in an attempt to mark her symbolically as his property. Even though the jade bangle is too small for Mei Kwei's wrist, Old Yoong refuses to change it with a larger but cheaper one. Being forced to wear small jade bangle, Mei Kwei at first sees it as an "invading", "hurting", and "hard gift" (Lim, 1998: 133)—showing that Mei Kwei sees it as a form of oppression. However, after she remembers the story of the Second Grandmother's lotus feet, she unconsciously submits to Old Yoong's oppression—which can be seen in the novel when her wrist finally gives up to the pressure coming from the jade bangle: "...the bones [in Mei Kwei's hand], finally reduced to a shivering helplessness, gave in at last" (Lim, 1998: 133). Mei Kwei still considers Old Yoong's action as an "assault" or an "onslaught" (Lim, 1998, p. 133), but she no longer struggles against it as she previously does. This indicates that the memory of the Second Grandmother's lotus feet has influenced Mei Kwei and prevented her from rebelling against Old Yoong's oppression. The memory about the Second Grandmother's lotus feet taught Mei Kwei that women should be inferior and submissive to men, and this knowledge is what forces Mei Kwei to submit unconsciously to Old Yoong's subjugation. This pattern can be found repeatedly in the text and it indicates that Mei Kwei's personal memory has become a means of internalising gender-oppressive social constructions into her mind.

1.2 *Memory as a defence strategy against gender oppression*

Living in a patriarchal community, Mei Kwei has been described as unable to rebel against gender oppression. As a result, Mei Kwei does not have any choice but to make an effort to cope with it by selecting certain positive memories that enable her to survive under the pressure. These memories are listed in the following table.

Mei Kwei's memory of her sexual intercourse with Father Martin is an example of how memory can be used as a defence strategy against gender oppression.

This sexual intercourse happens near the end of the novel when Mei Kwei has resided in Singapore. After being accused of cheating with Father Martin by her husband Austin Tong, Mei Kwei leaves Luping and works as a cabaret girl in Singapore. Old Yoong, hearing the news about Mei Kwei's staying in Singapore, offers her to be his mistress. In need of financial support to bring up her children, Mei Kwei finally agrees to Old Yoong's offer. A few years later, Father Martin—who was previously forced to go back to his homeland because of the scandal—goes to Singapore on his own initiative to look for Mei Kwei. They end up having a sexual intercourse as an expression of love to each other, even though they know they still cannot be together. Mei Kwei has to go back to Old Yoong as he has already promised her the financial support and protection for her children in exchange of her being his mistress, and Father Martin has to continue his service in Vietnam.

The previous discussion has shown that Mei Kwei lives in a Chinese community where the concept of male domination prevails. This concept results in the emergence of rules that tend to discriminate women, one of which is related to virginity. In the patriarchal Chinese community, women are expected to give their virginity only to their husbands, and violation to this rule will result in their being positioned as worthless animals. However, this rule does not apply to men. This means that the society views virginity as women's gift to their husbands.

Mei Kwei contradicts this rule by having a sexual intercourse with Father Martin. When the event takes place, Father Martin is still a virgin (considering his position as a pastor that has to lead a celibate life), while Mei Kwei is not. Mei Kwei's virginity is said to be "all gone and squandered" (Lim, 1998: 327); she has been married and has other sexual relations with her ex-fiancé Old Yoong and some other men when she works in a cabaret. In this context, virginity is Father Martin's gift for Mei Kwei—a man's gift for a woman. Mei Kwei's sexual intercourse with Father Martin can therefore be considered as an event that breaks the concept of male domination—a memory worthy enough to be selected.

The memory itself is selected as a defence strategy against Old Yoong's oppression. When the sexual intercourse happens, Mei Kwei is practically under Old Yoong's dominance as she is still positioned as his mistress. Living under Old Yoong's dominance means living under his gender oppression throughout her life. This realisation causes Mei Kwei to consciously create and select a valuable memory as a defence strategy. The novel explains that the memory of her sexual intercourse with Father Martin "would give a sublime rhythm to her life" (Lim, 1998: 327), which means that remembering the memory will strengthen and enable Mei Kwei to survive under Old Yoong's oppression. This indicates that memories can function as a defence strategy against gender oppression.

Table 3.   Memory used as a defence strategy against gender oppression.

Characters	Memory used as a defence strategy
Mei Kwei's father	Memory of a coin from her father. Memory of two kind foreigners (a Japanese soldier named Sato-san and an Indian worker).
Big Older Brother	Memory of Big Older Brother's kindness in their childhood. Memory of Father Martin's white handkerchief. Memory of a photo of Big Older Brother and Mei Kwei in their childhood.
Old Yoong	Memory of God King and Moon Maiden (the male and female protagonist in a Chinese opera Mei Kwei watches). Memory of Mei Kwei's sexual intercourse with Father Martin.
Austin Tong	Memory of the First Month Celebration clothes (representing two events: First Month Celebration and Mei Kwei's adoption).

1.3   *Memory as a form of female subjectivity*

The analysis in the previous parts of this study has shown that Mei Kwei's memories always consist of two parts: the negative and positive memories. The negative memories internalise gender-oppressive social constructions into her mind and thus prevent her from rebelling against the male characters' gender oppression, while the positive memories help her survive under the oppression. Like the two sides of a coin, these memories continuously take turns in influencing and shaping Mei Kwei's life. Nevertheless, as an individual, Mei Kwei always chooses to remember the positive memories and repress the negative ones. This indicates that Mei Kwei positions herself as a subject who has the right and power to choose what is good for herself in spite of everything that is thrown her way—thus performing her subjectivity as a woman. In addition, Mei Kwei's conscious choice of using personal memories as a defence strategy against gender oppression shows that memory can be seen as a form of female subjectivity.

2   CONCLUSION

The study shows that the female protagonist's memories have two roles in relation to the gender oppression she experiences. First, Mei Kwei's memories function as a means of internalising gender-oppressive social constructions into her mind. Through stories told by the other female characters and Mei Kwei's personal experiences, the concept of male domination is internalised and reinforced into her mind. Accordingly, these memories prevent her from getting out of the oppressive treatment of the male characters when she has grown up. Second, Mei Kwei's memories function as a defence strategy against gender oppression. By selecting the positive memories and at the same time repressing the negative ones, Mei Kwei is able to adapt and survive under the male characters' dominance.

Through these contradictory roles of memories, Lim as the author shows that memory is a sort of battlefield where two opposing forces meet: one is the gender-oppressive social constructions, and the other is the women's resistance against gender oppression, which can actually be seen as the manifestations of the aforementioned concept of male domination. On one hand, Lim shows that the now-established patriarchal system develops from something as simple and natural as memory. On the other hand, Lim also tries to raise the readers' awareness that women's resistance might start from the same thing: memory. Meanwhile, Mei

Kwei's decision to select positive memories and repress negative ones reflects how a woman might position herself as a subject that has the power and the right to choose for herself. This way, memory can also be seen as a form of female subjectivity.

REFERENCES

Anderson, M. C. & Levy, B. J. (2009) *Suppressing unwanted memories.* Association for Psychological Science. [Online] *18*(4), 189–194. Available from: http://memorycontrol.net/AndersonLevy09.pdf.

Lim, C. (1998). *The Teardrop Story Woman.* London: Orion.

Literature of Singapore (2016) Focus Singapore. [Online] Available from: http://www.focussingapore.com/information-singapore/literature/.

Millett, K. (1970) *Sexual Politics.* Chicago, IL: University of Illinois Press.

Patke, R. S. & Holden, P. (2010) *The Routledge Concise History of Southeast Asian Writing in English.* New York, NY: Routledge.

Qu, R. (2010) *The symbolic meaning of lotus flowers in Chinese culture.* China Culture. [Online] Available from: http://www.chinaculture.org/chineseway/2010-07/23/content_386723_4.htm.

Schacter, D. L. (1996) *Searching for memory: The brain, the mind, and the past.* New York, NY: Basic Books.

Schudson, M. (1995) Dynamics of distortion in collective memory. In: D.L. Schacter (Ed.), *Memory distortion: How minds, brains, and societies reconstruct the past.* Cambridge, MA: Harvard University Press. pp. 346–364.

Cultural Dynamics in a Globalized World – Budianta et al. (Eds)
© *2018 Taylor & Francis Group, London, ISBN 978-1-138-62664-5*

Women and corruption in Okky Madasari's *86* and Anggie D. Widowati's *Laras*: A feminist study

A. Akun
Department of English Studies, Faculty of Humanities, Bina Nusantara University, Jakarta, Indonesia

M. Budiman
Department of Literature, Faculty of Humanities, Universitas Indonesia, Depok, Indonesia

ABSTRACT: The aim of this study was to criticise the representation of women in two Indonesian women's novels about corruption by comparing the corruption behavioural construct of female characters during the Indonesian reformation era. In particular, it reveals how the relationship between female characters and corruption habitus is built through the narration, irony, metaphor, character and characterisation in both stories. Furthermore, the study aims at interrogating the text to reveal the extent to which the female authors are trapped in strengthening the biased male-dominating ideology by exposing the texts' absences, omissions and silences, as well as the story parts and social practices unspoken or thought of as being natural. The research method used was Pierre Macherey's text interrogation to expose the unspoken truth of the texts. Furthermore, Susan Faludi's backlash theory and Pierre Bourdieu's habitus and social arena were used. The result of the study shows that although the two Indonesian female authors while writing about corruption appear to represent women in seemingly dominating roles, they still cannot escape the patriarchal ideological trap as revealed by their ironic marginalisation of women as done by many writers.

1 INTRODUCTION

'There is a common sense in our society to properly assume that behind the corrupt men (husbands) there are greedy and demanding wives...just inversely true as behind the husbands' career success, there are great wives' (www.jurnalperempuan.org/bias-gender-dan-perilaku-korup.html). The seemingly influential position of women actually shows that they have socially been constructed as never being equal to men, either in good deeds or in bad ones. More specifically, women have disadvantageously been justified as men's reason or barrier for corruption as elaborated in Indonesian novels from the 1940s to the 1980s. Mariam, the first wife of the protagonist Bakir who fights corruption to the bitter end, and Sutijah, the kept woman who causes Bakir's corruption in Pramoedya Ananta Toer's novel *Korupsi* (Corruption, 1954) are examples of female characters in this category, just as Fatma, Dahlia and Hasnah in Mochtar Lubis' *Senja di Jakarta* (Twilight in Jakarta, 1970). Throughout the development of the Indonesian literature, there have been very few works discussing women corruption despite the massive corruption practices. The research on women and corruption is also very rare (Sarjono, 2012, p. 2). Thus, this topic is still challenging to be elaborated to see how women are represented and positioned in the authors' attempt to resent women's involvement in corruption cases.

Furthermore, men's involvement in corruption is well tolerated and broadly unquestioned as if it is men's habitus and arena. However, when women commit corruption, they are exposed not only in terms of their crime, but also in terms of their womanhood, female and feminine identity. This trend can be traced in the recent corruption cases committed by women as critically discussed by Mariana Amiruddin (2012) in her article "Dari Payudara

Melinda Dee, Rambut Ungu Miranda Goeltom, Hingga Tas 'Hermes' dan Kerudung 'Louis Vuitton' Nunun" (From Melinda Dee's Breasts, Miranda Gultom's purple hair to Nunun's Hermes Bag and Louis Vuitton Veil). She has concluded that there is an indication of gender bias when the media exposes female corruptors because the coverage of female corruptors has become a separate profile in news about their private life and lifestyle, which is not the case for male corruptors. For example, Sjoberg and Gentry, quoting Wight and Myers, have emphasised this point, 'when a woman commits an act of criminal violence, her sex is the lens through which all of her actions are seen and understood' (2007, p. 29). This thought has reduced and degraded women not only for their crime but also in the way they are discussed, '…women who are violent are highlighted, exploited, and fetishized… discussions of women's violence debase women and reduce them to their sexuality' (p. 46). Another research on corruption and women is Mashuri's *Goda dan Lupa: Wajah Korupsi dalam Novel Indonesia* (Temptation and Forgetting: Corruption's Face in Indonesian Novels), which elaborates the triadic relation of wealth–power–women as the root of corruption (Sarjono, 2012, p. 25).

This study aims to criticise the representation of women in two Indonesian women's novels about corruption by comparing the corruption behavioural construct of female characters during the Indonesian reformation era. *86* is a novel by Okky Madasari, a former journalist. It is about a young woman named Arimbi, a suburban girl and a university graduate from Central Java, who is directly employed as a civil servant typist in the corrupt district court in Jakarta under Bu Danti, a senior and corrupt court registrar. After four years of naivety, she finally joins the others (judges, attorneys, lawyers) in the banal corruption practice in the court. Meanwhile, *Laras* is a novel by Anggie D. Widowati, also formerly a journalist. The story centres on the female protagonist Laras who has an ambition to take revenge on her irresponsible biological father because he left Laras' mother during her pregnancy after a love affair and refused to accept Laras due to her being an unwanted child. Laras consciously takes revenge by becoming a mistress of Handoko, the finance minister, to gain material wealth, pleasure and status.

This study aims to reveal how the relationship between female characters and corruption habitus is built through the narration, irony, metaphor, character and characterisation in both stories. Furthermore, the study interrogates the text to reveal the extent to which the female authors are ironically trapped in strengthening the biased male-dominated ideology by disclosing the texts' absences, omissions and silences, as well as the story parts and social practices unspoken or thought of as being natural.

2 METHOD

This is a qualitative study that used Macherey's text interrogation to expose the promised truth (ideological project), the revealed truth (realisation) and the unconscious (unspoken truth) of the texts.

3 RESULTS AND ANALYSIS

3.1 *The unspoken and feminist backlash*

Macherey believed that the silence in the text speaks more about something that the author 'cannot' speak of, '…to know what the writer is saying, it is not enough to let him speak, for his speech is hollow and can never be completed at its own level…in its every particle, the work manifests, uncovers, what it cannot say. This silence gives it life…for in order to say anything, there are other things which must not be said…to reach utterance, all speech envelops itself in the unspoken…Speech eventually has nothing more to tell us: we investigate the silence, for it is the silence that is doing the speaking' (2006, pp. 93–96). The power of the silence and unspoken has given the possibility to a more productive feminist reading of a text because the ideological determinations can be revealed more through 'the silences, gaps and contradictions of the text' (Moi, 1995, p. 94) compared to its explicit statements. Eagleton clarified further the role of the silence and unspoken in relation to the disguise of ideology of the text and how an author is unconsciously forced to reveal a certain ideology:

'It is in the significant *silences* of a text, in its gaps and absences that the presence of ideology can be most positively felt. It is these silences which the critic must "speak". The text is, as it were, ideologically forbidden to say certain things; in trying to tell the truth in his own way, for example, the author finds himself forced to reveal the limits of the ideology within which he writes. He is forced to tell its gaps and silences, what it is unable to articulate. Because a text contains these gaps and silences, it is always *incomplete*. Far from constituting a rounded, coherent whole, it displays a conflict and contradiction of meaning; and the significance of the work lies in the difference rather than unity between these meanings...' (Moi, 1995, p. 94).

There is no complete text, and ideology can be exposed through the study of silences and contradictions in the text.

Feminists find this way of uncovering ideology fruitful because most ideologies work beyond the silences and unconsciousness, and this gives the critics space to study the (contradictory) textual construct more comprehensively and therefore 'enable the critic to link to a specific historical context in which a whole set of different structures (ideological, economic, social, political) intersect to produce precisely those textual structures. Thus, the author's personal situation and intentions can become no more than one of the many conflicting strands that make up the contradictory construct we call the text' (p. 94).

Overall, Macherey's method of analysis takes three instances: the ideological project (the 'truth' promised), the realisation (the 'truth' revealed) and the unconscious of the text (the return of the repressed historical 'truth') as the significant unspoken ideology (Storey, 2009, p. 76).

Faludi's backlash concept tries to critically make all aware of the fact that women are silently and ideologically blamed for their own unhappiness due to their own feminist struggles for equality. Women are made worried or scared that the freedom they have achieved enslaves them so that they seem to be convinced and constructed through discourses that their worst enemy is women themselves (not men) (Faludi, 2006, p. 2). Moreover, backlash works uncontrollably and unrealised among everyday dynamics, where the enemy can ironically be somebody calling herself feminist, as Faludi (2006, p. 13) states, 'The backlash is not a conspiracy, with a council dispatching agents from some central control room, nor are the people who serve its ends often aware of their role; some even consider themselves feminists'. The beauty myth as proposed by Naomi Wolf is also part of the backlash (beauty backlash) where women are psychologically terrorised through the play of a woman's image as naturally imperfect—causing fear, lack of confidence and embarrassment and labelled as fat or ugly *feminazis*—without having the idealised beauty promoted by the media and beauty industry (Wolf, 2012, pp. 2–3). To reach an awareness of this disadvantaged position, the study of the silences or unspoken is very crucial as it provides insight into the infiltrating ideology of the backlash.

Furthermore, to understand the banal practice of corruption and its connection to women, Bourdieu's habitus theory is needed. Bourdieu formally defined habitus as:

"the system of 'durable transposable dispositions, structured structures predisposed to function as structuring structures, that is, as principles which generate and organize practices and representations that can be objectively adapted to their outcomes without presupposing a conscious aiming at ends or an express mastery of the operations necessary in order to attain them. Objectively 'regulated' and 'regular' without being in any way the product of obedience to rules, they can be collectively orchestrated without being the product of the organizing action of a conductor... The habitus is sometimes described as a 'feel for the game'...The habitus is the result of a long process of inculcation, beginning in early childhood, which becomes a 'second sense' or a second nature" (Bourdieu, 1993, p. 5).

The social arena is the site where this habitus works well just as a field where any game is played. The habitus thus functions in two ways: as a structured structure and a structuring structure. The habitus requires no conscious obedience to rules and is practised naturally by relying on improvisation in the social arena, where the game is collectively played or orchestrated even without a leader or conductor.

3.2   *Social arena and corruption habitus*

The two novels represent women's involvement in corruption habitus in different social arenas. *Laras* pictures the protagonist Laras in the kept woman social arena, while *86* deals with the protagonist Arimbi in the district court arena. *Laras* depicts the main character as knowing her potentials and position as well as being smart, active and independent from the first-person singular point of view. Meanwhile, *86* pictures a woman as more independent and dominating than men from the third-person point of view. Both female protagonists enter the social arena where corruption has become a common practice. Laras consciously plunges herself into the corrupt arena to achieve her personal ambitions, while Arimbi needs four years of naive work before drifting into and finally accepting the corruption practices.

3.3   *The banality of corruption habitus*

The banality of corruption habitus is clearly seen through the presentation of no characters refusing the practice in social arenas. The metaphoric jargon *86* (meaning 'we all know') itself shows this intensity of corruption. Expressions of calling corruption results as 'blessing', 'prosperity', 'bonus' and so on indicate that the crime has been accepted as non-criminal income, even seemingly blessed by God. The use of these metaphors has even ironically veiled common place corruption practices from outsiders instead of uncovering everything clearly as assumed to be the real function of metaphor. Other metaphors such as the 'sand house' and 'instant noodles' all refer to an instant way of achieving materialistic dreams through corruption.

3.4   *The unspoken irony of women's representation*

Despite the fact that both stories are written by women—to some extent criticising women's marginalisation as well—it is revealed through the elaboration of the unspoken that the texts constitute backlash in their attempt to put women in a dominating position, but unconsciously represent women the other way around. The following discussion of the irony in these novels will clarify this standing.

There are five situations where women are represented as actually occupying marginalised position although presenting them as dominating on the surface.

First, women are represented as 'fighting' against other women, blaming or downgrading one another but never blaming or even forgetting that men are actually the real actors of the crime. In *Laras*, Bu Ayu (Rahayu, the legal wife of the minister Handoko) is made collide with Laras as well as the beautiful singer in terms of faithfulness, beauty and motivation, even though Handoko is the one who is unfaithful and commits bigger corruption. Laras keeps mocking Bu Ayu as lacking beauty or physical attraction and a source of her husband's corruption to fulfil her glamorous life style, thus justifying Handoko's infidelity to find other kept women. Laras—doing the same capitalising of her beauty and smartness—also mocks the beautiful singer as having only physical beauty without intelligence and hard work. These women fight one another to get the unfaithful and corrupt Handoko. This unspoken truth of the fight over beauty or body capital actually roots from the men's expectations or patriarchal ideology where women are pushed back to their traditional role as faithful wives or mothers with the second sex stigma, while men can be polygamous and have hegemonic control over women's body.

In *86*, the real corrupt actors are the lawyers, judges and attorneys. However, the story plots a dominant and never-ending conflict between Bu Danti, the senior registrar who arranges the corruption/bribery flow, and Arimbi, the clerk who takes small bribe in the flow of big legal cases. Another example is the fight between the maid Tutik and her master's wife when she is caught having a purely sexual affair with the master. The male master is depicted to slowly escape from the room as the two women harshly fight each other almost to death.

There was no further story about the cheating master. This silence or omission proves the author's unconscious reluctance to blame men, as she is supposed to from a feminist perspective. The story's plot turns to the conflict between Arimbi and Tutik instead, over their same sex relationship with its entire dilemma seen indirectly as a social and moral punishment.

Second, still connected to the first, women are clearly punished in both stories for their crime, but not men. In *Laras*, the unfaithful and corrupt men such as Handoko and Soedjatmiko (Laras' biological father) are both unpunished. The plot omits their legal and social punishment. Meanwhile, female characters, such as Laras and Bu Ayu, are socially and morally punished. Laras is psychologically and morally punished through her eventual moral conflicts after losing Hendra (her ex-boyfriend and idealised man who is murdered by Handoko). Bu Ayu is plotted as severely stressed and finally insane at the end and is tragically considered dead by her husband Handoko. Above all, women such as Bu Ayu and Laras have been unconsciously constructed as the real reason for men's corruption, implying that the true bad characters and the blame go to women.

It is also ironic that, from the law reinforcement point of view, men committing huge unlawful corruption crimes in *86* remain formally and socially unpunished. More ironically, the female characters who know less about the law and commit only small crimes get caught by KPK (Corruption Eradication Commission) and are formally and socially punished. Both Arimbi and Bu Danti are formally punished, while none of the law enforcement men are punished. The story even omits and is silent over their legal punishment.

Third, polygamy is silently tolerated and women are blamed or justified for men's unfaithfulness. This is related to some extent to beauty myth promoted by the male culture. Women overthrow other women when their male partners turn away to other women, especially when beauty and physical attraction seem to be the main reason. However, it is more complicated than that reason, although the bottom line is the same toleration of this unfaithfulness.

Handoko turns to Laras because Bu Ayu has no more bodily beauty attraction. This does not seem to be the real reason for the turning away. The reason is disputed because Tutik's master turns to her not because she is beautiful or has physical attraction (she is big, muscular, dark, ugly and with low social status). It is more about men's polygamous nature that has socially been reinforced and tolerated as being natural. This toleration takes its hidden and unspoken way through men's and women's ambiguous attitudes, for instance, toward faithfulness and virginity. This aspect of faithfulness and the moral values (burden) attached to female virginity have disadvantaged women beyond silent acceptance. In *Laras*, this ambiguity is clearly seen through the depiction of the wild and smart Laras, the naive commoner Widya and the unfaithful yet idealised Hendra. In her revenge to life's unfairness and poverty, Laras has decided to plunge herself into the high-class lifestyle of the minister's kept woman social arena for material wealth and bodily pleasure. This decision is taken partly due to her lost relationship and virginity to Hendra. He left her without any clear reasons and with no news for years. When she met the already married Hendra years later, she was in a dilemma whether or not to accept Hendra who kept tempting her to reunite their relationship. Hendra wished so much that Laras would give him another chance and accept him despite his married status to the commoner Widya. The story cleverly avoids giving direct answers to the dilemma but offers a resolution through Hendra's and Widya's death. However, the idea of letting Hendra to be polygamous is still reinforced as a resolution of the story when Widya lets Hendra 'have' Laras as his ideal wife through the religious use of a 'dream'. This is not a common dream because the author uses the religious dimension as the reaction to Laras' seemingly supernatural experience. The use of *arsy* and the reading of the *Surat Yasin* together after the supernatural dream strengthen the seriousness of the dream and thus the acceptance of this polygamy ideology. The bottom line is Hendra is allowed to have both Laras' wildness and Widya's domestication.

Fourth, men's double standards towards female virginity also justify their polygamous nature, especially when women unconsciously use man-made language in describing the experience. Both novels describe women's sexual experience through man-made language. Laras describes her wildness and sexuality concerning her virginity using the word *merenggut* (taken by force) by saying '*Hendralah yang telah kupilih untuk merenggut kesucianku*' (Hendra is the one I chose to take my virginity, *Laras*, p. 27). The above statement seems to imply woman's autonomy over her own body but using man-made language *merenggut*, which assumes a man's active control over a woman's body because it implies that man has the right and domination to take something with hard effort or by force. Moreover, both Hendra and Laras seemed to regret the experience. Laras kept condemning herself and showed self-hatred for her lost virginity and position as a kept woman. Meanwhile, Hendra kept showing his guilt over his taking Laras' virginity, by even saying that all men as husbands will expect a virgin wife (*Laras,* pp. 88–9). Laras is represented as labelling herself with the 'whore' image when she is not virgin and playing the game in the kept woman arena. The double standards are seen clearly in the fact that Hendra marries a commoner Widya (who is not wild, is obedient, virgin and domesticated), but keeps wishing to have Laras (who is wild and not a virgin).

In *86,* Arimbi and Tutik also use man-made language such as '*nyoblos*' (penetrate, *86,* p. 175) to describe a woman's sexual experience even from a woman's perspective. This penetration according to Dale Spender is man-made language showing domination and activeness, describing more physical activity rather than emotions, feelings or love.

Fifth, in corrupt social arenas, religion is used as merely a lifestyle or distinction in the penance mechanism over the result of corruption. Laras wears her veil when she contributes to the orphanage foundation, Islamic boarding school or woman's organisation. She calls herself a 'female Robin Hood'. Arimbi and Bu Danti also show similar attitudes in presenting their religious identity. In their corrupt life, Arimbi uses, for instance, the expression *Alhamdulillah* and Bu Danti uses *Natalan* (Celebrating Christmas) as religious signifiers. However, these are nothing more than a lifestyle because their corruption does not reflect their religious practices.

4 CONCLUSION

Following Macherey's three instances of the analysis method, this study concludes as follows. The ideological projection as the promised or spoken truth is the idea that women can dominate and commit corruption like men. This was clear enough from both stories written by two female authors where female characters were represented as the main and dominating ones. This ideological projection was realized by the authors by featuring female protagonists playing the game in different social arenas. They were depicted as having control over their own life as well as their corruption habitus. However, the unconscious (or the unspoken truth) of the texts revealed that the female authors were trapped in marginalising women by representing them as the real criminals of corruption, depicting women fighting other women, omitting punishment for male corruptors, tolerating men's double standards towards polygamy and women domestication and using man-made language to elaborate women's sexual experiences.

Although the two Indonesian female authors while writing about corruption appeared to represent women in seemingly dominating roles, they still could not escape the backlash of the patriarchal ideological trap as revealed by their ironic marginalisation of women as done by many writers.

REFERENCES

Amiruddin, M. (2012) Dari payudara melinda dee, rambut ungu miranda goeltom, hingga tas 'hermes' dan kerudung 'louis vuitton' nunun (From Melinda Dee's breasts, Miranda Gultom's purple hair, to Nunun's Hermes bag and Louis Vuitton veil). *Jurnal Perempuan,* 72, 97–108.

Bias Gender dan Perilaku Korup. (n.d). Retrieved from: www.jurnalperempuan.org/bias-gender-dan-perilaku-korup. html [Accessed 1st June 2015].

Bourdieu, P. (1995) *Sociology in Question.* London, Sage Publications Inc.

Bourdieu, P. (2012) *Pierre Bourdieu: Arena Produksi Kultural-Sebuah Kajian Sosiologi Budaya* (Pierre Bourdieu: Cultural Production Arena-A Cultural Sociology Study). Bantul, Kreasi Wacana.

Eagleton, T. (1991) *Ideology: An Introduction.* London, Verso.

Faludi, S. (2006) *Backlash: The Undeclared War against American Women.* New York, Three Rivers Press.

Habsari, S.U.H. & Haryono, A.T. (2014) Pemberitaan koruptor perempuan dalam perspektif gender (News about female corruptors in gender perspective). *Jurnal Dinamika Sains,* 12 (28), 68–91.

Hellwig, T. (1994) *In the Shadow of Change: Images of Women in Indonesian Literature.* Berkeley, The Regents of the University of California.

Hooks, B. (2000) *Feminism is for Everybody.* Cambridge, South End Press.

Lubis, M. (1970) *Senja di Jakarta* (Twilight in Jakarta). Jakarta, Badan Penerbit Indonesia Raya.

Macherey, P. (2006) *A Theory of Literary Production.* London, Routledge.

Mashuri. (2012) Goda dan Lupa: Wajah korupsi dalam novel Indonesia (Temptation and Forgetting: Corruption's Face in Indonesian Novels). *Jurnal Kritik,* 2, 25–49.

Moi, T. (1995) *Sexual/Textual Politics: Feminist Literary Theory.* London, Routledge.

Sarjono, A.R. (2012) Perkara korupsi dalam sastra Indonesia (Corruption in Indonesian literature). *Jurnal Kritik,* 2, 2–9.

Spender, D. (1985) *Man Made Language.* New York, Routledge & Kegan Paul.

Storey, J. (2009) *Cultural Theory and Popular Culture: An Introduction.* Harlow, Pearson-Longman.

Toer, P.A. (1954) *Korupsi* (Corruption). Jakarta, Balai Pustaka.

Wolf, N. (2012) *Beauty Myth: How Images of Beauty are Used against Women.* New York, HarperCollins e-books.

Cultural Dynamics in a Globalized World – Budianta et al. (Eds)
© *2018 Taylor & Francis Group, London, ISBN 978-1-138-62664-5*

Contesting representations in the gendered space of politics: Hillary Clinton's representations in *Living History*, *A Woman in Charge*, and *Her Way*

D. Hapsarani & M. Budianta
Faculty of Humanities, Universitas Indonesia, Depok, West Java, Indonesia

ABSTRACT: The aim of this research is to show the contesting representations of Hillary Clinton in her autobiography *Living History* (2003) and two biographies (2007): *A Woman in Charge* by Carl Bernstein and *Her Way* by Jeff Gerth and Don van Natta. The results of the analysis on *Living History* reveals that Hillary negotiates with the gendered writing convention of political autobiography and the double binds traps in building her image as a credible and capable world leader. Meanwhile, the study on the two biographies brings to light the consistent strategies to frame Hillary with gender stereotypes, double binds, as well as a negative and disturbing personality.

1 INTRODUCTION

The fact that political elections in the United States are deeply gendered has sparked a great deal of research on women studies to figure out the reasons why it is so difficult to convince the American people to vote for a woman's leadership. Falk (2010) confirms her assumption that the way media treated women candidates has not changed after a thorough study on the nine presidential campaigns involving women candidates in the United States from 1872 up to 2008. Waylen, et al. (2013) put politics both as practices and a study within a wider social context in which gender plays a very determining role, and they are convinced that in order to change the current, political science it should start to embrace gender perspective in their study.

With all this mushrooming research on gender and politics at the background, gender, once again, became a hot button issue in the 2016 political election as Hillary Clinton managed to win the presidential nomination from the Democratic Party. Josh Marshall (2016), a respected political journalist and blogger, dubbed the election as "the ultimate gendered election," while Bartash (2016) from The Market Watch predicted that the election will lead to "the biggest battle of the sexes in American political history" since Hillary has to face Trump, the embodiment of extreme masculinity. It is not surprising that Hillary is still experiencing offensive onslaught as in the 2008 election.

Offensive onslaught and negative representations are not new for Hillary who has started to become the target of attack ever since she assumed an active role in the administration of her husband's presidency back in 1993–2001. The following pictures display some negative stereotypes that are consistently and continuously attached to Hillary in the virtual media in 2008 and 2016. Hillary has been portrayed as the embodiment of a threat to the society.

Figure 1.   Negative stereotypes of Hillary Clinton found in 2008 and in 2016.

With negative representations in the media and strong association between presidency and masculinity, Hillary is compelled to manoeuvre in this gendered space in representing herself as a woman candidate for president. Following the campaign tradition, Hillary chose to construct her self-representation by writing an autobiography, *Living History*, years before her candidacy. Writing a political or campaign autobiography is problematic in itself as the writing convention of the genre brings forth masculine characteristics. Long (1999) identifies four elements of masculinity in the writing convention of autobiography: a linear plot which emphasizes success, a call or destiny in the public domain, a solitary hero displaying independence and self reliance, and avoiding emotional exploration by repressing the emotional experiences. These elements reflect the gendered space of politics and political elections which is dominated by masculinity and masculine stereotyping references. Therefore, it is interesting to see how Hillary makes use of references available in the gendered society to convince the readers of her capabilities and leadership.

Another interesting phenomenon was the publication of two unauthorized biographies on Hillary in 2007, six months prior to the election. *A Woman in Charge* was written by Carl Bernstein, a Washington Post veteran journalist famous for his investigative report (with Bob Woodward) on the Watergate scandal resulting in the resignation of President Richard Nixon. *Her Way: The Hopes and Ambitions of Hillary Rodham Clinton* was written by Jeff Gerth and Don Van Natta, two Pulitzer-Prize winning journalists from New York Times. Considering the background of the writers and the publication of the biographies, it can be assumed that the purpose of their publication was to challenge Hillary's self-representations in her autobiography. Both books relied and also commented on Hillary's story in *Living History*. This paper aims to show the contesting representations of Hillary Clinton and the barricades that a woman presidential candidate has to break through to gain the voters' confidence.

The strategy of representations in the three life narratives will be examined by investigating the framing used in telling Hillary's life story. Framing is "selecting and highlighting some facets of events or issues and making connections among them so as to promote a particular interpretation, evaluation, and/or solution" (Entman, 2004, p. 5). The practice of framing commonly used in media studies is applicable in studying life narratives. Life itself has no structure and no meaning. It is the writer who selects and highlights a particular aspect of the subject's life in order to unify the unrelated incidents and to interpret life experiences to make them meaningful and to give a sense of purpose. Framing can be identified by examining key words, metaphors, myths, concepts, symbols, visuals, ideas, and actions that are consistently and repeatedly used within the text.

This study uses a feminist framework to investigate Hillary's strategies in navigating her life story through the maze of gender stereotypes, gender bias, double binds, and double standards to build her leadership image. At the same time, by analyzing the two biographies, the study examines the strategies of the three biographers in re-establishing the conventional gender stereotypes and gender bias and in reinforcing the use of double binds and double standards in telling their story about Hillary.

2 LITERATURE REVIEW

Research on gender and politics has been growing abundantly in the U.S., as researchers and scholars attempt to discover the barriers that prevent women from winning the presidential election. The results of these studies indicated a progress. The earlier research confirmed the negative effects of gender stereotypes on women candidates (Huddy and Terkildsen, 1993; Kahn & Goldenberg, 1991), while the latest research found evidence showing that gender stereotypes do not negatively affect the electability of women candidates, for there is a complex interaction involving several elements, among others the gender of the candidate, the candidate's party affiliation, the gender of the voters, and the gendered perception of the political party in nominating their candidates (Sullivan, 2007; Shaver, 2013; Dolan, 2013; Bauer, 2014; Hayes and Lawless, 2015). Also large in number is the research on gender stereotypes in the media aiming to reveal the extent of media bias against women candidates. Kahn's analysis of newspaper coverage of senatorial and gubernatorial candidates in the 1980s (Kahn, 1991) shows gender bias in the media, as women candidates received less coverage and more negative coverage than their male counterparts. Though women candidates nowadays receive more coverage than before, they are still defined by the media through their gender.

A great deal of research focuses on Hillary's strategy in dealing with these barriers. In evaluating Hillary's failure in the presidential election of 2008, Lawrence and Rose (2010) identify three interlocking variables to analyse female presidential candidates: gender stereotypes, media routine, and candidate and her context. By evaluating Hillary's campaign strategy in navigating her path around and through the use of gender stereotypes, Lawrence and Rose draw conclusions on factors to consider for future woman presidential candidate in developing her campaign strategy. This research covers an extensive range of campaign media from campaign speeches, website, advertisement, to media coverage (blogs, newspapers, and television stations). However, the research does not cover Hillary's preliminary campaign strategy using memoir or autobiography. Kaufner and Perry-Giles (2007), seeing a political memoir as a campaign medium to build arguments for future office, analyse the contrasting narrative style in *Living History* and *Hard Choices* and identify Hillary's assumed political identities to position herself as a political leader. One interesting phenomenon with Hillary's candidacy that has been overlooked by the previous research is the contestation of representations within the genres of life narratives. This paper will focus on the contestation between Hillary's self-representation and other representations that challenges her representations.

2.1 *Hillary's representation in* Living History

Hillary's autobiography, *Living History*, shows that Hillary gives careful thoughts on how to present herself as a credible and capable world leader. In her strategy to evade the traps of gender stereotypes, gender bias, double standard, and double bind, she uses the strong points of both gender traits. She adopts some masculine traits to construct her representation as a competent political leader, while at the same time she also tries to balance the masculine qualities with selected feminine traits to soften her image without disrupting her leadership image.

To position herself as a qualified political leader, she uses frames that are her success as the embodiment of the national ethos, the self-made man, and thus positions herself as a true American. Her claim that she is equal to other male political leaders in the American Dream can also be read as her statement on the myth that has hitherto rejected women from attaining the highest leadership position in the country.

The patriotism frame is more complicated for women, as heroic acts are perceived as masculine performance. Unfortunately for Hillary, patriotism has taken a centre stage in presidential campaigns since 1988 and has become more salient after 9/11 and the ensuing war against terrorism. Hillary uses patriotism to frame her life during the Vietnam War period to justify her opposing standpoint as an act of nationalistic posture. The frame allows her to affirm the possibility of a woman who has no war experience to denote her patriotism by treating wars as a discourse into which a woman can enter. A similar frame is used in addressing the most controversial and embarrassing case, the Monica Lewinsky affair. Although Hillary shares her private feelings toward the scandal, she justifies her decision to stick to her marriage by framing it with her sense of patriotism and responsibility of a citizen to protect her country.

To strengthen her position as a strong political leader, Hillary consistently contrasts herself to women with traditional gender roles and builds an image that she is a non-conventional and progressive woman qualified to become a world leader and that she is the symbol of social change itself. Nevertheless, despite her efforts to put herself at the same level as male political leaders, she retains some feminine attributes to soften the masculine image. For example, she highlights the positive warm and congenial image of her staff, the Hillaryland which consists mainly of women and contrasts it with Bill Clinton's competitive and individualistic presidential staff in the White House. This is an important strategy considering the operation of the double binds against women in politics. If she appears too masculine, she will be doomed as a threatening woman, whereas if she appears too feminine, she will be perceived as weak and thus she will lose her leadership credibility.

2.2 *Representation of Hillary Clinton in* A Woman in Charge *and* Her Way

The study on the two biographies shows that both books use negative frames in telling Hillary's story. Hillary is represented as ambitious, manipulative, compromising, dishonest, and dominating. These negative representations challenge Hillary's self-representation in *Living History* by undermining her credibility and integrity as a trustworthy leader. Along with the negative qualities, the writers impart gender stereotypes and emphasize Hillary's dominant masculinity as she is portrayed as a dominating and aggressive woman.

However, although both books convey negative representations, they employ different framing strategies. Bernstein uses the frames of evolution, hypermasculine leadership, and narcissistic personality disorder, while Gerth and Van Natta use the frames of double standard, image making, and negative personality. With the frame of evolution, Bernstein portrays Hillary that looks positive on the outside as she is perceived of having the capability to change from a transitional woman (a progressive woman who is still trapped within the conventional values) into an independent female politician who has the courage to disconnect her leadership from her husband's leadership. However, behind the seemingly positive image, Bernstein reveals that Hillary's evolution is not the result of her hard work and independence, but the fruit of her manipulative strategy and compromise in presenting herself as a faithful traditional wife to draw people's sympathy and thus gaining her popularity. This representation challenges Hillary's self-representation as Berstein insists that Hillary's senatorship is an ascribed status, not an achieved status as portrayed by Hillary through the myth of the American Dream. Bernstein also relates the frame of Hillary's evolution with her moral degradation and lack of integrity as he emphasizes more on how Hillary compromises her idealism and moral values to attain her political standing.

Although Bernstein uses masculine traits as a frame to highlight Hillary's personality, his choice of masculine traits are different from the ones used by Hillary. While Hillary uses the positive traits of masculinity to strengthen her leadership capability, Bernstein challenges her self-representation by applying hypermasculinity traits as his frame. She is not only portrayed as an unruly woman, but also as an insatiable monster who is craving for power.

Another frame used by Bernstein is the frame of narcissistic behaviour indicated by an inflated self-esteem, a sense of self-entitlement due to self aggrandizement, lack of empathy toward others, and difficulties to accept failure. To build the foundation of this frame, Bernstein uses a different perception in telling Hillary's childhood, specifically in relation to her father. By claiming that Hillary has fabricated her childhood story to conceal the truth, Bernstein contests, denies, and invalidates Hillary's version of story and questions, the authenticity of Hillary's self-representation, as well as her integrity. Based on this deconstruction, Bernstein builds up his frame of narcissism to operate "the madwoman in the attic" metaphor. Thus, Hillary is claimed to be not only unreliable but also mentally unfit to become the president.

Gerth and Van Natta use a different set of frames in their biography. They operate frames that highlight the double-standard, image making, and negative personality starting from the prologue. The double-standard frame is set up by their statement that "[f]or years, Hillary has known that the standard for the first female president would be far higher than that for a man. A woman president would have to be super strong" (Gerth and Van Natta, 2007 p. 8). This statement of double standard is normalized with the quotation from Charlotte Whitton, a Canadian politician (1963) which states that "[w]hatever women do, they must do twice as well as men to be thought half as good. Luckily, this is not difficult" (Gerth and Van Natta, 2007 p. 340). The last sentence strengthens the normalization of the double standard by confirming that although it might not seem fair, the double standard is still achievable for women. Gerth and Van Natta also include Eleanor Roosevelt's answer in an interview on the possibility of a woman to become a president. She mentioned the conditions a woman should have if she wants to be a president: her capacity and the people's trust in her integrity and ability. These criteria are used as the measuring rod to evaluate Hillary and in the end justify their claim that Hillary has neither the capability nor the integrity and ability.

To convince the readers that Hillary does not meet the criteria set by Eleanor, Hillary's role model, the writers operate the frame of image making and negative characters. The image making frame sets Hillary as any ordinary politician who brushes up her image to conceal the reality. To stretch it even further, Gerth and Van Natta represent Hillary as a politician who does not know her true identity since she has been used to apply different types of personas for different situations. Overall, Hillary is portrayed as an ambitious, dishonest, compromising, and manipulative politician who is driven with the desire to dominate.

Although the two biographies, *A Woman in Charge* and *Her Way*, employ different frames, both of them construct counter representations to subvert Hillary's self-representation in *Living History*. Both biographies essentially confirm and support the negative representations about Hillary in the media which can be classified as a symbolic annihilation.

The research shows that gender inequality has placed Hillary in a less fortunate position as the result of the complexity of the negotiations a woman should comply in constructing her own political persona. Autobiography as one of the most effective media for political campaign due to its ability to manipulate and influence the audience becomes problematic when used by women presidential candidates. Although, in an autobiography, the writer has the authority to construct the narration of her own life and experiences without any mediation, in reality the writer is bound to the convention of autobiography dominated by myths, values, stereotypes, and characteristics associated with masculinity. Consequently, positioning becomes not only very important but also complex for women candidates in writing their autobiography. They have to negotiate the convention of autobiographical writing and gender stereotypes to avoid the trappings of the double binds. This complexity of gender positioning is a form of suppression toward women that curbs them from entering politics since they have to face greater, more complicated, and intimidating situations than male candidates. Despite the fact that people expect women candidates to have masculine characteristics, they are also expected to keep their femininity. If they appear too masculine, they will be perceived as cold and threatening, while on the other hand, the traits of femininity pose a risk to their leadership. In addition, if they choose to negotiate between masculine and feminine traits, they will be judged as inconsistent, image making, and compromising (see Figure 2). Whatever their choice is, the result is always negative.

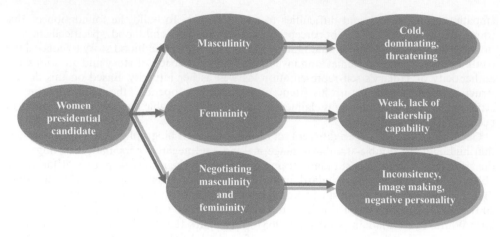

Figure 2.   The double binds traps.

The adversity continues on. Once a female presidential candidate constructed her political persona, it will then be scrutinized by the readers, which include the media. As indicated by the two biographies on Hillary Clinton, reinterpretation that produces counter-representations might occur in the form of a number of contesting representations.

The practices of repression is operated in doing the reinterpretation through the use of particular frames, such as gender stereotypes (hypermasculinity, "the mad woman in the attic" metaphor, the unruly woman), double standards (by using prominent women figures as the justification), negative personality, and image making. All of the frames are used for the same goal, which is to paralyze the character of the subject which can be classified as a form of symbolic annihilation (Gerbner & Gross, 1976).

Considering that the representations in the two biographies are not far too different from the ones circulating in the media prior and during the writing of the biographies, there are two possible conclusions to draw. First, the biographies' adoption of the frames that have been constantly used by the media indicates the powerful influence of the prevailing frames used to undermine women presidential candidates. Second, given that the three biographers are political journalists, there is a possibility that they are part of the institution that builds the news framing Hillary in the media. In that case, these political biographies are campaign tools used to voice the resistance or rejection toward women presidential candidates.

3   CONCLUSION

Based on the narrative strategies used in the three life narratives, gender still dominates politics in America as well as the narration convention of life writing. The concepts of gender roles, gender stereotypes, double standard, and double binds are being intensified in the writing of the political biographies of Hillary Clinton.

Lastly, the analysis reveals how gender asymmetry operates in the writing of campaign auto/biography of women presidential candidates. The gendered narration convention in autobiographical writing intensifies the complexity of positioning for women candidates. Since women presidential candidates are still an anomaly and due to the domination of masculinity in politics and campaign autobiography, the existing concepts and mental representations within the convention have not included women candidates. As a result, women candidates have to negotiate with the masculine concepts or mental representations while calculating other consequences that might arise due to the implementation of double standards and double binds. The writing of campaign biographies with women candidates as the subject contesting the self-representation constructed in the autobiography of the subject amplifies the challenges and complexity of the situation. It is undeniable that the complexity faced by Hillary Clinton cannot be separated from her liabilities; however, apart from Hillary's political baggage, the consistent use of gender stereotypes, double standards, and double binds as the strategy to resist Hillary's candidacy as president shows that politics is still a gendered space.

The results of this research affirm the research on the consequences of gender stereotypes conducted prior to the U.S. election of 2008. This research proves that gender stereotypes, double standards, and double binds are still believed to be a powerful and influential strategy to shape the way voters evaluate women presidential candidates. This research offers new possibilities to see the dynamics of the gender issue in politics through the formation and the contestation of representations in campaign auto/biographies. Since the gender issue does not only affect female presidential candidates, but also male candidates, gendered perspective research can also be applied to autobiographies written by male presidential candidates. Are male candidates free from the gender issue or do they also have to play with gender stereotypes in representing themselves?

REFERENCES

Bartash, J. (2016) *The 2016 election is shaping up to be the ultimate battle of the sexes*. [Online] Available from: www.marketwatch.com/story/the-2016-election-is-shaping-up-to-be-ultimate-battle-of-the-sexes [Accessed on 11th May].

Bauer, N. (2014) How partisans stereotype female candidates: Untangling the relationship between partisanship, gender stereotypes, and support for female candidates. *Presentation at the Visions in Methodology Conference*, 20–22nd May 2014 Hamilton, Canada, Mc Master University.

Duerst-Lahti, G., & Kelly, R. M. (1995) Gender power, leadership, and governance. Ann Arbor: University of Michigan Press.

Entman, R.M. (2004) *Projections of power: Framing news, public opinion, and U.S foreign policy*. Chicago, University of Chicago Press.

Falk, E. (2010) *Women for President: Media Bias in Nine Campaigns*. Chicago, University of Illinois.

Gerbner, G., & Gross, L. (1976) Living with television: The violence profile. *Journal of Communication*, 26 (2), 172–194.

Hayes, D., & Lawless, J.L. (2015) A Non-Gendered Lens? Media, Voters, and Female Candidates in Contemporary Congressional Elections. *Perspectives on Politics*, 13 (1), 95–118.

Huddy, L., & Terkildsen, N. (1993) The Consequences of Gender Stereotypes for Women Candidates at Different Levels and Types of Office. *Political Research Quarterly*, 46 (3), 503–525.

Kahn, K.F., & Goldenberg, E.N. (1991) Women candidates in the news: An examination of gender differences in U.S. senate campaign coverage. *Public Opinion Quarterly*, 5 (2), 180–199.

Lawrence, R.G., & Rose, M. (2010) *Hillary Clinton's race for the white house: Gender politics and the media on the campaign trail*. Boulder, Lynne Rienner Publishers.

Long, J. (1999) *Telling women's lives: Subjet, narrator, reader, text*. New York & London, New York University Press.

Marshall, J. (2016) *The Trumpian Song of Sexual Violence*. [Online] Available from: http://talkingpoints-memo.com/edblog/the-trumpian-song-of-gender-violence [Accessed on 24th May 2016].

Prentice, D.A., & Carranza, E. (2002).What women and men should be, are allowed to be and don't have to be: The content of prescriptive gender roles. *Psychology of Women Quarterly*, 26 (4), 269–281.

Waylen, G., Celis, K., Kantola, J., & Weldon G.W.K.C.J.K.L. (2013) *Gender and Politics: A Gendered World and a Gendered Discipline*. New York, Oxford University Press.

Cultural Dynamics in a Globalized World – Budianta et al. (Eds)
© *2018 Taylor & Francis Group, London, ISBN 978-1-138-62664-5*

Construction of woman in the Sundanese magazine *Manglè* (1958–2013): A corpus-based study of metalinguistic signs

S. Yuliawati & R.S. Hidayat
Department of Linguistics, Faculty of Humanities, Universitas Indonesia, Depok, Indonesia

ABSTRACT: This study examines the meaning of woman as constructed in the corpus of the Sundanese magazine *Manglè*, published between 1958 and 2013. Using corpus-based and Barthes's semiotic approaches, the study focuses on the usage of five Sundanese nouns denoting woman in the magazine, spanning four different eras: the Guided Democracy, the New Order, the Transition to Democracy and the Reform Era. The study regards the nouns as signs whose meanings are derived from Barthes's model of metalanguage. It argues that such meanings are the result of the extension of expression (E_2) in the secondary system of a semiological chain, which makes it possible to be analysed both qualitatively and quantitatively using corpus analysis. Therefore, the most or least frequently used signs and the meaning they carried can be explored by investigating the diachronic frequency and semantic preference of the nouns. This contributes to explaining the metalinguistic signs of women and gender construction from different perspectives. This study reveals that diachronically, the word *wanoja* was found to be the only sign used to denote women in *Manglè* with constantly increasing frequency. Moreover, it shows that women who were initially portrayed as dependent with regard to their traditional roles were becoming increasingly portrayed as independent in terms of their existence in the public sphere.

1 INTRODUCTION

Gender identity is the most important social category in people's lives that is given at birth, and thereafter, all social interactions are influenced by gender assignment (Weatherall & Gallois, 2003: 487). Gender identity is socially constructed using different individuals, groups and social institutions such as law, religion, education, cultural norms, beliefs, values and the media, which in the modern age have played a substantial role in the way people see, perceive, understand and construct gender (Ali & Khan, 2012: 343).

In the study of gender, language is one of the mechanisms to construe the notions of how gender is constructed because language is the fundamental means through which meanings are maintained, constructed, contested or resisted other meanings (Eckert & McConnell-Ginet, 2003). Two views related to language and gender that are prominently discussed are: (1) language is merely the reflection of society (language-as-mirror), so the social category based on gender is reflected in the patterns of language use and (2) language contributes actively in creating gender division (language-as-reproductive) (Talbot, 2001: 14). In spite of continuous debate on the two different views, it must be acknowledged that both of them can be regarded as a way to see how gender is constructed through language. Furthermore, Talbot (2001) argued that the most important task in studying the relationship between language and gender is to explore the multifarious roles of language in creating and maintaining gender division in society.

One of the linguistic phenomena found in the Sundanese society that contributes to the construction of women is the expression of *parawan jomlo*, a term typically used to describe a woman who has passed the usual age for marriage. The term obviously causes some discomfort to the woman, her parents or her family. Consequently, it is quite common for Sundanese parents to have their daughters marry early rather than let their daughters to stay single (Edi, 1984). Moreover, the position of women—with respect to their marital status—is aggravated even more by a Sundanese proverb *kawin ayeuna, isuk pepegatan* ('it is better to marry and then divorce, than to never get married at all'). This view is probably one of the causes why the divorce rates in some areas of West Java are higher than anywhere else in Indonesia (Jones, Yahya & Tuti, 1994). The other linguistic phenomenon found is in the mixed-gender personal binomials. On the one hand, women are often mentioned in the second position, such as in the phrases *salaki jeung pamajikan* and *laki rabi* ('husband and wife'). On the other hand, women are also frequently mentioned in the first position, such as in the phrases of *indung bapa, ema jeung bapa* ('mother and father') and *nini aki* ('grandmother and grandfather'). These reveal that it is common to mention women behind men when talking about women in relation to men; however, in relation to family, women are positioned ahead of men. Motschenbacher (2013) found that positioning of men and women in relation to the order principle mostly carries a harmful discourse about the normative positions and the roles of men and women in society. Consequently, mixed-gender binomials create a type of structure in which gender power differentials, gender binaries and gender differences have become firmly materialised. They, however, should not be understood as natural, but rather as discursively and linguistically mediated social constructions (Cameron, 1992).

Corpus linguistic that involves the analysis of large collections of computerised texts (known as corpora) representing certain language uses could provide some insights into the way gender is constructed by certain societies. By identifying the frequent linguistic patterns derived from the large samples of texts, the construction of gender can be studied through the combination of both qualitative and quantitative analyses. As argued in Baker (2005, 2010, 2014), language and gender studies in the past preferred to use qualitative analysis and small-scale data. However, in recent years, there has been a shift towards larger sketches of texts and combining the forms of qualitative and quantitative analyses. The objective of this study was to investigate the construction of woman diachronically, namely period 1 (the era of Guided Democracy, 1958–1965), period 2 (the New Order, 1966–1998), period 3 (the Transition to Democracy, 1999–2003) and period 4 (the Reform Era, 2004–2013), in the corpus of *Manglè*, a magazine published in Sundanese language, based on the usage of five nouns denoting women (*geureuha, mojang, pamajikan, wanita* and *wanoja*) using corpus linguistic and Barthes's semiotic approaches. By integrating the two approaches, a deeper understanding about the construction of woman in *Manglè* can be explored.

2   METHODOLOGY

A corpus from collection of texts taken from *Manglè* published between 1958 and 2013 was constructed for this study. The corpus was built from a sample of 92 editions of the magazine using the calculator size sampling[1]. Then, to determine the total number of words used to construct the corpus from period 1 to period 4, the technique of proportional systematic random sampling was employed. The size of the *Manglè* corpus was 2,940,537 words, consisting of 78,081 words from the corpus of period 1; 1,897,777 words from the corpus of period 2; 324,614 words from the corpus of period 3 and 641,065 words from the corpus of period 4.

To study the construction of woman in the corpus of *Manglè*, we use a mixed-method design in which quantitative and qualitative approaches are combined to provide a more complete understanding of the research topic than either approach alone (Creswell, 2014). The author uses three procedures of the corpus analysis to investigate the Sundanese nouns denoting women, that is, a frequency analysis, a collocation analysis to examine the strength of collocations statistically and the concordance contextual analysis to determine the semantic categories. To measure the strength of collocations, the study uses MI score of 3.00 or higher and the minimum frequency of 5, generated by the corpus software, WordSmith Tools 6.0. Subsequently, the corpus analysis results are used as a basis to interpret the way women

---

1. Available in http://www.surveysystem.com.

are mentioned in *Manglè* using Barthes's semiotic approach, specifically through the meta-language process.

From the perspective of Barthes's semiotic approach, the phenomenon shows that the society tends to regard something that are natural as the result of the process of significa-tion[2], and derived from the processes of metalanguage[3] – the extension of Expression/E_2 – and connotation[4] – the extension of Content/C_2. In other words, the signification moves in two directions: either towards the metalinguistic process as indicated by the synonymic phenomenon particularly in the case of verbal signs or towards the connotative process as indicated by the various concepts referring to the same form. Grounded on that theory, this study argues that in revealing the construction of women based on the words used to denote women in the *Manglè* corpus implies interpreting the signification of women through verbal signs. The five nouns denoting women, being the focus of the analysis, are regarded as the result of the metalinguistic process of the signification (the extension of E_2). Therefore, the construction of women here is specifically interpreted through that process.

3   ANALYSIS

On the basis of the list of word frequency of the *Manglè* corpus generated by WordSmith Tools 6.0., the frequencies of nouns denoting women from period 1 to period 4 could be identified. Table 1 presents a normalised frequency of the nouns in words per million found in every period that shows the diachronic change of word use frequency. There were sev-eral findings from the frequency analysis that may contribute quantitatively to the study of women's constructions in *Manglè*.

One of the important findings to note about the nouns in the *Manglè* corpus is the fact that while the frequency in the usage of *geureuha, mojang, pamajikan* and *wanita* was gradually decreasing from period 1 to period 4, *wanoja* was the only noun whose frequency was con-stantly increasing. The usage of *wanoja* began to rise significantly in *Manglè* during period 2 (the New Order, 1966–1998), whose occurrence had been found to increase by eightfold since period 1 (the era of Guided Democracy, 1958–1965).

In the first period, the noun *wanoja* was found only 13 times in a million words, but in the second period, it was found 111 times in a million words. In addition, observing the frequency of *wanoja* from period 1 to period 4, the considerable increase in frequency of the word could be identified. The usage of the word increased by 23-folds: from 13 words/mil-lion in the Guided Democracy period to 301 words/million in the Reform Era, resulting in a frequency of occurrence that was nearly the same as the occurrence of the most frequent noun, that is, *pamajikan* (321 words/million). Considering the frequencies of the other four nouns that continue to decline diachronically, it can be concluded that *wanoja* is the only noun that gained popularity to use for denoting women in *Manglè* – and its rise began in the New Order period (1966–1998).

---

2. Barthes defined a sign as a system that consists of an expression (E) in relation (R) to content (C) and identified two orders of signification: the primary system (E1 R1 C1) and the secondary system (E2 R2 C2). The secondary system can be resulted from the extension of expression (E2) known as the process of metalanguage or from the extension of content (C2) known as the process of connotation (Nöth, 1995).
3. In the metalinguistic process, users of a sign extend the expression in the secondary system of the semiological chain, so there are more than one expressions referring to the same content. For exam-ple, there are several words in Indonesian referring to a place for prisoners, such as *penjara, lembaga pemasyarakatan, hotel prodeo* and *kurungan* (Benny, 2014).
4. In the connotation process, users of a sign extend the content in the secondary system of the semi-ological chain, so there is more than one concept used to signify an expression. For example, every word used as a sign to signify a place for prisoners has a specific meaning in the case of connotation, such as the specific meaning of *penjara* is a place to punish criminals, while *lembaga pemasyarakatan* is a place to turn criminals into good citizens (Benny, 2014).

Table 1. Frequency of the word usage denoting woman in words per million.

Nouns	Period 1	Period 2	Period 3	Period 4
Geureuha	156	8	6	3
Mojang	545	244	166	164
Pamajikan	584	524	379	321
Wanita	298	146	105	59
Wanoja	13	111	179	301
Corpus size	78,081 words	1,897,777 words	324,614 words	641,065 words

Another interesting finding from the *Manglè* corpus is the usage of *pamajikan*. The usage of *pamajikan* demonstrates the opposite phenomenon from the usage of *wanoja*. Although *pamajikan* was the most frequent word encountered amongst all the five words in all periods, its usage continued to decrease over time. In fact, in the Reform Era (2004–2013), the occurrence of *pamajikan* was close to *wanoja*, which was the least-used noun in the Guided Democracy period (1958–1965). Figure 1 illustrates the trends in word usage in consecutive order, from period 1 to period 4. It shows that the frequencies of *pamajikan* and *wanoja* noticeably start from different positions in period 1 (*pamajikan* is at the top position of the word frequency axis, while *wanoja* is at the bottom), but then they move to almost the same point in period 4. It shows that the word *pamajikan* is getting less popular for denoting women, while the popularity of *wanoja* tends to rise as the noun of choice for denoting women in *Manglè*.

The last finding about the frequencies of the nouns for denoting women that is also significant to observe is the occurrence of *geureuha*. The usage of this word began to decrease dramatically in period 2 (the New Order, 1966–1998) and continued to decrease until period 4 (the Reform Era, 2004–2013). In the *Manglè* corpus of period 1 (the era of Guided Democracy, 1958–1965), the usage of *geureuha* was found to be 156 times in a million words, but in period 2, the usage of the word was only eight times in a million words, which means that its occurrence decreases by nearly 95%. Moreover, the occurrence of the noun *geureuha* decreased by 98% in the first to the fourth period, from 156 times in a million words to three times in a million words. On the basis of the frequency analysis, the usage of *geureuha* to denote women in *Manglè* will probably disappear soon as a result of its rare usage.

On the basis of the lexical meanings of the five nouns[5], the study shows that the editions of *Manglè* published between 1958 and 2013 (from the era of Guided Democracy to the Reform Era) mostly used the noun *pamajikan*, which is a word usually used by men to refer to the women they are married to and regarded as a low-level language (*Kasar*). According to Satjadibrata (in Anderson, 1993), *Kasar* is normally used by a speaker to a referent of a lower status than either the speaker/listener or to a referent of the same status when the speaker speaks to well-acquainted or younger listeners. Referring to the lexical meaning, it is likely that the women who are most frequently mentioned in *Manglè* during that period are married women who are regarded to have an equal or lower status than either the speaker/listener. On the contrary, the noun *geureuha* is a word of a high-level language (*Lemes*) and is used to refer to a married woman of a higher or equal status (when the speaker speaks to interlocutors with

---

5. The lexical meanings of the five nouns from three Sundanese dictionaries, KBS (2010), KBS (2009) and KUBS (1969), are as follows:

Geureuha is a high-level language (*Lemes*) used to refer to a married woman that is usually used by lower-status speakers to upper-status hearers.

Mojang is defined as a word used to refer to a young woman, but there is no explanation whether it belongs to high (*Lemes*)- or low-level language (*Kasar*).

Pamajikan is a low-level language (*Kasar*) used to refer to a married woman and is usually used by a man to address his married woman, which the antonym is *salaki* (a married man).

Wanita is simply defined as a word used to refer to a woman that is borrowed from Sanskrit—no further explanation about its speech level, whether it is a low (*Kasar*)- or high-level language (*Lemes*).

Wanoja, which is borrowed from *Kawi*, is the synonym of *wanita* that is used to refer to a woman, and the speech level of the word is also not explained.

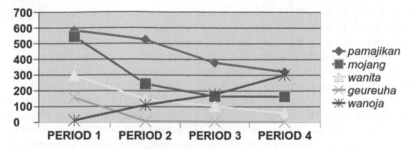

Figure 1.   Frequency of the word usage denoting woman in words per million.

a higher status and strangers) that is likely to disappear from the corpus of *Manglè*. The opposite phenomenon shown by the occurrences of both nouns in the diachronic corpus of *Manglè* demonstrates that women tend to be constructed through a low-level language (*Kasar*) rather than a high-level one (*Lemes*). Besides, there is a tendency that women with marital status are more interesting to be discussed in the magazine. To investigate how women are constructed semantically, the analysis looks further at the semantic categories of the five nouns referring to women based on their relationships with the co-occurring words using a semantic preference analysis.

On the basis of the analysis of semantic preference, the semantic categories of the nouns referring to women from period 1 to period 4 can be identified by observing their significant collocates, that is, words that statistically co-occur with the nouns in a significant number. On the basis of the analysis, semantic differences were generally found among the nouns. The noun *geureuha*, which can be identified only in the first period due to its infrequency in other periods, is associated with family because of its repeated co-occurrences with the word *putra* (child/children). Similarly, the noun *pamajikan* is closely related to family almost in all periods (from period 2 to period 4) because it co-occurs with the word *anak* (child/children). Moreover, the other co-occurring words that strongly connect *pamajikan* to the semantic category of family are *adi* (sister/brother), *bapa* (father), *barudak* (children), *budak* (child), *indung* (mother), *mitoha* (mother/father-in-law) and *indungna* (his/her mother) – and these are mostly found in period 2. The noun *pamajikan*, especially in period 2, is also associated with places, such as *imah* (house/home), *imahna* (his/her house), *suhunan* (roof/house), *dapur* (kitchen), *kamar* (bedroom) and *lembur* (hometown), which depict women's traditional places. Unlike *pamajikan* and *geureuha*, the nouns *wanita* and *wanoja* tend to co-occur with words whose semantic categories are occupational and organisational. The noun *wanita*, for example, co-occurs with the words referring to women's participation in organisations such as *anggota* (member), *dharma wanita* (women's association), PKK (the Supervisors of Family Welfare), *ketua* (chair), organisasi (*organisation*), *persatuan* (association) and *koperasi* (cooperative), and with the words referring to occupations, such as *karir* (career), *pelacur* (prostitute), *usaha* (business) and *wiraswasta* (entrepreneur). These significant collocates are mostly found in period 2. Slightly different from *wanita*, the significant collocates of *wanoja* whose semantic category is organisational are not connected with men/men's position, such as *dharma wanita* and PKK. The noun *wanoja* co-occurs with the words *organisasi* (organisation) and *tokoh* (figure) showing women's participations in the public sphere independently. Besides, *wanoja* is found to be the only noun that does not co-occur with its antonyms— words referring to men—and words referring to family (*pamajikan* and *salaki, wanita* and *pria/lalaki, mojang* and *jajaka, geureuha* and *putra*). Particularly in period 2 and period 4, *wanoja* also co-occurs with the words referring to places, but it is different from the semantic category of places for *pamajikan*. Here, the places refer to country or region, such as Sunda, Indonesia and Kuningan, which are commonly used to describe the women's places of origin. Moreover, the women referred to in this context are usually evaluated positively. Meanwhile, the noun *mojang* tends to co-occur with words referring to beauty, such as *geulis* (beautiful), body such as *leungeun* (hands) and *sora* (voice) and places such as Bandung and Jawa Barat

(West Java). From the categories, it shows that although in general the five nouns can be used to denote women, the nouns have their specific meanings.

From the perspective of Barthes's semiotic approach, the five nouns referring to woman (*geureuha, mojang, pamajikan, wanita* and *wanoja*) are regarded as signs that are necessary to signify. The signs were used by their users, that is, the writers in *Manglè*, to signify the concept of 'woman as a social category'. The fact that there is more than one sign used to signify the concept shows that the users of the signs extend the signification of woman through the extension of expression (E_2) in the secondary system of the semiological chain, known as the metalanguage (Barthes, 1991). As a result, the meaning of woman as a social category is signified by various signs, such as *geureuha, mojang, pamajikan, wanita* and *wanoja*. Referring to the corpus analysis, the signification process of woman through metalanguage in *Manglè* is demonstrated through not only the fact that there are various signs used to signify the concept of 'woman as social category' but also the sign users' tendency to use signs diachronically, as illustrated in Figure 1. The sign users in *Manglè* did not use a single sign to denote woman; instead, they were using five signs with varied levels of occurrence from period 1 (the era of Guided Democracy, 1958–1965) to period 4 (the Reform Era, 2004–2013). The sign most frequently used was *pamajikan*, because its occurrence in all periods was always the highest of all signs. Other signs that show a decrease in their use are *geureuha, mojang* and *wanita* – with *geureuha* as the one whose usage decreased the most. On the contrary, there is an interesting finding that shows that *wanoja* is the only sign whose usage persistently increased at all periods.

It could be argued that the analysis of the signification of woman through metalanguage in the corpus of *Manglè* is indicative of several points. First, the declines of almost all signs denoting women show that the degree of signification of women in the magazine was decreasing. Next, the finding on the considerable decline of the sign that belongs to the *Lemes* (formal language) category, and that contradictorily, the most frequent sign found in all periods belongs to the *Kasar* (informal language) category, reflects that women were mostly mentioned colloquially and were most probably regarded to have a lower status. Moreover, it generally indicates that the Sundanese language is becoming more egalitarian.

4 CONCLUSION

After analysing the construction of women in the corpus of the Sundanese magazine *Manglè*, several points can be summarized. First, a move towards the usage of *wanoja* among other signs to denote women reflects the changing trend in the word usage that plays a significant role in the construction of women in the magazine. Second, there is a dramatic shift in how women were portrayed in *Manglè*: in the New Order period (1966–1998), women were more frequently associated with family and domestic roles, whereas in the subsequent periods, women were more frequently portrayed in relation to organisations, occupations and other places that reflected their independency and participations in the public sphere. In other words, it can be concluded that the construction of women in *Manglè* tended to shift diachronically. Initially, women were portrayed as dependent, that is, their presence was frequently related to someone (*anak, adi, bapa, barudak, budak, indung, mitoha, indungna,* etc.) or something (*imah, imahna, suhunan, dapur, kamar, lembur,* etc.) that was closely associated with their traditional roles. However, later on, women were increasingly portrayed as independent, that is, their presence was associated with their existence in the public sphere. Referring to Cameron (1992), the findings should not be understood as natural, but rather as the results of the discursively and linguistically mediated social constructions of gender.

Third, we argue that the combination of Barthes's semiotic and corpus linguistic approaches used for studying the construction of woman in the Sundanese magazine provides a more thorough description. This is because the signification of woman can be explored not only qualitatively through the identification of a synonymic phenomenon (as the extension of the expression in the primary system), but also quantitatively through the analysis of sign occurrences diachronically. Therefore, the shift in the sign users' perspectives on women's issues can be investigated by analysing the way they used the signs diachronically. Thus, this study proposes a theoretical framework to study verbal signs using a mixed-method design, such as the combination of the metalanguage of Barthes's signification and corpus linguistic approaches.

REFERENCES

Ali, G. & Khan, L.A. (2012) Language and construction of gender: A feminist critique of sms discourse. *British Journal of Arts and Social Sciences, 4*(2), 342–360.

Barthes, R. (1991) *Mythologies.* USA: Twenty-fifth printing.

Benny, H.H. (2014) *Semiotika dan dinamika sosial budaya (Semiotics and socio-cultural dynamics).* 2nd edition. Depok: Komunitas Bambu.

Blackburn, S. (2004) *Women and the state in modern Indonesia.* UK: Cambridge University Press.

Butler, J. (1986) Sex and gender in Simone de Beauvoir's Second Sex. *Yale French Studies,* (72), 35–49. doi:10.2307/2930225.

Cameron, D. (1992) *Feminism and linguistic theory.* Basingstoke: Macmillan.

Creswell, J.W. (2014) *Research design: Qualitative, quantitative, and mixed methods approaches.* 4th edition. USA: Sage Publication.

Eckert, P. & McConnell-Ginet., S. (2003) *Language and gender.* UK & US: Cambridge University Press.

Ekadjati, E.S. (1984) *Masyarakat Sunda dan kebudayaannya (The Sundanese and their culture).* Jakarta: Girimukti Pusaka.

Hoey, M., Mahlberg, M., Stubbs, M. & Teubert, W. (2007) *Text discourse and corpora: Theory and analysis.* London & New York: Continuum.

Jones, G.W., Yahya, A. & Tutu, D. (1994) Divorce in West Java. *Journal of Comparative Family Studies, 25*(3), 395–416.

Motschenbacher, H. (2013) Gentlemen before ladies? A corpus-based study of conjunct order in personal binomials. *Journal of English Linguistics, 41*(3), 212–242.

Nöth, W. (1995) *Handbook of Semiotics.* USA: Indiana University Press.

Stubbs, M. (2002) *Words and phrases: Corpus studies of lexical semantics.* New Jersey: Blackwell Publishing.

Talbot, M.J. (2001) *Language and gender: An introduction.* UK: Blackwell Publishers Ltd.

Teubert, W. & Krishnamurthy, R. (2007) *Corpus linguistics: Critical concepts in linguistics.* 1st Volume. London and New York: Routledge.

Tognini-Bonelli, E. (2010) Theoretical overview of the evolution of corpus linguistics. In: O'Keeffe, A. & McCarthy, M. (Eds). *The Routledge handbook of corpus linguistics.* London: Routledge.

Weatherall, A. & Gallois, C. (2003) Gender and identity: representation and social action. In: J. Holmes & M. Meyerhoff. *The Handbook of Language and Gender.* Cornwall: Blackwell Publishing. pp. 487–508.

West, C. & Zimmerman, D. (1987) Doing gender. *Gender and Society, 1*(2), 125–151.

Cultural Dynamics in a Globalized World – Budianta et al. (Eds)
© 2018 Taylor & Francis Group, London, ISBN 978-1-138-62664-5

Metaphors of women in the Sundanese magazine *Manglè* (1958–2013): A corpus-based approach

A. Alkautsar & T. Suhardijanto
Department of Philosophy, Faculty of Humanities, Universitas Indonesia, Depok, Indonesia

ABSTRACT: This paper aims to see how women are described in the Sundanese language through the use of metaphors found in *Manglè* magazine, published from 1958 to 2013. The analysis is divided into four different periods: the Era of Guided Democracy, the New Order, the Transition to Democracy, and the Reform Era. By exploiting the corpus-based approach within the framework of Lakoff and Johnson's Conceptual Metaphor Theory, the study focuses on two Sundanese nouns denoting women: *awéwé* and *mojang*. A combination of quantitative and qualitative methods is used to identify the features of linguistic metaphors for women: collocational behaviour and syntactical behaviour, as well as its frequency of occurrence in different periods of time, thus providing a more complete description of women in the Sundanese society. The research finds out that as a metaphor, *awéwé* has been spoken in several different contexts: marriage (a wife, or a partner in an extramarital affair), companionship (a lover), practices of prostitution (a prostitute), and as an entity of weakness. Meanwhile, the metaphorical *mojang* has only been spoken in the context of beauty pageants (contestants or winners). Frequency-wise, the metaphorical senses of *awéwé* have experienced rises and falls in their usage throughout time, except for the sense of entity of weakness that keeps decreasing until the latest period. A rather similar fluctuation happened to nearly all the metaphorical senses of *mojang*, excluding the sense of contestants of beauty pageants, which sees a drastic increase in its usage.

1 INTRODUCTION

Lakoff and Johnson's Conceptual Metaphor Theory or CMT (1980) and several other publications marked the establishment of what Lakoff (1993) referred to as the contemporary view of metaphors. The basic idea of CMT lies in the notion that metaphors work on a thinking level, implying that human's cognitive mind is metaphorical by nature. If we look deeper into it, the cognitive mind is packed with such a wide variety of concepts amassed by the owner's experience, be it physical, social, or cultural (Lakoff & Johnson, 1999). Upon learning a new particular concept, human—the speaker of the language—would attempt to understand the new concept by looking at it through another existing concept the speaker has experienced before (Koller, 2011). Lakoff and Johnson state that inside the mind, a transfer process occurs between two different conceptual domains—from the source domain where the more concrete concept based on experience resided, to the target domain that contains the abstract, newfound concept to provide a plausible explanation to the speaker.

The classical view of metaphors puts a solid line between the non-literal (metaphors) and literal language. Metaphors are seen as simply the ornaments or additions to the literal language—the reason why metaphors have long been associated with poetry and literature. In contrast, one of the biggest consequences of CMT is the realisation that daily language in human communication—on the linguistic level—mostly consists of metaphors, including the words widely regarded as literal by their speakers, as these words are products of metaphors on a thinking level. To differentiate these 'metaphorical' words with the metaphors on a thinking level, Steen (1994) suggests the term *linguistic metaphors* is for the former, while Lakoff and Johnson introduced the term *conceptual metaphors* for the latter.

As Wikberg (2008) asserts, the development of corpus linguistics enables researchers to describe meanings of certain word combinations, either in its literal or non-literal sense. Sinclair (in Cheng, 2012) reported two important discoveries concerning corpus-based language studies, they are: (1) language is all about the creation of meaning, and (2) language has a tendency to be phraseological. The term *phraseological* refers to repeated word patterns co-selected by its creators—the speakers of the language—in order to create a specific meaning. McEnery and Hardie (2012) state that meaning is not attributed solely to a single word, but rather to a group of words whose uniqueness could be determined by how frequent the words appear together with other certain words (collocational patterns) or in certain structures (syntactical patterns). In other words, meaning is not isolated to only one word (*word-based*), but is distributed among the surrounding words (*phrase-based*) as chosen by the speaker of the language (Sinclair, 2004).

Deignan's (2005) corpus-based research about metaphors on the Bank of English corpus categorised the linguistic metaphors into four different types: 1) innovative metaphor; 2) conventional metaphor; 3) dead metaphor; and 4) historical metaphor. The second type, the conventional metaphor, is described as highly dependent on the core sense of the word and the context surrounding the word, which means that both the collocational and the syntactical patterns should be thoroughly examined. Additionally, Deignan states that conventional metaphors have the most examples in the corpus. As the Bank of English consists of texts from the media, which means it consists mostly of ordinary texts, what Deignan calls conventional metaphor is basically the CMT's area of focus.

The objective of this research is to investigate how women are described in the Sundanese language through the use of metaphors found in the long-running Sundanese magazine, *Manglè*, from 1958 to 2013. The approach of corpus linguistics within the framework of CMT is applied to identify and examine the metaphorical senses of two Sundanese nouns denoting women, *awéwé* and *mojang*. Based on the research done by Yuliawati (2016), the frequency of occurrence of *awewe* and *mojang* throughout 1958–2013stays on the top ranks among the Sundanese nouns denoting women, although in later periods, *wanoja* overtook *mojang* (see Table 1).

In the literal sense, the Sundanese dictionary *Kamus Basa Sunda* (Danadibrata, 2009) defines *awéwé* as *jelema nu saperti indung urang* (a person like a mother, referring to the female sex, or a woman), *mojang* as *parawan* (a virgin, or a young woman), and *wanoja* as *istri, wanita* (a woman, or similar to *awéwé*). The findings that *awéwé* and *mojang* occur most frequently show that both words are the most used in accordance with their respective

Table 1. The frequency of occurrence of nouns denoting women in *Manglè* (1958–2013).

| Noun | Frequency of occurrence (in percentage) | | | |
	Period 1 (58–65)	Period 2 (66–98)	Period 3 (99–03)	Period 4 (04–13)
awéwé	0.064%	0.064%	0.052%	0.052%
mojang	0.054%	0.024%	0.017%	0.016%
wanoja	0.001%	0.011%	0.018%	0.030%

(Source: Yuliawati, 2016).

definitions; hence, this becomes the rationale in determining *awéwé* and *mojang* as the focus of the research.

Inspired by Deignan's research, the conventional metaphors are the main focus in this paper and they are examined quantitatively through the frequency of occurrence and qualitatively through their collocational and syntactical patterns. However, seeing metaphors as products of the conceptual thought that is based on a physical, social, and cultural experience, the frequency of occurrence of metaphors with its specific patterns could indicate the norm of language use in society, alongside the patterns for literal senses of words (Teubert & Krishnamurthy, 2007).

2 METHODOLOGY

This research uses nearly all articles in *Manglè* (e.g. editorials, profiles, columns, stories, etc.), leaving out advertisements and quizzes. The required texts for the corpus were divided into several periods based on the context of Blackburn's gender ideology of the Indonesian state from 1958 to 2003 (in Yuliawati, 2016): 1) the Era of Guided Democracy (1958–1965), 2) the New Order (1966–1998), and 3) the Transition to Democracy (1999–2003). Another period was later added to cover the texts of *Manglè* publications after 2003: the Reform Era (2004–2013). This totalled to 2,455 editions of *Manglè*, spanning from 1958 to 2013. The magazine's texts had to be converted into electronic data by scanning them with an OCR (*optical character recognition*) tool or retyping them manually on the computer. The electronic texts were stored as *plain text* (.txt) files and uploaded to the software.

However, considering the limitation of the corpus software capacity compared to the abundance of the data, only samples from each period were used to build the corpus. The application of proportional cluster random sampling on each period reduced the number of samples to a total of 91 editions, proportionally distributed in each period. The details of the constructed corpus divided by each period can be seen in Table 2.

This research combines both quantitative and qualitative approaches (Creswell, 2014) for the data on the frequency and concordance of the corpus, respectively (Sinclair, 1994). The metaphors of the Sundanese words *awéwé* and *mojang* were identified and examined in the corpus by investigating the features of their linguistic metaphors, as well as the collocational and syntactical patterns (Deignan, 2005). Initially, the metaphorical nature of one sense could be detected by comparing it with the definition provided by dictionaries (i.e. the literal sense). The suspected metaphorical sense was then examined through how the most significant collocates on the left and right side of the word appeared together on a number of citations (i.e. the collocational patterns).

Different metaphorical senses could be unique with their set of collocates, or they share the same collocates, in which a wider look at the whole context was needed. Subsequently, another investigation was conducted by looking into the syntactical patterns—grammatical tendencies of metaphorical words in certain senses, e.g. word classes. The examined metaphors and their patterns, together with how frequently they occurred in the corpus, could

Table 2. The corpus size of *Manglè*.

	Period	Corpus size (words)
1	The Era of Guided Democracy (1958–1965)	78,081
2	The New Order (1966–1998)	1,897,777
3	Transition to Democracy (1999–2003)	324,614
4	Reform Era (2004–2013)	641,065
Whole corpus size (words)		2,941,537

show how *awéwé* and *mojang* were described, at least giving us one of the many descriptions of women within the Sundanese society from one period to another.

3 RESULTS & DISCUSSION

3.1 *Awéwé*

Table 3 presents the frequency of occurrences (shown by number of citations) of five metaphorical senses of *awéwé* from Period 1 to Period 4. It is interesting to note that while the frequency of other metaphorical senses was fluctuating between periods, the frequency of metaphorical sense #5 (entity of weakness) drastically dropped from Period 1 (8.82%) to Period 2 (1.59%) and kept decreasing persistently in later periods.

Another interesting note is that the reduced usage of sense #1 (wife) in Period 4 happened together with the rise of sense #2 (lover) in the same period. Its consequences, context-wise, will be discussed later.

Meanwhile, the word *awéwé* tends to appear together with its lexical collocates, as shown in Table 4.

Table 3. The metaphorical senses of *awéwé*.

| # | Sense | Number of citations (and its percentage) | | | |
		Period 1 (58–65)	Period 2 (66–98)	Period 3 (99–03)	Period 4 (04–13)
1	Wife (spouse)	3 (8.82%)	56 (4.44%)	17 (8.54%)	19 (6.03%)
2	Lover (companion)	0	8 (0.63%)	1 (0.5%)	11 (3.49%)
3	Partner in extramarital affair (cheater)	1 (2.94%)	21 (1.67%)	5 (2.51%)	5 (1.59%)
4	Prostitute (courtesan)	0	12 (0.95%)	12 (6.03%)	2 (0.63%)
5	Entity of weakness (not masculine)	3 (8.82%)	20 (1.59%)	3 (1.51%)	2 (0.63%)
	Total frequency of *awéwé* in the corpus	34	1260	199	315

Table 4. The lexical collocates of *awéwé*.

| # | Sense | Collocates | | | |
		Period 1 (58–65)	Period 2 (66–98)	Period 3 (99–03)	Period 4 (04–13)
1	Wife (spouse)	–	*salaki* *beda* *salakina* *pamajikan* *andalemi* *boga*	*Salaki* *cari* *aya* *urusan* *batawi* *ceuk*	*lalaki* *salakina* *tinande* *nasib* *budak* *boga*
2	Lover (companion)	–	*lalaki* *paduduaan* *setan*	–	*sejen* *neangan* *babarengan* *beda* *boga*
3	Partner in extramarital affair (cheater)	–	*sejen* *mimin* *ditaroskeun* *imah* *resep* *salaki*	*Sejen* *salaki* *raresep* *ngarandapan* *ditikah* *ditaroskeun*	*sejen* *nyelewer*

(*Continued*)

Table 4. *(Continued)*.

#	Sense	Collocates			
		Period 1 (58–65)	Period 2 (66–98)	Period 3 (99–03)	Period 4 (04–13)
4	Prostitute (courtesan)	–	*bangor* *ketengan* *sampeuran* *bondon* *beunang* *wts*	*tempat* *sauted* *bangor* *purah* *lalaki* *ngangkat*	–
5	Entity of weakness (not masculine)	*pangawakan*	*pangawak* *lalaki* *ngangsrog* *pangawakan* *tanaga* *ngalawan* *hengker*	–	–

Sense #1 (a wife) and sense #3 (a partner in an extramarital affair) are related to the context of marriage, since they share the same collocates *salaki* (husband) and *salakina* (her husband), as seen in the citations below.

Sense #1   *Sabab eta mah urusan salaki, lain urusan awéwé.*
        Because that is the husband's business, not the wife's.

        Di hareupeun salakina, eta awéwé teh kurulang-kuriling siga peragawati.
        In front of his husband, the wife walks around like a catwalk model.

Sense #3   *Contona, anu hayang tengtrem rumah tangga, anu embung salaki ngabagi asihna jeung awéwé sejen ku Haji Djedje dipalakiahan.*
        For example, anyone who wants peace in her household, who never wants her husband to share his love with another woman, will get treated by Haji Djedje.

        Mun teu nyelewer manggih deui awéwé sejen, tangtu lantaran salakina kabur ninggalkeun bari jeung teu puguh alesanana.
        Other than having an affair with another woman, there is no other clear reason for her husband to leave.

Most of the expressions for *awéwé* in sense #1 show how *awéwé* stands freely as a noun. However, a few other expressions also have a degree of fixedness to the point that they become decomposable, although they remain as a noun.

        Komo ayeuna mah sesebutan "awewe dulangtinande" teh geus teu merenah deui.
        Furthermore, nowadays the term *"awewe dulangtinande"* is not relevant anymore.

The fixed expression of *awéwé dulangtinande* (women are always driven by men) is idiomatic; thus, the metaphors could only be seen when all the words appear together. This is also the case for *awéwé sejen* found in citations of sense #3, albeit having a lower degree of fixedness. The word *sejen* itself means different.

Sense #2 (a lover) is more related to an intimate companionship or date—relationship between an unmarried man and a woman, as shown by the collocate *paduduaan* (together, restricted to two people only) or *babarengan* (together, common)", while *salaki* is absent. Contextually, both collocates are used when describing a date and they do not form a fixed pattern with *awéwé,* which remains a noun. There is an interesting finding showing that *awéwé* in this sense also collocates with *setan* (the devil) as shown in the citation below.

Tingali deui, hadits di luhur; netelakeun yen tiap lalaki paduduaan jeung awewe, pasti setan nu katiluna.
Check again the hadith above; it says that every time a man gets alone with a woman, the third company must be the devil.

This quote contextually refers to one of the teachings in Islam, which is widely embraced by the Sundanese people, as 95% of them are Muslims (Ekadjati, 2010). According to Jami' at-Tirmidhi: 2165 (translated by Abu Khaliyl), whenever a man is secluded with a woman, the devil (and all the temptations it brings) is accompanying them as the third person. This shows that an intimate companionship has always been less preferred than a legitimate marriage among the Sundanese, and sometimes the practice could go as far as allowing an early marriage or even a forced marriage (Marlina, 2006). However, the frequency of the sense's usage in Period 4 increased, while the frequency of sense #1 was the opposite. This could mean the Sundanese people were getting more resistant to marriage and its practices, yet also shows a more welcoming attitude towards intimate companionship as the time goes.

Sense #4 (a prostitute) tells us about a woman performing paid sexual services for customers (usually men). The expressions are mostly fixed and idiomatic with an immediate collocate on the right side of *awéwé* forming noun phrases, such as *awéwé bangor* (*bangor* means naughty), *awéwé ketengan* (*ketengan* means unit price), *awéwé sampeuran* (from the term *sampeur* or come over), *awéwé bondon* (*bondon* means prostitute), *awéwé panggilan* (from *panggil* or call), etc. Some of these phrases are shown on the citations below.

Jeung lain awewe bangor bae tapi loba mahasiswi-mahasiswi nu gampil diajak sare.
And not just prostitutes, so many female students are also easy to sleep with.

Awewe bondon lamun ceuk barudak ngora mah.
(She is) a prostitute, as the children said.

Kawasna moal jauh ti awewe panggilan, malah teu mustahil kelas profesional.
(She) seems like a prostitute, and it is possible she could be a professional.

Sense #5 (an entity of weakness) tells the opposite of women and men, mostly from the physical sense. Most of the expressions in sense #5 do not directly say that women are weak compared to men, but rather place one of the attributes of women together with a certain condition that is normally handled by men or attributed to men. One example of this is shown below.

Najan pangawak awewe Tiah henteu nolak pagawean lalaki.
Despite having a woman's body, Tiah never refuses to do men's work.

The word *pangawak* means body. Physically, women are weaker than men and most of the jobs (*pagawean* means job) done by men are commonly too heavy for women's physics to handle. Other expressions include *tanaga awéwé* (*tanaga* means strength), *mental awéwé* (*mental* means mental), *pangawak awéwé*, and *pangawakan awéwé* which are collocationally fixed as noun phrases with the word *awéwé* as the modifier of the phrases. Another way to express the weakness of women is shown in the citation below.

Tepi ka dina hiji peuting mah kuring bijil aral ngarasula, kuring ceurik nyegruk, kawas awewe.
Until one night, unable to hold myself from lamenting on how my life has been, I burst into tears like a woman.

The phrase *ceurik nyegruk* or 'burst into tears' is commonly attributed to women as it shows emotion. Therefore, crying men—who are commonly attributed to thinking and logic—are considered weak like a woman (*kawas awéwé*). The attribute of weakness to women stems from the fact that women have been positioned as lower beings than men in the Sundanese society since the feudal era (the late 19th century and the early 20th century) in Indonesia, as shown in the written texts of those times, such as *Wawacan Sajarah Galuh* and *Wawacan Carios Munada* (Stuers, 1960; Wiraatmadja, 1980 in Marlina, 2006). Not to mention that as a general trait, men have been always physically stronger than women. As Table 3 suggests, the persistent decrease of usage for sense #5 perhaps shows that the norms of the Sundanese society were moving to a more progressive direction, where men and women could be positioned more equally.

3.2   *Mojang*

A glance at Table 5 shows us that all the metaphorical senses for *mojang* are contextually related to each other, as they all spoke about beauty pageants. The only difference lies on the component of the beauty pageant, which is represented by the metaphorical *mojang*. Period 4 saw a drastic increase of usage of all senses (notably sense #2, the contestant of a beauty pageant), except for sense #1 (beauty pageant), which fell from 11.48% in Period 3 to 5.45% in Period 4. Nevertheless, despite the drop from Period 3 to Period 4, the usage of sense #1 in Period 4 was still much higher compared to its use in earlier periods.

The word *mojang* tends to appear together with the lexical collocates shown in Table 6.

The tendency of *mojang* to appear together with its counterpart *jajaka* (literally means young male) started in later periods, especially on metaphorical sense #1. Most citations have *jajaka* on the right slot from *mojang*, as seen below.

Sense #1   *"Mojang jeung Jajaka Jawa Barat 2001 di TVRI Bandung, Mang".*
           "(We are watching) Mojang and Jajaka Jawa Barat 2001 on TVRI Bandung, Uncle".

Sense #2   *Diantarana wae, Kota Bandung: Novi Kumala S (Mojang) jeung Padika M (Jajaka);*
           "Among them are, Bandung: Novi Kumala S (Mojang) and Padika M (Jajaka);

Sense #3   *Kalawan nangtukeun jeung milih Mojang Jajaka Pinilih...*
           By deciding to choose the 1st winner of Mojang Jajaka...

One citation, however, has *jajaka* appear on the left slot from *mojang*. This happens on a citation for sense #1.

   Nu teu bisa dipopohokeun mah, cenah, basa keur jadi Juara I sakaligus jadi Juara Favorit dina Jajaka/Mojang Parahyangan.
   But she could not forget, she said, when she was crowned both as the 1st winner and the Most Favourite in Jajaka/Mojang Parahyangan.

The collocational patterns for mojang and jajaka are mostly fixed, such as Pasanggiri Mojang Jajaka Jawa Barat and Mojang jeung Jajaka Priangan (sense #1), Mojang Jajaka Pinilih (sense #3).

For sense #2, the majority of citations show that *mojang* stands alone, although still appears together with *jajaka*, as shown in the citation below.

   Kab. Purwakarta: Ria Monika (Mojang) jeung Imam Ashari A (Jajaka);
   Purwakarta Regency: Ria Monika (Mojang) and Imam Ashari A (Jajaka);

Mojang also collocates with names of place, city, or region like Bandung, Parahiyangan, Priangan, Jawa Barat, etc. These collocates appear on the right side of *mojang* in every citation.

Table 5.   The metaphorical senses of *mojang.*

#	Metaphorical senses of *mojang*	Number of citations (and the percentage)			
		Period 1 (58–65)	Period 2 (66–98)	Period 3 (99–03)	Period 4 (04–13)
1	Beauty pageant	0 (0%)	6 (1.14%)	7 (11.48%)	6 (5.45%)
2	Contestant of a beauty pageant	0 (0%)	4 (0.75%)	1 (1.64%)	26 (23.6%)
3	Winner of a beauty pageant	0 (0%)	2 (0.75%)	0 (0%)	5 (4.55%)
Total frequency of *mojang* in the corpus		12	528	61	110

Table 6. The lexical collocates of *mojang*.

#	Sense	Collocates			
		Period 1 (58–65)	Period 2 (66–98)	Period 3 (99–03)	Period 4 (04–13)
1	Beauty pageant	–	*saembara* *Parahyangan* *bade* *alit* *jajaka*	*jajaka* *pasanggiri* *Barat* *Jawa* *Priangan*	*jajaka* *Jawa* *Barat* *pasanggiri* *kagiatan*
2	Contestant of a beauty pageant	–	*nomer* *hayang* *dibuka* *baheula*	–	*jajaka* *kota* *kab* *Tasikmalaya* *Sukabumi*
3	Winner of a beauty pageant	–	*parahyangan*	–	*jajaka* *milih* *wakil* *pinunjul* *barat*

Syntactically, all the obtained phrases by combining *mojang* and its collocates mostly do not undergo word class changes as they remain as noun phrases. Some examples of these noun-phrases are seen below. It should be noted that specific names (*Mojang* and *Jajaka* beginning with capital letters) are not translated to English.

Sense #1	*saembara mojang*	young woman contest
	kontes mojang	young woman contest
	Pasanggiri Mojang Jajaka Jawa Barat	West Java Mojang Jajaka Contest
Sense #2	mojang nomer 91	(young woman) contestant number 91
Sense #3	Mojang Jajaka Pinilih	1st Winner of *Mojang Jajaka*
	Mojang Jajaka Kameumeut	Most Favourite *Mojang Jajaka*

The increased usage of the metaphorical senses of mojang could be attributed to the hype of beauty pageants in Indonesia, especially the national-level pageants, such as Puteri Indonesia and Miss Indonesia. The chosen contestants in such pageants were the winners in lower-level competitions, such as a province-level contest. The usual path for a contestant would be competing in the city level, then competing in the province level, winning on the national level, and finally becoming the best in the world. As beauty pageants could pave one's way to bigger opportunities and fame, young people are enthusiastic to participate from the earliest stage of the competition, not excluding the people in Jawa Barat (West Java), where most Sundanese people reside. It could be said that Pasanggiri Mojang Jajaka, which is annually held in cities and regencies in West Java and in a higher, province level, is one of the starting steps to such opportunities and fame.

Contextually, all the metaphorical senses of *mojang* above are associated with beauty, which indeed serves as one of the key things in beauty pageants. The tendency that *mojang* often collocates with the name of a place (city, province, or region) shows these places take pride in the qualities of the females born in that place. In return, this could also mean that women are proud of the place where they come from. The nature of beauty pageants, not excluding Pasanggiri Mojang Jajaka itself, is "showing off" the finest qualities of its contestants to public. To win a prestigious event like Pasanggiri Mojang Jajaka could mean a higher prestige for both the winner and the place she represents, or will later represent when competing in a higher level.

Ekadjati (1984) stated that in the Sundanese society, women gained their significance after they are coupled with men through marriage, even going as far as the Sundanese proverb that says *kawin ayeuna, isuk pepegatan* (it is better to marry and divorce than never marry at all). Apparently, this issue of significance is unconsciously shown in Pasanggiri Mojang-Jajaka, as the final purpose of the beauty pageant is to find a couple—a young man and woman—as opposed to a single man or a single woman.

4 CONCLUSION

This research shows that the combination of the CMT framework and corpus linguistics method could provide a more complete description of women through the use of metaphors in Sundanese. Looking at it qualitatively, the significance of collocates, the collocational behaviour, and the syntactical behaviour show how words are chosen to deliver specific metaphorical senses in certain contexts. For the case of metaphorical *awéwé*, its senses are used when describing women in the context of marriage (a wife and as a partner in an extramarital affair), companionship (a lover), practices of prostitution (a prostitute), and an entity of weakness. Meanwhile, the senses of metaphorical *mojang* are limited to the context of beauty pageants (event, contestants, and winners).

Quantitatively, by examining the frequency of occurrence, we can see how often these identified metaphorical senses are used in different periods of time. This research reveals that most metaphorical senses of *awéwé* experienced their rises and falls in usage throughout Period 1 to 4. However, this is not the case for the sense of entity of weaknesses as it drastically dropped since Period 2 and never rose even slightly, which might indicate the change of view about the position of women in the present Sundanese society. Another interesting phenomenon is that in Period 4, the usage of the sense of wife decreased, while the usage sense of lover rose, which might show the change of attitude towards marriage (more resisting) and companionship (more welcoming). Meanwhile, nearly all metaphorical senses of *mojang* also experienced their rises and falls in usage through all periods, notably the sense of beauty pageant contestants that shows a very drastic increase. This increase of usage can be attributed to the hype of beauty pageants in Indonesia, especially among the Sundanese people.

REFERENCES

Cheng, W. (2012) *Exploring corpus linguistics: Language in action*. New York: Routledge.
Creswell, J.W. (2014) *Research design: Qualitative, quantitative, and mixed methods approaches*. 4th Edition. New York: Sage Publication.
Deignan, A. (2005) *Metaphor and corpus linguistics*. Amsterdam: John Benjamins Publishing.
Ekadjati, E.S. (1984) *Masyarakat Sunda dan kebudayaannya (The Sundanese and their Culture)*. Jakarta: Girimukti Pusaka.
Ekadjati, E.S. (2010) Islam, agama pilihan utama dan abadi orang Sunda (Islam, the eternal and main religion of choice of the Sundanese). In: M. Hasbullah (Ed.). *Studi Sejarah Islam Sunda (A Study of Sundanese Islamic History)*. Bandung: Fakultas Adab & Humaniora, Universitas Islam Negeri Sunan Gunung Djati.
Koller, V. (2011) Analysing metaphor and gender in discourse. In: F. Manzano (Ed.). *Unité et diversité de la linguistique: Cahiers du Centre d'Etudes Linguistique*. Lyon: Atelier intégré de publication de l'Université Jean Moulin. pp. 125–158.
Lakoff, G. & Johnson, M. (1980) *Metaphors we live by*. Chicago, IL: University of Chicago Press.
Lakoff, G. (1993) The contemporary theory of metaphor. In: A. Ortony (Ed.). *Metaphor and thought*. 2nd edition. Cambridge: Cambridge University Press.
Lakoff, G. & Johnson, M. (1999) *Philosophy in the flesh: The embodied mind and its challenge to western thought*. New York: Basic Books.
Marlina, I. (2006) Kedudukan wanita menak dalam struktur masyarakat Sunda: Studi kasus di kota Bandung (The status of prominent women in the structure of Sundanese society: A case study in the city of Bandung). *Sosio humaniora*, *8*(2), 185–205.

McEnery, T. & Hardie, A. (2012) *Corpus linguistics: method, theory and practice*. Cambridge: Cambridge University Press.

Perkawis Manglé (n.d.) Available at: http://mangle-online.com/pidangan/perelean/1354423703. Accessed on the 29th September 2016.

Sinclair, J. (1991) *Corpus concordances collocation*. Oxford: Oxford University Press.

Sinclair, J. (2004) *Trust the text*. London & New York: Routledge.

Steen, G. (1994) *Understanding metaphor in literature*. London: Longman.

Stuers, C.V. (1960) *The Indonesian woman, struggle and achievement*. Gravenhage: Mouton & Co.

Teubert, W. & Krishnamurty, R. (2007) *Corpus linguistics: Critical concepts in linguistics*. Vol. 1. London & New York: Routledge.

Wikberg, K. (2008) The role of corpus studies in metaphor research. In: N.L. Johannesson & D.C. Minugh (Eds.) *Selected papers from the 2006 and 2007 Stockholm Metaphor Festivals*. Stockholm: Department of English, Stockholm University. pp. 33–48.

Yuliawati, S. (2016) Profil semantis nomina perempuan dalam korpus majalah berbahasa Sunda (Manglè, 1958–2013) (*Semantic profiles of women's noun in the corpus of Sundanese language magazine*, Manglè, 1958–2013). Presented in KIMLI 2016, the 24th–27th August 2016. Denpasar, Bali.

Cultural Dynamics in a Globalized World – Budianta et al. (Eds)
© 2018 Taylor & Francis Group, London, ISBN 978-1-138-62664-5

The collocation and grammatical behaviour of two nouns denoting women in Sundanese: A corpus-based analysis of language and gender relationship

P.G. Bagasworo & T. Suhardijanto
Department of Indonesian Literature, Faculty of Humanities, Universitas Indonesia, Depok, Indonesia
Department of Linguistics, Faculty of Humanities, Universitas Indonesia, Depok, Indonesia

ABSTRACT: For many years, people have been discussing the potential relationships, intersections, and tensions between language structures and gender in different ways. This paper focuses on the collocational and grammatical behaviour of two nouns in Sundanese denoting woman as used in the corpus of *Manglè* magazine. The corpus comprises of texts taken from *Manglè* published between 1958 and 2013. This paper examines collocational and grammatical tendencies between a node and its collocate where the node acts as a subject or an object of a sentence. In this paper, the analysis only focuses on two Sundanese nouns denoting woman and view them as the node of collocational constructions. This study approaches Sundanese corpus data from the Romaine perspective that looked at sexism in language through collocational and grammatical evidence (Romaine, 2000). By taking this approach, this analysis allows us to reveal deeper information about sexism in language through collocation behaviour between a subject or an object and the surrounding words. The result shows that *pamajikan* tends to co-occur with words that are semantically related to family and women's submissive quality, while *wanoja* tends to co-occur with words that are semantically related to independence.

1 INTRODUCTION

The relationship between language structure and gender has been the subject of discussion for a long time, even since the days of the ancient Greek philosophers. Aristoteles, the Protagoras philosopher who coined the terms *masculine, feminine,* and *neuter* to classify nouns, can be regarded as the first philosopher relating language structures with gender (Tannen, 2007). Lakoff (1975) claims that a number of language characteristics are indeed based on gender and some of them are related to the structures of language, such as the uses of intensifiers in an excessive way, hypercorrect grammar, particular imperative words and adjectives, etc.

Research on the relationship between language and gender consists of three areas of study. First, the study of language and gender that focuses on the language variations associated with a particular gender. Second, the study that explores the norms and social conventions producing a gender-based language use. Third, the study that focuses on how gender is constructed and operated in particular ways that are local and context-specific. In relation to these three areas, the approach employed for the current research is closer to the description of the third approach. On the subject of this research is specifically the use of collocation that has an implication in gender.

Corpus-based research has been regarded as one of the characteristics of current linguistic research. Corpus, a collection of texts selected under a certain mechanism, becomes the foundation for linguists to reveal a language phenomenon and to attest language theories. In relation to the use of corpora, there have been numerous studies that examine gender-based

371

language on corpora. Kjellmer (1986) studied the frequency and distribution of masculine and feminine pronouns with the words man/men and woman/women in the 1961 Brown and LOB corpora. Kjellmer found that there was a tendency of more "masculine" items than "feminine" ones in both corpora. However, masculine bias is found more in the North-American Brown corpus than in the LOB corpus.

Sigley and Holmes (2002) also examined the comparative frequencies of man/men and woman/women in the Brown and LOB corpora, the Wellington Corpus of Written New Zealand English, the Freiburg-Brown Corpus of American English, and the Freiburg-LOB Corpus of British English. They found that the frequency of women in writing doubled between 1960s and the early 1990s. However, the frequency of references to women as individuals remained smaller than the references to men as individuals. It seems that adult males were more frequently referred to singularly, while adult women were more commonly referred to collectively.

Vigliocco & Franck (1999) investigated whether the system of language production, more specifically in French and Italian, used conceptual information about biological gender in the encoding of gender grammatical agreement between a subject and predicate. Both French and Italian have a gender category system that includes a distinction between the nouns reflecting the sex of the referent (conceptual gender) and the nouns for which gender does not reflect the sex of the referent (grammatical gender). Their findings implied that syntactical features that reflect conceptual features are retrieved from different sentence elements. This information ensures the accuracy of the sentence encoding and makes it more efficient.

Meanwhile, in line with Kjellmer (1996) and Sigley and Holmes (2002), Pearce (2002) studied the representation of men and women in the BNC by focusing on the collocational and grammatical behaviours of the noun lemmas WOMAN and MAN (including the nouns woman/women and man/men). In his study, Pearce explored the functional distribution of the target lemmas and revealed the structured and systematic nature of the differences in the way the words referred to adult males and females were patterned with other words in different grammatical relations. The collocational evidence of the lemma WOMAN shows that it is related to "sociological" discourses, i.e. women were presented as objects of sociological enquiry within a discourse that acknowledged their subordinations and tried to redress it in that context. According to Pearce, the finding was not surprising due to the construction of BNC that came from the collection of texts mainly taken between 1975 and 1994, known as the 'second wave' of feminism—the period of extensive discussion in academic circles about gender and the nature of women's oppression.

Moreover, according to Pearce, there were at least three observable patterns that had limited distribution to particular text domains. First, men were mainly portrayed to be aberrant in news reporting, and the distribution of the sequence adjective of "deviancy" + MAN (such as *dangerous, armed,* and *convicted*) seemed to confirm this. Second, the adjectives of "neuroticism" in the pattern of "neuroticism" + WOMAN (such as *distraught, hysterical,* and *silly*) occurred in prose fiction texts. Third, the social class markers (such as working-class and middle-class) premodified WOMAN over nineteen times more frequently per million words in the social science genre than they do in the overall BNC. It indicates that the "sociological discourse" was likely associated with this domain of genre.

The current study is rather similar to the study by Pearce (2002) in the case of examining words referring to women in Sundanese that occurred and are used in particular collocational and grammatical patterns in the corpus of Sundanese magazine *Manglè*, and the texts were collected between 1958 and 2013. In this study, the relationship between language structure and gender is revealed through the patterns of collocational and grammatical behaviour of two Sundanese words: *wanoja* and *pamajikan*. The rationale in determining the two words as the focus of the analysis is based on the study by Yuliawati (2016) about the frequencies of Sundanese nouns denoting women found in *Manglè* magazine. The two words were found to have an inverse frequency distribution from time to time (see Figure 1). For this reason, it will be interesting to analyse these two words further based on the corpus.

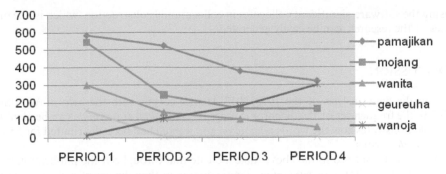

Figure 1.   The frequency of WOMAN in word/million (cited from Yuliawati, 2016).

The current research examines not only the grammatical behaviour, particularly in relation to the subject and predicate of a sentence, but also the collocational behaviour of the words *wanoja* and *pamajikan* in order to reveal the way a woman is constructed in Sundanese society. The collocational patterns may unveil the associations and connotations of words, and subsequently the assumptions they embody (Stubbs, 1996). Furthermore, Romaine (2000) shows how sexism in language could be demonstrated with collocational behaviour. According to her, in English there were several pairings of gender bias displaying various kinds of semantic and discursive asymmetry, e.g. *master* and *mistress*, *god* and *goddess*, *wizard* and *witch*, and *bachelor* and *spinster*. Besides, Romaine found in her study that the gender terms for females were associated negatively more frequently than the gender terms for males. Although the study does not include gender terms for males, the present research aims at revealing the associations and connotations embedded in the words *pamajikan* and *wanoja* by investigating their collocational and grammatical behaviours.

2   METHODOLOGY

To examine the collocational and grammatical behaviour of the words *pamajikan* and *wanoja*, the corpus linguistic approach is employed in this research. Referring to Jones and Waller (2015), a corpus provides information for researchers about the frequency of grammatical patterns in particular contexts. By using a corpus, information concerning which pattern is more frequently used may be obtained. A corpus, however, cannot demonstrate the reasons why such pattern is frequently used.

To reveal particular grammatical patterns in a corpus, several statistical test techniques are applied. They are frequency test, n-gram, and collocation test (Lindquist, 2011). However, according to Jones and Waller (2015), the statistical tests can only be applied to identify grammatical patterns. To examine further on why and how certain grammatical patterns are used, concordance lines displaying the use of a word or a group of words that we investigate can be employed.

The corpus used for the research is constructed from the collections of texts taken from *Manglè* published between 1958 and 2013. The corpus is divided into four periods: Period 1 (the Era of Guided Democracy, 1958–1965), Period 2 (the New Order, 1966–1998). Period 3 (the Transition Era, 1999–2003), and Period 4 (the Reform Era, 2004–2013). The total of the corpus size is 2,940,537 token. The number of words for each period is 78,081 tokens in the corpus of Period 1; 1,897,777 tokens in the corpus of Period 2; 324,614 tokens in the corpus of Period 3; and 641,065 in the corpus of Period 4.

The research uses corpus software, namely WordSmith Tools 7.0, to do a corpus analysis. The software helps corpus researchers to manage and explore the corpora they have. Some features in the WordSmith Tools 7.0 can provide word list, key word list, and concordance.

By using the software, significant collocates in collocational construction, which is one of the focuses of the research, can be easily generated.

3 DISCUSSION

Based on the data analysis of the 1958–2013 *Manglè* corpus with the help of WordSmith Tools 7.0, the frequencies of the diachronic usage of *pamajikan* and *wanoja* can be identified. Table 1 demonstrates how the words *pamajikan* and *wanoja* had been distributed for 55 years in the *Manglè* corpus.

The numbers displayed in the above table indicate that throughout the publication period from 1958 to 2013, the use of *pamajikan* is more dominant than the use of *wanoja*. In spite of that, in Period 2 (1966–1998), a significant change occurs, and it is indicated by the increased use of *wanoja* by more than 60% per period. The use of *wanoja* increases from Period 2 to Period 3 (61.2%) and from Period 3 to Period 4 (68.15%). Meanwhile, the use of pamajikan is decreasing by 10% on average, with the highest decrease from Period 2 to Period 3, which is 27.7% (see Table 1).

From the comparison of ten collocates co-occurring with the words *pamajikan* and *wanoja*, there are several interesting things to discuss as we examine Table 2. First, the significant collocates of the word *pamajikan* are *kuring* 'I/my', *anak* 'child', *boga* 'possess', *salaki* 'husband', and *imah* 'house'. The occurrences of *anak*, *salaki*, and *imah* as the significant collocates of the word *pamajikan* demonstrate that the word is semantically related to domestic sphere.

Furthermore, from a close examination of the most frequent collocate of the node *pamajikan*, which is *kuring*, it is found that the frequency of the collocate *kuring* is 15.09% of the total occurrences of *pamajikan*, and all collocates in the *Manglè* corpus. From the distribution, the collocation of *pamajikan* and *kuring* occurs 65.93% out of the total 91 texts in the corpus. The collocate *kuring* equally occurs either before or after the node *pamajikan* with the detailed percentage of 8.02% (on the left) and 7.07% (on the right). The interesting thing from the occurrences on the right as generated from WordSmith is that more than half of the occurrences of *kuring* are at the position of immediately following the node *pamajikan*. Thus, in other words, *pamajikan* occurs quite frequently in the phrasal combination of *pamajikan kuring* 'my wife'.

Based on the analysis shown in Table 2, another interesting collocate to study further is the verb *boga*. It can be seen that the verb occurs more frequently on the left (6.85%) than on the right (1.17%) of the node *pamajikan*. Then, if it is studied further, the analysis shows that among its occurrences, which are on the left of the node, the word *boga* occurs more than 75% immediately on the left of the node *pamajikan*. Thus, the words form the clausal construction of *boga pamajikan* 'possessed/possesses a wife.'

From the analysis of the words *kuring* and *boga* that co-occur with *pamajikan* in the corpus, there is a tendency that Sundanese women, in their role as wives, remain subordinate to men who become their spouse. In this relationship, the husband is the 'possessor', while the wife is the 'possessed'.

Among the ten significant words co-occurring with the word *wanoja*, there are three highest collocates that are interesting to discuss further: *kaum* 'group', *boga* 'possess', and *milu* 'follow'. The collocate *kaum* 'group' always co-occurs before the node *wanoja*. It indicates that the word *wanoja* is frequently used in a collective meaning. Meanwhile, the collocate

Table 1. The frequency of WOMAN in word/million.

Nouns	Period 1	Period 2	Period 3	Period 4	Total
pamajikan	584	524	379	321	1372 (74.25%)
wanoja	13	111	179	301	476 (25.75%)
Corpus size	78,081 words	1,897,777 words	324,614 words	641,065 words	2,941,537 words

Table 2. The collocates of *Pamajikan* and *Wanoja* in the *Manglè* corpus.

| | Pamajikan | | | | | Wanoja | | | |
Word	Text	Total	Total Left	Total Right	Word	Text	Total	Total Left	Total Right
KURING	65.93%	15.09%	8.02.%	7.07%	*KAUM*	13.92%	6.93%	6.93%	0.00%
ANAK	62.64%	9.11%	6.56%	2.55%	*BOGA*	10.13%	2.73%	0.47%	2.26%
BOGA	57.14%	8.02%	6.85%	1.17%	*MILU*	8.86%	1.47%	0.21%	1.26%
SALAKI	46.15%	5.61%	3.94%	1.68%	*RESEP*	8.86%	1.68%	0.42%	1.26%
IMAH	26.37%	2.41%	1.75%	0.66%	*KURING*	7.59%	1.26%	0.84%	0.42%
MILU	13.19%	0.95%	0.29%	0.66%	*PINTER*	3.80%	0.63%	0.00%	0.63%
LALAKI	9.89%	1.17%	0.66%	0.51%	*GEULIS*	3.80%	0.63%	0.21%	0.42%
RESEP	4.40%	0.29%	0.07%	0.22%	*ANAK*	2.53%	0.42%	0.21%	0.21%
PINTER	0.00	0.00	0.00	0.00	*IMAH*	2.53%	0.42%	0.21%	0.21%
KAUM	0.00	0.00	0.00	0.00	*SALAKI*	0.00	0.00	0.00	0.00

boga dominantly occurs on the left or in the position after the node *wanoja*, which indicates that *wanoja* is used in the contexts where Sundanese women are regarded to have the rights to possess or to have a will or wish. This implies that *wanoja* is used to denote Sundanese women who are regarded as powerful, and it is semantically positive.

Comparing the significant collocates between the node *pamajikan* and *wanoja* is another interesting thing to discuss. There are no collocates *pinter* and *kaum* that co-occur with the node *pamajikan*, as found in the node *wanoja*. It indicates that at least in the *Manglè* corpus, *pamajikan* has never been associated with both concepts. On the contrary, there is no collocate *salaki* that co-occurs with the word *wanoja*, as it is found in the node *pamajikan*. Nevertheless, the word *anak* is still found to be one of the significant collocates of the node *wanoja* (see Table 2). Therefore, *wanoja* is still associated with *anak*, even though it is not related with *salaki*.

The next discussion on the use of *pamajikan* and *wanoja* is in terms of grammatical behaviour from the perspectives of subject-object construction and phraseology. Table 3 displays information about words that co-occur with the nodes *pamajikan* and *wanoja*, side by side. In the table, the inclusion of subject and object is done by listing all the words within the span position 5:5 of the node *pamajikan*. From the list obtained, only the words that are in the position of one to the left and one to the right of the node are used.

In Table 3, the result shows that the occurrence of *pamajikan* is more frequent as the object of a sentence than the subject. There are several interesting things to discuss about some of the words shown in the table. One of them is the position of co-occurring words toward the node *pamajikan* from the perspective of sentence structures. For instance, the word *dibawa* that co-occurs after the node *pamajikan* is a passive form. Consequently, although in that construction the word *pamajikan* is the subject of the sentence, the collocate is semantically the patient or the participant that undergoes the actions, for example, "*Tapi ari kapangum-baraan mah pamajikan the papada teu dibawa*" ("*However, living outside the country, their wives were taken along*"). It demonstrates that semantic roles are important to consider when analysing the subject and object of a sentence.

Another interesting thing presented in Table 3 is the word *boga*, which in its every single occurrence makes *pamajikan* the object of the sentence. It is important to notice because it indicates, as explained previously, that *pamajikan* is often positioned as the 'possessed'. For example, in the sentence of *Dina Lain Eta*: "*Mahmud kacatur boga pamajikan, ari Neng Eha pepegatan jeung salakina, lantaran diteruskeun oge percumah matak baeja dimasiat lahir-batin.*" (In 'Not That Thing', it is told that Mahmud possessed a wife, while Mrs. Eha was divorced from her husband because continuing their marriage would be useless, yet bringing more physical and spiritual sins.")

In Table 3, the word *wanoja*, as examined from its function as the subject and object of a sentence, is less frequent than the word *pamajikan*. Besides, based on its co-occurrence

Table 3. The collocational behaviour of *Pamajikan* and *Wanoja as* the subject and the object of the sentence.

Pamajikan			Wanoja		
Word	Subject	Object	Word	Subject	Object
Balik	0	3	*bisa*	4	4
Bisa	0	12	*boga*	6	0
Boga	0	89	*milu*	5	0
Cek	30	3	*resep*	4	1
Ceuk	33	7			
Cul	0	5			
datang	0	6			
Hirup	0	5			
Inget	0	7			
Nyaho	0	5			
Ret	0	5			
tembal	5	1			
Make	2	1			
Balik	6	0			
Diuk	5	0			
Indit	4	1			
Cek	0	7			
hariweusweus	6	0			
Ceuk	0	6			
bangun	5	1			
dibawa	6	0			
ambek	5	1			
Bisa	10	0			
ceurik	11	0			
Milu	6	1			
Nanya	6	1			
ngajuru	9	0			
ngomong	4	2			
norojol	6	0			
Nyaho	4	0			
nyampeurkeun	5	0			
Nyarita	4	2			

with its co-occurring words as shown in the table, the word *wanoja* tends to function more frequently as the subject of the sentence. With the word *boga*, for instance, *wanoja* tends to function as the subject. In this case, *wanoja* is portrayed in the sentence as a woman who has willingness or desire.

In this current paper, the grammatical behaviour of the phraseological aspect is observed in the contexts when the words *pamajikan* and *wanoja* become an adjectval modifier and head in a phrase construction. As the adjectval modifier, *pamajikan* and *wanoja* modify their co-occurring words. The list of phrases in which *pamajikan* and *wanoja* are the head or modifier of a phrase can be seen in Table 4.

Some phrases in which *pamajikan* is the modifier, as shown in the table, can be classified into several groups of semantically related phrases. For example, the group of phrases relating to family are *anak pamajikan* (children and a wife), *salaki pamajikan* (husband and wife), *lanceuk pamajikan* (the sister/brother of a wife), *sodara pamajikan* (the relatives of a wife), *bapana pamajikan* (the father of a wife), and *kulawarga pamajikan* (the family of a wife). Then, there is a group of phrases relating to the possession of an object or quality, such as *sora pamajikan* (the voice of a wife), *imah pamajikan* (the house of a wife), *hate pamajikan* (the heart of a wife), *sieun pamajikan* (afraid of a wife), *harta benda pamajikan* (the property

Table 4. The phraseology of *Pamajikan* and *Wanoja*.

Pamajikan		Wanoja	
Adjectival modifier	Head of Phrase	Adjectival modifier	Head of phrase
sora pamajikan	pamajikan geulis		
hate pamajikan	pamajikan sieun	kaom wanoja	wanoja Sunda
sodara pamajikan	pamajikan babakti	kaum wanoja	wanoja ngarawat kulit
bapana pamajikan	pamajikan kolot	kuningan boga tokoh wanoja	wanoja Indonesia
beuteung pamajikan	pamajikan solat	tokoh wanoja	wanoja Bandung
duit ke pamajikan	pamajikan ceurik	pangarang wanoja	wanoja Kuningan
salempang ka pamajikan	pamajikan ka randa	organisasi wanoja	wanoja legislatip
imah pamajikan	pamajikan nitah	sora wanoja	wanoja meunang
nyiksa pamajikan	pamajikan ngora	caleg wanoja	wanoja Arab
sieun pamajikan	pamajikan nyigeung	kegeulisan wanoja	wanoja Bali
kulawarga pamajikan	pamajikan nyentak	majalah wanoja	wanoja desa
nitah pamajikan	pamajikan jebi	pahlawan wanoja	wanoja pasundan
lanceuk pamajikan	pamajikan muncereng	pemberdayaan wanoja	wanoja Aceh
anak pamajikan	pamajikan kabur	cinta ti wanoja	wanoja nu dilukis
salaki pamajikan	pamajikan ambek-ambekan	hak kaum wanoja	wanoja Jerman
harta benda pamajikan	pamajikan nuturkeun	mutiara wanoja	wanoja pamingpin
	pamajikan akang	pakean wanoja	wanoja nu ti keur SMA
	pamajikan satia	percaya deui ka wanoja	wanoja nu wani
	pamajikan keuheul	wakil wanoja	
	pamajikan panasaraneun		
	pamajikan sabar		

of a wife), *nitah pamajikan* (ordering a wife), and *nyiksa pamajikan* (torturing a wife). In the phrases showing the possession of *pamajikan*, some imply that *pamajikan* is represented as someone who is inferior, i.e. *nitah pamajikan* and *nyiksa pamajikan*, but there is also a phrase showing the superiority of *pamajikan*, i.e. *sieun pamajikan*. The three phrases are associated with the wife's spouse, so it can be said that the superior and the inferior in this case are *salaki* (husband).

Then, with regard to the use of *pamajikan* as the head of a phrase, the construction of the phrases tends to relate women semantically to their habit. The classification of phrases is also made here, and some of them indicate the behaviour and quality commonly possessed by a *pamajikan*, such as *pamajikan satia* (a loyal wife), *pamajikan babakti* (a dedicated wife), *pamajikan ambek-ambekan* (an angry wife), *pamajikan sieun* (a wife who is scared), *pamajikan kolot* (an old wife), *pamajikan nitah* (a wife ordered), *pamajikan nyentak* (a wife snapped), *pamajikan panasaraneun* (a wife was curious), and *pamajikan sabar* (a patient wife). In this case, the behaviour and the quality of *pamajikan* that tend to follow the husband might be related to the culture that prevailed in Sundanese society that adopts a patriarchy system. Nevertheless, there are some verbs showing *pamajikan* as the superior subject. For example, *pamajikan nitah* and *pamajikan nyentak*. This, however, does not occur for the spouse of *pamajikan*, which is *salaki*. This finding is supported by the recent statistic data from the National Commission on Violence against Women. It is stated that in 2015, there were 11,207 cases of violence against women on domestic level, and 60% of them (or 6,725 cases) were violence against wives (National Commission on Violence against Women, 2016). The number can finally represent a similar condition in the data corpus being studied.

Based on its position as an adjectival modifier as presented in the above table, *wanoja* and its co-occurring words can be categorised into several groups of semantically related phrases. The first group consists of phrases which are closely related with political terms, such as *kaom wanoja* (a group of women), *kaum wanoja* (a group of women), *tokoh wanoja* (female figures), *caleg wanoja* (female legislative candidates), *pahlawan wanoja* (a female hero), *organisasi wanoja* (the organisation of women), *pemberdayaan wanoja* (women's empowerment), *hak kaum wanoja* (women's rights), and *wakil wanoja* (women's representatives). It represents women who in recent decades have participated in the Indonesian government or parliament.

Considering the available corpus only covers data until 2013, we include the data on the representatives of women derived from the paper on the Policy of Women's Participation in Politics and Government, the United Development Programme, the House of Representatives of the Republic of Indonesia (DPR RI), and the Regional Representative Council of the Republic of Indonesia (DPD RI). The participation was observed from two general elections, i.e. the 2004–2009 election and the 2009–2014 election (UNDP, 2010). During these two periods of administration, women representation increased in two parliament chambers of Indonesia. The figure of women representatives rose from 22.58% to 26.25% in DPD RI, while in DPR RI, the figure jumped from 11.80% to 18.04%. Although incrementally the rise may not be largely significant, it can be used as a benchmark for a country that is still struggling for equal rights with regard to its policy makers.

We also have some additional data relating to the two parliament chambers in a smaller scope, i.e. women representatives from West Java. However, the data available is only for the 2009 general election. There were 22 women representatives or 24.18% seats filled by women in DPR RI, and 25% seats filled by women in DPD RI. The interesting fact is that the 22 women representatives have made West Java the province with the highest number of female delegates. Therefore, this could be the reason why there are many phrases where the adjectival modifier *wanoja* is closely associated to politics.

In the same table, there is also a group of phrases characterising *wanoja* as the representation of womanhood that is related to appearance and beauty. The group of phrases, for instance, is *sora wanoja* (women's voice), *majalah wanoja* (female magazines), *cinta ti wanoja* (love from women), *mutiara wanoja* (pearl of women), and *pakean wanoja* (women's clothes). If the five phrases are investigated further based on their contexts in sentences, it can be seen that women are frequently mentioned in relation to appearance and beauty. Thus, the two large groups of phrases with *wanoja* as adjectival modifiers demonstrate the tendency of using *wanoja* in the political sphere and in the topics related to appearance and beauty.

Regarding the occurrence of *wanoja* as the head of a phrase, the word tends to modify the origin or the identity of women. The group of phrases found, for example, are *wanoja Sunda* (Sundanese women), *wanoja Indonesia* (Indonesian women), *wanoja Bandung* (women from Bandung), *wanoja Kuningan* (women from Kuningan), *wanoja Arab* (Arabian women), *wanoja Bali* (Balinese women), *wanoja desa* (women from villages), *wanoja Pasundan* (women from Pasundan), *wanoja Aceh* (women from Aceh), *wanoja Jerman* (German women), and *wanoja nu tikeur SMA* (female students of a senior high school). The phrases demonstrate that the word *wanoja* in the corpus is associated with places of origin.

The last group of phrases discusses women in relation to superiority. The phrases are *wanoja pamingpin* (a female leader) and *wanoja nu wani* (a brave woman). Compared to *pamajikan* in the same column, the general character of *wanoja* is quite different. The difference is caused by its co-occurring words that show the tough and independent qualities of women.

4 CONCLUSION

The current research is about the collocational behaviour of two Sundanese words referring to women, namely *pamajikan* and *wanoja*. In Table 4, the collocational behaviour shows that *pamajikan* tends to co-occur with words that are semantically related to family and women's submissive quality. Additionally, the words that co-occur with *pamajikan* tend to involve only a few people in the process of communication. On the contrary, *wanoja* tends to co-occur with words that are semantically related to independence. The word *wanoja* is used to talk about women who show strong personality of free expression and toughness, and are associated with their places of origin, so the place where a *wanoja* is from or her identity can be seen.

From the sentence structures, compared to the word *wanoja*, the word *pamajikan* is more frequently used as the object of the sentence. Moreover, if it co-occurs with the verb *boga* (possess), *pamajikan* is always found to be the object of the sentence. However, the same thing does not occur with the word *wanoja*.

REFERENCES

Attenborough, F. (2014) Words, contexts, politics. *Gender and language special issue: Gender, language, and the media. 8*(2), 137–146.

Bari, F. & Venny, A. (2010) Perempuan di ranah politik (Women in politics). In: UNDP. *Partisipasi Perempuan dalam Politik dan Pemerintahan (Women's participation in politics and government)*. Jakarta: UNDP. pp. 4–6.

Cheng, W. (2012) *Exploring corpus linguistic: Language in action*. New York: Routledge.

Danadibrata, R.A. (2009) *Kamus basa Sunda (Dictionary of Sundanese language)*. Bandung: PT Kiblat Buku Utama.

Jones, C. & Waller, D. (2015) *Corpus linguistics for grammar: A guide for research*. New York: Routledge.

Komisi Nasional Anti Kekerasan terhadap Perempuan (National Commission on Violence against Women). (2016) *Lembar Fakta Catatan Tahunan 2016 (Annual Fact Report Sheets 2016)* [Online] Available from: http://www.komnasperempuan.go.id/lembar-fakta-catatan-tahunan-catahu-2016-7-maret-2016/. Accessed: 22nd September 2016.

Lakoff, R. (1973) Language and woman's place. *Language in Society, 2*(1), 45–80.

Panitia Kamus Lembaga Basa dan Sastra Sunda (The Committee on Sundanese Language and Literature). (1980) *Kamus umum bahasa Sunda (General dictionary of Sundanese language)*. Bandung: Tarate Bandung.

Pearce, M. (2008) Investigating the collocational behaviour of man and woman in the BNC using Sketch Engine 1. *Corpora, 3*(1), 1–29.

Satjadibrata, R. (2010) *Kamus bahasa Sunda (Dictionary of Sundanese language)*. Bandung: PT Kiblat Buku Utama.

Sigley, R. & Holmes, J. (2002) Looking at girls in corpora of English. *Journal of English Linguistics, 30*(2), 138–57.

Sinclair, J. (1991) *Corpus concordances collocation*. Oxford: Oxford University press.

Speer, S.A. (2005) *Introduction: feminism, discourse, and conversation analysis*. In: S.A. Speer. *Gender talk: Feminism, discourse and conversation analysis*. London & New York: Routledge. pp. 7–8.

Stubbs, M. (2002) *Words and phrases: Corpus studies of lexical semantics*. Oxford: Blackwell Publishing.

Tannen, D. (2006) Language and culture. In: R.W. Fasold & J. Connor-Linton. *An introduction to language and linguistics*. Cambridge: Cambridge University Press.

Tannen, D. (2007) *You just don't understand: Women and men in conversation*. New York: William Morrow Paperbacks.

UNDP. (2010) *Partisipasi perempuan dalam politik dan pemerintah (Women's participation in politics and government)*. Jakarta: UNDP Indonesia.

Vigliocco, G., & Franck, J. (1999) When sex and syntax go hand in hand: Gender agreement in language production. *Journal of Memory and Language, 40*(4), 455–478.

Yuliawati, S. (2016) *Profil semantis nomina perempuan dalam korpus majalah berbahasa Sunda* (Manglè, 1958–2013) (*Semantic profiles of women's noun in the corpus of Sundanese language magazine, Manglè*, 1958–2013). Paper presented in KIMLI 2016, Denpasar, Bali, 24th–27th August 2016.

The role of women in Javanese literature: A case study of *Serat Gandakusuma* SJ 194 NR 324 and SW 5 SB 47

W.P. Sudarmadji & A. Prasetiyo
Department of Area Studies, Faculty of Humanities, Universitas Indonesia, Depok, Indonesia

ABSTRACT: This study is a modified version of a dissertation. It analyzes the text entitled *Serat Gandakusuma*. Here, the focus of the analysis is the study of two texts of Gandaku-suma produced in two different scriptoria. The first one, which was edited philologically, is a text produced in the *keraton* (palace) peripheral. The second one is a text that was pro-duced in the coastal peripheral. We will analyze how the two texts that come from different peripherals present the roles of female characters with regard to the type of culture in each peripheral based on how each text formulates the story, which strongly involved the main female character, especially to support the main male character for achieving the true dignity.

1   INTRODUCTION

The development of literature in Java can be traced back at least to the 9th century A.D. Records indicate that, during that era, an ancient Javanese literary work entitled *Kakawin Râmâyana* was composed in 820–832 Çaka (Poerbatjaraka 1952, p. 2–5). Since then, the world of Javanese literature has continued to undergo development and innovation. The growth of Javanese literature cannot be separated from multifarious and intertwining social and cultural factors and phenomena. One of the most important factors in the development of the Javanese literature was the religious transition from Hinduism–Buddhism to Islam among the Javanese people. While some original features of the old faiths were retained, the arrival of Islam had inevitably introduced new features to the Javanese literature, especially when the new faith had taken deep root among the local populace.

The influence of Islam on the Javanese literature is clearly demonstrated in a literary work entitled *Menak*. The *Menak* tale is actually a Javanese adaptation of *Hikayat Amir Hamzah* (the Epic of Amir Hamzah) in Malay tradition (Poerbatjaraka 1952, p. 109). Pigeaud argued that the *Menak* tale has been known in Java since at least the 17th century. Since that time, it has continued to develop. The development was such that it eventually lost much of its original trace and became an independent and self-contained Javanese–Islamic romance (Pigeaud 1967, pp. 212–213).

In addition to the popularity of *Hikayat Amir Hamzah*, which led to its adaptation in the form of the *Menak* cycle, there was a great interest among the Javanese people in romances originated from the Arabic–Persian tradition. Those romances were then adapted by Java-nese poets or authors based on Malay manuscripts. With their strong Islamic tinge, these Javanese romances remained very popular in the northern coast of Java until the 19th cen-tury. They are *Joharmanik, Jaka Nastapa, Prantaka, Jatikusuma*, and *Jatiswara* (Pigeaud 1967, pp. 219–222).

2   SERAT GANDAKUSUMA

The text analyzed in this study is *Serat Gandakusuma*. This work displays the characteristic of a Javanese romance with an Islamic nuance. The data consist of two versions of the text, which, taken together, form a complete corpus for this research. The first text is kept in the

Manuscript Room of Universitas Indonesia Library with the code number SJ 194 NR 324, while the second text is kept in Sonobudoyo Museum Library Yogyakarta with the code number SW 5 SB 47.

Our full corpus of *Serat Gandakusuma* consists of 10 manuscripts, which are kept separately in Sonobudoyo Museum Library Yogyakarta and Universitas Indonesia Library's Manuscript Room. Text selection was performed based on the descriptions of *Serat Gandakusuma* manuscripts produced by us. However, some parts of the texts cannot be examined because they do not fulfill two criteria: completeness and availability in either of the two places where the manuscripts are stored—the former was the most important inclusion criterion in this work. Some sections of the texts have even been completely lost, so they had to be excluded from our analysis.

From the metrical system of *tembang macapat* (Javanese lyric poem), in which both versions of the text were composed, and the writing style of a number of words and the prolog of the text, it is clear that the two versions of *Serat Gandakusuma* are the products of two different scriptoria. The first one is the coastal scriptorium, which is marked by its characteristic diction, distinctive language style, and more metrical flexibility. The second one is the palace scriptorium, which is indicated by more metrical rigidity, better spelling consistency, more profuse and poetical word choice, and formal language.

The general plotline of these two texts is about the journey of Raden Gandakusuma, the crown prince of Bandaralim. One night, he was abducted and then killed by his half-brother, Prabu Jaka or Menak Tekiyur. His corpse was then flung into the sea. The presence of the corpse in the ocean generated a chaos in the nature. This condition forced the goddess Sarirasa, the ruler of Sirrullah in the depth of the ocean, to rise to the surface to see what had triggered the turmoil. Upon seeing a light emanating from Gandakusuma's corpse, the goddess Sarirasa approached it and revived it using her magical power. She then led Gandakusuma into a long and perilous journey, which concluded with a great reunion of the royal family of Bandaralim and Gandakusuma's coronation as a king, accompanied by Sarirasa as his queen. From the formulation of the tale, *Serat Gandakusuma* can be categorized as Javanese Sufism literature, which typically features a protagonist who has to undergo a long journey to achieve salvation.

It is interesting to examine *Serat Gandakusuma* because it emphasizes the role of a female character as both the medium and the goal of the protagonist, Raden Gandakusuma, in completing his spiritual journey. Indeed, women often play a prominent role in the Javanese Sufism literature. The main focus of this work is how the two texts, which were produced in two different scriptoria, reflect and position the role of its main female character, the goddess Sarirasa. We first show the difference between the dissertation as the original source of this work and this study. This study discusses the role of women as represented by the main female character shown through the formulation of plots, parts of dialogs, and the authors' point of view in both texts. It does not include in-depth discussion on the Sufi concept, which is symbolized through the elements in the structure of the story and the comparison between them.

3   AN OVERVIEW OF SERAT GANDAKUSUMA

This section outlines the general plotline of the epic as produced by two different scriptoria and discusses several important constituting elements of the epic. The version of *Serat Gandakusuma* with the code number SJ 194 NR 324 will be referred to in this study as "text A". This set of manuscripts contains 38 *pupuh* or chapters written in *tembang macapat* style. The version of *Serat Gandakusuma* with the code number SW 5 SB 47 will be referred to in this study as "text B", and it consists of 41 *pupuh* or chapters.

Both texts share the same general plotline, which consists of:

- The introduction of the characters
  Both texts begin with elaborating Gandakusuma's identity as the son of the king of Bandaralim.

- The abduction of Gandakusuma

  Both texts share this part of the tale in which Gandakusuma was abducted from the palace of Bandaralim. The abductor was one of Gandakusuma's own brothers, Prabu Jaka, also known as Menak Tekiyur, assisted by another character named Darba Moha (in text A) or Dremba Moha (in text B).
- The assassination of Gandakusuma

  After being abducted, Gandakusuma was assassinated by the same character, Prabu Jaka or Menak Tekiyur. His body was then thrown into the sea.
- The appearance of the goddess Sarirasa

  In both texts, Gandakusuma's death is described to trigger an extraordinary turbulence in the surrounding nature. This great turmoil was immediately felt by a supernatural being named the goddess Sarirasa, who lived at the bottom of the ocean, in a cave named Sirrulah. She then left her abode to search for the cause of the turmoil.
- The resurrection of Gandakusuma

  The goddess Sarirasa then found and revived Gandakusuma. Gandakusuma fell in love with her, but she did not accept his love directly.
- The deliverance of the goddess Sariraga and the marriage of Gandakusuma and the goddess Sariraga

  After a long conversation with the goddess Sarirasa, Gandakusuma agreed to help her release the goddess Sariraga who was captured by a king named Prabu Dasaboja or Prabu Dasabahu. After she was released, Gandakusuma was then married to the Goddess and then crowned as a king in Kakbahbudiman or Gabahbudiman, which was the home of the goddess Sariraga.
- The seizure of the magical heirlooms of Baginda Amir

  In general, these heirlooms would assist Gandakusuma in defeating his enemies, the heathen kings.
- The marriage of Gandakusuma and the goddess Sarirasa

  After seizing the entire set of magical heirlooms, Gandakusuma married the goddess Sarirasa. However, the goddess gave Gandakusuma one condition for marrying her: he had to be able to answer a riddle that she would give. The riddle was answered correctly, and the marriage took place.
- The defeat of Prabu Jaka

  At the end of the tale, Prabu Jaka or Menak Tekiyur was finally defeated. Bandaralim was reclaimed, and Gandakusuma rightfully assumed his throne.

4 THE MAIN FEMALE CHARACTER AND HER ROLE IN GANDAKUSUMA'S TALE

It can be deduced from the above plotline overview that even though Gandakusuma serves as the protagonist, there are many parts of the story which demonstrate that he is very much dependent on a female character named the goddess Sarirasa. First, Sarirasa revived Gandakusuma. Second, Sarirasa assisted Gandakusuma in releasing the goddess Sariraga from Prabu Dasaboja's grip. Third, she planned the strategy for waking Kanekaputra from his meditation. She also introduced the idea to recapture the magical heirlooms of Baginda Amir. Finally, during the final battle, she helped Gandakusuma to defeat his enemies.

Both texts evidently suggest that the goddess Sarirasa played a very important role in the development of the story. Nevertheless, due to the fact that the texts were produced in two different scriptoria, they are expected to display some differences in terms of character formulation and description of the goddess Sarirasa. Below are some of the differences.

4.1 *The identity of the character*

Both texts have a different way of describing who the goddess Sarirasa was. This is shown in the two passages below, translated from the original text.

Text A:

(Chapter 5: 2)

It was said that the princess was very difficult to describe. According to some people, she was actually the son of the Prophet Kilir. There were also some who said that she was a daughter who became the heart of water. Nobody knew except the Holiest One, who knew everything and possessed the characteristics of *sama'* and *bashar*.

Text B:

(Chapter VII: 1–4)

It was told that she was the one who sat on the throne, the adorable gem of the ocean, famous all over the world, and she was truly the woman whose beauty was incomparable. She was neither a genie nor a heavenly angel. She did not belong to the human race, nor did she become a descendant of the giants. She was in actuality the heart of water (the ocean).

She was the noblest essence (heart/mind). She was revered like angels. She united her soul at the depth of the sea, reigning over the ocean. Anything she wished, it could come true. She was highly skillful in both hard and fine tasks, and she could also adjust herself to both the young and the elderly. The blessing of a noble life remained with her. A life which was full of the purest perceptions.

Such perception was conceived in the real picture of Sang Sukma Maha Luhur (the Most Glorious Soul) in the disguise of the Princess Sarirasa, the essence of the earth. The beautiful light glowed faintly. The rainbow shone brightly, and the stars moved about. Sparkling when (one) beheld (it). Light fell like a full moon. Her origin was the same as that of the One who was the highest in the universe.

4.2   The resurrection of Gandakusuma

Both texts also give different accounts regarding how Gandakusuma was revived. Text B is somewhat more dramatic because it describes the process in more detail. Text A gives a briefer description.

Text A:

(Chapter 5: 13 and 15)

"God is never mistaken about the people who perform good things by giving help. They will certainly receive kindness. If they do evil, they will be punished. That is His sure promise. If this is so, it is better for me to give him) help. O, handsome body, I ask you to retreat to the seashore."

The goddess Sarirasa immediately took the medicine that was put in a container in the form of a small bottle made of diamond, which she put in her hair bun. She quickly opened the bottle and raised it over the wounds on the body three times.

Text B:

(Chapter IX: 4, 12–13)

The princess took the holy water. She then dropped some of the water over Raden Gandakusuma's head. All his wounds were healed in an instant.

Like a flower, the princess then lifted up her heart and said solemnly, "Gandakusuma, live, by His permission!" The prince did not wake. The princess said again, "O, Tambangraga, wake! Wake, by the word of mine; wake, by the word of the Highest One."

Still, the crown prince did not recover. The princess said again, "Live, o, pilgrim. Live, by my word, by my will."

4.3   The riddle of the goddess Sarirasa

As mentioned in the plotline overview above, the goddess Sarirasa gave Gandakusuma a riddle, which he had to answer correctly as the condition for marrying her. Surprisingly, both texts feature two different riddles, as shown by these two passages below.

Text A:

(Chapter 14: 24 and 26)

"During the full moon, the weaving loom is broken in one thrust. After the departure of the eastern star, where can it be found?" The King's heart was suddenly filled with joy upon hearing the princess's words. The king stood up immediately. He quickly approached the princess.

"Later, when we are in our bed, (o) my sweet one, I shall tell you." Then, His Majesty the King summoned the celebrant, preacher, and muezzin, no one was left uncalled. Prabu Anom at that time submitted to his celebrant. Thus, he was married to the goddess Sarirasa.

Text B:

(Chapter XXV: 31–33)

"… answer this riddle correctly: is the Sirullah cave at the bottom of the serene ocean?" The king answered solemnly,

"It is called the vast ocean, it is crystal clear and brightly brilliant. It is the palace of the grandest perception. Meanwhile, you take abode in the truest perception, the spirit of the life of the whole universe, it is the manifestation of the essential spirit, the king above all who live, just like Gandakusuma as the place where perceptions gather.

The Sirullah cave is the place of seclusion of the Living One, who lives without being vivified. There live you, I, and Sariraga. Therefore, all three must never separate. Humans and God. The earliest perception is marked by a true human being.

Below is an outline of the appearances and roles of the main characters in the epic.

1. The death of Gandakusuma

In both texts, the goddess Sarirasa made her first appearance after Gandakusuma was murdered. In order to prepare for the goddess' first appearance, both texts create a leading cause in the form of fierce natural phenomena triggered by Gandakusuma's death.

2. The resurrection of Gandakusuma

The goddess Sarirasa felt that the ocean was in a terrible turmoil, so she rose to the sea surface to find out the source of the trouble. After witnessing that the turbulence was being caused by Gandakusuma's corpse, Sarirasa moved to revive the prince. At this point, there are several subplots, which are relevant to our discussion, as follows:

– Both texts agree that the goddess Sarirasa had known the cause of Gandakusuma's death and felt pity on him.
– Text B recounts that, before the goddess Sarirasa revived Gandakusuma, she had been involved in a battle with a Chinese princess named the goddess Karsinah. The battle between those two female characters ended with Sarirasa's victory.
– The media that Sarirasa used to revive Gandakusuma was medicine, which was carried in a small diamond bottle and put in her hair bun (text A) or holy water (in text B). The water was poured on the injured parts of Gandakusuma's body (text A), while the medicine was slowly poured over his head (text B).
– After pouring the water/medicine, two different things happened. In text A, Gandakusuma immediately recovered and lived again. Meanwhile, in text B, Gandakusuma did not wake until the goddess Sarirasa said, "Live, o, pilgrim. Live, by my word, by my will".

3. The deliverance of the goddess Sariraga

In the next episode, the goddess Sarirasa gave a plan and directions to Gandakusuma on how to live his second life. The first mission was to release the goddess Sariraga who was captured by Prabu Dasaboja in the land of Kandhabuwana. The most important subplots are as follows.

– The goddess Sarirasa tested the faithfulness of Gandakusuma.
– The goddess Sarirasa gave Gandakusuma a condition for being united with her.
– Prabu Dasaboja was conquered and converted into Islam. This involved securing Prabu Dasaboja's commitment to being the foremost and most loyal ally of Gandakusuma.

In addition, this deliverance mission became very important because it would serve as a means by which Gandakusuma could assume the throne of Gabahbudiman, the home of the goddess Sariraga.

4. Supporting Gandakusuma's claim as a king

In this episode, the goddess Sarirasa assisted Gandakusuma in completing all the requirements that he would need to support his claim as a king. Those requirements are as follows:

- a skillful and powerful chancellor or prime minister;
- sacred heirlooms, which could augment the king's dignity;
- a queen (the goddess Sarirasa herself).

5. Adept commander in the battlefield

Finally, the goddess Sarirasa proved to be a very adept commander, as well as a brilliant warrior with supernatural powers. She was able to defeat the heathen kings. Of all heathen kings who had been killed in the battle, some were revived by her to be converted into Islam and also had to swear loyalty to Gandakusuma as his new allies.

5   THE GODDESS SARIRASA IN THE "COASTAL" TEXT AND THE "PALACE" TEXT

It seems that, based on the above textual comparison, the goddess Sarirasa played a significant role in enabling Gandakusuma to achieve his ultimate triumph at the end of the tale. Even though both texts exhibit a similar general plotline, there are still several differences in the way both texts depict the goddess' role. In text A, which was composed in the coastal scriptorium, the author and/or scribe tended to exercise more carefulness in describing the identity of the goddess Sarirasa. It seems that the author described only what is necessary and put more emphasis on the goddess' physical appearance.

On the contrary, text B, which was composed in the palace scriptorium, tends to be more dramatic in nature and explicitly represents the goddess Sarirasa as "the real picture of *Sang Sukma Maha Luhur* (the Most Glorious Soul)". With regard to the resurrection of Gandakusuma, text B clearly vests more power and strength in the goddess Sarirasa, which enabled her to revive Gandakusuma by the mere power of her words. This cannot be found in text A. In text A, Gandakusuma was revived by means of medicine in a container, which the goddess placed in her hair bun.

In spite of some minor differences, both texts also give the same general account of the division of role and power between Gandakusuma and the goddess Sarirasa after Gandakusuma became both the king and the husband of the goddess Sarirasa. As a general rule, ever since Gandakusuma was able to solve Sarirasa's riddle, Sarirasa's dominance over Gandakusuma's life and future plans, both as an individual and as a leader, had been gradually fading. Sarirasa no longer served as Gandakusuma's strategy and decision maker. However, she continued to be his companion as his queen, as well as his confidante, by employing all of her magical powers, especially her healing and resurrection powers.

6   CONCLUSION

Literature is an inseparable part of a society that has produced it. The two versions of Gandakusuma tales are the products of a society that flourished in a particular place and time. These spatial and temporal factors have inevitably exerted a great influence on how female characters, together with their roles and functions, are represented in the tale.

Text A, as the product of the coastal scriptorium, tends to show a higher degree of explicitness as well as carefulness when it comes to the depiction of the goddess Sarirasa as a female character. On the contrary, text B, as the product of the palace scriptorium, tends to further explore Sarirasa's character in terms of physical and non-physical qualities. In text A, Sarirasa is depicted as a woman with supernatural powers, but a constant effort is also made to avoid giving an impression that Gandakusuma, as the main character, was inferior to her. In text B, Sarirasa is depicted as having all kindness and strengths. Sarirasa's identity is established from the very beginning of her appearance and maintained until the end of the tale.

Our analysis of both texts demonstrates that the ancient Javanese culture, as reflected in *Serat Gandakusuma*, gives an equal treatment to both male and female in terms of room and power. Nevertheless, in spite of her powers, Sarirasa did not exert her influence by force. She is still depicted as an ideal Javanese woman; she might wield a great power and influence, but she also had to display all the qualities required of a typical Javanese woman: gentleness, patience, and calmness.

REFERENCES

Behrend, T.E. (1990) *Katalog Induk Naskah-Naskah Nusantara Jilid I Museum Sonobudoyo Yogyakarta (Main Catalogue of Manuscripts from the Indonesian Archipelago Volume I of Museum Sonobudoyo Yogyakarta)*. Jakarta, Djambatan.

Behrend, T.E. & Pudjiastuti, T. (1997) *Katalog Induk Naskah-Naskah Nusantara Jilid 3-A Fakultas Sastra Universitas Indonesia (Main Catalogue of Manuscripts from the Indonesian Archipelago Volume 3-A of the Faculty of Letters of Universitas Indonesia)*. Jakarta, Yayasan Obor Indonesia.

Damono, S.D. (1978) *Sosiologi Sastra Sebuah Pengantar Ringkas (The Sociology of Literature, A Brief Introduction)*. Jakarta, Pusat Pembinaan dan Pengembangan Bahasa Departemen Pendidikan dan Kebudayaan.

Endraswara, S. (2013) *Teori Kritik Sastra: Prinsip, Falsafah, dan Penerapan (The Theories of Literary Critiques: Principles, Philosophy, and Application)*. Yogyakarta, Caps.

Hamid, I. (1989) *Kesusastraan Indonesia Lama Bercorak Islam (Old Indonesian Literature with Islamic Nuances)*. Jakarta, Pustaka Al-Husna.

Handayani, C.S. & Novianto, A. (2004) *Kuasa Wanita Jawa (The Power of Javanese Women)*. Yogyakarta, LkiS.

Pigeaud, Th. G. Th. (1967) *Literature of Java Catalogue Raisonné of Javanese Manuscripts in the Library of the University of Leiden and Other Public Collections in the Netherlands Volume I Synopsis of Javanese Literature 900-1900 AD*. The Hague, Martinus Nijhoff.

Poerbatjaraka, R.M.Ng. (1952) *Kapustakan Djawi (Javanese Literary Bibliography)*. Djakarta, Djambatan.

Poerwadarminta, W.J.S. (1939) *Baoesastra Djawa (Javanese Dictionary)*. Batavia, J.B. Wolters'.

Prawiroatmodjo, S. (1981) *Bausastra Jawa-Indonesia (Javanese-Indonesian Dictionary)*. Jakarta, PT. Gunung Agung.

Cultural Dynamics in a Globalized World – Budianta et al. (Eds)
© 2018 Taylor & Francis Group, London, ISBN 978-1-138-62664-5

Posuo, space and women: Buton community's customary tradition and its preservation

I. Ibrahim & M. Budiman
Department of Literature, Faculty of Humanities, Universitas Indonesia, Depok, Indonesia

ABSTRACT: *Posuo* is a specific ritual intended for Butonese women, spanning from the menstrual period to the pre-wedding period. The purpose is to mark a woman's maturity as a sign that she is ready to enter a new phase of life in marriage. The advent of modernity and globalisation has brought about a change in *posuo*. Originally, it signified the transition of women's status from adolescence to maturity, but nowadays it has drifted away from *adat* (customs). In this study, we analyse the impact of such a change by first describing the structure of *posuo* and then looking at how today's community in Buton respond to it. Ethnographic approach is used to obtain as much data as possible from various resources. The next step is to answer how *posuo* survives despite the change, causing the change and the ways in which the Butonese community give meaning to such a change. There are three reasons for the existence of *posuo* until today: (1) as part of the local traditional customs, it is not considered to be against Islamic teachings, (2) it serves as one of the requirements for women who are about to enter marriage life and (3) and it has become an inseparable element of cultural identity.

1 INTRODUCTION

The Buton Island area covers the expanse of islands of Southeast Sulawesi and some parts of the mainland peninsula of Sulawesi Island. Buton was formerly a sultanate and the centre of the development of Islam in Southeast Sulawesi. The island is included in the administrative region of Southeast Sulawesi province.

In the 19th century, the government of the Butonese sultanate implemented Islamic sharia. *Tasawwuf* teachings developed by Sultan Muhammad Idrus Kaimuddin (Zahari 1977, 28) had a broad impact on various aspects of community life, including the terms of cultural and art practices. All forms of traditions and arts that were not in accordance with the teachings of Islam were 'cleaned' from the palace. In the end, a number of traditions and arts that are still maintained by the community can only be carried out outside the court. Some examples that can be mentioned here include the *maataa* in Cia-Cia or *pakande-kandea* in Baruta. In addition, *kabanti* as a literary performance was considered contrary to the moral ethics of the Sufis, and it was banned around the palace (Asrif, 2015, p. 2). Unlike *kabanti*, even though *posuo* tradition has the influence of Hindu in its practice, it is accepted in the palace territory. A different treatment that is given to *posuo*, in contrast to other traditions and art forms, is one of the reasons this research is conducted.

Posuo is a ritual performed specifically for women in Buton, marked by the first menstruation until maidenhood. The ritual is meant to mark the maturity of female members of the society, marking their entrance to a new phase of life and their readiness to foster their own household. The people of Buton believe that 'a female cannot be considered mature, despite signs of physical maturity, before she is inaugurated in *suo*'.

Etymologically, the word *posuo* originated from *po*, which means 'do', and '*suo*', which refers to a room located at the back of a house. It refers to the division of space in a traditional house in Buton, which is divided into three rooms: *bamba* (front), *tanga* (middle) and *suo* (rear). Thus, *posuo* can be defined as the activities carried out in a room behind a house. The rear room occupied by the participants is the room where the whole ritual takes place. Only females (the participants and *bisa*) are allowed to enter. *Bisa* is an older woman who has adequate knowledge about the customs and who acts as a mentor for the participants. Under the customary terms, *posuo* should be performed for eight days and eight nights, but the current circumstances have led to varying durations.

Even though Butonese people consider *posuo* as part of the Islamic tradition, in reality, pre-Islamic traditions are still apparent. Their influences can be seen in the series of the structure of implementation, conduct and myth incorporated in it. For example, there is a belief that if the content of the equipment bowl of the participant falls and spills, then their faces will be blackened or it will cause rainstorm in the village. Another belief is that, during the process of beating the drums, if one of the drums cracks, it is believed that one of the participants is no longer a virgin.

The strict customary rules of *posuo* have received various responses from younger generations, especially the female Butonese. Some of them considered that the ritual is no longer relevant, while others think that the ritual is still necessary. The two different perspectives reveal that the *posuo* tradition of Buton is facing a serious problem as, on the one hand, it can strengthen its tradition and, on the other hand, it threatens its own sustainability. On the basis of the issues discussed above, here we explain the changes that have happened to the *posuo* ritual and their implications to the culture of Buton. We aim to describe the causes and reasons for the changes, including revealing the reasons for the existence of *posuo* tradition until today.

2 METHOD

The research data were attained through the ethnographic method, by conducting in-depth interviews with the subjects and objects of the tradition. Furthermore, information was also received from 'secondary set of eyes', which refers to the help of female partners, due to the customary ban for males to be in the *suo* area.

3 ANALYSIS

Posuo is performed in three steps: *pauncura*, *baliana impo* and *matana karia*. The three steps are the main activities that the participants need to attend. Before the main activities are performed, however, there are two introductory activities called *maludu* and *malona tangia*. For further explanation, the steps for implementing *posuo* are described in Table 1.

Table 1 is based on *posuo* that is organised during the fieldwork in four different locations: at Mr Jafar's house, in the area of Buton palace, Melai Village and Baubau City. Mr Jafar held the *posuo* ritual for his daughter, together with his wife's relatives. The ritual was held for four days and four nights, from 31 August to 4 September 2012. The second *posuo* was organised in Laompo Village, Batauga Subdistrict, South Buton Regency, which was held in the family home of Lakina Burukene. The *posuo* was attended by 28 participants from the Burukene family, from Bautaga and family members from Papua who came for the ritual. Similar to the one held in Melai, *posuo* in Laompo was held for four days, from 1 August to 4 August 2014. Third, moose was held in Lipu, Betoambari Village, Baubau City, in the house of Parabelana Lipu. The ritual was attended by 12 participants, and it was held for 4 days, from 31 July to 3 August 2014. Fourth, *pahora'a* was held in the house of Mrs Wa Rosi in Takimpo, Pasar Wajo Sub-district, Buton Regency, which was held from 19 to 21 August 2013, as part of the Buton traditional party to celebrate Sail Indonesia 2013, attended by 12 participants.

Table 2 describes the changes in the *posuo* ritual.

According to Table 2, there are approximately five aspects that seem to experience significant changes in the implementation of the *posuo* ritual, namely the objective, time, location,

Table 1. Steps for the implementation of *Posuo* in Buton.

Name of the activity	Time	Activity
Maludhu	16.00	A group of women sing *barjanzi* verses with tambourines. The activity is held inside the *suo* in the late afternoon after *Ashar* prayer, with a break for *Maghrib* prayer, and it continues until an announcement is made that the participants can enter the *suo* room
Malona tangia	18.30	Invited guests are the family and friends who arrive and sit in the yard of the house. Guests, both men and women, gather for the opening night of *posuo*. Women who are close to the family or members of the family are instructed to go inside the house and sit in the living room. The family provides dinner called *haroa*, which is served in a *talang* and covered in *tudung saji* made from woven palm trees
	19.00	Participants of *posuo* who are assisted by *bisa*, their parents and an escorting girl as well as a functionary walk together going around the village towards the house where *posuo* is organised; they are welcomed by the guests. They climb the house stairs, the *bisa* steps up first, followed by the parents and the companions
	19.30	*Parika* (the leader of *bisa*) reads the names of *posuo* participants and explains the requirements that need to be fulfilled by the participants inside the *suo*. This activity is known as *malona tangia*, a sign that *posuo* is about to start. While crying, the participants kick their feet up to the wall, an activity called *ranca*. The louder the cry, the harder the kick to the wall. *Ranca* can only be done by the participants from the *walaka* class, while the participants' cries from the *kaomu* class are accompanied with drums and a gong
	20.00	Dinner is prepared by the family. After dinner, *malona tangia* is considered done, and the participants will sleep in the *suo* accompanied by a *bisa*
	04.00	During *shubuh* (early morning prayer), two (or more) boys are assigned the task of bringing water from the specified river. They take two bottles to be filled with water. The first bottle is called *uwe malape* (good water) and the other one is called *uwe madaki* (bad water). This task has to be done in secret, as if the boys were stealing, because their action should not be seen by other people
Pauncura	07.00	The *bisa* conducts *tuturangi*, which refers to the preparation of the water that is going to be used for bathing the participants. In front of the *bisa* are four bottles of water, and each participant will use two kinds of water as explained above.
		This activity starts with preparing *dupa* and *kemenyan* (incense and frankincense). The *bisa* tries to blow the bucket from the incense sticks, while another person opens the water bottle and gives it to a *parika*. Parika is assigned to read the *batata*, a kind of prayer, by putting her mouth closely to the mouth of the bottle, as if she were talking to the water. The *parika's* hand holds the bottle with four fingers, while her index finger points to something. Then, the *parika* blows the opened water bottle before closing it back again and replacing it with the next bottle. This continues until the fourth bottle
		After spells are cast onto all the bottles of the *posuo* participants, *parika* takes the frankincense and spreads it to the incense. The smoke of the frankincense covers the room, and one by one the bottled water is put inside the smoke and held by the four *bisas*. The bottles are moved in the count of seven times to the left and eight times to the right

(Continued)

Table 1.   *(Continued)*.

Name of the activity	Time	Activity
	07.30	The process of bathing the *posuo* participants (*pebaho*) is carried out after the bath water is ready. The participants enter the bathroom with the *bisa*. The participants squat facing east, and then the *parika* showers the water to the head of the participants, while the other *bisa* wash the participants' hair in the count of seven times to the left and eight times to the right. The participants change their seating position from facing the east to the rights, and then following the previous step, the *parika* showers the water and the *bisa* washes the hair.
		The sarong worn by the participants is taken off by tautening, and the replacement sarong is prepared. *Bisa* must cast a spell on it before it can be worn. The sarong worn for the bath should never be worn ever again according to the local belief. After taking a bath, the participants enter the *suo* with *bisa* to continue the process to *tuturangiana kiwalu*
	08.00	After the process of *panima* is completed, the *bisa* gives a mouthful of food, such as rice and a slice of egg to the participants. This activity is called *posipo* (feeding). *Posipo* is given only once to the participants, as they can continue on their own
	09.00	The *bisa* takes turmeric scrub that has been ground to be used by the participants. This activity is called *pomantomu*. The *bisa* only shows the participants how to use the turmeric scrub on the left and right arms once, and then the participants have to continue on their own, using the scrub on the face, neck, torso and legs. The scrubbing process goes on for up to three days before the *baliana impo* stage. After *pomantomu*, the *bisa* gets out of the *suo* room, and the participants remain in the room and they are not allowed to communicate with other people outside the *suo*, unless particularly when the *bisa* needs it
Baliana impo	06.00	The fourth day is known as *baliana impo* or a position change. The participants who originally sleep with their heads facing the east change the direction to the west. On the fourth day, the participants undergo a procession similar to the first day, except for *pebaho* and *pasipo*. Next, scrubbing, which originally uses turmeric scrub, is now done with cold powder made from rice called *pobura bhae*
		In the *suo*, they receive guidance from the *bisa* regarding manners, culture and customs, and they are asked to do *dhikr* and to always recite *istighfar*
Matana Karia	16.00	In the late afternoon before sunset, the participants go through the process of *pebaho* (shower) similar to the first day. Then, the participants put on their make-up and get dressed in traditional clothes called *kombo*. One by one each participant goes out of the *suo* room guided by the *bisa*, and then they sit in front of the guests. Then, the guests who are invited by the wife of *moji* or *muezzin* of the grand mosque of Buton palace, from the *walaka* class, symbolically make the process of trampling of the ground to each participant; this is called *palandakiana tana*. The soil is later stored in a container and then rubbed on the soles of the participants of *posuo*. This procession lasts until the evening, and the event in the evening is called *matana karia* or the main party

******Maludhu* and *malona tangia* are not considered as one full-day activity. Days are counted according to the three steps of *posuo*, which are *pauncura* (sitting down/benediction) on the first day, *baliana impo* (changing sleeping position) during the ritual (for the fourth day in an eight-day ritual and second day for a four-day ritual) and *matana karia* (the main event) as a closing. The description of each step is given in Table 1.

Table 2. Changes in *Posuo* tradition.

Posuo	Then	Now	Reason
Objective	Inauguration of a female in Buton, from *kabua-buatokalambe*	It is now a requirement for a woman who is going to marry. Although a girl has shown signs of physical maturity or reached a certain age to do *posuo*, if she does not have plans to get married, *posuo* is not urgent	There is a perception that time and age are not the main issue, because the focus is not on the timing of the ritual but the actual implementation of *posuo*
Time	Customary rules require *posuo* to be held for eight days and eight nights	Now, *posuo* is organised for four days and four nights, and sometimes it is held only for two days and two nights, or even one night	The community in Buton believe that as a custom, *posuo* still needs to be held. However, for efficiency, participant's conditions and other technical reasons, the length of the ritual needs to be adjusted. Some parts of the ritual, from *pauncura* to *matano karia*, are still implemented according to customs. The length of time, however, is shortened
Location	The word *suo* refers to the back part of a house in the structure of a traditional house called *banua tada* (house on stilts). *Posuo* must be done in that part of the house	Nowadays, *posuo* is often held in the front part of the house. This shift, however, does not change the name of *posuo* into *po-bamba* (the front part of the house) or *po-tanga* (the middle part of the house)	Because the house structure has changed into a house made out of concrete (although some still have their houses on stilts), *posuo* is held at the front part of the house. The same arrangement is also done in the palace with a house-of-stilts form, although generally the back of the house has become a kitchen made out of concrete, so *posuo* needs to be done in a room at the front part of the house
Participant	The participant has to be a teenage girl who has just had her first menstruation	The participant can be teenage girls to unmarried women	Cost and time adjustment
Bisa	In the Buton/Wolio palace, a woman has to be an old and widowed to be a *bisa*. In other places, such as Lipu, Takimpo and Laompo, to be a *bisa*, a woman must still be married, and she is usually the wife of the *modji* or the imam of the mosque	In the Buton/Wolio palace, a woman has to be old and widowed to be a *bisa*. In other places, such as Lipu, Takimpo and Laompo, to be a *bisa*, a woman must still be married, and she is usually the wife of the *modji* or the imam of the mosque	The different status of *bisa* is based on the customary agreement in each area

(Continued)

Table 2. (Continued).

Posuo	Then	Now	Reason
Steps	*Pauncura* *Malona tangia* *Baliana impo* *Matana karia*	*Pauncura* *Malona tangia* *Baliana impo* *Matana karia*	As for the steps, there has not been any change because in general, each step is considered as a fixed rule, the order of which has been arranged according to customary rules
Social Status	To make clear of social stratification during the *posuo* ritual, it is regulated that during the ritual, the *kaomu* class should be accompanied with drums, while those from the *walaka* class should not be accompanied with drums	In Lipu, those who are not from the *kaomu* class could still be accompanied with drum music during moose or the *posuo* ritual	The practitioners of *adat* in the *karie* area believe that the use of drums during the *posuo* ritual should not be an exclusive right for the *kaomu* class This understanding is a free interpretation made by the *adat* functionaries in the *kadie* area

participants and social status. Meanwhile, the aspect of *bisa* and the implementation steps do not change. In terms of the reasons for the change in each aspect, it seems that the dominant influence that changes the implementation of the *posuo* ritual is public perception. Besides perception, economic factors affect the implementation of the *posuo*.

Perception here refers to the changing perspective on *posuo*, which is now seen as an obligation. This is why the time chosen is usually before the wedding day. Because *posuo* is usually arranged three or four days before the wedding, it affects the structure of the ritual. The part that is reduced is actually the essential part of the ritual because during the mentoring nights, the *bisa* will provide advice to the participants about the patterns of act and behaviour directly related to the objectives of the ritual.

4 CONCLUSION

IAs mentioned in the Introduction, the Butonese people come from a community with a very strong maritime tradition. A question has been arising as to whether the sea is the main dimension of the Butonese people's life as represented in *posuo*. When observing the process of implementation and the use of cultural material, no idioms could be found related to the sea. How can a community with strong cultural roots in a maritime tradition ignore the presence of the sea in its cultural practice? This raises the question as to whether the 'sea' also takes part in *posuo* in this context.

Concerning the status and role of Butonese women, according to Schoorl (2003, 213), although they are influenced by the social class system, they never see the importance of understanding the differences between men and women. This is illustrated by the position of the empress or the queen who was essential in Sultan's decision-making process. According to oral sources, during the period of the sultanate, the selection of a new sultan depended on the eligibility of his wife to play the role of an empress. This is one record about the candidacy of an empress who had to be educated and to master mysticism or *ilmuu*. If these requirements were not met, a potential sultan was unlikely to be selected. There is a belief that Butonese women are matured in the *suo* to acquire good and firm knowledge in order to have a role and a position like an empress. However, in the present circumstances, it can be concluded that the value to be invested during the ritual process of *posuo* is devitalised because the objective emphasises more on the accessorial aspect than on the content. Thus, the short duration of *posuo* weakens its function because it is no longer present as an educational function intended to prepare the maturity of a Butonese woman.

REFERENCES

Anceaux, J.C. (1987) *Wolio dictionary (Wolio—English—Indonesia)*. Leiden, Foris Publication Holland.
Asrif. (2015) Kesusastraan Buton Abad XIX: Kontestasi Sastra Lisan dan Sastra Tulis, Budaya, dan Agama (19th Century Butonese Literature: Contestation of Oral and Written Literature, Culture, and Religion). Paper presented at *The Annual Meeting of the Butonese Cultural, Buton*.
Malihu, L. (2008) Buton dan tradisi maritim: Kajian sejarah tentang pelayaran tradisional dissss Buton Timur (1957–1995) (Buton and maritime tradition: Historical study on traditional sailing in East Buton (1957–1995)). [Thesis], Depok, FIB Universitas Indonesia.
Schoorl, P. (2003) *Masyarakat, Sejarah, dan Budaya Buton (Butonese Community, History, and Culture)*. Jakarta, Penerbit Djambatan.
Southon, M. (1995) *The Naval of the Perahu: Meaning, and the Values in the Maritime Trading Economy of a Butonese Village*. Canberra, ANU Press.
Zahari, A.M. (1977) *Sejarah dan Adat Fiy Darul Butun (History and Customs of Fiy Darul Butun)*. 1st, 2nd, and 3rd editions. Jakarta, Departemen Pendidikan dan Kebudayaan.
Zuhdi, S. (2010) *Sejarah Buton yang Terabaikan: Labu Rope Labu Wana (The Abandoned History of Buton: Labu Rope Labu Wana)*. Jakarta, Rajawali Press.

Cultural Dynamics in a Globalized World – Budianta et al. (Eds)
© 2018 Taylor & Francis Group, London, ISBN 978-1-138-62664-5

Cosmopolitan female Muslim travellers in *Berjalan di Atas Cahaya*

S. Hodijah & C.T. Suprihatin
Department of Literature, Faculty of Humanities, Universitas Indonesia, Depok, Indonesia

ABSTRACT: Travel writing is a medium to report about the world, reveal the traveller's self and represent other cultures. Problems related to a conflict of values, either cultural or religious, could be seen in the works of female Muslim authors. *Berjalan di Atas Cahaya* is a compilation of travel stories written by Hanum Rais, Tutie Amaliah and Wardatul Ula in Europe. They share their thoughts, experiences and feelings during their trips, as well as the various stereotypes of Islam centred on the position of women in both the public sphere and the global world. The awareness of female Muslim travellers as cosmopolitan characters is based on female perspectives. In this article, we examine the discourse of cosmopolitan female Muslims aimed at describing how female Muslim travellers find and address the conflict of values, cultures and religions and explaining the construction of Islamic cosmopolitanism by prominent female Muslim travellers. Furthermore, we argue that such a construction is carried out by characters, points of view and the tones of the writings examined.

1 INTRODUCTION

Travelling has been one of the many aspects that play an important role in the establishment of the Islamic civilisation, from the spread of the religion by the Prophet Muhammad S.A.W to the recent development of Islamic people. In *Popular Dictionary of Islamic Terms* (2012:117), travel is related to *hijrah*, which is 'the event in which the Prophet Muhammad S.A.W travelled from Mecca to Medina to safeguard himself from the oppression of the Quraysh tribe, a move to leave a sinful place or sinful deeds to a better place or deeds in order to devote himself to Allah'. On the basis of this definition, in Islam, travel is of two types: physical and spiritual.

Since the independence of Indonesia, travel, as a means of connecting people to the global world, can be seen through the works of Hamka, Usmar Ismail and Bahrum Rangkuti in writing their experiences of travelling to Muslim countries, such as Egypt and Pakistan (Salim, 2011:82). Muslim novelists, such as A. Fuadi, Agustinus Wibowo and Habiburrahman El Shirazy, have also described the tradition of travelling and writing the travel experience from the viewpoint of men.

In the context of travel, female Muslims have not been much exposed and discussed. However, in fact, Burhanudin and Fathurahman (2004:113) explained that since the mid-1990s, the discourse of Islam and women in Indonesia has become more open and an interesting topic to discuss, and even a Malaysian Muslim activist, Zainah Anwar, has considered Indonesia as the future standard regarding the issue of women and Islam because of its largest Muslim population.

Mernissi (1994:174) shared the same opinion regarding the fact that the future lies in the eradication of limitations through communication and dialogues to create a global prospect, in which all cultures can be unique. *Berjalan di Atas Cahaya* (*Walking above the Light*) described the travel of female Muslims in Europe and the impressions they had about the cultural differences. The stories are interesting as they were written by female Muslims showing their independence as well as stereotype problems related to veil, public–domestic role and social manner in Islam. Long been functioned as the foundation of civilisation through their roles as both wives and mothers, female Muslims are currently more involved in functioning leaders, as their viewpoints are listened to by the world, which responds too.

This study focuses on the difference between the cultural and religious values of female Muslims and the idea of cosmopolitan female Muslims built through the development of cosmopolitan zones. To analyse cross-culture interactions, we use the relationship between the cosmopolitan concept and the principles in travel writing. Thompson (2011) defined travel writing as the record and findings between one's self and new things or people through space travel as well as negotiation between similarities and differences. The principles of travel writing are reporting about the world, revealing the self and representing others.

Fussell and Thompson (2011:15) explained that travel writing emphasises personal factors and retrospective subjectivity, as it comes as a report on the experience of the author as a traveller, in order for the readers to experience his responses, impressions, thoughts and feelings. In addition, Thompson (2011:72) explained that the subjective description emphasises the discussion about the self and his/her emotional responses and opinion. In relation to the description, the analysis model proposed by Mills in *Discourse of Difference* (1991) analyses the subjects and objects from the perspective of the person who delivers and receives the information. In travel writing, the narrator, through the subject 'I', becomes important as the first person as well as the story teller of the world, places and people he/she met during the travel so that the readers depend solely on the narrator to obtain information.

Travellers always have the go-international idea. Fine (2007:4) explained that cosmopolitanism emphasises the universal idea and the rights of the people worldwide. The increase in the cosmopolitan quality in identity was analysed by Rantanen (2005:124) through the following five zones: media and communication, learning another language, living or working abroad or having family members living abroad, living with a person from a different culture and engaging with foreigners in your locality or across a frontier.

2 METHOD

By using the qualitative method and the feminist approach, the cosmopolitan ideas are analysed through the following positional and intrinsic components: characterisation, point of view and tone. During the analysis of the data, the special expressions of the characters' ideas and dialogues are then linked with Islamic perspectives and the development of the cosmopolitan zone. In this respect, this study tries to show the narrator's decision in revealing the roles of female Muslims and the formation of Islamic cosmopolitan values through the alternative discourses between the collision and the formation of cultural and religious values.

3 RESULTS AND DISCUSSION

Problems regarding the position of newcomers and settlers, discrimination and culture in a society can be observed only through conducting research on literary works with cross-cultural interaction. Kurnia (2000) described the traditional image of Indonesia in four German novels in the 20th century. Meanwhile, the discussion on Islamic cosmopolitanism was analysed by Budiman (2013) regarding the female main character in the novel *Geni Jora*. Finally, Mashlihatin (2015) analysed the novel *99 Cahaya di Langit Eropa* from the postcolonial perspective.

So far, research on travelling records has focused on male characters and male authors. In contrast to the studies conducted by Budiman and Mashlihatin, this study focused on *Berjalan di Atas Cahaya*. By reviewing the definition of the travel and the formation of the idea of Islamic cosmopolitanism, the research on the experiences and direct observations of three female Muslims is focused more on the problems and influence of value collision.

Berjalan di Atas Cahaya consists of stories written by the following three authors: Hanum Salsabiela Rais, Tutie Amaliah and Wardatul Ula. The similarity in the identity of the authors as female Muslims and the spatial difference influence the factors explored. Hanum, for example, provided a detailed description on the value collision between Eastern and Western families:

> I myself have not got the courage as those in Europe to let go of my children when they are 17 or 18... People coming from Asia and those from Europe are like two groups coming from two complete opposite poles on the issue of raising children. That is where religion takes place.... (p. 40).

The collision of values discussed earlier is related to the difference in mindsets that forges family relationship. The construction of house and family members is shown from the viewpoint of the narrator, which shows the different ways and styles through which European people let their children go. From a specific point of view, Europeans are seen to have a better value regarding the matter through the character 'I' as well as contrastive courage. The different ways of raising children are described as unique.

Although the character 'I' has lived and interacted with the European society, it prefers Asian or Eastern identity, as shown in her statements and acknowledgement of the family. Such an assumption indicates courage and openness of the European (Western) values, while wariness and not being open are considered as Asian (Eastern) values. 'I' considers religion as the point that differentiates the two cultures. An example is provided in Islamic cosmopolitan regarding how Markus, a *mualaf* (a person who converts to Islam) from Switzerland, was raised:

> 'I only learned from my mother who is a Catholic teacher. She raised me with religious values. When I told her about my choice to become a Muslim, it was the hardest moment for her. True, time heals. Time heals everything. Although now I have a different faith from her, she is proud of me because I also follow her example in raising my children by getting close to God', said Markus (p. 48).

Dialogues between Markus and Hanum show that religions, both Islam and Catholic, teach religious values in life. Through these dialogues and dramatic description, the point of view focuses more on the memory that Markus had about the figure of his mother. By maintaining a good parent–child relationship and applying religious values on the development of children, Markus emphasised that a role model deals with attitude as opposed to being limited to just the basic values. Such point of view shows that cosmopolitanism supports diversity in unity from the perspective of value equivalence. Involving God and religious values in parenting is a common practice in every religion, and Markus is successful.

In addition to the collision in the method of parenting, the collision of values is shown when dealing with old parents:

> I understand that most elderly in Europe are united in an organisation in the house for the elderly by their children. However, it does not mean that the children do not love them. It is just a common practice in Europe in dealing with the elderly... (p. 24).

Then, the collision is again described from the viewpoint that compares the society's mindset in taking care of family. The narrator's first point of view is based on the experience in Europe, which is then compared to that in Asia. The description of others is humble in Asia, whereas it is bolder and freer in Europe. 'I' uses the experience of 'knowing' and 'direct involvement' to describe the feeling of the other character. The explanation that Europeans tend not to take care of their parents is in contrast to that of Indonesians, who live together in a society in one big house. Europe is considered as a place with a slow growth phase, as well as its cycle of family life.

On the contrary, the collision towards Islamic points of view can be noted from the following dialogue between Hanum and Xiao Wei, a Chinese student in Austria who strongly believes in communism:

> 'I do not see any flaw in you as mentioned by people about Muslims. They all talked about things they had not seen before...I bring Indonesian food for you. You invite me to join your husband's events so that I can practice my English' (p. 74).

The above dialogue shows the development of the character's mindset towards female Muslims and vice versa. Through a comparison of Islamic values between Indonesia and Chinese communism, the readers are led to the understanding in comparing and concluding that the Chinese are communists, which is against Hanum's character. However, an empirical experience strategy was used by Hanum to show the awareness of both parties in terms of giving labels and stereotypes.

The collision between Islamic good values and negative images of Islam is described through the development of the awareness of Xiao Wei, whose statement shows that effort could change the awareness of non-Muslims about Islam, provided Muslim men and women portray the goodness of Islamic values to people. The relationship between Xiao Wei and Hanum reveals how social friendship as a medium could strengthen the understanding about each other. From the dialogue, it seems that they both build the awareness that not all Muslims are terrorists, but they show good attitude. Furthermore, the following three cosmopolitan zones are shown: media and communication, learning language and involvement to interact in one country.

Another cosmopolitan character in Hanum's travel is Nur Dann. In an interview, Nur is portrayed through the questions and direct answers about the choice of preaching through rap music:

> 'Changing perspective, people's point of view about veil as the symbol of being conservative in female Muslims. Veil is like when you wear a special hat for a rap and make it tilted. You feel comfortable if you wear it while doing rap. I said that I would feel comfortable when wearing veil....'
> 'This is how you preach, is it?'
> '*Ja!*' (p. 35)

What makes the dialogue interesting is that the collision between the cultural values in rap music and energetic conveyance does not prevent Nur from using a veil in festivals and associating the veil with a special hat for a rapper. From the viewpoint of the narrator, Nur's optimistic tone represents female Muslims in choosing the way of preaching and changing audience's point of view, which shows the quality of Islamic cosmopolitan.

Meanwhile, the collided character in Tutie's story focuses on the stereotype of female Muslims and the East. '*After you*. Please go ahead before me,' she briefly responded. 'Better, I come later. This black veil that covers the face will make you wait long,' she continued (p. 83).

The quoted phrase given above explains the difficulty faced by Layla, a female Muslim who wears a veil covering her face, towards airport checking. Although Tutie, being a Muslim woman, has different principles in wearing a veil covering the face every day, she still feels Layla as her sister. Wearing *hijab* and veil covering the face, female Muslims are seen negatively from the Western perspective. However, this does not stop female Muslims doing their regular activities, even in Western countries. Meanwhile, the point of view from European countries is represented by Tutie through the busy condition picturing the traffic of European passengers in an airport. 'They walk straight, while discussing business, entertainment, vacation, or music concert. In every of the steps, as if such noise appeared *is for money, money, money.....*' (p. 81).

The quoted phrase given above describes the difference and attitude of the European society. In a subjective description, through the expression 'does not care', the image of 'walking straight' and the discussion about 'business, vacation, music', Tutie assumes that Europeans tend to be individualistic or in group only when they have the same interest. This is in contrast with the appearance of Layla that makes Tutie regard her as being different from the rest of Europeans. The negative and subjective tone is described strongly in the mindset of Tutie.

In her travel as a graduate student in Austria, Tutie also experienced collisions, such as in terms of hangout culture, drinking wine and gossiping that she could not endure. The ability to socialise is obviously accompanied by the experience of being isolated by friends. In the end, Tutie gives a speech about her cosmopolitan idea in the graduation:

'...I grew up and was raised in a Muslim family... Yet, I'm very lucky I can continue my studies here, where the knowledge of West and East meets, and the border becomes unseen. Here, all people think globally, but still hold the principle and respect differences. I believe this is a powerful weapon to survive and become the champion in the global era' (p. 123).

This quoted phrase describes how Tutie sees Europeans objectively and positively. Tutie's cosmopolitan quality increases as she interacts with others in the campus. This shows that the position and judgement about female Muslims become more positive when interacting in a public area. Tutie also describes Europe as part of the world to develop cosmopolitan in her last quote. Through an oral explanation in the academic space, Tutie shows the success in her living zone and her studies abroad.

Last but not least, the character of Wardatul put more emphasis on the romance and problems that Muslim women from other countries face, especially those from countries with conflicts. This can be seen from the viewpoint of a Serbian Muslim woman:

> Dzelila comes from Sandzak, an area in Serbia. Islam has become a religion practiced by the minority group in a country once under the authority of Yugoslavia. Although born as a Muslim, Dzelila never tries to deepen her understanding about Islam.
> Their families are against them when they want to wear veils. (p. 150).

The abovementioned viewpoint of the narrator shows the difficulties faced by female Muslims abroad when they compromise their faith for the society. In the quote, Turkey as a meeting place of students from various countries becomes a space where everyone can be open about themselves. There is a point of view regarding Wardatul's luck of living in Indonesia. 'I wish I had been born in Indonesia, we would perhaps have understood Islam, just like you,' said Elma (p. 151). Through empirical experience, Indonesian female Muslim readers are placed in the same position as Wardatul's. Other female Muslims have an envious tone towards Wardatul because of her being at ease with practising religious activities in Turkey, which represents collisions experienced by female Muslims in the world.

In building a female Muslim cosmopolitan discourse, the characters also present cosmopolitan characters, women and men, Muslims and non-Muslims. Cosmopolitanism is built through the spirit of equality as humans with the respect for diversity, such as what Hanum says, 'In addition, those who help are not all Muslims. Mostly, they are not believers. Indeed, the brotherhood is not only with Muslims, but also with those of different religions. Sometimes Allah SWT sends them to help me through life' (p. 8).

Hanum's cosmopolitan quality is built through the aforementioned five zones. Her story with the characters in Europe begins with media and communication as well as learning a foreign language. Hanum presents more female characters, such as A Man, Xiao Wei, Heidi, Bunda Ikoy, Ce Siti and Nur Dann. A Man, Xiao Wei and Anne represent Europeans with their spirit of tolerance towards Hanum and Islam. Hanum focuses on reporting European hospitality and openness, a diaspora Muslim family in Europe, different mindset and lifestyle between West and East and the mission for Muslims to become representatives of Islam worldwide. Hanum reveals herself as a female Muslim and minority in Europe, and she regards the Europeans and immigrants in Europe as 'other'.

Meanwhile, Tutie's cosmopolitan quality is described through the three zones (ambivalent cosmopolitan), which are learning a foreign language, living abroad and interaction with foreigners through campus activities. Subjective description is more towards discussing discrimination issues in Europe, the progress in Europe and the difficulties faced by Muslim men and women in becoming the citizens of the world due to Islamophobia. The culture collision is described through issues concerning *hijab* and stereotyping of female Muslims in Europe. However, the success in coping with differences and stereotyping is achieved by overcoming obstacles and becoming the best graduate. Tutie reports the stereotype and discrimination, reveals herself as an Asian and a Muslim and represents the European and Middle-Eastern people as 'other'.

On the contrary, Wardatul shares her experience when studying in Turkey during her early life. 'I always bear Imam Al Ghazali's words that support my life. (He said) do *hijrah*, and do not be afraid of what you leave behind, as you will get better replacement, even more' (p. 158).

In positioning herself, Wardatul uses the authority of an Islamic figure, such as Al Ghazali, in relation to the goodness given when deciding to perform *hijrah* or move from one place to another. It becomes a principle and a spirit in life for her to be a traveller. The opportunity to succeed overseas at a relatively young age is something special, and Wardatul gains the full advantage of it. The collision occurring and at first only involving some issues, such as women and young age, does not stop her moving forward because her family, who are very local, always support her and the Islamic religious values support her in every step. Wardatul's cosmopolitan quality is described through the aforementioned three zones (ambivalent cosmopolitan).

4 CONCLUSION

Berjalan di Atas Cahaya builds a discourse and voice of independent female Muslims who are able to adapt with the environment and new culture. Such condition is very evident as being viewed from the improvement of the cosmopolitan quality throughout the story. The complete cosmopolitan quality is obviously seen in Hanum as she experienced five *Rantenan* zones and wrote more experience and assumption. Meanwhile, Tutie and Wardatul show three qualities of cosmopolitan zones.

A detailed subjective description is presented to show the independent vision and the character's success in coping with cultural differences with the local society. This exposition shows that the national cultural identity becomes the highest priority. Islamic values, on the one hand, being foreign for the local people, create problems for female Muslims; on the other hand, it improves their faith to change the negative perception by being actively engaged in social interactions. This resolution leads to the final discourse that cosmopolitan female Muslims are everywhere and that readers can position themselves like the three characters. As a whole, collisions occur on issues related to different mindsets of people from certain countries in seeing family relationship, women's roles and the image of *hijab*. All characters see the position of female Muslims as more empowering and refer to Islamic teachings as the guidelines in travelling. 'I' is described as an observer as well as the actor in the interaction.

Last but not least, the characters describe physical and spiritual travelling as a great opportunity for women to understand the world and the people, as well as to re-define the image of female Muslims within the global context. All characters highlight the images and stereotypes of *hijab* in the countries they visit. Accepting differences is considered as an obstacle to become cosmopolitan female Muslims. New and unaccustomed values are taken as comparison to the existing values. A final and emphasised attitude is to adopt good values from the world and maintain those values as local and religious values. Through physical and spiritual travel, female Muslim travellers build the discourse of travel as a bridge to understand and generate awareness of the 'other'. Globalisation is regarded as an opportunity to be more open to the world without losing their identity as Indonesian cosmopolitan female Muslims.

REFERENCES

Astuti, D. (2012) *Kamus popular istilah Islam* (Dictionary of Islamic terminology). Jakarta, Kalil.
Budiman, M. (2013) Geni Jora: Islamic cosmopolitanism. In: *Reimagining the archipelago*. Jerman, LAP LAMBERT Academic Publishing.
Burhanudin, J. & Fathurahman, O. (2004) *Tentang perempuan Islam: Wacana dan gerakan* (On Muslim women: Discourse and movement). Jakarta: Gramedia Pustaka Utama.
Fine, R. (2007) *Cosmopolitanism*. London & New York, Routledge.
Kurnia, L. (2000) *Citra Indonesia dalam empat novel Jerman abad kedua puluh* (Indonesia's image in the four German novels in the 20th century). [Doctoral dissertation]. Universitas Indonesia.

Mashlihatin, A. (2015) *Novel 99 Cahaya di Langit Eropa sebagai cerita perjalanan poskolonial* (99 Lights in the Sky of Europe: a novel about postcolonial journey). [Master's thesis]. Universitas Gadjah Mada.

Mernissi, F. (2003) The meaning of spatial boundaries. *Feminist postcolonial theory*, ed. Reina Lewis and S. Mills. USA and Canada: Routledge.

Mills, S. (1991) *Discourse of difference: An analysis of women's travel writing and colonialism.* London, Routledge.

Rais, H. S., T. Amaliah, & W. Ula. (2013) *Berjalan di atas cahaya* (Walking in the light). Jakarta, Gramedia Pustaka Utama.

Rantanen, T. (2005) *The media and globalization.* London, SAGE Publication.

Salim, H. (2011) Muslim Indonesia dan jaringan kebudayaan (Indonesian Muslims and cultural networks). In: *Ahli Waris Budaya Dunia: Menjadi Indonesia 1950–1965.* Bali, Pustaka Larasan.

Thompson, C. (2011) *Travel writing.* London & New York, Routledge.

Cultural Dynamics in a Globalized World – Budianta et al. (Eds)
© 2018 Taylor & Francis Group, London, ISBN 978-1-138-62664-5

The mosque as a space for Minangkabau women

H.F. Agus & M. Budianta
Department of English Studies, Faculty of Humanities, Universitas Indonesia, Depok, Indonesia

ABSTRACT: With regard to the routine activities in mosques in West Sumatera, in this article, we aim at elucidating how women position themselves in a segregated space. Through examining the forms of spatial management by observing and interviewing mosque congregations, we demonstrate the extent to which mosques can serve as a space that both limits and supports women participating in their communities on the basis of the social contexts and issues in the local environments. Furthermore, we show the contextual and cultural differences between two mosques located in different areas, Nurul Amri Mosque in Padangpanjang City and Nurul A'la Mosque in Tanah Datar district. We also show the complexity of hidden meanings of gender differences in a religious space that is interlinked with religious values and at the same time influenced by the socio-cultural life in Minangkabau.

1 INTRODUCTION

Mosques often attract tourists because of their unique architecture and history. The architecture of a mosque is influenced by the Middle Eastern cultures, foreign cultures that were brought during the colonial period and the local culture (Khan, 1997). As a Moslem from Sumatera, I experienced an awkward moment when I was trying to find the women's praying room in a mosque in West Java and accidently entered the men's praying room, which made me aware of the influence of local culture on the architecture of the mosque. This inspired me to conduct a study on the gendered spatial division in mosques in West Sumatera, my hometown, which shows a mixture of Islamic culture and Minangkabau's matrilineal tradition.

For the Muslim community, a mosque is both a symbol of quiddity and a centre of Islamic activities (Khan, 1997). In large parts of the society, mosques disseminate Islamic principles to both the local community and the global network through intra-mosque connections (Bano and Kalmbach, 2012). According to some women, mosques do not always provide equal space for them as men. A large number of research works have shown the efforts made by women in various Muslim communities worldwide to achieve equal access to mosques or to offer a different interpretation of a tradition that is dominated by men. Thus, in some areas where women are not allowed to participate in religious spaces, such as mosques, there have been significant changes, and those mosques have become alternative spaces for women to gain knowledge about Islamic principles and to play a role in their community. This shows how a mosque is not just a solid structure but can be more flexible and open to women, although this change still requires much effort.

As one of the provinces in Indonesia that is known to implement Islamic teachings in secular institutions (obligating *hijab* for female students and civil servants as well as participation in the Ramadhan *Pesantren* and *Wirid Remaja*), West Sumatera supports the local identity through mosques. In this study, we aim to investigate the extent to which the spatial management in mosques influences the role of women and how women use this space to redefine their position in the socio-cultural context of Minangkabau. In order to answer this question, contextual data will be collected through desk research and observation of the spatial management of mosques as well as through interviews. Furthermore, the concept of culture proposed by Stephen Greenblatt (1995) indicates that culture might have two opposite natures, namely constraint and mobility. Culture also functions as a controlling gesture of 'praise and blame' to achieve the ideal behaviour in the community, which will be used in this study to describe the relationship between culture as constraint and as mobility by using mosques as a religious space. The concept of space and gender in religion refers to the conceptualisation of Kim Knott in *The Location of Religion* (2005), where he demonstrates how space and religion are inseparable as "there are no places in which religion may not, in some sense or other, be found", in order to check how gender as one of the elements could be discussed in religious space. We also argue that the mosque is a public space, rather than a static space, which is not only influenced by the Islamic culture, but is also a space to interpret the role of women in customs and Islam.

2 SPATIAL DIVISION

Protests by Muslim women in various countries show the importance of providing equal access to physical space in mosques. Thus, physical space becomes one of the main considerations in studying the role of women in public spaces such as mosques. In this study, we observe the spatial division of two mosques located in different areas, namely Nurul Amri Mosque in Padangpanjang City and Nurul A'la Mosque in Tanah Datar district. Padangpanjang is the smallest city in West Sumatera, which is nicknamed *Serambi Mekkah*, that is, the Porch of Mecca. This nickname is related to the city's image as one of the places where Islamic schooling (*madrasah and pesantren*) developed in West Sumatera. Tanah Datar, on the contrary, is one of the districts where the *nagari* (village) way of living is still maintained in some parts and it consists of four tribes that have influential traditional leaders in decision-making processes in a societal life. These two cities are chosen because of their distinct characteristics in terms of customs and Islam.

In general, from the perspective of the positioning of the praying space, the spatial division for men and women in the two mosques does not display any difference. The spatial interior is similar to that in other mosques in West Sumatera, with the praying space for men at the front and near the rostrum and that for women at the back and separated only by a temporary divider (i.e. screen). The ratio of their praying spaces is 1:1 with an entrance located either facing one another or side-by-side.

Observation of the two mosques apparently shows the differences during Friday prayers. Nurul Amri Mosque in Padangpanjang provides a relatively smaller praying space for women than for men (though not 1:1), whereas Nurul A'la Mosque, which is located in the village, does not provide praying space for women during Friday prayers.

Figure 1. Nurul A'la Mosque (Tanah Datar District). The left image is the men's praying space, while the right image is the women's praying space (source: personal archive).

Figure 2. Example of spatial comparison while carrying out daily prayers.
(Nurul Amri Mosque and Nurul A'la Mosque models)
Notes:
1. Blue: men's praying space
2. Green: women's praying space
3. Red: rostrum
4. Black: divider.

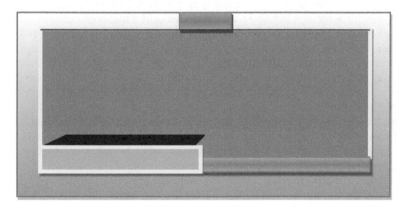

Figure 3. Example of spatial comparison during a Friday prayer in Nurul Amri Mosque.
Notes:
1. Blue: men's praying space
2. Green: women's praying space
3. Red: rostrum
4. Black: divider.

Female congregation is one of the factors determining the provision of praying space for women as it often facilitates the arrival of women to mosques (the same holds for the reason that women are not obligated to partake in congregational prayers or Friday prayers).

Under this condition, similar spatial management of mosques does not result in any difference in the treatment of women. However, the availability of equal space and facility supports the comfort and practicality for female congregation in mosques. As a result, there is no conflict related to spatial arrangement, even though it remains segregated with a divider. On the basis of the information provided by the caretakers of the two mosques, during daily prayers (other than Friday prayers), female congregation is larger than male congregation. The availability of space for women can support the congregation without feeling isolated or not facilitated.

The mosque, which represents the Islamic ideology, also accommodates customs, but in a different way. This is observed from the role of customs in decision-making processes related to social issues. In both the mosques, we find that the Minang women are projected as the *Bundo Kanduang,* a personification of a noble woman with strong leadership in the Minang matrilineal community. The figure of *Bundo Kanduang* is described in the Minangkabau literature to be a strong, wise, knowledgeable and independent character (Mina Elfira, 2007). *Bundo Kanduang* is a central character in Minangkabau customs. In relatively ethnically homogenous villages practicing customary laws, important affairs are settled by a meeting in an open field called *Medan Nan Bapaneh,* involving the four pillars of the Minang customs: *Bundo Kanduang, Ninik Mamak* (patriarchal leader), *Alim Ulama* (religious leader) and *Cadiak Pandai* (intellectual, cultural leader). The role of *Bundo Kanduang* is played by a senior and well-respected woman because of her position and knowledge about customs. In cities and villages with a more heterogeneous population, *Bundo Kanduang* does not refer to such actual representation of female body, but a collective norm about how a female leader should be. In cities, community affairs are settled by a formal state institution, KAN (*Nagari Custom Meeting*). In general, KAN in cities focuses more on *nagari* matters, such as the issue of *Tanah Ulayat* (the hereditary land in *nagari*).

Nurul Amri Mosque is located in an area that no longer implements decision-making processes in *Medan Nan Bapaneh* because it is composed of multicultural communities. Nurul A'la Mosque is located in *nagari*, which still uses the process of decision making by the existing traditional leaders. For both congregations of Nurul Amri Mosque and Nurul A'la Mosque, *Bundo Kanduang* remains a reference, but in different forms and interpretations.

The interviews conducted in the congregation of Nurul Amri Mosque in Padangpanjang show that the women understand the story and mythology of *Bundo Kanduang* as part of the identity of Minang women, which motivates them to study and play a larger role in their community. Even though this female figure is not present or represented in customary meetings, the mythical character of a brave, independent and knowledgeable woman is reflected in the way Minang women are positioned by the congregation in religious contexts.

This has a different impact on female congregation in Nurul A'la Mosque, which still has institutionalised *Bundo Kanduang* as the representation of women in the customs. The mosque becomes a more limited democratic space because of the tendency of the decision makers who are only monopolised by traditional leaders, especially *Alim Ulama* (including the aforementioned four pillars in customs). The result is that members of the congregation who do not hold position in the structure of the traditional society have limited voice compared to those in the city where the customary structure has been institutionalised into a more formal and centralised institution.

Even though the role of *Bundo Kanduang* can be debated, for a number of Nurul Amri congregation members, the memory of the *Bundo Kanduang* figure makes them more vocal towards their position in society. There is a need to show the inseparability of women and their Minangkabau identity. Open access for women to mosques is one of the factors that can support the balance of the bargaining process for the position of women in society. For female congregation, mosques become a place that facilitates not only studying the religion but also discussing and understanding the concept of women in Islam and customs.

In an interview conducted in the female congregation of Nurul Amri Mosque, some of the women quoted a hadith on women during the time of the Prophet Muhammad and related it to their current condition. They had the opinion that passivity is not the character of Minangkabau women. Eva (aged 60 years) held the opinion that the mosque is part of her daily life. According to her, the prohibition or appeal to carry out worship in the mosque is something that is absolute and cannot be debated. Similarly, Dian (aged 55 years) also related her life experience to her role in the mosque. Upon discussing the discrimination towards women and the prohibition of women going to the mosque in other communities, Dian expressed her opinion about the *mazhab* (school or teaching) and cultural differences as something that should be respected but should not confine women. For such reasons, these women showed a critical attitude towards religious interpretations and connected their interpretation with their daily experiences.

In the context of both congregations (as well as other congregations) relating the hadiths to the endeavours of women, they all refer to *Bundo Kanduang* as a reference for the ideal character of the Minang women. For these women, the *Bundo Kanduang* character is not just a traditional figure, even though she does not exist physically. On the basis of the explanations given by both Eva and Dian as well as other members of the congregation, there has been much effort devoted to elaborate the complexity of the interpretation of a hadith. For them, without context, a hadith cannot just be the reason to cease access to women, especially when it provides a means for worship and for the pursuit of religious knowledge. In all possible occurrences, Eva and Dian inserted phrases "our grandma the Minang woman", which implies how important is identity as a Minang woman for them to show the difficulties faced by them in everyday life.

Unlike the congregation of Nurul Amri Mosque, the female congregation of Nurul A'la Mosque was found to be more passive. The women who frequently visit the mosque focus more on worship and mosque-related activities set by its caretakers. These findings show that every district or smaller community unit could have different interpretations on the role of women. On the basis of the experience of Ayu (aged 30 years), a member of the female congregation of Nurul A'la Mosque, the management of the mosque is limited to some central characters in the local customs. Ayu believes that decision-making processes related to this mosque do not involve the congregation in general, but are decided only by the members of the committee with *Alim Ulama* as the leader, *Ninik Mamak* and *Bundo Kanduang*. The society only involves in making decisions without discussion. Ironically, in this context, the representation of *Bundo Kanduang* is also considered quite passive. In several customary decision-making processes, *Bundo Kanduang* also tends to follow the opinions of other leaders, so that the female congregation is not free to express their opinion. For the women, this tradition is very confining, as they realise that access to express opinion is crucial.

This is very contextual and can differ from one *nagari* area to another; however, these findings reveal that the role of women in tradition also affects religious and social spaces. In other words, the participation of women in mosques in the above two cases of mosques is parallel to the interpretation of the traditional role of women in customs. Greenblatt (1995) argued that culture has two opposing forces, namely constraint and mobility. Culture may become a constraint and an obstacle; however, it may also be a driving force for social change. In other words, there are two oppositions within culture. It can lead to the establishment of borders that serve as a controlling mechanism within the society as well as an opening space for changes.

4 WOMEN'S INITIATIVE IN EVERYDAY ACTIVITIES IN MOSQUES

Regardless of the fact that social institutions have a strong influence in determining the position of women, Minang women in their daily lives play an important role as an agent in the negotiation for wider access and larger opportunities. The efforts are not made through huge demonstrations, such as those carried out by the Muslim women who demonstrated in mosques, as well as through the discourse of gender equality. The efforts are visible from the self and through everyday experience and small, yet concrete proof of self-achievement. The everyday life strategy of Minang women in securing their place in religious activities is not a threat to other parties and hence it is more acceptable. Although small in scale, daily initiatives of women have widened their public roles.

The main members in a mosque's committee are usually men led by *imam* and *khatib*. The central position is limited because the role of *imam* and *khatib* can only be held by men. Almost all mosques adopt this tradition (even though some women activists try to offer new interpretation towards this practice), except for women's mosques. Despite the limitation from the central management, several mosques allow women to participate in other divisions (such as community and education affairs). Some women expand their role through their initiative by starting new programs or activities in the mosques. This simple discussion not only allows women to be more involved in social activities through mosques but also builds their interest to participate and to express their opinions in public space without feeling inferior (in relation to states, profession, education, etc.). Even though the scale of the activities is very small, such as monthly community meetings or weekly sermons and learning the interpretation of the Koran, these simple activities become events that help members in the society interact with each other.

Women holding many positions in the management structure of a mosque opens more possibilities for them to expand their role and knowledge in understanding Islam. However, this access will be ineffective without efforts from the society to support and develop the functions of mosques. Many women in Nurul Amri Mosque actively propose to rebuild the mosque for the younger generation. Ratih (aged 61 years), triggered by several incidents in her community (the use of drugs and bullying), suggested that the mosque's caretakers should revive the Teen Mosque program for teenagers to interact and be involved in discussions related to Islam and other pressing issues. Ratih began the program by disseminating the ideas from house to house in the neighbourhood and became a volunteer for other activities, such as Marching Band trainer to attract the interest of the teenagers.

Ratih's efforts show that women can hold *amanah* mandates in managing mosques. One of her major considerations as a member of the mosque management is the rules and behaviours that are appropriate in the religious space. According to her, going to a mosque means understanding the rules in it, the main one being not to disturb other members who are carrying out worship.

According to Greeblatt (1995), culture has a control mechanism within the concept of gesture, which is highly efficient in determining the criteria for inappropriate behaviour as well as models and ideals. In producing gesture that both inhibits and allows such behaviour, the technique of praise and blame is implicitly implemented in daily life. Through acts of respect, praising smiles and approving nods, one's act and behaviour are supported and accommodated; however, through sharp looks, satire and insults, certain behaviours are perceived to be bad and will be rejected. Hence, for the congregations, appropriate behaviours become one of the elements to prove their ability and awareness in managing the mosque.

In other various activities, such as *Majelis Taklim,* a joint *wirid* among groups in the same sub-districts, and celebrations of religious holidays and social activities, female congregation is very enthusiastic and willing to cooperate with men congregation. Often women propose the forms of activities, the person to deliver the sermon, as well as the donators. In general, these activities are commonly seen in mosques in other areas, except in larger mosques like *Masjid Negara, Masjid Besar* and *Masjid Agung.* These large mosques are managed professionally because those activities require bigger management to serve a larger society. In general, smaller mosques at the local community and sub-district levels are supported by the voluntarism of local members, motivated by the aspiration to study and find a solution of social problems in their community.

5   CONCLUSION

The mosque as a space can be made to both limit and empower women. Spatial negotiations in mosques for women are one way for them to gain access related to how the society interprets the role of women in a wider context. By observing the space in the two mosques as case studies, Minang women in general have flexible and friendly space. However, spatial management does not guarantee women freedom to involve in the decision-making processes in mosques. The challenge for women comes from the interpretation of the Islamic ideology and the practice of local customs. In a community that institutionalises a matrilineal custom in a more rigid way, the role of women in a religious space is more limited. On the contrary, in a community where the myth of female leader operates as a collective reference, women can develop thoughts more freely regarding their positions in Islam and customs. This shows that the interpretation of the role of women from one custom to another can be different.

This study also shows that the participation of women in activities in mosques also has a significant influence on the size of the space that is provided for them. The presence of women in the management structure of mosques is crucial in helping them to express their opinions and empower other women to be more confident, although there are limitations and segregation in mosques. Thus, space as a socio-cultural context and the active participation of women to increase agency in space are equally important.

This study is one of the early works conducted to develop in-depth research on the relationship among space, religion and women in West Sumatera. The complexity of other variables, such as the variance of traditions, city–village differences, government policies related to women and other elements, such as education and social status, can contribute to further research on religious spaces and gender role for Minang women.

REFERENCES

Bano, M. & Kalmbach, H. (2012) *Women, Leadership, and Mosques*. Brill, Leiden.
Greenblatt, S. (1995) Culture. In: Frank, L. & Thomas, M. *Critical Terms for Literature Study*. Chicago, University of Chicago.
Holod, R. & Hasan-Uddin, K. (1997) *The Mosque and the Modern World*. Thames and Hudson Ltd.
Knott, K. (2005) *The Location of Religion*. London, Equinox Publishing.
Mina, E. (2007) Bundo Kanduang: a powerful or powerless ruler? Literary analysis of Kaba Cindua Mato (*Hikayat Yang Muda Tuanku Pagaruyung*) (Fable of the Young Prince Pagaruyung). *Makara, Sosial Humaniora*, 11(1), 30–36.

Cultural Dynamics in a Globalized World – Budianta et al. (Eds)
© *2018 Taylor & Francis Group, London, ISBN 978-1-138-62664-5*

The realisation of metaphorical speech act in a religious themed text: Pragmatics analysis on a popular religious self-help book

S. Altiria & A. Muta'ali
Department of Linguistic, Faculty of Humanities, Universitas Indonesia, Depok, Indonesia

ABSTRACT: This research was based on Charteris-Black's (2004) notable argument about how metaphors in religious texts are basically utilized as tools for persuasion with affective value of a religious moral code. This research used the Critical Metaphor Analysis (CMA) approach to find out what persuasion strategy is used in the metaphorical speech act of a particular popular religious text regarding premarital romantic relationship. By using a descriptive qualitative research method and the combination of cognitive linguistic theories, this research finds that there is a polarization in metaphorical meaning between tenors in the data. The tenors also imply gender bias; men are portrayed as the source of problems, whereas women are viewed as the sole victim of premarital romantic relationship. Finally, the analysis on the speech act strategy reveals that it is delivered by using assertive and directive way of speech. The result here is in line with Charteris-Black's findings of didactic nature in sacred texts. This research might shed a light for further research in various disciplines on the study of religious texts in Indonesia.

1   INTRODUCTION

The modern expression of Islamic piety in Indonesia has caught the attention of global academics for years. An annual *Indonesian Update* conference held in the Australian National University in 2007 specifically discussed the particular topic with a number of scholars presenting numerous points of interest such as the marketing of religious preaching books and cassettes as a manifestation of customer's religious self-image, and Islamic law discussions on the Internet that led the public to become more close-minded and less tolerant (Fealy & White, 2012). Among all observations on piety expression, Hoesterey (2012, 2013, 2015) focuses on Abdullah Gymnastiar (Aa Gym), an iconic figure that has changed the trend of religious preaching in Indonesia with his approach on popular Islamic discussion, religion as the authority holder, Muslim's subjectivity, and public devotion to religion. Hoesterey highlights Aa Gym's unique passion to popularize and mobilize Islamic idioms as a singular answer to all problems, including economy and politics. Through his books, TV programme, seminars, and self-healing trainings, Aa Gym introduced to Indonesian Muslims a whole new horizon of popular preaching that is later followed by other celeb-preachers (Hoesterey, 2012).

After years of saturated market on popular religious preaching, a new popular preacher Felix Siauw emerged through social media with millions of followers and subscribers. One particular notorious topic brought up by him is the campaign to refuse any kinds of romantic relationship before marriage, through his book titled *"Udah Putusin Aja"* or *"Just Put an End to It"*. Here, he tries to convince millions of people who bought his book to despise any kinds of premarital romantic relationship. The book itself is a kind of a self-help book based on religious values. It was claimed that it could be used as a guide for youths to avoid romantic relationship before marriage and as a guide for parents to supervise their children misbehaviours.

The book ignited objections from other Islamic scholars. Akhiles (2014) asks Muslims to carefully think about and learn the basic of Islamic law used to support Siauw's argument. He was displeased by the way the book elaborated circumstances in a rigid fashion and is prone to subjective bias. He states that we need to perceive Islamic law with a fair perspective and based on good intention, rather than "forcing scriptures to reality without dialogue or manipulating a sacred text in order to win the debate" (Akhiles, 2004, p. 7). Fathoni (2015) also expresses his disappointment in his book titled "*Dear Felix Siauw*" that specifically questions Siauw's rigid claim on religion-based morality. These critics reflect a big burden for "*Just Put an End to It*" book that has potentially misled general readers in viewing reality and morality.

The style of argumentation and reasoning used in particular religious narratives has a unique pattern that presumably was not intended by the writer. Metaphors were frequently used in general religious preaching in social media, especially in Siauw's book, to express ideas. This is consistent with Stern's (in Gibbs, 2008) argument that metaphors do not only occur in classic literatures, but also in casual conversation. Metaphors in popular religious preaching are also reflected in Ortony's (1975) idea that metaphors are not only used to amplify the beauty of a language, but also to embed important values in it. Furthermore, Charteris-Black (2004) states that metaphors can adequately express religious messages because they deliver abstract concepts to concrete examples in real life.

Considering the significance of language expression in Siauw's book that also represents mainstream religion preaching expression in Indonesia, it is important to take a closer look on the topic through a question, "what kind of speech act strategy is being used to communicate the metaphorical text in the book?" This research used cognitive linguistic theories, consisting of semantics (metaphor-semantic cognitive) and pragmatics (speech act) to analyse the book's content comprehensively. In the pragmatic discussion, the researchers investigated the interaction between a linguistic approach and general issues that are related to the language. A semantic approach was used to set the boundary so that the observation and elaboration stay on the linguistic elements without expanding to other domains.

Multiple steps of analysis in this research were based on Gibbs' (1999) argument that metaphors in the linguistic, cognitive, and social context are different, so a careful analysis was needed to elaborate metaphorical expressions appropriately. Therefore, to answer the research question, three theories were needed, namely conceptual metaphor theory by Lakoff and Johnson (1980, 2003) to find the type of metaphors used in the data, relevance theory by Sperber and Wilson (1995) to find moral messages that are communicated in metaphorical speech acts, and Searle's (1975) speech act classification to classify metaphorical speech acts as well as to find speech act strategy used in communicating the particular moral messages.

1.1 *Research on metaphors in Indonesia*

There are quite a few numbers of studies on speech act of religion issues in Indonesia. One research by Rahyono, Sutanto, Rachmat, and Puspitorini (2005) observed the speech act in mass media focusing on language profile of mass media in post New Order era. Another research by Sitanggang (2009) also focused on the political nuance of the language in her thesis.

As for metaphors, the discussions are vastly available and dynamically elaborative on many important issues in public. Yin (2013) states that the scope of the metaphorical text analysis has extended to various sectors, including ideology (Goatly, 2006), political economy (Heradstveit & Bonham, 2007, and Cammaerts, 2012), education (Cameron, 2003), and religion (Charteris-Black, 2004). There was also specific research that observed casual daily conversation (Ponterotto, 2003) by using a cohesion approach on the speech/text of the conversation. Yet, there is still a little research on metaphors in religion-themed texts in Indonesia. One of the most notable studies can be found in Murtadho's (1999) investigation on the Indonesian translation of the Holy Koran. Most of the analyses on metaphors in Indonesia mainly focused on classic literatures, such as poems, song lyrics, and novels (Hasan, 2000, and Karnedi, 2011), but the metaphors in religion-based texts are not yet properly discussed in academic research.

1.2 *Conceptual metaphors*

The theory of conceptual metaphors was first introduced by Lakoff and Johnson (1980). According to them, metaphors involve two conceptual structures, which are the source domain and the target domain. Lakoff (in Cruse 2004:201) states that the source domain is concrete, while the target domain is abstract. Besides the two domains, there is also a set of mapping and accordance between the domains. The theory does not only focus on the linguistic expressions in metaphors, but also the semantic of conceptual metaphor or the speech's metaphorical expression. Taverniers (2002, p. 5) further explains that conceptual metaphor is part of cognitive linguistic discourse, which stretches on a dimension from the simplest to the most complex. This dimension can be seen from the structure of metaphor in speech and is divided into three kinds, which are orientational metaphor, ontological metaphor, and structural metaphor.

Orientational metaphor is the simple dimension metaphor. It consists of semantics' basic notion, such as time, quantity, status, changes, causes, and category members which metaphorically can be defined as the length of conceptual basic elements (e.g. space, motion, force) (Lakoff in Cruse, 2004, p. 204). The examples are: *argument is a container*; *happy is up*; *sad is down*. Ontological metaphor is the natural and persuasive metaphor in human cognition. It is a tool to understand about the experiences/sequences of human life. It can be used to conceptualize an abstract projection of a sentence into something more realistic (Cruse, 2006, p. 127). The entities of this metaphor can be derived from sentences which are related to one another. (Lakoff & Johnson, 1980, 2003). The examples are: *inflation is an entity* which is derived from these related sentences: *Inflation is lowering our standard of living* and *If there's much more inflation*. Structural metaphor is a concept which conceptualizes another new concept metaphorically (Lakoff & Johnson, 1980, 2003). It is a whole configuration formed by the mapping of metaphor domains (source domain & target domain) and it is built upon some related primary metaphors. Examples: the Structural Metaphor: *A purposeful life is a journey* from the primary metaphors: *People should have purpose in life*, and *Purposes are destination; People should act so as to achieve purpose*, and *Actions are motion* (Ning yu in Gibbs, 2008, p. 247).

This conceptual metaphor represents the basic concepts of an idea or description based on a set of metaphors found in the speech. This theory is still considered as a crucial approach in the studies on metaphors as found in Ponterotto (2014) decades after the initial conceptualisation of the theory. The research explored the metaphor of *"happiness is up"* as an emotional concept through the investigation on the verbs in the sentence.

1.3 *Relevance theory*

Relevance theory (Sperber & Wilson, 1995) views that a speech automatically creates an assumption that subsequently helps interlocutors to understand the meaning intended by the speaker. In relevance theory, people are perceived to efficiently process information; in this case, it means that there is an equilibrium on the effort to process information and the produced contextual effect (Cruse, 2004, p. 383). The processing effect means that the less effort given to discover the facts, the bigger the relevance of the facts. In other words, conspicuous facts need less effort to understand, if compared to the less obvious facts; direct inferences need less effort than indirect inferences. Subsequently, contextual effect means that the more contextual effect given, the stronger the relevance of those particular facts. This contextual effect includes additional information, the new information that either strengthens or weakens old information. New facts, which are irrelevant with the information, will not be processed.

This article agrees with the main premise of relevance theory, which states that in order to maintain a fluid communication, the speaker needs to understand the relevance principles, namely 1) that human cognition tends to be used to maximize relevance, and 2) that every demonstrative communication automatically communicates information with optimum relevance. Relevance theory in this article is used to find metaphorical meanings embedded in the speeches. This is based on the notion that the relevance between the source domain and the target domain can lead to certain meanings, which later can be used to form the reflector or determinant of illocutionary function in a metaphorical speech.

1.4 *Speech act taxonomy*

Searle (in Cruse, 2004) classifies speech act into five categories known as macro-functions. The first is representative/assertive, an illocutionary function that ties the speaker with the truth of expressed proposition, such as: stating, proposing, bluffing, whining, expressing opinion, reporting, imposing, suing. Next is the directive, an illocutionary function that aims to produce an effect of action performed by the speaker, for instance: ordering, asking, instructing, begging, forcing, giving advice. The third one is the commissive, an illocutionary function that makes speaker (more or less) tied to action in the future, such as promising and offering. This illocutionary functioning is to bring happiness and is less competitive because it does not facilitate speaker's needs, but rather the interlocutor's needs. The fourth function is expressive, an illocutionary function to express psychological attitude of the speaker toward situations implied in the illocution, such as gratitude, congratulation, forgiving, admiring, sympathizing, etc. Finally, declarative, an illocutionary function in the form of statements which have the influence to change one's status instantaneously, such as resigning, baptizing, firing, giving names, sentencing, ostracizing, promoting, divorcing, etc.

2 METHODS

This research uses a descriptive qualitative method. A mathematical calculation (e.g. sum, frequency, percentage) is only used to state numbers in order to support the data analysis. Critical Metaphor Analysis (CMA) is a metaphor research approach, underlined by the principle that metaphors are unique phenomena. These are important examples on the human experience reflected on language, especially about how pragmatic, context, speaker's diction, and semantic are seen as a unified linguistic system of language interpretation discourse (Charteris-Black, 2004). Metaphors by the definition of CMA can only be explained by considering the relationship among domains of cognition, pragmatics, and semantics

Table 1.   Source data of the research.

The original theme from book	Number of data	
	Sentence	Proposition
Rest Area (prolog)	8	15
Premarital romantic relationship is a sin	18	30
Just Let it Go! And the mentality of man who are involved in premarital of romantic relationship	29	54
Valentine's Day	20	47
Marriage as piety	4	9
Love as piety	12	19
Solution for not ready to marry	16	33
Move on	17	26
Love for the youth	17	36
Let's improve our self-quality	20	37
Epilogue	16	30
SUM	177	336

(Charteris-Black, 2004, p. 2). This approach has been proven capable of explaining metaphors in various kinds of discourses, such as politic, journalism, and religiosity.

The source of data in this research is the book entitled "*Udah Putusin Aja*" or "*Just Put an End to It*" by Siauw (2013). There are 180 pages in the book, and the data used in this research are the metaphorical sentences.

The data gathering started with the *documentation* process which included a) selecting the data that resembled social media statuses, b) rewriting all data by maintaining the themes based on the structures in the book, and c) performing codification to the metaphorical speech data to ease the exploration of statements. After documentation, the data were simplified so that irrelevant data could be eliminated. This *simplification* is based on macrostructure theory by Van Dijk (in Renkema, 2004), that is a) a screening and marking speech unit that can be eliminated without reducing or changing the contextual meaning (*deletion*); b) reconstructing speeches that have been screened without reducing or changing the contextual meaning (*re-construction*), and c) generalizing similar data with homogenous meanings without reducing or changing the contextual meaning (*generalisation*). Furthermore, to get the expected data, the researcher needed to find the proposition or sentence fragment expressing metaphors, so that the propositions without metaphorical expression were not used as data to be processed any further. After the proposition of expressed metaphors had been found, we conducted a *subjectification*, which is grouping propositions that have similar meaning to avoid overlapping topics. The subjects on the selected sentences would be further called as tenors or metaphor subjects.

Generally, CMA integrated the cognitive semantics and pragmatic process based on evidence gathered from the corpus linguistics analysis. The CMA approach consists of three steps of the metaphor analysis, which are 1) identification, 2) interpretation, and 3) explanation. Identification and interpretation started from exploration in the sentence subjects that were expressed with metaphors (tenor). The analysis and elaboration included the identification of source domain and target domain, and subsequently a further analysis was conducted to interpret the relationship between domains. This interpretation later produced a new concept into the speech tenors. At this stage, the data produced were sentences' propositions with metaphorical expressions, followed by speech codes where the proposition came from and a lexical indicator that formed the metaphors.

In the process of identification and interpretation of metaphor domains, there were two types of data, namely metaphor domains whose existence cannot be known unless by referring to special indicators in the form of lexical unit cohesion function (implicit metaphor) and metaphor domains that can be known directly (explicit metaphor/simile) (Steen in Cameron & Low, 1999, p. 82). The analysis of implicit metaphors in this article is based on Lakoff and Johnson's (1980, 2003) classification of metaphor types. These steps were used to form a "new concept", but for specific metaphor of simile supposition. This "new concept" in metaphorical texts can be directly identified through the referential accordance between metaphor domains.

The next step is the explanation of metaphors. The analysis conducted in this step is an analysis of illocutionary act in metaphorical speeches. Based on Charteris-Black's (2004) main premise, this analysis is intended to discover the speech act strategy that includes performative and illocutionary function of the speech in delivering the messages. The new concept gathered would be unified in a metaphorical speech meaning. This meaning is very useful to

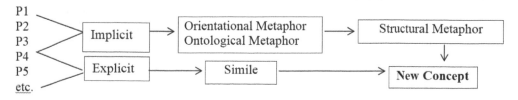

Figure 1.   The identification and interpretation process of the research.

417

THE CHOICES OF METAPHOR IN DATA
↓
PERSUASION
↓

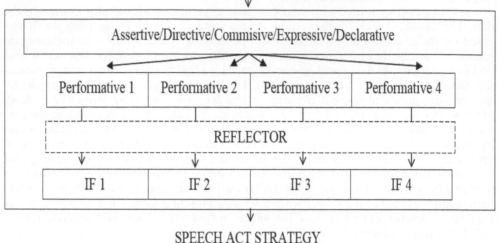

Figure 2.   The explanation process of the research.

state the contextual indicator (in this article, it is called *reflector*), as well as to bridge the gap of analysis among metaphorical speeches, performative speech act, and illocutionary function (IF) that underlines it.

3   RESULTS

3.1   *Metaphors identification and interpretation*

The investigation finds that there are two kinds of metaphors based on what the metaphor domains in the speeches are, namely implicit and explicit metaphors. The implicit metaphors in this research include these categories of conceptual metaphors: orientational metaphors, ontological metaphors, and structural metaphors. Implicit metaphors found in the source data are "premarital romantic relationship", "marriage", "love", "life", and "dignity", while explicit metaphors found in the source data revolve around "men" and ideas about love, and most of them are in the form of *simile* figurative language.

Based on the meaning of the particular metaphors, a meaning polarization is found. They are positive metaphors (+) and negative metaphors (−). This meaning polarization is found in the following tenors: "premarital romantic relationship" and "marriage"; "love" (in general) and "premarital love"; "men who are involved in premarital relationship" and "men who are not involved in premarital relationship". In the tenors of "life" and "dignity" there is no meaning polarization found in the source data, which means that both of them had neutral metaphorical meaning (n).

Based on the result above, these are metaphorical messages implied; 1) It is better to get married as soon as they are deemed appropriate because marriage is a source of happiness, and premarital romantic relationship is an unfavourable action which should be avoided because it will bring disadvantages, especially for women; 2) Men who are not involved in premarital romantic relationship are good/well-behaved, while men who are involved in premarital romantic relationship are bad; 3) Religion-based love relationship is a good deed, premarital love relationship is a bad deed, and love in general is neutral; 4) Men are portrayed as the source of problems in premarital romantic relationship, while also benefiting from it, and women are viewed as the sole victim of it.

3.2 *Metaphors explanation*

The analysis in this step focuses on speech's illocutionary action. This analysis is based on the findings on the previous step and it determines how these metaphors are communicated. Subsequently, the analysis method is laid upon Searle's (1979) classification of the macro-function in speech act.

The realization of metaphorical speeches in the source data can be seen as an occurrence of performative action, which means the action in the speeches can be determined (Austin, 1962). For instance, by stating that premarital romantic relationship is a sin, the writer of the book imposes bad value on the activity by relating the deed with unwise values. In this case, the source data deliver metaphors and at the same time intend to persuade teenagers to consider avoiding premarital romantic relationship. Performative actions found in the source data are 1) arguing, 2) directing, 3) persuading, 4) expressing, and 5) naming/labelling. Subsequently, the speech's illocutionary function analysis is performed to find out the strategy of delivering the message in the source data.

Renkema (2004, p. 17) states that researchers need to put relevant factors or adequate background information into account when determining an illocutionary action. For instance, from these following sentences; 1) *If you take the garbage out, I will give you beer,* and 2) *If you keep this up, you will have a nervous breakdown,* they were both valid as a speech of promise. However, the second sentence requires a further explanation on the idiom of *"a nervous breakdown",* a simple nominal metaphor of *"dangerous,"* so that the illocutionary function in the second sentence is no longer considered as a promise, but a warning.

Therefore, the illocutionary function analysis in this article is based on the investigation of metaphorical meaning from the tenors in Table 2. These metaphorical meanings form a contextual indicator (reflector). The analysis of illocutionary function by using reflector is done by Sitanggang (2009) on her research on implicatures of political discussion in Indonesian media. Aside from the use of reflector in determining illocutionary function, the analysis also needs to add an additional indicator that can be seen directly on the speech, for instance, the existence of performative verbs, instructive words, punctuation, adverbs, etc. Searle (1969) calls this as *illocutionary force indicating devices* (IFIDs). Regarding this consideration, this research uses two indicators in determining illocutionary function, namely reflector (contextual indicators) and IFIDs (additional indicators).

In the analysis of illocutionary function on metaphorical speeches in source data, there is a pattern that forms speech strategy from the relationship of metaphorical meaning and illocutionary function in the speech. In general, speech strategy used in source data in delivering moral messages are divided into two types, 1) speech strategy in tenor/sub-tenor considered as bad things (negative), and 2) speech strategy in tenor/sub-tenor considered as good things (positive).

Table 2. Metaphorical meaning of tenor on the data.

Tenor (−)	Metaphorical meaning		Tenor (+)
Premarital love relationship	short-term threat game misery lust violation	long-term guaranty serious thing happiness appropriateness obedience	Marriage
Love in premarital condition (valentine)	short-term depended on lust full of lies easy to get	commitment mandate, blessing certainty patience, sincerity expectation, struggle, a battle	Love in general
Men who are involved in premarital love relationship	Spoiled Egoistic Emotional	Though, Strong willed Aware to their capacity Full of preparation	Men who are not involved in premarital relationship

Table 3. Performative act and reflector of metaphorical speech act in data.

Types of Speech act	Performative act	Reflector
Assertive (45)	Perceiving	The value of the wisdom, information
Directive (19)	Ordering	A change, deed/action, idea
Commissive (15)	To influence	Consequence of previous action
Expressive (6)	To express	Hope, appreciation, anxiety (emotion)
Declarative*	Naming/labelling	The value of truth

N: 85 (metaphorical sentences)/*declarative speech acts are also found in assertive & directive.

Table 4. The result of illocution act analysis.

Tenor/sub-tenor (–)	Performative act	Reflector	Illocutionary function
Premarital love relationship	1. Perceiving	1. The value of the wisdom	1. Claiming
	2. Ordering	2. Deed; a change	2. Asserting
Men who are involved in premarital love relationship	3. To influence	3. Consequence of previous action	3. Prohibiting
	4. To express	4. Anxiety (emotion)	4. Ordering
	5. Naming/labelling	5. The value of truth	5. Threatening
Love in premarital condition			6. Insulting
			7. Judging

Tenor/sub-tenor (+)	Performative act	Reflector	Illocutionary function
Marriage	1. Perceiving	1. The value of the wisdom	1. Stating
Men who are not involved in premarital love relationship	2. Ordering	2. Deed; a change	2. Suggesting
	3. To influence	3. Consequence of previous action	3. Promising
	4. To express	4. Anxiety (emotion)	4. Appreciating
Love in general	5. Naming/labelling	5. The value of truth	5. Praying

The first speech strategy in negative tenors/sub-tenors consists of 1) claiming/asserting bad deeds, 2) prohibiting misbehaviours and ordering to change (demanding a change), 3) threatening by emphasizing the consequences of a sin, 4) insulting emotionally, and 5) judging a sin through elaboration of personal believed values. The second strategy is speech strategy in positive tenors/sub-tenors, namely 1) stating good deeds, 2) suggesting ideas, 3) promising beneficial consequences of good deeds, and 4) appreciating and praying through hope and prayers (see Table 3).

Another finding in the source data is the recurring discrimination between men and women in premarital romantic relationship. The speeches in the book frequently position men as the perpetrator of a sin and women as the victim of a sin. Careful observation in the source data does not find any speech that men experience disadvantages in premarital relationship. In fact, it implies a "judgment" towards men as the sole cause of women's sufferings in premarital relationship. It can be seen on this particular definitive statement in the source data, "premarital relationship is 100% disadvantageous for women and 100% beneficial for men, sooner or later" (Siauw, 2013, p. 38).

4   DISCUSSION

The result shows that the book is intended to shift the readers' mind-set on premarital relationship with a set of ideation to reconstruct the meaning of premarital relationship.

Subsequently, the book frequently mentions the needs to change, which later leads to speeches of prohibition to commit sin and reinforcement to do good deeds. This is consistent with the fact that persuasion speech act is frequently found in the source data. Rahyono, et al. (2005) state that illocutionary function found in metaphorical speeches in negative tenor/subtenor is mostly unwise illocutionary action, such as claiming, asserting, prohibiting, directing, threatening, and judging. From the 12 illocutionary functions found in the source data, 7 of them are unwise illocutionary actions, 4 are positive, and one is neutral. It can be said that the preaching narratives used in the source data tend to express negative illocutionary act that reflects unwise values.

Based on the finding in the analysis, it can also be seen that the preaching narratives in the source data mostly have assertive (52.94%) and directive (22.35%) tone; this is consistent with Charteris-Black's (2004) research where religions' sacred texts tend to be *didactic* or aim to give instruction and direction on religious morality. In religions' sacred texts, the concept of marriage, maturity, and heritage revolves around the values of obedience with the prize of heavenly advantages and punishments of hellfire for the deviants. However, the concept of hope, ambition, and energy are absent in the religion texts analysed by Charteris-Black (2004, p. 221). The concept of heavenly advantages and punishments of hellfire stated in Charteris-Black writing can be found in the source data on "influencing" speech act strategies, which are "promising" and "threatening" respectively. Referring to the metaphors of "light" and "darkness" in the Holy Koran (Charteris-Black, 2004), which states that "good is light" and "sin is darkness", the narratives in the source data generally position premarital relationship as "darkness", a sinful act that is mostly initiated by men to gain benefits, and thus, men involved in the particular relationship are considered evil/immoral, while women involved in the relationship are victims of men's sinful act.

5   CONCLUSION

The analysis on metaphorical speeches in the source data has answered the question stated in the beginning of this article, where speech strategies are used in delivering or communicating the messages. The investigation in this research is a representation of cognitive linguistic discussion (cognitive semantics and pragmatics), and in a more specific term this research is a discussion on metaphors of preaching texts in an Indonesian popular self-help book. In line with a psychological investigation, which suggests that religious sermons' narratives tend to repress the audiences' own thoughts and replace them with directive notions (Schjoedt, Sørensen, Nielbo, Xygalatas, Mitkidis, & Bulbulia, 2013). The findings in this research also suggest that preaching narratives tend to constrain the readers' own perceptions by shifting the idea of premarital relationship through assertive and directive metaphorical speeches.

We realize that the data used in this research do not represent the entire narratives on Islamic preaching and only represent preaches with a similar linguistic pattern. It is important for more advanced research to analyse various preaches from representative Muslim societies with a computerized corpus method to get the whole picture of how religion preaching is conducted in Islam. One potential data source might be religious accounts on social media that frequently promote religious teachings in daily life.

REFERENCES

Akhiles, E. (2014) *Do I have to let it go.* Yogjakarta, Safirah Publisher.
Austin, J. L. (1962) *How to do things with words* (2nd Edition). New York, Oxford University Press.
Cameron, L. (2003) *Metaphor in educational discourse.* Landon-New York, Continuum.
Cammaerts, B. (2012) The strategic use of metaphors by political and media elites: the 2007–2011 Belgian constitutional crisis. *International Journal of Media and Cultural Politics*, 8(2/3), pp. 229–249.
Charteris-Black, J. (2004) *Corpus approaches to critical metaphor analysis.* New York, Palgrave Macmillan.

Cruse, A. (2004) *Meaning in language: An introduction to Semantics and Pragmatics* (2nd Edition). Oxford, Oxford University Press.

Fatoni, M. S. (2015) *Dear Felix Siauw: Just a Correction, to clarify misperception*. Depok, Imania Publisher.

Gibbs, R. W. (1999) In Cameron and Low (Eds.). *Researching and applying metaphor*. Cambridge, Cambridge University Press.

Gibbs, R. W. (Ed). (2008) *The Cambridge handbook of metaphor: Metaphor and thought*. Cambridge, Cambridge University Press.

Goatly, A. (2006) Ideology and metaphor. *English Today*, 87(22/3), 25–39.

Hasan, D., C. (2000) *The translation English Metaphor into Indonesian Metaphor: Case study of metaphor translation in novel Absolute Power, Bloodline, and Rising Sun*. Depok, Universitas Indonesia.

Heradstveit, D., & Bonham, M. (September 2007) What the axis of evil metaphor did to Iran. *The Middle East Journal*, 61(2), 421–440.

Hoesterey, J. (2012) Morality marketing: the Rise, the Fall, and the Reconstruction of Aa Gym Brand. In G. Fealy and S. White (eds.). *Online Fatwa and Moral Commodification of Celebrity Preachers*. Jakarta, Komunitas Bambu.

Hoesterey, J. (2013) Shaming the state: Subjectivity and Islamic ethics in Indonesia's pornography debate. *Boston University CURA*. [Online]. Available from: www.bu.edu/cura.

Hoesterey, J. (2015) *Rebranding Islam piety, prosperity, and a self-help guru*. Stanford, Stanford University Press.

Hosen, N. (2012) Online *fatwa* in Indonesia: From *fatwa* shopping to scholars' googling. In G. Fealy and S. White (eds.), *Online Fatwa and Moral Commodification of Celebrity Preachers*. Jakarta, Komunitas Bambu.

Karnedi. (2011) *The translation of conceptual metaphor from English to Indonesian: Case study of the economic text book translation*. Depok, Universitas Indonesia.

Lakoff, G. (2008) The neural theory of metaphor. In R.W. Gibbs (ed.). *The Cambridge Handbook of Metaphor: Metaphor and Thought* (17–38). Cambridge, Cambridge University Press.

Lakoff, G., & Johnson, M. (2003) *Metaphors we live by*. Chicago, Chicago University Press. (First Publication 1980).

Murtadho, N. (1999) *The Al-Quran metaphor and its translation in Indonesian: The study of metaphor light, darkness, and some of God characters*. Depok, Universitas Indonesia.

Ortony, A. (1975) Why metaphors are necessary and not just nice. *Educational Review*, 2, 45–53.

Ponterotto, D. (2003) The cohesive role of cognitive metaphor in discourse and conversation. In Barcelona (ed.), *Metaphor and metonymy at the crossroads: A cognitive perspective* (283–298). Berlin, Mouton de Gruyter.

Ponterotto, D. (2014) Happiness is moving up: conceptualizing emotions through motion verbs. *Cognitive Linguistics Conference 4th UK* (265–283).

Rahyono, F. X. (2012) *The study of meaning*. Jakarta, Penaku.

Rahyono, F. X., Sutanto, I., Rachmat, R., Puspitorini, D. (December, 2005) The wisdom in language: A case study on pragmatics towards media language profile in post-New Era. *Makara Sosial Humaniora*, 9(2), 46–56.

Renkema, Jan. (2004) *Introduction to discourse studies*. Amsterdam, John Benjamins Publishing Company.

Schjoedta, U., Sørensena, J., Nielboa, K. L., Xygalatasa, D., Mitkidisa, P., & Bulbuliab, J. (2012) Cognitive resource depletion in religious interactions. *Religion, Brain & Behavior*. [Online] 3(1), 39–86. Available from: doi: 10.1080/2153599X.2012.736714.

Searle, J. (1969) *Speech Acts: an essay in philosophy of language*. Cambridge, Cambridge University Press.

Searle, J. (1975). A taxonomy of illocutionary acts. In *Language, mind, and knowledge* [Minnesota Studies in the Philosophy of Sciences 7]. Minneapolis, University of Minnesota Press.

Searle, J. (1979) *Expression and meaning: Studies in the theory of speech act*. Cambridge, Cambridge University Press.

Siauw, Felix Y. (2013). *Just put an end to it*. Bandung: PT Mizan Pustaka.

Sitanggang, Natal P. (2009) *Reflexivity of conversational implicature in political talk show at Indonesian television for ten months towards the legal campaign of General Election in 2009*. Depok, Universitas Indonesia.

Sperber, D., & Wilson, D. (1995) *Relevance communication and cognition* (2nd Edition). Oxford, Blackwell Publishers Ltd.

Taverniers, M. (2002). Metaphor. In: Jef Verschueren, Jan-OlaÖstman, Jan Blommaert and Chris Bulcaen (eds.) *Handbook of Pragmatics 2002*. Amsterdam, Benjamins.

Yin, Xiaojing. (2013) Metaphor and its textual functions. *Journal of Language Teaching and Research*, ISSN 1798-4769, 4(5): 1117–1125.

Cultural Dynamics in a Globalized World – Budianta et al. (Eds)
© 2018 Taylor & Francis Group, London, ISBN 978-1-138-62664-5

Imagery characters and ideology of Islamic fundamentalism: Adaptation of the novel *Les Étoiles de Sidi Moumen* and the film *Les Chevaux de Dieu*

D. Jinanto & J. Tjahjani
Department of Literature, Faculty of Humanities, Universitas Indonesia, Depok, Indonesia

ABSTRACT: In this article, we report the ideological theme of the novel *Les Étoiles de Sidi Moumen* and the film *Les Chevaux de Dieu* by conducting an adaptation study. An ideological text is expressed using a structural approach and a critical discourse analysis of the aforementioned works. The result of this adaptation study reveals different ideological texts, especially in the transformation of the portrayal of the fundamentalist character. The narrative and cinematographic elements of the film eliminate some of the ideas found in the narrative structure of the novel, particularly with respect to several emotional aspects of the character. The film's text relies on the interactions between characters to shape the fundamentalist figure. The result of this study is an ecranisation capable of depicting the image of the Islamic fundamentalist character. The image of the fundamentalist character can be observed as a form of self-awareness in human values, as found in the novel, or as an internalisation of sovereignty practices unbeknownst to the characters in the film.

1   INTRODUCTION

Francophone literature works, especially those originated from the Maghreb, often describe Islamic cultural values in the French language, with the purpose of presenting identities that are different from those of the Europeans or most Africans. The objective of Maghreb's francophone works is to incorporate Islamic identities in the characters without eliminating the elements of European culture, such as European fashion or football (Murphy, 2003). The complexity of these identities becomes the foundation for observing ideological text. This complex situation can be simplified by presenting actual phenomena related to identity issues, as shown in contemporary francophone works.

Terrorist attacks that have taken place since the early 2000s has created a bad impression on Islamic identity associated with the networks of radical groups. These attacks have tempted francophone writers to develop religion-based themes filled with terrorism, radicalism or fundamentalism. Fundamentalism is a tenet that attempts to defend something as the truth. The truth is a value believed to have been sent (probably through a scripture or spiritual belief) to influence the path of the human life (Dawkins, 2013, pp. 375–376). In some recent cases, fundamentalist groups have interpreted the tenet as a call to action to bring people back to religious teachings. Through the belief in the truth of religious teachings, fundamentalist groups often justify their violent acts.

The 11 September 2001 tragedy made radical fundamentalists the main topic of discussion in various media. Popular literature and scientific works have considered this case to have endless discussion, which prompted many writers to conduct studies with radical event background as a significant factor of creative process.

Fiction about the life of terrorists has become common nowadays. The following are some examples of terrorist attacks that have inspired fictional works: the 11 September 2001 tragedy in *Terrorist* (2006) by John Updike (the United States), Bali Bombing I tragedy on 12 October 2002 in *Conspiracy* (2005) by Muhammad Najib (Indonesia) and the serial suicide bombing incident in Morocco on 16 May 2003 in *Les Étoiles de Sidi Moumen* (*Stars of Sidi Moumen*) (2010) by Mahi Binebine (Maroko). The last work was adapted by a Moroccan director, Nabil Ayouch, in 2012 in his film titled *Les Chevaux de Dieu* (*Horses of God*). The novel *Les Étoiles de Sidi Moumen* and the film *Les Chevaux de Dieu* narrate the life story of a street kid named Yachine who decides to sacrifice himself together with his friends as suicide bombers at the end of the story.

Each medium has unique ideological texts related to the theme as the focus of the study. Different viewpoints from two media can provide different interpretations of the same fundamentalist story. Islamic fundamentalism stories are related to the depiction of the characters that play roles in the works. An adaptation study on the text *Les Étoiles de Sidi Moumen* and the film *Les Chevaux de Dieu* could show the difference in the depiction of characters in the context of Islamic fundamentalist ideology.

1.1 *Adaptation study*

An adaptation study is the one that shows the interpretation of the portrayal of ideas in the form of text. This study perceives media as vehicles that carry an idea. A different ideological text will change the idea due to the difference in the medium (Damono, 2014, pp. 13–14). This statement is similar to the idea proposed by Boumtje (2009) who states that currently films are a medium most commonly used to communicate a work. A film allows people to understand a work, a poem, play or novel, more easily. Therefore, the story of the novel *Les Étoiles de Sidi Moumen* (*Stars of Sidi Moumen*), which deals with the ecranisation with a different title, *Les Chevaux de Dieu* (*Horses of God*), can show different Islamic fundamentalist ideas in two different media.

The essence of finding meaning in the social context is the main focus of an adaptation study, as explained by Sapardi Djoko Damono in the book *Alih Wahana* (adaptation): '*What is contained in each text can be analysed to find a broader interpretation because a text is a compilation of other texts; therefore, the more texts there are in a work mean the more meanings can be found*' (Damono, 2014, pp. 15–16). Sapardi Djoko Damono's idea about adaptation complements the theory of adaptation proposed by Linda Hutcheon, according to which, any adaptation of a work is a form of repetition of interpretation expressed in a different way (Hutcheon, 2006, p. 7).

Damono (2014, pp. 118–119) asserted that a film often replaces the function of narrator with motion picture. In other words, the motion pictures in films can explain many interpretations thanks to cinematography. Therefore, Sapardi Djoko Damono emphasised that different media will create different meanings.

1.2 *Symbolic power*

Power is considered as a social practice resulting from the relationship between cultural participants and their role in an arena. Randal Johnson, in an introduction on Bourdieu's concept of social practice, defined arena as a space that is structured by rules and inter-agent relationships with a *habitus* to determine the position of the dominant agent. The position of the agent, which can be identified on the basis of the capital ownership of each agent, determines the pattern and structure of the arena (Bourdieu, 2010, p. xvi). The practice of power occurs with the presence of a more dominant capital in the inter-agent interaction.

Inter-agent conflict occurs when an agent tries to hold the highest position in an arena. The dominant agents will defend their capital to keep their positions stable, whereas the dominated agents will try to improve their positions by increasing their capital in order to become stronger agents (Widjojo, 2002, p. 43). Haryatmoko (2015) further stated that sometimes an agent unknowingly enters an arena of power conflict because of the strategy of a dominant agent to defend his/her capital in an arena. This can be understood as a reproductive nature of *habitus*, that is, when the dominant agent plays the same role when a new agent is entering the arena.

The practice of power may take the form of violence-inducing symbolic power unbeknownst to the agents in it. Bourdieu (1991) explained that symbolic power is used by the dominant agents to differentiate themselves from the dominated agents in order to be recognised. Symbolic power can be identified by anatomising the discourse implied by a certain social practice. The attention required in disclosing the symbolic practice of power is achieved by observing the recognition received from the dominated agents.

1.3 Structure of the novel Les Étoiles de Sidi Moumen

The novel *Les Étoiles de Sidi Moumen* has 18 untitled chapters. All of these chapters have Yachine as the narrator. The role of Yachine as the narrator-character is to narrate the complete story. The first-person viewpoint in this novel is useful for interpreting each occasion on the basis of the character's subjectivity.

Each story is narrated from the initial situation to the final situation. The transformation of characters is shown using the Greimas functional chart, which groups different events in a formula in order to see the whole story from the movements of situations (Zaimar, 2014, p. 41). Therefore, the indoctrination experienced by the characters of *Les Étoiles* can be understood as a transformation. The journey of *Les Étoiles*' group as a transformation is presented in Table 1.

Table 1 reveals a plot of the transformation of the *Les Étoiles* into a fundamentalist group in *Le Garage*. The initial situation of *Les Étoiles* in a poor neighbourhood led to the selection of this group as a fundamentalist group. Fundamentalist doctrines brought this group to the primary test, that is, the wish to become martyrs. The downfall test is experienced by this group during the bombing. The transformation of *Les Étoiles* results from the education and training in *Le Garage*. The death after the execution awakened Yachine, as a member of *Les Étoiles*. The reflection of *Les Étoiles* through the depiction of another street kid expressed Yachine's regret to explain the useless act of a sacrifice.

According to the functional chart presented in Table 1, fundamentalism can be instilled into an individual in a relatively short time. This inter-functional relationship shows several phases of indoctrination of Islamic values experienced by Yachine, which motivates him to make a change in life, that is, to get his family and himself out of poverty and to become a respectable person.

1.4 Structure of the film Les Chevaux de Dieu

The characteristic of a film as a medium that is different from a novel lies in the aspect of cinematography that reinforces the narrative aspect of a story. The film *Les Chevaux de Dieu* with 107 minutes of running time visualises the life story of Yachine chronologically from a child to the bombing executor. The film's narration uses an *objective-point-of-view* camera technique. This cinematic visualisation strategy focuses on the incidents that happened in the

Table 1. *Les Étoiles'* transformation functional chart.

| Initial situation | Transformation | | | Final situation |
	Screening test	Primary test	Victory/ downfall test	
Les Étoiles, a group of street kids who live in poverty.	Les Étoiles' participation as a member of a fundamentalist group in Le Garage.	Les Étoiles' wish to become a martyr after receiving fundamentalist education.	Les Étoiles' action as the executor of the bombing in Genna Inn, Casablanca.	A group of street kids with the potential to become new martyrs. Les Étoiles' reflection: Depiction of an unchanged situation.

Table 2. *Les Chevaux'* transformation functional chart.

Initial situation	Transformation			Final situation
	Screening test	Primary test	Downfall test	
Les Étoiles, a group of street kids who often fight, oppressed in the neighbourhood.	The involvement of Yachine, Hamid, Nabil and Fouad as students in a fundamentalist group.	The selection of Yachine and his friends to become martyrs, which is marked by the change of the name Les Étoiles to Les Chevaux.	Bombing in Casablanca with Les Chevaux group as the executor.	A group of Sidi Moumen street kids who watched the bombing from a far.

life of the character Yachine. Table 2 presents a functional chart that shows the storyline of the film.

Table 2 shows the transformation of Yachine's group and their friends who have a background as Sidi Moumen street kids. Their early life is full of violence and oppression being in slum and poor areas. The film's text visualises and justifies the fact that poverty is closely related to criminal activities. Fundamentalist groups work to protect Yachine and his friends from all sorts of trouble. The plot formed from this transformation explains the logical chronology of the selection of fundamentalist group to create martyrs.

Some name changes in the plot of *Les Chevaux de Dieu* mark the important events in the story. The transformation of the character Yachine is clarified in the film's text when the name Yachine is changed to Tarek. This name change shows the difference between Yachine outside *Le Garage* and Tarek inside *Le Garage*. Furthermore, the text emphasises the evolution of two different characters from one figure due to the Islamic fundamentalist doctrine.

1.5   *Ecranisation and symbolisation of fundamentalist ideology*

From the narrative structure, a change of information in the ecranisation of *Les Étoiles de Sidi Moumen* can be identified. On the basis of a statement by Damono (2014) that a change in an ecranisation may be an inclusion or exclusion of information from a narrative aspect; the study of adaptation can reveal ideas from each text. The novel *Les Étoiles de Sidi Moumen* and the film *Les Chevaux de Dieu* put forth the fundamentalist ideology through figures that can be divided into two categories, namely the fundamentalist and non-fundamentalist.

The martyr figure in the novel and the film is represented by the youngest fundamentalist figure. Poverty, hardship and the absence of hope for the future enable this type of figure to receive a doctrine easily. The indoctrination of fundamentalist values makes the street kids to play a role in their group. Further details about this youngest fundamentalist group can be seen from the differences between *Les Étoiles* and *Les Chevaux* presented in Table 3.

Description of their family backgrounds shows the interest of street kids in joining the fundamentalist group. The novel's text appears to place the family as an important element that influences the actions of *Les Étoiles*. In the novel, poverty in each family is narrated in detail as a problem, while the family aspect in the film seems to be ignored. The film's text depicts the fundamentalist figure as a character that does not have an emotional bond with his/her family. The family in the film appears to have values that are contradictory to the Islamic fundamentalist values. With a family value-erasing pattern, the Azzi figure is not included in the film, because Azzi's story in the novel's text prioritises family emotions, which is depicted through the character's conflict with his father.

Khalil's characterisations can also be compared between the two texts from different media. The similarity of the Khalil characters between the novel and film is that they have a life ambition. However, the difference is that the novel's text depicts a change in Khalil's ambition. Khalil in the novel undergoes a change of ambition after joining the fundamentalist

Table 3. Differences between *Les Étoiles* and *Les Chevaux*.

Screening test	*Les Étoiles*	*Les Chevaux*
Family background stories	Yes (story of each group member comes with a family background story)	No (family is only a supporting character in the story)
Number of members	6 (Yachine, Hamid, Nabil, Fouad, Khalil and Azzi)	5 (Tarek/Yachine, Hamid, Nabil, Fouad and Khalil)
Executors (martyrs)	Yachine, Hamid, Nabil, Khalil, Azzi	Tarek/Yachine, Hamid, Nabil
Figures not taking part in bombing mission	Fouad	Fouad and Khalil

group. On the contrary, Khalil in the film tries to achieve his life ambition and refuses to join the fundamentalist group. The story of Khalil in the novel explains that ideology indoctrination from the fundamentalist group has the potential to change the character's ambition. However, the story of Khalil in the film explains that violent fundamentalist indoctrination can be rejected by sticking to the worldly ambition.

Language shift also explains a different ideological text. The use of the French language in the novel shows that the Moroccan writer wants to convey some information to the French readers. This shows the attempt made by the francophone work in expressing its opinions of fundamentalism issues to the world. The character's regrets in the novel represent a criticism to the radical act that is associated as part of the Islamic culture. In the film, however, the use of Arabic in the dialogues between the characters conveys a message to the Arab people that fundamentalist groups are a threat to the Islamic culture. This indicates a criticism, like in the novel's text, but with a different focus. The criticism through the film's text attempts to warn the Arabic speakers of the symbolic power developed by the fundamentalist groups.

1.6 *Power and fundamentalism*

The position of *Les Étoiles* as training participants makes them a group dominated by figures that have a bigger capital in the *Le Garage* arena. The novel's text depicts the inability of *Les Étoiles* to resist the teachings of the emirs. Yachine's hatred towards the fundamentalist group leader, Abou Zoubeïr, appears at the end of the story as a form of the character's rejection of the violent fundamentalist ideology indoctrination.

The emirs aim to provide useful information to *Les Étoiles* on the basis of the Islamic fundamentalism doctrines, including Islamic philosophy, self-defence and technology. With the information provided in *Le Garage*, *Les Étoiles* are able to increase their cultural capital, which becomes the strategy of the fundamentalist group to recruit members. This also becomes a regenerative membership system of the fundamentalist group.

The figure who symbolically possesses the most dominant power is Syekh, who appears only once in *Le Garage*. The respect shown by all residents of *Le Garage* makes Syekh a sacred figure, and he has the highest power. Syekh's advice to *Les Étoiles* allows this group to justify all of their actions as the truth from the fundamentalist perspectives.

The activities in *Le Garage* can be considered as those of fundamentalists to make a person believe in a religious truth as a life guidance. This zeal could prompt the fundamentalist group to justify their actions to become martyrs. The activities in *Le Garage* affect the understanding of Islamic fundamentalist values as the defensive mechanism of Muslims. This forms the fundamentalist *habitus* that is practiced immediately by the *Les Étoiles* group.

The fundamentalist *habitus* is formed in *Les Étoiles* in a relatively short time by absorbing religious teachings intensively. When the agents in the lowest position in the arena are capable of judging a case by religious truth, the group leader can easily direct them to a certain act.

This direction creates zeal inside these agents to perform a radical action that they believe to be the truth. This demonstrates the influence of the fundamentalist logic that can change the mindset of a character, from being a street kid to a martyr.

The agents depicted in the *Le Garage* arena can be explained using Bourdieu's capital mapping, which is in turn used to show the influence of power in an arena. Bourdieu's capital mapping uses two axes. The horizontal line indicates the variations of opposing economy and cultural capitals, whereas the vertical line indicates the global capital volume. This global capital is related to the economic, social, cultural and symbolic capital possessed by an agent in a certain arena (Jourdain & Naulin, pp. 94–95). On the basis of the capital ownership in the *Le Garage* arena and the people in Sidi Moumen, the agent capital mapping picture (Figure 1) shows the position of the agent and the shift in the position of *Les Étoiles* as follows:

On the basis of the character capital mapping (Picture 1), the *Les Étoiles* or *Les Chevaux* group are the agents in the lowest position in two different arenas. The first arena (indicated by the blue line) is Sidi Moumen in general. This arena places *Les Étoiles* as the agents in the lowest position because of their economic limitations. The attempts of this group to gain economic capital by working do not change their position to the lowest. Something different happens when this group joins the *Le Garage* arena (indicated by the red line). *Les Étoiles* is provided with cultural capital to shift the position of the group.

The Islamic fundamentalist education changes the cultural capital ownership in the form of Islamic values that are physically learnt by the agents in the lowest position. This sort of indoctrination is considered capable of changing someone's *habitus* rapidly. A pseudo-shift is created (indicated by the dotted line) by providing a symbolic capital to the agents in the lowest position as an important agent in the arena. The symbolic capital is considered as false due to the lack of recognition from the people. This capital is only felt by the lowest agents. This sort of system places *Les Étoiles* or *Les Chevaux* as agents who are exploited to become martyrs.

On the basis of the mapping scheme, it can be understood that certain agents are stable, whereas other agents experience a change in capital. The stable agents tend to work to create martyrs. These agents can maintain their positions without becoming martyrs. In the lowest position, there are agents whose capital ownership can be shifted. The lesser the capital owned by the agents in the lowest position, the easier it is to deliver the fundamentalist indoctrination. Therefore, the task of becoming martyrs will be given to these agents. When agents in the lowest position are no longer available, the system will find new agents to fill the vacant position in the arena.

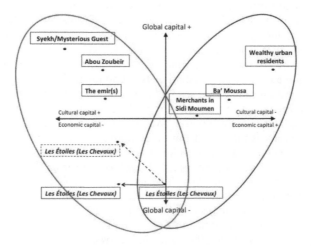

Figure 1.   Character capital mapping in Sidi Moumen.

2 CONCLUSION

On the basis of the storytelling that focuses on the idea of Islamic fundamentalism in two media, the novel *Les Étoiles de Sidi Moumen* and the film *Les Chevaux de Dieu* share a similar idea to display groups of uneducated youths from poor areas as fundamentalist figures. The characters that come from these areas were portrayed to have issues in dealing with hardships. It is in this crevice of life that the Islamic fundamentalist groups take the opportunity to establish new fundamentalist groups to work as martyrs. Therefore, both texts question the validity of the Islamic fundamentalist ideology that justifies radical actions.

These two texts of different media criticise the validity of the Islamic fundamentalist ideology through the disclosure of power practices. Both texts show symbolic forms of violence by different methods. The novel text shows that the decision to become a martyr is a damaging one. In other words, the narrator-character provides an understanding of the awareness of symbolic violence that occurs within the fundamentalist group. The film text presents the symbolic violence through visual interpretations that depict the fundamentalist group as a threat that places the *Les Étoiles* or *Les Chevaux* group as the victims of power.

Through their narrative exposure, both texts present character portrayals of Islamic fundamentalism values that tend to ignore human values. Adaptation explains the role of storytelling from the viewpoint of the first person. This study could describe the internalisation of fundamentalist values that makes the character directly unaware of power practices. The depiction of the martyr figure as the victim of power practice can be seen in the film text from the viewpoint of the third person by using the *objective-point-of-view* technique. The differentiation of interpretation from the viewpoint of the third person is a characteristic of the film that makes it an effective medium to convey a message about power practices.

Adaptation plays a role in providing different interpretation to the text. The fundamentalist figures turned into martyrs depicted in the novel become wasteful figures. The regrets depicted make the figures in the novel a reflection of humans who need emotional and social awareness before making a decision. On the contrary, the film text depicts the characters as associated with a threat of power practice based on Islamic fundamentalism. The depiction of characters as victims of power reveals a warning of practices that have the potential to create radical actions.

REFERENCES

Boggs, J.M. (2008) *Art of Watching Films* (Seventh edition). The United States of America, Mayfield Publishing Company.
Boumtje, M. (2009) L'impact du film en cours de littérature francophone. *The French Review*, 82 (6) 1212–1226. American Association of Teachers of French. Available from: http://www.jstor.org/stable/25613823.
Bourdieu, P. (1991) *Langage et pouvoir symbolique.* Paris, Éditions Points.
Bourdieu, P. (1993) *The Field of Cultural Production: Essays on Art and Literature (2010). Arena Produksi Kultural: Sebuah Kajian Sosiologi Budaya*, Yudi Santosa, translator). Yogyakarta, Kreasi Wacana.
Damono, S.D. (2014) *Alih Wahana.* Ciputat, Editum.
Dawkins, R. (2013) *The God Delusion* (Zaim Rofiqi, translator). Depok, Banana.
Murphy, D. (2003) Beyond tradition versus modernity: postcolonial thought and culture in Francophone sub-Saharan Africa. *Francophone Postcolonial Studies: A Critical Introduction* (Charles Fordsick & David Murphy, eds.), London, Arnold. pp. 221–230.
Haryatmoko. (2010) *Dominasi Penuh Muslihat: Akar Kekerasan dan Diskriminasi* (Domination Full of Deceit: Root of Violence and Discrimination). Jakarta, Gramedia Pustaka Utama.
Hutcheon, L. (2006) *The Theory of Adaptation.* New York, Routledge.
Jourdain, A & Naulin, S. (2011) *La théorie de Pierre Bourdieu etses usages sociologiques.* Paris, Armand Colin.
Widjojo, M.S. (2003) Strukturalisme konstruktivis: Pierre Bourdieu dan Kajian Sosial Budaya (Constructive Structuralism: Pierre Bourdieu and Socio-Cultural Studies). In: Irzanti Sutanto & Ari Anggari Harapan, (eds.). *Prancis dan Kita.* Jakarta, Wedatama Widya Sastra.
Zaimar, O.K.S. (2014) *Semiotika dalam Analisis Karya Sastra* (Semiotics in Literature Work Analysis). Depok, Komodo Books.

School-centric Islamic education in Indonesia: A deconstructive analysis of Muhammad Abed Al-Jabiri

N.S. Wibowo & Naupal
Department of Philosophy, Faculty of Humanities, Universitas Indonesia, Depok, Indonesia

ABSTRACT: Violent acts committed by Islamic fundamentalists at the ideological level in Indonesia have become frequent nowadays. In this article, we examine the extent to which Islamic education is implemented in Indonesia in relation to the impacts of global fundamentalism at large. The term "Islamic education" in Indonesia does not only mean formal Islamic education but also Islamic education in the form of religious guidance (*fatwa*), which is especially provided by *Majelis Ulama Indonesia* (Indonesian Ulema Council). Islamic education also comes from Islamic writing. In this study, we carry out deconstructive hermeneutics of Muhammad Abed Al-Jabiri to examine the origin of Islamic writing in Indonesia. We find that Islamic education in Indonesia is school-centric (*mazhab*-centric), resulting in the school receiving new ideas of fundamentalism aimed at triggering violent acts by the fundamentalists. The results show a correlation between the violent acts committed by the Islamic fundamentalists and the Islamic education in Indonesia. To solve this problem, schools must develop moral consciousness to refer to the teachings of Qur'an, instead of the school-centric ones.

1 INTRODUCTION

Initially, the diversity of Islam in Indonesia, or commonly called *Islam Nusantara,* did not cause any clash or physical conflict. However, in fact, ideological conflicts have penetrated into Islam Nusantara. The conflicts taking place are influenced by not only internal factors but also global factors, which play a significant role. One of the conflicts causing social turmoil is the conflict between *Sunni* and *Shia* that is worsened by the presence of the Salafi-Wahabi.

The conflict between Sunni and Shia in Indonesia can be traced back to 1984 when Indonesian Ulema Council (MUI) announced a *fatwa* against Shia teachings (Majelis Ulama Indonesia, 2015, pp. 17–18). In a short time, this conflict pervaded to issues regarding policies of regional government offices. Recently, in 2016, the Bogor regional government has banned Shia followers to celebrate Ashura Day. Earlier, in 2012, a conflict between the Sunni and Shia followers took place in Sampang, Madura, which culminated in the exile of the Shia followers from Sampang. They could only be allowed to return from the exile if they had acknowledged Sunni as the only official doctrine of Islam. Until 2016, the Shia followers were in exile in Sidoarjo, and this issue is the background for this study. This study was conducted using Muhammad Abed Al-Jabiri's deconstructive method, which analyzes the relationship between one Islamic text and another and that between one occasion and other past occasions to explain the present situation in the Islamic world.

1.1 *Muhammad Abed Al-Jabiri's deconstruction*

The deconstruction by Muhammad Abed Al-Jabiri is conducted by analyzing the habits of *mazhab* followers in Islam and their association with traditions, for example, pre-Islamic tradition that existed before Islam. In addition, Al-Jabiri's deconstruction is based on the intertextual studies between one *mazhab* and another to learn their influence on the other (Nughroho, 2007, p. 103) and forms the so-called Al-Jabiri's "dominating thought". Al-Jabiri stated in his work that the foremost task of people wanting to reform and reconstruct the thought (dominant thought) is to criticize the dominant thought (reason):

> ... all of it cannot be done but through the critique of a dominant reason, and the critique process must be done inside the reason itself by deconstructing the principles, making the effectiveness more dynamic, expanding, and also enriching the concepts and new knowledge that we gained from any places, even though it originated from the past, philosophers, or scientific thoughts. (Al-Jabiri, 2014, p. 30)

Using the hermeneutical deconstruction, Al-Jabiri categorized Islamic epistemology into three forms: *bayan, irfan,* and *burhan.* Furthermore, he described two different types of reasoning, namely active reasoning and dominant reasoning.

Active reasoning is defined as a "cognitive activity done by the mind when reviewing, analyzing, forming concept, and formulating basic principles" (Al-Jabiri, 2014, p. 28), and Seyyed Hossein Miri defined the reason as "constructive reason consisting of general principles, frameworks, concepts, methodologies, and all those elements which are, in some ways, able to influence a culture and produce a new discourse. Thus, this reason is characterized by creativity and productivity" (Miri, 2012, p. 44). Dominant reasoning is defined as "principles and rules that we make as a guideline in arguing (*istidlal*)" (Al-Jabiri, 2014, p. 28). Miri defined the reason as "the common and current reason along with the principles and rules institutionalized and accepted within a culture and historical period; in this framework, it is of absolute value" (Miri, 2012, p. 44). Precisely, active reasoning is one of the aforementioned three forms of Islamic epistemology, whereas dominant reasoning is the result of these epistemologies, which is called *ijtihad* or *ijma'*.

Bayan is the original thinking in Arabic Islam. The main idea of this epistemology uses different aspects related to the Arabic language, like *nahwu, sharaf,* and *balaghah,* in interpreting the main Islamic sources, Al-Quran and Sunnah. This epistemology is called "Arabic Religious Reasoning" by Al-Jabiri (Ibid, p. 495).

Irfan is *tasawuf* or *sufi* way of perspective that bases its interpretation on *"ilham"* (inspiration) believed to have come from God or God's Guardian in a close relationship with Muhammad bin Abdullah (AhlulBayt). The method is called *mukasyafah* or "cleavage", because it tries to divide the implicit meaning (spiritual meaning) of Al-Quran and the meaning of Sunnah.

Burhan is similar to Western philosophy, especially Aristotelian. This epistemology uses a concluding method through "empirical experience and rational concluding" (Ibid).

Although Al-Jabiri groups Islamic epistemology into three forms, according to him, they are still used in the religious field. In other words, the aim of these three forms is to interpret two Islamic sources, namely Al-Quran and Sunnah.

1.2 Mazhab *in Islam*

Mazhab, in the book of *Syiah Menurut Syiah* (2014), is considered as a "relative religion". In this book, religion is divided into two categories, namely absolute religion and relative religion. Absolute religion is believed to be passed on by God directly to prophets; relative religion is "an interpretation to sacred and absolute revelation" (Tim Ahlulbait Indonesia, 2014, p. 26). For further comprehension, I define *mazhab* as the school resulting from dissenting opinions of *ulemas* after all prophets left them. In Islam, there are some *mazhabs*, and the most embraced *mazhabs* by Muslims are Sunni and Shia. Sunni can be divided into *Hanafi, Maliki, Syafi'i, Hanbali,* and *Zahiri,* whereas Shia has two main *mazhabs,* namely *Ja'fariyya (ItsnaAsyariah)* and *Zaidiyya.* Furthermore, there are other *mazhabs* called *kalammazhab,* such as *Asy'ariah* and *Mu'tazilah,* and *ghulat* community, such as *Rafidha, Ahmadiyya,* and *Salafi-Wahabi.* I categorize Salafi-Wahabi as a *ghulat* community because of their extreme acts toward other Islamic groups with different opinions and faith. This *ghulatmazhab* is often miscalculated as the majority. The relationship between the *ghulat* community and the rise of other *mazhabs* and its implication in Islamic education in Indonesia will be discussed below.

According to Quraish Shihab, Sunni or *Ahlul sunnah* teachings originated when Abu Hasan Asy'ari built a *mazhab* called *Asy'ariah,* whose method is based on *fiqh* belief, which was developed by four great Imams of Islam, namely Muhammad bin Idris as-Syafi'i, Malik bin Anas, Ahmad bin Hanbal, and Abu Hanifah (Shihab, 2014, pp. 57-58). In reference to the perspective of Muhammad Abed Al-Jabiri, Sunni tends to use the *bayan* epistemology in developing its *mazhab*.

Upon further investigation, the Asy'ariah *mazhab* did not wholly use *bayan*. This is because Abu Hasan Asy'ari was the alumni of Mu'tazilah before he switched to the *fiqih* method. Sunni, since the era of Abu Hamid al'Ghazali, has experienced changes. After using *bayan* and *burhan*, it has now begun to syncretize both of them with *irfan*. This happened because Al-Ghazali is a Sufi practitioner in the mid to late part of his life. Thus, it was natural when he began to interpret Qur'an using the method of *mukasyafah* or over viewing; furthermore, he investigated it from the language and logical concepts.

Shia can be considered as contradictory to Sunni, with regard to the political concept and not the epistemological concept, because it refused Abu Bakr bin Abu Quhafah, 'Umar bin Khatthab, and 'Utsman bin 'Affan as caliphs. For Shia, the legitimate caliph was 'Ali bin Abu Thalib, as he was appointed directly by Muhammad bin Abdullah during the period of Ghadir Khum.

According to Muhammad Abed Al-Jabiri, Shia tends to use *irfan* in studying Qur'an and Sunnah. This is undeniable because Shia always refers to words of their *imam* that they acknowledge, such as Ahlul Bayt Muhammad, and they believe that the *imam* receives enlightenment directly from God. However, when we think of some Shia ulema like Mulla Sadra, we can find not only the *irfan* epistemology but also the *burhan* epistemology, because Shia refuses *qiyas* (analogy) as it is considered weak. Shia prefers the use of logics or a ratio as a whole than *qiyas* only.

The *Bayan* epistemology can also be found in Shia. This is due to their belief that *ushulfiqih* is important in reviewing Qur'an and Sunnah. This type of thought can be found in different works of Muhammad Husain Thabataba'i, especially in his exegesis work *Al-Mizan*.

1.3 The impact on Islamic education in Indonesia

The rise of *mazhab* in Islam resulted in the rise of the *ghulat* community. According to Ali Rabbani Gupaygani, from Syekh Mufid, *ghulat* originates from the word *"ghuluw"*, which means "... exceeding the limit and out of the stable border" (Gulpaygani, 2014, p. 371). The *Ghulat* community is often considered as a non-Islam group like Baha'i, Alawiyah, and Ahmadiyya Qadiani. However, the *ghulat* community could originally refer to those who still embrace Islam; however, because of their strong attitude, they are called the *ghulat* community or "extremists" similar to groups such as Lahore Ahmadiyya, Salafi-Wahabi, and Shia Rafidhah.

The *ghulat* community often say that the group only exists in Shia. However, if the meaning of *ghulat* is "extremist", then Sunni *mazhab* may also contain the *ghulat* community. In fact, there are more *ghulat* communities in Shia, such as Nausiriyah, Rafidhah, Alawiah, Bathiniyah, Druze, and Baha'i, as well as in Sunni. Examples of Sunni *ghulat* are Lahore Ahmadiyya and Salafi-Wahabi. In contrast to Qadiani Ahmadiyya, which claims that Mirza Ghulam Ahmad is the Prophet after Muhammad, they claim that Mirza Ghulam Ahmad is Imam Mahdi (*Ibid*, pp. 417-418).[1] Lahore Ahmadiyya can be considered as a Sunni, because they claim to follow Sunni teachings, especially in terms of *aqidah* of the pillars of faith and the pillars of Islam. We can refer to a book on the thought of Lahore Ahmadiyya in *Islamologi* written by Muhammad Ali. Salafi-Wahabi, which claims that they follow the teachings of Ibn Taimiyah and the literalism of Ahmad bin Hanbal. The rise of the *ghulat* community can then influence Islamic education in Indonesia.

Islamic education follows at least one of the *mazhabs*, because the focus of the teachings must be based on the purpose of educating about Islam. On the contrary, the education using the *mazhab* faces some challenges in relation to the sharpening conflict between the followers of Sunni and Shia in Indonesia. With the presence of the *ghulat* community, especially Shia's

---

1. See The Lahore Ahmadiyya Movement for the Propagation of Islam on site http://aaiil.org/text/qadi/intro/cmprsn.shtml

Rafidhah, many Islamic educational institutions, especially those from the Indonesian Ulema Council could not differentiate between Shia Rafidhah and Shia movements in general.

One of the causes of the conflict is the intensive and continuous *aqidah* teaching. From elementary school to high school, even in non-formal education institutions, it is always taught that the Islamic *aqidah* includes the six pillars of faith, (a) to Allah, (b) to the angels of Allah, (c) to Prophet and *Rasul*, (d) to the Holy Quran, (e) to the end of life, and (f) to the *qadha* and *qadar*, and five pillars of Islam, (a) *syahadah*, (b) *sholah*, (c) *alms*, (d) fasting, and (e) *hajj* (The Ministry of Religious Affairs of the Republic of Indonesia, 2014, pp. 5–6). In fact, two of the pillars are not purely from Islam, but an interpretation of Sunni over a *hadits* called Jibril Hadits. In Shia tradition, there are no such pillars. According to Quraish Shihab, Shia does not formulate this, because they believe Islam and Faith as one (Shihab, 2014, p. 92); therefore, it is impossible to separate or formulate them differently. However, in a book written by Indonesian Ulema Council (MUI), Shia formulates the five pillars of Islam as (a) sholah, (b) fasting, (c) alms, (d) hajj, and (e) *willayah*, as well as the pillars of Faith as (a) *tauhid*, (b) *nubuwwah*, (c) *imamah*, (d) Al-'Adl, and (e) Al-Ma'ad (Tim Penulis MUI Pusat, 2012, p. 85). However, for Gulpaygani, what is mentioned by MUI is not the two pillars in Islam, but the principles for the Shia *mazhab*, especially Shia Ja'fariyya (Gulpaygani, 2014, p. 160).

In addition to the different *aqidah*, MUI claims that all Shia followers tend to be labeled as *kafir* (heretics), except those who support Imam Ali as the legitimate caliph. This is evident in the statement issued by MUI, "...the revolutionary Iran leader, al-Khumaini says that 'Aisyah, Talhah, Zubair, Mu'awiyyah and so on, so even though physically they are not impure, they are worse and more disgusting than dogs and pigs" (Tim Penulis MUI Pusat, 2012, p. 54). By contrast, not all Shia followers are labeled as *kafir*, as stated by themselves (Tim Ahlulbait Indonesia, 2014, pp. 313-316). According to Gulpaygani, the insult to the Prophet's companions is directed to Shia Rafidhah. Even the word "*rafidhah*" is still debated (Gulpaygani, 2014, pp. 424-427). It is essential to know that Shia Rafidhah tends to label Muhammad's companions as *kafir* when they do not support Imam Ali.

In addition to Rafidhah, the minority Shia (*ghulat*) that is often mentioned as the majority Shia believes that Imam Ali is god or the true prophet. Furthermore, Shia *ghulat* is related to those who believe that the currently existing Qur'an is incomplete. In fact, for the majority Shia (Ja'fariyya and Zaidiyya), referring the term "*kafir*" to the Prophet's companions, considering Ali as god, and saying that Qur'an is incomplete are considered to be un-Islamic. Iran, whose population is mostly Ja'fariyya Shia, uses the same Qur'an as Sunni (Shihab, 2014).

From the perspective of Al-Jabiri, *mazhab*-centered Islamic education in Indonesia is repetitive and non-comparative, because it tends to use the dominant logic or *ijtihad* from elementary school up to high school without any revision. The Islamic education can be considered as non-comparative, because when examining an issue, the educators only look at it from one perspective (one *mazhab*). An example of this is their examination of the *aqidah* as described above. Islamic education in Indonesia tends to use the basic perspective of Sunni (*ijtihad*), so that an ideological conflict could arise when it encounters a perspective of Shia. This educational characteristic can create the third characteristic of Islamic education in Indonesia, namely generalization, which occurs when Islamic education in Indonesia has to encounter the perspective of *ghulat* community, which is a minority of all Islam communities. The *ghulat* community often represents themselves as a group like Shia because of a tendency of generalization that occurs within the Islamic education in Indonesia; therefore, the *ghulat* community is positioned at the same level as the present Shia group.

Another impact of the *mazhab*-centered education is that all followers are easily infiltrated by radicalism, such as by the Salafi-Wahabi movement, the followers of which claim that their teaching is the same as that of Sunni, which follows one *mazhab*, Hanbali. Their teaching is to eradicate all types of *bid'ah* that they claim as the teachings of Shia. The labeling of *bid'ah* (heresy) and the misinformation over the Shia teachings did not stop at the ideological stage, but worsened into the killing of the Shia people in Karbala (El-Fadl, Sejarah Wahabi & Salafi, 2015, p. 41). This act was then followed by most Indonesian Muslims, as mentioned in Introduction, by banning the celebration of Ashura Day in Bogor and making the Shia people into an exile in Sampang, Madura.

2 CONCLUSION AND SOLUTIONS

Mazhab-centered Islamic education in Indonesia has created several problems resulting from not only internal issues but also external affairs, especially from the Middle East. The foremost challenge for Islamic education in Indonesia is the existence of the *ghulat* community, which provokes generalization between the *ghulat* community and one of the *mazhabs*, the Shia. The second challenge is the advent of radicalism, including that of the Salafi-Wahabi movement, which instigates violence in Indonesia toward the Shia group, at the ideological level not only through the release of *fatwas* against the group but also through physical violence, similar to that in the Shia community in Sampang, Madura.

For this reason, Islamic education in Indonesia must undergo reformation. On the issue of generalization, Quraish Shihab stated, "One of the complaints from the Shia... is that writers... often times write negative opinion about Shia, even though they have perished as a group" (Shihab, 2014, p. 19). As mentioned above, Islamic education in Indonesia, especially that supported by MUI, tends to highlight the negative opinions on the Shia. For Shihab, the opinion of ulema from the majority Shia groups like Ja'fariyya and Zaidiyya should be the reference, not the minority (*Ibid*). In addition, one of the common points of Sunni and Shia can be viewed from the perspective of Shia Zaidiyya. This may happen because Shia Zaidiyya tends to use the Hanafifiqh in religious matters. Upon further analysis, the Sh'aJa'fariyya, whose fiqh is different from that of Sunni's *mazhab,* is much more different from that of Zaidiyya. The views of the five *mazhabs* of Sunni are based on Ja'farifiqh. This is clear because the Imams of the four *mazhabs*, Abu Hanifah and Malik bin Anas, are the students of Imam Ja'far ash-Shadiq (Ash-Shadr, 2014, p. 145) who was the sixth Imam of the Shia Ja'fariyya doctrine. The question of whether the Shia Ja'fariyya is part of Islam due to different views in seeing fiqh can be analyzed from the statement of Shia scholar named Muhammad Husain Kashiful-Githa, "Consider the libraries of the Shia's. The rows are well-stocked! Examine our own library as well!" (Kashiful-Githa, p. 4). Furthermore, Kasyiful-Githa denied the accusation that all Shia is not Islam by saying, "Their religion is pure '*tawhid*' (Oneness of God)" (*Ibid*, p. 15).

Amir Maliki Abitolkha states:

> ... Islamic education has to change its strategy and operational method. The changes in strategy and methods require overhauling the models as well as its institution, so that the education can be more effective and efficient in pedagogic, sociologic and, cultural aspects. (Abitolkha, 2014, p. 116)

An institutional overhaul is certainly needed, especially from the perspective of the institution, particularly if the institutions refer to MUI. According to Novan Ardy Wiyani, there are nine important points in developing Islamic education in Indonesia, some of them being fairness, respect for others, and tolerance (Wiyani, 2013, p. 75).

As the highest body of education in Indonesia, MUI should adopt a more comparative method, that is, it uses not only the internal Sunni methods but also other *mazhabs*, such as Shia Ja'fariyya and Zaidiyya, suggested by Shihab and Kashiful-Githa. Hence, I disagree with the fact that the school-based education proposed by Abitolkha refers to giving the full authority to the schools to create a curriculum. Even though "... it's open and inclusive to foreign resources of school environment..." (Abitolkha, 2014, p. 118), it is possible for the schools to have full authority in choosing the resources that support their *mazhab* only. MUI should still design the curriculum that must be continually improved. For that purpose, MUI cooperates with the Ministry of Religious Affairs to create a tolerant and non-*mazhab*-centered curriculum. After that, a supervisory body should be established to find schools still using *mazhab*-centered curriculum. The curriculum made by MUI and the Ministry of Religious Affairs should not be absolute, but a divider curriculum. The schools or similar institutions may deal with the rest. Hence, Abitolkha meant that there is still an interrelationship between the highest institutions and local resources.

Within MUI, there should also be non-Sunni ulemas. This is done to establish a relationship between Sunni and Shia, to restore each interpretation of the Qur'an and Sunnah using active logics or the aforementioned three Islamic epistemologies, and not only using the dominant logic or *ijtihad* that is considered absolute and unchanged. According to Al-Jabiri, these three epistemologies used in many *mazhabs*, especially by Sunni and Shia, refer to the two same sources: Qur'an and Sunnah. They use the three epistemologies because Qur'an and Sunnah do not consist of *bayan* (language), but *irfan* (spirit) and *burhan* (logics). It is widely believed that Islamic education in Indonesia can be more creative and comparative.

With the aforementioned improvements in Islamic education, Indonesia can tackle the issues of radicalism and fundamentalism. With the creative Islamic education, the society is expected to not only accept *ijtihad* carelessly but also examine the truth of the *ijtihad* and its legal sources. With the comparative Islamic education, the society is expected to be thoughtful in drawing conclusions by not only using one perspective of *mazhab*, but also examining a situation by acquiring knowledge on different *mazhabs*, so as to avoid *mazhab*-centered education. Furthermore, the comparative education can inform the society that Islam is not as rigid as it is thought to be, but it provides several alternatives in solving current problems.

REFERENCES

Abitolkha, A.M. (2014) *Problematika Penyelenggaraan Pendidikan Islam (Pendekatan Manajemen Berbasis Sekolah)* (Problematic Islamic Education Administration [School-Based Management Approach]). Tadris, 9 (1), 111–131.

Affan, H. (2013) Konflik Keluarga, Mazhab atau Politik (Family, Mazhab, or Political Conflict). [Online] Available from: http://www.bbc.com/indonesia/ [Accessed 2nd August 2016].

Ahmad, M.G. (2012) *Barahin-e-Ahmadiyya Part I & II.* London, Islam International Publication Ltd.

Ahmad, M.G. (2014) *Barahin-e-Ahmadiyya Part III.* London, Islam International Publication Ltd.

Ali, M. (2016) *Islamologi* translated by Kaelan, R. & Bachrun, H.M. Jakarta, Darul Kutubil Islamiyah.

Al-Jabiri, M.A. (2014).

NU Online. (2015) Kebijakan Walikota Bogor soal Syiah Dinilai Diskriminatif (The Bogor Mayor's Policy on Shia Considered Discriminative). [Online] Available from: http://www.nu.or.id/post/read/63162/kebijakan-walikota-bogor-soal-syiah-dinilai-diskriminatif [Accessed 2nd August 2016].

Nughroho, S.E. (2007) *Muhammad 'Abid al-Jabiri (Studi Pemikirannya tentang Tradisi (Turas)* (The study on his thought on tradition). Yogyakarta, Universitas Islam Negeri Sunan Kalijaga.

Ramdhoni, F. (2013) Di Balik Merebaknya Konflik Sunni-Syiah di JawaTimur (Behind the arising conflict of Sunni-Shia in East Java). [Online] Available from: http://www.nu.or.id/post/read/47029/di-balik-merebaknya-konflik-sunni-syiah-di-jawa-timur [Accessed 2nd August 2016].

Riswanto, A.M. (2010) *Buku Pintar Islam* (Islam Smart Book). Bandung, Penerbit Mizan.

Shihab, M.Q. (2014) *Sunnah-Syiah Bergandengan Tangan! Mungkinkah?* (Sunni-Shia working together? Is it possible?) (ed.). Tangerang, Lentera Hati.

Sobhani, J. (2001) *Doctrines of Shi'i Islam: A Compendium of Imamis Beliefs and Practices.* (R. Shah-Kazemi, Trans.) New York, I.B. Tauris & Co Ltd.

Thabataba'i, M.H. (1975). *Shi'ite Islam.* (S. H. Nasr, Trans.) New York, State University of New York Press.

Thabataba'i, M.H. (1983). *Al-Mizan: An Exegesis of the Qur'an.* (S.A. Rizvi, Trans.) Tehran, World Organization for Islamic Services.

The Lahore Ahmadiyya Movement for the Propagation of Islam. (n.d.). *A Comparative Study of the Beliefs of the Two Sections of the Ahmadiyya Movement (Lahore vs. Qadiani Groups)*.[Online] Available from: The Lahore Ahmadiyya Movement for the Propagation of Islam: http://aaiil.org/text/qadi/intro/cmprsn.shtml [Accessed 29th July 2016].

Tim Ahlulbait Indonesia. (2014) *Syiah menurut Syiah* (Shia according to Shia). Jakarta, Dewan Pengurus Pusat Ahlulbait Indonesia.

Tim Penulis MUI Pusat. (2012) *Mengenal & mewaspadai penyimpangan Syi'ah di Indonesia* (Knowing & Being Aware of the Shia Deviation in Indonesia). Jakarta, Nashirus Sunnah.

Wardah, F. (2013) Aktivis: Kekerasan Terhadap Warga Syiah Sampang Berpotensi Genosida (Activists: Violence towards Shia followers in Sampang having the potential for genocide to happen). [Online] Available from: VOA Indonesia: http://www.voaindonesia.com/a/aktivis-kekerasan-terhadap-warga-syiah-sampang-berpotensi-genosida/1737595.html. [Accessed 2nd August 2016].

Wiyani, N.A. (2013) Pendidikan Agama Islam Berbasis Anti Terorisme di SMA (The Anti-Terrorism Islamic Religion Education in Senior High School). *Jurnal Pendidikan Islam,* 2 (1), 65–83.

Developing English teaching materials based on Content-Based Instruction (CBI) approach for the Islamic management and banking students at IAIN Imam Bonjol Padang

Elismawati & Mukhaiyar
Faculty of Tarbiyah and Teacher Education, Imam Bonjol Islamic State Institute, Padang, Indonesia
Faculty of Language and Art, Padang State University, Padang, Indonesia

ABSTRACT: The lack of English teaching materials containing business topics is one of the factors leading to the students' poor grasp of English for Business Purposes. This research is designed to develop English teaching materials based on the content-based instruction approach for the students of the Islamic Management and Banking Study Programme at IAIN Imam Bonjol Padang. The method is R&D, adopted from the ADDIE model consisting of five phases: Analyse, Design, Develop, Implement, and Evaluate (ADDIE). The phases are structured to allow an exploration of the students' need, to develop the design and to identify its validity, and to verify the practicality of the developed learning materials. However, the evaluation phase is excluded due to the limited time and budget. The respondents of this research were taken from two classes of first-year students (79 students) and one class of third-year students (39 students). The results of the needs analysis show that the students preferred to learn English for Business Purposes as opposed to General English. The business topics that they wish to learn the most were related to subjects of recruitment, CV writing and application letters, job interviews, business and telephoning etiquettes, etc. In the implementation phase, it was found that the developed materials were considered as practical for students of the Islamic management and banking.

1 INTRODUCTION

English has become the most powerful and essential tool of communication in the era of globalisation. Almost no parts of the world—and only very few of our daily routines—that are not exposed to English, because as a means of communications, English permeates almost all aspects of our lives—from politics, economics, science, technology, education, and socio-cultural aspects. English becomes the language of many nations, and the language in every sector of the post-modern age. This means that they who reject English will eventually be excluded from the global community. As a result, the teaching of English should be given a high priority at all levels of the education system. Richards (2002) states that English is becoming the language of globalisation, which is highly needed in a business-based management for commerce and trade. In higher education in Indonesia, English is taught in every study programme, including in the Islamic Management and Banking (MPS) Study Programme at IAIN Imam Bonjol Padang.

Nowadays, people have started to consider the importance of learning English as a tool for global communications. This certainly has given English a somewhat special status (Crystal, 2003), and therefore English Language Teaching (ELT) has experienced continuous evolution and development (Varela, Polo, Garcia, & Martinez, 2010). Thus, it is no wonder that with globalisation and the role of English as an international language, the requirement of English competence is obvious (Astika, 2012). As a result, the enthusiasm to learn English has intensified. English has a very important position in Indonesia. This is reflected by several relevant policies issued by the Ministry of Education of the Republic of Indonesia (e.g. the Decree No. 060/U/1993, No. 020/U/2003, & No. 12/U/2012), which require English to be taught in schools, starting from lower secondary schools to university level.

The difficulties mostly faced by Indonesians who learn English as a foreign language, in fact, as admitted by the students, are that the globalisation era has significantly affected a number of sectors in Indonesia, such as education, economics, politics, and social (Rokhayani, 2012). This presents a new challenge for Indonesians to compete in the global arena with people from all over the world. To compete in this era, it is necessary to have human resources who are fluent in English.

Through interviews and observations conducted at MPS IAIN Imam Bonjol Padang, the author focused on several issues in English classes, especially those related to the course design, students, and learning materials. The course is designed independently by the lecturers. The institution, in this case the Faculty of Islamic Economic and Business (FEBI), does not determine the specific goals to be achieved from the course. The institution only expects that the students will be able to use oral and written English in business contexts, but such goals are too broad to be defined by the lecturers who design the courses.

With that in mind, the author carried out a study in developing English learning materials by applying the content-based instruction approach. This approach can be an alternative solution for at least three reasons. First, the content-based approach is appropriate for vocational schools or colleges, such as business schools in which the learners are trained to master English for academic and job-related purposes (Davies, 2003). According to Snow (in Villalobos, 2013), the content is subject matters that consist of topics or themes based on the interests or needs of the learners in an adult EFL setting, or it may be very specific, such as the subject that the students are currently studying. The second reason is that the content-based syllabus can be developed with the principles of ESP (Chen, 2015), in this case are English for business, academic, and job purposes. The third reason is that the content-based approach enhances the intrinsic motivation of the learners and empowers them to focus on the subject matters that are important to them (Brown, 2007).

Given the facts, this study is aimed to answer the following research questions: 1) What are the needs of the students at MPS Department IAIN Imam Bonjol Padang? and 2) How are the content-based learning materials for the MPS students designed?

2 LITERATURE REVIEW

2.1 English for Specific Purposes (ESP)

ESP is defined as an approach to language teaching in which all decisions with regard to content and method are based on the learners' reasons for learning (Hutchinson & Waters, 1987). ESP has a long history in the field of language teaching, which started in the 1960s, when General English courses could not meet the needs of language learners. There are three reasons common to the emergence of ESP courses: the demands of Brave New World, a revolution in linguistics and focus on the learners (Hutchinson & Waters, 1987).

2.2 The development of teaching material for ESP

The development of teaching material as one aspect in teaching English for Specific Purposes is aimed to analyse the students' needs and develop teaching materials that is synchronized with their needs. Byram (2000) argues that teaching material development is a process to identify the students' needs and to tailor-design the material to fulfil those needs. This means that the development of teaching material for ESP not only focuses on providing the required material, but also on analysing the students' needs. Hence, the material provided will prepare and support the students to fulfil their needs in workplaces.

2.3 Content-based instruction approach

Brinton, Snow, and Wesche (in Brown, 2007) define the content-based instruction (CBI) as an approach that integrates the content learning with language teaching aims. More specifically, they stated that CBI refers to the concurrent study of language and subject matter, with the form and sequence of language presentation dictated by the content material.

The principle of CBI is to ascertain the connection between CBI and developing English teaching materials. Richards and Rodgers (2001) claim that there are two central principles as the basis of CBI. Firstly, people learn a second language more successfully when they use the language as a means of acquiring information, rather than as an end itself. Secondly, CBI better reflects the learners' needs for teaching a second or foreign language. These two principles reveal that CBI leads to more effective language teaching and meets the student's needs.

2.4 *Needs Analysis (NA)*

Needs analysis has a vital role in the process of designing and carrying out any language course (Hutchinson & Waters, 1987). Similarly, Dudley-Evans and St John (1998) put NA as the first stage before moving to other stages, such as course and syllabus design, material selection (and production), teaching and learning, evaluation in the process of designing ESP courses, and works related to designing language instruction, including developing teaching materials.

3 METHODOLOGY

This study is a research and development (R&D) one. It focuses on developing English teaching materials based on the content-based instruction approach, in which the respondents are divided into two groups. The participants of this research are two classes of first-year students (79 students) and one class of third-year students (39 students).

4 DATA ANALYSIS AND DISCUSSION

4.1 *Data analysis*

Using the research and development method of the ADDIE model that includes analysis, design, development, implementation, and evaluation, this research analysed the data with the first four phases, excluding evaluation—through applying various instruments for different respondents. The analysis was based on descriptive statistics, such as frequency, percentage, and means. In the analysis phase, frequency and percentage were calculated to identify the students' needs from a set of questionnaires that covers aspects of target situation analysis and objective needs; wants, means, subjective needs analysis; present situation analysis; lack analysis; learning needs analysis; linguistics analysis; what is expected of the course; and means analysis. The highest percentage of the student's responses towards the analysis was a priority to be considered in the design phase. Before distributing the instrument to 39 third-year students, an expert of business content validated it.

The data from the designing and developing phases were analysed by an instrument called validation checklists to determine the validity of the product. There were three validation checklists to be completed by one business-content expert and two material development and language use experts. They evaluated the design and developed materials by scoring the content, language understanding, presentation, and writing mechanics. The scoring system referred to the rubrics of materials development from The Guideline of Materials Development, issued by the Department of National Education in 2008. The classification of the validity of the product was conducted with Ridgway's category (2005).

In the implementation phase, other descriptive statistics, such as means, percentage, and average scores were calculated to summarise the practicality of the products. The scores were obtained from the responses classified using a four-point scale: 1 for impractical; 2 for less practical; 3 for practical, and 4 for very practical. Two sets of questionnaires were used as the instruments. The first one was distributed to two classes of first-year students. The questionnaire was used to evaluate the students' perceptions regarding the compliance of time, the convenience/ease of use of the content-based English learning materials for business students, and the benefits of using the content-based English learning materials for business students. Another set of instrument was given to one Business lecturer and one English lecturer with business background, to gain the lecturers' perceptions on the accuracy of time, the convenience/ease of use, and the completeness of the components of the content-based teaching materials for business students.

The average scores from the two groups of 79 students and 2 lecturers were calculated separately to find out the final score regarding product practicality from each group. The data from each phase were then interpreted and discussed. The results in the analysis phase were obtained from the analysis of eight aspects. The first one was personal information about the learners and their attitudes toward English. It was revealed that 73.54% of the students realised that English is very important. More specifically, they agreed that English is important not only for academic, but also for professional needs. Therefore, 69.13% of the students preferred learning English for Business Purposes as opposed to General English. However, 74.21% of the students did not bother to make the extra efforts to improve their English. Consequently, for the second aspect, it was revealed that their proficiency in English and English for Business Purposes were mostly categorised as poor.

Furthermore, the third aspect, i.e. language information about a target situation, shows that the conditions were caused by the absence of the needs analysis before class and the lack of learning materials on business topics. For that reason, they could not improve their ability in using English for both academic and working purposes in the business world.

The fourth aspect of the analysis was related to the situation involving English in business situations. It was revealed that the highest percentage was business-related situation at 79.29%, followed by situations involving writing CV and job application letters at 78.12%, business reading etiquettes at 75.56%, banking administration at 71.12%, and giving instructions and directions at 64.11%. The rest of the situations—such as job interviews, requesting information by e-mail, telephone calls, explaining customer interactions, and giving suggestions and solutions—ranked the lowest, all combined to less than 25%.

The fifth aspect was related to the need for language learning that covers language skills and other learning activities that students need most. It was found that 82.61% of the students need to improve their speaking skills, followed by listening skills at 80.14%, reading at 79.53%, and writing at 21.78%. The learning activities they wanted the most was pair work (82.00%). Another aspect was the analysis related to the business topics the students wanted to learn the most. It was found that the students were interested in five topics, ranging from 84.44% to 71.11%. The findings show that the students need to improve their fluency in English for business—for both academic and working purposes at 90.14%. The last aspect was the means analysis. It revealed that 52.14% of the students would like to learn English with business content materials from the first year in the second semester, and they would like to be taught by a team of lecturers with business and English backgrounds.

Following the analysis, it was necessary to prepare a diagram as a guide in the design phase. The design of the developed materials is illustrated in Figure 1.

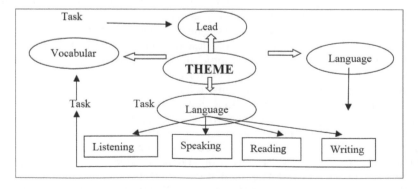

Figure 1. Design process.

Figure 1 briefly shows how the design process was initiated from the development of main themes that the students wanted to learn the most. The themes were developed from the data obtained from the previous phase. From the themes/main topics, 10 sub-topics were developed by a number of handouts used for one semester, with 14 meetings excluding mid-term and final tests. The themes also became the starting points to decide on the organisation and the contents of the developed materials. The organisation consists of four main parts, and they are: lead in, vocabulary, language focus, and language skills. Then, the language skills are divided into four parts covering listening, speaking, writing, and reading skills. Completing the contents and tasks of each part, the activities related to selecting an appropriate syllabus, setting up learning objectives, choosing the topics and sub-topics, collecting and choosing the appropriate texts, making decisions with regard to learning sources, and designing tasks were all documented as the blueprint of the developed materials.

Then, the expert of material development, who later recommended some suggestions, validated the blueprint. First, the expert suggested drawing a chart to outline the organisation of the materials. Second, the expert suggested including examples related to business contexts in the language focus section. After the design phase, the author prepared the blueprint as printed materials in a handout model. To get a better result and a functional product, the business content and material development experts validated the product that covers the aspects of contents, language understandings, presentations, and writing mechanics.

Based on the experts' evaluation and validation, the developed materials were rated as valid. However, they suggested revising some of the elements. The first validator recommended revising the front page of each handout. She also suggested revising the level of difficulty of the task in handout 2. Thus, the task was revised from writing a short paragraph to rearranging jumbled words into complete sentences. The second validator advised giving a clear explanation to prevent a misunderstanding and to ensure the students would know what to do first. The third validator revised the materials regarding business contents and their sequences. He suggested to include discussions about office administrations.

Based on the validators' suggestions, the author revised and printed the materials to be a product that would be applicable in a try-out session during the implementation phase. After the try-out session, the author found that the lecturers and the students agreed that the developed materials were practical for business students. However, the indicator of the compliance of time was rated at the lowest score. The highest score was obtained from the indicator of the benefits of the material. It scored 90.94 and was categorised as very practical.

5 DISCUSSION

In this part, the author discusses the four phases of research procedure that provide the answers to the research questions defined in Chapter 1.

5.1 *Analysis*

Before designing a learning material to be used in a learning process, it was necessary to conduct an NA first. Hutchinson and Waters (1987) clearly point out that the needs analysis has a vital role in the process of designing and carrying out any language courses. Furthermore, defining the students' needs is the way to find out the reasons of the students to learn English (Dudley-Evans & St. John, 1998). The data on the analysis phase shows that the students at MPS IAIN Imam Bonjol Padang needed English for Business Purposes, which would allow them to master business subjects, as well as to prepare them to use Business English for job purposes. However, the data revealed their masteries of English for Business Purposes were poor. Therefore, they prefer to learn topic-related business English.

In learning English for Business Purposes, the students mostly preferred to do paperwork for their classroom activities. In addition, the analysis also shows that they would like to be taught English for specific business purposes by a team of lecturers with business and English backgrounds.

5.2 The design

Based on the results of the NA conducted in the first phase, the author continued to the second phase i.e. the design phase. This phase was the part that inspired the author to find the answer to the second research question: How is the content-based learning material for the MPS students designed? Therefore, in this phase, the author decided to design a course syllabus based on the information collected in the analysis phase. Since the analysis shows that the students prefer learning English for business purposes as opposed to English for general purposes, the author decided to pick a content-based syllabus as a guide to develop the teaching materials. It is based on the premise first proposed by Richards (2001), who states that with a topical syllabus, content—rather than grammar, functions, or situations—becomes the starting point in the syllabus design. Therefore, the author listed all business topics in the questionnaire, the syllabus, and the materials that were based on the business language tasks and skills as defined in the business curriculum structure proposed by Hussin (in Orr, 2002). The language tasks and skills mentioned in the curriculum are explained in detail in Chapter 2. Based on the analysis of the language learning needs and the weakness analysis, it was found that the students' were weak in speaking skills. Therefore, they needed more speaking-related exercises than other activities. Thus, the author put more exercises in the speaking section. They covered not only questions and answers, but also role-playing exercises with business related topics.

5.3 Development

After the design phase, the author prepared the learning materials, followed by a validation process. The validation result shows that the learning materials developed on the content-based approach for students at MPS-IAIN IB Padang were categorised as "very valid". It was well expected as the author had strictly followed the procedure for developing learning materials adapted from Nunan's procedure (1991). In the procedure, the author outlined the activities to follow, such as selecting topics, collecting data, determining what the learners need to do about the texts, creating pedagogical activities/procedures, analysing the texts and activities to determine the language elements, creating activities focusing on language elements and learning skills/strategies, and creating application tasks.

5.4 Implementation

After the completion of the developed materials in a handout model, the author came to a try-out session to see the feasibility of the developed materials in an activity called the implementation phase. The result shows that in general, the students consider the developed materials as "practical". However, the indicator of the compliance of time received the lowest score. It was categorised as "fairly practical". The author assumed that such result was due to the overwhelming amount of subject matters discussed in each handout. Furthermore, the students found that the exercises were much more difficult as it was their first time using materials tailor specifically to business-related topics.

The highest score in the "very practical" category was obtained from the benefits of using the content-based English learning materials for business students. It means that learning materials that encompassed the topics developed from business contexts are very helpful to support the students' mastery of English for both academic and business related purposes.

In line with the students' perception, both lecturers and students also rated the compliance of time as "fairly practical" for the same reason. However, the indicator related to the completeness of the components of the content-based teaching materials for business students obtained the highest category, i.e. "very practical". It was quite reasonable because the product covers the four languages skills and is complemented with language focus and vocabulary sections, followed by relevant exercises.

6 CONCLUSION

This is a research and development study whose aim is to develop English teaching materials based on the Content-Based Instruction (CBI) approach for the Islamic Management and Banking (MPS) students. Specifically, the primary objectives of the study are to find out the real needs of MPS students, to design and develop the learning materials based on those needs, and to find out the validity and practicality of the developed materials. As a result, the final product of this study was a set of content-based syllabus and content-based handouts. The findings in the analysis phase revealed the real needs of the business students. They preferred learning English for Business Purposes as opposed to General English. The findings in the design and development phases show that the product was categorised as "very valid." Results in the implementation phase proved that the products were practical to use in business schools. However, the author realises that there might be some imperfections in the final product due to the limitations of the research.

REFERENCES

Astika, G. (2012) Teacher and child talk in PAUD English class. In: M. Syafe'I, H. Madjdi, & Mutohar (Eds.) *Proceedings of the 2nd National Conference on Teaching English for Young Learners in Indonesia*. Kudus, Indonesia: Muria Kudus University. pp. 46–57.

Brown, H.D. (2007) *Teaching by principles: An interactive approach to language pedagogy*. 2nd Edition. New York: Longman.

Chen, Y. (2015) Content-based business English course for EFL. *The Internet TESL Journal*, 16(1). Available from: http://iteslj.org/Lessons/Chien-BusinessEnglish.html.

Crystal, D. (2003) *The Cambridge encyclopaedia of the English language*. Cambridge: Cambridge University Press.

Davies, S. (2003) Content-based instruction in EFL context. *The Internet TESL Journal*, 9(2).

Dudley-Evans, T. & Jo St. John, M. (1998) *Developments in English for specific purpose: A multi-disciplinary approach*. Cambridge: Cambridge University Press.

Hutchinson, T. & Waters, A. (1987) *English for specific purposes: A learning centred approach*. Cambridge: Cambridge University Press.

Mario-Cal, V. (2010) *Current issues in English language teaching and learning*. Cambridge: Cambridge University Press.

Michael, B. (2000) *Teaching material development*. Cambridge: Cambridge University Press.

Ministry of Education of the Republic of Indonesia. (1993) Decision No. 060/U/1993.

Ministry of Education of the Republic of Indonesia. (2003) Decision No. 020/U/2003.

Ministry of Education of the Republic of Indonesia. (2012) Decision No.12/U/2012.

Nunan, D. (1991) *Language teaching methodology*. Cambridge: Cambridge University Press.

Orr, T. (2002) English for specific purposes. In: J. Burton (Ed.) *Case Studies in TESOL Practice Series*. Virginia: Teachers of English to Speakers of Other Languages, Inc.

Polo, F.J.F., Garcia, L.G. & Martinez, I.M.P. (2010) *Current issues in English language*. Newcastle: Cambridge Scholars.

Richards, J.C. (2001) *Approaches and methods in language teaching*. 2nd edition. Cambridge: Cambridge University Press.

Richards, J.C. (2001) *Curriculum development in language teaching*. Cambridge: Cambridge University Press.

Richards, J.C. (2005) Communicative language teaching today. *RELC Portfolio Series 13*. Singapore: SEAMEO Regional Language Centre.

Riduwan. (2005) *Belajar mudah penelitian guru, karyawan, dan peneliti muda (Easy learning research for teachers, workers, and junior researchers)*. Bandung: Alfabeta.

Ur, P. (1991) *A course in language teaching: Practice and theory*. Cambridge: Cambridge University Press.

Villalobos, O.B. (2013) Content-based instruction: A relevant approach of language teaching. *Journal of Innovation Education*, 17(20).

Cultural Dynamics in a Globalized World – Budianta et al. (Eds)
© *2018 Taylor & Francis Group, London, ISBN 978-1-138-62664-5*

The main ideas of Muhammad Mursi's speech at the 67th United Nations General Assembly: A critical discourse analysis

D. Lu'lu & A.T. Wastono
Department of Linguistics, Faculty of Humanities, Universitas Indonesia, Depok, Indonesia

ABSTRACT: A speech given by an orator contains certain ideas that need to be conveyed. These ideas also include power relations. This study aims to reveal the main ideas of Muhammad Mursi's speech at the 67th United Nations General Assembly and to explain the language strategies used to establish the power relations between him and his audience. The speech was delivered in Modern Standard Arabic. The method used in this research is a qualitative one. This study employs Fairclough's (2010) critical discourse analysis (CDA) as the core theory. As the core theory, it also encompasses other theories, such as van Dijk's macro rules (1980), Halliday's functional grammar (2014), Nida's components of meaning (1979), Yule's presupposition (2010), and the theory of power relations by Foucault (2008) and Fairclough (2015). The result shows that there are two main ideas behind the speech, namely Muhammad Mursi's identity as the President of Egypt after the revolution, and Egypt's new visions of itself and the world. Furthermore, the language strategies used to build the power relations were applied through the use of powerful dictions, epistemic modalities and interjectional particles, functional imperatives, emphatic particles and nouns, statement on the identity of the orator, knowledge, and deletion of several participants.

1 INTRODUCTION

A speech is a means of communication delivered by an orator. Through a speech, an orator conveys the idea(s) in front of an audience. On the 26th September 2012, Muhammad Mursi[1] had the opportunity to speak at the 67th United Nations (UN) General Assembly (GA). The theme was "Bringing about Adjustment or Settlement of International Disputes or Situations by Peaceful Means" ("FEATURE: What does the UN General Assembly do When the General Debate Ends?", 2012). The speech was delivered in Arabic. Meanwhile, the audience includes, among others, the President of the 67th UNGA, the UN Secretary-General, and the heads of the UN member states.

In his speech, Mursi presented some ideas that were manifested in several discourses. They were related to issues in the Middle East and the world, such as the condition in Egypt after the revolution and issues on Palestine, Syria, Africa, nuclear weapon, the international system, and the global economy. The speech deliverance showed that Mursi was engaged in a social interaction with his audience in an effort to bring about social change.

Regarding the delivered ideas, Mursi attempted to build power relations with his audience when he spoke. Power relations were apparent in the diction he used. For example, he used the verb أَدْعُو /ad'ū/, which means 'I invite' when he talked about Palestine. The word has the same root with دَعْوَة /da'wah/, which means 'a calling, an invitation, a lecture', which contains the concept of an invitation accompanied by an action. Before Mursi, no Arab leaders used that word when they were engaging with the international community to support Palestine's independence.

---

1. Muhammad Mursi was directly elected in a general election after the Egypt Revolution in 2011 as the President of Egypt. He served as a president from 2012 to 2013.

The diction usage indicated that Mursi used power to invite his audience, and that he was serious to end the Palestinian conflict. It showed that the power relations built by Mursi were the embodiment of the social struggle in the international community because power relations are always associated with the social struggle between community (Fairclough, 2015). Based on that, the objective of this paper is to reveal the main ideas of Muhammad Mursi's speech at the UN General Assembly. In addition, it also endeavours to explain the language strategies he used to establish power relations with the audience. Critical discourse analysis (CDA) as proposed by Norman Fairclough (2010) is used as the main theory.

2 METHODOLOGY

The paper uses a qualitative method. It is chosen to expose the reality so that the real empirical facts are assumed in a socio-cultural context. Trochim (2006) states that the qualitative method can provide results for an in-depth and detailed research on social phenomenon. Gubrim (1992) in Somantri (2005) states that the qualitative method in CDA puts a great emphasis on practice and context. In line with those statements, Fairclough's CDA has two dimensions of analyses, namely the textual dimension and the socio-cultural practice dimension that encompassed the situational, institutional, and social contexts. Both are linked with the dimension of practical discourse.

The source of data in this paper is Muhammad Mursi's speech at the 67th UNGA. The data were retrieved from http://www.c-span.org/video/?308405–2/egyptian-president-morsi-united-nations-general-assembly-address, which is a broadcast of his speech. Data collection was conducted by watching the video, preparing orthographic transcription and transliteration of the speech, and translating the text from Arabic to English. After that, the data source was classified into three sections: introduction, content, and conclusion. The content section, especially the propositions, was chosen as the data of Mursi's speech. It was codified by numbering each of the discourse, paragraph, and proposition.

3 CRITICAL DISCOURSE ANALYSIS AS THE THEORETICAL FRAMEWORK

Critical discourse analysis (CDA) as proposed by Norman Fairclough (2010) was used as the main theory in this study. The theory focuses on the formation of discourses as an attempt to create social change (Fairclough, 2010). Fundamentally, CDA aims to investigate the relations of causality and determination between discursive practices, events, and texts, as well as structures, relationships, and the broader social and cultural processes (Fairclough, 2010).

The objective was achieved by analysing the linking text at the micro level and the social context at the macro level. As a mediator, Fairclough created "the bridge" for both, namely the discursive practice. Accordingly, Fairclough designed a framework consisting of three dimensions: (1) the dimension of text, (2) the dimension of practical discourse, and (3) the dimension of socio-cultural practice (Fairclough, 2010). The dimension of text includes the linguistic description of the text's language. The dimension of practical discourse includes the interpretation of the relationship between the discursive process and the text. The dimension of socio-cultural practice includes an explanation of the relationships between the discursive processes and the social processes (Fairclough, 2010).

With regard to the dimensions of text, there are three kinds of analyses: analysis of the propositions, of the grammar analysis, and of the lexical analysis. Analysis of the propositions was conducted to obtain the macro-proposition in each discourse. It was obtained by using van Dijk's macro-rules theory (1980). There are four ways of implementing the macro-rules, i.e. the deletion [D] rule, the selection [S] rule, the generalization [G] rule, and the construction [C] rule (van Dijk, 1980). It should be noted, however, that the [S] and [D] cannot be performed on the same proposition because the proposition operated by the [S] is a proposition that is relevant to other propositions to build the macro-proposition (Renkema, 2004). Therefore, it cannot be eliminated.

Grammatical analysis was used to reveal the relations of social identities, representations, and the distribution of information (Fairclough, 2010) using Halliday's systemic functional linguistics (2014). Social relations refer to how participants interact through the language displayed in the text. Social identities are concerned with how participants identify themselves. Halliday refers to both elements as the interpersonal function. Meanwhile, representation is concerned with how individuals, groups, situations, or experiences are shown in the text. That element is in line with Halliday's view, which is referred to as the ideational function. On the other hand, the distribution of information is related to how the information is presented. In Halliday's view, that element is parallel to the textual function.

Next, lexical analysis is used to identify the common and diagnostic components that belong to the particular diction or phrases used. For that, the component analysis of meaning offered by Nida (1979) was used. Common components refer to the components of meaning that connect a word to others in the scope of semantics, while diagnostic components refer to the components of meaning used to distinguish one word to another.

The dimension of practical discourse requires intertextuality, which refers to the notion of presupposition by Yule (2010). Presupposition is defined as what an orator assumes to be true or known by the audience (Yule, 2010). It was used to interpret the language strategies used by the orator to establish specific power relations with the audience. This dimension also uses the theory of power relations proposed by Foucault (2008) and Fairclough (2015).

Foucault (2008) states that power can be realized through knowledge. Knowledge always has the effect of power. For instance, an orator who speaks to promote peace has more information and knowledge than the audience concerning the situations in a specific area that requires assistance from other countries to bring about peace in that area. His knowledge on the situations can be used as a tool to build power relations. It shows that power is an application of knowledge, which is manifested in a discourse.

In line with Foucault, Fairclough (2015) writes that power relations are always associated with struggles. Such struggles refer to the process that involves individuals, groups, or institutions with different interests. In a discourse, there are two aspects of power relations. First, power in a discourse that focuses on discourse as a medium to carry out and enact power relations. It is suggested that a discourse is the site of a social struggle. Second, the power behind a discourse that focuses on how the order of discourse is formed and constituted by power relations. Thus, the discourse is put as a stake in the social struggles. Based on that, the presence of power relations can be technically inferred in the discourse through dictions, phrases, grammar, or the order of discourse used by the actor of the discourse. In addition, an expression of affirmation and management of topics also constitute as indicators of power relations in a discourse.

The dimension of socio-cultural practice explains the result of discourse interpretation and its connection to the macro context (Fairclough, 2010). In this paper, the macro context was at the situational and social levels. Situational level is the process of text production at a typical condition, so it produces a text that is different from any other conditions. Discourse, as a result of social interaction, is considered as an act that responds to a condition. On the other hand, social level functions are about finding out how macro context affects the production of a discourse. Furthermore, the social level is also used to reveal how the discourse was produced.

4   CRITICAL DISCOURSE ANALYSIS OF MUHAMMAD MURSI'S SPEECH

The main ideas and language strategies were obtained after applying the three stages of Fairclough's CDA. At the first dimension, which is the description of the text, there are ten macro propositions.

Language strategies practiced by the orator was revealed through the multifunctional meaning, namely through the interpersonal, ideational, and textual functions. In the interpersonal function, the declarative mode was the grammatical mode that was used in all discourses. Its focus was the agent and the action that was taken. In the ideational function, the data implies partisanship of the orator on the issues concerning Egypt, the Palestinian, Syria, Sudan, Somalia, the elimination of nuclear weapons, criticism of the international

Table 1. Macro propositions of Muhammad Mursi's speech.

MP Code	Macro Propositions
I	وَنَسْعَى لِرُؤْيَةِ مَصْرِ الْجَدِيدَةِ إِنَّنِي مُنْتَخَبٌ دِيمُقْرَاطِيٌّ /innanī muntakhbun dīmuqrāṭiyyun wa nas'ā li ru'yati maṣr al-jadīdah/ 'I am indeed democratically elected and we strive for the new Egypt's vision.'
II	أَدْعُوكُمْ إِلَى دَعْمِ أَبْنَاءِ فِلَسْطِينِيٍّ لِجِهَادِ بِنَاءِ دَوْلَتِهِ الْمُسْتَقِلَّةِ /ad'ūkum ilā da'mi ḥallihā 'bnā'i l-filasṭīniyy li jihādi binā'i daulatihi al-mustaqillah/ 'I call you to support the striving Palestinians to establish the Independent State (of Palestine).'
III	لَابُدَّ أَنْ نَتَنَاوَلَ الْقَضِيَّةَ وَنَتَحَرَّكَ جَمِيعًا لِوَقْفِ الْمَأْسَاةِ الدَّمِيَّةِ فِي سُورِيَا /lā budda an natanāwala al-qaḍiyyata wa nataḥarraka jamī'an li waqfi al-ma'sāti ad-damiyyah fī sūriyā/ 'We must take the issue in hand and we move together to stop the tragedy in Syria.'
IV	يَحْتَاجُ أَشِقَّاءَنَا فِي السُّودَانِ إِلَى الدَّعْمِ بَعْدَ وِلَادَةِ دَوْلَةِ جَنُوبِ السُّودَان /yaḥtāju asyiqqā'unā fī as-sūdān ilā ad-da'mi ba'da wilādati daulati janūbi as-sūdān/ 'Our brothers in Sudan need support after the birth of the South Sudan State.'
V	إِنَّ هَذِهِ حَالَةُ شَعْبِ الصُّومَال وَإِنَّنِي أَدْعُو الْأُمَمَ الْمُتَّحِدَةَ إِلَى مُوَاصَلَةِ دَعْمِ جُهُودِ حُكُومَتِهِ /inna hāżihi ḥālatu sya'bi aṣ-ṣūmāl wa innanī ad'ū al-umama al-muttaḥidah ilā muwāṣalati da'mi juhūdi al-ḥukūmati/ 'This is the case of the Somalian people and I call for the UN to continue supporting the government's efforts.'
VI	تَتَّصِلُ مَبَادِئُ الْعَدْلِ وَالْحَقِّ بِتَحْقِيقِ الْأَمْنِ وَالِاسْتِقْرَارِ الَّذِي يُحَقَّقُ بِالتَّخَلُّصِ الْكَامِل مِنَ الْأَسْلِحَةِ النَّوَوِيَّةِ وَكَافَّةِ أَسْلِحَةِ الدَّمَارِ الشَّامِل /tattaṣilu mabādi'u al-'adli wa al-ḥaqqi bi tahqīqi al-amni wa al-istiqrār al-lażī yuḥaqqaqu bi at-takhalluṣi al-kāmil min al-asliḥati an-nawawiyyah wa kāffati asliḥati ad-damāri asy-syāmil/ 'The principles of justice and truth are related to achieving security and stability, which can be achieved by eliminating all nuclear weapons and weapons of mass destructions.'
VII	حُصُوصًا عَلَى قَارَّةِ إِفْرِيقِيَا وَإِنَّنَا جَمِيعًا مُسْتَعِدُّونَ لِلتَّعَاوُنِ مَعَ أَيِّ طَرَفٍ إِلَى فِي الْعَلَاقَاتِ الدُّوَلِيَّةِ تَمْتَدُّ مَفَاهِيمُ الْعَدَالَةِ وَالْحَقِّ وَالكِرَامَةِ مِنْ مَنْظُورِنَا مُسْتَقْبَلِهَا الْأَفْضَل /tamtaddu mafāhīmu al-'adālati wa al-ḥaqqi wa al-kirāmah min manẓūrinā fī al-'alāqāti ad-duwaliyyah khusūṣan 'alā qārrati ifrīqiyā wa innanā jamī'an musta'iddūna li at-ta'āwuni ma'a ayyi ṭarafin ilā mustaqbilihā al-afḍal/ 'The concepts of justice, truth, and dignity extend our perspectives in international affairs, especially on the African continent. We are all ready to cooperate with any parties for the better Africa in the future.'
VIII	الَّذِي ضَرُورَةُ الْإِصْلَاح وَيَجِبُ أَنْ يَكُونَ الْجَمْعِيَّةُ الْعَامَّةُ وَمَجْلِسُ الْأَمْنِ عَلَى قِمَّةِ أَوْلَوِيَّاتِهِ إِنَّنَا نَنْظُرُ إِلَى الْوَضْعِ الْحَالِيّ /innanā nanẓuru ilā al-waḍ'i al-ḥāli al-lażī ḍarūratu al-iṣlāḥ wa yajibu an yakūna al-jam'iyyatu al-'āmmah wa majlisu al-amn 'alā qimmah aulawiyyātihi/ 'We indeed look at the current situation, which requires reformation, and the General Assembly and the Security Council should be at the top of the reformation priorities.'
IX	إِنَّ النِّظَامَ الدُّوَلِيَّ يَسْتَقِيمُ طَالَمَا نَتَكَاتَفُ وَنَتَحَرَّكُ سَوِيًّا فِي التَّصَدِّي لِظَاهِرَةِ الِازْدِوَاجِيَّةِ /inna niẓāma ad-duwaliyy yastaqīmu ṭālamā natakāfu dan nataḥarraku sawiyyan fī at-taṣaddī li ẓāhirati al-izdiwājiyyah/ 'The International System is completely upright as long as we hold hands and move together in the face of double standard phenomenon.'
X	إِنَّ تَفَاقُمَ الْأَزَمَاتِ الْمَالِيَّةِ وَالِاقْتِصَادِيَّةِ يَجِبُ أَنْ يَدْفَعَنَا لِحَوْكَمَةٍ اقْتِصَادِيَّةٍ دُوَلِيَّةٍ جَدِيدَةٍ /inna tafāquma al-azamāti al-māliyah wa al-iqtiṣādiyyah yajibu an yadfa'anā li ḥaukamatin iqtiṣādiyyatin duwaliyyatin jadīdah/ 'The worsening of financial and economic crises should provoke us to build a new international economic governance.'

system that still applies the double standards that have an impact in the performance of the United Nations General Assembly and the Security Council, and to make developing countries as the basis of the global economic governance. Meanwhile, in the textual function, the propositions that are considered important by the orator and need attention from the audiences are signed with the emphatic particles to highlight the starting point of the messages.

Then, in the second dimension, there are 80 micro propositions, which include power relations between Mursi and his audiences. In the third dimension, especially in the situational level, the discourse on the international forum was delivered at the 67th UNGA. The language used was the Standard Arabic language. In the social level, there are three factors affecting the appearance of the discourse, namely peace and security in the Middle East and the world, the culture of double standard in the international system, and the global economic governance that is detrimental to the people.

5 CONCLUSION

There are two main ideas of Mursi's speech at the 67th UNGA: (1) the identity of Muhammad Mursi, and (2) the new Egypt's visions. Both of them were used by Mursi to bring about world peace that is based on justice (الْعَدْل/al-'adl/), truth (الْحَقّ/al-haqq/), freedom (الْحُرِّيَّة/al-hurriyyah/), dignity (كَرَامَة/al-kirāmah/), and social justice (الْعَدَالَة الإجْتِمَاعِيَّة/al-'adālah al-ijtimā'iyyah/). Furthermore, in the speech, Mursi used several language strategies to build power relations with the audience. The language strategies were applied using powerful dictions. For instance, in terms of nouns, there were the use of the words الْحَقّ/al-haqq/ 'truth' that collocates with the word الْعَدْل/al-'adl/ 'justice' to describe the new Egypt's visions, الْجُهْد/al-juhd/, in a single form, and the word الْجُهُوْد /al-juhūd/ 'the striving', in a broken plural form, to describe the persistence of the efforts made by the agents without time limit. Meanwhile, in terms of verbs, there were the use of the words يُنَاضِل/ yunāḍilu/ 'to struggle' that collocates with the word يُجَاهِد/yujāhidu/ 'to strive' to describe the persistence of the efforts made by the agents, the word أَدْعُو/ad'ū/ 'to invite; to call' that has the same root with the word دَعْوَة/da'wah/ 'an invitation; a calling; a lecture', which indicates that the call is accompanied by an action, the word نَتَنَاوَل /natanāwal/ 'to take in hand' which contains the concept of soon, and the word أُؤَكِّد/u'akkidu/ 'to emphasize', which indicates that Mursi had the discretion to use his power because he could control the information needed to get more attention from his audience. Moreover, Mursi used the declarative modes of functioning imperatives manifested in epistemic modalities. He also used interjected particles (أَلَا/alā/ and أَمَا/amā/) of functioning imperatives in the beginning of the propositions. The third strategy used by him was the use of emphatic particles (إِنَّ/ innal, أَنَّ/annal, لَكِنَّ/lākinnal, ـلَ/la-/, and قَدْ /qad/) and emphatic nouns (كُلّ/kullu/ and جَمِيْعًا/jamī'an/).

REFERENCES

Fairclough, N. (2010) *Critical discourse analysis: The critical study of language.* 2nd edition. London & New York: Routledge.

Fairclough, N. (2015) *Language and power.* 3rd edition. London & New York: Routledge.

FEATURE: What does the UN general assembly do when the general debate ends? (2012) Retrieved from: http://www.un.org/apps/news/story.asp?NewsID=43305&Cr=General+Assembly&Cr1=&Kw1 =general&Kw2=assembly&Kw3=#.Vx19OzB97IV. [Accessed on the 25th April 2016]

Foucault, M. (2008) *Ingin tahu: Sejarah seksualitas (Curiosity: History of sexuality).* (Translated by Rahayu S. Hidayat). Jakarta: Yayasan Obor Indonesia.

Renkema, J. (2004) *Introduction to discourse studies.* Amsterdam & Philadelphia: John Benjamins Publishing Company.

Somantri, G.R. (2005) Memahami metode kualitatif (Understanding the Qualitative Method). *Makara Seri Sosial Humaniora, 9*(2), 57–65.

Trochim, W.M.K. (2006) Qualitative measure. Retrieved from http://www.socialresearchmethods.net/ kb/qual.php. [Accessed on the 22nd April 2014]

van Dijk, T.A. (1980) *Macrostructures: An interdisciplinary study of global structures in discourse, interaction, and cognition.* Hillsdale, New Jersey: Lawrence Erlbaum Associates.

Yule, G. (2010) *The study of language.* 4th Edition. New York: Cambridge University Press.

Cultural Dynamics in a Globalized World – Budianta et al. (Eds)
© *2018 Taylor & Francis Group, London, ISBN 978-1-138-62664-5*

An ontology of violence based on moral teachings: A case study of *Salafi Wahabi* radicalism and the story of *Kresna Duta* (Ambassador Kresna)

Naupal
Department of Philosophy, Faculty of Humanities, Universitas Indonesia, Depok, Indonesia

ABSTRACT: This research begins with the problems of violence based on moral teachings. Monism in a moral view represents the face of exclusivity and intolerance. It is believed that there is an ideological motive behind a violent act emerging from moral teachings. Wahabi's teaching is considered as a radical view in interpreting religious texts. Meanwhile, from the cultural perspective the story of Kresna Duta indicates violence based on the assumption of the existence of non-egalitarian power relation among people with different social status. The method used in this research is Derrida's deconstruction and hermeneutics. Wahabi and *Kresna Duta* moral teaching texts will be deconstructed to find out the hidden elements causing the distortion of the texts' meaning. The method will also be used to show that moral monism and single interest-loaded interpretation will become the trigger of horizontal conflicts in Indonesia and that they can be solved by using dialogues in both rational and communicative acts by participants as said by J. Habermas. This research will be a contribution to the idea of rearranging the order of a tolerant Indonesian society.

1 INTRODUCTION

Indonesia is a country and nation composed of heterogeneous ethnic groups, religions, and cultures. In this era of globalization, when group identity is assimilating and flowing, religious and ethnic identity is factually important to pay attention to because there are still people who refer to their moral values on the basis of religion, ethnicity and culture. The faith to one moral belief can still be witnessed around us and is sometimes implemented through violence.

The issue of violence based on one moral teaching is important to be brought up as now, in the climate of democracy, there is a strong tendency of equality in ethnicities, religions, and classes in Indonesia. Even though we have entered the era of modernization, in reality the problem related to ethnic, religious and racial identity still exists. This is proven by the emergence of alliances of primordialism in Indonesia rooted in ethnic, religious and racial bonds. This does not only occur in Indonesia, but it also happens in global communities as said by Huntington (1997, p. 28), "In this new world, the most pervasive, important, and dangerous conflicts will not be between classes, rich and poor, or other economically defined groups, but between people belonging to different cultural entities."

This research becomes important because if we look at the Indonesian context, each of the religions, ethnicities, and cultures has a moral message that is considered ideal and correct; in this research the Salafi Wahabism moral teaching and the story of Kresna Duta will be the focus. The diversity is an empirical reality; a reality that continues to exist in Indonesia. This reality should be mediated seriously; otherwise, a continuous conflict will be real and present. This is the reason why it is the right time to reinvent and identify the ontology of violence coming from moral teachings by using the deconstruction method introduced by Derrida (1998) and to solve these problems using the ethical discourse approach by Habermas (2007).

1.1 Monism

A lot of violence occurs around us. The interesting part is that there is a similarity in their ontology, namely to send a moral message and the value of truth embraced by certain groups. One group claims themselves as being in the right side while considering other groups in the wrong side. Nevertheless, it is an obvious fact that different groups and societies understand and organize human lives differently. Clifford Geertz calls this a humanity based on group pride (Geertz, 1973).

Moral monism refers to a view that there is only one way of life that is fully humane, true, or the best, and that the others are not complete or not united (Farekh, 2000). The way of thinking is really dichotomous, "ego" and "other". "Ego" is emphasized, while "other" has to be eliminated. To show that one way of life is the best, the moral of monism should be based on something that is inter-cultural. For the followers of this belief, humans are basically the same, even though they exist within a different time and place. For the monism community crime, like mistakes, can take different shapes but good, like truth, only takes a single form.

In order to show that one way of life is the best, moral monism needs to spread it to all human beings. Moreover, moral monism refers to an idea that there is universality, uniformity, and objectivity everywhere. The fact is that this philosophy is confronted with other groups who have the same view or this philosophy will also be confronted with groups fighting for particularity, contingency, and subjectivity that believe that morality does not have to be based on the interpretation of truth, ideology, and domination. Moral monism is in the end no more than an ideological moral that leads to violent acts which in the end will ruin the unity and harmony in diversity.

1.2 Moral teachings in the movement of Salafi Wahabism in Indonesia

At the beginning, when Islam entered Nusantara, it penetrated well because it could adapt to the local culture that existed at that time like *Wayang* which was used as an instrument to introduce Islam (Riddle, 2001). However, these days the face of Islam in Nusantara has turned into a theological and fundamental Islam identified by the presence of a Salafi Wahabism movement. With its financial power, this movement has succeeded in establishing its foundation and educational institutions, and by using the label *ahlus sunnah wajamaah* in the context of Indonesia in which Islam is the majority and is also the dominant theological school, this movement has been able to draw the sympathy of the majority of Indonesian Muslims. This movement initially focused on purifying faith and the integrity of individual moral values. They implement this by interpreting religious texts literally and practicing the religious values formally and rigidly.

With a focus on these two issues, the Salafi Wahabism movement appears to be exclusive and takes the position that is in the opposite direction of the majority movements in Indonesia, either from the physical side (clothes) or spiritual side (teaching doctrine/ideology). Physically, they can be identified from the ornaments that they wear publicly in many big cities in Indonesia; Jogjakarta, Solo, Semarang, Bandung, Depok, Jakarta, and Makassar. The men wear *jalabiyah* (long robes), *imamah* (turbans), *isban* (long trousers up to the ankle), and *lihyah* (long beards) while the women wear *niqab* (long black dress covering the entire body) which distinguishes them from local Muslims. These clothes signify their difference from other Muslims.

Moreover, ideologically, the basis of their moral teachings come from literal interpretation of the religious texts, such as the teaching of *tauhid* (purifying the faith to Allah and the refusal of *tawasul* and pilgrimage to the graves), *al-wal wal-barra'* (the solidarity among followers, and refusing those who have different views), anti *hizbiyyah* (refusal of political preaching, which is considered as *bid'ah* and *thaghut*), *hijrah* (moving from digression to correctness), and *kafir* and *bi'dah* (guiding those who deviate from the genuine teaching).

On the one hand, the moral teachings of Salafi Wahabism considers the above functions as the ideology that imbeds social relations among the members of the group. Its moral teaching is also an identity that distinguishes them from others. From Bourdieu's perspective, this identity serves as a social capital (Bourdieu, 1994), as it functions as networking or resources because the group share the same doctrine. It can also be an adhesive that supports solidarity among the members. On the other hand, it can also become a tool to discriminate others. More importantly, Salafi Wahabism teachings can become an ethical legitimization of social relations. In other words, Wahabi moral teachings are true monism morality that will bring people to possess an attitude of exclusivity where they can be categorized into two groups only: right or wrong, God's followers or unbelievers, the chosen ones or the sinners. The effect of this dualism is the acts of violence as they start from a mental process when seeing 'others' as the different or the wrong one, then stigmatizing, humiliating, destroying, and finally killing them (Semelin, 2005).

Hafner explains that the moral teachings of Salafi Wahabism become the central topic of Islamic expansion and not as a religion anymore but as a political and social power. He adds that the changes in religious culture are related to "reconstruction in political bodies or parties in which some meanings are distributed but others are neglected" (Hefner, 1987, p.78.).

The Salafi Wahabism teachings can also be considered as neo-fundamentalism by Oliver Roy. According to him, Salafi Wahabism is an Islamic movement that tries to Islamize people at the grass root level (Roy, 1996). John O. Voll defines Wahabism as a prototype of rigid fundamentalism in modern Islam because this movement possesses harsh views in defining who can be considered as Muslims. This movement also strictly differentiates religious people from unreligious ones (Voll, 1994).

1.3 Moral teachings in the story of Kresna Duta

Puppet shows in Indonesia are closely related to the influence of Hinduism. According to Professor Poerbatjaraka, the Mahabarata epic was brought from India to Indonesia. The Mahabarata epic was translated into Javanese during the reign of King Dharmawangsa Teguh in Central Java from 913 to 929 Caka (991–1007 AD) (Poerbatjaraka, 1957) and this is the origin of the Wayang drama (Moelyono, 1975).

In the Mahabarata Epic, the figures of Pandawa in a puppet show are positioned on the right side while the figures of Kurawa are on the left. Pandawa is a symbol that represents 'virtue', while Kurawa is a symbol of 'evil'. At the peak of the show, there is a war between the two families, which means a war between good and evil. The good will win, while the evil will lose, and it is the message carried in the *wayang* show.

In the story of Kresna Duta, violence takes place between two related families, Kurawa and Pandawa. It is clear that violence has to be the choice as the three emissaries, namely Lady Kunti, King Drupada, and Kresna, fail to set a consensus through dialogues. The war of Baratayuda is a symbol of truth, justice, godliness against evil, injustice, and insubordination. The war that kills many lives must take place as a result of the failure of those carrying the mission of truth.

In the conflict between Kurawa and Pandawa, Kresna is the emissary of Pandawa. He carries an obligation to bring justice as an avatar who descends to the world. Truth and justice have to be guarded, while injustice and wrongdoings have to be destroyed. Duryudana is the one who does the arbitrary. His character in the play is very egoistic as he is concerned only about himself and his hunger for power. Karna does not defend Pandawa. Although he knows that the Pandawa brothers are his real brothers, he defends Duryudana instead.

The war has catastrophic effects to these puppet figures because the brothers kill each other. On Kurawa's side there are Bisma (Pandawa's eldest) (*Note: Bisma is the grandfather of Pandawa and Kurawa*), Durna (Pandawa's *guru*) and Karna (the eldest brother of Pandawa). Meanwhile in Pandawa, there is Salya (*Note: as far as I know Salya was tricked to defend Kurawa by Duryudana, so he is on the side of Kurawa although unwillingly*), the uncle of Nakula and Sadewa.

Puppet show is a fantasy and imagination in which there is a story known as the *lakon*. The lakon presents something that does not exist in reality (Amstrong, 2005). Although it is a fantasy and imagination, it has a lot of moral values. Javanese society believes that the *lakon* (story) of Kresna Duta contains a lot of moral teachings. The most important teaching is that the protection of the truth is very important. The absence of truth will cause someone to lose intuition that becomes a drawback (Duncan, 2008). That is why truth should be fulfilled and crystalized, even though violence and war happen like in the *lakon* above.

It can be said that there are some moral teachings in the story of Kresna Duta. The first one is that truth and justice must be upheld, while injustice and wrongdoings must be destroyed as shown in the role of Kresna. The second is that the promise of loyalty to one's duty should be fulfilled no matter what the condition is, even through self-sacrifice as shown in the role of Karna's loyalty to Duryudana. The third is that there is always a dilemmatic choice in life to keep the existing social order as shown in the role of Salya. Salya faces the dilemmatic choice between supporting Duryudana, his son-in-law, or supporting Nakula and Sadewa, members of the Pandawa, who are his own nephews.

Moreover, the moral teachings in the *lakon* of Kresna Duta tell us that there are always good and bad, right and wrong, in this life; and there is always the battle between good and bad or between right and wrong, and war is the chosen path to get rid of evil. This *lakon* also teaches that protection and defence of the right and truth is very important, even though they have to be upheld with violence and war. Behind this moral teaching, there is a dichotomous perspective; right and wrong, good and bad. The truth is given the legitimacy to take a violent action against the wrong.

1.4   *Reinventing the truth claim of moral monism*

From the two cases above, we realize that dogmatism and the spirit of being concerned only about ourselves will aggressively bring us to the acts of violence. At the same time, it also dulls our moral sensitivity. The idea that says that there is only one highest and supreme or most humane way of life is logically unacceptable (Berlin, 1969). It is a naive assumption that a single truth can form a unity without looking at space and time. This is due to the fact that not all social order has the same structure.

Moral monism suffers from defects and deviations. For Wahabism, non-Wahabism and even Muslims who disagree with the official interpretations of their central doctrines are all wrong and have no value, and Kresna Duta illustration of such opposition to hierarchy is so strong that it ends in war.

Moral monism makes a dangerous mistake of assessing other people's way of life, and this attitude can cause a hermeneutic disaster. This is caused by the shallow interpretation of texts and the notion that their group is the most ideal group. Essentially, texts are not just a single sign but a network of signs. This means that what we can find and know is the traces of the truth themselves and not the truth in itself. This is what is meant by the trace concept according to Derrida.

In his three books, *Of Gramatology*, *Writing and Difference*, and *Speak and Phenomenon*, Derrida refuses the claim of the standard truth because it is often used to construct a single meaning. Consequently, he closes himself and only accepts communities sharing the same understanding or meaning. Derrida is not interested in accepting the transcendental truth that is free of space and time. Through the deconstruction method, he tries to show the weaknesses of that way of thinking. There are always parts in the texts to be questioned, such as what is the ideal, the main and the original initiatives. Derrida shows the democratic model of thinking that is open to differences and alternatives. Thus, Derrida's concept of 'difference' becomes very important because it does not only refer to alternative understanding or difference "between...", but his concept of 'difference' also refers to the uncertain steps and directions. There is always something that is withheld in its certainty (Penolope, 2006).

Deconstruction is then implemented by putting aside the main idea of the texts to the edge and putting the present initiatives from the edge to the centre. It is subsequently confirmed that there is no model of binary opposition thinking, a model of thinking that prioritizes one side and neglects the other. Furthermore, there is no part of the text that is stagnant or permanent. There is no atom in the texts. This concept by Derrida is called undecidable category.

In his deconstruction model, there is a principle that is summarized by Derrida: *sans savoir, san voir, sans avoir* (does not know, does not have, and does not possess). *San savoir* (does not know) means that a text cannot always be interpreted by the interpreter in his totality, so he does not have the authority in his interpretation. *Sans voir* (does not see) indicates the limitation of the five senses that a man has about the truth; this can lead us to respect differences. *Sans avoir* (does not possess) reminds us that the truth is on the lap of the interpreter, but it is disseminated through other interpreters. Why did Derrida harshly oppose the opposition hierarchy and the claim of absolute truth and absolute moral? According to him, opposition hierarchy and the claim of absolute truth are no more than standardizing the meaning which is neither more nor less than personal emphasis in assignment opposition or authority. As a consequence, the comprehension is normative and those who differ are subject to sanctions. In other words, the special treatment of one entity over another is a product of cultural manipulation for the purpose of political interest in a dimension of power, and this needs to be deconstructed.

In that way, we can know the ideological content behind the texts of moral teachings, whether it comes from religion or culture. In the history of Islam, the alliance between religion and politics can be seen in the founding of the Saudi Arabian dynasty. In this case, Wahabism supports the authority and the government gives protection and support to Wahabism. The authority receives religious legitimacy from the Wahabism movement as it supports the kingdom, while Wahabism receives protection and political support from the kingdom. The result is an absolutism of truth which is a merge between absolute religious authority and absolute political authority.

In the case of Kresna Duta, opposition hierarchy is so strong that it ends in war. Each party, both Pandawa and Kurawa, are firm in their positions and, hence, war is unavoidable. In the story, war is regarded as the only solution to destroy evil. Even supporting evil to speed the ruin of evil is justified as shown by the figure of Karna. Karna is the eldest son of Lady Kunti, which means Karna is the blood brother of Pandawa. In the *lakon*/story, it is described that the reason of Karna's support for Kurawa is that he has a very genial attitude, namely defending evil does not mean supporting evil, but his taking side to evil means to smooth the Karma so that Baratayuda war can soon occur. In this way, evil can soon be eliminated. There is an absolute truth that needs to be defended even though many lives will be sacrificed.

1.5 *Constructive dialogue as a key to civilized society*

As described above, Indonesia is a heterogeneous country with many religions, ethnic groups, and cultures each with their own moral teachings. Violence will still be a reality if monism perspective remains in the mind of the majority of the Indonesian population. In regard to this matter, a constructed dialogue is expected to be the way out and the key to civilized society.

Habermas calls this ethical discourse. Ethical discourse is not an ethical base that can provide ready answers for the moral questions presented, but it is a way to ascertain the return of the norms of moral that become the questions (Magnis-Suseno, 2007). Ethical discourse attempts to answer universal problems. Ethical discourse does not stop at the phase of the implementation of universal norms, but it needs to be justified in rational discourse among participants through understanding to reach consensus.

According to Habermas, there are two principles in ethical discourse. The first principle is universality that depicts the moral consideration that I want to apply correctly if it is also wanted by the majority. Second, the universal truth should be discoursed to reach a consensus or mutual understanding. In this way, only the norms that have been confirmed in the practical discourse to which everyone agrees can be ascertained in their truth (Habermas, 2007).

By implementing the two principles in ethical discourse, it is expected that the acceptance of pluralism of moral teachings, coming from either religion or culture, can grow. Why does ethical discourse become important? It is a reality that there is violence in the name of moral teachings that come from religion and culture. Secondly, the binary opposition is intentionally used systematically by one group claiming the truth to corner other groups in a marginal position or to label them as valueless. Thirdly, the discrimination to 'others' smoothens the emergence of conflict and violence.

This is where ethical discourse tries to fulfil the basic needs of all followers of moral teachings that exist in Indonesia. Through ethical discourse, it is essential to describe that the formats of relations among moral teachings universally agreed upon by all parties are not decided by cultural capital owned by the dominants which is then to be forced as referred norms for everyone. More importantly, the moral teachings that have been agreed on are not meant to assimilate other groups into the general moral teachings through the economic, political, and social engineering as if the morality agreed by the dominant group will come into priority. Ethical discourse also teaches us to agree or disagree, as well as to be tolerant and ready for non-ending consensus.

2 CONCLUSION

Ethics are actually experience, so the practice of moral teachings will become more effective if it is done through experience. Moral teachings can only be obtained through less meaningful cognitive understanding; thus, it brings us to false interpretation. For example, the moral of monism that underestimates other groups can come. Values and norms should be a disposition that is incorporated inside every individual personality as a result of skills in practical behaviour developing in a democratic and plural social environment. Thus, it is urgent that every individual should interact socially with different religions, cultures as early as possible, so that our younger generation can have a paradigm that life is not bordered by religion.

The perspective of monism should be changed into pluralism through constructive dialogues. The purposes of pluralism are: first, as a citizen to actively participate in all social groups from different moral teachings; second, in relation to identity, to encourage the society to admit and accept cultural diversity so that the sense of belonging and commitment to society can grow; third, to meet the demand for social justice. The people can only enjoy peace and prosperity when the country can maintain social justice.

REFERENCES

Amstrong, K. (2005) *A Short History of Myth*. Great Britain, Canongate Books.

Berlin. (1969) *Four essays on liberty*. London, Oxford University Press.

Bourdieu, P. (1994) *Raison pratiques. Sur la theorie d l'action*. Paris, Seuil.

Derrida, J. (1998) Difference. In: Rivkin, J. & Ryan, M. (eds.) *Literary Theory: an Anthology*. USA, Blackwell Published Inc.

Duncan, I. (2008) *Right*. British, Avumen.

Farekh, B. (2000) *Rethinking Multiculturalism, Cultural Diversity, and Political Theory*. Houndmills, Macmillan Press.

Geertz. (1973) *"Thick Description" in The Interpretation of Culture*. New York: Basic Books.

Habermas, J. (2007) *Moral Consciousness and Communicative Action* Translated by Christian Lenhardt and Shierry Weber Nicholsen. UK, Polite Press.

Hefner, R. (1987) Islamic Conversion in Modern East Java. In: William R. (ed.) *Islam and the Political Economy of Meaning: Comparative Studies of Muslim Discourse*. London and Sydney, Croom Helm.

Magnis-Suseno, F. (1996) *Etika Jawa: Sebuah Analisa Falsafi tentang Kebijaksanaan Hidup Jawa* (Javanese Ethics: a Phylosophical Analysis of Javanese Life Wisdom). Jakarta, PT Gramedia Pustaka Utama.

Penolope, D. (2006) *How to Read Derrida*. New York, WW. Norton.

Poerbatjaraka. (1957) *Tardjan Hadidjaja: Kepustakaan Djawa (Tardjan Hadidjaja: Javanese Bibliography)*. Jakarta, Penerbit Djembatan.

Riddle, P. (2001) *Islam and the Malay-Indonesian: Transmission and Responses*. London and Singapore, C. Hurst and Horizon Books.

Roy, R. (1966) *The Failure of Political Islam*. Cambridge, MA, Harvard University Press.

Semelin, J. (2005) *Purifier et Detruire*. Paris, Seuil.

Voll, J.O. (1994) *Islam: Continuity and Change in Modern World*. Syracuse, NY, Syracuse University Press.

Cultural Dynamics in a Globalized World – Budianta et al. (Eds)
© *2018 Taylor & Francis Group, London, ISBN 978-1-138-62664-5*

The story of caliph Abu Bakr, Umar, Usman, and Ali to the battle of Hasan and Husain in Karbala: Islamic epics from Ambon

D. Kramadibrata
Department of Literature, Faculty of Humanities, Universitas Indonesia, Depok, Indonesia

ABSTRACT: In classical Malay Literature, *Hikayat Muhammad Hanafiyyah* (hereinafter abbreviated as HMH) is classified as an Islamic epic and is influenced by Shia doctrines. HMH is known as one of the oldest Malay manuscripts (approximately from the 16th century) and is adapted from Persian literature into Malay literature. HMH narrates the story of Muhammad Hanafiyyah who was fighting for Ali's descendants. The text of HMH is written in more than 30 manuscripts and spread all over the world. According to Brakel, there are three versions of HMH, namely (1) *versio simplex* (*x* version); (b) *versio ornatior* (*y* version); and (c) *versio orna-tissima* (*y1* version). HMH is very popular because some manuscripts describe Muhammad Hanaffiyah as *imam mahdi*. In Haruku Island, Ambon, there is one manuscript entitled "*Hikayat Khalifah Abu Bakar, Umar, Usman, dan Ali sampai Peperangan Hasan dan Husain di Karbala*" (The stories of Abu Bakar, Umar, Usman, and Ali up to the War of Hasan and Husain in Karbala) (abbreviated as HKAUUA). This manuscript contains the story of Muhammad Hanafiyyah and this belongs to y1 version. This paper discusses the characteristics of HKAUUA as an Islamic epic from Ambon.

1 INTRODUCTION

The spreading of Islam in *Nusantara* (Indonesian Archipelago) has had an impact on *Nusantara* literature. Malay literary works are enriched by literary translations from Arabic and Parsi languages. In the world of Malay literature, there are three renown Islamic epics (heroic tales). They are *The Story* (in Malay: *Hikayat*) *of Iskandar Zulkarnain, The Story of Amir Hamzah*, and *The Story of Muhammad Hanafiyyah* (hereinafter referred to as HMH) (Iskandar, 1996:124; Braginski, 1998:134). HMH emerged in Pasai in the 80s of the 14th century (Braginsky, 1998:128). In Malay society, HMH played an important role in spreading Islam as a religion. During the spreading of Islam, the stories in HMH served as models of bravery for the audience.

Brakel (1975) studied the literary works of Classic Malay. Based on Brakel's research (1975:56), HMH is categorized as the oldest Malay text (approximately in the 16th century) and it can be traced back to Parsi. Initially, the HMH text contained Shi'itic elements, but Brakel concluded that the more recent texts of HMH contain less Shi'itic influence.

Brakel succeeded to collect and register 30 manuscripts that tell the stories of HMH. One of the things that made HMH popular is the illustration of Muhammad Hanafiyah as the *mahdi* (Shia's followers' long awaited leader).

HMH tells the story of Muhammad Hanafiyyah's determination to defend the death of Husayn, who was the grandson of the Prophet Muhammad in Karbala. The first Malay readers who were exposed to the Shia *mazhab*, the character Muhammad Hanafiyah was considered as the true and the devout hero of Islam. He dedicated his whole life to fight against the disbelievers or *kafir*. He was also Ali's supporter, whom, according to the Shia *mazhab*, was the only family member who had the right to hold the position of the caliph (Braginski, 1998:134).

In HMH, the story ended with the disappearance of Muhammad Hanafiyyah in a cave after defeating his enemy. A number of groups within the Shia *mazhab* who revered Muhammad Hanafiyyah as *mahdi* believed that he did not die, but hid in the mountains waiting for the right time to return (Braginsky, 1998:134). Among the different Shia sects, Muhammad Hanifiyyah is the central character revered by the Kaisaniyah sect (Brakel, 1975:6).

However, the character of Muhammad Hanafiyyah, who is lauded by the Malays in HMH, almost does not have any connection with the Muhammad Hanafiyyah in historical records. As a historical figure, he is the son of the Caliph Ali and a slave named Khaula from the Hanifah tribe (Brakel, 1975:2). Muhammad Hanafiyyah lived during a period of upheaval due to the opposition in fighting for the position of Caliph between the followers of Ali and the Ummayah. In the dispute, however, Muhammad Hanafiyyah did not have any important influence (Braginsky, 1998:134).

The prototype of HMH Parsi consists of two parts. The first part falls within the Arabic-Parsi genre known as *maqtal* (murder stories). This part tells the story of the tragic death of the two of Ali's sons, who were revered as heroes within the Shia *mazhab*, i.e. Hasan and Husain. This part of the saga is read during special celebrations to commemorate the two characters. The second part falls in the Parsi saga genre. Based on the tragic death of Husain, the story of Muhammad Hanafiah's revenge to Yazid, who murdered Ali's sons, was later written (Braginsky, 1998:135).

In the beginning, the Shi'itic elements in the HMH text constituted the main part of the early period of Islam in Indonesia. However, more recently, due to increasing tendency of anti-Shi'itic elements in Malay literature, the congruity and tensions in the first part of HMH have disappeared, and they merely became a brief summary of the early history of Islam. Wieringa (1996) calls the removal of Shi'itic elements in texts as "de-Shiazation."

In a review on HMH, the narrative structure of HMY Malay consists of three parts (Brakel, 1975:16). The first part is the introduction containing the Prophet Muhammad's biography and the early days of his leadership. Part of this story is derived from *The Story of Nur Muhammad*. The second part consists of several episodes, which are childhood stories of Hasan and Husain; the biographies of the three Calips, Abu Bakr, Umar, Usman; the biography of Ali; and the death of Hasan and Husain in Karbala. The third part contains tales of war of Muhammad Hanafiyyah during the thirteen years he fought with his eight companions against Yazid and Marwan. The war ended with a victory for Muhammad Hanafiyyah, the burning of Yazid in a well that he himself had dug, and the appointment of Zainal Abidin, the son of Husain. The final part tells the story of how Muhammad Hanafiyyah suddenly disappeared from the battleground and entered into a supernatural cave. The mouth of the cave immediately closed upon him entering (read Brakel, 1975:18; Hadi, 2013: 98–100).

In his research, Brakel complied a *stemma* (genealogy) to determine the kinship of the HMH text in the Classic Malay literature. The HMH text is divided into three versions: (1) *versio simplex*; (b) *versio ornatior*; and (c) *versio ornatissima*. Later Brakel grouped the 30 HMH manuscripts into the three versions, as follows.

1. *Versio simplex*: **x** is contained within manuscripts of A, E, F, I, and M;
2. *Versio ornatior*: **y** is contained within manuscripts of B, G, K, L, N, O, P, Q, R, V, W, X, BB, CC, DD; and
3. *Versio ornatissima*: y_1 is contained within manuscripts of C, D, H, S, T, U, Y, Z, and AA.

Brakel provided a detailed description about the x and y versions. The y_1 version has not yet been described in detail. Upon explaining the relationship among the x, y, and y_1 versions, it becomes clear that there are a number of episodes that fall into the y_1 version, which are

1. The episode when Ali instructed his sons, Hasan and Husain to throw away the Zulkafar sword into the sea.
2. The Israel King helped Husain's family who were taken as hostages by Yazid. The Israel King then joined the war against Yazid;
3. The episode when Yazid's children, Said and Mahid, escaped from their father's army. They then assisted Muhammad Hanafiyah's army and died as martyrs.

One of the manuscripts that is part of the y_1 version, which is manuscript H, dated back to 1732. The y_1 version comes from Riau-Johor (Brakel, 1975:84–86). In the history of Malay literature, the Riau-Johor period took place during 1511–1779. During this period, adaptations of many Islamic literary works into Malay language took place (read Iskandar, 1996:274 and Collins, 2005:47–49). On that account, it can be concluded that the y_1 version dated back to the 18th century.

Brakel presented the HMH edition text based on the x and y versions that were closest to the archetype (Brakel, 1975:90). Most of the editions of the texts presented refer to the x version. The HMH edition text Brakel compiled uses a combination method.

The story of HMH consists of the following:

1. The history of the Prophet Muhammad's life until the early days of his Prophecy. Part of this story was written based on the Saga of Nur Muhammad;
2. Hasan and Husain's childhood stories;
3. The stories of the Caliphs Abu Bakr, Umar, and Usman;
4. The story of Ali;
5. The death of Hasan and Husain as martyrs;
6. The story of Muhammad Hanafiyyah's revenge;
7. The final story of Muhammad Hanafiyyah.

The HMH edition text edited by Braken began from the second to the seventh stories. Brakel presented the HMH edition text into two main parts, as follows:

The first part consists of 26 episodes, which are:

1. Episodes 1–9: Hasan and Husain's childhood
2. Episodes 10–16: the history of the first three Caliphs (Abu Bakr, Umar, and Usman);
3. Episodes 17–19: the story of Ali;
4. Episodes 20–26: the death of Hasan and Husain.

The second part consists of 22 episodes, which are:

1. Episodes 1–20: Muhammad Hanafiyyah's revenge;
2. Episodes 21–22: the final story of Muhammad Hanafiyyah (Brakel, 1975:18).

As aforementioned, Brakel had not worked yet on the manuscripts in the y_1 version in detail. According to Brakel, the manuscripts in the y_1 version are very complex due to the additional stories they contain.

This paper discusses "The Story of Caliph Abu Bakr, Caliph Umar, Caliph Usman, and Caliph Ali to the Hasan and Husain Battle in Karbala" (hereinafter referred to as HKA-UUA), a manuscript from Haruku Island in Ambon. Based on its contents, the HKAUUA text is part of the y_1 version.

2 "THE STORY OF CALIPH ABU BAKR, CALIPH UMAR, CALIPH USMAN, AND CALIPH ALI TO THE HASAN AND HUSAIN BATTLE IN KARBALA"

2.1 *The manuscript in Haruku Island, Ambon*

In Haruku Island, in Kabau Village, three manuscripts that are the results of Islamic literature were found. They are (1) The Story of Nur Muhammad and The Story of Nabi Bela Bulan, The Story of Nabi Bercukur and The Story of Nabi Wafat, and The Story of Haji and The Story of Saidina Umar; (2) The Story of the Prophet Muhammad; (3) and HKAUUA. The manuscripts are all written in *Jawi* script. The three manuscripts belong to *Bapak Wali Bangsa* Amanullah Ripamole, an *imam* and an Islamic teacher, who lives in Kabau Village in Haruku Island, Ambon.

HKAUUA drew attention because the title portrayed that the manuscript contained the biographies of the caliphs to the tragic stories experienced by Hasan and Husain. After being read, the manuscript tells the story of the Muhammad Hanafiyyah's determination to defend

Ali's descendants. In general, the plot of HKAUUA is the same with the plot of HMH. However, there are some differences.

The HKAUUA manuscript was copied by *Bapak Wali Bangsa* Amanullah Ripamole in 1999 because the original manuscript was damaged. The text is written in *Jawi* script and it has 143 pages. The text was copied on a drawing book with lines written on each page to make the writing neat. Nevertheless, there are a number of mistakes in the transfer, among which are missing words and mistakes that were made while linking the pages. In this manuscript, there are also drawings of red flowers decorating certain words, for example the name the Prophet Muhammad and other characters.

The study of HKAUUA is based on phillology principles (read Kramadibrata, 2015). After analyzing the text critically, the text is translated into Latin script and the content of HKAUUA is analysed. By tracing written sources about Ambonese culture, it appears that the characteristics of HKAUUA are closely related to the Ambonese culture.

2.2   *Analysis of HKAUUA*

There are a number of interesting findings in HKAUUA.

First is in terms of language. From the phonological aspect, in the HKAUUA text there are words commonly found in the Classic Malay text. For example, the writing of the word *menengar* for 'to listen'; the use of /k/ and /g/ is often mixed, such as the word *Kufah* that is written into *Gufah, kendi* is written into *gendi*; the addition of /h/ in a number of words, for instance *harang, mahu, hambah*; the removal of /h/ in a number of words, such as *amparan, mara*; the use of the glottal stop symbol (symbolized with *hamzah*) as a marker of the vocal series, for example *na'ik, ma'u, dimetera'i, pe'kerja'an*; the use of the glottal stop symbol at the beginning of a word with a vocal phoneme, such as *'oleh*.

In the text, the sound schwa (e) is symbolized with the vocals a, i, and u. For example, *benua* is written as *banua*, the word *tepi* is written as *tipi*, and the word *sembah* is written as *sumbah*. This kind of variation is also found in *The Story of Tanah Hitu* (Straver et al., 2004:28).

In the text, a number of archaic words, such as those from the Persian language, like *darwis* (Sufis who live in poverty) and *paighambar* (prophet), were found. Both words were written based on the writer's pronunciation. The word *darwis* became *durubasa*, while the word *paighambar* became *pangambur* or *pangumbar*. In addition, some words were written based on the pronunciation of the writer, who came from Ambon. For instance, the word *cincin* was written into *cincing, hutan* was written into *hutang*, or the word *pulang* was written into *pulan*.

From the morphological aspect, a number of characteristic affixations were found. For example, the formation of the word *diper-nya*, which has a perfective meaning is present in the sentence "*Syahdan maka Utbah dan Ubaidullah Ziyad pun diperikutnya oleh Masib Kaka.*" In addition, irregular assimilation is also found in the preffixation of *me*N-. There are affixes that are not assimilated, for instance in the word *mehela* for *menghela*. There are affixes that assimilate with a number of uncommon assimilated pattern forms, such as the word *memerdekakan* that becomes *mengerdehakan* and the word *mengerjakan* that becomes *mengerejakan*.

From the syntax level, a number of passive sentence patterns were found in the Malay language, which commonly appeared in the 18th and 19th centuries. In addition, sentence constructions influenced by Ambonese language were also found, among them being the sentence with the structure *dia punya*. For example is *Maka ada seorang Yazid punya hulubalang Syahrab/Zanggi namanya, tengah empat puluh akan tingginya (Yazid has hulubalang named Syahrab/Zanggi, and his height is fourty five)*.

Second of all, from the content aspect, the storyline of HKAUUA is similar to that storyline of HMH but with several differences. The story of HKAUUA began with Umar's death. After Umar died, Usman was appointed as the new caliph. Usman later was killed and Ali was appointed as the caliph.

During his caliphate, Ali repeatedly went to war against Muawiyah, who was Usman's brother. Due to Muawiyah's cunningness, Ali was killed by the man who took care of his horse. Afterwards, there was a dispute about who would be the next caliph. Ali's group (known in Islamic history as the Shia Ali sect) believed that Hasan, who was Ali's son, was entitled to become the new caliph. On the other hand, the Bani Umayyah sect, from Usman's group, wanted Yazid to be the new caliph.

The conflict triggered a war in Karbala. As predicted by a fortune-teller, Hasan, who was Husain's brother, was poisoned to death. Meanwhile, Husain had a tragic death—he was beheaded—in Karbala. After the death of Hasan and Husain, their brother, Muhammad Hanafiyyah sought revenge over the death of Hasan and Husain. With the help of Husain's spirit, Muhammad Hanafiyyah succeeded in defeating Yazid.

HKAUUA does not have the same story as the first, the second, and the third stories in HMH. In HKAUUA, a number of episodes from the y_1 version are as follows:

1. The story of Ali instructing his sons to throw away the Zulfakar sword;
2. The King of Israel helping "the content of the house of the Prophet";
3. The beheading Kasim's seven children to make up for the loss of Husain's head; and
4. The story of Said and Mahid leaving their father's troops.

In HKAUUA, an additional episode was found, which is about the beheading of Kasim's seven children to make up for the loss of Husain's head. This episode may have existed due to its connection with the tradition of beheading that occurred in Ambon in the past (read Subyakto, 2007:188). Meanwhile, the story of Sahid and Mahid is an intentional interpolation, which occurs because of changing times. The last part of the HKAUUA text does not contain the story of Muhammad Hanafiyyah's disappearance in the cave. Thus, HKAUUA only presents the heroic character of Muhammad Hanafiyyah as a supporter of Ali's family, and not as the *mahdi*.

Based on these elements, the HKAUUA text is included as part of the y_1 version. The y_1 version dated back to the 18th century. On that account, there is a possibility that the HKAUUA text is derived from texts dating back to the 18th century or later. There have been conjectures that the HKAUUA text has been copied multiple times, changed, and adapted to local conditions and culture.

2.3 *Muslim society in Haruku Island*

Muslim community in Haruku Island are known as Hatuhaha Muslim community. They believe that they are Ali's descendants (read Rumahuru, 2012, 2013). The HKAUUA text comes from Kabau Village in Haruku Island, an area which is believed to adhere to Shia *mazhab*. Initially, there were assumptions that the HKAUUA text showed characteristics of Shia thoughts. In the text, a number of elements related to Shia, such as the Asyura celebration, were indeed found. However, based on the analysis of the HKAUUA text, it can be concluded that the HKAUUA text does not reflect Shia perspectives, particularly the perspective that believes Muhammad Hanafiyyah is the *mahdi*. The final part of the HKAUUA text does not contain any story that claims Muhammad Hanafiyyah as the *mahdi*.

In the HKAUUA text, Muhammad Hanafiyyah is illustrated as a heroic character, who was committed to his duty and sought revenge for Husain's death. After accomplishing his duty, he immediately appointed Zainal Abidin as the king. In this matter, the removal of the part about the *mahdi* can be considered as part of the process of "de-Shiazation" in the story of Muhammad Hanafiyyah.

In relation to the belief that Hatuhaha Muslim community follows Shia *mazhab*, there is a text titled "The Karbala Sermon" that shows that Hatuhaha Muslim community does not follow Shia. They appreciate *ahl al-Bait*. Part of the praises in the text are addressed to the names of Muhammad, Ali, Fatimah, and Usman. There is a part of the text which condemns Yazid's act, but there are no parts that condemn the three caliphs before Ali. Therefore, it can be concluded that the influence of Shia in Hatuhaha Muslim community does not lie on the ideological level. Shia's influence can be observed in Hatuhaha Muslim community only on the cultural level, as seen in Aceh and Minangkabau.

In Hatuhaha Muslim community, Islam merged with local customs. The Muslim community in Hatuhaha is divided into two groups, Islam sharia and traditional Islam. The Islam sharia group performs Islamic rituals based on teachings of Islam, while the traditional Islam group performs Islamic rituals based on the local customs. This difference may lead to conflicts; however, there is one unifying tool, a traditional ceremony called *ma'atenu*.

Ma'atenu or *cakalele* is a ceremony that takes the form of a war dance which depicts the story of a courageous struggle against a tyranny. Such courage is demonstrated by insusceptibility the dancers have for sharp weapons. Based on its history, *ma'atenu* is a ceremony held to revere Ali who fought against the disbelievers. This ceremony is then linked to the historical event which occurred in Hatuhaha Muslim community who for several times fought against disbelievers. Thus, *ma'atenu* ceremony is not an event related to Shia tradition to commemorate Husain's death, but a ceremony to revere the ancestors of Hatuhaha Muslim community, whoare Ali's descendants.

Based on the elaboration above, it is shown that upon being introduced into Hatuhaha Muslim community, Islam was adapted into the local customs. The Islam that was introduced came with a Shia characteristic and was merged into *ma'atenu*. Along with the introduction of Islam, heroic Shia tales were also spread, as those told in HKAUUA.

3 CONCLUSION

Based on the elaboration above, from the aspect of language, it is shown that the HKAUUA text dated back to the 18th century. Due to repeated copying up to the 20th century, there are changes in the forms of words (affixation) and sentence structures in Malay language. From the phonological aspect, a number of words in Malay are written based on the pronunciation of the Ambonese language because the text was written by a person who speaks the language.

Based on the content, the HKAUUA text is not the same with HMH. According to the classification by Brakel, the HKAUAA text is part of the y1 version. In the HKAUUA text, there is the episode that tells the beheading of Kasim's seven children to make up for Husain's beheading. This episode appeared in relation to the tradition of beheading Ambonese's local culture in the past.

The HKAUUA text experienced "de-Shiazation" because it did not contain any illustration of Muhammad Hanafiyyah as the *mahdi*. For Hatuhaha Muslim community, Muhammad Hanafiyyah is a Muslim hero who could perform his duty well. HKAUUA only presents the heroic character of Muhammad Hanafiyyah as a supporter of Ali's family, but not as the *mahdi*.

Upon being copied repeatedly, the Shi'itic elements in the text continue to cease. Although the main Shi'itic element in the text has disappeared, HKAUUA is a symbol of a bond between Hatuhaha Muslim community and Ali. The copying of the HKAUUA manuscript by *Bapak* Amanullah Ripamole is to maintain this bond.

REFERENCES

Braginsky, V. (1998). *Yang Indah, Berfaedah, dan Kamal: Sejarah Sastra Melayu dalam Abad 7–19* (The beauty, the useful, and the perfect: Malay literature history in the 7th – 19th centuries). Translated by Hersri Setiawan. Jakarta, INIS.

Brakel, L.F. (1975). *Hikayat Muhammad Hanafiyyah: A Medieval Muslim-Malay Romance*. The Hague, Martinus Nijhoff.

Collins, J.T. (1974). *Catatan ringkas tentang Bahasa Melayu Ambon* (Brief records on Ambon Malay Language). In: Dewan Bahasa, April 1974, pp. 151–162.

Collins, J.T. (2005). *Bahasa Melayu Bahasa Dunia: Sejarah Singkat* (Malay Language the World Language: Brief History). Translated by Alma Evita Almanar with the foreword by Dendy Sugono. Jakarta, Yayasan Obor.

Hadi W.M.A. (2013). Jejak Persia dalam sejarah kebudayaan dan sastra Melayu (The trace of Persia in Malay culture and literature history), *Media Syariah* XVb(1), 89–103.

Iskandar, T. (1996). *Kesusasteraan Klasik Melayu Sepanjang Abad* (Malay Classical Literature throughout the century). Jakarta, Libra.

Istanti, K.Z. (2008). *Sambutan Hikayat Amir Hamzah dalam Sejarah Melayu, Hikayat Umar Umayah, dan Serat Menak* (The story remarks of Amir Hamzah in Malay History, Umar Umayah Story, and *Serat Menak*). Yogyakarta, Seksi Penerbitan Fakultas Ilmu Budaya Universitas Gadjah Mada.

Kramadibrata, D. (2015). *Hikayat Khalifah Abu Bakar, Umar, Usman, dan Ali Sampai Peperangan Hasan dan Husain di Karbala: Edisi Teks dan Kajian Latar Belakang Agama dan Budaya* (The Stories of Khalifah Abu Bakar, Umar, Usman, and Ali up to the war of Hasan and Husain in Karbala: Text Edition and the Background Study of Religion and Culture). [Dissertation]. Depok, Fakultas Ilmu Pengetahuan Budaya Universitas Indonesia.

Leirissa, R.Z. et al. (1999). *Sejarah Kebudayaan Maluku* (Malaku Culture History). Jakarta, Departemen Pendidikan dan Kebudayaan RI.

Radjawane, A.N. (1964). Islam di Ambon dan Haruku (Islam in Ambon and Haruku). In: Sidjabat, W.B. (ed.). *Panggilan Kita di Indonesia Dewasa Ini*. Jakarta, Badan Penerbit Kristen, pp. 70–85.

Rumahuru, Y.Z. (2012a). Dialog adat dan agama, melampaui dominasi dan akomodasi (Dialogue of Customs and Religions, exceeding domination and accomodation), *Jurnal Al-Ulum*, XII (2), 303–316.

Rumahuru, Y.Z. (2012b). *Islam Syariah dan Islam Adat: Konstruksi Identitas Keagamaan dan Perubahan Sosial di Kalangan Komunitas Muslim Hatuhaha di Negeri Pelauw (Syariah Islam and Customary Islam: Religion Identity Construction and Social Change in Hatuhaha Muslim Community in Pelauw Country)*. Jakarta, Kementerian Agama RI.

Rumahuru, Y.Z. (2013a). Agama sebagai fondasi perkembangan masyarakat dan perubahan sosial: studi kasus Orang Hatuhaha di Negeri Pelauw, (Religion as the foundation of community development and social change: a study case of Hatuhaha people in Pelauw Country) Maluku Tengah, *Harmoni*, 12 (1), 144–160.

Rumahuru, Y.Z. (2013b). "Kebudayaan dan tradisi Syiah di Maluku: studi kasus Komunitas Muslim Hatuhaha" (Shia Culture and Tradition in Maluku: A case study of Hatuhaha Muslim Community) in *Sejarah & Budaya Syiah di Asia Tenggara (History & Culture of Shia in Southeast Asia)*. In: Sofyan, D. (ed.). Yogyakarta, Sekolah Pascasarjana Universitas Gadjah Mada, pp. 255–270.

Soeratno, S.C. (1991). *Hikayat Iskandar Zulkarnain: Analisis Resepsi* (The story of Iskandar Zulkarnain: Reception Analysis). Jakarta, Balai Pustaka.

Straver, H., van Fraassen, C., van der Putten, J. (2004). *Ridjali Historie van Hitu: een Ambonse geschiedenis uit de zeventiende eeuw*. Utrecht, Landelijk Steunpunt Educatie Molukkers.

Subyakto. (2007). Kebudayaan Ambon (Ambon Culture). In: Koentjaraningrat (ed.). The 22nd edition. *Manusia dan Kebudayaan Indonesia*. Jakarta, Penerbit Djambatan.

Wieringa, E. (1996). Does traditional Islamic Malay literature contain Shi'itic elements? 'Alî and Fâtimah in Malay hikayat literature, *Studia Islamika, Indonesian Journal for Islamic Studies*, 3 (4), 93–111.

Islamic discourse in German online mass media: Intra-migrant Muslim perspective on refugee issues. Shift in the position of three Muslim individuals: Feridun Zaimoglu, Necla Kelek, and Navid Kermani

L. Liyanti
German Study Program, Faculty of Humanities, Universitas Indonesia, Depok, Indonesia

ABSTRACT: Since the late 80s, Germany has been dealing with new social, political, and religious issues brought in by the extended stay of its guest-workers who actually should have returned to their home countries once their employment contract ended. These guest-workers and their family are mostly from Turkey with Islam as their most prominent identity. The Islamic discourse has appeared in many aspects of German's social and political life since then. In 2006, The German Home Office initiated the First Deutsche Islamkonferenz (DIK) aiming to build a shared future with its Muslim citizens. Not only the Islamic large organizations but also ten individual Muslims from different backgrounds were invited to this conference. My previous research focuses on DIK, the problem of representation, and the position of the three Muslim individuals in proposing the term *deutscher* Islam. Meanwhile, in this paper, their position is examined based on the new emerging phenomenon raised by the arrival of millions of Muslim refugees in Germany. Van Dijk's Critical Discourse Analysis (Ideological Square) is applied to analyze the texts.

1 INTRODUCTION

After experiencing a massive destruction due to its defeat in World War II, Germany recovered and successfully achieved significant improvement on its economic and national development known as *Wirtschaftwunder* in the late 1950s–1960s. This led to the shortage of workers and forced Germany to get additional workers from various countries, including Muslim-majority countries, such as Turkey, Morocco, and Tunisia. These workers were called 'guest workers'/*Gastarbeiter* since they were invited to work for certain times and had to go back to their home countries after one year if the contract was not extended (Fetzer and Soper, 2005:99). However, these workers chose to stay and bring their families to Germany. In 1961, there were 65,000 Muslim workers in Germany. In 1989, it increased to 1.8 million, and in 2002 the number of Muslims in Germany was 3.4 million (Fetzer and Soper, 2005:102); the majority, around 1.8 million, came from Turkey (website DIK, 2009). Their presence has brought significant changes in the history of Germany. These Muslim migrants as stated by Martin Sökefeld are often seen as *doppelfremder*/double strangers and brought double problems: cultural and religious (Soekefeled, 2004). Aside from the religious and cultural differences, the readiness of the first generation of Muslim migrants to live together with the German was polemical. The guest workers mostly had poor educational and economic backgrounds. They did not have professional skills and could not speak German properly.

Since the presence of Muslim migrants' families, the discourse of German Islam has developed. The discussion about Islam has become tenser after 9/11. The German government realized this situation, and they held a dialogue between the State and Muslim representatives for the first time. During 2006–2009, the first stage of the DIK/Deutsche Islamkonferenz was in progress as a part of the political policy to create a future together with the Muslims living in Germany. The German government invited some Islamic organizations along with the German Muslims with a migration background. The main theme was *Deutsche Muslim, deutscher Islam*/German Muslims, German Islam, which brought about a heated discussion both in the conference and in mass media.

My previous research titled *Gibt es einen deutschen Islam? Intra-migran Discourse in Germany* questions the concept of German Islam/*deutscher Islam* proposed by three Muslim individuals who were invited to the conference and appeared in German mass media. These three Muslim individuals with a migration background are Feridun Zaimoglu, Necla Kelek, and Navid Kermani. Feridun Zaimoglu and Necla Kelek are the children of Turkish guest workers. They came along with their parents when they were children; Zaimoglu was one year old, while Kelek was nine year old. Navid Kermani was born in Germany from middle-class and educated Iranian parents. By questioning the concept of German Islam, the study covers the religious, socio-cultural, and political discourses. This is related to the concept of image, representation, and diaspora of Muslim migrants in Germany. In this paper, the foundation of my previous research is applied to see the development based on the current social and political situations: the arrival of refugees and asylum seekers from Muslim countries as the Syrian civil war broke in 2011.

The website of BAMF (Bundesministerium fuer Migranten und Fluechlinge Germany) published that throughout 2015, the number of asylum seekers was 476,649 people; this shows the highest number since the Syrian war in 2011.[1] In December 2015, there were 46,730 people who legally requested asylum, mostly from Syria (54%), Iraq (10.4%), and Afghanistan (9%). Their arrival in Germany was pushed by the political stance of the German Chancellor, Angela Merkel, who said, "*Wir schaffen es*"/"We can do it", and opened Germany to the refugees. The tragedies experienced by the refugees also took part in the acceptance of refugees and asylum seekers. One of them was the picture of Alan Kurdi, a 3-year-old child, who was drowned after the ship bringing him from the gulf of Turkey to Greece wrecked.

Although there has been a warm acceptance, there has also been refusal. The increase of crimes and the cultural differences as well as different religious values are the concerns of the people refusing the refugees and asylum seekers. This has led to many discussions and comments including those from German Muslims having a migration background: Nekla Kelek, Feridun Zaimoglu, and Navid Kermani.

Research Question. In this research, the position of Kelek, Zaimoglu, and Kermani in proposing their arguments on refugees coming to Germany was analyzed through their articles in German online media.

Corpus and Justification of Corpus. The articles chosen are from online media from 2015 (the peak of the arrival of the refugees and asylum seekers) up to mid–2016. After collecting all articles of those three German Muslims in German online media, one article with an intense discussion about the coming of refugees/asylum seekers was chosen.

Theoretical Framework. In this research, Van Dijk's ideological square theory is applied. Van Dijk defines ideology as socially shared cognitive resources and is fundamental for social practices, interaction, intra and intergroup relations (Van Dijk, 2001:731). He further argues that most ideologies are relevant in situations of competition, conflict, domination, and resistance between groups that result in a polarization on the basis of in-group and out-group differentiation, typically between 'us' and 'them'. Moreover, this often features the following overall strategies of what might be called the ideological square: Emphasize Our good things, Emphasize Their bad things, De-emphasize Our bad things, and De-emphasize Their good things (p.734). His point of the polarized differentiation of 'us' versus 'them' is stressed in this paper as this paper aims to expose the position of the three German Muslims with a migration background when talking about the refugees coming to Germany. I believe the position can be very dynamic as being Germans and (once) being Muslim migrants. Both

---

1. The number of asylum seekers in Germany based on year: 53.347 (2011), 77.651 (2012), 127.023 (2013), 202.834 (2014).

are parts of their identity: their position as German citizens on the one hand and their background as Muslim migrants on the other hand.

Van Dijk's categories of ideological discourse analysis—that he used in his example when analyzing some fragments from a debate in the British House of Commons on asylum seekers—are applied. Some of them are: *actor description [meaning], authority [argumentation], categorization [meaning], comparison [meaning], disclaimers [meaning], evidentiality [meaning, argumentation], generalization [meaning, argumentation]* (see, Van Dijk, pp. 735–739). To see the position of the authors, the Islamic discourse in their articles was analyzed. To define the Islamic discourse, I frame it as the opinion, argument, and thought regarding Islam used by the speakers to negotiate their position regarding the refugees.

2 CONTEXT

As mentioned above, this paper tries to see the position of three German Muslims with a migration background when proposing their arguments on refugees coming to Germany. This paper is a development from my previous research which has found several main theses as follows:

1. "The Discourse of Intra-Muslim German with a migration background in the context of *Deutsche Islamkonferenz (DIK)* is the very product of hybridity *within* Islam in the German diaspora, showing that the hybridization of understanding Islam in the diaspora is also influenced by the relation *within* the Muslim community in addition to the relationship with German and Western society or other migrant communities".
2. Stressing on the progress within the Muslim community, the DIK also shows the improvement of the Muslims' negotiation regarding their position in Germany as there has been a shift from talking *about* Muslims to talking *to* Muslims and *for* Muslims.
3. The Discourse of Intra-Muslim German with a migration background in the political framework of the DIK accentuates the problem of representation. Representations—as in many other contexts–have a tendency to reduce the complexity of reality. As Zaimoglu argues that the representation of Islam is dominated by the picture of liberal Muslims, his counterpart, Kelek (known as a liberal Muslimah), argues that she has no place anywhere. My finding shows a gap between the two, as the liberal Islam is more acceptable for the Western value, it is more embraced in the political discourse as well as in media discourse; however, since the representation of Islam in the grass-roots level seems to be more dominated by the "non-liberal" Islam representation, there is a stronger confrontation between the value of Islam and Western values.

From these three points, the second point is the basis of this paper's development, which is to see the position of 'us' and 'them' of the authors.

3 ANALYSIS

To begin this research, I collected all articles from Necla Kelek, Navid Kermani, and Feridun Zaimoglu related to the issue of refugees and asylum seekers dated from May 2015 to May 2016. The articles chosen are as follows:

3.1 *Necla Kelek*

- One short article titled *Fluechtlinge muessen sich aendern, wenn sie in dieser Gesellschaft ankommen wollen* (413 words) was published in FOCUS online[2] (29th September 2015).

---

2. http://www.focus.de/politik/deutschland/gastebeitrag-von-necla-kelec-fluechtlinge-muessen-sich-aendern-wenn-sie-in-dieser-gesellschaft-ankommen-wollen_id_4977927.html.

- Two long interview articles: *Merkel muss darueber nachdenken, was sie uns zumutet* (2367 words) was published in Die Welt online[3] (12th February 2016), and *Muslime brauchen dringend Aufklaerung* (4643 words) was published in Deutschlandradio Kultur online[4] (13th February 2016).

3.2 *Feridun Zaimoglu*

One short article titled *Da Kenne ich die Deutschen aber anders* (582 words) was published in Frankfurter Allgemeine Zeitung online[5] (7th September 2015).

3.3 *Navid Kermani*

Two long interview articles: first one titled *Schaffen wir das?* (4996 words) was published in Der Spiegel 23rd January 2016 and can be accessed online[6], and *Kulturgespraecht, Interview mit Kermani Einbruch der Wirklichkeit, Beobachtungen auf dem Fluechtlingtreck durch Europa* (1505 words) was published in SWR[7] (28th January 2016).

Necla Kelek. Kelek's short article titled *Fluechtlinge muessen sich aendern, wenn sie in dieser Gesellschaft ankommen wollen* (*Refugees Must Be Willing to Change If They Want to Be Accepted in This Society*) published in FOCUS online (29th September 2015) was chosen. In this article, Kelek discusses what needs to be done by the refugees if they want to be accepted in the German society. She begins with a statement of *problematic actor description* that the coming of these refugees has brought in the potential of ethnical, religious, and cultural conflicts (Z.4).[8] Then she provides *evidence* as she mentions the news of the dispute between the Muslims and Christians in the camp (Z.6).[9] Kelek also makes a *generalization* as she admits that although some of these refugees fled from those plagues; they are culturally still the same (with those). After making the generalization of all Muslims coming to Germany, she *compares* these Muslims with Germans. Islam is fundamentally different from the liberal freedom and social values in German society (Z.16).[10] Moreover, there is a *disclaimer* in her article as Kelek proposes that these coming refugees must learn to respect their and others' freedom and be willing to change if they want to be accepted in *this* society (Z.19).[11] It is clear that she is very critical towards the arrival of Muslim refugees. She highlights the dispute between the Muslims and Christians at the very beginning and continues to focus on arguing about the negative side of Islam in the rest of her article. Stressing these beliefs as her main argument, she comes up with the solution that the newcomers should change their attitude to be accepted by Germans.

Kelek is very critical not only towards the Muslim refugees but also towards the Islamic organizations in Germany. There is a *negative other-presentation* as she states that "We must

---

3. http://www.welt.de/politik/article152184546/Merkel-muss-darueber-nachdenken-was-sie-uns-zumutet.html.

4. http://www.focus.de/politik/deutschland/gastebeitrag-von-necla-kelec-fluechtlinge-muessen-sich-aendern-wenn-sie-in-dieser-gesellschaft-ankommen-wollen_id_4977927.html.

5. http://www.faz.net/aktuell/feuilleton/debatten/feridun-zaimoglu-zu-deutschlands-gastfreund-schaft-13788463.html.

6. https://magazin.spiegel.de/SP/2016/4/141826761/index.html.

7. http://www.swr.de/swr2/kultur-info/beobachtungen-auf-dem-fluechtlingstreck/-/id=9597116/did=16865244/nid=9597116/1uozh0 m/index.html.

8. *Z.4 Denn mit ihnen kaemen "etnische, religioese und kulturelle Konflikte".*

9. *Z.6 Die Nachrichten ueber Auseinandersetzungen zwischen Muslimen und Christen in den Erstaufnahmelagern machen nicht nur den Behoerde und freiwilligen Helfern Sorgen.*

10. *Z.16 Einige sind sicher auch von diesen Plagen geflohen. Aber alle kommen mit einer kulturellen Praegung hierher, die sich von dem libertaeren Freiheitsbegriff unserer Zivilgesellschaft fundamental unterscheidet.*

11. *Z.19 Sie muessen lernen, die eigene Freiheit und die der anderen zu respektieren. Sie muessen Gewohnheiten ablegen, sich aendern, wenn sie in dieser Gesellschaft ankommen wollen.*

ensure that mosque organizations do not spread conservative Islam in the name of human-ity" (Z.26).[12] The negative other-presentation is stronger as she argues that these organiza-tions integrate the migrants into their groups but not to the German society (Z.30),[13] and she warns the German society not to "let the fox guard the hen house" (Z.21–22).[14] In my previous research, her dispute with Zaimoglu over the issue of secular and conservative Mus-lims appears in this article as she stresses the danger of conservative Islamic organizations in Germany.

From this article, Kelek's position is clear. In the beginning, she tries to be neutral by mentioning both the refugees and the *aufnehmende Gesellschaft* using the word "*sie*"/they, but then she further uses the word "*wir*" and "*uns*" to identify herself with Germans (Z.26). All notes about the arrival of the refugees are connected to their religion and have a negative tone when comparing them to Germans. In this point, *the ideological square* from Van Dijk which emphasizes *their* bad things is clearly shown.

However, it is interesting to see how Nekla Kelek introduces (or is introduced) at the end of the articles: "German social scientist and writer Necla Kelek, born in Istanbul. She was a member of Deutsche Islamkonferenz".[15] All information highlights Kelek's identity as a part of Muslim that has a migration background. The strategy to show her identity as a Muslim and migrant can be seen as an attempt to strengthen her argument, which she really knows, because she is part of it. From another perspective, we can read her argument in criticizing Islam as her attempt to propose her definition of Islam that fits into the German society. Moreover, she is a fine example for this kind of Muslim.

Feridun Zaimoglu. Zaimoglu's article titled *Da Kenne ich die Deutschen aber anders* (*But I Know Germans Are Different*) published in FAZ online (7th September 2015) was chosen. Different from Kelek's article that stands as an independent article, Zaimoglus' article is an answer to the previous article written by Milos Matuschek *Warum macht unser Mitge-fuehl schlapp?* In his article, Matuschek argues that Germans do not have enough empathy for the refugees and states that German's *Willkommenskultur* (Welcoming Culture) existed only in the 2006 World Cup and ended there. Matuschek also argues that the refugees are mostly seen as a threat rather than an opportunity. Zaimoglu who does not agree with Matuschek, wrote this article as an answer. Like Zaimoglu, Matuschek, who came from Poland 33 year ago, is a German with a migration background. Having the same back-ground as a migrant, Zaimoglu gives another side of Germans—that he knows for sure—that German's *Willkommenskultur* is not only seen nowadays, but it has been known for a long time (Z.5),[16] and indeed the hospitality of Germans in welcoming people from all over the world is *das zweite grosse Wunder des deutschen Staates* (the second great wonder of Germany) (Z.20).[17] Furthermore, in his article, Zaimoglu criticizes what he calls as *Eli-tauslaender* or 'the one who always blames', and he is willing to teach the people who once accepted them (Z.27).[18]

In his final word, Zaimoglu concludes one very interesting point, "My country is hospita-ble, and in my country the guest will, in the future, be the host (Z.56)".[19] This emphasizes his important point. His voice is very positive and optimistic regarding the *wilkommenkultur* in

---

12. *Z.26 wir muessen auch strickt darauf achten, dass die Moscheevereine diese Menschen nicht- unter dem Deckmantel der Hilfe- in ihren alten Mustern des konservativen Islam bestaetigen.*
13. *Z.30 Mit den Islamverbaenden, die sich als Missionare von der Tuerkei, aus Kuwait, Katar und Saudi-Arabien finanzieren lassen, macht man den Bock zum Gaertner. Sie werden wie bisher die Einwanderer in ihre Gemeinden integrieren, aber nicht in dieses Land.*
14. *macht man den Bock zum Gaertner.*
15. *Die deutsche Sozialwissenschaftlerin und Publizistin Necla Kelek wurde in Istanbul geboren. Sie war Mitglied der Deutschen Islamkonferenz.*
16. *Z.5. Die herzliche Aufnahme wird bei uns nicht nur in diesen Tagen buchstabiert.*
17. *Z.20 Die Aufnahme fremder Menschen aus aller Welt das zweite grosse Wunder des deutschen Staates.*
18. *Z.27 Elitauslaender, kann nicht anders, als das Volk, das ihn aufnahm, geringzuschaetszen und erziehen zu wollen.*
19. *Z.26: Und mein Land ist gastfreundlich. Und in meinem Land werden aus Gaesten kuenftige Gastgeber.*

Germany. He believes that in German the guests are welcomed and live a good life as in the future they will also welcome and host others. Pointing many interesting and positive points towards the refugees coming to Germany, Zaimouglu does not bring up the Islamic discourse in his text but rather general consensus of migrants and refugees. This might be because he responds to the previous article that focuses on the argument of German's attitude and does not mention the Islamic discourse.

Zaimoglu's position of 'us' and 'them' in this article cannot be identified clearly as he seems to flow between these poles. On the one hand, he identifies himself as a German using the word "we" when writing about Germans and showing many examples of Germany's *Willkommenskultur*. On the other hand, he states a positive argument towards the arrival of migrants/refugees by providing a statement explaining that Germans with a migration background make Germany richer.[20] Zaimoglu also mentions that in the future migrants will welcome others coming to Germany, and this can be seen as a continuous process since in Germany "They" will always become "We". Thus, seeing from one perspective, *the ideological square* from Van Dijk cannot be used properly in this regard, as the position of 'us' and 'them' here is not clear in relation to Zaimoglu and the Muslim refugees, but the position of 'us' and 'them' here is between Zaimoglu and Matuschek. For the same reason, the categories of the ideological discourse analysis can only be understood in terms of doing a *positive self-presentation* when talking about both Germany and the refugees. Considering this argument from another perspective, the ideological square from Van Dijk can be seen in a point of "Emphasize Our good things".

The absence of Islamic discourse in Zaimoglu's text is in accordance with his profile information at the end of the text. Different from Kelek who states herself as a member of Deutsche Islamkonferenz, Zaimoglu concludes the information by highlighting only his migration background and promoting his book *Siebentuermeviertel:* born 1964 in Bulo Turkey, coming to Germany when he was one year old, a child of a guest-worker, raised in Munich, and publishing his roman, *Siebentuermeviertel.*[21]

It is interesting to see why Zaimoglu promotes his book in this article. As I searched the book, I came up with an interesting result. The book tells the story about a German man and his son who fled to Turkey in 1939 to avoid NS and got a shelter from a Turk man in Turkey. This is a reversal of what happens now in which Germans become a host to people (mostly Muslims) who flee from their county to avoid a dangerous condition.

Navid Kermani. Navid Kermani is very engaged in the refugee' issues. Since 2014, he has been doing a project with the refugees by joining their journey from Iraq to Europe and Germany via the Balkan route to picture their journey and understand everything much better. He wrote a report of his journey and then published it as a book titled *Einbruch der Wirklichkeit, Beobachtungen auf dem Fluechtlingtreck durch Europa.* He was awarded *Friedenspries des Deutschen Buchhandels* for these reports (Der Spiegel: 2016, interview article *Schaffen wir das?*). This project also attracts the media to write about him and his book, have him as a speaker, or offer him an interview session. Thus, it can be understood that during this time there was no article written by Kermani himself in German online media. From some coverage, I found two interview-articles with Kermani in online media. I chose one titled *Schaffen wir das?* Because this article focuses more on the issue of the refugees, Germany and Europe, while the others are concerned more about his book.

At the beginning of the text, there is a very interesting standpoint of Kermani as the interviewer writes: "The writer (Kermani) proposes that if we remain calm and be realistic, our life will be uncomfortable".[22] Furthermore, at the beginning of the interview, Kermani starts with a *comparative* statement that from now on Germany will be much more interesting

---

20. *Z.50–53 Es wird immer gesagt, dass herkunftsfremde Deutsche auf welchem Betätigungsfeld auch immer, ob in der Gastronomie oder in der Literatur, eine Bereicherung für das Land seien.*
21. *Der Schriftsteller Feridun Zaimoglu wurde 1964 in Bolu in der Türkei geboren und kam als Einjähriger nach Deutschland. Als Kind von Gastarbeitern wuchser in Muenchen auf. Soeben veröffentlichte er seinen Roman Siebentuermeviertel.*
22. *Der Schriftsteller schlaegt vor, cool zu bleiben. Und realistisch: Unser Leben wird unbequem.*

and lovely in comparison to monoculturally dull Germany in the 1950s (pp. 99–104).[23] He is realistic that the life could be uncomfortable, and the danger is real (pp.315–320), but he is also optimistic that all of those can be dealt as he quotes Hoelderlin's words "where there is danger, there will be a hero" (p. 436).[24]

In this interview, Kermani uses one interesting word *unsereiner* to define a person like him (a German with a migration background). This *unsereiner* would lose the most if something wrong happened in Europe, and pushing the *unsereiner* to leave Europe (pp.673–677).[25] The reaction of the interviewer is also interesting, saying Kermani is also a German citizen, just like he is (p.678),[26] and that Kermani belongs the *Volks* all these times, talks in the German Federal Parliament, and knows more about German literature than he does (pp.687–691).[27] Kermani responds in this context that the interviewer belongs to the *Volksgemeinschaft* that is ideal for Mr. Gauland and Mr. Hoecke, who want to build a "pure" Germany (even if the interviewer does not agree with that concept). Kermani argues, at the moment when the German people would be grouped between those *Volks* and all the newcomers or Islam, he will be placed as the others (pp.679–686).[28] This is because he believes that "if our *Gesellschaftmodell* collapses, then the ethnic root will matter" (pp.692–697).[29]

From these points, we can see how Kermani formulates his position in the context of newcoming migrants. He identifies himself as a German, using the word "we" and "our" when discussing about German and Germany (p.265), but shows a great empathy and hope for "them", showing they can add more colours to Germany and describing their suffering of becoming refugees (p.303). Furthermore, he admits that he is part of the *unsereiner*, a person who does not belong to the *Volksgemeinschaft* as he has an Iranian ancestor. In this context, we can see that he is aware of the fact that his position is not fixed. He is not only part of Germans but also part of the *unsereiner*. The ideological square from Van Dijk here is in the same context as in Zaimoglu's article, which is to "emphasize Our good things". However, after reading and discussing Kermani's text, I find the voice of Kermani different from that of Kelek (that is more pessimistic) and Zaimoglu (positive-optimistic). Kermani sounds clearly realistic-optimistic. He can see what Kelek argues as a potential conflict, and he also sees that these people could make Germany better, similar to Zaimoglu's argument.

Kermani's optimistic voice is even more solid because at the end of the interview, when the interviewer asked him about Merkel's slogan "*das schaffen wir*" and whether they can really do it, Kermani's answered, "I don't know, but at least we try".[30]

4 CONCLUSION

As the development of my previous research's findings, I proposed the question of Kelek, Zaimoglu, and Kermani's position of 'us' and 'them'. To answer this, I searched their focal

---

23. *Z.99–104 Moechte ich, mochten Sie zurueck zu einer Monokultur, gar zu einer homogenenen Volksgemeinschaft? Mir erscheint das Deutschland von heute spanneder, auch liebenswerter als, sagen wir, der Muff der Fuenfzigerjahre.*
24. *Z.436 Der Schriftsteller schlaegt vor, cool zu bleiben. Und realistisch: Unser Leben wird unbequem.*
25. *Z.673–677 Kermani: […] Und ausserdem: Unsereiner, also jemand wie ich, hat mehr zu verlieren als Sie. Wo soll ich denn hin, wenn es kein Europa gibt?*
26. *Z.678 Spiegel: Sie sind genauer deutscher Staatsbuerger wie ich.*
27. *Z.687–691 Spiegel: is das nicht ein bisschen kokett? Sie gehoeren laengst dazu. Sie reden im Bundestag. Sie wissen mehr ueber deutsche Literatur als ich. Die Bedrohung ist fuer uns dieselbe.*
28. *Z.679–686 Kermani: Aber Sie gehoeren nun einmal der Volksgemeinschaft an, auf die sich Herr Gauland und herr Hoecke berufen, auch wenn Sie das vielleicht doof finden. In dem Moment, da hier plotzlich jemand aufteilen will–die gehoeren zum Volk und all die Zugezogenen oder etwa der Islam nicht–, gehoere ich zu den anderen.*
29. *Z.692–697 Kermani: Ja, das stimmt. Dennoch: Wuerde unser Gesellschaftsmodell kippen, spielten ethnische Zugehoerigkeiten eine Rolle, fiele ich auch wieder heraus. […]*
30. *Z.999 SPiegel: Schaffen wir das also? Kermani: Keine Ahnung, aber versuchen wir es doch wenigstens.*

voice in their articles. I found that Kelek is more pessimistic, Zaimoglu is positive-optimistic, and Kermani is realistic-optimistic. Kelek's position of 'us' and 'them' is very clear (We = German, They = Muslim Refugees). Van Dijk's ideological square can be seen in a point of "Emphasize Their bad things" as she describes immensely how bad "They" are. In Zaimoglu's article, the position of 'us' and 'them' is not clear in the relation of Zaimoglu and the Muslim refugees, but the position of 'us' and 'them' here is between Zaimoglu and the previous writer Matuschek. In this text, we can see the shift between 'us' and 'them' (We = German and Muslim refugees) as he has drawn a positive self-presentation in talking about both Germany and the refugees. Van Dijk's the ideological square can be seen in a point of "Emphasize Our good things". The same is found in Kermani's article in defining his position (We = German and *unsereiner,* Muslim refugees). The ideological square from Van Dijk here is in the same context as in Zaimoglu's article, which is to "Emphasize our good things".

Using van Dijk's ideological square to analyze the texts in the context of intra-migrant context, I found that this is interesting as the people's position of 'us' and 'them' shifts from one side to another.

REFERENCES

Bundesamt fuer Migranten und Fluechtlinge. (2016) *Asylzahlen. Nationale Asylzahlen.* Retrieved from: http://www.bamf.de/DE/Infothek/Statistiken/Asylzahlen/asylzahlen-node.html;jsessionid=563972 F506C853 ADCE64DBB849E8913B.1_cid368 [accessed on 25th August 2016].

Bundesamt fuer Migranten und Fluechtlinge. *Das Bundesamnt im Zahlen. Asyl.* (2015) Retrieved from: http://www.bamf.de/SharedDocs/Anlagen/DE/Publikationen/Broschu-eren/bundesamt-in-zahlen-2015-asyl.pdf?__blob = publicationFile [Accessed on 25th August 2016].

Deutsche Islamkonferenz. (2009) *Geschichte der muslim in Deutschland in muslime in Deutschland.* Retrieved from: http://www.deutsche-islam-konferenz.de/DIK/DE/Startseite/startseite-node.html

Feridun, Z. (2015) *Da Kenne ich die Deutschen aber anders.* Retrieved from: *Frankfurter Allgemeine Zeitung* [Online] http://www.faz.net/aktuell/feuilleton/debatten/feridun-zaimoglu-zu-deutschlands-gastfreundschaft-13788463.html.

Joel S.F. & Soper, J. C. (2005) *Germany: Multiple Establishment and Public Corporation Status.* In *Muslims and the State in Britain, France, and Germany.* Cambridge, Cambridge University Press.

Kelek, K. (2015) *Fluechtlinge muessen sich aendern, wenn sie in dieser Gesellschaft ankommen wollen.* Retrieved from: FOCUS [Online] http://www.focus.de/politik/deutschland/gastebeitrag-von-necla-kelec-fluechtlinge-muessen-sich-aendern-wenn-sie-in-dieser-gesellschaft-ankommen-wollen_id_4977927.html.

Kermani, N. (2016) Schaffen wir das? In: *Der Spiegel.* Retrieved from: https://magazin.spiegel.de/ SP/2016/4/141826761/index.html.

Liyanti, L. (2010) *Gibt es Einen Deutschen Islam? Intra-Migrant Discourse in Germany.* Universitaet Tuebingen (Master's Thesis).

Sökefeld, M. (2004) *Das Paradigma Kultureller Differenz: Zur Forschung und Diskussion über Migranten aus der Türkei in Deutschland.* In: Sökefeld, M., (Ed.); Jenseits des Paradigmas kultureller Differenz. Bielefeld, Transcript.

UNHCR. (2015) *Global Trends Forced Displacement in 2015.* Retrieved from: http://www.unhcr.org/ sta tistics/unhcrstats/576408cd7/unhcr-global-trends-2015.html.

Van Dijk, T.A. (2001) Critical Discourse Analysis. In: Schiffrin, Deborah, Deborah Tannen, and Heidi E. Hamilton (eds). *The Handbook of Discourse Analysis.* Blackwell Publishers, pp. 352–371.

Van Dijk, T.A. (2006) *Politics, ideology and discourse.* In: Ruth Wodak, (Ed.), *Elsevier Encyclopedia of Language and Linguistics. Volume on Politics and Language,* pp. 728–740.

Cultural Dynamics in a Globalized World – Budianta et al. (Eds)
© *2018 Taylor & Francis Group, London, ISBN 978-1-138-62664-5*

Abuya Dimyathi: A charismatic mursyid of *tarekat* Syadziliyah during the New Order regime (1977–1982)

J. Syarif & Y. Machmudi
Arabic Studies Program, Faculty of Humanities, Universitas Indonesia, Depok, Indonesia

ABSTRACT: Before the Indonesian presidential election of 1977, *Kyai* Haji Muhammad Dimyathi, known as Abuya Dimyathi, was arrested for opposing the government's order for the Bantenese to support a party as he had already expressed in one of his Friday prayer sermons. His arrest, afterward, provoked strong reactions in various social classes and led to an outrage of violence in Banten, which the government could hardly control. However, the situation did not last long as Abuya Dimyathi handled it calmly. The purpose of this work is to study the life of Abuya and his role in the community, starting from his arrest during the New Order regime in 1977 until his death in 2003. It also includes the intellectual genealogy and the factors that motivated him to protest against the ruling regime. In this study, we use a historical method through narrative approach. Some theories on socio-religious movement and charismatic leadership were applied. Because of the non-availability of written sources, the data were mainly collected by conducting in-depth interviews with the people who witnessed the historical events and the present-day public figures.

1 INTRODUCTION

Banten has a long history spanning from the beginning of Banten Sultanate to the Unitary State of the Republic of Indonesia. With reference to local sources, Lubis (2003) noted that the first mention of the name Banten could be found in *Carita Parahyangan,* which was written in 1580 (Guillot, 1990; Guillot, 2008). Because of its strategic location in the west of Java, Banten attracted many traders from various parts of the world. According to the famous explorer Tome Pires (1530), Banten was a bustling port city in the Sunda Kingdom, and had been visited by people around the world long before the Banten Sultanate was founded in 1526. Ayatrohaedi revealed, as quoted by Lubis, that Banten had at least stood by the mid-10th century or even the 7th century.

In an Islam community like in Banten, *kyai* (Islamic teacher and leader) plays important roles in everyday life. *Kyai* forms an elite group, which plays not only a traditional role as a religious leader but also a role in the sociopolitical transformation in Banten. Therefore, a *kyai* is an important figure who has influenced the culture and history of the formation of this community. To date, *kyai* has formed sub-social groups that play an important role in Banten, although their traditional roles and positions have been "eroded" by modernization, and most of the roles have been shifted to professionals or other institutions (Hamid, 2011). However, these changes do not destroy all of their social positions and roles completely; a *kyai* remains a public figure who is respected by the Bantenese.

A *kyai* who leads a *tarekat* is called a *mursyid*. Initially, *tarekat* was defined as "the way to God". However, this term is commonly associated with the organization or congregation encompassing the Sufism method. In this study, the term "Sufi" could be interpreted as a *tarekat* teacher. They play a role in developing the teachings of Islamic esotericism that focus on the cleanliness and purity of heart by performing acts of worship to achieve a close relationship with God and for the fulfillment of His pleasure or approval. Their methods focus on meditation and typical deeds *(dhikr* and *wird)*. Through this development, the role of Sufism also extends to social activities through the *tarekat* movement. *Mursyid* is a term used in the *suluk* tradition and *tarekat* addresses a spiritual leader within a *tarekat* congregation. Three criteria need to be fulfilled by an ulema to become a *mursyid*. First, a *mursyid* needs to acquire full and comprehensive understanding of religion while being a spiritual leader. Second, a *mursyid* must understand the wisdom of those possessing *ma'rifat billah* in advance. Third, a *mursyid* must also learn the tactics and strategies applied by the rulers (king/political leaders) (Hadi, 2010). In Banten, a charismatic *mursyid* is called an Abuya. A *kyai* as a *tarekat* teacher must display and pioneer his/her role in spreading and developing the religion through the *tarekat* movement and, therefore, receive a positive endorsement from the society that drives the movement into a social power that dynamically responds to religious and sociopolitical challenges.

The influence of a *kyai* is ever-increasing, depending on his/her charisma. The influence of charismatic ulema since the Dutch colonial era has managed to annex the Banten Sultanate; hence, the position and role of *kyai* in the social system of the Bantenese community has also been questioned. The position, role, and social network of a *kyai* have been established through a very long historical process experienced by the Bantenese community, namely the establishment of the Banten Sultanate, the colonialism government period, and post-colonialism era. This historical journey has created a Banten society known as a devoutly, aggressive, and rebellious society (Kartodirdjo, 1966).

1.1 *Political instability before the 1977 public election*

The election fanatical attitude toward religion and rebellious spirit of Banten reemerged when *Kyai* (Abuya) Dimyathi, an ulema who led an Islamic Boarding School (*pesantren*) in Cidahu, Pandeglang, was arrested. This incident was related to the second round of general election during the New Order regime and negatively affected the political condition.

Abuya Dimyathi was detained by police officers after delivering a Friday prayer sermon for hindering the district and village heads from intimidating the Bantenese community and its surrounding to vote for a certain party, Golkar, in the upcoming general election. The government's pressure on the public authorities to support Golkar was a continuation of its efforts to stay in power by making the electoral system flexible for the government under the reign of President Suharto. From 1973, numerous political parties that were previously sovereign in Indonesia had been forced to join one of the three major parties. The government restructured those previous parties into only two political parties, leaving Golkar party alone. One of the two parties was *Partai Persatuan Pembangunan* (PPP), which is a collaboration of parties with Islamic ideology, such as *Nahdlatul Ulama* (NU), *Pergerakan Tarbiyah Indonesia* (Perti), *Partai Syarikat Islam Indonesia* (PSII), and *Partai Muslimin Indonesia* (Parmusi). The other party was *Partai Demokrasi Indonesia* (PDI), which involves parties with ideologies of nationalism and Christianity, such as *Partai Nasional Indonesia* (PNI), *Partai Kristen Indonesia* (Parkindo), *Partai Katolik*, *Partai Murba*, and *Ikatan Pendukung Kemerdekaan Indonesia* (IPKI). Abuya Dimyathi explained that the state was not Golkar, but the government and therefore people would not support Golkar in the second-round general election in the New Order regime (Dimyathi, 2009). On the basis of the data collected from the interview conducted on 20 January 2016 with *Kyai* Ariman, who has been Abuya Dimyathi's disciple since 1975 and is also a witness of Abuya's arrest, the arrest was unnecessary as Abuya Dimyathi did not have any intention to provoke the people to not vote for Golkar. Nevertheless, he emphasized that Golkar was essentially the same as the other parties, serving as a *"firqoh"* or road/channel to deliver the political aspiration of the society. Apparently, the government, however, regarded this as an act of defiance against it, which then led to the arrest of Abuya Dimyathi and detainment without a proper judicial process.

The detainment of Abuya Dimyathi caused strong reactions in the community. Bantenese from various classes provoked the arrest and demanded his immediate unconditional release. The situation in Banten worsened, resulting in a riot that caused casualties among the people of Banten. The government could hardly control the masses, although the situation did not last long as Abuya Dimyati handled it calmly. In an interview conducted on 29 July 2016,

H.M. Murtadlo mentioned that the efforts of Abuya Dimyathi to calm his followers originally received a huge reaction. However, they had already been out of control. According
to Murtadlo, a fistfight nearly broke out between Abuya and the authorities that seemed to
have the intention to cause turmoil (Dimyathi, 2009).

In fact, the imprisonment and detainment of Abuya Dimyathi were due to the strong
suspicion of the government against the *kyais*, especially those who politically opposed the
government. Before the government implemented a policy for the merging of political parties
in 1973, resulting in NU becoming part of the PPP, a prominent *kyai* known as the *mursyid*
of the *tarekat* Syadziliyah had not shown any active involvement in practical politics. Allegedly, he was one of the *kyais* who did not agree with this merging policy of the government,
which first showed a consolidation among political parties with Islamic ideologies, yet it was
part of the government's strategy to manipulate the political life of Muslims.

Nonetheless, Abuya Dimyathi did not show his opposition for the government openly,
even to his disciples. It seems that he faced a dilemmatic position. On the one hand, if he
had been actively involved in practical politics, it would mean that he agreed and was willing
to support the ill intention of the government to weaken the political life of Islamic people.
On the other hand, if he showed disapproval openly, it would negatively influence the well-
established communication among his fellow *kyais* that had been well maintained, especially
for those who supported the PPP. These considerations presumably made him to be more
focused on teaching and guiding his fellow disciples and the people around his *pesantren* in
the development of religious life as well as in the *tarekat* (Bruinessen, 1999, p. 279) through
his daily activities. Thus, it is evident that he did not support Golkar, because if he had done
so, the authority would not have detained him for frequently conducted religious lectures.

1.2 *Charismatic ulema and mursyid of tarekat Syadziliyah*

A charismatic leader is the most important characteristic of any religious movement. The
concept of "charisma" or authority can be used as a theoretical consideration of this study.
As proposed by Jackson (1990), together with the "traditional authority," charisma is a
power that emerges during the interaction among individuals or groups, and at a certain
level of interaction, an actor as a patron changes the behavior of the client. In fact, Abuya
Dimyati was renowned as a *kyai hikmah*, who can understand secrets (Ahmad, 2007). The
kyai is believed to have mastered the understanding of the mystical secrets veiled from the
understanding of ordinary people. Therefore, he was often visited by officials from the central government. Soekarno is said to have been seeking advice to stop invaders from entering Indonesia. However, it was considered more a folklore than a fact because it cannot be
verified.

Abuya Dimyati was a prominent figure in the highly respected *tarekat* Syadziliyah (Hadi,
2010), which was established by Shaikh Syadzili (died in 1258 AD). His full name was Sayyid
Abil-Hasan Ali bin Abdillah bin Abdil Jabbar Asy Syadzili Asy Syarif Al-Hasani. The
tarekat was attributed to the name of its founder, Abul Hasan Ali asy-Syazili (Gumara,
Tunisia, around 593 H/1196–1197, M-Hotmaithira, Egypt, 656 H/1258 AD). Within *tarekat*
Syadziliah, Syazili's lineage was associated with Hasan bin Ali bin Abi Talib. Thus, he also
had blood relationship with Fatimah, the daughter of the Prophet Muhammad. In Indonesia, Shaikh Shadhili was known for the practices of *aurad* (*wird-wird*) and *hizibs*, which were
widely practiced by the students of the *pesantren* and Muslims in general.

Abuya Dimyathi also practiced Qadiriyahwa Naqsabandyah. Historically, he received
ijazah Qadiriyah in Mecca and from Syadziliyah, who was one of the ulemas who became his
teacher when he was still learning from some of the greatest scholars in West Java, Central
Java, and East Java. *Tarekat* indicates a person who has received a series of teachings or
inheritance of a certain lineage from the Prophet, and is deemed capable of practicing and
teaching it to others in the future. He is known as a storehouse of knowledge, especially in
ilmu hikmah. There are many interesting stories to note about Abuya Dimyathi, especially
the one associated with *hikmah* or *karomah* throughout his life that cannot be rationalized
academically (Dimyathi, 2009). Since his early years, he had been known to be religious, and

he loved to sit and *tirakat,* which later made him a highly respected and great ulema. He was famous for his words every time he was asked about the *tarekat* he followed, *"Tarekat aing mah ngaji..."* (My *tarekat* is studying the Holy Qur'an) (Hadi, 2010). At the time of the revolution, Abuya Dimyathi was directly involved in an armed battle against the Dutch and the Allied Forces that attempted to reoccupy the recently independent Republic of Indonesia when he joined the Tanagara Front (Dimyathi, 2009). It was mentioned that Abuya who was still young became a member of the Hizbullah, a paramilitary unit that also took part in maintaining the independence of the Republic. As an ulema, he was one of the targets of the Dutch who had issues with a group of ulemas that were considered to have the potential to spread radical teachings.

1.3 The monographs of Abuya Dimyathi

Despite not being a *mualif* (author of books), Abuya Dimyathi could produce small monographs that he wrote in spite of his tight lecturing schedules, and he diligently practiced *wirid* throughout his life. His writing style in all of his monographs was considered strong and subtle, filled with *uslub* and *balaghah* as well as the nuances of brilliant literary, such as a small treatise entitled *Ashl al-Qadar.* This monograph contains a description of some important figures in the War of *Badr,* accompanied with the *munajat* poetised by Abuya Dimyathi. There is another treatise titled *Rasn al-Qasr,* containing messages for wanderers and the explanation of *Hizib Nasr.* Another small treatise is dedicated to his disciples, *ihwal sanad,* tarekat *Syadziliyah* characteristics, and the appropriateness for the *salik* in *taqarrub* (intimate terms) with God titled *Hadiyyah al-Jalaliyah.*

Once, his son asked for his permission to write a *Sharh* (explanation) of *Tafseer Al-Jalalayn,* but Abuya Dimyathi denied his request because of his strong belief that "spending the time reading the Holy Qur'an is the highest of all deeds". Furthermore, there are still several books written by *Salaf* classical scholars, which have not been read and examined.

1.4 Three intentions of knowledge

Abuya Dimyathi taught his disciples to be always *istighal* (self-occupied) and *tabahhur* (self-drown) in the oceans of knowledge. In several occasions, especially when teaching the principles of the Qur'an to his disciples, Abuya Dimyathi often explained that each disciple needed the "trilogy of intention" in pursuing knowledge. Those three intentions were that:

1. Seeking knowledge is a manifestation of gratefulness to God for *Ni'mah al-Aql* (grace of intelligence) and *Ni'mah Shihhat al-Badan* (grace of healthy body).
2. Seeking knowledge is to remove ignorance *(li izalah al-jahli)* and clean the deeds to be valid from all the things that shall be muddied and aborted.
3. *Ngaji* is an obligation in its nature, as it has been commanded *(masyru'iyyah).* Because it was a *masyru'iyyah,* the status of *ngaji* was different from *dhikr* and any other way of worship practice (daily prayers, fasting, etc.).

1.5 Establishing Raudhatul Ulum Islamic Boarding School (pesantren)

In 1965, Abuya Dimyathi established a *pesantren* in Cidahu, Tanagara village, Cidasari sub-district, after wandering in many places and studying for some time. Afterward, he settled and actively taught as well as fulfilled his role as a mursyid for *tarekat* Syadziliah. His intention in founding the *pesantren* needs to be explored even further; it could be solely because he had finished his study and practiced what he had learned in the *tarekat,* or it was possibly aimed to prevent communist influence, especially in Cidasari, Pandeglang, Banten. It is known that, Partai Komunis Indonesia (Indonesian Communist Party or PKI)'s insurgency broke out on 30 September 1965.

There were only 10 students at the beginning of the *pesantren* who studied in his modest hut. He started by teaching the principles of the Qur'an to his children, and to his surprise, it spread throughout the whole village. Students from all over the place came to study the religion (Islam), by building their own huts using wood and bamboos they carried from their homes. The number of students increased rapidly. They were from not only Pandeglang but also Banten and all over Indonesia. *Pesantren* Cidahu was formally established in 1965 with Abuya Dimyathi as its first Chairman. It was widely known that the *pesantren* taught religion in a *tarekat* way, an inheritable manner from *salafi* ulema (Hadi 2009). At present, the number of students at *Pesantren* Raudhatul Cidahu Ulum is approximately 1,000. In fact, some of the former students have set up their own schools.

Even though he was strict to his children (twisting their ears for coming late for the *majlis*), Abuya was known to be very gentle toward his students. Although treated with such gentle approach, the students of Raudhatul Ulum had a highly disciplined attitude. Abuya Dimyathi remained active leading *Pesantren* Cidahu until he passed away on 3 October 2003. At present, the schools are inherited to his two sons, *Kyai* Haji Much Murtadho Dimyathi and *Kyai* Haji Muhtadi. The *pesantren* is now widely known for its *tarekat* Syadziliyah congregation led by *Kyai* Muhtadi.

2   CONCLUSION

To understand the political attitudes Abuya Dimyathi that strongly opposed the rulers of the New Order regime in 1971, we should examine his role as a *kyai* as well as a leader of *tarekat* Syadziliyah. In the terminology of Sufism, essentially, a *tarekat* is a practical method run by *Sufis* in guiding their disciples to experience the nature of God. The pattern of teacher–student relationship is called the social community of the *tarekat* congregation. The disciples of a *tarekat* who usually come from different occupations show that, as followers of Sufism, they are in compliance and their role is dedicated to the *tarekat* movements.

The characteristics of socio-religious movements create a sociopolitical relationship, which eventually turns into a sociopolitical movement that takes place in a community through the social relations that serve the political movements internally as well as toward other communities. The social movement with a religious factor is always led by a charismatic leader, that is, through the concept of "charisma" or authority that can be used as a theoretical consideration of the socio-political movements.

"Traditional authority" as concluded by Jackson is usually held in a dyadic relationship (in a pair) that tends to be personal, spreading, soulful, and everlasting. Dyadic relationships are usually shown by the tendency of the elites (teachers) who require support from clients (students) and vice versa. The study of Abuya Dimyathi shows that political behavior is mainly based on one's constant position in the structure of traditional authority relationships. If the traditional authority figures (teachers) are involved in politics outside the community, the followers (disciples) would mobilize and provide their rights or even physical aids for the teachers and obey the establishment of the elected political figures embraced by the authority figures. The commotion that appeared in Pandeglang, Banten, in the midst of Abuya Dimyathi's arrest can be understood from the aforementioned perspective. The strong reaction displayed by the Bantenese society is a form of dyadic relationship, between Abuya Dimyathi and his disciples.

The position of religious leader in the socio-religious and sociopolitical movement seems so strong that they are usually at the connecting and supporting positions, even though they are not the only group with the access to the outside system. The mediator position in a complex society can also be achieved to fortify vulnerable spots in the local system with a whole wider system. The mediator also acts in mediating conflicting groups, maintaining the driving force, and maintaining the dynamics of the communities that are necessary for their activities. Although it is not always the same, the mediator can also be a cultural broker and agent of modernization actively trying to introduce new elements to public.

The government's actions against Abuya Dimyathi during the 1977 election can be theoretically considered as a regime with a strong suspicion that the *ulema* (in this case, Abuya Dimyathi) would become a mediator who would not support the regime. In fact, he turned out to be an agent with the potential to become harmful to the government's structural stability at that time.

REFERENCES

Ahmad, H.M. Athoullah. (2007) *Ilmu Hikmat di Bantren Studi Kasus Praktik Islam Mistik di Serang* (The Science of *Hikmat* in Banten: A Case Study of The Practice of Mystical Islam), Banten, Yayasan Bvani Hanbal (Yasbah).
Ali, Mufti. Interview with the Head of Bantenology IAIN Sultan Maulana Hasanuddin Banten (SMHB). (Personal communication 17th June 2015).
Bruinessen, M.V. (1999) *Kitab Kuning, Pesantren, dan Tarekat: Tradisi-Tradisi Islam di Indonesia* (*Kitab Kuning*, Islamic Boarding Schools, and *Tarekat*: Islam Traditions in Indonesia). Bandung, Mizan.
Dimyathi, H.M.M. (2009) *Manaqib Abuya Cidahu dalam Pesona Langkah di Dua Alam*, without publisher.
Guillot, C. (1990) *The Sultanate of Banten.* Jakarta: Gramedia.
Guillot, C. (2008) *Banten Sejarah dan Peradaban Abad X—XVII* (Banten: History and Civilization between the 10th and 17th centuries). Jakarta: KPG.
H.M. Murtadlo. Interview (Personal communication 29th July 2016).
Hadi, H.M. (2010) *Tiga Guru Sufi Tanah Jawa, Wejangan-Wejangan Ruhani* (Three Sufis in the Land of Java, Spiritual Teachings). Yogyakarta, Pustaka Pesantren.
Hamid, A. (2011) Pergeseran Peran *Kyai* dalam Politik di Banten Era New Order dan Reformasi (The Shifting Roles of *Kyais* in Politics in Banten in The Era of New Order and Reformation). *Jurnal Ilmiah Bidang Keagamaan dan Kemasyarakatan,* 28 (2), 340–363.
Jackson, K.D. (1990) *Kewibawaan Tradisional, Islam, dan Pemberontakan* (Traditional Charisma, Islam, and Upheaval). Jakarta: Pustaka Grafika.
Kartodirdjo, S (1966) *The Peasant's Revolt of Banten in 1888 tts Conditions, Course, and Sequel, an Ease Study of Social Movements in Indonesia.* Leiden, Martinus Nijhoff.
Kyai Ariman. Interview. (Personal communication, 20th January 2016).
Lubis, N.H. (2003) *Banten dalam Pergumulan Sejarah: Sultan, Ulama, Jawara* (Banten in Historical Contest: Sultan, Ulema, *Jawara*), Jakarta, LP3ES.

Regional identity and cultural heritage

Cultural Dynamics in a Globalized World – Budianta et al. (Eds)
© *2018 Taylor & Francis Group, London, ISBN 978-1-138-62664-5*

Foreword by Mikihiro Moriyama

In this section, we have 21 articles of a various theme on cultural heritage in and around Indonesia.[1] Most of materials discussed here is cultural assets that show local knowledge and regional identity. These articles based on careful research will enrich our knowledge on cultures in Indonesia and enhance our understandings on culture in general. In fact, Indonesian studies have greatly contributed to academia in the world. A large number of scholars have produced numerous articles and books. Ethnographic descriptions and findings about peoples and societies in Indonesia have been global assets since colonial times.

Such knowledge of local peoples in the colonies, however, was needed for self-interested reasons, namely managing the colonies and maximally profiting from the land and the people. T.S. Raffles, N.J. Krom, H. Kern, C. Snouck Hurgronje, C. van Vollenhoven and others wrote on the peoples. Likewise, language studies and multiple efforts toward creating dictionaries in the Archipelago were needed for propagation purposes by Christian missionaries since their arrival in the 17th century. Examples are B.F. Matthes on Buginese, Taco Roorda on Javanese, H.N. van der Tuuk on Batak, S. Coolsma on Sundanese. These research results comprise our basis for research on cultural heritage in Indonesia, regardless of what the original purposes were.

Not only descriptions of Indonesian peoples and cultures, but also theory and important concepts were discovered through research made by scholars from all over the world. For instance, Clifford Geerts made a wonderful contribution through his anthropological fieldwork in the islands of Java and Bali, and Benedict Anderson made an unprecedented contribution to the humanities with his high command of the language and deep attachment to local peoples.

By acknowledging such great contributions, here I raise a question that researchers on Indonesian studies are not completely free themselves from a colonial mindset. This problem seems to persist in academia both outside and inside Indonesian society. I notice that the term "indigenous studies" is recently used in academic circles in Indonesia. I will consider whether this term is adequate for research on cultural heritage in Indonesia or not. First of all, I retrace the definition of the term "indigenous" as it changes through the times—from the colonial period through the Japanese occupation and into the 21st century.

Rethinking the Term "Indigenous"

It is useful to begin the discussion of "indigenous studies" with a brief survey of the term "indigenous" since the Dutch colonial period. The words *inboorlingen* and *inlanders* were used by the Dutch to denote people in their colony. Perhaps these words can be translated as native rather than indigenous. Another word used in colonial times was *inheems*, which denotes indigenous with condescending nuance. For sure, these terms clearly show a disrespectful view of the people the Dutch encountered. The Japanese did the same from their arrival in the Archipelago in the late 19th century through the occupation period of World War II. They used the words *genjyu min* (indigenous people, disdainfully) – or, even worse, *dojin*, literary translated as "land man" but with a connotation of native or savage.

---

1. This preface was written on the basis of my presentation for the Asia-Pacific Forum for Research in Social Science and Humanities Conference (APRiSH) 2016, at Universitas Indonesia, Depok, West Java, Indonesia. The research was supported mostly by a grant from UI-RESOLV (Visiting Scholar Program) of the Faculty of Humanities Universitas Indonesia, and in part by Nanzan University Pache Research Sbsidy I-A-2 for 2017.

These terms denoting "indigenous" in the history of Indonesia mostly connotes looking down on others and distinguishing others from outsiders' higher position. Those who used the term did not see Indonesians as being in the same level as their own in terms of civilization. The "others" were considered as backward and uncivilized, so that their "masters", the colonizers, would enlighten them. It was quite an arrogant idea of the outsiders.

What happened here was that the newcomers found a culture and society different than their own, and did not think that every culture and society has its own values and philosophy. They evaluated foreign cultures with their own measuring standards. In short, the term "indigenous" is not free from such negative connotations because of the history of how the term was used in the colonial past: it has a discriminatory nuance when used to identify a group of people.

The next question is whom the term "indigenous" is aimed at in a global age. The term has tended to be used by arrivals, mostly Westerners, to a foreign island or continent to identify those who were already living there. This happened in the Archipelago too as we have seen above. This cannot be separated from colonial history either.

Then, we will see how the term is used in the 21st century in the following. The declaration on the Rights of Indigenous Peoples was adopted by the General Assembly of the United Nations in 2007. The aims of the declaration are to protect the rights of indigenous peoples around the globe. Article 2 reads as follows:

> *Indigenous peoples and individuals are free and equal to all other peoples and individuals and have the right to be free from any kind of discrimination, in the exercise of their rights, in particular that based on their indigenous origin or identity.*

An Indonesian translation read as follows:

> *Masyarakat pribumi dan tiap-tiap individu bebas dan setara dengan segala bangsa dan semua individu dan mereka mempunyai hak untuk terbebas dari segala macam diskriminasi, dan dalam pelaksanaan hak mereka, khusunya yang berdasar atas hak-hak mereka, khususnya yang berdasar pada asal-usul atau identitas mereka (Pasal 2 Perserikatan bangsa-bangsa 2007).*

Here the Indonesian translation for indigenous peoples is *masyarakat pribumi* (literally, "local society"). This translation does not seem appropriate because the term *pribumi* has been used to denote the Indonesian nation since the colonial period.[2] It has even been used to distinguish Indonesians as colonized from Europeans as colonizers. The people at whom the declaration aims is a different group. The report of the Indigenous People Alliance of the Archipelago reads as follows:

> *Indonesian laws use various terms to refer to indigenous peoples, such as masyarakat suku terasing (alien tribal communities), masyarakat tertinggal (neglected communities), masyarakat terpencil (remote communities), masyarakat hukum adat (customary law communities) and, more simply, masyarakat adat (communities governed by custom). Indigenous peoples live in forests, mountains and coasts. Some are nomadic and some are sedentary, and they are engaged in gathering, rotational swidden farming, agroforestry, fishing, small-scale plantations and mining for their subsistence needs. They traditionally live on their ancestral land and water. They depend on nature, as they believe the earth is a common property that has to be protected for its sustainability. They have their own knowledge about how to manage nature (Indigenous People Alliance of the Archipelago 2012: 1).*

In this passage, the term "indigenous people" is translated as isolated peoples such as *orang Baduy* (Baduy people in the Banten regency) or *orang Dani* (Dani people in the West Papua Regency). The United Nations declaration means that such isolated peoples in Indonesia should be differentiated from aborigines in Australia. Australian aborigines lived in the island before the Europeans came, while orang Baduy are Sundanese who chose to live in isolation when the Sundanese area was conquered by the Islamic-Javanese kingdom. Baduy people have distinctive customs and a unique culture ensuing from their own determination.

---

2. Sometimes it was used as a counter-term to *non-pribumi*, denoting mainly Chinese Indonesian.

My question is whether the English term "indigenous people" as stated in the United Nations declaration is used appropriately in the context of Indonesian society. It might mislead to differentiation from or even discrimination of fellow peoples from the majority of the Indonesian nation. Is it proper to label a group of people as indigenous because they still keep their traditional way of life with their own customs, *adat*? Indonesia is a nation that consists of a large number of ethnic groups with their own cultures, as stated and respected by the constitution of 1945.

Through history all peoples in the Archipelago were uniformly called indigenous by colonizers in various languages, despite noticing the great diversity among them. In this global age Indonesian city dwellers call people living in isolated locations "indigenous", as if they were masters from the colonial past. They seem to discriminate against those who live in an underdeveloped environment by not remembering the history of the term in their own colonial past. It seems appropriate to use a different term to denote such peoples instead of using the English term "indigenous people". As mentioned above, in Indonesian law there are Indonesian terms equivalent to the English ones: *masyarakat suku terasing* (alien tribal communities), *masyarakat tertinggal* (neglected communities), *masyarakat terpencil* (remote communities), *masyarakat hukum adat* (customary law communities) and *masyarakat adat* (communities governed by custom).

The same holds true for Indonesian studies. It is not necessary to use "Indigenous Studies" in English to identify ethnic studies such as Sundanese culture or anthropological studies on remote communities like orang Baduy. The term "Indigenous Studies'" in the Indonesian context reminds us of a colonial mindset and a discriminatory view. Is it more appropriate, then, to use the term "Indonesian Studies" the same way we use "Japanese Studies" for doing research on Japanese people, culture and society in a global age?

Respect for Local Culture

Articles in this section show their attachment for a local culture and careful research on their materials. This methodology and mindset should be respected for our further research on cultural heritage in Indonesia or Indonesian Studies. Neverthless, we know a different methodology based on comparison.

Comparative studies between cultures could clarify the meaning of cultural practices and possibly break through existing theories. In this method discipline plays a more significant role for analysis and interpretation of a local culture. For instance, an anthropologist will observe a local ritual and interpret it by comparing it with an equivalent ritual elsewhere. In this case the researcher does not become attached to a single place for a long time and learn the local language well. He/she will move from one place to another to test his/her hypothesis.

By contrast, another methodology consists of first learning the language of a research location and becoming involved in the society as well as deepening relationships with local people. Knowing a local culture deeply and finding a significant logic in the culture may lead to new theories. The direction of the vector is from specific to universal. For instance, Benedict Anderson found an idea of Imagined Communities through his Indonesian studies based on a deep understanding of the people and a thorough command of the Javanese and Indonesian languages. In the case of this highly talented scholar his research did not stop in Indonesia but covered the Philippines and Thailand too. In this process, the researcher sometimes develops an attachment to the people and they end up having a special place in his heart. Sometimes this becomes important to support our research motivation and adds extra meaning to our life.

Either way, a researcher's respect for the local people and an unpretentious attitude remains significant towards getting to know cultural heritage in this global age. We have more possibilities thanks to technological innovations for research even in the field of the humanities. At the same time, we have to be careful about our attitude and methodology for a deeper understanding of others.

Mikihiro Moriyama
Nanzan University, Japan

References

Aliansi Masyarakat Adat Nusantara. 2012. *Country Technical Notes on Indigenous Peoples' Issues, Republic of Indonesia*, International Fund for Agricultural Development (IFAD); Asia Indigenous Peoples Pact (AIPP).

Anderson, Benedict. 2116. *A life beyond boundaries*. London; New York: Verso.

Asian Development Bank, Environment and Social Safeguard Division, Regional and Sustainable Development Department. 2002. *Indigenous People/Ethnic Minorities and Poverty Reduction, Indonesia*, Manila: Asian development Bank.

Moriyama, Mikihiro. 2005. *Sundanese Print Culture and Modernity in 19th-century West Java*. Singapore: Singapore University Press.

—. 2012. "Regional Languages and Decentralisation in Post-New Order Indonesia: The case of Sundanese." In *Words in Motion—Language and Discourse in Post-New Order Indonesia*, edited by Keith Foulcher, Mikihiro Moriyama and Manneke Budiman, 82-100. Singapore: NUS Press.

—. 2013. *Semangat Baru: Kolonialisme, Budaya Cetak, dan Kesastraan Sunda Abad Ke-19*. Revised version, Jakarta: Komunitas Bambu.

—. 2015. "Bahasa Sunda dalam Berdoa", in Julian Millie and Dede Syarif ed., *Islam dan Regionalisme*, pp. 107-116, Bandung: Pustaka Jaya.

Semali, Ladislaus M. and Joe. L. Kincheloe eds. 1999. *What is indigenous knowledge? Voices from the academy*, New York; London: Falmer Press.

Sillitoe, Paul, Alan Bicher, and Johan Pottier eds. 2002. *Participating in Development: approaches to indigenous knowledge*, London; New York: Routledge.

United Nations. 2007. *United Nations Declaration on the Rights of Indigenous Peoples*,

Cultural Dynamics in a Globalized World – Budianta et al. (Eds)
© 2018 Taylor & Francis Group, London, ISBN 978-1-138-62664-5

Wawacan Samun: Between the convention and the creation of *Wawacan*

M. Holil & T. Pudjiastuti
Department of Literature, Faculty of Humanities, Universitas Indonesia, Depok, Indonesia

ABSTRACT: This research analyzes Wawacan Samun (WS) texts in Sundanese manuscripts with the *wawacan* genre. This research aims to show the tensions between convention and creation in the texts. To achieve this objective, philology and literary theories were used. The research found that out of five WS manuscripts, there are two story versions, i.e. *pesantren* scriptorium and *kabupaten* scriptorium. In terms of the tensions between convention and creation, it is clearly seen that the editors of WS texts still follow the conventional *pupuh* metrum writing with many errors found. The creation done by WS text editors focuses on narrative aspects.

1   INTRODUCTION

The text of Wawacan Samun (hereinafter referred to as WS) constitutes a *wawacan*, which is a literary genre known among the Sundanese after their exposure to Javanese culture (Mataram Islam) in the early 17th century (Adiwidjaja, 1950; Salmun, 1963; Ekadjati, 1983). The occupation of Sunda land by Mataram for around 50 years had a large influence on the Sundanese language and literature. Such influence can be observed, among others, from the literary manuscripts of *wawacan* and *guguritan* (Rosidi, 1995, p. 385). A *wawacan* is a long story or a saga written in a poetic style referred to as *dangding*. *Dangding* is a series of particular poems about particular matters. A *dangding* consists of several particular poetic type known as *pupuh*. Unlike *wawacan* which constitutes long epic stories, *guguritan* usually constitutes a short *dangding*, consisting of a few stanzas, and lyrical in nature (Rosidi, 2013, pp. 28, 34).

The *wawacan* genre was introduced throughout Sundaland via Priangan Residency. Historically, Priangan Residency was formed from the unification of two regions, namely Galuh and Sumedanglarang. These regions, along with Banten and Cirebon, constituted parts of Pasundan, which since the fall of Sundanese Kingdom (1579) had never been under one single authority. According to Ekadjati (1991), Priangan Residency was located in the southern part of West Java, stretching from Cianjur on the west to Ciamis on the east, which in a modern day covers the regencies of Cianjur, Bandung, Garut (formerly: Limbangan), Tasikmalaya (formerly: Sukapura), and Ciamis (formerly: Galuh).

The *wawacan* genre was introduced throughout Sundaland via two channels, namely the districts and Islamic boarding schools (*pesantren*). The *wawacan* spread throughout the districts and was often written in *cacarakan* characters. Meanwhile, the ones that spread throughout the boarding schools were often written in *Pegon* characters (Rosidi, 1966). In the context of Sundanese manuscripts, *cacarakan* letters were also known as Sunda-Javanese letters or Sunda-Mataram letters. This alphabet adopts the Javanese *hanacaraka* letters with some minor adjustments, such as the elimination of two letters (*tha* and *dha*) which are not used in the Sundanese language. Meanwhile, *pegon* letters refer to the Arabic letters used to write texts in Sundanese language.

This genre received a positive response in the Sundanese literary world. This was evident from the increasing number of the works of Sundanese writers within this genre. The *wawacan* genre even became very popular, particularly from the late 18th century to the early 20th century (Ekadjati, 1983; Rosidi, 1966; 1983; 1995; 2013; Rusyana, 1969; Moriyama 2005).

There have been a relatively large number of *wawacan* texts written. Among the 404 Sundanese manuscripts, which are part of the collection of the National Library of Indonesia, Jakarta, there are 72 manuscripts of *wawacan* genre. Meanwhile, among the 789 Sundanese manuscripts in the Library of Leiden University in the Netherlands, there are 114 manuscripts of *wawacan* genre. This number may increase if the collection of Sundanese manuscripts in the society is consulted. A number of Sundanese manuscripts in the society have been listed by Ekadjati (1983), Ekadjati and Darsa (1999). However, the current quantity of the listed manuscripts does not yet represent the estimated total number of Sundanese manuscripts in the society.

In its initial development within Sundanese literature, many *wawacan* were reproductions or adaptations of Javanese or Malay literatures. In the later phase, there was evidence that *wawacan* writers tried to re-explore the plethora of the "authentic" Sundanese literature genre.

Between the late 18th century and the early 20th century, *wawacan* genre had reached its peak of popularity. In that period, the popularity of this genre eventually motivated the authors of *wawacan* to create stories based on their own creativity. No longer simply translating or adapting stories from other literatures, the authors of *wawacan* started to create their own stories, as well as to embed the unique Sundanese characteristics in their works. One method that they employed was by examining and reviewing the plethora of Sundanese literatures from the previous eras. In this context, the emergence of WS can be perceived as the response of *wawacan* authors to changing times and the readers' horizons. The WS authors wanted to deconstruct the genre convention of *wawacan* by deviating from it, resulting in the clash between conventions and creativities within their literary works.

From the manuscripts search, it is known that there are WS manuscripts. Based on this, we can raise a question. Are those five manuscripts variants or versions? In connection with the peak of popularity of the *wawacan* genre, the question that is raised now is how is the attitude of the WS text writer in relation to *wawacan* genre convention.

This research is done to find out if these five WS texts are variants or versions and to know the attitude of the WS text copyist against *wawacan* genre writing convention. This research provides insights to the public and Sundanese culture researchers on the differences between WS texts that are produced in a duchy region (*kadipaten*) and a school of Koranic studies (*pesantren*). Furthermore, it seeks to show the attitude of Sunda literature writers against convention of the popularity of *wawacan* genre.

The WS texts studied in this research consist of old literary works written in manuscripts. Therefore, the principles of philological research are first applied to achieve a deeper understanding of WS texts, as suggested by Maas (1963) and Reynold and Wilson (1978). This research is based on old texts. The approach used for the text edition is through the philology research method, starting from manuscript inventory, manuscript description, text comparison, and text critique. Text critique is needed to explore texts that have long been processed (Reynold & Wilson, 1978, p. 187). In order to present the WS text edition, text criticism has been done with the method suggested by van der Molen (2011, pp. 82–84).

Next, WS texts are regarded as literary works. The meanings within the WS texts are explored using the theoretical framework within the literary field, particularly the theories that are related to the tensions between convention and creation which always exist in every stage of particular literature history (Culler, 1975, p. 1983).

2 WS MANUSCRIPTS

There are five manuscripts containing WS texts in the manuscript inventory list, namely two manuscripts from the collection of the National Library of Indonesia (SD. 26 and SD. 187).

Manuscript SD. 26 (hereinafter referred to as manuscript A), while manuscript SD. 187 (hereinafter referred to as manuscript B), and three manuscripts from EFEO Bandung collections (KBN-441, KBN-70, and KBN-148). Manuscript KBN-441 (hereinafter referred to as manuscript C), manuscript KBN-70 (hereinafter referred to as manuscript D), and manuscript KBN-148 (hereinafter referred to as manuscript E). Based on the characters or letters within the texts, the five manuscripts can be put into two categories: four manuscripts written in *Pegon* letters (A, C, D, and E) and one manuscript written in *cacarakan* (B). The *Pegon* letters used in the four WS manuscripts shared a number of similarities in terms of the shapes and the vocal signs. There is no visible difference among the letters used in the four manuscripts. Meanwhile, the *cacarakan* letters used in manuscript B, particularly in terms of the vocal signs, are not known nor used at present. The *cacarakan* alphabet used in manuscripts B is relatively old, and it had been used and became popular before the book of Coolsma was published in the early 20th century (1904). Based on that, the manuscript was estimated to be written in the late 19th century or the early 20th century.

Not all manuscripts are in good physical conditions. The order of the five manuscripts based on their physical conditions is as follows: B, A, D, E, C. The manuscript B has very good physical condition, clear and readable letters, and excellent cover and binding conditions. The manuscript A has good physical condition in general, clear and readable letters, good cover and binding conditions, yet some of the right parts of the manuscript were cut out during the re-binding process causing the loss of some letters. This impedes the reading process of the text, since it made several letters/syllables/words unreadable. The manuscript D is a photocopied text, so the letters are a little unclear. However, it is still readable. The manuscript E consists of both printed digital photos and photocopied texts. The one used for the reading is the latter. The photocopy of manuscript E is not as good as the photocopy of manuscript D. Although some parts of the text are readable, there are also many parts that are unreadable caused by the dark color of the paper. The manuscript C consists of prints out from microfilms and digital photos. In both forms, the text is no longer readable. Since the ink has faded, the letters are hardly recognizable. For this reason, manuscript C is not examined in this research.

Each of the four manuscripts examined has their own distinguishing characteristics based on the contents of the texts. Manuscript A constitutes the most complete, detailed, and oldest among the four. Manuscript B has a complete story from the prologue to the epilogue, without any colophon. In addition, there are a number of irregularities related to the logic of the story. Manuscript D and manuscript E contain incomplete stories since they only contain 25% of the full WS story.

A great deal of information has been obtained from the comparison between the four manuscripts. First, the four manuscripts share various similarities, particularly in the main points of the story. Therefore, the four manuscripts are assumed to have the same root. Secondly, manuscripts A, D, and E also share significant similarities. In some cases of reproduction, the choice of words in manuscript E is less complex, avoiding the mistakes of words in manuscript A, and correcting the mistakes of the writers of manuscript A in any case of deviation from the convention of *pupuh* writing. This becomes the supporting evidence for the notion that manuscript E might be a reproduction of manuscript A. Meanwhile, despite sharing the same chronological order of the story, the choices of words in manuscript D are very different from those of manuscript A. The one that is very different from the rest is manuscript B. In addition to different alphabets, manuscript B also has a different chronological order of the story. In other words, the WS text version in manuscript B is different from those in the other three manuscripts.

Based on such philological study, it can be concluded that there are two versions in the WS story, namely the Islamic boarding school version and the district version. The former can be found in three manuscripts, namely A, D, E, while the latter is only found in manuscript B. From the three manuscripts with the former version, only manuscript A contains the complete story. Therefore, only WS texts in manuscripts A and B are analyzed in the literary aspect analysis.

3 WAWACAN SAMUN: BETWEEN CONVENTION AND CREATION IN THE WRITING OF *WAWACAN*

The writing convention of *pupuh* in the context of Sundanese literature is influenced by Javanese literature. Therefore, a comparison between the writing convention of *pupuh* in Javanese and Sundanese literature contexts has to be drawn. From the comparison between the writing convention of *pupuh* in the Javanese literature context (Karsono, 2001) in the Sundanese literature context (Coolsma, 1985; Salmun, 1963), it is found that the writing rules or conventions of *pupuh* in both literatures share a number of similarities. Nevertheless, there are several differences in the repertoires and conventions of *pupuh* in both literatures. In Sundanese literature, there are two kinds of literature that are not known in Javanese literature, namely *Ladrang* and *Lambang*. There is a modification in the Sundanese version of Pupuh Balabak (Javanese), or also known as Balakbak (Sundanese) (Salmun, 1963). The rule of 12a + 3e in Javanese and Sundanese literatures (Coolsma, 1985) changed into 15e (the sum of 12 + 3) in the following periods. Meanwhile, the writing rules of *pupuh* Gurisa (Javanese: *Girisa*) and *Lambang*, according to Coolsma and Salmun, were "switched". However, the two varieties of *pupuh* are rarely used in the writings of *wawacan*.

The authors/writers of WS texts of both manuscripts A and B do not seem to show the characteristics of authors who are experts and skilled in composing WS texts. Two writers exhibited a high degree of carelessness. There are numerous mistakes in following the writing conventions of *pupuh*, particularly the number of *engang* (syllables) in each line in which both lacked or exceeded the ones stated in the conventions.

There are three types of deviation from the convention of *pupuh* in manuscript A. The first one is the excess of syllables, especially those which are more than one syllable. The number of syllables used in the WS lines exceeds more than what is stipulated in the convention. The second one is the lack of syllables, particularly those which lack one to two syllables. Lastly is the difference in *guru lagu*, which is the vowel sound at the end of each line. The last vowel sounds in a number of lines are different from those stipulated in the convention. From these three types of deviations, the first one happens more frequently compared to the second and third types. From 1910 deviations, 1378 cases constitute the first type of deviation. The second type of deviation is found in 348 cases, while the third type is found in 210 cases. In other words, of all the deviations, 72.14% constitute the first type of deviation, 18.21% constitute the second one, and 10.99% constitute the third type.

Deviations from *pupuh* convention in manuscript B also exhibit similar features with the ones in manuscript A. To begin with, there are also three types of deviation from *pupuh* convention. The first type is the excess number of syllables from the stipulated number in the convention, amounting to the excess of one to three syllables. The second type is the lack of syllables from the number stipulated in the convention, particularly those which lack of one to two syllables. The last type is the difference in *guru lagu*: the last vowel sounds in a number of lines are different from those stipulated in the convention. From these three deviation types, the first one happens more frequently compared to the second and the third types. From 1686 deviation cases in manuscript B, 1082 cases constitute the first type of deviation. The second type of deviation is found in 501 cases, while the third type is found in 103 cases.

Based on the data above, it can be seen that the deviations from the *pupuh* convention in the WS texts in manuscript A and manuscript B are relatively similar. These similarities include the types of deviations and the most frequent as well as the least frequent deviation types.

The deviation from the *pupuh* convention raises a number of questions: Are the numerous numbers of deviations from the *pupuh* convention really caused merely by the WS authors'/writers' ignorance, lack of skills, and carelessness? Or, are there any other possible factors?

The lack of skills and the carelessness of the authors/writers most likely play a part in the occurrence of the deviations. However, in the WS context, there seem to be several other factors.

First, there was a decline in the popularity of the *wawacan* genre at the end of the 19th century, which was the period when the WS texts were written. The year when manuscript A was written is 1299 H (1882 AD), and it is inscribed on the colophon of the manuscript. Meanwhile, manuscript B is assumed to be written at the end of the19th century or in the early 20th century. Moriyama (2005) states that at the end of the 19th century the *wawacan* genre began to be overshadowed by the prose genre introduced by the Europeans (the Dutch).

Secondly, the end of the 19th century also marked the period when many Sundanese authors began to look back at the Sundanese literature which had for long been neglected by their predecessors. The young authors at the time were motivated to explore the literature repertoire of their ancestors. They became aware that the *wawacan* genre had been created after and influenced by the "occupation" of a Javanese kingdom (Mataram Islam). This motivated the new authors to rediscover the old Sundanese literature genres. One genre of the old Sundanese literature that was studied by many of these authors was *carita pantun* (Rosidi, 1966; 1995).

Thirdly, the WS texts in manuscript A and manuscript B were made/reproduced to be performed. Such assumption is made since there is "interaction" with the "audience" in certain parts of the texts. This is also backed up by the fact that *wawacan* texts have often been used in various performances, such as reading out (*mamaca*) or singing the written texts without any instrumental accompaniment (*beluk* or *gaok*). The *wawacan* texts that are performed might vary in every region (regencies, sub-districts, or villages) and in every social event (weddings, seventh-month pregnancy ceremonies, or circumcision ceremonies), depending on the preferences and purposes. Apparently, WS texts have been very popular among the people of Kadipaten, Majalengka, and the surrounding regions and are performed in various social events (Interview with Rukmin in his house: Tarikolot, Kulur Village, Sub-district Sindangkasih, Majalengka Regency, West Java, on Saturday, 7th February 2015).

A consideration of these factors can help us understand why there are many deviations from the *pupuh* writing convention in WS texts. In addition to the authors' skills, it seems that the authors/writers of WS texts did not really give much thought to the deviations they made in the creation of the texts. The writers were more concerned with developing good and interesting narratives that would condition the "audience" to focus on the storylines. Interesting narrative styles, such as the one commonly used in the character descriptions in poetic stories, were used to arouse sympathies. Likewise, many humors inserted in the narratives adopt the humors commonly found in poetic stories.

It is obvious that in the creation of their works, the authors/writers of WS texts used their knowledge on poetic stories, known as *carita pantun*, as their source of inspiration. This includes the narrative styles that are considered attractive to the readers or listeners of the stories. In this matter, the authors or writers of WS texts seemed to free themselves from the more familiar writing convention of *wawacan*. However, they still applied some key features in the convention despite making many "mistakes" in their texts.

4 CONCLUSION

The tensions between convention and creation can be observed in the writing of WS texts. On one hand, the authors of WS texts often deviated from the writing convention of *pupuh*, yet on the other hand, the authors did not create any new patterns that might lead to the creation of a new convention. These deviations might be the result of several factors. First, they might be caused by the lack of skills or knowledge on the *pupuh* convention. Second, they might be the result of carelessness. Thirdly, the authors/writers might no longer put too much thought on the *pupuh* convention since at the time of the writing/reproduction of the WS texts the popularity of *wawacan* genre had been waning or declining. Thus, many of them turned their attention to old Sundanese literature genres, particularly the poetic stories of *carita pantun*. They chose to focus on the creation of well-written narratives with attractive language styles in order to be able to survive.

REFERENCES

Adiwidjaja, R.I. (1950) *Kasusastran Sunda II (Sundanese Literature II)*. Jakarta, J.B. Wolters–Groningen.

Coolsma, S. (1985) *Tata Bahasa Sunda (Sundanese Grammar)*. Translation from *Soendaneesche Spraakunst (1904)*. Jakarta, Djambatan.

Culler, J. (1975) *Structuralist Poetics: Structuralism, Linguistics, and the Study of Literature*. London, Routledge and Kegan Paul.

Ekadjati, E.S. (1983) Sejarah Sunda (History of Sunda). In: Edi S.E. (eds.). *Masyarakat Sunda dan Kebudayaannya*. Bandung, Girimukti Pasaka.

Ekadjati, E.S. (1983) *Masyarakat Sunda dan Kebudayaannya (Sundanese Community and Their Culture)*. Bandung, Girimukti Pasaka.

Ekadjati, E.S. & Darsa, U.A. (1999) *Katalog Induk Naskah-Naskah Nusantara Jilid 5A: Jawa Barat Koleksi Lima Lembaga (Main Catalogue of Indonesia Archipelago Manuscripts Edition 5A: West Java Five Institution Collection)*. Jakarta, Yayasan Obor Indonesia & Ecole Francaise d'Extreme-Orient.

Maas, P. (1963) *Textual Criticism*. Oxford: Oxford University Press.

Moriyama, M. (2005) *Semangat Baru: Kolonialisme, Budaya Cetak, dan Kesastraan Sunda Abad ke-19 (New Spirit: Colonialism, Printing Culture, and Sundanese Literature in 19th Century)*. Jakarta, KPG (Kepustakaan Populer Gramedia) in collaboration with the Resona Foundation for Asia and Oceania.

Reynolds, L.D. & Wilson, N.G. (1978) *Scribes and Scholars*. London, Oxford University Press.

Rosidi, A. (2013) *Puisi Sunda (Sundanese Poems)*. Bandung, PT Kiblat Buku Utama.

Rosidi, A. (1995) Kasusastraan Sunda (Sundanese Literature). In Ajip Rosidi, *Sastera dan Budaya: Kedaerahan dalam Keindonesiaan (Literature and Culture: Local Culture in Indonesian National Culture)*. Jakarta, Pustaka Jaya. pp. 376–411.

Rosidi, A. (1983) *Ngalanglang Kasusastran Sunda*. Jakarta, Pustaka Jaya.

Rosidi, A. (1966) *Kesusastraan Sunda Dewasa Ini (Sundanese Literature Nowadays)*. Jatiwangi, Tjupumanik.

Rusyana, Y. (1969) *Galuring Sastra Sunda*. Bandung, Gununglarang.

Salmun, M.A. (1963) *Kandaga Kasusastran*. 2nd ed. Bandung/Jakarta, Ganaco N.V.

Saputra, K.H. (2001) *Sekar Macapat.* Jakarta, Wedatama Widya Sastra.

Cultural Dynamics in a Globalized World – Budianta et al. (Eds)
© *2018 Taylor & Francis Group, London, ISBN 978-1-138-62664-5*

Piwulang teachings in *Serat Darmasaloka*

S. Rohman & D. Kramadibrata
Department of Literature, Faculty of Humanities, Universitas Indonesia, Depok, Indonesia

ABSTRACT: In this study, we discuss a classic Javanese manuscript entitled *Serat Darmasaloka* (SD). SD is a narrative poetry known as *tembang macapat*. On the basis of the content, SD is classified as *piwulang*. As a *piwulang* text, it contains several moral values and teachings in the story. The SD text provides a good model of the story that is filled with values of life, religion, tradition, and social system of the society. Here, we discuss the noble ideas of the Javanese society presented in SD.

1 INTRODUCTION

Serat Darmasaloka (hereinafter referred to as SD) is a Javanese manuscript in the form of poetry, or *tembang macapat. Tembang macapat* is a form of Javanese poetry that uses new Javanese language controlled by the poetic patterns of *guru gatra* (the number of stanzas), *guru wilangan* (the number of syllables in each stanza), and *guru lagu* (the last vocal in each stanza) (Karsono, 1992, p. 8).The SD text narrates the story of Siti Maryam, the daughter of Syeh Ngabdullah, who was banished to a forest due to allegations of adultery. The SD text contains several pieces of advice (*piwulang*) and teachings of Islam.

A total of three manuscripts are found under the title Sêrat Darmasaloka. The first, a manuscript titled *Serat Darmasaloka*, is kept at Keraton Surakarta library under the catalogue number KS 539 B. The second, a manuscript C titled *Pethikan saking sêrat Darmasaloka*, is kept at Keraton Surakarta library under the catalogue number KS 539. The third, a manuscript titled *Sêrat Darmasaloka*, is kept at Universitas Indonesia Library under the catalogue number PW 26/KS 77.

Judging from its colophon, the KS 539 B manuscript was completed on 29 Sura year Je 1814 AJ, 29 Muharram 1302 H, or 18 November 1884 AD. The KS 539 C manuscript was completed on Monday, 28 Ramadhan 1816 AJ or 20 June 1887 AD. Finally, the PW 26/KS 77 manuscript was completed on 25 Jumadilawal year Je 1846 AJ, or 30 March 1916 AD.

Following a careful comparison of both the readability and content by a philological study, the SD text in the KS 539 B manuscript was chosen as the basis of text editing and content analysis. This text was selected not only because of it being the oldest, but also because of the coverage of materials and the quality of readability, which refers to the clarity of the text and the least amount of deviation compared to other manuscripts. In this study, will not address text editing.

As a *piwulang* text, SD contains moral values and teachings for the readers. The research objective is associated with the moral values and teachings contained in SD. As a narrative poem, the structure of the SD text (plot, characters, and settings) has been analyzed (see Rohman, 2016). The storyline, conversation among characters, and settings reflect the moral values and teachings contained in SD. According to various written sources about the Javanese culture, some of the moral values and the main teachings in the text are revealed. In this study, we discuss the moral values and teachings contained in SD.

2 DISCUSSION

2.1 *Summary of the SD text*

The story begins with the introduction of the character Seh Ngabdullah, who lived in Mecca with his two children Abu Bakar and Siti Maryam. Seh Ngabdullah and Abu Bakar were about to set forth on a pilgrimage to Medina, but Seh Ngabdullah did not want to leave his daughter, Siti Maryam, alone at home. Therefore, he asked Kiai Sangit to stay at the house to protect her. A conflict arose when Kiai Sangit advanced to Siti Maryam to commit adultery, but the resolute Siti Maryam refused.

Kiai Sangit then met Seh Ngabdullah and claimed that Siti Maryam had propositioned him. A furious Seh Ngabdullah ordered Abu Bakar to kill Siti Maryam without knowing the truth of the accusation. Confused Abu Bakar decided not to kill his sister and instead banished her to a forest. He lied to his father, saying Siti Maryam was dead.

In the forest, Siti Maryam met an old widow named Ni Kalimah and went on to live with her. One day while shepherding goats, Siti Maryam met the King of Bagedad, who fell in love with her and wanted to marry her. King Bagedad asked Ni Kalimah for Siti Maryam's hand in marriage. Siti Maryam then married the king and became the queen of his kingdom. They lived happily and were blessed with two sons.

Another conflict arose when Siti Maryam was on her way to visit her father and brother in Mecca. During the journey, the King's aide, Ki Patih, performed treachery and attempted to rape Siti Maryam. Because she refused, Ki Patih killed her two sons. She managed to escape and disguised as a coffee merchant in the kingdom of Ngedhah-ngedhah. Ki Patih returned to Bagedad and told the king that his wife and children were dead after being preyed by tigers in the forest.

The King was stricken with grief upon hearing the news. He asked Ki Patih to show him the place where his wife and children died. Meanwhile, the story turns to Seh Ngabdullah, who asked his son Abu Bakar to show him Siti Maryam's grave. Seh Ngabdullah, Abu Bakar, and Kiai Sangit then went into the forest. In the middle of the forest, they met King Bagedad and Ki Patih.

Following the advice of a merchant from Ngedhah-ngedhah, the five men went to the kingdom of Ngedhah-ngedhah, where they met Siti Maryam, who was still in disguise. Siti Maryam told them a tale filled with occurrences taken from her own life experience. Seh Ngabdullah, King Bagedad, Abu Bakar, Kiai Sangit, and Ki Patih were astonished to hear a tale so similar to their lives. Siti Maryam then revealed her identity and disclosed all of the wrongs that Kiai Sangit and Ki Patih did to her, who were then punished by King Bagedad.

Siti Maryam and King Bagedad then had five more children. After their children were grown, King Bagedad suddenly left the kingdom because of a wish to devote himself to worship. After his departure from the kingdom, the royal family descended into chaos.

King Bagedad left a letter for Siti Maryam and their children, instructing his eldest son to ascend the throne in his place. However, the prince refused to become the new king before meeting his father. The prince went on a long journey in search of his father. During his journey, he met a hermit named Ki Darmasaloka, whom he asked for guidance. Finally, the prince met his father in Mecca. He returned to the kingdom afterward and became the new king of Bagedad. The kingdom of Bagedad became more prosperous under his rule.

2.2 *Values in the SD text*

From the above summary, it is evident that the names of the characters originated from Arab, for example, Seh Ngabdullah, Abu Bakar, King Bagdad, Siti Maryam, and Java: Kiai Sangit, Ki Patih, and Ki Darmasaloka. The story takes place in Mecca, Medina, Baghdad, and the forest. Thus, the SD text is a combination of the elements of Islamic and Javanese origin. This section analyzes the moral values and teachings delivered through the SD text. From the storyline and the conversation among the characters, the noble ideas contained in the SD text are as follows.

2.2.1   *Moral lessons for women*

Moral lessons for women are shown through the exemplary life of Siti Maryam, who was a prominent character in the story. Through her character, the text describes the attitudes deemed necessary for every woman. Siti Maryam is depicted as a very beautiful female character who is resolute, patient, obedient to her husband and father, and believes in destiny and the power of God. Her strong character becomes apparent when she rejected Ki Sangit's advances and did not concede to his threats when he propositioned her.

The SD text emphasizes that a woman should primarily possess the characteristics and qualities of Siti Maryam. This is shown when King Bagedad advised his sisters to set Siti Maryam as an exemplary character.

The text encourages women to possess a noble character as exemplified through Siti Maryam's behavior. Women are urged to be able to manage their feelings and to overcome and restrain vices, such as anger, betrayal, and pride. The text further encourages women to trust fully in destiny and the power of God, which in the end will bring awareness for them to always surrender to God's will.

In general, the moral values delivered through the SD text describe attitudes that women should have in life. The story urges women not to focus only on physical beauty, but they should also be able to control themselves against the negative aspects of human nature and live a life of faith and trust in order to be blessed with the grace of God. Several major values promoted in the SD text for women are determination, sincerity, patience and trust, obedience, and a strong belief in God's power and destiny.

2.2.2   *Tasawwuf teachings*

The teachings of *tasawwuf* or Sufism are shown in the conversation between the prince of Bagedad and Ki Darmasaloka. First, they discuss the characteristics of God's saints (*wali*). This can be understood as the efforts a person must make in order to become a noble human being. The SD text mentions the following six efforts: (1) withstanding sleepiness; (2) withstanding hunger; (3) not being embarrassed by a state of undress; (4) being very close to God; (5) not being afraid of death; and (6) not being afraid of one's fellow creatures.

In the Javanese culture, the efforts are called *laku tapa brata*, which includes the efforts made to achieve the perfect state of life based on the importance of controlling and suppressing carnal desires. These efforts are aimed at eliminating despicable natures and achieving perfection of life.

Laku tapa brata is done by sleeping, eating, and drinking less, suppressing lust, as well as managing one's feelings in order to achieve a state of being one with God or *manunggaling kawula Gusti*. The SD text metaphorically compares humans to boats in the middle of the ocean, saying that they must control themselves to prevent them from getting lost.

In addition to *laku tapa brata* that has to be performed to achieve perfection of life or *manunggaling kawula Gusti*, the SD text mentions the interpretation of perfection of life through the steps of *syariat, tarekat, hakikat,* and *makrifat.* These four steps are symbolized by the four levels or degrees of being a *wali*, as described in the following quote:

//Dene ingkang rumuhun/drajatipun ing wali puniku/badannira pribadi uning yen wali/tuwin janma liyanipun/inggih ugi sami wêroh//
//Dene ping kalihipun/badanira pribadi tan wêruh/janma liyan punika wikan yen wali/dene ingkang kaping têlu/janma liyan datan wêroh//
//Badane dhewe surup/drajat wali ingkang kaping catur/dhirinira priyongga datan udani/janma liyan tan sinung wruh/kang mirsa amung Hyang manon//(SD text, pupuh XI, stanzas 30–32)

Translation
//The first one/among the degrees of being a wali/is being self-aware that one is a wali/and so other people/are also aware of it//
//The second one/is not being aware that one is a wali/but other people are aware of it/while the third one/is other people not being aware of it//

//But the person is self-aware/while the fourth degree of being a wali/is not being aware on one's own/and others are also not aware/only God knows//

The level of *syariat* is symbolized by a person who is self-aware of being a *wali* or feels like one, and others are aware of it as well. The level of *tarekat* is symbolized by a person who is not aware or does not feel like a *wali*, but other people are aware of it. The level of *hakikat* is symbolized by a person who is self-aware or feels like a *wali*, but others are not aware of it. The fourth level or *makrifat* is symbolized by a person who is not aware of being a *wali*, and other people are not aware of it as well. The only one that knows is Allah.

The *tasawwuf* teachings present in the SD text are part of Javanese mysticism, which refers to the tendency of ascetics to avoid worldly matters in order to achieve purity and silence, perform *tapa-brata,* and live with limited means as a prerequisite to become a saint of pure heart.

2.2.3   *Javanese principles of harmony and respect*
A particular characteristic of the Javanese culture is its ability to respond to and receive an influx of cultural influences from outside while still maintaining its own authenticity. Javanese values cannot simply be erased as they are part of the characteristics of the Javanese culture, and this includes the values of harmony and respect, which have become the basis of the attitudes of the Javanese people. The principles of harmony and respect are essentially centered on one purpose, which is to avoid conflict in order to achieve a peaceful and harmonious situation.

The moral values that are part of the principles of harmony and respect in the SD text are apparent in the good relationship between kingdoms, whose people mutually help each other. An example of this is the willingness of King Ngedhah-ngedhah to assist King Bagedad when the latter was in the land of Ngedhah-ngedhah. King Ngedhah was willing to fetch King Bagedad and take him to the palace. The story describes a harmonious relationship between the kingdoms of Bagedad and Ngedhah-ngedhah, with both kings holding mutual respect for one another. King Bagedad was very thankful for the kindness his compatriot and the people of the court of Ngedhah-ngedhah showed him. He then encouraged that the good relationship between Bagedad and Ngedhah-ngedhah be maintained, reflecting an attitude of always striving to avoid conflicts. The following quote is the message King Bagedad gave to the people of the court of Ngedhah-ngedhah to always maintain harmony and avoid conflicts with each other:

… /Sri Bagêdad ngandika/eh ta Patih lan sagung para tumênggung/aku wèh slamêt mring sira/ lan bangêt panrimèng mami//
//Têtulungmu kabêcikan/ingkang tumrap marang sarira mami/tan kurang kapara langkung/ muga Hyang mahawasa/nglêstarèkna gyaningsun têpung sadulur/sumrambaha wadya bala/ lanang wadon gêdhe cilik//
//Bagêdad lan Ngedhah-edhah/aja ana ingkang sulayèng pikir/têrusa saturun-turun/haywa pêgat sagotran/di kang lali den elingna kang satuhu/kang luput den ngapuraa/mangkono traping pawong sih// (SD text, pupuh VII, stanzas 52–54).

Translation
…/King Bagedad said/dear Patih and all officials/I pray for all of you to always prosper/and my endless thanks//
//All the help and kindness/that you have given me/do not lessen but instead increase/hopefully God almighty/allows us to be family forever/down to all the soldiers/man or woman, big or small//
//May Bagedad and Ngedhah-edhah/stay far from disputes/stay harmonious from generation to generation/and never be separated/if some do wrong, remind them/for those who do wrong, forgive them/that is how people love one another//

This excerpt contains a message to always help, respect, and maintain harmony with one another. Any person, regardless of his/her wealth and strength, still needs the help of others, showing the human nature of being unable to live without other people.

The principle of harmony and respect can also be seen in the way King Bagedad always consulted or deliberated with his officials before making a decision. Deliberation is defined as a decision-making process by mutual consultation. Ideally, during the process of deliberation, all voices and opinions are heard and considered to be of equal importance in order to solve a problem. Deliberation aims to give everyone a chance to express their opinions, so as not to make a one-side decision. In this way, all parties can agree to the joint decision-making process (Suseno, 1983, p. 51).

In addition to mutual help and deliberation, the SD text includes a message of self-restraint that must be exercised by all people to avoid conflicts. In the text, Seh Ngabdullah was unable to control his anger when provoked by Kiai Sangit. Because of the lack of self-restraint, Seh Ngabdullah ordered his son Abu Bakar to punish Siti Maryam out of his anger. At first, Seh Ngabdullah felt satisfied and relieved to be able to vent his anger; however, in the end, he realized his mistake of punishing someone without knowing the truth. He became aware that his anger and ego rendered him unable to think of anything else or the result of his actions. Feeling remorseful and haunted by guilt, he asked Abu Bakar to lead him to Siti Maryam's grave.

In this case, awareness and vigilance, also referred to as introspection, are emphasized for self-control as some of the norms to prevent conflicts. A person must exercise self-restraint to prevent any conflicts, because emotions can easily result in negative things. In addition, self-control is necessary to perform actions or make decisions. This means that the decision-making process should not be rushed, and any positive or negative aspect as well as consequences must be carefully considered in advance.

These attitudes of mutually helping each other, being respectful, performing deliberation, and controlling oneself essentially teach us the moral value to maintain the harmony of life and avoid conflicts. This is in accordance with Geertz (1983), who identified the Javanese as a type of people who are straightforward and tend to avoid conflicts. According to Suseno, there are two most dominant rules determining the patterns of social association in the Javanese society. The first rule is that in every situation a person must behave in such a way that does not lead to conflicts. The second rule demands that a person's way of talking and carrying oneself must always show respect for others according to their degree and position. The first is called the principle of harmony, and the second is called the principle of respect (Suseno, 1983, p. 38).

2.2.4 Islamic teachings

The SD text contains one of the aspects of Islamic law, namely the law of *qisas*. *Qisas* is the law of equal punishment for perpetrators of premeditated murder or deliberate assault causing bodily injury. In the SD text, this legal system is reflected in the punishment given to Ki Patih, who was sentenced to death for killing two children of King Bagedad. The excerpt describing this is as follows:

//Khukumipun Ki Sangit punapa lampus/Sèh Dullah turira/Kajawi karsa Sang Aji/Kyai Sangit awit botên dosa pêjah//
//Lêrêsipun binucal kewala cukup/Winorkên wong kumpra/Sageda mulang agami/Pan pun patih sampun lêrês ukum kisas// (manuscript A, *pupuh* VIII, stanzas 37–38).

Translation
//Should Ki Sangit be given a punishment of death/Seh Dullah said/excepting the King's own wishes/Kiai Sangit is not guilty of murder//
//It is best that he is banished/to live with contemptible people/so that he will learn religion/for Patih a qisas punishment is justifiable//

Qisas is a form of punishment in the legal system commonly used in Islamic countries that imply Islamic penal law, such as Saudi Arabia. *Qisas* is enforced to (1) respect human dignity by allowing equal retaliation for the loss of life and bodily injury; (2) deter hostility and bloodshed, so that security and peace can be maintained; and (3) make people think twice about committing crimes. The Qur'anic verses explaining the implementation of *qisas* are *surah* (chapter) Al-Baqara *ayah* (verses) 178–179.

The SD text also contains a message that prohibits adultery. This is evident in the storyline that features two attempts of adultery against Siti Maryam by Kiai Sangit and Ki Patih. However, neither of them succeeds because of Siti Maryam's firm determination. The SD text depicts adultery as a great sin, and anyone who commits the act shall receive God's punishment.

495

Those deemed sinful are not only the ones who tempt or suggest committing adultery but also those who accept such advances. The Qur'anic verse forbidding adultery is *surah* Al-Isra *ayah* 32.

3 CONCLUSION

As a *piwulang* text, several moral values contained in the story are identified. These values include moral values for women, *tasawwuf* teachings, principles of harmony and respect, and Islamic teachings.

The content of the SD text reveals a combination of Javanese values and the teachings of Islam. The text is based on an Islamic story, but it also includes Javanese values, such as the mystical *Kejawen* teachings called *manunggaling kawula gusti* and the philosophy of true *raos* (sense/soul). These teachings are very well known among the Javanese as practices to achieve perfection of life.

Therefore, it can be concluded that the writer of SD tried to create a Javanese literary work by utilizing resources from the Islamic literature (the religion of the majority of Javanese at that time) while still maintaining Javanese values which were considered necessary to be preserved, as Islam further developed and strengthened in the Javanese society. The author believes that through this article the readers can further enrich their knowledge of Islam, while still preserving existing Javanese values that had been ingrained in the Javanese society.

REFERENCES

Baried, S.B. (a), et al. (1985) *Pengantar Ilmu Filologi (Introduction to Philology)*. Jakarta, Pusat Pembinaan dan Pengembangan Bahasa Depdikbud.

Baried, S.B. (b), et al. (1985) *Memahami Hikayat dalam Sastra Indonesia (Understanding Hikayat in Indonesian Literature)*. Jakarta, Pusat Pembinaan dan Pengembangan Bahasa Depdikbud.

Behrend, T.E. (1990) *Katalog Induk Naskah-Naskah Nusantara Jilid I Museum Sana Budaya Yogyakarta (Master Catalogue of Manucripts from the Indonesian Archipelago Volume I of Museum Sana Budaya Yogyakarta)*. Jakarta, Djambatan.

Behrend, T.E. (1995) *Serat Jatiswara: Struktur & Perubahan di Dalam Puisi Jawa1600–1930 (Serat Jatiswara: Structure & Changes in Javanese Poetry 1600–1930)*. Jakarta, INIS (Indonesian-Netherlands Cooperation in Islamic Studies).

Mulder, N. (1983) *Kebatinan dan Hidup Sehari-hari Orang Jawa (Spiritualism and the Everyday Lives in Java)*. Jakarta, PT Gramedia.

Pigeaud, T.G. (1970) *Literature of Java: catalogue raisonné of Javanese manuscripts in the Library of the University of Leiden and other public collections in the Netherlands. I: Synopsis of Javanese literature 900–1900 A.D.* The Hague, Martinus Nyhoff.

Riffatere, M. (1980) *Semiotic of Poetry*. London, Methuen & Co. Ltd.

Robson, S.O. (1994) *Prinsip-prinsip Filologi (The Principles of Philology)*. Jakarta, RUL.

Rohman, S. (2016) *Serat Darmasaloka: Edisi Teks dan Kajian Isi (Serat Darmasakola: Text and Content Analysis)*. [Thesis]. Fakultas Ilmu Pengetahuan Budaya, Universitas Indonesia, Indonesia.

Saputro, K.H. (1992) *Pengantar Sekar Macapat (Introduction to Sekar Macapat)*. Depok, Fakultas Sastra Universitas Indonesia.

Saputro, K.H. (2001) *Puisi Jawa: Struktur dan Estetika (Javanese Poetry: Structure and Aesthetics)*. Jakarta, Wedatama Widya Sastra.

Simuh. (1988) *Mistik Islam Kejawen Raden Ngabehi Ranggawarsita: Suatu Studi Terhadap Serat Wirid Hidayat Jati (Kejawen Islamic Mysticism of Raden Ngabehi Ranggawarsita: A Study on Serat Wirid Hidayat Jati)*. Depok, UI Press.

Simuh. (2003) *Islam dan Pergumulan Budaya Jawa (Islam and the Struggle of Javanese Culture)*. Jakarta, Teraju.

Suseno, F.M. (1983) *Etika Jawa Sebuah Analisa Falsafi tentang Kebijaksanaan Hidup Jawa (Javanese Ethics: A Philosophical Analysis on Javanese Wisdom)*. Jakarta, PT Gramedia.

Suseno, F.M. (1987) *Etika Dasar, Masalah-masalah Pokok Filsafat Moral (Foundations of Ethics, The Main Problems of Moral Philosophy)*. Yogyakarta, Yayasan Kanisius.

Van Luxemburg, J., Bal, M. & Weststeijn, W.G. (1987) *Tentang Sastra (On Literature)*. (A. Ikram, Trans.). Jakarta, P.T. Intermasa.

Cultural Dynamics in a Globalized World – Budianta et al. (Eds)
© *2018 Taylor & Francis Group, London, ISBN 978-1-138-62664-5*

Old Papuan manuscripts: A general review

T. Pudjiastuti
Department of Literature, Faculty of Humanities, Universitas Indonesia, Depok, Indonesia

ABSTRACT: Papua is one of the easternmost regions of the Republic of Indonesia. Islamic culture started to invade Papua in the 17th century through three pathways, namely the Sultanate of Tidore, Seram, and Raja Ampat. Written culture in Papua began since then. Thus, traces of the Islamic culture in the form of handwritten old manuscripts can be found in West Papua and the Raja Empat Islands. The major challenge is to find the exact number of such manuscripts and the genre of the texts. The objective of this research was to record, preserve, and inform about the existence of the texts and to explore about the content. The characteristic of the research is codicology with the manuscript as the object. Previous research has recorded and digitalized manuscripts and the properties of the aforementioned three cultures. It is known from the perspective of the context of the texts that the majority of the Papuan manuscripts contain Islamic teachings. The materials used are European paper and local paper called *koba-koba* leaves. The letters commonly found in the manuscripts are Arabic, Lontara, and Jawi in Arabic, Makassar, and Malay languages.

1 INTRODUCTION

Hand-written old manuscripts are a type of heritage that the Indonesian ancestors have passed on to the future. As a cultural property, old manuscripts are artifacts that explicitly present some inscriptions. According to Molen (1985), scripts play a very important role in the history of mankind, our daily life, science and knowledge, power, politics, and so on. Civilizations without and with inscription basically differ. From the perspective of cultural heritage, ancient manuscripts are authentic evidence of the advancement of a certain community. Papua is one of the ethnic groups in Indonesia with a cultural heritage in the form of inscription. This is not surprising because there is evidence that Papua has had a written culture since the arrival of Islam in the 17th century. On the basis on this matter, the research questions are: (1) where are the Papuan manuscripts located today? (2) how many are there? and (3) what is the genre of the texts?

The main objective of this study was to inform the society that the Papuan people have long been familiar with written culture, proven by their manuscripts. The other purpose was to preserve the heritage of the Papuan people from being extinct (due to damage) or sold. Furthermore, this study holds scientific interest of informing the manuscript data to scientists, which facilitate their studies based on each knowledge field.

This research is codicological with the manuscripts as the object of study. Therefore, all Papuan manuscripts involved in this research have been studied using the codicology approach. Every manuscript has been described and studied, and a detailed note has been taken, including its cover, size, genre, illumination (if available), and other aspects. Then, every page of the manuscripts is digitized (their picture is taken) to be kept in manuscript CDs.

2 PAPUAN ISLAMIZATION

According to Muridan (2013), Islam came to Papua during the expansion of the Sultanate of Tidore under the leadership of Sultan Nuku. However, according to some sources in Papua,

Islam came through other ways too, such as Seram, Banda, and the Hadramaut (the Arab peninsula). Islamic leaders and *petuanan* (kings) in Papua contend that Islam spread along the coastal areas from Sorong and Fak-Fak. With regard to the issues of Islamization in Papua, the following three things need to be explained: (1) the issue of the figures who brought Islam to Papua; (2) the route that Islam had taken to enter Papua; and (3) the heritage of Islam in Papua.

The figures responsible for bringing and spreading Islam in Papua can be divided into two groups. The first group includes those believed to have a line of descent to the Prophet Muhammad SAW, known as the sayyid. The second group includes those of Arab descent, known as the shaykh. During the Islamic proselytism period, both the sayyid and the syakh married the locals and reproduced Arab-Papua and Arab-Malay descendants. The descendants of the Islamic proselytism in Papua are recognized by their surnames, namely Al-Hamid, Al-Katiri, Assegaf, Said bin Agil, Bafadal, and Kabiran.

Islam is believed to have entered Papua through three routes, namely the Sultanate of Tidore, Raja Ampat, and the Archipelago of Seram-Banda.

1. The Tidore Route. From the stories told by their ancestors, Papuans believe that the Sultanate of Tidore had been present in Papua long before the arrival of colonists and the Christian missionaries in the hinterland of Papua. In fact, the Christian missionaries came into Papua escorted by some people from the network of Tidore Sultanate. According to Andaya (1993), the Papuans associated the Tidore Sultanate with Islam. This was illustrated through an event in 1705 when the Jogugu (Prime Minister), the sea captain Salawati, and Waigeo of the Raja Empat Islands received the Sultan of Tidore's envoys. Everyone present in the event said "Amen" when the Sultan of Tidore's envoy finished reading the Sultan's letter. In other words, Papuans respected the Sultan of Tidore similarly to Allah, despite the fact that only his letter was present.
2. The Raja Ampat Route. According to experts, the Archipelago of Raja Ampat was ruled by four kings (*petuanan*), namely King Salawati, King Waigeo, King Misool, and King Waigama. According to some resources from Raja Ampat, Islam arrived in Papua through Salawati, brought into, and proselytized by Islamic proselytizers from Hadramaut who came for the purposes of trading, seeking better sources of livelihood, or avoiding the social and political dynamics in their homeland.
3. The Seram-Banda Route. According to most of the resource persons from Sorong and Fak-Fak, Islam entered Papua through people from Seram-Banda, who came to Papua and settled there, and spread along the coastal areas, such as Sorong, Fak-Fak, and the Raja Empat Islands. While spreading Islam, some of them married the locals, eventually reproducing descendants of mixed origins of Maluku–Papua.

3 RESEARCH RESULTS

Find traces of the Islamic culture that had been recorded in the written tradition of Papua is a challenging task due to the following reasons:

1. The absence of figures who still recognize and keep the written tradition of their ancestors.
2. The influence of the social and political dynamics of the colonial era that has made the owners of the manuscripts unwilling to surrender their collection of manuscripts to strangers.
3. The confines of customs and fear and assumptions that the old manuscripts that they have are "sacred heritage" that is not to be read by ordinary people or is not to be taken out from its storage anytime.

Fortunately, in some enclaves assumed to have been strongly associated with the history of Islam in the Land of Papua, resource persons could still be found, willing to provide information regarding the legacy of the written tradition of Papua. As a result, a number of ancient Papuan manuscripts and historical objects that have been stored by people have been successfully recorded.

A total of 89 old manuscripts and three historical objects have been recovered through research conducted in 2013 and 2014 in Sorong, Fak-Fak, Kokas, Misol, Waisai, Salawati, Patipi, and Patimburak. The three objects are one dagger with an inscription in Arabic written in gold ink, one piece of talisman flag with an Arabic inscription on it, and one ceramic plate with Arabic scripts. The manuscripts and historical objects are held by 23 manuscript owners. The following are brief information about the 89 old manuscripts and the three historical objects, their owners, and the number of manuscripts in their collection.

The 89 old Papuan manuscripts recovered were written in Arabic, Jawi (Arab-Malay), and Lontara (Bugis-Makassar script) scripts using the languages of Arabic, Malay, and Bugis-Makassar.

The material used to write the text varies, namely European paper (Churchill, 1935; Heawood, 1950), lined paper, plain paper, *dluwang* or Javanese paper (Pudjiastuti, 1997), and the material used for writing the traditional Papuan manuscript is called *koba-koba* leaves. *Koba-koba* leaves are obtained from red fruit trees (Latin: *Pandanus Conoideus*). The tree is of the family *pandanus* (Latin: *Pandanaceae*), which usually grows in huddle in Papua. It can grow up to 3 m. The following is the information about how the *koba-koba* leaves are utilized as a surface to write: the leaves are cut into the desired size and dried in sunlight. Then, they are flattened and the surface is smoothened with *kuwuk* (sea snail shell). Finally, they are used as a surface to write a text or manuscript.

The Papuan texts consist of several genres, namely a history of some figures, prayers, language grammar, medicines, genealogy, tales, literature, legal treaties, and Islamic teachings, such as prayers, *tauhid, zikir* (dhikr), and the Koran.

The following are the information about the owners and the number of manuscripts and historical objects:

1. Abdurrahman Kastella

Abdurrahman Kastella from Ambon is the owner of Papuan manuscripts who lives in Sorong, Papua, for several decades. As a mosque imam, he has stored 10 old manuscripts that he inherited from his late grandfather. These manuscripts are mostly handwritten and partly in the form of lithographs (stone prints), which include *Risalah Hukum Jimak* (Jimak Legal Treaties), *Kitab Mujarobat* (The Book of Useful Medicines), *Kisah Nabi Muhammad SAW* (Tales of the Prophet Muhammad SAW), *Doa Tawasul* (Tawasul Prayers), *Zikir* (Dhikr), *Kitab Nikah* (the Book of Marriage), *Maulud,* and *Kumpulan Doa* (collection of Prayers).

2. Jafar Bugis

Jafar Bugis is a Bugis-Makassar descent who lives in Sorong, Papua. In addition to his position as a mosque imam, he is a trader. He has three old manuscripts, two of which were written in Arabic and Lontara scripts and Arabic and Bugis-Makassar languages, respectively, and the third manuscript was written using three scripts, namely Arabic, Lontara, and Jawi in Arabic, Bugis-Makassar, and Malay languages. They are *Kitab Bagan Zikir* (Dhikr Chart Book), *Kitab Zikir* (the Book of Dhikr and Prayer), and *Kitab Kumpulan Doa* (the book of collection of prayers).

3. Iman Latuconsina

Iman Latuconsina is a Papuan of Ambonese descent. He has lived in Sorong, Papua, for a long time. As a mosque imam, he only has one old manuscript in the form of a horizontal roll (*rotulus*), whose content is about *Khutbah Jum'at* (Friday sermons).

4. Muhammad Bafadal

Muhammad Bafadal is a Papuan of Arab descent. His great grandfather came to Papua from Hadramaut in the 17th century to proselytize Islamic religion and conduct trading simultaneously. As the imam of the mosque in Doom Island, he has only two old handwritten manuscripts in Arabic script and Arabic language. They are *Kitab Tasawuf* (the Book of Tasawuf) and *Kitab Masalah Agama Islam* (the Book of Islamic Religion Issues).

5. Ma'bud Wadjo

Ma'bud Wadjo is of Makassar origin who moved to Sorong, Papua, and has seven old manu-scripts, which he borrowed from his uncle. They are *Khatmal al Kajah, Maulid Barzanji, Wirid, Tashrifan* (Arabic grammar), *Rahasia Shalat* (the Secrets of Shalat), *Ibadah* (Wor-ship), and *Doa* (Prayer).

6. Dra. Nursia Salim, S. H.

Nursia Salim is a judge at the religious court of the city of Sorong. She has one manuscript that she borrowed from her late uncle, which is *Doa dan Zikir* (Prayer and Dhikr).

7. Hajj Abdurrahman al-Hamid ("Mister Teacher")

Abdurahaman al-Hamid, a Papuan of Arab descent, is from the Island of Misool at the Raja Empat Islands who lives in Sorong. In addition to his role as the mosque Imam, he is an English teacher and hence known as "Tuan Guru" (Mister Teacher). He borrowed the ancient handwritten manuscripts from his family living in the Island of Misool. He has five manuscripts in total, which are *Kitab Ajaran Islam* (the Book of Islamic Teachings), *Ilmu Siri dan Doa* (the Science of Siri and Prayer), *Doa dan Maulid* (Prayer and *Maulid*), *Wirid Abdullah Asy Syatari* (the Wird of Abdullah Asy Syatari), and *Catatan Kepemilikan* (Notes on Ownership).

8. Mohammad Nasib Baria

Mohammad Nasib Baria is a Papuan of Ambonese descent who lives in Sorong. He bor-rowed the following five manuscripts from his family: *Doa Kanz al-'Arsy* (Prayer of Kanz al-'Arsy), *Kumpulan Doa* (Collection of Prayer), *Kumpulan Doa Mustajab* (Effective Prayers), *Doa Istigfar Rajab* (Rajab Istighfar Prayer), and *Doa* (Prayer).

9. Haji Azis

Haji Aziz is a Papuan of Bugis descent who lives in Sorong. He has three manuscripts writ-ten in Arabic, Jawi, and Lontara alphabets in Arabic, Malay, and Bugis-Makasar languages, namely *Kitab Pengetahuan Agama* (Religious Knowledge Book), *Kitab Ahlak* (Book on Morality), and *Kumpulan Doa* (Prayer Collection).

10. Thalib Salim

Thalib Salim is Nursia Salim's younger brother who lives in Sorong. Thalib Salim is the Chairman of the MUI (Council of Indonesian Ulema) of Sorong branch. The three manu-scripts that he has are his family's property. They are *Kumpulan Doa* (Prayer Collection), Bagan *Shalat dan Doa* (Chart of Shalat and Prayer), and *Tahlil*.

11. Ahmad Iba bin Ismail Iba

Ahmad Iba bin Ismail Iba is an indigenous Papuan. He is a *petuanan* (raja) of the Island of Patipi at the Raja Empat Islands. Despite being the King of Patipi, he resides in Sorong. The ancient manuscripts in his collection are borrowed from his father and there are 11 of them, namely *Juz 'Amma*, six Quran, *Khutbah Idul Adha* (Eid al-Adha Sermon), *Akidah dan Kutbah Idul Adha* (Faith and Eid al-Adha Sermon), *Zikir* (Dhikr), and *Doa* (Prayers).

12. Ali Iha

Ali Iha is of the Mollucan-Papuan descent who lives in Fak-Fak. He borrowed four manu-scripts from his family. They are *Doa* (Prayer), *Maulid* (Maulid), *Keutamaan Doa* (the Virtue of Prayer), and *Kumpulan Doa* (Prayer Collection).

13. Abdillah At-Tamimi

Abdillah At-Tamimi is a Papuan who is a descendant of the Arab syaikh. He currently lives in Fak-Fak as a trader as well as a religious teacher. The four manuscripts that he has are borrowed from his ancestors. They are: *Silsilah dan Doa* (Genealogy and Prayer), *Kumpulan Doa* (Prayer Collection), *Kitab Fikih: Tata Cara Memandikan Mayat* (the Book of Fiqh: Procedures of Bathing Corpses), and *Doa Haykal* (Haykal Prayers).

14. Mohammad Taher Arfan

Mohammad Taher Arfan is not an indigenous Papuan. His ancestors moved from Tidore to the Doom Island. He has one manuscript, whose condition is very bad. The paper has been consumed by the ink, making it difficult to read. The sheets have stuck between one and another so much that they can no longer be released. The title of the manuscript is *Catatan Desa Salawati dan Doa* (Notes on Salawati Village and Prayer). In addition, he also has one piece of a talisman flag with the Arabic script in Arabic language written on it. Furthermore, he has one dagger on which the Arabic script in Arabic language was written in gold ink.

15. Ali Umbalat

Ali Umbalat is an indigenous Papuan from Misool Island at the Raja Empat Islands, who now lives in Sorong. He obtained two manuscripts from his family, namely *Kitab Hidayah As-Salikin* and *Kumpulan Hikayat Nabi* (Collections of Prophet's Tales).

16. Husen Umbalat

Husen Umbalat is Ali Umbalat's younger brother. He lives in Misool Island at the Raja Empat Islands. His collection of eight old manuscripts is borrowed from his family, namely: *Kumpulan Masalah Agama* (A Collection of Religious Issues), *Kutbah Id, Kutbah Jumat* (Eid Sermons, Friday Sermons), and five manuscripts concerning *Kutbah Idul Fitri* (Eid Al-Fitr Sermons).

17. Musa bin H. Husen Salim

Musa bin Haji Husen Salim lives in Saonek Island, the capital of the Raja Empat Islands. He is an indigenous Papuan, and the son of Saonek's mosque Imam, Husen Salim. The manuscript that he has is the only manuscript that could be saved from the "manuscript burial" event conducted by the local people of Saonek due to fear of keeping manuscripts that are considered hieratic. The title of the manuscript is *Kumpulan Doa Tarawih* (A Collection of Tarawih Prayer).

18. Jaelani Kuda

Jaelani Kuda is of a mixed Arab–Papua descent. He currently lives in Patimburak and Kokas in Fak-Fak district. His family is the traditional guardian and Imam of the Patimburak mosque that is located in Kokas district. The old manuscripts in his collection are borrowed from his family. Apart from the manuscripts, he keeps an old ceramic plate decorated with Arabic inscription in Arabic language. He actually has many manuscripts, but most of them are in bad condition, and only six of them are still readable, namely *Khifayatul Quran, Adabul Quran, Tauhid, Akidah, Shalawat,* and *Nazam Barzanji*.

19. Iksan Kuda

Iksan Kuda is the elder brother of Jaelani Kuda who also lives in Kokas district. He has only one old manuscript, namely *al Quran* (the Koran).

20. Muhammad Ali Fuad

Muhammad Ali Fuad is of Arab–Papua descent and lives in Kokas district. As the imam of Kokas mosque, he has four old manuscripts, namely *Irsyadul Iman, Hikayat Nur Muhammad* (Tales of Nur Muhammad), and *Kitab Mikraj Nabi* (the Book of the Prophet's Ascension).

21. Ibrahim Sagara

Ibrahim Sagara is of Arab–Papua descent. He lives in Patipi Island at the Raja Empat Islands. As the imam of the mosque in Patipi Island, he has four manuscripts of lithographed (stone print) Koran.

22. Saleh Barweri

Saleh Barweri is of Arab–Papua descent. He lives in Kokas district. He has only one manuscript entitled *Hikayat Nabi Bercukur* (Tales of How the Prophet Shaved).

4 CONCLUSION

On the basis of the above description, it can be concluded that the people of Papua have been advanced and more civilized and had possessed a written culture long before the invasion of colonialists.

From the perspective of resource persons and the heritage of Islamic culture in the form of old manuscripts and historical objects found in Sorong, Fak-Fak, and the Raja Ampat Islands, we can conclude that the majority of the Papuan manuscripts are concerned with Islamic issues.

According to Andaya, the Papuan people associated the Sultanate of Tidore with Islam and hence they responded by saying "Amin" when the letter of the Sultan of Tidore was read to the public in the 1705 event. Apparently, the expression "Amin", which was uttered by the Jogugu, the sea captain of Waigeo, and Salawati at that time, was not aimed at respecting the Sultan of Tidore similarly to the uttering of the name of Allah in the Koran; however, it was most probably because of the fact that the final part of the letter was a prayer.

From the perspective of the script and language used in the old manuscripts that were written in Arabic, Jawi, and Lontara scripts in Arabic, Bugis-Makassar, and Malay languages, it can be estimated that the Islamic culture invaded Papua not merely because of the expansion of the Sultanate of Tidore, Ambon, and Seram, but also because of the expansion of Bugis-Makassar.

The names of the manuscript owners, such as At Tamimi, Bafadlal, and others, show that they are the descendants of the sayyid or syaikh from the past who came to Papua to spread Islam.

The author believes that this study will be critical for scientists interested in conducting research in eastern Indonesia, especially for those who want to know about the history, culture, and the characteristics of the Papuan society in the past. Furthermore, this study will help complete the mapping of the Nusantara manuscripts throughout Indonesia.

REFERENCES

Andaya, L.Y. (1993) *The World of Maluku: Eastern Indonesia in the Early Modern Period*. Honolulu, University of Hawaii Press.

Churchill, W.A. (1935) *Watermarks in Paper in Holland, England, France, etc. in the XVII Centuries and Their Interconnection*. Amsterdam, Menno Hertzberger.

Heawood, E. (1950) *Watermarks: Mainly of the 17Th and 18th Centuries*. Hilversum, The Paper Publications Society.

Muridan, W. (2013) *Pemberontakan Nuku: Persekutuan Lintas Budaya di Maluku—Papua sekitar 1780–1810* (Nuku Rebellion: Cross Cultural Fellowship in Maluku-Papua around 1780–1810). Jakarta, Komunitas Bambu.

Pudjiastuti, T. (2006) Kertas Tradisional (Traditional Papers). In: Pudjiastuti, T. & Kartika, G.D. *Naskah dan Studi Naskah* (Manuscript and Manuscript Studies). Bogor, Academia.

van der Molen, W. (1985) *Sejarah Perkembangan Aksara Jawa in Aksara dan Ramalan Nasib dalam Kebudayaan Jawa* (History of the Development of Javanese Script in Javanese Literature and Prophecy). Translated by Soedarsono, Retna Astuti and I. W. Pantja Sunjata. Yogyakarta, Proyek Penelitian dan Pengkajian Kebudayaan Nusantara (Javanologi) Direktorat Jendral dan Kebudayaan.

Cultural Dynamics in a Globalized World – Budianta et al. (Eds)

Articulation and contestation of cultural identities in Riau Province: The case of 'Mandau Regency'

S. Basuki & M. Budiman
Department of Literature, Faculty of Humanities, Universitas Indonesia, Depok, Indonesia

ABSTRACT: In this study, we discuss the articulation of cultural identities in Riau Province, Indonesia, which—through fund transfers from the Central Government in the form of oil and gas revenue sharing—has become a success story of decentralisation in the post-Soeharto era. Decentralisation has provided opportunities for creating new regions, which Riau has made uses of since 1999. However, the wish of residents of Mandau district in Bengkalis Regency to form 'Mandau Regency' has yet to be fulfilled. Through the articulation of 'Mandau Regency', the idea about a new regency that is separate from Bengkalis has risen in the minds of the Mandau people. This idea is in line with the articulation that requires Mandau to stay within Bengkalis according to the 'native Malay population'. This contestation occurs in the middle of a decentralisation design within the arena of authority, which is called governmentality by Foucault. Here, we find that organisers in Jakarta still decide the fate of the Mandau people; however, this must be done through cooperating with local organisers in Riau. The idea of 'Mandau Regency' will continue to exist together with new articulations of cultural identities.

1 INTRODUCTION

Decentralisation in Indonesia began when Law No.22/1999 on Regional Governments and Law No. 25/1999 on the Balance of Finance between the central and regional governments were issued. The regulation on revenue-sharing has caused Riau Province to gain much attention as a success story. Riau has always ranked in the top five provinces with the highest per capita regional income in Indonesia. Bengkalis Regency, which has oil fields in Mandau district, has become one of the few regencies in Indonesia with the highest regional income and expenditure budget (APBD).

Decentralisation regulations also provided opportunities for creating new regions. In 1999, Bengkalis Regency facilitated the creation of three new regencies and cities, including Dumai City, Rokan Hilir Regency and Siak Regency. By 2009, Bengkalis Regency had already created another regency, namely Meranti Islands Regency. However, to date, the wish of the residents of Mandau to form a new regency has yet to be fulfilled.

The 'Mandau Regency' issue will be discussed using a cultural studies approach, which examines culture as a form of representation in the context of power within a society (Barker, 2011, p. 38). Culture is understood as a shared meaning on life that is vulnerable to power domination. From the viewpoint of cultural studies, this issue is closely related to the articulation of cultural identity. Articulation (Hall, 1996, pp. 141–142) is the temporary connection between two different elements within a certain context. The main principle is the delivery of an idea and the temporary connection between the idea and its initiators. This connection is related to a certain context, which in turn relates to the past as well as the future (Hall, 1990, p. 225). The 'Mandau Regency' idea may be deemed appropriate or inappropriate to be connected to the Mandau people and its surroundings, depending on who connects or does not connect them and what context is chosen by its initiators.

The temporary connection between an idea and its initiators becomes more crucial when the idea is a cultural identity. Hall (1990, pp. 223–225) described 'identity' as 'a matter of "becoming" as well as of "being"', which undergoes continuous transformation on the other. Hall (1990, p. 222) also explained that identity is not something transparent and simple; therefore, it is more appropriate to understand identity as a never-ending process.

The people of Mandau, the majority of whom are immigrants, try to define themselves as being one with the region where they live. The context that causes the emergence of the articulation of 'Mandau Regency' is the wish to separate from Bengkalis. Nevertheless, they also re-define themselves with Malayness as the main factor. The context is a movement driven by the 'newcomers' to form a new regency without prioritising Malayness.

We argue that decentralisation is in fact a form of governmentality. The concept of governmentality, as put forth by Foucault (1991), is a policy designed in such a way that the organisers can direct the behaviour of people. This situation of governmentality forms the background in the 'Mandau Regency' issue. The creation of new regions must follow the rules determined by the central government. We will discuss how cultural identities in Mandau are articulated amidst the governmentality of decentralisation and how they are contested, resulting in the fact that the 'Mandau Regency' idea has not been realised.

2 METHOD

The author collected data by interviewing informants in Pekanbaru and Bengkalis, including public figures, academicians, bureaucrats and journalists. Secondary data were collected by observing the related official documents published by the central government and regional governments. These two methods facilitated the observation of the articulation and contestation of cultural identities regarding 'Mandau Regency' and the complexities involved due to the existence of governmentality in Riau Province.

3 RESULTS AND DISCUSSION

The centre of the Bengkalis Regency government is in Bengkalis Island, but the regency's territory is spread out to Sumatra Island and several surrounding islands (Figure 1). The capital of the Mandau district is Duri. The Duri Oil Field has been exploited since the 1950s by PT Caltex Pacific Indonesia. Mandau district plays a strategic role in the economy of Bengkalis as approximately 80% of the regency's GRDP comes from the mining and excavation sector, which is located mainly in Mandau. The population of Mandau district in 2011 was approximately 200,000 people, or approximately 44% (the largest share) of the total population of

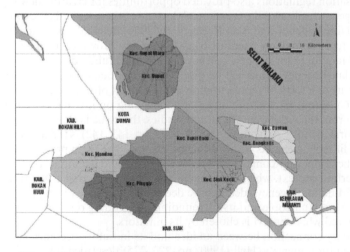

Figure 1.   Map of Bengkalis Regency (Bengkalis dalam Angka, 2012).

Bengkalis. Studies supporting 'Mandau Regency' typically focus on the increasingly higher population, the increasingly higher volume of economic activity and the workload of government officers. These factors are indeed the cause of the creation of new regions.

An ethnic Javanese community figure in Mandau said that the Mandau people 'do not have a shared cultural identity'. According to him, because of the presence of the Duri Oil Field, most of the Mandau district residents are newcomers. The similarity among Mandau residents only lies in the economic factor of earning a living with their professions. To this informant, Mandau people 'tend to maintain their own subcultures without undergoing acculturation'. Therefore, there is no specific cultural identity in Mandau. An ethnic Minang informant also said that 'culturally, there is no such thing as Mandau identity'. The Mandau 'identity' is only limited to economic pragmatism.

This study on the Mandau people is the establishment of a cultural identity, which is then articulated as an effort to support the creation of Mandau Regency. Hall (2007, p. 2) described that a similar view on many things and an ability to understand each other are actually qualities of people with the same culture. Therefore, a narration about cultural identity does not have to be related to the former glory of a kingdom or a tradition that has passed on through generations.

Meanwhile, according to Li (2000), a cultural identity can be articulated or unarticulated depending on the situation faced by the related community. According to the community figures who support 'Mandau Regency', this articulation of cultural identity emerged due to the 'threat' of being part of Bengkalis Regency, which is considered to have exploited Mandau's natural resources while neglecting the welfare of people. However, we are unsure of whether the formation of 'Mandau Regency' will improve the welfare of people.

The official stance of the Bengkalis Regency Government was to disapprove plans to create Mandau Regency. The reasons were that several requirements for the formation of Mandau Regency as stipulated by the law had not been met and that Bengkalis was considered to have had too many regions separating from it. The Riau Provincial Government had the same official stance.

During interviews with several ethnic Malay figures in Pekanbaru, the formation of Mandau Regency was perceived as a threat to the ethnic Malay identity in Bengkalis. They were concerned about the possibility of a regional leader that is non-Malay or non-Muslim if Mandau becomes a regency. The location of Mandau Regency near Rokan Hilir Regency, which borders North Sumatra, has raised concerns that a Christian figure might emerge as a leader in Mandau if it becomes a regency. These informants thought that there is financial support for the idea of a Christian Bataknese figure to lead Mandau. This is the formation of a supporting narration for an articulation, which is not always completely in accordance with the facts in the field.

The feeling of being threatened by a non-Malay potentially leading Mandau Regency is then responded with an articulation of '*anak jati* Melayu or a native Malay to lead Mandau'. An ethnic Malay academician mentioned the importance of the principle of 'Malay as the crown' derived from an expression implying that a regional leader in Riau must be of ethnic Malay descent. These explanations have prompted the author to detect a contestation of space in Mandau.

For Malay elites in Pekanbaru, Mandau Regency is considered a 'non-Malay element' threatening Riau as 'an ethnic Malay land'. Thus, they feel the need to relate Malay-ness with the leadership in Mandau. This feeling of being threatened was practically non-existent during the process of creating Rokan Hilir and Siak Regencies.

Regarding the issue of the creation of Mandau and Christianisation, an informant who is one of the Mandau figures supporting 'Mandau Regency' revealed that the Christianisation issue was fabricated by Pekanbaru and Bengkalis to hide their failure in delivering public service. This informant, who is a Muslim, believed that the new regency demand is motivated more by a reason of justice instead of prioritising a particular cultural identity. Two other Muslim informants generally agreed with this opinion, who considered the issue of cultural identity or ethnicity, religion, race and inter-group relation (SARA) in Mandau to be too 'high level' because residents are generally too busy with their economic activities. According to Mandau figures, the accusations of Malay figures in Pekanbaru about Christianisation are closely linked to the fact that ethnic Malays in Mandau are not adequately represented in the bureaucratic circles or the entrepreneurial world. If Mandau Regency is created, then ethnic Malay residents will get their access to power reduced. This opinion put forth by Mandau figures is a counter-narration to the one from Malay figures.

A meeting with a Muslim Bataknese informant provided an opportunity for the author to listen to the Bataknese group. According to him, there is no such conflict as stated by bureaucrats in Pekanbaru or Bengkalis. The informant also explained the system of the Bataknese kinship, which renders conflicts among them or with other residents in Duri to be an exaggeration. The informant inferred that the perception that 'SARA issues' exist in Mandau is caused by the jealousy of bureaucrats in Pekanbaru (many of whom are ethnic Malays) against non-Malays who dominate the economy in Mandau.

The Mandau informants who are pro-regency creators have the same opinion that in the era of decentralisation Mandau was discriminated by Bengkalis in various aspects. Mandau is considered to be raised by the newcomers, not by the 'pity' of Bengkalis. Meanwhile, an ethnic Malay informant, who is a former member of the Bengkalis Regional House of Representatives (DPRD), admitted that Bengkalis has long treated Mandau unfairly. The regional income of Bengkalis mostly comes from Mandau, especially from oil and gas revenue sharing. Furthermore, a waste retribution fee is imposed on Mandau residents, whereas no such fee is imposed in Bengkalis. Despite this, public facilities for Mandau residents have yet to be improved. According to this informant, he has raised this issue since about 2008 but received no response from the elites in Bengkalis.

This ethnic Malay informant suggested Bengkalis to release Mandau if it is not serious in solving the problems in Duri. The informant even said that if he were a Mandau resident, he would be at the forefront to demand Mandau to separate from Bengkalis. Everything has a fee in Duri, and according to the informant, the fees were even higher than those in Medan. The informant thought that Bengkalis should not mind if Mandau (as the source of the largest share of income for Bengkalis) separates itself, as long as large infrastructures in Bengkalis are built by the central government.

Decentralisation regulations have caused the creation of new regions in Bengkalis Regency to be managed by the Bengkalis Regent and DPRD, as well as the Riau Governor and DPRD. This authority was given to them through the Government Regulation No. 129 of 2000, which is a technical regulation on creating new regions as stipulated by the Law No. 22 of 1999. In line with the revision of the Law No. 22 of 1999 by the Law No. 32 of 2004, the Government Regulation No. 129 of 2000 was also revised by the Government Regulation No. 78 of 2007. Executive and legislative officials in Bengkalis and Riau are the decisive authorities on whether Mandau people's aspirations can be delivered to the central government and the DPR.

The author's opinion is that the issue of the creation of new regions is not simply a technical matter of whether the requirements in the regulation can be met or not. This issue is part of governmentality, where, for common prosperity, the behaviours of the people of Mandau, Bengkalis and Riau are directed in such a way according to the will of those who designed decentralisation in Jakarta. However, Foucault (1991, p. 102) reminded us that this situation is a 'very specific albeit complex form of power' that requires careful calculation in order to run smoothly. The complexity is apparent in the case of 'Mandau Regency'. Apart from the interactions between local actors, the creation of new regions also includes the design of interactional patterns between local actors and players in the DPR.

Although the official stance of the Bengkalis Regency and Riau Provincial governments stated that Mandau district had yet to fulfil the requirements to become a regency, the articulation of 'Mandau Regency' had a certain position of power. This position was then directed into political practice. Lobbies from Mandau figures to the DPR in the 1999–2004 period caused legislators to issue a Right for Initiative in proposing a Draft Law (RUU) on the Creation of Mandau Regency. Former President Susilo Bambang Yudhoyono even issued a Presidential Mandate (Ampres) No. R.01/Pres/2007 on 2 January 2007. Ampres is a cover letter for the representatives of the central government who are present in the DPR to deliberate a draft law, in this context on the Creation of Mandau Regency. This caused the rejection expressed by Bengkalis figures to Vice President Jusuf Kalla in January 2007.

The contestation of political lobbies has so far been won by the ethnic Malay figures in Bengkalis and Pekanbaru, resulting in the central government and DPR to discontinue the deliberation of the RUU on the Creation of Mandau Regency during the 2004–2009 and 2009–2014 periods. Until October 2013, the proposal to deliberate the creation of Mandau Regency still had not been included in the agenda of DPR. Several administrative requirements, including a recommendation from the Bengkalis DPRD (as the parent regency), had yet to be fulfilled.

The Riau Gubernatorial Election of 2013 aspirated the supporters of the formation of Mandau Regency, as several governor candidates supported the cause, including Annas Maamun, who was later elected Governor. During the campaign period, the 'Mandau Regency' idea was declared by the candidates to be a suitable element to be connected with the Mandau people. The connection ended when Governor Annas Maamun preferred Mandau district to become 'Duri City' rather than 'Mandau Regency'. Annas' concept is another counter-articulation against the 'Mandau Regency' idea, which has been expressed openly, even in various demonstrations.

4   CONCLUSION

A cultural identity does not have to be related to histories of kingdoms or century-old traditions. A shared view on societal issues and an ability to understand each other are qualities of people with the same culture. This is the basis that can be used to form a narration on cultural identity, which is then articulated. A number of informants from various ethnicities apparently see a common 'threat' if Mandau remains in Bengkalis; hence, they articulate the 'Mandau Regency' idea. At the same time, the ethnic Malay figures, especially in Pekanbaru, feel that 'Mandau Regency' will threaten Malay-ness; therefore, they articulate the idea of a 'native Malay to lead Mandau'. As a result, contestation occurs between the two articulations of cultural identities.

The case of 'Mandau Regency' shows that governmentality of decentralisation in the issue of creating new regions provides an opportunity for political practices that local elites in Riau are often unable to control. However, they also have the opportunity to cooperate with actors in Jakarta. The targets of governmentality are able to cooperate with their organisers in the central government. Governmentality by the central government has limitations, wherein cooperation with local organisers in the regency and provincial level is required so that Jakarta can direct the people in Mandau and Bengkalis. Although not directly expressed, the main reason for cooperation of organisers in the Mandau context is oil exploitation in Duri, which is practically under the control of the central government. Cooperation in governmentality between central and regional actors will still place the central government in the most advantageous position.

The idea about 'Duri City' may drive the emergence of contestation between articulations of cultural identities in new patterns. There can be a contestation of ideas between Mandau residents living in Duri and Mandau residents outside Duri. New actors may emerge in Duri as well as in Bengkalis, thus driving new cooperation patterns between the targets and the organisers of governmentality. The arrest of Annas Maamun by the Corruption Eradication Commission (KPK) caused the leadership in Riau Province to be shifted to Deputy Governor Arsyadjuliandi Rachman, who was called Acting Governor until 2018. Therefore, this results in an even wider possibility for contestation between articulations of cultural identities involving Mandau residents.

REFERENCES

Barker, C. (2011) *Cultural Studies: Theory and Practice*, 4th edition. London, SAGE Publication Ltd.
BPS Kabupaten Bengkalis. (2012) *Bengkalis dalam Angka 2012* (Bengkalis in Numbers 2012). Bengkalis, Badan Pusat Statistik Kabupaten Bengkalis.

Foucault, M. (1991) Governmentality. In: Burchell, G., Gordon, C. & Miller, P. *The Foucault Effect: Studies in Governmentality*. Chicago, The University of Chicago Press. pp. 87–104.

Grossberg, L. (1996) On Postmodernism and Articulation: An Interview with Stuart Hall. In: Morley, D. & Chen, K-H. *Stuart Hall: Critical Dialogues in Cultural Studies*. London, Routledge. pp. 131–150.

Hall, S. (1990) Cultural Identity and Diaspora. In: Williams, P. & Chrisman, L. *Colonial Discourse & Postcolonial Theory: A Reader*. London, Harvester Wheatsheaf. pp. 222–237.

Hall, S. (1997) *Representation*. London, Sage Publications and the Open University.

Li, T.M. (2000) Articulating Indigenous Identity in Indonesia: Resource Politics and the Tribal Slot. *Comparative Studies in Society and History* 42, 149–179.

Cultural Dynamics in a Globalized World – Budianta et al. (Eds)
© *2018 Taylor & Francis Group, London, ISBN 978-1-138-62664-5*

Utilization of the Senjang tradition as cultural identity of the people in Musi Banyuasin, South Sumatra

A. Ardiansyah & T. Christomy
Department of Literature, Faculty of Humanities, Universitas Indonesia, Depok, Indonesia

ABSTRACT: In this article, we examine the utilization of Senjang as the cultural identity of Musi Banyuasin (Muba), South Sumatra. The transfer of authority in the fields of culture and education facilitates strengthening the identity of each area. The district government of Muba seeks opportunities in the autonomous policy by utilizing cultural aspects to strengthen the identity of Muba through Senjang, which is one of the living oral traditions among the people of Muba district, South Sumatra Province, usually performed in special occasions, such as wedding ceremonies, thanksgivings, and local government's ceremonial agenda. It is called Senjang because the verses and music are sung and played alternately. Senjang is performed in Musi language, and performers of Senjang are called Tukang Senjang. The objective of this study is to investigate the utilization of Senjang as a means to understand the cultural identity of Muba. The study found some utilizations of Senjang by the local government and community as part of local political processes.

1 INTRODUCTION

In the beginning of 2010, after the district separation, the author had a chance to be part of a wedding reception in the city of Sekayu, Muba Regency, South Sumatra, during which, he witnessed a performance called Senjang by the Sekayu people. A man and a woman gave a unique Muba performance on the stage after they were called by the master of ceremony.

The man was wearing a high-collared suit or *beskap* and *tanjak* as the head cover, and the woman was wearing a modern *kebaya*, high-heeled shoes, and *tengkuluk* as the head cover. First, they bowed their heads to show respect, greeted the audience, and signaled the single organ player to start the music. Then, they danced by moving their hands and legs following the rhythm of the music. After the music stopped, the man started to recite Sampiran the Senjang verses, about four to five lines, and then stopped, and the music was played again. The music accompaniment and the verse recital were done alternately until the end of the performance.

The author also noted that the guests enjoyed and responded to the verses recited by the performers by saying *setuju* (agree) and *beno* (correct), accompanied by gestures of approval such as clapping, giving thumbs up, and shaking heads.

The show lasted for approximately 15 min and gained positive response from the audience. The performance had a special impression on the author, who is a non-speaker of the Musi language and not from Muba, which then leads to the following questions: What kind of art is that? What is the intention behind the performance of this art? Is this art the identity of Muba? Why did the master of ceremony say that this art was originally from Muba? The questions ultimately led the author to explore about this Muba art further.

In 2002, the Muba Regency was divided into two districts, namely the Musi Banyuasin district (Sekayu) and Banyuasin (Pangkalan Balai). The division did not only mean a political separation of the areas, but also affected those regions culturally. The political separation eventually led to contested cultural symbols. Competitions among different traditions also took place. Indirectly, there were competitions to determine an identity. This happened because all traditions could not represent a region simultaneously.

Muba district, before the separation, had some oral cultural expressions. A study conducted by Gaffar et al. (1989, 13–21) shows that those traditions were, among others, Senjang, Andai-andai Panjang, Mantra, Serambah, and Nyambai. The medium used in those traditions was the Musi language. Senjang is the only cultural expression that still exists today.

The main reason for the survival of Senjang is that it is included as a local content in the local literature course in the school-based curriculum of the 2004 Curriculum by the education policy makers of Muba district. Another reason is that the district has consistently held the annual Randik Festival, one of the most important agenda of which is the Senjang competition among the sub-districts in Muba district. The people also play a significant role in increasing the re-popularization of the Senjang oral tradition. The support of the people can be observed in public occasions, like wedding ceremonies, where Senjang is common, just like the one the author witnessed.

In addition to the original questions, finally, it is recognized that there are many issues behind the tradition. The presence of Senjang does not only fulfill the art in Musi Banyuasin, but is also used as a communal cultural identity in Musi Banyuasin.

With regard to this central question, the purpose of this study is to show whether Senjang can be used as a medium to understand and represent the cultural identity of Musi Banyuasin following the division, because the changes cannot be separated from political and cultural interests in Musi Banyuasin, given the different formation of cultural identity.

Here, we use a qualitative method with an ethnographic approach. During this study, the writer conducted interviews with informants, including Senjang artists, public figures, and bureaucrats. Observation–participation and interview were the main techniques used in this study.

2 RESULTS AND ANALYSIS

Muba community and culture have not attracted adequate attention of researchers worldwide. According to the author, Senjang tradition and Muba have been rarely mentioned in the literature, among them are the studies on the oral tradition of Muba conducted by Zainal Abidin Gaffar et al. (1989) titled *Struktur Sastra Lisan Musi* and by Ahmad Bastari Suan et al. (2008) titled *Sastra Tutur Sumatra Selatan: Peran dan Fungsinya dalam Masyarakat Ogan Komering Ilir dan Musi Banyuasin*.

The study conducted by Gaffar demonstrates the structure of the Musi oral literature. It focuses only on folk tales and their components. Gaffar et al. did not explain other kinds of oral tradition in Muba. It only documents some folk tales and analyzes the components that build the folk tales. Suan et al. (2008, pp. 97–101) explained some oral traditions in two districts, namely Ogan Komering Ilir district (OKI) and Muba district; however, the research has not explained traditions in Muba completely. Suan et al. described the oral traditions in Muba that still exist together with their functions and roles. One of them is the Senjang oral tradition. The article written by Oktaviany (2008, pp. 1–10) titled *Senjang: Tradisi Lisan yang Masih Eksis di Musi Banyuasin* discusses the topic specifically. Overall, the research about Muba community and culture so far, as shown by the above overview, has generally been beyond the social and cultural contexts that cover them, and has not been explored in depth.

Some existing cultural research focuses on the challenges faced by traditions and arts outside of Muba, which is related to the struggle of social and cultural groups and identity determination. The dissertation of Anoegrajekti (2006), for instance, discusses Banyuwangi Gandrung (a traditional dance from Banyuwangi). Anoegrajekti explained that Gandrung is being hegemonized by three powers—market, bureaucracy, and religion—so that the art represents a combination of all three powers with the domination of market. In such a context, identity is something that is not just created politically, but something that also moves forward. The dissertation also shows the powers that compete for identity, which are market gandrung, tradition gandrung, and religion gandrung.

Meanwhile, the unique role played by the local government in the formation of a region's identity is highlighted in the study conducted by Eri and Hasmi (in Ninuk 2008, p. 119), which concludes that if the elites in the region come from or are close to certain communities or ethnic groups, a kind of extra attention is given to the culture in their home areas. This attention can be in the form of allocating a large amount of budget.

To address the issues related to oral tradition, the author consults the results of the study conducted by Tuloli (1990). This study observes the relationship between the community and culture that encompasses it. According to Tuloli, all oral traditions that are maintained by their respective communities must have functions in them. A tradition having functions in the community will last longer, whereas the one with functions not in accordance with the wishes of the people will disappear.

States that function and appear are very much influenced by who the doers are, for whom the tradition is performed, and how much involvement do the participants or interpreters have in the performance. A traditional performance may have some functions as well, namely (1) to strengthen and attack political authorities, (2) to strengthen or be against tradition, (3) to satirize, (4) to be used as propaganda, (5) to fulfill daily needs, (6) to be proud, (7) to express beauty and love, (8) to complain, (9) to express issues that cannot be expressed openly, (10) to avoid difficult conditions, (11) to find oneself or one's identity, (12) to entertain, (13) to unite and separate people, and (14) to be used as religious guidance (Finnegan 1992, pp. 126–127).

The situation is the same with Senjang, which still exists and is guarded by the community because it still has functions. The community, especially the local government, sees Senjang not just as entertainment, but also as a means of propaganda, social criticism, collective solidarity, and cultural enhancement, which are all used to form the cultural identity of Muba.

The people of Muba make use of Senjang as a means of propaganda to campaign several things that are related to the programs conducted by the government of Muba district. The propaganda is very visible in the Senjang lyrics and verses that are performed in the Senjang competitions initiated by the government of Muba district. One example is shown in Figure 1.

The propaganda is seen clearly in the lyrics/verses of Senjang performed by Ema and Manto in the Sriwijaya Festival in Palembang in 2014. The delivery of the Senjang lyrics/verses in an occasion attended by people from all districts in South Sumatra proves that Muba district is considered different from other districts and cities in South Sumatra. In addition, in the text, Ema and Manto also mentioned the progress achieved by Muba and mentioned a superior program of their district, the Permata program, which was believed to lead the people to a better and more prosperous life.

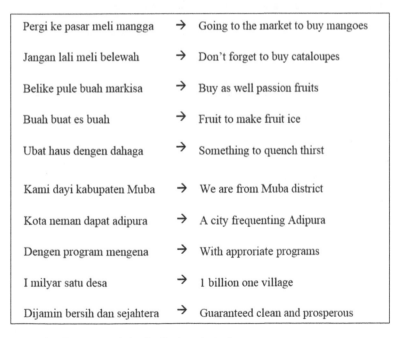

Pergi ke pasar meli mangga	→	Going to the market to buy mangoes
Jangan lali meli belewah	→	Don't forget to buy cataloupes
Belike pule buah markisa	→	Buy as well passion fruits
Buah buat es buah	→	Fruit to make fruit ice
Ubat haus dengen dahaga	→	Something to quench thirst
Kami dayi kabupaten Muba	→	We are from Muba district
Kota neman dapat adipura	→	A city frequenting Adipura
Dengen program mengena	→	With approriate programs
I milyar satu desa	→	1 billion one village
Dijamin bersih dan sejahtera	→	Guaranteed clean and prosperous

Figure 1. Example of propaganda in the Senjang lyrics/verses.

The One Billion One Village program is claimed by Muba district as one of its kind in South Sumatra. The following is an excerpt of Senjang that contains Permata, one of the programs of the pair of Pahri Azhari and Beni Hernandi, which was performed in the Randik Festival by the representatives of Sanga Desa sub-district (Figure 2).

The above-mentioned Senjang lyrics/verses contain the program conducted on by the government under the administration of Pahri and Beni (2012–2017), which is Permata Muba 2017. Permata Muda is the acronym of *Penguatan ekonomi kerakyatan, Religius, Mandiri, Adil dan Terdepan maju bersama* (the strengthening of public economy, religious, independent, fair, and being at the very front). In carrying out the Permata Muba, the government launched the One Billion One Village program in Musi Banyuasin. The program is claimed to be the first and foremost program and acts as a differentiator that separates Muba from other districts in South Sumatra. They call it the Pioneer of Village Development. Traditional art performance is quite effective in delivering messages, especially for the village people or the people in general.

In addition to being used to convey messages about development, Senjang is used as a medium to convey social criticism. This is interesting because Muba people still consider it taboo to criticize somebody directly, particularly when the person being criticized is a leader, a superior, or a relative, as well as taboo to criticize the current condition of the country.

As already explained above, Muba district has a vision called Permata 2017. The vision is then translated into missions, which are (1) the strengthening of the public economy on the basis of local resources and wisdom, which are independent, competitive, and religious; (2) developing centers of growth and creative service that are supported by information technology and communication; (3) enhancing the equity of sustainable development that is fair and environmentally friendly; (4) developing quality human resources and religious sociocultural environment; and (5) creating a trusted, clean, honest, professional, and democratic governance.

To explain the vision and missions through the Permata Muba program, the community recites the following Senjang lyrics/verses (Figure 3).

Kami rakyat selalu nukung	→	We people always support
Smangat Bupati memajuke Muba	→	The spirit of the district head to advance Muba
Permata Muba junjung besame	→	Permata Muba supported by all
Kitek harus Serasan Sekate	→	We have to be one negotiation one word
Kami sebagai generasi mude	→	We as young people
Siap untuk memajuke Muba	→	Ready to advance Muba

Figure 2. One Billion One Village program in Senjang lyrics/verses.

Dalam mengapai permata Muba	→	In reaching the Permata Muba
Parah murah dak begetah	→	Cheap rubber is sapless
harge sembako melambung tinggi	→	Prices of daily needs fly high
kirek ada Bapak Camat	→	If the head of district is here
Tolong telpon Pak Bupati	→	Please call Mr. Regent
Makmane carek solusi	→	How to find solution

Figure 3. Senjang conveyed in the 2013 Randik Festival by representatives of the Plakat Tinggi sub-district.

To the community, the Permata 2017 program can be achieved if the rubber price is high because, if the rubber price is low, their living is affected. Looking at such a condition, criticism can be done through Senjang. How will they achieve the Permata Muba if the rubber price is low and the rubber is totally sapless? Because the community needs a solution to the problem, the head of sub-district is asked to call the district head directly in order to find a solution to solve the problem of low rubber price and high prices of daily commodities. With Permata Muba 2017 as a way to create democratic governance, the head of a sub-district can call the first person in the Bumi Serasan Sekate so that he may find a solution in no time.

Beni Herniadi (Acting District Head of Muba) explained the author that it was easier for him to remember what the Senjang performers described because through that he finally understood what the Muba people wanted. A certain road that was damaged somewhere was repaired directly because it had never been reported by the related service. With the existence of Senjang, finally, the problem was being attended to be solved as soon as possible.

According to Beni, Senjang was very relevant for the current condition, especially for those who understood the words that were expressed, from humorous lyrics to lyrics laden with moral preachings. He added that Senjang was supposed to be used by people to express their aspiration so that they would not need to demonstrate, as criticism was sufficiently delivered through Senjang. For Uzer Effendi (the Chairperson of the Local Parliament for the period 2008–2014), the criticism expressed by Senjang performers was in accordance with the prevailing facts. If there was criticism, it must have been based on a real incident, especially on unfulfilled campaign promises.

The utilization of Senjang as a means to express criticism is common in Muba. Local officials understand well that there would be criticism expressed by Senjang performers if they are invited to perform. So far, there has been no report of any Senjang performer being arrested because they criticize the government. This is because the following format is followed by the performers: first, they ask for permission to start and apologize; second, they ask for permission to leave and apologize. Accompanied by music and dance during the performance, the criticism expressed by the Senjang performers was sensible. Senjang criticizes but does not hurt, and controls but does not dishonor the parties that are being criticized. Furthermore, it is agreed that is the performance of Senjang is usually relevant with the current condition of the community.

The Muba people wish for a secure, peaceful, and prosperous community. Therefore, Senjang is used to build collective solidarity of the people of Muba. The emergence of the collective solidarity is caused by the nature of Senjang that goes beyond differences in age, sex, and social status. Senjang is not only entertainment but also a means to keep the people's sense of solidarity and the bonds among people attending the performance.

Through Senjang, the people of Muba become close. This is proved when people from the city of Palembang responded to Senjang very enthusiastically. One reason was that the audience was mostly Muba people relocated to Palembang. Those people who left their hometown long ago felt entertained.

In addition, in some festivals, the chosen themes were those aimed at increasing Muba people's sense of solidarity, unity, and togetherness, such as in the 2014 Randik Festival with the theme of "Through the Randik Festival We Foster the Unity among Sub-Districts in Musi Banyuasin District". This is related to the motto in the symbol of Muba district, Serasan Sekate, which means always prioritizing on discussions to make one unanimous decision. The aim of the festival itself is to sustain and develop the local art and culture and enhance the creativity of art activism in Muba district.

With regard to the aim and the theme that have been planned in the festival, the position taken by Muba district in the effort to utilize traditional art becomes more obvious. To achieve that, the unity among people and among sub-districts in Muba district is also important because so far only Muba district had held Senjang competitions. Senjang is still performed in some areas that are in the border between Muba district and Musi Rawas district, which is Muara Rupit area, which borders Sanga Desa village. That is also the case in Banyuasin district that borders Muba district or in areas that still use the Musi language on a daily basis, such as in the Rantau Bayur area. However, Musi Rawas and Banyuasin cannot claim that Senjang comes from their areas because no Senjang festival had been held there.

The following is a Senjang text delivered by Ema with Irsan in a wedding reception (Figure 4).

Senjang performers who delivered the above Senjang lyrics/verses wanted to convey that the Senjang art is their culture. The word *kami* (we) in the text refers to Ughang Muba or the people of Muba. The word *kami* only refers to the speakers and does not include people whom they converse with. In the next line, it is emphasized by the words "cultural heritage, old culture". Our (kami) cultural heritage is still developing. This emphasis is also evident in the text that was delivered in the Senjang performance at a wedding reception in Sekayu, which can be read below.

On the basis of the text presented above, at the beginning of the Senjang performance, the Senjang performers asked the audience for permission to perform Senjang. Senjang is the original culture of Muba, and the use of the word original (*asli*) refers to something that is original, as it is not mixed. The word original can also mean something that is possessed from birth, not acquired. Therefore, because the art is originally from Muba, it belongs to the people of Muba, and it is presented in an occasion that is related to customs, like in wedding ceremonies. In wedding ceremonies, all relatives, friends, and guests bless to bring luck and throw away bad luck. The Senjang lyrics/verses show that this traditional art indeed belongs to Muba and is performed at occasions in relation to customs, such as wedding ceremonies and thanksgivings.

The Senjang tradition in Muba community is utilized to confirm the culture that belongs to them. This confirmation is also supported by the government, in this case, the Ministry of Education and Culture decreeing that Senjang is an intangible cultural heritage belonging to Muba. The decision was made in 2015. This decision proves that Senjang as the confirmation of culture is also supported by parties outside of the community. The decision was based on the fact that the tradition still exists and has maestros and supporters. In addition to being the mandate of the 1945 Constitution, Article 32, the decision has the aim to encourage governors, the heads of government in the provincial and district levels, stakeholders, and the general public to participate in sustaining culture, by taking care of, protecting, and managing the culture in the best possible way.

Para hadirin sekalian	→	Ladies and gentlemen
Cobo-cobo kami nak Senjang	→	We try to perform Senjang
Seni budaya asli Muba	→	Muba's original art
Pade resepsi pernikahan	→	In the wedding reception of
Kupik reza ngen kuyung Ade	→	Kupik Reza and Kuyung Ade
Sanak dulur merojong galek	→	Relatives and friends all having fun
Tamu undangan banyak datang	→	A lot of guests come
Ngunde tuah muang celake	→	Bringing luck, throwing away bad luck

Figure 4.   Senjang lyrics/verses for wedding reception.

Cobo-cobo kami baSenjang	→	We try to deliver Senjang
Senjang budaya kami Ughang Muba	→	Senjang is our culture people of Muba
Tinggalan Budaya, Budaya lame	→	Cultural heritage, old culture
Sampai mikak masih bakembang	→	Until now still developing

Figure 5.   Senjang lyrics/verses delivered by performers.

3 CONCLUSION

From the information provided above, it can be concluded that people from Muba use Senjang tradition as a tool for propaganda, social criticism, collective solidarity, and affirming culture. As the tool of propaganda, Senjang is used to campaign for something that is related to programs undertaken by the Government of Muba. Senjang is also used to convey a message of development, and the gap is also used as a medium to deliver the social criticism. Senjang is also used as a means of conveying criticism that has become a common practice in Muba. In addition, it is used to build the collective solidarity of people in Muba. The emergence of solidarity in the society in performing Senjang can neglect all differences in gender, age, and social level. Senjang is not only an entertainment but also a tool to maintain the sense of solidarity and the bonds of the community attending the show. Moreover, the attitude of consultation becomes the culture affirmation in this community. The use of Senjang tradition as a tool to validate the culture can be seen from the adoption of Senjang by the government in 2015 as an intangible cultural heritage belonging to Muba.

Thus, using a number of these types of tradition, we arrive at the formation of the cultural identity of Muba communities that is stated in Senjang lyrics/verses that have been well performed in the customary and traditional events.

REFERENCES

Anoegrajekti, N. (2006) *Gandrung Banyuwangi: Pertarungan Pasar, Tradisi, dan Agama Memperebutkan Representasi Identitas (Gandrung Banyuwangi: Market Competition, Tradition, and Religion Competing Identity Representation)*. [Dissertation]. Depok, FIB Universitas Indonesia.

Finnegan, R. (1977) *Oral Poetry*. Cambridge, Cambridge University Press.

Finnegan, R. (1992) *Oral Traditions and the Verbal Art. A Guide to Research Practices*. New York, Routledge.

Gaffar, Z.A. Subadiyono, Supriyadi, Yusuf, H. & Nurbaiti. (1989) *Struktur Sastra Lisan Musi (Oral Literary Structure of Musi)*. Jakarta, Departemen Pendidikan dan Kebudayaan.

Haris, Y. (2004) *Bumi Serasan Sekate dan Penduduknya (Bumi Serasan Sekate and its inhabitants)*. Pemerintah Kabupaten Musi Banyuasin, Sekayu.

Hastanto, S. (2010) Budaya Identitas Bangsa, Identitas Budaya Bangsa (Culture of National Identity, Cultural Identity of the Nation). In: Nurhan, K. (ed.) *Industri Budaya, Budaya Industri. Kongres Kebudayaan Indonesia 2008 (Cultural Industries, Industrial Culture. Indonesian Culture Congress 2008)*. Jakarta, Kementerian Kebudayaan dan Pariwisata Republik Indonesia.

Irawan, E. & Baharuddin, H. (2008) Jenggirat Tangi, Politik, dan Imajinasi tentang Perempuan (Jenggirat Tangi, Politics, and Imagination about Women). In: Kleden, N., Rohman, A.A. (eds.). *Etnografi Gandrung: Pertarungan Identitas (Gandrung Ethnography: Identity Battles)*, Depok, Desantara Press.

Isaacs, H.R. (1993) *Pemujaan terhadap Kelompok Etnis: Identitas Kelompok dan Perubahan Politik (Idols of the Tribe: Group Identity and Political Change)*. Jakarta, Yayasan Obor Indonesia.

Lord, A.B. (2000) *The Singer of Tales*. Cambridge, Massachusetts London, Harvard University Press.

Oktavianny, L. (2008) *Senjang: Tradisi Lisan yang Masih Eksis di Musi Banyuasin (Senjang: Oral Traditions that Still Exist in Musi Banyuasin)*. Palembang, Pusat Penelitian Balai Bahasa Palembang.

Ong, W.J. (1989) *Orality and Literacy: The Technologizing of the Word*. London and New York, Routledge.

Paeni, M. (2011) Melihat Kembali Nasionalisme Indonesia dalam Konteks Masyarakat Plural Melalui Perspektif Sejarah (Recalling Indonesian Nationalism in the Context of Plural Society through a Historical Perspective). In: Lan, T.J., & Manan, M.A. (eds.). *Nasionalisme Ketahanan Budaya di Indonesia: Sebuah Tantangan (Nationalism of Cultural Resilience in Indonesia: A Challenge)*. Jakarta, Yayasan Obor Indonesia.

Pudentia, M.P.S.S. (2008) *Metodologi Kajian Tradisi Lisan (Methodology of Oral Tradition Studies)*. Jakarta, Asosiasi Tradisi Lisan.

Sims, M.C. & Stephens, M. (2005) *Living Folklore: An Introduction to Study of People and Their Traditions*. Utah, Utah State University Press.

Suan, A.B., et al. (2008) *Sastra Tutur Sumatra Selatan: Peran dan Fungsinya dalam Masyarakat (Oral Literature of South Sumatra: Role and Functions in Society)*. Palembang, Dinas Pendidikan Provinsi Sumsel.

Teeuw, A. (1994) *Indonesia antara Kelisanan dan Keberaksaraan (Indonesia between orality and literacy)*. Jakarta, Pustaka Jaya.

Tuloli, N. (1990) Tanggomo: Salah Satu Ragam Sastra Lisan Gorontalo (Tanggomo: One of the Varieties of Gorontalo Oral Literature). [Dissertation]. Depok, FIB-Universitas Indonesia.

Cultural Dynamics in a Globalized World – Budianta et al. (Eds)
© *2018 Taylor & Francis Group, London, ISBN 978-1-138-62664-5*

Colophon in the *Hikayat Pandawa* manuscript

M. Buduroh & T. Pudjiastuti
Department of Literature, Faculty of Humanities, Universitas Indonesia, Depok, Indonesia

ABSTRACT: Since the beginning of the tradition of manuscript copying in Malay in the 14th century, the copyist or authors have often been unknown. In general, the manuscripts only contain information regarding the kings or their superior who ordered the copying process. And the copyist did not write their identity on the copied manuscripts. However, this situation has changed since the 19th century, as they included their identity on the copied manuscripts. This also happened in the tradition of manuscript writing in Betawi, as is evident from the identity of the authors found in the last part of these manuscripts. Nevertheless, this did not occur in Hikayat Pandawa manuscripts, which form part of the collection of the Republic of Indonesia National Library. Through investigating manuscript colophon, in this study, we will describe the identity of the authors and its association with the tradition of writing Malay manuscripts in Betawi.

1 INTRODUCTION

The colophon is an important part of the manuscript identity in the tradition of manuscript copying in Malay, through which a researcher can identify the author, the time of writing, and the method of writing of a manuscript. This information is important in the study of Malay manuscripts to know their history of copying, which facilitates the analysis. However, there has been less attention paid to the colophon on the study of Malay manuscripts. In other words, when there is only one indicator of the manuscript identity, it is considered without further consultation on any other source that may also exist in the colophon of the manuscript. Nevertheless, sometimes the colophon itself contains very limited information or sometimes no information regarding the colophon.

The exact time of emergence of colophons in the tradition of copying Malay manuscripts is not known. However, its form reveals that the tradition of writing colophons is presumably influenced by the tradition of copying manuscripts from Persia or the Middle East. The colophon on Malay manuscripts can be identified through the form of writing. It is usually located at the end of the manuscript, resembling an inverted triangle. In addition, there are words, such as "the end" or "this is the end", which mark the end of the story, followed by time and the name of the author.

Chambert-Loir (2014, 263) in his article "Malay Colophons" described colophons as the "last paragraph" written deliberately by the copyist. He further explained that the information found in the colophon is related to the copied work (i.e., the text), the author, the place and date of writing, the circumstance and the purpose of writing and copying itself (the manuscript), the date of copying (Islamic and CE; the name of the day; hour), the name of the copyist; the location of copying, the details of the manuscript (paper; format), the owner of the manuscript, and the source text (date, owner). Furthermore, Chambert-Loir revealed that, from the viewpoint of literature, colophons show the presence of texts in a particular place. In other words, colophons contain information about the place where the text is known and appreciated, and from the viewpoint of codicology, colophons illustrate the role of copyists in the Malay world (2014, 269). In order to address these problems, we will describe the colophon contained in a manuscript entitled Hikayat Pandawa coded Ml. 15 in the manuscript collection of the National Library of the Republic of Indonesia.

The Hikayat Pandawa manuscript contains part of the Mahabharata. The story begins with information about a king named Maharaja Bismaka ruling a country named Mandirapura. Maharaja Bismaka had three daughters, namely Dewi Amba, Dewi Ambawani, and Dewi Ambalika, for whom he conducted a contest for the princes who would like to marry them. Dewabrata, prince of Astinapura, won the contest. The story ends with information about Gatotkaca who flew to a country named Pringgodani, whereas the Pandawa gathered in heaven and had a good time there.

The manuscript was written in Malay language using Arabic–Malay letters with black and red inks. Scratches of black ink look thin with red rubrication. The manuscript contains 475 pages. European paper sized 31 × 20 cm was used for the writing. Two types of paper were used, as is evident from the two different paper stamps in the manuscript, namely a woman carrying a spear paper stamp (No. 50 with the year of production 1848) and a lion carrying a sword paper stamp (No. 102 with the year of production 1846) (Voorn 1960). The name of the author and the time of writing are not found in the colophon.

The colophon on the manuscript is a poem consisting of 15 stanzas. The writing of colophon in a poetry form is indeed a reminiscent of the shape of colophons in the manuscripts found in Pecenongan area. In the 19th century, this place was specially known for manuscript copying initiated by the Fadli family, in which there were four copyists and writers, namely Sapian bin Usman, Sapirin bin Usman, Ahmad Beramka, and Muhamad Bakir. Unfortunately, the colophon does not contain information on any of the four names. A similar finding was observed in the search for Pecenongan's collection of manuscripts that have been compiled in *Katalog Naskah Pecenongan Koleksi Perpustakaan Nasional, Sastra Betawi Akhir Abad ke-19* (National Library Collection of Pecenongan Manuscripts Catalogue, Betawi Literature End of 19th Century), and the manuscript of Hikayat Pandawa is not included in this collection. As the content of the manuscript is on Wayang stories, there are nine manuscripts that are contained in or sourced from the Mahabharata. The manuscripts are Hikayat Angkawijaya, Hikayat Asal Mulanya Wayang, Hikayat Gelaran Pandu Turunan Pandawa, Sair Perang Pandawa, Hikayat Agung Sakti, Hikayat Maharaja Garbak Jagat, Lakon Jaka Sukara, Hikayat Wayang Arjuna, and Hikayat Purusara.

The colophon also shows a problem of the manuscript borrowing. It is mentioned that the text should not be borrowed for a long time, and is limited to only 5 days, before which the borrower/reader should return it. It is quoted in the following lines:

> *bunga tunjung teram-teram*
> *bunga kesturi bunga delima*
> *jangan dipinjam terlalu lama*
> *a-n-t-h hari kelima*

> *ambil kaca di dalam pakan*
> *anak belibis segera kandangkan*
> *siapa yang membaca sembah pasukan*
> *jikalau habis segera pulangkan*

Lending or renting manuscripts was unusual in the tradition of copying Malay manuscripts. This is especially true in the tradition of copying Malay manuscripts in Betawi. Some texts, especially the works of Muhamad Bakir, reveal that the scripts he copied were also lent with the amount of time, with the amount of rent specified in the colophons. Unlike the Hikayat Pandawa text, the information about lending or renting the text does not mention the amount of rent to be paid by the borrower.

In addition to providing information related to manuscript borrowing, the manuscript contains a description of the origin of the copied story. The story in this Hikayat Pandawa manuscript originally came from Java. The story was "transferred" from Javanese language into Malay, which could be seen on the first page of the following manuscript: "Once upon a time there was a story of our ancestors which was told in Javanese language before transferred into Malay" (p. 1).

Information about the origin of the story is also confirmed in the colophon by mentioning the Javanese language elements in the story. Besides, the copyist admitted that the story was different from its rules. Here is an excerpt.

Telah dikarang ceritera Pandawa
bahasanya campur Melayu Jawa
Jikalau tuan membaca jangan tertawa
sebab aturannya janggal banyak kecewa.

Pandavas or Pandawa are basically the figures featured in the Mahabharata originated in India. The story started spreading in the 5th century CE in Asia and is still well known globally. Furthermore, the epic story can be found in various forms, such as movies, serials, and books. This is possible because the story changed during its spread because of the difference of cultural settings between the origin of the story in the Mahabharata found in Java and the form of kakawin, such as Kakawin Bharata Yudha. The story is about the battle between Pandavas and Kauravas or Kurawa and written in Old Javanese by Mpu Panuluh and Mpu Sedah in 1157 CE. It is written in the form of kakawin, which is different from the parwa form.

Kakawin is considered as the source of the copy of Hikayat Pandawa Jaya, the manuscript coded Ml. 91 from the collection of the National Library of Indonesia. The copying of manuscripts from Javanese language to Malay language might have happened around the second half of the 14th century. During the process of copying the manuscript from Javanese language to Malay language, a shape transformation occurred and new characters that were not featured in the Mahabharata text originating from Java emerged. Cultural transformation can also occur in the process of copying. For the Javanese, Mahabharata was sometimes considered to be sacred, whereas in the tradition of Malay literature, it was regarded as profane. In addition, the relationship between the characters in the story is different because the characters were often considered to have links with the real life in the Javanese tradition, which is achieved by attributing the genealogy of the kings into the life story of the Mahabharata, whereas in the Malay tradition, it is still regarded as not related to the kingdom. However, in terms of linguistics, the characters in Kakawin Bharata Yudha can be still found in Hikayat Pandawa Jawa (Tjiptaningrum 2004).

In the tradition of manuscript copying in Java, Kakawin Bharata Yudha was copied by a master craftsman, namely *empu*, referring to a person with expertise as a writer in the kingdom. However, the situation is different in Hikayat Pandawa, that is, the story was copied by a person who was not only a writer, as indicated in the following quote.

Cerita ini hamba terhentikan
sedang banyak kerja yang dipikirkan
bi dan kurang tuan maafkan
...ti yang kasih jangan ubahkan.

Another important issue in the colophon of this manuscript is writing thin scratches, which is also a form of writing. The thin scratches indicate technical problems in writing. Other problems were the limited availability of ink as a material for writing and the problems with writing instruments, such as a broken pen. This is stated by the author in the final stanza of the colophon as follows:

Bakar setanggi di dalam surat
pecahkan piring dalam pelita
apalagi hendak dikata
dakwatnya kering kalamnya patah.

In another part of the colophon, the author also suggested that the story should be read only under the light so that the reader can clearly see the script, as described in the following quote:

Pandawa susupan telah terlarang
membatu lirani sembarang-barang

dakwatnya putih hitamnya kurang
tuan membaca itu tempat yang terang.

Writing scratches are important for the form of writing in the manuscript because of their unique characteristic of being owned by a single copyist or a scriptorium. For example, in a script that was copied on the initiative of the Algemeene Secretarie in Batavia in the 19th century, the form of writing is unique based on the standards of the institution. Although they were copied by different people, the manuscripts had some similarities in terms of writing. Their time of copying and writers were recorded, which include Muhammad Cing Saidullah (1819–1828), Mohammed Sulaiman (1819–1830), Muhammad Hasan bin Haji Abdul Aziz (1834), and Abdul Hakim (1819–1826) (Rukmi 1997).

Given the form of writing scratches and the activities of manuscript copying in these institutions, it seems that the Hikayat Pandawa manuscript does not belong to the category of manuscripts copied by those copyists. On the basis of the time of copying, they generally worked as copyists in the early 19th century. Meanwhile, the paper stamp on the Hikayat Pandawa manuscript confirms the time after the aforementioned period. The investigation on the paper stamp used showed that the paper was produced between 1846 and 1848. Estimating that the paper arrived and was used in Batavia approximately 5–10 years later, the Hikayat Pandawa manuscript was possibly copied in the 1850s or the 1860s.

It can be predicted from the characteristic feature of the manuscript colophon that it was copied in another place, Batavia or Betawi. Especially, Pecenongan area was one of the manuscript-copying places, where several manuscripts were copied. This activity in this place was started by Sapirin bin Usman or Guru cit. He actively copied manuscripts from 1858 to 1885. He taught his expertise in copying to his son, Ahmad Beramka, and his nephew, Muhamad Bakir, who actively involved in copying and even produced works of new compositions, including the story that originated in the Mahabharata. One characteristic of the writing of Muhamad Bakir is that he included his name in the manuscripts he wrote or copied. Therefore, it is most unlikely that the Hikayat Pandawa manuscript, which does not include the name of the author, is the work of Muhamad Bakir.

Special attention is paid to the writing scratches in the Hikayat Pandawa manuscript, the writing form of which is similar to that of the woks produced by Muhamad Bakir. However, the size of the writings is smaller and more tightly written. One can see the striking differences between the forms of writing on the Hikayat Sultan Taburat (Ml.183 D) manuscript copied by Sapirin bin Usman and Muhamad Bakir by comparison. The form of writing of Hikayat Pandawa is more likely to be found on the manuscript written by Sapirin bin Usman or Guru cit.

3 CONCLUSION

Colophon in a manuscript is actually the information about its identity, which includes the name of the author, date of writing, and the method of writing. However, sometimes we cannot find all the information explicitly in the colophon. Therefore, it is required to have a supplementary section in the manuscript to explore the necessary information about the identity of the manuscript. By analyzing the form and content of the colophon, which is then combined with information about the form of writing and paper stamp used in the Hikayat Pandawa manuscript, the writer and the place of writing can be identified. It is probable that this manuscript was written by Sapirin bin Usman, the author of the early generation in Pecenongan area. Thus, the text is also one of the manuscript collections of Pecenongan, although it is not included in the catalog of Pecenongan manuscripts.

REFERENCES

Chambert-Loir, H. (2009) *Hikayat Nakhoda Asik (Sapirin Bin Usman) dan Hikayat Merpati Mas (M. Bakir) (Tale of Asik the Skipper (Sapirin Bin Usman) and The Tale of the Golden Pigeon (M. Bakir)*. Jakarta, Masup.

Chambert-Loir, H. (2014) *Iskandar Zulkarnain, Dewa Mendu, Muhammad Bakir, dan Kawan-kawan: Lima Belas Karangan tentang Sastra Indonesia Lama (Iskandar Zulkarnain, Dewa Mendu, Muhammad Bakir, and Comrades: Fifteen Essays on Old Indonesian Literature)*. Jakarta, Kepustakaan Populer Gramedia.

Fitzgerald, J.L. (1983) The Great Epic of India as Religious Rhetoric: A Fresh Look at Mahabharata. *Journal of the American Academy of Religion,* 51 (4), 611–630, http://www.jstor.org/stable/1462584. Accessed: 24th August 2010 at 01:05.

Hassan, T.F. (2004) Sastra Melayu Bersumberkan Kakawin Jawa Kuno (Malay Literature Based on Ancient Javanese Kakawin). In: Sedyawati, E., Sugono, D., Zaidan, A.R., Djamaris, E. & Ikram, A. (eds.). *Sastra Melayu Lintas Daerah (Cross-country Melayu Literature).* Jakarta, Pusat Bahasa, Departemen Pendidikan Nasional.

Iskandar, T. (1996) *Kesusastraan Klasik Melayu Sepanjang Abad (Classical Malay Literature Throughout the Ages).* Jakarta, Penerbit Libra.

Karim, N., Purwanto, D., Isyanti, D. & Nurita, Y. (2015) *Katalog Naskah Pecenongan koleksi Perpustakaan Nasional: Sastra Betawi Akhir Abad ke-19 (National Library Collection of Manuscript Catalogue: Betawi Literature End of 19th Century).* Jakarta, Perpustakaan Nasional RI.

Rukmi, M.I. (1997) *Penyalinan Naskah Melayu di Jakarta pada Abad XIX: Naskah Algemeene Secretarie Kajian dari Segi Kodikologi (Copying Malay Manuscripts in Jakarta in the 19th Century: Algemeene Secretarie Manuscript of Codicology).* Depok, Fakultas Sastra, Universitas Indonesia.

Cultural Dynamics in a Globalized World – Budianta et al. (Eds)
© *2018 Taylor & Francis Group, London, ISBN 978-1-138-62664-5*

Bismaprawa: An Old Javanese text from the Merapi-Merbabu tradition derived from *Adiparwa*

A. Kriswanto & D. Puspitorini
Department of Philology, Faculty of Humanities, Universitas Indonesia, Depok, Indonesia

ABSTRACT: *Bismaprawa* is a *codex unicus* originating from Merapi-Merbabu manuscripts written around 1669 AD. *Bismaprawa* is now kept at the National Library under the registration number 6 L 145 which is grouped into Merapi-Merbabu manuscripts. *Bismaprawa* is a manuscript of the Merapi-Merbabu tradition, and its story comes from *Adiparwa*. It is recognizable by the fragments of the story in it. However, it does not mean *Bismaprawa* always adheres to the framework of the source text because both are still different. The difference is caused by the presence of new elements that could be borrowed from other texts or elements derived from *Bismaprawa* itself, which is not contained in *Adiparwa*. The *Bismaprawa* and *Adiparwa* intertextual relationship can be traced from two aspects, namely the relationship related to character and the relationship associated with the event. Based on both relationships, it can be concluded that the new figures and new events that appear in *Bismaprawa* are not sourced from *Adiparwa*. *Bismaprawa* text creation can be seen as an attempt for the reader and writer to interpret *Adiparwa* as the source text. Interpretive effort by creating new elements is considered to represent a living tradition of text inheritance according to the situation of supporters.

1 INTRODUCTION

The *Bismaprawa* manuscript is currently maintained at the National Library of the Republic of Indonesia and is included as part of the library's collection of Merapi-Merbabu manuscripts. The Merapi-Merbabu manuscript collection is a set of manuscripts discovered on the western slopes of Mount Merbabu which was integrated into the collection of *Bataviaasch Genootschap* around 1852 (Bleeker, 1852, p.6). Mount Merbabu and its surrounding areas were the centre for the studies of Hindu-Buddhist literature and religion. However, this did not mean that the community shut itself from external influences. This is proven by the existence of Islamic texts which were created in these regions (van der Molen and Wiryamartana, 2001, p. 55). Based on date indicators found in several texts with colophons, it is evident that the texts produced by the Merbabu scriptorium were written or copied between the second half of the seventeenth century and the first quarter of the eighteenth century (Wiryamartana, 1993, p. 506). However, these texts do not provide clear indication about the time when the Merbabu scriptorium first came into existence; that is, whether the literary community was established not long after the ruin of Majapahit or long before the foundation of Mataram kingdom in Central Java (van der Molen and Wiryamartana, 2001, p. 55).

The Merapi-Merbabu manuscript collection consists of texts which were written in various Javanese traditional literary forms, such as *kakawin, parwa,* and *kidung*. *Bismaprawa* was written in old Javanese language in the form of *parwa*, similar to *Korawasrama, Tantu Panggelaran, Nawaruci,* and others (Poerbatjaraka, 1952, pp. 57–71). The story in *Bismaprawa* was inspired by the story in *Adiparwa*, an Old Javanese prose work originating from an Indian text (Ras, 2014, p. 143) which was written in the Old Mataram Kingdom around 991–1009.

Bismaprawa literally means "the story of Bisma," but *Bismaprawa* does not merely tell a story of Bisma because it is more related to the stories in *Mahabharata*, particularly those in

Adiparwa. The story recounted in *Bismaprawa* was inspired by the story recounted in *Adiparwa*. However, *Bismaprawa* does not seem to be adhering to the narrative framework of the source text because both texts are different in many respects. *Bismaprawa* is different from *Adiparwa* because it displays some distinctive features, such as borrowed elements from other texts or elements unique to *Bismaprawa* itself, which do not exist in its source text. Thus, *Bismaprawa* cannot be properly read without understanding the background of its source text, *Adiparwa*. However, readers must bear in mind that the most important thing is not the borrowed elements themselves, but the way they are integrated into the new narrative structure and whether their meanings remain the same or have undergone certain changes. This is a very interesting subject of investigation (Zaimar, 2014, p. 93).

The production of *Bismaprawa* as part of the Merapi-Merbabu tradition demonstrates that the Javanese society has also inherited the story of *Adiparwa*, albeit in a different literary form and plotline. This finding is significant because so far we only know *Adiparwa* from Balinese literary tradition (Juynboll, 1906, pp. III-VI). It is, therefore, obvious that the discovery of *Bismaprawa* must be considered as an important development in the studies of Javanese literature because it is through *Bismaprawa* that the traces of *Adiparwa* can be finally identified years after its alleged disappearance from the Javanese literary tradition.

1.1   *Relationship between Bismaprawa and Adiparwa*

The story of *Bismaprawa* was inspired by the story of *Adiparwa*. However, this does not mean that *Bismaprawa* always adheres to the narrative framework of *Adiparwa* because *Bismaprawa* has its own narrative framework. According to Julia Kristeva, this may happen because of productive reading, which means that *Bismaprawa*'s writer had his own reading materials as the sources for rewriting the story. Therefore, the relationship between *Bismaprawa* (as the resulting text) and *Adiparwa* (as the source text) may be considered as intertextual in nature. The intertextuality between *Bismaprawa* and *Adiparwa* as observed in this study is not constructed on the basis of similarities, but on the shifts or changes occurring during the production of *Bismaprawa* and the factors which cause such shifts or changes. This is in line with Julia Kristeva's argument that the intertextual relationship is built, among others, by a dual-face reality which consists of the process of reading and the process of writing. The shifts and changes which are indicated by the existing differences between *Bismaprawa* and *Adiparwa* aptly demonstrate that both reading and writing processes actually took place during the production of *Bismaprawa* as a form of intertextual relationship between the two texts.

The intertextuality of *Bismaprawa* and *Adiparwa* is observable through certain objective criteria, such as characters and events. The criterion of character may take two forms: (1) the introduction of new characters or places in the resulting text (*Bismaprawa*) and (2) any additions of or changes in the names of characters already existing in the source text (*Adiparwa*). Changes in the names of characters which only involve spelling or pronunciation modifications are not included in our analysis. An example of this is the modification from Durgandini (*Adiparwa*) to Drigandini (*Bismaprawa*). The criterion of event may take two forms: (1) events which affect the course of the narrative in certain fragments of the story and (2) new events which were added to already existing events in certain fragments of the story. If both criteria are present simultaneously, they should be regarded as a single criterion of an event because basically both criteria are inseparable in any narrative works.

The character-based relationship, the intertextual relationship between *Bismaprawa* and *Adiparwa,* generated new characters who originated from other texts or created by the writer of *Bismaprawa*. They are as follows:

1.2   *Citrasena*

In *Adiparwa*, the main character Bisma had siblings named Citranggada and Citrawirya. In *Bismaprawa*, Bisma's siblings were Citranggada and Citrasena. In other words, the character Citrawirya in *Adiparwa* was replaced by Citrasena in *Bismaprawa*. The name Citrasena is, however, can be found in other older texts, such as *Korawasrama*. *Korawasrama* is estimated to be written in around 1635 AD (Swellengrebel, 1936, p. 47). Below is a quotation from *Korawasrama* which mentions Citrasena:

> *gĕmuh kakayang ring pupu kiwa, palinggihanira sang Citranggada, gĕmuh kakayang pupu tengen, palinggihanira sang Citrasena*
> 'round and bent on the left thigh, [that was] the seat of Citranggada; round and bent on the right thigh, [that was] the seat of Citrasena.

The names Citrasena and Citranggada as mentioned in *Korawasrama* and *Nawaruci* are apparently not intended to refer to Bisma's younger siblings, but to certain gods or other heavenly beings because *Bismaprawa* also implicitly mentions that Citrasena actually bore the same name as that of a heavenly being, and this fact prompted the being to kill Bisma's younger sibling, as shown in the following passage.

> *Dan ucapĕn widadara sang Citrasena, ya ta ingilangakĕn antĕn sang Prabata ri gatining amamaḍa dewata.*
> 'And it is told that the gallant Citrasena, Prabata's younger sibling, was slain because of his god-like appearance.'

The word Citrasena as the name of a heavenly being may be traced back to old sources from Hindu mythology (Williams, 2003:101). Citrasena was a musician in heaven who played many roles in both Purana and epic literature. The name Citrasena in *Korawasrama* came from Hindu mythology known long before the production of *Korawasrama* itself. The appearance of the name Citrasena in the 17th century Old Javanese narrative literary work might have been due to the influence of *tutur*, another type of the Javanese literary text.

Thus, the first mention of Citrasena as the name of a heavenly being has been found, but the first mention of Citrasena as Pandu's son or Bisma's younger sibling in *Bismaprawa* did not take place until its appearance in the *Cantakaparwa* text as indicated in the quotation below.

> *Tan warnan lawasnira sang prabu makuren, ya ta masiwi kakalih sami jalu, ingaran sang Citranggada Citrasena.*
> It is not told how long the King had been married, but he had two sons named Citranggada and Citrasena.

Cantakaparwa is a narrative text whose manuscripts can be found in the Balinese tradition.

In terms of characters, *Bismaprawa* is related to *Cantakaparwa*, although it is unknown which text was produced first. Based on this intertextual relationship, it can be said that the name Citrasena did not originate from *Bismaprawa*, but from the same narrative tradition which also gave rise to *Cantakaparwa*. The change of the character's name from Citrawirya to Citrasena demonstrates that the intertextual relationship between *Bismaprawa* and *Adiparwa* is marked by the presence of new characters from other related texts.

Preta. In *Bismaprawa*, Preta is the name used by Kunti before she became a priestess following the death of her husband, Maharaja Pandu. This is mentioned in the following passage:

> *Sang Prĕta sira arĕp alabuwa sira api, ḍatang Bagawan Bisma Bagawan Dyuwasa tan pasung sira tumuta, akon sira wikĕn angganya, sangaskrĕtanira panḍita Baṭari Kunṭi.*
> Preta wished to jump into the fire, but Bagawan Bisma and Bagawan Dyuwasa prohibited her; they told her body to become a wiken (priestess), and her ordained name being the priestess Batari Kunti.

The name Preta and the sequence of events leading to its transformation into Kunti cannot be found in *Adiparwa*. In *Adiparwa*, Kunti is the name of a daughter of King Kuntaboja, a king in Kuntawisaya.

> *Ana ta anak sang Kuntaboja stri, inaranan sang Kunti, kanya sĕḍĕng ahayu.*
> 'It was told of a daughter of Kuntiboja named Kunti, a beautiful girl.'

According to a source from the Indian *Mahabarata*, Preta or Prita is another name for Kunti (Dowson, 1888, p. 42). The name Patah in *Cantakaparwa* or Preta in *Bismaprawa* actually came from India and was not created by the writers of both Javanese texts. The name Preta does not exist in other manuscripts from the Merapi-Merbabu tradition. The name Kunti appears as a replacement for the name Preta in *Bismaprawa* as the result of the death of Preta's husband, Pandu. This event does not exist both in *Adiparwa* as the source text and in *Cantakaparwa* as another text from the same literary tradition.

The fact that both *Cantakaparwa* and *Bismaprawa* give a similar treatment to *Adiparwa*—that is, that both texts feature the names Preta and Citrasena, which had previously appeared in older texts, proving that there is an obvious intertextual relationship between *Cantakaparwa* and *Bismaprawa*, even though both texts originated from different literary traditions.

1.3 *Nama Widura*

In Bismaprawa, it is told that Pandu had a step brother named Nama Widura. In Adiparwa, this Pandu's brother was simply called Widura, as shown below.

> *sang Widura makebu anakning mantri patih.*
> 'Widura, whose mother was a maid in the palace'

A character named Widura also appears in other texts, such as Korawasrama, as shown below.

> *sang Widura pwa makapurohita sang Pandawa.*
> 'Widura became Pandawa's teacher,

However, *Korawasrama* once refers to Widura by using the name *sang anama Widura*, even though the phrase may have been intended to mean 'the one with the name Widura':

> *antĕnira, sang anama Widura pinakatuwa-tuwa de sang Pandawa.*
> 'the younger brother, whose name was Widura, was made a leader by Pandawa.'

Even though the writer of *Korawasrama* did not intend to give additional words to Widura's name, the writer of *Bismaprawa* may have copied those words and intended them to be part of Widura's proper name. If this was the case, the transformation of the verb *anama* ('to name') into the noun *Nama* ('name') seems plausible. Such pattern of word class shifts was common in the transformation of other Merapi-Merbabu manuscripts. For instance, in the transformation of *Arjunawiwaha* into *Wiwaha Kawi Jarwa*, several patterns of shifts were generated out of the writers' creativity, such as the introduction of new characters and places whose names were derived from the names of other entities which are not characters or places, the provision of names for previously nameless entities, and adaptations in the form of conversations (Wiryamartana, 1990, pp. 229–232).

Based on the similarity of term usage found in *Bismaprawa* and *Korawasrama* and the similarity of word class shift patterns between *Bismaprawa* and other texts from the Merapi-Merbabu tradition, it can be concluded that the name Nama Widura in *Bismaprawa* comes from the phrase 'anama Widura' in *Korawasrama*. This is supported by the fact that several *Korawasrama* manuscripts were also discovered amongst the manuscript collection produced the Merapi-Merbabu scriptorium, the same literary society which gave rise to *Bismaprawa*.

2 EVENT-BASED INTERTEXTUALITY

2.1 *Events leading to the transformation of Bisma's name*

Events which led to the transformation of Bisma's name are related to the title of the text itself, even though Bisma only takes a prominent role in one fragment of the text. Interestingly, this "Bisma fragment" in *Bismaprawa* recounts a different story from the story recounted in *Adiparwa*.

In *Bismaprawa*, it is told that when Prabata won a competition and brought three princesses to be married to his younger siblings, he promised that if any of his younger siblings did not want to marry one of the princesses, he would marry her instead. Both of Prabata's two younger siblings chose Amba's two sisters to be their respective wives, and they did not wish to have any more marriages. Amba demanded Prabata to honor his promise and marry her, but he refused although his teacher, Bagawan Parasu, forced him to do so. Amba then always followed wherever Prabata went. One day Prabata tried to frighten her by stretching out a bow and an arrow in the hope that she would no longer follow him. Unfortunately, the arrow was shot by accident and struck Amba's heart, causing her to die. Before dying, Amba, at Prabata's own request, put a curse on Prabata that he would never die until the appearance of a princess who bore a bow. After her death, Prabata was later given the name Bisma.

The change of name from Prabata into Bisma in *Bismaprawa* is mainly driven by romantic factors because, in this text, Prabata had never taken a vow not to marry again (as is the case in *Adiparwa*), because he was able to make a promise to marry any princess whom his brother did not want to marry. However, it turned out that Prabata broke his promise and refused to marry Amba who was not chosen by his brothers.

It is also mentioned in *Bismaprawa* that Bisma was a hermit. After Amba's death, Prabata immediately went to Wirakanda to perform meditation. This episode does not exist in *Adiparwa*. In another part of the text, Wirakanda is described as a *mandala*, which is indicated in the phrase '*mandalaning Bisma ri Wirakanda*' (p. 175). *Mandala* was a village for religious people, which was commonly established in the middle of a forest (Sedyawati (ed.), 2001, p 102). In Wirakanda, Bisma meditated for a long time. He decided to lead the life of a recluse in Wirakanda because he was filled with deep remorse for accidentally killing Amba and for requesting Amba to put a curse on him. In this regard, *Bismaprawa* seems to put more emphasis on Bisma's status as a hermit or priest. It is also implied in another part of the text that Bisma had actually received the title of a priest (*bagawan*):

> *Sang Prěta sira arěp alabuwa sira api, datang Bagawan Bisma, Bagawan Dyuwasa tan pasung sira tumuta.*
> 'Preta wished to go into the fire, Bagawan Bisma and Bagawan Dyuwasa came and forbade her to die.'

Bisma's depiction as a hermit does not exist in *Adiparwa* or in any other texts which feature the character of Bisma. This shift of Bisma's characterization to a hermit signifies the image of the Merapi-Mcrbabu society as the abode of hermits, the place where *Bismaprawa* manuscripts were written. Mount Merbabu was in fact a centre of Hindu-Buddhist literature and religion. According to an Old Sundanese source, the Merbabu or Damalung area was the place for a religious community which was visited by Bujangga Manik for the purpose of studying religious teachings (Noorduyn, 1982, pp. 416–418). The status of a hermit was attached to Bisma as an attempt to represent the image the hermits or religious people who lived in the Merapi-Merbabu area, along with their respective backgrounds. Therefore, the use of the name Bisma in *Bismaprawa* is neither a coincidence nor a copying error, but serves as a kind of symbolic representation of a society.

2.2 Events leading to the end of Parasara's meditation

At the beginning of *Bismaprawa*, there is a fragment about Batara Pramesti and Batari Uma who transformed themselves into white sparrows to distract Parasara in his meditation. This fragment about Batara Pramesti and Batari Uma's transformation, which led to the meeting of Bagawan Parasara and Sayojanagandi, does not exist in *Adiparwa*. *Adiparwa* only narrates that Bagawan Parasara met Sayojanagandi because the former requested Sayojanagandi to take him across the river, as shown in the following passage.

> *Ana ta rěsi Bagawan Parasara ngaranira, anak Bagawan Sakri, putu de Bagawan Wasista. Sira ta mangaděg i pinggir ing Yamunatoya, mamalaku aliwatakna ring lwah.*
> 'There was a hermit named Bagawan Parasara, the son of Bagawan Sakri and the grandson of Bagawan Wasista. He was standing by the river Yamuna, asked to be taken across the river.'

In *Bismaprawa*, the meeting of Parasara and Sayojanagandi was preceded by an event as recounted in this following passage.

Dyan ucapĕn Baṭara Pramesṭi mangucap lawan Baṭari Uma, tumingal ring Bagawan Parasara. Ayun sira bancanaha, dadi ta sira prit putih ingaran sang Kawruhta, wadon sang Priyawati. Asusuh sira ring jaṭa, mangantiga manak alunga ameta mangsa, sang Priyawati mangiringi sang Parasara, lunga sang Priyawati lawan sang Kawruhta. Mangkana matĕlasan sang Parasara mara pya ri desa sunya angupayaha istri. Katingalan sira sang Drigandini, katrĕsna retanira tumiba ri siti, ingiringakĕn maring tĕlĕng muwah maring ya pĕtung.

And it was told that Batara Pramesti spoke to Batari Uma, (upon) seeing Bagawan Parasara. They wanted to tempt him, (so) they transformed themselves into a white male sparrow named Kawruhta and a female sparrow named Priyawati. They set a nest in jata, laid eggs that hatched, and then went for preys; Priyawati accompanied Parasara, and then Priyawati and Kawruhta went away. Thus Parasara ended his meditation; off (he) went to a quiet village to find a wife. As he saw Drigadini, he was aroused and his sperm fell on the ground, flowed to the base, and onto a bamboo.

This fragment is a new element which does not exist in either *Adiparwa* or other texts contemporary with *Bismaprawa*. The event of Parasara's arrival on the bank of the river Yamuna by the assistance of a god in the disguise of a bird is necessary for establishing a connection amongst other previously existing narrative elements, so that the logical relationship between new and old elements, as well as the natural orders of events in the text, can be maintained. Thus, we can conclude that events leading to the end of Parasara's meditation are actually new elements introduced by the *Bismaprawa*'s writer and do not exist in the source text.

3   CONCLUSION

The intertextuality between *Bismaprawa* and *Adiparwa* has given birth to new elements as the result of re-reading and re-writing processes. The intertextual relationship between the two texts is observable in two dimensions - the criterion of characters and the criterion of events. The criterion of characters consists of new characters and changes in or addition to the names of existing characters in *Adiparwa*. Based on the analysis of the criterion of characters, we can conclude that the names of new characters in *Bismaprawa* (1) originated from other texts related to the epic of the Pandawas and Korawas and (2) were created by the writer of *Bismaprawa* because they do not exist in any other texts which were derived from *Adiparwara*.

The criterion of events consists of new events which change or add to the existing plotline in certain fragments of *Bismaprawa*. Based on the analysis of the criterion of events, we can conclude that the new events recounted in *Bismaprawa* were created by the writer of *Bismaprawa* because they do not exist in any other texts which were derived from *Adiparwa*. These new events are actually the reflection of elements and circumstances beyond the text itself— that is, of the Merapi-Merbabu tradition in which *Bismaprawa* was composed.

On the one hand, one might argue that the changes introduced by the Merapi-Merbabu scriptorium may be considered as corruption of the original text. On the other hand, these changes may also be considered as part of a conscious attempt to preserve a literary work by adopting it to the real situations of the supporting society (Wiryamartana, 1990:198). Thus, the creation of *Bismaprawa* may be considered as an effort on the part of the readers, who in turn became the writers, to understand and interpret *Adiparwa* as a source text within the context of Merapi-Merbabu literary tradition. Therefore, the presence of these changes in *Bismaprawa* does not only reflect the writer's interpretation of the source text but is also a form of collective consciousness of the members of different communities which give a similar treatment to existing literary works.

REFERENCES

Acri, A. (2011) *Dharma Pātañjala: A Śaiva Scripture from Old Java: Studied in the Light of Related Old Javanese and Sanskrit Texts*. Gonda Indological Studies, Groningen, Forsten.

Allen, G. (2000) *Intertextuallity*. London & New York, Routledge.

Ash Shiddieqy, T.M.H., Kriswanto, A., & Mansyur, M. (2012) *Gita Sinangsaya: Edisi teks dan terjemahan (Gita Sinangsaya: Text and translation edition)*. Jakarta, Perpustakaan Nasional RI.

Bleeker, P. (1852) Verslag der werkzaamheden van het Bataviaasch Genootschap van Kunsten en Wetenschappen, van September 1850 tot April 1852, namens het bestuur des Genootschaps voorgelezen in de algemeene vergadering op den 27 sten April 1852. VBG 24.

Creswell, J. W. (2009) *Research Design: Qualitative, Quantitative, and Mixed Methods Approaches*. 3rd edition. California, Sage.

Culler, J. (1981) *The Pursuit of Signs: Semiotics, literature, deconstruction*. London & New York, Routledge.

Dowson, J. (2002) *A Classical Dictionary of Hindu Mythology and Religion, Geography, History and Literature*. London, Routledge.

Edmunds, L. (2001) *Intertextuallity and the Reading of Roman Poetry*. Baltimore and London: The Johns Hopkins University Press.

Ensink, J. (1967) *On the Old-Javanese Cantakaparwa and its tale of Sutasoma.* 's Gravenhage, Martinus Nijhoff.

Juynboll, H.H. (1906) *Ādiparwa, Oudjavaansch prozageschrift*. 's Gravenhage: Martinus Nijhoff.

Molen, W. van de. (1981) Aims and Method of Javanese Philology, *Indonesia Circle*, 26, 5–12.

Molen, W. van de. (2011) *Kritik Teks Jawa, sebuah pemandangan umum dan pendekatan baru yang diterapkan kepada Kunjarakarna (Javanese Text Criticism, a general view and a new approach applied to Kunjarakarna)*. Jakarta, Yayasan Pustaka Obor Indonesia.

Molen, W. van de. & Wiryamartana, K. (2001) The Merapi-Merbabu manuscripts: A Neglected Collection. *Bijdragen tot de Taal—, Land-en Volkenkunde*, 157(1), 51–64.

Noorduyn, J. (1982) Bujangga Manik's journeys through Java: topographical data from an Old Sundanese source. *Bijdragen tot de Taal—, Land-en Volkenkunde*, 138, 413–442.

Pigeaud, Th.G.Th. (1924) *De Tantu Panggĕlaran. Een Oud-Javaansch prozageschrift, uitgegeven, vertaald en toegelicht*. 's Gravenhage, Smits.

Poerbatjaraka. (1952) *Kapustakan Djawi*. Djakarta, Penerbit Djambatan.

Poerwadarminta. (1939) *Baoesastra Djawa*. Groningen, Batavia, J.B Wolters.

Ras, J.J. (2014) *Masyarakat dan Kesusastraan di Jawa (Society and Literature in Java). (Achadiati, Ed.)*. Jakarta, Yayasan Pustaka Obor Indonesia.

Reynolds, L.D. & Wilson, N.G. (1991) *Scribes and Scholars; A Guide to the Transmission of Greek and Latin Literature*. 3rd edition. Oxford, Oxford University Press.

Riffaterre, M. (1984) Intertextual representation: On mimesis as interpretive discourse, *Critical Inquiry*, 11(1), 141–162.

Robson, S.O. (1994) *Prinsip-prinsip Filologi Indonesia (Principles of Indonesian Philology)*. Jakarta, RUL.

Supomo, S. (1964) Sastra Jendra: Ngelmu yang timbul karena kakografi (Jendra Literature: Knowledge that comes from Kakografi). *Majalah Ilmu-ilmu Sastra Indonesia*, 2 (2), 177–186.

Swellengrebel, J.L. (1936) Korawāśrama, een Oud-Javaansch prozageschrift. *Uitgegeven, vertaald en toegelicht*. Santpoort, Mees.

Teeuw, A. (1986) Translation, Transformation, and Indonesian Literary History. In: Grijns and Robson, (eds.) *Cultural Contact a Textual Interpretation*. Dordrecht-Holland/Cinnaminson-USA: Foris Publications. [VKI 115].

The Old Sabhāparwa, a summary and reconstruction of its manuscript. (2005) In: *Seminar Internasional Jawa Kuna 8-9 Juli 2005*. Depok, Universitas Indonesia.

Tjahjani, J. (2013) *Ambiguitas Genre dalam Trilogi Les Romanesques Karya Alain Robbe-Grillet (Genre Ambiguity in the Trilogy of Les Romanesques by Alain Robbe-Grillet)*. [Dissertation]. Depok, Universitas Indonesia.

Williams, G.M. (2003) *Handbook of Hindu Mythology*. California, ABC-CLIO.

Wiryamartana, K. (1990) *Arjunawiwaha Transformasi Teks Jawa Kuna lewat Tanggapan dan Penciptaan di Lingkungan Sastra Jawa (Arjunawiwaha Transformation of Ancient Java Text through the Response and Creation in Javanese Literature)*. Yogyakarta: Duta Wacana University Press.

Wiryamartana, K. (1993) The scriptoria in the Merbabu-Merapi area. *Bijdragen tot de Taal, Landen Volkenkunde* 149, 503–9.

Worton, M., & Still, J. (Eds.). (1990) Intertextuality theories and practices. Manchester and New York: Manchester University Press.

Zaimar, O.K.S. (2014) *Semiotika dalam Analisis Karya Sastra (Semiotics in the Analysis of Literary Works)*. Depok, Komodo Books.

Zoetmulder, P.J. (1983) *Kalangwan Sastra Jawa Kuno Selayang Pandang (Ancient Javanese Kalangwan Literature at a Glance)*. Jakarta, Djambatan.

Cultural Dynamics in a Globalized World – Budianta et al. (Eds)
© *2018 Taylor & Francis Group, London, ISBN 978-1-138-62664-5*

Uttaraśabda in Java and Bali

A. Kurniawan & D. Puspitorini
Department of Linguistics, Faculty of Humanities, Universitas Indonesia, Depok, Indonesia

ABSTRACT: The aim of this research is to provide access to less well-known Javanese literary text, Uttaraśabda, which can be categorised as a *tutur* (utterance) text. It contains dialogues on religious speculation, which is also valuable for the study of religions in Indonesia. To make the text readable, textual editions should be arranged. However, the previously-necessary step taken is to explain the cultural background of all material evidence. It also stands as a consequence on the textual analysis. Uttaraśabda is documented in fifteen copied manuscripts: eight from Java and seven from Bali. If it is assumed that a manuscript is as an equal textual variant, then careful observation of each manuscript can be expected to give explanation on cultural-background peculiarities. Particularly, there are two manuscript groups from different cultural loci: Java and Bali; therefore, this research will apply a philological approach based on textology. The approach will provide comprehension about textual problems and the history of textual development. Thus, the understanding on the reception process to a *tutur* text, at least in two different cultural loci, can be dug up. Hence, this research wishes to contribute generally in the study of *tutur* which so far still has been polarised only on Balinese or Javanese traditions.

1 INTRODUCTION

There is an undeniable fact that we have not obtained a comprehensive portrayal about the history of Javanese literature which has been going on for more than a thousand years. There are a number of literary works which have been discussed before. One of them is Uttaraśabda. The previous scholarly works are just kept silent, except for some catalogues which contain information about its material evidence: manuscripts (Cohen Stuart, 1872; Poerbatjaraka, 1933; Pigeaud, 1967–70; Behrend, 1998; Setyawati et al., 2002). However, every catalogue only provides limited information. To make Uttaraśabda accessible, textual editions need to be prepared.

Is it worth making Uttaraśabda accessible? First, it is less known. Other materials which record its text are unfamiliar to the present-time readers. Meanwhile, research about this text will make an important contribution to the study of the religion development in Indonesia. Uttaraśabda is a pre-Islamic work. It can be categorised as *tutur*: a genre of a philosophical text on theological speculation, and it sometimes contains religious manuals for ritual (see Acri, 2009; Soebadio, 1971, pp. 3–4). Uttaraśabda contains dialogues between Uttaraśabda and his elder brother, Ajñānaśura, about those subjects. Analyzing Uttaraśabda and its textuality will give a comprehensive understanding about the dynamics of thoughts on a religious concept, particularly in the society where its text is produced.

Indeed, the efforts to make Uttaraśabda accessible began when a Dutch philologist, A.B. Cohen Stuart, was employed as a manuscript conservator in the Batavian Society of Arts and Sciences (Bataviaasch Genootschap van Kunsten en Wetenschappen, abbreviated as "BG") from 1862 to 1871 (Behrend, 1993, p. 423). With the assistance of Radèn Panji Suryawijaya and Radèn Mas Samsi (Suryawijaya's son), he made a copy of many palm-leaf manuscripts in a Buda script which was brought from Mt. Merbabu, Central Java, in the middle of the 19th century. They transliterated from Buda into the type of script which was used commonly for a Javanese text at that time: Modern Java script. In spite of that, there were no scholars interested. Uttaraśabda was considered as a minor literary work.

Pigeaud did a very limited analysis in the first volume of his catalogue that was published in the late 1960s. Uttaraśabda is classified as a Javanese *tutur* from Bali in this catalogue (Pigeaud, 1967, pp. 69–74). However, he had a reason for his classification because it was based on two copied manuscripts compiled by Proyek Tik (Tik Project) which were kept at Leiden University Library, the Netherlands. Is it true that Uttaraśabda was written in Bali as argued by Pigeuad? From the dating of the manuscripts, the Javanese materials are a few centuries older than the Balinese ones.

From this step, we uncovered that Uttaraśabda developed in two sets of traditions: Javanese and Balinese. These two different traditions share similar historical trajectories in literature, specifically pre-Islamic literature. Some colonial orientalists, such as Kern (1900) and Gunning (1903), argued that pre-Islamic Javanese literature should be depended on the text of Balinese tradition which has more superior quality than Javanese tradition itself. They positioned Bali as conservation space for Javanese culture before Islam established its domination along Java Island. However, this theory would be rejected later. Every tradition has potential and rules to grow in each scope. Thus, every textual artifact should be positioned as a variant which has certain functions in a specific historical phase. One manuscript: one variant. Hence, fifteen manuscripts which document Uttaraśabda text should be examined thoroughly.

This article will discuss the manuscript distribution in Java and Bali and when those manuscripts were produced. Those categories will be very useful in reconstructing the genotext (which could be called as cultural context) which encloses the production process of each phenotext (manuscript). The discussion has been limited on the palm-leaf manuscripts from those traditions. The copied manuscripts in Java script and from Proyek Tik are not included because their functions are different from the original function intended by the text. Information about where and when a manuscript was produced could be identified from the colophon or any external sources.

This is just a beginning step of the ongoing research. Its aim is to make a crude mapping connected to textual transmission and reception in every historical stadium. The comparison of each manuscript will be discussed in another occasion. We will start the discussion about the textuality of Uttaraśabda from this step. Furthermore, by this discussion, we can decide the most suitable edition model for this text.

2 MANUSCRIPTS FROM JAVA

Five palm-leaf manuscripts are of Javanese traditions, and they are included in the manuscript group named Merapi-Merbabu collection. This collection was discovered by authoritative colonials for the first time in 1822 through a survey initiated by Van der Capellen, a Governor-General who ruled from 1816 to1826. In 1852, thirty years after being discovered in the western slope of Mt. Merbabu, there was a report that those manuscripts existed in Batavia and became the first collection of BG library. In total, there were 400 palm-leaf manuscripts in Buda and Java scripts, which covered many genres of texts: *kakawin, kidung, tutur, pawukon,* and *mantra* (Van der Molen, 2011, pp. 141–148; Wiryamartana and Van der Molen, 2001, pp. 53–55). Five manuscripts which document Uttaraśabda text are PNRI 6 L 46, PNRI 7 L 49, PNRI 1 L 170, PNRI 1 L 225, and PNRI 86 L 334. Four manuscripts contain the complete text, but PNRI 7 L 49 is in a poor-conditioned bundle of fragments from several manuscripts. There are just five folios (*lempir*) which contain the text (cf. Setyawati et al., 2002, p. 38). All of them are heavily damaged. PNRI 7 L 49 will not be discussed in this article.

2.1 *The origin of the manuscripts*

Although excavated from Mt. Merbabu, information from the colophon shows that the text was copied (or produced) in several different locations. PNRI 6 L 46 and PNRI 1 L 170 were copied in Rabut Pamrihan, an ancient name for Mt. Merbabu (Noorduyn, 1982, p. 416).

More precisely, PNRI 1 L 170 was copied in the northwest slope (*imbang bayabya*), in the hermitage (*panusupan*) called Wanakarya. Other manuscripts, i.e. PNRI 1 L 225 and PNRI 86 L 334, were copied in Sang Hyang Giri Mandharagĕni (ancient name of Mt. Merapi) and Sang Hyang Giri Karungrungan (ancient name of Mt. Ungaran).

Terminologies such as '*rabut*' and '*panusupan*' can explain the status of the scriptorium. Both refer to a religious center or a place for spiritual education in the past. We can trace this back to a Middle Javanese prose work, Tantu Panggĕlaran (Pigeaud, 1924), or from an Old Sundanese poem, Bhujangga Manik (Noorduyn and Teeuw, 2009). In the last text, for example, one part of the story describes a main character named Bhujangga Manik who came back from the east area of Pasundan in West Java, after studying religions in Pamrihan located in Mt. Damalung (Noorduyn and Teeuw, 2009, p. 292). This brief explanation can clarify what kinds of environment in which Uttaraśabda text was produced and what its role was. Of course, it had a relation with religious activities in that society.

2.2   *Dating problems*

Among the four complete manuscripts, only two manuscripts have dating evidence: PNRI 1 L 170 and PNRI 1 L 225. It should be noted that the dating element which will be emphasised is the year. The colophon of PNRI 1 L 170 informs that the text was already written in 1585. Thus, through the chronogram "watu sina(m)bi hoyĕging wong" (the stone was accompanied by the trembling man), we know that the text in PNRI 1 L 225 was already written in 1611. Unfortunately, there is no definite conclusion from previous scholars about how to convert the Merapi-Merbabu calendar system to the common system (AD calendar). Van der Molen (2011, pp. 93–108) and Wiryamartana (1984) have tried to convert it by comparing it with the Tĕnggĕr calendar system, especially Pasuruan Tĕnggĕr. However, these efforts are not reaching the precise system yet, because the Tĕnggĕr calendar system is also still undecided (see the reconstruction by Proudfoot, 2007).

As stated by Wiryamartana (1984), even though there are a number of pre-Islamic texts from the Merapi-Merbabu collection, the dating evidence from the manuscripts indicates that they were actually from the later period, i.e. Mataram era, from the 17th to the 18th century. The sufficient evidence comes from PNRI 32 L 313 which also includes the Merapi-Merbabu collection. The Gita Sinangsaya text in this manuscript was copied in 1592 and represented in a variety of chronograms (*sĕngkalan*). The colophon also explains that the year 1592 was the same year in which the Prince of Madiun, Wiramanggala, was murdered and the ammunition warehouse (*paobatan*) was exploded. These incidents happened in 1670 AD as recorded in babad stories and colonial archives (Ricklefs, 1978, pp. 179–180). Thus, Gita Sinangsaya in PNRI 32 L 313 was written after the aforementioned incidents, perhaps a few months after. Thus, the year 1592 was in close proximity to the years 1585 and 1611.

Therefore, the time when the Uttaraśabda from the Merapi-Merbabu collection was copied— and also had a particular function in its society—was approximately during the second part of the 17th century, between the reign of Amangkurat I/Sinuhun Tĕgal Arum (1645–1677 AD) and Amangkurat II (1680–1702 AD). At that period, obviously, the Islamic influence had been spread out all over Java. However, the existing pre-Islamic text in Java in that period can trigger a question: what kinds of societies that became its proponents? Interestingly, in the Merapi-Merbabu collection, there are both Islamic and pre-Islamic texts, and they come from the same period.

Another interesting aspect which can be taken into account is the use of Buda script (for more explanation about this script, see Pigeaud, 1970, pp. 53–60; Van der Molen, 2011, pp. ...). At the period when Merapi-Merbabu manuscripts were written, Java script had already been well acknowledged massively. In several manuscripts of this collection, Java script was also used, even though it was not as much as Buda script. There was a possibility that the writer or copier of the manuscripts knew Java script very well despite the use of Buda script. Therefore, it is not surprising to find one or two Java scripts among the lines of Buda script in a manuscript. They might have made some mistakes in their work. Using the Buda script was their conscious choice, and it also had certain functions in their society.

However, the reasons behind making this conscious choice have not yet been known. Previous theories claim that there were particular societies that lived in exile after Islam spread out all over Java. They maintained their old way of living conservatively. Unfortunately, that theory can no longer be used because the Islamic influences had also infiltrated into the literary work in the Merapi-Merbabu collection. The society in this literary environment was more open than the conservative. The use of Buda script could be considered as having a function as an identity marker besides the written devices to make limited communication among this society.

2.3 *Manuscript materials*

Manuscripts of the Merapi-Merbabu collection were produced in a place high above the sea level. However, these manuscripts were made from the materials which could not be grown in such environment. The Palmyra (*Borassus flabellifer*) only grows in lowlands which is hot, dry, located about 500 meters above the sea level, and particularly in a coastal area (Ensikolopedi Indonesia IV, 1992, p. 2046). If the palm-leaf manuscripts were produced in highlands, where did they get the materials from? Van der Molen (2011, pp. 108–110) estimated that they got these materials from the coastal area in north Java where Palmyra plants can be grown well. It is important to take into account that the members of literary society chose palm leaves as the writing sheet rather than other materials which they were also familiar with, such as tree-bark paper, *gĕbang*, and so on. It is possible that palm leaves had social and religious functions, or other functions related to identity.

3 MANUSCRIPTS FROM BALI

The existence of Uttaraśabda manuscripts of Balinese traditions is related to Gedong Kirtya in Singaraja, which was established in 1928 at a convention held in Kintamani. Dutch scholars, brahmana, and local aristocrats attended this convention. Through this institution, palm-leaf manuscripts which contain Uttaraśabda of Balinese traditions—which were originally disseminated as personal collections—were able to be accessed by public.

Four Balinese manuscripts used in this research belonged to Gedong Kirtya. These manuscripts were coded with Roman number "III" which implies that they were included in the *tutur* or *wariga* group. This categorisation was made by a Dutch scholar, Roelof Goris, when he was working as a curator for this institution in 1931 (Swellengrebel, 1966, pp. 205–228). It could be used as the basis to identify what literary genre that Uttaraśabda of Balinese traditions should be included.

So far there are only two manuscripts that can be accessed which are (1) K 247 and (2) T/xxiv/5 DOKBUD (the second manuscript will be named as "lontar Pusdok" in this article). Three other manuscripts will only be identified through secondary sources that provide information about them. It is still ongoing research. It still needs more new data to accomplish our understanding on the function of Uttaraśabda in Balinese traditions.

3.1 *The origin of the manuscripts*

As a matter of fact, information about where each manuscript comes from could be identified from internal evidence (colophon). In the first folio of K 247, there is information in Malay written in Roman character: "dapat dibeli dari curator Kirtya di Moenggoe [Badoeng]" (could be bought from Kirtya's curator in Munggu, Badung). There is no other detailed information in this manuscript. According to a regular report from Kirtya, there was a pĕdanda (Balinese priest) from Munggu, Badung, who was employed as an administrator in Kirtya, i.e. Ida Pĕdanda Gde Pĕmaron Munggu (Palguna, 1999, p. 281). Perhaps this manuscript comes from him.

The colophon of lontar Pusdok provides more comprehensive information. This manuscript comes from Desa Dencarik, Kecamatan Banjar, in Kabupaten Buleleng. Someone named Putu Ngurah Mĕrta was considered as the writer. Unfortunately, we do not know who he is. The writer's profile can reveal in which social group the Uttaraśabda text became functional. These problems still wait for further research. Moreover, three other manuscripts' origin can be discerned from the catalogue.

The Palm-leaf manuscript which became the exemplar of K 6527 originated from Griya Pĕmaron, Munggu, Mĕngwi. K 6538 took its exemplar from the palm-leaf manuscript possessed by Makele Tarĕna, Puri Kawan, Singaraja. Its location is so close to Gedong Kirtya. K 1525 was a copy of a palm-leaf manuscript from Bakung, Sukasada, Buleleng. Hence, there were three manuscripts from North Bali (K 1525, K 6538, and lontar Pusdok), and two others from South Bali (K 247 and K 6527). Moreover, there were two scriptorium categories where the manuscripts of Uttaraśabada were produced: *griya* and *puri*.

To Balinese societies who recognise *warna* as the divisional system in social roles, the concept of *griya* has always been associated with the priesthood (*brahamana*). A *griya* is where priests (*pedanda* in Balinese) used to abide in, where any priesthood activities—especially literary—were held inside. On the other hand, a *puri* is associated with a place where *ksatria* or a group of noblemen used to live in. They also had a particular activity related to literature. This difference would definitely constitute the literary characters and genres which developed each scriptorium.

We can roughly assume that the types or genres of literary works developed in a *griya* were adjusted to the needs of the priest group. Textual genres, such as *tutur, mantra, babad*, or stories about a figure of priest, or even texts on ethical codes in priesthood, were developed in *griya*'s. Meanwhile, the literary genres, such as kakawin, kidung, laws, ethics, and babad or history about kings' genealogy, were developed customarily in *puri*'s.

This division was maybe significant to explain the functions of these texts. By considering the substance of Uttaraśabda which is about religious speculation and ritual manual, it would have been appropriate if such a text was developed in a *griya*. Its function can be related to priesthood religious activities. Nevertheless, that text was also living in a *puri*. The question remains: why did a *tutur* text, like Uttaraśabda, also exist in a *ksatria* circle? What was its necessity? This problem should be further investigated. This problem will be adjourned for a while by considering that the exemplars of K 6527 and K 6538 have not been physically accessed.

3.2 *Dating problems*

Uttaraśabda manuscripts of Balinese traditions are not as old as those of Javanese. The dates in these manuscripts could be acquired from internal information, which is the colophon. However, *lontar Pusdok* was the only manuscript that has dating information in its colophon. We have not found any such information from K 247 and also K 1525. These manuscripts were recorded in the list of the Kirtya Liefrick-Van der Tuuk manuscript collection which was published in 1948. Consequently, it could be predicted that both manuscripts were produced at the beginning of the 20th century. The copies of both manuscripts were also produced in the early 1960s through Proyek Tik which was conceived by Christiaan Hooykaas and Kĕtut Sangka.

Meanwhile, the Uttaraśabda text in *lontar Pusdok* was already completely inscribed by 19th November 1988, coincided with the birthday of Goddess Saraswati (piodalan Saraswati): Saniscara Umanis, wuku Watugunung, in 1910 Śaka. According to Hindu-Balinese traditions, any kinds of literary activities, either reading or writing, were not allowed from sunrise to sunset. During this time, people gave some offerings in front of literary works, such as palm-leaf manuscripts (*lontar*). The purpose of this ritual was to wish for the blessings of science from Sarasvati—who was believed as the protector of art and science—that would be descended into the literary works. Any literary activities would be permitted again after the sunset in the evening. Literary texts would be read during the night.

Thus, why was *lontar Pusdok* finalised during the celebration of piodalan Saraswati? When was the exact time it was finished if any literary activities were not allowed until the sunset? If the writers of this manuscript revered to the tradition, there is a possibility that the manuscript was finalised a day before the celebration. The expression "wus puput sinurat" (has already been inscribed) can be interpreted as on the celebratory day, and the manuscript was finalised and existed.

4 CONCLUSION: *TUTUR* IN TWO TRADITIONS

According to the previous explanations, besides evidence which has already been discovered, it could be concluded that Uttaraśabda of Javanese traditions was older than of Balinese traditions. In Java, Uttaraśabda was developed approximately at the end of the 17th century, in the religious circles living in the serene of a mountain area. Afterwards, the Uttaraśabda tradition seemed to have stopped, and its manuscripts were not found outside of Merapi-Merbabu scriptoria.

However, if we pay a closer attention to the text, there are a lot of elements which could also be found in *suluk* texts of Javanese Islam mysticism traditions. The use of simile expressions (*tamsil*) concerning God-and-human relationship can support these arguments. We may know that the depiction of the relationship between fish (humans) and water (God) in Uttaraśabda could also be found in Sěrat Sastra Gěnḍing by Sultan Agung of Mataram. Moreover, the relationship between shadow puppets (*wayang*) and its puppet master (*dalang*) was very popular in *suluk* texts (see Zoetmulder, 1991). It is no wonder that Raden Mas Samsi, the writer of the copied manuscript CS 78, also named Uttaraśabda as Suluk Lonṭang, even though these words do not exist inside the text. When we read it briefly, we would come up with an impression that the "horizon of expectation" of its copier was directed to Suluk Lonṭang. In his view, both works share many similarities. An intertextual analysis will potentially reveal these problems. The textual connection between two literary genres, i.e. *tutur* and *suluk*, will rise up as the gain of those attempts.

Meanwhile, in Balinese traditions, Uttaraśabda was developed in the early of the 20th century, although we could not really identify the precise time. Was the movement of this text from Javanese traditions to Balinese traditions related to Dhang Hyang Nirartha who went to Bali after the widespread of Islamic influence to eastern part of Java in the 16th century? There has not been any clue yet. However, when van der Tuuk was living in North Bali for a long period in the second half of the 19th century and collecting hundreds of palm-leaf manuscripts from many Balinese people for his lexicography projects, there was no palm-leaf manuscript entitled Uttaraśabda or Partha Ajñānaśura (see Brandes, 1901–26). In reality, three of five manuscripts come from North Bali. Therefore, it is possible that Uttaraśabda was brought into Bali, and obtained its proponents, at the beginning of the 20th century, or as late as the last 19th century. In present time, Balinese people recognise Uttaraśabda as the *tutur* text.

Thus, it is still ongoing research that needs more data to solve any problems internally. An in-depth analysis on this text—specifically on its dynamic structure, function, and dissemination—will give a thorough depiction about how a religious text was read and accepted in a specific society, specific time, and space. The understanding of its dynamics will open up new perspectives about how Indonesian people—especially in Java and Bali—comprehended the religious way and other related discourses of it.

REFERENCES

Acri, A. (2011) *Dharma Patañjala: A Saiva Scripture from Ancient Java, Studied in the Light of Related Old Javanese and Sanskrit Texts*. [Ph.D. Dissertation] Leiden University.

Barthes, R. (1987) Theory of the text. In: Robert Young (ed.) *Untying the Text: A Post-Structuralist Reader*. London & New York, Routledge & Kegan Paul.

Behrend, T.E. (1993) Manuscripts production in nineteenth-century Java. Codicology and the writing of Javanese literary history. In: *Bijdragen tot de Taal-, Land en Volkenkunde*, 149 (3), 407–37.

Behrend, T.E. (1998) *Katalog Induk Naskah-naskah Nusantara Jilid 4 Perpustakaan Nasional Republik Indonesia (The Main Catalogue of Archipelago Manuscripts of the 4th edition of the National Library of the Republic of Indonesia)*. Jakarta, Yayasan Obor Indonesia.

Brandes, J.L.A. (1901–26). *Beschrijving der Javaansche, Balineesch en Sasaksche Handschriften aangetroffen in de Natalenschap van Dr. H.N. van der Tuuk, en door hem vermaakt aan de Leidsche Universiteitsbibliotheek*, 4. Batavia, Landsdrukkerij.

Cohen Stuart, A.B. (1872) *Eerste vervolg catalogus der Bibliotheek en Catalogus der Maleische, Javaansche en Kawi Handschriften van het Bataviaasch Genootschap van Kunsten en Wetenschappen.* Batavia, Bruining & Wijt; s'Gravenhage, Nijhoff.

Ensiklopedi Indonesia, vol. 4. (1992) Jakarta, PT. Ichtiar Baru—Van Hoeve.

Goedenleer. 7 L 49. Palm-leaves, Buda script. Jakarta, Perpustakaan Nasional RI.

Gunning, J.G.H. (1903) *Bharāta-Yuddha: Oudjavaansch Heldendicht.* 's-Gravenhage, Nijhoff.

Kern, H. (1900) *Rāmāyaṇa, Oudjavaansch Heldendicht.* 's-Gravenhage, Nijhoff.

Molen, W. van der. (2011) *Kritik Teks Jawa: Sebuah Pemandangan Umum dan Pendekatan Baru yang Diterapkan kepada Kunjarakarna (Javanese Text Critiques: A General Perspective and New Approach Applied to Kunjarakarna)*, Achadiati Ikram (trans.). Jakarta, Yayasan Obor Indonesia.

Noorduyn, J & A. Teeuw. (2009) *Tiga Pesona Sunda Kuna (Three Charms of Ancient Sunda)*, Hersri Setiawan (trans.). Jakarta, Pustaka Jaya.

Oettarasabda. 1 L 334. Palm-leaves, Buda script. Jakarta, Perpustakaan Nasional RI.

Oettarasabda. 33 L 225. Palm-leaves, Buda script. Jakarata, Perpustakaan Nasional RI.

Palguna, IBM D. (1999) *Dharma Śūnya: Memuja dan Meneliti Śīwa (Dharma Śūnya: Worshipping and Researching Śīwa).* Denpasar, Yayasan Dharma Sastra.

Parimbon Arabisch, Javaansch en Soendasch. 6 L 46. Palm-leaves, Buda script. Jakarta, Perpustakaan Nasional RI.

Partha Ajñāṇaśura. K 247/IIIb/9. Palm-leaves, Balinese script. Singaraja, Gedong Kirtya.

Pemerintah Kabupaten Buleleng. (2011) *Buku Katalog Salinan Lontar UPTD Gedong Kirtya (A Catalogue Book of the Palm-Leaf Copy of UPTD Gedong Kirtya).* Singaraja, Dinas Kebudayaan dan Pariwisata Kabupaten Buleleng.

Pigeaud, Th. (1924) *De Tantu Panggĕlaran: Een Oud-Javaansch Prozageschrift, uitgegeven, vertald en toegelicht.* 's-Gravenhage, H.L. Smits.

Pigeaud, Th. (1967) *Literature of Java: Catalogue Raisonné of Javanese Manuscripts in the Library of the University of Leiden and Other Public Collections in the Netherlands, Volume I: Synopsis of Javanese Literature 900–1900 AD.* The Hague, Martinus Nijhoff.

Poerbatjaraka, R.Ng. (1933) Lijst der Javaansche handschriften in de boekerij van het kon. Bat. Genootschap, *Jaarboek Bataviaasch Genootschap* 1, 269–376.

Proudfoot, I. (2007) Reconstructing the Tĕnggĕr Calendar. In: *Bijdragen tot de Taal-, Land en Volkenkunde,* 163 (1), 123–33.

Ricklefs, M.C. (1978) *Modern Javanese Historical Tradition: A Study of an Original Kartasura Chronicle and Related Materials.* London, School of Oriental and African Studies University of London.

Sĕrat Oettara-sabda. 1 L 170. Palm-leaves, Buda script. Jakarta, Perpustakaan Nasional RI.

Setyawati, K., Wiryamartana, K., & Molen, W. van der (2002) *Katalog Naskah Merapi-Merbabu Perpustakaan Nasional Republik Indonesia (The Merapi-Merbabu Manuscript Catalogue of the National Library of the Republic of Indonesia).* Yogyakarta & Leiden, Penerbitan Universitas Sanata Dharma & Opleiding Talen en Culturen van Zuidoost-Azië en Oceanië Universitiet Leiden.

Soebadio, H. (1971) *Jñānasiddhānta.* The Hague, Martinus Nijhoff. [Bibliotheca Indonesica 7].

Swellengrebel, J. (1966) In memoriam Dr. Roelof Goris (with a bibliography by R.S. Karni). *Bijdragen tot de Taal-, Land en Volkenkunde,* 122 (2), 205–28.

Uttara Sabda Amrĕta. K1525/IIIb/36. Palm-leaves, Balinese script. Singaraja, Gedong Kirtya.

Uttara Sabda Amrĕta. T/xxiv/5/Dokbud. Palm-leaves, Balinese script. Denpasar, Pusat Dokumentasi Budaya Bali.

Wiryamartana, I.K. & Molen, W. van der (2001) The Merapi-Merbabu manuscripts: A neglected collection. In: *Bijdragen tot de Taal-, Land en Volkenkunde,* 157 (1), 51–64.

Wiryamartana, I.K. (1984) Filologi Jawa dan kuñjarakarna prosa (Javanese philology and kuñjarakarna prose), *Basis,* 33, 255–72.

Wiryamartana, I.K. (1990) *Ajunawiwāha: Transformasi Teks Jawa Kuna Lewat Tanggapan dan Penciptaan di Lingkungan Sastra Jawa (Ajunawiwāha: Ancient Javanese Text Transformation through Responses and Creation in the Javanese Literary Environment).* Yogyakarta, Duta Wacana University Press. [ILDEP].

Zoetmulder, P.J. (1991) *Manunggaling Kawula Gusti: Pantheisme dan Monisme dalam Sastra Suluk Jawa (Manunggaling Kawula Gusti: Pantheism and Monism in Javanese Suluk Literature)*, Dick Hartoko (trans.). Jakarta, Gramedia Pustaka Utama.

Cultural Dynamics in a Globalized World – Budianta et al. (Eds)
© *2018 Taylor & Francis Group, London, ISBN 978-1-138-62664-5*

Loloda in the world of Mollucas: The decline of the political entity in *Loloda* from *Kolano* to *Sangaji*

A. Rahman
Department of History, Faculty of Humanities, Universitas Indonesia, Depok, Indonesia

ABSTRACT: The objective of this article is to describe the decline of the Loloda Kingdom in terms of power in the regions of the North Mollucas during the 17th to 20th century, which was marked by the change of leadership from Kolano (King) to Sangaji (District Chief). We argue that the following four important factors caused the decline: (1) the weak control of the Loloda Kingdom over its territory; (2) the exploitation of its natural resources; (3) the steady decline of population due to political conflict, war, and migration; and (4) the political and military intervention of foreign (European) powers in the Mollucas, which directly affected the sovereignty of Loloda. These four factors contributed to the decline of the Loloda Kingdom in the Mollucas region. The research methodology used here is derived from the stage-wise history of science: heuristics (searching for, finding, and collecting data); criticizing the source internally and externally; interpretation of resources; and historiography (writing history).

1 INTRODUCTION

1.1 *The alliance of Moti and Moloku Kie Raha without Loloda*

Loloda was an ancient kingdom in the Moluccas World, but it was not included in the Moti Alliance (1322–1343). This alliance is well known as the Moloku Kie Raha (MKR), which means "the communion of the four mountain kingdoms of the Moluccas", including the kingdoms of Ternate, Tidore, Bacan, and Jailolo. The Moti Alliance, formed in the 14th century, was initiated by the Seventh King of Ternate, Sida Arif Malamo (1322–1332). If Loloda had been included in the alliance, then it would have become the Moloku Kie Romtoha (the five mountain kingdoms of the Moluccas).

However, the MKR concept was not consistent with the statement of the Governor of the Dutch East India Company (Vereenigde Oost-Indische Compagnie or VOC) in Moluccas, Robbertus Padtbrugge (1677–1682), in his memorandum of the handing-over to his successor, Jacob Loobs (1682–1686). Padtbrugge reminded Loobs that there were five major kingdoms in the Moluccas World with their own specific names, including "Loloda, toma ngara ma-beno" (royal ruler of the north gate of the Moluccas), "Jailolo, Jiko ma-Kolano" (royal ruler of the bays), "Tidore, kie ma-Kolano" (royal ruler of the mountains), "Ternate, Kolano of Maluku" (ruler of Moluccas), and "Bacan, Dehe ma-Kolano" (royal ruler of the south end of the Moluccas) (see also Andaya, 1993, pp. 51, 93, & 232; van Fraassen, 1987, pp. 18–24).

According to the local oral tradition as a mythology, Loloda is the oldest and largest kingdom in the region. There were two local traditions that described the supremacy and greatness of Loloda, namely the tradition of Biku Sagara and the Four Eggs of the Sacred Dragon (Galvao, 1544, in Jacobs, 1972: 80) and *Kroniek van het Rijk Batjan* (Chronicle of Bacan) by Coolhaas (1923: 480–481). Genealogically, the parent lineage of the Loloda kings is in line with the parent lineage of the MKR kings; however, geohistorically, the Loloda Kingdom had been "the ruler of the mainland and island of Halmahera" in the Moluccas since the 13th century. However, in reality, Loloda was then "eliminated and had disappeared" from the political arena of the Moluccas World with the decline of the political entities from being the Kolano (King) to becoming the Sangaji (District Head). This decline severely worsened after the territorial status of Loloda was also weakened during 1909–1915. Here, we attempt to address the reason for this decline, its mode of occurrence, and its consequences.

1.2   The king's weakness in the regional control system

The era of the Loloda Kingdom in the Moluccas World can be divided into the following several periods: (1) the mythological period (from the 13th century to the late 16th century); (2) the Portuguese–Spanish period (1512–1666); (3) the period of the Dutch East India Company (Verenigde Oost Indische Compagnie/VOC) (1607–1800); (4) the period of the transitional government from Dutch Government to the British Government (1801–1817); (5) the period of the transitional government from the British Government to the Dutch Government (1818–1908); and (6) the period of the decline of Loloda's sovereignty from the Kolano position (King) to the Sangaji leadership (District Head) (1909–1915).

The basic government structure of Loloda was the same as that of MKR, including important positions such as the Kolano (King), Jougugu (Prime Minister), Kapita Lao (Warlord), Bobato (Tribal Councils), Hukum (Judges), Sowohi (Supervisor of the Royal), and Alfiris (Secretary) (H.A. Usman, interview, 7 January 2016). This structure had not changed much until the 12th generation, since the period of Kolano Tolo Usman Malamo (1220) until Kolano Syamsuddin (1915). Despite the transformation by the Dutch to adopt a political structure of the sultanate of Ternate, Loloda retained its cultural freedom and moved dynamically to exit from the Ternate structural restraints (see also Susanto Zuhdi, 2010, p. 3).

Geographically, the administrative center of Loloda was based in Soasio, the capital city of Loloda, on the northwest coast of Halmahera with a territory covering the whole island of Halmahera and Morotai. The three main villages, namely Laba, Bakun, and Kedi (M.W. Hamad, interview, 7 January 2016) were the nearest villages to the center of the Loloda Kingdom, which had a very strong historical link that will later define Loloda's sovereignty in the period 1909–1915.

Laba and Bakun were the two main villages that became targets of the "attack and siege" strategy of the Dutch Army in 1909, because the people of these villages were involved in a rebellion under the leadership of a Loloda royal navy commander, Sea Captain Sikuru. Although not included in the MKR alliance, Loloda was an integral part of a group of local kingdoms in the Moluccas World, mythologically, genealogically, and geohistorically, which was the basis of its sovereign power. Unfortunately, such forces could not guarantee the continued existence of Loloda as a sovereign kingdom in the Moluccas World.

Geohistorically, Halmahera and Morotai Islands belonged to the Loloda Kingdom before they were conquered by the Ternate Sultanate. Nevertheless, since the late 19th century, only the King of Loloda had the power over the rest of the territory spreading from the Cape of Bakun to the Tobo-Tobo Bay, where the Loloda ethnic group was the predominant settlers. Restriction of the Loloda territory proved that these regions had been occupied by the Sultanate of Ternate and Tidore. The two kingdoms were always competing to gain control over the two Loloda regions because Loloda was the oldest kingdom and the true heir of Halmahera.

In the oral tradition of Ternate, the following poem describes the territorial boundaries of Loloda: "gudu-gudu tomamie; Susupu, Gamkonora, Tobaru se Tolofuo; Bantoli se Mandioli; Doitia se Doitai; Morotia se Morotai; Kolano Loloda o Gugumakuci" (Far away in the North; of Susupu, Gamkonora, Tobaru and Tolofuo; Bantoli, and Mandioli; Doi mainland and Doi overseas; Moro mainland and Moro overseas; King Loloda rules) (Hasan, 2001, pp. 172–173). This poem indicates that Ternate (with its own identity) was recognized on the Loloda territory, covering the mainland of Halmahera from the north to the west. Susupu, Gamkonora, Tobaru, Tolofuo, and Bantoli were in the west, whereas Doitia (Doi mainland), Doitai (Doi Islands), Morotia, and Morotai were in the north.

The aforementioned poem was supposedly to be presented before the British colonial government in the Moluccas during 1801–1817. The Loloda community indeed had several oral traditions using their own language; however, so far, not all of these traditions have been identified because of the history, culture, customs, and Loloda traditions (Y. Toseho, interview, 5 January 2016).

The end of the power of the Loloda Kingdom over these territories was estimated around the late 19th century (during the restructuring of the Dutch East Indies in 1866); however, during the VOC period (1607–1799), areas such as Susupu, Gamkonora, Tobaru, and Tolofuo were under the control of the Ternate Kingdom together with Loloda and Jailolo. During this period, the District Chief of Loloda still held the title of "Kolano" instead of "Sangaji" (see Leirissa, 1996, p. 64).

This situation is particularly relevant to the present-day situation in which the Loloda language is still used by the residents of these territories spreading from the northern part to the western part of Halmahera (Hueting, 1908; Masinambow, 2001, pp. 142–144). In the early 17th century (1606–1608), Loloda was attacked by Spain, which marked its weakening, and soon after that, it was conquered by Jailolo, as a new kingdom in North Halmahera (see also Lapian, 1948, p. 18). However, the sovereignty status of Loloda remained as a kingdom led by a Kolano.

From the 18th to the 19th century, Loloda was still actively involving in politics of the local inter-kingdom as the vassal of Ternate; however, the Kolano of Loloda was unable to maintain and govern the entire regions under his empire. In the early 20th century, Loloda experienced a total decline of its leadership, from Kolano to become Sangaji, together with the change of its status from a kingdom to merely a district under the pressure of the Ternate Sultanate and pacification politics of the Dutch as a form of its Pax Nederlandica in the Moluccas.

1.3   Natural resources and the dominance of Ternate

J.H. Tobias (1857) in the memorandum of the handing-over to his successor, C. Boscher (1859), wrote "Den inkomsten die Z.H. de Sulthan van Ternate wettig, van zijn gebied op Halmaheira trekt, zijn de volgende: Van Loloda: In waren: 6–10 katjes vogelnetjes in diensten..." (Sultan of Ternate, the ruler of Halmahera, always receives bounties from the Loloda people, such as 6–10 packages of bird's nest regularly). This shows that Loloda as the subordinate of Ternate has potential natural resources, such as bird's nest, which was considered precious at that time, at least in the opinion of the Sultan of Ternate (see MvO, from J.H. Tobias (1857) to C. Bosscher (1859), 1859 in ANRI, 1980, p. 183).

According to Baretta (1917: 80), Loloda was also included as a potential producer of sago and resin, whereas various types of fish, sea cucumbers, and other varieties of marine biota as shown by the various sea catch of the local fishermen were produced from the marine sector. The condition of water in Loloda was also described by C.F.H. Campen (1888, p. 155).

According to De Clerq (1890) and Van Baarda (1904), the economic life of the rulers of Loloda depends on the economy of the people. In the socioeconomic aspect, this means that the relationship between the Kolano (king) and their Bala (people) was built on the power authority and legitimacy of the people toward their leader. Therefore, the people as the infrastructure could strengthen the political system and the position of the Kolano. Moreover, a Kolano as a political superstructure ensured guarantee and economic protection for his people. In this context, people could manage economic resources on the territory of the kingdom (aha kolano) and the royal treasury (raki kolano).

Soerabaia Handelsblad (1908) mentioned the Residency of Ternate and its territory in the archives of 15th October to the end of November 1907. According to this report, there were many bamboo and sago forests along the Loloda River. These two types of plants also grew well on the muddy soil. Besides resin, several other forest products were also common, namely cane, sandalwood, ironwood, and teak. Sago and coconut trees grew wild everywhere. Clove trees also thrived in the region besides rice and corn. Bird nest caves were common in West Loloda. Over time, these commodities have been exploited by the people of Ternate and the orders of their Sultan. Apparently, the economic security of the natural resources of Loloda depended on the Ternate Sultanate, which was under the control of the Dutch.

1.4   Political conflict and demographical consequences

During the reign of Amir Al-Mukminin Hamzah Nazarun Minallahi Shah (Sultan Hamzah, 1627–1648), there were several rebellions in some territories of Ternate, such as in Seram,

Ambon, and the East Coast of Celebes. In addressing this problem, Hamzah had massively evacuated the people of Loloda from Halmahera Island to Jailolo since 1628, as Jailolo was abandoned by its own people as a protest against the liquidation of their kingdom by the Ternate Sultanate. The next mass evacuation of the Loloda people to Jailolo occurred in 1662, which led to a significant population decrease with only 200 people remaining in Loloda. The migration of the Loloda people was also due to the expansive policy of the Ternate Sultanate.

On the basis of a 17th-century folk tale, a Loloda prince named Sibu was believed to have lived, who escaped with his followers from Loloda to the east coast of Celebes, that is, to Manado and Bolaang Mongondow (Mapanawang, 2012: 133–134). From the story of the oral tradition of Loloda, Prince Sibu was King Loloda Mokoagow (1653–1689), who was well known and influential during his reign. Loloda Mokoagow (1653–1693) was the origin of the Manoppo dynasty in the Bolaang Mongondow Kingdom (Dunnebier, 1984, pp. 27–39 & Supit, 1986, pp. 70–115).

1.5   A determinant rebellion: "Sikuru of Laba" (1909)

As revealed from the local oral tradition, the rebellion of Sikuru of Laba village was believed to have started in 1908. However, from the source of written history, this event was discovered in early 1909. In the period 1909–1915, Loloda suffered a decline to the lowest degree of sovereignty and political status, from Kolano to Sangaji, as a result of the rebellion. A number of Dutch newspapers of that period reported that this social movement was led by Kapitan Laut (Kapita Lao) Sikuru of Laba in 1909.

As mentioned above, several Dutch newspapers had reported the revolt for the first time, two of which were *Nieuwe Amsterdamsche Courant, Algemeen Handelsblad*, Zaterdag, 13 Maart 1909; and *Nieuwe Amsterdamsche Courant, Algemeen Handelsblad*, Mandaag, 15 Maart 1909. These events had major implications for the sustainability and sovereign political entity of Loloda in the Moluccas World, which in turn further strengthened the Ternate on Loloda with support from the Dutch colonial government.

The Loloda rebellions in 1908–1909 led by Kapitan Laut Sikuru Laba of Laba Village caused a number of casualties. Three Dutchmen, a tax officer and two policemen, were killed, and eight others from Loloda were injured. A total of 30 fully armed Dutch soldiers were sent from Ternate to Loloda under the leadership of Lieutenant Meihuizen to end this rebellion, as described in the following newspaper report:

> "De resident van Ternate seinde den 9en Februari via Gorontalo: vier dezer bericht van posthouder Djilolo: Politie-adsistent Loloda 2 dezer in huiz Radja te Soa Sioe door bende Alfoeren kampong Laba overvallen en met twee politie opassers vermoord; militaire patroille, 30 man, onder Luitenant Meihuizen van Ternate gezonden, over viel nacht 5 op 6 dezer versterkte kampong Laba; kwaadwilligen vluchtten met achterlating van twee dooden, een geweer en veel blanke wapens; onzerzijds geen gewonden. Beide kampongs bezet...." (On 9th February, the Resident of Ternate via Gorontalo announced that four pieces of news have been delivered by the supervisors of Jailolo region, namely two police officers were suddenly killed in the palace of King Loloda in Soasio by a group of Alifuru people from the Laba village; as many as 30 military patrols were sent from Ternate under the leadership of Lieutenant Meihuizen, to blockade the Laba village for 5 to 6 nights; after killing two people, the criminals fled with many guns and weapons; from our side, no one was injured. We occupied two villages....) (See: *Nieuwe Amsterdamsche Courant, Algemeen Handelsblad*, Mandaag, 15 Maart 1909: 13).

This rebellion led to the dismissal and exile of Syamsuddin, the last Kolano of Loloda, in 1909, to Ternate, who was exiled together with his wife and children. The Sikuru with the other rebel members were punished by the Dutch. Within a period of 6 years until 1915, the sovereign status of Loloda as a kingdom was formally abolished and then integrated into the Kingdom of Ternate due to the Dutch colonial political pressure.

The Kolano title for the Loloda ruler was changed to Sangaji in 1909, and in 1915, Loloda was officially integrated under the reign of the Ternate Sultanate, coinciding with the death of Syamsuddin in 1915. The rebellion had actually been recorded in the Kolonial Verslag van Nederlandsch (Oost) Indie 1909, as reported below:

"......Lolode werden in Februari 1909, zonder bekende aanleiding, de ten laste van de lanschapskas in dienst genomen assistent an twee gewapende politiedienaren door de bevolking van Laba vermoord. Nadat de kampong door an militair detachement uit Ternate bezet was, waarbij op eenig verset werd gestuit, dat den aanvallers op 3 dooden en 8 gewonden te staan kwam, verbeterde de toestand spoedig. De gevluchte bevolking keerde terug en werkte, onder leiding van de nieuwent assistent, aan wegen en bruggen. Een onderzoek bracht aan het licht dat de radja, de djogoegoe en de kapitein Laoet van Loloda de hand in het gebeurde de moeten hebben gehad. Zij werden ontslagen en met de hofdschuldigen aan den moord naar Ternate gezonden" (.....in February 1909 in Loloda, due to an unknown cause, two armed policemen were killed by people from the Laba village. After that, the village was occupied by a military detachment sent from Ternate, the attack killed three people and injured eight others. The situation was soon amended, the displaced people went back to work under the guidance of a new assistant, to build roads and bridges. An investigation revealed that the king, Jougugu, and Kapitan Laut of Loloda were involved in the incident. They were fired and sent to Ternate for punishment as they were considered guilty of being involved in the killing) (See: KV (Oost) Indie, 1909, p. 77).

The aforementioned quote states that in February 1909, there was a sudden rebellion by the people under the leadership of Kapitan Laut (naval commander) of Loloda named Sikuru, which killed three people, a tax officer and two Dutch policemen, and wounded eight people inside the palace of Loloda, in front of Kolano Syamsuddin and his empress, Jou Boki Habiba, in Soasio. This event caused the dismissal of the Kolano (king), Jougugu (prime minister), and Loloda navy commander from their position and their punishment by the Dutch ((M. Mansyur, interview, 12 February 2015). Kolano Syamsuddin was also sent to Ternate to serve his sentence until his death in 1915.

Local oral sources mentioned that the revolt had already commenced in 1908 (H.M.J.B. Usman, interview, 4 January 2016) and no written sources reported it. It was stated that Sikuru, the naval commander who led the rebellion, was assisted by his two colleagues, namely Bagina and Tasa (Sulaeman, interview, 3 January 2016). This revolt was quelled by the Dutch on 15 March 1909. In fact, some Dutch newspapers reported the event; however, there were no details on the time and place of this insurgence.

The Loloda rebellion emerged due to the tax system (*belasting*) and the forced labor imposed by the Dutch on the Loloda people at that time. This event, which occurred in the period 1908–1909, marked a major historical movement in Loloda, which resulted in the decline of the Loloda Kingdom. The story of "Kolano Madodaga" or "the end of the Loloda Kingdom" is known as an expression in many local traditions that has been used still.

2 CONCLUSION

Five important steps had been taken by the Loloda Kingdom to maintain its sovereignty in the Moluccas World until the early 20th century, including fighting against the Spanish in the early 17th century that almost destroyed the kingdom and migrating due to political conflict. However, the roots of social, cultural, political, and geohistorical traces of Loloda in Halmahera were still upheld. People migrated from Loloda only to survive by building a new "political space" and "space politics". Loloda had built a new political entity in the eastern coast of Sulawesi, which spread in Manado and Bolaang Mongondow, by establishing an alliance with Nuku (Tidore), Jailolo, the British, and several other districts in Halmahera to resist the expansion of Ternate and the Dutch as the "enemies" in the northern part of Halmahera (1780–1810). The relationship pattern of the "allies and enemies" as an "instability settling" was also used by the authorities of Loloda, causing a rebellion by the people led by Kapita Lao Sikuru from Laba village (1908–1909) against Ternate and the Dutch in Soasio. The rebels were against the taxing policy of Ternate and the Dutch imposed on the people of Loloda. This is because they thought that both the Ternate Kingdom and the Dutch were external parties and that taxes could only be collected by, from, and for the Loloda people themselves, and they continued to maintain its traditional indigenous governance structures, even though it was under the authority of Ternate. This has been evident in every election of Jougugu (leader) in Loloda. It is possible that in addition to the rebellion other factors have also contributed to the deterioration of the sovereignty of Loloda; however, further research is required to determine these factors.

According to the study of the Moluccas World, Loloda has left no trace in history. However, a study conducted by the author using the latest historical resources of North Moluccas proves that Loloda has still "existed" as a kingdom with a political entity configured formally into the structure of the Ternate Sultanate. However, its existence has been forgotten and left unrevealed.

REFERENCES

Abdurachman, P.R. (2008) *Bunga Angin Portugis di Nusantara: Jejak-Jejak Kebudayaan Portugis di Indonesia (The Portugese Wind in Nusantara: Traces of Portuguese Culture in Indonesia)*. Jakarta, LIPI Press-Asosiasi Persahabatan dan Kerjasama Indonesia-Portugal dan Yayasan Obor Indonesia.

Andaya, L.Y. (1993) *The World of Maluku: Eastern Indonesia in Early Modern Period*. Honolulu, University of Hawaii.

Arsip Nasional Republik Indonesia. (1980) *Ternate: Memorie van Overgave (MvO), J.H. Tobias (1857) dan Memorie van Overgave, C. Bosscher* (1859). Jakarta, ANRI.

Baarda, M.J. van. (1904) *Het Loda'sch, in Vergelijking met het Galela'sch Dialect op Halmaheira*. BKI. 56 (1).

Baretta, J.M. (1917) *Halmahera en Morotai Bewerk Naar de Memorie van Den Kapitein van Den Generalen Staf. Mededeeling E.B.-Aflevering XIII*. Batavia, Javasche Boekhandel & Drukkerij.

Campen, F.C.H. (1888) *Beschrijving van de Westkust van Het Noorder Schiereiland van Halmahera*. BKI. 37 (1).

De Clercq, F.S.A. (1890) *Ternate the Residency and Its Sultanate (Bijdragen tot de kennis der Residentie Ternate)*, Translated from the Dutch by Paul Michael Taylor and Marie N. Richards. Washington, D.C., Smithsonian Institution Libraries Digital Edition, 1999.

Dunnebier, W. (1984) *Over de Vorsten van Bolaang Mongondow (Mengenal Raja-Raja Bolaang Mongondow) (Introducing the Kings of Bolaang Mongondow)*. Translation, Responses, and Review from R. Mokoginta and F.P. Mokoginta. Kotamobagu, DPRD Kabupaten Dati II Bolaang Mongondow.

Hasan, A.H. (2001) *Aroma Sejarah dan Budaya Ternate (The Historic and Cultural Aroma of Ternate)*. Jakarta, Antara Pustaka Utama.

Hueting, A. (1908) *Iets over de Ternataansch-Halmaherasche Taalgroep*, BKI, 60 (I), 369–411.

Jacobs, H.T.M. (1971) *A Treatise on the Moluccas (c.1544): Probably Preliminary Version of Antonio Galvao's Lost, Historia Das Moluccas*, Edited, Annotated, and Translated in to English from the Portuguese Manuscript in the Archivo General de Indies, Seville. Rome, Italy: Jesuit Historical Institute via dei Penitenzieri 20 00193, St. Louis University, St Louis, Mo. 63103, USA.

Lapian, A.B. (1994) Bacan and the early history of North Maluku. In: Visser, L.E. (eds.) *Halmahera and beyond: Social science research in the Moluccas*. Leiden, KITLV Press. pp. 11–20.

Leirissa, R.Z. (1996) *Halmahera Timur dan Raja Jailolo: Pergolakan Sekitar Laut Seram Awal Abad Ke-19 (East Halmahera and the King of Jailolo: The Rebellion around the Seram Sea at the Beginning of the 19th Century)*. Jakarta, Balai Pustaka.

Mapanawang, A.L. (2012) *Loloda Kerajaan Pertama Moluccas (Sejarah Kerajaan Loloda Maluku) (Loloda the First Moluccas Kingdom [The History of the Loloda Kingdom Maluku])*. Tobelo, Yayasan Medika Mandiri Halmahera.

Nieuwe Amsterdamsche Courant, *Algemeen Handelsblad*, Mandaag, 15 March 1909, No. 25827, editie van 4 uur, 82e jaar, on the part of Onze. Oost, Nederlandsche Mail, Java-blaaden, 13–17 Febr., p. 13.

Nieuwe Amsterdamsche Courant, *Algemeen Handelsblad*, Zaterdag, 13 March 1909, 825 editie, van 2 uur, Ternate deel, p. 3.

Perpustakaan Nasional Republik Indonesia. *Soerabaiya Handelsblad*, Tuesday, 21 January 1908, Third Sheet.

Supit, B. (1986) *Minahasa: Dari Amanat Watu Pinawetengan sampai Gelora Minawanua (Minahasa: From the Watu Pinawetengan to the Gelora Minawanua Decree)*. Jakarta, PT. Sinar Harapan.

Yakub A.L. (1985) *Bukti-bukti Kerajaan Loloda yang paling besar dan yang paling tua dan meliputi seluruh Halmahera (Proof of Loloda as the largest and oldest kingdom and covering the entire Halmahera). (The Rise and Fall of Loloda Kingdom)* (Manuscript). Ternate, Kantor Kebudayaan Kota Praja Ternate, 1980–1985.

Zuhdi, S. (2010) *Labu Rope Labu Wana: Sejarah Buton yang Terabaikan (Labu Rope Labu Wana: the Ignored History of Buton)*. Jakarta, Radja Grafindo Persada.

Cultural Dynamics in a Globalized World – Budianta et al. (Eds)
© 2018 Taylor & Francis Group, London, ISBN 978-1-138-62664-5

Votive tablets in Buddhist religious rituals in the *Nusantara* archipelago from the 7th–10th centuries AD

A. Indradjaja & W.R. Wahyudi
Department of Archaeology, Faculty of Humanities, Universitas Indonesia, Depok, Indonesia

ABSTRACT: This paper discusses the function of votive tablets found in many Buddhist sites in the *Nusantara* archipelago from 7th–10th AD. As a ceremonial object, a votive tablet's function is analyzed through specific analysis and context. The result shows that the function of votive tablets in relation to religious rituals, besides having the primary function as a media ceremony (offering), it is also used to "turn on" the sacred building (*stupa*) and as supplementary *Garbhapātra*. The function of votive tablets found in the religious ritual is known to be the result of hybridization between the Buddhist faith and the belief of local communities. In addition, the presence of votive tablets in religious ceremonies along with *stupika* and seal tablets is associated with the view that in any religious ritual, the figure of Buddha should be present along with every object associated with the Buddha such as votive tablet representing *rupakaya* (physical element) and mantras of the Buddha representing *dharmakaya* (spiritual element). It is the unification of the two elements (*rupakaya* and *dharmakaya*) which will turn on Buddha in the religious rituals.

1 INTRODUCTION

The intensive maritime trade in Southeast Asia brought a "blessing" to the spreading of Buddhism into the Indonesian archipelago (Poesponegoro and Notosusanto, 2009: 21–26). According to Claire Holt, a researcher of Indonesian arts, the period between the 7th–10th centuries AD is an important period for our knowledge of Buddhism and Buddhist art in Indonesia (2000: 35). Besides leaving a trail of monuments and statues of Buddha, some devout Buddhists left numerous relics/artifacts related to the religious activities at that time, including votive tablets.

Votive tablets are small-sized Buddhist icons usually made of clay and sometimes mixed with ashes, then printed with press technique, and subsequently burned or simply dried and used as ceremonial objects. According to Robert Redfield, an icon is defined as a sign to the object it represents. Icons are not always in the form of depictions or expressions as known in the Greek Orthodox Church, but they can also be an analogy (1971: 42; Sedyawati, 1985: 44). Icon in Buddhism is a form of visual object representing certain concepts, and to some extent, some icons are adapted from other traditions, for example the statue of Bodhisattva Avalokiteshvara is influenced by the figure of the Hindu god Siva (Sedyawati, 2009: 85).

The finding of votive tablets is often accompanied by *stupika* and a seal tablet. *Stupika* is a miniature *stupa* made of clay and a seal tablet is a short inscription containing Buddhist mantras (verses), which is also made of clay.

The findings of votive tablets in Indonesia were first reported in 1896 in the area around Yogyakarta and later published in *Oudheidkundige Verslag* (OV) around 1931–1935, which mentioned the existence of several votive tablets in Jongke Village, located around 7 km north of Yogyakarta.

The interest of researchers on the findings of votive tablet was first demonstrated by W.F. Stutterheim toward the votive tablet found in Pejeng in 1920 (Stutterheim, 1930: 4). Other researchers also wrote about the findings of votive tablets, among others A.A.Gede Oka

Astawa, Hariani Santiko, Issatriadi, Endang Sri Hardiati, and Peter Ferdinandus from the National Archaeological Research Center.

The research on the votive tablets can contribute to the understanding of the existence of Buddhism in the archipelago. In the context of Buddhism in the *Nusantara* archipelago, Edi Sedyawati (2009: 87) explained that Buddhism at that time had absorbed the Javanese culture into the universal Buddhism mixed with the local culture (Java). In terms of ceremonial objects, one of which is a votive tablet, this study attempts to explain the relationship between the function of votive tablets within Buddhist religious rites and their position in Buddhist religious rites in the *Nusantara* archipelago.

2 THEORY AND METHOD

According to Christopher Dawson, to understand the cultural influence of a society, one must first understand the religion professed by the community:

> "Religion is the key to history. We cannot understand the essence of social order without understanding the religion the community embraces. We cannot understand the results of their culture without understanding the religious belief into the background. In every age the main results are based on the culture of religious ideas and immortalized for religious purposes" (Zoetmulder, 1965: 327; Munandar, 1990: 2).

In other words, to understand the cultural objects closely tied to the religious aspect, it is essential to first know the prevailing concepts in the religion.

The word "religion" in anthropology also has a variety of meanings. Tylor describes the essence of religion as the belief in spiritual beings which can be interpreted as magic and that everything in nature has a soul. The belief in spirits then evolves into a belief in the gods of nature. Furthermore, the belief in the gods of the nature evolves into various ranks of gods from the lowest to the highest (Agus, 2006: 120–121).

According to Melford E. Spiro, religion is an institution consisting of culturally patterned interaction with culturally postulated superhuman being (Spiro, 1977: 96–7).

Interaction, according to Spiro, has two senses, namely: (1) the activities which are believed to carry out, embody, or be consistent with the will or desire of the superhuman beings of powers and (2) the activities which are believed to influence superhuman beings to satisfy the needs of the actors. The concept of superhuman beings refers to the belief in the entities that own supreme power and can do good or bad things, and have relationships with humans.

All institutions consist of (a) belief system, i.e. an enduring organization of cognitions about one or more aspects of the universe, (b) action system, an enduring organization on behavior pattern designed to attain ends for the satisfaction of needs, and (c) value system, an enduring organization of principles by which behavior can be judged on some scale of merit (Spiro, 1997: 97–98). The religious actors not only believe in the truth of propositions about superhuman beings, but also believe in these beings. They not only believe in the truth of religion propositions, but also – and more frequently—to certain practices (Spiro, 1997: 106).

According to Spiro, the existence of religion can be explained causally and functionally. Religion persists not only because it has functions, in which it does, or is believed to, satisfy desires; but also because it has causes, namely the expectation of satisfying these desires (1977:117).

According to Koentjaraningrat, there are five religious components that can always be found in all denominations. The five components can also be referred to as a religious identity of a religious group because the five components may differ from one particular religious group to other groups. The five components are: religious emotion, belief systems, system of rites and ceremonies, equipment of rites, and communities (Koentjaraningrat, 1980: 80–3).

Among the five components that have been outlined by Koentjaraningrat, the equipment of rites component, which in this case is the votive tablet, will be the focus of the discussion. Of course, with the assistance of other data, votive tablet functions in religious rites can also be reinterpreted.

The source of research data was obtained through data collection from the Segaran V, Batujaya Temple, Karawang, West Java; Borobudur Temple, Magelang; Kalibukbuk Temple, Bali; Gentong Temple, Mojokerto and Gumuk Klinting site, Banyuwangi.

The stages of research conducted in this study were adapted from the archaeological research method introduced by K.R. Dark (1995), where each artifact or structure (findings) can be seen as data containing archaeological information. However, the data only inform about their own existence and cannot necessarily be treated as archaeological evidence. The new archaeological data become archaeological evidence after being put into the framework of interpretation (1995: 36).

2.1 Archaeological data

The findings of votive tablets in the *Nusantara* archipelago can be classified into six groups. In the broadest sense, classification can be understood as to enter different units into a certain class or type that has not been previously determined. If a certain number of artifacts are categorized into one type, it means that such artifacts can be described as a group having the same attributes, which are different from other types. The classification or categorization of the findings of votive tablets is based on the opinions of Brew and Rouse, who stressed that the classification of a type was entirely based on the design made by the researchers, but it is not something that is already available in the data (Sedyawati, 1985: 22–8). The classification of the type of votive tablets is based on the depiction of relief. Broadly speaking, the depiction of relief on a votive tablet can be grouped into relief of figures and non-figures. The grouping of relief figures on the votive tablets follows the "divine arrangement" known in Buddhism. There are three divine arrangements namely, Tathagata (Dhyani Buddha), Dhyani Bodhisattva, and Manusi-Buddha. Beyond the three divine arrangements, there are also known other deities such as Tara. Identification of figures was done by analyzing the iconography, such as the postures, the attributes of that subject, and others that may be used for classification. The purpose of the identification and classification of characters in the votive tablet is to identify the characters depicted on the votive tablets and their position in the Buddhist pantheon. The research only found two non-figure depiction relief in the votive tablets, namely on the *stupas* and *kundika*, which then fell into their own category of types. Thus, the classification of types of votive relief tablets becomes Tathagata, Dhyani Bodhisattva, Manusi Buddha, Tara, Stupa, and Kundika.

In the context of archaeology, votive tablets were found in several places, among others in the base of the temple such as the one found in the Kalibukbuk Temple, in wells, or "*sumuran*", of the temple such as the one found in the Gentong Temple, under the base of the north-western side and the corner of the east wall of the temple such as the one found in the Segaran V, Batujaya Temple, and in the courtyard of the temple such as the one found at Borobudur Temple. In addition, there are votive tablets that were found in the grave in Gumuk Klinting Site.

2.2 Votive tablets as offerings

Votive tablets found in the courtyard of Segaran V Temple, Batujaya Temple, and the courtyard of Borobudur Temple are likely to be associated with religious activities conducted at the temples, i.e. the offering brought by the pilgrims to the holy places. This may be analysed by drawing a comparison with the practice of using tablets as a votive offering in Tibet. Votive tablets in Tibet are found around the *stupa* and always associated with the activity of pilgrims traveling to the holy shrines. They always bring a votive tablet mold and after arriving at the holy places, they will collect clay and then mold it. They always recite a mantra during the making of votive tablets until it is completed. Votive tablets have been created and used in religious activities during their stay around the shrine and then the votive tablets will be placed at the edge of the *stupa* as an offering and partly taken home as a souvenir (Li, 1995: 2).

The cultic practices of giving offerings in the area of the *stupa* have been performed continuously for a long time, making votive tablets deposit found in considerable numbers when they were rediscovered (Chipravati, 2000: 183).

In an archaeological context (the context of making, using, and disposing), their religious rituals to provide some offerings were conducted in the *stupa,* and the votive tablets were used as part of the offerings. Therefore, many votive tablets were found around the *stupa.* After the ceremony is completed, the object of this offering is cleared from the altar for worship. As an offering of a sacred object, it certainly must be carefully cleaned. Ethnographic studies conducted by Walker about ritual discard found after ceremonies, the entire ceremonial objects would be cleaned and then dumped or buried outside the ceremonial site (1995: 67–79; Wahyudi 2012: 219) and offerings that are difficult to dispose of will be deposited in one particular place.

Thus, the findings of votive tablets in a hole created around the base of the stairs and the walls of the Segaran V, Batujaya Temple and on the west side of the Borobudur temple courtyard could be the result of worshipping around the *stupa* area using votive tablets as offerings, which were placed not only in front of a *stupa* but in certain places around the *stupa.* Another possibility is that the collection of votive tablets is the result of the cleaning activity in front of the *stupa* after the ceremony is completed. Regarding the sacredness of votive tablets, to clean up the effigies of votive tablets would require specific places around the *stupa* to keep them.

2.3 *Votive tablets as* "Peripih"

The findings of votive tablets on the base and in the wells of a temple are associated with the founding ceremony of the *stupa* itself. It is also associated with the position of the votive tablets, *stupika,* and seal tablets as sacred objects. In India, the practice of establishing a *stupa* is a tradition that has lasted for thousands of years even before the time of the Buddha and is known as *caitya* or *dhātugarbha* (Dorjee, 2001: vii). The *stupa* was then intended to save physical relics (*dhatus*), which construction was ordered by the Buddha in the manuscript of Mahaparinibbana Sutta (Dorjee, 2001: viii).

The records on the regulations for erecting *stupas* are the Raśmivimala and Vimalosnīsa manuscripts which original version is drafted in Sanskrit language and then copied into Tibetan language (Dorjee, 2001: 23–4).

Raśmivimala states that votive tablets used to appear only after the pre-construction phase has been completed. Before a votive tablet is placed, the foundation hole, which is called *singgasana* or throne, is first created. In the middle of the throne, votive tablets are prepared together with other holy objects, such as precious stones, seeds, and soil (Dorjee, 2001: 41).

After votive tablets are placed in the throne, the hole is closed by a cover called the throne. A staircase is built on the top cover and its center is empty. The empty middle part is then filled with as many other holy objects as possible including votive tablets, *stupika,* and seal tablets. It is said that the number of votive tablets stored in the domes affect the levels of power and blessings of the *stupa* (Dorjee, 2001: 42).

After the *stupa* construction is completed, the ritual of purification is performed. The purpose of this purification ceremony is to embrace the wisdom of the Buddha through the power of meditation practice, the power of the ritual, and the loyalty of the followers. The lessons drawn are then consecrated into the Buddhist *stupa* and tied in the *stupa* by certain ritual procedures (Dorjee, 2001: 42).

The construction process of *stupa* which positions the *stupika* and votive tablets under the base of the ancillary Kalibukbuk Temple, could lead to the conclusion that the Kalibukbuk Temple seems to have the same procedure. This is due to the fact that votive tablets, *stupika*, and a seal tablet were found in certain positions arranged in the base of the Kalibukbuk Temple.

The position of the votive tablets at a slightly different base *stupa* was found in Gentong Temple, Trowulan. In this temple, votive tablets containing Buddhist mantra were discovered in the temple along with the *stupika* wells. The position of the votive tablets in the wells of the temple seems to have similarities with the tradition of Hindu temple construction in Java in the 8th–10th centuries AD.

In the Hindu religion, the temple construction is noted in a number of books of Vastusastra, among others, Mayamata, Manasara, Matsya-purana, and Bhuvanapradīpa (Anom

1997: 104). The process of pre-construction is followed by the laying of the foundation stone in the central hole (*ārdhāsila*). After the foundation hole (pitting) has been filled with soil up to ¾ of its depth, the *ārdhāsila* is topped with *nidhikalaśa* (container of wealth) made of stone or bronze. Adhārasila is then covered with the building of the temple (Kramrisch I, 1946: 110–12).

Apparently, the same approach is applied in Gentong Temple with the temple's wells/ *sumuran* at Gentong Temple which is filled with *stupika* and votive tablets. Other findings in these wells are ash, fragments of gold, and precious rocks. All of these findings may be considered as the content of *peripih* (*pendeman*) commonly found in Hindu temples. The similarity between the contents of *peripih* in Hindu temples with *stupika* and votive tablets and other findings (ash, fragments of gold, precious rocks) in Gentong Temple is that both are considered sacred objects that required to bring the gods.

As sacred objects in the wells of the temple, the purpose of laying *stupika* and votive tablets in a Gentong temple well is likely to be similar with the laying of a *peripih* at a Hindu temple. *Peripih* functions to "turn on" the temple building. Without *peripih,* sacred buildings cannot be used as a place of worship. Through the ceremony, the magical powers of the gods are gathered into one in the *peripih* seed and completes the establishment of the new temple (Soekmono, 1989: 217–8).

2.4 *Votive tablets as supplementary to Garbhapātra*

In Buddhism, a person's death is seen as a process to achieve freedom. The manuscript of *Maha Parinibbanasutta* elaborates how to handle the bodies of the Buddha who passed away, how to handle the fire, and how to put relics in a *stupa* (Blum, 2004: 205).

The common funeral rites before the presence of Hindu-Buddhism in the *Nusantara* archipelago included burying the bodies. In practice, the deceased is buried along with a number of everyday objects, which are called the burial gift. The belief in the survival in the afterlife encourages people to send the departed soul with burial gifts (James, 1957: Soejono, 2008: 83). In addition to everyday objects, the burial gifts in certain cases also include sacred objects such as a ceremonial axe. (Soejono, 2008: 104). Upon the arrival of Hindu-Buddhism in the archipelago, Javanese people were introduced to the funeral ceremonial procession of cremating the corpse (Bernet Kempers, 1970: 213).

The burial practice ceremony in Gumuk Klinting Site shows hybridization between Buddhism belief and non-Buddhism (local indigenous beliefs). The existence of Buddhist teachings is shown by the ashes found stored in a bronze *cupu*. The indication of this ash being the ash of a corpse is hinted by other findings such as a small knife, glass beads and hair, which in Buddhism are known as *paribhoga*—the objects owned by the figure of Buddha during his lifetime. The whole *paribhoga* is then put in a terracotta container called *garbhapātra* or ritual vessel.

In Gumuk Klinting Site, ash and *paribhoganya* stored in a *garbhapātra* are not placed in a *stupa* but buried. Then, *garbhapātra* containing the ashes and *paribhoga* are stockpiled by the arrangement of votive tablets, *stupika* and a seal tablet to a depth of two meters. The practice of burying the ashes with votive tablets, a seal tablet and *stupika* is totally unknown in Buddhist tradition, considering that votive tablets, *stupika* and seal tablets are seen as sacred objects. There is a possibility of inclusion of votive tablet, and a seal tablet and *stupika* in the graves of Buddha figures as a complement of *paribhoga*. It is hoped that buried figures will get the blessings of the Buddha, so that in the next life they obtain a good position.

The burial practices of the Buddha figure in Gumuk Klinting Site that includes *paribhoga* seem to have similarities with the traditions of the local communities that include objects used in everyday life and in the case of particular figures coming with sacred objects.

2.5 *The role of votive tablet in Buddhist religious ritual in* Nusantara

One role of the votive tablets that is considered important is being a medium to obtain merit easily on an unlimited scale. It appears that the votive tablet as a medium to obtain merit can fulfill the wishes of his/her supporters in order to prepare for a better life after death.

The presence of votive tablets became important and mandatory in every stage of the religious rites, including votive tablets that are considered to represent the Buddha himself. In practice, votive tablets are always accompanied by a *mantra*, and it can be understood that the votive tablets are considered as *rūpakaya* (physical manifestation) of Buddha and mantras as *Dharmakaya* (spiritual beings). The unification between *rūpakaya* and *Dharmakaya* is essential to invite "the presence of" Buddha who seems to be present in the religious rites where the rituals are performed.

The necessity to invoke the "Buddha" in any religious ceremony prompted the creation of votive tablets to represent the presence of "Buddha" in the ceremony. Thus, it can be said that the role of votive tablets becomes very important in any religious activities.

3 CONCLUSION

In *Nusantara*, votive tablets are known to have important roles in religious rites. Based on the archaeological findings of votive tablets, it is indicated that the votive tablets are used for religious rituals, where they are utilized as offerings placed within the area of enshrinement, or *stupa*, in order to gain merit. It can be seen from the findings of votive tablets around Segaran V, Batujaya Temple, and Borobudur Temple. Votive tablets are also used to fill the foundation of the temple as the ones found in the Kalibukbuk Temple, Bali and Gentong Temple, Mojokerto. It is also used to invoke the power and blessing of Buddha. Votive tablets are also used as a complement to *garbhapātra* for the deceased Buddhist figures.

The practice of placing votive tablets in the wells of the temple as *"peripih/ pendeman"* and the burial practices using votive tablets as a complement to *garbhapātra* indicate that the Buddhism that flourished in the archipelago was the Buddhism that had adapted to the ancient local tradition and intertwined with the local belief adjusted to the community needs.

REFERENCES

Agus, B. (2006) *Agama dalam kehidupan manusia: antropologi agama (Religion in the life of man: religious anthropology)*. Jakarta, Rajawali Press.

Anom, I.G.N. (1997) *Keterpaduan aspek teknis dan aspek keagamaan dalam pendirian candi periode Jawa Tengah* (studi kasus utama Candi Sewu) *(The synergy of technical and religious aspects in the construction of temples in the Central Java period) (primary case study of Sewu Temple)*. Dissertation. Yogyakarta, Universitas Gadjah Mada.

Blum. (2004) *Death in Encyclopedia of Buddhism*. In: Buswell & Robert (eds). New York, Macmillan Reference. pp. 203–210.

Chipravati, P. (2000) Development of Buddhist traditions in Peninsular Thailand: a study based on votive tablet (seventh to eleventh century). In: Nora A.Taylor (ed) *Studies in Southeast Asian Art: essay in honor of Stanley J.O.Connor*. New York, Cornell University. pp. 172–193.

Chutiwongs, N. (1984) *The Iconography of Avalokitesvara in Mainland South East Asia*. Leiden.

Damais, L.C. (1995) Agama Buddha di Indonesia (The Buddhist Religion in Indonesia). In: *Epigrafi dan sejarah Nusantara (The Nusantara Epigraphy and History)*. (pp.85–99). Jakarta, Pusat Penelitian Arkeologi Nasional.

Dark, K.R. (1995) *Theoretical Archaeology*. New York, Cornell University Press.

Djafar, H. (2010) *Komplek Percandian Batujaya: Rekonstruksi sejarah kebudayaan daerah pantai utara Jawa Barat (The Batujaya Temple Complex: Reconstruction of the cultural history of the north coast of West Java). [Dissertation]*. Bandung, Kiblat.

Dorjee, P. (2001) *Stupa and its technology* a *Tibeto-Buddhist perspective*. New Delhi, Motilal Banarsidass Publishers Private Limited.

Guy, J. (2002) Offering up a rare jewel: Buddhist merit-making and votive tablet in early Burma. In: A green and T.R. Blurton (ed.) *Burma, art and archaeology*. London, The British Museum Press.

Holt, C. (2000) *Melacak jejak perkembangan seni di Indonesia (Tracing the art development in Indonesia)*. Soerdarsono (trans.) Bandung, arti.line.

Koentjaraningrat. (1980) *Sejarah teori antropologi I (The History of Anthropological Theory I)*. Jakarta, Universitas Indonesia Press.

Kramrish, S. (1946) *The Hindu temple Vol. 1*. Calcuta, University of Calcutta.

Li, J. (1995) *Image of earth and water: the tsa-tsa votive tablet of Tibet*. Retrieved from: https: //www. asianart.com/li/tsatsa.html [Accessed on 21st June 2016].

Munandar, A.A. (1990) *Kegiatan keagamaan di Pawitra gunung suci di Jawa Timur abad ke 14–15. (Religious activities in the Pawitra of the sacred mountain in East Java during the 14th–15th centuries)*. Depok, Program Studi Arkeologi.Universitas Indonesia.

Oudheidkundig Verslag (1931–1935) *Oudheidkundig Verslag uitgegeven door den Oudheidku ndige Dient van Nederlandsch-indië 1931–1935*.Weltevreden, Albrecht;'s-Gravenhage, Nijh off.

Poesponegoro, Djoened, M., & Notosusanto, N. (2009) *Sejarah Indonesia II (Indonesian History II)*. Jakarta, PN. Balai Pustaka.

Sedyawati, E. (1985) *Pengarcaan Ganeśa masa Kadiri dan Sinhasāri: Sebuah tinjauan sejarah kesenian. (Depiction of Ganesha Statues during the Kadiri and Sinhasari: A historical review) [Dissertation]*. Depok, Universitas Indonesia.

Sedyawati, E. (2009). *Saiwa dan Bauddha di masa Jawa Kuna (Saiwa and Bauddha during the Ancient Javanese period)*. Denpasar, Widya Dharma.

Soejono. R.P. (2008) *Sistem-sistem penguburan pada akhir masa prasejarah di Bali (Burial systems during the prehistoric period in Bali). [Dissertation]*. Jakarta, Pusat Penelitian dan Pengembangan Arkeologi Nasional.

Soekmono, R. (1974) *Candi fungsi dan pengertian (The function and meaning of temples). [Dissertation]*. Jakarta, Universitas Indonesia.

Soekmono, R. (1989) Sekali lagi: Masalah peripih (Once again, the issue of *peripih*). In: *Pertemuan ilmiah arkeologi V (Vthe Scientific Archeological Meeting)*. Jakarta, Pusat Penelitian Arkeologi Nasional. (pp. 217–225).

Spiro, M.E. (1977) Religion: Problems of definition and explanation. In: *Anthropological approaches to the study of religion*. London: Tavistock Publications. pp. 85–125.

Stutterheim, W.F. (1930) *Oudheden van Bali (vol. 1)*. Singaradja, Uitgegeven Door De Kirtya Liefrink.

Wahyudi. W.R. (2012) *Tembikar upacara di candi-candi Jawa Tengah abad ke-8-10 (ceremonial ceramics in temples of Central Java during the 8th–10th centuries)*. Jakarta, Wedatama Widya Sastra.

Cultural Dynamics in a Globalized World – Budianta et al. (Eds)
© *2018 Taylor & Francis Group, London, ISBN 978-1-138-62664-5*

Tracing religious life in the ancient Bali period: An epigraphical study

N.K.P.A. Laksmi & W.R. Wahyudi
Department of Archaeology, Faculty of Humanities, Universitas Indonesia, Depok, Indonesia

ABSTRACT: In this study, we investigate ancient inscriptions written by kings in the ancient Bali period. Data in the ancient inscription were collected, selected, and interpreted in accordance with the following aspects: the models of holy places in the ancient Bali period, the relationships between the holy places mentioned in the ancient inscriptions and the ones that are still found in Bali, and the relationships between holy places and the society in the ancient Bali period. The ancient inscriptions show that the holy places in the ancient Bali period were built on peaks of mountains/hills, at riverbanks, near water springs, at lake banks, and around residential areas. They consist of main and additional buildings and some of them can still be traced from their physical existence, such as the Puncak Penulisan Temple at Kintamani, the Gunung Kawi Bathing Complex, the Tirtha Mpul, the Goa Gajah, the *Bhatara Da Tonta* at Trunyan, and the Kedharman Kutri Temple. The ancient Balinese society had very much associated with these holy places, especially in protecting, maintaining, and preserving them. Such functions are being continued by the modern Balinese society by performing *laku langkah* (working together), sending *bakat-bakat* (offerings), and performing religious rituals/ceremonies as stipulated in the ancient inscriptions.

1 INTRODUCTION

Ancient inscriptions not only show historical data, but also religious data. Traces of Hinduism and Buddhism can be found since the 8th century as found in religious ancient inscriptions. Nevertheless, pre-Hinduism traditions still remain to date. Holy places built in the ancient Bali period are physical evidence of the period having close relationships with the local communities.

We will first focus on the patterns of the holy places of the ancient Bali period. Then, we will also discuss the relationships that these holy places have with the communities surrounding them and their obligations in maintaining the holy places both physically and upholding the rituals.

2 DISCUSSION

2.1 *The patterns of holy places in the ancient Bali period*

2.1.1 *Types of buildings found at the holy places of ancient Bali*
In general, no clear or precise direction is available to be used as the basis for identifying religious holy places in the ancient Bali period. Nevertheless, there are some criteria that can be used for that purpose. The first criterion is designations, some of which are used in both single and mixed manners. The designations are as follows:

1. *Hyang/Sang Hyang/parhyangan*[1]: In some ancient inscriptions of the ancient Bali period, the holy places and/or gods the people worshipped had the title *Hyang/Sang Hyang/parhyangan* placed before the name. Examples are *Hyang Api, Hyang Tanda, Hyang Karimama* (Bangli Kehen Temple A), *Hyang Tahinuni* (Gobleg Batur Temple A), *Sang Hyang at Turuñan* (Turuñan AI), *Sang Hyang Wukir Kulit Byu* (Batur Abang A/Tulukbyu Temple A), *Sang Hyang Mandala ri Lokasrana* (Bangli Kehen Temple C), and *Sang Hyang Candri ring Linggabhawana* (Selumbung).
2. *Bhatara*[2]: Examples are *Bhatara Da Tonta* (Turuñan AI), *Bhatara Puntahyang* (Sembiran AI), *Bhatara ring Antakuñjarapāda* (Dawan), *Bhatara Dharma Hañar* (Sukawati A), *Bhatara Bukit Tunggal* (Gobleg Desa Temple AII), *Bhatara Kuñjarasana* (Tejakula), *Bhatara i Tuluk Byu* (Abang Abang Temple B/Tulukbyu B), and *Bhatara Partapan Langaran* (Langgahan). Similarly to *Hyang/Sang Hyang*, the designation of *Bhatara* is often used to refer to a designated holy place and/or god.
3. *Ulan*: This designation is only found in an ancient inscription of Sukawana AI (804 Ś), which is *ulan* at *Cintamani Mal* (a holy place at Kintamani Hill).
4. *Dangudu/Pangudwan*: This designation can be found in an ancient inscription of Bangli Pura Kehen A in "*dangudu kibhaktyan sang ratu di hyang karimama*" *(Goris, 1954a: 59)* meaning "a holy place worshipped by king at Hyang Karimama". The designation of *pangudwan* (*pangudwan bhatāra i baturan*) can be found in the ancient inscription of Batuan.
5. *Partapān/patapan*: Examples are the *Partapan* at *Bukit Ptung* (Babahan I), *Patapan i thani Latêngan* (Angsri B), and *Partapān Langgaran* (Langgahan).
6. *Tirtha*: The designation of *tirtha* is only found in an ancient inscription of Manukaya (884), namely the *tirtha in (water) mpul*.
7. *Katyagan*: The designation of *katyagan* is only found in an ancient inscription of Tengkulak A (945 Ś) and Tengkulak E (1103 Ś) in *katyagan ring amarawati*.

In addition to the aforementioned designations, there are some terms that are possibly used to refer to holy places/holy buildings, such as *sambar*[3], *prasādā*[4], *wihara*[5], *sanga*[6], *śalā*[7], *meru*[8], and *parpantyan*. However, these terms do not clearly refer to any specific location (toponymy).

2.1.2 *Locations/toponyms of the holy places in the ancient Bali period*

On the basis of the information obtained in the ancient inscriptions, some locations/toponyms of holy places from the ancient Bali period exist even today, such as:

Hills of mountains and hills. Bukit Cintamani, now known as Bukit Kintamani, is at Kintamani District, Bangli Regency, where the Pucak Penulisan Temple is located. The temple is also known as Tegeh Koripan Temple or the Panarajon Temple. Another hill is Bukit Karimama, which is located in Simpat Bunut Village, Bangli District, Bangli Regency. There is a temple on the hill called Kehen Temple. Meanwhile, Bukit Tunggal, also known as Gunung Sinunggal, is located in Tajun Village, Buleleng Regency, where a holy place known as the Gunung Sinunggal Temple is found. Finally, Bukit Kulit Byu, named as Abang Mountain, is located in Abang Village, sub-district of Kintamani, Bangli Regency.

---

1. *Hyang* = *dewa, betara*, holy soul, a designation for gods or holy souls; *parhyangan* = holy building, a place for worshipping gods or holy souls (Granoka, et al., 1985:44).
2. The word "*bhatara*" comes from Sanskrit, meaning (1) "god", (2) "king" who has passed away and purified or identified as the god; (3) the king who is still alive or active in ruling; (4) a site or holy place for worship (Goris, 1954b:223; Granoka et al.,1985:16; Mardiwarsito, 1986:125).
3. *Sambar* = temple (holy place) (Granoka et al., 1985:93).
4. *Prasādā* = prasada, meru (Granoka et al., 1985:82).
5. *Wihara* = temple, dormitory (Granoka et al., 1985:117).
6. *Sanga* = holy place for worshipping God (Granoka et al., 1985:93).
7. *Śalā* = temple, holy place (Granoka et al., 1985:92).
8. *Meru* is a type of buildings served as a place for worshipping as the symbol of Mahameru Mountain. The style prioritizes its beauty of the terraced roofs called *tumpang*. The roofs are always uneven in number as 3, 5, 7, 9, or 11 (the highest).

Riverbank. To date, 13 parts of the temple have been found on the banks of Pakerisan River, in addition to 3 temple buildings, 47 meditation caves, and 4 springs of holy water (Srijaya, 1996:97). The Goa Gajah site is located on the banks of Petanu River, and various holy buildings were found also on the banks of Penet River, such as the Balinese temples of Pucak Bon, Pucak Tinggan, and Taman Ayun, which contain several archaeological relics from both prehistorical and historical periods (Srijaya, 2010: 198–199).

Springs. In addition to riverbank areas, some holy places can be found near water springs, for example, at Tirtha Empul. The ancient inscription mentions that it is "the holy water in the water" (Goris, 1954a: 75), which may refer to Yeh Mengening Spring, in Tampaksiring Village, Tampaksiring District.

The edge of the lake. Information of ancient inscriptions of *Turuñan* AI, AII, B, and C proves that the villages around the Batur Lake were unified through the shared responsibilities or obligations toward the holy place of Sanga Hyang at *Turuñan*, in Terunyan Village. In addition to the Batur Lake, the bank of Tamblingan Lake consists of a holy place and residential area.

Around the residential area. The holy places or holy buildings located around residential areas are *Hyang Api* and *Hyang Tanda*, which were places of worship. Worshipping *Hyang Api* and *Hyang Tanda* is often mentioned in the inscriptions of the ancient Bali period. *Hyang Api* and *Hyang Tanda* could be located separately; however, in some residential areas, they are worshipped together besides other holy places.

2.1.3 *Buildings of holy places in the ancient Bali period*

Originally, holy places built in the ancient Bali period were not constructed together with other supporting buildings. There is an indication that the buildings were built gradually to meet the needs and considering the capacity of the worshippers. The use of organic materials has made the survival of some of the buildings impossible. Only the buildings made of andesite materials, stones, and bricks can relatively endure. In addition, the dynamic characters of the Balinese people, who still use such certain religious buildings, have caused much changes to the ancient holy buildings. Nevertheless, according to Nyoman Gelebet (2002:10):

> "traditional architecture is a reflection of space for doing human activities with repetitions on styles from generation to generation with little or no change at all that is caused by religious norms and based by local traditions and inspired by the condition and potency of the nature."

On the basis of the information of the ancient inscriptions and archaeological remains that can still be observed and supported by the present condition used as comparison, the condition in the ancient Bali period can be formulated. The buildings at the sites of the holy places in the ancient Bali period are categorized into two types, namely main and supporting buildings.

2.1.3.1 Main building

The main building is the representative building and/or *stana* that is "worshipped or purified" at the site of the holy place. This building is also known as the *pewayangan* (shadow) or *persimpangan* (temporal place to stay). On the basis of this concept, existence of various types of main building styles is possible, from simple buildings to larger, more beautiful, and complicated ones. It should also be noted that the buildings could either stand by themselves or complement other buildings. The buildings are as follows:

Bebaturan. After the spread of Hinduism and Buddhism in Bali, the belief system in the *perundagian* period that was based on worshipping ancestors still continued to persist in the ancient Bali period and even now. Various buildings used for the purpose of worshipping ancestors in the *perundagian* and megalithic periods still continue to develop dynamically. For example, construction of a stone altar consisting of one or more stone standings as the base and one or more laying stones as the place to sit without holders had developed vertically so as to become a terraced stone throne consisting of some stone layers, completed with holders. The people of Bali now recognize such simple construction as *bebaturan*. The stone throne previously used as the throne of the soul of ancestor in the megalithic society is considered as the early form of *padmasana,* the throne of the Sun God in the present society (Bernet Kempers, 1977:180; Sutaba, 1995: 226).

Lingga-yoni. *Lingga* can be defined with various concepts, such as a "sign" ("sign" male sex), "principle of maleness", and "symbol of fertility" (Liebert, 1976:152). In *Lingga Purana, lingga* is defined as a "sign", *"phallus"*, and "cosmos substance" (*prakerti* or *pradhana*) (Kramisch, 1988: 162). Meanwhile, *yoni* means "sex of female", *"vulva"*, and "sign of womanish" (Liebert, 1976: 355). In terms of iconography, *lingga* is depicted in the form of a cylinder; however, previously, *lingga* was depicted to have vertically arranged three parts, having a round form in the upper part, eight sides in the middle part, and four sides in the lower part. Furthermore, there are other variants, such as *lingga* face mask or *lingga* that is given certain decorations to resemble *karavista.* From iconographical aspect, *yoni* is a *lapik* or the base of a *lingga* with various shapes, such as rectangle, square, round, or octagon.

Arca. On the basis of their forms, *arcas,* or statues, found in Bali can be grouped into five categories as follows:

- *Arca* as the God Representations, which is an *arca* having certain *laksana* (sign or attribute) based on iconographies of Hinduism and Buddhism. The gods are the personifications of *Brahman,* and in their concrete forms, they are manifested in the forms of *arca.* Therefore, their presence also becomes the symbols of truth and is considered the ruler and creator of nature (Radha Krishman, 1971:38; Setiawan, 2008:50).
- *Arca* as the Representation of Priest or *Rsi,* which is an *arca* of a holy figure and is usually depicted as bearded men. Priest or *rsi* is a holy person who usually receives revelation from the God and teaches Hindu religion, such as Rsi Agastya.
- *Arca* as Representation of Ancestors, which is the *arca* of the soul of the ancestors or those who are respected and have been sacrificed. The *arcas* wear clothes and jewels similar to god *arcas,* but they do not have any attributes that can be connected with the *arca* of certain gods with two arms. In general, their hands hold flower buds or they are in meditation position and frozen as a dead body.
- *Arca* of Animal, which seems to be included as complementary *arca* or *wahana* for the gods or the figures of the main *arcas.*

Candi. In Bali, sites of *candi,* or temple, are built in the forms of *candi* carvings and *candi* buildings. Candi carvings are the *candis* built by carving stone cliffs at the edges of the rivers. Meanwhile, *candi* buildings are *candis* built using standard materials and the ideas of the builders.

2.1.3.2 Complementary buildings

Complementary buildings are those built with the purpose of supporting praying activities. The buildings can be temporal, semi-permanent, or permanent. In fact, it is difficult to describe them, because the artifact data that can be observed are still limited. However, in the ancient inscriptions, some buildings are identified as follows:

1. *Satra*: The form of this building is not known to date. It may be made of organic materials, such as bamboos and/or woods. It serves as the resting place and praying area. In the ancient inscription of Sukawana AI on page IIb line 4, it is mentioned:

   ..."paneken ditu di satra pyunyanangku, kajadyan pamli pulu, tiker pangjakanyan anak patikern anak atar jalan almangen"... (Goris, 1954a:53–54).
   which means:
   "the offering there at *satra* I present is made to buyer of *pulu* (rice container), mat, to be cooked (rice) by people who cook (camp), those who do travelling until late at night".

   Such information is described at ancient inscription 10 Duasa, Pura Bukit Indrakila AI on page IIa lines 4–6.

2. **Bantilan/Wantilan** is a large and open building usually containing terraced roofs used for public meeting (Granoka, *et al.*, 1985: 14). *Bantilans/wantilans* can still be found in Bali, but they may not come from the ancient Bali period. All *wantilan* buildings found today have newer styles. The function of the building is to hold public meetings, art performances, and cock fighting as part of temple ceremony.

The Patterns of Holy Places in the Ancient Bali Period are related to their kinds, their locations (toponymy), the buildings that possibly exist inside the holy places, and the relations of every building at the holy place site based ancient inscription data. Humans have skills and make efforts to use what nature provides for their lives. They also maintain balanced relationships with their fellow humans, their environments, and the spiritual realm (superhuman beings). Therefore, they build various buildings with different styles and functions, such as buildings dedicated to the superhuman beings believed to influence their lives.

Buildings are constructed at the places considered holy or sacred and they are made for the purpose of worshipping. Since the prehistorical period, Balinese people have considered hills and mountains to be the places where their ancestors live, and they constructed various buildings from large stones (*megalithic*), such as dolmen, menhir, stone room, stone grave, and terraced holy place (Soejono, 1984:205–208; Setiawan, 2002:203). After the introduction of the Indian culture to Bali in the 8th century, it seems that the practice of constructing religious buildings at the holy places on the peaks of mountains/hills still continued, although with new styles supported by the stories of the gods living in the mountains. In addition, the existence of water is also one of the important elements in choosing the locations of holy places, not too far from the residential area.

Relationships of holy places with ancient Balinese society. The ancient Balinese society has close relationships with holy places. The ancient inscriptions show that relationship can still be seen from the obligation of each community. The obligation of each community related to the existence of the holy place in the ancient Bali period can be described as follows:

The king. He is the one who holds authority with a capacity to hold the tradition with his knowledge on the holy books as the basis of his decisions. The king was obliged to solve any arising problem, including those related to religious matters. For instance, when there was population decrease due to some factors, which in turn affected the existence of the holy place and prevented people from performing ceremonies, the king had an obligation to maintain and protect the holy places in his area, including the related tradition. The king had the right to instruct his people and kingdom.

Religious leaders. In addition to the obligation to promote the religion, religious leaders had the capacity to lead and provide "religious services" at the holy place for the people. The in-depth knowledge possessed by religious leaders facilitated them to become teachers in various subjects besides religious matters, such as in government, architecture, and *arca* arts; however, theoretically, those were not their real fields.

Nonreligious leaders. Different from religious leaders, nonreligious leaders tend to play the role of executive. For a long time, these leaders had been obliged to leading the activities in the building or maintaining processes of the holy places. These leaders were responsible for promoting religious decisions to their subordinators and the society. They were also informed of the complaints from the society that were delivered to the king.

People (society). They constituted the majority group in the kingdom/the country. In the development and the maintenance processes of the holy place, the majority group had the obligation to provide goods, money, and actions given for the sake of the holy place. Those were carried out, for example, by performing *laku langkah* (working together), sending *bakat-bakat* (offerings), and performing religious rituals/ceremonies regularly as stipulated in the ancient inscriptions.

On the basis of the ancient transcription, it is known that the ancient Balinese society has close relationships with holy places around them. These relations can include the right to worship, both personally and collectively.

3 CONCLUSION

On the basis of the data obtained from the ancient inscriptions, it is concluded that, in the ancient Bali period, the religious holy places were built on peaks of mountains/hills, at riverbanks, near water springs, at lake banks, and around residential areas. The holy places consist of main and complementary buildings. Some of them can still be traced, such as Puncak

Penulisan Temple (Kintamani), Percandian Gunung Kawi Compound, Tirtha Mpul, Goa Gajah, *Bhatara Da Tonta* (Trunyan), and Kedharman Kutri Temple. People in the ancient Bali period had very close relationships with the holy places, because each society plays a very significant role to foster, maintain, and preserve the integrity of the holy places so that they can be in a good condition.

REFERENCES

Gelebet, I.N. (2002) *Arsitektur Daerah Bali* (Bali Area Architecture). Badan Pengembangan Kebudayaan dan Pariwisata Deputi Bidang Pelestarian dan Pengembangan Budaya Bagian Proyek Pengkajian dan Pemanfaatan Sejarah dan Tradisi Bali.

Goris, (1954b) *Prasasti Bali II (Bali Inscription II)*. Bandung, NV. Masa Baru.

Goris, R. (1954a) *Prasasti Bali I (Bali Inscription I)*. Bandung, NV. Masa Baru.

Granoka, I.W.O., *et al.* (1985) *Kamus Bali Kuno-Indonesia (The Ancient Balinese-Indonesian Dictionary)*. Jakarta, Pusat Pembinaan dan Pengembangan Bahasa Departemen Pendidikan dan Kebudayaan Jakarta.

Kempers, B.A.J. (1977) *Monumental Bali. Introduction to Balinese Archaeology Guide to the Monument.* Den Haag, Van Goor Zonen.

Kramrisch, S. (1946) *The Hindu Temple*. Calcuta, Calcuta University.

Setiawan, I.K. (2002) Menelusuri Asal-Usul Tempat Suci di Bali dalam rangka Pengelolaan Sumber Daya Budaya (Exploring the origin of holy places in Bali in order to manage the cultural resources). In: *Manfaat Sumberdaya Arkeologi untuk Memperkokoh Integrasi Bangsa (Advantages of archeological resources to strengthen national integrity)*. Denpasar: Upada Sastra.

Srijaya, I.W. (1996) *Pola Persebaran Situs Keagamaan Masa Hindu-Buda di Kabupaten Gianyar, Bali: Suatu Kajian Ekologi (The spreading pattern of religious sites during the Hindu-Buddhism Era in Gianyar Regency, Bali: An ecological study)*. [Thesis]. Program Studi Arkeologi Bidang Ilmu Pengetahuan Budaya. Universitas Indonesia.

Srijaya, I.W. (2010) Faktor-faktor yang Mempengaruhi Keletakan Situs Arkeologi di Kabupaten Badung (Factors influencing the location of archeological sites in badung regency), In: *Mutiara Warisan Budaya Sebuah Bunga Rampai Arkeologis (Cultural heritage pealr: an anthology to archeology)*. Arkeologi Fakultas Sastra Kerjasama dengan Program Studi Magister dan Program Doktor Kajian Budaya. Denpasar, Universitas Udayana.

Cultural Dynamics in a Globalized World – Budianta et al. (Eds)
© 2018 Taylor & Francis Group, London, ISBN 978-1-138-62664-5

Ancient water system and reservoir at the Muarajambi archaeological site

C.M. Arkhi & W.R. Wahyudi
Department of Archaeology, Faculty of Humanities, Universitas Indonesia, Depok, Indonesia

ABSTRACT: *Muarajambi is* an archaeological site that had experienced numerous events of archaeological ruins in the period of 9th–12th century AD. The structural ruins at this site showed the influence of Buddhism on both the layout and art in the structures, buildings and sculptures. On the basis of the influence of Buddhism on these premises, this site was considered a Buddhist archaeological site. This claim was supported by the latest theory, according to which the Muarajambi archaeological site was a Mahavihara, that is, a Buddhist teaching centre in Nusantara. The theory was developed on the basis of the archaeological ruins and the historical information collected from the journals written by I-Tsing in the 7th Century AD. I-Tsing was a Buddhist monk who travelled from China to India and lived in Muarajambi for 6 months. Thus, we can assume that the greater area of the Muarajambi site was also populated by quite a number of people and, therefore, it required some type of water resource to support the activities taking place inside the Mahavihara. The hypothesis that the site was located near a water resource could be true because of the presence of a river watershed near the site. In addition, there were traces of water channels, ponds and reservoirs on the site, all of which were likely built by the monks who lived there in the past.

1 INTRODUCTION

As an archaeological site, Muarajambi has a great and varied collection of archaeological findings that have been originated from the classical archaeology era of Indonesia. Some of the artefacts included beads, potteries, ceramics, bronze structures, stone statues, slabs, brick buildings or structures and earth-mounds, which may have hided the ruins of brick structures (commonly referred to as *menapo* in the local language). The most recent research conducted on the subject was presented in a dissertation by Agus Widiatmoko (2015) entitled "*Muarajambi Site as the Mahavihara from VII—XII AD*", which compares the ruined archaeological sites in Muarajambi with the archaeological sites in Nalanda and Wikramasila in India. Widiatmoko used a historical archaeology method and a historical analogy technique simultaneously by analysing a documented records of a Chinese monk named I-Tsing, written during his pilgrimage from China to India to study the true teachings of Buddhism in its birthplace. The author also used a similar approach for this study.

The information collected from the documents of I-Tsing mentioned a stopover in a town on the island of *Suvarnadvipa* (historians and archaeologist agreed that *Suvarnadvipa* is the Sanskrit name for the island of Sumatera in the past) before heading to India. The town was called Fo-Shi. I-Tsing mentioned that he sailed down a river to reach Fo-Shi, which may indicate that the town was located in the hinterland part of the island and there might be a watershed nearby. He also mentioned that the town was surrounded by a fortress and there were thousands of Buddhist monks practicing the teachings of Buddha as in Madhyadesa, India. It is evident from the documents of I-Tsing that thousands of Buddhist monks studied in Fo-Shi; therefore, it could be assumed that the town had facilities to support their lives and activities.

To date, several archaeological sites with Buddhist characteristics have been found in Sumatera, such as the Padanglawas site in North Sumatra Province, the Muaratakus site in Riau

Province and the Muarajambi site in Jambi Province. I-Tsing's further narrowed down to two archaeological sites from the classical era with a high concentration of archaeological ruins located near watersheds, the Batanghari River and the Musi River. However, more number of archaeological ruins were found in the site located near the Batanghari watershed, specifically in the Muarajambi area. Archaeological findings in this site showed strong Buddhist characteristics either in written objects, such as the Buddhist mantras inscribed on the golden plates, or in the form of a Buddhist figurine of *Prajnaparamitha*, and curved structures of stupas made of bricks that can be found in several archaeological structures. This fact assured earlier researchers and writers that the ruins in the Muarajambi site are the archaeological remains of Fo-Shi from the past.

1.1 *Problem formulation*

Muarajambi is an attractive landscape, surrounded by ponds, lakes and streams. In his documents, I-Tsing mentioned a tradition followed by the monks in Fo-Shi that resembles a tradition practiced in India. If this was true, then it was possible that the daily activities performed by the monk community in Fo-Shi were similar to the ones practiced in Nalanda monastery, India. Furthermore, he cited the routines performed by the monks in Nalanda monastery and the tradition of building bath ponds, which was used by the monks every morning upon waking up (Takakusu, 2014: 241–242). I-Tsing also mentioned other activities such as washing hands after a meal and screening insects from the water obtained from the water source. These assumptions were confirmed by the fact that many of the archaeological findings were located near rivers and other water sources.

In archaeology, different approaches are followed for handling or reviewing a particular landscape. Historical archaeologists are often divided between the natural environment and artificial environment. This fact is probably based on the positivist perspective that always considers the interactions between humans and the natural environment as a process of environmental adaptation. Later, from the viewpoint of hermeneutics, we also have a differentiated the concepts of 'space' and 'place'. 'Space' can be interpreted as the environment where people live, whereas 'place' is the space that has been interpreted by humans (Hoed, 2011, p. 110).

The proximity to the archaeological ruins and the landscape of Muarajambi provide a base for the assumption of associations between water and the presence of streams, ponds and lakes. From the viewpoint of hermeneutics, we aim to conceive the landscape as a 'place' that has significance through the memory and values constructed by the past inhabitants. To achieve this objective, the author proposed the formulation of the research problem in the following statements:

1. What is the significance of the proximity between the archaeological remains and the water channels and reservoirs in Muarajambi?
2. What is the significance of water with respect to religious activities of the Fo-Shi monks in the past in Muarajambi?

2   RESEARCH METHODS

In general, archaeology is the study of life in the past using documented archaeological evidence (Dark, 1995, p. 1). However, not all material evidence (artefacts) found in an archaeological research can be regarded as archaeological data. Only the evidence of archaeological material that has information related to the one found later can be regarded as archaeological data. Furthermore, archaeological data must first clear an interpretation framework to be considered as archaeological evidence (Dark, 1995, p. 36), that is, archaeological data in a study should be placed within a systematic framework so as to justify it scientifically. In this case, the new archaeological data can be interpreted only if it is placed in the right context, in a good spatial context, time context, as well as cultural context. By following this sequence, the research can be divided into four phases, namely:

1. Determination of data sources
2. Collection of archaeological information on the data source
3. Establishment of the archaeological evidence
4. Interpretation.

The first stage of determining the source of data was carried out by reviewing the available literature works on the subject, the findings from previous studies as well as the restrictions on the problems, the formulation of the coverage area of the research and the availability of data in the area of research. This was done to segregate data resources that can only be obtained as a set or in a singular form (datum), which would later be used in this archaeological research. This study included data on water channels, ponds, reservoirs and the ruins—found in the archaeological area of Muarajambi.

After determining the data sources, the relevant archaeological information was extracted from them through describing the location and the shape and size of the data unit. This description was made to obtain an important aspect of the research, namely shaping the context of the archaeological data from its provenience. As mentioned by Dark, context is important in understanding archaeological data. To obtain a context, an archaeologist should determine the significant relationship between the findings and the surrounding environment (Hodder, 1991, p. 43). Therefore, at this stage of the formation of archaeological evidence, we processed the archaeological information obtained from the data source and combined them with other related archaeological data. This was done by plotting the archaeological features found along the waterways. The subjects in this case are the ruins of ancient brick structures and the *menapo*. A similar procedure was carried out on the artefacts located along or near the ponds and reservoirs in the Muarajambi site. Therefore, by effective engineering, we could distinguish potential water sources from less potential water sources. After obtaining the archaeological data, the next phase is interpreting them as archaeological evidence.

One of the approaches used in the data interpretation process for this type of research is analogy, which is generally used to describe an unknown object by comparing it to another object with similar identifiable characteristics for reference (Sharer & Ashmore, 2005, p. 455). In other words, an analogy refers to a comparison of two identical objects, between the archaeological data and reference objects, by focusing on their similarities. The analogy process used in this study is direct historical analogy, the purpose of which is to describe the identity or the relationship between the archaeological data and the reference object used in the comparison by observing the common traits shared by the object. Furthermore, the particulars of the activity are recorded in the historical sources (Gamble, 2001, p. 87). In this study, the author compared the archaeological evidence and the description provided by I-Tsing in his documents.

2.1 *Environment description*

Administratively, the Muarajambi archaeological site is located in the district of Muarajambi, Jambi Province, at approximately 40 km from the city of Jambi. In 2013, this area was designated as the National Heritage Area by the Regulation of the Minister of Education and Culture (Kepmendikbud No.259/M/2013). Geographically, the region of Muarajambi is located on the banks of the Batanghari River with an altitude of 8–12 m above sea level. Astronomically, the region of Muarajambi lies at 1° 24' South Latitude–1° 33' South Latitude and 103.22' East Longitude–103.45" East Longitude (Geographic Coordinate System) or 48 M 346839 mT–356779 mT; 9833468 mU–9840765 mU (Universal Transverse Mercator). The site is located in the area of a natural dike (flood plain) where floods flow from alluvial deposits from the Batanghari River (BPCB Jambi, 2004, p. 7). The physiography of Muarajambi mostly consists of a gently sloping plateau (67.72%), followed by choppy and bumpy hills. Most areas of the plains in Muarajambi are at an altitude of 10–100 m above sea level (74.95%), and only a few areas (4.50%) have an altitude >100 m above sea level (BPCB Jambi, 2004: 9). The level of surface water is higher in the eastern part than the western part because of the topographic nature of the land; the eastern part of Muarajambi is a basin and a swampy area and therefore water does not get absorbed immediately by the soil and it flows as a runoff. The stream patterns in Muarajambi, depending on the intensity of the rainfall, is secondary to the risk posed by stream floods that occur when the main river cannot support the water tributaries (BPCB Jambi, 2004, p. 10). Therefore, we may conclude that, in general, Muarajambi is a lowland area that is prone to floods.

2.2 Data

Data obtained in this study can be grouped into the following three categories: water channels, ponds and reservoirs. The following is the description of the terms used in this article.

2.2.1 Water channels

In this study, a water channel is defined as a long basin of water flowing through the archaeological area of Muarajambi. The 17 water channels and their local names (see the list and map in the appendix) in the region are grouped into the following classifications of *parit*, *buluran* and *sungai*. The distinction in the local names is most likely associated with the width of the stream; however, obtaining the true meaning of their names is beyond the scope of this study and may require more studies involving toponomy or other similar approaches.

2.2.2 Ponds

In general, a pond refers to a broad alcove on the ground, which contains water. In this article, a pond is considered a niche or a recess on the ground, where a small amount of water is collected. In the area under study, there are 10 ponds (see the list and map in the appendix).

2.2.3 Reservoirs

A reservoir refers to a huge water storage used for satisfying human needs or managing water distribution. In this study, reservoirs refer to large water bodies located in the archaeological area, including *dano* (lake), *payo* and *rawa* (local terms). In the study area, there are six reservoirs (see the list and map in the appendix).

2.2.4 Features

As previously mentioned, Muarajambi is an archaeological site where several evidences of archaeological ruins have been found, including some large structures. In general, the archaeological structures with ruins found in the area were made of bricks. To date, approximately 12 archaeological structures and 69 *menapo* have been found and registered by the BPCB (*Balai Pelestarian Cagar Budaya/the Agency for the Preservation of Cultural Heritage*) Jambi (see the list and map in the appendix).

2.3 Data association

The framework of interpreting the historical analogy was obtained by understanding the value of and information found in the archaeological evidence. For this purpose, the author used relevant historical sources, including the travelogues of I-Tsing, which contained some information related to the tradition of Buddhism practiced in his motherland, China, countries of the southern island, Fo-Shi, as well as India. Among the many traditions observed by him, I-Tsing claimed that the traditions practiced by the monks in Fo-Shi matched those practiced in India. He noted that the monks in Fo-Shi maintained conformity with the sutras (*a type of religious literature*) similar to that of India, which is evident from his documents. Regarding water, I-Tsing mentioned that there were two types of water for daily activities, namely water for drinking (*Kundi*) and water for rinsing (*Kulasa*) (Takakusu, 2014, p. 130). Both could be obtained from wells, ponds or streams, according to the rules written in the Book of Discipline or the *Vinaya Sutra*. The synchronic phenomenon was demonstrated by combining the archaeological data and the description shown by the historical data taken from the documents of I-Tsing. Therefore, they can be elaborated in the following discussion.

2.4 Association of water channels with the features

The first step in limiting the geographical scope of processed data in this study was grouping the water channels with potentially stronger linkages with the archaeological features. This was done by re-grouping the remaining water channels located at a distance of ± 100 m from the banks of the waterways. This argument is based on the assumption that human activities related to water could not be carried out too far from the water source. The total number of features after re-grouping is listed in Table 1.

The highest number of features for a water channel indicates that the particular channel has potential association with human activities related to water in the past. The channels with the highest number of features are Sungai Jambi, Sungai Parit Sekapung and Parit Johor. Thus, the next set of water channels which also have potential association with human activities directly intersect with the aforementioned three channels. They also have a number of features distributed along the vicinity (see the map in the appendix), including Sungai Terusan, Sungai Melayu, Buluran Paku and Buluran Dalam. There are also other channels with connecting drains to the trenches that surround a feature, such as Parit Candi Sialang (which is connected to the trench around Candi Sialang) and the Parit Koto Mahligai (connecting Sungai Terusan Channels to the trench of Candi Koto Mahligai). Moreover, the water flow in some channels shows an anomaly of a reverse flow at some points, such as the inflows in

Table 1. Total number of features.

No.	Water channel	Total number of features in the range of ± 100 m
1	Buluran Paku	7
2	Parit Bukit Perak	1
3	Parit Candi Sialang	1
4	**Parit Sekapung**	**14**
5	Sungai Amburanjalo	2
6	Sungai Berembang	0
7	Sungai Danau Ulat	0
8	Sungai Kemingking	5
9	Sungai Medak	0
10	Sungai Selat	2
11	Sungai Selat II	0
12	Sungai Seno	0
13	Sungai Terusan	1
14	**Parit Johor**	**13**
15	**Sungai Jambi**	**34**
16	Sungai Melayu	2
17	Buluran Dalam	1

Further details can be found in the appendix.

Table 2. Interconnected water channel(s).

No.	Reservoir in the Muarajambi area	Interconnected water channel(s)	Total number of interconnected water channel(s)
1	Rawa Buluran Kli	Parit Johor	1
2	Dano Kelari	Sungai Jambi and Sungai Selat	2
3	Dano Ulat	Sungai Selat II	1
4	Payo Rimbo Terbakar	Sungai Melayu	2
5	Payo Rawang Bato	Parit Sekapung	1
6	Payo Terjun Gajah	–	0

Further details can be found in the appendix.

Sungai Selat from the Batanghari River. The reverse flow also occurred in the intersections of Sungai Terusan, from Parit Sekapung to Sungai Jambi channel, and it might indicate human intervention or human engineering in the past.

Another category of the water channel in the Muarajambi area which the author suspects to have an association with human activities is the reservoirs. In general, the region of Muarajambi is a swampy area, which is prone to floods; therefore, if the Batanghari River overflowed during heavy rainfall, then the water will simply go to the watersheds. Every reservoir in the area has a connection to one or two water channels, as illustrated in Table 2.

Because of this phenomenon, reservoirs found in the Muarajambi area might be naturally formed or artificially constructed to address similar issues. Nevertheless, compared to the results of earlier analysis on the waterways combined with data of the features, hypothetically, there should be a network of dams in the Muarajambi area intersecting with the water channels, forming a potential association with community activities in the past.

2.5   Association of features with ponds

The presence of ponds around Muarajambi is also of interest for archaeological findings. Some of the features were found within a radius of ± 100 m around the ponds. The use of the ± 100 m range as the perimeter in the analysis of features–ponds association is related to the earlier analysis. The results are presented in Table 3.

The above-presented recapitulation results show that some ponds have potential associations with the features due to their proximity, namely Kolam Parit Johor I, Kolam Parit Johor II, Kolam Telago Rajo, Kolam Sangkar Ikan, Kolam Sungai Jambi, Kolam Candi Tinggi and Kolam Pemandian Ayam 2. Another argument to support the author's hypothesis of the presence of humans around these ponds is the aspects related to their size, length and width, which seemed to support the fact that any activity requiring water would be carried out as close as possible to the abundant water sources. This conviction was also supported by the findings from an earlier hydrological study conducted by Sutikno, Aris Poniman and Maulana Ibrahim revealing that the base of Telago Rajo ponds (Kolam Telago Rajo) were cut crossing the layer of groundwater surface, so the overflow of groundwater would be directed to the pond. This finding provides strong evidence that Kolam Telago Rajo is manmade (Sutikno, 1992: 117). After reviewing the study cited above, we might surmise that a pond or *kolam* with a similar shape and size to Kolam Telago Rajo might also be artificially made in the past.

The documents of I-Tsing about the routines of the monks in Nalanda monastery, India, mentioned a tradition of building a bath pond of the size of 1 *yojana* (15 km); there were about 20–30 ponds with the size of 1 *mu* (0.84 m²) to 5 *mu* (4.2 m²) (Takakusu, 2014: 241–242). The monks would go to these ponds every morning with a bath cloth (because nudity is prohibited in Buddhism). On the basis of an analogy about traditions performed in Fo-Shi and India with

Table 3.   Association of features with ponds.

No.	Ponds in muarajambi	Amount of features in the range of ±100 m
1	Kolam Parit Johor I	4
2	Kolam Sungai Selat	0
3	Kolam Sungai Jambi	2
4	Kolam Candi Tinggi	2
5	Kolam Parit Johor II	4
6	Kolam Pemandian Ayam	0
7	Kolam Pemandian Ayam 2	1
8	Kolam Sangkar Ikan	2
9	Kolam Telagorajo	4
10	Kolam Tanjaan Galah	0

Further details can be found in the appendix.

regard to the liturgy, we may surmise that the reason for adjacent arrangement of ponds as seen in the features in Muarajambi was to follow the traditions of pond-making in India.

3 CONCLUSION

Historical data in the form of travelogues from I-Tsing, a Chinese scholar, have provided the information and reference in the research on the archaeological ruins at the Muarajambi site. By using the hermeneutics approach that had previously been referred to in the introduction of the archaeological environment, the archaeological site is considered to have not only a 'space' but also a 'place', which support the values of the monks in the past. Therefore, on the basis of this fact, we may draw the following conclusions.

The waterways, ponds and reservoirs in Muarajambi were of significance for the monks who lived at the time, because they provided the means for performing their religious activities and routines.

The water channels that were closely associated with the activities of the monks in the past were Sungai Melayu, Buluran Paku, Buluran In, Parit Temple beehive and the branching river channels to the temple features of Koto Mahligai.

Reservoirs intersecting the water network of channels in the second point are closely associated with the function of the water channels. In addition, reservoirs in the region support the management system in overcoming the outburst of floods during heavy rainfall in Muarajambi.

Clearly, Kolam Telago Rajo is an artificially constructed pond, the shape and size as well as the historical data of which imply the presence of other similar artificial ponds in the area.

The Muarajambi site, which has been designated as a National Heritage Area, is one of the major references in the history of Buddhism in the archipelago, as demonstrated by the numerous archaeological ruins in the area. The author believes that this study will lay the foundation for further studies in this area, and using scientific tools, we will learn more about the reversal of water flow and the integrated water networks in the Muarajambi site.

REFERENCES

BPCB Jambi. (2004) *Naskah Master Plan Wilayah I Situs Muarajambi, Kecamatan Marosebo, Kabupaten Muarojambi, Propinsi Jambi. (Master Plan Area I of the Muarajambi Site, In Sub-District Marosebo, District of Muarojambi, Province of Jambi)* Jakarta, Ministry of Culture and Tourism.

Dark, K.R. (1995) *Theoretical Archaeology.* New York, Cornell University Press.

Gamble, C. (2001) *Archaeology: The Basic.* London, Routledge.

Hodder, I. (1991) *Reading the Past, (Second Edition).* Cambridge, Cambridge University Press.

Hoed, B.H. (2011) *Semiotik dan Dinamika Sosial Budaya. (Semiotic and Dynamics of Socio-Culture)* Depok, Komunitas Bambu.

Sharer, R.J. & Ashmore, W. (2005) *Fundamentals of Archaeology.* California, The Benjamins/Cummings Publishing Company.

Sutikno, Poniman, A., Ibrahim, M. (1992) Tinjauan Geomorfolog-Geografis Situs Muarajambi dan Sekitarnya. (A Geomorphological-Geographical Review on the Muarajambi Site and its Vicinity) *Proceeding: Seminar on Ancient Malay History* Jambi, 7–8th December 1992. Jambi, Province of Jambi.

Takakusu, J. (2014) *A Record of the Buddhist Religion as Practiced in India and the Malay Archipelago (A.D 671–595).* Jakarta, Kemendikbud. (Ministry of Education and Culture).

Cultural Dynamics in a Globalized World – Budianta et al. (Eds)
© 2018 Taylor & Francis Group, London, ISBN 978-1-138-62664-5

Association of Borobudur Temple with the surrounding Buddhist temples

N.A. Izza & A.A. Munandar
Department of Archaeology, Faculty of Humanities, Universitas Indonesia, Depok, Indonesia

ABSTRACT: Borobudur Temple is a masterpiece exemplifying the excellence of architecture in the Indonesian civilization in the past. Unsurprisingly, it is recognized by UNESCO as a tangible World Cultural Heritage Site. Several studies have been conducted by scholars and experts, and according to one of them, Borobudur Temple is associated with other temples, such as Pawon and Mendut Temples. In this study, we offer a new interpretation on Borobudur and its surrounding temples as a cohesive unit for sacred procession. Factors such as location, religious background, ornaments, and statues are common in Borobudur and Ngawen Temples. Here, we use an archeological method, and data were collected from Borobudur and three other temples. The facets of location, religious background, ornaments, and statues of these temples were studied in depth to collect the necessary data. The collected data were then applied in the context of the Mataram Kuno (Ancient Mataram) period using a religious framework. The last step encompasses interpretation of the data. The author believes that this study will provide a new interpretation on the roles of Borobudur and the surrounding Buddhist temples as monuments for sacred procession in the ancient times in addition to the role of the former as a World Cultural Heritage Site.

1 INTRODUCTION

Borobudur Temple is a masterpiece of the ancient Indonesian people. It was recognized as a World Cultural Heritage Site by UNESCO in 1991 (Ramelan et al., 2013, p. 28). In addition to its historical and cultural value, Borobudur was once the center of religious rituals of Mahayana Buddhism, which was corroborated by the existence of other temples with Mahayana Buddhism around it (see Map 1).

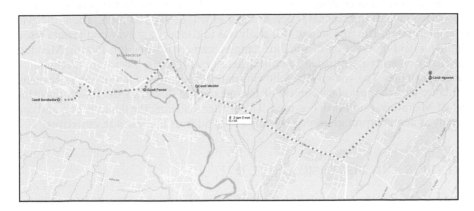

Map 1. Four temples from left to right (source: www.maps.google.com).

The first study on Borobudur was conducted during the Dutch East Indies era by Van Erp and N. J. Kroom, which coincided with the temple's restoration project (Ramelan et al., 2013, p. 27). The study indicated an association between Borobudur Temple and two other temples located nearby, namely Pawon Temple and Mendut Temple. This was based on the similarities with regard to the architectural style and ornamentation of the three temples, indicating that they were built in the same period, that is, the Sailendra dynasty era (Moens, 2007, p. 2). The next study was conducted by J. L. Moens in the 1950s (2007, pp. 93–99), which connected the three temples with Banon Temple, a Hindu temple located near Pawon Temple. Furthermore, it shows that Borobudur, Pawon, and Mendut Temples were all ritual centers of Mahayana Buddhism, whereas Banon Temple was a place for the followers of Siwa-Siddhanta. Another study conducted by IGN Anom imaginarily connected Borobudur, Pawon, and Mendut Temples, showing that the three temples were built along a straight line (Anom, 2005: 28). Totok Roesmanto also conducted a study on the location of Borobudur and the other temples surrounding it (2011: 99–120), which shows that the three temples are positioned along a single straight line, which was organized during the construction of Mendut Temple. It is also shown that the imaginary line connecting the three temples is linked to Mount Merapi.

Studies on the temples surrounding Borobudur show a similarity with regard to the period of construction, which is the era of Mataram Kuno (Ancient Mataram), as well as their religious affiliation, that is, Mahayana Buddhism, which excludes Banon Temple as it is filled with statues of Hindu Gods (Moens, 2007, pp. 94–98). These studies led to an interpretation that Borobudur Temple is highly associated with Pawon and Mendut Temples located in the east. The association between Borobudur and the two surrounding temples also identifies that the three temples were the centers for religious rituals in the past.

Geographically, Ngawen Temple is located in the east of Borobudur Temple. However, no study has been conducted revealing the association between Borobudur, Pawon, Mendut, and Ngawen Temples in the past. To further observe the association between the four temples, this study will focus on their location, religion, ornaments, and statues. The author believes that this research would provide a new interpretation of Borobudur and the surrounding Buddhist temples as monuments for sacred procession in the past and as a world heritage in the future.

2 METHODOLOGY

This study is conducted to explore the association between Borobudur and the other surrounding temples, especially Ngawen. On the basis of the framework established by K. R. Dark (1995, pp. 37–38), ensuring the availability of data resources in accordance with further discussion and the possible generation of data from the observation are crucial. These data will then be considered evidence for the context of the research. In the process of adhering the collected data to the context, relevant theories will be used as tools. Then, the evidence will be interpreted. The first step of this study was to conduct a survey on the data resources, including a field observation of the four temples, namely Borobudur, Pawon, Mendut, and Ngawen. The survey is intended to collect data regarding the association, religion, ornaments, and statues. The second step following data collection was to use them as evidence, in accordance with the periodical context of the research. In this stage, the theory regarding the belief of Mahayana Buddhism was used. Third, after matching the data, theory, and context, the analysis and interpretation phase took place to address the research objectives.

3 RESULTS AND DISCUSSION

3.1 *Association of the positioning of Borobudur Temple with the four nearby temples*

Borobudur Temple is located in the west of Elo River (see Map 1). The temple possesses several meanings related to the belief of Mahayana Buddhism. Moreover, in the past,

Borobudur had served as the center of other sacred buildings surrounding it (Huntington, 1994, p.136). Within a distance of 5 km around the temple, there are three other temples affiliated with Mahayana Buddhism, among which are Pawon Temple (1,150 m from Borobudur) and Mendut (2,900 m) (Kaelan, 1959: 122). Borobudur, Pawon, and Mendut Temples are located in the west of Elo River, and Ngawen is, in fact, located in the east side of the river, which is, in turn, 4 km away from Borobudur (see Map 1).

According to previous studies, Borobudur, Pawon, and Mendut Temples are positioned on a straight line and they form a triadic (a group of three) of sacred buildings affiliated to Mahayana Buddhism. However, according to Totok Roesmanto (2011, pp. 99–120), the imaginary axis connecting the three temples is not a straight line, and it is interpreted that they were the centers of religious rituals and processions in the past. Furthermore, it is suggested that the three temples were closely associated with Mount Merapi. Nevertheless, further examination of the map (Map 1) shows an addition temple called Ngawen Temple, from which a parallel imaginary axis can also be drawn, connecting it to the other three temples. Thus, on the basis of this fact, it can be interpreted that, in the past, the procession of the religious rituals might begin in Ngawen Temple and end in Borobudur.

3.2   Religious associations of Borobudur Temple with other nearby temples

Discussions on structures built during the Hindu–Buddhist era are highly associated with religious context. Revealing the religious background of a structure requires an observation of the components of the building. According to Soekmono (2005), temples in Indonesia can be classified in two major groups, namely Hindu and Buddhist temples. One of the main features of Buddhist temples is the existence of the stupas. A stupa is a bell-shaped structure of the shrine, which is a unique feature of Buddhist temples. Nevertheless, to explore more about the religious affiliation of a specific structure, we need to focus on the statues, reliefs, sketches, and other ornaments of structures.

The existence of stupas in the main body and reliefs of the temple (Photos 1 and 2) shows that Borobudur is a Buddhist temple. Stupas are also found in Pawon, Mendut, and Ngawen Temples.

In Pawon and Mendut Temples, the stupas are located on the top of the structures (Photos 3 and 4). The arrangement of the stupas is similar to that in Borobudur (Photo 5). Although the top part of Mendut Temple is not intact, stupas can still be found. Ngawen Temple, on the contrary, does not have a roof, but the ruins clearly show many stupa motifs (Photo 6).

Two major schools, namely Mahayana and Hinayana (Theravada), are found in Buddhism. Mahayana Buddhism is described as the "great vehicle", in which a holy man stays

Photos 1 and 2.   Stupas on the *Arupadhatu* level of Borobudur Temple (left) and stupa motifs on the reliefs of Borobudur Temple (right) (source: N.A. Izza, 2016).

Photos 3 and 4.   Pawon Temple with stupas (left) and stupas on the roof of Mendut Temple (right) (source: N.A. Izza, 2016).

Photos 5 and 6.   Stupas at the *Langkan* fence of Borobudur (left) and stupa ruins in Ngawen (right) (source: N.A. Izza, 2016).

Photos 7, 8, and 9.   *Bodhisattva* in Pawon Temple (left), Mendut Temple (center), and Ngawen Temple (right) (source: N.A. Izza 2016).

on the Earth, rather than going to heaven, in order to be able to help others (Irons, 2007: 17). Moreover, in Mahayana Buddhism, it is believed that a savior visits the Earth in the future, whereas Hinayana Buddhism or Theravada is described as a "small vehicle", in which the Buddha is merely the Buddha himself, without the presence of Bodhisattva.

In Borobudur Temple, there are relief panels about the life journey of Buddha, one of which is derived from the sacred text of Sutra Gandavyuha. This shows that Borobudur is a temple of Mahayana Buddhism (Irons, 2008; 17 & 57). In Pawon, Mendut, and Ngawen Temples, many of the relief panels depict the Bodhisattva (Photos 7, 8, and 9). This implies that these four temples are affiliated with Buddhism and follow the Mahayana school. In other words, the four temples located on the same imaginary axis are associated with one another on the basis of their religious affiliation, which leads to an interpretation of the four being the locations for sacred religious rituals in the past.

3.3 *Association of the ornaments found in Borobudur Temple with other nearby temples*

The ornaments include *kala-makara*, *kala*-reliefs, pillar ornaments, sculptures, stupa ornaments, antefixes, and other types of carving that decorate the temples. In this section, we will discuss from the smallest to the biggest temples because more ornaments can be found in bigger temples.

A staircase is found in Pawon Temple, whose left and right sides are adorned with reliefs, most of which, unfortunately, have faded. On the right side of the staircase, reliefs with *kala-makara* motifs are found (Photo 10). At the base of the staircase, most of the reliefs have also faded. The ruins of the reliefs show an image of *kinara-kinari* (heavenly creatures in the form of half-human half-bird) and some spiral plants. The relief panels are restricted to the pillar ornaments. In the body of the temple, two types of relief panels can be found with narrow *pradaksinapatha*, without balustrades. The first is a relief of the *Bodhisattva*, whereas the second relief is a depiction of the tree of life, with *apsara-apsari* (angels) at the top and *kinara-kinari* at the bottom (Photo 11). A long square hole is found on the second panel. On both relief panels and on the surface of the hole, pillar motifs can be seen. Inside the temple, there is a chamber for placing statues and recesses containing statues with the head of *kala* (Photo 12). On the upper section of the relief, ornaments (*quirlande*) depicting square-shaped flowers can be found. The antefix ornament with a *jaladwara* is found on the top in each of its four corners. At the very top, there is a composition of small stupas with bigger stupas placed in the middle (Photo 3).

Ngawen Temple is an installation consisting of five temples of the same size facing the same direction. Ngawen Temple was buried 1.5 m below the ground. Four of the temples' structures are in ruins, with only their base or foot visible, except for Ngawen II, which has been successfully restored. Each temple has a staircase with two sides containing ornaments depicting the *makara* with the same pattern of the *kala-makara* reliefs in Pawon Temple. At the base of the stairs, the reliefs of *Bodhisattva* are found with the tree of life in the middle, below which there is a jagged seam ornament surrounding the temple's structures (Photo 13). On the four sides of the temple foot, there are lion statues, each with a distinguished feature

Photos 10, 11, and 12. *Kala-Makara* in Pawon Temple (left), the tree of life reliefs in Pawon Temple (middle), and recesses of statues in the chamber of Pawon Temple (right) (source: N.A. Izza, 2016).

Photo 13. Jagged seam ornament surrounding the Ngawen Temple (source: N.A. Izza, 2016).

Photos 14, 15, and 16. Lion statues in Ngawen Temples I and V (left), in Ngawen Temples II, III, and IV (middle), ribbon garland motifs in Ngawen Temple II (source: N.A. Izza, 2016).

depending on its location; these lion statues can be found on temples I and V, as well as on temples II, III, and IV.

(Photos 14 and 15). In the body of the temple, there are recessed spaces filled with statues on both sides of the entrance, which is adorned with a carving of *kala* head containing pillars on each side. The upper part is adorned with ribbon garland motifs (Photo 16). Inside the chamber of Ngawen Temple II and in the ruins of Ngawen Temple V, there is a statue of Buddha in *padmasana* position without head and hands (Photo 17). The upper part is decorated with an antefix, whereas the roof of the temple has collapsed. The stupa is assumed to contribute to the ruins of the temple's roof (Photo 6).

Mendut Temple consists of a large main temple and smaller adjoining temples, only the bases of which are visible at present. The base of the temple is decorated with relief panels depicting a golden womb; the tree of life, surrounded by heavenly creatures; and jagged seams, surrounding the structure, which is the same as the ornaments found in Ngawen Temple (Photos 11 and 13). The reliefs on Mendut Temple follow a zigzag pattern between flowers and the tree of life.

The staircase in the main temple is adorned with *kala-makara* ornaments, which are bigger on Mendut than on Borobudur. The base of the relief panels is decorated with flowery carvings. On the first terrace, a balustrade can be seen. On the outside of the chamber located in the body of the temple, three large relief panels covering the three sides of the wall are found. The pattern of the panels on Mendut is the same as that of the panels on Pawon, which differ only in size. Relief ornaments on Mendut show the figure of Bodhisattva and the tree of

Photo 17.   Statues in the main chamber of Mendut Temple (source: N.A. Izza, 2016).

life, whereas the reliefs outside the temple's body are unreachable to human hands. The right and left sections of the main chamber of Mendut are adorned with the depiction of *yaksa* (Moens, 2007, p. 88). The reliefs are located on both sides of the entrance and are reachable to hands. The entrance of the main chamber has recessed niches without statues. The remaining statues are Bodhisattva statues on both sides as well as a Buddha statue in the middle. The Buddha statue in Mendut is the focal statue that depicts Buddha in *Dharmacakramudra* position (Photo 17). The front of the statue consists of an engraved ornament of a wheel between two deer. This is an illustration of the *dharma* wheel and the deer garden in the life story of Buddha (Moens, 2007, p. 88). The rooftop of Mendut Temple is partly in ruins; however, it shows that the top part of Mendut was once decorated with stupas as its peak with antefixes around it (Photo 4).

Borobudur does not have any chamber similar to that in the other temples. Moreover, the stupas in the other three temples are unreachable to human hands, with no statues inside, whereas those in Borobudur are arranged in different ways. According to A.J. Bernet Kempers (1976, p. 142), the stupas in Borobudur have a distinguished structure and arrangement that differs from that of other temples in Indonesia. The base or foot section of the temple is adorned with *Bodhisattva* figures. However, on the next terrace, the sculptured reliefs depict the life story of Buddha.

On both sides of the staircase located in all four sides of the temple, *kala-makara* ornaments are found. The reliefs on Borobudur are also dominated by engravings depicting the tree of life and heavenly creatures. The space between one panel and the next is adorned by pillars with shapes similar to the ones found in the other three temples. Antefixes in Borobudur are also similar to the ones found in the other three temples. Borobudur has much decoration along its circumference, as well as the decoration similar to the ones found in the other three temples.

Temple ornamentation may provide a clue for the identity of the temple. Agus Aris Munandar (2011) stated that from the viewpoint of architecture and style of the ornamentation, temples in Java may be grouped into two categories, namely temples built in the Old Classic era (including Buddhist temples in Central Java) and the ones built in the Modern Classic era. Furthermore, a study conducted by Marijke J. Klokke (2008) reveals that Buddhist temples that are the relics of Sailendra dynasty in Central Java show a distinct ornamentation characteristic. Earlier discussion on the ornamentation in Pawon, Ngawen, Mendut, and Borobudur Temples shows a similarity between one type of ornamentation and another with respect to the shape of the *kala-makara*, *kala* heads, antefixes, and reliefs of the tree of life and the decoration along the circumference. This similarity, according to Marijke J. Klokke (2008, pp. 159–161), shows that the four temples are Buddhist relics of the Sailendra dynasty. This ornamentation similarity confirms that the four temples were built in the same era with the same religious affiliation, namely Mahayana Buddhism.

3.4 *Iconography association of the statues in Borobudur Temple with other nearby temples*

One important aspect to be considered when observing temples is the existence of statues. According to Edi Sedyawati (1980, p. 209), a number of entities will be defined as the same characteristics of iconography, some of them are being created in the same period and on the same geographical location, having the same surroundings, belonging to the same religion, and being created under the same authority and by the same artists. Because Pawon Temple no longer has statues, the discussion in this section will be limited to Borobudur, Mendut, and Ngawen Temples.

Borobudur Temple comprises *Tathagatha* and lion statues. The *Tathagatha* statues in Borobudur are similar to the ones found in Ngawen Temple. The only difference lies in the backrest of the statues (*Prabhamandala*) found in Ngawen Temple because the statues found in Borobudur do not have *Prabhamandala* (Photos 18 and 19). In addition to the *Tathagatha* statues in Mendut, there is a statue of Buddha in sitting position with hands in a *Dharma-cakramudra* position and two *Bodhisattva* statues.

(Photo 17). Statues in Mendut were made of stone types different from the ones used in the making of Borobudur and Ngawen statues. Statues in Mendut were made of colorful stones with soft pores and are bigger than those in Borobudur and Ngawen, which are about the size of an adult human. Statues on Mendut are depicted as wearing jewels, whereas the ones in Borobudur and Ngawen are not.

Lion statues on Borobudur are depicted as standing with three legs and one of the front legs raised (Photo 20). Lion statues on Ngawen Temples I and V are depicted in a standing position with front and hind legs bending; meanwhile, on Ngawen Temples II, II, and IV, they are illustrated in a standing position with their hind legs and front legs supporting the temple (Photos 14 and 15). Although they were crafted to be in different positions, lion statues on Borobudur and Ngawen have the same facial expression and body shape.

Borobudur supposedly had two Dwarapala statues located around Dagi Hill (Picture 1), one of which is now in the Bangkok National Museum (Miksic, 2012: 18). Because the other statue has not been found, the Dwarapala statue was not included as the observation object of this study. In addition to the statues of Buddha and lion on Ngawen Temple, we found statues of Nandi, identical to those found in Hindu temples; therefore, it is assumed that the statues are an original feature of Ngawen Temple. This may be related to a series of disasters that destroyed Ngawen and buried it 1.5 m below the ground.

The study results of associations among the four temples from the perspectives of their positioning, religion, style of the ornaments, and iconography of the statues are presented in Table 1.

Photos 18 and 19. Buddha statue in Borobudur without *Prabhamandala* (left) and Buddha statue in Ngawen with *Prabhamandala* (right) (source: N.A. Izza, 2016).

Photo 20. Lion Statue on Borobudur Temple. (Source: Nainunis Aulia Izza, 2016).

Picture 1. The Painting of Dwarapala Statue in Dagi Hill near Borobudur (Source: H.N. Sieburgh in Miksic, 2012: 18).

Table 1. Summary of similarities among Borobudur, Pawon, Mendut, and Ngawen Temples.

Temple's name	Placement	Religion	Ornament	Statue iconography
Borobudur	Farthest west	Buddha Mahayana	Ornaments existing in other temples can be found in Borobudur	Tathagatha and lion statues
Pawon	1.15 km east of Borobudur	Buddha Mahayana	Kala heads, kala-makara, pillar, Bodhisatva, tree of life, heavenly creatures, and stupa summit	–
Mendut	2.9 km east of Borobudur	Buddha Mahayana	Kala heads, kala-makara, pillar, Bodhisatva, tree of life, heavenly creatures, stupa summit, jagged seam	Tathagata with soft manufacturing
Ngawen	4 km east of Borobudur	Buddha Mahayana	Kala heads, kala-makara, pillar, Bodhisatva, roof in shape of stupa summit, jagged seam	Tathagata, and lion statues in two styles

4 CONCLUSION

To conclude, Borobudur, Pawon, Mendut, and Ngawen Temples are located adjacent to each other and connected by an imaginary axis. Ornamentation style in Ngawen Temple shows a similarity among the four temples. The shapes of statues in Borobudur and Ngawen are also similar. Therefore, we may conclude that Borobudur is not only associated with Pawon and Mendut, but also correlates with Ngawen Temple in terms of religious affiliation, positioning, ornamentation style, and iconography as well as the structures that were used as locations for sacred Buddhist rituals during the Ancient Mataram era.

In addition to the association of Borobudur and the other temples, the association between Borobudur and Mount Merapi and the hills surrounding the temple is observed. More studies need to be conducted on Bubrah, Lumbung, and Sewu Temples. The similarities and the different patterns of association among Bubrah, Lumbung, Sewu Temples as well as the association between the Buddhist temples around Borobudur and those around Sewu Temple need to be studied.

ACKNOWLEDGMENTS

The author is grateful to Universitas Indonesia for providing a grant for this study through the indexed *Hibah Publikasi Internasional* program for the final projects of the students of Universitas Indonesia in 2016, without which the publication of this study would be impossible. Furthermore, the author thanks LPDP for providing a scholarship for the Master's program and also thanks all the lecturers at the Department of Archaeology, Universitas Indonesia, as well as the Borobudur Conservation Center for providing references and research permits.

REFERENCES

Anom, I.G.N. (2005) *The restoration of Borobudur*. Paris: UNESCO Publishing.

Dark, K.R. (1995) *Theoretical archeology*. New York: Cornell University Press.

Huntington, J.C. (1994) The iconography of Borobudur revisited: The concepts of Slesa and Sarva [Buddha] Kaya. In *Ancient Indonesian sculpture*. M.J. Klokke and P.L. Scheurleer. Leiden: KITLV Press. pp: 136–150.

Irons, E.A. (2008) *Encyclopedia of Buddhism*. New York: Infobase Publishing.

Kaelan. (1959) *Petundjuk tjandi: Mendut Pawon Borobudur* (Temple guidelines: Mendut, Pawon, Borobudur). Yogyakarta: Tjabang Bagian Bahasa, Djawatan Kebudajaan Departemen PP & K.

Kempers, A.J. (1976) *Ageless Borobudur Buddhist mystery in stone*. Servire: Wassenaar.

Klokke, M.J. (2008) *The Buddhist temples of the Sailendra dynasty in Central Java. Art Asiatiques*, tome 63, 2008. pp. 154–167.

Map Source: www.maps.google.com

Miksic, J. (2012) *Borobudur: Golden tales of the Buddhas*. New York: Tuttle Publishing.

Moens, J.L. (2007) *Borobudur, Mendut, and Pawon and their mutual relationship.* (M. Long, Transl). Retrieved from www.borobudur.tv.

Munandar, A.A. (2011) *Catuspatha: arkeologi Majapahit* (Catuspatha: Majapahit archeology). Jakarta: Wedatama Widya Sastra.

Ramelan, W.D.S. (ed.). (2013) *Candi Indonesia: Seri Jawa* (Indonesian Temples: Javanese Series). Jakarta: Direktorat Pelestarian Cagar Budaya dan Permuseuman (Directorate of Conservation of Cultural Heritage and Museums).

Roesmanto, T. (2011) *Keletakkan Candi Borobudur dan candi sekitarnya* (The positioning of Borobudur Temple and the temples surrounding it). Magelang: Balai Konservasi Borobudur (Borobudur Conservation Hall).

Sedyawati, E. (1980) *Pemerincian unsur analisa seni arca* (The elaboration of the elements of the statue art analysis). Pertemuan Ilmiah Arkeologi I (the First Scientific Archeological Meeting). Jakarta: Proyek Penelitian dan Penggalian Purbakala Departemen P dan K (Research and Ancient Excavation Project, Department of Education and Culture).

Soekmono, R. (2005) *Candi, fungsi, dan pengertiannya* (Temples, functions, and understanding about them). Jakarta: Penerbit Jendela Pustaka.

Forms and types of Borobudur's stupas

A. Revianur
Department of Archaeology, Faculty of Humanities, Universitas Indonesia, Depok, Indonesia

ABSTRACT: Candi Borobudur (Borobudur Temple) is the world's largest Mahayana-Vajrayana Buddhist temple, which is located in Magelang, Central Java, Indonesia. It was built by the Sailendra dynasty between the 8th and 9th centuries A.D. The temple was built with 10-step pyramid terraces, which are decorated with 2,672 relief panels, 504 Buddha statues, and 1,537 stupas. In this study, we aim to examine the forms and types of the stupas of Borobudur, which are composed of 1,536 secondary stupas and 1 primary stupa compared to those of other stupas found in Java and Bali islands. The stupas at Borobudur located from the second to ninth terraces are called the secondary stupas, whereas the one located on the tenth terrace is called the primary stupa. They are symbolic stupas, which consist of a base (*Prasadha*), a bell-shaped body (*anda*), a top support (*harmika*), and a top (*yashti*). The stupas are divided into four types, namely plain stupas, hollow space-square stupas, hollow space-diamond stupas containing the Dhyani Buddha *Vairocana* that represents the turning wheel of the dharma and the single main stupa that becomes the centre of Borobudur Temple reflecting Sailedra art-style. Here, we use a qualitative method, which is based on field observation and historical sources. The objective of this study is to provide a comprehensive description of the stupas in Borobudur from the perspective of historical archaeology.

1 INTRODUCTION

Candi Borobudur (Borobudur Temple) is known as a Mahayana-Vajrayana Buddhist archaeological site located in Magelang, Central Java, Indonesia. Borobudur was built between the 8th and 9th centuries by the Sailendra dynasty and was re-discovered by Sir Thomas Stamford Raffles in 1814 (Soekmono, 1976). Opinions of historians and archaeologists on the real name of Borobudur differ greatly. Raffles (1830) named it Boro Bodo, which means 'the name of the district' (Boro) and 'ancient' (Bodo). Thus, Boro Bodo can be interpreted as the ancient Boro (Raffles, 1830). The manuscript *Negarakrtagama* canto 77 (LXXVII) written by Mpu Prapanca in the 14th century mentions the domain of a *Vajradhara* or Vajrayana-Buddhism sect named *Kabajradaran Akrama* in Budur. This place is most likely the present-day Borobudur (Pigeaud, 1962; Soekmono, 1976). De Casparis (1950), on the basis of *Dasabhumika Sutra*, argued that Borobudur was built with 10 stages or *dasa bhumi*, and is named in relation to *Kamulan i Bhumisambharabhudhara*, which is mentioned in the inscriptions of Sri Kahulunan, dating back to the mid-8th century. *Kamulan* is a Sanskrit word, which means a root, origin, sacred place or shrine to worship ancestors. *Bhumisambhara* is interpreted as a barrow, hill or the level of the building that is identified with the temple called Borobudur. On the basis of the statements of de Casparis, *Kamulan i Bhumisambharabhudhara* can be interpreted as the temple of the Sailendra dynasty people located in Borobudur, which was built with 10 stages. Nevertheless, the original name of Borobudur is still under debate among scholars.

Borobudur Temple was built on a hill surrounded by several mountains, such as the Menoreh, Sumbing, Sindoro and Merbabu Mountains and Mount Merapi, and bordered by the Progo and Elo rivers. Borobudur could be linked to *Kunjarakunjadesa* in South India. Sands and andesite fractions lay the foundation of the hill where the Borobudur Temple is located (Moertjipto and Prasetyo, 1993). The temple was built as a stepped pyramid or *punden berudak* structure with six-square terraces and topped with three circular courtyards (Chihara, 1996).

The structure and function of Bodobudur Temple have been described in the literature. Borobudur could be regarded as a *Vajradhatu–mandala* and is, in turn, related to Mendut Temple, which is identified as a *Garbhadhatu-mandala*. The pair of mandalas is called *Dharmadhatu-mandala* (Chihara, 1996, Prajudi, 2009). According to a study conducted by Hoenig (as cited in Coomaraswamy, 1965), Borobudur was built as a temple. It was built with a nine-stepped structure, which served as a place of worship. Parmentier (1924) proposed that Borobudur was built as a monument crowned with a great stupa. However, because the large stupa structure could not withstand the huge weight, it was disassembled to attain the present-day structure of Borobudur (Chihara, 1996). Soekmono (1979) explained that Borobudur probably was not built as a temple, but as a place of pilgrimage where ancient Javanese Buddhists could gain knowledge. Magetsari (1997) argued that Borobudur was not established for the commoners, but only for a *Yogin*. Borobudur has a special place in the southwest area for the commoners and priests to accumulate a virtue. It could be seen that the spirit of the Monastic movement in India was materialised in this structure, which is influenced from the eastern school of India and the architecture of Bengal. Furthermore, Borobudur represents not only the creativity of Javanese geniuses but also one of the world's greatest constructional and artistic masterpieces (Brown, 1959).

Borobudur Temple has three levels representing the three worlds in the universe, namely *kamadhatu* or the world of desire, *rupadhatu* or the world of appearance and *arupadhatu* or the world without visual existence. At the level of *rupadhatu*, a man has left his desires but still has ego and resemblance. *Arupadhatu* is a world where ego and resemblance no longer exist. At this level, a man has been released from *samsara* and decided to break his affiliation with the mortal world. At Borobudur, *kamadhatu* is found at the foot of the structure, the five steps above it are described as *rupadhatu* and the third round terraces are described as *arupadhatu* (Stutterheim, 1956; Soekmono, 1974).

Similar to the Mahayana and the Tantric Buddhism, Vajrayana was practiced in Borobudur during ancient times, whose evidence could be found in the statues of *Pancatathagata* at Borobudur Temple, and it is related to *Guhyasamaja-tantra*. Another evidence is the teaching of *Paramita*, which is embodied in the relief of *Lalitavistara*, *Avadana* and *Jataka*. The implementation of *Yogācāra* and *Pramitayana* with Tantric philosophy is depicted by the relief of *Gandavyuha* and *Bhadracari*. The ability to integrate the philosophy of Tantric or Vajrayana and Mahayana through reliefs and sculptures in the temple indicates the high intelligence of Borobudur's architect. This is a unique feature of Borobudur (Magetsari, 1997).

The main component of a Buddhist temple, including Borobudur, is the stupa. The stupa, in the period before Buddha, would have served as a tomb and later became the symbol of Buddha's life. It was originally built to bury the relics of Buddha shortly after his body was cremated. In its further development, a stupa was used to store not only the relics of Buddhist monks but also Buddhist objects (Coomaraswamy, 1965; Soekmono, 1974). A stupa, which describes the concept of Buddhism, has several sections, namely the basis of the stupa (*Prasadha*), the parts of the ball (*dagob*) or bell (*genta*) and the top or crown (*yashti*) (Dehejia, 1972; Moertjipto and Prasetyo, 1993). The stupa was also decorated with parasols (*chattra*) at the top of the *yashti* (Fogelin, 2015). Kempers (as cited in Mentari, 2012) showed that Borobudur Temple is covered by stupas on its terraces. The stupas in Borobudur have a form different from that of other stupas in Indonesia. Borobudur, as the legacy of the Sailendra dynasty, has 1,537 stupas, which could be subdivided into 1,536 buffer stupas and 1 main stupa.

Many research studies on the stupas of Candi Borobudur have been carried out by various scholars. Academic discussions on the meaning of the stupas at Borobudur are integrated by Magetsari (1997), who stated that the stupas of Candi Borobudur represent the *Parinirvana* and emphasise the philosophy of Vajrayana and Mahayana Buddhism. The mini-thesis research conducted by Mentari (2012) describes the classification of the forms and types of Borobudur stupas. It intends to complement the research on the stupas of Borobudur by expanding it with historical archaeology studies as the framework. The stupas of Borobudur are compared to those of other Buddhist temples such as Mendut, Pawon, Ngawen, Kalasan, Sari, Lumbung and Sewu Temples as well as the Ratu Boko archaeological site and Pura Pegulingan in Bali.

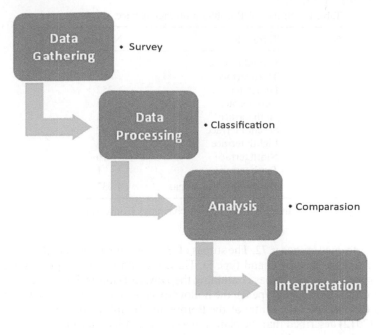

Figure 1.   Stages of research on the stupas of Borobudur.

2   METHOD

In this study, we focus on the stupas at Borobudur Temple and their comparison with stupas in other archaeological sites such as Mendut, Pawon, Ngawen, Kalasan, Sari, Lumbung and Sewu Temples as well as Ratu Baka archaeological site and Pura Pegulingan in Bali. We use a qualitative method, which is based on archaeological research such as the stages of data gathering, data processing, analysis and interpretation (Ashmore & Sharer, 2010).

The data were gathered from the sites in Borobudur, Mendut, Pawon, Ngawen, Kalasan, Sari, Lumbung and Sewu Temples; the Ratu Boko archaeological site and Pura Pegulingan; furthermore, and an archaeological survey approach was applied. This stage was followed by data processing. In this stage, the stupas were observed, described, classified and assessed by comparing them with others. The classification of stupas in Borobudur is based on the research conducted by Mentari. In the third stage, the meaning and function of stupa were analysed. In the final stage, the stupas at Borobudur Temple and in other archaeological sites were interpreted on the basis of the comparison of each stupa.

3   DISCUSSION AND DATA ANALYSIS OF BOROBUDUR'S STUPAS

Borobudur has 1,537 stupas located from the second-level terrace to the tenth-level terrace. The number of stupas on each terrace varies, as shown in Table 1.

Table 1 shows that the number of stupas on each Borobudur terrace is different. The third terrace has the most number of stupas (416). On the basis of this evidence, it can be concluded that the number of stupas on each terrace is the multiples of 8, except on the second, fifth and tenth terraces. The perforated stupas are erected from the seventh to the ninth terrace (Balai Konservasi Peninggalan Borobudur, 2004).

Mentari (2012) classified the stupas of Borobudur into four types. The author suggests that there are two forms and four types of stupas in Borobudur. The forms are plain and perforated. The plain stupas can be found from the second to the sixth terrace and in the tenth terrace, where the great stupa is located. The number of plain stupas is 1,465, whereas the

Table 1. Stupas of Borobudur on each terrace.

No.	Terrace level	Number of stupas
1	Second terrace	116
2	Third terrace	416
3	Fourth terrace	352
4	Fifth terrace	316
5	Sixth terrace	264
6	Seventh terrace	32
7	Eighth terrace	24
8	Ninth terrace	16
9	Tenth terrace	1
	Total number of stupas	1,537

Source: Center for Borobudur Archaeology Conservation (2004).

number of perforated stupas is 72. The stupas of Borobudur can be classified into four types, namely type A, type B, type C and type D. These four types are the plain stupas (type A), the hollow space-diamond stupas containing the Dhyani Buddha *Vairocana* that symbolises the turning wheel of *dharma* (type B), the hollow space-square stupas (type C) and the single main stupa that becomes the centre of the Borobudur Temple (type D).

Mentari (2012) described that the plain stupas or type A have a *Prasadha* with ornate lotus seams (*dalla*) and a semi-circle (*kumuda*), solid *Anda*, rectangular *harmika* and basic circle-shaped *yashti*. This type of stupa is the smallest in Borobudur. There are 1,464 stupas of this type, which are located from the second to the sixth terrace on the ledges, niches and roofed-gates or *paduraksa*. The hollow space-diamond stupas or type B stupas are characterised by a *Prasadha* with a flat seam (*patta*), lotus (gentha-side), ornate lotus (*dalla*) and semi-circle (*kumuda*), and the hollow space-diamond stupas are characterised by an *Anda*, a rectangular *harmika* and a basic circle-shaped *yashti*. There are 56 stupas of this type, which are located on the seventh and eighth terraces. The type B stupas contain a statue of Dhyani Buddha Vairocana with the *mudra* or hand gesture of *Dharma Chakra Parvatana*. The hollow space-square stupas or type C stupas are characterised by a *Prasadha* with a flat seam (patta), lotus (gentha-side), ornate lotus (*dalla*) and semi-circle (*kumuda*), and the hollow space-square *Anda*, by an octagonal *harmika* and basic octagonal-shaped *yashti*. There are 16 stupas of this type, which are located on the ninth terrace. This type of stupas contains no statues. The main stupa or type D is characterised by a *Prasadha* with a flat seam (patta), a lotus (gentha-side), an ornate lotus (*dalla*) and a semi-circle (*kumuda*). The *Anda* is a solid, rectangular and octagonal *harmika*, with basic octagonal-shaped *yashti*.

The *Yashti* on the main stupa has not been fully restored since the discovery of the temple. The reconstruction of the *yashti* was carried out on the basis of a picture showing that it previously contained three parasols (*Chatra*). However, the reconstruction was disassembled because there were many wrong interpretations, and the original stone of the *Chatra* was not suitable for reconstruction (Soekmono, 1976). There is a main stupa, with a belt adorned with vines, which is located on the tenth terrace. It has been suggested that the main stupa should be stylised with a parasol. The types of stupas and their position on each terrace are described in Table 2.

It has been found that the plain stupas (type A) are located from the second to the sixth terrace. In the author's opinion, the plain stupas on these terraces are probably used as the boundary between the ledge (*Vedika*) and the floor (*pradaksinapatha*). The hollow space-diamond stupas are located on the seventh and eighth terraces, and a statue of Vairocana inside them holds a religious meaning. According to the text of the *Vairocanabhisambodhi Sutra* (2005), Vairocana is placed in the centre of the mandala. It is suggested that the circular structure of Borobudur, which contains Vairocana statues, is part of the centre of mandala. Magetsari (1997) explained that Vairocana with *Dharmacakra-mudra* hand gesture meditates in the *Mayotama-samadhi* position. It is suggested that Vairocana has reached Buddha.

Table 2. Forms and types of stupas and their position on each terrace.

No.	Terrace level	Stupa form	Number of stupas	Stupa type
1	Second terrace	Plain stupas	116	Plain stupas
2	Third terrace	Plain stupas	416	Plain stupas
3	Fourth terrace	Plain stupas	352	Plain stupas
4	Fifth terrace	Plain stupas	316	Plain stupas
5	Sixth terrace	Plain stupas	264	Plain stupas
6	Seventh terrace	Perforated stupas	32	Hollow space-diamond stupas
7	Eighth terrace	Perforated stupas	24	Hollow space-diamond stupas
8	Ninth terrace	Perforated stupas	16	Hollow space-square stupas
9	Tenth terrace	Plain stupas	1	Main stupa
Number of plain stupas	1,465			
Number of perforated stupas	72			
Total number of stupas			1,537	

Type A   Type B   Type C   Type D

Figure 2.   Images of the four types of Borobudur's stupas.
(Source: Aditya Revianur (2016)).

Thus, Buddha has carried out his activities through his body as an intermediary. This argument is visualised in the temple in the form of type C stupas or the hollow space-diamond stupas. The Dhyani-Buddha Vairocana statues placed in this position show the ambiguity between the being and nothingness or *maya*. Buddha was visualised in the *arupadhatu* stages, but he is still able to carry out his activities. The activities presented in *arupadhatu*, which teaches *Dharma*, finalise and liberate all beings. They are manifested from the seventh to the ninth terraces. The Vairocana Buddha statue is placed in this stupa in order to demonstrate his activity; Buddha teaches all beings and to all directions.

The hollow space-square stupas, which do not contain a Buddha statue, hold a higher position than the hollow space-diamond stupas. They have become a symbol of the last level *arupadhatu*, eventually reaching a Parinirvana stage, which is symbolised by the main stupa. According to *Parinirvana Sutra* (as cited in Magetsari, 1997), Buddha of *Kamadhatu* went to the top, and after passing through various levels in *arupadhatu*, he entered the level of *arupadhatu* to eventually reach the level where feelings no longer exist. Then, Buddha went down from the top to the lowest level of *rupadhatu*. Then, he again reached the highest level of *rupadhatu* to finally enter *Parinirvana*.

3.1   Comparison between the stupas of Borobudur and other temples

Stupas are also found in other Mahayana-Vajrayana Buddhist temples, such as Mendut, Pawon, Ngawen, Kalasan, Sari, Lumbung and Sewu Temples, as well as the Ratu Boko archaeological sites and Pura Pegulingan. The form of plain stupas found in all of these sites is similar to that of Borobudur. Meanwhile, the form of perforated stupas can only be found in Borobudur. Table 3 shows a comparison of the form of plain stupas in Borobudur and in other temples and sites.

Table 3 shows that plain stupas can be found in Mendut, Pawon, Ngawen, Kalasan, Sari, Lumbung and Sewu Temples. It has been found that the plain stupas in Sewu Temple are similar to those of Borobudur, which are placed on the ledge. They reinforce the opinion about their function at Borobudur as a boundary between the ledge (*Vedika*) and floor (*pradaksinapatha*). The Sewu Temple was built in the late 8th century having *Vajradhatu-mandala* structure, with a great number of Dhyani Buddha figures (Suleiman, 1981, Chihara, 1996). However, there was limited information on the essence of Boddhisattva who was worshipped in the temple because the inscription of Kelurak, which was found at Sewu Temple, provided information only about a temple named Manjusri-grha or the house of Manjusri, and it could be built to worship Manjusri (Magetsari, 1981). On the basis of the similarities found, it can be suggested that Borobudur and Sewu Temples are most probably correlated and were erected at the same time. Both temples also represent Sailendra art and Mahayana and Vajrayana sects, such as stupas on the ledge and *Vajradhatu-mandala* structure.

The author found the main stupas or type A stupa at Pawon, Kalasan, Sari, Lumbung and Sewu Temples, as well as at the Ratu Baka archaeological site and Pura Pegulingan. The main stupas at Mendut and Ngawen Temples were built during the ancient times. The main stupas in both temples have collapsed or been damaged and cannot be reconstructed, as well as the

Figure 3.   Vairocana statues inside a hollow space-diamond stupa.
(Source: Aditya Revianur (2016)).

Table 3.   Stupa forms of Borobudur and other Buddhist temples.

No.	Temples/sites	Stupa forms of Borobudur and other Buddhist temples	
		Plain stupas	Main stupa
1.	Mendut Temple	√	
2.	Pawon Temple	√	√
3.	Ngawen Temple	√	
4.	Kalasan Temple	√	√
5.	Sari Temple	√	√
6.	Lumbung Temple	√	√
7.	Sewu Temple	√	√
8.	Ratu Baka archaeological site		√
9.	Pura Pegulingan		√

main stupa at Kalasan Temple, but several parts of the *Prasadha*, the slotted-square *Anda*, *Harmika* and parts of the damaged *Yashti* can still be seen. The main stupa at Pura Pegulingan in Bali, which is similar to a miniature of stupa, can be found at these sites.

The miniature of stupa at Pegulingan is probably related to the statues of *Pancatathagata*, which itself is related to the Vajrayana Buddhism doctrine, and it could be seen on the Dhyani Buddha statues that are placed on the four corners of the stupa (Astawa, 1996). The Ratu Boko Temple was built as a vihara and named *Abhayagirivihara*; it still preserves the legacy of Buddhism (Magetsari, 1981). The author found the type of the main stupa at Ratu Boko, but it has been reconstructed ever since its discovery. Meanwhile, Ngawen, Mendut, Pawon, Kalasan, Sari, Lumbung and Sewu Temples have one main stupa surrounded by plain smaller stupas. The main stupas at these temples are the symbols of *Parinirvana*. It has been found that these main stupas in each temple were not decorated with parasols or *chattra*.

4 CONCLUSION

The stupas of Borobudur Temple have two forms and four types. The two forms are plain and perforated stupas. The plain stupas are located from the second terrace to the sixth terrace and serve as the boundary between *Vedika* and *Pradaksinapatha*. A similar function is also found at Sewu Temple. The perforated stupas and the main stupas of Borobudur Temple are related to the teachings of both the Mahayana and Vajrayana Buddhism sects. These stupas, which are located on the seventh and eighth terraces, serve as a preparation stage to enter the *Parinirvana*, which is symbolised by a single main stupa. The Dhyani-Buddha Vairocana statues that are placed inside these stupas indicate the symbols of the Vajrayana sect. According to *Parinirvana Sutra*, the hollow space-square stupas symbolise the last level of *arupadhatu*, which eventually leads to the Parinirvana stage that is manifested by the main stupa. The stupas of Borobudur Temple and those found in other Buddhist sacred sites in Java represent the art of the Sailendra dynasty, which dates back to the period between 7th and 10th centuries. No parasols of the stupas have survived, and because of the limited archaeological evidence and references, it is impossible to reconstruct the parasols of the stupas. Borobudur itself was erected in the late 8th century on the basis of the comparison of its stupas with those at Candi Sewu. The stupas of Sailendra are related to the development of the Mahayana and Vajrayana Buddhism that were practiced not only at Borobudur but also in other Buddhist temples in Indonesia between the 7th and the late 14th centuries, according to *Negarakrtagama*, which mentions Vajrayana temples in Java and Bali.

REFERENCES

Ashmore, W. & Sharer, R. J. (2010). *Discovering our past: A brief introduction to archaeology*. New York: McGraw-Hill.

Astawa, A.A.G.O. (1996). *Agama Buddha di Bali kajian artefaktual* (Buddhism in Bali: An artifactual analysis). [Thesis]. Depok: Fakultas Sastra Universitas Indonesia.

Balai Konservasi Peninggalan Borobudur (Borobudur Heritage Conservation Hall). (2004) *Data ukuran bagian-bagian candi Borobudur: Ukuran dalam meter* (Measurement data of Borobudur temple parts in meter). Magelang: Balai Konservasi Peninggalan Borobudur.

Brown, P. (1959). *Indian architecture, Buddhist and Hindu periods*. Bombay: D. B. Taraporevala Sons & Co. Private Ltd.

Chihara, D. (1996). *Hindu-Buddhist architecture in Southeast Asia*. Leiden: Brill.

Coomaraswamy, A. K. (1965). *History of Indian and Indonesian art*. New York: Dover Publications, Inc.

De Casparis, J.G. (1950). *Inscripties uit de Cailendra-tijd: Prasasti Indonesia I*. Bandung: A.C. Nix & Co.

Dehejia, V. (1972). *Early Buddhist rock temples*. Ithaca, New York: Cornell University Press.

Fogelin, L. (2015). *An archaeological history of Indian Buddhism*. New York: Oxford University Press.

Giebel, R.W. (2005). *The vairocanabhisambodhi sutra*. Berkeley, California: Numata Center for Buddhist Translation and Research.

Magetsari, N. (1981). Agama Buddha Mahayana di kawasan nusantara (The Buddha Mahayana religion in the Indonesian archipelago). In Ayatrohaedi (Ed.) *Seri penerbitan ilmiah 7*. Jakarta: Fakultas Sastra Universitas Indonesia. pp. 1–28.

Magetsari, N. (1997). *Candi Borobudu rekonstruksi agama dan filsafatnya* (Borobudur Temple: Religious reconstruction and the philosophy). Depok: Fakultas Sastra Universitas Indonesia.

Mentari, G. (2010). *Bentuk dan tata letak stupa di Candi Borobudur* (The forms and positions of stupas at Borobudur Temple). [Mini-Thesis]. Depok: Fakultas Ilmu Pengetahuan Budaya Universitas Indonesia.

Moertjipto & Prasetyo, B. (1993). *Borobudur, Pawon, dan Mendut* (Borobudur, Pawon, and Mendut). Yogyakarta: Kanisius.

Parmentier, H. (1924). Dv A. Hoenig: Das formproblem des Borobudur. In *Bulletin de'Ecole Francaise d'Extreme-Orient*, 24, 612–614. Retrieved from http://www.persee.fr/doc/befeo_0336–1519_1924_num_24_1_3038.

Pigeaud, T. G. Th. (1962). *Java in the 14th Century: The Nagara-Kertagama by Rakawi Prapanca of Majapahit, 1365 A. D.* Leiden: Koninklijk Institut voor Taal-, Land- en Volkenkunde.

Prajudi, R. (2009) Perjumpaan dengan budaya India dan Cina (The meeting with the cultures of India and China). In Gunawan Tjahjono (Ed.). *Sejarah kebudayaan Indonesia: Arsitektur* (The cultural history of Indonesia: Architecture) (pp. 159–235). Jakarta: Rajagrafindo Persada.

Raffles, T. S. (1830). *History of Java vol II*. London: John Murray.

Soekmono. (1978). *Candi Borobudur* (Borobudur Temple). Jakarta: PT Dunia Pustaka Jaya.

Stutterheim, W.F. (1956.) *Studies in Indonesian archaeology*. The Hague: Martinus Nijhoff

Suleiman, S. (1981) *Monuments of ancient Indonesia*. Jakarta, Pusat Penelitian Arkeologi Nasional.

Cultural Dynamics in a Globalized World – Budianta et al. (Eds)
© 2018 Taylor & Francis Group, London, ISBN 978-1-138-62664-5

Variety of distinct style scripts in inscriptions found in Mandalas of the late Majapahit era: An overview of the paleography to mark religious dynamics

N. Susanti

Department of Archaeology, Faculty of Humanities, Universitas Indonesia, Depok, Indonesia

ABSTRACT: The late Majapahit era is found to mark the decline of Saivaism and Buddhism. In the late 15th century, Buddhism started to decline, whereas Saivaism was still developing, despite experiencing a decline. Trailokyapuri II and Trailokyapuri III inscriptions, which were issued by the King, mentioned names of figures and gods other than Siva, namely Sang Rsiswara Bharadhwaja, Bhatara Wisnu, Bhatara Yama, and Bhatari Durga. Literature works have also suggested life perspectives and "indigenous" religiousness as well as the establishment of religious buildings with mountain top features (*punden berundak*) and pyramidal architecture, such as the constructions in the slopes of Mount Penanggungan and Mount Lawu. Distinct script styles found in brief inscriptions in several sites that served as centers of religious activities (*mandala*) may confirm the assumptions regarding the religious life lead in the late Majapahit era. The content of the inscriptions provides hints of holy places/*mandala*, names of figures and gods, and moral teachings of the time. Paleographic analyses carried out using a dynamic method for the distinct script styles found in brief inscriptions from the late Majapahit era may provide information regarding the types of the currently existing *mandala*. Previous studies show that every *mandala* normally possesses a unique script style together with its diacritical symbols. The analysis of the content of the inscriptions reveals the figures, gods, and moral teachings of the era. Ultimately, the variety of distinct script styles may contribute to the information related to the life outside the palaces of the late Majapahit era, especially the religious life.

1 INTRODUCTION

The late Majapahit era has marked the decline of two religions, namely Saivaism and Buddhism. In the late 15th century, the role of Buddhism as an official religion started to decline, whereas Saivasim sustained and flourished up to the late 15th century, despite experiencing a decline. The inscriptions and literature works of that time have also revealed the existence of figures and gods belonging to neither Saivaism nor Buddhism.

For instance, the content of Trailokyapuri II and III inscriptions issued by King Girindrawarddhana Dyah Ranawijaya mentions the existence of other figures and gods, such as Sang Rsiswara Bharadhwaja, Bhatara Wisnu, and Bhatari Durga. Literature works have also suggested views of life pertaining to an "indigenous" religion and the establishment of religious structures, including those with mountain-like and pyramidal shapes, such as some constructions found in the slopes of Mount Penanggungan and Mount Lawu.

When Hinduism and Buddhism were introduced in Nusantara, they grew in an environment that had developed its own faith system, namely the worship of the spirit of the ancestors. Together with the development, these three systems of faith influenced each other, which was reflected in not only the system of the ideas but also the ritual activities and the material manifestation of the objects used to support such activities (Sedyawati & Djafar, 2012, p. 286).

Centers of religious development from the early stages were distinguished into three categories based on the environment where the religious teachings were developed, namely the palace area, the hermitage area, and the village area. Two groups of religious leaders were involved in palaces, namely the court priests and the high officials in charge of religious affairs. Hermitage sites that were later also referred to as *mandala* are religious activity sites located somewhat away from the living areas, normally deep in the forest, in mountain slopes, or in caves. The caves, for example, which were situated in the high plateau of Dieng and Ratuboko, are assumed to be the oldest centers of worship in Java. The word *Dieng* or *Dihyang* (ancient Javanese language) would possibly indicate the dwelling place of the spirit of the ancestors or *hyang*. The proximity of such caves and the oldest Hindu religious building compound and the finding of prehistoric ritual equipment, such as copper drum (*nekara*), possibly indicate the area having been used since prehistoric time by a group of people (Sedyawati & Djafar, 2012, pp. 286–295).

The description of hermitage life was obtained later, especially from the ruling era of King Airlangga and the era of the Majapahit Kingdom through the content of inscriptions and literature works conducted in the two eras.

On the basis of manuscripts of the Janggala-Kadiri era, namely Bhomantaka and Sumanasantaka, and those dated from the Majapahit era, such as Nagarakrtagama, Rajapatigundala, Arjunawijaya, Tantu Panggelaran, and Pararaton, some terminologies are indicated to relate to the naming of such religious centers, namely *mandala, kuti, dharmasala, karĕsyan, wanaśrama/asrama, patapan,* and *kadewagurwan.* Santiko (1990, p. 159) assumed that *wanasrama* or dormitory refers to a sacred place of hermits or *karĕsyan,* whereas *mandala* is another term for *kadewagurwan* (Sedyawati & Djafar, 2012, p. 292).

Soepomo (1977, pp. 66–67) distinguished *patapan* from *mandala* and stated that *patapan* is a quiet place of retreat that a person visits and stays for a certain period for a certain intention, whereas *mandala* is a compound of housing for hermits that is more of a permanent base and the occupants are referred to as *tapaswi* and *tapi* (Sedyawati & Djafar, 2012, p. 293).

Religious centers that flourished during the ancient Javanese era as well as served as places of learning and teaching are referred to as *widya gocara* as mentioned in the book Sutasoma. The book Sarasamuccaya describes a place of hermitage or *patapan* as the one that served as the dwelling place of a master (*Mpu*) and his students (*sisya*). Furthermore, it is described as being surrounded by a fence made of solid walls and having a gate, a pond, and a large banyan tree, under which the students played (Sedyawati, 2012, p. 302). Several types of other religious centers are also described in the literature works, for instance, *kadewagurwan* is described as being located at the base of a mountain and was occupied by a family of religious leaders or *dwija* (Brahmana) and holy women.

The educational system followed was similar to that of *gurukula,* with a single student being directed by his teacher at a hermitage. Literature works denote an interactive learning–teaching process, involving direct verbal communication between the teacher and the students. The materials taught were related to religion and the literature. The book Pararaton lists the subjects taught to the students, which included scripts, knowledge on the use of vocals, consonants, and change of sound of scripts or *candrasangkala;* elements of calendar system; and so on. It also includes the teaching of good characters, truth, and proper behaviors (Gonggong, 1993, pp. 101–117).

Further studies on the inscriptions of the late Majapahit era would find that there are several inscriptions that have distinct characters, indicated in their size and shape. These inscriptions are not as large as other inscriptions in general, which have a height of >1 m, and the shape is less regular. Examples include inscriptions made on natural stones that are relatively small. Scripts carved on these inscriptions are short, and their form has specific motifs different from those used to record king's announcements. The language used is ancient Javanese of Middle dialect. Unlike that of inscriptions issued by the royal palaces, the content of these inscriptions is not related to king's announcements, the elements of which are arranged chronologically. These inscriptions contain names of places, gods, and priests, as well as moral teachings, which leads us to assume that these inscriptions were made and issued by people who lived in *mandala*, comprising priests, students, and sages (Susanti, 2008; 2011, pp. 3–4; Rahayu, 2016, p. 13).

On the basis of several previous studies, it is assumed that there were religious centers, such as *mandala, kadewagurwan, kuti,* and *patapan,* during the late Majapahit era, as mentioned in the manuscript of Nagarakrtagama. By linking the sites where these inscriptions were found, the information obtained from Nagarakrtagama, and the journey of Bujangga Manik, the locations of those religious centers can be deduced. The chronological data of the *mandalas* may be concluded from the hints found in the content of the inscriptions and the development of the shape of the scripts.

The development of the script style can be observed by two methods, namely static method and dynamic method. According to the static method, introduced by de Casparis, the script is a hierarchy of lines. Thus, in analyzing, observing the scripts one by one is sufficient. The simplest script is the latest (de Casparis, 1975, p. 66). According to the dynamic method, introduced by Jean Mallon, the script is not just one line, rather a result of hand movements. The script consists of real and unreal elements. For example, in order to make a real line to attract a specific direction, the hand needs to be lifted first to make a real line on the different sides. The script changes can be seen from the combination of real and unreal elements (*duktus*). Increasingly complicated *ductus* indicates the younger characters of script style (van der Molen, 1985, pp. 9–10).

Bujangga Manik was a Sundanese Hindu priest who lived in the late 16th century. He went wandering overland from his place of origin in Sunda region to the eastern tip of Java and to Bali, and went back to West Java. He wrote his experience in a kind of travelogue (Noordijn, 1984, p. 1).

Anton Wibisono (2006) studied the development of scripts of different styles using the dynamic method, that is, by studying scripts on the basis of their shapes, script typeface (*ductus*), corners, size, and thickness. Inscriptions discussed at the time were Gerba, Widodaren, 1371 Śaka Damalung inscriptions, and Pasrujambe inscriptions. This research concluded that the oldest scripts with distinct styles found in various sites in East Java and Central Java were those of the Sukuh group of inscriptions.

From the temple compound of Sukuh, 10 inscriptions and statues as well as independent reliefs were found. According to the experts, the inscriptions and construction elements were of the same era, namely between 1361 Śaka, as mentioned in one of the reliefs, and 1363 Śaka, as written on the rear part of the statue of Bhima. In the travel journal of Bujangga Manik, he mentioned that during the westbound trip he went past villages in the southern part of Mount Wilis and Mount Lawu. Despite the names of the villages mentioned here being not recognized as the Sukuh temple compound is situated in the slope of Mount Lawu, there would most probably be scriptoria/*mandala* near the temple compound (Susanti, 2010, p. 6). Figure 1 shows the shapes of the scripts of inscriptions found in the compound of Sukuh temple.

More recent *mandalas* were believed to be located near Malang based on the finding of Gerba inscription in the village of Gerba in the regency of Malang and the finding of Widodaren inscription in the village of Widodaren. The names of these two inscriptions come from the villages where they were found. The shapes of the scripts used in these two inscriptions are the same, and so is the language used, namely ancient Javanese with Middle dialect. The content of these inscriptions is about moral teaching. Gerba inscription deals with sincerity in marriage life that is described as having the weight of heaven and earth and balance. Widodaren inscription addresses religious teachings and marriage life. Bujangga Manik shows that, on his return from the east, he went past the east–west part of East Java, east from Panarukan that covers the Ijen mountain compound. The journey expanded to Mount Raung in the area (*lurah*) of Telaga Wurung and straight to Balungbungan, where he stayed for more than 1 year to meditate (Noordijn, 1984, p. 26). Kasturi, Kukub, and Sagara are the names of three *mandalas* located near Mount Mahameru (Semeru) as mentioned in Nagarakrtagama (Pigeaud, 1924, p. 33) as well as in Batur inscription that is made of metal. The deciphering of Batur inscription reveals that the *mandala*s in Kasturi, Kukub, and Sagara indicate the same school of religious teachings (Hardianti, 2015). Is there any possibility that Gerba,

Sukuh inscription IV   Facsimile of Gerba inscription   Widodaren inscription

Figure 1.   Figure 2.

Widodaren, and Batur inscriptions were issued by the *mandala* located in the slope of Mount Semeru? Figure 2 shows the shapes of the scripts of Gerba and Widodaren inscriptions.

Both Gerba and Widodaren inscriptions do not bear any year. However, on the basis of the paleographic analysis using the dynamic method, it can be estimated that the scripts of the two inscriptions originated from a period between Sukuh inscription and Damalung inscription.

Damalung inscription was found in the village of Ngadoman near Salatiga, Central Java. This inscription bears the year 1371 Śaka. It mentions that the word *Damalung,* which implies a mountain, is assumed to be Mount Merbabu (Wiryamartana, 1993, p. 1), one of the *mandalas* visited by Bhujangga Manik to deepen his religious knowledge (Noordijn, 1984, p. 7). Several *mandalas* were found in Mount Damalung and were spread in the base, slope, and top of the mountain. These *mandalas* continued their existence as proven by the finding of 400 palm leaf manuscripts containing literary work. Unfortunately, the Damalung inscription was the only inscription found at the site. The shape of the scripts indicates their unique characteristics (Figure 3), and the language used is Ancient Javanese with Middle dialect. The content of Damalung inscription included praises to goddess Saraswati and moral teaching on recommended human characters (Susanti, 2010, pp. 6–7).

Pasrujambe inscriptions were found in the village of Pasrujambe, the regency of Lumajang, comprising 19 inscriptions with distinct styles. The content of these inscriptions consists of names of Figures (2) and names of gods and goddesses (6). Three of the inscriptions mention names of holy sites (*rabut pĕthak, rabut walang taga*); seven contain brief sentences; two bear the year of 1391 Śaka; and one of the inscriptions contains advice on life. In his travel journal, Bhujangga Manik revealed that he took a route along the southern beach of Java through Padangalun and reached Mount Watangan that faced the island of Barong (Nusa Barong). From there, he reached Sarampon, mentioned in Nagarakrtagama as Sarampwan, a place where King Hayam Wuruk stopped over while he was visiting Sadheng for several days (1359 Śaka). Afterward, Bhujangga Manik went past the village of Cakru on the beach south of Lumajang and reached the *lurah* of Kenep (not yet confirmed) and the area of Lamajang Kidul. Lamajang could probably have been the present Lumajang. It is assumed that there were several *mandalas* used as religious centers in Lamajang, judging by the numerous brief inscriptions found in this location. Figure 4 shows the shape of the scripts of Pasrujambe inscriptions.

The four groups of inscriptions mentioned above would most probably describe religious centers in the late Majapahit era as they all bear similar characters. This assumption is supported by the information contained in the literary works of the same era, at least in Nagarakrtagama and the travel journal of Bhujangga Manik. Nevertheless, the precise locations of several other inscriptions with scripts of distinct styles that were found sporadically in East Java, namely Tempuran, no. D 62, and Condrogeni inscriptions, as well as several brief inscriptions of distinct style scripts that are housed in the Majapahit Museum in Trowulan remain unknown.

Condrogeni inscription is part of the collection of Jakarta National Museum and has not been clearly deciphered. The scripts are of ancient Javanese with a unique style, and the language used is ancient Javanese of Middle dialect. The content of the inscription is believed to consist of religious teachings and advice that include praises to goddess Bathari. It was found in the village of Condrogeni, the regency of Ponorogo. Other discoveries in the proximity of the inscription include pieces of the statue of goddess Durga and stone inscription that bears the year of 1376. Figure 5 shows the shape of the scripts of Condrogeni inscription .

The place where D 62 inscription was found remains unknown. It has a quasi-linga shape and is part of the collection of the Majapahit museum in Trowulan. This inscription has distinct style scripts and uses ancient Javanese language of Middle dialect. It contains several words implying the existence of religious rituals and is assumed to have dated in 13 (12) Śaka.

Several brief inscriptions of a rectangular shape with unique style scripts and measuring approximately 20 cm × 60 cm can be seen at the Majapahit museum. These inscriptions bear the names of places such as *prapyanga, rĕksaguna,* and *rĕsi bhalajangidhwang.* Mount Penanggungan or *Pawitra* (holy) is mentioned in Sukci inscription (929 M) and in the ancient Javanese manuscripts of Nagarakrtagamaand Tantu Panggĕlaran. According to these sources, the Hindus and Buddhists treated the mountain as a sacred place. Numerous archaeological artifacts were found in the mountain constituting constructions and bathing facilities (*petirtan*), caves (*ceruk*), and hermitage (Aris Munandar, 1990). Hence, such brief inscriptions that constitute part of the collection of the Majapahit museum would most probably have originated from the sites of Pawitra or Mount Penanggungan.

The different shapes of the scripts in these inscriptions may have resulted from several factors, such as automatic intrinsic changes, or changes taking place in the writing instruments or materials used to make them, intentional factors such as the creativity of the writer in manifesting the values developed during certain eras, with the innovation made by a person or a group of persons who created new scripts, and the creation of new variations in the shape of the scripts offered by a group of people living far away from the center of the kingdom (the coastal communities or the *mandala* communities). According to a study conducted on the writers of Ulu manuscripts in Bengkulu, in every scriptorium or center of literary work, writers pay special attention to the creation of the shape of scripts and the punctuation marks in their manuscripts. As such, it would be highly possible that a person who understood the script of Ulu would yet face difficulties in deciphering and distinguishing such differences.

Damalung inscription

Figure 3.

Pasrujambe inscriptions

Figure 4.

Condrogeni inscription

Figure 5.

D62 inscription

Figure 6.

On the basis of the content of inscriptions with unique style scripts that originate from the *mandalas* of the late Majapahit era, a conclusion can be drawn on the different functions of the inscriptions. In general, they served as markers or reminders of names of places or indicators of the years when the inscriptions were written. Some also contain the names of gods to remind the readers (individuals or communities) of the gods people worshipped at the time. The names of gods also served as a means of concentrating in performing yoga exercise. Inscriptions such as Gerba, Widodaren, and Damalung served as reminders as they contain advice (Rahayu, 2016, p. 20).

Various unique style scripts in inscriptions originating from the late Majapahit era may contribute to the reconstruction of religious activities and religious centers located outside the royal palaces. For instance, a study conducted by Andriyati Rahayu on brief inscriptions containing unique style scripts in the temple compound of Sukuh describes the inauguration ceremony and religious rituals, namely the inauguration of Bagawan Gangga Sudhi to become a better person. The manuscript of Rajapatigundala mentions an obligation for conducting such ritual, whereas the manuscripts of Tantu Panggelaran and Calon Arang mention that, in order to become a priest, a person would possess an inauguration robe, an umbrella, and a pair of earrings (Rahayu, 2016, p. 18).

A close observation of the Pasrujambe inscriptions led to the following three assumptions. First, the existence of three places had different functions in terms of religious life, namely *rabut* that might have been the residence of religious community, *samadi* that served as a place for individual meditation, and *paṇyaṇḿan sarga* that was an object of certain worship. Second, artifacts found at the Pasrujambe site consisting of ritual instruments, such as jewel boxes, bowls, bells, plates, pots, and trays, also serve as indicators of the existence of religious activities at the site. Third, on the basis of the content of the inscriptions and the mentioning of names of figures such as Sang Kurusya and Sang Kosika, it can be assumed that the religious concept observed by the community was of the Pasupata school because the above-mentioned figures were students of Lakulisa, the founders of the Pasupata school (Rahayu, 2016, p. 20).

Studies of the inscriptions that have unique style scripts in terms of a variety in scripts and the language as well as the content play important roles in supporting the reconstruction of religious life that existed at a certain era, particularly the Majapahit era. The spread out locations of the inscriptions with unique style scripts may also provide us with the description of the identification and chronological data about the sites and enrich us with information about the religious life at that time far from the center of authority that remains rarely discussed to date.

REFERENCES

De Casparis, J.G. (1975) *Indonesian Palaeography, a History of Writing in Indonesia from the Beginnings to C A.D. 1500*. Leiden/Ǩln, E.J. Brill.

Fic, V.M. (2003) *From Majapahit and Sukuh to Megawati Sukarnoputri*. New Delhi, Shakti Malik Abhinav Publications.

Gaur, A. (1979) *Writing Materials of the East*. The British Library.

Gonggong, A. (editor). (1993) Sejarah *Kebudayaan Jawa* (The History of the Javanese Culture). Jakarta, The Indonesian Ministry of Education and Culture.

Munandar, A.A. (1990) Kegiatan Keagamaan di Pawitra: Gunung Suci di Jawa Timur Abad ke-14–15 (Religious Activities in Religious Sites: Holy Mountains in East Java in the 14th – 15th centuries). Unpublished Master's thesis for the Department of Archeology, Graduate Program, Post-graduate Program of Universitas Indonesia, Depok.

Pigeaud, Th.G.Th. (1960–1963) *Java in the 14th Century a Study in Cultural History: The Nagarakertagama by Rakawi Prapanca of Majapahit, 1365 AD*. Volume I-V. The Hague, Martinus Nijhoff.

Prasodjo, T. (1991) *Kajian Paleografis terhadap Prasasti-Prasasti Candi Sukuh* (Paleographic Study on Inscriptions at the Sukuh Temple), Research Report, Faculty of Letters, Universitas Gajah Mada, Yogyakarta.

Rahayu, A. (2016) *Kehidupan Kaum Agamawan Masa Majapahit Akhir: Tinjauan Epigrafis* (The Life of Religious People of the Late Majapahit Era: An Epigraphic Review). [Unpublished Dissertation] Faculty of Humanities, Universitas Indonesia, Depok.

Santiko, H. (1986) *Mandala (Kadewagurwan) pada Masyarakat Majapahit* (Mandalas in the Majapahit Society). A paper presented at the Scientific Workshop of Archaeology IV, Cipanas.

Sedyawati, E & Hasan, D. (2012) *Indonesia dalam Arus Sejarah* (Ed. Vol. 2). (Indonesia in the Flow of History). Bandung, PT. Ichtiar Baru van Hoeve.

Sedyawati, E. (2001) Masalah Pusat dan Pinggiran dalam Sastra Jawa (Central and Outskirt Issues in the Javanese Literature, in Sedyawati et al. (ed) *Sastra Jawa, Suatu Tinjauan Umum* (The Javanese Literature, A General Review). Jakarta, Indonesia Language Center and Balai Pustaka.

Sidomulyo, H. (2007) Napak *Tilas Perjalanan Mpu Prapañca* (Tracing Back the Journey of Mpu Prapañca*),* Penerbit Wedatama Widya Sastra in collaboration with Nandiswara Foundation of the Department of History, the Faculty of Social Sciences, UNESSA.

Susanti, N & Agung, K. (2006) Damalung: Skriptoria pada Masa Hindu-Buddha sampai dengan Islam (Scriptoria from the Hindu-Buddha Era up to the Islamic Era), Proceeding at the International Symposium of Manuscript Community of Nusantara (Manassa), Bandung.

Susanti, N. (2010) Skriptoria Masa Majapahit Akhir: Identifikasi Berdasarkan Persebaran Prasasti (Scriptoria of the Late Majapahit Era: Identification Based on the Distribution of Inscription), Proceeding at Seminar on Guided Chronology in Archaeology, Kuala Lumpur.

Susanti, N. (2012) Prasasti dari Desa Widodaren, dalam M.Suhadi (editor), *Aksara dan Makna: Membaca dan Mengungkap Kearifan Masa Lalu* (Scripts and Meaning: Deciphering and Revealing the Past Wisdom). Jakarta, The Indonesian Ministry of Education and Culture.

Susanti, N. (2016) Script and Identity, Prosiding 10th International Conference on Malay-Indonesia Relations; A Decade Promoting Bilateral Prosperity, Faculty of Arts and Social Sciences, University of Malay, Kuala Lumpur.

Susanti, N., Titik, P. & Trigangga (editors). (2015) *Inscribing Identity, the Development of Indonesian Writing Systems*. Jakarta, The National Museum of Indonesia.

Wiryamartana, I.K. & W. van der Molen. (2001) The Merapi-Merbabu Manuscripts: A Neglected Collection, *BKI*, 157–1, pp. 51–64.

Wiryamartana, I.K. (1993) The Scriptoria in Merbabu-Merapi Area, *BKI,* 149, pp. 503–509.

Cultural Dynamics in a Globalized World – Budianta et al. (Eds)
© *2018 Taylor & Francis Group, London, ISBN 978-1-138-62664-5*

Religious communities in the late Majapahit period at Pasrujambe site, Lumajang

A. Rahayu & N. Susanti
Department of Archaeology, Faculty of Humanities, Universitas Indonesia, Depok, Indonesia

ABSTRACT: There are 24 short inscriptions discovered in PasruJambe, Lumajang, East Java. The scripts used in these inscriptions are different from the scripts used in inscriptions issued by the king. The inscriptions are very short and only mention one word such as the names of gods (*batharamahadewa, batharamahisora,* and *bathariprtiwi*), holy places (*rabut macan pethak,* and *rabut walang taga*), saints (*bagawancaci* and *bagawancitragotra*), the rsi (*sang kurusya* and *sang kosika*) and moral teaching. Those characteristics indicate that the inscriptions were not issued by the king but produced by religious communities. This paper discusses the site of PasruJambe, where the short inscriptions were found. The method used is the epigraphy method, which includes collecting data, making transliteration of the inscriptions, and interpreting the data. This paper also uses manuscript and archeological remains as supporting data. The result is that PasruJambe was once a place of a religious community, where clergymen and their pupils learned about religious texts and issues. This kind of community was plenty in Majapahit era, and PasruJambe was only one of them.

1   INTRODUCTION

Inscriptions are among the types of data in archaeological research. According to Boechari (2012, p. 4), inscriptions are historical sources that are originated from the past and were inscribed on stone or metal media. They were the king's announcements about certain events that were inscribed on stone or metal media (Boechari, 2012, p. 4; Susanti, 1996, p. 6).

Inscriptions are very important for Indonesian ancient history because they are the primary sources which can provide rich information about the social and cultural conditions in the past, such as structures of kingdoms, bureaucracy, religions and beliefs, communities, economic matters, and traditional customs in ancient Indonesian societies (Boechari, 2012, pp. 6–7).

Most of the contents of the inscriptions from the Old Javanese period assign an area as *sīma* (Susanti, 1996, p. 20; Boechari, 2012, p. 6). *Sīma* inscriptions are those that mention a king's or a ruler's declaration to change the status of an area in relation to building maintenance, maintenance of religious buildings or public facilities, as well as a king's gratitude/compensation to an individual or a group of people. Examples of *sīma* inscriptions are among others Kudadu and Sukamerta inscriptions (Boechari, 2012, pp. 13–14).

Information in parts of the inscriptions, if studied meticulously, can give very interesting portrayals about the structures of kingdoms, bureaucracies, societies, and economy as well as religions, beliefs, and traditional customs of ancient Indonesian communities (Boechari, 2012, p. 25).

In PasruJambe, about 24 stone inscriptions were found in six different locations. From the date, which is 1371 Śaka (Suhadi, et al., 1996/1997, pp. 16–17), it is known that the stone inscriptions originated from the Majapahit period. However, the inscriptions of Pasru Jambe have different characteristics from the other inscriptions from the Majapahit period. The differences are from the aspects of physical appearance and content. An example of physical difference is that Pasru Jambe inscriptions were carved on un-worked natural stones while Majapahit inscriptions were carved on natural stones that had been worked on. Second, the scripts used in Pasru Jambe inscriptions have different shapes from those used in Majapahit ones in general. Third, Pasru Jambe inscriptions only consist of several lines, so they are called short inscriptions. Fourth, Pasru Jambe inscriptions contain the names of gods, names of religious figures, and religious advice while Majapahit inscriptions in general contain kings' statements/decrees. It is these differences that have led to the assumption that Pasru Jambe inscriptions were issued by religious communities outside the central kingdom. This is supported by the artefacts found in Pasru Jambe sites, such as lidded-boxes, bowls, trays, pots, plates, and stones with *yantra* picture on them, which are thought to be a series of ceremonial implements.

Those short inscriptions are the main sources to understand how life during the Majapahit era was from the point of view of communities outside the royal court. The contents of the inscriptions can provide information about life around the several places where they were issued, which in turn can give us new understanding about Majapahit during that period and more data to lead to a more complete picture of life history during the Majapahit era.

There have been many studies on communities in the Indonesian archipelago in the past, among others about the bureaucracy, laws, economy, and so on. However, those studies have focused on the communities within the royal court and their cultures. Studies on the cultures of communities outside the royal court area have rarely been done despite the fact that the cultures of the Indonesian archipelago as a whole were established not only by the communities who lived within the centres of kingdoms but also by those who lived outside them.

1.1   Problems

This paper will discuss the short inscriptions found in Pasru Jambe site and the reconstruction of the communities that issued those inscriptions.

1.2   Research methods

The research methods of this study used approaches proposed by Robert J. Sharer and Wendy Ashmore in *Fundamental of Archeology* (1979, pp. 114–120). The following are the phases of the research, which include among others:

1.2.1   Data collection

During this phase, all data regarding this study, which are inscriptions from the Majapahit era, were collected. First, references and literature connected to the research objects, transliterations, and translations of inscriptions from the Majapahit era, as well as records and experts' opinions about those inscriptions, were pulled together. The bibliographical data focus on articles about PasruJambe inscriptions. In addition, secondary data in the form of transliterations and translations of manuscripts related to the research aim were also used.

Afterwards, a field study was carried out to obtain the data, particularly the short inscriptions of PasruJambe site. During this phase, surveys were conducted to know the locations of the short inscriptions as the main source of the data. The inscriptions of Pasru Jambe are now being kept at MpuTantular Museum in East Java.

1.2.2   Data processing

The results of the data collection, which are bibliographical data and field data, were then processed to be used as the material of data analysis. The processing of bibliographical data was done by classifying resources relevant to the research aim. The next steps were transliteration and translation processes. Meanwhile, critical analyses were also done to determine whether there were anachronism or inconsistency with the period and whether or not the written sources were original.

The method to do the processes was comparing the inscriptions used as the data of this research with other inscriptions from the same period. In other words, every element, both a physical element and content, of the inscriptions used as the main data was compared with those of the other inscriptions from the same period to see whether or not there was a significant difference. If there is no significant difference, it can be assumed that the inscription is from the same period while if there are many significant differences then the originality or authenticity of the inscription is questionable.

1.2.3 Data analyses

A contextual analysis was done by observing the pattern found in the inscriptions and their relations to other artefacts. During this phase, observation was also made to see whether there were supporting artefactual data, for instance whether or not there were artefacts found at the location where the special-typed inscriptions were found. It was also during this phase that it was known that the data to be used in the reconstruction of the religious group's life were the inscriptions from Pasru Jambe.

Analyses on the pattern characteristics and contents of the inscriptions that are the objects of this research were also carried out. Furthermore, the contents were analysed, particularly the contents of those that have a different pattern from the inscriptions from the centre of the kingdom. The content analysis would show who issued the inscriptions and how their culture was. Furthermore, the factors that might have caused the differences were also scrutinised to see the reasons behind the characteristic differences among the inscriptions. From the results of the analyses, the life of the communities that issued the inscriptions was reconstructed.

1.2.4 Data interpretation

Data interpretation was conducted with the help of other textual and artefactual sources, namely manuscripts and artefacts. The manuscripts used were particularly those that contain information about the life of religious groups, among others *Nāgarakṛtāgama*, *Rajapatiguṇḍala*, *Calon Arang*, *Arjuna Wijaya*, and *Tantu Panggelaran*. Manuscripts greatly helped in the reconstruction of the life of the religious groups because data from the inscriptions are limited. Translating the manuscripts was as much as possible done by the authors although there were some that were translated by other people.

The artefactual finds that were used in the data interpretation are religious buildings and ceremonial implements. Those finds are very important in the process of interpretation as evidence of religious life at the location. Based on the analyses, a depiction about the religious environment during the period was made with the help of data from manuscripts and artefacts. The results of the interpretation provided a conclusion based on the contents of the inscriptions, manuscripts, and archaeological finds at the location where the inscriptions were found, which in turn can provide a portrayal of the life of religious communities during the Majapahit era. The following is a diagram of the research framework.

2 RESULTS

The following are the results of transliteration and translation of Pasru Jambe inscriptions:

No	Name of inscription	Transliteration	
1	Pasru Jambe 1	*Walěriṅa babadwoŋ samadi*	that (is) the border of a place where people meditate
2	Pasru Jambe 2	*Saṅ = anawakr n) dha*	The Bearer of Cepuk (lidded-box)
3	Pasru Jambe 3	*hyaṅakasa*	The Sky god
4	Pasru Jambe 4	*BathariPrtiwi*	The Earth goddess
5	Pasru Jambe 5	*°isaka 1371*	The Śaka year of 1371
6	Pasru Jambe 6	*Dhudhukuna*	Go to the healer
7	Pasru Jambe 7	*°isaka 1371*	The Śaka year of 1371
8	Pasru Jambe 8	*dhudhukuna*	Go to the healer

(Continued)

595

No	Name of inscription	Transliteration	
9	Pasru Jambe 9	Rabutmacan pĕthak	A holy place called macan pethak
10	Pasru Jambe 10	Bathara mahi sora	The god Mahisora
11	Pasru Jambe 11	Batharamahadewa	The god Mahadewa
12	Pasru Jambe 12	BagawanCaci	A holy man named Caci
13	Pasru Jambe 13	Bagawan Citra Gotra	A holy man named Citra Gotra
14	Pasru Jambe 14	Saŋku rusya	Sang Kurusya
15	Pasru Jambe 15	RabutLita	A holy place named Lita
16	Pasru Jambe 16	Paŋyaŋñan Sarga	Object of worship (to god) (for) common people[1]
17	Pasru Jambe 17	Rabut walaŋ ta ga	A holy place named WalangTaga
18	Pasru Jambe 18	Saŋko sika	Sang Kosika
19	Pasru Jambe 19	ikipañestu yaŋmami guru guru yen arabi den kadibotiŋñakasalawanpṛtiwi papa kabuktihi	This is a blessing (from) god, we (the) teachers If having a wife (you are) supposed (to be) like the weight of the sky and earth Misfortune (will) be proven (for those who disobey)
20	Pasru Jambe 20	Sakā bala sariwu	Come Troops or forces A thousand
21	Pasru Jambe 21	Riŋmaja loka	at maja loka
22	Pasru Jambe 22	Bathara Wisnu	The god Wisnu
23	Pasru Jambe 23	nilapaksi	Blue bird
24	Pasru Jambe 24	Rabut sida	A holy place named Sida

Based on the scripts being used, the inscriptions of Pasru Jambe are assumed to be from the same period. Although the places where they were found are different, the distance between one location and the others is not very far. Thus, it can be concluded that those inscriptions are from the same spatial and temporal contexts.

There is some information that in 1986 in the trench near the location where the inscriptions were found, a set of ceremonial implements made of bronze and ceramics were found. They are ceremonial bells, plates, boxes with a cover, a lid of a flask, bronze bowls, pots, and trays (Suhadi, M. et al., 1994/1995, p. 10).

Based on the function of each artefact found, the artefactual finds at the site of Pasru Jambe are assumed to be a set of ceremonial implements. This is closely related to the fact that one of the contents of Pasru Jambe inscriptions is *Sañanawakṛndha*, which means 'the *cepuk* (box with cover) bearer', and one of the artefactual finds at Pasru Jambe site is a box with cover known as *cepuk*. Besides a set of ceremonial implements mentioned above, at Pasru Jambe a stone with the picture of *yantra* on it, which is 39 cm long, 28 cm wide, and 7 cm thick, was also found. *Yantra* is a type of implement to help in meditation.

Information obtained from the contents of Pasru Jambe inscriptions provides names of places, gods/goddesses, and figures, as well as advice.

In the inscriptions of Pasru Jambe, three places, each with a different function, are mentioned.

Rabut. According to the Old Javanese Dictionary, *Rabut* is a sacred place or a place that has supernatural power (Zoetmulder, 2006, p. 897). There is no further information about the term *Rabut*.

A Place for *Samadi*. A place for *samadi* is mentioned in Pasru Jambe 1 inscription as *walĕrikababadwoŋsamadi*, which means the border of a newly cleaned place for people to do *samadi* (meditate). Probably in the past this location was used as a place for people to meditate; a special place to meditate individually (alone).

Paṇyaṇñan Sarga. The word *paṇyaṇñansarga* means a worshipping place for many people. Maybe in that location there used to be an object that was worshipped by many people. At the same location was found an inscription that mentions the *kosika* and an inscription that contains advice. Unfortunately, only a few other artefacts were found. There is only a report about the discovery of a set of ceremonial implements, such as ceremonial bells, plates, boxes with a cover (*cepuk*), bowls, pots, and trays made of bronze and ceramics. There was no *lingga* or other features that could have served as an object worshipped by many people. It is possible that the object was made of perishable material.

From the entire inscriptions of Pasru Jambe there is information that the establishment of religious places was carried out in the same year, which is 1371 Śaka, but there is no additional information or more exact date regarding the establishment of those religious communities or religious places that were dispersed at several locations at Pasru Jambe. Information from several inscriptions only mentions years, which all refer to 1371 Śaka. Presumably all the religious places were built at the same time or within the same short period of time.

From the mentioning of the names of figures in the contents of Pasru Jambe inscriptions, it can be concluded that there was a hierarchy of holy figures within the religious environment, which can be seen in the following diagram.

Diagram of Figures within the Religious Community of Pasru Jambe

*Bathara*s had the highest position as the worshipped gods; the Pasru Jambe environment worshipped Mahadewa, Wisnu, and Mahisora (an epithet for the gods, Śiwa or Visnu). In the second position were goddesses or *batari*s as the companions (*sakti*) of the gods, but only one goddess that is mentioned in the inscriptions, which is *Pratiwi*. In the third position were *sang Kosika* and *sang Kurusya*, who were two of *Lakulīśa*'s disciples. Lakulisa is the founder of Siva Pasupata, a school of Sivaism. In the fourth position was *dhudhukuna* (dukun) or shamans/witchdoctors, who are no ordinary human beings because in the inscriptions there is advice to go to a shaman (witchdoctor). In general, a shaman's function in Javanese communities is to cure illnesses. Meanwhile, there are other more important roles of shamans in Javanese communities, for example in Tengger where shamans lead purifying ceremonies during the *Kasadha* festival (Smith-Hefner, 1992, pp. 239–240). In the fifth position were *Bagawan*s, who were holy people in a high level, usually kings who left their worldly life to dedicate themselves to religious life and became *wiku*s (Zoetmulder, 2006, p. 185); they can also be categorized into *rṣi*.

The short inscriptions of Pasru Jambe which mention the names of places indicate that there were religious locations dispersed within the area. One of the contents of the inscriptions mentions the opening of a landscape to be made into a place to do *samadi* (meditation), which means that it was the initial opening of a landscape to be made into a place to meditate for an individual or several people in a small group. They were established in separate locations, but still within the same area, which is Pasru Jambe. Thus, Pasru Jambe was probably a religious community that consisted of several religious places, each with its own function. From the information obtained from the contents of Pasru Jambe inscriptions, we learned that there were three places, namely *Rabut*, *Samadi*, and *paṇyaṇñansarga*. It is unfortunate that there is no further information from the inscriptions that provide information about those places.

One of the contents of Pasru Jambe inscriptions, *"walĕrikababadwoṇsamadi"*, gives us information that there was work to clean a place to be used for *samadi* (meditation). Ascetics (hermits) usually choose certain locations. Rsis tend to do it in *grotto*s especially for meditating or other places that are considered suitable for their purpose and worship the god Śiwa in their minds, meditate, and do other religious activities at holy bathing places or terraced structures on mountain slopes with altars but no statue. It can be assumed that the number of holy bathing places (*patirthan*) was quite plenty, and they were usually scattered along mountain slopes that still retained dense forests or on other high places (Santiko, 1993, p. 16; Munandar, 2003, p. 122). The area where Pasru Jambe inscriptions were found and which is the hamlet of Munggir is located around a water spring with very clear water named Sumber Rawa. A small river that flows near the water spring is also called Kali Rawa (Rawa River). It seems as though in the past Sumber Rawa, besides being used for daily activities, was also considered as the source of water of life (*air amṛta*). It is located at the foot of Mount Semeru, which was believed to be a holy mountain and part of Mount Mahameru in India that was moved to the island of Java (Atmodjo, 1986, p. 40), thus supporting the assumption that the area where Pasru Jambe inscriptions were found was a place to carry out religious activities by a group of people who had left their worldly life.

3 CONCLUSION

Studies of Indonesian ancient history require data from various sources, both written sources and artefacts left by people in the past. Among the written sources for Indonesian ancient history are inscriptions. The inscriptions found in Indonesia are mostly about designating an area as *sīma*, as a king's bequest to a group of people or community or functionary that was praiseworthy to the king or a king's charity for religious purposes. However, the discovery of the short inscriptions from East Java has provided a new outlook regarding the content of an inscription. The inscriptions have different characteristics from the other inscriptions found thus far.

Until now there have been about 24 inscriptions with special characteristics found in Pasru Jambe. Based on the dates, it can be assumed that those inscriptions were from the late Majapahit era.

The conclusion based on the analyses of the contents of the inscriptions as well as the artifactual finds in Pasru Jambe area shows us a picture that during the Majapahit Era there was a religious community in that area, who lived together and had a significant role within the society, as seen by the presence of a set of ceremonial implements, which led to the assumption that the religious community also performed sacred ceremonies for the public. The entire inscriptions of Pasru Jambe provide information that the establishment of religious places was carried out in the same year, which is 1371 Śaka.

The religious concept adopted by the community at Pasru Jambe was possibly Pāśupata because there is an inscription that mentions *Sang Kurusya* and *Sang Kosika*. They were the disciples of Lakulīśa, the founder of Pāśupata sect. Pasru Jambe inscriptions also mention the names of gods and goddesses, which are *Māheśwara, Wisnu, Mahadewa,* and *BatariPratiwi.* Other figures mentioned in the inscriptions are *Hyaṅ Akasa, Dhudhukun, Bagawan Citragotra,* and *Bagawan Caci.*

From the mentioning of the names of figures in the contents of Pasru Jambe inscriptions, it can be concluded that there was a hierarchy of holy figures within the religious environment. *Bathara*s had the highest position as the worshipped gods. In the second position were goddesses or *batari*s as the companions (*sakti*) of the gods. In the third position were *sang Kosika* and *sang Kurusya*, who were two of *Lakulīśa*'s disciples. In the fourth position was *dhudhukuna* (dukun) or shamans/witchdoctors, who were believed to be able to communicate with the gods and to summon spirits of the ancestors to come to certain ceremonies. In the fifth position were *Bagawan*s, who were holy people in a high level, usually kings who left their worldly life to dedicate themselves to religious life and became *wiku*s.

At PasruJambe site was also found artefacts like *cepuk* (a box with a cover/lid), *bokor* (a bronze bowl), *genta* (a bell), plates, pots, and trays. Those artefacts are thought to be ceremonial implements. A manuscript titled *Rajapatiguṇḍala* explains that during ceremonies the religious community distributed holy water to the commoners' in a flask. Another manuscript *Atharvaveda-pariśiṣṭa*, mentions that one of the worshipping practices of the Pasupata followers is, smearing their bodies with holy ashes (*bhasma*), which was intended to ward off evil.

REFERENCES

Atmodjo, M.S. (1986) Mengungkap Masalah Pembacaan Prasasti (Revealing Inscriptions Deciphering Problems). *Berkala Arkeologi,* VII, pp. 39–55.

Boechari (2012) *Melacak Sejarah Kuno Indonesia Lewat Prasasti* (Tracking the Ancient History of Indonesia through Inscriptions). In: N. Susanti, H. Djafar, E. Wurjantoro, & A. Griffiths, Eds. Jakarta, Kepustakaan Populer Gramedia.

De Casparis, J.G. (1975) *Indonesian Paleography: A History of Writing in Indonesia from The Beginning to c. A. D. 1500.* Leiden/KOLN, EJ. BRILL.

Media Hindu. Pancamahabhuta. July 2016. (http://www.mediahindu.com/ajaran/panca-mahabhuta-sebagai-anasir-dasar-penyusun-alam-semesta.html).

Munandar, A.A. (2003) Candi dan Kaum Agamawan: Tinjauan Terhadap Jenis Candi Masa Majapahit (Abad ke-14–15) (Temples and Clergy: A Review of Temples in Majapahit era in the 14th and 15th Centuries). In: A. A. Munandar, *Aksamala.* Bogor: AKADEMIA. pp. 111–141.

Santiko, H. (1993) Penelitian Awal Agama Hindu-Siwa pada Masa Majapahit, (Early Study of Hindu-Shiva Religion during Majapahit era) paper presented in: *Simposium Peringatan 700 tahun Majapahit, tanggal 3-5 Juli 1993. Trawas Mojokerto, Jawa Timur.*

Sharer, R. & Ashmore, W. (1979) *Fundamentals of Archaeology.* California, Benjamin/Cummings Publishing Company, Inc.

Smith-Hefner, N.J. (1992) Pembaron: An East Javanese Rite of Priestly Rebirth. In *Journal of SEA Studies* Vol. 23 No. 2. Department of History National University of Singapore. pp. 237–275.

Suhadi, M., *et al.* (1994/1995) Laporan Penelitian Epigrafi di Wilayah Provinsi Jawa Timur (Epigraphic Research Report in East Java province). In *Berita Penelitian Arkeologi* No. 37. Proyek Penelitian Arkeologi Jakarta. pp. 1–38.

Susanti, N. (1996) *Prasasti sebagai Data Sejarah Kuna* (Inscriptions as Historical Data), Laporan Penelitian Proyek DIP-OPF Fakultas Sastra Universitas Indonesia. Jakarta, Fakultas Sastra, Universitas Indonesia.

Yogamag. Bhasma. May 2016. http://www.yogamag.net/archives/2006/lnov06/bhasma.shtml)tp://www.ibiblio.org/gautam/hind0003.html.

Zoetmulder, P.J. (2006) *Kamus Jawa Kuna Indonesia* (Indonesian Old Javanese Dictionary). (Daru Suprapta & Sumarti Suprayitna, trans.) (5th ed.) Jakarta, Gramedia Pustaka Utama.

REFERENCES

Cultural Dynamics in a Globalized World – Budianta et al. (Eds)
© *2018 Taylor & Francis Group, London, ISBN 978-1-138-62664-5*

Particles *pwa* and *ta* in the Old Javanese language

D. Puspitorini
Department of Linguistics, Faculty of Humanities, Universitas Indonesia, Depok, Indonesia

ABSTRACT: In this paper, we aim to explain the function of particles *ta* and *pwa* of the Old Javanese language on a syntactic level. Although the function of the particle *ta* as sentence (syntactic) constituent markers has been explained by experts, the difference between *ta* and *pwa* as the clause marker necessary in a discourse (a topical clause) has not been studied in detail. The data source used here is the Old Javanese narrative prose titled *Adiparwa*, which is estimated to be written in the 10th century. The explanation of the function of particles *ta* and *pwa* facilitates in-depth understanding of the aspects of the Old Javanese language, particularly for translating Old Javanese language texts to other languages.

1 PARTICLES IN THE OLD JAVANESE LANGUAGE

Old Javanese (OJ) has particles that occupy the second position in the order of sentence constituent, namely *pwa* and *ta*. This is illustrated in the following examples:

1. a. *Lunghā **ta** sang Uttangka*   [Ad 15:1]
      'Uttangka went'
   b. *Lunghā **pwa** sang Garuda*   [Ad 39:6]
      '*Garuda went*'

2. a. *Kita **ta** prasiddha mijil saking hatingku*   [Ad 84:1]
      'You were really out of my heart
   b. *Kita **pwa** magawe tapa mangke* [Ad 117:26]
      'You are performing the asceticism now'

The above examples show that both the particles *ta* and *pwa* that occupy the second position separate and modify two functionally different sentence constituents. Therefore, particle *ta* only appears in one sentence and not in either the beginning or the end of the sentence.

In example (1), the particles *ta* and *pwa* follow the element of the sentence that is considered important, namely the predicate (1) and subject (2). The element of the sentence that is considered important appears at the beginning of the sentence. Uhlenbeck used the formula A-*ta*-B to indicate the position of *ta* in a sentence and its structure. Particle *ta* separates the predicate that lies in position A of the subject, which is in position B (3):

3. *Měnga / ta / sang hyang prětiwi* [Ad 15:9]
          A   B
   'Sang Hyang Pertiwi (the earth) opened'

This formula also applies for particle *pwa*. Particle *pwa* separates the predicate that lies in position A of the subject that lies in position B (4):

4 *Atuha / pwa / ya.*   [Ad 126: 1]
   A   B
   'They grew up'

In general, the researchers of BJK agree that *ta* and *pwa* are similar particles that are emphatic (see Zoetmulder, 1954; Teselkin, 1972; Uhlenbeck, 1970; Hunter, 1988; Harimurti and Mardiwarsito, 1984). The explanation of the two particles as emphatic particles is based on their communicative function. The constituent that is emphasised in a sentence occupies the first position and is followed by particle *ta* or *pwa*.

Explanation concerning the function of the use of particle *ta* has been provided by Uhlenbeck (1970, 1987), Hunter (1988), and Hoff (1998). The occurrence of the use of *pwa* is approximately one-eighth of that of the particle *ta* (On the basis of the author's calculation, the particle *ta* appears as often as 2,299 times, whereas the particle *pwa* appears only 266 times.) Because they are considered similar, previous researchers have not studied particle *pwa* on its own or separately from *ta*. Consequently, there is less understanding on its function and use. A reasonably clear distinction between particles *ta* and *pwa* is illustrated with the probability of the presence of the two particles in one sentence structure:

4. *Katon **pwa** mahurip sang Kaca de nikang daitya, prihati **ta** ya amet upāya.*
   'When it is seen by that *daitya* (that) Kaca is still alive, they are worried (and therefore) finds (other) means.'

Zoetmulder also stated that *ta* and *pwa* are the same particles. However, Zoetmulder (1954) recognised the relationship between *pwa* sentence and *ta* sentence that appear in sequence as illustrated in example (4). The sentence containing *pwa* comes first, whereas the one containing *ta* comes second. The two sentences are closely related. Zoetmulder's statement facilitated a study on interclause relationship meaning in *pwa* sentences.

In this paper, the analysis of particle *pwa* is conducted with (i) a sentence structure that contains *pwa*; (ii) the interclause semantic relation within the sentence that contains particle *pwa* and (iii) a comparison between particle *pwa* and particle *ta*. The sentence containing *pwa* (hereinafter referred to as a *pwa* sentence) will be observed through clauses because, in principle, a clause is the biggest grammatical unit that is covered by particle *pwa*.

A clause containing particle *pwa* may appear as either an independent clause or a dependent clause. The A-*pwa*-B formula is used to explain the *pwa* sentence structure. The following sentences have a simple structure:

5. *Wruh pwa bhagawān Wasistha.*   [Ad 94:28]
   'Bhagawan Wasistha knew.'

6. *Sinambutnira pwa ya kalih.*   [Ad 40:22]
   'They were both greeted by him/her.'

7. *Datĕng pwa ya.*   [B 26:2]
   'He came'

In the above three sentences, A is filled with a single word in the form of an actor-focus verb and a goal-focus verb, whereas B is filled with a noun or a pronoun. A filler is a constituent that functions as a predicate, whereas B functions as the subject. This is in line with the explanation provided by several experts who state that the Old Javanese language has a main sentence structure in the P + S pattern that is usually followed by an emphatic particle, among others particle *pwa* (see Harimurti, 1984; Teselkin, 1972). However, some of the data that the author has observed also show a number of A-*pwa*-B sentence structures, where A is the subject (8), complement (9), or conjunction (10). For example:

8. *Kita pwa magawe tapa mangke.* [Ad 117:26]
   'You are performing the asceticism now'

9. *Mangke pwa kita turung mânak.* [Ad 117:2]
   'Now you don't yet have a son'

10. *Yāwat pwa kita kĕna śāpa de mami, tāwat kita tan pangguha ngkawijayān.* [B 68:2]
    'As long as you are under my curse, you will never win.'

1.1 The topic of the second discussion concerns the interclause relation in pwa sentences

Pwa sentences are often in the order of two clauses that have a meaningful relation. Structurally, a *pwa* clause is in the first order, followed by a clause that contains *ta* or a clause that does not have a particle. Example (11) illustrates a cause–effect relationship that is stated in the second order of a clause in the *pwa* sentence:

11. *Ahirĕng pwa warnany awaknya, inaranan ta sira dewī Krsnā.* [Ad 153:6]
    'Her body was black in colour; hence, she was named the goddess of Krsna'

Research regarding particle *pwa* is complemented by highlighting *pwa* in relation to particles of a similar type, namely *ta* and *ya* and the comparison between particles *pwa* and *ta* and their combined forms *ta pwa, pwa ya,* and *pwa ya ta*. In addition to being particles, *ta* and *ya* serve as pronouns (12) and parts of compound particles (13):

12. *Katon pwa ya de bhagawān Sthûlakeça.* [Ad 21:2]
    'He was seen by bhagawan Sthulakeça.'

13. *Kita pwa ya huwus krtayaça.* [Ad 203:14]
    'You are already famous.'

In sentence (13), *ya* is not a third-person pronoun because there is already a second-person pronoun *kita* to act as the subject. If *ya* together with *kita* is considered as a subject filler, then *pwa* must be considered as a coordinative conjunction. So far, there has been no proof to show that *pwa* has ever been used as a coordinative conjunction. In addition, the context of the story illustrates only *kita* who has a subject filler constituent. Therefore, *pwa* and *ya* are subject filler constituents. Therefore, *pwa ya* in example (13) must be considered as a compound form.

2 SENTENCE STRUCTURE

A clause is the largest grammatical unit that can be covered by particle *pwa*. In the A-*pwa*-B pattern, both A and B are constituents that are functionally different:

14. *Wruh pwa bhagawān Wasistha* [Ad 94:26]
    'Bhagawān Wasistha knew'

15. *Maturu pwa sira muwah* [Ad 100:11]
    'He slept again'

In a clause that begins with a predicate, the position of A is filled with predicate (P), and B is filled with subject (S). The presence of *pwa* relates to the subject filler, which is a topical pronoun. Hunter (1988) stated that *ta* as well as *pwa* are topic marker particles. The topic meant by Hunter (1988) is a pronoun that is grammatically compatible with the verbal predicate. BJK verbal predicates have certain characters that determine the semantic relationship between the verb and the argument. Such characters are associated with verbal affixes. Affixes such as *mang-* and *-um-* mark subjects having a role as agents, whereas the *-in-* affix marks the subject that has the role of the patient. Subject filler arguments in a form of a pronoun always appear as a topical pronoun, regardless of the role. Topical pronouns appear as an independent word, for example, *sira* (16), whereas non-topical pronouns appear enclitic, for example *-ira* (17):

16. *Dinudut pwa* **sira**   [Ad 32:12]
    A   B
    'He was pulled'

17. *Inungang**nira** pwa ikang sumur* [Ad 78:18]
    'The well was seen by him'

In example (16), B is filled with a subject that plays the role of a patient in a form of the third pronoun *sira*. The pronoun is a topical pronoun as it is compatible with affix *-in-*, which focuses on the patient. In example (17), the actor argument exists as a non-topical pronoun (enclitic *-ira*) that is attached to verbs with affix -in- because it is not a subject.

Particle *pwa* must be present when the subject is filled with an anaphoric topical pronoun and lies in position B (16). If it is filled with another category, then the presence of a particle is not necessary. We compare the following two sentences:

18. *Sumahur Bhagawan Bhisma.*
    'Bhagawan Bhisma answered.'

19. *Sumahur pwa sira.*
    'He answered.'

In example (18), the S filler is a noun; therefore, *pwa* is not necessary to be present. On the contrary, the S filler in sentence (65) is a topical pronoun; therefore, particle *pwa* must be present. The necessary presence of particle *pwa* in a sentence with a topical pronoun filling position B is further clarified with the following examples:

20. *Sundul ring ākaça pwa ya.*   [B 21: 14]
       A   C   B
    'It (that tree) is soaring to the sky.

21. *Tumĕmpuh pwa ya i bhagawān Bhīsma.* [B 78: 3]
         A   B   C
    'It (that spear) lunged towards the direction of bhagawan Bhisma.'

In sentence (20), the predicate *sundul* 'soar' needs a complement in the form of a prepositional phrase *ring ākaça* 'to the sky'. Similarly, *tumĕmpuh* also needs a complement in the form of a prepositional phrase *i bhagawān Bhīsma* (21). In the two sentences, the complements occupy different positions. In sentence (20), the complement (*ring ākāca*) immediately follows the constituent that occupies position A, so the pattern is A-C-*pwa*-B. In sentence (21), the complement (*i bhagawān Bhîsma*) occupies the position following B, so the pattern is A-*pwa*-B-C. The similarity between the two sentences is that the topical pronoun lies in the immediate position that follows *pwa*, namely in position B.

3   INTERCLAUSE SEMANTIC RELATIONSHIP

A *pwa* sentence often consists of two consecutive clauses without a conjunction (see also Zoetmulder, 1954:86). The order of the two clauses creates a relationship of meanings. In such case, the clause containing particle *pwa* always lies in the first position (hereinafter referred to as a *pwa* clause*)*, followed by another clause. The order is fixed:

22. *Katon **pwa** dewī Kuntī matanghi de sang Hiḍimbī, masö ta ya manĕmbah, mājarakĕn hyunya ri sang Bhīma*   [Ad 144:5]
    'It is seen by Hidimbi (that) goddess Kunthi woke up. She (Hidimbi) entered (then) bowed (and) expressed her desire for Bhima.'

The combination of a *pwa* clause with other clauses shows a sequential time relationship and the same time relationship. In the sequential relationship, two consecutive events occur in a very short time difference or almost simultaneously (23).

23. *I sĕḍĕng ning yajña ginawe, hana ta Sārameya, śwāna milu manonton yajña nira.. Katon pwa ya de sang Śrutasena, pinalu nira ta ya ikaug asu si Sārameya.*

'When the sacrifice ceremony was taking place, a dog (namely) Sarameya took part in watching the sacrifice ceremony. When it was seen by Sang Srutasena, the dog was hit by him.'

The order of the *pwa* clause and its pair are iconic because those illustrate the sequence of events as happening beyond language. In this case, the *pwa* clause that occupies the first position refers to the first event, followed by the other clause, especially the *ta* clause, which states the following event. In relation to the sequential time, the event that is marked by a *pwa* clause is an action that occurs in a short time, and then the second event occurs almost simultaneously, as stated by the following clause.

The second-time relationship is **simultaneous time relationship** (24). The *pwa* clause becomes the time background and the occurrence of an event that is stated in the next clause. Therefore, in the simultaneous time relationship, the matter or event that is expressed by the second clause is not a separate part of the event that is stated by the *pwa* clause:

24. *Sĕdĕng śānta **pwa** sang Prabu, mākanak i sira inaranan ta sang Śantanu.* [Ad 92:31]
    'When the King was in a state of calm, he got a son, therefore he named (his son) Śantanu

The explanation concerning the time relationship stated by the order of *pwa* clause and *ta* clause is in accordance with the result of Uhlenbeck's research on particle *ta*.

A combination of a -*pwa* clause with other clauses also states a cause–effect relationship. The *pwa* clause that states the cause is in the first order, followed by the other clause that states the effect. This order is fixed. Therefore, the relationship is iconic:

25. *Ring dwīpa pwa sirān wijil, ya ta sang Dwaipāyana ngaran ira.* [Ad 63:16]
    'He was born in the island; therefore, his name is the Dwaipāyana.'

26. *Katon pwa sira ahayu, mahyun tikang raksasa si Duloma.* [Ad 18:22]
    'She (the Puloma) looks beautiful, (therefore) Duloma, the giant desires her.'

4   COMPARISON OF PARTICLES *PWA* AND *TA*

There are three similarities between particles *pwa* and *ta*. **First**, both particles exist in a construction in the form of a clause. Therefore, the scope of the two particles is one clause. In a clause, *pwa* and *ta* play the role of modifying two constituents that are functionally different. Thus, the A-*pwa*-B pattern applies similarly to particle *ta*. Both can be paradigmatically related (see again examples 1(a) and 1(b).)

Second, both particles must be present when position B is a subject in the form of a topical pronoun, namely *sira* or *ya*:

27. *Katon pwa ya de bhagawan Sthulakeśa.*   [Ad 21:2]
    'He was seen by bhagawan Sthulakeśa.'

28. *Katon **ta** ya manglayang ring ākāça*   [Ad 57:14]
    'He is seen (currently) flying to the sky.'

In sentences that start with the verb *hana*, both *pwa* and *ta* are markers to introduce a new figure:

29. *Sira ta sumilih ratu ri Hāstinapura, ri lunghā sang Pāndawa nusup ing alas muwah, lawas-nira siniwi nĕmang puluh tahun, ndan kadi mahārāja Pāndu, sira śakta ring gunāburu. Asing wukir alas paranirāmet mrga. Hana pwa kidang tinutnira, anghel ta sira denya, ahyun anginuma wwe sira.*   [Ad 49:9]
    'He (Parikesit) replaced the king in Hastina, when Pandawa went back into the jungle; he had been a king for sixty years, like Pandu, he likes hunting. There was a deer he followed, he (became) tired; therefore, he wanted to drink water.'

30. *Hana sira brāhmaṇa, bhagawān Dhomya ngaranira, patapanira ry Âyodhyāwiṣaya. Hana ta śiṣyanira tigang siki, ngaranira sang Utamanyu, sang Ârunika, sang Weda.* [Ad 8:20–22]
'There was a *brahmana* called bhagawan Domya, whose hermitage is in the country of Ayodhya. There are three of his students, namely sang Utamanyu, sang Arunika, sang Weda.

The similarity between particles *pwa* and *ta* lies in the presence of the two particles in a simple sentence. **First,** the difference between the particles *pwa* and *ta* lies in their distribution in a compound sentence that consists of two clauses. The particle *pwa* is present together with particle *ta*. Both have an interclause relationship. In such case, the *pwa* clause that is semantically related to the *ta* clause always occupies the first place in the order. Therefore, the pattern is A-*pwa*-B, A-*ta*-B. In such sentence construction, particles *pwa* and *ta* are not paradigmatically related. The order cannot be reversed to become A-*ta*-B, A-*pwa*-B. The order of *pwa* and *ta* clauses mainly states the relationship of the meaning of time. This is in accordance with the rule defined by Uhlenbeck (1970, 1987), that is, a construction with *ta* is often preceded by a time clause. The particle *ta* is never present in such first clause.

Second, the particle *pwa* is not present in a prohibition sentence that begins with the word *haywa*, and an interrogative sentence is preceded by an indeterminate *pira,* whereas *ta* may be present in the two sentence types. For example:

31. *Haywa ta kita sangsaya*   [Ad 28:29]
    'Don't you worry!'

32. *Pira ta kwehning nāga mati?*   [Ad 23:19]
    'How many dragons that died?'

In contrast, particle *ta* did not exist in an equative sentence (32) and in sentences beginning with a *matangnyan* conjunction (33). However, although this is only supported by one datum, *pwa* may exist in the two sentences, like in the following quotes:

33. *Tanti pwa ngaraning pasamuka nikang lĕmbu*   [B 41:19]
    'The name of the herd of cattle is Tanti.'

34. *Matangnyan **pwa** mahāraja Krĕsna, tan ahyun nghulun i kawijayan tan kapengin ing rājya-wibhawa.* [B 41:19]
    'For the reason of king Kresna, I do not want victory, do not want the country and power.'

Uhlenbeck (1970, 1978) explained that the absence of particle *ta* in an equative sentence is because of the pattern A-*ta*-B, where the constituent is A ≠ B.

Third, in the expanse of the A-*ta*-B pattern, particle *ta* is never present when C occupies the position following A. For example:

35. *Katon / de sang wiku ring Śataśrĕngga / sira kabeh*   [Ad 121:19]
    A   C   B
    'They were all seen by the priest in Śataśrĕngga.'

Example (34) is a passive clause. In a passive clause, the position of C is normally taken by an actor that is preceded by particle *de*. Particle *ta* is never present if constituents A and B are interrupted by C. Particle *pwa* may be present in such pattern (35). However, the constituent that fills B, which is usually filled by a subject that plays a role of a purpose, is not present in the sentence:

36. *Pinahayu / pwa / de sang Basuki*   [Ad 28:28]
    'was greeted by Basuki'

37. *Katon / pwa / de sang Garuda, pinahalitnira tekāwaknira, sakawĕnanga masuk i sĕla ning cakra.*   [Ad 53:24]
    'When seen by the Garuda, his body was shrunk to be able to enter the wheel gap.'

5 CONCLUSION

The observation of particle *pwa* has provided new insights concerning the particle. Most of the previous researchers believed that *pwa* and *ta* are similar particles. However, we show that the similarity between the two particles lies only in their function as a modifier of two functionally different constituents. Hence, the scope is in one clause.

However, in the extra-sentence relationship, particle *pwa* has a function that is not similar to that of particle *ta*, namely to mark the meaning of time relationship and cause–effect relationship. Grammatically, particle *pwa* has an interclause scope in one compound sentence. The order of the *pwa* clause and the other clause that is related to meaning is fixed. The order is iconic because it illustrates the sequence of events as occurring beyond language.

REFERENCES

Adeelaar, A., & Himmelmann, N P. (eds.). (2005b). Javanese. In *The Austronesian Languages of Asia and Madagascar*. London, New York, Routledge Language Family Series.
Bahasa Parwa II: Tatabahasa Jawa Kuna (Old Javanese Grammar). (1993) Yogyakarta, Gadjah Mada University Press.
Becker, A.L, & Oka, I.G.N. (1974) Person in Kawi: An exploration of an elementary semantic dimension. *Oceanic Linguistics, 13*, 229–255.
Becker, A.L. (1982) Binding Wild Words: Cohesion in Old Javanese Prose. In Kridalaksana, H. & Moeliono, A. M (Eds.), *Pelangi Bahasa*. Jakarta, Penerbit Bhratara Karya Aksara.
Creese, H. (2001) Old Javanese Studies. A Review of the Field. In van der Mollen and Helen Creese (ed.) Old Javanese Texts and Culture. Bijdragen tot de Taal-, Land—en Volkenkunde, KITLV. *Journal of the Humanities and Social Sciences of Southeast Asia & Oceania, 157*, 1.
Dixon, R.M.W. (2010a) *Basic Linguistic Theory: Grammatical Topics (Vol. 1)*. Oxford, Oxford University Press.
Dixon, R.M.W. (2010b) *Basic Linguistic Theory: Grammatical Topics (Vol. 2)*. Oxford, Oxford University Press.
Givón, T. (1983) Topic Continuity and word-order pragmatics in Ute. In T. Givon (ed.), *Topic Continuity in Discourse: a Quantitative Cross-Language study, Typological Studies in Language 3*, 141–214. Amsterdam & Philadelphia, John Benjamins.
Givón, T. (2001a) *Syntax Volume I*. Amsterdam/Philadelphia, John Benjamins Publishing Company.
Givón, T. (2001b) *Syntax Volume II*. Amsterdam/Philadelphia, John Benjamins Publishing Company.
Gonda J. (1936) *Het Oud-Javaansche Bhīsmaparwa*. BJ 7, Bandoeng.
Gonda, J. (1959) On Old-Javanese Sentence Structure. *Oriens Extremus 6*, 57–68.
Hellwig, C.M.S., & Robson, S.O. (1986) Clitic, suffix, and particle: some indispensable distinctions in Old Javanese grammar. In Hellwig, C.M.S and S.O.Robson (eds.) *A Man of Indonesian Letters, Essays in Honour of Professor A. Teeuw*. [Verhandelingen van het Koninklijk Instituut voor Taal-, Land—en Volkenkunde 121] Dordrecht, Holland/Cinnaminton, USA, Foris Publications, 334–341.
Hoff, B.J. (1998) Communicative Salience in Old Javanese. In Janse, Mark, Verlinden, Ann (eds.), *Productivity and Creativity: Studies in General and Descriptive Linguistics in Honor of E.M. Uhlenbeck, Trends in Linguistics; Studies and Monographs, 116*. (pp. 337–47). Berlin & New York, Mouton de Gruyter.
Hunter, M.T. Jr. (1988) Participant Marking in Old Javanese. In *Balinese Language: Historical Background and Contemporary State. [Dissertation]*, 57–117. The University of Michigan.
Irrealis in Old Javanese/Irrealis dalam bahasa Jawa Kuno. (2005a) [Seminar] *Seminar Internasional Jawa Kuno*. Universitas Indonesia.
Jacobson, R., & Kawamoto, S. (eds). (1970) Position and syntactic function of the particle ta in Old Javanese. In R. Jacobson and S. Kawamoto (eds.) *Studies in General and Oriental Linguistics presented to Shiro Hattori*, 648–658. Tokyo, TEK Corporation for Language and Educational Research.
Juynboll, H.H. (1906) *Ādiparwa. Oud-Javaansch Prozageschrift*. 's-Gravenhage, Martinus Nijhoff.
Juynboll, H.H. (1912) *Wirātaparwa. Oud-Javaansch Prozageschrift*. 's-Gravenhage, Martinus Nijhoff.
Kalangwan Sastra Jawa Kuno Selayang Pandang (A Brief Study on Old Javanese Literature). (1985) Jakarta, Djambatan.
Kajian Morfologi Bahasa Jawa (Morphology of Javanese Language). Seri ILDEP 4. (1982) Jakarta, Djambatan

Kamus Jawa Kuna Indonesia (Jawa Kuna-Indonesian Dictionary). (2006) Jakarta, KITLV/Gramedia.

Lokesh, C (ed). (2000) The Old Javanese Word *de*. In Chandra, Lokesh (ed.), *Society and Culture of Southeast Asia. Continuities and Changes*, (pp. 179–190). New Delhi.

Mardiwarsito L & Kridalaksana, H. (1984) *Struktur Bahasa Jawa Kuno* (Old Javanese Language Structure). Ende, Flores, Nusa Indah.

Muslim, M.U. (2003) *Morphology, Transitivity, and Voice in Indonesian*. Dissertation La Trobe University.

Ogloblin, A.K. (1991) Old Javanese verb structure. In: Chandra, Lokesh (ed.), *The Art and Culture of South—East Asia*, (pp. 245–257). New Delhi.

Old Javanese-English Dictionary. (1982) Gravenhage: Martinus Nijhoff. (online version <sealang.net/ojed>).

Shibatani, Masayoshi (ed). (1988) Voice in Indonesian: A Discourse Study. In Masayoshi Shibatani (ed.) *Passives and Voice. Typological Studies in Language*, *TSL Volume 16*. Amsterdam, John Benjamins.

Sutrisno, *et al* (eds). (1985) The concept of proportionality: Old Javanese morphology and the structure of the old Javanese word kakawin. In Sutrisno, Sulastin, Darusuprapta., Sudaryanto (eds.). *Bahasa, Sastra, Budaya: Ratna Manikam Untaian Persembahan Kepada Prof. P.J. Zoetmulder*, (pp 66–82). Yogyakarta, Gadjah Mada University Press.

Sentence Pattern in the Old Javanese of the Parwa Literature. (1987) In Laycock, Donald C. Winter, Werner (De.). *A World of Language: Papers Presented to Professor S.A Wurm on Hie 65th Birthday* (pp. 695–708). *Pacific Linguistics, C-100*.

Purwo, B.K. (1986) Strategi Pemilihan *men-* dan *di-* (Strategies in Choosing men—and di-). *Wacana Bahasa Indonesia*. Masyarakat Linguistik Indonesia, Year 4 No.8.

Purwo, B.K. (ed.). (1989) *Serpih-serpih Telaah Pasif Bahasa Indonesia* (A Brief Study on Indonesian Language). Yogyakarta, Kanisius.

Teselkin, A.S. (1972) *Old Javanese (Kawi)*. Itacha, New York: Modern Indonesia Project Southeast Asia Program. Cornell University.

Uhlenbeck, E.M. (1964) *Critical Survey of Studies on the Language of Java and Madura*. Royal Institute of Linguistics and Anthropology, Bibliographical Series 7.

Van Valin Jr., R.D., & LaPolla, R.J. (1999) *Syntax: Structure, meaning and function*. Cambridge, Cambridge University Press.

Van Valin Jr., R.D. (2005) *Exploring the Syntax-Semantics Interfa*ce. Cambridge, Cambridge University Press.

Verhaar, J.W.M. (1980a) *Asas-Asas Linguistik Umum* (General Linguistic Principles). Yogyakarta, Gadjah Mada University Press.

Verhaar, J.W.M. (1980b) *Teori Linguistik dan Bahasa Indonesia* (Linguistic Theories and Indonesian Language). Yogyakarta, Yayasan Kanisius.

Zoetmulder, P.J. (1950/1983) *De taal van het Adiparwa: een grammaticale studie van het Oudjavaans*. Dordrecht, Foris Publications.

Zoetmulder, P.J. & I.R. Poedjawijatna. (1992) *Bahasa Parwa I: Tatabahasa Jawa Kuna* (Parwa Language I: Old Javanese Grammar). Yogyakarta, Gadjah Mada University Press.

Taylor & Francis Group image not present.

Cultural Dynamics in a Globalized World – Budianta et al. (Eds)
© *2018 Taylor & Francis Group, London, ISBN 978-1-138-62664-5*

The role of *Maṇḍalas* in the old Javanese religious community in *Tantu Panggĕlaran*

T.I. Setyani & T. Pudjiastuti
Department of Literature, Faculty of Humanities, Universitas Indonesia, Depok, Indonesia

ABSTRACT: This paper studies maṇḍalas in the *Tantu Panggĕlaran* (TP). The general objective of this paper is to contribute to the study of Ancient Javanese literature in the old Javanese civilization context in the late 15th century. Moreover, the specific objective is to understand the role of maṇḍalas in the old Javanese religious community in TP. The research is qualitative one, which uses the interpretation text theory. The research mainly aims to illustrate how important the role of maṇḍalas was to the old Javanese religious communities in TP. The specific highlight is that maṇḍalas are a cultural heritage left by the Indonesian ancestors. We paid close attention to TP since it contains important information in relation to maṇḍalas, which can be used as a data source for further research. Temporarily, the conclusion is that a maṇḍala in TP had very important roles in the old Javanese religious communities as a learning centre and also as a representation of its learners' achievement.

1 INTRODUCTION

As a religious institution, a mandala once held a very important position in Indonesia, particularly during the Old Javanese era. Different studies have produced different conceptions of mandala. Archaeologists proposed that a mandala was a place of education for any individual who wanted to improve his religious knowledge, and each mandala was led by a *dewaguru* or "head hermit" (Sedyawati, 1985, 2001, & 2009; Santiko, 1986; and Munandar, 1990 & 2008), Jung (1987). Munandar (presented on 7th February 2014) argued that mandalas can also be considered as an *ekagrata*—a tool that can facilitate meditation.

Another scientific study by Jung (1987) found that mandalas also represent the most exceptional magical circle of the human self. In his study of Hindu and Buddhist traditions and cultures, Soeroso (1998/1999) pointed out that mandalas stand as a symbol of the universe, while their cosmological configurations represent the plotting or hierarchical positions of the deities. Similar description was also offered by Zoetmulder (1983:442) and Kartika (2007:30) who proposed an idea that the cosmos itself is fashioned in the form of a cosmic circle or mandala.

Another opinion from Ras (2014:214) maintains that an unprotected retreat can be called a mandala. In a similar vein, Supomo (1977:66–67) and Hariani Santiko (1986: 112–114) argued that a *patapan* 'retreat' and mandala are not essentially different because both terms refer to a meditation retreat. The only difference is that a patapan was a temporary retreat, which a retreatant could use for a certain period of time until his purpose was fulfilled, while a mandala was a permanent retreat complex in which the hermits resided.

As gathered from various available studies, research results, and expert opinions, such conceptions of mandala serve as the basis for constructing our own definition of a mandala. This paper uses the conception of mandala as depicted in the ancient literary text of Tantu Panggĕlaran (hereinafter abbreviated as TP).

Ras (2014: 214) even argued that TP could be used to investigate the pattern of life in a mandala, as well as into the minds of its supporting community. Ras' opinion was based on

TP's content, which includes a brief cosmogony and therefore bears some resemblance to the *purāna*. According to Ras (2014:189), cosmogony is related to the creation of the world from a primaeval substance, whereas purāna is a collection of books consisting of eighteen scriptures concerning "matters related to ancient times." Based on Ras' explanation, it can be said that the concept of mandala, according to TP, has certain connections with the creation of the world out of some primaeval substances and with the religious landscape of the Old Javanese. In short, mandalas are indirectly related to the religious communities at that time.

This paper uses the 1924 redaction of TP by Theodoor Gautier Thomas Pigeaud, which appeared in his dissertation titled *De Tantu Panggĕlaran Een Oud-Javaanch Proza-Geschrift, Uitgegeven, Vertaald en Toegelicht* (Tantu Panggĕlaran, an Old Javanese Prose, Redaction, Translation, and Annotation) as the primary data source for investigating mandalas. The full TP text, written in the Old Javanese language, can be found from pages 57–128 of the dissertation. For the purposes of this paper, any reference to annotations or quotations pertaining to TP is made directly to Pigeaud's redaction (1924:57–128). Any TP quotation in this paper is made by mentioning the abbreviation of the book's title, the surname of the author, the year of publication, and the page(s) from which the quotation is taken. For instance: (TP Pigeaud, 1924:57–128). The full TP consists of seven narrative chapters that recount major events as follows.

- Firstly, the island of Java was in the state of imbalance because there were no humans or pillars (*lingga*) that could support it. Bhatārā Jagatpramanā sent Sang Hyang Brahma and Wisnu to create humans after going through a meditation (yoga) with Bhatārī Parameçwari in the island of Yawadipa (Java). Men and women were created and married so as to produce offspring. Then, these first human beings were introduced to the *katatwapratista* lesson ("the principles of the reality of life"), the ways to earn a livelihood, and various tools for survival.
- Secondly, Sang Hyang Mahameru (Mount Mahameru the Great) was relocated to bring balance to the island of Java.
- Thirdly, the cosmic balance between human beings and the island of Java was preserved by installing mountains and sacred retreats.
- Fourthly, the foundational balance of the island of Java was perfected by establishing mandalas, appointing a dewaguru or "head hermit" for each mandala, filling the sacred retreats, and defining levels of religious accomplishment, attributes of holiness, and lessons in meditation.
- Fifthly, Sang Hyang Mahameru was firmly installed, and the island of Java can finally be in the state of perpetual balance.
- Sixthly, the foundational balance of human beings in the island of Java was continuously improved.
- Seventhly, the completion of the mandala constructions was carried out.

Not only after the narrative reaches chapter four is the word mandala mentioned in TP, the word regularly reappears only in chapter five and chapter seven. In spite of this, the fact that no fewer than forty mandala names are mentioned in TP demonstrates that mandalas were really considered as an important concept. In general, this study is intended to examine, understand, and explain the importance of mandalas in TP. This study of mandalas, particularly the ones that flourished towards the end of the fifteenth century, is performed as part of our contribution to Old Javanese literature study.

In this paper, the concept of mandalas is analysed by applying the interpretative theory and the objective approach. The interpretative theory was applied to understand the text. The text interpretation focuses on the concept of mandala as depicted in TP. In this approach, understanding is considered to be based on interpretation, and this is achieved by expressing one's perception of, opinion of, or perspective on any entity associated with the interpretation of mandala. Interpretation is an attempt to uncover the original intention of a text based on what is written therein. In the objective approach, TP is seen as an autonomous literary work that can be used to examine the importance of mandala.

2 DISCUSSION

We have explained that the word mandala is first mentioned in chapter four of TP and continues to appear in chapter five and chapter seven, while there are forty mandala names mentioned throughout TP. Those forty names are Sarwwasiddā, Sukawela, Sukayajñā, Mayana, Guruh, Hahāh, Gĕrĕsik, Çūnyasagiri, Kukub, Tigaryyan-parwwata, Tigapatra, Jala-parwwata, Nangka-parwwata, Panasagiri, Ḍingḍing-Manuñjang, Labdawara, Maṇḍala in the island of Kambangan, Andawar, Talun, Wasana, Warag, Sanggara, Sagara, Trisamaya, Purwa-dharma-kasturi, Kasturi-purwa-dharma Bapa, Sang Hyang Maṇḍala Kasturi Selagraharong, Kasturi Gĕnting/Ghra-rong, Dupaka, Wariguh, Bhulalak, Kasturi Sri-manggala, Arggha-tilas, Jawa, Rebhalas, Kasturi Harĕng, Jiwana, Sukayajñā-paksa-jiwana, Panatmaku, and Layu-watang or Panatan.

According to the year recorded in TP's colophon, those forty mandalas were established in 1557 Çaka (1635 A.D.). This year signifies that the mass establishment of mandalas took place during the late Majapahit era. At that time, the Old Javanese civilization was still dominated by Çiwa-Hinduism and Buddhism.

Information about the generality of Çiwa-Hinduism and Buddhism in Java can be found in TP page 109 as follows.

> *Kahucapa tāmpu Mahāpalyat, mantuk ta sira maring nūsa Jawa. Pinalihnirā ta çariranira matmahan ta çaiwa sogata, mangaran sirāmpu Barang, sirāmpu Waluh-bang. Sirāmpu Bārang çewapaksa, sirāmpu Waluh-bang sogatapaksa* (TP Pigeaud, 1924: 109).
>
> "It is said that Mpu Mahapalyat returned to the island of Java. He divided himself into a follower of Çiwa religion and a follower of Buddhism, under the names of Mpu Barang and Mpu Waluh-bang. Mpu Barang was a follower of the Çiwa religion, Mpu Waluh-bang was a follower of Buddhism."

This demonstrates that information about the establishment of mandalas found in TP can serve as a representation of Old Javanese life during the late Majapahit era, a transition period marking the end of Çiwa-Hinduism and Buddhism era and the beginning of Islamic era.

All explanations in TP regarding the importance of mandalas are based on the role that this institution played in the Old Javanese communities. The role of mandalas is strongly related to the function and purpose of their formation or establishment. Chapter four mentions that mandalas were established to achieve two purposes: general and specific purposes. Its general purpose is indicated in this following passage.

> *... bhatāra Parameçwara tumulusakna magawe tantu praçista ri Yawadipa. Makāryya ta sira maṇḍala, makulambi magundala sira, tumitihi surangga patarana. Tinhĕr manganaknadewaguru sira, umisyani sira patapan kabeh...* (TP Pigeaud, 1924: 81).
>
> "Bhatāra Parameçwara perfected the foundational balance of the island of Java. He established mandalas using *makulambi* and *magundala* [The word magundala is mostly written as either makundala or kundala (Pigeaud, 1924)] and sat on a red cloth. He then appointed dewaguru ('head hermit') and filled all the sacred retreats."

From the quotation above, it can be deduced that mandalas were established by Bhatāra Parameçwara for the purpose of perfecting the foundational balance of Java Island. Moreover, the establishment of mandalas was accompanied by the establishment of sacred retreats within their vicinity. This might consist of various facilities, such as retreats for men, women, and hermits, as well as housing complexes for abĕt (a special religious community) and village religious officers (Pigeaud, 1924:81).

When building a mandala, Bhatāra Parameçwara displayed certain attributes of holiness: *makulambi* '(special) attire', *magundala* '(special) ring or earrings', and a red cloth on which he sat. The display of such attributes of holiness suggests that a certain ritual had to be performed when building a mandala. This is very likely since attributes of holiness were only donned when performing certain activities or actions, especially religious ceremonies or rituals, as well as the ordainment or inauguration of a dewaguru or "head hermit." The newly inaugurated dewaguru would receive a *payung*, *kundala*, and *kulambi* as his own attributes of holiness.

The specific purpose of a mandala's establishment is implied in this following passage.

Mojar ta Bhatāra Guru:

> *Kapan ta kang manusa limpada sakeng pañcagati sangsara? Dwaning makāryya mandala panglpasana pitarapāpa. Antukaning manusa mangaskara hayun wikuha; matapa sumambaha dewata, dewata suměngkaha watěk hyang, watěk hyang suměngkāha siddrāsi, siddārsi suměngkha watěk bhatāra. Lena sakerikā hana pwa wiku sasar tapabratanya; tmahanya tumitis ing rāt, mandadi ratu cakrawarthi wiçesa ring bhuwana, wurungnya mandadi dewata...* (TP Pigeaud, 1924:83)

Bhatāra Guru said:

> "When will humans be able to release themselves from the *pañcagati sangsara* ('five levels of incarnation in the cycle of rebirth')? The purpose is to make mandalas (as) places of salvation for the ancestors. A human being can achieve this by being purified first and becoming a *wiku*; by performing meditation until he becomes a *dewa*; after attaining the status of dewa, he will then be raised to the rank of *hyang*; the hyang will then be raised to the rank of *siddrāsi*; the siddrāsi will then enter the group of *bhatāra*. However, a different fate awaits those wikus who perform their meditation in a wrong way; their reincarnation will take place in this world, becoming mortal kings as the highest rulers on earth, who have failed to or do not become dewas ..."

This passage indicates that a mandala was established to serve as a place for the ancestors to attain salvation, so that they would be able to release themselves from the pañcagati sangsara. The way to achieve this salvation was to become a wiku (hermit). To become a wiku, one had first to undergo a purification and to perform deep meditation to reach the level of a dewa (deity) and then all the way up until one reached the level of a bhatāra. In other words, a mandala served as a place where people purified themselves to become wikus and perform meditation to reach the highest level of freedom from the pañcagati sangsara.

Nevertheless, a mandala did not only function as a place where people purified themselves and performed meditation. In TP, the word mandala is frequently used to refer to a much larger sacred complex, which did not only contain retreats. A mandala also contained dewaguru's residences, *ashrams* or boarding schools and/or training places, an education complex consisting of retreats for men and for women, a housing complex for abět (a special religious community), and a housing complex for village religious officers. This arrangement is indicated in this following passage.

> *Makāryya ta sira mandala, ... Tinhěr manganaknadewaguru sira, umisyani sira patapan kabeh, kadyana: katyagan, pangajaran, pangubwanan, pamanguywan, pangabtan, gurudeça....* (TP Pigeaud, 1924: 81).
> "He (Bhatāra Parameçwara) created a given mandala, ... Then he would appoint the head hermit, fill all the retreats with, among others, *katyagan* (common retreats), *pangajaran* (ashrams or training places), *pangubwanan* (retreats for women), *pamanguywan* (retreats for hermits), *pangabtan* (a housing complex for abět) and *gurudeça* (village religious officers), ..."

This passage suggests that a patapan 'retreat' was located in a mandala, which implies that patapan is a smaller part of a mandala. Furthermore, almost all patapans are mentioned with their respective owners' name attached to them, so it can be concluded that a patapan is a private retreat allocated for specific individuals. For instance, there are Bhatāra Guru's patapan, Bhatāri Uma's patapan, Bhatāra Içwara's patapan, Sang Gana's patapan, and Sang Kumara's patapan. In many cases, the owner's name became even more popular than the patapan's proper name. For instance, some of Bhatāra Guru's patapans actually had their own proper names: Ranubhawa (see TP Pigeaud, 124: 70), Gĕgĕr-katyagan (see TP Pigeaud, 1924: 71–72), Tandĕs (see TP Pigeaud, 1924: 72), and Kayutaji.

TP mentions the names of the founders of the forty mandalas who also acted as their respective *dewagurus* or "head hermits", but these names are not indicated as their owners. Some mandalas experienced a succession of *dewagurus*, who were given the responsibility for passing on the teachings of their predecessors. For instance, during the timeframe of TP, Sukayajñā mandala experienced five successions of leaders: Bhatāra Guru, the first leader, ordained Bhatāra Wisnu to replace him, Bhatāra Wisnu was replaced by Bhatāra Içwara, Bhatāra Içwara was replaced by Bhagawan Karmandeya, Bhagawan Karmandeya was replaced by Bhagawan Agāsti, and Bhagawan Agāsti was then replaced by Bhagawan Trnawindu, the last Sukayajñā dewaguru mentioned in TP.

This order of succession of *dewagurus* was in line with their respective duties. Bhatāra Guru created Sukayajñā mandala and purified the people who wished to be *wikus* and become his disciples there.

> *Akweh manūsa harĕp wikuwa, yata sinangāskaran de bhatāra Guru. Tambehaning mangaskarani bhagawan Wrhaspati, kaping kalih bhagawan Soma, kaping tiga bhagawan Budda, kaping pat bhagawān Çūkra, kaping lima bhagawan Raditya, kaping ĕnĕm bhagawan Saneçcara, kaping pitu bhagawan Hanggara. Samangkana kwehning çiksa bhatra Guru duk ing Sukyajñā* (TP Pigeaud, 1924:82).

> "Many people wished to become *wikus*, so they were purified by Bhatāra Guru. Firstly, he ordained Bhagawan Wrehaspati; secondly, Bhagawan Soma; thirdly, Bhagawan Budda; fourthly, Bhagawan Sukra; fifthly, Bhagawan Raditya; sixthly, Bhagawan Saneçcara, and seventhly, Bhagawan Hanggara. Such was the number of Bhatāra Guru's disciples when he was in Sukayajña."

Next, Bhatāra Guru ordained Bhatāra Wisnu to replace him as the dewaguru in Sukayajña, while Bhatāra Guru proceeded to establish the Mayana mandala. Bhatāra Wisnu took over Bhatāra Guru's duty to purify people who wished to become *wikus*. He then ordained Bhatāra Içwara to replace him as the next *dewaguru*.

Bhatāra Içwara took over Bhatāra Wisnu's duty to purify people who wished to become wikus. In addition to this responsibility, Bhatāra Içwara also bestowed the attributes of a Çiwa priest and gave ordination names to mortal men who had been purified and appointed as Çiwa priests. At the end of his tenure, Bhatāra Içwara ordained his successor, Bhagawan Karmandeya, as the next dewaguru.

Bhagawan Karmandeya then ordained his successor, Bhagawan Agāsti, who immediately performed his responsibility as a teacher of yoga and *samadhi* (meditation). In addition to that, Bhagawan Agāsti was also responsible for protecting the summit of Mount Kelaça. He blessed, purified, and gave names to Bhagawan Trenawindu and Bhagawan Anggira, a pair of twins who had been abandoned by their mother Galuh Sri Wiratanu, a daughter of the king of Daha. He was also entitled to teach *sang hyang kawikun* or 'a lesson on becoming a wiku', which was a kind of secret knowledge. Bhagawan Agāsti then ordained Bhagawan Trnawindu as the next *dewaguru* and returned to the heavens.

The duties of each dewaguru in Sukayajñā mandala can serve as the basis for identifying the main function of a mandala as a place where people became *wikus*, acquired the knowledge of sang hyang kawikun, learned yoga and samadhi, and even became Çiwa priests to teach other people to attain the level of a bhatāra.

The nature of sang hyang kawikun lesson was expounded in the Dingding-Manuñjang mandala. It was a secret lesson on understanding the meaning of becoming a *wiku* without having to learn from a teacher or donning a hermit's garb. Upon mastering sang hyang kawikun, an individual was entitled to purify his own wife and children, to be exempted from offering sacrifice, to be exempted from the obligation to worship and meditate, to be exempted from religious lessons, not to be known by any other people, and to be addressed as a bhatāra. In other words, in order to achieve the level of bhatāra, a *wiku* had to master sang hyang kawikun.

In Çünyasagiri mandala, Bhatāra Guru introduced sang hyang Hastitijati, which was a lesson on identifying one's true guide or teacher. This lesson was taught to enable people to purify themselves, to emulate the behaviours of the deities, to eradicate the stains of the world, to offer sacrifices, and to create a harmonious, peaceful, and prosperous society.

Kukub mandala was a place where all men and women from all classes and backgrounds could have the blessed opportunity to become hermits, as well as to improve and perfect their meditation in order to attain an eternal life in heaven. One way to achieve this divine life was by exterminating the evil within one's own self. The attainment of perfect meditation was exemplified by Resi Sidawangsitadewa who succeeded in mastering sang hyang kawikun and acquired the ability to pass through sang hyang sima brata by means of meditation. Such perfection is described in this following passage.

> ... sirā mangunkna tapa, malāghna monaçri, tan kalangkahan dening rahina wngi, tan panuwukning pangan turu. Kalinganya: tan hana hinistinira, tan bhuwana, tan swargga, tan bhatāra, tan kamoksan, tan kalĕpasĕn, tan suka, tan duhka; tan hana hinalĕmnira, tan hana keliknira; yata sinangguh tapa ngaranya... (TP Pigeaud, 1924: 92).
>
> "... he performed meditation, struggled quietly, never faltered both day and night, had neither rest nor sleep. Meaning: he wanted nothing, not the world, nor the heavens, nor bhatāra, nor kamoksan (perfection), nor freedom, nor happiness, nor sadness; he praised nothing, was bored with nothing; that is what is called the ultimate meditation skills..."

3 CONCLUSION

Our discussion above reveals that mandalas were established for the general purpose of perfecting the foundational balance in Java Island. However, they also served a more specific function as places where our ancestors could attain salvation, so that they could free themselves from the pañcagati sangsara (five levels of incarnation in the cycle of rebirth).

In general, all activities in a mandala can be considered as religious in nature. Each mandala was built to represent its important function and role within its religious community and especially in the development of the Old Javanese religions as mentioned in TP.

According to TP, the principal role of mandalas in Old Javanese religious communities was to serve as places where our ancestors could achieve salvation by freeing themselves from the pañcagati sangsara. Such an achievement could be made by completing all levels of education at a mandala under the tutelage of a *dewaguru*. This education was acquired in several levels, which started when an individual became a wiku, performed a profound meditation, mastered the sang hyang kawikun lesson, and eventually reached the level of a bhatāra. When all people were able to make such an achievement, the foundational balance of Java Island was assumed to take place.

According to TP's descriptions, a mandala evidently played important roles in the Old Javanese religious communities as both a centre for religious education and a representation of its disciples' religious achievements. Therefore, it can be concluded that the mandalas served as the centres of balance in the life of the Old Javanese communities.

REFERENCES

(2008) *Ibukota Majapahit, Masa Jaya dan Pencapaian (The Capital of Majapahit, Its Golden Age and Achievement)*. Depok, Komunitas Bambu.
(2009) *Śaiwa dan Buddha di Masa Jawa Kuna. (Śaiwa and Buddha in Ancient Java)*. Denpasar, Penerbit Widya Dharma.
Jung, C.G. (1987) *Menjadi Diri Sendiri, Pendekatan Psikologi Analitis. (To be Oneself, an Analytical Psychology Approach)* A. Cremers (trans.). Jakarta, Gramedia.
Kalangwan, (1983) *Sastra Jawa Kuno Selayang Pandang. (A Brief Look at the Ancient Javanese Literature)*. Jakarta, Penerbit Djambatan.

Kamus Jawa Kuna Indonesia. (Dictionary of Ancient Javanese) (2006) Jakarta, PT Gramedia Pustaka Utama.

Kartika, D.S. (2007) *Budaya Nusantara: Kajian Konsep Mandala dan Konsep Triloka/Buana terhadap Pohon Hayat pada Batik Klasik.* (The Culture of Nusantara: A Study on the Concept of Mandala and the Concept of Triloka/Buana on the Tree Life in Classical Batik). Bandung, Rekayasa Sains.

Munandar, A.A. (1990) *Kegiatan Keagamaan di Pawitra: Gunung Suci di Jawa Timur Abad 14–15.* [Thesis]. *(Religious Activities in Pawitra: Holy Mountain in East Java 14–15th Century).* Depok, Fakultas Pascasarjana. (Post Graduate Faculty)

Munandar. Consultant. (Personal communication, 7th February 2014).

Pigeaud, T.G.T. (1924) *De Tantu Panggělaran Een Oud-Javaanch Proza-Geschrift, Uitgegeven, Vertaald en Toegelicht.* [Dissertation, Rijksuniversiteit te Leiden]. 's—Gravenhage, Nederlandsche Boek en Steendrukkerif voorheen H L Smits.

Ras, J.J. (2014) *Masyarakat dan Kesusastraan di Jawa.* (*Community and Literature in Java*). Ikram A. (trans). Jakarta, Yayasan Pustaka Obor Indonesia.

Santiko, H. (1986) *Mandala (Kedewaguruan) pada jaman Majapahit.* {*Mandala (Kedewaguruan) in Majapahit Times*}. [Meeting] *Pertemuan Ilmiah Arkeologi ke IV (The 4th Archeological Science Meeting). 3th–9th March 1986, in IIb part; Aspek Sosial Budaya (Socio-Cultural Aspect), 149–168.*

Sedyawati, E. (1985) *Pengarcaan Gaṇeśa Masa Kadiri dan Singhasāri: Sebuah Tinjauan Sejarah Kesenian. (The Statue of Gaṇeśa in Kadiri and Singhasāri Times: A Review on Art History).* [Dissertation]. Jakarta, Universitas Indonesia.

Soeroso. (1998/1999) Jantra dan Mandala dalam arsitektur candi. (Jantra and Mandala in temple architecture) *Berkala Arkeologi Sangkhakala No. III/1998–1999. (Sangkhakala Archeology Periodical No. III/ 1998–1999).* Medan, Pusat Penelitian Arkeologi Nasional—Balai Arkeologi Medan, 41–57.

Supomo, S. (1977) *Arjunawijaya.* The Hague: Martinus Nijhoff and KITLV.

Tantu Panggělaran dan Manikmaya: Bandingan Kosmogoni. (Tantu Panggělaran and Manikmaya: A Cosmogony Comparison) (2001) [Seminar] *Seminar Jawa Kuno (Seminar on Ancient Java)*, Faculty of Humanities, Universitas Indonesia.

Zoetmulder, P.J. & Robson, S.O. (1982) *Old Javanese-English Dictionary.* The Hague, Martinus Nijhoff.

Cultural Dynamics in a Globalized World – Budianta et al. (Eds)
© *2018 Taylor & Francis Group, London, ISBN 978-1-138-62664-5*

Signatures in *Hikayat Sultan Taburat's* manuscripts: The existence of scribes in the 19th century

R.A. Suharjo & T. Pudjiastuti
Department of Library and Information Sciences, Faculty of Humanities, Universitas Indonesia, Depok, Indonesia

ABSTRACT: Most of the colophones of Malay manuscripts do not reveal the identities of scribes. However, in the late 19th century, most of the literature works transcribed in Pecenongan reveal their identities. Furthermore, some of them even have their signatures. One of the manuscripts copied in Pecenongan is *Hikayat Sultan Taburat* (HST). In total, there are nine HST manuscripts, seven of which have the signatures of the scribes. The objective of this study is to explain the significance of the signatures in HST manuscripts in relation to their placement and shape. In addition, we examine HST texts, scriptorium in Pecenongan, and the relationship between Muhammad Bakir and Tjit. According to previous studies, the signatures found in HST manuscripts serve as a sign of not only ownership but also the text reading and an aesthetic aspect to the writing.

1 INTRODUCTION

In the writing tradition of Malay manuscripts, display of the self-existence of scribes was initially prohibited. Therefore, detailed data on the identity of the scribes could not be found in most colophones of Malay manuscripts. Nevertheless, some scribes intentionally revealed their identities clearly in a few texts by not only including their full names but also engraving their signatures.

The Malay manuscripts containing the signatures of the scribes are those coded Ml.181, Ml. 179, Ml. 384, and Ml. 363. In the manuscript coded Ml. 181, entitled *Hikayat Jaran Kinanti Asmaradana,* the signature reads "Saidan"; in Ml. 24, *Hikayat Bunga Rampai,* the signature reads "R. Redjodiepoero"; in Ml. 384, *Hikayat Si Miskin,* the signature reads "Jusuf"; and in the manuscript coded Ml.363, entitled *Hikayat Pendita Raghib,* there are two signatures, whose meanings have yet to been found. Similarly, the signature present in the manuscript coded Ml.179, entitled *Hikayat Jaran Sari Jaran Purnama,* has yet to be deciphered.

Signatures can also be found in some of the *Hikayat Sultan Taburat* (hereinafter referred to as HST) manuscripts, in which, a number of signatures belonging to Tjit (Syafirin bin Usman bin Fadhli) and Muhammad Bakir bin Syafian bin Usman bin Fadhli have been discovered. On the basis of the content of the texts, these two persons were the scribes of the HST manuscripts. Furthermore, the commonly found signature is of Muhammad Bakir, whereas Tjit's signature was only found in one manuscript.

In addition to the signatures found in the HST manuscripts, Chambert-Loir and Kramadibrata (2013) asserted that the signatures of Muhammad Bakir can be found in other manuscripts present in a scriptorium owned by the Fadhli family in Pecenongan in the late 19th century. These manuscripts include *Hikayat Merpati Mas and Merpati Perak* (1887), *Hikayat Nakhoda Asyik* (1890), *Hikayat Raja Syah Mandewa* (1893), *Hikayat Syahrul Indra* (1893), *Syair Siti Zawiyah* (1893), *Hikayat Indra Bangsawan* (1894), *Hikayat Begerma Cendra* (1888), *Seribu Dongeng, Hikayat Asal Mulanya Wayang* (1890), *Hikayat Gelaran Pandu Turunan Pandawa* (1890), *Hikayat Maharaja Garbak Jagat* (1892), *Lakon Jaka Sukara* (1894), *Hikayat Sri Rama* (1314 H), *Hikayat Wayang Arjuna* (1897), and *Hikayat Purasara.*

The objective of this study is to discuss the significance of the signatures of Muhammad Bakir and Syafirin (Guru Tjit) found in the HST manuscripts.

This study is a codicological approach toward the physical manuscripts, which was conducted by describing and analyzing the placement and shape of the signatures as well as exploring the relationship between Muhammad Bakir and Tjit.

2 SIGNATURES IN THE HST MANUSCRIPTS

In this section, we will explain the various aspects of the signatures found in the HST manuscripts. The layout, form, and function of the signatures will be examined in relation to the HST texts and the scriptorium owned by the Fadhli family. In addition, the relationship between Tjit and Muhammad Bakir will be revealed. The purpose of this section is to trace the origin of a variant of Muhammad Bakir's signatures that include two names, "Tjit" and "Muhammad Bakir", written as one.

2.1 *The scriptorium owned by the Fadhli family in Pecenongan*

One of the manuscript scribal and rental places in Batavia is *Langgar Tinggi, Pecenongan*, which is owned by the Fadhli family, in which supposedly Fadhli came to Batavia and established Langgar Tinggi in Pecenongan (Saidi, 2000, p.133).

Although Fadhli sounds like an Arabic name, Chambert-Loir (2009, p. 9) stated that al-Fadhli is actually not a name of an Arab family. By 2011, in Arrahman Mosque (formerly known as Langgar Tinggi Mosque), there were two heirs of the family who became the head and the guardian of the mosque, namely Mr. Arnadi Mukti (deceased) and Anton Haris. Furthermore, Arnadi Mukti, who was the heir from Syafirin's family, served as the head of *Takmir* of Arrahman Mosque. Meanwhile, Anton Haris came from Muhammad Bakir's family. At that time, they had a letter containing the family tree of the Fadhli family with information written on a piece of paper. On the basis of the family tree obtained from his heirs, Fadhli had another name, that is, Tegal Arum. He originated from Mataram, Yogyakarta. However, the content of the letter needs to be explored.

There were four of descendants of Fadhli who worked as scribes, namely Syafirin bin Usman bin Fadhli, Ahmed Mujarrab bin Syafirin bin Usman bin Fadhli, Muhammad Bakir bin Syafian bin Usman bin Fadhli, and Ahmad Beramka bin Syafirin bin Usman bin Fadhli. Syafirin (Master Tjit) and Muhammad Bakir, in addition to being scribes, were Quran tutors, and both of them once served as the *Takmir* chairman of Langgar Tinggi Mosque, in which the staff has been taught by the family for generations.

In addition to serving as a place for worship and learning the Qur'an, according to the information obtained from the heirs, Langgar Tinggi Mosque used to serve as a center for manuscript rental service. However, the mosque no longer serves that purpose. The classic Malay manuscripts that were once stored in the mosque no longer exist. At present, the majority of the manuscripts are stored in the National Library in Jakarta. Some other texts, which mainly contain religious teachings, are kept by the families of the heirs.

Of the four aforementioned scribes, Muhammad Bakir was the most productive scribe between 1858 and 1908 (Mulyadi, 1994, p. 58). Meanwhile, Muhadjir (2002, p. 8) stated that Muhammad Bakir actively worked as a scribe from 1884 to 1906.

Chambert-Loir (2009, p. 246) stated that the manuscripts from the scriptorium owned by the Fadhli family in Pecenongan were written during the transitional period of two cultural realms. Furthermore, Chambert-Loir (2009, p. 246) stated that the family, as providing manuscript rental service, and their customers were considered part of the generation that still enjoys the fantastic adventures of the characters as well as questions and doubts the values expressed through that particular literature.

2.2 HST texts and manuscripts

From the investigation of several Malay manuscripts and local text catalogs, the HST texts can be found in nine manuscripts written using *Jawi* characters and Malay language. These manuscripts are currently stored in the Indonesian National Library, and they are coded Ml.183 A, Ml. 183 B, Ml. 183 C, Ml. 183 D, Ml. 183 E, Ml. 257 A, Ml. 257 B, Ml. 258, and Ml.259.

None of the HST manuscripts contain the complete text from the beginning to the end of the story of the *Hikayat*. Of the nine HST manuscripts, eight contain concatenated texts that form a complete story. They are ordered as follows: Ml. 257 A, Ml. 259, Ml. 183 D, Ml. 258, Ml. 257 B, Ml. 183 B, Ml. 183 C, and Ml.183 E. The HST manuscript coded Ml. 183 A contains texts similar to the manuscripts coded Ml. 257 A and Ml. 259.

The oldest HST manuscript is Ml.183 D. On the basis of the colophon, the scribing process of this manuscript was completed in January 1885. The second oldest HST manuscript is Ml. 183 B, whose scribing process was completed in November 1885. Meanwhile, the third older manuscript is Ml. 183 C, whose scribing process was completed in December 1885. Manuscript coded Ml. 183 E was completely copied in January 1886, whereas the manuscript coded Ml. 183 A did not have clear information on the time of the copy process, and only writings read "1887". There is no much information on the date in this manuscript. However, on the basis of the writings, it can be predicted that this manuscript was copied around that date. The sixth manuscript of HST is coded Ml. 258, whose scribing process was completed in April 1893. The next manuscript is coded Ml. 257 B, whose scribing was completed in October 1893. The eighth manuscript is coded Ml. 257 A, which was completely copied in January 1894. The last manuscript is coded Ml. 259, which was completely copied in May 1894.

2.3 Placement, shape, and function of the signatures in the HST manuscripts

A signature is a characteristic of the manuscripts of Muhammad Bakir and Syafirin. In the HST texts, there are several signatures of both scribes. Most of the signatures in the HST texts are engraved with the words "Muhammad Bakir", "Bakir Tjit", and "Tjit Muhammad Bakir".

The oldest signature can be found in the manuscript coded Ml. 183 D. On the basis of the information in the colophon, the scribing of this manuscript was completed on 14 January 1884. The results of the examination of the use of paper and characters in the manuscript show that it was probably written by two scribes, namely Muhammad Bakir and Syafirin. However, in the text, there is only one signature that belongs to Syafirin. In the manuscript, Tjit (Syafirin) engraved his name resembling an ornament. Some of them look more similar to a signature.

Most of the HST manuscripts have signature of either Syafirin or Muhammad Bakir. The signature of Muhammad Bakir is engraved in the HST manuscripts coded Ml.183 B, Ml. 183 C, Ml. 183 E, Ml. 258, Ml. 257 A, and Ml. 259. Meanwhile, the signature of Syafirin is engraved only in the HST manuscript coded Ml. 183 D. In essence, the signatures are engraved on both the body and outside of the body of the text (the colophon and poetry). Furthermore, there are signatures found in the middle of the binding, which is between two pages of the manuscripts, and there are also signatures located on the right- and left-hand sides of the lines of the text (flanking the lines of the text). The beginning of the text also bears signatures. Besides, there are signatures that lie between two parts of the text and located at the end of the text.

Almost all of the HST manuscripts contain signatures. There are only two manuscripts without any signature, namely manuscripts coded Ml. 257 B and Ml. 183 A. The manuscript coded Ml. 257 A has 25 signatures, most of which are engraved with "Muhammad Bakir". The manuscript coded Ml. 259 has 23 signatures, most of which are engraved with "Bakir Tjit".

In the manuscript coded Ml. 183 D, there are 22 signatures, most of which are engraved with "Tjit". Meanwhile, none of the signatures in that text bear the name of Muhammad Bakir.

Moreover, the manuscript coded ML. 183 B has 12 signatures, most of which are engraved with "Muhammad Bakir". Furthermore, each of the manuscripts coded Ml. 183 C and Ml. 183 E contain six signatures, most of which are engraved with "Tjit Muhammad Bakir".

The signatures in the HST manuscripts are interpreted in many ways, namely as a sign of ownership, as a marker for the identity of the manuscript owner, as an ornament, as a punctuation mark, and as a cover for parts in the manuscript that seem hiatus.

The signatures used as ornaments, for example, are present in the manuscript coded Ml.183 D on page 320. At the beginning of the text, the signatures of Muhammad Bakir are engraved to form the word "Syahdan". Some of the signatures in the HST manuscripts that function as a full stop (a punctuation mark) can be found in the manuscripts coded Ml. 259 and Ml. 257 A. The signatures in the middle of the text (on the binding) can serve as a catchword. Those signatures can be found in the HST manuscripts coded Ml.258 and Ml.259. The signatures in the HST manuscripts can also serve as a cover for text passages that are hiatus. Some parts of the body of the text with hiatus are covered with the signatures so as to make the right- and left-hand sides of the page look even. These signatures are contained in the HST manuscript coded Ml. 183 D.

The HST manuscript with the oldest signature is the one coded Ml. 183 D. This manuscript was written on the 15th Mulud 1301, containing 22 signatures only belonging to Syafirin.

On the basis of the analysis performed, it can be concluded that Syafirin is the first scribe who started the tradition of engraving a signature in the HST manuscripts. The subsequent manuscripts scribed by Muhammad Bakir were engraved with signatures that read "Muhammad Bakir", "Tjit Muhammad Bakir", or "Bakir Tjit".

Some of the signatures belonging to Muhammad Bakir that appeared in the HST manuscripts have inconsistent forms. However, the signatures that are commonly used by Muhammad Bakir, including the etcher on his tombstone, are engraved with "Bakir Tjit".

2.4 "Tjit" and "Bakir"

There are several signatures engraved in the HST manuscripts using a combination of two names, namely "Tjit" and "Muhammad Bakir". Several forms of Muhammad Bakir's signatures include the word "Tjit". Tjit (Master Tjit) is Syafirin, who is the uncle of Muhammad Bakir. These signatures imply the close relationship between Muhammad Bakir and Syafirin.

Syafirin bin Usman bin Fadhli had another name, *Guru* Tjit (Master Tjit). The title "Guru" is related to his profession as "a teacher in Islamic studies" and the *takmir* chairman of Langgar Tinggi Mosque, Pecenongan. Langgar Tinggi Mosque was established by the Fadhli family.

The family managed the mosque for generations and held positions from the *Takmir* chairman to the caretaker of the mosque (*marbot*). The edict issued by the mosque stated that the first *Takmir* chairman was Syafian bin Usman bin Fadhli, who was the father of Muhammad Bakir, and the second *Takmir* chairman was Syafirin bin Usman bin Fadhli, the younger brother of Syafian. Syafirin was the uncle of Muhammad Bakir, who became the third *Takmir* chairman.

The *Takmir* chairman of Langgar Tinggi Mosque would only be replaced if the official in question was deceased or unavailable. It can be assumed that Syafian bin Usman bin Fadhli had passed away before the death of Syafirin bin Usman bin Fadhli. During this time, Muhammad Bakir is believed to be still young and therefore he was then raised by Syafirin. This can be inferred from the conclusion of the text in the manuscript coded Ml.259 scribed by Muhammad Bakir as follows:

> "...dibilang yang saya tiada ada bekerja. Dari kecil menumpang makan dan pakai dari saya punya mamak. Maka itu saya minta kasihannya yang sewa ini buat sehari semalam
> sepuluh senadanya"

The manuscript coded Ml.183 A contains a concluding poem as follows:

> "Hal apa yang baca sudi terima. Sebab saya baru belajar. Lagi tintanya banyak malar. Ada yang halus ada yang kasar. Lagi yang menulis ada kurang sabar. Lagi umur saya belum berapa. Sementar ingat sementar lupa. Dengan tulisan ayah anda tiada serupa. Tulisannya hina terlalu papa."

There are two possibilities that can be inferred with regard to the poem. First, Muhammad Bakir learned the process of scribing from his father Syafian. Second, Muhammad Bakir learned the process from his uncle Syafirin. The word "father" in the poem may refer to Syafirin because Muhammad Bakir considered Syafirin as his father.

3 CONCLUSION

The signature of Tjit is only found in the HST manuscript coded Ml. 183 D. Except in the two manuscripts coded Ml. 183 A and Ml. 257 B, the signature of Muhammad Bakir can be found in the HST manuscripts copied from November 1885 to October 1893. The number of signatures found belonging to Muhammad Bakir increased in 1894, which is much higher than that of the previous years. In the HST manuscripts, signatures serve not only as a sign of ownership, but also as a sign of text reading and an aesthetic support to the writing.

Presumably, the manuscript copier in Pecenongan had been damaged and hence it was thought that self-existence was an important issue. This is why Muhammad Bakir and Tjit placed their signatures in the copied manuscripts in addition to having their names.

This work contributes to the study of codicology of Malay manuscripts, particularly the meanings and functions of signatures of the scribes in the Malay Pecenongan manuscripts in the late 19th century.

REFERENCES

Bakir, M. (1885) *Hikayat Sultan Taburat* (The Tale of Sultan Taburat), *Ml. 183 B.* Manuscript collection in the National Library of the Republic of Indonesia (PNRI).

Bakir, M. (1885) *Hikayat Sultan Taburat* (The Tale of Sultan Taburat), *Ml.183 C.* Manuscript collection in the National Library of the Republic of Indonesia (PNRI).

Bakir, M. (1886) *Hikayat Sultan Taburat* (The Tale of Sultan Taburat), *Ml. 183 E.* Manuscript collection in the National Library of the Republic of Indonesia (PNRI).

Bakir, M. (1893) *Hikayat Sultan Taburat* (The Tale of Sultan Taburat), *Ml. 257 B.* Manuscript collection in the National Library of the Republic of Indonesia (PNRI).

Bakir, M. (1894) *Hikayat Sultan Taburat* (The Tale of Sultan Taburat), *Ml. 257 A.* Manuscript collection in the National Library of the Republic of Indonesia (PNRI).

Bakir, M. (1894) *Hikayat Sultan Taburat* (The Tale of Sultan Taburat), *Ml. 259.* Manuscript collection in the National Library of the Republic of Indonesia (PNRI).

Chambert-Loir, H. & Kramadibrata, D. (2013) *Katalog Naskah Pecenongan Koleksi Perpustakaan Nasional* (Pecenongan Manuscript Collection in the National Library). Jakarta, Perpustakaan Nasional Republik Indonesia.

Chambert-Loir, H. (2009) *Sapirin bin Usman, Hikayat Nakhoda Asik [dan] Muhammad Bakir, Hikayat Merpati Mas dan Merpati Perak* (Sapirin bin Usman, Hikayat Nakhoda Asik [and] Muhammad Bakir, The Tale of Merpati Mas and Merpati Perak). Jakarta, Masup Jakarta.

Dain, A. (1975) *Les Manuscrits.* Paris, l'association Guillaume Bude.

Hikayat Bunga Rampai, Ml.24. Manuscript collection in the National Library of the Republic of Indonesia (PNRI).

Hikayat Jaran Kinanti Asmaradana, Ml. 181. Manuscript collection in the National Library of the Republic of Indonesia (PNRI).

Hikayat Jaran Sari Jaran Purnama, Ml.179. Manuscript collection in the National Library of the Republic of Indonesia (PNRI).

Hikayat Pendita Raghib, Ml.363. Manuscript collection in the National Library of the Republic of Indonesia (PNRI).

Hikayat Si Miskin, Ml.384. Manuscript collection in the National Library of the Republic of Indonesia (PNRI).

Hikayat Sultan Taburat, Ml. 258. Manuscript collection in the National Library of the Republic of Indonesia (PNRI).

Hikayat Sultan Taburat, Ml.183 A. Manuscript collection in the National Library of the Republic of Indonesia (PNRI).

Muhadjir. (2002) *Bunga Rampai Sastra Betawi* (Betawi Potpourri Literature). Jakarta: Dinas Kebu-dayaan dan Permuseuman Propinsi DKI Jakarta.

Mulyadi, S.W.R. (1994) *Kodikologi Melayu di Indonesia* (Malay Codicology in Indonesia). Jakarta, Lembaran Sastra Universitas Indonesia Fakultas Sastra.

Saidi, R. (2000) *Warisan Budaya Betawi* (Betawi Cultural Heritage). Jakarta, Lembaga Studi Informasi Pembangunan.

Syafirin. (1885) *Hikayat Sultan Taburat* (The Tale of Sultan Taburat), *Ml. 183 D*. Manuscript collection in the National Library of the Republic of Indonesia (PNRI).

Cultural Dynamics in a Globalized World – Budianta et al. (Eds)
© 2018 Taylor & Francis Group, London, ISBN 978-1-138-62664-5

Management of cultural heritage sites: A case study of Perkampungan Adat Nagari Sijunjung

F. Amril
Department of Archaeology, Faculty of Humanities, Universitas Indonesia, Depok, Indonesia

ABSTRACT: *Perkampungan Adat Nagari Sijunjung* is a representation of the township and the matrilineal culture of Minangkabau society. Located in Sijunjung, which lies between two rivers, the Batang Sukam and the Batang Kulampi, in an area covered by forests, the village was designated as a Cultural Heritage Site by a decree of the Sijunjung Head District number 188.45/243/KPTS-BPT-2014. Therefore, anything related to the management of the settlement should conform to the cultural heritage preservation principles. In general, the principles can be effectively applied to a protected site, even though it was uninhabited or already in ruin when discovered. However, Nagari Sijunjung is an inhabited locale where people have been living there for generations. The issue focuses on how to manage a living cultural heritage site while at the same time allowing rooms for the community to grow and expand. This paper endeavours to provide a management model for a cultural heritage site that functions as a living monument.

1 INTRODUCTION

1.1 *Background*

Perkampungan Adat Nagari Sijunjung is one of the indigenous settlements in West Sumatra. Located between two rivers, the Batang Sukam and the Batang Kulampi, this village has fertile soil good for farming. The forest surrounding the village also adds to the beauty of the Nagari Sijunjung landscape. In 2014 this village was designated as a Cultural Heritage Site by a decree of the Sijunjung Head District number 188.45/243/KPTS-BPT-2014. The designation of Nagari Sijunjung as a Cultural Heritage Site means that this site is now bound by the cultural heritage preservation principles.

According to Law of the Republic of Indonesia No. 11 of 2010 regarding Cultural Heritage, a cultural heritage site is defined as "a unit of geographic space that has two or more cultural heritage sites, which are adjacent and/or exhibit distinctive spatial characteristics". If we take the wording of the law above, then the Perkampungan Adat Nagari Sijunjung can be categorized as a Cultural Heritage Site for its distinctive spatial characteristics. Nagari Sijunjung does not share the typical spatial characteristics, which are common with other villages in Minangkabau. In this village, the traditional Minangkabau large houses or *rumah gadang* are built to line up following the contour of the road as opposed to other villages that generally allow a more loose composition with regard to the layout of their houses.

Perkampungan Adat Nagari Sijunjung, however, still retains common characteristics of traditional villages in Minangkabau, by having the following, *basosok bajurami* (village borders), *balabuah batapian* (a main road and a common bathing place), *barumah batanggo* (living quarters), *basawah baladang* (rice fields), *babalai bamusajik* (a meeting point and a place of worship), and *bapandam bapakuburan* (local cemetery). Besides those seven elements which are the manifestation of the culture's tangible characteristics, it also has two *jorong*s (a smaller unit of a *nagari*), namely Jorong Koto Padang Ranah and Jorong Tanah Bato, whose inhabitants are still practicing a series of traditions, such as *batoboh, bakaul,* and *membantai adaik,* among others. In addition, there are rules that people of Koto Padang Ranah and Tanah Bato should follow, including the one that prohibit them from building a new house parallel to the *rumah gadang* (Ministry of Tourism and Creative Economy of West Sumatra, 2015, p. 24).

Unlike plants or animals, cultural resources, whether it is tangible or not, is not something renewable that can reproduce itself. This has become the main reason to issue the principles of the management of cultural heritage (Darvill, 1987, p. 4). Perkampungan Adat Nagari Sijunjung is an example of a cultural resource that possesses both tangible and intangible properties. It has the potential to develop, but there are threats coming from natural processes and destructive human activities, such as vandalism (Grant, 2008, pp. 341–344). Even though it has been declared as a Cultural Heritage by the Sijunjung government, the possible threats still exist since there is yet a system to manage the Perkampungan Adat Nagari Sijunjung.

Byrne (2008, p. 155) argues that the preservation of cultural heritage that belongs to a minority group would require more efforts. Nevertheless, this does not presuppose that the preservation of a cultural heritage of the majority group will not face any problems. In this case, the Perkampungan Adat Nagari Sijunjung has issues related to its management system. Darvill (1987, p. 25) mentions that the management of a cultural heritage region in a rural area should include three main objectives, namely maintaining the diversity of archaeological resources in the original landscape where it is located, to make the archaeological resources meet the community's needs in terms of its management, and to resolve the conflict and competition for the use of land which contains the archaeological resources. Based on these objectives, it is pivotal to conduct a study to determine a management model for the Perkampungan Adat Nagari Sijunjung that will meet those three categories.

Cultural heritages can be classified into two categories. The first one is the dead monument, a cultural heritage that was already in ruin when it was discovered. The other type is known as the living monument and it refers to items or properties that continue to be of use to date (Explanation of the Law Number 11 of 2010, Point I General Provisions, Paragraphs 5–6). Obviously, there are distinct differences in terms of preservation as the dead monument has ceased to function, while the living one still serves a purpose. Dealing with a dead monument is relatively easy, especially in terms of its ownership. In contrast, the living monument still belongs to a group of people. Unfortunately, there are no explicit or clear guidelines regarding the management of living monuments in Indonesian laws and regulations. Hence, there is a need to carry out a study about the cultural heritages that fall under the category of living monuments, since it is still inhabited by the community, which still practices their tradition.

This research uses a descriptive qualitative approach. This approach allows an analysis of the qualitative data, which are obtained through observations in the field as well as through a review of available literatures on Perkampungan Adat Nagari Sijunjung. This study examines the cultural resources found in Perkampungan Adat Nagari Sijunjung through the Cultural Resource Management study, i.e. a study by Pearson and Sullivan (1995, as cited in Saptaningrum, 2007, p.17), which is an integrated cross-sectoral work designed to accommodate a variety of interests.

1.2 *Issues*

Perkampungan Adat Nagari Sijunjung is one of the traditional villages whose landscape and culture are still intact and well preserved. The village maintains the traditional Minangkabau village structures and its complements, such as the *rumah gadang,* rice fields, places of worship, and buildings for ceremonial activities. From the cultural perspective, the villagers still practice certain traditions, such as *batagak gala, nikah kawin, basiriah tando, membantai adat, batobo kongsi,* and many others. These traditions are some of the characteristics of Nagari Sijunjung, which runs on a matrilineal system adopted by most Minangkabau clans. This is an important aspect of the community that is still intact, although the site has long been designated as a cultural heritage area.

Unfortunately, there is still no proper management system specifically set up to manage the site. Granted, that there are several parties responsible for the preservation of this village, such as the local government (local government in Sijunjung and the West Sumatra governments), the Central government represented by the Institute for Preservation of West Sumatra Cultural Heritage, and most importantly, the local supporters from the region, including the *ninik mamak* (the village elders) and their followers. Considering the diversity of the stakeholders involved, it is crucial that they share the same point of view with regard to the site management.

Perkampungan Adat Nagari Sijunjung is a living cultural heritage due to the fact that the community is still residing on site and actively practicing their tradition. However, since the issue on cultural heritage site management has not been addressed properly, this site is urgently in need of its own management model. Based on the statement above, the researcher has prepared the following questions.

1. What is the significance of Perkampungan Adat Nagari Sijunjung?
2. What kind of management does a living cultural heritage need to preserve the site properly and to provide economic benefits to the local community?

2 DISCUSSION

2.1 *Perkampungan Adat Nagari Sijunjung*

Minangkabau culture is known for its traditional large-sized house called the rumah gadang, or sometimes *rumah bagonjong*. The name is derived from its spired, multi-tiered roof structure called *gonjong* in the local Minang language. There are some opinions regarding the origin of this shape. One mentions that the roof is purposely crafted to resemble a buffalo's horn; it is based on a mythical race between big buffalos from Majapahit and small buffalos of Minangkabau. Some even compare the roof with a stack of betel leaves. Betel leaves (*daun sirih*) are the common ingredients for chewing pan served at traditional ceremonies, and they are believed to prevent various diseases because of its antiseptic properties. Others claim that it resembles a boat shape, associating it with the tradition of *merantau* (a common practice among Minangkabau men to go to other region in search of a better life) in Minangkabau (Hasan, 2004, pp. 6–7).

In total, Perkampungan Adat Nagari Sijunjung has 76 rumah gadang arranged along the road in Jorong Koto Padang Ranah and Jorong Tanah Bato. Perkampungan Adat Nagari Sijunjung is the only village that is still survived from the historical Pagaruyung Kingdom from the 14th century. At that time, this village was inhabited by six tribes, namely Caniago, Malay, Panai, Tobo, Piliang, and Malay Tak Timbago. Nowadays, many rumah gadang still standing in the area, even though most of them are in a state of disrepair.

The existence of a rumah gadang is one of the requirements before setting up a family in Minangkabau (*barumah batanggo*). Another tradition is known as *balabuah batapian*, which manifests as the main road separating the village from the river that primarily serves as both washing and bathing facility. There is another term known as *basawah baladang* that refers to the paddy fields as the village agricultural areas. Fields in this village are well preserved and located at the back of rumah gadang. As a village with their own government, Perkampungan Adat Nagari Sijunjung has their own meeting place and a place of worship to perform traditional ceremonies. These are essentially parts of a Minangkabau's village layout called babalai *bamusajik,* consisting of specific buildings, such as a village hall and several mosques. It also has *bapandam bapakuburan* or the local cemetery. All the elements mentioned above are still existed in this village.

Rumah gadang, rice fields, roads and rivers, mosques and village halls, as well as the burial place, are the physical characteristics that are still visible in Perkampungan Adat Nagari Sijunjung. Both the tangible and intangible attributes are the proofs that the village's culture remains intact. Perkampungan Adat Nagari Sijunjung in Koto Padang Ranah and Tanah Bato is said to be the centre of the village in which all activities related to the culture and customs take place, namely *batagak gala, basiriah tando, penyelenggaraan mayat, bakaua adat, mambantai adat, musyawarah tobo* or *batobo, turun mandi,* and *manapati mamak (mangaku mamak).*

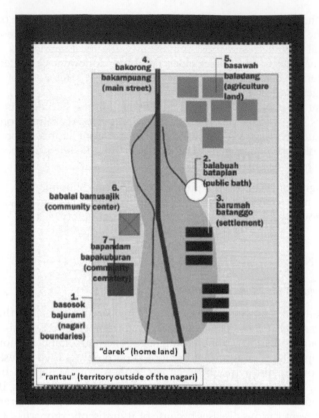

Figure 1.   Nagari conception in Tambo Alam Minangkabau (Source: Joni Wongso).

Figure 2.   *Batoboh*, one of the traditions still practised today (Source: Disparenkraf Sumbar).

Batagak gala is an act of awarding a title to someone who is appointed as the *sako* holder; unlike other forms of legacy, *sako* is usually a title ascribed to a person chosen to be the next tribe leader. Meanwhile, *basiriah tando* is a local proceeding to put someone on trial. There is also an event called *penyelenggaraan mayat*, or a communal discussion (*musyawarah*) to appoint a replacement for a *sako* holder when the current one has passed away. *Bakaua adat* is an event served to express gratitude for a good harvest. The event preceded by slaughtering a buffalo and offering a prayer for the abundance of the future harvest. Still related to an agricultural activity, a number of *Tobo Konsi* (groups of farmers) have been living in this village for a long period of time. Tobo's activities do not only include farming, but also perform tasks, such as building houses, fences, and other related tasks. Meanwhile, the ritual of *manapati mamak* or *mangaku mamak* is held to welcome newcomers to the village.

All the remains in Perkampungan Adat Nagari Sijunjung, either in the form of objects (tangible) or in the form of non-objects (intangible), are heavily imbued with the matrilineal views of the Minangkabau people. West Sumatra is said to host the largest matrilineal Muslim society in the world (Hadler, 2010, p. 3). In matrilineal societies, men have no rights to own land or to inherit a property; a married Minangkabau man will stay at his wife's home at night, and he is still part of his mother's house during the day (Kato, 2005 in Nurti, 2013, p. 74). In Minangkabau, there is a term *saparuik* (literal meaning: one womb) that means siblings from the same mother. Basically, they make up a family occupying a rumah gadang in the region (Mansoer, 1970, p. 6). Therefore, in addition to observing the importance of preserving the cultural heritage in Perkampungan Adat Nagari Sijunjung the emphasis is also on preserving the matrilineal system in Minangkabau as this way of life has been increasingly eroded over time.

2.2 *A proposed management model for Perkampungan Adat Nagari Sijunjung from the perspective of the stakeholders*

Stakeholders, according to Freeman (1984, p. 25), are any group or individual who can affect or be affected by the achievement of corporate goals. Freeman, who talks about stakeholders from the perspective of management strategy in a company, describes that a typical company usually has many stakeholders, including the owners, competitors, customers, employees, governments, local communities, and others. This certainly is not much different from the stakeholders involved in the effort to preserve a living cultural heritage site.

As mentioned before, Perkampungan Adat Nagari Sijunjung is a traditional village consisting of a series of rumah gadang, along with other essential elements that complement the traditional practice of running an indigenous village. The tradition is usually guided by the ninik mamak from one of the tribes living in the village. There are several tribes and each of them is led by the eldest member who is appointed as leader of the group. Each tribe will basically voicing their own interest. Nevertheless, these are always related to the area preservation. A mutual agreement on a plan to preserve the area was signed on 28 September 2013. All parties, including the chief and ninik mamak, have agreed to comply with the following rules and regulations: 1) to preserve Perkampungan Adat Nagari Sijunjung as a world heritage, 2) to cease adding permanent buildings and changing the form and function of the rumah gadang in the village, 3) to preserve the culture and the traditional customs as part of the matrilineal Minangkabau tradition, 4) to participate in maintaining, managing, and developing the area as a leading and genuine site for tourism, 5) to wear a traditional costume inside the village borders.

The agreement shows that the local community, as one of the stakeholders, is already aware of the importance of preservation. They have even gone as far as commencing an agreement to support the area to become one of the world cultural heritages. In this case, public awareness is not an issue at all. As mentioned by Al Najjar (1997, p. 41), raising public concern is beneficial to preserve cultural heritage. However, to coax public concern will not generate any greater impact without any support from both the Central and Local governments.

Governments have a role in terms of establishing the rules as well as imposing sanctions for violations, setting the policy and standard operating procedures for the preservation's implementation and utilization. Every level of the governments must work together to preserve and manage the heritage area. The government's point of view is certainly different from the community's in addressing this particular issue. The government, in this regard, consists of the Central government as represented by the Ministry of Education and Culture. For West

Sumatra, the responsibility is in the hands by the Institute for Preservation of Cultural Heritage (BPCB) for West Sumatra. As the technical implementation unit in the area, their main task is to preserve the cultural heritage that includes protection, development, and utilization. Protection includes the act of rescuing, securing, zoning, maintaining, and restoring the cultural heritage; development relates to research, revitalization, and adaptation. Lastly, utilization is the efforts to utilize the cultural heritage so that it will contribute to the welfare of the community (the Law of the Republic of Indonesia No. 11 of 2010).

For the provincial government, the role is assumed by the Department of Tourism and Creative Economy. It seems that the provincial government aims is to focus on the perspective of tourism and creative economy rather than culture so that this task is performed by the Ministry of Tourism and Creative Economy rather than the Ministry of Education and Culture. As for the local government, this responsibility is given to the Department of Tourism, Arts, Culture, Youth, and Sports. There is still no consensus in the nomenclature of the structure of regional governments at provincial and district levels. Hence, this causes a different focus in each of these fields, which later will affect the budget allocations in supporting the conservation action. Therefore, it is crucial for every stakeholder to share a similar view. To establish this condition, regular meetings between stakeholders are expected to foster a common vision and avoid possible conflicts that might occur between them. It can also enhance public participation in the decision-making process so that the community, as the rightful owner of culture, will not feel left behind (Al Najjar, 1997, p. 42).

Meetings between the stakeholders are also important to devise the most suitable management model for Perkampungan Adat Nagari Sijunjung. As mandated by the Law No. 11 of 2010, the management of the heritage area should be carried out by the responsible parties that consist of all stakeholders. In this case, they are the Central and Local governments, the indigenous community, and the business partners. The management models should also be tailored in accordance to the condition of the village. Quade (1982) argues in Situmorang (2002, p. 12) that a proper model is formulated from all relevant factors or elements resulting in a certain situation. Each element will have relation to one another so that the model will depend on the situation at hand and on the purpose of its creation.

In its management, a cultural heritage should be able to fulfil and appeal to the academic, ideological, and economic interests. Academically, a cultural heritage will allow researchers to gain more knowledge, whether in archaeology or any other subjects. In terms of ideology, it is expected that the local inhabitants will have a sense of pride and passion for their cultural identity or the nation in general. Lastly, a cultural heritage should benefit the community's economy; it should be able to encourage local artisans to develop their artworks and to attract tourists to help the community reap some economic benefits. (Rahardjo, 2013, p. 8–9). In relation with the academic interest, a study conducted by the researchers from various disciplines should be submitted to the management of cultural heritage so that all policies issued are in accordance with the interests of all stakeholders (Ramelan, 2012, p. 197).

The management model with the Agency of Area Management (*Badan Pengelola Kawasan*) is meant to become a forum for stakeholders from various groups. Hence, all stakeholders will have a similar position and interest, namely preserving the area's traditional custom. However, this does not mean that every stakeholder can do as they wish without acknowledging the principles of preservation (Sulistyanto, 2008, p. 302). Therefore, this agency must be able to accommodate all means needed to preserve the Perkampungan Adat Nagari Sijunjung. The preservation does not only protect the tangible properties, such as *rumah gadang,* mosques, or the village hall, but it should also include non-physical culture such as traditions and the local performing arts, which are still alive and run by the local inhabitants.

Thus, this agency will consist of several divisions that will manage the physical aspects of the region, including the maintenance of *rumah gadang,* places of worship, and the community centre. Furthermore, they also manage the rules concerning the management of rice fields and rivers as they are integral parts to this area. Meanwhile, there will be other departments dealing with the traditions and other intangible heritages in the area, such as *batobo* rituals, *batagak gala, turun mandi,* and others. In the end, the agency would cover the needs of preservation consisting of protection, development, and utilization of the Perkampungan Adat Nagari Sijunjung.

3 CONCLUSION

The key to successfully implement the preservation of Perkampungan Adat Nagari Sijunjung largely depends on the shared perception of each stakeholder. With the existence of a management board consisting of all stakeholders, it will result in effective conservation programs for the village. The joint parties will be able to devise a series of plans suitable to the needs of society and avoid the possibility of any violations to the cultural heritage preservation principles. Unnecessary clash between parties will be avoided because plans are the results of a joint agreement based on a mutual understanding. Monitoring and evaluating programmes are also expected to run smoothly to create better plans in the future.

Therefore, establishing a good management system and involving the locals are expected to improve the overall economic condition of the people. Furthermore, developing the village as a tourist attraction is an effective way to conserve not only the physical, but also the non-physical attributes that have constantly been threatened by outside influences. The involvement of the local inhabitants and various stakeholders in managing the Perkampungan Adat Nagari Sijunjung will also reduce a certain party from dominating the effort, in particular, the government. Thus, the management is directed to fulfil the need of all parties and benefit the villagers as well.

REFERENCES

Al Najjar, M. (1997) The Benefit of Developing Heritage Sites for Living Communities: The Petra Experience. In: Wiendu Nuryanti (Ed.). *Tourism and Heritage Management*. (pp. 36–43). Yogyakarta, Gadjah Mada University Press.

Anonym. *Undang-undang Republik Indonesia Nomor 11 Tahun 2010 tentang Cagar Budaya. (Law No 11 of the Republic of Indonesia – 2010 on Cultural Heritage)* Jakarta, Kementerian Pendidikan dan Kebudayaan (The Ministry of Education and Culture).

Byrne, D. (2008) Heritage as Social Action. In: Fairclough, G., Harrison, R., Jameson Jnr., J.H., & Schofield, J. (Eds.), *The Heritage Reader* (pp. 149–172). London & New York, Routledge.

Darvill, T. (1987) *Ancient Monuments in the Countryside an Archaeological Management Review*. London, Historic Buildings and Monuments Commission for England.

Dinas Pariwisata dan Ekonomi Kreatif Provinsi Sumatera Barat. (2015) *Naskah Rekomendasi Penetapan Satuan Ruang Geografis Perkampungan Adat Nagari Sijunjung Kabupaten Sijunjung Provinsi Sumatera Barat Sebagai Kawasan Cagar Budaya Peringkat Nasional. (Recommendation on the Designation of the Geographical Entity of Perkampungan Adat Negeri Sijunjung, Regency of Sijunjung, West Sumatra Province as a National Heritage Site)* Padang, West Sumatra.

Dinas Parsenibudpora Kabupaten Sijunjung. (2016) *Laporan Kegiatan Delineasi dan Zonasi Perkampungan Adat Nagari Sijunjung (Report on Delineation and Zonation of Perkampungan Adat Negeri Sijunjung)* Sijunjung, West Sumatra.

Freeman, R.E. (1984) *Strategic Management: A Stakeholder Approach*. Boston, Pitman Publishing.

Grant, J., Gorin, S. & Fleming, N. (2008) *The Archaeology Course Book an Introduction to Themes, Sites, Methods, and Skills* (3th ed.). London and New York, Routledge.

Hadler, J. (2010) *Sengketa Tiada Putus, Matriarkat, Reformisme Agama, dan Kolonialisme di Minangkabau (Unresolved Dispute, Matriarchal, Religious Reformation and Colonialism in Minangkabau)* Jakarta, Freedom Institute.

Hasan, H. (2004) *Ragam Rumah Adat Minangkabau: Falsafah, Pembangunan, dan Kegunaan (The Variety of Minangkabau Traditional Houses: Philosophical Meanings, Construction and Uses)* Jakarta, Yayasan Citra Pendidikan Indonesia.

Nurti, Y. (2013) *Perubahan Budaya Makan Pada Orang Minangkabau: Suatu Kajian Kasus Pada Acara Baralek Gadang di Kota Padang. (The Changing Eating Culture of the People of Minangkabau: Case Study during the Baralek Gadang Event in Padang)* [Dissertation]. Depok, FISIP, Universitas Indonesia.

Mansoer, M.D, Imran, A., et al. (1970) *Sedjarah Minangkabau. (The History of Minangkabau)* Djakarta, Bhratara.

Rahardjo, S. (2013) *Penelitian Arkeologi dan Pemanfaatannya bagi Kepentingan Publik. (Archaeological Research and Its Uses for Public Interest)* In: Saringendyanti, E., & Syarief, Y.I. (Eds.). *Prosiding Seminar Nasional dalam Rangka 100 Tahun Purbakala. (pp. 6-16). (Proceeding for the National Seminar on the Centenary of Antiquity)* Bandung, Balai Arkeologi Bandung.

Ramelan, W.D. (2012) *Permasalahan Pengelolaan Cagar Budaya dan Kajian Manajemen Sumberdaya Arkeologi. (Challenges in Management of Cultural Heritage and Studies on Archaeological Resources Management)* In: Rahardjo, S. (Ed.). *Arkeologi untuk Publik (Archaeology for Public)* (pp. 186–199). Jakarta, IAAI.

Saptaningrum, I. (2007) *Pengelolaan Kawasan Arkeologi di Kota Magelang. (The Management of Archaeological Site in Magelang)* [Thesis]. Yogyakarta, Pascasarjana UGM.

Situmorang, S. (2002) *Model Pembagian Urusan Pemerintahan antara Pemerintah, Provinsi, dan Kabupaten-Kota. (The Model on Division of Tasks between Central, Province and Regency-Municipality Governments)* [Dissertation]. Depok, Universitas Indonesia.

Sulistyanto, B. (2008) *Resolusi Konflik dalam Manajemen Warisan Budaya Situs Sangiran. (Conflict Resolution in Sangiran Cultural Site Management)* [Dissertation]. Depok, FIB, Universitas Indonesia.

Cultural Dynamics in a Globalized World – Budianta et al. (Eds)
© *2018 Taylor & Francis Group, London, ISBN 978-1-138-62664-5*

Nusantara philosophy: The study of meanings based on Indonesia's local wisdom in East Java and East Nusa Tenggara

L. Tjahjandari, T.I. Setyani & L.H. Kurnia
Department of Literature, Faculty of Humanities, Universitas Indonesia, Depok, Indonesia

ABSTRACT: This study specifically examines the role of philosophy, which can still be found in ethnic communities living in Indonesia. Each region has its own cosmological and philosophical treasury that becomes the guidance for the community to act, speak and behave. As a way of life, the Nusantara philosophy materialises in human understanding on the complexity of their natural surroundings. This study was carried out in East Java and East Nusa Tenggara by observing the communities' way of life, which revolves around solving a problem together and mutual assistance. This study also focuses on the communities' traditional systems, which preserve certain patterns of mutual cooperation, and provide new solutions to actual problems. This research is based on the cultural complexity, the philosophical view of life, and particular ethnic and philosophical advice, which are widely used today and for generations to come. Thus, the conception of the philosophy of living in a society and mutual cooperation can be explained.

1 INTRODUCTION

Indonesia's diversity is built upon its cultural richness. Each ethnic group in Indonesia has its own cosmological and philosophical treasury, which serves as the guideline for living and solving problems. Social ties are strengthened by the foundation of collective consciousness, helping collective problems to be solved with collective consciousness and action. A form of collective sense of sharing is known in Indonesia as *"gotong royong"*, which means solving various problems together.

The diversity of perspectives and the complexity of problems faced by each ethnic community have shaped unique patterns regarding collective actions. Local wisdom is a conceptual idea that lives inside a community, continuously grows and develops inside the community's consciousness, and regulates the community life. Local wisdom is a conception related to the life strategy of each community, which is manifested in certain patterns of activity and is highly influenced by the community's worldview. It is also a complexity of knowledge in form of norms, values, beliefs that underlies the society's behaviour and sometimes is expressed in the myths and traditions (Keesing, 1981).

Haba (2007) mentions six functions of local wisdom. The first is an identity marker of a community. The second is a unifying element (cohesive aspect) of people from different religions and beliefs. The third is that local wisdom is not coercive or top down because its cohesive power is more effective and lasting. The fourth is that local wisdom adds a sense of togetherness to the community. Fifth, local wisdom changes the mindset and mutual relation of individuals and groups by putting it on a common ground. Sixth, local wisdom can act as a collective mechanism to avoid various possibilities that may reduce or even damage communal solidarity believed to originate from and grow in the collective consciousness of an integrated community.

Gotong royong as a form of social solidarity is formed because of others' help, either for the sake of personal or communal interests, which makes it carry loyal attitude from every community member as a whole. Regarding this, Parsons (1951: 97–98) expresses that,

Loyalty is, as it were, the non-institutionalized precursor of solidarity, it is the "spilling over" of motivation to conform to the interests or expectations of alteration beyond the boundaries of any institutionalized or agreed obligation. Collective orientation on the other hand converts this "propensity" into an institutionalized obligation of the role expectation. Then whether or not the actor "feel like it", he is obligated to act in certain ways and risk the application of negative sanctions if he does not.

The life of an integrated community can be seen from the solidarity among its members through helping others without the others having to reciprocate, such as during a disaster or when helping someone in need. *Gotong royong* can be said as a feature of Indonesia, especially of those who live in rural areas where it has been practiced for generations, hence creating real social behaviours that shape the value system of social life. The existence of such value helps *gotong royong* to thrive in the community life as a cultural heritage worthy to preserve. Relating to gotong royong as a cultural value, Bintarto (1980: 24) explains that such value in the cultural system of Indonesians contains four concepts:

1. Humans do not live alone in this world, but they are surrounded by their community, society, and the universe. In this macrocosmic system, they feel themselves as merely a small element carried by the movement process of the universe.
2. Thus, humans are essentially dependent in all aspects of life on others.
3. Therefore, they have to keep trying as much as possible to maintain good relations with others, driven by the spirit of equality, and
4. Keep trying as much as possible to conform, doing the same thing as what their community do, driven by the spirit of equality.

The existence of such value system makes *gotong royong* always maintained and needed in various aspects of life. Thus, *gotong royong* will always exist in many forms, adapted to the cultural condition of each community.

In relation to such matters, this study discusses the conception of social life, which correlates with actual problems faced by each ethnic community, and how the conception of "*gotong royong*" is understood and implemented by referring to the local wisdom. Specifically, this study will discuss the conception of "*gotong royong*" in the Komunitas Belajar Tanoker (the Tanoker Learning Community) in Jember, East Java, and the *Arisan Pendidikan* (the Educational Gathering) in Manggarai, East Nusa Tenggara. The study focuses on how the worldviews and collective problems in both regions are contextually understood according to the current condition and how "*gotong royong*" is practiced in everyday life.

1.1 Nusantara philosophy and meaning of Gotong Royong

1.1.1 Tanoker learning community, Ledokombo in Jember, East Java

Jember is generally known as one of the largest suppliers of migrant workers in East Java Province. In the District of Ledokombo alone, it is estimated 70% of women leave their villages to work as migrant workers, not only abroad but also in regions outside Jember. There is a trend of dropouts among children and unemployment among the working age population. Social problems coming from outside, such as drug addiction and HIV/AIDS cases, only worsen the social condition there.

In Jember Regency, the District of Ledokombo is synonymous with poverty, given that the area is categorised as lacking, both in natural resources and human resources. The majority of the population is ethnically the Madurese, who often become the victims of negative stereotypes, such as being unruly, selfish, stubborn, difficult to cooperate, and envious with others' success. As a result, the perception that emerges is that it is impossible for the district to improve its prosperity and catch up with others.

Out of this concern, on 10th December 2009, a learning community named *Komunitas Belajar Tanoker* (Tanoker Study Community or Komunitas Tanoker in short) was established by a married couple, Supohardjo and Farha Ciciek. Supohardjo (Supo in short) is an alumnus of the doctoral program in sociology at the Faculty of Political and Social Sciences, Universitas Indonesia, while Farha Ciciek is a female activist who has worked much with the issues of gender equality. Supo is a native of Ledokombo. The name "Tanoker" was chosen by the children of the community. It is a word in Madurese, which means cocoon. They created a motto: "Friendly, happy, learning, creating". The persistence of Supo and Ciciek, who believe that all the negative stereotypes about this place are wrong and that the people of Ledokombo actually want progress, eventually gained positive feedback. Komunitas Tanoker, which was initially rejected by the local people and the teachers, now is increasingly visited by the children of Ledokombo.

The members of this playgroup are generally the children of migrant workers, agricultural labourers, small merchants, teachers, public servants, and domestic workers who live in the area. The central activity of this community revolves around the revitalisation of the traditional game of stilts. For Supo and Ciciek, stilts have a lot of meanings and strategic cohesive functions. The name Tanoker and stilts together become the icon of social transformation aspired by their community. Tanoker or cocoon, normally understood as isolation from the outside world, is given a more positive meaning, which is the embryo of a butterfly that shall fly high and far through a process of transformation. Meanwhile, the stilts, as stated in their website, are a symbol of great leaps forward and self-growth. These two simple yet meaningful philosophies become the spirit of empowerment for the children and their facilitators.

In our interview with an ex-migrant worker who had worked in Saudi Arabia and Taiwan, we found that initially, not all people of Ledokombo supported the effort made by Supo and Ciciek. The community, whose main orientation is playing, was opposed by some teachers who were worried that their students would prefer playing in Tanoker to taking extra classes in the afternoon. However, in the process, this community gained the acceptance of the local people, as evidenced by the increased participation of the locals in the activities of Tanoker. Parents who were initially suspicious of the motivation and purpose of Supo and Ciciek slowly became curious and enrolled their children in Tanoker to learn reading and mathematics. This opportunity was seized by Ciciek to mobilise the mothers in empowering activities, such as cooking and making handicrafts from simple local materials.

Teachers and principals were also embraced since the couple were fully aware that the support of the teachers had to be won in order to realise the vision of children empowerment. Komunitas Tanoker occasionally invited English and mathematics trainers to provide enrichment and improve the ability of the local teachers. Today, some activities aimed at developing teachers' capacity are held in cooperation with parents, teachers, religious leaders, and local figures, including programs, such as Inspiring Teachers and Creative Parents Training, Jaritmatika (counting fast using fingers), GASING (easy, fun, not confusing) method of mathematics, and English for Community, with the help of Jaritmatika Foundation in Salatiga, Surya Institute in Jakarta, and Yasmin Learning Centre in Jakarta.

An important aspect observed by the research team while visiting Komunitas Tanoker is that around one hundred children of the community did not show any sense of inferiority or backwardness in terms of general knowledge. They generally showed high enthusiasm, strong confidence, updated knowledge of the world, friendliness, and the ability to cooperate well among themselves. The attitude of these children totally is opposite to the negative stereotypes imposed all this time on the people of Ledokombo. Instead of meeting the stereotypical children from disadvantaged villages, during the observation the team gained a strong impression that they are children with progressive thinking and mentality, who are comparable with children from the cities.

Regarding the history of the stilts game in Tanoker, Supo and Ciciek have a firm belief that stilts can be an agent of social change in Ledokombo. It is written on the website of Tanoker that:

> Stilts can trigger social change. And that is what happens in Ledokombo District, Jember Regency. Stilts are a traditional game for children, in the form of extended feet made of bamboo that can make the player look taller. In urban areas, such a game is almost unseen. But in rural areas such as Ledokombo, stilts are part of the tradition and social life. During the previous flooding in Ledokombo, people used stilts as a means of transportation.

The decision to choose stilts as the centre of life in Komunitas Tanoker is not only determined by the fact that it is a local traditional game, but more than that; there is a history showing that stilts are closely related to the lives and survival of Ledokombo people, especially during a life-threatening crisis.

The stilting activity by the children of Tanoker is done not only in their place. They also participated in shows such as *Indonesia's Got Talent* and also *Jakarta Percussion Festival* in March 2013. All of this has helped building the confidence of the children who Ciciek describes as "socially orphaned." The existence of the Komunitas Tanoker is indeed aimed at avoiding a "lost generation" tragedy among the children of Ledokombo, which may be caused by the absence of proper care and education by their parents who are too occupied with their poverty or miss their children's most important growth period because they have to work abroad as migrant workers.

1.2 Arisan Pendidikan *in Manggarai, East Nusa Tenggara*

Manggarai is a region located on the western side of Flores Island. Flores, especially Manggarai, is known by most of the people in East Nusa Tenggara as a region that cares the most about education. This is reflected in their various traditional processions, philosophy, and thinking that are deeply ingrained in the Manggarai society. The Manggarai language has an old expression passed down through generations, *"Uwa haeng wulang, langkas haeng tala"*, which means "Grow and soar high until you reach the moon and stars in the sky".

This expression is a hope for the children to be successful persons in the future. The success is supported by several factors, one of which is education. Therefore, the dropout rate in Manggarai is almost non-existent, though most of the people there are poor. A form of support for educational success is shown through the traditional ceremony of *wuat wuai*, where people gather and contribute a sum of money for education. The practice is also known as *Arisan Pendidikan*. According to Thomas Harming (31), a local of Pagal, Manggarai, *Wuat Wa'i* also means supplies, especially supplies for a child who is about to pursue higher education. This value about the importance of education is implanted in every family in Manggarai through the philosophy passed down from generation to generation in accordance with the expression *"Uwa haeng wulang, langkas haeng tala"*. This expression has encouraged the people of Manggarai to pursue higher education. However, in fact, not all people in Manggarai can afford it. Financial problems and living costs have often barred them from achieving their dream. In this situation, *Wuat Wa'i* appears as the solution.

It is unique compared to other traditional concepts of *gotong royong* commonly found in other regions in Indonesia, which arise from ceremonial events of harvesting, planting, hunting, building traditional house, birth, wedding, and death ceremony. *Arisan Pendidikan* or *Wuat Wa'i* is a form of *gotong royong* in Manggarai, which supports modern education but is wrapped up in a traditional ceremony, becoming a tradition for generations. *Wuat Wa'i* in the Dictionary of Manggarai, as quoted by Adi M Nggoro (2006), is a cultural rite to bid farewell to someone who is leaving their hometown, either to pursue higher education or a better life (migrating).

Wuat Wa'i is usually held in modestly inside a house, or by setting up a small tent outside a house. The participants are usually close relatives within the clan, colleagues, and other relatives in the same village. The invitation of *Wuat Wa'i* is spread orally and face to face by parents who hold *Wuat Wa'i* for their children. They will say something like, "Sir, my child is going to Jakarta to pursue higher education. Let's gather in my home, we will support the departure of my child."

The people of Manggarai have a collective understanding regarding the implied message in this invitation. Without having to be asked explicitly, they already understand that the inviter expects a form of financial support or money. The amount also has a minimum standard which has been discussed and decided together before. At the location, the guests will be greeted in the local language one by one by the host and is offered traditional alcoholic beverage or beer, "Thank you for coming, this is the symbol of greeting from my family." The ritual is called *Tuak Kapu*. A basin or another container is placed near the host who is offering the beer, where the guests can put their contribution. They usually know the price of the beer offered, for example IDR 36,000. They will then pay the drink at least at twice the price, or usually IDR 100,000.

After the guests gather and receive the drink, the ceremony is continued with *Tura Manuk Bakok* which is a ritual of prayer chanting in the indigenous religion of Manggarai. The prayer is a plea to the Supreme Being (*Mori Kraeng*) to protect the journey of their child who will pursue higher education. *Bakok* in the phrase means white chicken, symbolising sincerity and safety. Besides *Tura Manuk Bakok,* there is also a ritual called *Teing Hang*, which means feeding the ancestors. The ritual is intended to invoke the blessing and feed the ancestors, held in form of the slaughtering of a white chicken and the offering of alcoholic beverage.

After the prayer, the ritual of *Wuat Wa'I* is completed, and after that, they will eat together. The food offered is very modest, since the main purpose of this ceremony is not to be extravagant. After the meal, the host will offer cigarettes and coffee for male guests. Just like the alcoholic beverage at the beginning, the cigarette and coffee also come with a basin for money. The guests will pay at least at twice the price of coffee and cigarette they enjoy. Before the guests leave, they are also expected to give contribution money for the lunch they had (usually around IDR 50,000). Thus, in one *Wuat Wa'i,* there are three contributions given by the guests to the host. The first is for the appetizer of alcoholic beverage. The second is for the dessert of coffee and cigarette. The third is for the main course that is given before leaving the location. The total amount of contribution from each guest (or family) ranges from IDR 250,000 to IDR 350,000.

There is a special sentence in the culture of Manggarai that is repeatedly told by parents to their children as advice:

"Lalong bakok du lakom, lalong rombeng koe du kolem."

And the translation is as follows:

Lalong means cock, *Bakok* means white chicken, *du lakom* means when you leave hometown, *rombeng* means colourful (such as adult cock which has red, black, and orange colours), and *du kolem* means when coming back.

The whole sentence means:

"When you leave your hometown, you are like a white chicken. When you return, you will be colourful."

The meaning of this advice is a child who is leaving to pursue higher education is like a white chicken which symbolises innocence and lack of experience. When the child returns, he/she is expected to have transformed into an adult cock with colourful feathers, rich in new experience and skills. A child who pursues higher education outside Manggarai is expected to return as a successful person.

No social sanction, such as seclusion or excommunication, is imposed by the people of Manggarai on those who do not attend the event. The willingness to participate arises out of the attitude of caring for others, and also because everyone will take their turn to hold it. If someone rarely comes to such an event, the only consequence they bear is probably not many will come when they hold the same event. There is also no obligation for the child to return. However, there is a sense that everyone who comes from Manggarai has a big obligation to dedicate their experience to their hometown. During early 1980s–1990s, *Arisan Pendidikan* was mostly held for boys. The impression that education was only for young men was still very strong. Even in the expression, the word *Lalong* refers to cock, implying a gender claim that it was only for young men. Nevertheless, today there has been a shift in thinking. Many young women now also pursue higher education, and their family also hold *Wuat Wa'i* ceremony for them.

1.3   *The meaning of* Gotong Royong *to the Tanoker learning group and* Arisan Pendidikan *Manggarai*

Based on the field observation and secondary data on the forms of *gotong royong*, both activities have unique characteristics shaped by the understanding and knowledge on local wisdom that is absorbed by the society. Thus, the pattern and implementation of *gotong royong* in each region has its own uniqueness.

As seen from the intention to share, and referring to the cognitive theory on the intention to cooperate with others, even with those who they do not know at all, these activities are based more on solidarity bounded with local identity. The tradition held in Ledokombo to promote their regional uniqueness, founded and initiated by local leaders with high motivation to develop their regions, is able to move other elements in the society to support their thinking. Together they strengthen this collective movement, which has become the basis for their local identity development. A form of *gotong royong* in this village has succeeded in showing a new colour, full of innovation yet still rooted from the local values. A form of *gotong royong*, which is rooted in local tradition for generations and still held widely in the society, can be found in Manggarai with their ceremony of *Wuat Wa'i*, which has been practiced for hundreds of years. The tradition is also closely related to their spiritual and religious values.

From the above analysis, it can be seen that *gotong royong* concepts found in both places also have developed dynamically over the years. Local communities now face challenges and problems arising from global pressure. New solutions to solve social problems are developed, such as the one found in *Wuat Wa'i* tradition, which now includes young women in Manggarai, and local leaders emerge to provide solutions, such as the one found in Ledokombo. New forms of *gotong royong* are created as a result of construction of ideas, which also involves elements of local tradition and wisdom.

What keep these *gotong royong* activities run well are the high solidarity and the sense of importance among the community members regarding education and children development for a better future. Such a perception then binds the people to participate in *Komunitas Tanoker* and *Arisan Pendidikan*. This bond has become a communal sense in the community, with no force or formal rule to coerce them. Individual interest becomes communal interest.

The value of sacrifice can also be observed in the activity of *Arisan Pendidikan*. The guests usually come from the lower class population, such as farmers, yet they are willing to contribute up to IDR 300,000 for the host. In Komunitas Tanoker, it can be seen in the initial sacrifice of Supo and Ciciek who were willing to return to their hometown to contribute to the development of their hometown. The attitude of caring for others and sacrificing personal interest signifies individual consciousness towards their group (group consciousness). Collective understanding and sense that exist without having to be explicitly said or standardised are a form of local wisdom that needs to be protected and preserved.

REFERENCES

Ayatrohaedi. (1986) *Kepribadian budaya bangsa* (local genius) (The nation's personality and culture (local genius)). Jakarta, Pustaka Jaya.

Bintarto, R. (1980) *Gotong royong: Suatu karakteristik bangsa Indonesia* (Gotong Royong: A characteristic of Indonesian people). Surabaya, Bina Ilmu.

Haba, J. (2007) *Revitalisasi kearifan lokal: Studi resolusi konflik di Kalimantan Barat, Maluku dan Poso* (Revitalising local wisdom: A study of conflict resolution in West Kalimantan, Maluku and Poso). Jakarta, ICIP and the European Commission.

Keesing, R.M. (1981) *Cultural anthropology: A contemporary perspective*. Boston, Holt, Rinehart, and Winston.

Nggoro, A.M. (2006) *Budaya Manggarai selayang pandang* (The Manggarai culture: an overview). Ende, Nusa Indah.

Parsons, T. (1951) *The Social System*. New York, Free Press.

Tanoker-Ledokombo. *Profil*. January, 2014.

Cultural Dynamics in a Globalized World – Budianta et al. (Eds)
© 2018 Taylor & Francis Group, London, ISBN 978-1-138-62664-5

Silpin's artistic freedom in the creation of Tathagata statues on the Rupadhatu level at Borobudur

G. Mentari & A.A. Munandar
Department of Archaeology, Faculty of Humanities, Universitas Indonesia, Depok, Indonesia

ABSTRACT: In this study, we investigate the art form created by Buddhist monks (*Silpin*), known as the Tathagata sculpture. The term *Tathagata* refers to a person who reaches the transcendental truth in life. Tathagata manifests itself in the form of the Buddha sculpture with some unique characteristics. Here, we examine 20 Tathagata sculptures located on the Rupadhatu level at Borobudur Temple. In this analysis, we use the E. B. Vogler (1948) concept on "bounded art" (Gebonden kunst), which is related to the creation of Hindu–Buddhist arts. The Tathagata art and its creative process are used as a parameter to achieve the goal. In this study, we find that the binding rules related to the creation of sculptural arts in general were not always observed in the creation of Buddhist sculptural arts. Moreover, the result of this study also reveals that *Silpin* had a widespread influence and innovations in the creation of the beautiful Buddhist arts, which is evident in Tathagata sculptures on the Rupadhatu level at Borobudur Temple.

Keywords: Silpin, Buddhist arts, Tathagata, Buddhist sculpture, Borobudur

1   INTRODUCTION

The existence of statues and idols indicates that the medieval Hindu–Buddhist community is more concerned about creating artworks made of stone, bronze, wood, or other metallic materials (Magetsari, 1997, p. 1). By definition, statues also include objects created by humans for a specific purpose. A statue is sometimes created to fulfill a specific need or is tailored for serving a certain purpose (Sedyawati, 1980, p. 213).

To date, several studies, both descriptive and philosophical, have been conducted on the statues of Buddha. However, they have often excluded detailed descriptions of the artistic style of Tathagata statues found at Borobudur. Therefore, in this study, we focus on the artistic style of Tathagata statues at Borobudur, which possess unique features similar to other statues in temples worldwide.

Statues are seen as a commemoration marking the success of *silpins* (artists) in transforming ideas into archaeological artifacts. To assess the intelligence of our local medieval artists, we need to conduct an in-depth analysis of their works. That being said, this study relies on the data collected specifically to gain a deep understanding of Tathagata Buddha statues at Borobudur. Furthermore, we use the hypothesis proposed by E.B. Vogler (1948) to assess the level of creativity of the artists. Thus, the objective of this study is to discuss the creativity of the artists by examining the Tathagata statues on the Rupadhatu level at Borobudur.

2   RESEARCH METHODS

The research on the relationship between the creativity of the artists and the artistic style of Tathagata statues began by collecting the relevant data. The data collection process involves literature reviews and field observations to obtain data related to the statues of Tathagata. Field observation is considered the most appropriate method for data collection as it allows direct access to the research object.

In conducting field observation, the specific parts of Tathagata statues on the Rupadhatu level at Borobudur were also measured. The measuring process was carried out by the maintenance technicians of Borobudur. The purpose of this step is to get the detailed dimensions of Tathagata statues. The dimensions of the statues' head and body, which were the objects of observation in this study, were taken. Later, the collected data were analyzed. In the data analysis step, data collected from both literature reviews and field observation were processed. A further analysis was conducted by synthesizing the artifactual data. Furthermore, the analysis was also carried out using Vogler's hypothesis about the nature of bound art. Ultimately, the results of this study are expected to improve our understanding on the nature of Tathagata statues on the Rupadhatu level at Borobudur.

2.1 *Tathagata statues on the Rupadhatu level at Borobudur*

At Borobudur, the Buddha statues, which are found on top of the temple's balustrades, are called Tathagata. According to Encyclopedia Britannica, the term *Tathagata* is derived from Sanskrit, which refers to Siddhartha Gautama as Buddha. In general, Tathagata in Buddhism means someone (*tatha*) who departs or arrives. It refers to the history of Buddha, about his past and future life, and how one can achieve enlightenment. *Tathagatagarbhasutra*, a Mahayana Buddhist scripture, contains a brief explanation of Tathagata, describing it as the very important from Buddha's perspective of life. In the development of Mahayana Buddhism, the term and concept of Tathagata becomes an embodiment of Buddha, which is hidden inside every person (Zimmerman, 2002: 39).

Tathagata, which serves as a solid manifestation of Buddha, exhibits detailed information on the experiences and achievements of Buddha. These experiences and achievements demonstrate a series of rituals practiced by Buddha in order to gain enlightenment and to share his insights to the human race. In the process, the concept of Tathagata is later manifested in the form of statues (Zimmerman, 2002: 40). This concept is confirmed by the presence of Tathagata statues on the upper level at Borobudur. Tathagata is designed to show the experiences of a *yogin* as dictated by Buddha (Coomaraswamy, 1998: 58). All of these experiences are embodied in five Tathagata statues.

The Tathagata statues are spread and cover almost the entire area of Borobudur. The first level of the temple is an empty court without any statues. Statues began to appear on the first terrace located on the first balustrade of the temple. Furthermore, the Tathagata statues are also found on the balustrades from the second to fifth level of Borobudur.

In total, 504 Tathagata statues are found in Borobudur, 432 of which are located on the Rupadhatu level, and the rest are found on the *Arupadhatu*. As the scope of this study is limited to examining statues located on the Rupadhatu level, we restrict observation up to the first to fifth terraces of Borobudur.

The first and second balustrades have 104 statues, while the third balustrade has 88 statues. The numbers of statues on the fourth and fifth levels are 72 and 64, respectively.

It is indeed acceptable to refer to the statues in all four directions of Borobudur, namely North, East, South, and West. On the basis of this fact, it is possible to assume that every side has the same number of statues. There are 26 statutes on each side of the first and second terraces. The next two terraces have 22 and 18 statues, respectively, on each side. Finally, the fifth terrace has 16 statues.

However, during our observation, we found many of the Tathagata statues to be damaged slightly or missing even their entire body parts (head, shoulders, and limbs). Therefore, it was decided to categorize these statues on the basis of their condition: intact, damaged (headless), and lost.

Referring to Table 1, in this study, we focus only on statues that were fully intact, as they are free from any damage that can complicate the analysis process. Here, we expect to provide some restoration ideas for the damaged Tathagata statues. The state of "wholeness" of a statue can also be measured by referring to the basic teachings of *Panca Tathagata* (*Five Tathagatas*) in Buddhist teachings. *Panca Tathagata* is also known as the concept of the five manifestations of Buddha's radiance, which also serves as the basis of this study. Therefore, the object of this study is limited to the statues that are categorized as "whole" on the basis of the *Panca Tathagata* and their locations.

In this study, we use an approach based on the journey of Buddha as depicted in the relief panels at Borobudur. According to records, this approach does not use the usual cardinal directions as a reference. On the basis of the journey of Buddha, Borobudur is divided into

Table 1.   List of the damaged and undamaged (intact) statues on the Rupadhatu level.

Number of statues based on their condition

| Statue condition(s) | Estimated number of statues on every balustrade | | | | | |
	I	II	III	IV	V	Total
Intact	37	37	47	46	42	209
Damaged (headless)	59	59	35	23	18	194
Lost	8	8	5	3	4	29
Total	104	104	87	72	64	432

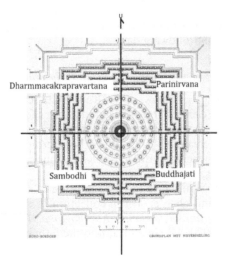

Figure 1.   Four-sector arrangement of Borobudur.

four sectors, namely the *Buddhajati* (the southeastern part of Borobudur), the *Sambodhi* (the southwestern part of Borobudur), the *Dharmmacakrapravartana* (the northwestern part of Borobudur), and the *Parinirvana* (the northeastern part of Borobudur). The journey of Buddha can be observed from the relief panels engraved on the walls of Borobudur (Munandar, 2012: 68).

Agus Aris Munandar (2012: 68), in his book *Proxemic Relief Candi-candi Abad ke-8–10 (The Relief Proxemics of Eighth Century Temples)*, explained that the life of Siddhartha Gautama is illustrated by the four sectors of Borobudur. In the *Buddhajati* sector, there is a *Buddhajati* symbol that represents the living phase. This sector is located at the southeast side, between the east and west stairways of Borobudur. Meanwhile, the period in which Gautama received his enlightenment can be seen in the *Sambodhi* sector, located at the southwest side, the area between the south and west stairways of the temple. *Dharmmacakrapravartana* represents the first teaching experience of Gautama. This sector is located in the northwest area, between the west and north stairways. The last sector, *Parinirvana,* depicts Buddha as he reaches Nirvana. The *Parinirvana* is located in the northeast area, between the north and east stairways.

The sector division itself is vital in the way the reliefs are arranged at Borobudur. It is possible to observe other parts of Borobudur by observing the phases of the journey of Buddha. By examining the arrangement of these sectors, it is interesting to observe the Tathagata statues, from which we may deduce that the artistic style of Tathagata statues is closely related to the journey of Buddha, as depicted in all the four sectors at Borobudur. According to Munandar (2012), all the sectors narrating Buddha's pilgrimage are related to the story told in the *Lalitawistara* relief panels, which is very much possible. Therefore, investigating the aforementioned four sectors of Borobudur is crucial in supporting the analysis of Tathagata statues in order to determine how the journey of Buddha is depicted and embodied in the Tathagata statues at Borobudur.

2.2 *Form of artistic freedom in sculpting the Tathagata statues on the Rupadhatu level at Borobudur*

In general, the depiction of Tathagata statues on the Rupadhatu level is almost identical to that of Tathagata statues found in other temples. The position of a Tathagata statue, especially its hand position or *mudra*, at every sector exemplifies Buddha's pilgrimage in accordance with the Buddhist teachings and the life story of Buddha. This shows that the creation of a statue follows a specific religious teaching, Mahayana Buddhism in this case.

Nonetheless, as it is realized, the sculpting of Tathagata statues at Borobudur disproves the theory of unbound nature (*gebonden*) of an artwork as suggested by E.B. Vogler. This argument is supported by the following data.

2.3 *The difference in the iconometric measurement of the statues' heads*

The embodiment of Tathagata statues on the Rupadhatu level has a different iconometric measurement on the head element from that of the other four sectors of the temple. This different iconometric measurement has an impact on the radiant sense or the *bhawa* that serves as a reference in the creation of a sculpture. Furthermore, each statue, with a different iconometric element, will generate a different *bhawa*, thus giving a different impression to observers and connoisseurs of statues.

A holy book was used as guidance for the sculpting of the statues, namely *Sipasastra,* as mentioned in the previous chapter. It is a collection of several holy books of different sizes, namely *Cilparatna, Amsubhedagama, Karanagama, Vaikhasagama, Brhat Samhita,* and *Pratilaksanam* (Rao 1920: 44). In addition to the holy book of *Manasara-Silpasastra,* other holy books are believed to contain the description of the Borobudur temple, namely *Sang Hyang Kamahayanan.* This holy book contains information about the religious background of Borobudur. To date, it has been a practically difficult task to determine the major source of reference regarding the iconography and iconometry in the sculpting of

Tathagata statues at Borobudur. Therefore, many experts are still conducting research on this particular topic.

Two elements with different iconometry measurements are always involved in the sculpting of Tathagata statues on the Rupadhatu level of Borobudur. The difference in the measurement is evident in the head of the statue. This difference inevitably results in the different impression and aura projected by the statue. This is related to one of the *sad-darsana* conditions, which is in turn related to the *bhawa* or the radiant sense. This iconometric difference can be seen in the element of *ushnisa* and the length of the lips of the statue. It is interesting to note that this iconometric difference is shown in all statues at every sector (Borobudur Conservation Hall (2013). *Kearsitekturan Candi Borobudur (Architecture of Borobudur Temple)*. Yogyakarta: Balai Konservasi Borobudur).

One of the sectors shows an exact iconometric size for each statue; however, it is different from the statues at the other sectors. In the *Buddhajati* sector, the Tathagata statues are assembled in the terraces from the first to fourth levels with the highest *ushnisa* to other statues in *Buddhajati*. The height of *ushnisa* decreases at the *Samboddhi* and *Dharmmacakrapravartana* sectors; however, it increases at the *Parinirvana* sector. Increased *ushnisa* height in the *Parinirvana* statue emphasizes the beauty of the statue, which in this sector displays the position of *abhaya-mudra*. The measurement of the *ushnisa* elements in the sculpting of the statues at the aforementioned four sectors, from the first to fourth terraces, is believed to be closely related to the tale of the journey of Buddha. The height of the *ushnisa* will impart a certain impression on the observers, so that they will come to know the lessons taught by the journey of Buddha.

The iconography of the statues on the fifth terrace is different from that of the statues from the first to fourth terraces in the four sectors. There is a significant decrease in the *ushnisa* size compared to that of the statues from the first to fourth terraces. This decreased may be due to a technical problem related to the structure of Borobudur, which becomes narrower with the passage of time to reduce the load on the temple.

Another unique element on each of the statues at each sector at Borobudur is the length of the lips. The longest lips can be found in the statues at *Parinirvana* sector. The shift in size of statues is similar to that in the *ushnisa* statues. The relatively long lips can be found at the *Buddhajati* and *Parinirvana* sectors; meanwhile, the length of the lips begins to decrease at the *Samboddi* and *Dharmmacakrapravartana* sectors. Relatively longer lips can impart a sense of enjoyment to the observers. This is related to the journey of Buddha at the *Buddhajati* and *Parinirvana* sectors. The length of the lips of the statues at the *Sambodhi* and *Dharmmacakrapravartana* sectors gives the impression of stiffness and deep concentration the statues are involved in.

Table 2.   Measurement of statue heads at Borobudur Temple.

	Tathagata statues					
Head element	Akhsobyat the Buddhajati Sector		Akhsobyat the Buddhajati Sector		Akhsobyat the Buddhajati Sector	
1. Ushnisa height	7.5 cm	1. Ushnisa height	7.5 cm	1. Ushnisa height	7.5 cm
2. Ushnisa circumference	37 cm	2. Ushnisa circumference	37 cm	2. Ushnisa circumference	37 cm
3. Distance between the eyes	4 cm	3. Distance between the eyes	4 cm	3. Distance between the eyes	4 cm
4. Width of the eyelids	5 cm	4. Width of the eyelids	5 cm	4. Width of the eyelids	5 cm
5. Length of lips	7 cm	5. Length of the lips	7 cm	5. Length of the lips	7 cm

It is very interesting to note that even the slightest difference in iconometry can give a very different impression. The provision of the iconometry difference with regard to the element of the statues is found in the set of rules governing the creation of Tathagata statues. Therefore, the pattern of the iconometric difference in the Tathagata statues at *Borobudur* shows the freedom in deciding the expression of statues, which will impart the desired impression of the journey of Buddha to the observers.

2.4 *The six conditions in the creation of a statue that does not have to be realized according to the holy book*

The embodiment of Tathagata statues on the Rupadhatu level at Borobudur does not always follow the guidelines with regard to the artworks of the sculpture. In fact, this study reveals that a condition known as the *warnikabhangga* was not implemented in the creation of Tathagata statues at Borobudur. Other conditions, such as *rupabedha*, *sadrsya pranana*, *bhawa*, and *lawanya*, were implemented in the Tathagata statues on the Rupadhatu level at Borobudur.

Sad-darsana is a criterion that needs to be fulfilled in the making of statues. This is because *sad-darsana* is an esthetic principle from India that is used in the making of Hindu and Buddhist artworks. Rupabedha requirement is clearly shown in the *mudra* of Tathagata statues on the Rupadhatu level at Borobudur, which helps observers in differentiating statues in every sector. It is the same as the *sadrsya* principle, which frames the rules regarding similarity and vision. It is evident that the statue is depicting a man sitting with crossed legs and doing the three hand poses. The rules of *pramana*, or the conformity of measurement, are also shown in the statues. The proportion of many elements is so clear that every observer can observe the proportion of the statues. Another condition is the *bhawa*, the radiant sense, or the aura, which is evident in the Tathagata statues on the Rupadhatu level at Borobudur. All statues project a profound impression to the viewers. On the contrary, an additional requirement in creating a statue is called the *lawanya* condition, which is related to the beauty aspect in statues. Profound impression from the *bhawa* is complemented by the *lawanya*. The presence of *lawanya* gives an appeal, which will allow the observers to enjoy the beauty of Tathagata.

Several aforementioned conditions are seen on Tathagata statues on the Rupadhatu level of the temple. Nevertheless, the *warnikabhangga* condition is not shown on the statues. The statues did not undergo the coloring process, in contrast to statues in India. This proves that, in the sculpting of statues, the condition with regard to the creation, that is, *sad-darsana*, does not have to be followed, and the case in point was the coloring of Tathagata statues at Borobudur. Coloring in this case is applying common colors (primary and secondary colors such as red, yellow, green, and blue). This condition was not implemented maybe due to a technical or some other reason. Therefore, it can be concluded that the condition with regard to statue making, despite being mentioned in the holy book, does not have to be followed.

However, the *warnikabhangga* condition may have to be explored further. The condition *Warnikabhangga* governs the coloring of statues that is not implemented in the Tathagata statues at Borobudur. Nevertheless, colors are assumed to be used on the statues from the play of lighting and the sunlight passing through Borobudur. The play of colors on Tathagata statues can be observed from sunrise to sunset. In any case, this study does not cover the *warnikabhangga* condition as used in the Tathagata statues at Borobudur any further. Therefore, a specific research on the subject may be required in this aspect.

2.5 *Statues that possibly went through changes*

Vogler (1949) stated that the changes occurring in artistic styles in the development of art in Java–Hindu are due to style mixing (*stijlvermenging*). Each component undergoes a unique development stage, and it does not specifically occur in one artwork. Each identifiable artistic style is an amalgam of its basic components with added variations. This statement needs to be verified with regard to a unique Tathagata statue at Borobudur.

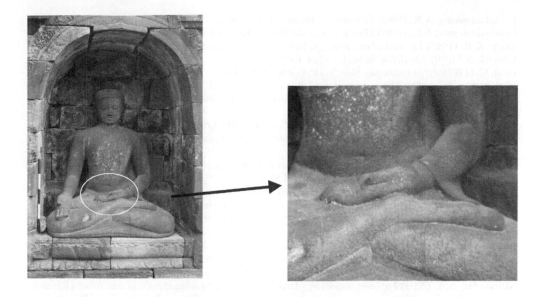

The reason behind the creation of this unique statue is still unclear and being studied, which may be related to a religious background or the result of forcing an artist. The reason behind the difference in parts of the statue cannot be found in the holy book. Therefore, the mere existence of the unique statue proves that the character of an artwork is not bound as previously stated by E.B. Vogler.

Throughout its development in many areas, uniqueness or individualistic style can be clearly observed in example statues in other areas. The shapes of the statues, in which their iconography cannot be identified clearly, may be because the teachings of Hindu and Buddhism in Indonesia had absorbed the traditional belief of the people. Well-known Hindu Gods were given features and characters that are not mentioned in conventional religious scriptures. These additional gods are referred to as *Gramadewata*, whose statues show that although the original feature of the statue was taken from the holy book, it was evolving in the later development; it was adjusted to meet the demands of the ancient Javanese society, especially the need of the *silpin* or the artist who created the statue artworks.

3 CONCLUSION

This study shows that the free and unbound characteristic of an artwork is shown through the difference of the iconometry in the elements of the heads of the statues, that is, the iconometric dimensions of the *ushnisa* and the length of the statues' lips. This iconometric difference influences the *bhawa,* which, being created from every statue, possesses a different iconometric element that gives a different impression to those who enjoy the artwork of the statues. The difference between these two iconometries is shown in all statues in every sector. This difference reveals a certain pattern that needs to be interpreted. The skill of statue making and the patterning clearly highlighted the creativity of the *silpin* that created the Tathagata statues at Borobudur.

REFERENCES

Balai Konservasi Borobudur (Borobudur Consertvation Hall). (2013) *Kearsitekturan Candi Borobudur* (Architecture of Borobudur Temple). Yogyakarta: Balai Konservasi Borobudur.
Battacharyya, B. (1958) *The Indian Buddhist iconography.* Calcutta: Firma K.L. Munkhopadhyay.
Bernet Kempers, A.J. (1959) *Ancient Indonesian art.* The Netherlands: C.P.J. Van Der Peet.

Coomaraswamy, A.K. (1998) *Elements of Buddhist iconography*. New Delhi: Munshiram Manorhalal.

Coomaraswamy, A.K. (1999) *Hinduism and Buddhism*. New Delhi: Munshiram Manorhalal.

Dark, K.R. (1995) *Theoretical archaeology*. Ithaca: Cornell University Press.

Glubok, S. (1969) *The Art of India*. London: Collier.

Holt, C. (1967) *Art in Indonesia*. New York: Cornell University.

Huntington, J.C. (1994) The iconography of Borobudur revisited. In Klokke, M.J. & Pauline L.S. (Eds.) *The concept of Slesa and Sarva (Buddha) Kaya in ancient Indonesian sculpture*. The Netherlands: KITLV.

Kar, C. (1956) *Classical Indian sculpture: 300 B.C. to A.D 500*. London: Alec Tiranti LTD.

Krom, N.J. (1927) *Barabudur: archaeological description, Volume 2*. India: The Hague Martinus Nijhoff.

Magetsari, N. (1982) *Pemujaan Tathagata di Jawa Tengah pada abad ke-9* (The Tathagata worshipping in Central Java in the 19th Century). Depok: FSUI.

Maulana, R. (1987) *Hiasan badan pada masa Hindu Buddha di Jawa pada estetika dalam arkeologi Indonesia* (Body decoration in the Hindu Buddha era on the aesthetics in the archeology of Indonesia), Archeological Scientific Discussion II. Jakarta: Pusat Penelitian Arkeologi Indonesia (Indonesia's Center for Research in Archaeology).

Maulana, R. (1996) *Perkembangan gaya seni arca di Indonesia dalam laporan penelitian FIB UI* (The statue art style development in Indonesia in the FIB UI research report). Depok: FIB UI.

Munandar, A.A. (2003) *Aksamala*. Bogor: Jaskarindo Pandukarya.

Munandar, A.A. (2003) *Laporan penelitian: Arca Prajnaparamita sebagai perwujudan tokoh* (Research report: Prajnaparamita Statue as the realization of a figure). [Thesis] Fakultas Ilmu Pengetahuan Budaya: Universitas Indonesia, Depok, Indonesia.

Munandar, A.A. (2012) *Proxemic relief candi-candi abad ke-8–10* (The temple relief proxemics of the 8–10 Centuries). Jakarta: Wedatama Widyasastra.

Rao, T.A.G. (1920) *Talamana or iconometry*. Calcutta: Superintendent Government.

Sahai, B. (1975) *Iconography of minor Hindu and Buddhist deities*. New Delhi: Abhinav.

Sedyawati, E. (1980) *Pemerincian unsur dalam analisa seni arca dalam penelitian ilmiah arkeologi* (The elaboration of elements in the statue art analysis in archeological scientific research). Jakarta: Pusat Penelitian Arkeologi Nasional.

Sedyawati, E. (1981) *Pertumbuhan seni pertunjukan* (The growth of performing arts). Jakarta: Sinar Harapan.

Sedyawati, E. (1985) *Pengarcaan Ganesha masa Kadiri dan Singhasari: Sebuah tinjauan sejarah kesenian* (Ganesha statuing during the Kendiri and Singhasari eras: A review of art history). Depok: FSUI.

Vogler, E.B. (1949) *De monstercop uit het omlijstingsornament van tempeldoorgangen en-nissen: in de Hindoe-Javaanse bouwkunst*. Leiden: E.J. Brill.

Zimmerman, M. (2002) *A Buddha within: The Tathagatagarbhasutra, the earliest exposition of the Buddha-nature teaching in India*. Tokyo: Soka University.

Cultural Dynamics in a Globalized World – Budianta et al. (Eds)
© *2018 Taylor & Francis Group, London, ISBN 978-1-138-62664-5*

Spatial narrative in a traditional Sundanese village

T. Christomy
Department of Literature, Faculty of Humanities, Universitas Indonesia, Depok, Indonesia

ABSTRACT: One of the most important strategies employed by indigenous people when dealing with change is assigning cultural markers to the concept of space. This process creates a symbolic demarcation between what can be accessed by 'outsiders' and what remains to be the privilege of insiders. Such cultural mechanism is highly effective for some indigenous people (*masyarakat adat*). In addition to helping the community fulfill the obligation to continue the legacy of their ancestors, the mechanism also serves as a means for managing tensions brought about by change that may weaken the customary foundation of the indigenous community. In reality, this process that the community undertakes may be challenged by various interests, including those coming from the in-group, especially from the young generation that has been initiated to these changes. To understand this issue, I will discuss the narrative of space as one of the most important strategies related to the identity awareness and the spatial construction of Sundanese indigenous communities.

1 INTRODUCTION

Although modernization has affected various aspects of Sundanese people, there remain some communities that still preserve ancient traditions in various ways, such as those found in Baduy, Kampung Naga, Cigugur, and Pamijahan. This paper concentrates on the cultural narrative of space and place of Cigugur and Pamijahan communities. These two kampongs share a similar mythical landscape because of their strong historical connection with the ancient West Javanese kingdoms of Sunda, Galuh, Galunggung, and Pajajaran. Despite their effort to preserve their traditional ways of life, these communities have been frequently challenged by various influences, forcing them to continuously reconstruct their position as part of *Pasundan* (Tatar Sunda). Cigugur, a peaceful and beautiful area on the slopes of Mount Ciremai, is well-known as the birthplace of Sunda Wiwitan. This community experienced various encounters with political forces during the colonial era, the 1960s, and the New Order era; the latter took the form of various prohibitions imposed on them by the government as part of the re-signification process of their local faiths. Meanwhile, Pamijahan, which originally served as the centre of *Tarekat Shattariyah* (the Shattariyah congregation) and its teachings (Christomy, 2008), has now become a destination for pilgrims from all over Indonesia who constantly flock the area. Development over the years has compelled the Pamijahan community to deal with changes coming both from outside and around their kampong. One of the most important strategies which the community frequently adopts when faced with inevitable changes is strengthening or reestablishing cultural markers attached to the concept of space. This measure is taken to establish a kind of symbolic demarcation between what can be accessed by outsiders and what remains the privilege of the insiders. In some indigenous communities (*kampung adat*), such cultural mechanism is highly effective and meaningful. In addition to helping the community to fulfill the obligation to continue the legacy of their ancestors, such mechanism also serves as a tool for managing tensions brought about by change which may weaken the customary foundation of the community in the long run.

Indigenous communities and their own conception of cultural space are often challenged by the existence of different conceptions of space. For instance, they have to face the presence of other small political units which, as part of the official government administration system, are superimposed on the existing cultural landscape. In such a case, coexistence is created between the indigenous societies (with all of its accompanying traditions and systems) and local government units, such as villages (desa), community associations (*rukun warga* or RW), or neighborhood associations (*rukun tetangga* or RT), which are compulsorily added to their cultural landscape. Ambiguities and disputes might arise when members of the community are compelled to adhere to a different conception of space in a different context.

A community's signification of space often has to face challenges, not only in the form of external influences, but also in the form of internal mechanisms. This mechanism involves the young generation that have somewhat lost access to their forebears' cultural competence and knowledge because of various factors, such as migration, the ineffectual process of cultural inheriting, and various challenges posed by information technology which provides the young generation with new perspectives on the future. In fact, there are several indigenous communities in West Java that seem to remain unable to establish their own version of tourism industry (a modern concept initially introduced by the regional government) and to devise their own ways to deal with this development. Such decisions are related to whether they will fully welcome this concept of tourism or establish a clear demarcation between "external" and "internal" realms, which must not be allowed to overlap at all costs. Such critical measure with regard to the "internal" realm was taken in 2006 by the Kampung Naga community, which refused to welcome large groups of tourists because they thought that such visits would disturb the community members in general and would only benefit the regional government in the form of parking fees, among other factors. There is a growing consternation as to whether they can really open their community to the general public as part of the larger industry, an idea that has never been part of their traditional signifying order. Moreover, there also develops an atmosphere of contestation and competition amongst neighboring indigenous communities over access to symbolic assets available within their territory as a consequence of these changes.

Such situation leads to the question regarding the role of space narrative in relation to the cultural landscape of indigenous communities. The answer to this question will provide us with a deeper understanding of how space is translated into a mnemonic device that, in Fox's terms (1997), can be preserved, transformed, and memorized. To answer this question, I shall examine how a cultural landscape is constructed within the awareness of the Sundanese people in two indigenous communities: Pamijahan and Cigugur.

Several cultural landscapes in West Javanese traditional kampongs are strongly related to a certain mythical space which has long become a source of legitimacy for a number of regional leaders. Space serves as an extension of the existence of ancestral figures recorded in traditional narratives and reflected in the form of various spatial metaphors. Due to their sacred nature, the elaborations of these metaphors demand certain competence on the part of local elders. Moreover, several anthropological studies in the eastern Indonesia show that space is never a fixed entity but is always dependent on the context where it operates and from which perspective it is signified (Fox, 1997; Parmentier, 1987).

The study of space is categorized into two general divisions. The first one sees space from the Cartesian paradigm, and the second is sees space as an inseparable part of the signification process. In the latter perspective, space is considered as an entity which is socially constructed and personally experienced at the same time (Fox, 1997, p. 1). This paper adopts the second approach supported by the application of relevant theses and theories from anthropology, especially those related to the key concepts of "landscape", "space", and "place" as proposed by Fox (1997, p. 1). He argues that cultural categories of space can be used to explain how human knowledge is "preserved", "transformed", and "memorized." He states, "Their impact is to stress the multiplicity of ways in which places are socially constructed and personally experienced" (Fox, 1997, p. 3). Fox starts his argument by explaining the relationship between the "genealogy" and "itinerary of natural entity within the conception of space as prevailing in Roti Island (1997)." Of his many important findings, the most relevant to this paper is the concept of "precedence," which can explain why a community's privilege and access to a particular space may be resented by other neighboring communities because those communities adhere to a different metaphor or "trajectory of the path," in spite of the fact that they share the same cultural root. Moreover, each of the kampongs may even have more than one space-based cultural reference, which in practice can be used randomly and potentially be manipulated by external influences. Invented by Fox, foxtopogeny has become an instrument to record a community's genealogy in the form of space. This is based on the assumption that genealogy relies on personal names and topogeny on place names. In both, "points of origin and termination are critical" (Fox 1997, 101). In other words, the historical narrative, which has been reflected in the "genealogy" can be synchronically perpetuated in the form of "topogeny".

What was discovered by Fox in Roti Island (1997) is in line with what had been previously discovered by Parmentier (1987) in Belauan Island through his concept of "semiotic of trajectory." Parmentier explains that progression of place can form a network, which will eventually form a space (Parmentier, 1987).

Another conception of space which is relevant to this study is that proposed by Pannel (1997). He continues Fox's argument (1997) and incorporates it into de Certau's conception of space within the perspective of cultural geography. In his opinion, space is considered as an individual's or a group of individual's arena of social practice, while place is considered as an "ordered representation," with both "place" and "space" occupying the same landscape (Pannel 1997, 63). In my previous study, I defined space as an area of experience which is constructed from each individual's journey of life, which can be analogized to a pathway: "A criss-crossing of the path and texts creates a space which is also open to interpretation" (Christomy, 2008, pp. 65–66).

2 THE SACRED LANDSCAPE

The spatial arrangement of pre-Islamic Sundanese indigenous communities is generally influenced by their cosmology, observing the world as consisting of three layers, which are the upper world, the middle world, and the netherworld. Human beings occupy the middle world, which is an imaginary space where humans interact with each other and perform activities. This space is sometimes considered as a "neutral" space, albeit full of traps and challenges.

In spite of this general cosmology, not all indigenous communities strictly adhere to such spatial arrangement. In several places, there are communities whose development forms an integral part of the development of Islam or is at least strongly related to Islam, such as Pamijahan. Moreover, there are also indigenous communities whose cultural landscape is influenced by both Islamic and Sundanese cultures. Cigugur can be considered as belonging to this category because the community's ancestors were once part of the syncretistic Islamic Javanese/Sundanese culture, even though its Sundanese features have become more prominent over time.

3 METAPHOR OF CLOSENESS

In Sundanese communities, cultural landscape is constructed symbolically from various narratives. One of the most important narratives for constructing a strong cultural landscape is the one which is related to genealogy. Cigugur Kampong was established by Madrais, an ancient figure who is believed to be a descendant of a Sultan of Cirebon but had to travel from place to place in his youth in search of knowledge. Based on the testimony given by Prince Djatikusuma (Madrais' grandson) during my field work, Madrais was obliged to learn and acquire the "paramount knowledge," as it was the custom amongst the Sultan's descendants. This "paramount knowledge" could be obtained by firstly learning Islam at various *pesantren* (Islamic boarding schools). Upon completion of their religious study, these royal descendants, by their own free will and effort, had to search for more knowledge in the real world by visiting places associated with Sundanese culture and values. These places were considered haunted and frightening places by many people at that time. Madrais, however, is said to have succeeded in seeking knowledge, completing his journey safely, and settling in a region which was to become Cigugur Kampong.

For Madrais' followers, his narrative of the journey is believed to be a pre-established fact and constitutes part of the community's collective conciousness. In his study of Amaya and Wulur communities, Pannel identifies a strong relationship between "landscape" and "pathway" whose intersection gives rise to what he calls "topostories," which are "the unique journeys of each of the beings and places where they stopped, served to define a number of discrete topographic estates, which today are linked to the different 'houses' in Amaya" (1997, 164).

In this context, Pannel wants to demonstrate that a "place" is not simply "established," but its establishment must be somewhat connected to a certain power which does not come from the figure performing the journey itself, but from "above", which, in Islamic terms, refers to the power of destiny. Such "divine" power is considered to operate in Madrais' journey narrative. Pannel's findings also apply to our Cigugur case in many respects, with a little variation in detail. For instance, the figure in Pannel's case is a supernatural being, not a real human as in our Cigugur case. Moreover, according to the local narrative, the young Madrais left the Islamic boarding school and embarked on his journey for knowledge of his own volition, solely because he was encouraged to do so by the "divine inspiration," a power greater than his. The Cigugur narrative places a great emphasis on the figure of Madrais as the main actor who tried his best to interpret the divine inspiration he received. Such travel tropes are common in congregational communities (tarekat or tariqa) that associate a search for knowledge with a journey. However, when I asked Prince Djatikusuma about this analogy, he diplomatically replied, "I'm sorry. I don't know anything about it (tariqa). I only know 'Sundanese ways'." In spite of this answer, it is clear for me that the journey has served as the foundation of Cigugur community on the slopes of Mount Ciremai. It is also clear that, as shown by the Cigugur case, when a particular place is strongly related to a narrative, it may serve as a cultural reference. This is in line with Pannel (1997, 165), "...stories and the landscape constitute map; they encode or embed that which is perceived as customary and cultural."

In this way, Cigugur is present as a cultural reference for the supporting community of the Sunda Wiwitan faith. Even though many members of the Cigugur community have migrated to various places in Java Island, the existence of Cigugur is still an important part of Sunda Wiwitan, a collective knowledge which is reinforced annually and passed on through an annual ceremony which can only be appropriately performed in Cigugur.

Cigugur's landscape has changed, and the changes affect the community's signifying order. One such change is the overlap between the mythical and historical narratives of the society. Prince Djatikusuma said that his ancestor did not come from Cigugur, but from Gebang, and as it has been described above, this ancestor had to travel far to seek knowledge and ended his journey in Cigugur. From the Prince's statement, it can be assumed that Madrais belongs to three prominent cultural backgrounds, which are Islamic, Javanese, and Sundanese. This point will prove important in understanding the Cigugur community.

Islam was introduced to Pasundan by figures who had strong affinity with the Javanese culture, and they went to West Java through Cirebon all the way down to southern Sunda or through Banten to the hinterlands. Sundanese kingdoms and their kings soon had to accept the fact that other kings in Javanese coastal areas who had converted to Islam became more superior than them in terms of military and political powers. The Sundanese addressed this situation by augmenting their pride and emphasizing the premise that Sunda could never be defeated by Islamic forces. Because of this, in the Sundanese oral tradition, the King of Pajajaran is never described as being killed in a battle, but experiencing *nghyang* (turning into a supernatural being). To this day, his fervent followers still believe that his spirit can still be seen in certain places in southern West Java. In my opinion, the Sundanese people must have found it difficult to accept Islam and admit the defeat of their Sundanese ancestors at the same time. Their position concerning this unpalatable truth remains inconclusive.

In such situation, the compromise is to mix various cultural features which may not necessarily be compatible with each other but still operate within the context of Cigugur landscape. In the Cigugur case, such ambiguity is addressed by adopting a more flexible position with regard to external influences and by opening access to other belief systems, such as Catholicism, which has also penetrated into Cigugur cultural landscape (Christomy, 2013). Forced by an escalating political tension in 1960s and the imposition of being categorized into one of the five state-recognized religions, Cigugur customary leaders decided to convert into Catholicism, along with a number of their followers. However, this situation has been reversed by their descendants. The new generations of Cigugur customary leaders encourage those who have converted into Catholicism to return to their original faith. In a short-term fieldwork in 2013, I was warmly welcomed by Prince Djatikusuma, an important figurehead of Sunda Wiwitan. According to him, patrilineal lineage is very important in the Sunda Karawitan faith. Because of this, in his opinion, daughters may be permitted to convert into Islam, Catholicism, or any other religions, but sons have to preserve their ancestral faith.

Visitors when visiting Cigugur will find a palace-like building that has is a reception hall called *Paseban*. This is Cigugur's main attraction and the centre from which Madrais' teachings are developed and learned. *Paseban* has been designated as a central coordinate or point of direction to which the community members may return and from which they reproduce a new signification system. *Paseban* serves as a collective knowledge or signifying order which has to be preserved. *Paseban* is also intended as a landmark to which community members who have taken the road to churches or mosques can return.

The second traditional community is Pamijahan, which is located approximately 60 km south of Tasikmalaya. Pamijahan is now growing into a popular destination for pilgrims who regularly visit sacred places associated with the Nine Apostles (Wali Sanga), other influential figureheads of ancient Islamic congregations throughout Java. No fewer than ten large buses enter Pamijahan every day, carrying devout pilgrims who come not only from Java, but also from Sumatera.

Pamijahan Kampong was once called Safar Wadi Kampong (Christomy, 2008). The name "Safarwadi" was also mentioned in a report by de Haan who described the kampong as part of Karang area which was quite notorious amongst the Dutch colonizers at that time as a rebel territory (Haan, 1910). Shaykh Abd al Muhyi, one of the kampong's ancestors, is estimated to live between 1660 and 1715 (Krauss, 1995, p. 112), in the same era as Shaykh Yusuf (Rinkes, 1910).

Manuscripts found during my fieldwork and in other places testify to the position of Shaykh Abd al Muhyi as an important figure behind the development of the Shattariyah congregation, not only in Pasundan, but also on Java Island in general (Christomy, 2008). Even though no original writing of Shaykh Abd al Muhyi has been found, there are many manuscripts written by his disciples which refer directly to Shaykh Abd al Muhyi as their teacher. There is some doubt about the existence of Shaykh Abd al Muhyi (Krauss, 1995), but, in my previous investigation, I was able to uncover many old manuscripts which can provide evidence for his existence (Christomy, 2008).

Following the death of Shaykh Abd Muhyi, the area where he established his congregation grew into a place of pilgrimage. Right now, the place is more commonly known as Pamijahan than as Safarwadi, its original name. A large number of pilgrims come to this site to pray at Shaykh Abd al Muhyi's tomb and to visit several places believed to be the birthplace of the Shattariyah congregation. Pilgrims usually come to Shaykh Abd al Muhyi's tomb before continuing their journey to Pamijahan Cave, which is known as a place of education for the congregation members at that time. A number of serious pilgrims will also visit other places around Pamijahan, such as Panyalahan Kampong, which is still genealogically related to Pamijahan. The kampong's rapid growth caught the interest of the regional government, which then decided to build tourism infrastructure such as roads to provide easy access to Pamijahan Kampong, a parking area to accommodate large interprovincial buses, and the parking fee collection.

The increasing number of pilgrims brought its own problems. Conflicts arose, both within the village and in other neighboring villages. There were questions regarding who was entitled to manage Shaykh Abd al Muhyi's grave, Gua Keramat (Sacred Cave), and other important artifacts. It culminated with an official legal investigation that was never completed. In the midst of such contestation, appropriate strategies and responses to cultural narratives can play an important role.

In both Pamijahan and Cigugur, blood ties have become an important metaphor that becomes the point of reference where the coordinates of the places are formed. In the Cigugur case, the reference point is Madrais; as an essential part of a collective memory, his existence is symbolized by specific landmarks in the form of the Paseban and the palace. As tangible evidence that becomes central in the life of the community, the Paseban and the palace must, in turn, be supported by the presence of male descendants who are expected to fully understand all of the teachings of their predecessors. Therefore, Cigugur's cultural landscape is not only marked by an abstract "map," but also by a physical "space" in Cigugur. This physical space is considered as a point of destination which can lead the "lost" community members back to their true and original faith of Sunda Wiwitan. The establishment of such a spatial reference point is very important. As a cultural landscape, Cigugur has to cope with various external influences. In addition to their traditional rivalry with Islamic groups, they now also have to deal with Catholicism. Although Christianity in general has existed in Pasundan ever since the colonial era, it had not been considered as a material problem in the Cigugur community until the mass conversion of a large part of its members into Catholicism. In Cigugur, there is even a growing tradition of Catholic pilgrimage to Gua Fatimah Sawer Rahmat, a local shrine dedicated to the Virgin Mary, with a pilgrimage track built after the fashion of Via Dolorosa in Jerusalem (Christomy, 2013).

It is important to emphasize that Prince Djatikusuma has succeeded in reconverting many members of the community into the Sunda Wiwitan faith. In a short conversation, he expressed his belief that religion is only a "way," while the essence of all religions is all the same. What is even more important to reflect on is the Sundanese proverb *"someah ka semah"* which literally means "be friendly to guests." For Prince Djatikusuma, this proverb means that his community is open to other beliefs as proven by its cultural landscape that shows the existence of those beliefs in the society. For instance, Islam, which is aptly demonstrated by the peaceful existence of many mosques for centuries, and Catholicism and Hinduism, which are marked by the establishment of several temples and meditation centres around Prince Djatikusuma's palace. What the Prince intends to do is to construct a narrative which enables elements of various belief systems to enter their community and, by doing so, strengthen the position of the Sundanese faith in its centre as an overarching power which can embrace those other major religions. In relation to this, Pannel, who elaborates de Certau's concepts of "space" and "place", argues that "Space is thus the actualization, by the operation of historical subjects, of what has already been created—place" (Pannel, 1997, p. 167). In the eyes of his followers, Djatikusuma and his *Paseban* are considered as the supreme power which has been able to "tame" other major religions and ideologies operating within the community.

This observation confirms that the Cigugur community feel a great need to address the influx of major religions by establishing a cultural narrative in the form of a place (the "paseban" hall) and a space (the *"hade ka semah"* proverb) which are intended to place Sunda as the centre of everything. However, the same strategy is not found in Pamijahan. Pamijahan is more concerned with the contestation over the landscape of pilgrimage sites and with the question of the legitimacy to be the centre of all aspects of the pilgrimage—not only matters pertaining to religious rituals, but also their impacts on the economy of the village. In both Cigugur and Pamijahan cases, "trajectory of ancestors" is found to provide a clear demarcation as what can be designated as the imaginary "map", "coordinate", or "place" in de Certau's terms. However, the difference lies in the fact that the pattern of "space" narrative in both villages is more directed inward, which might be due to several factors, such as the limitation of the area of the villages, an increasing number of population, the community's relationship with neighboring villages, and the regional government.

4 CONCLUSION

Historical narratives tend to be represented by markers that can serve both synchronic and symbolic functions. These markers enable a community living in a particular era (synchronic) to feel that they are able to give a signification to diachronic entities from their own perspective. Such symbolic situation can bring about transformation to the previously established signification through the narrative of "sediment trace of activity." Contestation at the level of "space" is balanced by the construction of a "place" narrative, which can function as a "map" directing the community members back to their point of origin. It is interesting to see how the Cigugur community adopts a more flexible conception of "space" in order to place their trajectory narrative at the centre of everything, precisely when external factors begin to affect their overall cultural life. Furthermore, the narrative of origin plays an important role in both communities because it greatly determines an individual's position within a particular space. Pamijahan has successfully amalgamated the Sundanese culture into the narrative of the Nine Apostles, while Cigugur attempts to return to its original Sundanese conception of space as part of the community's effort to cope with changes.

REFERENCES

Christomy, T. (2008) *Signs of Wali: Narratives at the Sacred Sites in Pamijahan Tasikmalaya.* Canberra, Research School Pacific and Asian Studies, the Australian National University.

Christomy, T. (2013) Narrating the Paths: The Case of Fatimah Sawer Rahmat, Cigugur Kuningan. Paper presented at the *5th International Conference on Indonesian Studies, 13th–14th June 2013, Yogyakarta.*

de Certau, M. (1984) *The Practice of Everyday Life.* Berkeley and Los Angeles, University of California Press.

de Certau, M. (1997a) Place and Landscape in Comparative Austronesian Perspective." In: Fox, J.J. *The Poetic Power of Place: Comparative Perspectives on Austronesian Ideas of Locality.* Canberra, Research School Pacific and Asian Studies, the Australian National University. pp. 191–195.

De Haan, F. (1910) *de Preanger-Regenschappen onder het Nederlandsch bestuur tot 1811.* Batavia, Bataviaasch Genotschap vn Kunsten enWeenschapen.

Fox, J.J. (1997b) *The poetic power of place: Comparative perspectives on austronesian ideas of locality.* Canberra, Research School Pacific and Asian Studies, the Australian National University.

Krauss, W. (1995) An Enigmatic Saint: Sheykh Abdulmuhyi of Pamijahan (1640–1750). *Indonesian Circle* 23, 65.

Pannel, S. (1997) From the Poetic of Place to the Politics of Space: Redefining Cultural Landscapes on Damer, Southeast Maluku." In: Fox, J.J. *The Poetic Power of Place: Comparative Perspectives on Austronesian Ideas of Locality.* Canberra, Research School Pacific and Asian Studies, the Australian National University. pp. 163–171.

Parmentier, R.J. (1987) *The Sacred Remain: Myth, History, and Polity in Belau.* Chicago, University of Chicago Press.

Cultural Dynamics in a Globalized World – Budianta et al. (Eds)
© *2018 Taylor & Francis Group, London, ISBN 978-1-138-62664-5*

Symbolic meaning of the *Śrāddha* ritual in the Merapi-Merbabu *Putru Kalĕpasan* text

I.M. Suparta & T. Pudjiastuti
Department of Literature, Faculty of Humanities, Universitas Indonesia, Depok, Indonesia

ABSTRACT: In this study, we examine the *Putru Kalĕpasan* text, which is one of the religious texts in the *Sastra-Ajar* scriptorium in the Merapi-Merbabu manuscript. This text is compiled with four palm-leaf manuscripts, namely L 42, L 222, L 271, and L 322, in the manuscript collection of the National Library, Jakarta. The text contains detailed information on the procedures of the ritual *Śrāddha,* known at the Old Javanese era. Here, we apply the method of textual criticism to reveal the *Putru Kalĕpasan* texts. In addition, we apply religious and content analysis techniques to understand the expression of thoughts and images of the ancestor worship ritual called *Hambukur* ceremonies. The following conclusions are drawn from this study: (1) the expression of offerings in *Putru Kalĕpasan* is an ongoing Old Javanese religious thought passed on by the *Sastra-Ajar* tradition in the 16th century; (2) the slaughter of animals used for offerings symbolizes purification of the soul of *daśamala* stains. Thus, essentially, *ruwatan* in the *Hambukur* ritual is intended for the ordination of *pitara* to become *Dewa Pitara* (Sanskrit: *Pitṛideva*), a sacred ancestral spirit who "becomes a god" and is considered to reach *kalĕpasan*, so that the sacred spirit of the ancestor can return to heaven.

1 INTRODUCTION

Putru Kalĕpasan (hereinafter referred to as PK), the text examined in this study, is classified as one of the genres in *tutur* – one of the four palm-leaf manuscripts (L 42, L 222, L 271, and L 322) that contains the PK text with colophon and *candrasangkala* (which reads "*Kuda gu(mulin-gu) lin siti*", and is equivalent to number 7551) that is estimated to be written in the Śaka year 1557 (1635 AD). The four PK texts are included in the collection of the National Library, Jakarta.

This article is categorized as a religious literature work that is considered sacred and used as a practical guide for a sanctification and ancestor worship ceremony known as the *Śrāddha* ritual in the Old Javanese era. The word *putru* is not recorded in OJED (Zoetmulder, 1982), but is referred to as *putrĕhu* (Sanskrit) in KBNW, a spelling that is found in Old Javanese (v.d. Tuuk, 1912). In Monier-Williams (1994: 627), this word form is recorded as *pitĕi-hū, "invoking the Pitṛis"*, or *pitṛ ihūya*, which is in line with the form *devahū* or *devahūya. Mamutru* or *amutru* tradition is the reading manner of "rhythmic pronunciation" of the *putru* text in the purification of ancestral spirit that is called *Mamukur* (= *Śrāddha*), which is meant for guiding the journey to paradise.

Kalĕpasan (<*lĕpas*) is an Old Javanese terminology ("authentic" from the Archipelago) used to describe a condition of inhibition, or freedom. The meaning of the term *lĕpas* is equivalent to that of the word *mokṣa* (Sanskrit); one of the meanings of the word *mokṣa* directly refers to "*kalĕpasan*" (Zoetmulder, 2011). The terms *lĕpas, kalĕpasan, mokṣa,* and *kamokṣan* are more intertwined with Yoga tradition, particularly related to *aṣṭangayoga* and *laku tapa-brata.*

2 AIMS AND BENEFITS

The purpose of this study is to specifically discuss the first part of the "content" of the PK text that is compiled in the L 222 manuscript. Thus, it aims to examine the description of the rites of offerings in the tradition of the *Śrāddha* ritual, which also refers to *Hatiwa-tiwa* and *Hambukur* rituals.

In addition, this study aims at revealing the symbolical meaning of the offerings of feeding the ancestors as well as describing it in the PK text in the Merapi-Merbabu manuscript. Thus, this study contributes to accomplish the historical data of the *Śrāddha* ritual known in the Majapahit era on the basis of the religious manuscripts from Merapi-Merbabu tradition. In fact, the contribution of this study is also important to understand a similar tradition of the *Śrāddha* as far as it has been continued in Bali.

3 METHODS

Textual criticism is the main method used in this study to reveal the texts of *Putru Kalĕpasan* (Robson, 1988). Besides, this study uses religious and content analyses to understand the expression of thoughts, symbolical meaning, and images of ancestor worship ritual called *Hambukur* ceremonies (Purwita, 1992). In that sense, the exposition of the aforementioned rituals involves the discussion of *Śrāddha* that is practiced in the Old Javanese era (9th–15th centuries AD) as revealed in several inscriptions and literary works produced in that period. Merapi-Merbabu *Putru Kalĕpasan* text is a textual heritage that records and explains the rites of *Śrāddha* (purification of ancestral spirits), which are related to the Indic society who lived outside the Palace in the Old Javanese era as recorded and salvaged by the tradition.

Because *Putru Kalĕpasan* texts are written in Merbabu scripts or the so-called *aksara Buda,* the first step taken was to transliterate it to Latin scripts for easy access. To understand the meaning of the "content", this analysis is completed by a translation into English. The "content" of the text used as the source of thoughts and knowledge on religious issues is classified into four parts, namely (1) the description of the type of animals to be sacrificed for *Hambukur or Śrāddha* rituals (pp.1v-3v); (2) the description of *gunung pitu* that is believed to be undergone by the *préta* and *pitara* of *"duruŋ lĕpas"* (pp. 3v-7r); (3) the depiction of the journey of *Saŋ Lĕpas* or *the Pitara* God to paradise and the depiction of the 29 levels of *kahyaŋan* (pp. 7v-15r); and (4) the closing of the text containing the teachings of *Saŋhyaŋ Dharma* meant for the descendants to start practicing *Karya Hayu* (pp. 15r-15v).

4 DEPICTION OF THE *ŚRĀDDHA* RITUAL IN THE OLD JAVANESE ERA

The information on the ancestor worship ritual or *Śrāddha* in the Old Javanese can be found, among others, in the *Karang Tĕngah* inscription from the *Śaka* year 746 (824 AD). The historical information in the inscription explains the sculpting procedure in the ritual that was prepared 12 years after the death and the cremation of King Indra or Ghananātha (King Awan). According to Soekmono (1977), King Indra mentioned in *Kelurak* inscription (782 AD) died in 812 AD. Thus, the 12-year gap after his death would be in 824 AD, as mentioned in the inscription. The myth of number 12 (12 days, 12 months, or 12 years) that is seen as the perfect moment to conduct the *Śrāddha* ritual (both in the Old Javanese era and in Bali) is symbolically related to the 12 *tattwa Śiwa*, which is explained in *Bhuwana Kośa* scripture III.7–11 (Mirsha, 1994).

The next written information related to *Śrāddha* is revealed in the *Śiwagṛha* inscription from the *Śaka* year 778 (856 AD), in which it is mentioned that the *dharma* built by King Rakai Pikatan- a placed that is called *parhyaŋan* – specifically called *Śiwalaya* or *Śiwagṛha*, according to Soekmono (1977), is an object meant for deceased kings who have reached the status of the gods (*"tĕwĕk bhaṭāra ginawe sinaŋaskāra weh…. ya ikana sīma puput mare bhaṭāra"*).

In the next development, the *Śrāddha* ritual procession is revealed in the *Jiu* I and III inscriptions from the *Śaka* year 1408 (1486 AD). The notes in these inscriptions explain that Dyah Ranawijaya held the *Śrāddha* ritual to commemorate the 12th anniversary of the death of *Śri Paduka Bhattara riŋ Dahanapura Saŋ Mokta riŋ Indrabhawana*; he was Bhre

Pandan Salas or Śri Adi Suraprabhawa, whose maiden name was Dyah Suraprabhawa Singhawikramawardhana (Soemadio, 1984; Djafar, 1986).

In addition to the data obtained from the inscriptions, the description of the *Śrāddha* ritual is revealed in an interesting manner in the *kakawin Nagarakṛtāgama,* from *pupuh* LXIII to *pupuh* LXIX (Pigeaud, 1960). The literary work explains how King Hayam Wuruk held the *Śrāddha* ritual in the year Ś 1284 (= 1362 AD) as a tribute to his grandmother named Rajapatni. The ritual was held 12 years (*dwidaśawarṣa*) after the death of Rajapatni (Pigeaud, 1962, IV).

Śrāddha ritual from the Majapahit era is in fact continued to be practiced regularly in the *Pitrayajña* ritual by the Balinese as well as the Hindu-Těnggěr societies (Pigeaud, 1962, IV). Besides, the ritual is also deeply recorded within the collective memory of the Javanese society as is evident in the life and belief of *Kejawen.* The term *Kejawen* or *Agami Jawi* itself is defined as a set of Hindu–Buddhist beliefs and concepts with mystical tendency that is syncretized and acknowledged as part of Islamic tradition in Indonesia (Koentjaraningrat, 1984).

The basic value of the *Śrāddha* ritual commonly known by the term *Nyadran* or *Nyěkar* (from: *Sěkar* or *Sěkah*) continues to be passed down and practiced regularly even at present. *Nyadran* (*N-* + *Śraddha* + *-ěn*) tradition for the followers of *Kejawen* is normally conducted annually in the month of *Ruwah* (from the word *Arwah* or spirit). *Nyadran* and *Nyěkar* (possibly derived from the word *Sěkah* changed into *Sěkar* or *Nyěkar*) are a form of continuation from the *Śrāddha* tradition in the Old Javanese era. *Sěkah* is a symbolic manifestation of ancestral spirit that takes the form of flowers or *sěkar,* termed as *puspa-sarira* in the *Mamukur* ceremony.

5 *ŚRĀDDHA* RITUAL IN *PUTRU KALĚPASAN*

Religious knowledge related to traditional processions and offerings in the *Śrāddha* ritual recorded in the PK text depicts several important aspects, including (1) *hatiwa-tiwa* ritual and (2) *hambukur* ritual or *Karya Hayu.* The exposition on these two ritual aspects will be explained below.

5.1 *Hatiwa-tiwa ritual*

The use of the term *atiwa-tiwa* that can be found in the PK text is not a new discovery. That term is often found in both Old Javanese and Balinese manuscripts. In *kakawin Rāmāyana* (Poerbatjaraka, 2010), it is explained that after the death of King Daśaratha, the Sang Bharata ordered the people to perform a cremation ceremony. Information about *Atiwa-tiwa* in *kakawin Rāmāyana* can be read in the following quote:

> "Huwus nira wěruh ri hetu nira saŋ narendrār pějah[...]/Kinonira ta saŋ balādhika tumunwa saŋ bhūpati/Maśoca ta maweh tilěm sira rikaŋ tilěmniŋ wulan/[...]" (III.27a, 32a-c)

> (Translation: "After knowing the cause of the death of the king [...]. It is instructed to the mighty people to burn the (deceased) body of the king (Daśaratha). To purify (the body), to hold the ceremony (tilěm) at the night of the dark moon (tilěm))"

The cremation ceremony of King Daśaratha revealed in *kakakwin Rāmāyana* is termed as *Tilěm,* which is conducted at *Tilěm* night (dark moon). The use of this term is also revealed in the *Ādiparwa* scripture 102.20–22 (Juynboll, 1906), which states:

> "[...] Māti ta sira sěděŋ yowana. Atyanta duḥka Saŋ Gandhawatī, malara sirānaŋis. Tan ucapakna lara Saŋ Ambalika [...] (tan pahiŋan). Ginawe ta ŋ pretatarpaṇa, saha widhi widhana, wineh śocāŋkěn tilěm, katěka ri kaparikramaniŋ śuddha śrāddha."

> (Translation: "He (the Citrawirya) passed away when he was young. The Gandhawatī was heartbroken, crying, and in perpetual suffering. The sadness of the Ambalika

too was beyond words [...] endless. He held a ceremonial offering to the ancestral spirit (*pretatarpaṇa*) with a holy ritual every *Tilĕm* (dead moon), until the time of the *Śrāddha* ritual for purification arrived).

A better description about this *Atiwa-tiwa* ritual in the Old Javanese can be found in the *Wirataparwa* scripture that came from the same period as *Ādiparwa* that was written in the era of King Dharmawangśa Tĕguh in the Śaka years 913–929 (991–1007 AD). In the *Wirataparwa* scripture, it is written: "[...] inikĕtakĕnya ri śawa saŋ Kīcaka, irikaŋ kāla pituŋ tabĕh tumūta yātriy**atiwa-tiwa**, umuśi deśa nikaŋ śmaśāna [...]" (Juynboll, 1912). Additional information related to *atiwa-tiwa* is also available in several texts compiled under the four manuscripts of PK, namely *pilĕpasi setra* (18r-18v), *ŋĕntas-ĕntasan* (18v-19r), *tĕbus petra-sasaje nira* (20r-24v), *puja pitara* (24v-25v), *hĕntas-ĕntasan-kalĕpasa nira Aji Dhaŋḍaŋ Gĕṇḍis* (28r-28v), *palilimbaŋan* (54r-55v), and *puja setra* (55v-56r).

5.2 Hambukur ritual or "Karya Hayu" (Śrāddha)

The term *Hambukur,* also known as "*Karya Hayu*", is recorded in the PK text in the expression "*ri sampun kita mukti* **hambukur** *bhūmi śayana*" (PK 76–89). In fact, its definition is not explicitly mentioned. The term *Hambukur* has been used in *Nagarakṛtāgama* 63.4 and 66.1 written as *buku-bukuran* in the *Śrāddha* ritual for the spirit of Tribhuwana Tunggadewi. It is assumed that the term *puspaśarira (= Sĕkah* or *Sĕkar)* in the *Mamukur* ritual in Bali is also derived from *Nagarakṛtāgama* 67.2, particularly from: *saŋhyaŋ puspaśarīra* (Pigeaud, 1960).

However, several centuries before *kakawin Nagarakṛtāgama* (1365 AD) was written by Mpu Prapanca, the word *bukur* had been used in *kakawin Sumanasāntaka* written by Mpu Monaguṇa from the era of King Kadiri, which mentioned the name Warṣajaya (= Jayawarṣa) that indicates a relative of the king (Worsley, *et al.*, 2014). *Kakawin Sumanasāntaka* (Worsley et al., 2014) stated *"aluŋguh irikaŋ bukur paŋaraŋanya nĕhĕr asidĕhāŋuregĕluŋ"* (2.4c), *"ri tiŋhalira riŋ bukur saŋ araras winulatanira maŋliga-ligā"* (4.1c), and *"Bukur pamiḍaran ndatan milu gĕsĕŋ kadi rinĕŋwan iŋ kilat"* (9.2c).

The definition of *Bukur* in *Sumanasāntaka* that is translated as *"Bukur* hall" contains at least three layers of meaning, namely (1) *Bukur* hall is the place where Bhagawān Tṛṇawindu practiced yoga and wrote literary works; (2) in the same *Bukur* hall, *widyādhari* Harinī (assigned with the duty of seducing a hermit), descending from Indra paradise, sat with her hair loose and was seen topless by the *bhagawān*; and (3) in that *Bukur* hall, the body of Dyah Harinī was laid and burned by the fire of a thunder cloud, the manifestation of the dreadful curse from Bhagawān Tṛṇawindu.

On the basis of the short depiction and information from *kakawin Sumanasāntaka* (2.4c, 4.1c, and 9.2c) and *kakawin Nagarakṛtāgama* (63.4 and 66.1), it can be concluded that *Bukur* in its relation with *Hambukur* (PK 76–89) refers to a particularly sacred hall connected to a sanctified figure and specifically used as the *sthana* for ancestral spirits. In OJED, *bukur* is defined as "*a small building (probably with meru-shaped roof)*" (I) (Zoetmulder, 2011).

The meaning of *Bukur* in the *Hambukur* or *Mamukur* ritual is still recorded and maintained well in the *Śrāddha* tradition in Bali. *Bukur* or *Madhya* is a tower with meru-shaped roof that functions as the site to place *Sĕkah* (*puspaśarira*), which will be carried to the sea. *Hambukur* ritual is also known as "*Karya Hayu*" (considered as sacred as *Dewayajña* ritual), which is a ritual dedicated to sanctify (ordain) the spirit from *pirata* to *pitara* or *the Pitara* God (Sanskrit: *Pitṛideva*).

In the ancestor worship tradition in Bali, an argument based on a source from *lontar* explains how *Mamukur, Tilĕman,* and *Maligya* in essence refer to the same meaning, which is the purification of *ādhyātmika* element, because human beings are formed from the *wāhya-ādhyātmika* element (body-soul). Therefore, the manifestations of *Tri Aŋga* are (1) *Maligya,* which is *uttama*, (2) *Tilĕman/Panilĕman,* which is *madhya*, and (3) *Hambukur* or *Mamukur/ Ngaroras,* which is *niṣtha*.

Hambukur ritual in the Balinese tradition is usually followed by the *Maligya* ritual (this term is not recorded in OJED), a ritual for the third-level purification of ancestral spirit that is meant to "unify" the holy ancestral spirit in *Pura Padharman*, and it would then become *Dewa Pitara*, which is the "deified" ancestral spirit. The procedures of the *Maligya* ritual are elaborated in the *Yajña Baligya Panilĕman* manuscript (*Kal* 79 Bali Cultural Documentation Center), as well as painted in the *Kiduŋ Karya Ligya* or *gĕguritan Padĕm Warak* (K 6313/Ivd), and *Kramaniŋ Karya Baligya/Mamukur* (K 6528 Ic).

In the aforementioned manuscript, the information about *Mamukur-Maligya* is considered as one unity. If such is the case, it can be concluded that the same perspective is applicable to analyze *Hambukur* ritual in the Merapi-Merbabu PK-text. In regard to this assumption, the PK text claims:

"[...] Ri sampun kita mukti **hambukur** bhūmi śayana, pusaḍi minakādinya, wehana ta kita susur, kase, kuramas, ndan ana kipat, suri, lĕŋa, burat. Ri tlas i macāmana, Ucapĕn ta kita muwuh skul, saṇḍaŋan den ana saṇḍiŋana wwe cāmani, sampiranta mareŋ swargga, [...] (PK 76–89).

"After you enjoy Bukur as a resting place, particularly in the podium, (next), first you will be offered a toothpick, perfumed ointment, shampoo, and also "kohl" (eye shadow), hairbrush, perfume and perfumed powder. After purifying yourself by washing your mouth (gargling), we will tell you (that) you can add supplements with rice, a set of clothes, and water to wash your mouth, (and then) put on your shawl to leave for heavens."

The above quote elaborates on the elements of offerings that are given to the purified ancestral spirits, including toothpick, perfumed ointment, shampoo, "kohl" (eye shadow), hairbrush, perfume, perfumed powder rice, a set of clothes, water for washing the mouth, and shawl. This offering, known as "*karya hayu*", is also denoted by the term *pitṛtarpana*, which refers to water and other types of offerings to the ancestors. It is clearly explained in *Sarasamuccaya sloka* 280 that "*...pitṛtarpana, ya śrāddha ŋaranya...*" (Pudja, 1981). Within the Indic tradition, as described by Monier-Williams (1994), *pitṛtarpaṇa* is "*the refreshing of the pitṛ's (with water thrown from the right hand), offering water etc. to the deceased ancestor*".

On the basis of that conceptualization, it can be understood that the forms and types of offerings as *pitṛtarpaṇa* in the Merbabu tradition are very particular as they are the results of interpretation of and deep contemplation on religious thoughts from the Old Javanese. This perspective can be clearly outlined in the following quote (apparently, the PK text = PK-Bali text, LOr 5132):

"[...] Saŋ mahurip kuminkina, haweha **bubur sasuru**, ŋūni kotamani **dadhi, nasi putih** hawak dharmma, **nasi baŋ** subhaga ni rāt, **kuniŋ ika nasi** wṛddhi putra phalanya, **nasi hirĕŋ** wṛddhi ikaŋ mas. Puŋpuŋ rare hulunanta! Sarwwa wija sarwwa haywa, yateka cihnaknanta glisakĕn aywa suwe denta baŋun Karyya Hayu [...] (PK 9–21).

It is evident from the above quote that there is an intertwinement in the concept and thought from the author of the PK text with the offerings in *Śrāddha* ritual. However, the element of water is not explicitly stated. This might be because water has regularly been used in the said ritual.

6 SYMBOLIC MEANING OF ANIMAL OFFERING IN THE *ŚRĀDDHA* RITUAL

It is explained that the offerings and details about the types of animals used for the offerings include (1) mackerel and snapper; (2) grouse, green pigeon, dried fish, kaliliŋan, wild chicken, wuru-wuru, turtle dove, and putĕr; (3) hunted animals like deer, antelope, and mouse deer; (4) buffalo meat, ducks, and chicken; (5) hedgehog, rasé, monitor lizard, iguana, and wṛtalan; (6) goose and freshwater fish; (7) tortoise, turtle, and shrimp; (8) wild boar; (9) buffalo; (10) milk and honey; (11) honey oil; and (12) rhino: meat, skin, and bones.

It is very challenging to comprehensively depict the types of animals that are offered as elements of upakāra-upacāra in the rites of sanctification (from the daśamala elements). It is implied in Wratiśasana that, according to the religious norms used as the theological foundation, the slaughtering of animals does not violate the ahimsa-dharma doctrine ("killing or hurting living beings is forbidden"). In this regard, the siddhanta brata is allowed to conduct himsa karma ("animal slaughter") if it is meant for the purposes of (1) dewa pūja, (2) athiti pūja, (3) pitra pūja, (4) walyākrāma pūja, and (5) dharmā wigāta, which is atmaraksa, and that covers dharma yuddha ("killing animals on the battlefield") and killing because of an attack from poisonous animals (Rani, 1961). The slaughtering of animals as sacrificial offerings that is in lieu with the above purpose is also based on the doctrine that is conveyed through mythology in the Purwa Bhumi Kamulan and Bhūmi Tuwa manuscripts, which explain how pañca-dewa (Iśwara, Brahma, Mahādewa, Wiṣṇu, and Śiwa) offered different types of animals such as eka-pada, dwi-pada, catur-pada, bahu-pada, and rahi-pada (Hooykaas, 1974).

Among the animals offered, apparently, *warak* (*Rhinoceros sondaicus/Rh. Indicus/Rh. unicornis*) has the most symbolic meaning. It is outlined that meat, skin, or bones from *warak* that are rubbed on whetstone and mixed in the offerings can bring happiness to the *pitara* and therefore this will keep the *pitara* in the paradise forever (*agěŋ pawehku nikaŋ sukha, salawasta haneŋ swargga*).

The name *warak* is indeed mentioned in several Old Javanese literary works, including *Ramāyāna* II.26 (Poerbatjaraka, 2010) "*...aŋhiŋ warak juga warěg rumuhun rikaŋ rwi*", XVII.14 "*...hana moŋ warak wwara ta siŋha pañjaran...*," as well as *Brahmāṇḍapurāṇa* 49.11 (Gonda, 1932), *Korāśrama* 32.13 (Swellengrebel, 1936), and *Nawaruci* 79.13 (Prijohoetomo, 1936) "*...warak upamanira Aŋkuspraṇa....*". However, the name *warak* mentioned in these literary works is not associated with the holding of a ceremony. A mythological saying can be found in the *Agastyaparwa* manuscript about the origins of *warak*. This can also be linked with the use of *warak* in *Śrāddha*. The mythological saying about rhinos in the *Agastyaparwa* scripture is elaborated as follows: "[...] *Kunaŋ anak bhagawān Pulaha i saŋ Mṛganya ŋ **kidaŋ**, śaśa, kañcil, waraŋkakan; nahan tānaknira saŋ Mṛgā. Kunaŋ anaknira i saŋ Mṛgamandā nyaŋ kuwuk, luhak, **tiŋgaluŋ**, **sapi**, **kubwa**, śarabha, uṣṭra, **warak**; nahan tānaknira i saŋ Mṛgamandā [...]*" (Gonda, 1936).

The use of *warak* meat for offerings in the sanctification of ancestral spirit apparently is not only known to be found in Merbabu and Bali PK texts. The information on the superiority of *warak* in offerings for the *Śrāddha* ritual has been doctrinally written in *Manawa Dharmaśastra* III.272. The following is a quote from the *Manawadharmaśastra* scripture III. 272 (Pudja, 1996):

"Kalaśākammahāśalkāḥ, **khaŋga** lohāmisammadhu,
Ānantyāyaivakalpante, munyannānicasarwaśaḥ."

(Translation: "The vegetables that go by the name Kalaśaka, fish by the name Mahasalka, **rhino meat** and red-goat meat, and all kinds of foods eaten by the hermits in the woods will satisfy them for eternity (in paradise).

In fact, the above quote is indicated as an intertextual relation of *rhino* in offerings for the *Śrāddha* between Indian ritual (as well as mentioned in the *Manawadharmaśastra* III. 272) and Old Javanese tradition at the end of the 16th century.

7   CONCLUSION

Several conclusions can be drawn from the above discussion, such as: (1) the text of *Putru Kalěpasan* from Merapi-Merbabu is a text within the scope of classic Javanese literature that records and explains in great detail the procedures for *Śrāddha* ritual, a prominent ritual in the Old Javanese era (9th–15th centuries); (2) the exposition of the rites of ancestral worship known as *Hatiwa-tiwa* and the *Hambukur* ritual in the PK text is a form of continuity derived from the Old Javanese religious thought that is still passed on and sustained in the *Literature-teaching* or *Sastra-Ajar* tradition around the Merapi-Merbabu mountains from the 16th century to the early 17th century; (3) animal slaughtering that is used as a sacrificial offering in the *Śrāddha* ritual is meant to symbolically purify the ancestral spirits from any impurity caused by the *daśamala* elements. Thus, essentially, the purification in *Hambukur* ritual is meant to ordain *pitara* to become *Pitara God*, (Sanskrit: *Pitṛdeva*), a worshipped ancestral spirit that is considered to have reached *kalĕpasan* so that the spirits may return home to the *kahyaŋan* of the gods or to unite with their avatar.

REFERENCES

Darmosoetopo, R. (1985) Pandangan Orang Jawa terhadap Leluhur (Tinjauan berdasarkan Data Tertulis), [Pertemuan] *Pertemuan Ilmiah Arkeologi III (PIA III),* Ciloto 23rd —28th May, pp. 519–529. Translation: Javanese View towards Ancestors (A Study on the Basis of Written Data), [Meeting] *Archaelogy Meeting III (PIA III),* Ciloto 23rd —28th May, pp. 519–529.

Djafar, H. (1986) Beberapa Catatan mengenai Keagamaan pada Masa Majapahit Akhir Several Notes on Religosity in the End of Majapahit Era, [Meeting] *Pertemuan Ilmiah Arkeologi IV,* Cipanas 3rd to 9th March, pp. 252–266. Translation: Several Notes on Religosity in the End of Majapahit Era, [Meeting] Archaelogy Meeting IV, Cipanas 3rd to 9th March, pp. 252–266.

Endraswara, S. (2015) *Agama Jawa, Ajaran, Amalan, dan Asal Usul Kejawen.* Yogyakarya, Narasi. Translation: *Religion,* Teaching, Practice of Java and Origin of Kejawen. Yogyakarya, Narasi.

Flood, G. (2003) The Śaiva traditions. In: Flood, G (Ed.) *The Blackwell companion to Hinduism.* USA, Blackwell Publishing.

Gonda, J. (1932) *Het Oud-Javaansche Brahmāṇḍa-Purāṇa, Proza-tekst en Kakawin. Bibliotheca Javanica,* No. 5. *Koninklijk Bataviaasch Genootschap van Kunsten en Wetenschappen.*

Gonda, J. (1935–1936) *Agastyaparwa, Uitgegeven, Gecommenteerd en Vertaald. BKI* No. 90, 329–419; No.92, 389–458; No. 94, 223–285.

Hooykaas, C. (1974) *Cosmogony and Creation in Balinese Tradition.* The Hague, Martinus Nijhoff.

Istari, T.M.R. (1997) Upacara Sraddha di Jawa dan Bali, *Jurnal Arkeologi Siddhayatra,* No. 1/II/28th— 35th May. Translation: Sraddha Ceremony in Java and Bali, Siddhayatra Journal of Archaeology, No. 1/II/28th—35th May.

Juynboll, H.H. (1906) *Adiparwa. Oud Javaansch Prozageschript Uitgegeven.* S-Gravenhage, Martinus Nijhoff.

Juynboll, H.H. (1912) *Wirātaparwwa. Oud Javaansch Prozageschript Uitgegeven.* S-Gravenhage, Martinus Nijhoff.

Kaler, I.G.K. (1993) *Ngaben, Mengapa Mayat Dibakar?* Denpasar, Yayasan Dharma Narada. Translation: Ngaben, Why Bodics are Cremated? Denpasar, Yayasan Dharma Narada.

Kamus Jawa Kuna-Indonesia. (2011) Jakarta, PT Gramedia Pustaka Utama. Translation: Ancient Java-Indonesia Dictionary. (2011). Jakarta, PT Gramedia Pustaka Utama.

Koentjaraningrat. (1984) *Kebudayaan Jawa.* Seri Etnografi Indonesia No.2. Jakarta, Balai Pustaka. Translation: Javanese Culture. Indonesia Ethnography Series No.2. Jakarta, Balai Pustaka.

Linus, I.K. (1986) Pemujaan Roh Leluhur di Bali: Suatu Pendekatan Tradisi Hindu, [Meeting] *Pertemuan Ilmiah Arkeologi IV (vol. IIb),* Cipanas 3–9 March, 258–268. Translation: Worship of Ancestral Spirits in Bali: An Approcah to Hindu Tradition), [Meeting] Archaelogy Meeting IV (vol. IIb), Cipanas 3rd —9th March, 258–268.

Mirsha, I.G.N.R. (1994) *Buana Kosa, Alih Aksara dan Alih Bahasa.* Denpasar, Upada Sastra. Translation: Lexicons, Transfers of Scripts and Languages. Denpasar, Upada Sastra.

Monier-Williams, M. (1994) *Sanskrit-English Dictionary.* Munshiram Manoharlal, Publisers Pvt Ltd.

Pigeaud, T.G.T. (1960, 1962, 1963) *Java in the 14th Century,* A Study in Cultural History (vol. I—IV). Koninklijk Institut voor Taal-, Land—en Volkenkunde; Translation Series 4, 1–4. The Hague, Martinus Nijhoff.

Pigeaud, T.G.T. (1967, 1968, 1980) *Literature of Java* (vol. I—IV). The Hague, Martinus Nyhoff.

Poerbatjaraka, R. Ng. (2010) *Rāmāyana Djawa Kuna.* Jakarta, Perpustakaan Nasional Republik Indonesia Translation: *Rāmāyana* of Ancient Java. Jakarta, National Library of the Republic of Indonesia.

Prasad, R.C. (1997) *The Śrāddha: The Hindu Book of the Dead, a Treatise on the Śrāddha Ceremonies.* Delhi, Motilal Banarsidass Publishers Private Limited.

Prijohoetomo. (1934) *Nawaruci. Inleiding, Middel-Javaansche Prozatekst, Vertaling.* Groningen: Bij J.B. Wolters.

Pudja, G. & Sudharta, T.R. (1996) *Manawa Dharmaśastra.* Jakarta, Hanuman Sakti.

Purwita, IB Pt. (1992) *Upacara Mamukur (Mamukur Ceremony).* Denpasar. Upada Sastra. Translation: Mamukur Ceremony. Denpasar. Upada Sastra.

Rani, Mrs. Sharada. (1961) *Wratiśāsana, A Sanskrit Text on Ascetic Discipline with Kawi Exegesis.* New Delhi, the Arya Bharati Mudranalaya.

Robson, S.O. (1988) *Principles of Indonesian Philology.* Dordrecht-Holland, Foris Publications.

Santiko, H. (1986) Maṇḍala (Kadewaguruan) pada Masyarakat Majapahit, [Meeting] *Pertemuan Ilmiah Arkeologi IV (vol. IIb), Cipanas 3rd—9th March, 149–169.* Translation: Maṇḍala (Kadewaguruan) in Majapahit Society, [Meeting] Archaelogy Meeting IV (vol. IIb), Cipanas 3rd—9th March, 149–169.

Sayers, M.R. (2008) *Feeding the Ancestors: Ancestor Worship in Ancient Hinduism and Buddhism.* Dissertation on Doctor of Philosopgy. USA, the University of Texas at Austin.

Setyawati, K., et al. (2002) *Katalog Naskah Merapi-Merbabu Perpustakaan Nasional Republik Indonesia.* Yogyakarta, Universitas Sanata Dharma-Univ. Leiden. Translation: Catalogue of Merapi-Merbabu Manuscripts National Library of the Republic of Indonesia. Yogyakarta, Univesitas Sanata Dharma-Univ. Leiden.

Soekmono. (1977) *Candi: Fungsi dan Pengertiannya.* Jakarta, Departemen Pendidikan dan Kebudayaan. Translation: Temples: Functions and Understanding. Jakarta, Department of Education and Culture.

Soemadio, B. (ed.). (1993) Sejarah Nasional Indonesia II. (ed. ke-3). Jakarta, Balai Pustaka. Translation: National History of Indonesia II. (3rd Edition). Jakarta, Balai Pustaka.

Swellengrebel, J.L. (1936) *Korawāśrama een Oud-Javaansch Proza-Geschrift, Uitgegeven, Vertaald en Toegelicht.* Proefschrift. Leiden, Rijkuniversiteit te Leiden.

van der Tuuk, H.N. (1912) *Kawi-Balineesch-Nederlandsch Woorden Boek* (vol. IV). Batavia, Lands Drukkerij.

Worsley, P. et al. (2014) *Kakawin Sumanasāntaka. Mati karena Bunga Sumanasa Karya Mpu Monaguna. Kajian sebuah Puisi Epik Jawa Kuno.* Jakarta, Yayasan Pustaka Obor Indonesia. Translation: *Kakawin Sumanasāntaka. Death Due to Sumanasa Flower by Mpu Monaguna. A Study on an Epic Poem of Ancient Java.* Jakarta, Yayasan Pustaka Obor Indonesia.

Zoetmulder, P.J. (1982) *Old Javanese-English Dictionary.* 'S-Gravenhage, Martinus Nijhoff.

APPENDIX 1

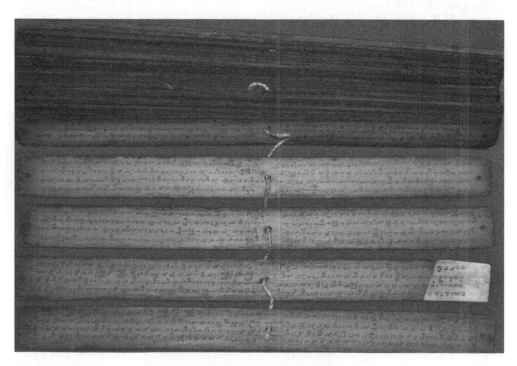

Photo: L 222 P10, the Merapi-Merbabu collection of National Library Jakarta.

Cultural Dynamics in a Globalized World – Budianta et al. (Eds)
© *2018 Taylor & Francis Group, London, ISBN 978-1-138-62664-5*

Ancient religious artworks in Central Java (8th–10th century AD)

A.A. Munandar
Department of Archaeology, Faculty of Humanities, Universitas Indonesia, Depok, Indonesia

ABSTRACT: In Central Java, Hindu-saiva and Buddha Mahayana thrived between the 8th and 10th centuries AD. The main hypothesis of this study is that Central Java was governed by two different dynasties in that period, namely the "Sailendravamsa" dynasty, which adopted Buddhism, and the "Sanjayavamsa" dynasty, which adhered to Hinduism. In this study, we discuss the ancient religious artworks related to Hinduism and Buddhism developed between the 8th and 10th centuries AD. Iconography and the architecture of sacred structures and the technique of making the fine arts are the tools used to explore the religious art style. We also discuss the theory proposed by Dutch archeologist E.B. Vogler, namely the theory of "the tied art" from the ancient Javanese society. Several conclusions are drawn from the interpretation achieved in this study, which is aimed to gain a deeper understanding of the ancient religious artworks in Central Java. The author believes that this study could provide further academic explanation on the origin of the supporters of these old artworks in Central Java.

1 INTRODUCTION

The form of art discussed in this study is limited to the fine artworks supported by available data, which could be retrieved anytime. The artworks selected in this study are those that are devoted to religious and ritual needs for worship or complementing the ancient sacred structures. Therefore, the forms of art chosen for this study include the following three aspects, namely the works of religious architecture, sculpture art, and reliefs. In the central part of Java, based on the dated inscription, there is evidence of existence of a civilization that had thrived under the principles of Hinduism and Buddhism. The Hinduism discussed in this study is the Hindu-saiva, in which Siva is the god. Meanwhile, the Buddhism discussed in this study is from Mahayana-ism, which recognizes pantheon in its system of belief.

Interestingly, according to the existing data, Hinduism and Buddhism had developed well in the central part of Java region only from the 8th to 10th centuries, and no archeological evidence was found in that region after the 10th century. In the study of Indonesian archeology, the period of development between the 8th and 10th centuries is often called the Old Classic period. In the late 10th century, apparently, the central powers of the royal government shifted from the central part to the eastern part of Java; thus, the cultural activities of the ancient Javanese society also shifted to the eastern part of Java region. The ancient Mataram Kingdom initially had developed in the central part of the Java region, and according to the Canggal inscription dated 732 AD, the kingdom was ruled by King Sanjaya, who was the first king to adhere to Hindu-saiva (Poerbatjaraka, 1952, pp. 49–58). From the perspectives of political history and archeology, it is crucial to identify the causes of the frequent relocation of the capital town of Mataram. According to the inscriptions, the capital of Mataram was named *Medang* or *Mdang*; however, in some places, it was known as *Mdang i Bhumi Mataram, Mdang i Poh Pitu,* and *Mdang i Mamratipura.* Yet the exact location of *Kedaton Mdang* is unknown to date. The inscriptions also show that the kings ruling Mataram were the family members of King Śailendra or Śailendravamsa. This dynasty is often related to Buddha Mahayana, because the term Śailendravamsa was used for the first time in the Kalasan inscription (700 Saka/778 AD), and King Śailendra apparently adopted the principles of the Buddha Mahayana. The inscription also describes King Tejapurnnapanna Panangkarana,[1] named *Sailendravamsatilaka* ("the pearl of Śailendra dynasty"), who was ordered to build some type of sacred structure to worship Goddess Tara, named *Tarabhawanam*. Tara is a common name for any goddess that is considered *sakti* (enchanting) in many Buddha levels in the Buddha Mahayana religion.

Another inscription named Kelurak inscription (in 782 AD) mentioning *Śailendravamsa* was found in the west of the Sewu Temple complex. It states that a sacred structure used for worshipping under the Manjusri belief was built and hence it was called *Manjusrigrha* in the inscription, which is the current Sewu Temple. *Sailendravamsa* is also mentioned in the *Abhayagirivihara* inscription (792 AD), which is found in the archeological area of Ratu Baka (Boechari, 2012, p. 198). Here, we agree with the opinions of experts from previous studies that rebutted the theory of the two dynasties, "Sanjayavamsa" and "Śailendravamsa" existing in the central part of Java from the 8th to 10th century. Nevertheless, only the Śailendravamsa dynasty existed in this period, and some members practiced Hindu-saiva, whereas others practiced Buddha Mahayana. This study was conducted to gain in-depth knowledge about the forms of religious fine arts during the old Mataram period in Central Java and to understand its characteristics, nature, and art style developed by the artists.

As stated above, the artworks developed during this period were used for religious purposes. Thus, it is assumed that the artists were highly associated with religious activities under the Hindu-saiva or Buddha Mahayana belief and not with any other purposes. It was certain that the adherents of both religions needed a sacred place to hold religious ceremonies, place the statues of gods as a symbol of worship, and display the reliefs depicting the teachings and systems of religious belief. Meanwhile, many other artworks, such as those of the performing arts, could not be studied because of the difficulty in obtaining the relevant data.

1.1 *Sacred ancient structures*

The sacred structures of Hinduism and Buddhism do not have a specific term, such as *grha, bhavanam, candi, vihara, kuti,* and *katyagan*; however, a common term that has always been used to refer to these structures is "temple". In addition, the term *patirthan* denotes a sacred compound in the form of a sacred water source; in Buddhism, there are terms like stupa and *caitya*. The sacred structures of Hindu-saiva in the central region of Java are Gedong Songo (in Ambarawa), Gunung Wukir, Sambisari, Kedulan, Morangan, Merak, Pendem, Lumbung, and Asu Temples, and the largest complex is Prambanan Temple, whose original name is *Sivagrha*. Meanwhile, Buddhist temples include Kalasan Temple (*Tarabhavanam*), Sari, Lumbung Temples, Sewu Temples (*Manjusrigrha*), Bubrah Temple, Plaosan Lor, Plaosan Kidul, Sajiwan, Banyunibo, Ngawen, Mendut, and Pawon, and the grandest Buddhist stupa is in Borobudur Temple.

Furthermore, it can be concluded that the differences between Hindu-saiva and Buddhist temples are attributed to their differences in belief systems and ritual tools. However, an interesting finding is that the temples belonging to those two religions have several similarities. The data on the similarities and differences between the architectural components of these two religions are presented in Table 1.

On the basis of the data presented in Table 1, similarities between the Hindu or Buddhist temples are described as follows:

1. Two types of materials, namely stone blocks (outer structures) and bricks (central structure) are used in the structures of the temples.
2. The temples have a three-level roof shape with a *kemuncak* (the top of the roof), which has the shape of either a *ratna* (in Hindu temples) or a *dagob* (a small stupa in Buddhist temples).
3. The edge of stairway is known as the *ikal-lemah* (weak-bend form) with the *Makara* end, and its base is adorned with the head of *Kala*.
4. The frames are flat (*patta*) with variations, namely semicircles (*kumuda*) and bell shape (*padma*).

Table 1. Similarities and differences in the architectural components of Hindu and Buddhist temples.

No.	Components compared	Similarity	Difference
1.	The temples use two materials, namely stone block and bricks (as well as other materials that quickly weather)	√	–
2.	Their architecture shows religious symbols	–	√
3.	Hindu temples are equipped with 3 ancillary temples (*perwara* temples) in front of the main temple, whereas the *perwara* of Buddhist temples are built surrounding their main temple	–	√
4	The temples have a three-graded roof shape with one *kemuncak*	√	–
5.	The temples have the edge of stairs called "ikal-lemah" with a Makara end, and their base is adorned with the head of Kala	√	–
6.	The yard of Hindu temples is not paved with stone blocks, whereas the surface of a Buddhist temple yard is covered with stone blocks	–	√
7.	The frames used are flat (*patta*) with variations, namely semicircles (*kumuda*) and bell sides (*padma*)	√	–
8.	The outer walls of Hindu temples have niches, whereas Buddhist temples do not have niches placed on their outer walls	–	√
9.	The entrance to the chambers of the temples and niches (*parsvadevata*) is adorned with the head of Kala above	√	–
10.	The temples are known to have a fence structure (*vedika*) on the edge of *pradaksinapatha* floor around the body of Hindu-saiva or Buddhist temples	√	–

5. The entrance to the chambers of the temples and niches (*parsvadevata*) is adorned with the head of *Kala* positioned above the entrance.
6. The temples are known to have a fence structure (*vedika*) on the edge of the *pradaksinapatha* floor in the body of Hindu-saiva or Buddhist temples.

Furthermore, the significant differences between the Hindu-saiva and Buddhist temples are as follows:

1. Their architecture shows religious symbols, with the roof component in the shape of a *ratna* for the Hindu temples and a *dagob* (a small stupa) for the Buddhist temples.
2. Hindu temples are complemented with three ancillary temples (*pervara* temples) in front of their main temple, whereas the *pervara* of the Buddhist temples is built surrounding their main temple.
3. Hindu temples have a niche on their outer wall, whereas Buddhist temples have no niches in their outer wall.
4. The surface of the courtyard of Buddhist temples is covered with stone blocks,[2] which are absent in the Hindu temple compounds.

The similarities in the structures of the Hindu and Buddhist temples show that the architectural art form is developed and supported together by the *silpin* (*artisan*) from both religions. Considering the similarities in the architecture of the temples, it can be assumed that the characteristics of the Hindu and Buddhist sacred buildings originated in the Old Classic period in Central Java. With the development of the ancient Mataram Kingdom in that period, it can be concluded that the Hindu temples have the same architectural characteristics as the Buddhist sacred buildings during the Mataram period under the reign of the royal family of *Sailendravamsa*.

1.2 Statues of gods

Apparently, there is a significant iconographic difference between the statues of the Hindu-saiva and Buddha pantheon, which is not necessarily discussed further. This difference attributes to the differences found among Brahma, Visnu, Siva, Parvati, and other Hindu statues and Panca Tathagata, Bhoddhisattva, and other statues of the Buddha pantheon. It is interesting to note that the same components are found in the statues of Hindu-saiva and Buddha pantheon. This similarity can be due to the characteristic features of the statue art developed in the central region of Java from the 8th to 10th century AD.

On the basis of the observation of several statues of the Hindu and Buddhist gods in the central region of Java, many similarities can be concluded as follows:

1. The pedestals of the statues are always in the form of a blooming lotus (*padmasana*) in both standing and sitting positions.
2. The shape of the backrest for the statues' body is called the *Prabhamandala,* and there is a circle named *Sirascakra* behind the head of the statues, which is considered as the symbol of god's divine light.
3. The ornaments in the Hindu and Buddhist statues have the same pattern, which is a series of jewels (*ratna*). A large jewel is in the center, which is surrounded by small jewels.
4. The statues are depicted as wearing clothes and accessories without many jewels. The accessories not only complement the appearance of the statutes but also contribute to their elegance.
5. The accessories include crown, ear accessories (earrings), *upavita* (a string of caste), brace-let, necklace, arm ring, belt, and ankle bracelet.

Through an in-depth study, many other similarities can also be examined, such as the size (height) of statues, which seems to be almost similar in both religions, or the working process of the surface, which depends on the materials used for making statues (e.g., stone or metal). The similarities in the appearance of those statues can thus be assumed to be due to the similarities in their working process, which are the characteristics of the Hindu and Buddhist statue arts found in Central Java during the Old Classical era (8th to 10th century AD).

1.3   *Characteristics of narrative relief art in the Old Classical era*

The characteristics of narrative reliefs developed in the Old Classical era (8th to 10th century AD) in Central Java are as follows:

1. The figures are described realistically or with a naturalist shape.
2. The faces of human and animal figures are always oriented toward the observers.
3. On the panel, there is an empty space, but there are no accessories at all.
4. Almost half of the media's depth (e.g., stone) is used for the relief sculpture.
5. A *haut relief* is carved into the stone surface (Bernet, Kempers 1959, pp. 45–46, Munandar 2003, p. 28).

After careful observation, it can be concluded that there are several characteristics closely related to the description of story reliefs in the Old Classical temples. First, each human fig-ure is depicted in a dynamic position, not in a static position, describing a movement. Statues in standing person always bear a smile, depicted as talking or meditating. It means that the impression of such a story relief is alive, teaching a lesson to the observers. Furthermore, the scene depicted is always the front ground, and not background; a far second ground or third background is found in the relief depiction in the temples during the Majapahit era (14th to 15th century) found in Central Java.

Every work of art is always produced using a creative process of its artists. They process the idea and use it in different art media and expressions, such as stone, metal, wood, body movement, and voice. An artwork would have no value if it is not appreciated by society. Therefore, the society becomes an important element in the artistic process. The elements of art are shown in Chart 1.

In the ancient Javanese society, the artworks of Hinduism and Buddhism have five ele-ments, as shown in the chart. However, their idea and concept always come from the lessons of the two religions. A *silpin* that carves the statue of Avalokistesvara from a stone borrows the concept of Bhoddisattva from the lesson of *Buddhadharma Mahayana.* Then, after com-pleting the sacred idea expression and the working process that are equivalent to medita-

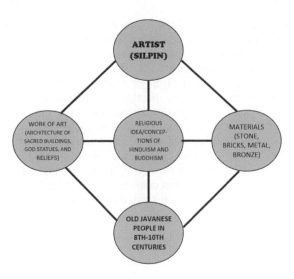

Chart 1. Elements of artwork that are interrelated (Source: Munandar et al. 2009).

tion, the statue is presented to the people for worshipping. The statue is then placed by the adherents of the Buddha Mahayana in a chamber of the temple accompanying the statue of Siddharta Buddha or others. For example, the statue of Avalokitesvara is appreciated by the people when it is used as one of the ritual tools in the worship of *Buddhadharma*. This process of religious artwork creation, according to E. B. Vogler (1948), is called "bound and used art", which developed in Central Java from the 8th to 10th century. "Bound" means that the art is subject to the rules and concepts of Hinduism and Buddhism, and "used" means that it is accepted by the society as a ritual facility, because the work of art produced has to be in accordance with the religious concepts used as the reference of its creation. According to Vogler, "bonded and used" art is difficult to change. Such art can experience some changes if there are several strong influences, such as any forces from outside of the artists. In fact, the artists are actually a group of clergy that is bound to the religious belief they adhere to. To examine the changes of "bound art", a different study needs to be conducted.

Claire Holt (2000), an expert in Indonesian cultural history, found that the art of Hinduism and Buddhism developed in Central Java from the 8th to 10th century had unique features. The statue art in Buddhism and the description of the stories in the relief of Borobudur Temple show the intricate smoothness and composure exhibited by the artists on their artworks. The artworks do not show any tension or dynamic quality on the reliefs in the *Rupadhatu* part of the temple, and all the images are kept in harmony. Nevertheless, the relief scenes on the covered base part (*Kamadhatu*) show a more dynamic movement (Holt 2000, pp. 49–50). Thus, Borobudur Temple shows two different impressions of beauty, and the artists showed their religious idea through their artworks. At Prambanan Temple, the reliefs carved are in fact more dynamic in their composition and more dramatic from the perspective of expressed feelings. It can be concluded that a real stylization symptom at Prambanan is more apparent, which is evident from its relief panel rooms that are more dominant than those of Borobudur Temple.

According to Holt, those who have brought a greater impression will then lead the way to glory (2000, p. 63), which is especially true based on the existing archeological data. The impression resulted from seeing the form of fine art at Borobudur Temple is "a gentle flow of calmness". The calmness itself can be seen from the building of its temple, which is in fact a giant stupa decorating the peak of a hill in the midst of natural scenery and surrounded by mountains. All statues of Tathagata that decorate the niche also show a flow of smoothness leading to *sunyata*. This impression can also be observed in the description of the story reliefs. Although the panel rooms are filled with many figures described, their presence does not appear to be crammed or crowded. In fact, those figures complement each other in a silent narration. Meanwhile, all of the forms of fine art in *Sivagrha* temples have the same objective, that is, to show glory and victory. The three main temples (Temples of Siva, Brahma, and Visnu) are complemented with the temples for worshipping other gods, and 224 *pervara* (ancillary) temples are added surrounding them. The statues are depicted in dynamic and "live" impressions, in the position of either meditation or welcoming the worshippers entering their sacred chambers. The relief portrays several meanings. The *Ramayana* is an epic story presenting the victory of Rama as the good side that beats the evil, which is symbolized by the figure of Rahwana. The story is full of heroism of the god-incarnated knights.

It can be concluded that the main characteristics of the ancient Mataram art created by the *silpin*s (artisans) from the Sailendra dynasty are "devotion and victory". Artworks are a full devotion to god by humans (people and other kingdoms). Both concepts could be maintained until the beginning of the 10th century; however, in the second half of the 10th century, they abandoned Central Java and migrated to the eastern part of Java region thereby bringing a different concept of art.

NOTES

1. The name of King Tejapurnnapanna Panangkarana, according to the Mantyasih inscription (829 Saka/907 AD) issued by King Balitung, was possibly Rakai Panangkaran, who ruled Mataram after King Sanjaya.
2. Because the courtyard of the temples is covered with stone blocks, from an archeological viewpoint, it can be concluded that several activities were held in that yard in the era of the Buddhism. Possibly, these Buddhists followed a kind of ritual of circling the main temple in a *pradaksina* motion. Therefore, in order not to damage the yard by the numerous pilgrims (especially during the rainy season), it had to be paved with the reinforcement structure of stone blocks. In the Hindu-saiva temple complexes, the yard of the temples was not covered with stone blocks. It can be concluded that the religious activities of the Hindu society in that era were not as active as those of the Buddhist society in their temples (Munandar 2016, pp. 23–24). Examples of Buddhist temples paved with stone block structure include Sewu Temple, which has a stone block structure in its first yard that is the closest to the main temple, and the Banyunibo, Plaosan Lor, and Lumbung Temples.

REFERENCES

Boechari. (2012) *Melacak sejarah kuno Indonesia lewat prasasti* (Tracing the ancient history of Indonesia through inscriptions). Jakarta: Kepustakaan Populer Gramedia.

Holt, C. (2000) *Melacak jejak perkembangan seni di Indonesia* (Tracing the art development footsteps in Indonesia). (R.M. Soedarsono, Trans.). Bandung: Masyarakat Seni Pertunjukan Indonesia (Society of Performing Arts in Indonesia) & arti. line.

Kempers, A.J.B. (1959) *Ancient Indonesian art.* Amsterdam: C.P.J.van der Peet.

Munandar, A.A, et al. (2009) *Lukisan Basoeki Abdullah tema dongeng, legenda, mitos dan tokoh* (Basoeki Abdullah's paintings with the themes fairytales, legends, myths, and famous figures). Jakarta: Museum Basoeki Abdullah.

Munandar, A.A. (2003) *Aksamala: Bunga rampai karya penelitian* (Aksamala: Research work anthology). Seri Kajian Arkeologi (Archeological Studies Series). Bogor: Akademia.

Munandar, A.A. (2016) Candi-candi Buddha di Jawa: Bukti perkembangan agama Buddha Mahayana dalam masyarakat Jawa kuno (Buddhism Temples in Java: Evidence of Buddha Mahayana religion development in ancient Javanese society). [Symposium] *Muara Jambi and Sriwijaya Saturday, 14 May 2016 for Festival Waisak for Indonesia 28th April—29th Mei 2016, NAS Building, Fl. 2, Pasar Seni, Taman Impian Jaya Ancol, North Jakarta.*

Poerbatjaraka, R.M.Ng. (1952) *Riwayat Indonesia I* (History of Indonesia I). Djakarta: Jajasan Pembangunan.

Global economy, urbanization and social change

Cultural Dynamics in a Globalized World – Budianta et al. (Eds)
© 2018 Taylor & Francis Group, London, ISBN 978-1-138-62664-5

Foreword by Abidin Kusno

The essays in this section are put together under the theme of "global economy, urbanization and social change." Readers however will notice that they deal with a variety of subject matters characterized by the disciplines within which they are embedded. They are nevertheless linked together, discursively, by their shared interest in humanistic approaches, and perhaps more importantly, in their profound interest in issues facing Indonesia today. All of their concerns are related to Indonesia even though some of the cases look foreign or transnational (such as code-switching in French language or epistemology in economics). They are also characteristically urban as represented explicitly in some case studies, but in most cases the urban is an implicit framework of thought. We don't see much a discussion on agricultural or rural Indonesia, but we see a strong interest in social dynamics especially those that are related to the social and technological change associated with the city life.

All the essays indicate that Indonesia is worlding for much of its existence. Finally, it is more appropriate than before to say that the percentage of the population living in cities is greater than ever. The country is encountering an urban generation that recognizes the full potential of the city as the crucible of social change, the locus of whatever is dynamic, and the playground of the global economy as well as the site for state disciplinary practices. The essays collected here could be said as representing such an urban generation. They are written by scholars interested in modernity and cosmopolitanism of Indonesia which they seek to preserve and develop in their scholarship.

The (Under)Currents

There is thus a number of essays that focus on the maritime cultures of Indonesia, and they all seek to capture the oceanic feeling of the archipelago, when the surrounding seas integrate rather than isolate islands. From a historical and archeological perspective, they offer a maritime history of Nusantara before European hegemony. For students of urbanism, such maritime history could be seen as an attempt to revise Eurocentric scholarship which claims that European colonialism laid the foundation of contemporary cities. The linguistic toponymic approaches tease out evidences that shows the power of indigenous sea trade which contributed to the formation and transformation of early port cities. The focus on Eastern maritime of Indonesia is also interesting considering the domination of Java and its land-based polity. Writing from the perspective of the waterway in the outer islands where currents move back and forth beyond Java, the profile of Nusantara appears to be larger than Indonesia.

In most of the essays, the agency of sea people is emphasized to show how local social institutions nurtured trade in their engagement with the world-economy; how statecraft intertwined with resource management. The essay on Bugis Makassar's boat tradition, for instance, shows how the outside world had long been part of the islands and how the fishermen community took advantage of technology and economic management in order to gain prosperity. The driving force behind the will to modernize was the fishermen's interest in material wealth, as represented in their investment in gold. Such a specific narrative of progress and modernity could be said as an attempt to write an autonomous history of the region even though, as all the essays indicate, Indonesian maritime cultures are part of a network of broader global economic enterprises.

Colonial and Postcolonial Cities

The voyage cultures of Indonesia also offer a ground to understand Indonesian cosmopolitanism and the colonial reactions to it, which take the form of spatial ordering of colonial "plural societies" and inter-imperial maritime treaties. One of the most dynamic approaches to the study of colonial and postcolonial cities has been the concept of governmentality. The interface between the norm and the form of the city produces a kind of government that seeks to organize populations and territories that seem unconnected to formal political institutions. How the materialities of space have consequences for political processes are recognized by authors in the essays dealing with colonial and postcolonial cities.

To govern a city is to create a map for the city, one that is not only to connect places, but also to control the flows of transports and people. The modernization of the commuter train for instance is ultimately about the disciplining of the urban population but it also involves in the "creative destruction" of communities and settlements that stand in the way of the network. The technology of ticketing is also part of urban pedagogy in which the state makes its appearance. The emphasis on governmentality however does not neglect questions of resistances. A number of articles pays attention to the impact of the change in the physical environment on the social and political consciousness of both commoners and elites. Others show that governmental control can never be total as shown in the case of a more decentralized popular transports such as *ojek* and *uber* which often use cognitive mapping to go around the city (and the *kampung*). The case of *uber* also shows the city as a space of social conflict which at once complicates class formation. Taxi drivers are workers who do not own car, but riding a taxi is expensive and affordable only for the upper-middle class.

The authors on colonial cities emphasize the agency of the colonized people by recognizing them as citizens and activists. The colonized were denied membership in the municipal council but they managed to engage in city politics at the community levels. Finally, it is important to point out that postcolonial cities are not always more liberating. An essay in this volume shows just how a community (in the domain of civil society) could be formed not only by class but also by explicit religious affiliation. As indicated in the essay on Islamic housing, the Qoryatus-salam Sani, one of the several Islamic housing real estates in Depok at the outskirt of Jakarta, has become the site for the reproduction of a religious ideology on which the exclusivity of the community is based. Here, residents are expected to conform to the norms spatialized in the housing complex. The authors show how such practices are resisted by some inhabitants but they raise concerns over the course of religious branding that is not simply motivated by property marketing but by place-making that is at once religious and political.

Language and Representations

One of the characteristics of the urban generation is their interest in the doubling of a language and the modern medium by which it is conveyed. Doubling here means the material qualities that seem to be more important than the functioning of the language as a speech act. Much like the urban that is not only a container of activities, but as a stage for performing status update, social media allows code switching that liberates linguistic code from the propriety of language. Several essays in this collection explore precisely the interface of language and status upgrade via social media. They look at how Facebook plays out the doubling of language that allows a particular language to move in and out its appropriate domain. The Indonesia-English code-switching, for instance, rearranges the hierarchy of the social world. The result is a performance of one's identity by way of identification with other languages all in and through the technology of Facebook.

The move between the technology and the social is central to the essays on virtual communications. The technology of Instagram for instance brings not only people, but also objects and feelings together. The essay on culinary branding shows precisely how the marketing of cuisine forms identity that is at once modern and traditional. Central to the study is the way traditional cultural practices that link food to kinship and family come to represent the possibility of being modern. Such possibility is linked to Instagram which integrates space and time and by doing so enable culture to be imagined as something that is both traditional and modern. This theme of cosmopolitanism is further explored in a series of essays on Facebook as the aesthetic of linguistic practices that moves a particular language beyond its provincial domain. French language (the elegance of which is in fact based on its parochialism), for instance, undergoes an alteration and expansion in order to be part of the world language.

Some of the essays consider the cultural aspects of linguistic turn, such as how a foreign language is localized to appear Indonesian. And such cultural refashioning has also shaped approaches to data and social science approaches: "Indonesians like the data talk for themselves in order to avoid a potential conflict and hence to maintain harmony between the writer and readers." An essay on the unit of analysis in social sciences goes as far as to reveal the social and moral assumptions of scientific approaches. It raises the possibility of a humanities approach to overcome the scientific pretension of the discipline of economics. Another dimension to speech and textual expressions offers a way to think about the mental appreciation and bodily reaction to lived reality.

Some of the essays investigate the domain of somatic experience (beyond language) as in the case of children with autism and the use of cohesive devices in deaf children as a way to understand their comprehension of the world. Their findings offer insights into the role of the visual, and by implication the visuality of the city in helping deaf children to self-construct their world.

Collecting Cultures

The storage of knowledge is not a cabinet of curiosity. Much like the invention of tradition, knowledge is curated, catalogued and organized in a manner that it became possible to see and understand the world. Library serves such function of knowledge construction. Its cataloguing of materials visualizes certain knowledge while exclude or marginalize others. In the context of the world after Google, however, it is no longer clear if the library still retains the aura of the embodiment of knowledge. The last set of essays raises such a question. There is something particularly Indonesian in the choice of the topics, for Indonesians today read more than before partly because they like to read social media. The world of social media has changed the culture of reading and it thus poses a question of whether library has a future. Yet, as some essays indicate, the inter-university competition is ironically taking the form of library building as the icon of the university. One could just consider the investment in the architecture of the library of the Universitas Indonesia.

Library while has becoming less authoritative continues to stand for what it means to be a university. Furthermore, as an essay in this collection indicates, university politics are increasingly organized around the roles and performances of library. The inter-library competition between universities is after all a manifestation of neoliberal idea that has penetrated the university system. Some of the essays here point to the influence of neoliberalism in Indonesian education as stemming from the late Suharto era which continues to the present. The tension between nationalism and neoliberalism, we could add, is part of urbanization which demands the education sector to take certain position in the market field. President Jokowi's emphasis on nationalism in education is part of an attempt to restore selfhood and community, one that would not be subsumed by the demand of the neoliberal urban.

Jokowi's nationalism is in some way a restoration of self and community, one that he envisions in terms of cosmopolitanism: the capacity to transcend "ethnicity" to achieve "cultural" identity. A set of essays in this collection deals with just such question of identity and difference, of ethnicity and culture. The revival of rituals in the multi-ethnic Bagan Siapi-api and the display of regional ornaments in restaurants, much like the images of cultures in the Instagram are all part of the attempts to apprehend what it is to be modern, mutable and progressive that is, how to become Indonesians, and how to re-worlding Indonesia.

Abidin Kusno
York University, Canada

Cultural Dynamics in a Globalized World – Budianta et al. (Eds)
© *2018 Taylor & Francis Group, London, ISBN 978-1-138-62664-5*

A toponymy based study on waterway trade in the ancient Mataram era (9th–11th century AD)

G.A. Khakam & N.S. Tedjowasono
Department of Archaeology, Faculty of Humanities, Universitas Indonesia, Depok, Indonesia

ABSTRACT: The ancient Mataram Kingdom in Central Java is one of the Hindhu–Buddhist kingdoms that had several written inscriptions. Most inscriptions about the management of water transportation were issued by King Balitung (9th century), King Sindhok (10th century) and King Airlangga (11th century). These inscriptions contained information about the development of patterns of trade conducted through river ways, mainly Solo and Berantas Rivers, as the main trade lanes. Archaeological and geographical evidences found in the areas surrounding the aforementioned two major rivers indicate attempts made to improve the economy by trading via waterways that had been explored at that time. Evidences found around the two rivers indicate existence of small port areas in the ancient Mataram era. In addition, on the basis of the toponymy of the areas around the harbours, this study reveals the functions of the ports. The findings of this study reveal the methods of using tracks and patterns of waterway transportation in trading at that time.

1 INTRODUCTION

Trading has always been contributed to the existence of kingdoms in the Indonesian Archipelago (Nusantara), especially in the ancient Mataram era that set its centre in Central Java and East Java (Soejono & Leirissa, 2010). Evidences of the long-lasting trade in the forms of written inscriptions and artefacts indicate that large rivers served as vital trading routes. In order to manage the trading routes, it was necessary to extend the power to regions that could develop prosperous trade on both regional and international scales, not to mention that Java as a transitional area was located in the region where commodities from the East and the West met (Meilink-Roelofsz, 2016). Evidence of the extension of power of the kingdom can be found in the inscriptions describing a conquest of an area developed by the king or a high-ranking official in charge of the area. For example, there was a position mentioned in the Kubu-Kubu inscription, stipulated by King Balitung and dated 827 Śaka or 905 AD, to specifically administer the areas under the king's colony, commonly referred to as *Juru ni Majajahan.*[1] As the king invaded to expand his colony, it was necessary to hold high-ranking officials responsible for the colonies, so that the kingdom's political and trading activities remain unaffected.

It is common for a kingdom to invade other areas outside its territory. In this case, trading was the main reason for the invasion by the kingdom, especially when the ancient Mataram era was still prevalent in Central Java, and the kingdom left a heritage of evidences in the form of inscriptions found in East Java (Soejono & Leirissa, 2010). On the basis of the data from the inscriptions, efforts to expand the colony, including the economic expansion, started during the period of Sri Maharaja Rakai Watukura Dyah Balitung Sri Dharmmodaya Mahasambhu, who had left written evidence in the form of inscriptions found in East Java, to be exact, behind the statue of Ganesha in Ketanen village (dated 826 Śaka) and in Kinewu village

---

1. The word *juru* means an intellectual person in charge of a task (Zoetmulder, 2006), while *majajahan* means a territory that has become the colony of a state or kingdom (Zoetmulder, 2006).

(dated 829 Śaka). He was later followed by King Daksa, Dyah Tulodong, Dyah Wawa and finally Pu Sindhok, who ruled the ancient Mataram in East Java (Boechari, 2012).

Many scholars opined on the displacement of the ancient Mataram from Central Java to East Java. According to de Casparis (1958), the displacement was due to the invasion of the Sriwijaya Kingdom. However, according to Schrieke, the displacement was due to the following two reasons: (1) the fact that the people were forced to build religious buildings in a relatively short time and (2) the strategic potential of the Berantas River area as an international trading area (Rahardjo, 2002; Boechari, 2012). Boechari (2012) suggested that the displacement of the centre of the kingdom was due to a volcanic eruption or probably an earthquake that had destroyed it. Another reason might have been an economic factor. Natural disaster often occurred in the territory of the kingdom, as described in the Rukam inscription dated to 829 Śaka, which also included information about the awarding of *sima* status to the king's grandmother in Rukam village due to the natural disasters that often occurred there (Nastiti, et al., 1982; Darmosoetopo, 2003; Soejono & Leirissa, 2010). Agus Aris Munandar suggested that the displacement of the centre of the kingdom to East Java was due to its adjacency to a sacred site near Mount Penanggungan, which was regarded as *pawitra* (Susanti, 2008).

Although scholars differ in their opinions, there is also a common factor, namely the economic or trading factor. The strategic location of Berantas River as part of an international trade zone as suggested by Schrieke, the invasion of the Sriwijaya Kingdom that aspired to manage the trade route as suggested by de Casparis and the economic factor as suggested by Boechari confirmed that the control over the trade route was the major reason behind the displacement of the centre of the ancient Mataram Kingdom from Central Java to East Java.

2 ISSUES AND OBJECTIVES

Several opinions have been provided regarding the displacement of the centre of the ancient Mataram Kingdom from Central Java to East Java; however, the economic factor mentioned above has been the most significant. The expansion towards the eastern part of Java had started since the period from the era of King Balitung to the displacement brought about under the leadership of King Pu Sindhok. Control over trade route in the Berantas River zone was a must, and several written inscriptions had suggested such considerations along the way to the area, and it was mentioned in the names of the villages where inscriptions were found. Names of villages and trade-related facilities can also reveal the attempts made to control the trade route as early as the reigning era of King Balitung. In this case, toponymic study can help identify the names of places written in the inscriptions as evidence of attempts to rule the area in East Java in order to gain control over the trade route at the time.

The objective of this study is to determine the reasons for the displacement of the centre of the ancient Mataram Kingdom from Central Java to East Java during the ruling period of King Balitung, through toponymic approach.

3 RESEARCH METHOD

Similarly to other archaeological studies, according to Deetz (1967), in this case, there are three supporting stages, namely data collection, data processing and data interpretation. The objective of the data collection stage is to collect all data, including those gathered from the field and taken from relevant literature works. The objective of the data processing stage is to find elements that suit the theme of the research. The data interpretation stage aims at discussing the results obtained from the previous two stages, data collection and data processing, and providing a context according to the research theme in the form of a conclusion.

3.1 *Data collection*

Data collection is intended to find all aspects related to the theme. The two main targets of this stage are field data and literature. Collecting data related to the inscriptions originated from the ancient Mataram Kingdom from the 9th to 11th century, especially during the period of King Balitung, which remained in Central Java until the ruling period of King Pu Sindhok, who displaced the centre of the kingdom to East Java was the objective of data collection stage.

The data collection and literature study involve collection of written sources from scholars or reviews related to the theme of research. Review of literature includes reviewing data collected from the ancient Mataram Kingdom regarding the displacement of the centre of the kingdom, trading, transportation and the names of the villages under the ancient Mataram Kingdom.

3.2 *Data processing*

In general, data processing involves searching elements related to the theme of the study. In this study, the elements used as data are the names of villages mentioned in the inscription, means of transportations in the trading activities of the kingdom and notes regarding the place where the inscriptions were found.

After the target elements are found, the next steps were to analyse the names of the villages and to study the current names of the villages that are assumed to be the trading centres in the ancient Mataram era, using the toponymic approach.

3.3 *Data interpretation*

After data collection and data processing takes place data interpretation. The objective of data interpretation was to provide a description or discuss about the data obtained from the previous stages and use them as additional data regarding the cultural history of the ancient Mataram era, particularly regarding the displacement of the centre of the kingdom to gain access to the strategic location for trading.

4 DISCUSSION

Inscription is a form of information written on stone or metal (Boechari, 2012). On the basis of the context, several types of inscriptions were available. For example, Jayapattra is an inscription that records matters regarding court rulings.[2] Normally, these inscriptions contained information about the protest of people against tax collectors as found in the Luitan inscription (dated 823 Śaka) and in the Palêpangan inscription (dated 828 Śaka) or about the disputes among people living in a village as found in the Guntur inscription (dated 829 Śaka) (Soejono & Leirissa, 2010). Another type of inscriptions is in the form of a long manuscript or those containing the year and the *śima* status. The *śima* type of inscription describes the awarding of the *perdikan* (special) status to a land or a village as a gift from the king for providing assistance or benefitting the kingdom through supporting sacred construction.[3]

The awarding of *śima* status to a village can provide a description about the village and the villages surrounding it and about the main materials of the commodities produced by the traders subjected to taxes. This was written in the *śima* type of inscription. The description of villages and their surrounding areas was included in the awarding of *pasek-pasek* within the *śima* awarding ritual, and the recipients of *śima* award invited officials of the surrounding villages to the event (Wurjantoro, 1981). In the inscription, some villages are indicating by their original names, whereas some others with the altered names.

4.1 *Waterway trade during the ancient Mataram era*

As mentioned above, trading in the ancient Mataram era was divided into two types, namely overland trade, which involved the role of markets, and trading activities through waterway access, which took place near ports. Trade by waterways, as stated by Nastiti, used streams of two major

---

2. Boechari, *op.cit. p. 4.*
3. Boechari, *op.cit. p. 6.*

675

rivers, namely Bengawan Solo and Berantas.[4] As such, water transportation means were required for trading via waterways in the ancient Mataram era. Several inscriptions issued in that era mentioned the use of boats for trade. These data are provided below:

4.2 Sangguran inscription

This inscription was issued by Rakai Sumba Dyah Wawa in year 850 Śaka. It mentions the use of a boat in trading, as specified below:

> ... parahu 1 ma(s)u(ŋ)hara (3) tanpatundāna.... (Wurjantoro, 2011: 256)

Translation:

> one boat [parahu 1] having three masts [ma(s)u(ŋ)hara (3)], having no deck [tanpatundāna]. (Wurjantoro, 2011: 263)

4.3 Gulung-Gulung inscription

This inscription was found in the village of Singosari, Malang. Information about the type of boat used for trading is provided in this inscription, as mentioned below:

> ... hajinya samangkana ikanang barahu pawalijān 1 ma suŋhara 2 tanpa tundāna.... (Wurjantoro, 2011: 256; Trigangga, 2003: 11)

Translation:

> ... the King, so was this traders' boat [samaṅkana °ikanaŋ barahu[5] pawalija 1] of 2 masts [ma su haarā[6] 2] without any deck [tanpa tundana]... (Trigangga, 2003: 42)

4.4 Sarangan inscription

This stone inscription was found in Mojokerto, East Java. Information about the type of boat used for trading is provided in this inscription, as mentioned below:

> ... lan °undahagi satuhan parahu masuhara 3 tanpatundāna.... (Prihatmoko, 2011: 41)

Translation:

> ... and the woodcraft artisans (the limit) 1 leader/chief, a river boat with mast (the limit) 3 without any deck.... (Prihatmoko, 2011: 41).

4.5 Linggasuntan inscription

This inscription was found in the village of Lawajati, Malang, East Java. It mentions the use of a boat in trading, as specified below:

> ... parahu 1 masuŋhara 3 tan patundāna .... (Wurjantoro, 2011: 270)

Translation:

> ... 1 river boat with mast (the limit) 3 without any deck.... (Trigangga, 2003: 52)

4.6 Turyyan inscription

This inscription was found in the village of Tanggung, Blitar, East Java. It mentions the use of a boat in trading, as specified below:

> ... parahu 1 magalaha 3 tanpa tundana .... (Casparis, 1987: 45)

---

4. Nastiti, op.cit. p. 48.
5. The word barahu should be parahu.
6. The word ma su haara is written in the inscription as masuṅhara.

Translation:

... one boat with 3 masts that have no deck [parahu magalaha 3 tanpa tuṇḍana....
(Prihatmoko, 2011: 43)

4.7 Jeru-jeru inscription

This inscription was found in the village of Singosari, Malang, East Java. It mentions the use
of a boat in trading, as specified below:

... parahu pawalijan 1. masuṅhar 2 tanpat uṇḍana.... (Wurjantoro, 2011: 286–287)

Translation:

... river boat with mast (the limit) 2 without any deck.... (Prihatmoko, 2011: 44)

4.8 Wimalasrama inscription

This inscription was found in the delta of Berantas River, Sidoarjo, East Java. It mentions
various traders who used boats as their means of transportation:

... maramwan tlung ramwan, parahu 6, masunghara 6 kunang, ikang hiliran 6, aki-
rim agong, akirim tambatamba 6, amayang 6, amuket kakap 6 amuket krp 6, atadah
6 anglamboan 6 amaring 6, anglam, 6 amuntamunta 6, 6 puket dago, 6 kirm dwal
baryyan 6, kirim panjang 6, anglahan 6, añjala 6, anjalawirawir 6, anjala besar 6,
amuwumuwu 6, amintur 6, anjaring balanak 6, jaringkwang-kwang 6, karing kakab 6
amibit 6, waring sugu 6, waring tundung 6, waring tadah 6, anghilihili 6, i kata kabaih
tan kaknana ya soddhara haji, kunang ikang kangkapan, wlah galah 6, kalima tundan,
parahu panawa kalima tundan, parahu pakbowan sawiji kapat tundan, parahu jurag 5,
parahu panggagaran 5, parahu pawalijan 5, parahu pangngayan 5, mwangapadaganga,
kunang wwitaning padagang bhataran pamli bha.... (OJO CXII: 245).

Translation:

... traders using raft/rowing boat (the limit) 3, merchants, boats (the limit) 6, riv-
erboat (with a mast?) (the limit) 6, trader in river downstream (the limit) 6, boat
transporting large size goods (the limit) 6, boat transporting medicine (the limit) 6,
fishermen with seine (the limit) 6, fishermen catching snappers (the limit) 6, fisher-
men catching grouper (the limit) 6, atadah (the limit) 6, fishermen with lambo boat
(the limit) 6, amaring (the limit) 6, anglam (the limit) 6, amuntamunta (the limit) 6,
puket dago (the limit) 6, boat transporting crockeries/chinaware (the limit) 6, anglaha
(the limit) 6, fish catcher using fish net (the limit) 6, fish catcher using jalawirawir
(the limit) 6, fish catcher using large fish net (the limit) 6, fish catcher using wuwu (the
limit) 6, crab catchers (the limit) 6, kwangkwang catchers with fish net (the limit) 6,
sugu catcher using waring (the limit) 6, tundung catcher using waring (the limit) 6,
tadah catcher (the limit) 6, anghilihili (the limit) 6, everyone are not subjected to tax
by the hajj, for all groups of traders using boats with oar or pole (the limit) 6 group,
(the limit) 5 with a deck, large boat (the limit) 5 with a deck, cattle (buffalo) trans-
porting boat 1 (group) (the limit) 4 with a deck, boat led by the owner (the limit) 5
panggagaran boat (the limit), trader's boat (the limit) 5, pangngayan boat (the limit)
5 and trader who initially traded for the king.... (Prihatmoko, 2011: 47).

Table 1 presents data regarding all of the seven inscriptions described above.

The above-described seven types of inscriptions that served as sources of data regarding
boat as a means of transportation for trade show that trading activities during the ancient
Mataram era were centred in East Java through the use of waterways. As indicated in Table 1,
inscriptions were found in areas through which Berantas River flowed, covering the regencies
of Nganjuk, Tulungagung, Malang, Blitar, Sidoarjo, Mojokerto, Jombang, Probolinggo and

Table 1. Names of inscriptions mentioning the means of water transportations in the ancient Mataram era.

No.	Name of the inscription	Type	Year	Place of discovery
1	Palebuhan	Copper	849/927	Goreng Gareng, Madiun
2	Sangguran	Stone	850/928	Ngendat, Malang
3	Gulung-Gulung	Stone	851/929	Malang
4	Sarangan	Stone	851/929	Mojokerto
5	Linggasuntan	Stone	851/929	Malang
6	Turryan	Stone	851/929	Blitar
7	Jeru-Jeru	Stone	851/929	Malang
8	Wimalasrama	Copper	–	Sidoarjo

Source: Early Tenth Century Java from the Inscriptions (Jones, 1984).

Table 2. Names of villages mentioned in the inscriptions of the ancient Mataram found in East Java.

No.	Name of the village	Inscription	Present toponymy
1	Sangguran	Sangguran	
2	Sepet	Sangguran	
3	Tugaran	Sangguran	
4	Kajatan	Sangguran	
5	Kdikdi	Sangguran	
6	Wungkaltihang	Sangguran	
7	Wungawunga	Sangguran	
8	Papanahan	Sangguran	
9	Tampur	Sangguran	Tempursari
10	Palebuhan	Palebuhan	
11	Garung	Palebuhan	Garung
12	Gulung-Gulung	Gulung-Gulung	
13	Kusala	Gulung-Gulung	
14	Himad	Gulung-Gulung	
15	Batwan	Gulung-Gulung	Batuan
16	Curu	Gulung-Gulung	Curung Rejo
17	Air Gilang	Gulung-Gulung	Gilang
18	Gapuk	Gulung-Gulung	
19	Kanuruhan	Gulung-Gulung	Kanjuruhan, Kejuron, Malang
20	Sarangan	Sarangan	Sarangan
21	Bangbang	Sarangan	Bambang
22	Kancing	Sarangan	
23	Linggasuntan	Linggasuntan	
24	Sumari	linggasuntan	
25	Wurakutan	Linggasuntan	
26	Mling-Mling	Linggasuntan	
27	Talijungan	Linggasuntan	
28	Pangawan	Linggasuntan	
29	Kanuruhan	Linggasuntan	Kanjuruhan, Kejuron, Malang
30	Turyyan	Turyyan	Turen
31	Kanuruhan	Turyyan	Kanjuruhan, Kejuron, Malang
32	Gurung-gurung	Turyyan	
33	Kanuruhan	Jeru-Jeru	Kanjuruhan, Kejuron, Malang
34	Tampuran	Jeru-Jeru	
35	Pangawan	Jeru-Jeru	
36	Tugaran	Jeru-Jeru	
37	Waharu	Jeru-Jeru	
38	Jeru-Jeru	Jeru-Jeru	Jeru, Turen, Malang

Lumajang and the cities of Surabaya, Sidoarjo, Malang, Blitar, Kediri and Pasuruan (Taqy-uddin, 2016). On the basis of this fact, it can be concluded that the displacement of the ancient Mataram capital from Central Java to East Java would most probably be based on the control of waterways for trading.

4.9 Port cities in the ancient Mataram era

As discussed above, trading during the ancient Mataram era had focused on the waterways, with boats as means of transportation. The above-described seven inscriptions mentioning boats as means of transportation would possibly be near rivers, which facilitated trading activities. The following are names of places mentioned in the inscriptions, that is, villages that participated in the awarding of *śima* status.

The toponymic study that focuses on the names of places (Rais, 2008) plays an important role in positioning a village or a location in the map. As presented in Table 2, 38 names of villages were found in the inscriptions that mention the use of boats for trading. Of the 38 names of villages, only 8 names could be identified according to the current toponymy. This was possibly due to the alteration of names or the merging of hamlets or villages. However, these 38 villages would likely be visited by traders who used boats for transportation.

As specified in the inscriptions used as sources of data for this study, several types of boats were used for transportation, including (1) boats without mast and deck, (2) river boats and (3) large boats. This information shows that trading activities in the ancient Mataram Kingdom was concentrated in East Java, through waterways or rivers in surrounding areas.

5 CONCLUSION

Previous studies have indicated that the control over the West Java area began in the ruling era of Rakai Dyah Balitung, which left six inscriptions in the East Java area. This was followed by other kings up to the displacement of the capital of the ancient Mataram to East Java by King Pu Sindhok. The transfer aimed at controlling the trade routes in the area surrounding Berantas River, which was highly strategic for international trade. The evidence of significantly rapid trading activities reveals the mentioning of various types of boats used to transport different sizes of commodities, as mentioned in the Wimalasrama inscription during the ruling period of King Pu Sindhok.

REFERENCES

Boechari & Wibowo, A.S. (1986) *Prasasti Koleksi Museum Nasional Jilid I*. Jakarta: the National Museum. Translation: Collection of Inscriptions of the National Museum Volume I. Jakarta: the National Museum.
Boechari. (2012) *Melacak Sejarah Kuno Indonesia Lewat Prasasti*. Jakarta: Kepustakaan Populer Gramedia. Translation: Tracing Indonesia's Ancient History through Inscriptions Jakarta: Kepustakaan Populer Gramedia.
Casparis, J.G. de. (1958) *Airlangga*. Presented in the inauguration as a Professor of the Old Indonesian History and the Sanskrit at the Pedagogic Institute of Universitas Airlangga, Malang, 26th April 1958.
Darmosoetopo. (2003) *Śima dan Bangunan Keagamaan di Jawa Abad IX–X TU*. Jogjakarta: Penerbit Prana Pena. Translation: *Śima* and Religious Buildings in Java in IX-X Centuries TU. Jogjakarta: Penerbit Prana Pena.
Deetz. (1967) *Invitation to Archaeology*. New York: The Natural History Press.
Jones, A.M.B. (1984) *Early Tenth Century Java from the Inscriptions*. USA: Foris Publications.
Meilink-Roelofsz, M.A.P. (2016) *Persaingan Eropa & Asia di Nusantara*. Depok: Penerbit Komunitas Bambu. Translation: Competitions of Europe & Asia in Nusantara. Depok: Penerbit Komunitas Bambu.
Nastiti, T.S., et al. (1982) Tiga Prasasti Dari Masa Balitung. Jakarta: Proyek Penelitian Purbakala. Kementerian Pendidikan dan Budaya. Translation: Three Inscriptions from Balitung Era. Jakarta: Proyek Penelitian Purbakala. Ministry of Education and Culture.

Nastiti. (2003) *Pasar Di Jawa Masa Mataram Kuna Abad VIII–XI Masehi.* Jakarta: PT Dunia Pustaka Jaya. Translation: *Markets in Java in the Era of Ancient Mataram VIII–XI AD.* Jakarta: PT Dunia Pustaka Jaya.

Rahardjo, Supratikno. (2011) *Peradaban Jawa Dari Mataram Kuno sampai Majapahit Akhir.* Depok: Komunitas Bambu. Translation: Civilisation of Java from Ancient Mataram to End of Majapahit. Depok: Komunitas Bambu.

Soejono, R.P. & Leirissa, R.Z. (2010) *Sejarah Nasional Indonesia II: Zaman Kuno.* Jakarta: Balai Pustaka. Translation: National History of Indonesia II: Ancient Era. Jakarta: Balai Pustaka.

Susanti, N. (1991) *Raja dan Masalah Perpajakan: Suatu Analisis Simbolik Integratif Jaman Raja Balitung (899–910 Masehi).* Jakarta. [Thesis] Universitas Indonesia. Translation: King and Taxation Problems: An Integrative Symbolic Analysis in King Balitung Era (899–910 AD). Jakarta. [Thesis] Universitas Indonesia.

Susanti, N. (2008) *Perpindahan Pusat Kerajaan Mataram Kuno dari Jawa Tengah ke Jawa Timur.* Translation: Displacement of the Centre of Ancient Mataram Kingdom from Central to East Java. Presented in a Seminar and Book Launching commemorating the One Century of the National Resurgence, Jakarta, 27th to 29th May 2008.

Wurjantoro. (1981) *Wanua i Tpi Siring: Data Prasasti dari jaman Balitung.* In: *Seri Penerbitan Ilmiah 7.* Depok: Fakultas Sastra, Universitas Indonesia. Translation: *Wanua i Tpi Siring*: Data of Inscription from Balitung Era. In: Scientific Publication Series 7. Depok: Faculty of Letters, Universitas Indonesia.

Zoetmulder. (2006) *Kamus Jawa Kuna Indonesia.* Jakarta: PT Gramedia Pustaka. Translation: Dictionary of Indonesia's Ancient Java. Jakarta: PT Gramedia Pustaka.

Cultural Dynamics in a Globalized World – Budianta et al. (Eds)
© 2018 Taylor & Francis Group, London, ISBN 978-1-138-62664-5

Development of Mandar's maritime trade in the early twentieth century

A.R. Hamid
Faculty of Humanities, Universitas Indonesia, Depok, Indonesia

ABSTRACT: In this paper, we explore the rise and fall of the Mandarese maritime network in the Makassar Strait in the early twentieth century. The objective of this study is to address the following three questions using a historical method: (1) how did the Mandarese establish their shipping network? (2) what commodities did the Mandarese trade and how were they distributed? and (3) why the maritime trade declined in the early 1940s? The results show that the establishment of the Mandarese maritime network was supported by cultural, geographical, and economic factors. The Pambauwang sailors mainly carried out shipping activities around the Pambauwang and Majene ports. The four main commodities traded were copra, rice, woven fabric, and dried fish, which were exchanged for other commodities from other ports in the archipelago. The ability to manage the cultural and natural resources was crucial in the growth of their maritime trade. In the early 1940s, Mandar's maritime network declined due to safety issues for ships sailing at sea, which affected their ability to buy and sell commodities. In this context, maritime safety is vital for the survival of Mandar's maritime network.

1   INTRODUCTION

Shipping and trade are inevitable for Indonesia, as more than half of its area is covered by sea with thousands of islands scattered along the archipelago, where many seafarers use boats and ships to transport goods. Some of the famous seafarers from Indonesia are those from the Mandarese community. However, they are often referred to as seafarers from Bugis, Makassar. In fact, their spirit is shown in their collective memory in the following expression: "Later the black eye will be separated from the white, then the sea, the boats, and the Mandarese will split (Rahman, 1988, p. 78)." The Mandarese who lived on the coasts were efficient traders (Nooteboom, 1912, p. 528) as well as Sulawesi's best sailors from the past (ENI, 1918, p. 664).

Unfortunately, Mandarese maritime history has been rarely studied. Some of the early studies on the subject include the studies conducted by Lopa (1982), Liebner (1996), and Alimuddin (2005; 2009). Lopa examined the legal aspect of sailing, whereas others discussed the boat technology in *sandeq* sailboat, which was smaller (weighing up to 3 tonnes) than the Mandarese trading boats in the first half of the twentieth century, which weighed approximately 30 tonnes. This study provides information about the maritime historiography of Eastern Indonesia.

The Mandarese were the people dwelled on the eastern part of the Makassar Strait, Sulawesi, that is, from the west coast of the Gulf of Mandar in the north and to the Gulf of Mamuju in the south. The main source of their income was shipping and trading. They built a maritime network, which covered areas of Sulawesi, Kalimantan, Java, Nusa Tenggara, the Moluccas, Irian, Sumatera, and the Malacca Strait, and the main hub of their network was in the Makassar Strait. The question of how they built the maritime network between 1900 and 1941 remains unexplored. Therefore, in this study, we attempt to address this question as well as the following three questions:

1. How did the Mandarese establish their shipping network?
2. What commodities did the Mandarese trade and how were they distributed?
3. Why the maritime trade declined in the early 1940s?

1.1 Establishment of the maritime network

The establishment of the Mandar maritime network was supported by cultural and geographical factors. The coastal areas of Mandar, particularly Pambauwang, produced no less than hundred boats of various sizes annually, ranging from small boats with an outrigger and loading capacity of 2–4 tonnes (*pakur* and *sandeq*) to large boats without an outrigger and loading capacity of 10–30 tonnes (*padewakang, palari, lete,* and *lambo*). Most of the ships were made by artisans from Mandar and other boat makers were from Ara and Bulukumba. The boats were used for not only trading but also transporting cargo (Mededeelingen, 1909, pp. 673–674; Nooteboom, 1940, p. 33).

The location of the Mandar settlement on the west coast of Sulawesi was very beneficial for utilizing the onshore and wind for sailing. The navigation system of boats was determined by their position, distance, and time. The boats usually sailed close to the mainland (island) using onshore wind, which blew from evening to morning. It is also known as *pelayaran pesisir* (coastal shipping). The boats sailing toward interisland passages depend on the wind from the sea that blows during the day; it is also known as *pelayaran antar pulau* (interisland shipping). In contrast to the onshore wind, which blows throughout the year, there is a steady sea wind known as monsoon, which blows from one direction from July to September (east/southeast wind) and from January to February (westerly wind). The knowledge of this system is well known to the Mandarese sailors, as is recorded in the *lontar pattappingang* from Pambauwang.

Two important ports of Mandar, with regard to the location and wind systems, were Pambauwang during the east monsoon and Majene during the west monsoon (Vuuren, 1920, p. 208). The boats sailed via Pambauwang port to explore the west coast of Sulawesi to the north toward the ports of Mamuju (Lariang, Karossa, Lariang, Budong-budong, and Pasangkayu) and continued to the west to sail through the Makassar Strait to the east coast of Borneo. In addition, some boats from Pambauwang sailed directly to south Kalimantan. Ships entered several ports there, including Pegatan, Banjarmasin, Sea Island, and the surrounding islands, to either sell or buy commodities or to take water and shipping supplies. Boats that entered the east coast sailed to the south coast and further into the eastern part of the Java Sea to the port of Surabaya (Kalimas), Gresik, Probolinggo, Banyuwangi, and others (*PewartaSoerabaia,* 1935). From there, most of the boats continued sailing along the northern coast of Java to Semarang and Batavia and to the west coast of Sumatra (Padang) and the Malacca Strait (Singapore). Sailors who used this route were known by the names of their destination, namely *Paboroneo* (Kalimantan), *Pajawa* (Java), *Papadang* (Sumatra), and *Pattumasik* or *Passala* (Singapore) (Lopa, 1982, p. 50).

Boats from the port of Majene sailed toward south to the ports of Pare-Pare and Makassar. In this voyage, the boats dropped anchor in the coastal areas of western Sulawesi (Pinrang, Barru, and Pangkep) and the small islands around them. After arriving in Makassar, they continued their voyage to the east. On entering the waters of the Selayar islands, the shipping lanes were divided into two paths. One path continued to the east into Banda Sea islands passing Iron Works (now Wakatobi) to the Moluccas Islands and to the west coast of New Guinea (West Papua). Ambon and Ternate were the main bases for boats from Mandar. Therefore, the sailors of this route, including Irian, were called *Paabung* and *Pattaranate*. The second route continued to the south into the Flores Sea up to Nusa Tenggara islands and the main port in Salaparang. The traders in the area were known as *Pasalaparang* (Rahman, 1988, p. 77).

In anticipation of the east or west monsoon, the Mandarese sailors had to quickly complete the preparation to return to Mandar. The boats were docked in different places when they leave and return, as the geography, monsoons winds, and the waves affected the safety of the ships. Boats departing from Pambauwang will return to Mandar and stopped in Majene, and vice versa. This in turn has made the two ports centers for shipping and commerce in Mandar (Sailing Direction, 1937, p. 466).

Mandarese sailors from Pambauwang, especially Bababulo and Luaor, were most active in shipping and commerce (*Pemberita Makassar,* 1924). *Rijsdijk* (1935, p. 9) noted that there were indigenous people who were good at running lucrative businesses of commercial shipping. Statistical data of the number of boats in Mandar in 1938show that there were 946 boats consisting of 437 units (46%) in Pambauwang and 323 units (34%) in Majene. The rest were in Mamuju and Polewali. Then, in 1939, there were 477 units (38%) in Pambauwang and 397 units (32%) in Majene, from 1,250 boats in Mandar (*Statistiek*, 1939, p. 60; 1940, pp. 61–62). Each boat in Pambauwang, especially in Bababulo and Luaor, brought in a commercial capital of 30,000 to 40,000 guilders in one seasonal cruise (*Rijsdijk*, 1935, p. 9).

In addition to the captain and boat crew, there were others, called *punggawa*, who also sailed with them in the voyage, who were under the control of the captain. Each merchant was required to report the amount of capital to the captain (Vuuren, 1917b, p. 333). The captain in Bababulo, Lolo Baharuddin, said that the amount of capital usually differed, but the cargo space and fares (5% of profits) were the same for all traders. Fares did not apply to traders who did not earn.[1]

1.2   *Commodities and distribution*

The Mandar commodities were of three types, namely (1) agricultural products, such as copra, rice, coconut oil, resin, rattan, and cotton; (2) marine products, especially dried fish and sea cucumbers; and (3) handicrafts, mainly woven fabric, boat ropes, screen weaving, rattan mats, and wicker baskets.

With regard to copra trade, two resident assistants of Mandar, Rijsdijk (1935) and Leyds (2006), noted that the whole coastal plain in Mandar consisted of coconut plantation. An estimate suggests that there were approximately 500,000 coconut trees; hence, Mandar was called "the coconut island of Sulawesi". Coconut was processed into copra and coconut oil. Copra was exported to Pare-Pare, Makassar, Java, Singapore, and Borneo, whereas coconut oil was used for local consumption. From 1932 to 1933, the price of copra in Mandar was between f 5 to f 5.5. In 1933, copra exports from Pambauwang reached 90,000 kg.

During the rice harvest season, boats Bababulo and Luaor to Pare-Pare and Pangkajene bought or brought paddy to the region, which was processed into rice by the local women. Furthermore, the rice was sold in markets in Pambauwang, Majene, Tinambung, and Balanipa (Leyds, 2006, p. 73). According to Los (2004), the *Onderafdeeling* head of Pangkajene, most of the rice was transported to Mandar and Borneo and some was taken to Makassar by the Mandarese boats (*Pemberita Makassar,* 1923). From Borneo, the boats returned to carry wood to be sold to the residents in Pangkajene, Pare-Pare, and Mandar. In Mandar, the wood was used to make boats and build houses.

In addition to rice, women produced woven fabric. The high-quality weaving adorned with European crystal patterns on the bottom part of the cloth came from Majene, and the low-quality woven was produced in Sendana (Mededeelingen, 1909, pp. 665–682). The fabric was brought in large quantities to the Moluccas Islands (Ternate, Sula, Bacan, Ambon, Aru, Kei, and Tanimbar) and was exchanged for cash crops such as copra, cloves, nutmeg, resin, and sea products, including sea cucumbers, scallops, and tortoiseshell. Commodities were sold to the Chinese and European traders in Makassar (*De Sumatra Post,* 1918; 1936). In addition to fabric, firearms and ammunition were purchased in Singapore and exported from the Dutch East Indies to the west coast of Irian (*Bataviasch Niewsblad,* 1939). Some Pambauwang traders bought fabric from the market in Balanipa, Pambusuang, Campalagian, and Mapili to sell in Padang and Singapore.

The next commodity was dried fish, caught by the Mandarese fishermen in the Makassar Strait and the eastern part of the Java Sea. Typically, the dried fish processed in the boat for 1 or 2 weeks was sold in nearby ports in East Java and South Kalimantan. In this endeavor, the Mandar fishermen interacted with the fishermen from Madura in sea for the exchange of information on the location of fish and shellfish that they needed. Before returning, the Mandarese fishermen bought household goods at the port and then carried mostly dried fish to sell in the market in Pambauwang and Majene. The dried fish brought from the western coast of Sulawesi was sold in Pare-Pare and Makassar.

Commodities from Mandar were exchanged for industrial commodities such as manufactured goods, grocery items, pottery, glassware, gloves, home appliances, and fishing lines

---

1. Interview with Baharuddin Lolo (88 years old) in Bababulo Majene, 4/2/2016.

brought by boats from Java, Makassar, and Singapore to Borneo, Sulawesi, the Moluccas, and Irian under the barter system. When the boats arrived at a local port, the captain allowed the traders, including the crew, to sell the commodities they carried and purchase new commodities to be sold in the next cruise destination. This commercial commodity management mechanism was run by the Mandarese seafarers.

1.3   The decline of the maritime trade

The decline of the Mandarese maritime trade was due to the disruption in shipping lanes and the difficulty of acquiring and distributing trade commodities as a result of World War II (1942–1945). The Makassar Strait was a battleground between the Allied forces and the Japanese forces, so it was not safe to sail. Warships were passing along the west coast of Sulawesi. Similarly, aircrafts were flying low over settlement areas or only slightly above the palm trees. Therefore, every house had a bunker hole as a hiding place when an aircraft or ship was passing by their settlement.

The boat makers from Mandar were deployed to make boats for the Japanese Army in Tonyaman Polewali. During this period, they no longer made boats for themselves or were properly paid to make boats for the boat owners. Everything was done by force and no fee was paid. The boats sailed across the Makassar Strait in the interisland shipping, except along the beach (coastal shipping) of western Sulawesi, from Polewali to Mamuju and vice versa, to avoid warships at sea.

"The last time my parents went to Singapore was in 1941. After that, they never went there again. That was the time of distress": that was the testimony of Baharuddin Lolo on commercial shipping in Bababulo. Before World War II, in 1942, his father, Lolo, a boat owner and captain sailed to Singapore and Ambon to trade each year on a *palari* boat, named *Parendeng*, with a loading capacity of 20 tonnes. Commodities brought from Singapore could not be sold because the boat could not sail to the Moluccas. Some traders sold them around Sulawesi; however, there were not many buyers, because crops such as copra and rice and crafts such as woven cloth, a medium of exchange in trade at the time, were difficult to find. The Mandarese maritime heyday, known as the "Age of Singapore and Ambon", ended in 1941.

Before 1942, the people of Bababulo produced rice and woven fabric. According to M. Jafar,[2] who sailed with his father when the boat went to Ambon, many people (women) brought fabric to be sold by the boat crew. All of the fabric was recorded by the captain on Bugis Lontar as "Kain Bugis" and sold by the boat crew when the boat arrived at the destination. When the boat returned, the captain gave the proceeds of the sales to the owners in the form of daily staple items purchased in Makassar when the boat dropped anchor there. Meanwhile, the local women, when their husbands went sailing, planted paddy and converted it to rice. This activity lasted from morning to afternoon and even continued until late in the evening. These activities stopped when the Japanese troops arrived in 1942. As a result, no commodity was available for trade boats.

Other commercial commodities that were difficult to obtain were copra and dried fish. The coconut and fruit trees were not abandoned by the farmers; however, during this period, the Japanese banned the residents from igniting fire, which was needed to produce copra through smoking process. Then, when farmers processed oil into kerosene, the activity was limited, and the Japanese seized most of the product. Similarly, although fishermen were not prohibited from fishing, their catch was seized by the Japanese.

The scarcity of the trading commodities had implications on the shipping business. The captain, his crew, and the *punggawa* faced capital constraints because it was difficult to make profit from the trade. The end of the "Singapore and Ambon" era became a turning point of the maritime trade. The critical role of the boat captain was declining: from the leader of shipping and commerce to the leader of shipping only. Shipping lanes were adjusted to the interests of merchants, as users of transport services, which were previously determined solely by the captain.

---

2. Interview with Muhammad Jafar (83 years old) in Bababulo Majene, 22/2/2016.

2   CONCLUSION

The ability of the Mandar people to manage the cultural and natural resources was crucial in establishing a maritime network in the Makassar Strait. This activity was mainly carried out by the Pambauwang, especially Bababulo and Luaor, based in the ports of Pambauwang and Majene. They brought local commodities to exchange for other commodities from other areas on each shipping season. In that process, different trends emerged. The first trend was obtaining commodities from the eastern region commonly done by barter system, and the type of commodities was limited to natural products and very few domestic industrial products such as woven cloth. The second trend was obtaining commodities from the western region, such as industrial products, through cash payment. The trend was managed well and brought prosperity to the Mandarese sailors.

The maritime activities of Mandar began to decline in the early 1940s when World War II broke out (1942–1945). The sea around the archipelago became unsafe for shipping boats. Similarly, the distribution of commodities, as part of the maritime trade, suffered a serious decline. Thus, the maritime safety was an important aspect for the sustainability of shipping and trades for the Mandarese in the Makassar Strait in the first half of the twentieth century (before 1942).

REFERENCES

Alimuddin, M.R. (2005) *Orang Mandar orang laut. (The Mandarese, The Seafarers).* Jakarta, KPG.

Alimuddin, M.R. (2009) *Sandeq: perahu tercepat Nusantara. (Sandeq, The Fastest Boat in Nusantara).* Yogyakarta, Ombak.

Bataviasche Nieuwsblad (7/10/1939).

*De Sumatra Pos*t (2/2/1918; 16/4/1936).

Leyds, W.J. (2006) *Memori Asisten Residen W.J. Leyds. (Memoar of the Resident Assistant W.J.Leyds). (diterjemahkan oleh Hanoch Luhukay & B.E. Tuwanakotta, 2006). (translated by Hanoch Luhukay and B.E, Tuwanakotta, 2006).*

Liebner, H. (1996) *Beberapa catatan tentang pembuatan perahu dan pelayaran di daerah Mandar, Sulawesi Selatan (Laporan penelitian) (Some notes on the boat making and sailing in Mandar, South Sulawesi, a Research Report).* P3MP YIIS, Universitas Hasanuddin.

Lopa, B. (1982) *Hukum laut, pelayaran dan perniagaan. (Law of the Sea, Shipping and Trade).* Bandung, Alumni.

Los, M.A. (2004) *Memori van overgave onderafdeling Pangkajene* (translated). Makassar: Badan Arsip dan Perpustakaan Daerah Sulawesi Selatan. (*Archive and Library Agency of South Sulawesi*).

Mededeelingen. (1909) *Mededeelingen betreffende eenige Mandarsche Landschappen.* Bijdragen tot de Taal,-Land-en Volkenkunde van Nederlandsche Indie, 62, pp. 646–746.

Nooteboom, C. (1912) *Nota van Toelichting bettrefende het Landschap Balangnipa. TBG,* XLI edition.

Nooteboom, C. (1940) *Vaartuigen van Mandar. TBG,* 80, pp. 22–33.

Pemberita Makassar (12/10/1923; 29/12/1924).

Pewarta Soerabaia (9/8/1935).

Rahman, DM. (1988) *Puang dan daeng: kajian sistem nilai budaya orang Balanipa Mandar (Puang and Daeng: Studies in the cultural values of Balanipa Mandarese)* [Dissertation] Ujung Pandang, Universitas Hasanuddin.

Rijsdijk, L.C.J. (1935) *Nota betreffende het landschap Pembauang.* MvO Serie 1e Reel. No.33. Jakarta, ANRI.

Sailing Direction. (1937) *Sailing direction for Celebes, Southeast Borneo, Java, and islands east of Java.* Washington, Government Office, United States.

Statistiek. (1939) *Statistiek van de scheepvaart in Nederlandsch-Indie over het jaar 1938.* Batavia, Drukkerij F.B. Smits.

Statistiek. (1940) *Statistiek van de scheepvaart in Nederlandsch-Indie over het jaar 1939.* Batavia, Drukkerij F.B. Smits.

Vuuren, L. van. (1917a) De prauwvaart van Celebes, *Koloniale Studien,* pp. 107–116.

Vuuren, L. van. (1917b) De prauwvaart van Celebes, *Koloniale Studien,* pp. 229–339.

Vuuren, L. van. (1920) *Het Gouvernement van Celebes.* Deel 1. Den Haag, Encyclopaedisch Bureau.

Cultural Dynamics in a Globalized World – Budianta et al. (Eds)
© *2018 Taylor & Francis Group, London, ISBN 978-1-138-62664-5*

The golden age of the East Indonesian economy during the NIT era (1946–1950)

L. Evita & Abdurakhman
Department of History, Faculty of Humanities, Universitas Indonesia, Depok, Indonesia

ABSTRACT: Federation model is a governance model that aims at developing economic sovereignty and national integration. If the decline of the Majapahit Kingdom was due to the weak economic power in the periphery that facilitated invasion of Muslim traders, then how can national integration be achieved in Indonesia through economic empowerment from the periphery? In this study, we will discuss the economic sovereignty in eastern Indonesia after its independence. The State of East Indonesia (Negara Indonesia Timur/ NIT),which was established in 1946, implemented a comprehensive system of economic openness and a two-way cooperation in all fields, thereby providing financial support to the government of the Republic of the United States of Indonesia (Republik Indonesia Serikat/ RIS). The emergence of the regional economic power in East Indonesia was influenced by the presence of foreign and local companies, including Mandeers, Seeman & Co, Company Insulinde Makassar, Moreoux & Co., Coprafonds and South Pacific Trading Company, and Manado, which flourished the economy in eastern Indonesia. Commodities available at that time were being sought for by the world. Through the disclosure system, the State of East Indonesia revived its glory in the kingdom of Makassar as the pivot of the world maritime economy until 1950. Here, we apply historical methods using archival sources and contemporary newspapers.

1 INTRODUCTION

In this study, we will describe the economy of eastern Indonesia during the NIT era. The fact that eastern Indonesia consists of many small islands is well known. One of the most significant tribes in the region is the Bugis-Makassar, which is famous for its sailors. In addition, they were experts in making boats connecting all islands in Indonesia. In 1947, there were boats measuring $>20m^3$ in NIT, 92% of which came from South Sulawesi. The boats produced by Bugis-Makassar are one of the largest contributors to the Indonesian economy.

Approximately 50% of the boats were destroyed by bombs and bullets during the independence war, thereby stagnating the trade and economy. The following are the efforts made by the government to revive the economy that made NIT experience a golden age.

1.1 *Establishment of fonds of South Sulawesi's boats*

Several attempts were made by the NIT government to revive the economy and trade activities. One of the successful attempts was the establishment of the Fonds of South Sulawesi's Boats agency. It was proposed by Najamuddin Daeng Malewa, the then Prime Minister of NIT. Najamuddin was also a trader from the native Makassar region; therefore, this agency was based in Makassar.

The purpose of the establishment of this agency was to provide loans to repair and make new boats, regardless of whether they were automatic or manual. The following are other purposes of the establishment:

1. Providing loans to makers of boats or motorboats for coastal shipping.
2. Providing loans directly or by intermediaries to finance the goods loaded by boats or motorboats.
3. Providing loans directly or through the items stacked, which would be loaded by boats or motorboats.
4. Establishing programs to educate Indonesian students as well as providing trainings for technicians, sailors, and employees.
5. Advising the government regarding designing the rules concerning sailing between the islands of Indonesia.
6. Establishing the *Veem company*, to which shipbuilding and other companies related to ships and sailing were affiliated by the Republic of the United Indonesian States (Negara Baroe, Monday, 20 October 1947: 2).

This agency received financial assistance amounting to f1,200,000 from the government. In order to revive the past glory of the economy, Najamuddin recommended improving the shipping transportation sector. He proposed that the transportation route, related to both shipping transport and private boats, should be supported by the government as well as the economic development in the fisheries sector. To achieve these targets, Najamuddin further requested the facility aids, such as motor equipment and machinery, to be given to the fishermen, through which he believed that these efforts could increase the number of cruise operations in Indonesia. He expected to increase the people's income by enhancing the shipping facilities.

Source: Collectie Moluks Historisch Museum.

Many measures were taken to increase the number of boats in South Sulawesi, one of which was to establish a company intended specifically to make boats in Tampoena village, in Boetoeng Strait. This area used to be the site of a Japanese airline company, which then became a shipbuilding area due to its suitable land condition. In addition, access to timber and labor was also easier, close to the regional center. Labor costs were still low, and the public facilities such as *pangreh* pradja and hospitals were within reach.

The following are the ship companies that participated in reviving the shipping activities in NIT (Harsidjo, Saleh Sarif, et al. 1953: 378):

1. *N.V Maskapai Kapal Sulawesi Barat*, established in July 1947 in Makassar. The company operated six ships serving the route of Makassar–Raha, Makassar–Selayar, Bone Bay, Makassar–western coast of Sulawesi to Toli-Toli (north).
2. *N.V. Nocemo in Manado,* established in July 1947. The company operated eight ships. It served the route of Manado–north coast of Sulawesi to Toli-Toli, Manado–Tomini Bay, Manado–Sangir/Talaud, and Manado–North Moluccas.
3. *N.V Lokumij in Gorontalo*. It served the route to Tomini Bay. The number of vessels were three.

4. N.V. Pelayaran Nasional in Gorontalo, established in March 1951. It had three ships serving the route to Tomini Bay as well.
5. *N.V. Perkapalan and Pelayaran Indonesia*, Gorontalo, established in April 1947. It operated one boat with a route in Tomini Bay.
6. *N.V. Maskapai Kapal Makmur, in* Gorontalo, established in December 1951, operating one boat.
7. *Al Ichwan Firm in* Gorontalo, having one boat with transporting 100 tonnes in Tomoni Bay.

In order to avoid competition among carrier ships in Gorontalo, according to the instruction of the People's Business Organization Department in Makassar in May 1952, Sulawesi's Combined Shipping was established. In addition, for the purposes of sea transport, seaports were built along the coasts of Sulawesi. One of them was the Makassar port, which is the economic center in the eastern region of Indonesia. In addition, there are seaports built in Pare Pare Bay, Kendari Bay, and Kolonedale Bay, all of which are natural ports frequently visited by large ships.

Makassar Port became the center of the economy in NIT. It is frequently visited by not only KPM ships but also overseas ships from Japan, the Philippines, and Australia (Ibid. p. 378). Table 1 presents a list of destinations of Makassar copra exports in 1949.

Journalists who visited Makassar in 1946 reported that the city was making progress. South Sulawesi had already experienced a new stage of development. The construction was in progress; therefore, a good infrastructure was inevitable. In addition, import of medicines, textiles, building materials, and others was needed. During his meeting with Van Mook in an effort to establish political cooperation, Najamuddin Daeng Malewa requested necessary facilities to expedite the development in South Sulawesi. He wanted to increase exports of copra, rattan, resin, coffee, timber, and nickel from South Sulawesi that could amount to approximately 30 million gulden, intending to cover the cost of imports. In addition, the oil fields located in Buton and Bone could be explored to support the economy. One of the sectors that could be used as well was the agriculture sector. South Sulawesi had a surplus of food (rice), approximately 60,000 tonnes, which could be used to cover the shortage of food throughout eastern Indonesia (Ide Anak Agung Gde Agung, 1985:86).

NIT is slightly different from other areas in terms of the economic pattern. In Java, for example, most of the area was occupied with various industries. On the contrary, in Sulawesi, people were still highly dependent on nature, such as agriculture, plantation, and marine products.

Table 1.   Destination of Makassar copra export.

Destination of exports	Total payload (tonnes)
Dutch	108,038
Germany	8,548
USA	13,310
England	14,731
Belgium	4,064
Czech	2,438
Canada	3,709
South Africa	2,540
Poland	505
Swiss	1,016
Singapore	2,134
Japan	7,112
Total (tonnes)	168,148

Source: A. Rasyid Asba, Kopra Makassar Perebutan Pusat dan Daerah Kajian Sejarah Ekonomi Politik Regional di Indonesia: 208.

Eastern Indonesian seas are rich with fish, with its fishermen inhabiting small villages on the coasts. The production of all seawater fishery in 1946 was no less than 15 million kg with a value of f30 million (Verslag van het Agentschap te Makasser over het jaar 1948: 44). It was realized when marine fisheries were of utmost importance for the people of eastern Indonesia, and the fish stocks were enough for the whole population of Indonesia. It was also important to focus more on the possibility of mechanization in fishing and the application of scientific methods. It is necessary for the marine fisheries department of the state to buy boats in large numbers and provide them to the fishermen.

1.2 Agriculture in NIT

During the NIT era, the availability of food resources in some places, such as Sumba, Flores, and Lombok, was quite sufficient. Rice, corn, and milk powder were delivered to prevent people from starvation. The price of rice in November 1949 remained the same as in previous months. Nevertheless, paddy harvest was not really satisfactory. A prolonged drought and invasion of pests were the reasons behind the harvest failure.

Rice fields in NIT occupied 680,000 hectares, comprising approximately 60% of South Sulawesi. Land planted with dry-field rice was estimated to be approximately 250,000 hectares. The annual production of rice reached 1 million tonnes, usually exceeding the needs of the population. Famous surplus regions were South Sulawesi and Bali-Lombok, which could produce about 75,000 tonnes of rice for both internal consumption and areas outside of NIT before the war. However, there were some rice-deficit areas in NIT, such as Manado and South and North Moluccas (Verslag van het Agentschap te Makasser over het jaar 1948:44).

Comparison of the development of rice production in NIT in the last two years of the golden age with the current development—after a slight setback during the Japanese occupation—shows a great improvement. In the first years, there was an increase in the surplus of rice due to the irrigation schemes as well as the improvement in farming techniques.

Indeed, the focus of the government was shifted to promote rice production for several years. For example, the government realized the opening of rice nursery centers, where the types of rice and a second plant could be chosen by the farmers to be distributed by seed growers. The Sadang project in South Sulawesi during 1947 and 1948 increased the irrigated land by 10,000 hectares (Verslag van het Agentschap te Makasser over het jaar, 1948: 66). After completing approximately 60,000 hectares, it could technically be irrigated. The main focus was on better cultivation methods in paddy fields and fertilization, and the provision of agricultural equipment and rat poison was taken into consideration as well. To stimulate production, agricultural weeding was also practiced. The intervention of the NIT Agricultural Bureau from the Department of Eastern Indonesian Economic Affairs was successful.

No exact figures were reported with regard to corn production. The only area with a surplus of corn was South Sulawesi, which had exported 50,000 tonnes before the war. Corn production before the war in eastern Indonesia reached 456,000 tonnes. At that time, farmers irrigated corn crops with water from the lake. On the contrary, the war that had broken out caused a decline in the production of crops such as nutmeg, coffee, and cloves, whose areas of production also decreased from 24,000 (1940) to 16,500, 28,000 to 25,000, and 3,400 to 3,300 hectares, respectively. The export of nutmeg and mace in 1939 reached 3,335 and 587 tonnes respectively, which dropped to 1,551 and 383 tonnes in 1947 and increased to 2,084 and 473 tonnes in 1948.

The volume of corn harvest in the rainy season in South Sulawesi was estimated to reach 200,000 tonnes, of which 30,000 tonnes was exported. If the "1949 harvest" during the normal dry season had prevailed, then in the following year, there would have been 50,000–70,000 tonnes of corns available for exports compared to zero in 1948. In addition to corn, fruit nuts such as peanuts, soybeans, and mung beans were important produce in eastern Indonesia, as they were could generate surplus for exports, too. Surplus of nuts were usually exported to South Africa (about 5,000 tonnes per year), whereas soybean and mung beans (about 5,000 and 4,000 tonnes per year, respectively) were usually sold interisland.

In 1949, corn crops in South Sulawesi alone expanded to 11,400 hectares, and the same also happened in Minahasa. Wetland and farmland of eastern Indonesia in 1941 reached approximately 680,000 and 250,000 hectares, respectively. The average annual harvest amounted to approximately 1,050,000 tonnes. The surplus of annual rice exports reached approximately 20,000 tonnes. This surplus was a result of various irrigation projects such as the Sadang project located in the northern part of South Sulawesi and the one near Tonsea Lama in Minahasa. Thus, the Sadang project in eastern Indonesia resulted in a surplus of 100–150,000 tonnes in 1953 (Ibid.).

In 1948, a study on the possibility of massive cultivation of beans was conducted in eastern Indonesia. A commission visited the area near Kendari, Flores, and Timor. Another important crop for food supply in NIT was cassava, as it was considered that there may have been opportunities to export tapioca. To summarize, the development of agro-food in the last two years of the golden age in NIT reached the prewar production capacity. Nonetheless, there were food shortages, for example, in Flores and Sumba. Because of the low income of people in that area, government had to provide food for the people. Basically, the food supply in NIT was becoming more satisfying in the last two years of the golden age.

The most important trade crop in NIT was coconut copra, with about 500,000 hectares of land and potential production of approximately 500,000 tonnes, approximately 80% of which were exported. Figure 1 shows the progress of the export of copra in eastern Indonesia during 1946–1949.

Figure 1 illustrates that the volume of copra exports in the NIT increased from 1946 to 1949, as well as in Makassar, although not as significantly as the eastern Indonesia exports. Some increase was also noted in other regions such as Manado.

In addition to copra, other annual crops in NIT were coffee, spices, kapok, and cocoa. Coffee did not grow well due to the Japanese imperialism. The Arabica coffee plants in Sulawesi and Bali also experienced bad harvest. To recover from the crop failure, a large number of nurseries for coffee seeds and seedlings were established. In 1940, the entire coffee plantation area in NIT covering 28,000 hectares (19,000 hectares in Bali and 5,500 hectares in Sulawesi) dropped to 25,000 hectares.

During these years, coffee proved to be a weak export product as its domestic value was higher than the export value. As a result, export of coffee dropped. The plantation areas for cloves also declined from 3,400 hectares before the war to 3,300 hectares. Cloves were also considered as a weak product. Nutmeg plantation also suffered greatly during the war. Many of the plantations were turned into food crop plantations. The areas of plantation declined from 24,000 to 16,500 hectares. While exports in 1939 amounted to 3,335 tonnes of nutmegs and 587 tonnes of mace, in 1947, only 1,551 and 383 tonnes of them were exported,

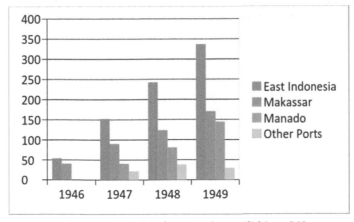

Source: *Sejarah Sulawesi Selatan Jilid 2. p. 243*

Figure 1.   Volume of copra exports in the NIT from 1946 to 1949.

respectively. Before 1948, these figures reached 2,084 and 473 tonnes, which showed a significant rise. Restoration of the plantation was still possible; however, they had to wait for 7–8 years before harvesting.

Cotton crops suffered severe losses during Japanese occupation. After the war, almost nothing was exported from NIT, considering that people had adequate income at that time. Therefore, they were not motivated to pick cotton. One experiment was conducted in this sense, yet it was unsuccessful. On the contrary, cocoa proved to be a profitable commodity. Although cocoa had been grown in eastern Indonesia minimally, it apparently had the potential to be a successful commodity, especially in South and Southeast Sulawesi. The State Agricultural Bureau for Eastern Indonesia attempted to stimulate the planting of cocoa in the country. Having a potentially high demand, with the world supply being low, this product showed good prospects. The first cocoa nursery was opened in Minahasa, Ambon, South Sulawesi, and Timor, and Flores would follow afterward.

In addition to cocoa, tobacco plants yielded good prospects, especially in South Sulawesi, Bali, and Lombok. During this time, in South Sulawesi, experiments were conducted by planting Virginia tobacco as a second crop in open fields. Tobacco plant production before the war was very high in NIT.

Besides corn and nutmeg, wood production in eastern Indonesia could be taken into account. Firewood was considered important as it was used as fuel for transport, as coal was not yet available due to the war. Figure 2 shows the annual production of wood.

Figure 2 shows that the total production of teak wood and firewood increased from 1946 to 1948:

- 1938   20,507 m³ teak wood and 19,477 m³ firewood
- 1946   32,238 m³ teak wood and 22,970 m³ firewood
- 1947   67,075 m³ teak wood and 56,711 m³ firewood
- 1948   72,012m³ teak wood and 71,092 m³ firewood

The increased production of wood was parallel to the rise in the market demand for timber. Timber was used as a substitute for coal as fuel for transportation. Coal became scarce due to the war and hence coal production was impaired.

1.3   Industry in NIT

Before the war, a nickel mine, mostly exporting to Germany, was opened in Kolaka. After independence, the mine was managed by the Oost-Borneo-Mij and Smelting Co. (USA), which continued operating the mine (Negara Baroe, Saturday, 11 October 1947: 2). An important issue was the energy resources to operate the factories. Therefore, to support industry in NIT, power plants were opened in order to provide power to operate the machines. Electricity was majorly generated from runoff water in the NIT region, Sadang, and Teppo near Pin-

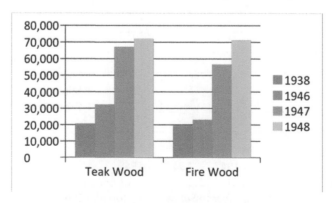

Figure 2.   Production of wood.
Source: *Kedaulatan Rakyat, Tuesday 20th May 1947:1.*

rang. This hydropower would drive the motors to generate electricity power. In addition to Sadang, there were power stations in the cities of Pinrang, Pare Pare, Rappang, Pangkajene, Tanroe Tedong, Sengkang, and Watangsoppeng. There was also one in Minahasa built in Tondano River, which served as a power source for industry.

Another important export product from eastern Indonesia was copra. Approximately 75% of the population was directly involved in the production, export, and trading of copra. A total of 315,000 from 530,000 tonnes of copra exported by the Indies in 1939 came from this region, which was located within the boundaries of NIT. In the same year, copra covered 80% of the total volume of exports and 60% of the total export value of Makassar.

This attracted the attention of the Western countries in the economic sector, and a number of Directors from Lever Brothers & Unilever Limited—a large industrial company in the United Kingdom—visited the region to set up a soap factory to meet the demands of soap in NIT and Kalimantan. NIT, producing more than 320,000 tonnes of copra annually, was a gateway to open a new economic sector in Indonesia. In addition, it facilitated the expansion of other industries, such as garments, tobacco, and animal skin (Negara Baroe, Saturday, 11 October 1947:2).

2 CONCLUSION

The State of East Indonesia (1946–1950) had considerable potentials and strengths due to their economic and finance sectors. As a state of the Republic of the United States of Indonesia, NIT was able to emerge as an economic power from the periphery. NIT was able to fulfill the need for food and agricultural products of its people and export the excess goods to Borneo and Java. The golden period of the NIT economy was facilitated by the government empowering all potential natural and human resources in it. Supported by dynamic policy rules, NIT attracted foreign investments; thus, a network of international economies was formed to enable the products from NIT to be enjoyed by the community and also be exported. In other words, this success contributed to the golden period of the economy of RIS at that time.

REFERENCES

Agung, I.A.A.G. (1985) *Dari Negara Indonesia Timur ke Republik Indonesia Serikat (From the East Indonesian Country to the Republic of the United States of Indonesia)*. Yokyakarta, Gajah Mada University Press.
Arsip Verslag van het Agentschap te Makasser over het jaar 1948.
Harsidjo, Sarif, S., et al. (1953) *Republik Indonesia Propinsi Sulawesi (Sulawesi Province of the Republic of Indonesia)*. Makassar, Djawatan Penerangan RI Propinsi Sulawesi.
Surat Kabar Berita Indonesia. (Friday, 6th January 1950) *Hasil-Hasil Perkebunan dan Sawah NIT (The plantation and paddy field results of NIT)*.
Surat Kabar Berita Indonesia. (Wednesday, 4th November 1949) *Politik Ekonomi RIS (RIS economy politics)*.
Surat Kabar Kedaulatan Rakyat. (Tuesday, 20th May 1947) *Kayu Bakar Mendapat Perhatian (Burning woods getting attention)*.
Surat Kabar Kengpo. (28th March 1950) *Perkembangan Ekonomi di NIT (Economic Development in NIT)*.
Surat Kabar Negara Baroe. (17th July 1947) *N.V Maskapai Kapal Celebes Selatan (Ship Company of South Celebes)*.
Surat Kabar Negara.Baroe. (Friday, 3rd October 1947) *Kekayaan Indonesia Timur (The wealth of East Indonesia)*.
Surat Kabar Negara Baroe. (Monday, 20th October 1947) *Usaha Kementrian Perekonomian NIT di Dalam Memenuhi Kebutuhan Perahu-Perahu (The effort of the NIT Ministry of Economy in fulfilling the need of boats)*
Surat Kabar Negara Baroe. (Saturday, 13th September 1947) *Pemandangan tentang Perekonomian Indonesia Timur (Perspective on the economy of East Indonesia)*.

Surat Kabar Negara Baroe. (Saturday, 31st May 1947) *Perguruan Tinggi Ekonomi di Makassar (The economic higher education in Makassar)*.

Surat Kabar Negara Baru. (Saturday, 11th October 1947) *Kemungkinan perindustrian di NIT (Industry possibility in NIT)*.

Surat Kabar Suluh Rakyat. (Thursday, 5th January 1950) *Pertanian NIT (NIT Agriculture)*.

Surat Kabar Utusan Indonesia. (Friday, 16th June 1950) *Sebab dan Akibat Naik Turunnya Harga Karet (Causes and effects of the price fluctuation of rubber)*.

Surat Kabar Utusan Indonesia. (Tuesday, 30th May 1950) *Rakyat Makassar Sambut Hatta "Industri Kayu dan Kopra Ditinjau" (The people of Makassar welcoming Hatta: "Wood and copra industry reviewed")*.

Cultural Dynamics in a Globalized World – Budianta et al. (Eds)
© *2018 Taylor & Francis Group, London, ISBN 978-1-138-62664-5*

Food culture and land use in ancient times

Taqyuddin & N. Susanti
Department of Archaeology, Faculty of Humanities, Universitas Indonesia, Depok, Indonesia

ABSTRACT: Discovery of ruins or artifacts of ancient times may lead to the discovery of their food culture, land use traditions, and agricultural system. Analyzing from different perspectives and using numerous theories as well as the findings from previous studies on ancient food culture and land use tradition, we could generate new approaches that can enrich and develop archeology and other fields of sciences and technology. From the analysis of the spatial archeology of the location of a temple built during ancient times, we might surmise that people who lived in that location may have had a certain agricultural system. This assumption, however, needs further research. Through epigraphic, toponymic, and geographical analyses; spatial archeological analysis of the locations described in the inscriptions; and by further observing and studying the ruins of the material culture, we could deduce the tradition and culture as well as the food habits at that time, including their preparation and processing methods using the available technology. Although data about the ruins of material culture and associated inscriptions are primarily used to support these findings, study of the environment of the area where these people lived, including its climate, geographical landscapes, and land and soil quality, is also considered as an influencing factor. We study the methods of maintaining, storing, processing, and consuming food to investigate our current tradition and culture, the results of which may serve as an evidence to deduce the past tradition and culture. In this study, we investigate whether the ancient food and land use cultures still exist today and whether they are being properly identified in the island of Java.

1 INTRODUCTION

In this archeological research, we carry out a spatial study on food culture and land use for agricultural purposes across the island of Java; this study needs to be accompanied by other studies on ancient political and economic systems derived from the ancient artifacts found in Java.

Here, we focus on the ancient food culture and the culture of land use and explore how agricultural activities produced food in the ancient Mataram era by studying its culture, knowledge on land use, and spatial identity with regard to agricultural activities.

We also refer the findings of several studies, which generally reaffirm the fact that the power and political stability of the ancient Mataram Kingdom depended more on its agricultural orientation than on other sources of income, such as trading. A few of the artifacts from that era provided the evidence of land use for agriculture, and various archeological studies suggested a strong governmental influence in the management of agricultural land use activities.

A close examination of the numerous written evidence found through many studies gives an indication of food culture and agriculture-related development (land use) as well as how the onsite existing evidence reflects the food sustainability and the culture of land use in the ancient Mataram era.

This study is based on a thorough examination of some ancient writings found in different regions of Mataram (Sumatera, Java, Madura and Bali), which were compiled by L.CH. Damais and published in 1952. A total of 290 partial and 81 complete transcripts and translations were methodically examined.

Spatial verification has also been carried out with regard to the territories of the Mataram Kingdom that encompassed the central part of Java to the eastern part of the island and matched the time line or the chronology of the ruling kings at the time. The facts were verified using inscriptions from the Airlangga Kingdom during the 11th century that encompassed the neighboring territories of Sidoarjo, Jombang, Kediri, to Malang located in the present province of East Java.

Typonomic indications, as mentioned in the inscriptions, have been confirmed and spatially plotted by field studies, which show that the current land and agricultural use reflects the past activities. By using this method, we could better understand the food culture and land use culture in the ancient Mataram era.

1.1 Theoretical view

A theory constitutes a set of thoughts containing general principles that can be used to assume, predict, explain, and complete a scientific analysis. Archeological theories can also be used to examine the food supply, food production methods, and land use in ancient times.

In particular, epigraphic theory is used to describe the culture and livelihood in ancient times on the basis of a valid, complete, and thorough analysis of ancient inscriptions from the Airlangga period. Some of these ancient stone inscriptions were found intact, and others were broken and hard to decipher. To overcome these problems, we employ the service of an epigraphist who can infer the lost data using Sanskrit Mantras.

After obtaining a full set of scripts, we have to translate these ancient scripts to our modern language to reveal their content. In this particular text of the Airlangga Period, we found a record of land autonomous status being awarded as a gift and a guideline about tax exemption policy in the kingdom. The document also contained a magical curse intended for those who disobeyed the king's edict. Time stamps containing year, month, and day were written in full, including a list of names of the king and members of the royal court. The inscription also provides information about the period of the king's reign and the territory under his rule.

In addition to studies, we rely on the theory of location, a geographic setting analysis to indicate the location where certain economic activities were carried out. We would plot the geographical location of the places (toponymy) indicated in the translated texts or inscriptions from the Airlangga Period on a modern-day map. By comparing the toponymy of modern names, we could indicate a particular location down to the lowest level such as a village.

We next discuss about the spatial theory. First, it helps to understand the change and dynamics on the earth surface and then connects physical geographical condition to the changes in human dynamics, cultural behaviors, names of places, and the dynamics of food culture and land use throughout the region.

1.2 Bureaucracy theory

Max Weber Theory of Bureaucracy is important in any archeological study, as it helps to understand the dynamics of duties in a society and the role of government in both ancient and modern eras. Other scholars, including Peter M. Blau and Marshall W. Meyer, described the following six main characteristics of bureaucracy: (1) having a formal job description, (2) a hierarchical structure of authority, (3) execution of tasks and decision-making based on a consistent system of regulation, (4) officials shall carry out their duties professionally, (5) working in a bureaucratic organization constitutes a career level, and (6) administrative staff will play major roles.

Previous studies show that ancient and traditional kingdoms in the Airlangga era applied a type of bureaucratic system, albeit a simple one as viewed by Max Weber (1971: 18–23; cf.

Blau 1970: 141–143), and a solid governmental structure with the best approach and policies in the use of land and water resources.

A cultural viewpoint would lead this study into a living culture, in which people try to fulfill their needs, ancient culture would be discarded by modern society, and old practices would be forgotten because of the development of the culture or buried as a lost civilization due to catastrophic disaster. In most cases, the lost culture and its associated civilization lay the foundation of archeological studies, which would provide a better understanding of the dynamics of our living culture.

A close examination of the manuscripts and artifacts would provide invaluable information with regard to the way of use and method of production of various resources as well as the tools used in agricultural activities, either in its basic form or as a product of an advanced technology. These parameters may be used to determine whether a particular culture is dependent on land-based resources or maritime resources.

2 RESEARCH METHOD

2.1 Data collection method

In this study, we used inscriptions from the Airlangga era discovered in East Java province as the research object, translated by experts in the field. The data, including toponymy, were analyzed to find evidence that can support the study hypothesis on food culture, land use, and their association with the surrounding natural resources.

The toponymy of names of places in the inscription was cartographically plotted on a modern-day map of East Java to determine their locations and positions, in which some of the names could not be identified.

2.2 Study on food culture and land use culture in ancient Java

Identifications of words found on the inscriptions revealed a certain food culture encompassing the raw ingredients used in food, type of food, methods of processing, and utilization. The spatial perspective also includes the attributes of the location based on the thematic distribution map.

2.3 Study on land use in ancient Java

The study of texts in the inscriptions provides clues and identifies the words containing information about land use in ancient times, especially for agriculture use, tool use, mystic rituals confirming the land use for agricultural purposes, the bureaucratic system, policies on taxes, and other related activities. It also provides geographical attributes for a specific location. Then, we will prepare a spatially developed thematic distribution map of land use.

2.4 Study on the utilization concept

Overall, data regarding the utilization of land and other important resources related to agriculture mentioned in the inscription, such as river network, basin, and topographic and geomorphologic aspects, were plotted in the thematic map, assuming that the overall condition has not changed much from the 11th century to the present day.

2.5 Distributional and overlay analysis

Spatial distributional analysis (SDA) was used to provide in-depth details necessary to view the structure of land use for agriculture as described by the inscriptions, including the dimensions of the location, morphometry, and other numerical measurements. Maps of various themes were shown by SDA to transform the information from the inscriptions into a spatial analysis.

Overlay analysis, based on the distributional map of various spatial themes, was conducted to provide more information about specific geographical settings according to the inscriptions. The findings from the overlay analysis were later used to find the pattern of food culture and the culture of land use during the Airlangga era (the ancient Java).

A number of relevant information were provided, including the subjects, tools used, model experiment and design, data collection technique, variables to be measured to support the spatial distributional and overlay analysis.

3 RESEARCH FINDINGS

The data source for this study was inscriptions from the ancient Mataram era and particularly those from the Airlangga era that began from year 941 Saka (1019 AD) to 959 Saka (1037 AD), and consisted of the followings:

1. The Pucangan inscription of 941 Saka (1019 AD) describing the inauguration of Airlangga as the king succeeding Dharmawangsa Teguh who died during a battle against King Wurawari.
2. Cane inscription of 943 Saka (1921 AD) mentioning the king awarding an autonomy and tax exemption status of the people of Cane village (Sima) as the kingdom's western stronghold and the title bestowed upon King Airlangga himself as the Mahapurusa, making him equal to God Vishnu.
3. Kagurukan inscription of 944 Saka (1921 AD) containing information on the granting of Sima status to the family of Dyah Kaki Ngadu for showing their high dedication to the king.
4. Baru inscription of 952 Saka (1030 AD) mentioning the granting of Sima status to Baru Village for providing boarding for the royal troops.
5. Pucangan inscription (in Sanskrit) of 954 Saka (1032 AD) containing the confirmation of the title bestowed to Airlangga as God Shiva using another name of the god, which is Sthanu (line 3).
6. Terep inscription of 954 Saka (1032 AD) containing information about King Airlangga leaving his Wwatan Mas Palace to travel to Patakan.
7. Kamalagyan inscription (Kelagen) containing commemoration about the construction of a dam at Waringin Sapta and the granting of Sima status to the village managing the dam.
8. Turun Hyang inscription of 958 Saka (1036 AD) mentioning the bestowing of Sima status for Han Hyang village.
9. Gandakuti inscription of 964 Saka (1042 AD) stating the status of King Airlangga as the Cakravarttin or the umbrella of the world, which is mentioned in Pucangan, Turun Hyang, and Kamalagyan inscriptions. This inscription (Gandakuti) describes the status of King Airlangga as a priest, indicating that he had renounced his throne.
10. Pamotan inscription of 964 Saka (1042 AD) stating that the center of the Airlangga Kingdom was at Dahana village. The inscription was found in the area of the present Pamotan village, the subdistrict of Sambeng, the district of Mojokerto, East Java.
11. Pandan inscription of 964 Saka (1042 AD) containing information of the granting of Sima status to Pandan village as a reward to the village officials.
12. Pasar Legi inscription of 965 Saka (1943 AD) mentioning the return of Airlangga as a king after chaos had befallen the kingdom.
13. Turun Hyang B inscription of 966 Saka (1944 AD) mentioning about the battle between King Garasakan and King Panjalu and the split of the Airlangga Kingdom into two kingdoms.

The toponymy of locations from the Airlangga era in the lowland and near the seashore starting from the villages of Waringin Sapta, Pamwata, and Kamalgyan is presented in Table 1.

Table 1. Toponymy in Airlangga inscriptions.

No.	Inscription	Year (Saka)	Toponymy found in the inscription	Toponymy at present
1	Pucangan inscription	941	Pugawat	Pucangan
2	Cane inscription	943	Cane	Baru/Cane
3	Kakurugan inscription	944		
4	Baru inscription	952		
5	Pucangan inscription (Sanskrit)	954		
6	Garumukha inscription	954		
7	Terep inscription	954	**Wwatan Mas**, Patakan	Terep
8	Turun Hyang inscription	958		
9	Pucangan inscription (ancient Java)	963		Pucangan
10	Kamalagyan inscription (Kelagen)	964	**Kamalagyan, Waringin Sapta**	Kelagen/Klagen
11	Gandakuti inscription	964	Kembang Sri	
12	Pamwatan/Pamotan inscription	964	**Pamwatan**	Pamotan
13	Pandan inscription	964		Pandan
14	Pasar Legi inscription	965		Pasar Legi
15	Turun Hyang B inscription	966?	**Garaman**, Watak Air Thani, Garun	

Information regarding the contents of inscriptions related to food ingredients is described in Table 2. Food ingredients and beverages include grains, eggs, meat of medium-sized and large animals, fowls, vegetables, tubers, sea fish, and freshwater fish. Drinks consisted of alcoholic drinks, tamarind juice, cane juice, coconut juice, and drinks made from leaves and flowers.

Tools and utensils identified from the inscriptions are found, including stabbing tools, pincers/pliers, needle for manual sewing, cutter, chisel, cooking utensils, eating and drinking utensils, lighting utensils, carpenter tools, plowing tools for working the land, containers, and nail cutter.

Information found is regarding land use for paddy fields, moors, and settlements. A more recent suggestion has been made regarding paddy field with irrigation, water use management, and land use in coastal areas and seashore.

The following is a brief description of the bureaucracy charged with land use management; the officials are assigned to manage the rice barns, regulate water distribution, harvest, and supervise forest conservation. On the basis of data, in early 1387 AD, many officials (*mantri*) were appointed in coastal regions, indicating land utilization near the coastal area.

At that time, people earned their living by working as traders, tools makers (for catching birds and fish), raw food sellers, sugar and beverage makers, cattle raisers, coal miners and sellers, and betel leaf making and selling in complete set with the lime. These professions are characterized by activities related to life on land and no profession has been found mentioning sea activities.

3.1 *River basin and topography in Central to East Java, and their relationship with the ancient agricultural setting*

The great river basins in Central to East Java, namely Bengawan Solo, Brantas, and Progo, uphold the dynamics of the agriculture area and distribution of kingdoms. As indicated in Figure 2, the civilization dynamics associated with the three river basins is shown by the distribution of discovered inscriptions, the distribution of temples associated with ancient Java civilizations (7th to 14th century), and the toponymy derived from the inscriptions found to date. Everything seemed to begin from the river basin of Progo and the surrounding area before moving toward the river basin of Brantas and passing through the river basin of Bengawan Solo in an eastward movement. A tight grouping is seen in the river basin of Progo and the surrounding area, which grows tighter in the river basin of Brantas compared to the river basin of Bengawan Solo.

Table 2. Identification of food ingredients from the ancient Mataram inscriptions.

No.	Food ingredients as mentioned in the inscription	Known name at present	Category of food and drinks
1	*Atak pīhan*	A type of nut	Nut
2	*Wuku*	A type of grain	Grain
3	*Skul/sgu*	Rice	Grain
4	*Hnus*	Squid	Sea animal
5	*Huraṅ*	Shrimp	Sea animal
6	*Gtam*	Crab	Sa animal
7	*Taṅiri*	Spanish mackerel	Sea animal
8	*Kura/capacapa*	Tortoise	Marsh/river animal
9	*Dlag*	Snake head murrel	Marsh/river animal
10	*Kawan*	Grouper	Marsh/river animal
11	*Hangsa*	Goose	Fowl/poultry
12	*Hayam*	Chicken	Fowl/poultry
13	*Tetis*	Squeezed food	Food
14	*Dūh ni nyūn*	Coconut juice	Natural drink made of coconut
15	*Jnu*, pandan (*puḍak*), *bunga/ skar campaga, bunga/skar karamān.*	Raw material used to make wine other than palm sugar/coconut	Drink made of leaves and flowers
16	*Kila/kilaṅ*	Fermented sugar cane	Drink made of cane juice
17	*Ciñca/kiñca*	Tamarind juice	Drink made of tamarind juice
18	*Jātirasa, madya, mastawa, pāṇa, siddhu, tuak/twak*	Drink, usually alcoholic	Alcoholic drink
19	*Biluṅluṅ/bijañjan, halahala, duri, kaḍawas, kaḍiwas, kaṇḍari, layarlayar, prah, rumahan, slar, wagalan*	Types of fish	Types of fish
20	*Haryyas/hinaryyasan/āryya*	Banana stem used as vegetable	Vegetable
21	*Atah atah/ḍuḍutan*	Uncooked vegetable salad	Vegetable
22	*Kuluban/kulub*	Cooked vegetable salad	Vegetable
23	*Rumwarumwah/rumbarumbah*	Vegetable salad	Vegetable
24	*Tahulan*	Fish bone	Food remains
25	*Celeṅ/wök*	Boar	Large game as source of meat
26	*Hadahan/kbo*	Buffalo	Large cattle as source of meat
27	*Kidaṅ/knas*	Deer	Large game as source of meat
28	*Wḍus*	Goat	Goat as source of meat
29	*Hantiga/hantrīni/hantlu*	Egg	Egg
30	*Suṇḍa*	Roots/tubers	Tubers

On the basis of the trend of river basin of Progo, the inhabitants of ancient Mataram apparently settled in the mountainous areas or central part of Java. It continued to the east by people settling in the central region of the island, and they only occupied the lowland areas and even getting closer to the seashore as they entered the river basin area. However, the distribution pattern of the evidences is mostly linear, in line with the river networks, except in part where the river basin of Bengawan Solo is in mountainous areas.

4 DISCUSSION

In this study, we used data sources encompassing inscriptions from the ancient Mataram era, which described the food culture, the type of food and beverages, and food and cooking

Table 3. Names of tools mentioned in the inscriptions (interpretation of the meanings: courtesy of Zoetmulder, Mardiwarsito).

No.	Name of the tool	Present name	No.	Name of the tool	Present name
1	*aṅkup*	Pincers or pliers	19	*paŋhatap*	Not available
2	*Dāŋ*	Copper vessel to hold steamer used to cook rice, or used as a cooking pot	20	*papañjutan*	Lantern/torch/lamp holder
3	*Dom*	Needle	21	*patuk*/patuk patuk	Small axe
4	*Dyun*	Cooking pot	22	*rimbas*	Axe for chopping wood
5	*Gulumi*	Three-pronged spear	23	*saragi °inuman*	Drinking utensil
6	*gurumbhagi/ karumbhagi/ kurumbhāgi*	Knife	24	*saragi pagaṅan inuman*	Eating and drinking utensils
7	*Hampit*	Weapon	25	*saragi pagaṅanan*	*saragi ketel* meaning copper pot or *saragi pagaṅanan* meaning eating utensils
8	*Kampil*	Bag	26	*saragi pewakan*	Pot to keep fish
9	*Kris*	Traditional dagger (keris)	27	*siku siku*	Carpenter's square
10	*kukusan*	Steamer	28	*tahas*	Metal tub or tray
11	*laṇduk*	hoe, spade	29	*tampilan*	Metal utensil
12	*liṅgis*/liŋgis	Crowbar	30	*taraḥ*	Metal tools, small axe
13	*Lukai*	Machete/cleaver/ chopping knife	31	*tarai*	Copper plate/bowl
14	*nakaccheda*	Nail cutter/clipper	32	*taratarah*	Metal tools, small axe
15	*padamaran*	Lamp holder	33	*tataḥ*	Chisel
16	*paliwta/ paŋliwetan*	Rice cooker	34	*tēwēk punukan*	Sharp stabbing Weapon
17	*pamajha*	Present name unknown	35	*waduŋ*	Axe
18	*paṅinaṅan*	Eating utensil	36	*wakyul*	Hoe

utensils used by those people (Tables 1, 2, and 3), providing a strong indication of an agriculture-based economy.

The socio-politics and the system of governance at that time also suggested a developed agricultural system, as can be seen by the large number of government officials (bureaucracy) assigned to the agricultural sector to provide management and administrative services. The established agricultural system also provides people with the infrastructure to earn their livelihood and pursue their professions (Tables 4 and 5).

Using the inscriptions from the Airlangga era, the data from the inscriptions and a geographical data overlay enable us to determine how food ingredients were distributed, land use, and toponymy of the ancient Mataram era. The major agricultural civilization was found along three biggest river basins of the Central and East Java, namely the Bengawan Solo, Progo, and Brantas. These patterns were in fact closely related to water sources, having sustained civilizations on the island of Java from ancient times to the present day.

The dynamic of the ancient Mataram civilization shifted slightly to the east of Java as opposed to the west; however, it was still within the enclosure of the river basin of Bengawan Solo as the main water source and the surrounding lowland that support the paddy fields. A small kingdom was ever established in the eastern part of Java, although the river basin of Brantas was seen as a cradle of civilization giving birth to the Airlangga Kingdom and its advance civilization proposed by this study.

Table 4. Land use as mentioned in the inscriptions.

No.	Year AD	Quote	Present name
1	824	*ika tanī C.A.3*	Village/paddy field
2	840	*lmaḥ C.IIa.1*	Land (Zoetmulder, 2011:583)
3	840	*riŋ piṅgir siriŋ. mwaŋ thani kaniṣṭha. C.VIIb*	Village/paddy field
4	873	*manususk sīma lmaḥwaharu*	Land (Zoetmulder, 2011:583)
5	873	*maṅaran bukit C.Ia.1*	Hill
6	878	*lmaḥniŋ kbu°an karamân °i mamali C.Ia.2*	Land in garden/small field
7	878	*lmaḥsu(kat) C.Ia.1*	Prairie (Boechari, 2012:294)
8	879	*ikanaŋ tgal C.Ia.2*	Open field; unirrigated dry field (Zoetmulder, 2011:1229)
9	879	*sinusuk gawayan sawah maparah śīmā ikanaŋ prāsāda C.Ia.3*	Paddy field (Zoetmulder, 2011:1084)
10	879	*manusuk lmaḥma nīma C.Ia.1*	Land (Zoetmulder, 2011:583)
11	879	*i kwak watak wka tga(l) C.Ia.1*	Open field; unirrigated dry field (Zoetmulder, 2011:1229)
12	881	*manusuk tgal C.Ia.1*	Open field; unirrigated dry field (Zoetmulder, 2011:1229)
13	881	*dadya sawaḥ tampaḥ 2 śīmā niŋ parhyaṅan C.Ia.1*	Paddy field (Zoetmulder, 2011:1084)
14	882	*panusukna lmaḥ C.Ia.3*	Land (Zoetmulder, 2011:583)
15	882	*(a)las dadyakna sawaḥsīmā nya. ikanaŋ lmaḥi ramwi watak halu C.Ia.4*	Forest cleared for paddy field
16	901	*manusuk lmaḥkbuan C.Ia.2*	Garden/small field land
17	901	*muaŋ sīmānya sawaḥlamwit 1 sampuŋ suddha C.VIIa.3*	Paddy field (Zoetmulder, 2011:1084)
18	902;903	*sawaḥkanayakān tampaḥ 7 C.Ia.3*	Paddy field (Zoetmulder, 2011:1084)
19	902;904	*lmaḥrāmanta i paṅgumulan C.IIIb.10*	Land (Zoetmulder, 2011:583)
20	902;905	*muaŋ sawaḥiŋ panilman C.III.10*	Paddy field (Zoetmulder, 2011:1084)
21	907	*wanu°a °i mantyāsiḥwiniḥni sawahnya satū. C.A.2*	Paddy field (Zoetmulder, 2011:1084)
22	907	*(a)lasnya °i muṇḍuan C.A.3*	Forest (Zoetmulder, 2011:23)
23	907	*pasawahannya ri wunut kwaiḥni winiḥnya satū hamat C.A.3*	Paddy field (Zoetmulder, 2011:1084)
24	907	*sawaḥkanayakān. C.A.3*	Paddy field (Zoetmulder, 2011:1084)
25	907	*mu°aŋ °alasnya °i susuṇḍara °i wukir sumwiŋ. C.A.3*	Forest (Zoetmulder, 2011:23)
26	908	*sawaḥ haji lān C.Ib.2*	Paddy field (Zoetmulder, 2011:1084)
27	908	*lu°a/luah/lwah*	River
28	909	*(u)misi anan lēbak gunuŋ tumut upan C.Xa.1*	Valley, lowland (Zoetmulder, 2011,581)
29	909	*Marawairawai C.Xa.1*	Marsh (Zoetmulder, 2011:931)
30	919	*sumusuk ikanaṅ alas C.Ia.2*	Forest (Zoetmulder, 2011:23)
31	929	*inanugrahan lmaḥ C.A.4*	Land (Zoetmulder, 2011:583)
32	929	*sira sawaḥi turyyan mamuat paṅguhan C.A.5*	Paddy field (Zoetmulder, 2011:1084)
33	929	*sawaḥpakaruṇan C.A.9*	Paddy field (Zoetmulder, 2011:1084)
34	931	*nikaŋ lēmaḥi warahu C.Va.2*	Land (Zoetmulder, 2011:583)
35	939	*ikaṅ lmah waruk ryy ālasantan C.Ia.3*	Land (Zoetmulder, 2011:583)
36	940	*rin pomahan kēbuan kēbuan pamli iriya kā 12 C.Ia.4*	Garden (Zoetmulder, 2011:480)
37	941	*dumual ikaṅ lmaḥrāma ryy ālasantan sapasuk banua C.Ia.4*	Land (Zoetmulder, 2011:583)

(*Continued*)

Table 4. (*Continued*).

No.	Year AD	Quote	Present name
38	944	*ikanaŋ lmaḥ C.A.7*	Land (Zoetmulder, 2011:583)
39	945	*irikaŋ luaḥ (lwah) C.A.8*	River
40	1053	*inugrahan sumima thāninya. mantĕn matahila drabya haji riŋ paknakna C.IIb.2*	Village, paddy field
41	1317	*sesiniŋ (saisi niŋ) gaga C.XIIb.3*	Unirrigated paddy field where rice grows in dry field (Zoetmulder, 2011:263)
42	1318	*sesiniŋ sagara C.XIIb.3*	Sea
43	1395	*halalang i gunung lejar. C.A.5*	Tall grass/ weed (Zoetmulder, 2011:328)
44	1396	*alas kakayu C.B.1*	Wood forest
45	1395	*hatuku latĕk. Luputa C.Ia.3*	Swampy, muddy land (Zoetmulder, 2011:576)
46	1386	*tiraḥ C.Ia.1*	Sea shore (Zoemulder, 2011:1260)
47	1387	*karaŋe patiḥ tamba C.Ia.2*	Fish pond (Zoetmulder, 2011:1190)
48	1388	*paŋananewetan sadawata anutug sāgara C.Ia.2*	Flat land (Wurjantoro, 2006:14)
49	1389	*peŋanane kulon babatane C.Ia.3*	Dry field (Wurjantoro, 2006:14)
50	1379	*Tāmbak C.Ia.2*	Fish pond (Zoetmulder, 2011:1190)
51	1379	*sawaḥ C.Ia.2*	Paddy field (Zoetmulder, 2011:1084)
52	1379	*Tgal C.Ia.2*	Open field; unirrigated dry field (Zoetmulder, 2011:1229)
53	1379	*Lĕmbah C.IIa.1*	Valley/flat land (Zoetmulder, 2011:584)
54	1379	*ing tāmbak C.IIa.3*	Fish pond (Zoetmulder, 2011:1190)

Table 5. Officials as found in the inscriptions.

No.	Śaka	AD	Name of the position	Source in the inscriptions	Functions
1	693	771	*pakalangka(ng)*	*D1.A.9*	Official in charge of rice barn
2	762	840	*lĕblĕb*	*D2.IIIb.3*	Official in charge of paddy field irrigation
3	762	840	*pakalaṅkaŋ pakaliṅkiŋ*	*D2.IIIb.3*	Official in charge of rice barn
4	762	840	*puluŋ paḍi*	*D2.IVa.1*	Official in charge of rice affairs
5	795	873	*puluŋ paḍi*	*D3.Ib.2*	Official in charge of rice affairs
6	795	873	*hulu wras*	*D5.IIIa.2*	Official in charge of harvest yield
7	823	901	*huru wras*	*D7.Ib.4*	Official in charge of harvest yield
8	824	902	*saŋ huluwras*	*D8.IIIb.12*	Official in charge of harvest yield
9	827	905	*lĕb ĕlĕb*	D9.IVb.2	Official in charge of paddy field irrigation
10	827	905	*makalaŋkaŋ*	D9.IVb.2	Official in charge of rice barn
11	829	907	*makalaṅkaŋ*	*D10.SiMuk.8*	Official in charge of rice barn
12	837	915	*pakalangkang*	*D11.SiMuk.11*	Official in charge of rice barn
13	851	929	*ḷbḷb*	*D13.SiMuk.17*	Official in charge of paddy field irrigation
14	851	929	*kalangkang*	*D13.SiMuk.20*	Official in charge of rice barn
15	851	929	*pulung padi.*	*D13.SiMuk.20*	Official in charge of rice affairs
16	851	929	*ḷbḷb*	*D14.SiMuk.17*	Official in charge of paddy field irrigation
17	851	929	*kalangkang*	*D14.SiMuk.17*	Official in charge of rice barn
18	851	929	*pasukalas*	*D14.SiMuk.20*	Forest guard
19	851	929	*ḷbblab*	*D15.SiMuk.10*	Official in charge of paddy field irrigation
20	851	929	*kalang(kang)*	*D15.SiMuk.10*	Official in charge of rice barn
21	851	929	*pakaluṅkuŋ*	*D12.Ia.13*	[N/A]
22	851	929	*puluŋpadi*	*D12.Ia.14*	Official in charge of rice affairs

(*Continued*)

Table 5. (*Continued*).

No.	Śaka	AD	Name of the position	Source in the inscriptions	Functions
23	852	930	*kalaṅkang*	D16.SiMuk.7	Official in charge of rice barn
24	853	931	*pakalaṅkaŋ. pakaliṅkiŋ*	D17.IVa.7	Official in charge of rice barn
25	861	939	*lĕblab*	D18.Ia.10	Official in charge of paddy field irrigation
26	861	939	*kalangkaṅ*	D18.Ia.10	Official in charge of rice barn
27	975	1053	**pakalaṅkaŋ**	D19.IIIa.5	Official in charge of rice barn
28	1022	1100	*lĕbalĕb*	D20.IIa.11	Official in charge of paddy field irrigation
29	1022	1100	*pakalaŋkaŋ*	D20.IIa.12	Official in charge of rice barn
30	1245	1323	*pakalaŋkaŋ-pakaliŋkiŋ*	D21.VIb.2	Official in charge of rice barn
31	1245	1323	*puluŋ paḍi*	D21.VIb.3	Official in charge of rice affairs
32	1309	1387	*mantriŋ tiraḥ*	D21.A.1	Official (*mantri*) in coastal area (*tiraḥ*)

Table 6. Professions as found in the inscriptions.

No.	Name of the profession	Meaning
1	*°ahikana*	Food vendor
2	*abakul wwawwahan.*	Fruit vendor
3	*adagang wuṅkuḍu*	*Mengkudu* (noni fruit) vendor
4	*amisaṇḍun/mamisaṇḍu(ŋ) manuk*	Bird trap maker
5	*anĕpis.*	[N/A]
6	*aṅgula/magula/manggula*	Sugar maker
7	*añjariŋ./mañjariŋ*	Animal catcher using net/net maker
8	*añulan/pañulaŋ aṇḍaḥ. añulaŋ kbo. sapi. Celaŋ,wḍus*	Duck, buffalo, cow/ox, boar, goat vendor
9	*Anwaŋ*	[N/A]
10	*Bawaŋ*	Garlic/shallot vendor
11	*haṅapu*	Lime maker
12	*majari*	Animal catcher using net
13	*makala kalā/manuk*	Animal/bird trap maker
14	*maluruŋ/maṅluruṅ/manglurung*	Castor oil maker
15	*mamĕlut/mamulaŋ wlut*	Eel vendor
16	*mamukat*	Fish catcher using net
17	*manahab manuk*	Bird catcher
18	*manaŋkêb.*	Trap maker (usually for birds)
19	*Manawang*	Net maker
20	*Manawaŋ*	[N/A]
21	*Manghapū*	Lime (for betel) maker
22	*manghapū.*	Lime (for betel) maker
23	*manūla w)uṅkuḍu/anulaŋ wuṅkuḍu*	Processing of noni fruit
24	*mañulaŋ haḍaṅan sapi wḍus aṇḍaḥ*	Vendor of food made of beef, buffalo, duck
25	*maŋhapū*	Lime (for betel) maker
26	*maŋharĕŋ*	Coal vendor
27	*pabṛsi/pambṛsi/pamṛsi/mamṛsi*	Maker of a type of drink
28	*pucaŋ sērĕḥ.*	Betel vendor
29	*wli hapū*	Lime (for betel) buyer

The notion about an advanced Airlangga civilization was supported by the morphology of the wide and deep Brantas River, thus enabling it to be used as waterways and major transportation routes in the kingdom; later, it would play an important role in the interaction between the kingdom and foreign civilizations, such as India, China, and Arab during the Airlangga era.

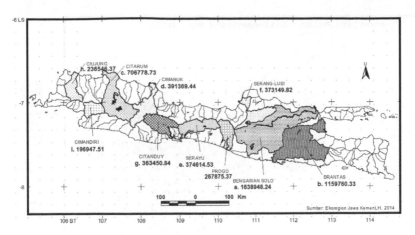

Figure 1.   Ten largest river basins (DAS) in the island of Java.

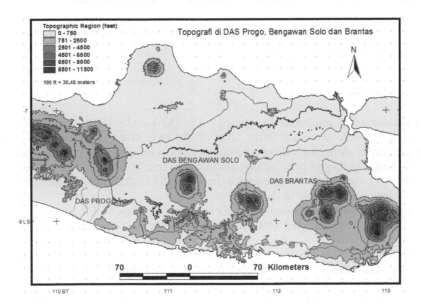

Figure 2.   Topography.

From a much broader perspective, the inscriptions and various analyses conducted on them and the geographical setting indicate that a basic agricultural civilization would progress with river transportation that will result in technology advancement and a full range of economic benefits.

The ancient Mataram civilization in Central Java up to the Airlangga era is viewed from the spatial perspective, particularly its natural resources that encompassed the topography, river networks, river basins, and the geomorphology aspects. In particular, the kings of the ancient Mataram era were adept in managing their natural resources. It would be seen that the crucial aspect in maintaining a great kingdom relies on the ability to provide for its people, to make them prosperous. Some efforts were made to be adaptive to a region, particularly the river basin area or the vast plains and fertile valleys that had sufficient supply of water.

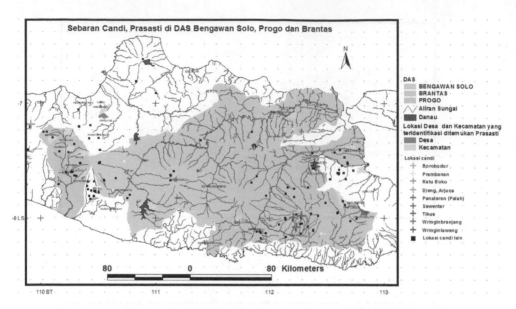

Figure 3.   Association between locations of temples and inscriptions discovered along the river basins of Bengawan Solo, Progo, and Brantas.

5   CONCLUSION

During the Airlangga era before the golden era of the kingdom, the king proposed several measures designed to dominate fertile areas of the river basin of Brantas by conquering smaller kingdoms along the valleys of Brantas to provide sustenance for his people. Subsequently, King Airlangga believed that controlling these areas would guarantee the social stability of his kingdom; he then bestowed special status to these areas, which he considered as loyal subjects and valuable to the kingdom.

Such privileges were also bestowed to officials, their families, and the population of some villages. The kingdom was ruled with a central bureaucracy system, and it developed its economy by using the existing natural resources, as well as by constructing dams to protect the people from floods, especially from Brantas River, and awarding special status to villages that guarded the dams and to those in charge of regulating the drainage system as indicated in the announcement about the appointment of such officials.

Unsurprisingly, the kingdom's agriculture was productive and able to fulfill the need of the people. Judging by the size of the lowland areas and the river water management system using dams, the agricultural productivity of the kingdom was sufficient. The kingdom also exported its agricultural products as well as organized other activities for the people in the period before harvest time. The center of power of the kingdom was close to the estuary of Brantas River, which is strategic and enabling the kingdom to control trading activities on the waterways using boats, as indicated by the ruins of ancient docking facilities in some parts of Brantas.

The culture of land use in the river basin area of Brantas was quite lavish due to its vast lowland plains and river valleys, in which the river basin has a leaf-vein pattern indicating the existence of forest absorbing the rainfall on the surface and preventing it from directly flowing in an accumulated mass to the main river, thereby preventing a drastic amplitude of tidal wave. The case would be different in rivers with an elongated shape and having a short distance between the lengthwise ridges and the main river accommodating the water flow, such as Bengawan Solo (with a tendency of making a parallel pattern) and the Progo River, which has a large gradient (the angle formed by the horizontal line and the highest vertical line), resulting in a strong current. Farmers were benefitted with the advantage of working on flat ground in the lowland with a thick layer of soil and a fine texture that contained volcanic nutrient element and an abundant supply of water. The land use culture was not strong and did not require more energy despite the modest tools available, compared with farming in hilly areas with strong sliding water, thin layer of soil, and coarse to rocky textures, where only certain plants could strive.

The spatial pattern found throughout this era indicates that the people in the Airlangga era utilized the areas in a linear position parallel to the river basin of Brantas on the riversides. It is apparent in the distribution of names in the inscriptions found in the proximity of Brantas River.

REFERENCES

Berger, P. & Luckman, T. (1979) *The Social Construction of Reality: A Treatise in the Sociology of Knowledge*. Penguin Book, New York.

Brandes, JLA. (1913) *Oud Javaansche-Oorkonden, Nagelaten Transscripties. VBG deel LX (Verhandelingen Genootschap van Kunsten en Wetenschappen)* M.Nijhoff,'s Hage.

Branson, J. & Miller, D., & Bourdieu, P. (2002) In: Beilharz, P. *Teori-Teori Sosial: Observasi Kritis terhadap Para Filosof Terkemuka (Social Theories: Critical observation toward the prominent philosophers)*. Yogyakarta, Pustaka Pelajar, pp. 43–57.

Bunasor. (1989) *Diversifikasi dan Program Pembangunan Pertanian. Hasil Konpernas X PERHEPI dalam Diversifikasi Pertanian (Diversification and Agricultural Development Program: The results of Konpernas X PERHEPI in Agricultural Diversification)*, 1995. Jakarta, Pustaka Sinar Harapan.

Daniel, R. & Giddens, A. (2002) In: Peter Beilharz. *Teori-Teori Sosial: Observasi Kritis terhadap Para Filosof Terkemuka (Social Theories: Critical observation toward the prominent philosophers)*. Yogyakarta, Pustaka Pelajar, pp. 191–200.

Dant, T. (1991) *Knowledge, Ideology and Discourse: A Sociological Perspective*. London, Routledge.

Danusubroto, A.S. (2008) *RAA Cokronagoro I (1831–1857) Pendiri Kabupaten Purworejo (The founding fathers of Purworejo Regency)*. Purworejo, Diterbitkan atas prakarsa HR Budi Sarjono BRE.

Ekadjati, E.S. (1988) *Naskah Sunda (Sundanese Manuscripts)*. Kerjasama Lembaga Penelitian Pajajaran dan the Toyota Foundation. Bandung Forum Kajian Komuniti Pesisir, Depok, FISIP UI.

Gittinger, J.P. (1986) *Economic Analysis of Agricultural Projects*. Second Edition. Jakarta, UI-Press.

Hardjowigeno, S. (2003) *Klasifikasi Tanah dan Pedogenesis (Land Classification and Pedogenesis)*. Jakarta, Akademika Pressindo.

Hardjowigeno, S. (2007) *Evaluasi Kesesuaian Lahan dan Perencanaan Tataguna Lahan (Evaluation of Land Suitability and Land Use Planning)*. Yogyakarta, Gadjah Mada University Press.

Herman, C. (1985) *Culture, Health, and Illness*. Bristol, Wright.

Kastaman, R., Kendarto, D.R. & Aji, A.M. (2007) Model Optimasi Pola Tanam pada Lahan Kering di Desa Sarimukti Kecamatan Pasirwangi Kabupaten Garut (Planting Pattern Optimizing Model in Dry Land in Sarimukti Village, Pasirwangi District, Garut Regency). *Jurnal Teknotan*, 1 (1), 1–12.

Koentjaraningrat. (1987) *Sejarah Teori Antropologi I (Anthropology Theory History I)*. Jakarta, UI Press.

Koga, Y. (1988) Farm Machinery Vol. II. Tsukuba International Agricultural Training Centre. JICA.

Kuntowijoyo. (2006) *Raja Priyayi dan Kawula (The Noble King and the common people)*. Yogyakarta, Penerbit Ombak.

Maryoto, A. (2009) *Jejak Pangan, Sejarah, Silang Budaya dan Masa Depan (The trace of food, history, cross culture, and the future)*. Jakarta, Penerbit Buku Kompas.

Metutia F.S. (1985) *Makanan Kelompok Lanjut Usia dan Konteks Budaya, Khasiat Makanan Tradisional (The food for the elderly group and cultural context, the benefits of traditional food)*. Jakarta, Kantor Menteri Negara Urusan Pangan RI, pp. 99–109.

Mubaryato. (1989) *Pengantar Ekonomi Pertanian (The Introduction to Agricultural Economy)*. PT. Jakarta, Pustaka LP3ES Indonesia, anggota IKAPI.

Munandar, A.A. (2011) *Majapahit: Kerajaan Agraris-Maritim di Nusantara (Majapahit: The Agrarian-Maritime Kingdom in Indonesian Archipelago)*. Depok, Departemen Arkeologi, FIB UI.

Munandar, A.A. (2012) *Pentingnya Arkeologi dalam Membangun Budaya Bangsa menurut R.M. Ng. Poerbatjaraka (The importance of archeology in building a nation culture according to R.M. Ng. Poerbatjajraka)*. Depok, Departemen Arkeologi, Fakultas Ilmu Pengetahuan Budaya, Universitas Indonesia.

Munandar, A.A. (2012) *Peradaban Majapahit: Data dan Masalah Interpretasinya (Majapahit Civilization: Data and its interpretation problem)*. Depok, Departemen Arkeologi, FIB UI.

Munandar, A.A. (2013) *Tidak Ada Kanal di Kota Majapahit (No canal in Majapahit cities)*. Depok, Departemen Arkeologi, FIB UI.

Nainggolan, K. (2005) *Pertanian Indonesia Kini dan Esok (Indonesian Agriculture: Now and Tomorrow)*. Jakarta, Pustaka Sinar Harapan.

Nganro, N.R. (2009) Dukungan Kebijaksanaan Pemerintah dalam Pengembangan Komoditas Pertanian yang Mendukung Ketahanan Pangan Nasional (Government Policy Support in Agricultural Commodity Development in Supporting National Food Endurance). SEMILOKA *Pengembangan dan Penerapan IPTEK dalam Mendukung Ketahanan Pangan dan Energi (Science and Technology Development and Application in supporting food and energy resilience)*. Organized by Kedeputian Bidang Dinamika Masyarakat, Kementerian Negara Riset dan Teknologi. 10 November 2009, Jakarta.

Pakpahan, A. (1989) *Refleksi Diversifikasi dalam Teori Ekonomi. Hasil Konpernas X PERHEPI dalam Diversifikasi Pertanian (Diversification Reflection in Economic Theory: The results of Konpernas X PERHEPI in Agricultural Diversification), 1995.* Jakarta, Pustaka Sinar Harapan.

Pigeaud, Th.G. (1967) *Literature of Java, vol. I-III.* The Hague, Martinus Nijhoff.

Rahardjo. (1999) *Pengantar Sosiologi Pedesaan dan Pertanian (Introduction to Rural Area Sociology and Agriculture)*. Yogyakarta, Gadjah Mada University Press.

Sandy, I.M. (1977) *Penggunaan Tanah di Indonesia (The Land Use in Indonesia)*. Jakarta, Publikasi No.75. Direktorat Tata Guna Tanah, Ditjen Agraria, Departemen Dalam Negeri.

Sandy, I.M. (1995) *Republik Indonesia Geografi Regional (The Regional Geography of the Republic of Indonesia)*. Depok, Department of Geography, Faculty of Mathematics and Natural Sciences, Universitas Indonesia.

Santiko, H. (1986) *Mandala (Kadewaguruan) pada Masyarakat Majapahit (Mandala (Kadewaguruan) in Majapahit Society)*. Paper presented at the Pertemuan Ilmiah Arkeologi IV Cipanas.

Setiawan, R.P.A. (2001) *Research Report on Development of Variable Rate Granular Applicator for Paddy Field. Laboratory of Agricultural Machinery.* Kyoto University, Research Report.

Siswoputranto, P.S. (1987) *Komoditi Ekspor Indonesia (Indonesian Export Commodity)*. Jakarta, PT. Gramedia.

Srivastava, A.K., Goering, C.E. & Rohrbach, R.P. (1993) Engineering Principles of Agricultural Machines. *ASAE Textbook Number 6*, American Society of Agricultural Engineers.

Sumodiningrat, G. (1989) *Aspek Sosial Ekonomi Diversifikasi Sektor Pertanian Pangan. Hasil Konpernas X PERHEPI dalam Diversifikasi Pertanian (The Diversification Economic Social Aspects of the Food Agricultural Sector: The results of Konpernas X PERHEPI in Agricultural Diversification)*, 1995. Jakarta, Pustaka Sinar Harapan.

Susanti, N. & Kriswanto, A. (2008) *Damalung; Skriptoria pada Masa Hindu-Buddha sampai dengan Masa Islam (Damalung: Scriptoria during the Hindu-Buddhism Era until the Islamic Era)*. Paper presented at the Simposium Internasional Manassa, Bandung.

Susanti, N. (2003) *Masa Pemerintahan Erlangga (During Erlangga Government)*. [Thesis]. Depok, Department of Archeology, Faculty of Humanities, Universitas Indonesia.

Susanti, N. (2008) *Bahasa dan Aksara di dalam Prasasti dan Naskah; Wujud Kekuasaan Simbolik pada Masyarakat Jawa Kuno (Language and Letters in Inscriptions and Manuscripts; the symbolic power realization in ancient Javanese society)*. Paper presented at the Kongres Kebudayaan, Bogor.

Taqyuddin. (2004) *Pengelolaan Bangunan Air di DAS Ciliwung Masa Kolonial (Water Building Management in Ciliwung DAS during the Colonial Era)*. [Thesis]. Depok, Department of Geography, Faculty of Mathematics and Natural Sciences, Universitas Indonesia.

Weber. (2015) *In Weber's Rationalism and Modern Society*. Edited and Translated by Tony Waters and Dagmar Waters, New York, Palgrave Macmillan.

Zimmerer, K.S. (2007) Agriculture, livelihoods, and globalization: The analysis of new trajectories (and avoidance of just-so stories) of human-environment change and conservation. *Journal of Agriculture and Human Values*, 24 (1), 9–16.

Zoetmulder, P.J. (1995) *Kamus Jawa Kuna-Indonesia (Ancient Javanese-Indonesian Dictionary)*. Jakarta, Penerbit Gramedia Pustaka Utama.

Cultural Dynamics in a Globalized World – Budianta et al. (Eds)
© *2018 Taylor & Francis Group, London, ISBN 978-1-138-62664-5*

The Treaty of 1855: Early American shipping, commerce, and diplomacy in the Indonesian archipelago (1784–1855)

Y.B. Tangkilisan
Department of History, Faculty of Humanities, Universitas Indonesia, Depok, Indonesia

ABSTRACT: The Treaty of 1855 is an agreement between the United States and the Netherlands concerning the opening of consular representatives in the Dutch overseas possessions, including the Dutch East Indies. The interests of the American international shipping and commerce in Asia laid the foundation to this treaty. The Americans faced several difficulties with such a mercantilist rule in European countries, like the Netherlands, which closed their possessions from the international affairs. The Gibson case (1852–1854) led to the agreement. According to the Americans, the Dutch East Indies, which became the independent nation of Indonesia on 17 August 1945, was a stopover point for their broader interests in the global world.

1 INTRODUCTION

At present, the relationship between Indonesia and the United States shows an unequal circumstance. Indonesian economy depends on the US economy, but the reverse is not true. The United States has played an important role in the Indonesian economy, especially after the global crisis of 1997/1998, from which it has not yet been fully recovered. The United States is still Indonesia's main export destination.

Moreover, the United States is the highest investor in Indonesia. Therefore, the role and influence of the United States in Indonesian domestic development are very significant.[1] America's exposure toward the Indonesian archipelago is deep-rooted. The United States went through similar phases immediately after the American Declaration of Independence that ended the British colonialism. It began with the commercial motives to build foreign economic networks.

With regard to the economy, the recent development of Indonesian historiography has resulted in increasing interests in the fields of economic history, together with the maritime and area history. The concerns are growing with the increasing number of publications on such studies. The international commerce as a subject of research lies between these disciplines. However, it seems that the subject is still neglected by most Indonesian historians.

2 PROBLEMS

The primary objective of this study is to reveal the policies and actions of the early American government to support their private shipping and commerce in other countries through

---

1. The involvement of the United States in the domestic affairs of Indonesia has begun since independence struggle of Indonesia, which resulted in the Round Table Conference in 1949. It ended the dispute between the Republic of Indonesia and the Dutch over the Indonesian archipelago. Thereafter, the United States involved in the rebellion of PRRI/Permesta. The next interferences of the United States were in the Papua New Guinea dispute, the Confrontation of 1963, and the political transition of 1965/66.

some diplomatic measures. Then, it aims to contribute to economic history and the history of relationship between Indonesia and the United States. This study focuses on the commercial history during the Dutch colonialism in Indonesia. The colonial rule introduced modern shipping and trading networks, using steamships, regular lines, and harbor systems. The colonial government issued some regulations for those activities in the early 19th century.

In this study, we deal with the efforts made by the United States to breach the mercantilist policy of the Dutch East Indies that hampered its shipping and trading activities in the Indonesian islands in the 19th century. On the route to Canton, the islands became a stopover point for American ships to restore their supplies and find commercial goods. However, in the Dutch East Indies, their activities were restricted by certain regulations. Therefore, they tried to solve the problems through diplomatic channels, although there were some imperialistic motives behind such efforts.

In this study, we focus on the treaty referred to by Gerlof D. Homan (1978) as the Treaty of 1855, which, according to various American sources, was about the consuls. In the *Staatblaad* (State Gazette) 1855 no. 65 (ANRI Collections) of the Dutch East Indies, it is stated that "*Overeenkomsttusschen Nederland en de Vereenigde Staten van Amerika, tot regeling der Voorwaarden op de Consulaire Agenten van die Staten in de Voornaamste Havens der Nederlandsche Overzeesche Bezittingen zullen worden Toegelaten* (the Agreement between the Netherlands and the United States of America, concerning the permits for opening the Consulate Agency of the United States at the Main Ports of the Dutch Overseas Colonies)". The American version sounds as "Consuls" (Bevans 1972: 28–33). The main point is concerning the regulation for opening the American Consular Office in the islands of the Dutch East Indies. It meant that the American shipping and commercial interests had been protected under a legal basis. The American shipping activities and commerce in the islands began in 1784 and 1786, respectively.

For the United States, the agreement had a close connection with its political doctrine and philosophy, which is the Freedom of the Seas or *Mare Liberum*, formulated by Hugo Grotius in 1607 and widespread in 1633. In the international commerce, the United States has adopted the doctrine of Free Market or Free Trade, which originated from the work of Adam Smith in 1776. In the United States, this thought was popularized by Benjamin Franklin in 1774. In the field of diplomacy, the United States was considered the most favored nation (Eckes 1995: 5).

The treaty resulted from the bilateral talks between the Dutch and the US governments since their first agreement in 1782 concerning the opening of diplomatic representative offices in both countries. The United States was interested in establishing representative offices in the Dutch overseas possessions, especially in the Dutch East Indies, because it needed such a commercial network to support its trade with Canton, China. Furthermore, the colony's waters constituted part of the route to the destination, which Commodore Galbraith Matthew Perry used to reach Japan. On the contrary, according to the Dutch, the agreement was a commitment of the US government to recognize its dominance over the area. Since the beginning of the 19th century, the Dutch had launched colonial expansion to all islands in the archipelago area (*Nusantara*).

3 METHODOLOGY

This study was conducted using historical methods that consist of heuristics, criticism, interpretation, and writing (Gottschalk 1980; Kuntowijoyo 1987). The main source is the written documents available in the archival depositories and online sources, including qualitative and quantitative data, such as statistics, tables, and graphs. The documents from the Dutch file collections of *Arsip Nasional Republik Indonesia* (ANRI), such as *Verslag van den Handel, de Scheepvaart en de Inkomende en Uitgaande Regten op Java en Madoera 1823–1869* (Batavia: Landsdrukkerij, 1825–1869), are treated as the primary sources. We also explore the American document *A Compilation of the Messages and Papers of the Presidents* (1911/1917).

To analyze the findings, the next step is using certain approaches from different fields of social science, such as politics, sociology, international relations, and economics. From the viewpoint of classical economics, commercial transaction is significantly influenced by the market mechanism, namely demand and supply. Adam Smith (1776) stated that the mechanism works as an invisible hand. David Ricardo (1817) put it into an international economic context by explaining the significance of the comparative advantages, which encourage the specialization of national production.

The idea of Free Market was challenged by Alfred Chandler (1977). He referred to a modern business corporation that acts as an exchange function that has previously been conducted by the market. He described that the mechanism depends on the role of economic actors arising from the development of business corporations. With the development of economic activities, which can be caused by several aspects such as immigration and expansion, business owners could no longer run the companies by themselves, and they need a more effective and better management system. The growing business needs an intensive and careful treatment conducted by salaried executives called managers.

Moreover, Chandler suggested eight propositions to answer the question of why the visible hands of the management replace the invisible ones. The first three propositions explain the initial performance of modern business corporation, from when, where, how, and why it begins. The other five propositions deal with the sustaining growth associated with where, how, and why a company grows and maintains the growth. The eight propositions are as follows:

1. The rise of modern multiunit enterprise
2. The establishment of managerial ranks
3. The improvement of efficiency in administrative coordination and profits
4. The success of managerial ranks in running the function of administrative coordination
5. The increasingly technical and professional performance in the career of salaried managers
6. The increasing separation of management and ownership
7. The commitment of managers toward long-term stability and the growth of the enterprise
8. The rise of large enterprises controlling the economics and causing the economic structural change

Before the rise of modern managers, which, according to Chandler, started in 1840, the business actors, especially in the distribution sector of trading, had been influenced by the market mechanism. The rate of the demand played an important role in the calculation and decisions of business actors. The commodities brought from the United States often were not welcomed by the market, such as in Canton, China. It happened when they miscalculated the market. Therefore, they had to find other commercial goods that were in high demand in other places, such as the Dutch East Indies. However, following further development, the American merchants tried to avoid such mechanism and began to affect the market, by either political support of the government or innovation.

International economists such as Paul R. Krugman and Maurice Obstfeld (2003: 4–10) offered seven subjects to discuss, namely (1) commercial advantages, (2) commercial patterns, (3) protectionism, (4) terms of trade, (5) exchange value, (6) international policy coordination, and (7) international capital market. The theory of commercial policies covers taxation on some international transaction or tariffs, export substitutions, import restrictions, voluntary export restraints, local content requirements, export loans, national procurements, red tape barriers, and free trade. In the 19th century, some commercial policies, from both the American government and the Dutch East Indies, showed several indications in line with this theory, especially regarding the transaction, tariff, protection, red tape barriers, and free trade.

International trade is closely related to international diplomacy. According to Harold Nicholson (Roy 1991: 3), diplomacy has at least five meanings, namely foreign policy, negotiation, mechanism of the negotiation, foreign office branches, and international negotiation skills. After observing some definitions of diplomacy, including that from Carl von Clausewitz, S. L. Roy (1991: 4–5) stated that the main aspect of diplomacy is negotiation, which is undertaken to achieve the national interest of a state.

Peaceful diplomatic actions are intended to maintain and encourage the national interest as long as possible. Furthermore, diplomacy deals with the international political agenda of a country. The modern diplomacy is closely related to statecraft and is inseparable from the representatives of the state. Diplomacy is manifested in the form of agreements, conventions, and treaties.

In the practice of diplomacy, a state possesses some instruments. The economic instruments cover the international trade and aid as a means to secure peace. The important role of the increasing international commerce is its potential impact on diplomacy. The steady increase in industrial growth increases the strength and position of investors and merchants, who can force the government to open new markets and gain trading contracts (Roy 1991: 120–121).

In addition to diplomacy, the frame of analysis uses the theory of imperialism proposed by James W. Gould (1972: 306). He stated that imperialism is the attempt of a state to dominate or to impose its will on a foreign area. The following the features of imperialism: (1) the actor represents the state or its apparatus, and is not limited to him/herself; (2) the action is an attempt, and its outcome is not important; (3) the goal is to dominate, rule, and impose its perspective, and permanence is not essential; and (4) the target is the foreign people or area. Therefore, imperialism could be used here to describe an attempt of a state to dominate or to impose its will upon any part of other areas or people.

Then, liberalism is studied versus mercantilism. The doctrine of free trade originates from the liberalism restriction of the role of the state or the government in the economic fields. By contrast, mercantilism stresses on the role of the state or government in commerce. Diplomatic efforts have been taken to bridge this gap.

In order to show the interaction between two ideas or policies from two countries, we need to discuss the theory of structuralism. From the perspective of sociology, Anthony Giddens (2004) proposed the theory of structuration. Giddens reconstructed the concept of structure and described the role of the agents or agencies of change during a social change.

The agency and structure do not undergo dualism and hence they cannot be separated; however, they interact over time or under a duality. From the perspective of realism, Christopher Lloyd (1993) developed a similar insight. Individuals belong to a structure, and they have the potential and ability to initiate changes like agencies. The interaction between the agency and the structure in social changes results in two factors, namely transformation and reproduction. In Indonesian historiography, structuralism and structuration were introduced and developed by R.Z. Leirissa (2002; 2003; 2006).

Structuralism does not neglect the theories of structuralists and individualists. It tries to connect the two aspects and minimizes the bipolarity by mentioning that they are not barriers for individuals and vice versa. Lloyd mentioned historian Emanuel Le Roy La Durie from the France's Annales Historical School famous for Structuralism who applied this theory. Moreover, structuralism suggests the use of some relevant theories under such a combination to explain the dynamics of the past.

3.1 *The early American voyage to Asia*

The United States involved in the international trading networks immediately after independence. However, the British, its former colonizer, did not open its ports to American shipping, neither did some other European nations, including France and the Netherlands. Therefore, the US government used a diplomatic strategy to overcome those barriers by making treaties.

Together with foreign policies, the US government took several preparation steps, one of which was Plan 1776. It contains a draft or blueprint that guided the making of international treaties. (www.americanforeignrelations.com/E-N/Freedom-of-the-Seas-Origins-of-the-conc...). In addition, the development of navigation and shipbuilding technology supported the effort to enter the international voyage and commerce. Moreover, the American commercial interests were disturbed with the breach of its former colonial era trading policies. Therefore, its main concern was to rebuild the networks. Canton in Asia, also known as Indie, was one of the main destinations of American overseas shipping (Dennett 1922).

More opportunities resulted when Europe was experiencing wars due to the French Revolution in 1789. In 1799, Napoleon Bonaparte led the spread of the revolution to other European countries. As a result, ships from countries that did not take part in the revolution like the United States took an advantage to serve the shipping lines between Europe and the colonies. Despite in insecurity in the lines, American shipping increased rapidly (Fichter, 2002).

The first American ship to cross the Atlantic was "the Harriet", which reached the Cape of Good Hope in 1784. In the same year, the ship "Empress of China" arrived at Canton, which passed by and stopped at some points of the Indonesian islands; however, it did not make any commercial transaction. In the following years, the islands became the stopover points in the American shipping route as a part of the Cantonese commercial network.

During the voyages, American ships encountered some barriers and threats, ranging from piracies, including the attacks from European countries that were under siege, to the natural and navigational factors that hampered the shipping. However, the main barrier was the application of the mercantilist policy, including the protective policy used by the Dutch East Indies (*Staatsblad,* 1818, no. 58; *Regeering Almanak,* 1823; *Arsip Statistiek,* no. 2. ANRI).

The Dutch ships were given the first priority to load the commercial goods. Therefore, American ships often were forced to purchase similar products at a higher price from the Dutch East India harbors, such as Batavia and the Java northern coast ports, or there would be no available commodities to load. The Dutch authorities also restricted trade in volume or value. Other difficulties included health issues and diseases. The crew of American ships suffered from dehydration under sundry and humid climate, leading to death.

3.2 *American commerce in the Indonesian archipelago*

The American trade in the Indonesian archipelago began with the arrival of the ship "the Hope" in Batavia in 1786. Samuel Shaw, the American consul for Canton, was also a passenger on the ship. In his second journey, he did not find any problem in obtaining the cargoes to be brought to Canton. Meanwhile, ship "Rajah" under the command of Captain Jonathan Carnes sailed to the northwestern Sumatran coast in 1789. He found a pepper cargo that increased the profit up to 700% when sold at home (Dennett 1922: 31). The Salem pepper trade with the Sumatrans developed since then.

The trade progressed well despite competition with the French and British traders. In addition to such competition, the American ships experienced two assaults from the local people of Kuala Batu and Muki in 1831 and 1838. The Americans retaliated by sending naval military operations considered as a form of punishment (Long, 1973; El Ibrahimy, 1993; Tangkilisan, 1997b). In the 1840s, the Salem–Sumatran pepper trade began to decline (Reid, 1973: 7; Malone, 1960: VII 338; Turnbull, 1972: 160).

In Batavia, coffee trade was carried out by the Rhode Island traders. They gained huge profit during the Napoleon wars in Europe. However, there were shortcomings, including unsecured shipping due to piracies, foreign attacks, natural difficulties, rivalry with British or Dutch traders, and the mercantilism policies. Related to mercantilism, the idea of appointing consulate officers who would secure the American commerce in the Dutch East Indies was implemented in Washington D.C., which did not receive the expected response. The idea emerged after the Gibson Affairs in 1853. The trading activity of Rhode Island in Java ceased in 1836 (Ahmat, 1965: 95).

3.3 *American overseas trading networks*

The pattern of American international networks consists of export and import activities, re-exports, carrying trade, and trade chain (Fichter, 2002; Hacker, 1982; *Verslag,* 1830–1850). Export referred to the flow of goods from the American domestic production sector to abroad, and import referred to the reversed flow, which was carried out by shipping. Re-export trade occurred when the imported goods did not reach the domestic markets, but were sold again to other foreign destinations. The goods originated mainly from Asia, especially from Canton and the Sumatran West Coast. Tea, coffee, pepper, silk, and Chinese wares were mainly traded. However, the balance of American trading from 1790 to 1855, including the export, import, and re-export, declined.

Trade chain referred to the trade from one port to another under a certain commercial network centered in Canton. This was different from the well-known trade pattern of commerce in Asia, which was concerned with the flow of goods from one place to another, but not the traders and their ships. This type of networking trade demanded the sailing traders to move around under a rhythm of market drives. The next pattern was the carrying trade, through which America made considerable profits, especially from the cargo rents (Hacker 1982: 236). However, the quantitative numbers of both patterns were yet to be found. The trade between the United States and the Dutch East Indies continued despite experiencing deficit because of these types of trade. The United States compensated the deficit from other profitable terms.

The main problems of trade chain are related to the issues regarding commercial goods, payment methods, and mercantilism policies. The American merchants had to monitor the highly demanded goods in the markets. Problems in the method of payment arose because the American currencies had not yet been accepted in international transactions. Therefore, they had to find and carry the Spanish silver and gold currencies, used mainly as their trading currencies in the Dutch East Indies. However, the main difficulty was the lack of trade protection, which gave rise to diplomatic efforts, namely to recognize and place consuls (Livermore, 1946).

3.4 *American diplomacy*

The US foreign affairs followed a doctrine of the most favored nation. Under such doctrine, the US government made various amities, navigations, and trade agreements. However, the efforts were not enough to give the American ships such freedom to sail and trade in the international economic sphere, especially in the colonized areas. It was because the colonial rulers did not open access to their possessions for other countries. Therefore, the American government used a diplomatic strategy by making a consular convention that allowed them to open a consulate office, which enabled the provision of protection for the American citizens and shipping and commercial activities in a colonial possession (*A Compilation Messages and Paper of the Presidents,* vol. XIX, tt).

The making of such a convention between the United States and the Netherlands was due to several factors. First, the decline in the Salem pepper trade in Sumatera and the trade with the Rhode Island traders in Java. Second, the British had taken a hold of Canton through the Opium War in 1842, and as a result, the Americans turned their focus to Japan. The interest of the United States in Japan was marked by the termination of the isolation politics (*Sakoku*) by Commodore M.C. Perry in 1854 triggered by the Gibson case (Locher-Scholten, 1994/2007).

Walter Murray Gibson was an American adventurer who arrived at Palembang, South Sumatra, in 1852. His sympathy to the local people trapped between a conflict among the Jambi Sultanate, and the Dutch put him in the middle of the conflict. Therefore, the Dutch suspected and then arrested him. His ship and cargo were seized. He was imprisoned in Weltevreden, from where he escaped and reached his homeland. He asked the government to support him in suing the Dutch government for his loss. The US Ambassador, August Belmont, responded the plea and took this opportunity to make the Dutch to open its overseas colony. He was inspired by the actions taken by his father-in-law, Commodore M.C. Perry, in the case of Japan (Curti1962: 34; *Readers Digest Family Encyclopedia of American History,* 1975: 1288; Johnson, 1964: 170).

3.5 *The Treaty of 1855*

Belmont failed to win the Gibson case. However, he was asked by the Dutch government to begin a discussion of establishing a consular convention, after some pressure. The result was the agreement of 22 January 1855, known as the Treaty of 1855. This was brought to both governments to be ratified. For the Dutch East Indies, it was announced in the *Staatsblad* 1855 no. 65.

The following are the results of the discussion held in a friendly atmosphere and under the full authority of both sides (the *Staatsblad* 1855 no. 65. ANRI):

a. The American consular representatives, namely the Consul Generals, consisting of Consuls and Vice Consuls, had a formal recognition and acceptance in all Dutch colonial harbors as the "most favored nation" (points 1 and 14).
b. The representatives had to hand the letter of confidence to the Dutch King (point 3).
c. The consular representative acted as an agent and protector for the trade activities of its citizens, including in any incident of shipwreck, and held an immunity privilege against the seizing according to the existing rules, but no diplomatic immunity (points 3, 4, 5, 6, and 14).
d. The quarrels that could not be settled at the consular level would be referred to the diplomatic representative in The Hague (point 6).
e. The two countries were able to appoint vice consuls at the harbors under their jurisdiction, and it was possible to appoint foreigners as consuls (point 7).
f. The consular office issued passport or other papers needed to travel or to reside in areas under its duties (point 8).
g. The consuls were obliged to support the authorities in any criminal case involving American citizens (points 10 and 12).
h. The duty also included to support the matters of inheritance or testimony of American citizens who passed away in the colony (point 11).

The representatives who signed this agreement were A. Belmont, the American Ambassador to the Netherlands, A. van Hall, as well as C. Pahud of the Netherlands. The Dutch King signed it on 8 April 1855 and the US President, Franklin Pierce (1853–1857), on 5 March 1856. On 25 May 1856, it was exchanged in Washington D.C. The Dutch published it in the State Gazette on 5 July 1856, and for the colony, on 29 October 1856. The treaty allowed the Americans to open consular offices in the Netherlands and India.

4 CONCLUSION

The United States is a liberal country. However, in international economics, the US government played a significant role in breaking the mercantilist restriction to its trade. The Treaty of 1855 was a prominent example. The factors include liberal trade doctrine, stopover point, mercantilism, and the opening of Japan. The Gibson case was the triggering factor. The agreement, on the one hand, permitted the United States to open consular representatives in the Dutch overseas possessions, especially the Dutch East Indies. On the contrary, the Dutch gained an advantage to secure its expansion in Sumatra, a prosperous mining island. In 1872, the Dutch declared a war against the Kingdom of Aceh. The plea for support from the kingdom was rejected by the US consul in Singapore. The United States had a diplomatic agreement with the Dutch that prevented such interference.

Moreover, the Indonesian archipelago was a stopover point for the US international commerce. In such context, the American commercial networks grew. This commercial pattern was the reason for the non-termination of the American shipping and commerce with the decline of American terms of trade with the Dutch East Indies, as the United States compensated the deficit from this trade.

To date, the United States has considered Indonesia as a stopover point for its global interests. In the Cold War era, Indonesia was a frontier to contain the Communist influence (*the Containment*). In terms of economy, Indonesia still depends much on the United States. It means that the United States has a rather significant influence on and plays an important role in the development of Indonesia.

REFERENCES

A Compilation of the Messages and Papers of the Presidents vol. 3, 1911/1917.

Ahmat, S. (1965) Some problems of the Rhode Island traders in Java 1799–1836. *Journal of Southeast Asian History*, 6, (1).

Bevans, C.I. (1972) *Treaties and other international agreements of the United States of America 1776–1949*, 10. Washington, Department of State Publication.

Burnett, E.C. (1911) Note on American negotiations for commercial treaties 1776–1786. *The American Historical Review*, 16 (3).

Chandler, A.D. (1977) *The visible hand: The managerial revolution in American business.* Cambridge, Mass, The Belknap Press of Harvard University Press.

Curti, M.E. (1962) Young America. In: *The American Historical Review, 32(1), Oct.*

de Hullu, J. (1988) On the rise of the indies trade of the united states of America as competitor of the east indie company in the period 1786–1790. In: MAP Meilink Roeloefs *et al. Dutch Authors on Asian History.* Dordrecht, Holland & Providence, USA, Foris Publication.

de Roo, L.W.G. (1919) Amerikaansche Contracten In: *Documentenomtrent Herman Willem Daendels Gouverneur-Generaal van Nederlandsch-Oost Indie*, 2 jil.,'s Gravenhage, Martinus Nijhoff.

Dennett, T. (1963) *Americans in eastern Asia: A critical study of United States' policy in the far east in the Nineteenth Century.* New York, Barnes & Nobles, Inc.

Eckes, A.E. (1995) *Opening America's market. U.S. foreign policy since 1776.* Chapel Hill, The University of North Carolina Press.

Fichter, J.R. (2006) *The United States, Britain, and the East Indies 1773–1815.* [Dissertation]. The History Department, Harvard University, Cambridge, Mass., August.

Gibson, W.r M. (1855) *The prison of Weltevreden: And a glance at the east Indian archipelago.* New York, J.C. Riker, 1855.

Giddens, A. (2004) *The constitution of society: Outline of the theory of structuration (Teori Strukturasi untuk Analisis Sosial).* Pasuruan, Pedati.

Gould, J.W. (1961) *Americans in Sumatra.* The Hague, Martinus Nijhoff.

Gould, J.W. (1972) American imperialism in Southeast Asia before 1898. *Journal of Southeast Asian Studies*, 3, (2).

Krugman, Paul R, Obstfeld M. (2003) *Ekonomi internasional: Teori dan kebijakan.* (International Economy: Theories and Policies). Jakarta, Raja Grafindo Perkasa.

Leirissa R.Z. (1999) Strukturisme dalam ilmu sejarah suatu alternatif. (Structuralism in the history of science is an alternative), *Makara Jurnal Penelitian Universitas Indonesia*, no. 3 Seri C.

Lloyd, C. (1993). *The structures of history.* Oxford, Blackwell Publisher.

Locher-Scholten, E. (1994) *Sumatraans Sultanaat en Koloniale Staat. De Relatie Djambie—Batavia (1830–1907) en het Nederlandsche Imperialisme.* Leiden, KITLV Uitgeverij.

Locher-Scholten, E. (2008) *Kesultanan Sumatra dan negara kolonial: Hubungan Jambi-Batavia (1830–1907)* (trans.). (The Sultanates of Sumatra and the Colonial States: The Relationship of Jambi-Batavia (1830–1907)). Jakarta, Banana & KITLV Jakarta.

Long, D.F. (1973) Martial Thunder: The first official American armed intervention in Asia. *Pacific Historical Review*, XLII, May, no. 2.

Quincy, J. (1847) *The journals of Major Samuel Shaw: The first American consul at Canton.* Boston, W. Crosby & H.P. Nichols.

Reeves, J.S. (1917) Two conceptions of the freedom of the seas. *The American Historical Review*, 22, (3), April.

Regeringsalmanak 1823, 1824, 1826, 1829, 1859.

Spector, R. (1972) The American image of Southeast Asia 1790–1865, A preliminary assessment. In: *Journal of Southeast Asian Studies,* 3, (2).

Staatbladen 1818, 1853, 1855, 1864.

Verslag van den Handel, de Scheepvaart en de Inkomende en Uitgaande Regten op Java en Madoera 1823–1869 (1825–1869) Batavia, Landsdrukkerij.

Cultural Dynamics in a Globalized World – Budianta et al. (Eds)
© 2018 Taylor & Francis Group, London, ISBN 978-1-138-62664-5

Individual unit of analysis in the debate on the methodology of economics as a social science

H.S. Pratama

Department of Philosophy, Faculty of Humanities, Universitas Indonesia, Depok, Indonesia

ABSTRACT: The development of theories in social sciences and humanities are influenced by debates on the unit of analysis. One key of the debate is about whether scientific investigation should be based upon a micro-level analysis (individual unit) or a macro-level analysis (*sui generis*). This methodological debate originated in the contraposition between Max Weber and Emile Durkheim in the late 19th century. The debate centred on the atomistic methodology vis-à-vis holistic methodology. In contrast to theory as a system of description that contains truth/falsity parameter, methodology is more about an open choice in scientific inquiry. Hence, the chosen methodology determines the possible theoretical outputs. This paper will use an analytical approach to discuss the unit of an analysis in social science methodology. It aims to demonstrate the theoretical implications that may result from the debate.

1 INTRODUCTION

There is a historical debate among social scientists on the nature of a society and the best method to understand it. On one hand, there are those who look at a society as an aggregation, a collection of individuals. On the other hand, there are those who see society as a collectivity. In this sense, the first group perceives a social phenomenon in terms of individuals and the interactions among them. Meanwhile, the second group perceives a social phenomenon as a social whole. In the history of methodology, the first perspective is referred to as methodological individualism, while the second as holism (Udehn, 2002).

The methodological individualism is a kind of atomism that uses an individual as its unit of analysis. An individual is similar to an atom in a physical reality. In order to generate an explanation of a social phenomenon, a scientist begins by explaining an individual. In this methodology, the characteristics, nature, or disposition of an individual must be known. Furthermore, this methodology develops a model of an individual who is rational. Rationality is considered as the essence of an individual. In turn, this methodology is related to the rational choice theory. Methodological individualism looks at a social whole as constructed by the individuals and the relations among them (Udehn, 2002).

Methodology and theory are different. Theory is, in an empirical term, a descriptive statement, which can be true or false. However, methodology is a metatheory, a metalanguage, which cannot be true or false, but fruitful or not. In this sense, methodology is a choice made by a scientist to ensure the fruitfulness of his/her research. The success of a methodology lies in the application of a methodology in enriching the research. The use of a methodology is initiated by Hobbes when he tries to explain how we can understand the social by understanding the psychology of an individual. The use of methodology reached its culmination in the Austrian School of Economics with Carl Menger, Joseph Schumpeter, Ludwig von Mises, and Friedrich von Hayek as its members.

The debate of methodology in that era was between holism (Emile Durkheim) and methodological individualism (Max Weber). Following Max Weber, they place rationality as the main feature of an individual, and make an individual as a unit of analysis in economics. The debate also focuses on whether this rationality is an empirical fact or only an ideal type. Von Mises, for example, sees rationality as factual. Weber and Hayek, on the other hand, believe that rationality is an ideal type and that scientists are able to make comparison between how the actual man behaves and how the ideal type conducted their actions. Through the application of methodological individualism, social scientists are able to examine social issues using an individual unit of analysis. The debate between methodological individualism and holism raises a fundamental question on how scientific social science is similar to economics. This paper will concentrate on this question and relate it with the problem of modelling in social science, especially in economics.

The rise of social sciences in the modern era has created some philosophical questions on how scientific social sciences are. These questions can be divided based on the systematic of philosophy by looking at what social sciences are (Ontology), how social sciences develop their theories and what the justifications are (Epistemology), and by looking at the values they have (Axiology).

Since the method of induction was reinvented by Francis Bacon in the 16th century through his book, *Novum Organum (or New Method)*, modern science has had a new primary tool to produce theories. Bacon's new method was used to replace the previous deduction method introduced almost 2,000 years before by Aristotle in his book *Organon,* which literally means method.

Induction has enabled scientist to observe a phenomenon in order to build a generalization. This generalization could then be treated as a causal explanation. In other words, induction is a kind of method that enables scientists to predict some future unobservable facts based on some observable facts in the past. As a result, explanation and prediction become integral parts of scientific activities. This method has developed from Bacon to logical positivism in the 20th century. Simply put, an induction or a generalization is in line with verification and confirmation in the late methodological debate.

Natural sciences fit in with the inductive method and the method continued to grow rapidly after Bacon. By using the induction method, natural scientists are able to find a pattern, regularity, and formulate them in historical statements or theories. A statement expected to be a theory must be historical. Historical components must be eliminated so that the historical statement is able to explain and predict certain event. To some degree, this historical statement is best represented with a mathematical language since this language is 'neutral' and it contains no historical component.

Theories that are simultaneously confirmed would be recognized as natural laws. By finding the natural laws, a natural scientist is able to give account to another phenomenon, making explanation and prediction as well. Carl Hempel names this process the Deductive-Nomological Model (DN-Model) referring to observations treated as "law-like statements", which in turn would be a major premise or a *nomos* (rules). Based on this logic, a scientific activity is considered as an attempt to translate the *nomos* into certain circumstances, condition, or data.

The development of social sciences has come a long way. In the past, there was a belief that the only science was natural science. This is a strange phenomenon since the development of social sciences tends to imitate that of natural science. August Comte's Positivism Project attempting to establish sociology as a social physics is a good example of how social sciences have followed the footsteps of natural sciences. The nature of social science and the mechanism behind it continue to be perceived as intriguing and philosophical questions.

The same issues emerge in economics. At the beginning, economics was formulated as a rule to manage households. Although *oikos-nomos* was not perceived as a science, the concept became the origin of economics. This is very interesting since the nature (or the origin) of economics is seen as a technique, not an explanation or a prediction as what we now know. Furthermore, the very nature of economics is not a scientific attempt (to explain), but a moral attempt (to do something). In this modern era, this view of economics as a moral attempt is open for debate.

The modern economics began with the publication of Adam Smith's book, *The Wealth of Nations,* in 1776. In the book, Smith tries to solve some problems of the society by proposing a market system as an exchange system. Moreover, a market functions as a moral system. Smith focuses on unintended consequences of human actions, not on moral intention. Market is perceived as a reliable system that enables society to do well by proposing an individual self-interest. Smith denies the importance of a moral intention as the basis for action. Instead he focuses on unintended consequences. In a market, every person is encouraged to pursue his or her self-interest. This selfish motive would lead to a harmony by ensuring that a perfect market is established. A perfect market is based on three prerequisites: self-interest, freedom, and competition. When these prerequisites exist in a market, equilibrium will take place. In other words, equilibrium is the unintended consequence or a condition, which seems to be regulated by an 'invisible hand'.

The development of modern economics relies on the normative agenda to create order in a society. The task of economics is redefined by Amartya Sen through his work, *On Ethics and Economics*. Sen argues that there is an irreducible task of economics. The work of economics shapes politics, moral, and engineering aspect (Sen, 1991). This means economics should be about how to gain social achievement, how to promote moral conducts, and how to create efficiencies. The three tasks of economics contain normativity or anything that must be done simultaneously and not partially.

As a discipline, economics holds a vital role in formulating a public policy. Policy makers take economics principles in ensuring social conduct, such as distributing advantages, taxing people, raising the rate of interest, building infrastructures, budgeting, and so on. This is possible because economics is considered as an established social science similar to natural science in the use of methods (Popper, 2005). This article will discuss how the normative (moral) agenda of economics is replaced by a scientific function and identify the discipline's capitalistic economic system by its use of a modelling method.

2 METHODOLOGY

This article is a qualitative research that focuses on a conceptual examination of economics through a philosophical perspective. Several steps must be taken in order to do this. The first step is a conceptual analysis. To explicate the philosophical problem in an economic modelling, a researcher needs to analyse the key concepts such as reductionism and model-dependent realism. The next step is abstraction. Philosophy is different from other fields of studies in its capacity to develop speculative abstraction beyond the empirical data. In this sense, a philosophy researcher tries to abstract an actual model to arrive at a more abstract understanding about the model. This is called the abstract concept of modelling. This concept is different from other disciplines such as economics that takes a model as it is. The third step is questioning. Almost every scientific field uses a model in their research, but there is a very tiny room for them to question the ontology of model they use. In this article, the researcher tries to question the use of a model in economics beyond its empirical significance. A philosophical approach in looking at the issue involves philosophical systematism, which includes ontology, epistemology, and axiology. It questions what something is, how we know it, and how we value it. The last step is making judgment about the subject. It involves making conclusions and explicating the implication, especially the philosophical implication.

In developing ideas and generating philosophical problems on economic modelling, the researcher focuses on exploration in philosophical texts. Hermeneutics is used to approach these texts. Hermeneutics is a branch of philosophy studying texts in a critical manner to expose the basic assumption, prejudice, and context (Baggini, 2010). The researcher examines several primary texts, which include Julian Reiss, 2013, *Philosophy of Economics: a Contemporary Introduction* (New York: Routledge) and Lars Udehn, 2002 [2001], *Methodological Individualism: Background, History, and Meaning* (New York: Routledge).

Economics is similar to natural science or an exact science based on its ability to generate statements that are verifiable and falsifiable with empirical data and avoiding speculations that exceed empirical observation. In this sense, economics works like positive science. On the other hand, philosophy is more speculative, and it looks at the ontology and the axiology of an issue (especially, ethical issues) (Reiss, 2013). This argument leads to a rigorous distinction between the positivistic side and the normative side of economics. The normative side belongs to philosophical discussion. In addition, economics has isolated itself from observing a phenomenon and generating a social pattern.

Through an induction method, economics attempts to find general laws in an economy by observing a particular phenomenon. This method expects to find universal synthetic statements. It is believed that we can find a universal statement from a singularity. According to Carl Hempel, this "general laws" should be discerned between a statement, which is "necessarily true", and that which is "accidentally true". Necessarily true statements are true laws, while accidentally true statements are statements that are true only by coincident (Fetzer, 2014). If we observe multiple phenomena in a wide setting of time and place, we will be convinced that a general law can be discovered. Unfortunately, David Hume states that epistemologically we can only see the singulars (or particulars, not universals). Everything that we see is a singular observation, a singular statement, and this is a kind of *phenomenalism*. We cannot find the universals because it means we will need to collect all phenomena, both actual and possible, in all kinds of time and place settings (Popper, 2002). This leads to a criticism on the reliability of an induction method because no one can accumulate all possible data and this means our best theory of anything is not certain.

Can economics find the general laws in economics and have it accepted and recognized as a positive science? Positive science consists of organized scientific statements, which describe and explain phenomena as it is. Positive science primarily was a philosophical movement initiated by August Comte in the 19thcentury to establish sociology as social physics (Udehn, 2002). In this project, Comte considered science as value-free activities by making rigorous dichotomy between fact and value. Science takes only facts. In this sense, science is characterized by the method of reduction and naturalism. Reductionism is an idea that its smallest unit or element should be analysed in every complex phenomenon. In physics, scientists reduce reality into physical matters, particularly an atom or a particle. In biology, scientists reduce a living being into a gene. In economics, an economic phenomenon is reduced to an individual action. Particles, genes, and individuals are the atoms of every science. Naturalism means that we can explain reality by its natural mechanism. This is the heritage of a Newtonian view dominant in the modern age. It assumes that reality follows some "exact mechanisms" or general laws (Reiss, 2013).

Economics do two things simultaneously. On the one hand, it generates scientific statements, which are the result of facts finding, generalization, prediction, and model development. This leads to some observable and measurable reality. On the other hand, economics needs to take into accounts moral goals, such as welfare, equality, and justice.

In its development, economics tries to imitate the natural science method found in reductionism. Based on atomism, reductionism is an idea that a complex reality must be divided into its smallest unit. Every positive science adopts reductionism.

In the context of economics, reductionism can be seen from an epistemological and an ontological perspective. From an epistemological perspective reductionism enables a scientist to make an explanation or prediction. Nevertheless, this is not the case for an ontological perspective. Ontology of reality is still beyond a scientific explanation. In this sense, a scientific theory is not equal to the reality. Theory is only a representation of (some) realities.

To help with their explanations, economists use models. Modelling is a scientific activity that enables scientists to explain their hypothesis. Models must be distinguished from theories, since theories are descriptive statements about reality, whereas models are hypothetical statements about reality. Economists, like other scientists, use models in explaining economic reality (Morgan, 2013).

In general, a model is a theoretical construct. There are some familiar definitions of models. First, a model is an instrument to see reality. A model is built to help economists in observing an economic phenomenon. By using a model, reality is stabilized and regulated in certain forms (Morgan & Morrison, 1999). Second, a model is a causal representation. A model in this sense is a form of representation of causal relations that occur in certain economic events. Third, a model is an idealized phenomenon. Economic phenomenon is diverse and random. A particular phenomenon is idealized using an abstraction procedure by eliminating accidental and attributive aspects. Through this procedure, an economist is able to build an ideal construction of an economic phenomenon characterized by regularity and universality. Fourth, a model is an imagination (Krugman, 2013). This means that a model is a kind of an imagination used by an economist to develop a narrative about an economic world. In other words, a model is a kind of a mental entity developed to narrate the world.

The aims of a model are to help economists generate explanations of the economic world (or economic reality) (Harford, 2006). An explanation has to meet the truth-likeness criteria (verisimilitude) derived from two components. The first is a specific phenomenon. If there is a causal relation between two things in a specific economic phenomenon, economists see a causal relation. The second is a general phenomenon. This component relies on a great quantity of a specific phenomenon; economists see a causal relation among those phenomena and come to understand the causal mechanism that shapes every single economic event (Reiss, 2013).

By developing and using models, economists should be able to understand the economic reality. However, the process is not as simple especially when we take into account what happened in the world in the past decades. Financial crisis occurred several times, particularly in 1997 and 2008. These crises raised philosophical questions about the nature and the activity of economics. Financial crisis is an independent reason for why the philosophy of economics appears to have a significant role once again both academically and in practice (Reiss, 2013). There are some challenging questions, especially one related to the scientific status of a theory of economics. Does economics fail as a science?

Some believe that the economic crisis was caused by banks that did not follow the government regulations. However, methodologically, a crisis is a fact caused by the failure of a mainstream economic model used to oversee and overcome the economic reality. As Joseph Stiglitz states, mainstream economics uses a model based on unrealistic assumptions of information, perfect competitions, and perfect markets, which lead to negation of regulations (Stiglitz, 2007). A similar thesis is written by Paul Krugman who states that economic mainstream focuses too much on bad theories, which make unrealistic assumptions about the market. These closed-minded theories fail to recognize and overcome the catastrophic crisis. Those bad theories focus only on mathematical truth, but they fail to give some valuable public policy recommendations (Zreiss, 2013). Economists should rethink the basic assumptions behind their theories and models.

Since economics mostly focuses on generating models and a mathematical analysis of a certain phenomenon, the discipline becomes a formal study. Formal economics seems to dominate almost all economists practice nowadays. These formal characteristics are distinguished from the substantive characteristics. A study by Karl Polanyi states that economics has transformed itself from a substantive study to a formal study. At the beginning of its conception, economics was about the day-to-day household economies. Economics had a normativity to manage a household, distributing welfare, and understand how people run their economies. This was long before modern economics postulates market as the only economic system. After adopting the scientific tools, economics is more similar to other (exact and natural) sciences that focus on explaining and predicting some events. This is a formal character of economics, since it deals with, at some level, historical statements (theories). Economists today use mathematical modelling to explain economic events. In addition, this modelling is not a neutral and value-free entity because it relies on some philosophical commitment.

The practice of economics as a discipline carries within itself philosophical issues concerning epistemology, ontology, and axiology. By ontology, we can find debates on elementary concepts, such as what object is, what property is, what an individual as an economic agent is, what natural law and social law are, and whether there is causality in economic reality or not. By epistemology, we can raise questions about what kind of methodology is used by an economist and why, how about measurements, and how to properly conduct observations. Meanwhile, by axiology, we can see axiological consequence of economics theory and activities, such as welfare, wellbeing, virtue, and theory of justice. Thereby, economics as science must realize that economics has two sides, both positive and normative. In this context, economics must develop and rethink its own methodology, models, and assumptions.

The main framework of mainstream economics is a rational choice. The assumption of a rational choice is widely known in economic methodology development. Mainstream economics is all about making a rational choice. This rational character of an economic agent is debatable. Some economic schools use the concept of rationality as an "ideal type", a kind of heuristic tools to explain the real economic behaviour. This concept is influenced by Max Weber and the Austrian School (Udehn, 2002). To put it simply, rationality is not about the reality of an economic activity. In the real world, people can behave rationally or irrationally. However, economists will interpret those real activities by an ideal type of rationality. This means, if we find an irrational action in the real world, we can say that it is a deviation from the ideal type of rationality. By this methodological procedure, economics makes a simple move to understand economics reality. Nevertheless, this advantage also contains disadvantages. By using this very simple model of rationality, economics cannot observe the real complex phenomena of economy. In a series of experiments, for an example, psychologist Dan Ariely concludes that human being is predictably irrational (Ariely, 2009).

Finally, in developing a model, there are some requirements that need to be fulfilled. First, a model has to be logically permissible. Every model must have a perfect coherence. Second, a model has to have an empirical domain that can be verified or falsified. Third, every model should be precise in describing reality to fulfil its reliability (Frigg & Hartmann, 2012). By contrasting a model with the reality, we might understand if this model is reliable or not, since a model can fail completely (Frigg & Hartmann, 2012).

When we build, develop, and adopt a model to see the reality, paradoxically at the same time the ontology of the world fits to the model. This is called a problem of a model-dependent realism. A model-dependent realism is an idea that reality is determined by a model that is used to apprehend it (Hawking, 2011). Similar to Kantian constructivism, reality is dependent on mental categories. Whenever economics is modelling an individual as a rational agent, economic reality is quickly corrected and fitted to the model. In fact, financial crisis and inequality are not compatible with the model because the real economic activities are much more complex and cannot be reduced to a rational interpretation. Modelling becomes an integral part of scientific activities because this enables scientists to explain a phenomenon in simple way.

The fact is economics uses models to explain reality and it is the same way to say that economics is constructing the economy reality. We have been talking about the nature of economic science as a normative science (aside from the positive quality), which follows the normative conclusion that we should take this model to construct (and explain) an economic phenomenon. It is an epistemic reason to propose the use of a normative model to promote justice and equality as the normative agenda of economics. The same reason has been motivating the report of the Commission on the Measurement of Economic Performance and

Figure 1. Relation between model and reality.

Social Progress in 2009. The commission questions the use of gross domestic products as a welfare indicator, which has been used by the world for decades (Sen, Stiglitz, & Fitoussi, 2009). They find that GDP is not a reliable indicator to measure a country's welfare.

To some degree, the world economy today has challenged economists to reflect on their activities. This article has shown how this discipline has transformed from its origin, which focuses more on normativity, to the current style, which imitates exact and natural sciences that focus on how to explain and predict a future economic event. In a broader sense, economics today does what Polanyi explains as 'economistic fallacy' or identifies economy with a market system (Polanyi, 2001). In reality, people do a different kind of economy, from the past to the present. Market system is only one type of economy. Therefore, focusing on how to explain and predict people's behaviour in a market system with some modelling assumptions about rationality is a hasty practice.

4 CONCLUSION

In conclusion, economics since its modern inception, as found in Adam Smith's title of his masterpiece, *The Wealth of Nations*, has had a normative agenda. We can paraphrase it as justice, equality, welfare, or wealth. This is a fact that economics has two sides: positive and normative. Isolating economics only in terms of positivistic is a foolish idea since economics have a moral agenda. Moreover, as a science, economics should rethink and develop its methodology and models to be better so that it can help explain the world, overcome the obstacles, and comply with the moral (normative) agenda.

The first step to do this is to reflect on the origin of economics as an attempt to manage households and their economy. The practice of economy varies as found in multiple historical and anthropological studies. In this sense, identifying economy with a market system is an economic fallacy that must be avoided by every economist. The scientific function of economics in explaining and predicting economic events must not be limited to the market system and rationality modelling. Economists should be open to think out of the box and realize that their activities rely on some untested assumption about what economics means. By adopting the normative modelling, economics as a discipline would be more humanistic and realistic.

The use of methodological individualism in placing an individual unit of an analysis at the centre of social sciences (like economics) activities is the simplest way to understand social reality. However, the methodology reduces humans and the society to an individual level. This reduction is seen in the model developed in the social science. It may sound simple, but it is not that simple. The philosophical problem in the use of the methodology has an implication on the type of theories built and the models made. Furthermore, the use of this methodology produces a commitment to the market system.

REFERENCES

Ariely, D. (2009) *Predictably Irrational: The Hidden Forces That Shape Our Decisions*. New York, HarperCollins.
Baggini, J. (2010) *Philosopher's Toolkit: a Compendium of Philosophical Concepts and Methods 2nd Edition*. New York, Wiley-Blackwell.
Fetzer, J. (2014) Carl Hempel. *Stanford Encyclopedia of Philosophy* (2014). <http://plato.stanford.edu/entries/hempel/#AnaDis> [Accessed 30th January 2015].
Frigg, R. & Hartmann, S. (2012) Models in Science. *Stanford Encyclopedia of Philosophy* (2012). http://plato.stanford.edu/entries/models-science/. [Accessed 30th January 2015].
Granovetter, M. (1985) Economic action and social structure: The problem of embeddedness, *American Journal of Sociology 91*.
Hawking, S. (2011) *The Grand Design*. New York, Bantams Book.
Krugman, P. (2013) *End This Depression Now!* New York, WW Norton & Company.
Morgan, M.S. (2013) *The World in the Model: How Economists Work and Think*. New York, Cambridge University Press.

Morgan, M.S. & Morrison, M. (editor). (1999) *Models as Mediators: Perspectives on Natural and Social Science.* New York, Cambridge University Press.

Polanyi, K. (2001) *The Great Transformation: The Political and Economic Origins of Our Times.* Boston, Beacon Press.

Popper, K. (2002 [1959]) *The Logic of Scientific Discovery.* New York, Routledge.

Popper, K. (2005) *Unended Quest.* London. Taylor and Francis e-Library.

Reiss, J. (2013) *Philosophy of Economics: A Contemporary Introduction.* New York, Routledge.

Ross, D. (2014) *Philosophy of Economics.* New York, Palgrave Macmillan.

Sen, A. (1991) *On Ethics and Economics.* New York, Wiley-Blackwell.

Sen, A., Stiglitz, J. & Fitoussi, J-P. (2009) *Report by the Commission on the Measurement of Economic Performance and Social Progress.*

Stiglitz, J.E. (2007 [2006]) *Making Globalization Work.* New York, WW Norton & Company.

Stiglitz, J.E. (2013) *The Price of Inequality: How Today's Divided Society Endangers Our Future.* New York, WW Norton & Company.

Turner, S.P. & Roth, P.A. (editors). (2003) *Companion to Philosophy of Social Science.* Oxford, Blackwell Publishing.

Udehn, L. (2002 [2001]) *Methodological Individualism: Background, History, and Meaning.* New York, Routledge.

Cultural Dynamics in a Globalized World – Budianta et al. (Eds)
© *2018 Taylor & Francis Group, London, ISBN 978-1-138-62664-5*

Colonialism and segregation: An analysis of colonial policies in the Strait of Malacca 1795–1825

T.R. Fadeli
Department of History, Faculty of Humanities, Universitas Indonesia, Depok, Indonesia

ABSTRACT: This paper examines the Dutch and the British colonial policies related to trade and societies in the Strait of Malacca focusing on Malacca and Penang. Covering the period from 1795 to 1825 – the period of transition and transfer of power—this paper attempts to observe colonial policies that resulted in social segregation. It begins with the description of the Strait of Malacca, i.e. its historical background, followed by the trade and societies in the region. It then assesses the British and the Dutch methods of ruling and their policies that result in social segregation in the Strait of Malacca's societies. Using both primary and secondary materials collected from the British Library and the SOAS Library, this paper concludes that some policies set up by the colonizers grant privileges to a particular ethnic group.

1 INTRODUCTION

For more than a thousand years, the Strait of Malacca has been an important gate to the international trade route for traders from Europe, Asia, and also Africa. The Strait of Malacca, also known by its cities, Malacca, Penang, and Singapore, had its origins in the early fifteenth century when the region witnessed the overlapping influences of the Thai kingdom of Ayutthaya and the Javanese kingdom of Majapahit (Sandhu & Wheatley, 1983, p. 3). Borschberg (2008) claims that the Strait of Malacca then rose as an area with diverse ethnic groups and at the same time became a region considered as one of the most strategic areas for trading in the Southeast Asia.

In recent times, there has been a growing interest in the studies of the area of the Southeast Asia, particularly the Strait of Malacca. Most of the studies are to learn about its societies and their activities, yet they encourage a more comprehensive history of the Strait of Malacca. Historical studies of economic and trading activities in the Strait of Malacca have also gained extensive attention in recent years. Some of the recent work includes the research of Donald B. Freeman, Nicholas Tarling, Leonard and Barbara Andaya, and Nordin Hussin, who have written on this topic extensively (Andaya, 1962). Basu (1985) points out that the debate on this topic has been discussed amongst historians, but rather than only considering the history of commerce in the Strait of Malacca, many scholars believe that studies should be expanded to an interdisciplinary analysis including the roles of traders, influential institutions, organizational factors, and political and economic problems that have been influenced by commerce.

The importance of understanding various issues from a historical perspective is increasingly recognized nowadays, not only by academics but also by other non-academic groups in the society. As a result, problematizing races and ethnicity becomes one of the most important themes in the study of the history of the Southeast Asia (Andaya, 2008, p. 1). With the presence of colonizers in the Strait of Malacca, the political, economic, and social life of the Strait of Malacca was transformed.

This paper analyses the social segregation demonstrated by colonial policies set up by the British and the Dutch colonial governments in the Strait of Malacca during the late eighteenth until early nineteenth centuries, focusing on Malacca and Penang as port cities.

This paper also compares the British and Dutch colonial legacies by focusing on their ruling methods. Covering the period from 1795 to 1825 – the transitional period between the British and Dutch administration in the region (De Witt, 2008, p. xiii) – the paper argues that there were several colonial policies and trading systems set up by the colonial governments to grant privileges to certain groups in the Strait of Malacca. The fact that there was an extraordinary role of the Chinese in the Strait of Malacca raised a question about whether the British and Dutch colonial governments had established policies that gave privileges to some particular ethnic groups, in this case, the Chinese. This research is important because it aims to provide an explanation of the policies set up by the colonial governments, which significantly affected the nation-building process in the region.

This research uses both primary and secondary materials as sources. The primary materials from the pre-colonial and post-colonial period are often scarce and often discovered not in a good condition. Using archived documents from the *Strait Settlements Records*, *Archives and Special Collection* from the library of the School of Oriental and African Studies, and the *Melaka Records* from the India Office Records at the British Library, this paper examines the economic and social conditions and analyses the economic policies and trading systems adopted by the British and Dutch in the Strait of Malacca during the chosen period. The limitations of the sources are that they are mostly written in English and Dutch languages; as a result, examining texts written only in these languages means neglecting the primary sources written in Malay and other vernaculars. However, the high degree of discipline possessed by the colonial governments of the British and Dutch then, completeness of the information can be obtained.

In the first part of this paper, the background of the Strait of Malacca, including a brief overview of the geography of the region, and the methodology of research will be discussed. In this part, the importance of the Strait of Malacca as an international trade route will be explained. The discussion is not only limited to the geographical context, but will also include a brief historical background of the politics, economy, as well as society and ethnicities in the Strait of Malacca.

The second part will focus on the economy, trade, and society in the Strait of Malacca. This part will concentrate on the trading commodities and also the trading patterns in the Strait of Malacca. From the discussion, it can be seen that natural conditions, such as weather, seasons, and geographical locations are very influential in the trade and social lives in the Straits of Malacca.

The third part will attempt to analyse colonial policies set up by the British and the Dutch colonial administrations. In this section, problems of social segregation become relevant to the discussion. There were policies set up by the colonial governments that created social isolation and ethnic-based work classification in the Strait of Malacca. This part will be followed by the conclusion.

1.1 Trade and society in the Strait of Malacca

Located on the east coast of Sumatra and the west coast of Thailand, the Malaysian Peninsular, and Singapore, the Straits of Malacca is in a very strategic location. Currently, the four countries bordering the Straits of Malacca are Indonesia, Malaysia, Singapore, and Thailand. The Indonesian archipelago is located in the south-eastern part of the Strait of Malacca; there are over ten thousand islands in the Indonesian archipelago and most of these have been destinations for international trade over the past centuries.

Reid (2015, p. 70) argues that the strait is located on one of the busiest trading routes in the Southeast Asia. There are hundreds of islands in the Malacca Strait, which are part of the Malay-Indonesian Archipelago. The Archipelago covers most of the areas in the Malay Peninsula and various other islands. Sau Heng (1990, p. 23) discovered that many settlements had emerged as ports or trading areas around coastal regions and in the riverside area. Some small ports that still exist have grown to be paramount for trade. According to Hussin (2007, p. 1), "Although not all of these settlements and ports became important trading centres, some of them emerged as important regional exchange ports or entrepots." The strategic location had helped many ports in the Strait of Malacca to grow and become places for global commercial trading activities.

Numerous ports in the Strait of Malacca emerged dramatically over a period. These ports were filled by various traders from the third century when they sought particular commodities, such as herbs for medical use, spices, and gold, which were distributed to this area from other areas of Indonesia and the Malaysian archipelago (Kathirithamby-Wells, 1990, p. 1). Some of the traders were from India, Arabia, Persia, and also China. Hussin (2008, pp. 2–3) states, "Chinese traders from South China had also been trading with the ports in the Straits and beyond, although since the sixteenth century they rarely ventured west of the Straits." The traders from China and India used the Strait of Malacca as a hub linking the trade between the South China Sea and the Indian Ocean.

Due to the Strait of Malacca's strategic location, several port cities emerged and became significant in Southeast Asia. According to Masselman (1963, p. 223), some important port cities are Malacca, Riau, Siak, Palembang, Kedah, and Aceh. These port towns played a major role in the trade in the archipelago. Many immigrants settled in these cities both for short and long periods of time. They played a vital role in the growth of the port cities in the Strait of Malacca.

Port cities in the Strait of Malacca have a multicultural and multi-ethnic population. Settlers who lived in the Strait of Malacca were not only traders who came from Asia, but also from Europe and Africa. Various groups with different languages, religions, and races came for trading. Hussin (2008) and Kaur (2004, p. 5) make clear that traders from all around the world, who came to many port cities in the Strait of Malacca, could be separated into two groups: traders from Europe, Indian subcontinent or China, and traders who came from the region of the archipelago. He adds that the two groups could be classified into four categories. Firstly, the European traders and merchants from large trading companies, such as VOC (*Vereniigde Oostindische Compagnie*), EIC (East India Company), and many traders from the regions of Europe such as French, Spanish, Portuguese, Danish, and English traders, who belonged to some of the many companies that traded in the past. The second group is Indian traders who came from the Indian subcontinent to buy and sell goods in the Strait. This group consists of Gujarati, Moors, Hindus, and Chulia Muslims. The third group was the Chinese traders from Mainland China. Finally, the merchants and traders who traded in the Strait of Malacca coming from the Archipelago comprised the fourth group.

Trading activities that exist between the Eastern and the Western worlds had also been established long ago in the Strait of Malacca. Many Westerners who came to the region to buy and sell goods carried out trading activities with Easterners. In the middle of the seventeenth century, there were merchants who came from European countries, such as Portugal and the Netherlands (represented by *Vereenigde Oost-Indische Compagnie* officials), followed by Britain (represented by East India Company officials), France, and Denmark (Chauduri, 1985, p. 21, see also Souza, 1986, p. 228, and Eang, 1996, p. 28).

There were several types of goods that were traded in the Strait of Malacca. The commodities came from the Indian subcontinent, and then continued to be transported to Southeast Asia and the South China Sea. The merchants in the Strait of Malacca traded five main categories of commodities: manufactured goods, mineral ores, forest products, human slaves, and products for food (Hussin, 2008, p. 12, see also Reid, 1983). The basic needs of life, such as sugar, salt, and clothing, were the types of commodities most frequently traded by merchants from the Indian subcontinent through Southeast Asia to the South China Sea. The manufactured products traded in the region were cotton textiles, porcelain and glass, silk, precious stones, and jewellery (Chauduri, 1985, pp. 19–20). Raw materials for industries were also traded in the region. Some of the raw materials were produced locally, but demand was then also high for imported materials.

Merchants from other parts of the world were looking for other commodities in the Strait of Malacca. Gum resins and aromatic woods were some of the many commodities traded there. According to Villiers and de Matos (as cited in Hussin, 2008, p. 13), "spices such as cloves, nutmegs, and mace were in great demand among the Arabs, Persians, and Indians, but not by the Chinese, who preferred pepper." At certain periods only, there was an increase in demand from European traders who sought Chinese tea. Predominantly, the commodities traded in the Strait of Malacca were mainly raw materials and food items. Meanwhile, the traders imported manufactured items from India and China.

The trends and patterns of trading commodities did not change from the seventeenth to late nineteenth centuries. As Cortesão (1944, p. 265) perceptively states (as cited in Andaya, 2008), "Melaka's economic success could be measured by the fact that it had not one but four *syahbandar*, the officials appointed to handle all matters dealing with foreign commerce at the port." This indicates that there was relatively large-scale trading for various commodities in the Strait of Malacca. It must, therefore, be recognized that the Strait of Malacca had been crucial to the trading world of Southeast Asia. Its strategic location benefited the region economically. Thus, the cities in the region had a lot of opportunities to grow and develop rapidly.

1.2   *Social segregation in the Strait of Malacca*

From the first half of the seventeenth century to the late eighteenth century, the territories of the Strait of Malacca that were claimed by the Dutch spread widely. The Dutch government claimed jurisdiction over a region spanning about two thousand and three hundred square kilometres (Harrison, 1986, p. 8). By 1795, more than 14,000 people (Extract from Vol. 705 Home Miscellaneous Series, 441, p. 34) representing many ethnicities including Malays, Chinese, Indians, Arabs, Eurasians, some Javanese, and hundreds of soldiers, civilians, and employees of the Dutch administration filled the area.

The Chinese had settled in the region since the sixteenth century. Most of the Chinese were contractors, merchants, artisans, small traders, and shopkeepers. They married local women from various places in Indonesia and settled in the Strait of Malacca. According to Harrison (1986, p. 9), the Chinese were described as hard workers, enterprising, and acquisitive. Some of the Chinese were relatively wealthy and had a culture of respect towards elders and, as Harrison (1986, p. 9) adds, "the older families retained great pride in their own cultural tradition."

Apart from the Chinese, there were also Indians who settled permanently in Malacca. The Indians in Malacca who consisted of several sub-ethnic groups took ordinary jobs in the region. The proprietors held their lands on a sort of feudal contract with the Dutch government, having the right to exact a rent amounting to one-tenth of the land from the farmer, in return for an agreement to open the land up to maximum cultivation and also to maintain infrastructure. In fact, most of these proprietors were absent and lived in the town pleasantly, using land agents as collectors.

The Dutch administrated the region and controlled the population through their direct process of ruling. A small group of the VOC (United East India Company) officials directly ran the administrative control in the Strait of Malacca. One of the administrative systems introduced by the Dutch was indirect taxation. As Harrison (1986, p. 9) explains, this system "...was a method of indirect taxation by which revenue was raised through 'farming out' certain monopoly rights or privileges to individual entrepreneurs or 'farmers'."

The Dutch not only implemented the aristocratic system, but also segregated ethnic groups. The Dutch classified ethnicities in order to ensure the welfare of each group as well as the government's administration. On the other hand, to manage society in Malacca, the British implemented an administrative system that was not much different from the Dutch's. Despite the similar administrative systems, the British administration implemented an associative rather than assimilative approach.

In spite of all methods of ruling by the colonial governments, the British and the Dutch colonial governments tried to segregate ethnic groups to maximize economic activities in the region. The colonial governments assumed that each ethnic group had different capabilities so that distribution of work had to be done prudently. Each ethnic group had activities related to the location of their residence. For example, the Europeans settled in the west or the north side of Penang. Such areas were mainly occupied for administrative matters. In the same way, the Chinese and Indians resided in the central business area in the town. This was merely because the daily activities of the Chinese, Indians, and some Malays required them to stay in the area where they worked.

Since economic motive was one of the main purposes of colonialism, the colonial administrations built organized institutions to extract natural resources from their colonies (Hrituleac, 2011, p. 28). To achieve this goal, the colonisers usually had different strategies and approaches to administer their colonies. According to Calvin (2011), "The British had another motivation in segregating the division of labour: keeping wages low by suppressing collective action. By ensuring the distribution of labor along racial lines, the colonial authorities were able to keep wages low, and therefore, maximize profits for the Empire."

Other issues on policy that needed to be considered were the privileges given to some particular ethnic groups in the region. Several commercial privileges, which were the exclusive privileges of preparing opium for smoking and for retailing prepared opium; retailing Asiatic liquor and American rum; distributing liquor; retailing paddy within the settlement of Malacca; retailing *sirih* (betel) leaves within the town and port of Malacca and its suburbs; keeping stalls in government market; and keeping pawnbrokers' shops in the town and suburbs of Malacca (Public Notice Regarding the Farming by Auction of Various Commercial Privileges at Malacca, 15th April 1830) were granted to the Chinese traders in Malacca by the Dutch. Most of the traders in Malacca were Chinese and Indians; apparently, this illustrates the inequality that existed and hampered the economic development of Malacca.

The above privileges were granted to the Chinese people because they had certain characteristics. The nature of the Chinese described by William Skinner (as cited in Mearns, 1983, p. 140) is "displaying extreme industriousness, willingness to labour long and hard, steadiness of purpose, ambition, desire for wealth and economic advancement, innovativeness, venturesomeness, and independence." The transition period of 1795–1825 saw growing Chinese involvement in trade and economic activities in Southeast Asia. The colonial government that administrated the port town expected the arrival of Chinese traders. Moreover, Chinese settlements in the port town were greatly encouraged. The colonizers believed that the Chinese people were important in the process of economic development of a port town.

Other trade-related privileges were also provided to the Chinese by the British colonial government during their administration in Malacca. The rules for the Office of Registrar of Imports and Exports explained that every merchant vessel had to report to the office as soon as possible after arrival, and the Office would levy a duty of six Spanish dollars per pound of opium. Article 11 states that the rules may apply to Europeans and Natives (Rules for the Office of Registrar of Imports and Exports, Malacca, 20 April 1825). Therefore, the Chinese were exempt from paying taxes to the Office. Evidently, the British administration granted privileges to the Chinese to keep them settled and help the economy of Malacca. Undoubtedly, the Chinese gained privileges from the British and the Dutch colonial governments in the Strait of Malacca.

2   CONCLUSION

It has been explained that the British and the Dutch were the two greatest colonial powers in the Strait of Malacca. During the Dutch administration, they used an indirect system of rule. The system was also utilized by the British to set up colonies in Malacca and Penang. Also, the Dutch used the assimilative approach in governing Malacca, while the British government preferred the associative one.

The segregation in the colonial cities was common in the form of patterns of settlement. In Malacca, the Dutch residential segregation policies applied to ethnic groups based on the work performed. The British implemented the same policy when administering Penang. In an attempt to segregate the society, the British colonial government granted privileges to the Chinese. This attempt is evidenced by the formation of government bodies to protect the Chinese people. Chinese people had a greater chance to develop their economy because they were seen as a hard-working group and were the key to economic development, not only in the Strait of Malacca but also in the Southeast Asia.

It can be said that the Strait of Malacca was a perfect example of the location where the competition between two European colonial powers took place. The fact that the Dutch and the British granted privileges to the Chinese led to the segregation in the Strait of Malacca, and it was undoubtedly a significant influence in the nation building process in the region. The granting of privileges could have also been one of the many factors in the emergence of inter-ethnic conflicts that would occur in the Southeast Asia in the future.

The assimilative administration system implemented by the Dutch colonial government created a gap between societies and economic inequality, which was inherited over time. Without economic equality, inter-ethnic tensions will emerge, social jealousy will develop, and conflicts are inevitable. For this reason, this research is important because today's conflict originated from the colonial policies in the past.

REFERENCES

Andaya, L.Y. (2008) *Leaves of the Same Tree: Trade and Ethnicity in the Straits of Malacca*. Honolulu, University of Hawai'i Press.

Appendix to report giving geographical and racial a/c.

Basu, D.K. (1985) *The Rise and Growth of the Colonial Port Cities in Asia*. London, University Press of America.

Borschberg, P. (2010) *The Singapore and Melaka Straits: Violence, Security, and Diplomacy in 17th Century*. Singapore, NUS Press.

Calvin, E. (May 2011) The march of racial progress in Malaysia, in *The Progress of Memory*, Issue 4.

Cortesão, A. (1944) *The Suma Oriental of Tomé Pires*. London, Hakluyt Society.

De Witt, D. (2008) *History of the Dutch in Malaysia: A Commemoration of Malaysia's 50 Years as an Independent Nation and Over Four Centuries of Friendship and Diplomatic Ties Between Malaysia and the Netherlands*. The Netherlands, Nutmeg Publishing.

Drabble, J. (2000) *An Economic History of Malaysia, C.1800-1990: The Transition to Modern Economic Growth*. Hampshire, Macmillan Press.

Eang, C.W. (1996) *The Hong Merchants of Canton*. London, Curzon Press.

Elson, R.E. (1997) *The End of Peasantry in Southeast Asia: A Social and Economic History of Peasant Livelihood, 1800–1900s*. New York, St Martin's Press.

Extract from Vol. 705 Home Miscellaneous Series, 441, p. 34.

Extract of Regulations of the Governments of Batavia and Malacca Regarding Private Trade on Company's Ships etc. October 1790–November 1799.

Fong, M.L. (1989) The social alignment patterns of the Chinese in nineteenth-century Penang. In: *Modern Asian Studies Vol. 23, No. 2*. Singapore, National University of Singapore.

Harrison, B. (1986) *Holding the Fort, Melaka under Two Flags, 1795–1845*. Kuala Lumpur, the Malaysian Branch of the Royal Asiatic Society.

Heng, L.S. (1990) Collecting centres, feeder points and entrepots in the Malay Peninsula 100 B.C. – A.D. 1400. In: J. Kathirithamby-Wells and John Villiers, (eds.). *The Southeast Asian port and polity rise and demise*. Singapore, Singapore University Press.

Higgot, R. (1999) The political economy of globalization in East Asia: The salience of region building. In: C. Olds, P. Dicken, P. Kelly, L. Kong, and H.W. Yeung, (eds.) *Globalization and the Asia-Pacific*, edited by. London and New York, Routledge.

Hoyt, S.H. (1993) *Old Malacca*. Kuala Lumpur, Oxford University Press.

Hrituleac, A. (2011) *The Effect of Colonialism on African Economic Development*, Aarhus University, Business and Social Science.

Hussin, H. (2008) *Southeast Asian Religion, Culture and Art of the Sea*. Kuala Lumpur: Institute of Ocean and Earth Sciences (IOES). University of Malaya.

Hussin, N. (2007) *Trade and Society in the Straits of Melaka: Dutch Melaka and English Penang, 1780–1830*. Singapore, NUS Press.

Hutterer, K. (1977) *Economic Exchange and Social Interaction in Southeast Asia: Perspectives from Prehistory, History, and Ethnography*. Ann Arbor: University of Michigan.

India Office Records and Private Papers.

Kathirithamby-Wells, J. & Villiers, J. (1990) *The Southeast Asian Port and Polity Rise and Demise*, Singapore, Singapore University Press.

Kathirithamby-Wells, J. 'Introduction: An Overview:' In: J. Kathirithamby-Wells and John Villiers, (eds.). *The Southeast Asian port and polity rise and demise*, Singapore, Singapore University Press.

Kaur, A. (2004) *Wage Labour in Southeast Asia since 1840: Globalisation, the International Division of Labour and Labour Transformations*. London, Palgrave Macmillan.

Malacca Historical Society. (1936) *Historical Guide of Malacca*. Singapore: Printers L, G. (1963). *The Cradle of* Colonialism. New Haven, Yale University Press.

Mearns, L. (1983) Formal association with the Chinese community of Melaka. In: Sandhu. K.S & Wheatley, P (eds.) *Melaka: the transformations of a Malay capital c. 1400-1980 Vol. II*. Kuala Lumpur, Oxford University Press.

Milne W. Report describing life and work at Malacca.

Morse, H.B. (1966) *The Chronicles of the East India Company Trading to China, 1635–1834. Vol. I.* Taipei, Ch'eng-Wen.

Newbold, T. J. (1839) *Political and Statistical Account of the British Settlement in the Straits of Malacca, Pinang, Malacca, and Singapore*. London, John Murray.

Public Notice Regarding the Farming by Auction of Various Commercial Privileges at Malacca 15th April 1830.

Purcell, V. (1968) *The Chinese in Malaya*. New York, Oxford University Press.

Reid, A. (2015) *A History of Southeast Asia*. West Sussex, John Wiley & Sons.

Rules for the Office of Registrar of Imports and Exports, Malacca, 20th April 1825.

Sandhu, S.K. & Wheatley, P. (1983) *Melaka: The Transformations of a Malay Capital c. 1400-1980 Vol. I*. Kuala Lumpur, Oxford University Press.

Sandhu, S.K. & Wheatley, P. (1983) *Melaka: The Transformations of a Malay Capital c. 1400-1980 Vol. II*. Kuala Lumpur, Oxford University Press.

Schoff, W. (1912) *The Periplus of the Erythraean Sea*. London, Longmans Green.

School of Oriental and African Studies Library.

Souza, G.B. (1986) *The Survival of Empire: Portuguese Trade and Society in China and the South China Sea, 1630–1754*. Cambridge, Cambridge University Press.

Tarling, N. (1962) *Anglo-Dutch Rivalry in the Malay World, 1780–1824*. Cambridge, Cambridge University Press.

Tong, C-K. (2010) *Identity and Ethnic Relations in Southeast Asia: Racializing Chineseness. Singapore*, Springer.

Turnbull, C.M. (1983) Melaka under British colonial rule', in Sandhu. K.S & Wheatley, P (eds.) *Melaka: The transformations of a Malay Capital c. 1400-1980 Vol. I*. Kuala Lumpur, Oxford University Press.

Ultra Ganges—Malacca Incoming Correspondence Box 1, 1815–1820.

Webster, P. (1987) The elementary monsoon. In: J.S. Fein and P.L. Stephens. *Monsoons*. New York, Wiley.

Winstedt, R. (1962) Malaya and Its History, 6th Edition. London, Hutchinson.

Woodman R. (1997) *The History of the Ship*. London, Conway Maritime Press.

Cultural Dynamics in a Globalized World – Budianta et al. (Eds)
© *2018 Taylor & Francis Group, London, ISBN 978-1-138-62664-5*

The impact of economic modernization on the lifestyle of the Palembang society in 1900–1930

N.J. Utama & L. Sunarti
Department of History, Faculty of Humanities, Universitas Indonesia, Depok, Indonesia

ABSTRACT: This article attempts to explain the social changes that occurred in Palembang at the beginning of the 20th century, when economic modernization also brought changes in social life. The main issue discussed in this article is how the impact of economic modernization has influenced the changes in the society of Palembang, especially the changes in lifestyle and its orientation in the period of 1900–1930. At the beginning of the 20th century, the people of Palembang were indulged by economic progress, especially in trade and plantations. There was a kind of consumerist pattern after economic development, such as the widespread and ownership of imported goods. In addition, the economic modernization also brought an impact to the orientation of people's life, from river to land. This was caused by the increase of the large number of cars and bicycles. Yet, the river transport was an obstacle. Although there are a lot of studies about social changes, only few discuss the change about Palembang. One example is a study by Masyhuri who conducted an investigation about the social changes in Palembang, but the study discusses the main factor that was caused by the political development in the 19th century. This paper, on the other hand, talks about the impact of economic modernization on the social conditions in the 20th century. It is fascinating to discuss the pattern of changes and its impact on the society in Palembang itself using a historical method. This research is also supported by relevant sources such as archives, newspapers, journals, and secondary sources.

1   INTRODUCTION

Palembang is a city that has a great reputation as a port and trade city. The primary basis for this opinion is because this region is very strategic. The consequence of the status of this city is that Palembang is open to all cultural influences brought by foreign traders. Since the period of the Kingdom of Sriwijaya, Palembang has been famous as a multicultural city (Wolters, 2011). In addition, as a trade city, Palembang can easily experience changes in social, economic, and cultural issues. Recorded until the 20th century, this region was controlled by three different powers and political cultures, namely the Kingdom of Sriwijaya, the Palembang Sultanate, and the Dutch colonial government (Reid, 2011, pp. 2–3, pp. 8–11).

The 20th century was the golden era for the economy of Palembang. Around the first quarter of the 20th century Palembang was at its most prosperous period compared to previous periods. Financially, most of its inhabitants experienced an economic increase in the early 20th century. Moreover, economic developments such as this brought about not only changes in the physical construction (especially the construction of the city), but also social changes in Palembang. In reality, the type of changes that occurred in Palembang was due to the external influence of a structural nature coming from outside of Palembang (Soekanto, 1983, pp 26–27), and one of the most dominant was the world's rubber commodity prices.[1]

---

1. During the first two decades of the 20th century, there was massive planting of rubber in Palembang. This was due to the increasing demand for rubber for industrial raw materials, especially for tires, so that rubber prices soared (Purwanto, 2002: 205–206).

There are, as yet, few studies on social changes in Palembang. One study that has sparked a discussion on the study of social change is the research conducted by Masyhuri (1993). The study portrays the era when a political transition between the Palembang Sultanate and the Dutch colonial government occurred. This political transition caused changes especially in the structures of the economy in Palembang. Unfortunately, the study does not clearly depict the social changes in the society such as changes in lifestyle.

The main issue discussed in this paper is how economic modernization has influenced the changes in the society of Palembang, especially the changes in lifestyle and its orientation between 1900 and 1930. The argument to answer that question in this paper is that there were changes (especially changes in lifestyle) in Palembang from 1900 to 1930, driven by economic development. This research uses a historical method (Sjamsudin, 2007), which first stage includes the collection of data. The second stage is the source's critique of the collected materials. The third stage is the interpretation of the gathered data. The last stage is the writing of history or historiography, which is the writing of the interpreted data into an article. Meanwhile, to support this paper the writers also use several concepts and theories, especially in explaining the issue of social changes. This article uses the concepts of social change, specifically from Piotr Sztompka that argues that social changes as a result of economic development threatens the existence of old traditions. Several forms of the old tradition are influenced, reshaped, or even swept away by the government and even by the communities themselves (Sztompka, 2011, p. 43).

1.1 Palembang's condition in the early 20th century

Geographically, Palembang was in a strategic position in relation to two major ports at the end of the 19th century, Batavia and Singapore. In addition, nine large watersheds from the Musi River, which was the longest river, also fed this area (Faille, 1971, p. 16). Because of its location which is slightly protruding into the inland areas of Sumatra (+/- 100 km), the river had an important role in the distribution of individual and economic activities in Palembang. Therefore, transport on Musi river was also dependent on the water flow and the condition of the river. For example, during the dry season, the river's water level would be low, making mobilization to the inland difficult and time consuming. The best time for traveling through the river in Palembang would be the rainy season (Sevenhoven, 1971, p. 11).

The area was also divided into two sub-cultures (i.e., upstream and downstream areas) with differences that distinguished not only the geographic conditions of both areas, but also their social, cultural, and economic conditions. The two sub-cultures were often referred to as *Iliran* and *Uluan* (Irwanto, et al., 2010). *Iliran* refers to those who lived in the downstream area of the river or near the city of Palembang. Geographically, this was the lowland area that was dominated by swamps. *Iliran* itself was often interpreted as a region of city people, and then associated as a modern society that was close to civilization. The majority of citizens in this area were traders who used to transport agricultural and plantation commodities from the rural areas to the city of Palembang.

Uluan refers to the inland areas in Palembang: the upstream areas around the Bukit Barisan Mountains, Ranau, and Ogan Komering. The natural condition of this region was mainly fertile plateau. Therefore, this region was the production centre of plantation products in demand by domestic and international markets, such as rubber, coffee, tea, hard wood, resin, and others (Abdullah et al., 1985, pp. 14–17). In terms of socio-economic conditions of the population, the majority of the region's inhabitants were plantation growers. *Uluan* people were also often regarded as less-developed people (in terms of education and religion). However, this condition seemed to change when Palembang entered the 20th century: extensive land holdings and estates owned by the majority of the *Uluan* population apparently transformed them to become a modern society.

As a major trading city, Palembang experienced a quite significant population growth, especially after the discovery of several new economic resources such as rubber and the discovery of mining sources which absorbed a lot of the local workforce as well as laborers and porters coming in from Java. The population of Palembang in 1900 totaled about 600 thousand, and it increased to reach 1 million in 1930 (Zed, 2003, p. 66; Stibbe, 1919, p. 264). The composition also included foreign ethnic communities in Palembang, which were dominated by the Chinese and the Arabs who had inhabited the area since the period of the Palembang Sultanate (Utama, 2015, pp. 6–13). The diverse population and the rapid mobility were the driving factors toward social change in Palembang in the early 20th century. Yet, the most important factor which had a significant effect was the process of commercialization of agriculture and plantation, leading to a rapid advancement of economy. Furthermore, since the area of uncultivated land was so vast, the colonial government was unable to manage the entire land in Palembang, and the local people opened new plantations, especially for high-priced commodities such as rubber (Zed 2003, pp. 85–88).

1.2   *Economic modernization*

In the early 20th century, Palembang was known as one of the biggest exporters of natural resources for the Netherlands Dutch Indies, with commodities from the agricultural sector (rubber) and mining sector. In fact, the early 20th century was not only the era when Palembang was known as the biggest supplier for commodities for the market, but it was also known as a port city for transit. Palembang was not only considered as strategic in terms of geography, but it also produced natural resources as early as from the Sriwijaya Kingdom era to the Palembang Sultanate era. The commodities included gold, ivory, woods, cattle, tin, pepper, etc. (Vlekke, 2008, pp. 44–45). However, in the early 20th century, the most dominant plantation commodities in Palembang were rubber and coffee. Although the rubber plantations were dominated by small-scale holders, the management was not traditional. The method of recording and cultivation of the plantation applied by these small-scale holders were the same as that used by the Europeans (Furnivall, 2009, pp. 338–339). Rubber is actually not a native plant of Palembang and it was said to be originally from the Malay Peninsula.[2] Interestingly, in Palembang, there was a type of sap plant known as *"rambung"*, but this plant could not be cultivated extensively in Palembang. Then at that time, the rubber tree (*Hevea Brasiliensis*), which had more economic value, was introduced. Apparently, the climate of Palembang was also suitable for planting rubber trees, making this plant quickly become the target plant of many farmers in Palembang (Purwanto, 2002, p. 203).

As for coffee, although the plantation area and the profits were not as extensive as that of rubber, coffee was, however, widely grown in the inland and upland areas of Palembang. Nevertheless, the production of coffee and rubber by the local people should not have been underestimated, since the production of both commodities cultivated by the local people had even exceeded the production of the companies belonging to the colonial government and the private sector (local coffee production reached a total of 90% of the total coffee production in the Palembang *Uluan* region) (Zed, 2002, p. 300, Peeters, 1997, p. 123).

Interestingly, there was a difference in the management style of the plantations in Palembang compared to that of Java and east Sumatra, which was mostly controlled by the government and private companies. In Palembang, there were no agrarian conflicts, especially related to land disputes between the people and the government. The vast area of vacant lands in Palembang enabled small-scale plantation holders and the government to live peacefully. Besides that, the government seemed too focused on the mining sector, as in the early 20th century, mining commodities were discovered in Palembang in significant quantities. Therefore, there was no overlap between commercial plantation lands and farmlands, and there was no friction stemming from the interests of their management (Zed, 2003, pp. 88–90). This was unlike the condition in some regions in Java where the paddy fields were an integral part of the sugar cane plantations. A different picture could also be observed from the welfare of farmers;

---

2. Rubber tree (Hevea Brasieliensis) was first developed in the region of the Malay Peninsula circa 1876–1877. It is estimated that a lot of rubber seedlings planted in Palembang originated from this area. The spreading of the rubber tree was introduced by the Hajj pilgrims who transited in the region. One of the first rubber plantations in Palembang was established in Sugihwaras in 1910–1912. (Drabble, (unknown year): 570).

during the "rubber boom", farmers in Palembang became prosperous, but farmers in Java were less successful in the era of agriculture commercialization.[3]

In addition, having a good network of traders and commodity brokers, the farmers were able to establish a strong collaboration between the plantation sector and the trade sector. The inland communities (*Uluan*) were the actors in the plantation sector, while the urban communities (*Iliran*) established the trade relations and product marketing of the estates. The network brokers in Palembang in the early 20th century were controlled by the Chinese traders. One important factor that made the Chinese traders dominate the trade, ranging from small to large scale, in Palembang was their mastery of the river transport business in Palembang and because most "wheel ship hulls" that operated in Palembang until 1920 were controlled by Chinese families (Zed, 2003, pp. 94–96. *Kemoedi* 3rd July 1926).

The economic progress that occurred in Palembang in the early 20th century had a significant impact on the economy of the people. Until the 1920s, the collaboration between the local people and the estate-owned mines had impacts on the colonial government and made Palembang one of the important export regions in the Dutch East Indies (Linblad, 2000, pp. 344–347).

1.3 Social and orientation change in Palembang society

The development and economic growth that occurred during the two decades in the early 20th century brought significant changes in the lifestyle of the people of Palembang. In the 13 years of the planting and rubber tapping period, the total income of the rubber farmers in Palembang reached 95 million Gulden, not to mention the profit calculated by the brokers and traders who marketed these commodities (Purwanto, 2002, p. 218). Interestingly, besides becoming one of the largest exporting regions in the Dutch East Indies, Palembang also became one of the main destinations for importers of luxury goods. Significant transformation during the period of economic development in Palembang could be recognized by the changes that include the physical development of the city, the large number of people going for the Haj pilgrimage, progress in education, the purchase of imported goods (particularly cars and bicycles), and the adoption of new habits that were considered modern.

The physical development of the city was regarded as a symbol of prosperity in Palembang. Between the years 1900 to 1930, Palembang was transformed into a city with a European style and culture in many corners. Construction of facilities was done in this city such as filling the rivers to build roads, and building courtyards/parks, post offices and telephone lines, pawn shops, churches, schools, theaters, hotels, meeting halls, and the construction of the most famous building of those days, the water tower (Irwanto, 2011, pp. 50–52). Yet, unfortunately, most of the facilities and buildings established by the colonial government were used for the advantage of the colonial government and the faction itself, such as the real estate development (*Talang Semut*) which was to become the future elite housing for Europeans (Ali and Sujiwati, 2015, p. 8).

The significant development not only occurred in Palembang city, but also in the suburb region in Plaju. In this region, a large oil refinery was built, and in addition to the oil refineries, the company also built various facilities supporting its employees ranging from housing, sports facilities (courtyards and swimming pools), clubs, hospitals, and shopping stores. Such transformations in Plaju are in contrast with the condition of Plaju before the 20th century, which was one of the marginal, backward region and separated from the city center. However, in the midst of late modernity and development in 1920–1930 Plaju could easily compete with Palembang (Tanjung, 2015, pp. 303–302).

Next, the economic development in Palembang also had an impact on the society in terms of performing the Hajj Pilgrimage.[4] After the rise in rubber prices, the number of

---

3. As illustrated by Clifford Geertz on sugar cane farmers in Java at the same time also became laborers and factory workers (Geertz, 1971: 327).
4. Hajj is one of the obligatory rituals for Muslims who are physically and financially able to perform it.

people performing the Hajj from Palembang also increased significantly, especially in the years when the price of rubber was expensive. Prior to 1920, the average number of the Hajj pilgrims from Palembang was about 1,400 people, but not including the data from the years of 1915 and 1916. The significant increase began after 1920, especially in 1927, which represented the culmination of the rubber trade in Palembang. The records show that there were 7,000 pilgrims from Palembang at that time. Interestingly, when the price of rubber fell, the demand for the Hajj pilgrimage also dropped. For example, in the post-Malaise period, Hajj enthusiasts in Palembang declined to the number of 500 people (Peeters, 1997, p. 151).

Changes also occurred in the consumption and lifestyle of the urban community. There were striking differences between the periods before and after the economic modernization that occurred in Palembang. As previously mentioned, the large influx of rubber farmers in the early by 20th century was reflected by the fact that luxury goods were no longer difficult to buy. The consumption pattern also changed with the number of entry-imported goods, including food. Surprisingly, before the rise of rubber planters, rice production in Palembang was sufficient to meet the needs of the region, but after the expansion of rubber plantations, rice had to be imported from outside of Palembang. In addition to rice, other goods were also imported on a large scale such as fabric, sewing machines, Western cigarettes, and others (Purwanto, 2002, pp. 218–219).

Imports of foreign cultures also occurred during this period. Many urban people in Palembang started to adopt modern cultures. The government facilitated these habits, and the public valued them. Cinemas, which were built in Palembang, not only served the foreign (white) audience but also the local people. Several groups of young people were also interested in forming a kind of assemblage of art and sports that was considered modern. This association was named "*Wy Ontwaken*" that a few days later was renamed to "*Madjoe Adil Setia*", embracing the activities of young people, namely in music, *toneel* (theater), and football (*Kemoedi* 3rd July 1926, 10th July 1926).

However, the most important import items that changed the orientation of Palembang were land vehicles such as cars and bicycles. Having an increased income the people could afford to buy bicycles and cars that were then considered super luxury goods and were complementary goods for the rich in Palembang. There were about 300 cars in Palembang in 1922; this figure then increased to about 1,400 cars in 1924. Meanwhile, for bicycles in Palembang, the number reached 19,000 and they were imported between the years 1910–1929 (Purwanto, 2002, pp. 219–220). Even a year later in 1925, the number of cars in Palembang increased sharply to more than 3,000 cars (Peeters, 1997, pp. 135–136).

Between 1900 and 1920 these land vehicles had no effect on the existing river transportation. Indeed, the cars at that time were only the size of a cart, but the superiority of these vehicles lays in their speed. Nevertheless, before 1920, river transportation was still in demand because the operational costs were much cheaper than that of cars which needed infrastructure such as roads and bridges (Zed, 2003, pp. 110–112).

However, the rampant spread of cars after the 1920s then brought a negative impact on the river transportation in Palembang. The traders and farmers in Palembang apparently had enjoyed the economic progress of Palembang and were quite content in those years. Therefore, these farmers and traders did not consider developing their business by investing more to expand water transportation. In contrast, many people in Palembang spent more on cars rather than buying or repairing ships. As a matter of fact, the commodity trading business was very strategic to be further developed with its important function to support the distribution of goods, although they mostly relied on foreign groups. Meanwhile, cars were more widely used as a symbol of prestige for leisure or just driving around town (*Pertja Selatan* 15th July 1926).

In the end, the economic growth in Palembang during this period (1900–1930) ended with the onset of the Great Depression (*Malaise*). The crisis in 1929 hit the industrial sector and greatly weakened the rubber business that was "booming" in Palembang. Great losses as a result of the decline in rubber prices struck the region in 1928. The economic boom and the large money in-flow were not put to good use, so that when the Great Depression struck, the people of Palembang had no money. Most of the money was spent to meet the consumptive desire to buy cars for leisure, and was even used for the pilgrimage journey as a social prestige rather than for religious purposes (*Moesi* 11th May 1928). As explained above, most of the people in Palembang did not have any intention to develop their commodity business. In subsequent years, the price of rubber continued to rise and the consumptive pattern between 1900 and 1930 continued to repeat itself, even to this day.

2 CONCLUSION

Major economic development in Palembang in the beginning of the 20th century brought changes not only in the economic sector, but also in the social aspect of the society in Palembang. The rise of rubber prices in the world market led to significant improvement in the standards of living in Palembang during the period of 1900–1930. Modernization was apparent in the city's significant development. The city's progress was mainly around *Talang Semut* area as indicated by the European-style buildings such as public service buildings, theatres, housing complex, roads, and bridges. In addition, the economic development also brought significant social changes in the community, such as the rise in the number of people going to the Hajj pilgrimages in line with the growth of the economy. Furthermore, western culture was also adopted, such as watching films in cinemas and establishing art clubs and sports clubs. Another impact of this economic boom was the increasing consumption pattern among the people of Palembang. Imports of goods also increased, ranging from imported food up to luxury goods. Cars, as luxury goods, seemed to be used only as a prestige of that era rather than being utilized for its actual function. Besides that, the use of cars also began to cause a decline in the river culture. Apparently, the Palembang people were happier to travel to places using a car rather than going by boat or ship. This last point brought an unfavorable impact for the development of businesses which were mainly owned by natives/locals of Palembang as they spent more money for consumptive purposes and for prestige rather than for business development. These issues show how developments in the economic sector had affected the people's lifestyles and also changed the public perception on certain values, such as a shift in the meaning of the river.

REFERENCES

Abdullah, M., et al. (1985) *Kota Palembang Sebagai Kota Dagang dan Industir (The city of Palembang as a Trade and Industrial City)*. Jakarta, Depdikbud.
Ali, N. & Sujiwati, M. (2015) Pembangunan Kota Palembang dengan Konsep Tata Ruang Hijau pada Masa Hindia-Belanda (The Construction of Palembang City with a Green Environmental Concept during the Ditch-Indies Era). *Journal Tamaddun (UIN Raden Fatah Palembang)* Vol. XV No.1.
Colombijn, F. & Coté, J. (2015) *Cars, Conduits, and Kampongs, the Modernization of the Indonesian City 1920–1960*. Leiden, KITLV.
Dalton, G. (1971) *Economic Development and Social Change*. New York, American Museum Sourcebook in Anthropology.
Faille, P.D. (1971) *Dari Zaman Kesultanan Palembang (Since the Era of the Palembang Sultanate)*. Jakarta, Bhratara.
Furnivall, J.S. (2009) *Hindia Belanda, Studi Tentang Ekonomi Majemuk (The Dutch-Indies, A Study on Plural Economics)*. Jakarta, Freedom Institute.
Irwanto, D., et al. (2010) *Iliran dan Uluan (Iliran and Uluan)*. Yogyakarta, Eja Publisher.
Irwanto, D. (2011) *Venesia Dari Timur: Memaknai Produksi dan Reproduksi Simbolik Kota Palembang Dari Kolonial Sampai Pasca Kolonial (Venice of the East: Understanding the Symbolic Production and Reproduction of the City of Palembang from Colonial to Post-colonial Times)*, Yogyakarta, Penerbit Ombak.
Linblad, J.T. (2000) *Sejarah Ekonomi Modern Indonesia Indonesian (Modern Economic History)*. Jakarta, LP3ES.
Linblad, J.T. (2002) *Fondasi Historis Ekonomi Indonesia (The Historical Foundation of Indonesian Economy)*. Yogyakarta, Pustaka Pelajar.

Mashyuri. (1993) *Perdagangan Lada dan Perubahan Sosial di Palembang 1790–1825* (Pepper *Trade and Social Change in Palembang 1790–1825)*. [Thesis]. Departemen Ilmu Sejarah, Universitas Indonesia, Indonesia.

Newspaper *Hanpo* 1926.

Newspaper *Kemoedi* 1926.

Newspaper *Moesi* 1928.

Newspaper *Pertja Selatan* 1926.

Peeters, J. (1997) *Kaum Tuo-Kaum Mudo, Perubahan Religius di Palembang (The Old and The Young, Religious Change in Palembang) 1821–1942*. Jakarta, INIS.

Rahim, H. (1998) *Sistem Otoritas dan Administrasi Islam, Studi tentang Pejabat Agama Masa Kesultanan dan Kolonial di Palembang (Islamic Authority and Administration System, A Study on Religious Officials of the Sultanate and Colonial Era in Palembang)*. Jakarta, Logos.

Reid, A. (2011) *Menuju Sejarah Sumatra (Toward Sumatran History)*. Jakarta, Yayasan Obor Indonesia.

Van Sevenhoven, J.L. (1971) *Lukisan Tentang Ibukota Palembang (A Description of the Capital City of Palembang)*, Jakarta, Penerbit Bharata.

Soekanto, S. (1983) *Teori Sosiologi Tentang Perubahan Sosial (Theory of Sociology on Social Changes)*. Jakarta, Ghalia Indonesia.

Stibbe, D.G. (1919) *Encylopaedie van Nederlansch-Indie*. Gravenhage-Martinus Nijhoff Leiden, N.V.E.J. Brill.

Sjamsudin, H. (2007) *Metodologi Sejarah (Historical Methodology)*. Yogyakarta, Penerbit Ombak.

Sztompka, P. (2011) *Sosiologi Perubahan (Sociology of Change)*. Jakarta, Prenada.

Utama, N.J. (2015) *Perebutan Ruang Sungai di Palembang (The Conquest for River Space in Palembang)*. Presented at Third Graduate of History Seminar, Universitas Gadjah Mada.

Wheatly, S. (No year) *Malaka*. Kuala Lumpur, Oxford University.

Wolters, O.W. (2011) *Kemaharajaan Maritim dan Perniagaan Dunia Abad III-VII (The Maritime Empire and World Trade in the 3rd–7th Century)*. Jakarta, Komunitas Bambu.

Vlekke, B.H.M. (2008) *Nusantara Sejarah Indonesia (Nusantara, the History of Indonesia)*. Jakarta, KPG.

Zed, M. (2003) *Kepialangan Politik dan Revolusi Palembang 1900–1950 (Political and Revolutionary Brokerage of Palembang 1900–1950)*. Jakarta, LP3ES.

Cultural Dynamics in a Globalized World – Budianta et al. (Eds)
© *2018 Taylor & Francis Group, London, ISBN 978-1-138-62664-5*

The impact of modernization on the economy for fishermen in Makassar City

P. Rifal & L. Sunarti
Department of History, Faculty of Humanities, Universitas Indonesia, Depok, Indonesia

ABSTRACT: This paper discusses the impact of modernization on the economic development of fishermen in Makassar. In general, fishermen are considered as a group of people who are poor with a subsistence oriented economy. However, the fishermen in Makasar tend to show a more commercially oriented economy compared to those in other regions in Indonesia. More advanced living began to emerge as they started to use modern technology, such as actuator boats, fish preservation (cold storage), and trading networks. With modernization, the economic life of those fishermen developed significantly as seen from their income per capita, which is higher than that of the workers in the field of construction and those working at sugar and rice mills. With higher incomes, every family of the fishermen could afford to buy a 'golden ringgit' as a symbol of wealth in the Bugis-Makassar tradition. This research is a historical study using a qualitative approach. The data were obtained through interviews with fishermen and entrepreneurs of the local fisheries. Furthermore, the data were processed using the historical method, which includes heuristic, criticism, interpretation, and historiography.

1 INTRODUCTION

In general, fishermen are considered as poor and marginalized people. BPS recorded that in 2011, the number of poor fishermen in Indonesia reached 7.87% or 25 million people, and 14% of the national total poor population reached 31.02 million people (*Koran Sindo*, the 4th April 2016 edition). From the record compiled by *Koran Sindo* (2016) there were 16 million fisherman households, which had revenues of 1.1 million rupiahs per month. From an economic perspective, most fishermen are known to be more oriented towards subsistence economy rather than commercial fishing. However, the fishermen in Makassar tend to adopt a market economy such as those in Singapore. This economic orientation began to appear after they started to use modern technology in fishing. At the macro level, in 1973 they had a surplus of 2000–3000 tons of catch. Cooperating with Japan, companies like Sendid and Bonecom successfully exported 400 tons of catch (*Pedoman Rakyat*, edition XXVII No. 1. Saturday, the 5th May 1973, and *Pedoman Rakyat*, edition XXVII No. 77, Friday, the 1st June 1973). At the micro level, most fishing households in Makassar would possess several "golden ringgits," which symbolize success in the tradition of the Bugis-Makassar people. The annual income per capita was Rp.69,800.000, with a monthly income of Rp.5,816.000 (*Harian Fajar*, the 11th October 1973, and *Pedoman Rakyat*, edition XXVII No. 77, Friday, the 1st June 1973). Their income was higher than that of workers in other sectors. Based on the data from the Statistical Pocketbook 1968 (as cited in Booth, A. and Sundrum, 1981, p. 263), the income of sugar factory workers and those working at Gilligan rice mills was Rp.2,179,000/month, while the highest salary received by professional and technical workers in the private sector was as much as Rp.4,479.00/month. Therefore, the fishermen's income was twice as much as that of sugar factory workers, and in fact it was even higher than that of workers in the engineering field.

Prior to modernization, such a phenomenon would have been almost hard to find. With fishing areas located along the coast, the fishing production was greatly influenced by climatic conditions. During the western monsoon between October and April, it was difficult to catch fish, since the fishing boats could be used only in the area near the coastline. The study of modernization and its socio-economic impacts on fishermen was carried out by Sumardiati (1999). The findings show that modernization does not necessarily help improve the fishermen's livelihood. Instead, it may cause conflicts between the modern and the traditional fishermen. However, in the case of Makassar fishermen, modernization does not seem to inflict such conflicts.

According to de Jonge (1989), modernization will attract new investors. However, as the operational costs increase due to modernization, fishermen will have to borrow money from investors. In addition, their catch is usually monopolized, since investors will determine the selling price of the catch, which leads to further dependence on the investors. This would create an unfavorable condition for the fishermen. Nevertheless, Makassar fishermen do not experience this problem. The income of the *sawi* (small fishermen) is higher compared to the income of workers in other sectors. Social mobility occurs vertically upward among fishermen. From the above statement, it is necessary to further understand the factors related to the economic progress among Makassar fishermen and the impact of modernization on their livelihood.

In view of the above circumstances, this article aims to analyze the impact of modernization on the lives of Makassar fishermen. The "golden ringgit" owned by most fishermen households, which have a relatively large income per capita, serves as an entry point in analyzing the factors of economic development. Progress in science is always correlated with advances in technology, which will eventually alter the production means used by fishermen. According to Schoorl (1980, p. 1), modernization is a process of transformation and a change in the community in all aspects. The real modernization should be understood not only as a change from traditional practice to modern techniques, but it also changes the means of production such as marketing and preservation of fish and transforms the mindset of the society (Ahmadin, 2009, p. 29). Thus, modernization includes changes in technology, fish preservation, marketing, and people's mindset.

2 RESEARCH METHODS

This research is a historical study using the qualitative approach. The data sources of this study are archives that include documents related to the economic activities of fishermen, taken from the Library and Archives Board of South Sulawesi, which was in the form of Municipality Government Archives of Ujung Pandang for the period of 1926–1988. In addition to these documents, there were also newspapers such as *Harian Fajar* and *Pedoman Rakyat* published around 1970s and also used as data sources. Upon reviewing these documents, we conducted an interview with Haji Abdul Rahman Baddu, one of the resource persons for historical data, who received direct benefits of technological changes in fishing boats. The information retrieved from the documents and interviews was then analyzed chronologically in a dialogue, which is separated from experience, using historical imagination (Kartodirdjo, 1992, pp. 90–92). As a result, the historical fact obtained through criticism and interpretation was written in historiography.

2.1 *The living condition of Makassar fishermen*

Most fishermen of Makassar live in the coastal areas and the islands around Makassar. According to Sutherland (2009, pp. 100–101), the Makassar indigenous people living in the coastal areas have long worked in the fishery sector and have trade skills like those of the Chinese. Using *wangkang* boats, the fishermen would catch sea cucumbers for the local market in Makassar, which also serves as the center for financial transactions and commercial activities. This happened around the 1720s until the 1840s, but in the subsequent years these economic activities became rare. Based on the records, the trade export of sea cucumbers had decreased, and in fact the fishing activities for export became obsolete.

Entering the beginning of the independence era, fishing activities were still very much influenced by weather conditions. The types of boats used in the 1960s or before modern technology were *Pa'jala, Paterani, Balelang, Sampan, Seppe-seppe, Lepa-lepa,* and *Perahu Pemancing* (Archive of the Municipality of Ujung Pandang Reg. 1179). The best type of boat used by fishermen was the canoe. This type of boat is made from a large piece of wood that can fit two or three people. Riding this boat was relatively easy and it was only used for fishing near the coastline. Therefore, these fishermen from Makassar in that era were considered to be traditional fishermen. From an economic perspective, these fishermen were more oriented towards a subsistence economy since their catch would mostly be intended to fulfil their own daily needs. If they were able to catch more fish, they would share them with their neighbors who were also fishermen. When they sold their catch, they would make a profit. Once they managed to collect the capital return from the sale, they would give the remaining catch to their needy neighbors. This indicates a subsistence economy rather than one oriented for profit.

After the modernization, the area of the catch expanded to the offshore area. Having boats with high-tech engine, they could overcome big waves when fishing. Therefore, fishermen were now able to fish in farther areas that had once been unreachable when using traditional boats. These modern boats enabled the fishermen to catch more fish as they expanded their fishing areas.

2.2 *Boat modernization*

Technology is one of the driving forces of human creativity in a competitive world. In this case, technology can transform input into output with a high economic value. As a result, it may help to enhance the welfare of the community. Thus, technology plays an important role in poverty alleviation (Masyhuri, as cited in Bondan et al., 2009, p. 58). The use of advanced technology in the modern era is unavoidable where efficiency is required. When selecting the appropriate technology, it is necessary to avoid the extensive use of advanced technology before considering the impact or benefits for the society (Thee Kian Wie, 1981, p. 63). With the changes in technology such as the use of wind power, sails with mechanical power become a positive trend for the Makassar fishermen. The distribution of fishing boats and sailboats in Indonesia is presented in the following table:

The below table shows that Sulawesi has the largest number of motor boats and sailboats. The change of the actuator from sailboats to motorboats is quite significant. As recorded in 1970–1971, the rise of the number of motor boats in Sulawesi reached 93.81%, while other regions, such as Java and Kalimantan, only increased by 4.35% and 8.83% respectively. Thus, apparently the Sulawesi fishermen are more open for advancement in technology compared to their counterparts in other regions, especially in Java and Kalimantan. Their openness to technology became more apparent after they had established cooperation between local entrepreneurs and Japanese companies. The year 1968 marked the beginning of the development of fishing technology. At that time, the company chairman, Abdul Rahman Baddu,

Table 1.   Distribution of motorboats and sailboats for fishing 1970–1973.

Area	Motorboats				Sailboats			
	1970	1971	1972	1973	1970	1971	1972	1973
Java	736	768	1197	1200	41558	42481	44027	44.000
% Increase		*4.35*	*55.86*	*0.25*		*2.22*	*3.76*	*−0.18*
Borneo	940	1023	1856	1900	19.216	18012	19802	20000
% Increase		*8.83*	*81.43*	*2.37*		*−6.22*	*9.88*	*1*
Sulawesi	97	188	278	300	119664	103954	102.640	102000
% Increase		*93.81*	*47.87*	*7.91*		*−13.11*	*−1.26*	*−0.62*

Source: ANRI, Ministry of Agriculture Reg 361.

entered into a collaboration with Japan (Interview Baddu Abdul Rahman, the 16th April 2016). In the 1960s, cooperation with other countries was widely open, especially after the enactment of foreign investment laws. Through Bonecome, the company that managed the fishery, Japanese businessmen provided subsidies in the form of machinery and capital for boat building.

Abdul Rahman Baddu (2016) stated that he then asked for the machinery for catching shrimps offshore, so that the boats could even face bad weather conditions where they had to deal with big waves. After he gained a lot of profit from his fishing business, he bought houses, a piece of land, and a car. With huge profits, in 1970, he again asked for a Brand Yanmar engine with the boat size of 33 PK and the capacity of 10 tons. It was a very large boat, and when compared to the largest boat today, there is a 7-ton difference between them. After that, the number of boats and their size increased. Similar to Makassar fishermen, the increasing demand for machines encouraged the government of Ujung Pandang Municipality (now Makassar) on the 12th August 1970 to provide Marcuri 60 PK outboard motor with the engine number 1106048 Mark 751 model through the Fisheries Cooperative of Management Municipality of Ujung Pandang (Municipality Archives of Ujung Pandang, Reg 102). Having new machine tools, the fishermen were able to reach areas that had previously been unreachable. The fishermen of Makassar, especially those in Kampung Baru and Kampung Gusung, were able to reach up to Australian waters, and even local authorities (Police XVIII Sul-Selra) were overwhelmed in supervising them (Municipality Archives of Ujung Pandang, Reg 1190).

In the early days of the mechanization of fishing, the fishermen were not able to operate the machines properly. The most common problem among them was the damage to the actuator of the engine. According to Baddu (2016), who became Chief Executive of Bonecom, in the process of installing engines on fishing boats, his company immediately asked for assistance from a mechanic specialist from Japan. The boats were made in Ara, BiraBulukumba, and some smaller boats were made in the islands of Makassar such as Kodingareng and BarangCaddi. However, the rapid technological development did not necessarily alienate the traditional fishermen. They tried to use the technology that they thought would bring economic benefits for them. Smaller boats were modified to adjust to the engine, but the shape of the old boats did not change much. The only change was the tonnage of the boat. The more fish they caught, the bigger their desire to enlarge their boats. Thus, the presence of motorized boats among the fishermen affected their economic status.

2.3 *Marketing of fishermen's catch*

According to the Accountability Report Management Board and Supervisory Fisheries Cooperative Insan PPI Paotere Makassar, in 2015 there were 120 group leaders who joined the cooperative, and every leader was responsible for "guiding" as many as three to five boats. To run the boat, the crew usually consists of one *juragan* (captain) and three to six *sawi* (fishermen laborers). The number of group leaders and fishermen who are members of cooperatives are 2206 people (LPJ Insan Fisheries Cooperative, 2016, pp. 28–78). Each cooperative serves to bring together economic perceptions between the head of the group and his fishermen. If any problems occur between the two, or between them and other parties, the cooperative is there to provide solutions to the problems.

In the context of Makassar fishermen, the local entrepreneurs have taken part in the fishing activities. Most of them later would become the heads of the group providing the capital and marketing the catch (Rifal and Linda Sunarti, 2016, p. 10). In the 1970s, the trading networks of the catch expanded. Prior to that, the trading network was limited to the local market. For the local consumption, the fish were sold at nearby markets, or sold by *papalele* (distributor/seller) in every house *sambalu* (subscriptions) in the community. In the 1960s, the marketing area was still around nearby markets, such as *Pasar Sentral, Pasar Pongtiku,* the outskirts, *Pasar Maricaya, Pasar Cidu, Pasar Kalimbu, Pasar Sawah,* and *Pasar Terong* (Municipal Archives of Ujung Pandang, Reg. 523). Markets were usually open early, and by riding a bicycle *singkking,* the *papalele* brought the fish to the market. In addition, *papalele* also sold fish directly to the people who mostly came from the Paotere Fishery Port. Besides selling to nearby markets, they also sold fish to areas such as Sungguminasa (Gowa) and Maros that led directly to nearby countries. A few years later, after the rapid growth in transport technology, cars began to be used to transport fish. Since the use of cars, the marketing network has spread to farther areas, such as Sidrap, Enrekang, and Palopo.

During the period of international trade, the development of fishing industry in Makassar was more complex. To facilitate the development of trade networks, in 1973 two exporting companies, namely Serdid and Bonecom, were established. Based on the records, until May 1973 Sendid could export 178 tons of catch, which exceeded their expectation. The target was actually 400 tons, but after entering the month of May the export increased by 50%. The target could be achieved with a surplus. Meanwhile, the company Bonecom, on behalf of Abdul Rahman Baddu and Supu in May 1979, was able to collect a catch (especially shrimps) totaling 32,210.6 tons, some of which were exported (*Pedoman Rakyat*, edition XXVII, Saturday the 5th May 1973, and Municipal Archives of Ujung Pandang. Reg 1182). The fishing industry in the city of Makassar was profitable for both the local traders and fishermen. Industrialization also benefited the local government as the revenue increased through taxation, market distribution, and exports. Mariso Marauni, B.A. (Head of District) delivered the message of the Mayor of Ujung Pandang, "The majority of the fish caught for our people come from marine fisheries, stretching from the District Tallo Utara Kota to the coastal region of Tamalate District. The fishermen continue to improve techniques of fishing, because fish are an important source for rich nutrients (if consumed), and therefore those can support regional development" (Marauni, *Pedoman Rakyat* edition XXXII No. 32. Friday the 7th April 1978).

2.4 *Fish preservation system (cold storage)*

Besides the development of technology for the boat, the local government in cooperation with PT Astra branch of Ujung Padang on the 30th May 1972, provided cold storage facilities in every fish auction in Makasar, especially the fish auction in Paotere (Archive of the Municipality of Ujung Pandang, Reg 279). The auction is a fish trading center in the city of Makassar, a meeting place for merchants and fishermen. The cold storage facilities could also produce ice blocks for the fishermen to preserve their catch. The catch would be stored and covered with shards of ice blocks, which could last for days. The more ice was used, the longer the fish could be kept fresh. This was the first time the production of fresh fish could be implemented by the majority of the fishermen. Before using cold storage facilities, fishermen relied on salt to preserve their fish; as a result, many people consumed dried fish instead of fresh ones. They used nets to dry the fish under the sun on their boat. The fish were cut in half or *fengka* and then dried. The price of fresh fish was usually higher than that of dried ones since fresh fish are more nutritious.

The new knowledge became something needed by the fishermen. Without proper knowledge, it was difficult to develop expertise in the fishing business. With the potential for sizeable fishery accompanied by the development of more advanced technology, it was necessary for fishermen to improve their knowledge. On the 7th April 1978 a course was held for the fishermen. This course was conducted for 1 month, from the 6th April to the 6th May 1978. It was attended by fishermen from 23 areas in South Sulawesi (*Pedoman Rakyat*, edition XXXII. No. 31. Friday, the 7th April 1978). The course focused on improving fishermen's understanding in the use of technology, such as engines and cold storage (cooling devices), and in anticipating bad weather. Through this course, the fishermen could compare their prior knowledge with the new knowledge they learned from the course. For example, they could combine their skills in forecasting weather using astrology with weather-forecasting machines. With more advanced knowledge and technology, they could overcome all sorts of constraints.

3 CONCLUSION

Modernization has an important role in developing the economy of the fishermen of Makassar. The increase in catch was able to improve the lives of the fishermen, who previously could not afford to buy gold. After modernization, many fishing households could afford to buy gold as an investment instrument. Per capita income of fishermen was higher than that of those working in the technical fields and factories. Traditional boats such as canoes and *seppe-seppe* gradually transformed into sophisticated modern boats with a larger size. Therefore, fishermen could expand fishing areas farther from the coast. The houses of fishermen made of wood in the coastal cities and islands of Makassar were renovated with concrete. This shows that modernization could change the economy of the fishermen of Makassar and bring them closer towards prosperity.

REFERENCES

Ahmadin. (2009) *Ketika lautku tak berikan lagi* (When my sea no longer provides). Makassar: Rayhan Intermedia.

Arsip Kotamadya Ujung Pandang (Archives of Ujung Pandang Municipality). (the 12th August 1970) Kebutuhan alat angkutan laut bagi tugas pengurus pusat koperasi perikanan/nelayan Kotamadya Makassar (The need for means of transport on the sea for the central administrators of the fishing/ fishermen cooperative in Makassar Municipality to conduct their duties). Reg 102.

Arsip Kotamadya Ujung Pandang (Archives of Ujung Pandang Municipality). (the 15th February 1975) Adanya perahu-perahu pelayan yang sering mengadakan operasi penangkapan ikan di perairan Pantai Australia (The fishermen boats often catching fish in the waters near the Australian coastlines). Reg 1190.

Arsip Kotamadya Ujung Pandang. (Archives of Ujung Pandang Municipality) (the 30th May 1972) Adanya cold storage pelelangan ikan Kotamadya Ujung Pandang atas kerjasama antara Pemerintah Daerah Kotamadya Ujung Pandang dengan PT. Astra, Cabang Ujung Pandang (Cold storage in the fish auction place in Ujung Pandang as a result of the cooperation between the Local Government of Ujung Pandang and PT. Astra, Ujung Pandang Branch). Reg 279.

Arsip Kotamadya Ujung Pandang. (Archives of Ujung Pandang Municipality) (the 5th June 1979) Laporan realisasi pembelian/pengumpulan udang di Kotamady Dati II Ujung Pandang, Bulan Mei 1979 (The report of shrimp purchase/catching in Dati II Municipality of Ujung Pandang, May 1979). Reg 1182.

Arsip Kotamadya Ujung Pandang. (Archives of Ujung Pandang Municipality) (September 1966) Jumlah ikan yang ikirim keluar melalui Pelabuhan Makassar untuk bulan Juni s/d September 1966 (The number of fish exported through Makassar Port from June to September 1966). Reg 1179.

Arsip Nasional Republik Indonesia. (National Archives of the Republic of Indonesia (the 15th June 1974) *Tabel penyebaran kapal motor dan perahu layar tangkap 1970–1973* (Table of the motorboat and sailing boat distribution 1970–1973). Reg 361.

Baddu, H.A.R. (the 16th April 2016) Ketua Koperasi Insan Perikanan Pelabuhan Perikanan Paotere (The Head of Fishermen Cooperative of Paotere Fishing Port). *Wawancara Pribadi* (Personal Interview). Kota Makassar.

Booth, A. & Sundrum, R.M. (1987) *Distribusi pendapatan* (Distribution of income). In A. Booth & P. McCawley (Eds.), *Ekonomi Orde Baru* (The Economy of the New Order regime). Jakarta: LP3ES.

Harian Fajar. (the 11th October 1973) Kesejahteraan kaum nelayan saat ini cukup baik (The welfare of the fishermen is good enough nowadays).

Jonge, de H. (1989) *Madura dalam empat zaman: Pedagang, perkembangan ekonomi dan Islam* (Madura in four eras: Traders, economic development and Islam). Jakarta: Gramedia.

Kartodirdjo, S. (1992) *Pendekatan ilmu sosial dalam metodologi sejarah* (The social science approach in historical methodology). Jakarta: Gramedia Pustaka Utama.

Koperasi Insan Perikanan. (2016) Laporan pertanggungjawaban Badan Pengurus dan Pengawas tahun buku 2015 dan rencana kerja tahun buku 2016 (The accountability report of the Management and Supervision Staff of the 2015 and the work plan of the 2016). Makassar: PPI Paotere Makassar.

Koran Sindo. (edition of the 4th April 2016) Ironi nelayan di negeri surga maritim (The irony of fishermen in the maritime heaven country).

Masyhuri. (2009) Iptek dan dinamika ekonomi nelayan (Science and technology and fishermen economic dynamics). In Bondan Kanumoyoso, et al. (Eds.), *Kembara bahari: Esai kehormatan 80 tahun Adrian B. Lapian*. Depok: Komunitas Bambu.

Pedoman Rakyat. (XXVII edition. No. 1 Saturday, the 5th May 1973) Ekspor udang Sulsel terus meningkat (the South Sulawesi's shrimp export keeps increasing), p.1.

Pedoman Rakyat. (XXVII edition No. 77. Friday, the 1st June 1973) Produksi udang Sulsel dapat dipasar-kan sampai ke luar negeri (the South Sulawesi's shrimp production can be marketed abroad), p. 1.

Pedoman Rakyat. (XXXII edition No. 32. Friday 7th April 1978) Nelayan punya andil besar dalam pembangunan (Fishermen having a big role in the development), p. 1.

Pedoman Rakyat. (XXXII edition No. 31. Friday, the 7th April 1978) Motorisasi kapal pegang peranan penting (Boat motorization having an important role), p. 8.

Rifal & Sunarti, L. (2016) Local enterpreneurs of Makassar fishermen 1954–1998. A paper presented on the *10th Internasional Conference on Malaysia-Indonesia Relations*, Universitas Malaya, 16–18th August 2016.

Schoorl, J.W. (1980) *Modernisasi pengantar sosiologi pembangunan negara-negara sedang berkembang* (Modernization: Introductiomn to the sociology of development of developing countries). Jakarta: Gramedia.

Sumardiati, S. (1999) *Perikanan dan usaha nelayan di Kecamatan Muncar Kabupaten Banyuwangi 1969–1983* (Fishing and fishermen's effort in Muncar District Banyuwangi Regency of 1969–1983). The thesis of Program Studi Ilmu Sejarah Program Pascasarjana Bidang Ilmu Pengetahuan Budaya Universitas Indonesia.

Sutherland, H. (2009) Teripang dan perahu wangkang perdagangan Makassar dengan China pada abad ke-18 (kl. 1720-an -1 840-an) (Dried sea slugs and *wangkang* boat, trading between Makassar and China in the 18th century (from 1720s to 1840s). In Rogel Tol, Keesvan Dijk, Greg Acciai-oli (Eds.), *Kuasa dan usaha di masyarakat Sulawesi Selatan* (Power and business in South Sulawesi Society). Ininnawa: Makassar.

Wie, K.T. (1981) *Pemerataan, kemiskinan, ketimpangan* (Equality, poverty, disparity). Jakarta: Sinar Harapan.

Cultural Dynamics in a Globalized World – Budianta et al. (Eds)
© *2018 Taylor & Francis Group, London, ISBN 978-1-138-62664-5*

Colonial modernity, Indonesian nationalism, and urban governance: The making of a colonial city, Surabaya (ca. 1890–1942)

A. Achdian
Department of History, Faculty of Humanities, Universitas Indonesia, Depok, Indonesia

ABSTRACT: Studies on the socioeconomic problems experienced by the local people in the built environment of Indonesia have always provided information about the practice of late colonial politics from the 19th to the early 20th century. However, the engagement of nationalist politics to shape and influence the direction of the policy-making process on a city's governance is rarely considered in systematic ways. By focusing on Surabaya in the early 20th century and its importance as an intersection of eastern and western parts of the Indonesian archipelago, this study highlights the political articulation of nationalist politics at the local level. It also shows the sustained and persistent efforts made by Indonesian nationalists to offer unique versions of city governance as well as propose a substantial challenge to the structure of colonial power at the local level.

1 INTRODUCTION

Toward the end of January 1931, tensions rose during the regular session of *Gemeenteraad van Soerabaja* (Surabaya's City Council). Council member J.K. Lengkong, the leader of an indigenous fraction, started a debate regarding the issue of equitable representation in the council. He criticized the overrepresentation of European members despite their minority in the city compared to the native Indonesians. Lengkong wanted to increase the number of seats for the representative of indigenous people in the council, at least equal to that of European members. He also proposed the removal of taxes and property prerequisite in the voting system that barred the natives to elect their representatives to the council. Surabaya's mayor and council's chairman, Mr. W.A.H. Fuchter, suspended Lengkong's proposals to the next session in the first week of February.[1]

During the discussion of Lengkong's proposal, Mayor Fuchter opened the session by calling it as irrelevant because "the distribution of seats was not determined by local economic conditions alone", but "other" factors beyond the council's decisive authority.[2] He said that the current practice was "sufficient" and there is no need of such a discussion. However, Fuchter's attempt to halt the discussion was to no avail. Lengkong insisted that his fraction remains with their position to table the motion, even if the Chinese fraction—their devoted allies—would not lend their support. The Chinese members Tan Tjiang Ling and Ong Swan Yoe, however, took a neutral stance by reminding that the increase of the number of seats of the council was not within the domain of the council's authority, but within that of the Volksraad or National Council.

The *Indo-Europeesche Vereeniging* (Indo-European Union) and *Vaderlandsche Club* (Fatherland Club) as the major fractions in the council rejected the proposal. Council member Guldenar, the speaker of the IEV fraction, stated the reason for him rejecting the

---

1. "De Stadsgemeenteraad van Soerabaja." *Soerabajasch Handelsblad*, 29 January 1931.
2. "Staadsgeementeraad van Soerabaja." *Soerabajasch Courant*, 3 February 1931. Overall relevant descriptions of the debate that occurred as a response to the motion of Lengkong were rooted in this article.

proposal was the city council was an institution exclusively designed for Europeans or those assimilated as Europeans under the jurisdiction of *Gemeente van Soerabaja* (Surabaya's city administration). He further said that although the majority of the members of the council were Europeans, it did not mean they overlooked the interest of the native population. He reiterated how Europeans in the council vehemently worked for the advancement of native Indonesians in the city. Councilmember Jan Verboom, a spokesperson from the conservative party, the *Vaderlandsch Club* (Fatherland Club), shared Guldenar's view by adding a racial tone representing a grim fact of Surabaya as a colonial city. He said that the council was "a western enclave in the eastern society" and further argued that the development of Surabaya as a major commercial city was unquestionably the result of western capital and European labors; hence, it was logical if they then dominated the council.

The heated debates about the natives' political rights and their participation in the local political body in Surabaya at the time marked an interesting feature of a neglected theme in the historiography of nationalist movement: the articulation of nationalist politics about the city. The current historiography has several limitations in its view about the historical significance of Indonesian nationalist politics. First, it viewed the nationalist practice as only a movement of organizations, ideologies, ideas, and the paramount importance of nationalist leaders. Second, it overlooked the repertoire of collective actions and public performances of Indonesian nationalist to reclaim and reterritorialize the city in line with their political project about the city. Third, it came under the orthodoxy of reading the sources and dynamics of nationalist politics from the viewpoint of colonial central administration in Batavia, thus misrepresenting the rich texture of political contestation that pitted the nationalist vis-à-vis private Europeans as dominant social and economic forces in the colonial city. Drawing on empirical evidence from Surabaya as a major colonial city at the turn of the 20th century, this study is an attempt to provide an overlooked feature about the nature of nationalist politics in the city and their efforts to influence the city's governance to cause a nationalist practice.

1.1 *Surabaya and the making of colonial urban governance*

From the late 19th to the early 20th century, Surabaya rose into prominence as the center of commerce and industry in the Dutch colony of the Netherlands–Indies. The opening of the Suez Canal in 1869 and the introduction of colonial liberal policy in 1870 accelerated the waves of European capital, enterprises, institutions, technology, and migrants, which doubled the number of European residents in Surabaya from 4,500 in 1870 to 10,000 in 1890 (Von Faber, 1931). In the early 19th century, European settlements were more characterized by the presence of colonial soldiers and bureaucrats inside the fortress, whereas, by the mid-19th century, lawyers, doctors, bankers, engineers, journalists, book holders, clerks, and white-collar workers working for European companies—who call themselves *orang particulier sadja* (literally private citizens) – dominated urban life.

By 1930, the population of Surabaya had increased to 341,000, making the city the second highest populated area after Batavia as the center of Dutch colonial administration, which had a total population of approximately 533,000. The natives constituted the majority with 271,255 people or 79.6% of the city's population, followed by the Chinese and Foreign Orientals comprising approximately 44,245 people or 13% and Europeans with 25,900[3] people or 7.6% of the total population (Volkstelling, 1930). A diverse racial background of the city's population and the dominant position of European citizens in the modern sector of economic development made Surabaya a typical colonial city like any other major cities under European colonialism in Asia and Africa from the late 19th to the early 20th century (Yeoh, 2003).

An important event that occurred during the so-called "imperial globalization" period (Kidambi, 2007) was the emergence of what official city historians at the time called

---

3. Among the Europeans, the Dutch was the majority with 87%, followed by the Japanese (2.4%), British (1.2%), and Armenians (1%). There were also another 3% of those legally equated as Europeans (Dick, 2003: 125).

de middenstanden, (the middle class), that is, "a group of people who were not too rich but saw themselves as above ordinary citizens" (Von Faber, 1931). Together with their ascendancy as the dominant social forces in the late 19th century, the white European middle class became the patrons of burgeoning public culture, artistic, and cultural development in various European clubs, associational life, and newspapers that thrived in Surabaya. The two major newspapers, the *Soerabajasch Handelsblad* and *Nieuwe Soerabajasch Courant,* were keen to publish various opinions highlighting the population demand, at least among Europeans, for an autonomous local body to manage the city's affair. The poor condition of the infrastructure, health, and cleanliness of the people had become a persistent subject of critical discussions against the colonial state among the European citizens, as well as criticism to the centralized colonial bureaucracy and the incompetence of local colonial officials. Especially during periods of plagues,[4] criticism turned into public resentments and indignant protests as exemplified by petitions sent to Queen Wilhelmina in the mother country. The petitions, signed by hundreds of European citizens, expounded complaints about the ignorance of colonial officials to their long-awaited proposal for the construction of modern water supply in the city rampaged by cholera plagues. The depth of exasperation of the citizens to the work of the colonial bureaucracy and local officials might best be illustrated through the words of Adriaan Paets tot Gonsayen, a staunch liberal and successful lawyer, who openly stated in public meetings that the *ambtenaar*s and their obsession with careerism in Indie were a "disaster".

The hopes for the autonomous local body, however, had to wait a decade until the initiation of the Dutch's Ethical Policy and the commencement of the Decentralization Law in 1903 (Decentralisatie Wet 1903), which provided the legal bases to furnish the participation of citizens (*medezegenschap*) in local affairs. On 1 April 1906, *De Gemeenteraad van Soerabaja* (Surabaya's Council) was formally established, albeit far from satisfying the hopes of European citizens in Surabaya. The total of 15 seats reserved for the European members of the council were filled with 8 officials appointed by the government and 7 others elected by registered Surabaya city-dwellers. As an effort "to educate" the natives and other groups to the practice of European "democratic" institution, the government appointed three native officials in the colonial government and three others from Chinese and Arab community leaders. The government also amended the 1854 Colonial Constitution by removing the legal restriction that prohibited any political activities in the colony. The establishment of the city council as a key political institution in major colonial cities, therefore, marked the beginning of a modern urban governance in Surabaya at the turn of the 20th century.

1.2   *The political articulation of Indonesian nationalists*

Despite the 1930 census showing that half of the total number of native Indonesians who settled in Batavia, Bandung, and Surabaya were born in urban areas and the rest were born in provinces with large cities (Volkstelling, 122–123), they were not regarded as city residents. This feature was most clearly reflected in the statistics record system of the city, which merely displayed the citizen profiles of European, Chinese, Arabic and other foreign orientals and excluded the native population from the category in the statistical calculations (Verslag van den Toestand der Staatsgemeente Soerabaja, 1937). This statistical registration model remained until the end of Dutch colonialism in Indonesia (*Statistische Berichten der Gemeente Soerabaja*, 1930).

The absence of native Indonesians as a social category in the official demographic calculation throughout colonial cities in the Netherlands-Indies did not mean an absence of their roles and concerns toward urban environment. Since the beginning of the formation

---

4. Throughout the 19th century, the city experienced five cholera outbreaks that took the lives of natives as well as Europeans. The poor conditions of the city's sanitation infrastructure and water supply became the main theme for European private citizens to blame the colonial officials and the overcentralization of colonial bureaucracy that was considered as the main problem of the governments' irresponsiveness to the needs of local population.

of autonomous cities as well as the establishment of city councils to become representative bodies for the citizens in line with the decentralization policy of the colonial government, the educated native Indonesians from various political groups have continued to speak up their insistence and demand for improved social conditions as well as their political rights as citizens. Similarly to many European citizens, they turned into persistent spokespeople in raising issues related to the life of natives as equal citizens, such as their political rights, sanitary conditions, and healthy neighborhoods, or even the effort to improve the life conditions of native Indonesians as the major population of the cities. In the case of Surabaya, the formation of political contestation symbolized not only their resistance toward colonial political practices represented by the colonial state but also their opposition to the European citizens as the dominant social power in the colonial city of Surabaya.

As portrayed in several Indonesian nationalist publications in Surabaya during the second and third decades of the 20th century, they could provide indications on how "city politics" appeared to feature the activities of Indonesian nationalists. An article published in the newspaper Proletar, again, showed the indication to how the patterns of actions were formed in an urban environment:

> Thousands of Proletar bulletins were spread throughout Surabaya city, to go against the ice companies union which asked for monopolies to the government.
>
> Proletar considered not to wait until the proposal had been approved by the government. Yet, Proletar previously intended to refuse the proposal since the government is not supposed to help capitalists. As what had already happened by now... the proposal of ice companies union was refused. This means we prevailed.
>
> However, Surabayans, please do not lose yourselves in the euphoria, because tomorrow or the day after, such a proposal can be on the table of officials. Thus, let's strengthen the people and labor unions, as a political movement that will be against all the capitalist politics.[5]

For readers of the current generation, the description as quoted above seems to be close to the contemporary experience of Indonesians at present. The collective actions undertaken by some educated student activists, labor unions, journalists, and others that have influenced the current Indonesian political dynamics, in the end, is no longer a new narrative within the historical experience of Indonesian society. Through their publications, the anticolonial activists conveyed how collective actions will affect the direction and orientation of public policy within their urban environment.

In contrast to communist groups, which stood outside the political mainstream by not being involved in urban governance, other groups in Surabaya who joined the Indonesian Study Club under the leadership of Dr. Soetomo featured different methods by participating in the city council. It is indubitable that their involvement was limited by a representation system that provided an unbalanced proportion within the representative body, as the majority of the seats were occupied by European and Indo-European representatives. European's view on cooperative native Indonesians in colonial politics and their engagement to aspire the future development of the city could be seen from the notes in *Locale Belangen*, showing an optimism of a few Europeans toward the political participation of native Indonesians.

> Lately appeared a small group of native Indonesian intellectuals who started to be actively engaged in their role as modern citizens. Every beginning is difficult for certain. Despite the fact that the concern is either still limited, or neglected because of their passive sentiment of *non-cooperative*, after all, we are not supposed to be affected by such kind of attitude...
>
> We can see that there is a significant development within the society of the Netherlands-Indies that is still growing...Urban life in the Netherlands-Indies demonstrates the way of progress, which becomes the beginning of a new phase:

---

5. "Kemenangan Kita", Proletar, 01/07/1925.

a disconnection from the past that paves a way for the development of modern citizens.[6]

However, a positive outlook toward Indonesian nationalist participation in the political realm had to overcome the challenges from Europeans who were worried about this political progress. As mentioned in the beginning, the political mechanism of colonial policy was retained to ensure the dominance of Europeans in the city council. Together with those limitations, the activists of the Indonesian national movement ran their political agendas from compromises to confrontations. This was captured in political maneuvers taken by Soetomo's group in the city council. In April 1925, the members of the city council decided to appoint Mr. Van Gennep as the council representative for executive affairs, as Dr. Soetomo and three colleagues resigned from their positions as the representatives of Indonesians, which then triggered political turmoil over the city and forced the colonial government to appoint any native Indonesians who were willing to replace Dr. Soetomo and colleagues in the city council.

The political attitude of Soetomo and his colleagues received a positive response from other political powers outside the colonial political institutions, such as PKI. A daily newspaper Proletar reviewed Soetomo's resignation as follows:

Dr. Soetomo Standing on Two Boats
We have just heard that regarding his engagement in the popular movement and his resignation from the city council, Mr. Soetomo has been warned by his superior. It is clear that a threat to Dr. Soetomo will come if he remains constantly engaged in the movement, and then he will be imposed with certain rules.

In an attempt to make a decision due to this case, we have no capacity, though, as we are not yet able to make sure whether the news was actually true or not. According to what we have known, the proof which reveals the truth of the news has not been acknowledged. Yet, would it be impossible that Dr. Soetomo will encounter such threats.[7]

Proletar, in this way, described the "pressure" that might have arisen as a result of the political attitudes of Soetomo and his colleagues as they resigned from the city council. On the next reviews, it was mentioned that the impact of their political maneuver within the city council was quite effective to influence the way and development of politics in the city council related to the interests of Indonesian natives:

Then the strike of government officials occurred, the government seems to like this hassle. For the village, the government doesn't mind spending f100,000 in 1925.

If the government actually does, let's clean up the villages which are dirty or have no ditches.

And strangely enough, why was the money recently disbursed while the village has long been abandoned? Why after the strike at the city council?

If we relate it to the strike of those four members, it costs f100,000 or f25,000 per person. Amazing. It is what appears at a glance, and further, it is the result of political contestation, although it doesn't seem visible, it is precious. Shortly said a multiple of thousands as the fruit of the actions mentioned above.[8]

Considering the above review by Proletar, changes in a policy within the Surabaya city council in line with the action taken by Soetomo and his colleagues provided an interesting insight about the articulation of the city politics amid the political maneuver of Indonesian nationalists. It indicated that the pace of modernity in the colony through the growth of colonial cities and political institutions had influenced the perspectives and actions of Indonesian nationalists on the formulation of the urban policy. Their political practice created a contestation arena in

---

6. "A Tudor. "De stedelijke traditie in Nederlandsch-Indie", in *Locale Belangen*, 15e Jargaang, 1 September 1928.
7. "Pemerintah dan Pemogokan Empat orang lid Gemeenteraad Soerabaja." *Proletar*, 30 May 1925.
8. "Tjiamah Tjiung, Goemintah", *Proletar*, 30 May 1925.

urban governance that served the interests of not only Europeans as the dominant social group but also the native Indonesians as the major population in the city of Surabaya.

2 CONCLUSION

A picture of Indonesian nationalist contribution in the city council has provided interesting clues in some aspects. First, it is related to their concern for city politics, which remains neglected in history studies of the last decades of Dutch colonialism in Indonesia. The direction and orientation of the studies on Indonesian nationalism, which emphasizes more on ideas, figures, institutions, and organizations as well as their confrontation with the colonial state, have inevitably missed an interesting overview of everyday political practices in characterizing the history of anticolonial movements.

Surabaya as a colonial city in the Netherlands-Indies has demonstrated an important role of Indonesian nationalists in city politics through a series of their open actions and political articulations within formal institutions that offered a different form of "governance" beyond the social and political dominant power of the white Europeans. As citizens, activists of the anticolonial movement were also sensitive to issues in the city they lived in the scope of the discourse of anticolonial politics. In addition to the popular conception of the nation as an "imagined community" (Anderson, 1983), it is necessary to mention that Indonesian nationalism is a particular product of urban community as the center of colonial modernity in the 20th century in the Netherlands-Indies.

Through the development of cultural hybrid in the colonial cities, new categories in enriching the studies on British colonialism in India are withdrawn to the debate concerning the practices of the colonial power at the local level, racial segregation that overlapped with the discourse on class and identity politics, as well as colonial power structure that was filled with uncertainty and inconsistency compared to the nature of power, which was monolithic and final within the old historiography of British colonialism in India. A review on the development of Surabaya's urban history has also revealed a similar picture. The development of urban environment from the late 19th century to the early 20th century portrays that the firm boundaries of racial and ethnic segregation that formed the typology of the colonial city as presented by the king did not entirely appear.

REFERENCES

Anderson, Benedict. R'OG. (2006) *Imagined Community*. 1st Edition. London, Verso.
Basundoro, P. (2012) *Sejarah Pemerintah Kota Surabaya (The History of Surabaya City Government)*. Yogyakarta, Elmatera.
Departement van Economische Zaken, Afdeeling Nijverheid: 1949.
Dick, H. (2003) *Surabaya. City of Work. A Socioeconomic History, 1900–2000*. Singapore, Singapore University Press.
Elout, C.K. (1930) *De Groote Oost*. NV. Den Haag, Boekhandel.
Laporan-Laporan tentang Gerakan Protes di Jawa pada Abad ke-XX (Reports on Protest Movements in Java in the 20th Century). Arsip Nasional Republik Indonesia. Jakarta. 1981.
Politiek-Politioneele Overzichten Van Nederlandsche-Indie. Deel III. 1931–1934.
Ricklefs, M.C. (2012) *Islamisation and Its Opponents in Java*. Singapore, NUS Press.
Shiraishi, T. (1990) *An Age in Motion. Popular Radicalism in Java, 1912–1926*. Ithaca, Cornell University Press.
Statistical Pocketbook of Indonesia. Batavia Landsdrukkerij. 1941.
Tudor, A. De stedelijke traditie in Nederlandsch-Indie, dalam *Locale Belangen*, 15e Jargaang, 1st September 1928.
Volkstelling 1930. Volume I. Batavia, Landsdrukkerij, 1936.
William, F. (1989) *Pandangan dan Gejolak. Masyarakat Kota dan Lahirnya Revolusi Indonesia (Views and Turbulence of City Society and Birth of a Revolution in Indonesia)*. Jakarta, Gramedia.

Efforts to overcome the problems of food demand in South Sulawesi in the Guided Democracy period (1959–1965)

A.A.A. Mulya & M. Iskandar
Department of History, Faculty of Humanities, Universitas Indonesia, Depok, Indonesia

ABSTRACT: Indonesia is no different from other developing countries of the world when it comes to facing problems related to meeting food demands of the people. Unfortunately, in South Sulawesi, strategies to overcome food demand, especially demand for rice, resulted in not only increased rice production but also increased smuggling activities by the emergence of separatist movements of Darul Islam/Tentara Islam Indonesia (DI/TII) Kahar Muzakkar. The aim of this study is to reconstruct the efforts made by the government in the Guided Democracy period in South Sulawesi to overcome the problems. It is crucial to explain the efforts made by the current government as well as the challenges faced by during 1959–1965. Both qualitative and quantitative data are presented in this study. Methods used in this study were heuristic, critics, interpretation, and historiography.

1 INTRODUCTION

Rice has long been the staple food of Indonesians. According to some experts, Indonesians have known rice since 1000 BC (Khudori, 2008). However, the exact period of its emergence as staple food remains unknown. Nevertheless, it is hypothesized that rice production has been widely practiced since the 8th century in Java.

Studies about rice are interesting to be investigated, especially those with regard to its role among the people of South Sulawesi. At present, when we talk about the Bugis-Makassar tribe that resides in the south peninsula of Sulawesi, their maritime aspect attracts our attention. This is true because people of the tribe are widely known as great seamen. Nonetheless, this tribe should also be recognized for their agricultural practices. Cristian Pelras, in his book *Manusia Bugis* (The Bugis People) stated that the Bugis-Makassar tribe had been known for their agricultural practices long before they were known for their maritime activities. This is evident in their customs, such as *Mappadendang*, in which the members celebrated their successful rice harvests by organizing parties. Furthermore, the book Mattulada, *Lingkungan Hidup Manusia* (Human Environment) mentions one of the three main aspects of the great kingdoms of Bone, Luwu, and Makassar in the South Sulawesi peninsula and their vast and wide agriculture land (*padangmaloang*). This shows that agriculture played an important role in the past glory. Unfortunately, studies about food demands, particularly of rice, in South Sulawesi, are scarce, as most of them address problems related to rice in Java or in Indonesia in general. Therefore, the aim of this study is to review the problems of food demand, particularly of rice, in South Sulawesi in the Guided Democracy period (1959–1965). Several prominent problems in this period were the increase of price, pest attacks that obstructed rice production, and peace intrusion by Darul Islam/Tentara Islam Indonesia (hereinafter abbreviated as DI/TII).

1.1 *Background of the study*

South Sulawesi[1] is one of the main rice producers in Indonesia, apart from Java. Production areas include Pinrang, Sidenreng Rapang (Sidrap), and Bone Regencies. South Sulawesi receives regular interchanging annual rainfall, which is advantageous for agricultural practices, especially the production of rice. In Sulawesi, the agriculture sector is divided into two groups, namely foodstuff and plantation sectors. The foodstuff sector includes rice, corn, yam, and bean cultivation. The varieties of rice cultivated in South Sulawesi are rice paddy (*Oryza sativa*), *gadu* paddy, and *gogo* paddy. Meanwhile, the plantation sector includes coffee, chocolate, and coconut production.

Rice, on the one hand, holds an important position in the Bugis-Makassar culture. This is evident in a widely known cultural tradition called *Mappadendang*, in which the society celebrates successful rice harvests to show gratitude to God. The name *Mappadendang* is derived from the practice of singing (*dendang*) traditional songs by the women during rice pounding. Because this tradition is considered very sacred, the women usually wear *bajubo-do*.[2] The women of the Bugis-Makassar tribe wear traditional clothes, whereas the men wear a coiled headband, black clothes, as well as black *sarong* around their bodies.

Meanwhile, covering approximately 20% of the province's total area, paddy fields are considered one of the largest contributors in the island of Sulawesi (Southeast Sulawesi and North Sulawesi). This is also evident since South Sulawesi has become a rice producer in Indonesia, particularly in Pinrang, Sidrap, and Bone regencies. The rice cultivation areas are supported by four distinct irrigation systems, namely technical irrigation, half-technical irrigation, village irrigation, and rain-fed irrigation, of which the village irrigation system is mostly used by the farmers in South Sulawesi. Meanwhile, Sidrap and Pinrang regencies, whose cultivation areas are the largest, used the technical irrigation system. These activities are supported by the construction of dam *Bendungan Benteng*, which has been active since the period of Dutch colonialism in Indonesia.

The spacious cultivation area of South Sulawesi is proportionally in line with its number of farmers. It is presumed that, 60% of South Sulawesi population were farmers in 1906, which increased to approximately 71% in 1930. Furthermore, there was no sign of job switch for the past 20 years in the South Sulawesi society (Harvey 1989).[3]

The farmer society in South Sulawesi, especially the Bugis-Makassar tribe, is divided into three structural levels, namely kings and royal families (*bangsawan*), liberate people (*to maradeka*), and servants (*ata'*) (Abdullah, 1985).[4] In general, noblemen own the cultivated lands and employ the liberate people or servants to manage their lands. Before Indonesia's independence, these workers were only given food and drinks as payment for their labor. This is seen as a special tribute for the landowners because of their high societal status. In addition, the landowners managed an authority to solve problems related to the workers within the society.

The agrarian law in Indonesia during the period of 1959–1965 was closely related to the condition in Indonesia that put the Guided Democracy into practice. This was first marked by President Soekarno's decree on 5 July 1959, according to which all government policies were sourced from the president. The policies applied to all sectors, including the agrarian affairs. One of the agrarian policies applied in this period was the Basis Agrarian Law (also referred to as UUPA) number 5 of 1960. This policy was founded on the basis of the spirit of anti-colonialism and feudalism as mentioned by the then Agrarian Minister Mr Sadjarwo that "the struggle to reform our agrarian laws is firmly connected with Indonesian history of struggle to be free from the grasps, influences, and other remnants of invasion,

---

1. In the beginning of its establishment in 1960, Southeast Sulawesi Province was combined with the South Sulawesi Province. The area was then called as Level I Area of South-Southeast Sulawesi. In 1964, the province was divided into Level I of Region Province into South Sulawesi.
2. The traditional clothes of Bugis-Makassar women.
3. Harvey, B. *Pemberontakan Kahar Muzakkar, dari Tradisi ke DI/TII*, p. 59.
4. Abdullah, H., *Manusia Bugis Makassar, Inti Idayu Press*. 1985. p. 109.

particularly the struggles of peasant society to free themselves from the feudal systems of land and exploitation of foreign capital" (Wibowo, 2009).

1.2 The condition of South Sulawesi's agriculture in the beginning of the Guided Democracy period

During the early periods of implementing the newly practiced policy, Indonesia experienced higher rainfalls (in 1960) compared to the previous years. This condition lasted throughout the year, which then contributed to the increase of harvests in South Sulawesi. In Pinrang regency (Level II Region),[5] the rainfall recorded was 2,086 mm with an average of 115 rainy days annually. In addition, 41,856 Ha of total cultivated paddy fields was verified in the regency, larger than that in the previous year (32,542 Ha). An increase in land utilization for paddy cultivation of 9,314 Ha is evident from these data.[6] Out of the total 41,856 Ha of the land, 20,000 Ha was irrigated using the technical irrigation system, 3,700 Ha used the half-technical irrigation system, 13,400 Ha used the traditional irrigation system, and the remaining 4,756 Ha depended on rainfall for its irrigation.[7]

Different types of rice were cultivated in these areas. The farmers in Pinrang regency planted three varieties of rice, including rice paddy (*Oryza sativa*), *gadu* paddy, and *gogo* paddy. However, the most common species was rice paddy, which occupied 39,574 Ha of the land. Rice production from such large areas reached 49,467 tons, which rapidly increased in 1959 to 48,993 tons from just 32,981 Ha of cultivated lands. On the contrary, *gadu* paddy and *gogo* paddy were produced on 5,519 and 665 Ha of cultivated lands, achieving harvests as much as 7,392 and 399 tons, respectively. In line with the increase of the areas of cultivated lands, the *gadu* paddy harvest also increased by 1,693 tons from the previous year, while *gogo* paddy did not experience such increase.

In regions of Bone regency, the data of the rainfalls in 1960–1961 from the Department of Agriculture in Bone show that the rainfall distribution in the first three months of 1961 was identical to that of the last three months in 1960. This high rainfall facilitated rice cultivation. However, there was minor land damage due to flash floods that flooded as much as 16 Ha[8] in the first three months of 1961.

In Bone regency, *gadu* paddy cultivation also developed in terms of land occupied. This is apparent from the data obtained from the first three months of 1961, according to which 3,010 Ha of land was used to produce this rice variety, slightly larger than that in 1960 (2,292 Ha). On the contrary, land occupied by *gogo* paddy fields decreased. In the first three months of 1960, it was 2,180 Ha, which decreased to 1,729 Ha in the same period of 1961, a decline of 460 Ha. The total volume of rice harvested throughout the South Sulawesi region in 1960 was 515,000 tons with 435,000 Ha of cultivated lands.

The increase of land use for cultivation is believed to be due to the increasing demands of food in Indonesia, particularly in Java (Mubyarto, 1989),[9] which continuous increase of population. On the contrary, the population of South Sulawesi did not grow as rapidly as that of Java. Essentially, according to the 1961 data, the food demand of Indonesia was 90 kg/year. This high demand urged the government to maximize cultivation lands outside Java.

1.3 Problems of food demand in South Sulawesi in the Guided Democracy period

Despite satisfactory outcomes in the agriculture sector in South Sulawesi, particularly its rice production, efforts were not easy. Data from 1960–1961 show increase of rice production;

---

5. Level II Region is equal to Regency/City. Maschab, M. *Politik Pemerintahan Desa di Indonesia*, p. 99.
6. The Yearly Report of Level II Region Department of Agriculture of Pinrang in 1961.
7. *Opcit.*
8. Archives of Department of Agriculture of Bone in 1961.
9. Annual increase of agricultural production in Java was only 2.9%. Mubyarto, *Pengantar Ekonomi Pertanian.*, p. 31.

however, in 1962, pest attack was a major threat to paddy fields. The pest is called *sundep*[10] (a type of caterpillar that eats young leaves of paddy plants) in the district of Sidenreng Rappang (Sidrap). Meanwhile, the separatist movement of DI/TII also caused other problems in South Sulawesi regions. Besides, high prices of all staple commodities, especially rice, became a hurdle to agriculture in South Sulawesi.

Data obtained from *Marhaen*'s daily report on 14 January 1960 show that there was about 30% increase in the cost of rice, from Rp 3.5 to Rp 5 per kilogram. (Marhaen Daily, 1960)[11] The data also included an interview with the head of Indonesian Rice Mill and Traders Society (IPPBI) of Makassar, Mr M. Daud Latif, according to whom, this increased price of rice was due to the absence of licensed rice distributors among the farmers who milled rice (millers).

A report from Sidenreng Rappang district submitted to the Governor of Southeast Sulawesi shows that approximately 2,147 Ha of land was damaged due to *sundep* attack. The attack occurred in a large scale and led up to 100% destruction of the contaminated paddy plants. This attack occurred because the plants were not sprayed with pesticides. Meanwhile, in Pinrang district, there were mice attacks, damaging hundreds of acres of paddy fields.[12]

The threat from DI/TII also aggravated the problems related to agriculture in South Sulawesi. Sidrap Regent's Report described the situation of rice harvest in mid-1962, in which DI/TII group demanded the farmers to hand over 20% of their rice harvests. They also forced the farmers to pay Rp 5 for every paddy they gave.[13] Furthermore, in Pinrang regency, the *gadu* paddy farmers were chased away from their rice fields by the DI/TII members. They did not only intrude the farmers' work but also forced them to hand over their rice harvests. They also threatened to beat the farmers and prohibit them from cultivating their lands.[14] In fact, a farmer named Manrung Padjalele from Patampanua district of Pinrang regency was kidnapped and was held as a captive from August 1964 by DI/TII, which ended in his death.[15]

1.4 *Efforts to overcome the problems of food demand in South Sulawesi*

It has been mentioned in the previous section that the high price of rice, pest attacks, and threats from DI/TII are the three main problems in the government's efforts to improve the agriculture sector and resolve the problems of food demand in South Sulawesi. Therefore, Level 1 Region (Province) of South Sulawesi cooperated with the Level 2 Region (Regency) to solve the problems.

In order to decrease the high price of rice, the government provided several solutions. First, miller farmers were allowed to sell their rice directly without any agents. Then, a shop was established specifically for provision of staple commodities. These efforts were made to control and stabilize the rising price of rice and other foodstuff within the province.

Meanwhile, to improve the agriculture sector and increase the rice production, the government of Sidenreng Rappang regency on 9 January 1963 instructed the use of high-quality seeds in Maritengngae district. The rice varieties used for this purpose were *Bengawan* and *Si Gadis*. It also instructed and supported the rice cultivation by using high-quality rice plants, making good use of land, and fertilizing plants by spraying, weeding, and exterminating pests. The regency government even launched a 1-month cultivation course facilitated by officers of the agriculture office to improve the harvests.[16]

In addition, an organization named South Sulawesi Agricultural Development Foundation was officially established on 8 February 1961 to carry out agricultural development plans and provide recommendations and technical supports to not only individuals but also agricultural

---

10. *Sundep* is a kind of caterpillar pests that attack the paddy field areas. It usually attacks young leaves, causing the plant to wilt and then die.
11. *Marhaen Daily*, 14 January 1960.
12. Damage caused by mice in Pinrang regency was not mentioned in detail in the Archives of the Department of Agriculture of Pinrang.
13. Report of Regent of Sidenreng Rappang, on 28 June 1962.
14. Report of Head of Patampanua District, Pinrang Regency on 11 November 1963.
15. Report of Regent of Pinrang, Andi Djalante Tjoppo on 26 August 1964.
16. Instruction of Regent of Sidenreng Rappang, Andi Sapada, on 28 November 1962.

companies when needed. Ibrahim Manwan was appointed as the head of the foundation. It was highly expected that this foundation would succeed in increasing South Sulawesi's food production, which is the mission and vision of the foundation.[17]

Alternatively, in Pinrang regency, the efforts to increase its agricultural production and prevent land destruction from pest attacks included fertilization of *gadu* paddy plants as instructed by the head of the Department of Agriculture of Pinrang. This instruction was mandatory to all *gadu* paddy farmers.[18]

In order to solve the problems and security threats caused by DI/TII, the head of the foundation urged the farmers not to feel intimidated by such threats. Andi Muhammad Dalle, the Commander of Mattiro Walie District Military Command, instructed all merchants who wanted to buy and sell rice to submit a report to the district military command. This was in order to prevent and block DI/TII from disrupting the economy. In addition, some security personnel were appointed to monitor and provide security to the farmers so that they could continue cultivating their lands without feeling threatened by DI/TII soldiers.[19] Eventually, the security disruption caused by DI/TII soldiers reduced over time as they disunited. Some of the heads and members quit and even surrendered voluntarily.

2   CONCLUSION

The measures taken for increasing the agricultural production in South Sulawesi in the Guided Democracy period began in 1960. The situation in the beginning of this period, particularly in South Sulawesi's regencies of Pinrang, Sidrap, and Bone, showed good outcomes between 1960 and 1961. It was evident from the increased harvests compared to the previous year. Climate and high rainfall was a contributing factor for this. Some cultivated fields were damaged due to the high rainfall. However, it did not have any impact on the volume of harvest.

Unfortunately, the efforts were not always easy. The two main hurdles to the increase of food production were the attack of a pest called *sundep* and security interference by DI/TII soldiers. In Sidrap regency, *sundep* attacked approximately 2,000 Ha and inflicted 100% damage to the paddy plants. In addition, the farmers were threatened by the demands from DI/TII soldiers to hand over 20% of their harvest and pay Rp5 for every bunch of rice they took.

The South Sulawesi government made significant efforts to overcome these problems. The first instruction was to inform the farmers to start cultivating high-quality paddy seeds. Besides, they were told to irrigate their fields, fertilize their crops, and perform weeding. Furthermore, the government initiated a forum of agriculture extension carried out by the officers of the agriculture office to help the farmers improve their harvests.

Meanwhile, to help the farmers from the security interference caused by DI/TII, it was instructed that the farmers must submit a report on the goods they sold to the District Military Command. This was done to block DI/TII's economy. The security personnel also provided the farmers with security so that they could work without worrying about the threats from DI/TII.

It is interesting to note here the shift of a patron–client pattern in the South Sulawesi society, which was made possible by the nation's interference in the agricultural practices of the society. It has been mentioned that the traditional patron–client pattern of this society was centered in the noblemen (*arung/karaeng*) as the landowners and in *maradeka* as well as *ata* as the workers. Nevertheless, this position began to shift when the state took part in becoming the patron for the farmers. The shift, however, was even more evident during the New Order Period with Sidrap farmers' tradition of *Tudang Sipulung*.[20] In the past, this tradition

---

17. Report of the Board of Patrons of South Sulawesi Agricultural Development Foundation on 8 February 1961.
18. Head of Pinrang Department of Agriculture instruction, Abdullah Lawi, on 26 October 1965.
19. Instruction of Commander of Mattiro Walie District Military Command, A.M. Dale, on 13 October 1962.
20. *Tudang* is referred to sitting, while *Sipulung* means together. The literal meaning of *Tudang Sipulung* tradition is to sit together.

was held by the farmers to determine the patterns of rice cultivation. In the New Order period, the government of Sidrap regency turned the tradition of *Tudang Sipulung* into an agenda, which served as a means to deliver the government's agricultural programs. This agenda was then led by the Regent of Sidrap.

REFERENCES

Abdullah, H. (1985) *Manusia Bugis Makassar* (Bugis Makassar People). Jakarta, IntiIdayu Press.

Greg A. (1994) What's In A Name? Appropriating Idioms in the South Sulawesi Rice Intensification Program. *The International Journal of Social and Cultural Practice*, 35, 39–60.

Harvey, B. (1989) *Pemberontakan Kahar Muzakkar, Dari Tradisi Ke DI/TII* (The Rebellion of Kahar Muzakkar: From Tradition to DI/TII). Jakarta, Pustaka Utama Grafiti.

Instruction of Commander of Mattiro Walie District Military Command Vak. B/ Dim 1406/, AM. Dale, dated 13th October 1962. Archives of Level II Region of Sidenreng Rappang.

Justus M.K. (1963). Indonesia's Rice Economy: Problems and Prospects. *The American Journal of Economics and Sociology*, 22 (3), 379–392.

Khudori. (2008) *Ironi Negeri Beras* (The Irony of 'Rice Country'). Yogyakarta, INSIST Press.

Lindblad, T., et al. (1998) *Sejarah Ekonomi Modern Indonesia* (The History of Indonesia's Modern Economy). Jakarta, LP3ES.

Marhaen Daily dated 24th November 1959.

Marhaen Daily dated 13th January 1960.

Marhaen Daily dated 14th January 1960.

Marhaen Daily dated 20th February 1960.

Marhaen Daily dated 6th February 1960.

Maschab, M. (2013) *Politik Pemerintahan Desa di Indonesia* (The Politics of Rural Governance in Indonesia). Yogyakarta, Penerbit Pol Gov.

Mattulada. (1994) *Lingkungan Hidup Manusia* (Human Environment). Jakarta, Pustaka Sinar Harapan.

Mubyarto. (1989) *Pengantar Ekonomi Pertanian* (Introduction to Agricultural Economics). Jakarta, LP3ES.

Pelras, C. (2006). *Manusia Bugis* (The Bugis People). Jakarta, Penerbit Nalar.

Report of the Board of Patrons of South Sulawesi Agricultural Development Foundation regarding 'Establishment of Yayasan Pertanian Sulselra' dated 8th February 1961. Archives of Level II Region of Bone.

Report of the Head of Mattiro Bulu District, Muhammad Arsjad, regarding 'Hundreds of Acres of Lands Attacked by Mice,' dated 8th May 1964. Archives of Level II Region of Pinrang.

Report of the Head of Patampanua District, Muhammad Musjtari, dated 11th November 1963. Archives of Level II Region of Pinrang.

Report of the Head of Pinrang Agriculture Office regarding 'A Comprehensive Report of O.G.M,' for January 1963, dated 5th March 1963. Archives of Level II Region of Pinrang.

Report of the Head of Pinrang Agriculture Office, Abdullah Lawi, regarding 'Plans on Food Production Increase,' dated 26th October 1965. Archives of Level II Region of Pinrang.

Report of the Regent of Pinrang, Andi Djalante Tjoppo, regarding 'A Kidnap of P. Manrung Kp. Padjaleleby Group of Men,' dated 26th August 1964. Archives of Level II Region of Pinrang.

Report of the Regent of Sidenreng Rappang, Andi Sapada, regarding 'Investigation and Examination of people's Rice,' dated 3rd April 1962. Archives of Level II Region of Sidenreng Rappang.

Report of the Regent of Sidenreng Rappang, Andi Sapada, regarding 'Twenty Percent (20%) Extortion of Rice and so on from People by Individuals on Behalf of Organisation ex. DI/TII and Act of Tax Collection on Rice Brought from Villages to Cities,' dated 28th June 1962. Archives of Level II Region of Sidenreng Rappang.

Report of the Regent of Sidenreng Rappang, Andi Sapada, regarding 'Agriculture Course to All Rice Field Officers in Level 2 Region of Sidenreng Rappang,' dated 28th November 1962. Archives of Level II Region of Sidenreng Rappang.

Wibowo, L.R., et al. (2009) *Konflik Sumber Daya Hutan dan Reforma Agraria: Kapitalisme Mengepung Desa (Conflict of Forest Resources and Agrarian Reform: Capitalism surrounding the villages)*. Yogyakarta, Alfamedia Citra.

Cultural Dynamics in a Globalized World – Budianta et al. (Eds)
© *2018 Taylor & Francis Group, London, ISBN 978-1-138-62664-5*

A study on how Uber seizes the transportation space in Jakarta

D.T. Gunarwati & L. Kurnia
Department of Literature, Faculty of Humanities, Universitas Indonesia, Depok, Indonesia

ABSTRACT: In this study, we address the development of the application-based transportation service organization Uber, which has caused a direct disturbance to the taxi company Blue Bird, which had dominated for decades in Jakarta. Uber is considered illegal by the government due to its noncompliance to the existing regulations. Uber uses informational ways and forms a network society, which is a concept proposed by Manuel Castells (1996). Uber seizes the consumption space of taxi users and opens up the employment production space in Jakarta's transportation industry. The data were obtained by ethnographic approach (field study), including in-depth interviews with the drivers and users of Uber, and literature review on the topic of Uber controversy in mass media. The result of the research shows the presence of loopholes, which are not in accordance with the pattern of the network society and informational capitalism proposed by Castells. Uber's informational capitalist system has initiated a discussion on the shifting forms of capitalism, which is being responded by the state.

1 INTRODUCTION

Jakarta has been the capital city of the Republic of Indonesia since the Indonesian Proclamation of Independence on 17 August 1945. Traffic congestion has been one of the concerning issues in Jakarta for many years. This is due to the significantly uneven proportion of private vehicles and public transportations in the streets of 98% and 2%, respectively. The population of Jakarta was 10.2 million in 2013 (Damardono, 2013, p. 9). In the absence of reliable public transportations, the urban community prefers to use private vehicles or taxis. Despite their status as public transportation, taxis, being more convenient and faster than other means of public transportation, are often associated with people of the upper middle class.

The popularity of taxis declined after the increase of taxi tariff, due to the increasingly severe traffic congestions and fuel price increase. At the same time, technological advances led to the development of online mobile applications that offer transportation services similar to taxis. Uber, which officially began to operate in August 2014, has rapidly become a popular alternative among Jakarta urbanites. However, Uber, which is available in 68 countries, has caused controversies, such as lacking the official license to operate unlike regular taxi companies. The increasing popularity of Uber among Jakarta urbanites incites issues with the authority. The Provincial Government (*Pemprov*) of DKI Jakarta and the Land Transportation Organization (*Organda*) of the Special Capital Region (*DKI*) of Jakarta have considered Uber illegal due to the use of black license plates for their vehicles, which signifies their status as private vehicles, allowing them to evade commercial vehicle tests required by the Law of Road Traffic of 2009.

The operational expansion of Uber and other application-based companies created conflict with the providers of conventional public transportation, particularly taxis. Public transportation drivers staged a large-scale demonstration on Tuesday, 22 March 2016. The demonstration organized by the taxi drivers allegedly was organized due to the disturbance caused by the operation of Uber to the business of taxi companies, such as Blue Bird Group, the biggest taxi network in the country. The government later issued a regulation allowing Uber to operate only after meeting some requirements.

In this study, we aim to show how the development of Uber has changed the transportation pattern of Jakarta urbanites, the pattern of employment, and the relationship between employers and workers. Although its operation is in contrast to the existing regulations, Uber has received support from the community. Hence, the government has to negotiate confronting this global company. This study focuses on the changes in urbanites' consumption patterns and the employment pattern in the transportation business, as well as the government's position in law enforcement to confront this global public transportation company.

The theoretical framework used by the authors is informational capitalism proposed by Manuel Castells, a professor of technology and community communication at Annenberg School of Communication. In his book trilogy, Information Age: Economy, Society, and Culture (Castells, 1996; Castells, 1997; Castells, 1998), he observes the emergence of new, and even revolutionary, types of society, culture, and economy. This new inclination first developed in the United States, which has been quite visible since the 1970s, namely the emergence of information technology marked by the dominance of television, computer, and the likes. The revolution progressed, and since the 1980s, a phenomenon described by Castells as informational capitalism has emerged, which has simultaneously led to the formation of the informational society.

In the case of Uber's operation in Jakarta, the authors assume that a cultural transformation process has taken place due to the operation of a global network company (Uber) practicing informational capitalism. This has a pervasive effect on human civilization as Uber users form a global community, and it slowly gains control over transportation in many countries.

Here, we use a cultural studies approach from the perspective of Manuel Castells' informational capitalism, which is argued to be promoted by certain actors, namely the network company (global company) represented by Uber. The authors further argue that the negotiation, contestation, and resistance between the older capitalism, represented by taxi companies, and the newer capitalism, represented by Uber, have seized the consumption space of the urbanites and the labor production in Jakarta's transportation sector.

In addition to literature review, this study collects relevant data using ethnography methods, which are carried out through in-depth interviews, directive interviews, nondirective interviews, and quoting an Uber user from a private blog. The field study requires the authors to take some trips using Uber to get the first-hand experience as Uber passengers. Meanwhile, to obtain information from other stakeholders, the authors utilize selected mass media publications in both printed and electronic forms.

1.1 *Seizing the transportation space in Jakarta*

The city of Jakarta is well known for traffic congestion. The *TransJakarta* Bus system introduced in 2004 to resolve Jakarta's traffic congestion has yet to serve its purpose effectively. The urban society prefers to use private vehicles as a more convenient means of transportation. Taxis, as part of public transportation, have a different characteristic and market share compared to other means of public transportation. Taxis are considered as an exclusive form of public transportation. The market share of taxis in Indonesia's large cities is rather high, particularly among passengers with a high level of mobility or those who opt for a particular kind of convenience and safety (Sjafruddin, Widodo, & Kuniati, 2001).

Blue Bird Group is the taxi company that dominates the market share in the country until today. It is the extension of industrial capitalism ruling the transportation industry in Indonesia. Industrial capitalism is determined by the amount of capital deposited by the company to establish their businesses. As stated by David Harvey (1992 in Scott, 2012, p. 2), capitalism, at its core, is a system of production involving the search for profit by means of investment in physical assets and the employment of workers to produce sellable goods and services.

The domination of Blue Bird taxis in Indonesia has started to cease when the taxi tariff was increased in December 2014 due to the increasingly severe traffic congestions in Jakarta. With regard to Uber, through its mobile application, users of Uber can estimate the cost of a journey. Uber operates on a business model that stems from the "sharing economy", as mentioned by Julyet Schor (2015), and allows people from the lower-economy class to use certain assets that they do not have. One of the Uber users interviewed by the researchers, RS, 31, mentioned that one of the reasons for her to choose Uber was the feeling of having a personal driver.

> "Actually, I also use Uber as it makes me feel like I have my own driver. Aside from that, Uber does not require payment in cash. They use credit cards. So the payment is easier. I do not need to worry if I do not have enough money when the street is heavily congested, or when I do not bring a lot of cash with me." (Interview with RS was conducted on 7 April 2015)

Unlike any other forms of public transportation, Uber offers the sensation and feeling of having a personal driver without necessarily having one. Furthermore, Uber provides a more competitive price compared to taxis. By using Uber, the consumers enjoy the feeling of being part of the "upper class", indicated from the feeling of "as if" having a personal driver, as quoted above. This has been made possible by Uber, which uses cars with black license plates, meant for private vehicles.

Even so, as an extension of informational capitalism, Uber service can also be exploitative toward its users. The rising popularity of Uber among Jakarta urbanites has led to the increasing demand of their service, which may trigger the increase of surcharge from the current basic price. The imposing of surcharge occurs when the number of Uber vehicles in a location is limited and the demand is high. For example, on weekends, the tariff of Uber can be double that of the normal price, and will only decrease when the demand for Uber service drops within the next 30 to 50 minutes.

As a network company, Uber, which has operated in Jakarta since August 2014, allows the formation of a more open system in labor absorption. According to Castells, the key factor in a global international economy and network society is the network company, that is, work individualization through aspects such as more flexible time. Nevertheless, essentially, the transformation process varies between countries (Ritzer, 2012, p. 970 in Rahayu, 2015, p. 20).

Uber allows the formation of a more liquid employment pattern. LM, of age 41 years, said that aside from being an Uber driver, he also works as a driver for Grab Car. He does this to compensate for daily operational costs. According to him, even though he works for two different mobile application-based companies, as long as he meets the required targets, there are no problems.

Researcher:	How is your experience of being an Uber driver to date?
LM:	Not bad. It's because I handle two jobs, one with Uber, one with Grab. Since Uber gives weekly payment, we (the drivers) struggle with the (operational) costs. Weekly payment is a daily struggle. With Grab, we get daily payment.
Researcher:	What are the differences?
LM:	It's quite the same. But for money, Grab is faster. Grab has higher tariff compared to Uber.
Researcher:	Then, the car is yours?
LM:	Yes, it's my own car.
Researcher:	*Pak*, aren't there any legal regulations stating you cannot be a Grab driver if you are already an Uber driver, and the other way around?
LM:	There are no problems with that.

(Interview with LM was conducted on 16 February 2016)

What LM does by becoming a driver for Uber and Grab Car service can be viewed as an act of negotiation within the existing working system. LM, temporally and spatially, can be a driver for both Uber and Grab Car. This is made possible by the informational capitalism business model performed by Uber that creates what Castells proposed as a space of flows in which space has turned into a world dominated by processes instead of physical locations. Nevertheless, the physical locations continue to exist in reality.

As previously explained, the recruitment mechanism for drivers is not too complicated. However, the absence of a written contract between the drivers and network companies like Uber allows the company to conduct exploitative action toward the workers. An informant interviewed by the researchers, SP, 36, confessed that working in Uber is like being under a massive exploitation system.

> "I got minus (earnings) working as taxi driver, but here with Uber, at least I got plus (earnings), although it really wears me out, since the meter is half of the taxis'. So you see my work here is the same with me driving two taxis. The (tariff of) taxi from here to the airport is IDR 100,000, but they (Uber) only charge IDR 59,000. It's like almost half the taxi tariff, around 40%. But everywhere I go, there will be orders. My previous passenger was still in the car when I got another order notification. When I was at Ciracas, it was the same; the notification went off for the order I just said." (Interview with SP was conducted on 20 March 2016)

SP's disclosure builds on the argument of Uber being the embodiment of informational capitalism, which exploits labor. Referring to Castells' informational capitalism, Uber's exploitation resembles the characteristic of the previous capitalism, which creates an exploitative production relation by marginalizing the social structure and other culture. Orders are given regardless of the working shifts or working hours for the drivers. Rejecting orders leads to unfavorable ratings for the drivers and consequently the suspension of their accounts. When a driver's account is suspended, the driver will no longer be able to go online and receive any orders from passengers.

Uber, which is now also available in 68 countries, has caused controversies due to the absence of legal business licenses unlike regular taxi companies. Similar to other countries, Uber, which has gained more popularity among Jakarta urbanites, has raised issues with the authority. Prohibition toward Uber has been done a couple of times, by the provincial government (*Pemprov*) of DKI Jakarta as well as the *Organda* of DKI Jakarta, the Department of Transportation of DKI Jakarta, and the Governor of DKI Jakarta, Basuki Tjahaja Purnama. Despite being banned by the provincial government of DKI Jakarta, this global application company has not terminated their operations in it. Uber services can still be accessed by the users. The spokesperson of Uber claims that Uber is a technology company and not a transportation company.

On 17 December 2015, the Indonesian Government, this time represented by the Ministry of Transportation, officially banned application-based transportation services in Indonesia. The prohibition is stated in the Notification Letter No. UM.3012/1/21/Phb/2015 dated

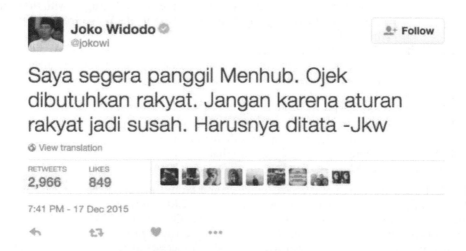

Figure 1.   Jokowi's tweet @jokowi on 17 December 2015.

9 November 2015 signed by the Minister of Transportation, Ignasius Jonan. The decision immediately caused a controversy among the netizens. Just a day after the prohibition, President Joko Widodo requested the ban to be lifted through his Twitter account. Although the president's directive is informal, the Transportation Ministry later revoked the prohibition order.

The President's order as stated in the tweet was sufficient enough to revoke the regulation issued by the Ministry of Transportation.

The indecisiveness of the government, both the central government and the provincial government of DKI Jakarta, in banning Uber's operations shows that the state has started the disposition in countering Uber as a form of foreign informational capitalism. Implementing Castells' framework (1996), the state has become weaker in a world of new economic globalization, and its dependency to the global capital markets will increase simultaneously having less bargaining power. In the case of Uber in Jakarta, the state's position also keeps changing due to different opinions among the stakeholders.

The state's vacillating attempt to counter this global multinational company and the failure to ban the transportation application led to a large-scale demonstration organized by public transportation drivers, particularly taxi drivers, on Tuesday, 22 March 2016. The protest ended in anarchy. As the situation was worsening, the management of Blue Bird Taxi, which had the most drivers demonstrating on the streets, later organized a press conference. The Commissioner of Blue Bird Plc., Noni Purnomo, explained the company's corporate stance as the provider of taxi transportation service, in which they claimed that there was no real competition with the online transportation services. She stated that they had released a statement 2 days before the massive demonstration mentioning that they did not condone their drivers to join the protest.

Even so, the anarchical actions of the taxi drivers were video-recorded and widely circulated through social media, drawing criticism from the netizens. According to one netizen, Nuniek, in her private blog, the anarchical demonstration smeared the reputation of taxi companies.

> "In the press conference held today, the Blue Bird management stated they are not held accountable for what happened today. They released a statement 2 days ago that they *do not* encourage their drivers to join mass demonstration. But no one can push aside the fact that the demonstration today was dominated by taxi drivers wearing their uniform, driving their fleets. The polite, honest, reliable taxi drivers has transformed into "hangry" people. Hangry is a special term made by my 8-year old daughter to describe someone who is *hang* (read: error) because s/he is hungry and angry at the same time. The 24-hour free ride offer didn't get good sympathy from netizens."

Nuniek in her blog deplored the response of the company's management stating that they were not responsible for the demonstration conducted by their drivers. The previous image of Blue Bird drivers being honest, polite, and reliable instantly changed into a group of hungry and angry people.

Castells in *The Informational City* mentioned the shifting from the structural logic to the flow logic, or known as the flow hegemony. In the informational capitalism era, those who control the network are the dominant issue ruler. This is what has made Uber, with their network power, to gain great sympathy from the society and to receive wide support to proceed with their operations. On the contrary, Blue Bird was ridiculed and considered to have failed to innovate and improve their working system in the face of emergent competitors equipped with information technology excellence.

After the mass protest of the public transportation drivers, the government set 31 May 2016 as the deadline for online application-based transportation services to comply with the existing public transportation regulations. During the transition, they were allowed to operate but prohibited to expand their business. One of the mandatory terms is the establishment of an official business entity of their own or cooperating with operators or business entities that have transportation operational licenses. The vehicles are also required to take a commercial vehicle test to warrant the eligibility of passengers' safety (Simbolon, 2016).

2 CONCLUSION

Through information technology, Uber has rapidly expanded its service in more than 400 cities in 68 countries in just about 6 years. Uber, which uses new vehicles owned by their partners and provides services with lower price compared to conventional taxi, has changed the consumption pattern of the urbanites, including in Jakarta. The existence of "space of flows" as proposed by Castells (1996), which is defined as remotely communicated simultaneous social practices material support, has allowed Uber, which is based in the United States, to operate online across the world.

In the case of Uber's operation in Jakarta, the researchers conclude that Uber's network power has established a network society, which has now developed into a global system with new characteristics in the current globalization era. Even so, the findings in the research also show certain loopholes that are not in accordance with the patterns of the network society and informational capitalism proposed by Castells.

Uber is mostly used by the middle-class community to feel the experience previously reserved for the "upper class community", which is having personal drivers and private vehicles. The class issue in transportation is something characteristically Indonesian. Together with the urban development starting from Batavia to Jakarta and recently the concept of Smart City, the idea of owning private vehicles and personal drivers has always been thought to raise the prestige of the passengers. However, there are still a number of issues faced by some Jakartans who are less technologically literate, such as the surcharges applied by Uber application during rush hours, which is often beyond their understanding.

Uber has also changed the existing employment system by opening opportunities for workers outside the range of productive ages to work as drivers and earn income. The technology-based companies have also reduced the power relations between the workers and employers. The partners and drivers of Uber are still willing to work with the regulation stipulated by Uber, despite only working through the application system on their gadgets. Uber has no longer highlighted interpersonal relations. Instead, Uber accentuates human and technology relations. However, in Jakarta, there are also Uber drivers who resist the system, for example, by also working as Grab Car drivers.

Uber has faced several challenges in many countries, including Indonesia, due to its non-compliance toward the regulations on investment return shares between the company and the state. However, the state's resistance toward Uber has been responded differently by the community. Uber, which develops their operations through a global Internet network, has created a network society. This causes the state from initially being repressive toward Uber to slowly being more accommodative in order to gain benefits from the operations of the informational capitalism company.

REFERENCES

Ajidarma, S.G. (2004) *Layar Kata (Words Screen)*. Yogyakarta, Bentang.

Arvirianty, A. (2016) *Saham BIRD, dan TAXI Fluktuatif (BIRD and TAXI Stock Values Fluctuate)*. Media Indonesia: page 19.

Belk, R. (2014) You are what you can access: Sharing and collaborative consumption online. *Journal of Business Research*, [Online] 67 (8): 1595–1600. Available from: http://www.greattransition.org/publication/debating-the-sharing-economy [Accessed 16th December 2015].

Castells, M. (1996) *The Rise of the Network Society: The Information Age: Economy, Society, and Culture Volume I*. Oxford, Blackwell Publishers.

Castells, M. (1997) *The Power of Identity: Information Age: Economy, Society, and Culture Volume II*. Oxford, Blackwell Publishers.

Castells, M. (1998) *End of Millenium: Information Age: Economy, Society, and Culture Volume III*. Oxford, Blackwell Publishers.

Damardono, H. (2013) *Ayo Lawan Kemacetan (Let's Fight Congestion)*. Jakarta, PT Kompas Media Nusantara.

Kusno, A. (2009) *Ruang Publik, Identitas dan Memori Kolektif: Jakarta Pasca-Suharto (Public Space, Identity and Collective Memory: Jakarta Post-Suharto Era)*. Yogyakarta, Penerbit Ombak.

Kusumawijaya, M. (2004) *Jakarta: Metropolis Tunggang Langgang (Jakarta: Helter-Skelter Metropolis)*. Jakarta, Gagas Media.

Rahayu, Wahyono, B., Rianto, P., Kurnia, N., Wendratama, E., Siregar, A.E. (2015) *Menegakkan Kedaulatan Telekomunikasi & Penyiaran di Indonesia (Defending Telecommunication and Broadcasting Sovereignty in Indonesia)*. Jakarta, Yayasan Tifa and PR2Media.

Rahayu, Wahyono, B., Rianto, P., Kurnia, N., Wendratama, E., Siregar, A.E. (2016) *Membangun Sistem Komunikasi Indonesia: Terintegrasi, Adaptif, dan Demokratis (Establishing Indonesia Communication System: Integrative, Adaptive, and Democratic)*. Jakarta, Yayasan Tifa and PR2Media.

Rhea, A. (2015) *Defining the Digitally Networked Sharing Economy: A Rhetorical Analysis of Airbnb.* [Thesis] Departments of English and Economics, Texas Christian University, Forth Worth, Texas.

Saputra, I., Simbolon, C.D. & Arvirianty, A. (2016) *Taksi Daring Siap Penuhi Persyaratan Pemerintah (Online Taxis Agree to Government's Conditions)*. Media Indonesia: page 1.

Schor, J. (2015) *The Sharing Economy: Reports from Stage One.* [Online] The Boston College. Available from: https://www.bc.edu/content/dam/files/schools/cas_sites/sociology/pdf/TheSharingEconomy. pdf [Accessed 15th January 2016].

Scott, A.J. (2012) *A World in Emergence: Cities and Regions in 21th Century.* Celthenham, UK, Northampton, MA, USA, Edward Elgar Publishing Limited.

Shaheen, S. & Chandari, N. (2014) Transportation Sustainability Research Center (TSRC) University of California Berkeley. App-Based, On-Demand Ride Services: Comparing Taxi and Ridesourcing Trips and User Characteristics in San Francisco. [Online] Available from: http://tsrc.berkeley.edu/ node/797 [Accessed 28th October 2015].

Sjafruddin, A., Widodo, P., & Kurniati, T. (2001) Demand rate and elasticity of the urban taxi service based on the stated preference data case study in Bandung, Indonesia. *Journal of the Eastern Asia Society for Transportation Studies,* 4 (3), 177–189.

Suwardiman. (2015) *Penetrasi Internet Belum Merata (Internet Penetration has not been Evenly Distributed)*. [Online] Available from: http://print.kompas.com/baca/2015/07/21/Penetrasi-Internet-Belum-Merata [Accessed 30th March 2016].

Yuliani, P.A. (2014) *Tarif Taksi Naik Operator belum Untung (Rising Taxi Tariff, Operators are yet to Gain Profits)*. Media Indonesia: page 8.

Commuter line e-ticketing system and the disciplining of urban citizens

A. Sandipungkas & L. Kurnia
Department of Literature, Faculty of Humanities, Universitas Indonesia, Depok, Indonesia

ABSTRACT: The e-ticketing system and its following policy have managed to change the culture of urban society in using public transportation. The e-ticketing system has changed the perception of train stations from a dirty, crowded, unorganized, and unsafe environment to a cleaner, safer, more comfortable, and more systematic environment. This research examines the disciplining practices in train stations and the Commuter Line trains in relation to the implementation of the e-ticketing system. The purpose of this research is to understand how the state political culture contributes to the process of disciplining and educating the people through the new system.

1 INTRODUCTION

Since 2013, there has been a significant change in the *Jabodetabek* (the areas of Jakarta, Bogor, Depok, Tangerang, and Bekasi) Commuter Line train service with the implementation of the E-Ticketing system. The system has altered the area of train stations from a dirty, crowded, unorganized, and criminal-filled place to a cleaner, safer, more comfortable, and more systematic environment (BeritaSatu.com, 2016). Since its implementation, it has managed to change the culture of urban society in using public transportation (Antaranews, 2013). This research examines the disciplining practices that occur in train stations and the Commuter Line in relation to the implementation of the e-ticketing system. The purpose of this research is to understand how the state political culture contributes to the process of disciplining and educating the people through the automation of public transportation service.

One of the previous studies addressing the e-ticketing system is the research carried out by Grace Ng-Kruelle, et al. entitled *E-Ticketing Strategy and Implementation in an Open Access System: The case of Deutsche Bahn* (2006). The research analyses the advantages and challenges surfacing from the implementation of the e-ticketing system for the train service business in Germany, as well as its developmental strategies in utilizing the open access ticket system, which gives users direct access to tickets (Loh, Kramer, & Kruelle, 2005; Kruelle, 2003 in Kruelle, 2006, p. 7). In Indonesia, Singgih Rahadi, in his thesis entitled *Proses Inovasi Layanan Sistem e-ticketing pada Kereta Commuter Jabodetabek (The E-ticketing System Innovation Process in Jabodetabek Commuter Trains)* (2014), describes the innovation process of the e-ticketing provided by PT KAI Commuter *Jabodetabek*. The research addresses the motive behind the innovation, the dynamics in developing the innovative idea, as well as the implementation strategies and the benefits of the e-ticketing system. Another study focusing on the Commuter Line station following the implementation of the e-ticketing system is carried out by Ratna Kusmiati, in her thesis entitled, *Analysis of Commuter Line Passengers Queue toward the Exit Door of Tanah Abang Central Jakarta* (2014). Kusmiati finds that the long queues in Tanah Abang train station have not only been caused by inadequate facilities and infrastructure, but also by problems emerging from the implementation of the e-ticketing system. Meanwhile, Abidin Kusno, in Chapter V of his book entitled *Ruang Publik, Identitas, dan Memori Kolektif: Jakarta Pasca-Suharto (Public Space, Identity and Collective*

Memory: Jakarta Post-Suharto Era) (2009), examines the public transportation policy and its relation to the politics of urban culture. He finds that the Busway Project, as a form of populist policy of the administration of Jakarta's former governor Sutiyoso acts not only as a 'shock therapy' to resolve Jakarta's traffic congestion issue, but also as a form of disciplining technique in Jakarta. The Busway Project serves as an apparatus used by the city government to restore its authoritative power before the people after diminishment following the end of Suharto's regime. Kusno's research on the Busway Project is used as the model of analysis for this research.

1.1 *E-ticketing: The innovation of commuter line service*

In 2008, PT KAI (*Kereta Api Indonesia*) created a subsidiary company called PT KAI Commuter Line (hereinafter referred to as PT KCJ) in accordance with the President's Instruction No. 5/2008 and the SOE Ministry Letter No. 5-653/MBU/2008. PT KCJ is established to improve the focus and quality of the train service in the *Jabodetabek* area (krl.co.id, 2013), with a scope of works ranging from public and cargo transportation businesses, maintenance of infrastructures, support in the concession of train infrastructures and facilities, and performance based on the general purpose of implementing the principles of a limited liability company (The Deed of Establishment of PT KAI Commuter Jabodetabek number 457:15, 2008). Previously, the Urban Transport Division of Jabodetabek, under the management of PT KAI, operated the Commuter Line service. Since 2013, one of PT KCJ's strategies to achieve its main purpose is by implementing the e-ticketing system (The Internal Report of the National Train Revitalization Technical Team: 30, 2008).

The e-ticketing, or electronic ticketing, system is an electronic card-based system in which a computer chip containing information, such as balance, fare, and travelling plan, is planted to replace paper-based tickets. The e-ticketing system was first implemented in American business flights in the 1980s, and it revolutionized the conventional ticketing system through its efficiency and practicality. The implementation of the system requires a set of operational tools, which include electronic cards, card dispensers, and automatic gates for tapping in/out. Generally, the benefits of the e-ticketing system are easier access for train passengers, improvement of the service quality of the ticketing system, and optimization of revenue (Ng-Kruelle, Swatman, and Kruelle, 2006).

The quality of the Commuter Line service and the conditions of the train stations before the implementation of e-ticketing were problematic and far from convenient. The stations were dirty and unorganized due to overcrowding. There were a great number of unsupervised and unregulated illegal activities. The Commuter Line service experienced recurring delays and technical problems. The ineffectiveness of the conventional paper-based ticketing system also caused many problems, such as the so-called *atapers*, long queues, and pickpockets in the trains and stations. The term *atapers* refers to passengers who sit on the roof, or *atap* in Indonesian, of the trains during the ride. Some stated that the *atapers* are usually associated with free riders, while in fact there are some people who buy tickets but still prefer to sit on the roof. Thus, the management of PT KCJ considers the e-ticketing system as the appropriate solution to resolve all of these issues, while at the same time improving the service quality. This system is believed to be capable in improving the service system, while at the same time changing the culture of the unruly passengers that have been going on for a long time (Rahadi, 2014). Preparation for the implementation of the e-ticketing system in *Jabodetabek* Commuter Line started with the sterilization process of the stations from peddlers and illegal kiosks both within and around the stations, the closings of illegal access to the stations, and the rearrangement of the stations in order to install the operational tools for the e-ticketing system. The implementation of the e-ticketing system in the *Jabodetabek* Commuter Line results in the usage of single and multi-trip electronic cards replacing paper-based tickets, the usage of automatic gates for the passengers to check in and out replacing the ticket inspector officers, and the implementation of the progressive fares. In other words, the e-ticketing system means the automation of the stations. The term 'automation' is derived from the word

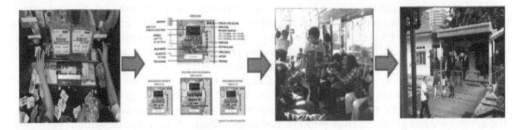

Figure 1.   The procedure of purchasing a conventional ticket with flat rate: the same rate is applied for both short and long-distance trips IDR 8,000. (Mechanism: purchase the ticket at the counter—obtain the paper ticket with the destination information—ticket inspection process inside the Commuter Line—leave the station freely).

Figure 2.   The procedure for the e-ticketing system with the progressive rate: the rate for the first five stations is IDR 2,000 and increases by IDR 1,000 for every 3 next stations. Since April 2015, the progressive rate is based on the travel distance of the commuter line in order to make it fairer. The first 25 km is IDR 2,000, and it increases by IDR 1,000 for every next 1–10 km (Mechanism: Purchase ticket on the counter—Obtain card ticket from dispenser—Tap the card on the automatic gate—Tap again on the automatic gate to leave the station).

Grundrisse by Marx which refers to a time when the social structure of the capitalist society depends heavily on the development of technology (Marx in Mulyanto, 2013).

1.2   *Marginalization from e-ticketing*

Before the implementation of the e-ticketing system in stations, PT KCJ had begun the sterilization process of the stations by evicting street peddlers and illegal residents and closed illegal access for almost 7 months since 2012. This was followed by the rearrangement of the stations, namely the installation of the operational tools required for the e-ticketing system. In carrying out the sterilization process, PT KAI as the owner of the station area collaborated with security forces to clear out the kiosks on the train platforms and the stations' parking areas. The eviction process faced great resistance especially from the merchants and people who had peddled and lived around the station areas for a long time. They felt that their presence did not violate any laws, especially because they had already paid rent money to the station administration. In certain areas, resistance even led to violent frictions between protesters and the securities deployed by PT KCJ. Moreover, in some stations such as Pondok Cina, Depok, an altercation occurred between the passengers and the protesters because the Commuter Line passengers felt that the protesters' action to close some parts of the Commuter Line railway interfered with their travel.

Although encountering resistance, PT KAI remained adamant in carrying out the eviction process for the passengers' convenience. Ignasius Jonan as the KAI director at the time said that the eviction was done for convenience and safety reasons, as the number of Commuter Line passengers were increasing. Jonan considers the eviction legitimate, since the traders' rent contract in the station had gone up. This case is similar to the eviction targeting street peddlers and a number of trees in certain roads in Jakarta to improve the development of the Busway Project in 2005. For the purpose of public convenience, the lower-class citizens were forced to be evicted in the process (Kusno, 2009). This phenomenon can be commonly found in urban areas where the lower-class citizens are marginalized for the sake of public convenience. According to David Harvey (2006), urban restructuration, which includes the restoration of the transportation infrastructure, must go through the 'creative destruction' phase with the marginalization of lower-class citizens. This also includes those who are marginalized from political power as a result of the dispossession of their living spaces. The term 'creative destruction' refers to the Schumpeter theory popularized by David Harvey (2006), which means that to build a progressive capitalistic production system the non-capitalist system must be left to die or be destroyed first. This is then followed by technological advancements, increased labour productivity, as well as increased efficiency of social relations and the distribution of work for production/circulation to accumulate maximum capital (Harvey, 2006). The practice implemented by PT KAI through PT KCJ is a form of accumulation by dispossession (Harvey, 2006) to implement the e-ticketing system and to improve the quality of the Commuter Line service as the backbone of public transportation and as a significant source of future income. The DKI Jakarta Regional Regulation No. 5 of 2014 Article 141 on Commuter Line states that Commuter Line service is expected to be the backbone of urban transportation in the *Jabodetabek* area. The Presidential Decree No. 14 of 2015 on the Master Plan for the Development of National Industry 2015–2035 also states that trains are expected to be capable of sustaining the national economic growth (the Jakarta Regional Regulation 5: article 141 of 2014 and the Presidential Decree 14 of 2015).

1.3   *E-ticketing as a disciplining technique*

Jacques Ellul in the introduction of his book entitled The Technological Society states, "In our technological society, technique is the totality of methods rationally arrived at and having absolute efficiency in every field of human activity" (Ellul, 1964, p. XXV). Ellul explains that, the "techniques in the technological society blanket all human activities, and abolish individuality in a network that transforms or obliterates the qualitative" (Ellul, 1964, pp. 284, 287). Techniques are needed by the state in order to exercise power, while simultaneously achieving the maximum quality of the institutional functions under its control, which include disciplinary and social-economic functions.

Before the implementation of the e-ticketing system, the spaces in the station and the train platforms were almost without supervision and control of power. Criminal activities in the stations were even considered as normal occurrences. The exits from the station were too lenient and led to a lack of control in the calculation of the passengers who actually paid for the tickets. It was also difficult to control peddlers inside the stations who even used the stations as their illegal settlements. Consequently, train stations and their platforms became dirty, unorganized, and unruly. The root of this problem was the feeling of freedom of not being under any supervision inside the stations and the Commuter Line trains. The e-ticketing system and its apparatus deprive them from the feelings of freedom and being unsupervised.

By purchasing and possessing the e-ticket, a person consciously registers themselves to a system that forces them to submit to a set of rules reinforced by the system. The e-ticket also requires a minimum nominal value that has to be paid as an insurance for every person who uses the train service and can be reimbursed when they tap-out upon arriving at their destination. This implies that without paying a certain amount of money a person does not have the right to enter the station. This qualification becomes the minimum standard in disciplining train passengers following the implementation of the e-ticketing system, which is in line with Deleuze's argument,

> The disciplinary society is composed of enclosed spaces, and individuals are continually trapped in one or another of those spaces (family, school, factory, army, prison, and hospital) that have moulded them (Deleuze, 1992, pp. 3–5).

In these spaces, people are being supervised and controlled so that arbitrary activities that contradict or oppose the interest of the state's exercise of power can be minimized or even eliminated. Because of the implementation of the e-ticketing system on every Commuter Line stations, we can no longer find any street peddlers selling their merchandise freely, performing street musicians, people smoking cigarettes freely, brawls between students, even demonstrations against the government's policy staged inside the stations and the trains.

Commuter Line service users, who are mostly citizens from satellite areas commuting to and from Jakarta, can notice that today there are security officers patrolling and stationed in various places inside the stations as well as inside the trains. They are aware that CCTV cameras are installed in every train station and platform. They will hear the voice from the loudspeakers both in stations and Commuter Line trains reminding them not to smoke and litter, and informing them about the safety line on the platform, the waiting lounge, and the special carriage exclusively for women. When those restrictions are broadcasted, people will feel that these messages are directly addressed to them personally. If there is someone who happens to litter in the stations or in the Commuter Line trains while those messages are being broadcasted, other people will feel obliged to remind the person to stop doing so. The feeling of being watched and the ability of the stations and the train developers to immediately handle the offenders make every person in the stations and the trains watch themselves and others. They are conditioned to constantly be alert and not to let one self or another violate the rules and become the target of punishments, such as being reprimanded, expelled, and fined.

1.4 *E-ticketing and the presence of the state*

Before the implementation of the e-ticketing system for the Commuter Line passengers, the control for entering and exiting the stations and trains, especially the economy-class trains, were loosely enforced. Illegal kiosks spread everywhere in the stations. Thousands of passengers without tickets were commonly seen in the stations and the trains. People could freely carry any kind of objects inside the trains, such as sacks, cardboards, bicycles, vegetables, and even farm animals like chickens or ducks. They occupied the spaces in the trains which significantly reduced the space for other passengers. These occurrences, which happened on a daily basis, became an enormous detriment for PT KCJ to provide quality service and caused them as the operator of the Commuter Line service experienced a great financial loss.

The ideal purpose in implementing the e-ticketing system is to improve the service quality and increase the income for PT KAI through PT KCJ and to resolve the issues that have prevented PT KAI from gaining potential and maximum financial profit from every train passenger. By filtering the people entering the station area, PT KAI through PT KCJ ensures that only those who have purchased the ticket can gain access inside the station, and only those who possess official licenses are allowed to operate their businesses inside the stations. The filtration through licensing and automatic gates has eliminated the potential losses for PT KAI as happening in the past. The implementation of the e-ticketing system, along with its regulations and supporting apparatus, ensures that the Commuter Line passengers and the station space users are only those who are rightfully entitled to be there. In other words, every party involved must pay to PT KAI through PT KCJ. As a result, the *atapers* and other free-riders have ceased to exist. The same goes for the peddlers, both in the stations and in the train carriages, as well as illegal residents. Most importantly, there is also a decrease of potential criminal activities in the station areas and inside the train carriages after the implementation of the e-ticketing system (kereta-api.info, 2013).

By implementing the e-ticketing system, the state has exercised its maximum function in controlling the source of income through the disciplining of its citizens, in this case the passengers of the Commuter Line and the parties that utilize the station areas for commercial purposes. At the same time, the state is present through PT KCJ and exercises its function as a protector and public servant by optimally providing comfort for those who obey the existing regulations and mechanisms. In the end, it is clear that harmonious and regulated cooperation from every party involved could bring benefits for all, not only for the state as the operator of mass transportation, but also the citizens who are in need of a public transportation system that is safe, comfortable, and free from traffic congestion.

REFERENCES

Antaranews.com. (2016) *Mengubah Budaya Bertransportasi Warga Ibukota (Changing the Transportation Culture of Jakarta Citizens)*. [Online]. Available from: http://www.antaranews.com/print/383155/mengubah-budaya-bertransportasi-warga-ibu-kota [Accessed 23th May 2016].

Beritasatu.com. (2016) *Direvitalisasi, Stasiun KRL Kini Megah dan Cantik (Revitalized, Train Stations are Now Luxurious and Beautiful)*. [Online] Available from: http://www.beritasatu.com/megapolitan/341823-direvitalisasi-stasiun-krl-kini-megah-dan-cantik.html [Accessed 20th May 2016].

Deed of Establishment of PT KAI Commuter Jabodetabek No. 457 of year 2008 dated 15th September 2008.

Deleuze, G. (1992) *Postscript on the Societies of Control*. (M. Joughin, Trans.). October 59 (Winter): 3–7. Cambridge, MIT Press.

DKI Jakarta Regional Regulations. (2014) Number 5 Article 141.

Ellul, J. (1964) *The Technological Society*. New York: Vintage Books.

Harvey, D. (2006) Neo-liberalism as creative destruction. *The ANNALS of the American Academy of Political and Social Science*. 88 B(2),145–158.

Kusmiati, R. (2014) *Analisis Antrean Penumpang Commuter Line Jabodetabek Menuju Pintu Keluar Stasiun Tanahabang Jakarta Pusat (Analysis of Commuter Line Passengers Queue toward the Exit Door of Tanah Abang Central Jakarta)*. Depok, Universitas Indonesia.

Kusno, A. (2009) *Ruang Publik, Identitas, dan Memori Kolektif: Jakarta Pasca-Suharto (Public Space, Identity and Collective Memory: Jakarta Post-Suharto Era)*. Yogyakarta, Ombak.

Merdeka.com. (2013) *Bentrok penggusuran pedagang Stasiun Duri, 11 luka-luka (Clash during Peddlers Eviction Process in Bukit Duri Station, 11 Injured)*. [Online]. Available from: http://www.merdeka.com/jakarta/bentrok-penggusuran-pedagang-stasiun-duri-11-luka-luka.html [Accessed 12th June 2016].

Ng-Kruelle, G., Swatman, P.A. & Kruelle, O. (2006) *e-Ticketing Strategy and Implementation in an Open Access System: The case of Deutsche Bahn*. Germany.

Presidential Decree. (2015) Number 14. On Master Plan for the Development of National Industry 2015–2035.

PT. KAI Commuter Jabodetabek. *Tentang Kami (About Us)*. [Online]. Available from: http://www.krl.co.id/#m_tentangkami [Accessed 23th May 2016].

Rahadi, S. (2014) *Proses Inovasi Layanan Sistem e-ticketing pada Kereta Commuter Jabodetabek (E-ticketing System Innovation Process in Jabodetabek Commuter Trains)*. Depok, Universitas Indonesia.

Rosa, D.F. (2013) *KAI Sukses Halau Atapers Berkat E-ticket (KAI Succeeds in Driving Away Atapers with E-Ticketing)*. [Online]. Available from: http://kereta-api.info/kai-sukses-halau-atapers-berkat-e-ticket-1516.htm [Accessed 23rd June 2016].

Salim, H.J. (2013) *Pembongkaran 83 Kios di Stasiun UI, Mahasiswa dan Polisi Bentrok (Eviction of 83 Stalls in UI Station, Students and Polices Clash)*. [Online]. Available from: http://news.liputan6.com/read/599058/pembongkaran-83-kios-di-stasiun-ui-mahasiswa-dan-polisi-bentrok [Accessed 2nd June 2016].

Seruu.Com. (2013) *Penggusuran 82 Kios di Stasiun Pasar Minggu Berujung Bentrok (Eviction of 82 Stalls in Pasar Minggu Station Ends in Riot)*. [Online]. Available from: http://indonesiana.seruu.com/read/2013/04/18/158334/penggusuran-82-kios-di-stasiun-pasar-minggu-berujung-bentrok [Accessed 29th May 2016].

Tim Teknis Revitalisasi Perkeretaapian Nasional. (2008) *Laporan Interim Tim Teknis Revitalisasi Perkeretaapian Nasional (Interim Report of Technical Team for the National Trains Revitalization)*.

The dynamics of production and consumption in Islamic housing

J.D. Shanty & M. Budianta
Department of Literature Studies, Faculty of Humanities, Universitas Indonesia, Depok, Indonesia

ABSTRACT: This paper aims to show Islamic identity construction in Qoryatussalam Sani, one of the new Islamic housing estates in Depok. Established in 2010 next to the KSU road in the Sukmajaya Regency, Depok, West Java, this Islamic estate promotes the concept of the New Madinan Society, with an underlying conservative Tarbiyah ideology. In marketing the product, this housing estate does not focus merely on the physical dimensions of the housing, but more on the Islamic community building. This orientation can be seen in the promotional brochures and advertisements, which stress the proposed religious activities. The paper argues that within the increasingly strong hegemonic paradigm of the Tarbiyah ideology, there is room for negotiation and contestation.

1 INTRODUCTION

The commodification of religion in the housing business has become rampant in Indonesia since the fall of the New Order regime in 1998. The process of democratization after this Reform Era has opened spaces for competing life styles and freedom of expression. Hedonism and religiosity compete in the consumer society. As a response to the fear of globalization and capitalism, many housing developers market their products with the label "Islam/Muslim/Syar'i/Syariah" attached to the concept of their housing projects. The public interest in housing with such labels was considerably high (Zainab, 2015). Even though some scholars have argued that this is just a new kind of capitalism wrapped in religiosity, Islamic housing is clearly still high on demand in urban and suburban areas in Indonesia, especially in Java.

As one of the satellite towns around Jakarta, Depok also has many Islamic housing clusters in various locations. With a large number of the working-age population (according to Central Bureau of Statistics in 2010, nearly 70% of the population in Depok is at the age 20–44, and a majority of them are Muslims), Depok is a potential site for Islamic housing business. This idea is supported by the mayor of Depok who continues to promote the existence of Islamic housing (Nurdiansyah, 2016). The demand for this Islamic kind of housing has made the housing developers expand their business from the center to the remote areas of Depok, ranging from the eastern part (Cimanggis) to the western part (Sawangan) of Depok, and from the northern part (areas near Universitas Indonesia) to the southern part (Citayam) of Depok. It shows that almost all areas in Depok are tinged with Islamic housing clusters.

For the developers whose orientation is merely on business and profit, commodification of religion on housing business in Depok seems quite promising due to the large and increasing number of urban middle-class Muslims in Indonesia. According to Lukens-Bull (2008), the idea of infusing religious values into commodities is a strategic entry into the market of the pious. However, such an assumption could be taken as a simplification of the matter because there are some housing projects developed not to gain profit but to spread certain ideology, such as creating Islamic spaces in the midst of urban social life whose values have become increasingly distant from what is idealized (al-Laham, 2012). This ideological contestation underlies the construction of residential Muslim housing as a social space that can control and perpetuate a form of power (Lefebvre, 1991).

Qoryatussalam Sani follows the model of other Islamic housing development by placing the mosque, the house of worship, as the focal point that marks the public and social facilities of the usual Islamic housing. Nevertheless, this physical aspect has never been the primary focus in marketing its product. It is the community building that brings the target market to this housing estate. This can be seen from the written concept on how to construct a New Madinan Society at residential community level in its brochures which are more prominent than the designs of the houses offered. Moreover, the advertisements of this housing also feature an *ulama* figure, Ustadz Amang Syafruddin, Lc., as the conceptor of the New Madinan Society development for the residential community. Through an advertisement on YouTube, Ustadz Amang explained further the concept of the housing that represents an idealized Islamic community. In its effort to realize the community, Qoryatussalam Sani has managed to entice *da'wah* activists to buy the property and help build the community.

By studying the production and the consumption of an Islamic housing cluster, this research aims to investigate the process of place-making through the development of Islamic housing with the concept of "the New Madinan Society," used as the tagline by the developer. It will examine how the ideology of Tarbiyah is formulated and disseminated by the housing developer of Qoryatussalam Sani through advertisements and how the concept of the Islamic residential community is practiced by the dwellers. The data gathered for this research consist of written texts from the brochures as well as the training manual authored by the developer management team, and ethnographic descriptions based on a four-year fieldwork I conducted in Qoryatussalam Sani from 2012–2016. This paper has been developed out of a concern to understand the ideology behind the place-making and structured in two subtopics: the first one deals with issues relating to the production of the concept, and the second one discusses the consumption continuing with the implementation phase or reproduction of the concept.

1.1 The Tarbiyah movement in the concept of the New Madinan Society

Towards a New Madinan Society is not just as a mere tagline attached to Qoryatussalam Sani, but also the underlying marketing concept. In Islamic discourse, the concept of a Madinan Society is a guideline that functions as a benchmark for the progress of a civilization. It is also a representation of the hopes and dreams of Muslims in order to realize Islamic revivalism. This concept refers to the social order city-state of Medina during the reign of Prophet Muhammad (PBUH) in 622 AD (Azra, 2004: 9–11). Adhering to such a notion, the developer, as a cultural intermediary, intended to build a community as exemplified by the Prophet in its housing products. The concept then was developed based on what is believed to be an Islamic way of life in accordance with the guidance of the Qur'an and hadith. It has proved to be a quite powerful marketing strategy that can be evidenced by the number of consumers who have bought houses based on considerations of the concept (16 out of 19 informants interviewed stated that the housing concept was an important consideration).

The desire to construct the imagined Madinan Society is represented through the content of the brochures and YouTube video advertisement. Unlike any other housing projects intended for the middle-class society, Qoryatussalam Sani does not use names commonly used for housing estates in Indonesia, such as *Griya, Graha,* Estate, or Residence. The name 'Qoryatussalam' is taken from Arabic meaning 'Kampung Damai' or peaceful village. In Indonesian the term

Figure 1.   Head of the various official letters and advertising brochures of Qoryatussalam Sani housing.

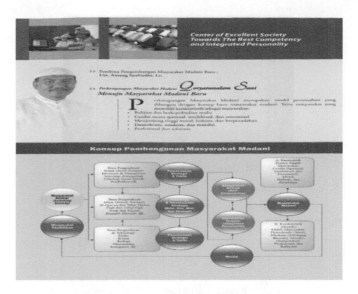

Figure 2.   Page of the brochure that features notions on how to build a New Madinan community featuring Ustad zAmang Syafruddin, Lc.

'kampung' (village) has a connotation that does not represent middle-class residence, but it is often associated with a sense of warmth and humility. Meanwhile, the word 'Sani,' according to some informants, means 'Yang Kedua' or 'The Second' (in Arabic 'tsani'), assuming that the first peaceful village is the hereafter. It turned out that the word 'Sani' might have referred to the name of a person who had an important position in the development of this housing, Mr. Herman Sani. In addition to using Arabic for its name, the representation of Islam in this housing is reinforced by the colour green, complete with the picture of a mosque with its minarets framed by an octagonal star as seen in its logo. The concept of a Madinan Society intended for the housing community is further explained by a public figure from the Justice and Prosperity Party (*Partai Keadilan Sejahtera*), the rising political party after the reform. The figure of Ustadz Amang Syafruddin, Lc, who is active in the field of Islamic education, appeared both in the brochures and YouTube video advertisement.

In the brochures, the concept of a New Madinan Society is presented in the form of a concept map containing notions on how to build a community based on the example of the Prophet's society. It is explained that the intended New Madinan Society originates from a 'Learning Society'. Learning and education are the key words that appear most frequently in the brochure. The importance of education is also the final goal of the concept. Therefore, in building this community the presence of educational forums becomes significant. The families expected to support this community are "religious and noble" families with "spiritual, intellectual, and emotional capacities" to uphold the noble values of morality, civilization, law, democracy, and independence. Two other terms mentioned in the brochure are the quality of being "moderate" and "democratic." The text on the brochure is then narrated through a video advertisement on YouTube. The last two words which suggest openness and tolerance of differences seem at odds with Ustadz Amang's overall concept of an Islamic way of life.

In the YouTube video advertisement uploaded on 16th March 2011, Ustadz Amang stated, "This is how we prepare for a generation, a new generation, the generation of the Quran. It's a generation who loves learning, a generation who understands law, even ethics, and morals. They will grow up in an atmosphere of the reconstruction of Islamic personality...." The religious leader also said, "Each week they (the residents) will also be involved in various training activities, for self-development, family development, and community development." He invites the audience (i.e. prospective consumers) to construct "various competencies" toward "a great and noble civilization." In other words, this concept represents the ideals and

Figure 3.   Qoryatussalam Sani Youtube video advertisement.

expectations which can only be achieved through congregational living with the love of learning to form an Islamic personality that leads to a Quranic (holy) generation. A strong desire to
make the values of Islam as a way of life to revive the glory of Islam is clearly expressed in the
message. The vision in building the community through activities corresponds to the function
of Tarbiyah as discussed by Hasan al-Banna (al-Banna, 2004, in Machmudi, 2010: 2).

Tarbiyah, a revolutionary socio-religious movement allegedly changed the face of Islam in
Indonesia in the 1980s and enriched the classification of Islam that originally focused only on
traditional Islam and modern Islam (Machmudi, 2008). Tarbiyah, in essence, is a reflection
of the Western concept of education (Al-Attas, 1980: 28). In the context of Indonesia today,
the term Tarbiyah is better known as the label of an Islamic movement which flourished in
secular campuses (Machmudi, 2010). This movement was then transformed in 1998 into a
political party named the Justice Party, which later turned into the Justice and Prosperity
Party (PKS) in 2003 (Machmudi, 2008: 217), a party that has won the regional elections in
Depok since 2005. The idealized concept of this housing seems to be in tune with the spirit
of the Tarbiyah movement that focuses on education.

1.2   *Tarbiyah ideology dissemination through trainings*

The concept of Madinan Society subsequently entered the stage of socialization through
trainings conducted by the developer for its first consumers. From the website page created
by the developer team, it is reported that the initial training was carried out on 29th August
2010 attended by 20 participants who were among the first occupants. This training confirmed that the developer had serious intentions in initiating a community. As an initiator of
the community, the developer did not organize the community developed in the housing, but
rather gave advocacy to the occupants who were expected to build the community themselves
(Stoecker, 1997:11). The occupants were supposed to be the backbone of the community.
Hence, trainings were needed to disseminate the underlying concept of the community.

According to the manual, to construct a New Madinan Society, two approaches are
required, namely cultural and structural approaches. It is mentioned that the orientation of
the cultural approach is to raise awareness and develop a habit to establish an Islamic personality, while the structural approach aims to construct a structured society whose terminal

Semua tahapan proses ini dilakukan dengan dua pendekatan:

Pertama: Pendekatan kultural yang berorientasi membentuk kesadaran dan habit (kebiasan) hidup yang selalu interaktif dan kondusif dengan Islam. Di tataran individu pendekatan ini lebih difokuskan pada terbentuknya kepribadian Islami (*Syakhshiah Islamiyyah*), sedangkan di tataran masyarakat dan negara diorientasikan ke arah terbentuknya peradaban Islam (*Hadlarah Islamiyyah*).

Ke dua: Pendekatan struktural yang berorientasi membentuk sebuah masyarakat yang terstruktur dengan sebuah otoritas dan konstitusi yang lebih berdaulat. Dalam proses selanjutnya masyarakat ini diarahkan kepada sistem yang lebih mandiri dalam kehidupan berbangsa dan bernegara dengan terbentuknya Khilafah Islamiyyah.

Figure 4.   Approaches to construct a New Madinan Society. (Source: Training Manual, page 7).

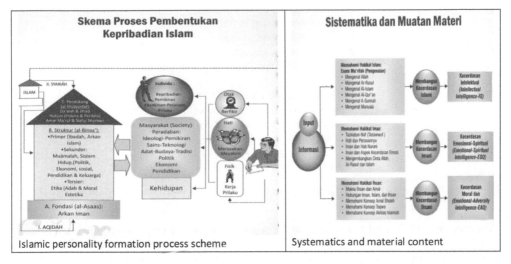

Figure 5.   Invocation to transform to a state of surrendering only to God. (Source: Training Manual, pages 31 and 35).

objective is to build the civilization of Islam. The cultural approach is further elaborated in the manual as personality development becomes the basis for disseminating the concept.

Programs with regards to moral issues and thoughts are some of the main concerns of the Tarbiyah movement (Machmudi, 2010). Mental and moral formation is the foundation of the fundamentalist concept of individual militancy, where life is kept simple by referring to religious teachings. The emphasis on the spiritual (*ruhiyah*) aspect continually is expected to strengthen their faith and make the people surrender only to God.

Having faith in religion involves feelings, thoughts, and behavior. Placing a religious concept over a community can be interpreted as an attempt to homogenize and naturalize the way of feeling, thinking, and behaving as "common sense." When religion dominates, those who feel, think, and act outside the given guidelines might indirectly feel excluded from the community. Therefore, religion serves as a discourse in which meanings are constructed (Hall, 1996), while at the same time it functions as a means of tightening community ties.

In order to make religion become the living culture of a community, the developer proposed daily, weekly, and even annual agenda for the community members. These schedules are designed with a reference to rituals of Islamic worship. These individual and collective routines unconsciously construct experiences which influence the thinking and actions of the people in the community. Culture in everydayness disguises the ideology underneath, so it

No.	Waktu	Kegiatan	Keterangan
1	03.30 – 04.30	Qiyamullail	1. Waktu shalat disesuaikan dengan jadwal yang berlaku
2	04.30 – 05.30	Shalat Shubuh Berjama'ah dan Wirid Ma'tsurat	
3	05.30 – 06.30	Kajian Tadabbur al-Qur'an	
4	07.00 – 12.00	Aktifitas Rutin	
5	12.00 – 12.30	Shalat Dzuhur Berjama'ah	2. Aktifitas rutin sesuai dengan tugas dan agenda masing-masing
6	13.00 – 15.30	Aktifitas Rutin	
7	15.30 – 16.00	Shalat Ashar Berjama'ah	
8	16.00 – 17.00	Kajian Islam Ibu-Ibu	
9	17.00 – 17.30	Wirid Ma'tsurat	
10	17.30 - 18.15	Shalat Maghrib Berjama'ah	
11	18.15 - 19.00	Kajian Islam dan Diskusi	
12	19.00 – 19.30	Shalat 'Isha Berjama'ah	
13	19.30 - 21.00	Bersama Keluarga	
14	21.00 – 03.30	Istirahat	

Figure 6. The proposed daily agenda that suggests religious activities in public and private sphere. (Source: Training Manual, page 40).

Pertama : Penyusunan konsep kaderisasi.
Kedua : Pendidikan dan pelatihan.
Ketiga : Rekomendasi kader-kader
formal dan informal leaders.

Pembentukan Masyarakat Pendidikan:
• Pembentukan kelompok-kelompok Pembelajaran (*Learning Communities*).
• Pembangunan Masyarakat Pembelajaran (*Learning Society*).

Figure 7. Propagating the ideology through learning groups. (Source: Training Manual, page 18).

looks natural (Hall, 1996). The ideology represented in the concept is wrapped in a religion which regulates everydayness which makes it difficult for the members to resist. The strategy that regulates the everydayness of the community gives structure to frame and control the everyday practices of its members (de Certeau, 1984).

Broadening the spectrum of the ideology to a wider environment is also the expectation in establishing this community. Groups are formed in small study circles (*halaqah*) that consist of 5–10 members each, scouted continuously in competence-based levels. From those circles, networks can be developed and widened. This community is to become a pilot project of the New Madinan Society development that is expected to grow and affect the surrounding areas. The tendency to share what is believed to be the truth is the reason to disseminate the ideology (Dant, 1991).

Establishing an idealized community in commercial housing such as Qoryatussalam Sani can be considered as an act of place-making where the intention to create public spaces that promote religious life is capitalized (Schneekloth & Shibley, 1995). This is a political act that encourages disciplines in pursuit of the intended religious qualities.

1.3 *Implementation of the concept through the consumption of meaning*

The message conveyed by the developer through the training can be seen as a discourse that is decoded further into a new one by the consumers. Approximately 60% of the trainees (about 23 out of the 38 families) took the dominant-hegemonic reading position where the meaning of the concept was received as intended, while about 12–13 families were on negotiated position and even fewer kept their oppositional position. Those who took the negotiated position recognized the legitimacy of the dominant reading, but could not fully accept what was intended, whereas those who took the oppositional position understood the literal and connotative meaning of the discourse, yet were in opposition to the dominant discourse (Hall, 2001: 172–173). Therefore, the consumption of this concept faced many challenges in its implementation. There were suspicions and fears that hindered the concept from finding its concrete form for a number of reasons. First, the colour selection of one type of the housing resembled the official colours of the Justice and Prosperity Party. Second, the brand ambassador of Qoryatussalam Sani, Ustadz Amang Syafruddin, L.c., was affiliated with the same party. Lastly, the models of the community activities offered are similar to the ones in the party.

In the first half of 2014, there was an incident caused by the resentment of the residents toward the developer who was negligent in providing the public infrastructure and facilities of the housing. Ustadz Amang Syafruddin who was in the front row of the director staff became the target of their exasperation. The hatred turned to criticism of the background and ideology of the developer. On the other hand, there were people who felt that Ustadz Amang was the figure who was not supposed to be blamed because he was the only one who could bridge the residents with the developer. One of the informants interviewed saw how the conflicts escalated due to differences in understanding the daily practices and certain rituals in Islam. He was suddenly aware of the ideology infiltrated in the neighborhood, although he also believed that this ideology would not get a positive response if imposed. Nevertheless, this conflict split the residents into two. Although the conflict was admitted as a mere misunderstanding and has long been considered resolved, the trauma of the dispute is still clearly visible on both sides.

The disagreements due to the colour of the political party in this housing became the dynamics of the early formation of the community. It ended with a "cold war" that actually exacerbated and obstructed the communication in the neighborhood management. For a while, the activities to build the community stopped due to various disagreements among the residents. Not long afterwards, one of the informants, a mother of six, said that she could not remain silent and do nothing. She decided to move on and reorganized the community. She was also one of the few mothers who decided to enliven the mosque with religious activities. They were former campus activists who used to be engaged in student organizations, social events, and political parties. They have proven to consistently develop the small circles in and outside the community with active Islamic study groups. These community activists became important social actors who mobilized the community, constructed daily practices, and educated the members of the community (Ledwith, 2011). They acted as models for other members to determine the direction and objectives of the community by disseminating the dominant ideology embedded in our cultural attitudes, defining what is normal and acceptable (Thompson, 2007). The way they dress, speak, and behave becomes a common symbol system shared with the community, as an indication that they are part of the community. In other words, it is difficult for other community members to ignore the dominant ideology brought by these activists.

The community activists become a model for the inhabitants and indirectly produce what McMillan and Chavis (1986) called the boundaries of the community membership. How these community activists present themselves is seen by its members as a representation of *kaffah* (true; complete; perfect). They helped establish the community norms, direct the members to the "right" ways to frame their identity (Thompson, 2007). They always look simple and flexible with their unassuming long, loose veils in soft colours. Their demeanor is one of humility enhanced by polite, focused, and well-chosen vocabulary. They are hardly ever seen to raise the intonation of their voices. They consistently work hand in hand in organizing weekly religious studies (*majelis taklim*), and take turns in giving talks on light and easy-to-digest materials related to everyday household routines. Gradually the other members of the community follow and appropriate the attitude of these activists.

To strengthen the community ties, the need of similar perspectives of the world is constructed through the frequency of interaction (Nisbet and Perrin, 1977). Therefore, the community activists endeavoured to invite the inactive members by involving them in the communication

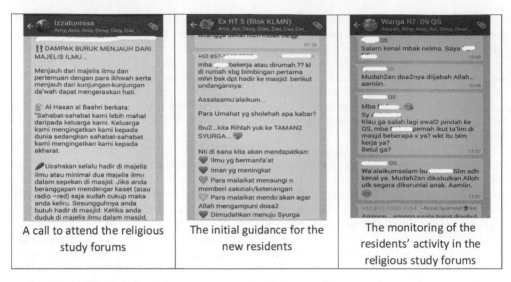

| A call to attend the religious study forums | The initial guidance for the new residents | The monitoring of the residents' activity in the religious study forums |

Figure 8. Community interaction using *Whatsapp Groups*.

network built through *Whatsapp group* media and assist them in forming the mentality that suits the desired character of the community. The use of *Whatsapp group* has expanded from announcements and warnings to the socialization of their programs and the monitoring of the residents' activity. Through the interaction in this forum, the rest of the inhabitants are aware that their participation is monitored; thus, the actual supervision is conducted by the fellow members of the community. The effort to influence other members has made this forum effective as a panopticon in establishing the hegemony of the Tarbiyah ideology.

2 CONCLUSION

The flourishing of Islamic housing in Depok has become a marker of the rise of a new Islamic middle class in Indonesia. By using Qoryatussalam Sani as an illustration, this study offers a critical perspective on the use of physical space to create a new social arena in which political and religious ideologies overlap. Although the grandiose community development as idealized by Ustadz Amang has not yet materialized, it has generated an organic community building embodied in the process of the place-making. It is made possible by the space creation that invites the participation of members in the community activities. The two words mentioned by Ustad Amang in the advertisements, "moderate" and "democratic," were put to test in the process. The pressure to conform to the intended Islamic way of life shows that the utopian concept of a New Madinan Society was far from being moderate and democratic. Although met with some resistance, the strategy used by the residential activists in building the community has in a way been successful in reproducing the ideology. Whether or not this residential community will become a model in spreading the Islamic community requires further research.

REFERENCES

Al-Attas, M. (1980) *The Concept of Education in Islam.* Kuala Lumpur, ISTAC.
Al-Lahham, A. H.-D. (2012) Housing Communities between Islamic Freedom and Capitalist Kaleidoscope. *Architecture & Planning,* 24, 99–116.
Azra, A. (2004) *Menuju Masyarakat Madani* (Toward Peaceful Society). Bandung, PT Remaja Rosdakarya.

Dant, T. (1991) *Knowledge, Ideology, and Discourse.* London & New York, Routledge.

de Certeau, M. (1984) *The Practice of Everyday Life.* (S. Rendall, Trans.) Berkeley, Los Angeles, & London, University of California Press.

Hall, S. (1996) The problem of ideology: Marxism without guarantees. In S. Hall, D. Morley, & C. Kuan-Hsing, *Stuart Hall: Critical Dialogue in Cultural Studies* London: Routledge. pp. 24–46.

Hall, S. (1997) *Representation: Cultural Representation and Signifying Practices.* London, Sage.

Hall, S. (2001) *Media and Cultural Studies: Key Works.* In: M. Durham, & D. Kellner, (eds.). Malden and Oxford, Blackwell.

Ledwith, M. (2011) *Community development: A critical approach.* Bristol Policy Press.

Lefebvre, H. (1991) *The Production of Space.* Cambridge, Mass., Blackwell.

Lukens-Bull, R. (2008) Commodification of Religion and the 'Religification' of Commodities: Youth Culture and Religious Identity. In: P. Kitiarsa (ed.), *Religious Commodifications in Asia: Marketing Gods* London, Routledge. pp. 220–234.

Machmudi, Y. (2008) *Islamizing Indonesia.* Canberra, ANU E Press.

Machmudi, Y. (2010) Islamism and Political Participation: A Case Study of Jemaah Tarbiyah in Indonesia. *The International Workshop on Islamism and Political Participation in South-East Asia: Global Context and Trends.* Sydney, Lowly Institute for International Policy.

McMillan, D.W. & Chavis, D.M. (1986) Sense of community: a definition and theory. *Journal of Community Psychology (14)1*, 6–23.

Nisbet, R. & Perrin, R. (1977) *The Social Bond.* New York, Knopf.

Nurdiansyah, R. (2016) *Walikota Depok Dukung Keberadaan Perumahan Bernuansa Islami* (Depok Mayor Supporting the Presence of Islamic Nuance Housing). Retrieved from: http://nasional. republika.co.id/berita-/nasional/jabodetabek-nasional/16/07/26/oaxcpf365-wali-kota-depok-dukung-keberadaan-perumahan-bernuansa-islami [Accessed on 15th August 2016]

Schneekloth, Lynda H. & Shibley, R.G. (1995) *Placemaking: The Art and Practice of Building Communities.* New York, Wiley.

Stoecker, R. (1997) The CDC model of urban redevelopment: a critique and an alternative. *Journal of Urban Affairs, 19*(1), 1–22.

Thompson, N. (2007) *Power and Empowerment.* Lyme Regis, Russell House.

Zainab, W.O. (2015) *Di Bawah Bayang-bayang Labelisasi Syariah* (Under the Shadow of Syariah Labelling). Retrieved from: https://islamindonesia.id/kolom/di-bawah-bayang-bayang-labelisasi-sya riah.htm [Accessed on 10th August 2015].

Cultural Dynamics in a Globalized World – Budianta et al. (Eds)
© *2018 Taylor & Francis Group, London, ISBN 978-1-138-62664-5*

Typical French linguistic process on Facebook

F.G. Junus & M. Laksman-Huntley
Department of Linguistics, Faculty of Humanities, Universitas Indonesia, Depok, Indonesia

ABSTRACT: The widespread use of Facebook (FB) as social media greatly affects the development of the world's languages. Many new forms of language have emerged as an impact of that medium. Crystal (2006) stated that conversations using the Internet as a medium tend to use a language variation, 'netspeak,' or spoken-writing. FB users prefer to write their conversations in the way they speak. The French language, which is known for its very complicated relationship between grapheme and phoneme, faces the same phenomenon. This paper is based on a research on the use of the French language on FB. The aim of the research is to describe what linguistic process occurs in the use of the French language on FB. The data used were taken from the status blocks of 111 accounts. From each account, we took five status blocks, so there are 555 files as corpus data. 'Antconc,' an application to determine the frequency of words in a corpus, was used in processing data. The result illustrates that there are some linguistic processes, such as abbreviation, morphophonemic, and a new phenomenon that are typical processes on FB.

1   INTRODUCTION

Language as a communication tool is continually growing and changing in human life. Over the past decade, communication technology has grown rapidly and has become more sophisticated. Communication between individuals or groups in a community is becoming faster and traversing geographic boundaries, crossing not only the island but also the continent. Technological developments have changed the way people communicate. Hellen Watt (2010), in her research on the influence of modern communication technologies on language, as cited in Crystal, called this new era the era of written speech and spoken writing that uses communication technology such as the Internet or social networks, which tend to be rid of oral communication. Baron (2008) claimed that the Internet is dramatically changing the way people communicate. This phenomenon also affects language change, both locally and globally (Thangaraj & Maniam, 2015).

Crystal, in his book *Language and the Internet* (2004), stated that the use of the Internet greatly impacts language; he used the word 'netspeak' to explain the use of language on the Internet. According to Crystal, 'netspeak' is a language variation that occurs in graphic, orthographic, grammatical, lexical, and discourse features. Moreover, for Crystal (as cited in Baron, 2003) the suffix 'speak' in Netspeak refers not only to 'speak' but also to 'write' activities, including the receptive elements that are 'listen and read'.

Communication technology using the Internet constantly evolves. In the history of its development, communication modes, such as iDegrees.com., LiveJournal, Friendster, LinkedIn, MySpace, and Hi5, have already existed since 1997 and become pioneers of social networking, such as Facebook and Twitter (Boyd & Ellison, 2007). According to Ellison, Steinfeld, and Lampe (2007) in the *Journal of Computer Mediated Communication*, social networking sites have an orientation that ranges from contextual work (LinkedIn.com), exploration of romantic relationships (which is the original purpose of Friendster.com), a common interest in music and politics (MySpace.com), to campus student population (the original intention of the establishment of Facebook.com).

Unlike other types of social media networks, Facebook (FB) appeals enormously to the social networking users. According to statistical data (http://expandedramblings.com/index.php/by-the-numbers-17-amazing-facebook-stats/), there are over 1.65 billion monthly active FB users, which is a 15 percent increase year over year. With the number of users per month of 1.65 billion (as of the 27th April 2016), the number of FB users is far beyond that of other social media such as WhatsApp which has a billion users (http://expandedramblings.com/index.php/whatsapp-statistics/), Twitter with its 310 million users (http://expandedramblings.com/index.php/march-2013-by-the-numbers-a-few-amazing-twitter-stats/), and Line with its 218 million users (http://www.statista.com/statistics/327292/number-of-monthly-active-line-app-users/). Nowadays, FB is the most popular among social media (https://zephoria.com/top-15-valuable-facebook-statistics/). FB users spread all over the world, and all people with different backgrounds, ethnic groups, and languages, use the site as an alternative to build social relationships with each other. Today, statistical data (http://www.statisticbrain.com/) states that FB is available in more than 70 languages.

The number of FB users in France until September 2015 was around 30 million. Laurent Solly (Chantrel, 2015), the Director General of the French FB, said that today, nearly one in two French people use FB to share many subjects. In other words, FB has become the most popular social media in France, defeating Snapchat, Tumblr, Pinterest, Twitter, and others. Children and teenager social network users in France most frequently use FB; they prefer using FB than Snapchat (http://www.emarketer.com/Article/Young-Social-Network-Users-France-Love-Facebook-Snapchat/1013682#sthash.X00LQmDf.dpuf). This is why we have based our research on the use of the French language on FB.

The data used in this study have been taken from http://www.facebook.com/. To create corpus data, we picked out the status and its comments that were uploaded in 2015. From the corpus of 555 files or conversations, 220,946 total word tokens, 19,105 word types, there are 184 words to be analyzed. The data have been compared with the words, phrases, and clauses compared to written language convention to find the linguistic processes that occur. For phonetic transcription of every data, we refer to the *Dictionnaire de la prononciation française dans son usage réel* (Martinet & Walter, 1973) and French *International Phonetique Alphabet* in Bechade (1992, pp. 14–15).

1.1 *Language use on FB*

As a highly efficient mode of communication in the era of technology that needs speed, FB becomes the first choice of many people. FB users not only share news, but are also able to speak directly, such as conducting face to face conversation with others. This is the reason that makes people prefer to communicate using social media like FB rather than to communicate with their neighbors. It is evident everywhere that people prefer to establish communication by using their gadget with other people who are physically distant from them rather than with people around them.

The use of FB as a mode of communication not only reduces oral communication, but has also brought a change in the use of language, especially written language. In French, for example, we can find many linguistic changes that have occurred in the writing of words, phrases, and clauses created by the French FB users. Changes in writing graphemes have become very interesting. French is known as a language which is very complicated in its relation between spelling and pronunciation, while FB communication needs simplicity in writing conversations because of space and writing time limitation. Therefore, the FB comment space becomes a lexical creativity space for French speakers. They must be smart in producing a written conversation which is proper and communicative for reasons of language economy.

Language economy reason has no longer been relevant since 2011 when FB increased the capacity of space from only 420 characters to a maximum of 63,206 (Protalinski, 2011). Likewise, the providers of technological devices have made improvements by adding communication device features that enable users to write correctly; nevertheless, this lexical creation in writing on FB conversations is still conducted by FB users and is becoming increasingly varied.

1.2 Linguistic processes in French conversations on FB

In their FB conversations, French language users do some lexical creation as follows:

1.2.1 Abbreviation and truncation

One of the lexical creations always done by FB users is abbreviation. It is a process of shortening words using only the initial letters of the word (Krautgartner, 2003). One type of abbreviation eliminates the end of a word or syllable known as apocope. In our corpus there are 20 data, for example, *mdr (mort de rire)* 'laugh out loud' and *bg (beau garcon)* 'handsome boy,' *pv (privé)* 'private,' and *cc (coucou)* 'hello.' However, not only at the end of syllables have FB users eliminated letters, but they have also eliminated the segment in the middle of a word or syncope. There are 23 data, such as in *pr (pour)* 'for,' *vs (vous)* 'you,' and *dc (donc)* 'so.' In the process of abbreviation, the only thing occurring in the creative process of writing is that the abbreviated form is not pronounced letter by letter, but is read based on the words that compose it. An interesting phenomenon of the abbreviation performed by FB users is that they abbreviate a word, phrase, or clause by using consonants which have the same pronunciation. For example, there is the replacement of the grapheme <q> with grapheme <k>, both of which have the same pronunciation, such as /k/ as in *tkt* for *t'inquiète* /tɛ̃kjɛt/ 'don't worry,' and *pk* for *pourquoi* /puʀkwa/ 'why,' or the replacement of the grapheme <c> with grapheme <s> which has the same pronunciation /s/ in the *dsl* for *de cela* /dəsəla/ 'thereof,' or replacement of both <q> and <s> in *psk* for *parce que* /parsəkə/ 'because.'

Another lexical creation occurring in French writing on FB is truncation. According to Bouzidi (2009), truncation is a lexical creation process consisting of deletion or removal of one or more syllables of a word. The truncation process occurs because the word seems too long. Unlike abbreviation, the result of truncation is read as a word, not the basic word. In FB conversations, the users conduct some truncation processes, such as aphaeresis or omission of segments at the beginning of the word as in the word *ti (petit)* 'little.' This case refers to what French speakers do in a verbal conversation, for example the phrase *un petit peu* /œ̃pətipø/ 'a little bit' which tends to be uttered as /œ̃tipø/. Furthermore, in FB conversations the removal of a segment at the end of the word or apocope (22 data) can also be found, as in the word *fort (fortement)* 'strongly,' *deg (degueulasse)* 'disgusting,' *pro (prochain)* 'next,' *dispo (disponible)* 'available,' and *perso (personellement)* 'personally.'

In some cases, FB users add some grapheme after doing truncation (4 data) as in *frero (frère)* 'brother,' and *bise (bisous)* 'kiss.' They also replace the grapheme <s> with <z> like in *biz (bisous)*. The purpose of these additions or replacements is that the final consonant in the truncated word can be pronounced or pronounced differently. However, there is also an addition on truncated words that does not change the pronunciation of the word, as in *choux (chouette)* 'great,' which is a modified writing that does not alter the meaning of the word. Moreover, FB users also reduplicate words that have been truncated (5 data), such as *dodo (dormir)* 'sleep,' *mimi (mignon/e)* 'cute,' and *tata (tante)* 'aunt.'

1.2.2 Morphophonemic

FB users also seem to write their conversations with alternative words that have the same pronunciation as the words or phrases to which they refer, or they write the conversations on FB as if in verbal conversations, whereas the trend in a verbal conversation is that the speakers frequently change phonemes. Phoneme change occurs mainly when there is a meeting of two or more morphemes, a phenomenon that is called morphophonemic. In French, omission, or the process of removing a segment of the phonetic context, often occurs. This usually occurs because of the emphasis on segments that follow, for example, the removal of the last vowel of a word when the following word also begins with a vowel. Another example is the removal of the vowel /y/ which is represented by a grapheme <u>. It can be seen from the change of *tu es* /ty e/ 'you are' to *tes* /te/, or the clause *tu as* /ty‑a/ 'you have' to *tas* /ta/. The last is still experiencing another removal of a non-pronounced grapheme <s> and becomes *ta* /ta/.

The omission that is the most common in verbal conversations is the omission of schwa /ə/. This omission occurs in two cases. First, the omission occurs on the first syllable of a phrase or clause due to the emphasis on segments that follow. Second, when a clause contains multiple schwas, then the first schwa will be eliminated. In FB conversations, any omission in the first case is illustrated by the removal of grapheme <e>, which represents the pronunciation of schwa as in *dla (de la)* 'of the,' and *jsui (je suis)* 'I am.' The second case can be seen in the writing of the clause *jme (je me)*.

1.2.3 *Other linguistic processes in FB conversations*

In FB conversations, users seem to arbitrarily write a word, phrase, or clause. They only rely on the pronunciation of those words because when the words are detached from the context of the utterance, they would not have the same meaning as referred to in the context. There are 18 data, for example *sa* 'his/her' for *ça* 'it,' *mes* 'my' for *mais* 'but,' and *foi* 'faith' for *fois* 'time.' They also write words that have other grammatical forms (3 data) such as *quel* 'which' (masculine) to replace *quelle* (feminine), because both have similar pronunciation. FB users freely use some words that do not exist in French (10 data), as in *ke (que)* 'that,' 'who,' *kan (quand)* 'when,' and *koi (quoi)* 'what.' Moreover, they do not write letters that are unpronounced like some final consonants such as *s*, *p* and *t* in the words (6 data) *ouai (ouais)* 'yeah,' *tro (trop)* 'too,' and *vien (vient/viens)* 'come' or unpronounced graphemes in verbal conversations such as <l> and <ls> as in the subject pronoun *il* or *ils*. In other cases, these two subjects are also replaced by <y>. These two cases occur when the pronoun subject is followed by a verb with a consonant initial, such as *i prend* 'he takes', *i sont* 'they are,' *y parle* 'he speaks,' and *y va* 'he goes.'

Another form created by FB users is replacement where they replace a word or clause with a single alphabet (6 data) which has the same pronunciation as the word or the clause referred to, such as in *a (as)* 'have,' *c (c'est)* 'it is,' and *g (j'ai)* 'I have.' They also replace nasal final vowels in the interjection *ben* /bã/ with oral vowels such as *ba* /ba/ and *beh* /be/. The phenomenon of written oral conversation in French FB conversations is also seen in the presence of several dialects, such as the dialect of Southern France *nan (non)* 'no' and *oue (ouais)* 'yes' or the dialect of Northern France which is known as Chtimi, *ch* that refers to *c'est*.

Another interesting thing in written conversations on FB is the *siglaison* or *sigle* formation which is categorized by Bouzidi (2009) as an acronym. Unlike abbreviation, acronyms are not spelled letter by letter but are pronounced as a word. Acronyms are made not only by taking the initial letters of each word, but also other elements of the basic word, such as the syllable. The acronymization (Beard, 2001; Bouzidy, 2009) takes place in order to ease pronunciation. FB users apply some acronymization that has similar pronunciation with a phrase or referral clause. The differences are minimal and often overlooked in verbal conversations. An example is in *ct* /sete/ which replaces *c'était* /setɛ/ 'it was'; or at *tmtc* /teemtese/ which replaces *toi même tu sais* /tɯɑmɛmtysɛ/ 'you, yourself know'; and in *oklm* /okɑelem/ that replaces *au calme* /okalmə/ 'be calm.'

Also found in FB conversations as frequently as in verbal conversations is ellipsis. Ellipsis is the omission of certain elements in a sentence. In verbal communication, the speakers often omit the negation *ne* in their utterance. An example is *c'est pas* instead of *ce n'est pas* 'it is not,' In FB conversations, the acronym for this ellipsis is seen in *jpp* /ʒəpepe/ which stands for *je peux pas* /ʒəpøpɑ/ 'I cannot' or *jmp* /ʒiɛmpe/ stands for *j'aime pas* /ʒɛmpɑ/'I do not like.' The ellipsis of the subject *il* occurs in *ya (il y a)* 'there is' and *fo (il faut)* 'it must.' The latter example replaces the digraph <au> with <o> which is homophonous, and a the omission of final consonant <t> which is not pronounced in oral communication.

FB users during some conversations often create an atmosphere as if they are communicating orally or face to face. It is seen from the abbreviation which shows some action such as *mdr (mort de rire)* 'laugh out loud' (7 data), and *ptdr (pété de rire)* 'laugh so hard' (4 data). Moreover, in writing some of these expressions, they also make some repetition to give the impression of long duration, as in *mdrrrrr, mdddr,* and *ptdrrrr*. The repetition also performed on the final vowel in a word gives the impression that the duration of "saying" the words is very long as in *merciiiiii (merci)* 'thank you' or *ouiii (oui)* 'yes.'

People sometimes use argot in verbal conversation. According to Dubois (1999, p. 105), argot is all words, phrases, or grammatical forms used by people from the same social group or profession to distinguish them from other groups. FB users tend to use argot (8 data) to give a different impression in their talk. For example, the word *kiffer (aimer)* 'love,' *taffer (travailler)* 'work' (verb), *taffe (travail)* 'work' (noun), *gosse (enfant)* 'child,' *niquer (faire l'amour)* 'make love,' and *wesh (salut)* 'hi.' All of these argots are loaned from Arabic, while from English the argot is *oki (d'accord)*. Another argot known as verlan is also used by FB users. Verlan is an argot in the French language featuring inversion of syllables in a word, and it is very common in youth language, for example: *meuf (femme)* 'woman' and *tof (photo)* 'photo.'

As in verbal conversation, FB users also perform 'code switching' in their conversation. It was found in the data that they switch the French language into several other languages, such as English, Spanish, Portuguese, and Arabic. FB users do implement code switching into English only on special occasions, such as while saying "Happy birthday" or "I love you" and also when answering comments from friends who speak English. Meanwhile, FB users use code switching into Spanish and Portuguese when they comment on the status or photos or reply to their friends' comments written in those languages. For example, in Spanish "*Eres un Fenomeno cariño*" 'you are very remarkable, dear' or, in Portuguese "*obviamente*" 'of course.' Therefore, it can be concluded that code switching into Spanish and Portuguese is situational. Likewise, the use of Arabic occurs in the use of argot as *kiffer, taffer*, etc. In some online dictionaries, such as Reverso, Linternaute, Exionnaire, those argots have been entered as a new vocabulary in French, but in Word Reference and Larrouse online, only *taffe* and *kiffer* have become the entries.

However, the number and frequency of foreign languages used in FB conversations are very small compared to the use of the French language itself. Based on the interviews conducted, French FB users also use a language other than French with friends who cannot speak French, for example, by using English, Spanish, *créole marocaine*, or French Patois, Japanese, and Indonesian. In situations when they have to use foreign languages, they prefer to use English which allows them to communicate with friends from around the world. Thus, it can be said that French FB users are more proud in using French than other languages because they speak more in French than other languages, and they switch from French to another language merely on some occasions.

2   CONCLUSION

Creative processes of writing by FB users are abbreviation, morphophonemic, and other linguistic processes. The first two processes are commonly performed in general writing, while the latter process is typical in written FB conversations. In writing and understanding conversations, users rely heavily on an understanding of the context because the words which are written sometimes have another meaning or no meaning at all when they are removed from the context. A typical thing in FB conversations is FB users trying to animate the situation of a verbal conversation and package it in a variety of abbreviations as an activity record during a conversation. They also do creative writing with the aim of reinforcing the impression that the conversation is 'exactly' reported as it is.

This is in line with Barron (2008) and Crystal (2004), who explain that communication on the Internet, although in a written form, feels more like talking than writing. It also confirms what is stated by Crystal (2004) that netspeak also arises at the level of grapheme. The phenomenon of language use and their change on FB as we found in this research not only confirms what is said by Crystal about netspeak but also indicates the tendency of French people to simplify French orthography. This phenomenon may show the aspiration of the French speakers to change the spelling (orthography) of French— which is renowned as being very complicated—to be easier and simpler. However, it still needs further research.

Although this phenomenon seems to show that French FB users want an alteration in French orthography, they remain proud of the French language, which can be seen from the frequency in the use of French as being much higher than that of other languages in their conversations. In addition, from the interviews conducted, French FB users said that they used other languages in their conversations just in order to build communication with people from all over the world. These facts may also indicate that French FB users are now more open or have a more positive attitude towards foreign languages.

REFERENCES

Bamman, D., Eisenstein, J. & Schnoebelen, T. (2014) Gender identity and lexical variation in social media. *Journal of Sociolinguistics*, 18(2), 3–46.

Baron, N.S. (2003) The language of the Internet. In Farghali, A. *The Stanford handbook for language engineers.* Stanford: CSLI.

Barron, N.S. (2008) *Always on: Language in an online and a mobile world.* New York: Oxford University.

Beard, R. (2001) Derivation. In Spencer, A. & Arnold, Z. M. *The handbook of morphology* (pp. 44–65). Massachusetts: Blackwell Publishers.

Bechade, H.D. (1992) *Phonetique et morphologie du Français: Moderne et contemporain.* Paris: Presses Universitaires de France.

Besner, D. & Smith, M.C. (1992) Basic process in reading: Is the orthographic depth hypothesis sinking? In Frost, R. & Katz, L. (Eds.) *Orthography, phonology, morphology and meaning* (pp. 45–66). Amsterdam: Elsevier Publisher B.V.

Bouzidy, B. (2009) Créativité lexicale par réduction en français contemporain. *Synergies Algérie, 5,* 111–117.

Boyd, D.M. & Ellison, N.B. (2007) Social network sites: Definition, history and scholarship. *Journal of Computer-Mediated Communication.* 13(1), 210–230.

Burki, A., Fougeron, C., & Gendrot, C. (2007) On the categorical nature of the process involved in schwa elision in French. *Interspeech*, 1026–1029.

Chantrel, F. (2015) *Facebook dépassse les 30 millions d'utilisateurs actifs en France.* In Blog du Moderateur. Retrieved from http://www.blogdumoderateur.com/facebook-france-30-millions/.

Coëffé, T. (2016) *Les 50 chiffres à connaître.* In Blog du Moderateur. Retrieved from http://www.blogdumoderateur.com/50-chiffres-medias-sociaux-2016/.

Crystal, D. (2004) *Language and internet.* Cambridge, Cambridge University Press.

Dubois, J. (1999) *Dictionnaire de la langue française: Lexis.* Paris: Larousse-Bordas.

Durrand, J. (2009) L'alphabet phonétique international. In Herrenschmidt, C., Mugnaioni, R. M.-Savelli, J. & Touratier, C. *Le monde des écritures.* Paris: Gallimard. 1–19.

Fagyal, Z., Kibbee, D., & Jenkins, F. (2006) *French a linguistik introduction.* New York: Cambridge University Press.

Fromkin, V., Rodman, R. & Hyams, N. (2003) *An introduction to language.* USA, Wadsworth.

Krautgartner, K. (2003) Techniques d'abréviation dans les webchats francophones. *Linguistik online,* 15(3/03), 47–67.

Lampe, E. C., & Steinfield, C. (2007) Benefits of Facebook "friends": Social capital and college studenst' use of online social network sites. *Journal of Computer-Mediated Communication*, 12(4), 1143–1168.

Léon, P. (1992) *Phonétisme et prononciation du français.* Paris: Nathan.

Martinet, A., & Walter, H. (1973) *Dictionnaire de la prononciation française dans son usage réel.* Paris: France Expansion.

Philips, S. (2007) *A brief history of Facebook.* Retrieved from http://www.theguardian.com/technology/2007/jul/25/media.newmedia.

Pooley, T. (1996) *Chtimi: The urban vernaculars of Northern French.* Clevedon: Multilingual Matters Ltd.

Protalinski, E. (2011) Faceook has increased the status update box character limit to 63.206. That's more than a 12-fold increase the last limit of 5.000 characters. Retrieved from http://www.zdnet.com/article/facebook-increases-status-update-character-limit-to-63206/#!.

Smith, C. (2016) *By the numbers: 200+ Amazing Facebook User Statistic.* Retrieved from http://expandedramblings.com/index.php/by.the.numbrs-17.amazing-facebook-stats/.

Thangaraj, S. & Maniam, M. (2015) The influence of netspeak on students' writing. *Journal of Education and Learning,* 9(1), 45–52.

Watt, H.J. (2010) How does the use of modern communication technology influence language? *Contemporary issues in science communication and disorders* (144–148).

Cultural Dynamics in a Globalized World – Budianta et al. (Eds)
© *2018 Taylor & Francis Group, London, ISBN 978-1-138-62664-5*

Indonesian-English code-switching on social media

M. Aulia & M. Laksman-Huntley
Department of Linguistics, Faculty of Humanities, Universitas Indonesia, Depok, Indonesia

ABSTRACT: In most Indonesian urban cities, code-switching is not related to the use of regional languages but to the use of English, which is taught in schools. In addition, many city dwellers also receive exposure to the language by contact with English speakers, which may occur extensively in daily conversations. This research aims to investigate the way speakers communicate on the internet, particularly through social media. Even though virtual communication is expressed through the medium of writing, it still has features of a speech act. Preliminary research was conducted in 2015 to describe the code-switching used by radio announcers. Previous study showed that intrasentential switching occurred mostly in conversations, and it relates to the speaker's ability to speak both languages. In this research, the authors highlight the frequency of code-switching on social media by examining 30 Facebook status updates of a radio broadcaster. The research uses a qualitative method by analyzing the topic of the statuses and its correlation with the language used. The result shows that the participant uses English to express herself, while Indonesian is used only when she is triggered by something in Indonesian or about Indonesia, and to promote products.

1 INTRODUCTION

Speaking two or more languages within an utterance is common in multicultural countries such as Indonesia. This is based on people's need to communicate across cultures leading to language contact. The contact may take the form of bilingualism relating to code-switching (Appel & Muysken, 2005). However, in most Indonesian urban cities, code-switching is not related to the use of regional languages but to the use of English. This phenomenon occurs extensively not only in daily conversation but also in mass media, such as radio.

If we listen to most conversations on radio shows in Jakarta, the broadcasters often use a mix of Indonesian and English. This kind of language use occurs frequently in almost every radio programme on most radio stations. This language mixing may reflect the binguality of both the speakers and the listeners.

Research about code-switching in radio has been conducted before (Aulia, 2015), and the results show that there were 725 code-switching in the recorded conversations. According to Poplack (1980), code-switching is divided into three types: intersentential switching, intrasentential switching, and tag switching.

As a continuation of the study by Aulia, this research examines how one of the radio broadcasters, whose conversations were investigated in the earlier study, uses her language. The investigation focuses on the use of Indonesian and English, and how they are used through the social media, particularly on her Facebook status updates. This starts with a question: How does the radio broadcaster use her language on Facebook? In other words, does she use Indonesian-English code-switching when she writes as she does when she speaks? Crystal (2001) states that language used on the internet has features of spoken discourse, although it is written. Therefore, by conducting this research, the authors can identify if there is a difference between code-switching used by the broadcaster when she speaks and when she writes a written form of a spoken language on Facebook.

Using a qualitative method, this research analyzed 30 status updates of a radio broadcaster on her social media account, Facebook that were collected in May 2016. The analysis identifies the pattern of language use, particularly code-switching between Indonesian and English, on the status updates. The factors of the occurrence of code-switching are also investigated to identify the speaker's motivation to switch languages.

2 CODE-SWITCHING ON FACEBOOK

Poplack (1980) defines code-switching as the alternation of two languages within a single discourse, sentence, or constituent. She divides code-switching into three types: intersentential switching, intrasentential switching, and tag switching. In this research, the language switch found in the data is separated mostly between discourses. Intersentential and intrasentential switching occurs only at the level of words and phrases, as well as tag switching.

The result shows that language switch often occurs on the broadcaster's Facebook status updates. Thirty status updates were investigated, 18 are in English, four are in Indonesian, and 8 updates are a mix of both languages. It was found that there are certain patterns in choosing the language she uses when she writes her status updates. The following are some examples found in the data.

Status updates in Indonesian

"Saya bersama Yuyun."
('I am with Yuyun.')
(attachment: a picture inscribed with the words 'Saya Bersama Yuyun #Nyala UntukYuyun', with a long caption written in Indonesian about a girl named Yuyun)

"Terharu pagi2:')"
('Still early but I am touched.')

(Attachment: a video of two Indonesian citizens taking part in a talent show in England, with a caption written in Indonesian)

Indonesian is used to write a status if it is shared together with an attachment (pictures, videos, articles) relating to issues about Indonesia or the Indonesian language. In other examples found in the data, Indonesian is also used to promote some products. Therefore, status updates in Indonesian occur only if the language user is triggered to use it.

– Status updates in English

"Hosted the launching of Fitflop new collection for Spring Summer 2016. This is definitely the most comfortable I've been MCing an event. Thanks to the footwear." (Attachment: a picture of herself)

"Lotsa stuff on FB (or internet in general) can turn you into a depressed cynic. Thanks Chewbacca Mom for bringing back good ol simple fun to my timeline. Go get em!" (Attachment: an English article about a person dressed in Chewbacca costume, with pictures)

From the topics and the attachments included in the status updates above, English is used to express herself. English is also used to talk about her personal life. This shows that personal thoughts and reflections are mostly written in English.

– Statuses in Mixed Languages

There are two types of language mixing in this category. The first type is status updates in Indonesian with a slight occurrence of English as seen in the example below.

> *"Hari ini Tiga Mami Kece ngobrol-ngobrol dengan majalah Mother & Baby, mengenai me-time, persahabatan, dan peran ibu. Seru banget!*
> *Tunggu cerita lengkapnya di Mother & Baby Indonesia edisi Juni."*
> ('Today Three Cool Mommies chit chat with Mother & Baby magazine about me-time, friendship, and mother's role. So exciting! Wait for the full story in Mother & Baby Indonesia June edition.')
> (Attachment: a picture of herself with her two friends)

The second type is status updates in English with a slight occurrence of Indonesian, for example as below.

> "That's right girls. The only BS you need to have in your lives are Bags and Shoes.
> *Camkan."*
> ('Note that.')
> (Attachment: a picture of her two children holding a purse)

The occurrence of one language when talking in a different language is found in the data, as shown in the examples above. The broadcaster's reason to choose which language to use, based on the authors' observations of the base language used to write the statuses, remains the same as previously described. Indonesian is used to promote some products and English is used to talk about personal life. The occurrence of the other language is significantly lower compared with the base language used in the whole discourse of the status updates, which only occurs at the level of words, short sentences, or taglines.

2.1 *Language and identity*

As we mentioned above, from 30 Facebook status updates observed, 18 were in English and three were in English with a slight occurrence of Indonesian. Thus, it can be said that there are 21 statuses that were written in English (the occurrence of one language in the other one is discussed on (10) in 2.2), which shows that English is used more than Indonesian.

In these 21 status updates, the subject mostly talks about herself, her family, and her friends. The topics also focus on expressing her thoughts, daily activities, or trips abroad with friends. By using a particular language to talk about all of these, it is expected that there are reasons for her language choice.

Fishman (1965) states that one of the factors of language choice is the desire to be identified as a member of a group. Guy (1988) also states that language relates to identity. This means that a person can show his/her identity from the language he/she uses.

In Indonesia, the use of English has a particular effect in the society. A person who can use English is considered as someone who has higher status than those who can only use Indonesian. As a fact, English language learning can be accomplished only through education, therefore, being educated means that a person is more likely to be able to move to a higher social status. Another reason that has more to do with a phenomenon in the mindset of the Indonesian society, is that by embracing all "Western" features, one would be regarded as being modern and superior. Then, a person will be considered or consider him/herself inferior if he/she cannot adapt to modernity. This relates to what Edwards (2009) explains about Asian societies. They often think that European languages are superior, and this is not based on linguistic reason but on the experience of those language societies. This mindset about superior-inferior is the reason why English is considered a social status determinant.

Therefore, the 21 status updates written in English can be related to the subject's identity. Below is another example:

> "Our next vlog series will be on Sydney, an urban wonderful of glamour and natural charm. We saw and cruised the beautiful harbor, enjoyed the dynamic cultural

landscape and innovative fashion scene, proved its gastronomical reputation, and experienced for ourselves an outdoor lifestyle more vibrant than anywhere else. Subscribe to our Youtube channel and join us as we explore one of the world's most iconic cities, and see for yourselves why we fell madly in love with Sydney." (attachment: a link to a vlog about her trip to Sydney)

Here, she talks about her experience exploring Sydney, a city in Australia, and the activities she did in that city. By using English, there is an impression that she has a particular social status making her different from the others. Readers also get the impression that she reflects something that is perceived to belong to a higher class by the society.

English structure is used to write her English status updates. Indonesian structure in those statuses is not found. This means that she can distinguish both structures so that there is no language interference or the use of one language pattern in another.

2.2   *Language and situation*

There are four status updates written only in Indonesian, while there are five status updates written in Indonesian with a slight occurrence of English. It can be said that there are nine status updates written entirely in Indonesian. The authors found that she uses Indonesian only if she is triggered by pictures, articles, or videos in Indonesian or about Indonesia. See an example below.

> "Buat Ahok!"
> ('For Ahok!')
> (Attachment: a picture of herself holding her identity card, standing beside a banner inscribed with the words 'KTP Gue Buat Ahok' ['my identity card is for Ahok'])

Fishman (1965) states that one of the factors of language choice in multicultural societies is a situation. He stated that there are some particular situations that can make a person choose one language over the other, and that one thing relating to a situation is his/her surrounding.

In the example above, Indonesian is used because of a situation. The status is shared with a picture of a banner inscribed with Indonesian words. It can be said that the banner is the trigger for her to use Indonesian.

The use of Indonesian to promote something is also found in the data.

> *"Buibuk, kalau mau cari buah-buahan segar berkualitas main2 aja ke Citos, di atrium lagi ada Melbourne Market yang jual buah2an aneka rupa dan sayur-mayur segar yang semuanya produksi Victoria, Australia. Gampangnya sih kapan aja bisa belanja ke Hypermart dan Foodmart, selalu tersedia berbagai pilihan buah-buahan segar, termasuk buah2 berkualitas terbaik Victoria ini.*
> *Live healthy, eat healthy, eat Australian fruits!"* ('Moms, if you want to find good quality fruits just go to Citos. In the atrium, Melbourne Market is selling various fresh fruits and vegetables produced in Victoria, Australia. You can find them easily at any time in Hypermart and Foodmart, they always provide various choices of fresh fruits, including these good quality fruits from Victoria. Live healthy, eat healthy, eat Australian fruits!') (attachment: a picture of herself hosting an event)

The example above can also be related to a situation. The status update is shared with a picture of herself hosting an event. By using Indonesian to promote a product, in this case Australian fruits, and writing in Indonesian to update her status, she repeats what she said earlier as the host of the event. The combination of the picture and the topic can be considered as a situation triggering the use of Indonesian.

We can also see an English sentence in the example. Nevertheless, the occurrence does not affect the examination of the whole status because it is only used as a tagline for the event. Taglines generally are said to fulfill the need to communicate the values of a product; it is not something that a person chooses to say about what he/she thinks.

Fishman (1965) also states that another factor of a language choice is topic. He said that one of the things that relates to topic is the need for specific terms that can be used to make a conversation more effective and easier to understand. A person will use a particular language if he/she thinks that it has a more precise word than in any other languages as seen in the example below.

> "My latest blog is about Lilou's Jedi Camp Birthday party last month, and how our family's collective love of Star Wars will be one of our shining lights to guide us through the galaxy of mess and troubles in life. Lebay? Maybe a little bit. If you love Star Wars, read this, you must. If you're just stuck in traffic, read this, you can."
> ('exaggerating') (Attachment: a link to her blog, with a picture of her daughter)

In the data above, she used the Indonesian word *lebay* as a language switch. There is an equivalent word in English, which is 'exaggerating', but *lebay* is considered more powerful to express her feelings, so she chose to switch to Indonesian. The example above is what we have mentioned earlier in 2.1. in which the occurrence of one language in another can also be found in the data. Another example of this discussion is found in the status update (5).

2.3 *Language on the internet*

Crystal (2001) states his view about the use of language on the internet. He used the term netspeak to name the conversations that occurred on social media. He states that although it is written, language that is used on the internet is much more like spoken language. This means that people are likely to use a spoken style rather than a written style when they communicate via the internet.

There are various forms of netspeak found in the data. In example (2), netspeak can be identified with the use of '2' as a reduplication form and a face symbol to reflect self-expression. In example (4), netspeak can be identified in the form of lotsa (lots of), ol (old), and em (them). These compressions occur so that the writing process can be done more quickly, similar to other examples found in the data such as omg (oh my god), BS (bull shit), and congrats (congratulations).

From 30 status updates posted on Facebook, netspeak only occurs in nine updates in which five are written in English and four are in Indonesian. Netspeak in English occurs in the form of a compression style, while netspeak in Indonesian occurs in spoken style of written words and the use of '2' as a reduplication. The few occurrences of netspeak in the data are most likely caused by the medium of the language: the status updates. In social media, a status is a one-way communication, where a person does not need to converse with another person, like chatting. Unlike chatting, which needs more speed and simplicity in replying messages, status writing provides more time for the writer to express what he/she wants to express. Therefore, most status updates that are posted are still in a written style.

If we relate this to the previous study, this study attempts to compare the occurrence of code-switching in a particular speaker, i.e. if the large quantity of code-switching occurs in his/her spoken language it will also appear in his/her written language as netspeak. The result shows that the occurrence of a language switching in a status update generally varies depending on the topic. In other words, code-switching when speaking is not likely to occur in writing. As it was mentioned before, these status updates are in a written style and English netspeak only occurs in the form of compression.

3 CONCLUSION

Based on this study, we can conclude that the speaker has a language preference when updating her statuses on Facebook, which refers to her identity and the situation. She uses English to write about her personal life and to express herself, and this relates to her desire to create a particular identity to be seen by society. Meanwhile, she uses Indonesian when triggered by something in Indonesian or about Indonesia and to promote some products. This means that she needs particular situations in which to use Indonesian. Nevertheless, this conclusion is made based only on analyzing the data. A further study about this case needs to be done by interviewing the subject to understand her exact language attitude.

Netspeak is found in the data, although it is not frequent. It is presumably because a Facebook status is a one-way communication, so the time taken to write it is slightly longer. It is different from a two-way communication, like chatting, that does not provide a long time to reply to messages.

The theory about netspeak expects the researcher to analyze the subject's language switch in her writing, and compare it with the results found in the previous study (Aulia, 2015). However, the language style found in the data are mostly written not spoken, so there is no code-switching as when a radio broadcaster talks orally during a radio show. The only thing that occurs is a patterned language switch based on what the radio broadcaster is talking about. English structure is also used in the radio broadcaster's status update in English. In other words, those status updates are not literal translations of Indonesian causing language interference.

REFERENCES

Appel, R. & Muysken, P. (2005) *Language Contact and Bilingualism*. Amsterdam, Amsterdam University Press.

Aulia, M. (2015) Bentuk-bentuk Alih-kode di Radio. *Prosiding Seminar Nasional Sosiolinguistik–Dialektologi* (Forms of Code-switching in Radio. *Proceedings of The National Seminar on Sociolinguistics Dialectology), FIB UI, 9–10 November 2015, 264–269*.

Crystal, D. (2001) *Language and the Internet*. Cambridge, Cambridge University Press.

Edwards, J. (2009) *Language and Identity: An Introduction*. Cambridge, Cambridge University Press.

Fishman, J.A. (1965) "Who speaks what language to whom and when?" in: Wei, L. (eds.) *The Bilingualism Reader*. London, Routledge. pp. 82–98.

Guy, G.R. (1988). "Language, social class, and status" in: Mesthrie, R. (eds.) *The Cambridge Handbook of Sociolinguistics*. New York, Cambridge University Press. pp. 159–185.

Poplack, S. (1980) "Sometimes I'll start a sentence in Spanish y Termino en Español: toward a typology of code-switching". in *Linguistics*. 18 (7/8), 581–618.

Cultural Dynamics in a Globalized World – Budianta et al. (Eds)
© *2018 Taylor & Francis Group, London, ISBN 978-1-138-62664-5*

Use of English hedges by 12 learners in academic writing: A case study of pragmatics

Y. Widiawati & F.X. Rahyono
Department of Linguistics, Faculty of Humanities, Universitas Indonesia, Depok, Indonesia

ABSTRACT: The main purpose of academic writing is to inform other researchers about the writers' findings in certain research. In this case, the writers will propose claims. For non-native English speakers like Indonesians, this is a tough task to do. L2 learners find difficulty in writing academic texts or making claims. One of the strategies that L2 learners use is using hedging devices. This study aims to identify the hedges in academic writing produced by Indonesian researchers or writers. According to Levinson (1987) in his theory of FTA (Face Threatening Act), those words mostly function as a tool for speakers or writers to make them comfortable and save negative face. It means that the writers should choose the appropriate words to achieve their communicative goals. The data were taken from 5 dissertations written in English. The method used is a descriptive-qualitative analysis. The study focuses on 2 kinds of hedging strategies proposed by Hyland (1996), namely, **writer-oriented hedges** and **reader-oriented hedges**. The first strategy consists of (1) passive voice, (2) dummy subjects, and (3) abstract rhetors. The latter consists of (1) personal attribution and (2) conditionals. The results reveal that writer-oriented hedges, such as: passive constructions and dummy subjects, are the most frequent hedging devices used by Indonesian researchers. The conclusion of this study is that Indonesian researchers frequently used passive constructions and modality (can, may, might, should). It means that Indonesians like to let the data talk for themselves in order to avoid a potential conflict and hence maintain a harmony between the writer and readers.

1 INTRODUCTION

For most Indonesians writing in a foreign language is difficult. It is a difficult task for L2 learners of English, especially when they have to write academic texts or writings. As members of a particular discourse community or researchers or writers, they wish to publish their works or findings in international journals. To reach this goal, the use of hedges in their writing will be important as hedges can be used in conforming to academic writing standards (Banks, 1996).

It was Weinrich (1966) who first introduced the word "hedge". He called these devices "metalinguistic operators". A few years later, Lakoff (1972) in his article entitled *Hedges: A Study in Meaning Criteria and the Logic of Fuzzy Concepts* made this concept more popular and it has created the greatest initial impact. Lakoff defined this concept as "words whose functions are to make meanings fuzzier or less fuzzy". He said that "sort of" is an example of a hedge. The following year Fraser (1975) introduced the hedged performative. He states that a hedged performative is based on the use of modality, such as: *will, can, must* or semi modality like *want to, would like to, wish to.*

Hedges are pragmatic features that the speakers or writers use to strengthen their assertions, tone down uncertain or potentially risky claims, emphasize what they believe to be correct, and appropriately convey collegial attitudes to the listeners (Hyland, 1996). Myers (1989) also states that hedges can be used to mitigate propositions. Hedges will help reach "the optimal relevance" (Sperber and Wilson, 2001) between speaker and listener or writer and reader. The writer should make some choices in strategy and linguistic forms in order to adapt to his or her intentions. Hedges are often chosen to achieve goals. Brown and Levinson define hedges as particles used to mitigate propositions to less strong. Indonesians are well known as friendly people who like to keep a low profile. Most of them can easily make friends with others, both local and foreign people. They show intimacy and warmth to people around them (Maryanto, 1998).

Furthermore, Brown and Levinson's theory of politeness (1987) states that an FTA is a violation of the speakers' or writers' privacy and freedom of action, for which hedges could provide a possible compensation. Hence, negative politeness enables the speaker or writer to go on record, but with redress, which means that the speaker or writer makes an effort to minimize the imposition of his/her claims.

Hedges may also influence writers when they write texts, especially academic texts. This argument is supported by Hyland (1996) who states that academic texts are full of hedges. Hedges (particles, lexicons, and clausal hedges) are pragmatic markers that attenuate or weaken claims. Academic texts or scientific texts are not only content-oriented and informative, but also seek to convince and influence their audience. There is an increasing number of research studies about the use of hedges in a variety of disciplines (for example: Hyland, 1994, 1996, 1998, 2000; Salager-Meyer, 1991, 1994, 1998; Skelton. 1997; Meyer, 1997). In addition, Myers (1998) examined a corpus of biology research articles.

In scientific writing, vagueness can be seen as a motivating factor for the use of hedges, for example in the case of where some exact data is missing or if precise information is irrelevant in the preliminary results. Hedges will protect writers from making false statements by indicating either lack of commitment to the truth value of propositions or a desire not to express that commitment categorically. In contrast, Salager-Meyer (1994, 151) asserts that hedges are "ways of being more precise in reporting results". She adds that by doing this, the credibility of a statement may be increased.

Along with the argument of being vague, Joanna Channel (1994) states that the language system permits speakers to produce utterances without having to decide whether or not certain facts are "excluded or allowed by" them. Hedges, however, are sometimes required to capture the probabilistic nature of reality and the limits of statements (Toulmin, 2013). In fact, the use of hedges is typical of professional writing to make absolute statements more accurate (Hyland, 1998). Moreover, hedges play a critical role in academics' presentations of their own work (Hyland, 1998).

Hedging is an important interactional strategy used in communication. This strategy can make the communication go more smoothly. Therefore, to become an effective communicator, a speaker should be able to know how and when to use hedging devices in different processes of communication. Hedging devices here mean the verbal propositions employed by participants of communication (both speaker and listener) to prevent conflict, to avoid being blunt, to weaken or strengthen the illocutionary force and protect the face (Brown & Levinson 1987; Stenstorm 1994, Salager-Meyer 1994).

Leech (1983) proposes six maxims of politeness principles (PP), which are tact, generosity, approbation, modesty, agreement, and sympathy. The tact maxim regulates the operation of directive speech acts and addresses the dominant type of politeness that can be measured on a cost-benefit scale. The more costly the action, the less polite it is. Brown & Levinson (1987) claim that in any social interaction participants devote much of the time to face-work. They argue that "face" is something that concerns human beings universally, and it is divided into negative and positive faces. The first one deals with a negative politeness strategy, which gives freedom to individual actions and a desire to be unimpeded. In other words, it is called a strategy of independence or *deference politeness strategy*. Meanwhile, the positive face deals with a positive politeness strategy, which attempts to save the listener's face. This strategy is also called the strategy of involvement or *solidarity politeness strategy*. The following is an example of this: *I really **sort of** think/hope/wonder....* (Brown & Levinson, 1987, 116).

Being polite means being a considerate conversational partner. In terms of negative politeness, being polite means choosing the right words to express communicative messages that might be felt to be face-threatening for the addressee, such as refusals, criticisms, or claims in order to prevent conflicts. In written communication, researchers present their own findings or claims by using pragmatics markers, or hedges. Hedging devices are the critical tools to prevent potential arguments and save FTA. Look at the example: *close the window if you can* (Brown & Levinson, 1987, 162). By using "if", this sentence of command is weakened or hedged.

1.1 *The importance of scientific Hedging*

Hedging devices are mostly used to mitigate propositions or claims. As Hyland (1996) states in his article *Nurturing Hedges in the ESP Curriculum*: "Hedges, therefore, have an important role in a form of discourse characterised by uncertainty and frequent reinterpretation of how natural phenomena are understood" (Hyland, 1996, 478). Furthermore, he adds that academic discourse involves interpretative statements because cognitions are variably hedged. Writers offer an assessment of the referential they provide, rather than being factual and impersonal in order to alert readers to their opinions.

In addition to this, hedges are used to prevent conflicts in order to avoid humiliation of both speakers and listeners or writers and readers. In the context of academic writing, authors tend to mitigate the force of their scientific claims by means of hedging devices in order to reduce the risk of opposition and minimize the face threatening acts (FTA) involved in making claims. This argument goes along with Hyland who states that one of the functions of using hedges is to allow writers to anticipate possible negative consequences of being wrong (Hyland, 1996, 479). Academics seek agreements for the strongest claims they use for their evidences, as this is how they gain their academic credibility, but they also need to cover themselves against the embarrassment of categorical commitment to statements that later may be shown to be inaccurate. Hedges also help writers to develop and maintain their relationship with the readers, addressing affective expectations in gaining acceptance for claims. Although academic writings try to persuade and convince the readers, they can be rejected. Thus, they should use the strategy of preventing it by using the hedges.

Following Hyland's theories about hedges from the perspective of relationship between writers and readers, I conducted investigations on *Writer-oriented Hedges* and *Reader-oriented Hedges* in dissertations. I am interested in investigating the relationship between writers and readers because this has not much been explored by researchers, especially hedges used by L2 learners in academic writing. *The Writer-oriented hedge* is a kind of strategy that facilitates the communicative strategy by which writers can get his readers to see the real world from their points of view. The writers persuade readers to accept their claims by seeing the evidences through three subcategories: (1) passive voice, (2) abstract rhetors, and (3) dummy subjects (Hyland, 1996). Meanwhile, *Reader-oriented Hedges* try to involve the readers in the writers' claims to minimize the uncertainty from the readers, which might happen. The subcategories belonging to this are (1) personal attribution and (2) conditionals.

2 METHODOLOGY

The corpora for this data were taken from 5 dissertations written in English by Indonesian student writers who were studying at the Faculty of English Applied Linguistics in one of the prominent private universities in Indonesia.

The research methodology used was the qualitative descriptive method. I employ this method because my intention was to obtain insights into the strategies used by postgraduate student writers of English Applied Linguistics. I studied a relatively small number of dissertations. This is in accordance with the main characteristics and spirit of the qualitative approach which states that what stands out in a qualitative study is the depth and breadth of the analysis, not the number of subjects studied. A qualitative study has nothing to do with statistical significance; rather it seeks to pursue a profound understanding of a particular phenomenon by utilizing all resources, data, observation, and even subjective interpretation.

Corpus Selection. In this study, I chose to analyse the Discussion section of the dissertations for two reasons:

a. The Discussion section is the section where postgraduate student writers put in their claims of their research;
b. This section is the most important and crucial part of their dissertations.

Data Collection Techniques. In collecting data, I listed all propositions found in the Discussion sections that contain hedges. When investigating such hedges, I categorized them by the subcategories being determined. This section will include the contexts, graphs, tables and interpretations of the results.

3 RESULTS AND ANALYSIS

I used both theoretical and empirical perspectives to make sense of what happened, as well as, the context that caused it to happen. After that I interpreted the collected data to seek answers for my research questions.

The steps that I undertook were:

a. I identified hedged words, phrases, and clauses based on the indicators provided by Hyland (1998). Those are found in the Results and Discussion section of every dissertation;
b. I classified those hedges into the subcategories being determined;
c. I put the hedged units in the tables to show how they are distributed in percentage. By calculating the frequency of hedges, it would be easy to look at the tendencies of the hedging strategy used;
d. Finally I interpreted the data in relation to the strategy being used.

3.1 *The data*

For this study, I investigated the Results and Discussions section in 5 (five) dissertations written in English by student writers who were studying at the Faculty of English Applied Linguistics. I chose the names randomly, but I preferred to pick them based on the year when these dissertations were written, which ranged from 2011–2013. The names are kept in initials to keep them confidential. The data are as follows:

Table 1. Topic selection.

No	Writer (initials)	Title of dissertation	Number of page of results & discussion section	Number of hedged units in results & discussion section
1	CH	The production and recognition of english word stress: an auditory word priming study	124	137
2	YY	Verbal communication of emotions: A case study of obama-mccain presidential debates	122	142
3	IID	EFL learners' metaphor competence english proficiency, english exposure and learning style	89	101
4	HT	English collocational mismatches in second language writing	178	191
5	SS	The construction of self in academic writing: A qualitative case study of three indonesian undergraduate student writers	84	114

I was interested in investigating the Results and Discussion section because this section is a very important part of the dissertation as this is the part where the student writers make their claims of the results of their research. When making claims, student writers try to persuade readers who come from their academic community. In scientific writing, hedges are effective and propositional functions that work in rhetorical partnership to persuade readers to accept knowledge claims (Myers, 1985).

4 RESULTS

Table 2. Realization of hedges.

No.	Initial	Writer-oriented	Reader-oriented	Others
1	CH	PV: 33.23% DS: 12.34% AR: 24.21%	PA: 8.54% C: 2.35%	Others: 19.33%
2	YY	PV: 38.68% DS: 19.54% AR: 12.30%	PA: 10.37% C: 1.56%	Others: 17.55%
3	IID	PV: 36.25% DS: 18.64% AR: 14.45%	PA: 6.35% C: 1.75%	Others: 22.56%
4	HT	PV: 40.25% DS: 16.56% AR: 21.32%	PA: 4.92% C: 1.55%	Others: 14.4%
5	SS	PV: 28.35% DS: 19.68% AR:15.54%	PA: 17.65% C: 2.24%	Others: 16.54%

Notes:
– PV: Passive Voice;
– DS: Dummy Subjects;
– AR: Abstract Rhetors;
– PA: Personal Attribution;
– C: Conditionals.

5 ANALYSIS

5.1 *Writer-oriented Hedges*

From the results obtained, it can be seen that the student writers often use passive construc-tions (PV). This is because they want to avoid being blamed for making errors in presenting their claims (Brown & Levinson, 1987, 194). The sentences below are a few examples of this:

1. Verbal communication *can be expressed* literally (YY, p. 51).
2. The results *are summarized* in the following table (CH, p. 135).
3. Metaphors *were produced* most by low English proficiency (IID, p. 103).

From the examples above, the absence of agency (Hyland, 1996b, 444) is central to the characteristics of *Writer-oriented Hedges*. Those sentences (1, 2, and 3) are the first subcate-gory of this strategy. In connection with agency, as an agent in the process of producing a piece of scientific knowledge, a scientific writer seeks to place their discoveries in a wider community.

Meanwhile, the second subcategory of this strategy is the dummy subject (DS). For gram-marians like Quirk et al. (1985), dummy subjects are considered to be expletive, meaning that "it" in English is regarded as an "empty" subject as can be seen in the sentences below:

1. *It* seemed that learners from the high English proficiency were more serious (IID, p. 162).
2. *It* can be seen that their vision and mission were presented in series of words (HT, p. 112).

The above sentences show that "it" is used as a dummy subject as it is the most neutral of the pronouns and thought to be an "impersonal subject". The dummy subject 'it' can be used with other reporting verbs like *"seem"* as shown in example no. 4 above (Sinclair, 1990, 331).

The last subcategory in this strategy is abstract rhetors (AR). Scientific writing can perhaps be used to challenge the theory of classical rhetoric. Within classical rhetoric, humans are recognised as the only actors that can speak (Myers, 1996, 22). The term 'rhetor' can simply be defined as an orator, and therefore has been understood to be a practitioner of the art of using language skilfully for persuasive purposes. However, scientific writing can now be used to imagine a rhetorical situation in which everything is a possible rhetor, including non-humans, such as: a piece of a research. Look at these sentences below:

1. *The table* above shows that words with final stress produced much less correct... *(CH, p. 154)*.
2. *The results* show that on average learners knew 71% of the relationship... (IID, p. 104).

Usually humans are subjects of the various verbs of saying. However, in scientific writing humans are not always the subjects of the actions that people are talking about. The data that the researchers find can be said to be stating arguments. These can be regarded as an explanation.

5.2 *Reader-oriented Hedges*

The second strategy is the reader-oriented hedge, which tries to involve the readers in the writer's claims. These hedges will help a scientific writer make sure that his/her research is reliable. Personal Attributions (PA) are the substrategy that a writer uses not only to convey information but also serves as a professional attitude about the readers and their negotiation of knowledge claims (Hyland, 1996b, 446). Personal perspective can be attributed to scientific claims. This can be observed in the following sentences:

1. We can infer that the students will get benefit more from repetition (CH, p. 143).
2. Our interaction partner expects that we will feel this way (YY, 142).

The second subcategory of this strategy is conditional (C). Alternative opinions such as conditionals are commonly used as personal views. Alternative conditionals are widely discussed by the grammarian Quirk et al. (1985) who argue that they may be used for open or hypothetical conditions. On the one hand, open conditionals are neutral; they leave unresolved the question of nonfulfillment of the condition. On the other hand, a hypothetical condition conveys the writer's belief that the condition will not be fulfilled and hence the probable or certain falsity of the proposition. Look at the claims below:

1. If we understand the words' meaning, we can see the speaker's feeling.. (YY, p. 151).
2. The result shows that if 20 items were used, then reliability of the instrument would be low... (IID, p. 87).

6   CONCLUSION AND IMPLICATIONS

Hedging devices are often used by Indonesian student writers because such devices help them conceptualize the claims that they are going to convey. Moreover, such devices will assist the student writers to communicate with their readers and academic community because the claims will be accepted by the readers if the writers successfully communicate them to them. It can be said that hedges are communicative tools to negotiate with the potential readers. A claim can be presented by using PV, DS, and AR, which belong to writer-oriented hedges. Meanwhile, reader-oriented hedges give the maximum degree of visibility to the writers. The writers try to develop a relationship with their readers. Reader-oriented hedges make the content of the claims more tentative. In conclusion, the use of writer-oriented hedges is more preferable because the empirical evidences are able to explain and describe what the findings of the research are. Thus, the student writers do not need to say much about them.

The findings of the study have important implications to the study of academic writing as one of the compulsory courses taught in Indonesian universities where English is still not given much attention. In universities, English is only a minor subject in non-English Departments. It has been indicated that the writers' awareness of the use of hedging in writing is essential because the ability to use hedging devices helps writers craft their statements appropriately to produce credible, rational, and convincing claims.

Hedging is also important to maintain the writers' academic credibility. Furthermore hedging will help writers protect their reputation as scholars and minimize possible damage. In science, writers may hedge because of small samples, preliminary results, uncertain evidence or imperfect measuring techniques (Hyland, 1996:479).

REFERENCES

Brown, P. and Steven C.L, (1987) *Universals in Language Usage*. Cambridge, Cambridge University Press.

Channell, J. (1994) *Vague Language*. Oxford, Oxford University Press.

Cherry, R.D. (1988) Politeness in written persuasion. *Journal of Pragmatics*, 12 (1), 63–81.

Fraser, B, (1975) *Hedged performative. In Peter Cole and Jerry L. Morgan* (eds.), Syntax and Semantics 3: Speech Acts. New York, Academy Press, 187–210.

Fraser, B, (1980) Conversational mitigation. *Journal of Pragmatics*, 4 (4), 341–350.

Fraser, B, (1990) Perspective in politeness. *Journal of Pragmatics*, 14 (2), 219–239.

Holmes, J. (1984a) Modifying illocutionary force. *Journal of Pragmatics*, 8 (3), 345–365.

Holmes, J, (1984b) Hedging your bets and sitting on the fence: Some evidence for hedges as support structure. *The Relo*, 24 (3), 47–62.

Hubbler, A. (1983) *Understatement and Hedges in English*. Amsterdam, John Benjamins Publishing Company.

Hyland, K. (1996a) *Talking to the Academy: Forms of Hedging in Scientific Research Articles*. Written Communication 13 (2), 251–281.

Hyland, K. (1996b) Writing without conviction? Hedging in science research articles. *Applied Linguistics*, 17 (4), 433–454.

Johnson, B. and Christensen, L. (2008) *Educational Research: Qualitative, Quantitative, and Mixed Approaches*. Thousand Oaks, CA, Sage Publication.

Lakoff, G. (1972) The Pragmatics of Modality. *Chicago Linguistics Papers*, 8, 229–246.

Lakoff, G. (1973) *The Logic of Politeness: or, Minding Your p's and q's*. Papers from: Regional Meeting of the Chicago Linguistics Society 9.

Leech, G.N. (1983) *Principles of Pragmatics*. London, Longman.

Lyons, J. (1995) *Linguistic Semantics: An Introduction*. London, Longman.

Maryanto. (1998) *Hedging Devices in English and Indonesian Scientific Writings: Towards a Sociopragmatic Study*. [Thesis] Jakarta, Atmajaya University.

Myers. (1985) The pragmatics of politeness in scientific articles. *Applied Linguistic*, 10 (1), 1–35.

Nikula, T. (1997) Interlanguage view on hedging. In: Markannen R. and H. Schoder (eds.). *Hedging and discourse: Approaches to the analysis of a pragmatic phenomenon in academic texts*. Berlin, Walter de Gruyter, 188–207.

Prince, E.F., Joel F., and Charles B. (1982) On hedging in physician discourse. *Proceedings of the Second Annual Symposium on Language Studies, 83–96*.

Skelton, J. (1988) The care and maintenance of hedges. *English Language Teaching Journal*, 42 (1), 37–48.

Sperber, D. and Wilson, D. (1995) *Relevance: Communication and Cognition* (2nd edition), Blackwell, Oxford.

Thomas, J.A. (1983) Cross-cultural pragmatic failure. *Applied Linguistics*. 4 (2), 91–112.

Cultural Dynamics in a Globalized World – Budianta et al. (Eds)
© *2018 Taylor & Francis Group, London, ISBN 978-1-138-62664-5*

"You are what you eat ... and post": An analysis of culinary innovation and cultural branding in *Panggang Ucok's* Instagram account

S.M.G. Tambunan & M.R. Widhiasti
Department of English Studies, Faculty of Humanities, Universitas Indonesia, Depok, Indonesia

ABSTRACT: Culinary practices in our everyday lives have been analyzed in a great deal of researches arguing that food and drinks are essential parts in identity construction. The main function of culinary practices is to establish a sense of belonging in a particular environment. One of the tools in disseminating information about culinary practices whilst constructing one's cultural identity through culinary practices is social media, in this case Instagram. Through a Cultural Studies perspective, this research investigates how Panggang Ucok, a culinary business in Jakarta, relies heavily on culinary innovation mixing localities from all over Indonesia to sell its products. At the same time, through its Instagram account, there are multiple strategies used to construct personalized images by emphasizing on the owner's family as well as close and friendly relationships with the employees and customers unlike other Instagram pages owned by other culinary businesses, which focus solely on the products. In other words, in the dynamic relationship between the business owner and the 'imagined audience' of the Instagram page, Panggang Ucok is constructing an ideal image of a traditional-modern, commercial-private, and local-global articulation of identity.

1   INTRODUCTION

Social media has become a space where we document and curate our everyday lives, including the meals that we eat daily. People are snapping photos of their meals before they eat either in a fancy restaurant or a food stall on the side of a busy road. This has transformed a cultural practice as a necessary daily routine to a competitive sport between social media users. They will hunt for the newest and most trendy/happening culinary gems to elevate their own personal status. Eating has always been a focal point in socializing, and our culinary practices reflect an array of social and cultural status, such as our class, gender, and to which generation we belong to (Rousseau, 2012). Furthermore, the existence of social media in its connection to culinary practices has destabilized the dissemination of food information through old media without the presence of a centre of information flow. The driving force of social media is a conversation opening new spaces to talk about food and culinary practices. Instagram, for example, has become a tool for today's consumers to learn about products based on the experiences of other consumers instead of experts or food critics, or even food reviews by journalists in the old media, namely printed newspapers or magazines.

Marketing via social media in the culinary business has experienced a significant development in the last few years. "Unlike traditional media such as company websites or paid outdoor advertisements, social media create a two-way street between businesses and their core consumer segments" (Bui 2014, 6). Even though, on one hand, owners of culinary business make use of social media as a promotional tool, consumers are also able to give feedbacks to the business owners through their social media posts. As mentioned earlier, conversations created by social media between the actors are tools to construct particular images and build brand awareness for new consumers. In the process of branding, business owners need to continuously innovate, even though their first priority is to gain profit. However, at the same time, they need to construct images that will become the main characteristic of that culinary business.

For culinary business owners selling traditional menus, innovation marketed through social media could also be seen as a representation of identity. As a case study, this research investigates a culinary business, *Panggang Ucok*, which was established in August 2015, selling *Babi Panggang* or Roasted Pork from North Sumatra. However, *Panggang Ucok* (which will be abbreviated into PU throughout this article) is not only selling its Bataknese or North Sumatra identity. This particular ethnic identity is somehow obscured since there are constant innovations repackaging the food they sell inspired from different localities in Indonesia. The research focus is on how *Panggang Ucok* as a representation of 'traditional yet modern' culinary business owners utilizes social media through strategies of innovation and collaborative yet emotional branding. By conducting a thorough textual analysis on its Instagram account, the main objective is to investigate how social media is constructing the desired images, creating conversations between culinary actors, and articulating a dynamic identity formation through culinary practices.

1.1 *"Make yourself feel at home": PU's strategic branding*

PU has used its Instagram account (@panggangucok) to promote the restaurant's products by constantly posting the pictures of the products and alluring testimonials from their consumers. There are several narratives that are used to signify the culinary business' distinct features. One of the main narratives being used is the discourse of family. Among the photos posted on @panggangucok instagram page, almost half of them are photos portraying Bang Ucok (the owner)'s family or employees, who are depicted as part of the Panggang Ucok's family. Most of the pictures are depicting his family or his employees' daily lives because as business grows strong, the family's centre of activities is at the restaurant. By posting pictures related to his personal life, Bang Ucok uses @panggangucok Instagram account more like his personal account, even though it is used to sell roasted pork. The account is exclusively used as a business account since it has been created for business purposes. The strategy to manage the account like a personal account creates amity between Panggang Ucok and its consumers. This is a strategy that is rarely used by other online culinary business owners who mainly use impersonal or very business-like posts to sell their products.

In general, the pictures (and captions) of family members and employees posted in @panggangucok reflect a sense of affection and gratitude towards the people who work for the business. When this account gained 1000 followers, for example, @panggangucok posted a picture of its employers holding a cake with a caption "All credits go to these humble people behind every pork bowl we have enjoyed." Furthermore, before Eid Mubarak, @panggangucok posted a picture of Ucok sending his employees away to go *mudik* or go back to their hometown to celebrate Eid Mubarak. By showing these pictures and from the captions, PU is expressing an identity of kinship that has become the foundation of the business. Moreover, PU relies mostly on online transportation apps, such as GoFood from Gojek or GrabFood from Grab, for the order-delivery system. In @panggangucok, we can see a lot of posts depicting PU's gratitude towards the drivers from Gojek or Grab. These drivers, from the posts, are considered as part of the family because of their contribution to the business. As an example, PU posted pictures of Gojek drivers having a free meal or a corner in the restaurant with free drinks for the drivers who come to pick up orders.

By emphasizing on the discourse of family and kinship reflecting the restaurant as a 'home,' PU strategically uses the converse nature of modern culinary practices, which relies on business-like features of convenience, casualness, and speed. Moisio, Arnould, and Price (2004) argue that in modern times, particularly in an urban setting, food consumption has been altered to fit the fast-moving lifestyles. The consumers' culinary habits have been changed, for example, the development of fast-food industry altering family life and consumption rituals. PU is reassessing the meaning of eating out or even eating at 'home' as relying on its family and kinship identity to create a more 'feel at home' or 'be part of the family' nuance, not only from the food they sell but also from their business features depicted in their social media.

1.2 *Panggang Ucok's ambiguous Bataknese identity*

In the past, PU's was an online culinary business selling its product by mainly posting the products on Instagram. Customers could order the food via instant messaging services and it would be delivered to the customer's house by Gojek, the online transportation service. The restaurant used to be the main kitchen in which Ucok and his staff cooked and prepared the food before being delivered to the customers. Even though their main menu is Bataknese roasted pork, PU is different from a regular traditional Bataknese restaurant, which is called *Lapo*. If someone wants to eat Bataknese roasted pork or other Bataknese delicacies in Jakarta, they will immediately think about *Lapo*. The word *Lapo* actually refers to a tavern or a space in many areas in North Sumatra where people (mostly men) gather to drink coffee, tea, or *tuak*, a traditional alcohol fermented drink, and some Bataknese food. In its origin, *Lapo* is not actually a place to have big meals as it mainly sells drinks and a few choices of food. The main function is as a space to get together and engage in conversations. However, in Jakarta, *Lapo* is, first and foremost, a restaurant selling Bataknese food for the Bataknese diasporic communities in the capital city.

PU portrays its restaurant as the 'modern' *lapo*, but the only thing that signifies the restaurant as lapo is simply because it sells Bataknese roasted pork as the main menu. The interior or even the types of food and drinks they sell are completely different from the regular *lapo*. However, by referring to *lapo* in constructing their images on Instagram, PU is associating itself with the image of a Bataknese restaurant. On the other hand, not only do they sell food that one could never find in a regular *lapo,* the way they serve the traditional Batanese roasted pork is also distinct. In a *lapo*, roasted pork is served in a plate separated from the rice. Other dishes will also be served on separate plates. However, in PU, the roasted pork is served in the same plate with the rice and other delicacies, such as the Medanese anchovies and chilli. This style of serving roasted pork is actually quite similar with the Chinese *nasi campur* (mixed rice dish) in which rice is served with multiple pork dishes in one plate.

Besides the menu, the interior design in the restaurant is different from a regular *lapo* as it adds a modern and trendy touch to the setting of the restaurant. On the walls of the restaurant, the owner has put some posters with inspiring yet also funny quotes about eating (or eating pork). This type of interior design is often found in many modern/contemporary/urban cafés in Jakarta. Some examples of the writings on the wall are: "people who love to eat are always the best people" or "eat lots of pork when times get rough, and eat even more when times are great". This modern touch of the interior design is parallel with the packaging PU used to serve their menus, which is a combination of the 'old' and the 'new.' For delivery and take away, for example, PU serves the roasted pork in a rice bowl. This is uncommon for a *lapo*, which usually prepares the takeout in food wrapping papers. Through this practice, PU once again makes itself distinct from a regular *lapo*, therefore distancing itself from a Batakense identity, but at the same time uses other Bataknese identity features. The name Panggang Ucok is the main signifier. Among the Bataknese people, the word *panggang* (literally means roasted) refers to *babi panggang* (roasted pork), although there are other kinds of roasted meat. In other words, when someone hears the word *panggang*, they will not associate it with roasted chicken or beef because it has been associated with roasted pork. Moreover, Ucok is a widely popular Bataknese name, which actually means a young Bataknese man, and in PU, Ucok refers to the owner, Sahat Gultom.

Another signifier of PU's Bataknese identity is the way it interacts with the customers on Instagram. When PU responds to the customer's comments or in the caption, it often uses Bataknese language, or the content is related to the Bataknese identity. For example, in one of the posts, it says: "*Mauliate*, sold out! Thank you." It was actually an announcement that they no longer have any roasted pork they could sell that day. *Mauliate* in English means thank you, and by using this Bataknese word to show his gratitude, PU articulates its Bataknese identity and at the same time shows respect to his Bataknese customers as the ones who can understand the language. This does not mean that PU's customers are mainly Bataknese people because nowadays PU has been very successful and the customers come from a variety of ethnicities. Even so, it still uses Bataknese language to often communicate with its customers implying a strong association with its Bataknese identity.

This indefinite representation of identity through PU's culinary practices is a result of constant negotiation between continuity and transformation reflecting the dynamics of culture. Roland Barthes (1997) in his article "Toward a Psychosociology of Contemporary Food Consumption" states that food and culinary practices need to be seen as "a system of communication, a body of images, a protocol of usages, situation, and behaviour" (21). Food has become signifiers and the meanings are produced in the process of making the food or the way we consume it. "Techniques of preparation" and "habits" of consumption, or their contexts—how, why, when, where, and by whom the units of signification are prepared and eaten" (22) are the foundation of the process of identity formation in culinary practices.

1.3 *Re-inscribing culinary practices through innovation*

To explore the implication of PU's ambiguous representation of its Bataknese identity, the analysis will look at one of the most renowned menus, which is the Roasted Pork Burger. It looks like any regular burger but besides the roasted pork, it also contains anchovies from Medan, North Sumatra, and *andaliman* chilli, a traditional Bataknese chilli. The anchovies and chilli are known as Bataknese food, but through innovation, Ucok, the owner of PU, has made these two types which were originally eaten in supplementary with rice now eaten as part of a burger. Innovation is done by transforming the way of serving the dish in order to attract the consumers. On one hand, the changes in the culinary practices have made roasted pork and Medanese anchovies (and *andaliman* chilli) famous or more renown because now one does not need to eat them with rice. On the other hand, PU's roasted pork has also been considered as less Bataknese compared to the 'original' one. One of the reasons is because the roasting process is not the same since PU uses modern ways of roasting the meat unlike the traditional way of roasting the pork directly on top of firewood. Original roasted pork will have black marks from the firewood roasting, which you will not see in PU's roasted pork.

PU menus are incorporating elements of Bataknese traditional cuisines with local dishes coming from other parts of Indonesia disarticulating the Bataknese identity. Menus, such as *Babi Gepuk Sambal* (Chili Pork), exemplify innovation and the mix of culture between Bataknese and Sundanese (*Gepuk* is a Sundanese dish) culinary practices. *Gepuk* is actually made out of beef because in West Java where this disc comes from, the dominant population is Muslim. Therefore, it is unimaginable to have *gepuk* to make out of pork. However, in North Sumatra, most of the populations are Christians and most of the dishes are pork-based meals. A mixing of culture in this sense does not only come from two different ethnicities, but it is also heavily influenced by the religious beliefs, which in turn affect the culinary practices in the two different locales. Besides from Indonesian localities, PU also creates meals inspired by American or European food, such as the aforementioned Pork Burger and Panacota Pudding, an Italian inspired dessert. It could be concluded that through constant innovation in the menus, PU is articulating 'traditional yet modern' culinary practices, which are also reflected in the setting and design of the restaurant.

Culinary practices are spaces where individual and collective identities are constructed and articulated. According to Gabaccia (1998) in *We Are What We Eat: Ethnic Food and the Making of Americans*:

> ".... the production, exchange, marketing, and consumption of food have generated new identities—for foods and eaters alike ... eating habits changed and evolved long before the rise of a modern consumer market for food. Human eating habits originate in a paradoxical, and perhaps universal, tension between a preference for the culinary familiar and the equally human pursuit of pleasure in the forms of culinary novelty, creativity, and variety." (Gabaccia, 5–6)

Changes and evolution of culinary practices, just like what is being done by PU, should not only be seen as an effect of the changing market or merely as an economic practice. There is a cultural dynamics reflecting constant negotiation between the familiar, the food we are comfortable with, and the new, creative, and innovative ones. Most of the time, the need to innovate comes from limitations; for example, for diasporic communities, when the first generation, like Ucok's mother and father, left their hometown, they carried their culinary knowledge and experiences that became the basis of their culinary practices in the new place (Tan 2011). The need to reinvent new food through localization and invention bring about the reproduction of familiar food and taste from existing local ingredients and resources, which might be completely different from what they have in their hometown.

2 CONCLUSION

To sum up, PU exemplifies its identity through its Instagram account by emphasizing on narratives of family, ambiguous identity, and innovation. In relation to identity, Panggang Ucok is at the same time Bataknese food and not Bataknese food as it is both modern and traditional. Innovation is an important factor, as recipes are constantly reinvented to try to find familiar flavours from the past whilst creating new ones. These narrations on identity and innovation could be concluded as part of the identity formation of the culinary actors, in this case for the owner of Panggang Ucok. It has become a space where narrations could be conveyed and where culinary actors could make meaning out of their fluid and complex identity. The exploration of the re-inscribing of culinary practices does not only inform us about how food is constantly transforming, but also problematizes the dynamics of culture and its implications.

REFERENCES

Bui, T.V. (2014) *Social Media on a Stick: A Uses and Gratification Approach toward Helping Mobile Food Vendors Engage Consumers on Instagram.* Retrieved from: http://hdl.handle.net/11299/166761, the University of Minnesota Digital Conservancy.

Barthes, R. (1997) Toward a Psychosociology of Contemporary Food Consumption. *Food and culture: A reader, 2,* 28–35.

Bhabha, H.K. (1994) *The Location of Culture.* United Kingdom, Psychology Press.

Dalessio, William R. (2012) *Are We What We Eat? Food and Identity in Late Twentieth-century American Ethnic Literature.* Amherst, NY, Cambria.

Gabaccia, Donna. (1998) *We Are What We Eat: Ethnic Food and the Making of Americans.* Cambridge, Harvard University Press.

Hall, Stuart. (1997) *Cultural Identity and Diaspora.* pp. 222–237.

Lai, A.E. (2010) *The Kopitiam in Singapore: An Evolving Story about Migration and Cultural Diversity.* Asia Research Institute Working Paper Number: 132.

Moisio, R., Arnould, E.J. & Price, L.L. (2004) Between Mothers and Markets: Constructing Family Identity through Homemade Food. *Journal of Consumer Culture,* 4(3), 361–384.

Rousseau, S. (2012) *Food andSocial Media: You are What You Tweet.* Rowman Altamira.

Sidharta, M. (2011) The Dragon's Trail in Chinese Indonesian Foodways. In: C. B. (Ed.). *Chinese food and foodways in Southeast Asia and beyond.* Singapore, NUS.

Tan, C.B. (2011) Cultural Reproduction, Local Invention and Globalization of Southeast Asian Chinese Food. In: *Chinese Food and Foodways in Southeast Asia and Beyond* (pp. 23–45). NUS Press, Singapore.

Cultural Dynamics in a Globalized World – Budianta et al. (Eds)
© *2018 Taylor & Francis Group, London, ISBN 978-1-138-62664-5*

The use of cohesive devices in deaf and hearing children's writing

Novietri & B. Kushartanti
Department of Linguistics, Faculty of Humanities, Universitas Indonesia, Depok, Indonesia

ABSTRACT: This research aims at exploring the use of cohesive devices by deaf and hearing children, 15–19 years of age. The sources of the data are their handwriting. The instrument is a story picture. It is found that both the deaf and hearing children groups show similar tendencies towards the use of ellipsis, hyponyms, meronyms, and collocations. Similarities are also found in conjunctions—except adversative and temporal markers, which are used more productively by the hearing children. The deaf children tend to use repetitions and antonyms, whereas the hearing children tend to use references and synonyms. These similarities and differences in using cohesive devices are influenced by several factors, amongst which are the teaching method, children's ability, family, children's environment, and the sign language used by the deaf. The findings deal with the strategies on learning at school.

1 INTRODUCTION

In the first five years of their life, hearing children learn to interact verbally with adults, which includes telling stories, both in verbal and written forms. In comparison, deaf children do not have access to hearing speech in order to understand or to react. Therefore, there are some delays in their language development, including their production (Bernaix, 2013, p. 5). They need some treatments in order to communicate with others. The way deaf and hearing children interact with others influences their capacity for using cohesive devices in writing.

Cohesive devices are important tools for understanding and interpreting a discourse (Halliday and Hasan, 1976, p. 4). Halliday and Hasan (1976) divide cohesive devices into grammatical and lexical cohesions. Grammatical cohesion is a relation between elements in discourse, which can be found in reference, substitution, ellipsis, and conjunction (Halliday and Hasan, 1976; Halliday and Matthiessen, 2014). Reference is a semantic relation indicating that information has to be retrieved from elsewhere (Halliday and Hasan, 1976). There are three types of reference: personal, demonstrative, and comparative. Substitution is a replacement of a certain constituent by word or sentence. Ellipsis is an omission of a word or part of a sentence. Conjunction is a relationship indicating how a subsequent sentence or clause is linked to the preceding or the following (parts of the) sentence (Renkema, 2004). Halliday and Hasan (1976, p. 238) distinguish four categories of reference, namely additive, adversative, causal, and temporal references.

Lexical cohesion is a relation between lexical elements in a discourse, which can be found in repetition, synonym, hyponym, meronym, antonym, and collocation (Halliday and Hasan, 1976; Halliday and Matthiessen, 2014). Repetition is a relationship in which one lexical item restates another (Halliday and Matthiessen, 2014). Synonyms are different phonological words that have the same or very similar meanings (Saeed, 2003). Hyponym is a relation of inclusion; the meaning of a more general word (Saeed, 2003). Meronym is a term used to describe a part-whole relationship between lexical items (Saeed, 2003). Antonym is a relation of lexical items which are opposite in meaning. Collocation is a relation of two lexical items having similar patterns (Halliday and Matthiessen, 2014).

The paper aims at comparing the use of cohesive devices in writing production by hearing children and deaf children. This is part of the first author's thesis (Novietri, 2016). In this study, we mainly apply Halliday and Hasan's (1976) and Renkema's (2004) theories on cohesive devices to analyze the children's writing. We also examine factors that influence the use of cohesive devices by both groups.

Participants in this research are 15–19-year-old high school students. Two groups are compared: hearing and deaf children. Each of the group consists of 10 children. All of them come from middle class families. In general, they were born in Jakarta. Participants in the hearing group are those with a good academic report, as suggested by their teachers. Participants in the deaf group are those who have 90 dB of deafness (for characteristics of deafness, see Rodda and Grove, 1987).

The sources of the data are the children's handwritten stories, based on a task of storytelling composition. We used a pictured story titled *Gara-Gara Usil* (Dwi and Sabariman, 2014) as an instrument. It was taken from *Bobo*—a children's magazine. After the handwriting was obtained, we transcribed and typed it in a written document format using the Microsoft Word application. Once the orthographic transcription was done, we grouped the data sentence by sentence and numbered each of the sentences. The cohesive devices were identified and classified. Once they were classified, the use of each type and its appropriateness in text and context was then analyzed. We also counted the occurrences of each type in order to examine the tendencies found in both groups. To have an overall illustration on the use of cohesive devices by the deaf children, we also interviewed their teachers. Based on these steps, we have analyzed the influencing factors on the use of cohesive devices in children's writing.

2 DISCUSSION

It is found that both groups use various types of cohesive devices, except the substitution. Both use grammatical and lexical cohesions. We find references, ellipses, and conjunctions (the grammatical cohesion), and repetitions, synonyms, hyponyms, meronyms, antonyms, and collocations (the lexical cohesion) in the children's writing. Nevertheless, we find that there are some differences between both groups, as shown in Table 1.

2.1 *The use of references*

We find that deaf children tend to use the singular first person pronoun *saya* or 'I',—a pronoun which is used mostly to show formality (for more detail, see Table 2). They also tend to use proper names to refer to the characters in the story. They did not use the pronoun *aku* or 'I'—an informal

Table 1. The use of cohesive devices in deaf and hearing children's writing.

Cohesive device	DC										HC									
	1	2	3	4	5	6	7	8	9	10	1	2	3	4	5	6	7	8	9	10
Reference	√	√	√	√		√	√	√	√	√	√	√	√	√	√	√	√	√	√	√
Ellipsis	√	√	√	√	√	√	√			√	√	√	√	√	√	√	√	√	√	
Conjunction	√	√	√	√	√	√	√	√	√	√	√		√	√	√	√	√	√	√	√
Repetition	√	√	√	√	√	√	√	√	√	√	√	√	√	√	√	√	√	√	√	√
Synonym				√				√			√	√		√					√	√
Hyponym		√						√			√		√			√	√			
Meronym								√						√						
Antonym		√	√		√		√	√		√				√				√	√	
Collocation	√	√	√	√	√	√	√	√	√	√	√	√	√	√	√	√	√	√	√	√

DC = deaf children; HC = hearing children
Number refers to individual participant.

Table 2. The use of references.

Reference	DC										HC									
	1	2	3	4	5	6	7	8	9	10	1	2	3	4	5	6	7	8	9	10
Saya	√	√					√	√		√	√									
Aku											√	√		√			√	√		
Kami		√	√								√						√			
Kita			√			√					√			√			√	√		
Kamu	√	√					√				√	√		√	√		√	√		
Ia			√								√	√	√	√	√	√	√	√	√	√
Dia	√			√		√		√												√
Nya		√		√		√	√	√			√	√	√	√	√	√	√	√	√	√
Mereka	√			√			√	√			√	√	√	√	√	√	√	√	√	√
Itu	√		√				√		√	√	√	√	√	√	√		√	√	√	√
Tersebut			√								√	√	√	√	√	√	√	√	√	√

DC = deaf children; HC = hearing children
Number refers to individual participant.

variant of *saya* 'I', which is used mostly in casual or informal situations—in all contexts. Meanwhile, hearing children use *aku* in informal contexts, or in quoted conversation, besides *saya*.

Based on our interview with their teachers, the deaf children learn language mostly based on daily conversations. To interact with these children, the teachers use the standard Indonesian from which the children learn how to use the references. Therefore, they opt for *saya*, instead of *aku*. The deaf children are also used to using *saya* in daily conversations.

It is important to briefly discuss the Indonesian sign language, its users, and its dictionary, in relation to the use of references. Deaf children use mainly sign language in conversations. The pronouns *aku* and *saya* are also included in the Indonesian sign language system. *Kamus Sistem Isyarat Bahasa Indonesia* (2001) (hereinafter referred to as KSIBI; *Sistem Bahasa Isyarat Indonesia*, or SIBI, is one of the Indonesian sign language systems), an Indonesian sign language dictionary, distinguishes the visual signs of *aku* and *saya*. Meanwhile, *aku* and the first person possessive personal pronoun *–ku* are not distinguished. In comparison with SIBI, according to *Bahasa Isyarat Jakarta* (2014) (hereinafter referred to as BIJ), a Jakarta-Indonesian sign language dictionary, the visual sign of *aku* and *saya* is the same. BIJ is a variant of *Bahasa Isyarat Indonesia* (hereinafter referred to as BISINDO), another Indonesian sign language system. At school, the deaf children learn SIBI. They talk and use SIBI in classroom, while they learn BISINDO from their deaf friends or family.

We find that deaf children use the first person plural pronoun *kami* (exclusive) and *kita* (inclusive) in both appropriate and inappropriate contexts. The similarity between *kami* and *kita*, in terms of some semantic components, may cause inappropriate use; hearing children are no exception. *Kami* and *kita*, in some regional languages, are not distinguished, as well.

Kamu or 'you', the singular second person pronoun, in children's writing occurs when there is a dialogue in the story. When there is no dialogue, the pronoun is not used. There are only a few *kamu* in the story, as only a few of participants use dialogues in their story.

The hearing children tend to use the singular second person pronoun more frequently than their counterpart group. Nevertheless, the differences between both groups are not striking. *Kamu* can be used to address either an older to younger and familiar person, a person whose status is higher than the interlocutor, or a person in an intimate situation (Alwi et al., 2000, p. 253). Both groups use the pronoun in the appropriate context.

Deaf children use more kinship terms or proper names to address the second person. The finding correlates to the technique of conversation, which is learned at school, as it may influence the use of the address terms. Usually, the teachers ask the children to write down what they have been told. Therefore, the deaf children's writing rarely uses dialogues in their writing.

The third person pronoun tends to be used frequently by hearing children. It occurs in stories which consist of more than two characters. Meanwhile, it is rarely used by deaf children.

For deaf children, it is easier to do a dyadic conversation than a triadic or more. They need face to face interaction with their interlocutor. In daily conversations, they tend to point or mention the person's name instead of using pronouns. For them, it makes their intention clearer. For example, when deaf children use the pronoun *dia* or 'he/she' to refer to one of their friends, they need to point the person they refer to. The interlocutor and the pointed friend should pay attention to his mouth or gesture. Many of the deaf children often make mistakes in using third person pronouns in writing a story. Therefore, they tend to avoid using them and opt for repeating the character's name.

In comparison with their counterpart group, the deaf children tend to use less the demonstrative *itu* or 'that.' This finding is in line with the interactions with their teachers and friends in class. When deaf children intend to mention unknown or unnamed objects, they will point at them through gesture or picture. In an ongoing conversation, deaf children tend to mention the objects they know about without using *itu*. Therefore, in a written story, demonstratives are rarely used to refer to the objects they discuss as they think the objects are already known by both the writer and the reader.

The hearing children use more the demonstrative *tersebut* or 'the mentioned' than their counterpart group. It is influenced by their written language and daily conversations. Even though *tersebut* and *itu* are similar in terms of their semantic element, both have to be distinguished in terms of their visual sign. There is a visual sign for *itu* in KSIBI (2001), but no visual sign for *tersebut*. *Itu* may occur at any situation and at any kind of books. Meanwhile, *tersebut* mainly occurs in books. As mentioned by Purwo (1984, p. 121), *tersebut* is mainly found in newspapers, scientific papers, or speech texts. Therefore, books read by children influence the children's competence on the use of *tersebut*. Deaf children begin to learn how to write from their spoken language. The absence of *tersebut* in the spoken language may influence its frequency in the written form.

2.2   *The use of ellipses*

In our data, it is found that hearing children use ellipses more frequently and variously than deaf children do, but the difference is not striking (for more details, see Table 3). Deaf children can use ellipses in an appropriate context, but many of them opt for repetition in order to make the readers understand. This is also in line with the teaching method in class where the teachers use more repetitions in conversations. The children often write down what they have talked about, and write down the repetitions. In turn, the teachers often ask the deaf children in order to make sure that they understand their own writing.

2.3   *The use of conjunctions*

Both hearing and deaf children used various kinds of conjunctions. They use different types of conjunctions: additive (*dan* or 'and', *lalu* or 'and then'), adversative (*tetapi* or 'but,' *sedangkan* or 'whereas,' and *namun* or 'however'), and temporal (*ketika* or 'when,' *saat* or 'when,' *seraya* or 'while,' *sembari* or 'while,' and *setelah* or 'after'). Nevertheless, we find that hearing children and deaf children tend to use various types of conjunctions respectively. The hearing

Table 3.   The use of ellipses in deaf and hearing children's writing.

	DC										HC									
Ellipsis	1	2	3	4	5	6	7	8	9	10	1	2	3	4	5	6	7	8	9	10
Word	√	√	√	√			√	√		√	√		√	√	√	√	√	√	√	
Phrase					√			√			√			√	√	√			√	
Clause						√							√					√		

DC = deaf children; HC = hearing children.
Number refers to individual participant.

814

children tend to use adversative and temporal conjunctions more frequently than their counterpart group. They use *tetapi, sedangkan, namun, ketika, saat, seraya, sembari,* and *setelah.* The deaf children tend to use additive conjunctions. They use *dan* and *lalu* more frequently than hearing children.

In KSBI (2001), not all conjunctions are found. Therefore, not many deaf children know that actually there are more conjunctions in the written language. The use of temporal conjunctions by deaf children depends on the number of books they read. The more often deaf children read books, the more various temporal conjunctions are used in their writing. The use of conjunctions is summarized in Table 4.

2.4 *The use of repetitions*

As discussed before, the deaf children use repetitions more frequently than their counterpart group. Using repetitions indicates that they confirm what they mean, or they want to show some degree of importance of things in their writing. It is already discussed in the previous section that the teachers use more repetitions in conversations. The teaching method also influences deaf children in writing. Moreover, repetitions influence children's memory. Since they lose their hearing, deaf children rely on what they see in order to interact. It is reflected in their written language. Repetitions are also used as a strategy to make readers understand their writing and to make it more comprehensive. In other words, deaf children use repetitions in order to make coherent writing. The use of repetitions is summarized in Table 5.

Table 4. The use of conjunctions in deaf and hearing children's writing.

Conjunction	DC 1	2	3	4	5	6	7	8	9	10	HC 1	2	3	4	5	6	7	8	9	10
Dan	√	√	√	√	√	√	√	√	√	√	√	√	√	√	√	√	√	√	√	√
Lalu		√	√				√		√	√	√			√	√					
Tetapi					√													√	√	√
Tapi						√						√								
sedangkan											√					√				
Padahal				√																
Namun											√	√	√		√			√		
Ketika															√					√
Saat											√	√				√		√		
Seraya											√	√								
Sambil		√		√	√	√	√	√			√	√			√			√	√	
Sembari														√		√				
Sebelum																			√	
Setelah						√								√	√	√			√	

DC = deaf children; HC = hearing children.
Number refers to individual participant.

Table 5. The use of repetitions in deaf and hearing children's writing.

Repetition	DC 1	2	3	4	5	6	7	8	9	10	HC 1	2	3	4	5	6	7	8	9	10
Word	√	√	√	√	√	√	√	√	√	√	√	√	√	√	√	√	√		√	√
Phrase		√	√		√	√		√	√	√				√				√	√	
Clause		√			√										√					

DC = deaf children; HC = hearing children.
Number refers to individual participant.

2.5 The use of synonyms

Synonyms are used more frequently by the hearing children than by their counterpart group. It is found that deaf children do not use various lexical choices. As already discussed, deaf children prefer to use repetitions. According to their teachers, the deaf children do not fully understand the concept of synonyms. For example, they understand the concept of *bangku* or 'seat' and *kursi* or 'chair,' but in sentences they do not use those alternately. Once they choose a lexical form, it is possible that they will not substitute it with other lexical forms that have the same semantic component.

2.6 The use of hyponyms

It is found that both the hearing and deaf children groups have the same tendency to use hyponyms, especially in verbs. The use of hyponyms is influenced by the instruments, which are pictures composed as a linear story. They use some similar words referring to the objects in the pictures. The use of hyponyms indicates the children's intention to tell specific objects in order to be understood by the readers. An example is the use of the words *olahraga* or 'to do sport' and *lari pagi* or 'jogging.' They are aware that *olahraga* has a general meaning that needs to be specified in its varieties. Therefore, they add *lari pagi*. Some use one of the options. The finding on hyponyms would have been different if they had been given a theme without pictures to elicit the story.

2.7 The use of meronyms

As in hyponyms, it is found that both the hearing and deaf children groups have the same tendency to use meronyms, especially in nouns. The reason is also similar in regard to how the pictures, being the instruments, influence their lexical choices. It is also found that the children's competence in writing is dependent on their ability to imagine and develop the plots.

2.8 The use of antonyms

Antonyms tend to be used more frequently by the deaf children than their counterpart group. The deaf children tend to use verbal antonyms in their story, especially *tanya* or 'ask' and *jawab* or 'answer.' In fact, many of them use dialogues in their writing. They use direct sentences like *Dia bertanya, "Ada apa...?"* or 'He asked, "What happened...?"' and *Dia menjawab, "Ada..."* or 'She answered, "There was...".' Some of the hearing children use this kind of sentence, but the pairings are inappropriate, such as *tanya/balas* or 'ask/reply,' *tanya/kata* or 'ask/tell,' *ucap/jawab* or 'say/answer,' and *ujar/jawab* or 'state/answer.' Therefore, it leads us to an assumption that deaf children are more competent in using antonyms. It is also shown that deaf children can describe pictures more variously by using antonyms, such as *kiri/kanan* or 'left/right' or *kurus/gemuk* or 'thin/fat.' We find one deaf child who uses a pair of adverbial antonyms *sudah/belum* or 'already/not yet.' The pair is used in tag questions, which are mainly found in an ongoing conversation. The use of a tag question indicates that his spoken language influences his writing.

2.9 The use of collocations

We found that the deaf children use collocations more frequently than their counterpart group. Nevertheless, the difference is not striking, as both groups use the same instrument. They use words relating to the pictures, from which children use lexical choices related to sport, kinship terms, emotion, communication, house, and its parts. The analysis on collocations shows that both the hearing and deaf children groups can develop a story using their imagination, even though there are certain things that are not illustrated in the instrument. The use of collocations is summarized in Table 6.

Table 6. The use of collocations in deaf and hearing children's writing.

Collocation	DC										HC									
	1	2	3	4	5	6	7	8	9	10	1	2	3	4	5	6	7	8	9	10
Sport	√	√	√	√		√	√			√	√		√		√		√		√	√
kinship term		√		√	√	√		√	√		√	√	√	√				√		
Emotion	√	√		√	√	√	√		√	√	√	√	√	√	√					
Place											√	√	√							
communication	√	√	√	√	√	√	√	√	√	√	√	√		√	√	√	√	√	√	
Character			√																	
Food							√													
House	√	√	√	√	√	√	√	√	√		√	√	√	√	√		√	√	√	√

DC = deaf children; HC = hearing children.
Number refers to individual participant.

3 CONCLUSION

It has been shown that both the hearing and deaf children groups can use cohesive devices in their writing. An interesting finding in this research is on the use of references. Both groups use various kinds of references. Nevertheless, it is found that the use of certain references in the deaf children's writing is influenced by the corresponding visual signs. This particular finding has to be explored further in order to have a full picture on how deaf children learn both written and spoken languages through sign language.

It has also been shown that there are some differences between the groups in terms of their competence in composing a story. We have also discussed that the differences are influenced by the teaching methods used—especially for deaf children. Yet, some evidence that supports this conclusion is still needed for further investigation.

REFERENCES

Alwi, H., Dardjowidjojo, S., Lapoliwa, H., & Moeliono, A.M. (2000) *Tata bahasa baku Bahasa Indonesia* (The standard grammar of Indonesian language) (3rd ed.). Jakarta: Balai Pustaka.

Bernaix, N.E. (2013). Oral and written narrative production in children who are deaf and hard of hearing. *Independent Studies and Capstones*, 1–15. St. Louis: Washington University School of Medicine.

Departemen Pendidikan Nasional (Department of National Education). (2001) *Kamus sistem isyarat Bahasa Indonesia* (Indonesian sign language system dictionary) (3rd ed.). Jakarta: Departemen Pendidikan Nasional.

Dwi & R. Sabariman (2014) Paman Kikuk, Husin, dan Asta: Gara-gara usil (Uncle Kikuk, Husin, and Asta: Due to being naughty). *Bobo: Teman bermain dan belajar* (Bobo: Play and learning mates), 27, 30–31.

Halliday, M.A.K. & Hasan, R. (1976) *Cohesion in English*. London: Longman.

Halliday, M.A.K. & Matthiessen, C.M.I.M. (2014) *Halliday's introduction to functional grammar* (4th ed.). New York: Routledge.

Novietri. (2016) *Penggunaan alat-alat kohesi dalam tulisan anak tuli dan anak dengar* (The use of cohesive tools in the writing of deaf and hearing children). [Unpublished Thesis]. Depok: Universitas Indonesia.

Purwo, B.K. (1984) *Deiksis dalam Bahasa Indonesia* (Deixis in Indonesian language). Jakarta: Balai Pustaka.

Renkema, J. (2004) *Introduction to discourse studies*. Amsterdam: John Benjamins Publishing Company.

Rodda, M. & Grove, C. (1987) *Language, cognition, and deafness*. New Jersey: Lawrence Erlbaum Associates.

Saeed, J.I. (2003) *Semantics* (2nd ed.). Oxford: Blackwell Publishing.

Tim Produksi Bahasa Isyarat Jakarta (Jakarta Sign Language Production team). (2014) *Bahasa isyarat Jakarta: Kamus pendamping untuk buku pedoman siswa 1 tingkat 1* (Jakarta sign language: Supplementary dictionary for student guidelines I level I). Depok, Universitas Indonesia.

Cultural Dynamics in a Globalized World – Budianta et al. (Eds)
© 2018 Taylor & Francis Group, London, ISBN 978-1-138-62664-5

Acoustical analysis of pitch contour in autism spectrum disorder

L. Roosman, T.W.R. Ningsih, F.X. Rahyono, M. Aziza & Ayesa
Department of Linguistics, Faculty of Humanities, Universitas Indonesia, Depok, Indonesia

ABSTRACT: This study is aimed at examining a pitch contour of Indonesian prosody in children with Autism Spectrum Disorder (ASD). The sample data are taken from utterances of ASD children sitting on the third, fourth, fifth, and sixth grade of an Indonesian elementary school. The PRAAT program was used to analyse acoustic data in the pitch contour. The stimuli used are declaratives and interrogatives modes contrasted to get information of the pitch contour differences between the two modes of sentences. The results of the research revealed that there are no significant differences between the two modes in the F0 mean, pitch movement, and pitch range. The ASD children show a low F0, flat pitch movement and a narrow pitch range; therefore, they are significantly different from the control speakers. The boundary tone and pitch variation are not found in the pitch contour of ASD speakers. In most cases, the results show that the pitch contour of ASD is flat, while the TD speakers could produce the pitch contour of statements and questions differently.

1 INTRODUCTION

First-language acquisition during childhood proceeds with apparent ease. Children will learn to speak their language with a perfect pronunciation and perfect command of grammatical rules. Intonation is an important aspect of language that seems to be easily, if not automatically, acquired by children in L1 (Chun, 2002: xiii). However, for children with autism syndrome, this is not the case.

Autism is a pervasive developmental disorder and is defined as abnormal development, in terms of abnormal social interactions, communication, certain behaviour, stereotype, and repeated actions (Wing and Gould, 1979; Aitken et al., 1998). Recent research in autism focuses on three autism disorders domains: socialization constraints, language, and imagination deviation. In language disorder, many autistic children encounter mute conditions or difficulties in acquiring the language as well as using the language in certain styles, such as infrequent initiation of starting conversations, inability to do voice imitation, recurring repetitions or parroting words and sentences, and having odd intonations or vocal rhythms.

Baltaxe and Simmons (1985) state that one of the communication disorders of autistic individual is prosody deviation. The prosody of ASD (Autism Spectrum Disorder) speakers tend to be more atypical especially shown in a pragmatic affective aspect (Shirberg et al., 2001). They also have difficulty in understanding the changing of tones, modulation, or accentuated words from people's speech (Koning and Magil-Evans, 2001 in Attwood, 2008: 218). This shows that speech sound perception has an implication towards the speech sound production. Hubbard and Trauner (2007) found that the difficulty in prosody imitation of ASD speakers is because they have a wider pitch range than the affective prosody imitation. It means that the children who were subjects of this study exaggerated the intonation they heard. Moreover, they did not use duration signs so their responses did not show certain emotional features accurately.

Baltaxe et al. (1984) analysed the intonation of five ASD children compared to Aphasia and TD (Typical Developed) children of 4–12 years old. The spontaneous controlled speech was recorded and analysed acoustically in terms of pitch range, declination, pitch contour, and the influence of frequency declination and intensity. One of the findings shows that the pitch contour of ASD is wider than a child with aphasia, but the difference from TD children is insignificant. However, ASD speakers have a varied pitch range more than others. They can speak either very fast or very slow; there is no average pitch range information provided to show their specific intonation patterns.

Prosody is the properties of speech sounds that can be found beyond sequence segments. The properties occur in language units vary from syllables to longer units as sentences and paragraphs. Noteboom (1997: 640) provides some examples of prosody, such as the pitch of speech, stretching and shrinking duration of utterances, loudness, and intonation. The parameters of prosody can be arranged in two structures: melodic and temporal structures. The melodic structure of utterances consists of pitch parameters that are high and low pitch movements, and the intonation. The temporal structure contains the rhythm of speech, a combination of short–long utterances, and pauses.

Prosodic functions operate in different domains. Prosody has a lexical function on a word domain. At the word level, tone in tone languages is distinctive* similar to stresses in stress languages like English and Dutch (Cutler & van Donselaar, 2001). Other functions of prosody work in different domains. As a boundary marking, prosody can demarcate word boundaries, phrase boundaries, and sentence boundaries. Speeches are usually prolonged and fall with a lower pitch in the final position. Prosodic characters also define communication differences. A statement usually ends with a falling pitch, while a question with a rising tone (Ladefoged and Johnson 2011: 24). An accentual focus is carried out louder, with some pitch movements, and sometimes a longer duration (Beckman, 1989; Eefting, 1991; Eefting & Nooteboom, 1991).

Pitch is one of the acoustical parameters that play an important role in the speech melody. Pitch is investigated by measuring a vocal fold vibration and fundamental frequency (F_0) (Nooteboom, 1999: 642). The more the vocal folds vibrate, the higher the fundamental frequencies are. Combinations of pitch movements form the intonation of the sentence.

A previous study has shown that children with autism cannot differentiate declarative and interrogative sentences in their speeches. Therefore, we investigated whether such a case also occurs in Indonesian autistic children. Simple declarative sentences can be realized as interrogatives if they are spoken with a certain degree of a rising tone at the end of a sentence.

2 METHODOLOGY

This experimental phonetics research uses production experiments to investigate the intonation of the sentences. The production data are digitized, and the intonation contours are extracted. Perception experiments in this paper are not performed. Perception analyses are performed by the researchers to elicit the "close-copy of" the intonation contours.

The current phonetic researches use computer programs to analyse the speech sounds. This prosodic research uses a specific computer program called PRAAT (Boersma & Wenink, 2015), to depict the pitch contours and to measure the fundamental frequencies (F_0 in Hz). In particular, we measured the mean F_0 of all sentences and calculated the pitch ranges by measuring the F_0 maximum and F_0 minimum. Furthermore, to investigate the occurrence of declination in the pitch contour of the declaratives, we measured the initial F_0 and finale F_0.

The subjects are five elementary school children (grade 3 and higher) with autism syndrome disorder. They were asked to produce two sentences in two modes: declarative and interrogative modes. The data for this study are two sentences (1) *Dia membaca buku* 'He/she reads a book' and (2) *Dia bermain bola* 'He/she plays football/soccer' in two different modes. To illustrate the differences in the intonation contours between the declaratives and interrogatives, below are examples of the sentence *Dia bermain bola* spoken by a normal TD child of the same age delivered in two modes: declarative and interrogative one.

---

* In Mandarin, the domain of tone could be a bound morpheme (Yip 2002).

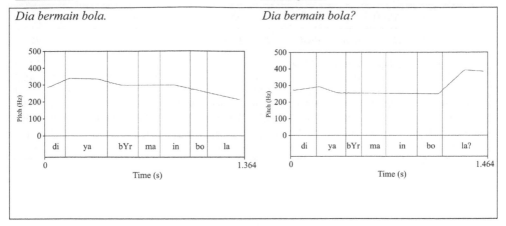

Figure 1. The pitch contours of *Dia bermain bola* in declarative and interrogative modes.

Figure 1 shows that the declarative sentence has a declination (55 Hz), which ends with a lower pitch. An interrogative usually ends with a rising tone. English-speaking children aged 10 normally have a pitch range between 200 and 500 Hz (Gut, 2009: 172). However, previous studies indicate that the pitch range is language specific (Van Bezooijen 1995, Mennen 2007). Indonesian children might have a wider pitch range or the opposite of a narrower pitch range. Our TD children have a pitch range of ca. 130 Hz (F_0 mean 291.5 Hz) in the declarative sentence and ca. 293 Hz (F_0 mean 271.1 Hz) in the interrogative sentence.

The data recording were performed by reading passages. The sentences were typed on a sheet of paper and the subjects were asked to read them aloud for four times. However, during this study, we only selected one with the best quality. As a comparison, a simple imitating experiment was performed in which the ASD children had to imitate their speech therapist.

3   RESULTS AND DISCUSSION

The results of the pitch measurements listed in the table are the mean pitch of the sentence (F_0 mean), the initial pitch (F_0 start), the final pitch (F_0 end), the lowest pitch (F_0 min), and the highest pitch (F_0 max). By taking the F_0 min from the F_0 max, we are able to have the pitch range of the sentence. The table is made separately for each sentence mode.

3.1   *Declarative sentences*

From Table 1 we can see that four of the five ASD children have a relatively high F_0 mean (266.3 Hz on average), but one of them has a lower F_0 mean (ca. 179 Hz). On the other hand, nine of the ten sentences have a relative small pitch range (ca. 70 Hz), only ASD3 has a relative wider pitch range (ca. 159 Hz), while reading "*Dia bermain bola.*" It seems that our ASD subjects read the sentences in a rather flat manner; they did not make adequate pitch movements. The following contours are from ASD1 and ASD3 reading "*Dia bermain bola*"

It is noticeable that the intonation contours of the declarative sentences do not always imply a clear declination. The contour of ASD3 from Figure 2 has, for instance, a declination of 27.66 Hz. ASD1 has no declination; it even looks more like an inclination. The two contours differ from each other. The pitch contour of ASD1 is rather flat, while the pitch contour of ASD3 indicates occurrence of pitch movements.

From Table 1 we can see that not all sentences end with a lower pitch. In other words, our ASD children are inconsistent in repeating a declining declarative sentence.

Table 1. Pitch parameters of declarative sentences read by 5 ASD children.

	ASD1		ASD2		ASD3		ASD4		ASD5	
	Book	*Ball*	*Book*	*Ball*	*Book*	*Ball*	*Book*	*Ball*	*Book*	*Ball*
F_0 mean (Hz)	183.88	179.44	258.74	265.40	267.9	271.46	295.43	290.12	236	237.78
F_0 start (Hz)	175.8	166.3	274.38	261.86	239.3	221.98	297.9	258.65	240.75	245.2
F_0 end (Hz)	142.1	190.1	228.49	255.66	284.7	194.32	313.2	286.52	230.93	205.4
F_0 min (Hz)	142.11	159.85	209.93	252.06	231.95	194.32	257.59	258.08	191.23	193.78
F_0 max (Hz)	202.69	196.41	301.17	294.73	326.61	353.53	340.87	311.96	265.46	269.93
Pitch range (Hz) (F_0 max–F_0 min)	60.58	36.61	91.24	42.67	94.66	159.21	83.28	53.88	74.23	76.15

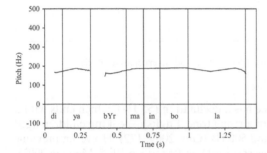 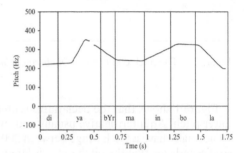

Figure 2. Intonation contours of declarative sentences "*Dia bermain bola*" spoken by ASD1 (left) and ASD3 (right).

3.2 *Interrogative sentences*

The F_0 mean (248.88 Hz) is rather lower than that of normal children. If we compare the pitch parameters in Table 2 with those in Table 1, we might see that the differences are small. The ASD subjects are indeed insensitive to the differences between declarative and interrogative sentences. The following is the intonation contours of ASD3 reading "*Dia bermain bola*" in the declarative sentence (straight line) and interrogative sentence (dot-dash line).

However, the pitch ranges of the interrogative sentences (82.49 Hz) are higher than the declarative sentences. They have attempted to make a rising tone somewhere in the sentence. Below is the intonation contour of the interrogative sentence "*Dia bermain bola*" read by ASD4. In Figure 4, the final pitch of the interrogative (267.25 Hz) is lower than the initial pitch (288.97 Hz). However, if we observe the overall intonation contour, the sentence ends with a final rising tone. It has a declination contour (dashed line) towards the final position and a rising tone on the last constituent, the object of the sentence "bola." Below is a comparison with another interrogative sentence with a rising tone at the end of the sentence "*Dia membaca buku*" as read by ASD3.

From Figure 5 we can see that the contour starts to rise on the predicate "membaca", and rises higher on the object "buku". This interrogative melody differs from the interrogative contour from the same speaker in Figure 3.

The melodic structures of declarative sentences of the ASD subjects indicate atypicality. Their mean pitches are relatively high, but the pitch ranges are small. It means that they often speak with higher tones, but do not make adequate pitch movements. In other words, their speeches are relatively flat. Moreover, when they were asked to read the interrogative sentences, they could hardly produce a representative melody. In most cases, the interrogative sentences were read with almost the same contour as the declarative sentences. However, the interrogative sentences have somewhat wider pitch ranges than those of the declarative sentences. In a few cases, they could make a final rising tone. It is noticeable that the ASD children have a prosodic problem in reading the sentences.

822

Table 2. Pitch parameters of interrogative sentences read by 5 ASD children.

	ASD1		ASD2		ASD3		ASD4		ASD5	
	Book	Ball	Book	Ball	Book	Ball	Book	Ball	Book	Ball
F₀ mean (Hz)	182.8	172.45	256.59	254.98	272.74	272.19	284.83	261.16	242.18	271.79
F₀ start (Hz)	179	188.6	276.5	287.4	238.28	239.30	270.98	288.97	251.9	288.4
F₀ end (Hz)	131.1	175	252.4	255.2	273.93	219.09	265.03	267.25	191	307.5
F₀ min (Hz)	131.1	134.51	227.16	201.94	232.03	219.09	264.72	209.90	190.37	223.86
F₀ max (Hz)	200.9	197.8	301.01	303.06	321.35	343.81	334.05	300.97	323.86	316.84
Pitch range (Hz) (F₀ max–F₀ min)	69.8	63.29	73.85	101.12	89.32	124.72	92.54	91.07	133.49	92.98

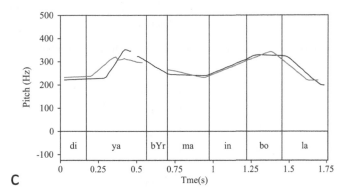

Figure 3. Intonation contours of *"Dia bermain bola"* in declarative (straight line) and interrogative read by ASD3.

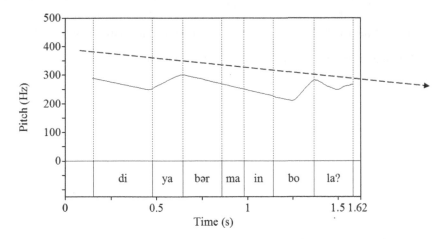

Figure 4. Intonation contour of interrogative sentence "Dia bermain bola" read by ASD4.

Furthermore, when the ASD children were asked to say the same sentences by imitating their speech therapist, their intonations to some extent are more adequate. The therapist said aloud the same sentences as in the reading task, and the ASD subjects had to imitate him immediately. Let us take a look at the following illustrations.

The pitch contour of the declarative sentence of ASD1 resembles the pitch contour of the therapist. Both contours have slight declinations and gently rise on the subject of the sentence "Dia" and the object "bola". The therapist has a pitch range ca 51.30 Hz, and the speaker ASD1 has a pitch range circa 53.92 Hz.

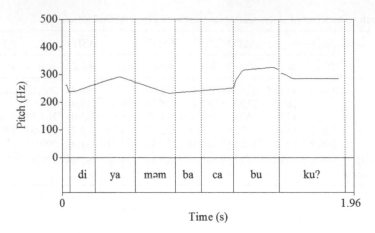

Figure 5. Intonation contour of the interrogative sentence "Dia membaca buku" read by ASD3.

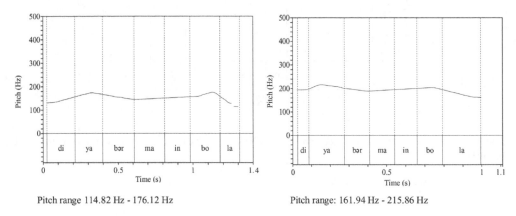

Pitch range 114.82 Hz - 176.12 Hz   Pitch range: 161.94 Hz - 215.86 Hz

Figure 6. Declarative sentence spoken by therapist (left) and imitation spoken by ASD1 (right).

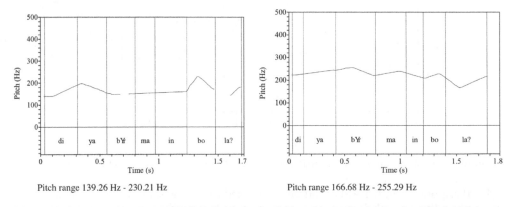

Pitch range 139.26 Hz - 230.21 Hz   Pitch range 166.68 Hz - 255.29 Hz

Figure 7. Interrogative sentence spoken by therapist (left) and imitation spoken by ASD1 (right).

When the therapist produced the interrogative sentence and the ASD subject imitated the therapist, we have the following contours.

The pitch contours as illustrated in Figure 6 do not look similar to each other; it is noticeable that the ASD subject could produce the interrogative sentence with a final rising tone.

The pre-final syllable "bo" is realized in both utterances with a pitch point, although the point made by the therapist is much sharper than that of the ASD speaker. It suggests that the ASD speaker can produce, to some extent, pitch movements. Furthermore, the pitch range of ASD1 (88.61 Hz) resembles the pitch range of therapist (90.95 Hz).

4 CONCLUSION

This experimental research reveals that the ASD children have difficulty in differentiating interrogative from declarative sentences when they were asked to read word sequences ending with a question mark and the same sequences ending with a full stop. They spoke with a relatively high tone, but with a narrow pitch range. Almost all of the sentences were read without obvious pitch movements. The declarative sentences are relatively flat, declination occurs infrequently, and the boundary tone has in most cases no remarkable fall. The pitch parameters of the interrogative sentences are almost similar to the pitch parameters of the declarative sentences. One of the ASD children realized the interrogative sentence with a similar melody to the declarative sentence. Nevertheless, the interrogative sentences have, on average, wider pitch ranges than the declarative sentences. A few interrogative sentences indicate the occurrences of pitch movements. These sentences have a rising tone on the final constituent.

The imitating task apparently was better performed than the reading task. The experiment with the therapist as role model shows that ASD children are familiar with imitation therapy. The autistic children were able repeat the intonation of the therapist adequately. The imitation task reveals a declarative pitch contour with a declination and a falling boundary tone, and the interrogative sentence is realized with a noticeable pitch rise at the end of the sentence.

REFERENCES

Aitken, K. et al. (1998) *Children with autism diagnosis and intervention to meet their needs*. London, Jessica Kingsley.

Attwood, T. (2008) *A complete guide to Asperger's Syndrome*. London, Jessica Kingsley Publishers.

Baltaxe, C. (1984) The use contrastive stress in normal, aphasic and autistic children. *Journal of Speech and Hearing Research*, 27, 97–105.

Baltaxe C.A. & Simmons, J.Q. (1975) Language in childhood psychosis. *Journal of Speech and hearing Disorders*, 40(4): 439–458.

Beckman, M.E. (1986) *Stress and non-stress accent*. Dordrecht/Cinnaminson, Foris.

Bezooijen, R. van (1984) *Characteristics and recognisability of vocal expression of emotion*. Dordrecht, Foris.

Boersma, P. & Weenink, D. (2015) *PRAAT, doing phonetics by computer, version 5.4.18*, [www.praat.org].

Chun, D.M. (2002) *Discourse Intonation in L2: From Theory and Research to Practice*. Amsterdam, John Benjamin.

Cutler, A. & van Donselaar, W. (2001) Voornaam is not (really) a homophone: Lexical prosody and lexical access in Dutch, *Language and Speech*, 44, 171–195.

Hefting, W. (1991) The effect of 'information value' and 'accentuation' on the duration of Dutch words, syllables, and segment, *Journal of the Acoustical Society of America*, 89, 412–424.

Eefting, W. & Nooteboom, S. G. (1991) The effect of accentedness and information value on word durations: a production and a perception study. *Proceeding of the 22nd International Congress of Phonetic Sciences* (pp. 302–305), Aix-en-Provence.

Gut, U. (2009) Introduction to English phonetics and phonology: [including CD]. Frankfurt, M: Lang.

Hubbard K. & Trauner, D.A. (2007) Intonation and emotion in autistic spectrum disorders, *J. Psycholinguist Res Mar*; 36(2), 159–173.

Ladefoged & Johnson (2011) *A Course in Phonetics. International Edition* (with CD-ROM), Boston, Wadsworth, Cengange Learning.

Mennen, I. (2007) Phonological and phonetic influences in non-native intonation. In: Trouvain, J. & Gut, U. (eds.) *Non-native Prosody: Phonetic Descriptions and Teaching Practice (Nicht-muttersprachliche Prosodie: phonetische Beschreibungen und didaktische Praxis)* (pp. 53–76). Mouton De Gruyter.

Nooteboom, S. (1997) The Prosody of Speech: Melody and Rhythm. In: W.J. Hardcastle & J. Lavers (eds) *The Handbook of Phonetic sciences*. Oxford, Blackwell. p. 640–673.

Shriberg, L.D., Paul, R., Mc Sweeny, J.L. Klin, A., Cohen, D.J. & Volkmar, F.R. (2001) Speech and prosody characteristics of adolescents and adults with high functioning autism and asperger syndrome. *Journal of Speech, Language, and Hearing Research*, 44, 1097–1115.

Wing, L. & Gould, J. (1979) Severe impairments of social and associated abnormalities in children: Epidemiology and classification. *Journal of Autism and Developmental Disorder*s, 9, 11–29.

Yip, M. (2002) *Tone*. Cambridge, Cambridge University Press.

Cultural Dynamics in a Globalized World – Budianta et al. (Eds)
© *2018 Taylor & Francis Group, London, ISBN 978-1-138-62664-5*

Characteristics of word duration in children with autism spectrum disorder

T.W.R. Ningsih, F.X. Rahyono & Ayesa
Department of Linguistics, Faculty of Humanities, Universitas Indonesia, Depok, Indonesia

ABSTRACT: This study focuses on the prosodic aspect of Autism Spectrum Disorder (ASD) associated with characteristics of the disorder. The subjects are male ASD speakers aged 9 to 12 years old. The data was collected using reading and imitation techniques and subsequently processed using the PRAAT program. The results show that there are variations in duration among ASD speakers. The duration from the reading technique tended to be longer in comparison to the imitation technique, and the size of the differences varied greatly. Both techniques, reading and imitation, produced longer durations for interrogative sentences than declarative sentences. Interrogative sentences produced by ASD speakers tended to be longer on the final syllable of the object constituent. This indicates that ASD speakers have difficulties in detecting the end of a speech boundary. Meanwhile, the short duration produced by ASD speakers on a subject constituent indicates that they have a tendency to fail to understand the function of a subject constituent in a sentence.

1 INTRODUCTION

Duration can be defined as a series of articulators and the time dimension of an acoustic signal. Lehiste (1970) argues that duration can be associated with quantity if it functions as an independent variable in the phonological system of a language. Duration is related to the determination of an articulator's movement and its measurable channel. The speech duration can be influenced by several phonetic factors, while the duration of a segment can be determined by the nature of the segment itself, namely articulation point and behavior. According to Sugiyono (2003b), duration is the time span required for the realization of a sound segment that is measured in milliseconds. If the segment is in the form of a sentence, the time span is usually called temporal. Structural temporal, which is also known as duration, is a set of rules that determines the duration of a speech pattern.

Autism is a pervasive developmental disorder whose symptoms appear before the age of three. Children who show autism symptoms (ASD) often have a very slow speech development. Symptoms that can be identified include an underdeveloped vocabulary, for example the ability to give meaning to an object is not developed. A verbal symptom especially prominent among ASD speakers is their difficulty to understand the meaning of a word during a conversation. Incorrect pronunciation is another symptom. Even if the vocabulary is well developed, it is often not used for communication. Word repetition, echolalia, and incoherent grammar are several other symptoms. In addition, children with autism show qualitative disturbances in social interaction, play, and communication that are often accompanied with repetitive behaviors and stereotyped patterns (American Psychiatric Association, 1994; Klin, et al., 2000).

ASD speakers are also inept in social communicative functions. Researchers suggest that the disorder also affects the speech prosody. McCann et al. (2007) believe that the accuracy level of prosodic production is related to the language skill level. Peppé *et al.* (2010) also state that prosodic functions are related to the level of communicative function. ASD speakers may have linguistic abnormalities (Ploog et al., 2009), which eventually will affect the perception and production of their prosody. Pecora (2009) discovered that children with ASD tend to speak in higher tones and with longer durations. This finding is also supported by research that was conducted by Bonneh et al. (2010). They proved that children with ASD have a wider tone range and greater variations. Diehl (2012) investigated ASD speakers' difficulties in producing appropriate durations, responding and mimicking the prosody, and having prosodic perception.

Speech production disorders are associated with speech reflective motor differences particularly stress and duration. Some researchers reported that ASD speakers pronounce words and sentences with different stresses (Paul et al., 2005; Shriberg et al., 2001). This is perhaps related to ASD speakers' difficulties to understand and process the theory of mind (McCann & Peppe, 2003). Other speech disorders involve speech duration, frequent pauses, and the length of the pauses (Zajac et al., 2006). Paul et al. (2005) developed a hypothesis arguing that ASD speakers are unable to produce normal prosody and have social skill impairment and poor communication.

The ASD speakers' prosody disorder is identified as a communication disorder (Baltaxe & Simmons, 1985). However, this disorder is often overlooked, despite the fact that prosody impairment is considered a main obstacle to social group acceptance (Shriberg et al., 2001). Moreover, as prosody disruption can occur at an unpredictable duration of time, by then other disorders will have already been resolved. This is supported by Paul et al. (2005) who state that if ASD speakers have difficulties in adapting to social environment and social imitation, the prosody they produce is atypical.

Based on previous studies, one of the ASD symptoms is a deficit in the production of duration. The duration produced through a spontaneous speech is declared unstable. The duration produced by ASD speakers using reading techniques is also unstable, and there is a tendency to expand the duration of the final syllable. This trend is associated with production deficit of final speech (Shriberg et al., 2001). However, so far there has been no detailed explanation about the indications of the disorder. Several tests using certain techniques need to be conducted in order to identify the indications of the disorder.

This study used two techniques of data collection, namely reading and imitation techniques. Both of these techniques enabled a comparison of speech production among ASD speakers. As already known, imitation is a technique used by therapists to provide speech therapy to ASD speakers. Similarly, the echolalia symptom is shared among most ASD speakers. Researchers suspect that the imitation technique yields more significant results than the reading technique. It is expected that through these two techniques, the capabilities of ASD speakers to perform receptive and productive speech can be revealed.

2 RESEARCH METHODS

The subjects were male ASD speakers aged 9 to 12 years old. They were diagnosed with autism disorder when they were 3 years old and received treatment in a rehabilitation center for children with special needs at an early age. During data collection, the subjects were students in the same elementary school but at different grade levels. Subjects A (9 years old), B (9.7 years old), and C (9.8 years old) were 3rd grade students, while subject D (11.4 years old) was a 5th grade student, and subject E (12 years old) was a 6th grade student.

This study is a phonetic experimental study using the subjects' utterances as data. Utterances produced by each subject were recorded using a special recording device in a sound-proof room. The recorded utterances were saved as wav files and then processed using the PRAAT program (Boersma & Weenink, 2001). This study applied the IPO approach (Instituut voor Onderzoek Perceptie) ('t Hart Collier & Cohen, 1990; Rahyono, 2003). IPO approach includes three main activities, namely (1) experimental speech production, (2) the speech acoustic analysis, and (3) the experimental speech perception test.

To obtain the data, two techniques were used, namely reading and imitation techniques, and the following instruments were employed.

Instruments 1:	1.a.	*Dia* (he)	*membaca* (reads)	*buku* (a book). (Declarative sentence)
		subject	verb	object
	1.b.	*Dia* (he)	*membaca* (reads)	*buku* (a book)? (Interrogative sentence)
		subject	verb	object
Instruments 2:	2.a.	*Dia* (He)	*bermain* (plays)	*bola* (soccer). (Declarative sentence)
		subject	verb	object
	2.b.	*Dia* (He)	*bermain* (plays)	*bola* (soccer)? (Interrogative sentence)
		Subject	verb	object

2.1 Data screening techniques:

1. The reading technique

The subjects were asked to read the text (research instruments) three times using the reading technique. The data was screened three times considering that it was selected based on the degree of speech naturalness. The utterances were recorded using a high-quality recorder.

2. The imitation technique

The subjects were accompanied by a therapist to carry out imitation. The therapist read the text, and the subjects were asked to imitate. Similar to the reading technique, the imitation was carried out three times. The number of utterances obtained from the reading technique is 45 ($3 \times 5 \times 3$) utterances of declarative sentences (instrument 1 and instrument 2), and 45 utterances of interrogative sentences (instrument 1 and instrument 2). The imitation technique generated 45 utterances of declarative sentences (instrument 1 and instrument 2) and 45 utterances of interrogative sentences (instrument 1 and instrument 2).

2.2 Data selection test

The data selection test was conducted in order to obtain a natural primary contour of declarative and interrogative utterances. The selected data were 45 utterances produced by 5 research subjects for each instrument. The data selection test provided information of the score of each utterance. The data consisted of two declarative sentences and two interrogative sentences for each subject. The imitation technique generated the same number of data.

2.3 Acoustic analysis

After the data selection test was conducted, the primary contour of declarative sentences and interrogative sentences was obtained. The data was then processed in the acoustic analysis stage. The first phase is the segmentation of the utterance. The segmentation is useful to determine the boundaries of the constituencies to be analyzed, and it is useful as a basis for measuring the duration of each constituent.

3 RESULTS AND DISCUSSION

Based on the results of the acoustic analysis, the values of F0 and duration were obtained. A sample of duration calculation using the PRAAT is presented in Figure 1.

Figure 1 illustrates the acoustic analysis and duration using PRAAT, '*Dia membaca buku?* (He read the book?)'. It also illustrates the respective value of the duration of each syllable. Syllable '*di*' has a duration value of 180 ms, while syllable '*ya*' (94 ms), syllable '*mem*' (174 ms), syllable '*ba*' (174 ms), syllable '*ca*' (140 ms), syllable '*bu*' (248 ms), and syllable '*ku*' (331 ms). The total duration in Figure 1 is 1444 ms.

Based on the selected data, the values of F0 and speech duration produced by each subject using two different techniques were obtained.

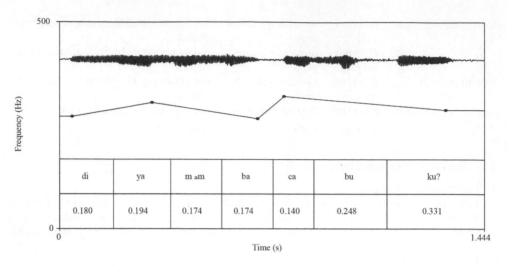

Figure 1.   Duration measurement in PRAAT.

3.1   *The reading technique (Instrument 1)*

The speech durations produced by the subjects were analyzed acoustically. The results of duration measurements in the reading technique are depicted in Table 1 and Table 2. Table 1 shows that the longest duration in declarative sentences was produced by subject A at 2052 ms, and the shortest duration was produced by subject E at 1219 ms. In interrogative sentences, the longest duration was produced by subject B at 2408 ms. The lowest duration was produced by subject E at 937 ms. Based on the analysis of the durations of declarative and interrogative sentences, subject E produced the shortest duration, and it was consistently on the subject constituents. The shortest durations on the subject constituent, i.e., syllable 'di' and syllable 'ya', were produced by subject E with the subject constituent being shorter than the other constituents.

The duration of the instrument *'bola/ball'* shows that declarative sentences had the longest duration, which was produced by subject B at 1896 ms on the subject constituent, and the shortest duration was produced by subject E at 945 ms on the object constituent. In interrogative sentences, the longest duration was produced by subject A at 1975 ms on the predicate constituent, and the shortest duration was produced by subject E at 236 ms on the predicate constituent.

Based on the duration analysis of instruments 1 (*buku*/book) and instruments 2 (*bola*/ball), subject E shows the same characteristics. E produced durations which were shorter than the durations produced by the other subjects. In the second instrument (*bola*/ball), subject E produced the shortest duration on the predicate constituent, which is at the beginning of the syllable 'ber'.

3.2   *Imitation technique*

The results of duration measurement in the imitation technique are shown in Table 3 and Table 4.

The data show that the longest duration on declarative sentences (*buku*/book) using the imitation technique was produced by subject A at 1313 ms on the object constituent. The shortest duration was produced by subject E at 521 ms on the subject constituent. In interrogative sentences, the longest duration was produced by subject A at 2569 ms on the object constituent, and the shortest duration was produced by E at 52 ms on the subject constituent. Subject E produced the shortest duration on the subject constituent.

In a declarative sentence mode, the shortest duration was produced by subject A at 1611 ms on the object constituent, and the shortest duration was produced by subject E at 605 ms

Table 1. Duration with the reading technique (Instrument 1).

Subject	Age	Instrument DCL/INT	Syllables						
			di	ya	mem	ba	ca	bu	ku
A	9	DCL	285	279	201	207	188	326	566
		INT	233	270	268	173	219	312	499
B	9.7	DCL	477	179	217	237	228	202	406
		INT	447	525	273	182	243	437	301
C	9.8	DCL	222	285	223	146	267	193	234
		INT	197	284	117	303	689	258	359
D	11.4	DCL	109	178	167	86	181	171	361
		INT	97	109	235	79	134	197	261
E	12	DCL	159	53	420	65	114	145	263
		INT	94	112	88	87	162	135	259

Table 2. Duration with the reading technique (Instrument 2).

Subject	Age	Instrument DCL/INT	Syllables						
			di	ya	ber	ma	in	bo	la
A	9	DCL	169	394	137	273	296	232	307
		INT	169	557	159	267	211	204	408
B	9.7	DCL	456	255	180	142	260	187	416
		INT	477	165	213	213	160	227	256
C	9.8	DCL	247	208	141	205	217	210	522
		INT	346	304	167	260	137	231	565
D	11.4	DCL	125	287	152	141	152	134	474
		INT	97	109	235	79	134	197	261
E	12	DCL	104	99	55	71	137	132	347
		INT	128	880	99	112	113	137	236

Table 3. Duration with the imitation technique (Instrument 1).

Subject	Age	Instrument DCL/INT	Syllables						
			di	ya	mem	ba	ca	bu	ku
A	9	DCL	250	270	197	192	146	233	147
		INT	40	147	207	174	122	288	303
B	9.7	DCL	20	189	208	185	175	226	217
		INT	183	168	182	193	122	275	243
C	9.8	DCL	289	139	160	229	158	292	401
		INT	197	284	117	303	689	258	359
D	11.4	DCL	148	133	169	127	140	123	392
		INT	50	121	179	128	152	245	298
E	12	DCL	62	100	68	89	146	385	226
		INT	52	38	81	127	105	129	106

on the subject constituent. In interrogative sentences, the longest duration was produced by subject A at 2858 ms on the object constituent, and the shortest duration was produced by subject E at 953 ms on the predicate constituent. The results of duration measurement show that subject E produced the shortest duration on the two types of sentences.

The overview of the total duration comparison between the reading technique and imitation technique using instrument 1 is presented in Figure 2.

Table 4. Duration with the imitation technique (Instrument 2).

Subject	Age	Instrument DCL/INT	Syllables						
			di	ya	ber	ma	in	bo	la
A	9	DCL	278	254	120	174	216	271	298
		INT	205	103	172	196	631	262	289
B	9.7	DCL	124	270	395	101	216	173	261
		INT	161	118	153	196	156	183	375
C	9.8	DCL	245	116	163	241	173	213	458
		INT	305	355	164	275	142	219	438
D	11.4	DCL	80	190	133	132	117	133	317
		INT	129	305	294	232	224	223	422
E	12	DCL	58	97	97	49	136	84	84
		INT	97	113	134	90	146	112	261

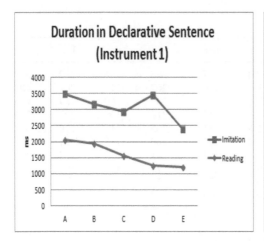 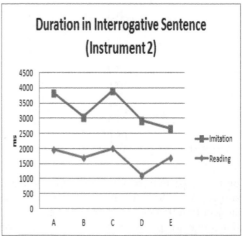

Figure 2. Duration in instrument 1 'buku' (book).

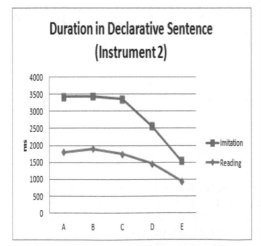

Figure 3. Duration in instrument 2 'bola' (ball).

The comparison of the total duration produced by each subject using instrument 2 is presented in Figure 3.

The results of the speech duration calculation are presented from Table 1 to Table 4. Each table describes the durations of utterances produced by the subjects using two techniques on declarative sentences and interrogative sentences. The results show that subject E produced the shortest duration on both reading and imitation techniques. There was no significant difference between the duration of the declarative sentence mode obtained from the reading technique and that from the imitation technique. In interrogative sentences, the highest duration was produced on the object constituent. In the second instrument, the results show that the longest duration in both reading and imitation techniques was on the object constituent, while the shortest duration was on the subject constituent.

This study shows that the measurement results of contrastive durations using reading and imitation techniques show different results. The difference in durations was calculated based on the duration per syllable on each constituent, namely subject, predicate, and object, and the total duration. Based on the four instruments, the longest duration was on the interrogative sentence mode, which is characterized by extending the vowel in the final syllable. Declarative sentences are produced in a shorter duration than that of interrogative sentences. The results of the duration measurements reveal abnormality in the length of the duration which indicates a production problem among ASD speakers.

4 CONCLUSION

The durations produced by ASD subjects show various results, especially from the reading technique. In this technique, the duration produced by one of the subjects is considered to be very short, especially when producing the subject constituent. This indicates that the subject did not focus on the subject constituent. Meanwhile, another ASD subject produced utterances with a longer duration on the object constituent. This also indicates that the ASD subject has difficulties to determine speech boundaries. Based on the measurement results, it can be concluded that the duration production of each subject varies widely, which is presumably because the subjects did not have enough knowledge about how to produce declarative sentences and interrogative sentences. Another probable explanation is the emotional instability of ASD speakers in producing speech that leads to instability in their speech duration production.

REFERENCES

American Psychiatric Association. (1994) *Diagnostic and Statistical Manual of Mental Disorders (4th ed.)*. Washington, DV: Author.

Boersma, P. & Weenink, D. (2001) PRAAT, a system for doing phonetics by computer. *Glot International*, 5 (9/10), 341–345.

Bonneh, Y.S., Levanon, Y., Dean-Pardo, O., Lossos, L., & Adini, Y. (2011) Abnormal speech spectrum and increased pitch variability in young autistic children. *Frontiers in Human Neuroscience*, 4, 1–7. doi:10.3389/fnhum.2010.00237.

Diehl, J.J. & Paul, R. (2012) Acoustic differences in the imitation of prosodic patterns in children with autism spectrum disorders. *Research in Autism Spectrum Disorders*, 6, 123–134.

Hart, J., Collier, R. & Cohen, A. (1990) *A Perceptual Study of Intonation: An Experimental-Phonetic Approach to Speech Melody (1990)*. Cambridge, Cambridge University Press.

Klin, A. & Volkmar, F.R. (2000) Treatment and intervention guidelines for individuals with Asperger syndrome. In: Klin A, Volkmar FR, Sparrow SS, eds. *Asperger syndrome*. New York, Guilford Press; pp. 340–366.

Lehiste, I. (1970) *Suprasegmentals*. Oxford, England: MIT Press.

McCann, J. & Peppe, S. (2003) Prosody in autism spectrum disorders: a critical review. *International Journal of Language & Communication Disorders*, 38(4), 325–350. doi:10.1080/1368282031 000154204.

McCann, J., Peppé S., Gibbon, F.E., O'Hare, A., & Rutherford, M. (2007) Prosody and its relationship to language in school-aged children with high-functioning autism. *International Journal of Language & Communication Disorder*. Nov–Dec; 42 (6), 682–702.

Noterdaeme, M., Mildenberger, M., Minow, F. & Amorosa, A. (2002) Evaluation of neuromotor deficits in children with autism and children with a specific speech and language disorder. *Europe Child Adolescent Psychiatry* 11, 219–225.

Paul, R., Augustyn, A., Klin, A. & Volkmar, F. (2005) Perception and production of prosody by speakers with autism spectrum disorders. *Journal of Autism & Developmental Disorders*, 35, 205–220.

Paul, R., Shriberg, L. D., McSweeny, J., Cicchetti, D., Klin, A., & Volkmar, F. 31(2005b) Brief report: Relations between prosodic performance and communication and socialization ratings in high functioning speakers with autism spectrum disorders. *Journal of Autism and Developmental Disorders*, 35(6), 861–869. doi:10.1007/s10803-005-0031-8.

Pecora, L. (2009) *Acoustic and Prosodic Characteristics of Spontaneous Speech in Autism*. Unpublished Senior Honors Thesis, University of Massachusetts, Amherst, MA.

Peppé, S., Cleland, J., Gibbon, F., O'Hare, A. & Castilla, P. M. (2010) Expressive prosody in children with autism spectrum conditions. *Journal of Neurolinguistics*, 24, 41–53.

Ploog, B.O., Banerjee, S. & Brooks, P. J. (2009) Attention to prosody (intonation) and content in children with autism and in typical children using spoken sentences in a computer game. *Research in Autism Spectrum Disorders*, 3(3), 743–758. doi:10.1016/j.rasd.2009.02.004

Rahyono, F.X. (2003) *Intonasi Ragam Bahasa Jawa Keraton Yogyakarta (Kontras) Deklarativitas, Interogativitas, dan Imperativitas* (Disertasi) (Intonation of Keraton Yogyakarta Javanese Language (Kontras) Declaravity, Interogativity, and Imperativity (Dissertation). Jakarta, Universitas Indonesia.

Shriberg, D.L., Paul, R., McSweeny, L.J., Klin, A., Cohen, J.D. & Volkmar, F.R. (2001) Speech and prosody characteristics of adolescents and adults with high-functioning autism and Asperger Syndrome. *Journal of Speech, Language, and Hearing Research*, 44, 1097–1115. doi:10.1044/1092-4388(2001/087).

Sugiyono (2003) *Pedoman Penelitian Bahasa Lisan: Fonetik* (Guidance of Oral Language Research: Phonetics). Jakarta, Pusat Bahasa.

Zajac, D.J., Roberts, J.E., Hennon, E.A., Harris, A.A., Barnes, E.F. & Misenheimer, J. (2006) Articulation rate and vowel space characteristics of young males with fragile X syndrome: Preliminary acoustic findings. *Journal of Speech, Language, and Hearing Research*, 49(5), 1147–1155.

Cultural Dynamics in a Globalized World – Budianta et al. (Eds)
© 2018 Taylor & Francis Group, London, ISBN 978-1-138-62664-5

Experimental approach of perception toward autism spectrum disorder intonation

F.X. Rahyono, T.W.R. Ningsih, Ayesa & M. Aziza
Department of Linguistics, Faculty of Humanities, Universitas Indonesia, Depok, Indonesia

ABSTRACT: This study aims at finding intonation patterns of utterances produced by speakers with Autism Spectrum Disorder (ASD). The acoustic units used to mark the intonation pattern are: 1) pitch movement, 2) final intonation, and 3) baseline slope. The data were uttered sentences in declarative and interrogative modes produced by 11-year-old children with ASD and TD (Typical Development). This is an experimental research using an IPO approach and a PRAAT program to analyse acoustic data. Based on real contour utterance patterns produced by the speakers with ASD and TD, the synthesis and substitutions of pitch movement between the ASD and TD intonations were made. The stimuli of pitch movement synthesis and substitution were then tested to know the meaning of acceptability through perception test and statistics using the Likert scale. The perception test results show that the speakers with ASD did not produce different utterance intonation between declarative and interrogative modes. The pitch movement, in both declarative and interrogative modes tends to be flat and show a pitch range under three semitones.

1 INTRODUCTION

Each language speaker has the ability to use the language in order to communicate with each other. Such a speaker has certain conventional sound systems, both in the segmental and the supra-segmental systems called prosody, which they collectively agree with. On a certain limit, even speakers of a young age have mastered the prosody of their mother tongue. Baltaxe and Simmons (1985), as cited by Azizi, state that prosody is an important variable from a children's language acquisition perspective. The prosody ability, which consists of emotional, and pragmatics utterances might be less stable for autism children compared to children with no learning difficulties since the latter are able to improve their awareness and prosodic abilities (Zahra Azizi, 2016). Peppe et al. (2007) state that this knowledge has improved the understanding of communication aspects, such as utterances filled with emotion, behaviour, and social signs.

Mozziconacci (1998) explains that prosody characteristics provide information about one's identity, physical and emotion conditions, attitudes, as well as utterances and act situations. The problems arise when the prosody variation that significantly deviates from the conventional prosody system is found. According to Mozziconacci, the speaker's health and emotional conditions are important factors to determine the prosody profile produced. A speaker with ASD is an example of those who produce deviant prosody characteristics from acceptable or natural prosody characteristics. The deviant prosody characteristics occur due to cognitive disorders suffered by the speakers with autism which cause intentional and comprehension constraints as well as other cognitive disorders (Baron-Cohen, Leslie, and Frith (1985); Frith (1989); Happ' (1993); Baron-Cohen, Tager-Flusberg, and Cohen (1993)).

Fay and Schuler (1980) describe a subset of autistic individuals who used a singsong mode to communicate rather than in a flat pattern. Goldfarb, Braunstein, and Lorge (1956) and Pronovost, Wakstein, and Wakstein (1966) found unusually high fundamental frequency levels in autistic speakers. Other voice disorders, such as hoarseness, harshness, and hyper nasality, have also been identified (Pronovost et al., 1966). Aitken et al. (1998) explain that autism is a pervasive development disorder (PDD), which incorporates abnormal development in social interactions, communication, certain behaviours, stereotypes, and repeated actions. Furthermore, autism starts from 36 months old (DSM-IV 1994), and nowadays social failures are seen as the main symptom of autism by many people (Baron-Cohen 1989).

Peppé (2009) explains that there are two expressive prosody disorders in speakers with ASD: the 'function' prosody disorder and 'pattern' prosody disorder. The 'function prosody' disorder causes the speakers with ASD to be unable to differentiate utterance intonation when only prosody is used. Based on the research, Peppé (2009) used declarative sentences contrasted with interrogative sentences. The result (Peppé, 2009) shows that speakers with ASD have failed to differentiate the contrast of the two sentence modes.

This study focuses on intonation, which is among one of the prosodic expressions. Intonations are ensembles of pitch variations in utterances caused by varying periodicity in the vibration of the vocal cord (t'Hart, Collier, and Cohen, 1990). The intonations are controlled by intonation sub-components (called prosody), in which the realizations are the forms of pitch, accent, loudness, duration, rhythm, and sound quality. These prosody features form intonation patterns. In a language intonation research, intonations serve as supra segmental; phonetic features are used to deliver lexical meanings structurally (Ladd, 1996). Based on Pierrehumbert (1980) who states that intonations are parts of structured linguistic prosody, and in American English, the intonations basically consider pitch stress, sentence accent, and pitch range.

Referring back to the prosody variation problems stated above, a language has productive characteristics. Rising and falling pitch movements that represent intonation have consequences to the variations of pitch contours. The sentence mode differences spoken have resulted in different intonation contours. Therefore, every speaker can produce different intonation contours. Pike (1972) states that variations of pitch movements, which build the intonation contour of a sentence, are not random variations. The variations of patterns, as well as their changes, are set in certain orders. The problem arises here when people with ASD produce the variations.

The debatable arguments of ASD speakers' different prosody production compared to normal speakers are interesting to be studied. Therefore, the researchers conducted an experiment to find out the intonation indicators or parameters produced by speakers with ASD and the prosody produced by normal speakers. This study was conducted to analyse two important things in the intonations: production of acoustic analysis, and utterance perception of informant with ASD and normal speakers. The intonation features as the product of the utterances, used PRAAT, which is F_0 score, pitch variation, excursion, and pitch ranges from the two informants with different verbal development ability. A perception test was conducted to assess the utterance production acceptability by speakers with ASD.

This study focuses on measurement of the intonation components consisting of: 1) pitch movement; 2) final intonation, and 3) baseline slope. Therefore, this study aims to answer the following research question: "How is the acceptability intonation of speakers with ASD as natural utterances?" Acoustic parameters used to mark the ASD intonation patterns were: pitch movements, pitch variations, and baseline slope intonation contours.

2 METHODOLOGY

This is an experimental study on the intonation characteristics in the speakers with ASD. This study used *IPO* scale (*Instituut voor Perceptie Onderzoek*), which is a phonetics experimental model, which examines signal acoustic analysis to statistics analysis of the utterance acceptability ('t Hart et al., 1990). The main procedures are utterance production, acoustic utterance analysis, hypothesis making and experiment design, stimulus making, perception test, and statistics analysis.

The data of this study are uttered sentences produced by speakers with ASD and TD. Both speakers were requested to say the same sentences: 1) *Dia membaca buku*, and 2) *Dia bermain bola.* Each sentence was stated in two modes: in declarative and interrogative modes. The data of both sentences were produced by a reading technique. In speakers with ASD, the utterance production also used an imitation technique.

The acoustic analysis step is a process of digitizing the utterances, which consists of two activities; utterance segmentation and utterance synthesis. The acoustic analysis was conducted using PRAAT* software program. In this step, the data acoustic collected are in the forms of fundamental frequency scores, pitch, duration, as well as pitch movement features and contour patterns.

The perception test started from designing stimulus based on the hypothesis and experiment design. The respondents of this study consisted of 219 normal adults (students) with TD. The main point of this method is the perception test. The intonation patterns, which are the point of this, study are collected from data utterances that have been checked through the perception test. The perception or acceptability test was conducted to find out whether the hypothesis works on the production results of real utterances and re-synthesis utterances.

The perception test of stimulus acceptability was conducted using the Likert scale and counted based on the descriptive statistics analysis. Next, the perception test was counted using the statistics analysis by software SPSS program ver. 17. The analysis was done with the Wilcoxon test.

3 RESULTS AND DISCUSSION

The result of the experimental production seems to indicate that utterance intonation of declarative modes shows a contour pattern, which significantly differs between ASD and TD utterance contours. Figure 1 shows that the ASD utterance on the sentence "*Dia membaca buku*" was marked by a fundamental frequency (F_0) variation that is relatively flat in the whole utterance.

The TD utterance is marked by a decline—in that the fundamental frequency (F_0) variation tends to decline from the beginning to the end of the utterance. Perceptually, the initial pitch of TD utterance at 19.79 st declines to 13.2 st at the end of the utterance. Based on the pitch produced, ASD utterance starts at 9.77 st and ends at 6.09 st in the final syllable /ku/. However, the final pitch is insignificant as a referent to form the baseline slope. The final pitch contour in the syllable /ku/ creates a declining pitch movement from 11.6 st to 6.09 st. Overall, the baseline slope that generates the flat characteristics was taken from the syllable /di/ in *dia* at the pitch 9.77 st and the syllable /bu/ in *buku* at the pitch 9.66 st. Hence, the baseline slope in ASD contour utterance is flat.

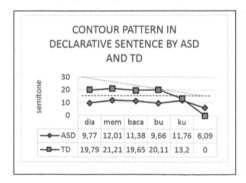
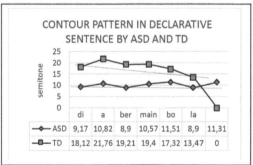

Figure 1. Contour pattern in declarative mode. Figure 2. Contour pattern in declarative mode.

---

*PRAAT is software designed by Paul Boersma and David Weenink (*Praat, a system for doing phonetics by computer, version 3.4.*). University of Phonetics Sciences of the University of Amsterdam, Report 132. 182.

Figure 2 below shows that ASD utterance (*Dia bermain bola*) is also marked by the fundamental frequency (F_0) variation, which is relatively flat throughout the utterance. Similar to the sentence *Dia membaca buku*, the TD contour is marked by a decline, from 18.12 st, it declines to 13.47 st at the end of the utterance. The ASD pitch starts at 9.17 st and ends at 8.9 st in the final syllable /la/. The pitch range is 0.18 st and it does not significantly change the flat characteristics of the baseline slope. In contrast with the contour in the sentence *Dia membaca buku*, the final contour of *Dia bermain bola* creates an inclining pitch movement of the syllable *la* from 8.9 st to 11.31st. This data production shows inconsistent contour pattern of ASD.

Based on the four selected contour patterns of declarative mode produced by ASD and TD speakers (cf. Figures 1 and 2), the hypothesis built on this study is: "intonation contour pattern on declarative modes of ASD utterance tends to be flat". The first experiment was conducted to test the acceptability intonation of the declarative utterance of the ASD compared to the acceptability intonation of TD utterance. The stimuli perception test was designed from the manipulation contour as follows:

1. The contour of the ASD was replaced by the contour of TD;
2. The contour of the TD was replaced by the contour of ASD.

Table 1 shows the result of experiment 1. Stimuli (1) and (5), the original contour of ASD for the sentence *Dia membaca buku* and *Dia bermain bola* resulted in a continuously a low score at 2.14 ('not good') and 1.89 ('not good'). Stimuli (2) and (6), the original contour of TD on the second sentence, scores continuously 3.6 ('good' to 'very good') and 3.3 ('good'). Stimulus (3), the contour of the ASD whose pattern has changed into the TD contour resulted in a mean score of 2.31. It means that the score increased from 2.14 to 2.31. Although this score increase is insignificant to change the level from 'not good to good', at least there is an increasing acceptability stimulus of ASD utterance after the contour pattern has been changed into a TD contour pattern. Stimulus (7), the sentence *Dia bermain bola* resulted in a mean score of 1.84. In this stimulus, there is no increase in acceptability score, even if it was lower by 0.05, since this was deemed insignificant.

Stimulus (4), the contour of the TD, which was replaced by the ASD contour pattern has a decreasing acceptability score, yet it insignificantly changes from 'good' to 'less good.' In the stimulus (8), the acceptability score decreased significantly from 'good' to 'not good.'

Experiment 2 was conducted to test the manipulation result of the F_0 slope. The modification experiment of this baseline slope was proposed to modify the intonation contour from the flat characteristics to the intonation contour, which has declining characteristics. The intonation contour on predicate and object declined to 1, 2, and 3 semitones. There are four stimulus utterances in this experiment, they are: 1) original utterance, 2) modification utterance declined to 1st (min 1st), 3) modification utterance declined to 2 st (min 2 st), and 4) modification utterance declined to 3 st (min 3 st).

Table 1. Mean value of declarative stimulus.

Number	Speaker	Sentence	Stimulus	Mean value
(1)	ASD	*Dia membaca buku*	Original contour	2.14
(2)	TD	*Dia membaca buku*	Original contour	3.6
(3)	ASD	*Dia membaca buku*	The contour of the ASD replaced by the contour of TD	2.31
(4)	TD	*Dia membaca buku*	The contour of the TD replaced by the contour of ASD	3.14
(5)	ASD	*Dia bermain bola*	Original contour	1.89
(6)	TD	*Dia bermain bola*	Original contour	3.3
(7)	ASD	*Dia bermain bola*	The contour of the ASD replaced by the contour of TD	1.84
(8)	TD	*Dia bermain bola*	The contour of the TD replaced by the contour of ASD	1.98

The perception test toward the four stimulus utterances resulted in the mean scores as shown in Table 2 below.

The stimulus of original utterance, marked by a flat characteristics in the baseline slope, which is the pitch line dragged from the initial utterance in the syllable [di] from the word *dia* 'he/she' and the syllable [bu] on *buku* 'book', received the lowest mean score (2.23, meaning 'not good'). Continuously, the baseline slope showed a decline with a score acceptability of 2.43, on the stimulus min 1st, 2.63 on the stimulus min 2 st, and 2.26 on the stimulus min 3 st. The highest mean score from the stimulus utterance min 2 st indicates that the decline on this utterance will be ideal if the end of the utterance is 2 st. If the final decline is more than 2 st, on the stimulus min 3 st, then the utterance will be unacceptable.

Experiment 3 was conducted to test the intonation acceptability of the interrogative ASD utterances by comparing them to TD utterances. The stimuli perception test was designed from the following manipulation contour:

1. The contour of the ASD was replaced by the contour of TD;
2. The contour of the TD was replaced by the contour of ASD.

The result of the experimental production seems to indicate that utterance intonation of the interrogative modes shows the contour pattern is significantly different from ASD to TD utterance contour. Figures 4 and 5 show that ASD utterance is marked by a fundamental frequency (F_0) variation that is relatively flat throughout the utterances. This contour pattern indicates that the flat pattern of a baseline slope in the declarative utterance is the same as the interrogative utterance. In the interrogative sentence, though insignificant, there is a tendency of a declining pattern. The TD contour pattern of interrogative modes is marked by an inclination, which is a tendency of the fundamental frequency (F_0) variant that increases from the beginning to the end of the utterance. However, from the data, the inclining pitch is found only in the final contour. This inconsistency appears in the final syllable of ASD pitch contour in the interrogative modes. In the sentence of *Dia membaca buku?* The pitch

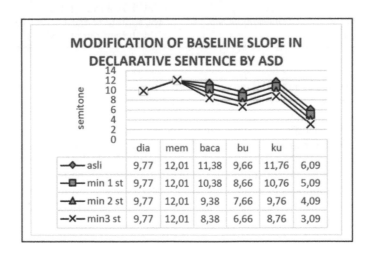

Figure 3.   Modification of the baseline slope.

Table 2.   Mean values of baseline slope modification.

No	Stimulus utterance	Mean
1	original utterance	2.23
2	minus 1 st	2.43
3	minus 2 st	2.63
4	minus 3 st	2.26

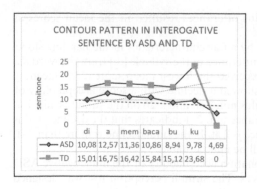

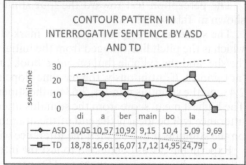

Figure 4. Contour pattern in interrogative mode. Figure 5. Contour pattern in interrogative mode.

Table 3. Mean value of interrogative stimulus.

Number	Speaker	Sentence	Stimulus	Mean value
(1)	ASD	*Dia membaca buku*	Original contour	1.9
(2)	TD	*Dia membaca buku*	Original contour	3.2
(3)	ASD	*Dia membaca buku*	The contour of the ASD replaced by the contour of TD	2.46
(4)	TD	*Dia membaca buku*	The contour of the TD replaced by the contour of ASD	1.57
(5)	ASD	*Dia bermain bola*	Original contour	2.05
(6)	TD	*Dia bermain bola*	Original contour	3.38
(7)	ASD	*Dia bermain bola*	The contour of the ASD replaced by the contour of TD	2.46
(8)	TD	*Dia bermain bola*	The contour of the TD replaced by the contour of ASD	2.02

movement in the final syllable /la/ moves down, but in the sentence of *Dia bermain bola?* The final pitch movement of the syllable/la/ is up and down.

Table 3 shows the result of experiment 3. Generally, the deviation of the ASD contour pattern in both declarative and interrogative modes appears in the fundamental frequency (F_0) variation which creates a flat baseline slope and pitch movement in the final contour. In the ASD interrogative contour, although insignificant, the decline can be seen. In the TD utterance, there is a significant difference between the contour pattern of declarative and interrogative modes marked by a variation of F_0 patterns. The contour of declarative modes is indicated by a decline, while in the interrogative modes it is marked by an inclination. Table 2 shows that the stimulus utterances of ASD, both real and modified, resulted in low scores.

Figures 6 and 7 below are the contour patterns of declarative and interrogative modes created by the experimental production with the imitation technique. The pitch movement throughout the contour indicates significant differences between the contours of declarative and interrogative modes. It means that speakers with ASD may actually realize the different intonation of declarative and interrogative modes through the imitation. Yet, the flat baseline slope patterns, as well as the inconsistent pitch movement patterns, appear in the contour patterns. Therefore, the utterances of the speakers with ASD tend to be flat with uncertain pitch variations.

Table 4 below shows the mean values acceptability contour produced by the imitation technique. The highest score (2.9 – 'very close' to 'good') resulted from stimulus no 3, which is the contour interrogative mode in the sentence of *Dia membaca buku*. The contour pattern

840

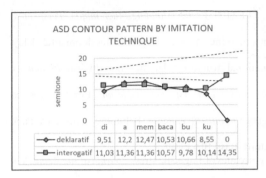
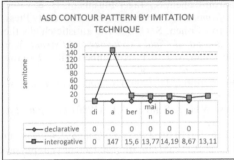

Figure 6. Contour pattern of declarative and Interrogative mode produced by imitation technique.

Figure 7. Intonation of contour patterns of declarative and interrogative modes produced by imitation technique.

Table 4. Mean values of baseline slope modification.

Number	Mode	Sentence	Mean
(1)	Declarative	*Dia membaca buku*	2.0
(2)	Declarative	*Dia bermain bola*	2.6
(3)	Interrogative	*Dia membaca buku*	2.9
(4)	Interrogative	*Dia bermain bola*	1.8

in the stimulus 3 is similar to that in the interrogative mode produced by normal speakers in Figure 4.

4 CONCLUSION

Communication disorder of speakers with ASD has given some effects toward intonation patterns, in both declarative and interrogative modes. Their intonations in both modes do not show any significant contrasts to mark the differences in such modes. The pitch ranges produced from the utterances are less than three semitones. Therefore, the pitch movements produced by speakers with ASD indicate a rather flat intonation contour. Besides, the pitch range that is relatively low, and the baseline slope of the ASD utterances from the speakers with ASD in both modes have flat patterns as well. Inconsistent contours are productively found in the final pitch movements. In the utterance production of imitation, the speakers with ASD also produced inconsistent final intonation contours. Based on the deviant findings of speakers with ASD, the hypothesis proposed for this study is that unstable emotion of the speakers with ASD could result in inappropriate final utterances. In other words, speakers with ASD may complete their expressions or utterances, but they are not able deliver the messages to properly.

REFERENCES

Aitken, K.J. (1998) Behavioural phenotypes in developmental psychopathology. *Association for Child Psychology and Psychiatry Occasional Papers Series*, 5–19.

Azizi, S.S.Z. (2016) The Tilt Model Acoustic Survey of intonation in Children with Severe Autism. *International Journal of English Linguistics*, 78–86.

Baltaxe, C.A.M. & Simmon, J.Q. (1985) Prosodic development in normal and autistic children. *Communication Problem in Autism*, 95–125.

Baron-Cohen, S., Leslie, A.M. & Frith, U. (1985) Does the autistic have a "theory of mind"? In: *Cognition, 21(1)*, 37–44.

Baron-Cohen, S., Tager-Flushberg, H. & Cohen, D.J. (1993) *Understanding other minds: Perspective from autism.* Oxford, Oxford University Press.

Baron-Cohen. S. (1989) The autistic child's theory of mind: a case of specific developmental delay. *Journal of Child Psychology and Psychiatry, 30*, 285–297.

Beckman, M.E. & Pierrehumbert, J.B. (1986) Intonational structure in Japanese and English. *Phonology Yearbook*, 255–309.

Bogdashina, O. (2005) *Communication issues, in autism and Asperger syndrome.* London, Jessica Kingsley Publishers.

Fay, W.H. & Schuler, A.L. (1980) *Emerging Language in autistic children.* Baltimore, University Park Press.

Firth, U. (1989) *Autism-explaining the enigma.* Oxford, Basil Blackwell.

Fry, D.B. (1955) Duration and intensity as physical correlates of linguistic stress. *Journal of the Acoustical Society of America*, 765–768.

Fry, D.B. (1958) Experiments in the perception of stress. *Language and Speech*, 126–152.

Fry, D.B. (1965) The dependence of stress judgments on vowel formant structure. *Proceedings of the 5th International Congress of Phonetics Sciences, Munster* (pp. 306–311). Basel & New York, S. Karger.

Goldfarb, W., Braunstein, P. & Lorge, I.A. (1956) A study of speech patterns in a group of schizophrenic children. *American Journal of Orthopsychiatry*, 544–555.

Happe, F. (1993) Communicative competence and theory of mind in autism: A test of relevance theory. *Cognition*, 48, 101–119.

Ladd, R.D. (1996) *Intonational Phonology.* Cambridge, Cambridge University Press.

Mozziconacci, S. (1988) *Speech Variability and Emotion: Production and Perception.* Eindhoven, Printers of the University of Eindhoven.

Peppé, S.J. (2009) Why is prosody in speech-language pathology so difficult? *International Journal of Speech-Language Pathology*, 258–271.

Peppe´, S., McCann, J., Gibbon, F., O'Hare, A. & Rutherford, M. (2007) Receptive and expressive prosodic ability in children with high-functioning autism. *Journal of Speech, Language, and Hearing Research*, 1015–1028.

Pike, K.L. (1972) *General Characteristics of Intonation.* England, Penguin Books.

Pronovost, W., Wakstein, M.P. & Wakstein D.J. (1966) A longitudinal study of speech behaviour and language comprehension of fourteen children diagnosed atypical or autistic. *Exceptional Children*, 19–26.

t'Hart, J., Collier, R. & Cohen, A. (1990) A perceptual study of intonation: an experimental-phonetic approach to speech melody. Cambridge, Cambridge University Press.

Cultural Dynamics in a Globalized World – Budianta et al. (Eds)
© *2018 Taylor & Francis Group, London, ISBN 978-1-138-62664-5*

Conflicts of interest among Indonesian university libraries in developing intellectual capital

Laksmi & L. Wijayanti
Department of Library and Information Science, Universitas Indonesia, Depok, Indonesia

ABSTRACT: The purpose of this study is to identify potential conflicts of interest among university libraries in their efforts to develop intellectual capital. To achieve this objective, world-class universities must maintain excellent library facilities. However, the funding and assistance given to libraries can give rise to various types of conflicts of interest. By using a qualitative approach and cultural-study method, the aim of this study is to identify how conflicts of interest in university libraries are efficiently and effectively addressed in their efforts to develop intellectual capital. The findings suggest that conflicts are usually triggered by the rectors, librarians, and other stakeholders who tend to treat libraries as convenient locations to discuss conspiratorial ideas for potential shifts in ideology and to contest bureaucratic power. The efforts to develop intellectual capital lack human and structural integrity which causes the end users to become a low priority. In regards to conflict resolution, the 10 libraries reviewed rely mainly on communication, compromise, accommodation, and cooperation.

1   INTRODUCTION

Conflicts of interest among university libraries are by and large conditions that may arise in the process of supporting the objectives of the university (Chowdhury, *et al.*, 2008, p. 27; Saifuddin, 2005, p. 340). Universities under globalisation are required to compete internationally to maintain their standing as world-class universities (Corrall, 2014, p. 21). Such conditions can engender a spirit of optimism among the parties who are involved in supporting libraries, but they can also lead to the emergence of certain interests that are unfavourable to the development of intellectual capital. Some college administrators are even involved in the industrialisation and commercialisation of education for their own benefits (Toha-Sarumpaet, 2012, p. 13; Ibrahim, 2007, p. 281; Armando, 2012, p. 118).

University libraries in Indonesia assume risk when dealing with conflicts, particularly when they operate in high-ranked universities. Based on the 2015 Decision by the Indonesian Ministry for Research, Technology, and Higher Education (hereinafter abbreviated as "Menristekdikti") on university rankings, the top 10 universities from 3,320 universities in Indonesia are ITB (Institut Teknologi Bandung), UGM (Gadjah Mada University), IPB (Bogor Agricultural University), UI (Universitas Indonesia), ITS (Institut Teknologi Surabaya), UNIBRAW (Universitas Brawijaya), UNPAD (Universitas Padjajaran), Airlangga (Universitas Airlangga), UNS (Universitas Sebelas Maret Surakarta), and UNDIP (Universitas Diponegoro) (Ristekdikti, 2015).

Based on this classification, the research question is how to address potential conflicts of interest within the top 10 Indonesian universities to minimise any obstructions in the development of intellectual capital. The significance of this research is that it identifies how conflicts of interest are efficiently and effectively handled by 10 academic libraries in the development of intellectual capital. The findings intend to provide input for university libraries, so that the resolutions for conflicts of interest can have positive results.

2 LITERATURE REVIEW

2.1 *Conflicts of interest in university libraries and intellectual capital*

A conflict of interest is a common symptom that occurs in an organisation due to differences between two or more interest groups (Trice & Beyer, 1993, p. 226). These differences include differences in beliefs, values, and understanding. They cause resentment toward or envy of others and generate dilemmas in professional loyalty, particularly when the general code of conduct is ignored (Stueart & Moran, 2007, p. 377). Meanwhile, the concept of conflict of interest in various fractions indicates the methods adopted to acquire specific benefits either for individuals or groups acquired through power relations (Saifuddin, 2005, p. 341).

Conflicts of interest in university libraries have become particularly salient under globalisation and are triggered by international rankings on the quality of higher education. The ranking system requires universities to constantly improve the quality of their facilities, including libraries (Helm-Stevens, *et al.*, 2011, p. 129). In order to develop intellectual capital, libraries require the support of three types of capital, namely human, structural, and consumer (Yazdi & Chenari, 2013, p. 4183). The mechanism adopted in resolving mutual disputes between parties is usually avoidance, compromise, competition, accommodation, or cooperation (Stueart & Morin, 2007, p. 379). Generally, leaders have a critical role in the mediation process that is necessary for immediate conflict resolutions and accurate handling of situations.

2.2 *Academic libraries*

Menristekdikti recently published a book providing guidelines for universities which designated academic libraries as technical implementation units. This publication advised heads of universities to adopt the Tri Dharma University principles of education, research, and community service in choosing, collecting, processing, treating, administering resources, and particularly in contributing to society in general. The university is an entity that houses libraries, which serve the conservation of knowledge, teachings, research, publications, and interpretations, and a place for the creation of knowledge, through research, and knowledge preservation (Helm-Stevens, *et al.*, 2011, p. 129). Furthermore, librarians should strive to create a scientific community that is just, honest, truthful, and open (Sedyawati, 2012, p. 480).

Intellectual development requires libraries to provide suitable locations to facilitate freedom of thoughts and opinions, expressions of ideas, and prevention of plagiarism. To support the university's objectives, the general purpose of university libraries is to meet the informational needs of all the academic members by providing reference materials and suitable facilities, lending services suited for different types of users and students of all academic levels, and active information for university environment and local industries (Kemdikbud, 2004, p. 3).

3 RESEARCH METHODS

A qualitative approach aims to understand interaction processes that occur from conflicts in daily life, such as behaviour, perception, motivation, and action in a holistic manner, using natural methods (Powell, 1997). This method seeks to explain a person's comprehension and subjectivity. Cultural studies are used to provide a critical framework to describe such processes among group members in an organisation and the production of conflicts through their actions, including their manners and deductive processes (Bradigan & Hartel, 2014, p. 8). The informants were determined by using purposive sampling and encompassed the principal or staff of the university library, the leader or staff of the university, and other stakeholders. Data collection was gathered by observation, interviews, and document analysis.

4 CONFLICTS OF INTEREST IN 10 MAJOR UNIVERSITY LIBRARIES IN INDONESIA

4.1 *Activities in the development of intellectual capital*

The aforementioned top 10 university libraries have developed intellectual capital through various types of programs. In developing intellectual capital, they all employ similar methods and most of them emphasise the importance of developing human-resource quality and improving facilities and collections.

In the development of intellectual capital, the quality of human-resource development is considered vitally important to ensure the best possible provision of services that enable academicians to expand intellectually. UGM library, UI library, and a number of other libraries provide the opportunity for librarians to deepen their knowledge through further education and various training courses (Universitas Gajah Mada, 2016). UI library increases the number of subject specialists by actively recruiting graduates from various subjects who are then given the opportunity to continue their studies in the library and information science. Subject specialists are expected to competently meet the information needs of the users. The libraries in UI, UGM, UNAIR, and UNDIP hold writing competitions for scientific articles on librarianship directed at academicians. Librarians can also enhance knowledge through FPPTI (Indonesian Higher Education Library Forum), which regularly holds forums on librarianship knowledge.

A number of other programs that aim to improve the facilities have also become a major preoccupation in most libraries. Space, both real and virtual, has become the main tool for academicians to share knowledge. Besides improving facilities, parent institutions can also gain additional operational funding through these endeavours. The development of collections is also one of the programs sought to develop intellectual capital. Under globalisation, the 10 libraries have been developing both conventional and electronic collections. UI library also provides open-access repositories under an agreement with UI students to make their final projects available to the general public.

4.2 *Conflict triggers*

Within the 10 university libraries, most conflicts are triggered by personal interests. Conflicts of interest involve three parties, namely university rectors, heads of libraries, and other stakeholders. Figure 1 shows the involvement of the participants in more than one program (b) to establish joint cooperation so as to achieve the goal of intellectual capital development that enables a university to become a world class. (c). Head librarians and their staff are the most vulnerable parties because libraries are the central units that manage the intellectual capital of universities. The library can also be an arena that gives research to various types of conflicts of interests. The entire program is structured to implement the Tri Dharma University principles, ranging from collection development, services, planning, and the overall operational control, including cooperation with external parties.

Conflicts of interest among university libraries usually emerge due to the participants' lack of understanding of their roles, responsibilities, performance, and environmental demands (d). Head librarians rely very much on the university's policy. In 2000, university autonomy was implemented and included in several libraries parallel to faculties (Prasodjo, 2012, p. 45). At that time, UGM library strategically made effective and independent decisions to initiate further studies for the staff, while UI library took the opportunity to build a magnificent new library building, which is considered to be the largest university library in Southeast Asia (Armando, 2012, p. 113) (f). However, since the Decision was repealed in 2012, libraries are now considered to be technical service units. This resulted in the scaling down of the authority of the head libraries and their incapacity to initiate programs freely.

A number of differences were also seen in the establishment of various information service corners funded by foreign countries, such as the United States and Saudi Arabia. Most librarians interpret the information provided as having hidden agendas or propagandas from the respective countries, and this is considered unethical. However, most of the libraries,

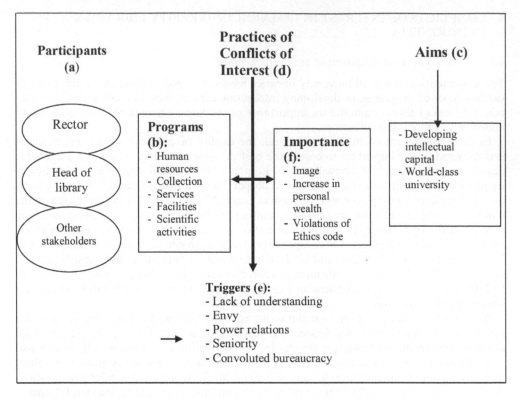

Figure 1. Practices of conflicts of interest.

such as the one at UGM, reasoned that the University is indeed biased toward the United States of America, so it is only natural that they have readily accepted the establishment of the American corner. The American Embassy, as the sponsor of the corner, also donates and funds scholarships that enable Indonesian librarians to study in the United States. However, there is a doubt that such an offer would be accepted by the head of the library, as they invariably believe that these offers would create a precedent for other agencies to ask for accommodation in the library. When this type of situation occurs, it can produce conflicts between the head of the library and the university rector.

Some university libraries have conflicts that occur among library staff, possibly due to position, seniority, or other negative perceptions that tend to be outside the context of intellectual capital development (e) (Trice & Beyer, 1993, p. 241). Certain library positions provide many advantages, and in some universities such assignments become a political posting. Therefore, many of the staff who aspire to these positions are discouraged, and some also experience their reputation being undermined through unfounded innuendos or defamatory anonymous letters to freely enable the political appointee. Regarding seniority, generally the older group, both in terms of length of work and age, are consistently favoured over their younger and perhaps even more knowledgeable and skillful co-workers. Seniors often make continuing their education and travelling on business trips either abroad or to other facilities their first preference. They also take advantage of certain conditions to improve their financial situation (f). Whether the junior staff would have been considered for the next jaunt or not is mostly immaterial, as programs inevitably change, and this engenders resentment and envy among the younger staff.

Conflicts with customers were not identified in any of the 10 libraries appraised. The difference between librarians and users has been understood and well addressed by the management of governance. UI library has merged the integrated services of the entire library of the faculties, which has resulted in some positive and negative consequences. Librarians consider the library's centralised management to be more propitious, and it will help reduce both efforts and costs (Chowdhury, et al., 2008, p. 28). However, this model has caused a sense of discomfort among many professors. They complain that centralisation causes the location and collection management to become less efficient, and the library facilities create an atmosphere reminiscent to that of a crowded mall. Thus, as most of the users are students who have lifestyles that tend to be hedonic, they enjoy the library not so much as a place to read, but as a place for discussions, romantic liaisons, relaxing, and socialising. It must also be understood that all the facilities provided by libraries intend to provide convenience and comfort for the users and entice them to frequent the library.

Facility expansion and improvements of some libraries were prohibited by the limited availability of land. At UNAIR, a cafeteria was recently set up in the library, and many consider it the reason why the library has in some measure turned into a slum. The head librarian is placed in an invidious position with no alternative but to play a game of cat and mouse with the university authority. Although both sides maintain their respective opinions, the final decision remains with the university rector. In addition, the collections and procurement process are often adversely affected by bureaucracy. For a considerable amount of time, almost all the libraries have made inappropriate purchases for their respective collections due to interference by the rectors.

4.3 Conflict resolution strategies

The 10 university libraries handle conflicts of interest using communication strategies, namely compromise, accommodation, and cooperation. These strategies involve human capital and reveal an interaction pattern among the participants (Yazdi & Chenari, 2013, p. 4183). Communication strategies are intended to maintain good relations internally and externally. As a technical service unit, libraries do not have any authority in policy change. Thus, if there is a policy that contradicts the library's rules, the head of the library must handle the situation using communication strategies through more personal approaches.

On a regular basis, the head of UGM library hosts a private dinner party at home in attempt to build mutual trust with university leaders, submits progress reports, and proposes strategic plans. The heads of the libraries in UI, UNPAD, and UGM hold undercover operations to investigate those who are presumed to have authority in order to identify and rectify problems that may arise within their ranks. In addition, these operations seek to eliminate prevalent obstacles and expunge any issues that could be detrimental to the process of building social networks with parties considered important. The head of Airlangga library generally addresses the participants as 'bapak', which means father and is a customary address for adult males in Indonesia. The use of the term 'bapak' gives the impression of a close relationship compared to more formal terms such as professor.

Even so, not all management decisions and improvement suggestions by staff or users are necessarily approved by the librarian. When conflicts arise in the hierarchy, the head of UI library will accept solutions that are considered normal but reject countermeasures that are perceived to violate the code of ethics. Discords triggered by conflicts of interest are a violation of the professional librarian ethics. In managing knowledge and resolving conflicts, librarians are not permitted to take sides. However, when certain influential people persistently push for setting up external information corners in the library, the head of the library may accept this decision on the basis of maintaining good relations with external partners.

To resolve conflicts among the staff, the head of UNAIR library creates activities that aim to engender a 'feel-good factor' and include talks on motivation during morning briefings. The head of the library in UGM emphasises that he is cordial with everyone, ranging from the janitors to the rector. Staff members who are unfriendly are counseled and encouraged by the head of UNPAD library to be more cooperative. Almost all the staff in the 10 libraries were provided with opportunities for further studies, involved in meetings with the rector, extended invitations to evening meals, and so forth. Conflicts are resolved through communication with users. UGM often assumes the position of a user to discover how the service is likely to be perceived by the real users. UNPAD puts great emphasis on service to the users and supervision. UNAIR holds competitions that allow students to submit proposals on the improvement of library services, so they feel that they are involved.

5 DISCUSSION

5.1 Library as social space in the growth of intellectual capital

Basically, a library is a unit that facilitates education, research, development, cultures, and also provision for recreational functions. In fact, four of the library's functions that should be directly involved in promoting intellectual capital development are steered away from the primary objective by individual and group interests. However, libraries must not be perceived as the panacea or the front of all knowledge. Therefore, when required, students should be encouraged to perform research elsewhere to augment their knowledge.

5.2 Library as a place for ideology shifts

Libraries should be the center for the creation, use, and dissemination of knowledge. Moreover, in an era of globalisation, libraries should function as the supporting factor in enhancing universities' international rankings. UI conceptualises the importance of a library's role in science as the epitome and standard bearer of knowledge (Universitas Indonesia, 2016). The library at UI is impressive with its magnificent architecture and equipped with numerous amenities (Toha-Sarumpaet, 2012, p. 118). UI library is adjusting to its role, as symbolised by the designation of the library as "crystal of knowledge," and does not focus merely on infrastructure development but the development of intellectual capital (Universitas Indonesia, 2016). Some libraries are still regarded by many as knowledge depository and not as a site of knowledge elaboration for academicians. UNPAD library is designated as *kandaga* which in Sundanese language means jewelry box or a box to store valuables. They also consider the library as a storage space.

Upturns in the ideologies of education also provide benefits for the economy. As libraries are usually in close proximity to various types of businesses, such as bookstores, cafés, restaurants, and grocery stores, they have become a tool to generate income, particularly in developing countries where people are under more financial pressure, and education tends to develop capitalism, pragmatism, and encourage the removal of irrationality (Zaini, 2015). Thus, libraries generally encourage economic advancements. In addition, some library facilities have been commercialised, and even academicians who wish to use a room, auditorium, or classroom must pay fees (Toha-Sarumpaet, 2012, p. 118).

5.3 Library as a place for power contestation

Before the centralisation of libraries was implemented in 2012, the policy of UI was not to provide external information. Historically, the library has also never accommodated information corners from foreign countries, such as the United States and South Korea. However, some of those that were sanctioned have been subsequently closed as a result of conflicts between the university authority and the head librarian. The library staff consider these contributions to be a double-edged sword with certain subscribers wishing to spread ideologies that do not necessarily support the intellectual capital development. In such conflicts, the power relations are dominated by the university leaders rather than the heads of libraries who should have full authority to balance out decisions and potential benefits in accordance with a library's function.

5.4 Practices of conflicts of interest in intellectual capital development

To become a world-class university, a library must be considered as the most strategic and enabling asset. Thus, rectors and university leaders will inevitably endeavour to construct libraries that are quintessentially representative with the most impressive architecture available. These goals intend to facilitate the needs of users, namely academicians in implementing the Tri Dharma University principles. Other stakeholders can provide assistance, either in cash or provision of intellectual documents, for example collections, but usually under certain conditions, it can be installment of their logos in the library, which can be in the form of a flag, a state emblem, or a brand. Many universities provide areas suitable for the American, Australian, and Sampoerna information corners. Accommodation is also given for prestigious cafés and restaurants, and some libraries even provide banking facilities. Many stakeholders consider academicians as a specialist resource to generate profit, as they are considered to have the ability to provide trustworthy and reliable scientific opinions. However, for the majority of academicians, this is regarded as a covert propaganda.

Heads of libraries are empowered to accept or reject proposals from stakeholders, but the rectors still hold the highest executive authority to formulate, initiate, and enable library policies. Generally, when a rector wishes to make a decision regarding a library's stakeholders, the rector would consult and request recommendations from the head of the library. Nonetheless, for certain matters the rector will deliberately not contact the head of the library, particularly if there are potential conflicts. However, if the head of a library refuses to accommodate and set up a corner on the grounds of bias, as the head believes a library should be neutral, this is supported by the code of ethics. Authority, such as the rector, can cause a bottleneck, which is unavoidable due to the source of library leadership. On the one hand, a library may not be considered a commercial enterprise, but on the other hand, there is awareness that funding is required to provide the knowledge base required by academicians. The centralisation of library management opens opportunities for certain parties to utilise funds illegally (Armando, 2012, p. 120). In addition, students today tend to have a hedonistic lifestyle, so they need a place to socialise, such as cafés, parks, and places with access to Internet connection. This situation was identified by stakeholders as a means to take commercial advantage of the opportunities presented and provide superficial facilities that do not necessarily support the development of intellectual capital. Practices of conflicts of interest are also prevalent in libraries, regarding their function and accessibility as repositories. As a knowledge repository, libraries need to be available for institutions to access research, but they must also be accessible to the general public. However, library leaders have objections, as they consider collections as purely internal assets. Due to this interpretation of freedom of information, the policy thus provides a limited kind of freedom (Koran SINDO, 2015).

6 CONCLUSION

Conflicts of interest among the three participants, namely the rector, the head of the library, and other stakeholders, in developing intellectual capital show that the conflicts were in fact disputes over structural capital. Conflicts tend to make the library a site of ideology shifts and disagreements of bureaucratic power, which include the images they wish to project to the international community and self-enrichment. Heads of the libraries and their staff in all 10 universities have programs in accordance with their capacities as librarians in developing intellectual capital. However, they remain hampered by complicated bureaucratic processes. The areas of problem also include staff resistance as a result of envy generated by the perception of not having realistic opportunities for further studies and advancement.

However, conflicts of interest in a library can be resolved when addressed properly by the head of the library through good communication and sympathetic approaches. These facts indicate that human capital is already well-established but stuck in a rut due to misunderstandings of the roles and functions of the library. Heads of libraries across the universities handle almost all of the conflicts of interest through sound communication strategies by way of compromise, accommodation, and cooperation.

ACKNOWLEDGEMENT

This work was supported by PITTA Grant 2016 from the Directorate Research and Community Engagement of Universitas Indonesia.

REFERENCES

Armando, A. (2012) Perpustakaan Pusat UI: boros, tak berguna, dan bocor (UI Central Library: wasteful, useless, and leaking out). In Toha-Sarumpaet, R.K., Budiman, M., & Armando, A. (eds.) *Membangun di atas puing integritas: belajar dari Universitas Indonesia (Building on the collapse of integrity: learning from Universitas Indonesia)*. Jakarta, Gerakan UI Bersih dan Yayasan Pustaka Obor Indonesia.

Bradigan, P.S. & Hartel, L.J. (2013). Organizational culture and leadership: exploring perceptions and relationships. In Blessinger, K. & Hrycaj, P. (eds.) *Workplace culture in academic libraries, the early 21st century*. Oxford, Chandos Publishing. pp. 7–20.

Corrall, S. (2014) Library service capital: The case for measuring and managing intangible assets. *Libraries in the Digital Age (LIDA)*, 16–20.

Chowdhury, G.G., *et al.* (2008) *Librarianship: an introduction*. London, Facet Publishing.

Helm-Stevens, R., Brown, K.C., & Russell, J.K. (2011) Introducing the intellectual capital model interplay: advancing knowledge frameworks in the not-for-profit environment of higher education. *International Education Studies* [Online] 4 (2). Available from: www.ccsenet.org/ies/ [Accessed 17th August 2016].

Ibrahim, I.S. (2007) Wajah dunia pendidikan di balik hegemoni media: penguatan atau pemiskinan wacana demokratik dalam ruang publik (The face of the education world behind the media hegemony: strengthening or impoverishing democratic discourse in public space). In Ibrahim I.S. (ed.) *Budaya popular sebagai komunikasi: dinamika popscape dan mediascape di Indonesia kontemporer (Popular culture as communication: dynamics of popscape and mediascape in contemporary Indonesia)*. Yogyakarta, Jalasutra. pp. 271–285.

Koran SINDO. (2015) Membangun Riset Perguruan Tinggi (Building academic research). [Online] Available from: http://nasional.sindonews.com/read/1064574/18/membangun-riset-perguruan-tinggi-1448481986 [Accessed 17th August 2016].

Prasodjo, I.B. (2012) Peran pendidikan tinggi dalam meningkatkan kualitas ilmu, karakter, dan kebangsaan (The role of higher education in improving the quality of science, character, and nationality). In Irianto, S. (ed.) *Otonomi perguruan tinggi: suatu keniscayaan (The autonomy of higher education: a certainty)*. Jakarta, Yayasan Obor Indonesia. pp. 42–77.

Powell, R.R. (1997) *Basic research methods for librarians*. 3rd edition. Greenwich, CT, Ablex Publishing Corporation.

Ristekdikti (2016) SK Klasifikasi dan Pemeringkatan Perguruan Tinggi di Indonesia Tahun 2015 (The Letter of Decision on the Classification and Ranking of Higher Education in Indonesia in 2015). [Online] Available from: http://ristekdikti.go.id/sk-klasifikasi-dan-pemeringkatan-perguruan-tinggi-di-indonesia-tahun-2015/ [Accessed 11th July 2016].

Saifuddin, A.F. (2005) *Antropologi kontemporer: suatu pengantar kritis mengenai paradigma (Contemporary anthropology: a critical introduction on paradigm)*. Jakarta, Kencana.

Sedyawati, E. (2012) Masyarakat ilmiah dalam pembangunan bangsa (Scientific society in nation development). In: Toha-Sarumpaet, R.K., Budiman, M. & Armando, A. (eds.) *Membangun di atas puing integritas: belajar dari Universitas Indonesia (Building on the collapse of integrity: learning from Universitas Indonesia)*. Jakarta, Gerakan UI Bersih dan Yayasan Pustaka Obor Indonesia. pp. 479–482.

Stueart, R.D. & Moran, B.B. (2007) *Library and Information Center Management*. 7th Edition. Westport, Connecticut, Libraries Unlimited.

Toha-Sarumpaet, R.K., Budiman, M., & Armando, A. (2012) *Membangun di atas puing integritas: belajar dari Universitas Indonesia (Building on the collapse of integrity: learning from Universitas Indonesia)*. Jakarta, Gerakan UI Bersih dan Yayasan Pustaka Obor Indonesia.

Trice, H.M. & Beyer, J.M. (1993) *The Cultures of work organization Englewood Cliffs*. New Jersey, Prentice Hall.

Universitas Indonesia. (2016) Profile. [Online] Available from: http://lib.ui.ac.id/profil.jsp?hal = 3 [Accessed 8th July 2016].

Yazdi, K.H. & Chenari, H. (2013) Intellectual capital and technological advances in knowledge society: How are these concepts related? *African Journal of Business Management*. [Online] 7 (40) Available from: http://www.academicjournals. org/AJBM [Accessed 17th August 2016].

Zaini, A.H.F. (2015) *Pendidikan dalam masyarakat kredensial (Education in credential society)*. Jakarta, Kompas.

Cultural Dynamics in a Globalized World – Budianta et al. (Eds)
© 2018 Taylor & Francis Group, London, ISBN 978-1-138-62664-5

Changes in the virtual reference services at Bina Nusantara University to meet users' information needs

M. Septiana & T.A. Susetyo-Salim
Department of Library and Information Sciences, Faculty of Humanities, Universitas Indonesia, Depok, Indonesia

ABSTRACT: Technological developments in the information age have transformed the face of library services, one of which is the reference services. The transformation is from a direct communication (face to face) to an online communication (virtual reference services). This transformation also affects the reference services at Bina Nusantara University library to meet its users' needs. Changes have been implemented from the beginning of 2009 until now in separate locations. Bina Nusantara University has campuses in three different locations with different online reference services. The purpose of this research is to analyse the transformation process and the effects it has on library users. The objective is to analyse the process of interactions between the librarians and the users following the transformation of library reference services at Bina Nusantara University library. The method incorporates a documentation study, observations, and interviews using a qualitative descriptive approach. The finding of this study shows that technology plays an important role in the transformation of reference services.

1 INTRODUCTION

Library development should involve flows of information, which are more sophisticated these days. Reference services should serve as a filter for explosion of information. The fact that students are turning to the Internet for research assistances may mean that they are computer literate, but this is not the same as being information literate (McDonald, 2004; Messineo & DeOllos, 2005; Goodfellow, 2007; Rowlands, Nicholas, & Huntington, 2007, and Gurganus, 2015). The Internet carries some advantages and disadvantages for learning processes. Some of the problems related to the internet are: (1) poor organization of documents, hence the difficulty in retrieving information leading to unsatisfactory results; (2) inadequate search tools for facilitating a fast information retrieval; (3) documents on the internet being transient; (4) overwhelming mass of updated documents causes the users to lose track of the original sources; (5) restricted access or denied access to information often encountered by the users (as cited in Ming, 2000; Shachaf & Snyder, 2007). Librarians should consider these problems as great concerns and find solutions for the sake of the users.

The transformation of technology creates changes in reference services. The transformation from a traditional (face-to-face) to electronic services or digital services and 24/7 access enable libraries to be responsive to the users' needs. This may serve as a motivation for libraries to find solutions to the limited access to information. Many academic libraries provide their services to the users by offering appropriate reference services through the available technology, such as telephone, e-mail, instant messaging, WhatsApp, Line, and other online applications. Lewis and DeGroote (2008) as cited in Gurganus (2015) discovered that the more access points were added to the library and institution website, the more likely the students will use the online services.

For the purpose of this paper, virtual references, also known as digital references, are defined as exchanges of information between the library reference staff and the users.

The virtual references are formed from traditional reference services and aimed to provide appropriate information to the users.

2 LITERATURE REVIEW

2.1 *Virtual reference services*

Virtual reference services are not different from answering questions face-to-face, but there is a lack of connection in the communication since no tones and gestures are displayed from the librarians to the users and vice versa. The reference services librarians may not be able to determine whether there is good communication with the users. Misunderstandings sometimes occur through virtual reference services. The media used sometimes cannot describe what the users feel. The emphasis here is that virtual reference services should employ some sort of communication standards in line with the prevailing communication pattern in the community.

Green (1876) defined reference works as personal assistances given by librarians to individual readers who seek information. As Margaret Landesman observed, "The ability to search full text ... turns every collection of online texts into a reference collection and provides an automatic concordance for every title (Bopp & Smith, 2011)." Nowadays, titles can no longer be tidily separated into 'reference works' and 'general collection.' The previous definition refers to the traditional method of classifying references works and books in libraries to assist the users in obtaining their needs. More and more libraries have developed programmes to promote reference works as their main services. In order to obtain the highest satisfaction from users, libraries are taking some steps to meet the users' needs with the use of technology. Rubin adopted, from the American Library Association Reference and User Service Association (RUSA), a guideline on the ethical codes for information providers, which defines many aspects of the elements listed and reflected in modern reference services: (1) a service-oriented perspective, (2) responsibility to select and organize collections in order to maximize effectiveness, (3) strong knowledge of the subjects, principles, and practices in the fields, and (4) commitment to either answering questions or referring to sources capable of answering (Bopp & Smith, 2011).

Cassell and Hiremath (2009) argued that virtual references are often underused simply because their existence is unknown to the users. It is important that libraries clearly mark the services and make them visible on all pages on their websites. Meanwhile, RUSA defined a reference transaction as information consultations in which library staff recommend, interpret, evaluate, and/or use information resources to help others to meet particular needs of information. Cassel and Hiremath (2009) also argued that virtual references librarians should aim to be approachable in the way they word their responses to the users. When librarians provide references to the users, they should think of a faster way to solve problems faced by the users. Librarians should know what is best for their users. RUSA Guidelines describe the ethics of a virtual references librarian because one has the right to get information one needs regardless of one's personality. Virtual reference services have various means to provide the users with what they want. Nowadays, e-mail and other popular means of communication are readily available. Therefore, those in the field of education must have e-mail addresses for communication. Chat applications, such as WhatsApp, Line, Instant Messenger, are known among young generation. Media applications are commonly used by the young generation to perform better communication. This type of communication method is used mostly because it is faster than traditional means of communication. The users can use these applications through their mobile devices using internet connection and Wi-fi network to access a vast collection of virtual references titles. The new forms of virtual collections are also emerging, combining electronic versions of both primary source materials and related reference source materials (Bopp & Smith, 2011, p.389).

2.2 *Information needs*

Information needs are basically derived from the two words "information" and "needs", meaning that humans need information to perform their activities. Need itself means something that humans should fulfil. Psychologists divide the concept of needs into three categories: (1) physiological needs, such as food, water, and shelter; (2) affective needs, which are related to emotions and feelings; (3) cognitive needs, which are related to skills, plans, and so on. From the concept of needs above, we can define information needs as affective and cognitive needs, a requirement for something that drives the lives of human beings so that they may live better and improve their knowledge.

When we talk about the information needs of users, we should bear in mind not to have some notion of the fundamental, innate, cognitive, or emotional 'need' for information, but a notion of information (facts, data, opinions, advice) instead as one of the means toward the end of satisfying such fundamental needs (as cited in Wilson & Wansgsh, 1981). Information needs have something that is indescribable, clearly, because information is performed by human cognitive and affective feelings.

Some of the research conducted on students' information needs at the university level is as follows:

> Susana Romanos de Tiratelas, as cited in Tamara (2012) proves that, "*This percentage is higher in the humanities (66%) than in the social sciences (49.8%). Although 32.5% in social sciences registered a frequency of less than once a week, humanities measured 12.1%. A total of 5.3% in social sciences did not know or answer how often they visited the library against 18.9% in humanities who go three or four times a week. Humanities scholars exhibit a more intensive use of the library.*"

> Heting Chu, as cited in Tamara (2012) states, "*Electronic books: viewpoints from users and potential users.*" *The respondents apparently preferred to use e-books that could be obtained from libraries or the Web without incurring any direct expense to themselves. They were unwilling to pay for e-books as they do for print books.*

> Broadus in Boone as cited in Tamara (2012) found, "*published several articles giving the results of a survey of the material used from 1982 through 1984 by humanities scholars at the National humanities centre, Research Triangle Park, North California. Broadus concluded that "materials relevant to humanistic investigations are even more widely dispersed in library collections" than previously thought. Broadus also found that thirty-one per cent of the requests were for journals and that twelve percent of the tides satisfied almost half of those request.*"

It can be concluded that information needs play an important role to promote students' skills and capabilities. Information-seeking behaviour of each student from the different major is related to their interests. It was shown that students prefer to find information in e-books from the libraries than to pay for e-books or any related online documents themselves.

3 METHODOLOGY

This research uses a qualitative approach with in depth-interviews, observations and a documentation study.

3.1 *Data collection*

Data collected for this research was derived from the reference services of the campus libraries of Bina Nusantara University in the form of reports. These reports contain monthly e-mail transactions and monthly reports from January to April 2016. The library transactions based on the monthly reference reports in 2016 are:

1. Monthly reports of library transactions through online media (WhatsApp, Instant Messenger and e-mail) and face-to-face interaction.
2. Transactions that were classified into questions on the subject forms and uses of social media.

Interviews with the reference services librarians in charge of the reference services.

3.2 Data analysis

The data were analysed by making some categorisation on the reference services, such as: (1) pattern of communication and transactions between the librarian and users at the reference services; (2) pattern of formal and informal communication in two different locations.

4 RESULTS AND DISCUSSION

The pattern of communication differs based on situations and conditions. The explanation of the data analysis is shown below. The result of the communication transactions based on media (face-to-face) and online (via email, IM, and WhatsApp) is shown in the following figure.

Figure 1 shows the differences between the two types of media in two locations. The users in Area 1 (Anggrek, Kijang, and Alam Sutra campuses) were more interested in coming and meeting the librarians to ask questions for their information needs. They came and talked directly about what they would like to know. In Area 2 (JWC and FX campuses), the users preferred to use online media communication.

Based on the pattern of information exchanges in the two locations, it is obvious that some users prefer to use technology to fulfil their need for information. These days, technology has become a cultural element in society. Technology is a human attempt to conserve their energy (power) and thought by using nature (Hoed, 2014, p. 229). The communication pattern involving the technology significantly changes the way we communicate. It is noticeable that technology access in the second area facilitates users to ask questions about references to the librarians. E-mail and Instant Messenger used by the librarians make transactions related to references efficient and effective. As cited in Cassell and Hiremath (2009), the problems with virtual reference arguably relates to the disadvantage of not having a face-to-face interaction where the users' tone of voice, facial expressions and body language will help the librarians to judge whether they communicate well with the users. On the other hand, Cassell and Hiremath (2009) stated that librarians in charge of virtual references could develop peer-reviewing systems to help one another in order to improve the quality of their work. Transcripts should be reviewed on a regular basis to ensure that the best possible service is provided.

Based on the interviews with the librarians of the reference services in two locations, the users sometimes asked various redundant questions, such as access to the library's collections, ways to find books and book borrowing systems, and errors on access in a website's system.

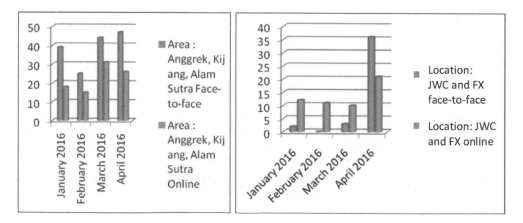

Figure 1. Number of visits by reference service users in Area 1 (Anggrek, Kijang, Alam Sutra) and area 2 (JW and FX).

4.1 Virtual reference services at the Bina Nusantara University Library

Bina Nusantara University Library is located in Jakarta, Indonesia. It is known as the Binus Library. Bina Nusantara provides multiple degree programmes (Bachelor, Master's, and Doctorate), international and regular classes with approximately 25,000 students (full and part-time). The campus libraries are located on the five different campus locations (Anggrek, Kijang, Senayan, FX, and Alam Sutera Campuses) with approximately 35 staff members, 15 of whom are librarians. The Binus Library has combined the traditional reference services with the virtual reference services since 2009. The Library uses social media to serve the users because it would like to provide the best services to the users. In the Anggrek campus, the reference services librarians have a reference desk to help the users find the information they want. The librarians will perform some in depth-interviews until they know what the users want. The Binus Library has many reference programmes in their online services, such as consultation about research materials, and explanations related to library collections. The users visit the library to find information for their research, especially from the thesis and e-book collections. In the event that they do not know exactly the information they need, they will ask and talk to the reference services librarians who will interview them and the users must repeat the subjects they need. On JWC and FX campuses, the users mostly use the online media to ask for help to find e-journals, e-thesis and e-books. The users will communicate virtually through e-mail and instant messengers. The reference services librarians will ask and answer their questions via e-mail. This programme is changed annually and the Binus Library has transformed the services gradually. Initially, the Binus Library used a chat application with a pop-up chat application. It also transformed the use of the social media in numerous ways. Facebook has been used to cover the information about events and activities in the campus libraries. The libraries have made efforts to attract the users through the use of various social media, such as Line, WhatsApp, Yahoo Messenger, IM, and Viber to facilitate users when they need some help. Also, the libraries have assigned a staff to answer users' questions through mobile device.

4.2 Obstacles

There are some problems in the implementation of the virtual reference services programme. First of all, the users sometimes ask redundant questions during face-to-face and virtual meetings. In Anggrek Campus, they keep visiting the library to make sure that the reference services librarians get their questions. Most users approach the reference desk and ask for help to the reference services librarians. This culture is still present until now. Meanwhile, the users of JWC library ask some reference questions through e-mail. This different culture shows changes in the information-seeking behaviour.

Secondly, the librarians of the reference service sometimes are not attentive due to the long questions in the interviews and other users had to wait for their turn. The third is the short conversations via e-mail and WhatsApp cannot cover what the users would like to know. The Anggrek campus library users usually use WhatsApp application to query books status and operational hours. Meanwhile in the JWC campus, the library users usually ask for the library's e-collections. From the interviews with the librarians, the data of the reports does not show the details on the questions asked by the users. The librarians also do not perform in depth reference interviews in the face-to-face and virtual services communication.

4.3 Recommendations

The Bina Nusantara Library should be aware of the obstacles due to communication barriers between the users and librarians. As cited in Westbrook and DeDecker in Katz (1993) who argued that the mechanism for references and instructions used by librarians in an academic setting should create a more integrated and complete approach for welcoming and validating their users' information needs. This structure consists of guidelines for three specific areas:

1. **The Facilities** should be comfortable and provide ease of use to the users: the facilities should be warm, with good lighting, and cater to the users' physical safety and special needs, such as those with disabilities.

2. *Service*: Librarians at the desk reference services should be proactive with regard to their interview skills, offer access to private areas in the library and be sensitive to the different learning styles, approaches, and depth-interviews about the users' information needs.
3. *Staff*: Library staff and librarians should be aware of the needs to welcome, develop and share the means to effectively collect messages, encourage staff involvement in setting up policies and procedures, and emphasize awareness of local and institutional demographics (p. 49).

Based on the analysis of this research, the Binus Library should perform the following activities to promote the reference services:

1. Understand the users of the Binus Library and the culture embraced by them. It is important to know the users' culture in order to approach them. The librarians, through their nonverbal actions such as turning away from the users, may inadvertently indicate that interview is over (as cited by Ross and Dewdney, 1999; and Cassell and Hiremath, 2009).
2. Analyse monthly reports in order to find out exactly what is needed by the users. The librarians should simultaneously consider the following three steps to avoid scattershot searches: (1) categorizing the answers (2) visualizing how the final answers will appear (3) testing the waters to check if the answers head towards the right direction (as cited by Cassell and Hiremath, 2009, p. 36).
3. Empower the reference services librarians by providing the necessary trainings on managing and resolving problems related to information searches. The librarians should evaluate the search results in order to provide answers effectively and efficiently. The evaluations the search results comprises (1) authority: placing authorship is a traditional way of establishing authority on any piece of information; (2) reliability: reputable print resources gain stature because of their credentials, authority, and transparent documentation of publishing pedigree; and (3) currency: related to printed resources, internet references are most valued for providing the kind of currency that print formats are unable to supply.

The virtual reference services should be made as continuous programmes and has to be developed further in order to meet the users' information needs. The library plays an important role in promoting the university's aim, which is to deliver excellent results in education.

REFERENCES

Boone, R.W. (1994) The Information Needs and Habits of Humanities Scholar. *RQ*. [Online] 34 (2), 203–215. Available from: http://www.jstor.org/stable/20862645?seq = 1&cid = pdfrefe-rence#references_tab_contents [Accessed 21th August 2016].

Bopp, R.E. & Smith, L.C. (2011) *Reference and Information Services-An Introduction*. 4th edition. Santa Barbara, Libraries Unlimited.

Gurganus, A.S. (2015) *Virtual reference in a community college library: patron use of instant messaging and log in chat service*. Malibu, Pepperdine University.

Katz, B. (1993) *Reference service expertise*. New York, The Haworth Press, Inc.

Shachaf, P., & Snyder, M. (2007) The Relationship between Cultural Diversity and User Needs in Virtual Reference Service. *The Journal of Academic Librarianship*, 33 (3), 361–367.

Tomic, T. (2012) *Evaluation of e-mail reference service in university Donja Gorica Library*. University of Ljubljana. Faculty of Arts. Department of Library and Information Science and Book Studies.

Wilson, T.D. & Wangsh (1981) On user studies and information needs. *Journal of Documentation,* 37 (1), 3–15.

Cultural Dynamics in a Globalized World – Budianta et al. (Eds)
© 2018 Taylor & Francis Group, London, ISBN 978-1-138-62664-5

Students' information behavior and the use of library reference service at Surya International School, Jakarta

R.R. Romadon & T.A. Susetyo-Salim
Department of Library and Information Sciences, Faculty of Humanities, Universitas Indonesia, Depok, Indonesia

ABSTRACT: The Internet has changed the information behavior of students in learning activities, especially in doing homework. The role of school library in providing reference service has been undermined by online search engines like Google, thus limiting staff size, collections, and opening hours. Even so, the richness of information resources on the Web presents counterproductive effects in students' ability to construct and articulate information on the basis of their understandings. In this study, we analyze the information behavior of students and its effects on the use of school library. Here, we use a qualitative approach to collect data through focus group discussions conducted at Surya International School, Jakarta. Findings from this study show that the digital native background of students has a significant effect on their information behavior and use of school library. The result of this study shows that the school library should adapt to students' information behavior and cooperate with other libraries to provide homework-assistance service both in person and online. This article is expected to provide insight into the information behavior of students for the evaluation and development of the reference service provided by the school library.

1 INTRODUCTION

The digital era has significantly changed the way people access information. Together with the increasing popularity of the Internet and the mass use of new mobile devices, access to information has become possible anywhere, anytime, and at the touch of our fingertips. The latest survey conducted by APJII (2016) indicates that 51.8% (132.7 million people) of Indonesia's total population use the Internet. It also indicates that 25.3% (31.3 million people) of Internet users are information seekers and 67.8% (89.9 million people) of Internet users use smartphones for information search. The result of this survey shows that people tend to access information through online resources from mobile devices rather than printed materials.

This phenomenon has also changed the information behavior of students in learning activities at school, especially on assignments. APJII also stated that, on the basis of profession, students are the second largest group of Internet users in Indonesia (69.8%). The National Library of Indonesia also stated that more than 70% of public library users are students (National Library of Indonesia, 2013).

Most young people who are known as digital natives are very comfortable with ICT, and this increases their expectations toward the services provided by school libraries, especially the reference service. In general, they use school libraries to complete homework assignments in time. However, in order to meet their increasing expectations, students will rely heavily on search engines, thereby diminishing the use of the library's reference service. School libraries, which are known for their comprehensive collections, also face problems, such as the limited staff size, collections, and opening hours. These limitations influence students' perception that school libraries can be substituted by search engines. However, the rich information resources on the Web have a counterproductive effect on the ability of students to construct and articulate information based on their understandings. Thus, rather than approaching this as an obstacle, it should be construed as an opportunity for the school library to maintain its role as the center for information.

For this purpose, we discuss identifying the information behavior of students and its effects on the use of school library reference service.

2 LITERATURE REVIEW

The Internet has changed the information behavior of students in learning activities, especially assignments or homework, by providing easy online information search. Although a wide disparity still occurs in access and skills based on socioeconomic status, most young people are very comfortable with technology, which leads to their increased expectations of library services. Nevertheless, the richness of information resources on cyberspace presents counterproductive effects in the ability of students to construct and articulate information on the basis of their understandings (Carr, 2008, Gunter, 2009, Jamali, 2011).

Even though students have their own personal information needs and interests, they generally use the library reference services for completing their homework assignments (Cassell & Hiremath, 2009). Information behavior includes selecting, searching, and using relevant information to complete an assignment.

In this section, we reviewed the existing literature that includes aspects of digital natives, information behavior, especially information capture and related theories as well as evaluation of library reference service.

2.1 *Digital natives*

The term "digital natives" was coined and popularized by Marc Prensky in 2001. It refers to young people who have grown accustomed to digital devices in their daily life (Dingli, 2015). Similar but more popular terms that are used interchangeably to refer to this group are "Google Generation", "Net Generation", and "Millennials".

In general, digital natives reside in major cities in which the use of technological and communication devices has increased rapidly and has a cultural impact on the community. Rowlands (2008) and Gunter (2009) defined this generation as mostly those born after 1993, since when the Internet has begun to develop as a source of information and communication media for public use, unlike its previous usage, which was restricted to military and academic environments.

Rowlands et al. (2008) claimed that although young people demonstrate an apparent effortlessness and familiarity with ICT devices, they rely heavily on search engines to view than to read, and they do not possess the critical and analytical skills to assess the information found on the Web.

2.2 *Information behavior*

Wilson (as cited in Pijpers, 2010) stated that information behavior comprises two key activities, namely information seeking and information use. Information-seeking behavior is defined as the purpose of seeking information to achieve certain goals. While seeking for information, the person may also interact with manual information systems, such as a library and an information center, or with ICT-based information systems, such as the Internet. Meanwhile, information-use behavior is defined as the physical and mental activities involved in incorporating information found in a person's existing knowledge base.

One of the IB theories that suits students or young people is the berrypicking theory developed by Marcia J. Bates in 1989. This theory shows the impact of the search process in the results. During the information-seeking process, the questions are refined gradually until they meet the needs of the information seeker. This theory is indicated through the trial-and-error stage in the process of using search engines.

Table 1. Berrypicking theory and information capture concept.

Berrypicking theory	Information capture
• Typical search queries are not static but evolving; • Searchers commonly collect information in bits and pieces instead of in one grand best-retrieved set; • Searchers use a wide variety of search techniques that extend beyond those commonly associated with bibliographic database; • Searchers use a wide variety of sources other than bibliographic databases.	• Handwriting copy of information that is encountered; • Photocopying a printed source, such as book chapters or magazine articles as a whole or partially; • Printing out an electronic document, such as Web pages, eBooks, and online journals; • Copying and pasting material from electronic sources into another file, often a Word document, which a student may be preparing for assignment purposes.

Information capture is another theory similar to this concept (Shenton, 2010). It involves the representation of materials, such as facts, opinions, and interpretations, in a form that renders them useful for some later activity.

2.3 The use of library reference service

In a digital environment, the interaction between librarians and users will no longer be truly face-to-face. Whitlatch (2001) stated that reference librarians must shift from their reference desks and develop new strategies to connect library materials to its users. Librarians in school libraries can develop outreach programs for students, as well as library instructional and consultation service, or homework-assistance programs (Intner, 2010). All of these efforts must be made for the library reference service to adapt, change, survive, and prosper in the new digital age.

3 METHODS

This study uses a qualitative approach with the focus group discussion (FGD) technique at Surya International School, Jakarta. We conducted a semi-structured and face-to-face interview for collecting data. FGD is a group discussion often consisting of 6–12 participants and guided by a facilitator or researcher to gain an understanding of their attitudes and perceptions that are relevant to the topic (Gorman, 2005). We used two methods of data recording from discussion, namely tape-recording and immediate note-taking after the discussion, to evaluate the library service. It also allows the participants to describe and express their behavior, beliefs, viewpoints, and preferences in their own words (Beck, 2004). The remaining of this section describes the preliminary interview with teacher-librarian, participants, interview structure, and interview analysis.

3.1 Preliminary interview with teacher-librarian

Before starting the FGD session with the students, we interviewed the teacher-librarian to get a first impression of the library itself, characteristics of its users, and problems in the services of the school library. The information obtained from this preliminary interview session is used to construct the quality questions of the study. In order to acquire sufficient background to develop the questions, we must have thorough knowledge about the study setting. We use this interview as a "springboard" to facilitate the discussion and build a list of possible questions for the next session with the students as the participants.

3.2 Participants

From the preliminary interview with the teacher-librarian, we learn the characteristics of the users of the school library and the low use of its service. Interviewing the students directly was not easy as most of them were not comfortable talking to strangers. In order to help us collect the data, we sought assistance from the teacher-librarian to persuade the students to let us conduct the interview in their leisure time at the library. We interviewed seven students from a junior high school, Surya International School, Jakarta. It is interesting to learn that all of the participants have learned in the library or acquired literacy skills since grade 9. They have weekly library time or information literacy session in the class.

3.3 Interview structure

The interviews were loosely structured and started by asking breaker questions such as how much they depend on digital devices. In the next session, we asked the students about their reasons and perspectives for using Google rather than existing library materials. Finally, we continued asking about their experiences, feelings, and opinions regarding the library. The interviews were conducted in the library room at Surya International School, Jakarta (SISJ) when the students were on their break time, which they would usually spend studying, doing assignments, and waiting for the next class session.

3.4 Interview analysis

The interview results were transcribed and interpreted by coding, and the data were linked by connecting using relevant codes to construct the data, which were to be displayed graphically. For the internal validity of the study, we used triangulation by sharing the categories or themes with the participants and asked them whether the conclusions match reality. We also used the external validity to examine how the study could represent the problem in general and whether its results or conclusions could be applied to similar studies in the future.

4 FINDINGS

Our interview session results in the identification of the students' digital native background, deadlines of their assignments, the lack of library collections, their false assumptions, their information behavior, and the use of the library reference service.

4.1 Digital native background

Most of the students in SISJ are digital natives. On the basis of their age, we can perceive that they were all born after 1993 (Rowlands, 2008, Gunter, 2009). This is indicated by the familiarity in using mobile devices or gadgets for their daily activities both inside and outside the classroom. They use these devices for three main reasons, namely communication, entertainment, and online search, as explained by interviewees 1, 4, and 5:

P1: We use cellphones and laptops every day.
P4: We use it for studying and accessing social media, but mostly for communication.
P5: I bring it [cell phone] into the class. I never go to school without it.

During the interview session, we observe that students always bring their mobile devices or laptops for completing their assignments through Google search.

4.2 Assignment deadlines

The deadlines for assignment completion also increase students' need for information. They want the information to be available to them quickly, easily, anytime, and anywhere, as expressed by interviewees 1, 5, and 7:

P1: Searching information in those books must take a long time.
P5: To meet the deadline, we just copy-paste because we don't have time to read it all at once.
P7: We prefer searching eBooks. We already have the textbooks.

Because of their digital native background, the deadlines set for the students' assignment sled to their high expectations toward the library to fulfill their needs. However, the school library's limited collections make them search for the fastest and easiest platform as an alternative solution, namely Google.

4.3 Lack of collections

Furthermore, there have been some complaints regarding the library's limitations, such as lack of collections, room size, computer units, and inadequate Internet connection.

P5: Maybe it is because of the lack of collections here, so we would rather search on Google instead.
P2: Add the number of collection, at least the books that we have not bought yet.
P5: Bigger rooms, I hope.
P3: I wish there were a lot of computers here.
P2: Sometimes the Internet connection is not available here.
P1: Actually the connection is fast, but many students connect to the network causing the connected to slow down.

The school library has limited collections, staff, and opening hours. The lack of collections has led the students to the false assumption that the library is now outdated and obsolete (Herring, 2014). They even think that everything they need is available on the Web (Rowlands, 2008).

4.4 Students' false assumptions

The students, with their digital native background, are disappointed with this condition and feel they no longer need to use the library, as everything is available on the Web (Rowlands, 2008). This is expressed by interviewees 1, 2, 3, and 5:

P1: We don't rely on it [library] anymore for assignments. Searching information in those books must take a long time.
P2: On Google, everything is available.
P3: It [library] doesn't help at all.
P5: Yeah, we don't need it [library] anymore.

From these statements, we found these false assumptions to be caused by the students' digital native background. The Web is never a substitute to the library, but rather an additional tool for information seeking (Herring, 2014). Therefore, the students' false assumptions are due to their digital native background, assignment deadlines, and lack of library materials.

4.5 Students' information behavior

There are two major findings regarding the students' information behavior:

1. The students still experience problems in formulating keywords for search engines that are suitable for their needs. Even though they have been taught literacy skills at previous levels, namely elementary and junior high, the abilities of students are still limited when articulating their information needs into words or terms that are adequate for online search:

   P6: *The "super-specific" topics are very hard to find. As it is, right now I am working on my paper in the area of psychology. It's so hard to find anything. I have difficulties formulating the right keywords for the search. But sometimes, it suddenly appears on my mind, even though the process takes time.*

P7: *To find the keywords, we do trial and error. Through this process, we can gradually get what we are searching for.*

Most of the students start the search by phrasing the questions vaguely from which thousands of search results are generated. This process is repeated by refining the keywords until they find the most suitable sources for them. This process is in line with the principles of the berrypicking theory (Bates, 1989).

2. The students tend to copy-paste the information from the online sources found on the Internet without analyzing it or having a comprehensive understanding of it. They are not even sure whether the obtained information is reliable or not:

P7: *All we need to do is just search and copy-paste the information to the paper. That's all.*

P5: *Well, it depends on what kind of materials is needed for the assignment, or which websites we want to use. If the source is unreliable, such as Wikipedia, we can't use it as a reference. But to meet a deadline, we just copy-paste it, because we don't have time to read it all.*

P2: *Unfortunately, we didn't have any session on library skills last semester. But when we were in elementary school, we were already taught about how to make citations.*

This copy-paste habit is one of the manifestations of information capture. Information capture is the key element in information behavior (Shenton, 2010). However, it does not only deal with the transformation from one format to another. It must be engaged with the transformation of information on the basis of the individual's interpretation and understanding.

4.6 *The use of library reference service*

Information behavior of students has an impact on the use of library reference service as described below:

P1: We don't rely heavily on the library's collections for our assignments, because there is an Internet connection here, right? Searching for those books must take a long time.

P4: They [collections] don't help us at all.

P7: We prefer searching for eBooks on the Web. We already have the tutorial books.

P5: Yeah, we don't need them [collections].

P2: On Google everything is available.

P5: Maybe it is because of the lack of collections here, so we prefer to search on Google instead.

From the transcript, we discover that there are both direct and indirect impacts of the information behavior of students on the library reference service. One of the direct impacts is

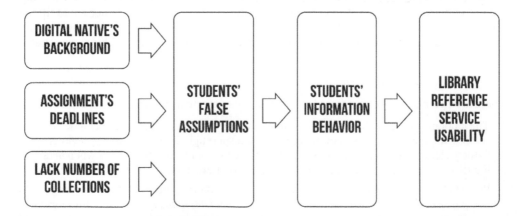

Figure 1.   Findings from the FGD session.

their assumptions that they no longer need to use the library reference service for completing their assignments as everything is available on the Web (Rowlands, 2008). Meanwhile, one of the indirect impacts is their preference to use the library room because it is the most quiet and comfortable place to study and discuss with their friends and teachers:

P4: We use this room [library] just for studying, for using our laptops only because it is quiet here.

P5: We can't use the student lounge, because it's too noisy.

P1: This is a good spot for hanging out with friends and teachers.

From these findings, we discover that the students' digital native background, their assignment deadlines, and the lack of library collections are the main factors behind their information behavior and false assumptions that everything is available on the Web. Finally, the information behavior has both direct and indirect impacts on the library reference services undermined by search engines. The analysis results are described in Figure 1.

5 CONCLUSION

We conclude that the information behavior of students arises because most of them are digital natives and have been accustomed to carrying out online search using their digital and mobile devices. Consequently, the students' assumption that everything can be found on the Web is reinforced by the limited number of library collections that can help them. Even though the students have been taught library literacy skills, they still need guidance and consultation from the librarian.

The identified information behavior of students includes: (1) the problems in articulating and formulating the appropriate keywords during the information-seeking stage and (2) the habit of copying and pasting the information found on the Web to their assignments without analyzing the level of reliability and validity.

The information behavior of students also has direct and indirect impacts on the use of library reference service. The direct impact is that they no longer use the library reference service and collections but prefer to search for information through search engines. The indirect impact is that the library becomes a quiet and suitable place for the students to study, discuss, or spend their leisure time with friends, librarian, or teachers. On the basis of these impacts, the librarian must evaluate and formulate new strategies in the development of the reference service in order to adapt with the information behavior of students.

5.1 *Recommendations*

The library can develop an in-house outreach program for students and change the term *reference service* to *homework assistance,* so that it can be adjusted to fulfill the information needs of the students in the digital era (Pfeil, 2005, Sonntag & Palsson, 2007, Dresang & Koh, 2009, Intner, 2010, Gratz, 2011). The homework-assistance program can assist the students in areas where the Internet cannot help. This program can also illustrate how the students can benefit from the library with all the available sources of both printed and online materials regardless of setting. This is why the school library must also maintain good coordination and collaboration with other school libraries, academic libraries, local public libraries, and even the national library.

On the basis of the conclusions drawn, we find that the new important role of the reference librarian at school is to develop cognitive skills on information-seeking behavior for students (Mokhtari, 2013). To overcome the limitations of school libraries, such as those related to collections, staff, and opening hours, the library must make good use of the current popular communication and information delivery tools, such as Webpages, databases, the Internet, Facebook, and Twitter. The result expected from this outreach program is the increase in the ability of the school library to provide better reference services and library consultation both in person and online.

REFERENCES

Asosiasi Penyedia Jasa Internet Indonesia (APJII). (2016) *Penetrasi dan Perilaku Pengguna Internet di Indonesia* (Penetration and Behavior of Internet Users in Indonesia). Retrieved from http://www.apjii.or.id/survei2016.

Bates, M.J. (1989) *The Design of Browsing and Berrypicking Techniques for the Online Search Interface.* Retrieved from University of California website: https://pages.gseis.ucla.edu/faculty/bates/berrypicking.html.

Beck, S.E. & Manuel, K. (2004) *Practical Research Methods for Librarians and Information Professionals.* New York, Neal-Schuman.

Bopp, R.E. & Smith, L.C. (2011) *Reference and Information Services: An Introduction.* 4th ed. Englewood, Colo., Libraries Unlimited.

Carr, N. (2008) *Is Google making us stupid?: What the internet is doing to our brains. The Atlantic.* Retrieved from http://www.theatlantic.com/magazine/archive/2008/07/is-google-making-us-stupid/306868/.

Cassell, K.A. & Hiremath, U. (2009) *Reference and Information Services in the 21st Century: an Introduction.* 2nd ed. London, Facet.

Dingli, A. & Seychell, D. (2015) *New Digital Natives: Cutting the Chord.* Heidelberg, Springer.

Dresang, E.T. & Koh, K. (2009) Radical Change Theory, Youth Information Behavior, and School Libraries. *Library Trends*, 58(1), 26–50. Retrieved from http://search.proquest.com/docview/220466124?accountid=17242.

Gorman, G.E. & Clayton, P. (2005) *Qualitative Research for the Information Professional: a Practical Handbook.* 2nd ed. London, Facet.

Gratz, A. & Gilbert, J. (2011) Meeting student needs at the reference desk. *Reference Services Review*, 39(3), 423–438. doi: 10.1108/00907321111161412.

Gunter, B. (2009) *Google Generation: are ICT Innovations Changing Information-seeking Behaviour?* Oxford, UK, Chandos Publishing.

Herring, M.Y. (2014) *Are libraries obsolete? An argument for Relevance in the Digital Age.* Jefferson, North Carolina: McFarland & Co., 2014.

Intner, C.F. (2010) *Homework Help from the Library: In Person and Online.* Chicago, American Library Association.

Jamali, H.R., & Asadi, S. (2009) Google and the scholar: the role of Google in scientists' information-seeking behaviour. *Online Information Review*, 34(2), 282–294. doi:10.1108/14684521011036990.

Mokhtari, H., Davarpanah, M., Dayyani, M. & Ahanchian, M. (2013) Students' need for cognition affects their information seeking behavior. *New Library World*, *114*(11/12), 542–549. doi: 10.1108/NLW-07-2013-0060.

National Library of Indonesia. (2013) *Indonesia Province and District/Cities Public Libraries Profiles.* Jakarta, National Library of Indonesia.

Palfrey, J. & Gasser, Urs. (2008) *Born digital: understanding the first generation of digital natives.* New York, Basic Books.

Pfeil, A.B. (2005) *Going places with youth outreach: smart marketing strategies for your library.* Chicago, American Library Association.

Pijpers, G. (2010) *Information overload: a system for better managing everyday data.* Hoboken, NJ, John Wiley & Sons.

Rowlands, I., Nicholas, D., Williams, P., Huntington, P., Fieldhouse, M., Gunter, B. & Tenopir, C. (2008) The Google generation: the information behaviour of the researcher of the future. *Aslib Proceedings*, 60(4), 290–310. doi:10.1108/00012530810887953.

Shenton, A. (2010) Information capture: a key element in information behaviour. *Library Review*, 59(8), 585–595. doi:10.1108/00242531011073119.

Sonntag, G. & Palsson, F. (2007) No longer the sacred cow—No longer a desk: transforming reference service to meet 21st-century user needs. *Library Philosophy and Practice.* Retrieved from http://www.webpages.uidaho.edu/~mbolin/sonntag-palsson.pdf.

Whitlatch, J.B. (2001) Evaluating reference services in the electronic age. *Library Trends*, 50(2), 207–217. Retrieved from https://www.ideals.illinois.edu/bitstream/handle/2142/8400/librarytrendsv50i2_opt.pdf?sequence=3&isAllowed=y#page=50.

Cultural Dynamics in a Globalized World – Budianta et al. (Eds)
© *2018 Taylor & Francis Group, London, ISBN 978-1-138-62664-5*

Transformation from conventional to virtual library reference services at Surya International School, Jakarta

D.P. Prabowo & T.A. Susetyo-Salim
Department of Library and Information Sciences, Faculty of Humanities, Universitas Indonesia, Depok, Indonesia

ABSTRACT: The school library as a learning center has an important role in reference service for students working on their assignments. However, various gadgets, the emergence of the Internet and popular search engines, and online encyclopedias drive students to search information from cyberspace therefore abandoning the library reference service. Nevertheless, the richness of information resources from the Internet has counterproductive effects on students' ability to construct and articulate information based on their understanding. This study analyses how reference service transformation is beneficial for the students' learning process at school. It uses a qualitative approach with in-depth interview and non-participant observation to examine the school library at Surya International School Jakarta (SISJ). Findings from this study show that students would rather search information online from their own gadgets than go to the library to work on their assignments. Thus, the school library is required to transform its reference service from conventional to virtual-based with the newest ICT commonly used by students. Chat To Librarian, which is a gadget-based reference service, can be an alternative solution for students. This paper is expected to provide a new perspective on school library reference services in a digital era.

1   INTRODUCTION

With the digital revolution, libraries are expected to adapt to changes in order to stay relevant to the developing era and the demands of users, especially with regard to services at school libraries. As a learning center, the school library plays an important role in helping students with assignments at school. Overwhelming information search on the Internet has affected school libraries directly and indirectly. Directly, there is a diminishing number of students' visits to school libraries. Therefore, many library collections and services within school libraries are used less optimally by students. Users nowadays prefer to search for information on the Internet via smartphones, because it is easier and faster than having to visit the library. Indirectly, the library as a source of information is gradually undermined in the eyes of users by sophisticated search engines like Google. These search engines are capable of generating thousands to millions of search results based on search text input within seconds. If we compare it to conventional library information service, we are not going to find similar information results which are as much or fast as Google. Therefore, in order to maintain existence and credibility in the digital era, libraries must transform their reference services by reforming and innovating library services to be able to fulfill the needs of users today.

The reason is that this service requires both physical and virtual interactions with users, unlike most of the services at libraries. The reference librarian is considered the 'spearhead' of library service. In other words, in addition to the quantity and quality of its collection, the overall library service quality depends on its reference service. Based on the problem previously stated, this article supports the idea of the transformation of reference service at school libraries into virtual service.

2 LITERATURE REVIEW

2.1 *Reference service*

Reference service is provided by librarians tasked to assist the selection and obtainment of information based on user request (Bishop, in Genz, 1998). Another definition is given by the Reference and User Service Association (RUSA) in Bopp and Smith (2001) in which reference service is defined as library information service and consists of various activities, including personal guidance, directory, information provision selected from reference sources, and access to digital information for library users. The term 'reference service' was first coined by Samuel Green in 1876. Green was a librarian at Worcester Public Library in Massachusetts, USA. The term developed from his opinion on the need for librarians to help users select books based on their needs.

2.2 *Reference service type*

According to Bopp and Smith (2001, p. 6), reference service consists of three basic types, namely information provision, library information resource guidelines, and information resource selection guidance. These three basic types of service are further expanded on by Cassell (2010) as follows:

• Library information service provides information service to users based on their requests and fulfills their specific needs;
• Learning service is the ability of the librarian to provide e-learning service on over-selection of information and easy access;
• Guidance service is similar to learning service. However, it provides more direct instructions and guidance to those who need guidance.

2.3 *Teenagers and reference service at school library*

Reference service has grown well along with technological development and the needs of the users they serve (Godfrey, 2008). The virtual reference world is still in its early stage of development, and many studies need to be conducted to determine its actual effectiveness (Forrest, 2008). Many studies have not been conducted on library service in the virtual world. Children growth rate is a hope for the virtual world, because today's children 'live' in the digital world and have the ability to explore virtually (Lou, 2010). It is important that the library explores how service can be provided in the virtual world, so that they are better equipped for the future.

Rapid electronic resource development as well as its availability in large amount causes libraries to change the vision and mission of reference service (Cassell, 2010). The growing rate of Internet usage also contributes to the need for libraries to change. Librarians and users are regularly bombarded with a variety of information whose amount is so significant that it requires an evaluation regarding authenticity and accuracy. The school library as part of a learning center at school also experiences change with the coming of technology (Beheshti, 2009). The large amount of digitized information has made people to prefer to search for information using computers and from the Internet. The rise of the Internet indulges people with free information resources that can be obtained more easily and faster. Reference service has to respond to this challenge by providing a combination of service that is personal in nature and based on users' requests.

School children today from pre-school to university have used digital equipment and information as an important part of their daily lives (Druin, 2008). Children (age 13–17 years old) have a very positive point of view towards technology, and technology is an important part of how they define themselves (Neuman, 2003). In the development of new technology and library service for the young generation, children's ideas may not be the first ideas to be considered. When visiting libraries, adults usually speak on behalf of children but seldom to them regarding resources and service planning or resources evaluation. Children and teenagers are potential users of virtual reference service, because they are familiar with computers and the Internet. Therefore, virtual service, such as Chat, videoconferencing, co-browsing, instant messaging (IM), Voice over Internet Protocol (VoIP) and e-mail, has to be implemented at libraries, especially school libraries (Cheng, 2008).

2.4 *Virtual reference service*

Many studies have been conducted on reference service and the evolution of virtual reference (see Lankes et al, 2000; Su, 2001; West, 2004; Sloan, 2006). However, a brief inspection on virtual reference will help determine its existence and place in the context of reference service. The separation of library service from the means of providing reference has changed and evolved in line with the needs of library users. Face-to-face reference service and reference interview are among the first to be developed. As technology progresses, telephone reference service has become an addition to reference service. The Internet has a deep impact on reference service and has delivered virtual or digital reference. Meanwhile, key features such as reference service are still in several different forms. The role of the reference librarian has changed, as users are able to perform a simple search by themselves.

Joint (2008) and Lankes (2000) indicate that traditional reference service provides mediation between users and collections. As the collections are transformed into online formats, assistance is required to locate the information. Virtual reference service develops as a means to provide assistance including through e-mail and chatting services. The first e-mail reference service can be traced back to the 1980s, while synchronization of chatting reference service began at the end of the 1990s (Stormant, 2007). Chatting and IM reference began as commercial projects, and many advantages were offered by co-browsing. Although virtual function is a good idea, it is often obstructed by technical problems, such as browser compatibility and problem authenticity. As a result, several libraries have shifted to IM delivery through existing IM service, such as AIM, Hotmail, and Meebo. Variation in the delivery method has also changed. Every new method provides a reference service that is in line with the previous service. IM reference service does not replace the traditional or virtual reference service. There is no single method exclusively appreciated by all library users and no replacement of one type of service with another.

3 RESEARCH METHOD

This study uses a qualitative approach with a case study method. The object of the study is the school library at Surya International School Jakarta (SISJ). Qualitative research aims to interpret the view point of the librarian and the students concerning the transformation of library reference service from conventional into virtual.

This study uses a purposive sampling technique in determining the respondents, and the chosen respondents are one librarian and three students at SISJ. According to Gorman, research data is collected through three basic activities, namely observation, interview, and document analysis. In this study, the author uses two activities, namely interview and observation (Gorman, 2005). Data analysis process in this research is conducted through test processing and analysis of the collected data before formulation of the conclusion based on theory. The steps in the analysis include data reduction, coding, presentation, and data interpretation, as well as the conclusion.

4 RESULTS AND DISCUSSION

Surya International School Jakarta (SISJ) is a well-known private international middle school in Jakarta with an international-based curriculum where every student is required to combine theoretical knowledge, a practice method in the field, and information technology usage. In addition, students are required to conduct studies at an early level, so it is necessary for them to be information literate. According to Cassell (2010), information literacy can only be developed at libraries through reference service. The function of the library at SISJ is to support the learning process at school. It provides a reference service called *Ask Librarian,* which can be accessed in person ore-mail.

During the school's early years, SISJ already had a school library, although at the time it did not provide reference service. The following is the SISJ librarian's statement:

> *"School library was founded together with this school. Its collection was not much, nor was its service. Reference service did not even exist. However, despite the limitations, we were still able to help students with their assignments." (Librarian, 14th September 2016).*

This is in line with the school library's function as a learning center for education and a simple research center enabling students to develop their creativity and imagination. The establishment of a reference service at SISJ school library is perceived to be the right decision, as it clarifies the librarian's role in providing and directing information service needed by students. At that time, reference service was carried out more often in person. Students would ask questions related to school assignments, such as how to search for the appropriate reading materials.

> *"The launching of the reference service is intended for us, librarians, so that in helping students we can be more directed." (Librarian, 14 Sept. 2016).*

Information technology development demands libraries to be able to fulfill all the needs of its main users, namely the students. Students' assignments are becoming more complex, so they need assistance in searching for literature in the library, as stated by the SISJ librarian below:

> *"ICT develops at a fast pace. As a consequence, the students' assignments are becoming more difficult and complex. Many of them look for literature in the library and ask directly to the librarian. It is part of the teacher's role to direct students to visit the library." (Librarian, 14 Sept. 2016).*

Information technology exists along with the development of smart mobile devices. Almost all students at SISJ use these devices as a means of communication to help those complete assignments in school. Their opinion of the smart mobile devices is given below:

> *"Laptop, tablet, cell phone. These three devices; it's like everywhere [I take them]. Especially the cell phone" (Rosa, 14th Sept. 2016).*

Rosa's statement clearly shows that smart mobile devices are part of her daily activities. In addition, these devices can be used to study and search for information related to the assignments in school.

> *"Well, it's fifty-fifty. Sometimes it helps me study, like searching for information, enhancing knowledge, but sometimes I use the cell phone for communication." (Rosa, 14th Sept. 2016.)*

This phenomenon matches the thesis statement from Druin and Neuman in which nowadays students at school access and use digital equipment and information as an important part of their daily lives (Druin, 2008). Children (age 12–16) have a very positive point of view towards technology, and technology is an important part of how they define themselves (Neuman, 2003).

The development of information technology through smartphones is marked by easy access to the Internet. It has made students prefer to search for information from the Internet through smartphones, as it is easier and saves time compared to a visit to the library. Indirectly, the library's function as a source of information is gradually undermined in the eyes of users by sophisticated search engines like Google. These search engines are able to generate thousands to millions of information search results based on search text input within seconds. Compared to the conventional library information service, the results will not be as many and as fast as through Google.

> *"Of course, I'd prefer to use Google rather than [visit] the library because the information found in the library is not complete. Google is more complete and faster" (Dana, 14th Sept. 2016).*

Usually, conventional reference service is provided by a librarian at a reference desk whose task is to help and guide library users search for information through reference collections, newspaper clippings, and indexes. Unfortunately, this traditional reference service has many deficiencies, such as the librarian's lack of reference in terms of quantity and quality, limited acquisition and update on reference collections, as well as the promotion of the reference service itself. As a result, the library reference service is less optimal in service and less used by library users than it used to be. In order to survive in today's era, the reference service has to adapt to changes by adopting information and communication technology supporting its service expansion in order to reach users virtually. At the moment, the existence of sophisticated technology and the Internet in the digital era drives the transformation from conventional reference service to virtual reference service.

According to the Reference and User Services Association (RUSA) of ALA, virtual reference service is reference service that is provided electronically, just like any other real-time service in which users use computers or other Internet-based equipment to communicate with a reference librarian without having to be present physically at the library building (Kern & Lankes, 2009). Based on the definition above, we can conclude that virtual reference service uses intermediate reference librarians to answer various reference questions in the virtual world. The communication channels that are generally used are online chat, web messenger, video conference, voice over IP, co-browsing, e-mail, and instant messaging.

Library users nowadays have understood the method to search for information instantly from the Internet using search engines, such as Google, as demonstrated by a survey released by Google Indonesia which stated that 61% of the urban society in Indonesia prefers to access the Internet through smartphones in a total of 5.5 hours per day (Jcko, 2015). However, even with easy access to the Internet and sophisticated technology, the abundance of search results requires guidelines and validity check. Therefore, guidelines and assistance from reference librarians are still in demand because the users need to filter and interpret the search results from various sources on the Internet. Up to this moment, libraries are still relied on for valid and trustworthy information. This is a great opportunity that must be grasped in order to survive in the digital era.

At present, the reference service at SISJ library is called *Ask Librarian*. With this service, users can ask questions related to assignments in school to the librarian. *Ask Librarian* is carried out through e-mail and the telephone. The users of *Ask Librarian* are not only students, but also teachers and staff members. According to the SISJ librarian, after the service was launched many students have been interested to use it. They even asked things related to their subjects through e-mail.

> *"Usually students ask about how to write bibliography, footnote, and all the things related to the writing format of scientific articles. They also ask about the use of reference collections, such as encyclopedia, how to cite from them, and what websites provide free articles to be downloaded. As the students at this school are already introduced to assignments related to research, they need specific reading materials, such as online international journals." (Librarian, 18th September 2016)*

The statement above shows that although students are adept at using Google as their information search tool, they still need the library and the librarian to explain to them how to write bibliography as part of their research assignments. This demonstrates the relevance of the reference service at the library.

"They can use Google by themselves to search for information. They are adept at that. The task of the librarian here is to direct them to reliable information sources on the Internet. For this task, the librarian cooperates with the teachers, as the students will submit their assignments to them. The teachers together with the librarian will direct the students to information that is relevant. We also teach them about plagiarism, because of the anti-plagiarism principle that we have here, so any source cited has to have its copyright examined. This way they know how to cite from the Internet. The point is that we can't just take everything from the Internet." (Librarian, 18th September 2016).

A search strategy is the most frequently asked question from students through *Ask Librarian*. Sometimes students experience difficulties to obtain the information that they need from the search engine because of an incorrect search strategy. This is where the users can benefit from the librarian's guidance.

"Usually they ask the questions during class. Sometimes they ask about the keywords to search for information for their assignments. Then they search for information using a certain keyword, but if they can't find the result they ask a librarian about it, for example, the synonym of the keyword. Well, that's what happened, more or less. Sometimes in thesaurus, we can find this term, main term, related term, and what they are. So, sometimes they don't know about a particular term to use as the keyword for a certain subject. The librarian also teaches them the Boolean logic so that the search results in relevant findings. Because sometimes they just input any keyword, but if they use the Boolean logic, there will be codes, like 'File type: pdf', so the search results will all be in PDF format. Another example is the combination of two keywords using 'AND' or one keyword using 'OR'" (Librarian, 14th September 2016).

Information search according to Gash (2000) is important because the "life" of information service in an information unit or library is in fulfilling information needs demanded by users, finding information needed by users, and providing the "way" for users to obtain the information needed. The information search process is important to yield relevant, accurate, and precise findings or information. A precise process and equipment use will generate precise information as well. A series of field studies show that the SISJ school library has transformed its conventional service into a virtual one through e-mail. This media can be accessed from various gadgets in order to make it easier for students to contact the librarian when they face difficulties with their assignments.

5 CONCLUSION

The library reference service remains relevant and is still needed by users in the digital era. In order to adapt to current developments, today's reference service has developed into a virtual reference service. This service, besides presenting valid and trustworthy information search results, also provides understanding to users, especially SISJ students about filtering and interpreting information obtained from various sources both printed and electronic. This is the distinguishing point and opportunity for libraries to survive in the digital era. Although users nowadays tend to use gadgets to access the Internet rather than to visit libraries, the library's role is not going to be replaced any time soon by sophisticated search engines like Google. Libraries are expected to find new ways to promote and present services that are in line with users' needs in the 21st century and the virtual reference service.

REFERENCES

Beheshti, J. (2012) Virtual environments for children and teens. *In tech.* 20 (3), 271–287.
Bopp, R.E. & Smith, L.C. (2011) *Reference and Information Services: An Introduction.* 4th ed. Englewood, Colo., Libraries Unlimited.

Cassell, K.A. (2011) *Reference and Information Services in the 21 st Century: An Introduction.* New York, Neal Schuman Publisher, Inc.

Cheng, Y.L. (2008) Virtual reference service. Bulletin of the American Society for Information Science and Technology. 34 (2), 6–7.

Collins, J.W. III. & Kasowitz, A.S. (Eds). (2000) *Digital Reference Service in the New Millennium: Planning, Management, and Evaluation,* New York, NY, Neal-Schuman Publishers.

Druin, A. (2003) *What Children Can Teach Us: Developing Virtual Reference for Children with Children.* Maryland, University of Maryland.

Forrest, L. (2008) Overvaluing the virtual, *American Libraries,* 39 (3) 11.

Gash, S. (2000) *Effective Literature Searching for Research.* 2nd ed. Hampshire, Gower.

Godfrey, K. (2008) A new world for virtual reference, *Library Hi Tech,* 26 (4), 525–539.

Gorman, G.E. (2005) *Qualitative Research for the Information Professional: A Practical Handbook.* London, Facet Publishing.

Joint, N. (2008) Virtual reference, Second Life and traditional library enquiry services, *Library Review,* 57 (6), 416–23.

Lankes, R.D. (2000) *Introduction: the foundations of digital reference.* In: Lankes, R.D.

Lou, L. (2008) Reference service in second life: an overview", *Reference Services Review,* 36 (3), 289–300.

Neuman, D. (2003) Research in school library media for the next decade: Polishing the diamond. *Library Trends,* 51(4), 503–524.

West, J. (2004) *Digital versus Non-digital Reference: Ask a Librarian Online and Offline.* Binghamton, Haworth Press.

Cultural Dynamics in a Globalized World – Budianta et al. (Eds)
© *2018 Taylor & Francis Group, London, ISBN 978-1-138-62664-5*

Empowering students with information literacy education as one of the library reference service functions at Surya International School, Jakarta

E. Cahyani & T.A. Susetyo-Salim
Department of Library and Information Sciences, Faculty of Humanities, Universitas Indonesia, Depok, Indonesia

ABSTRACT: Information plays an important role in our lives. However, information saturation makes it difficult to determine the accurate and reliable information. In this case, it is important to know how to search, obtain, and evaluate information from various sources in the current digital era. This circumstance has posed a new challenge to libraries as one of the institutions providing information. In this study, we examine the information literacy education program as an effort made by the library of Surya International School, Jakarta, to overcome the present challenge. Furthermore, we investigate the operation of the program in relation to the reference service in the library. We use a qualitative approach with a case study methodology. Data were collected through in-depth interviews and nonparticipant observations. Findings from this study show that information literacy skills help students in searching information in a library or the Internet efficiently. Students are able to obtain the exact information they need to complete their assignments and meet other purposes. In this study, we underline the importance of information literacy education in the early years of a student's life and its benefits on finding the right information.

1 INTRODUCTION

Library is not only about books, but also about providing services, which enable users to find a variety of reading materials for their entertainment or their research projects. This type of service is called reference service. The service began in the late 19th century when public libraries started to develop, and many people "did not know how to use the library" (Tyckoson, 2001, p. 186). One of the services offered is a library instruction. This means that through the reference service, librarians provide information and teach users how to use the facilities provided by the library.

With the advancement of technology, libraries face new challenges. In order to overcome these challenges, libraries have expanded their collections and facilities, allowing users to find information with ease. Moreover, with the rapid growth of information communication technology (ICT), there is a great demand for libraries to apply technology to cope with the increasing number of collections as well as the need for information.

At present, users are more familiar with electronic gadgets than in the past. The younger generation may have been introduced to gadgets in a very early age, which makes them fully conversant on electronics, including the Internet. Those belonging to such generation are called digital natives, a term coined by Marc Prensky. He used the term to describe those born in the last decade of the 20th century, the time when digital technology gained its momentum. "They have spent their entire lives surrounded by and using computer, video games, digital music players, video cams, cell phones, and all the other toys and tools of the digital age" (Prensky, 2001, p. 1). According to Alvermann (as cited in Selwyn, 2009, p. 365), they are perceived as "being literate in media and ICTs in ways that exceed what many [adults] know or even consider worth knowing." However, those who are not "born into the digital world but have, at some later point in their lives, become fascinated by and adopted many or most aspects of the new technology are digital immigrants" (Prensky, 2001, pp. 1–2). Therefore, library users of today comprise not only digital immigrants but also digital natives. At present, the challenges for librarians who are mostly digital immigrants are more complicated due to the rapid development of ICT. These librarians should know how to use the technology and understand the various challenges faced by users in using the sophisticated technology at the library.

Although ICT enables easy access to information without having to visit a library, users are prone to gain false and misleading information from the Internet. According to Bopp and Smith (2011, p. 223):

> "Whereas prior generations may have struggled with a scarcity of information, the challenge for today's user is finding what is needed in an environment of information abundance. In today's world, the information that is needed may be obscured by the sheer volume of resources available or the distraction users face as related but not quite 'on topic' information draws attention from one's intended search task."

Moreover, students of this generation tend to rely on the Internet for information related to their daily activities and school assignments. They "...do not distinguish between search engine results and, scholarly, authoritative databases" (Galvin, 2005, p. 354), because they often have "no concepts of how information is organised...do not understand indexing and are unaware of controlled vocabulary...and do not necessarily understand the distinction between Internet resources and information available from electronic databases" (Warnken, 2004, p. 153). Keen later described (as cited in Selwyn, 2009, p. 368) that they are "taking search-engine results as gospel, thus fostering a younger generation of intellectual kleptomaniacs, who think their ability to cut and paste a well-phrased thought or opinion and makes it their own".

Information on the Internet is supposed to be filtered to find the accurate one out of the many found from using a search engine, because

> "...information comes to individuals in unfiltered formats, raising questions about its authenticity, validity, and reliability. In addition, information is available through multiple media...and these pose new challenges for individuals in evaluating and understanding it (ALA.org)."

Filtering information is therefore important, especially for academic purposes where every piece of information needs to be verified and accountable. Library users, such as scientists, students, and professionals, should be very careful in searching information so as to find the reliable source of information. In general, the intended information is stored on certain Websites or in electronic journals requiring users' access to obtain the information needed. The role of librarians is to educate users the much needed skills, as stated by Galvin (2005, p. 354):

> "It is the librarian's responsibility to be attentive to the reference encounter and the opportunities for teaching presented in it, to guide the student to appropriate and valid online resources, and to teach the student to evaluate information found online."

Mastering this skill is crucial, especially for those whose works involve research. With this skill, one can find the data needed effectively and efficiently from reliable sources and combine them with other information to generate credible research.

2 LITERATURE REVIEW

2.1 *Information literacy*

The meaning of the term literacy has changed drastically. The current rapid development phase of the world demands everyone to be literate in all aspects of life. In the present era of digital information, in which abundant information is generated using ICT, everyone is required to be literate in terms of information. This requires searching, selecting, and using the information needed effectively in order to obtain true information for performing daily activities.

According to the Association of College and Research Libraries (2000), information literacy is "a set of abilities requiring individuals to 'recognize when information is needed and have the ability to locate, evaluate, and use effectively the needed information'. Therefore, an information literate individual is able to:

- Determine the extent of information needed;
- Access the needed information effectively and efficiently;
- Evaluate information and its sources critically;
- Incorporate selected information into one's knowledge base;
- Use information effectively to accomplish a specific purpose;
- Understand the economic, legal, and social issues surrounding the use of information and access and use information ethically and legally" (pp. 2–3).

2.2 Reference service

According to Tyckoson (as cited in Bopp and Smith, 2011), libraries perform three basic functions in their daily activities: (1) selecting, collecting, and preserving information for users; (2) organizing the information within its collections; (3) assisting users in their search and retrieval of information. The third function of libraries is known as reference service and has become a valid term worldwide as "the librarians who helped readers tend to use the books that were located in the reference collection; they gradually became known as reference librarians" (Bopp and Smith, 2011, p. 8).

Reitz (as cited in Bopp and Smith, 2011) further defined reference service as:

"including but not limited to answering substantive questions, instructing users in the selection of appropriate tools and techniques for finding information, conducting searches on behalf of the patrons, directing users to the location of library resources, assisting in the evaluation of information, referring patrons to resources outside the library when appropriate, keeping reference statistics, and participating in the development of the reference collection (p. 13)."

Samuel Green, the first person to discuss about the concept of library reference service, popularized this new concept in the late 19th century. He proposed the following four functions of a reference librarian (as cited in Tyckoson, 2001 and Bopp and Smith, 2011):

- Teaching people how to use the library and its resources;
- Answering the questions of library users;
- Aiding the library users in the selection of resources;
- Promoting library within the community.

Many changes have taken place in libraries since Green's statement in 1876. Libraries have gradually developed over time and applied various technologies to adapt to the ever-changing environment, the growing number of collections, and the needs for more complex information. However, reference service is always relevant and in high demand. Reference librarians have existed since the era of Samuel Green, and their function have not changed much according to Thomas Galvin (as cited in Bopp and Smith, 2011), which are:

- Assistance and instruction (formal or informal) in the use of the library;
- Assistance in the identification and selection of books, journals, and other materials relevant to a particular information need;
- Provision of brief, factual information of the "ready reference" variety (p. 12).

Reference service in libraries can be performed in many forms, especially with the current rapid development of technology. Some of the very early forms are still relevant to date, and some have been modified using ICT. The following list proposed by Bopp and Smith (2011) includes some forms of reference services offered in libraries around the world:

- Readers' advisory
- Ready reference

- Research consulting
- Subject specialist
- Bibliographic verification and citation
- Interlibrary loan and document delivery
- Instruction
- Literacy program
- Outreach and marketing (pp. 14–18).

3 METHODOLOGY

Data required for this study were obtained through interviews with 1 librarian and 13 randomly selected students. Seven students from grade 12 were interviewed face-to-face in a focus group discussion, and the other informants are six students from grade 10 who were interviewed via e-mail.

Questions related to their activities at the school library, methods of literature search for their school assignments, and the difficulties in searching for information from the Internet were asked to the students in the interviews. The librarian was asked to explain about what she teaches during an information literacy class, the type of reference service provided to the students, and the difficulties in teaching. The school library was observed to check the activities performed by the librarian and the students at the library and the types of reference services used by the students.

3.1 *Surya International School*

The school applies two curricula, namely national and international, known as the IB (International Baccalaureate). Students need to pass final examinations in both curricula to advance to a higher grade. There is no difference in terms of school assignment and teaching method between the two curricula. The learning process has an international standard. In this school, teachers encourage students to participate in class discussions and conduct research by searching information from various sources.

Students are welcomed to contact teachers if they face any difficulties. They can communicate with their teachers about their assignments via an application called Manage Bac outside the classroom. This application is used in international schools around the world for "efficient curriculum planning, assessment, and reporting while eliminating paperwork and enhancing communication" (managebac.com). Class schedules, students' attendance, class materials, school assignments, and conversations between students and teachers regarding assignments are displayed in this application. Therefore, students who could not attend a class can still keep up with the progress by reading the display in Manage Bac. Teachers, students, and librarians actively participate in this application to monitor the learning process and progress of students. Parents are also given the access to this application to view the learning progress of their children.

Most students studying in this school intend to continue their studies abroad. In this case, the school is seen as the place to equip the students with proficiency in a foreign language and to prepare students to be engaged in a learning process with an international standard. Therefore, the learning atmosphere in this school is different from that in state schools. Students are given the task to write essays, in which they are required to be creative and critical as well as observant of the environment around them.

3.2 *Findings*

To enhance their knowledge about the library and its use, students are directed to attend an information literacy program regularly. This program is perceived as useful to support the learning process of students in every subject taught in this school. This program is called library time and is not part of the school curricula.

Library time is a course in which students are taken to the school library to receive materials about the library and how to use the facilities in it. An application records students' attendance, materials, and assignments given, just like any other subjects of the curricula. However, their achievement in this course does not determine their advancement to the next grade. The class started last year and it is held once a week for 45 min. It is purposely intended for grade 10 students, because they have many essay assignments. Therefore, lesson materials given in the library time focus on using reference collection, writing bibliographies and abstracts, citing from printed books and electronic database, and obtaining information by exploring electronic databases that provide free e-books and e-journals.

Subjects taught in this class vary at each meeting. All students are required to bring their laptops or smartphones to practice the lesson. In the first session, they are given materials covering the basic knowledge about the library in general and its rules. They also learn about the school's library website, books, call number, and the arrangement of books. The course material also covers a more advanced level of information, such as how to explore the Web to find information from reliable websites for their essays' citation, how to use the principles of Boolean logic to find specific information, and how to write citation in MLA format. As for the next meeting, they are required to practice the lesson taught in the previous week. They can also apply this knowledge to their other assignments.

In fact, the course materials for this class are directly applicable and arranged in such a way to help students with areas or skills they find most difficult to master because

> "It is critical that students be taught to evaluate materials they find on the Web so they can make informed decisions about what would be considered reliable and appropriate for their research...this also means understanding the ownership of intellectual property as well as recognizing and accepting the responsibility to credit others for words and ideas in whatever format—text or graphic, print or electronic-they appear (Warnken, 2004, p. 154)."

Although students are familiar in Google search, "most often they do not consider the type of site or the validity of the information and are generally unaware of the concept of evaluation criteria" (Warnken, 2004, p. 153). Therefore, the librarian will direct and introduce them to credible websites providing accessible e-books and e-journals, such as EBSCO and Proquest by the National Library of Indonesia, because the school has not yet subscribed to electronic journals and books.

The librarian is always involved in the curriculum arrangement so that she can prepare for the next library collection development, which supports the learning process in this school. She communicates with other teachers to find the information needed for every class so that she can prepare the materials for the students visiting the library seeking those materials. Moreover, the librarian can work together with other teachers in assigning students certain essays that require them to search for information from not only electronic sources but also printed books. This exercise helps the students to recognize other sources and understand the activities of the library, including the organization of call numbers.

The relationship between reference service and the information literacy program depends on the frequency of students using the service. The more students need information to complete their school assignments, the more they will use reference service as stated by Galvin (2005, p. 354):

> "...students are more likely to use information skills when they learn these skills at the point of need. Reference librarians report that when patrons are aware of the wealth of resources provided by databases and other Internet sources, the reference desk can be busier than ever. Librarians do spend more time teaching those patrons who do come to the reference desk to use electronic resources."

This program allows students to ask and consult the librarian anytime. Although students are digital natives in nature, they still use the traditional method of consulting the librarian face-to-face. Some students use e-mail to contact the librarian; however, most of them feel more comfortable asking the librarian face-to-face. Moreover, they like to spend time studying or working on their assignments at the school library together. In this way, they can reach the librarian easily during any trouble.

3.3 *Obstacles*

The following are some of the difficulties found during the operation of this program. First, the perception of students about library and its contents prevents them from knowing the importance of library in supporting their life-long learning process. They are less interested in printed books and all issues associated with books. They consider them less important because printed books are outdated and the current trend is all about electronic. They prefer other types of materials related to gadgets and ICT as they can gain access to information faster and more easily. In addition, they tend to think they have found enough sources for their school assignments in the form of e-books and e-journals, which are available free of cost. Therefore, they think that library does not help much in terms of finding literature for their essays. This perception of students in which they "consider their searches successful when short search strings on popular Internet search engines yield many results,...are satisfied when they identify multiple sites" (Warnken, 2004, pp. 152–3) need to be corrected.

Their reluctance to learn about call numbers resulted in them getting lost in a larger library during material search. They are unaware of the fact that college or university libraries have several books, dictionaries, and encyclopedias and they have to find the information through their call numbers. The skill to identify call number is particularly useful when they need to search for information in such places. They do not know the importance of call number because of their school library being relatively small.

Because of their nature as digital natives, students are already accustomed to the use of gadgets in their daily activities. From the interviews with the students, it is obvious that they are familiar with Google search since their early years. However, sometimes they find difficulties finding the information they need because they do not type the precise keyword based on the principles of Boolean logic or they do not understand the vocabulary. Moreover, slow Internet connection seems to impede their efforts to find the right information.

4 CONCLUSION

In this paper, we showed the importance of information literacy skills for high-school students. These skills are especially helpful for academic purposes when the students have to obtain particular information to complete their assignments. Although students are familiar with Google search, they still need to be aware of the reliability, accuracy, and authenticity of the information they find through it. This awareness is built through the information literacy program that encourages students to be careful when searching for information in terms of credibility of the Web-based information.

It is expected that this program be continued to be taught in this school and in other schools, especially because students need to be prepared to attend colleges and universities. Being information literate is not only helpful in terms of academic life, but also that those people will be relatively more powerful than others. Schools and particularly school libraries should provide students with information literacy skills for them to obtain information efficiently and effectively.

REFERENCES

Bopp, R.E. & Linda, C.S. (editors). (2011) *Reference and Information Services—An Introduction* (4th edition). Santa Barbara, Libraries Unlimited.

Galvin, J. (2005) Alternative strategies for promoting information literacy. *The Journal of Academic Librarianship*, 31 (4), 352–357. Retrieved on 24th September 2016 from http://remote-lib.ui.ac.id:2076/10.1016/j.acalib.2005.04.003.

Information Literacy Competency Standards for Higher Education. (2000) [Brochure]. Chicago, Association of College and Research Libraries. Retrieved on 10th September 2016 from http://www.ala.org/acrl/sites/ala.org.acrl/files/content/standards/standards.pdf.

Prensky, M. (2001) Digital natives, digital immigrants. *MCB University Press*, 9 (5). Retrieved on 25th September 2016 from http://www.marcprensky.com/writing/Prensky%20%20Digital%20Natives,%20Digital%20Immigrants%20-%20Part1.pdf.

Selwyn, N. (2009) The Digital Native-Myth and Reality. *Aslib Proceedings New Information Perspectives,*. 61 (4), 364–379. Retrieved on 25th September 2016 from http://dx.doi.org/10.1108/00012530910973776.

Tyckoson, D.A. (2001) What is the best model of reference service? *Library Trends*, 50 (2) Fall 2001, 183–196. Retrieved on 20th February 2016 from https://www.ideals.illinois.edu/bitstream/handle/2142/8398/librarytrendsv50i2d_opt.pdf?sequence=1.

Warnken, P. (2004) The impact of technology on information literacy education in libraries. *The Journal of Academic Librarianship*, 30 (2), 151–156. Retrieved on 24th September 2016 from http://remote-lib.ui.ac.id:2076/10.1016/j.acalib.2004.01.013.

Cultural Dynamics in a Globalized World – Budianta et al. (Eds)
© *2018 Taylor & Francis Group, London, ISBN 978-1-138-62664-5*

Reconstructing Joko Widodo's national education ideology: A critical discourse analysis on the policies in Indonesia

R.M.P. Silalahi, N.J. Malik & E. Mulyajati
Department of Linguistics, Faculty of Humanities, Universitas Indonesia, Depok, Indonesia

ABSTRACT: Joko Widodo, the 7th president of Indonesia and one of the most popular politicians in Indonesia, is highly recognized for his controversial choices. Nomination of alleged corruption figures, cabinet reshuffles and death penalty of seven foreign drug convicts are some examples of Widodo's controversial choices. In an educational context, the discontinuation of the National Educational Curriculum 2013 and the abolition of the National Final Examination are a number of controversies undertaken during Joko Widodo's administration. In regard to the controversial choices made by President Joko Widodo in the education sector, this research aims to reconstruct Joko Widodo's ideologies of national education. In order to achieve this objective, the writers conducted a descriptive qualitative research on Joko Widodo's policies (which were reflected through the presidential speeches on education, legislation, and presidential and government regulation). Fairclough's three dimensional frameworks include 1) the linguistic description of linguistic elements, 2) the interpretation of discursive practice, and 3) socio-cultural explanation applied to answer the research question. To support the analysis, the writers analysed the linguistic elements by using Halliday's Systemic Functional Linguistics (SFL), incorporating three linguistic meta-functions (ideational, interpersonal, and textual). This research is expected to contribute to the improvement of discourse analysis studies and the Indonesian education system.

1 INTRODUCTION

Whether education is conducted toward oneself or toward other people, humans are always constantly involved in the process of education (Sukardjo & Komarudin, 2009, p. 1). This reality is the logical consequence of humans as cultural beings. The cultures build a social construction in which people/man must always involve themselves in the process of teaching and learning.

The Law Number 20 of 2003 on the National Education System defines education as a conscious effort to develop intelligence, competences, morals, and characters of the people for the sake of the country. This concept of education emphasizes the complexity of education which does not focus only on development of knowledge but also the development of moral and spiritual aspects. Therefore, the task and responsibility to improve education do not only lie on educational institutions but also on the entire Indonesian society.

The education in Indonesia is still in poor condition. In the survey conducted by Political and Economic Risk Consultancy (PERC), it is indicated that Indonesia is one of the countries with the worst educational system in Asia (retrieved from: http://www.kompas.com / kompas-cetak/0109/05/dikbud/mend09.htm). The poor quality of education in Indonesia is the result of instant education policies that are unable to meet the people's needs of education and to improve the development of the Indonesian people.

The current government of Indonesia led by President Joko Widodo is expected to provide a solution for the improvement of national education. A range of actions has been conducted by the current government in order to improve the education system in Indonesia. The discontinuation of the 2013 National Curriculum and the plan to discontinue the National

Examinations are among President Joko Widodo's controversial attempts to improve the quality of national education.

Upon recognising the poor condition of the national education system in Indonesia and after reviewing the efforts of President Joko Widodo to improve the national education, this research aims to reconstruct President Joko Widodo's ideologies of national education which are reflected through his political speeches during his term of governance. The process of analysis was grounded/based on the model of Critical Discourse Analysis by Fairclough (2013) which covers analysis on text dimensions, discourse practices, and socio-cultural context.

2 THEORETICAL FRAMEWORK

Fairclough (2013) states that Critical Discourse Analysis is conducted through three dimensional analyses, namely description of text, interpretation of discourse, and explanation of socio-cultural context. In this paper, the analysis on text dimensions was directed toward two elements of Systemic Functional Linguistics (SFL), namely lexicalization (the usage of lexical elements) and transitivity (syntactic classification at the clausal level).

Lexicalization is related to the usage of words or lexicons in a text. The usage of certain words or lexical configuration reflects a certain ideology that the creator of the text wishes to communicate and promote through a text. For example, the word *pembangunan* 'development' dominantly appeared in the speeches by former President Soeharto. This means that he emphasized development as the ideology and as the main goal in his political communication.

Transitivity is a construction of syntactic elements in a clause. It is used to represent reality through the elements of lexicogrammar. Halliday and Matthiessen (2014) classify transitivity into several processes in accordance with the functions of verbs in a clause, which are:

a. Material process, which refers to verbs that describe physical action, such as the verb 'make' and 'do';
b. Mental process, which refers to verbs that involve the five senses, such as the verb 'hear' and 'feel';
c. Verbal process, which refers to actions conducted using articulators, such as the verb 'say' and 'state';
d. Relational process, which is related to the action of 'defining,' 'symbolizing' and so on;
e. Behavioral process, which relates the doer with certain actions, such as the verb 'behave';
f. Existential process, which is related to verbs that describe the existence of some entities, such as the verb 'be'.

The interpretation of the discourse is related to the process of production and consumption of a text. Through this dimensional analysis, the information on the background and the implementation of power through text can be acquired. Explanation on socio-cultural context refers to the relation between the text and the social context outside of the text that influences communication process. This dimension covers social, political, and cultural aspects.

The result of this research unveils the ideology in the speeches by President Joko Widodo. Therefore, the term 'ideology' has to be defined. Fairclough (2013) defines ideology as a socially-shared belief communicated through text and geared toward certain social actions. It is assumed that the transcription of speech by Joko Widodo contains educational ideologies. The president as the leader of a nation promotes educational ideologies which will determine the direction of national education.

3 METHODOLOGY

This research used the model of Critical Discourse Analysis (CDA) by Norman Fairclough (2013) to examine speeches by President Joko Widodo. The data cover thirty seven (37) speeches that were delivered from 2014 (since Joko Widodo was officially elected as the president of Indonesia) to 2016. It was assumed that these speeches contained constituents that reflected the concepts of education. By following the design of CDA by Fairclough (2013), the data were analyzed in three levels of analyses, i.e. description of text, interpretation of text, and explanation of socio-cultural context.

4   ANALYSIS

President Joko Widodo emphasized the importance of aspects of character building, nationalism, and competences in his speeches. These three dimensions were clearly reflected in the lexical configuration explained below.

From the explanation above, it can be seen that President Joko Widodo gave emphasis on issues related to education and educational services. In addition, he explained other aspects which are related to:

1. character building
2. education on human values
3. civic education
4. two-way model of education
5. vocational education and training
6. educational equity

On the other hand, from the aspect of transitivity, President Joko Widodo utilized material process in his education discourse as can be seen in Table 2.

In the table above, it can be seen that constituents related to education acted as the entity that accepts material processes conducted by the actors or the doers of the action. This indicated that the actions conducted by the government in the context of education were directed toward the development of national education. Referring to the explanation above; it can be seen that education in President Joko Widodo's point of view was directed toward the issues below:

1. improvement of national education (no. 1)
2. priority on education sector (no. 2, 5, and 6)
3. cooperation in education sector (no. 3)
4. the citizens' access to education (no. 7, 8, 10, 13, and 16)

Table 1.   Lexicalization.

Lexicalization	Total	No.	Configuration	Total
Pendidikan 'education'	23	1	Bidang pendidikan 'education sector'	5
		2	Berpendidikan 'educated'	2
		3	pelayanan pendidikan 'educational services'	3
		4	pendidikan karakter 'character building'	1
		5	pendidikan kewarganegaraan 'civic education'	1
		6	pendidikan tentang nilai-nilai kemanusiaan 'education on human values'	1
		7	fase pendidikan 'phases in education'	1
		8	sistem pendidikan 'educational system'	1
		9	pembangunan pendidikan 'development of education'	1
		10	fasilitas pendidikan 'educational facilities'	1
		11	masalah pendidikan 'problems in education'	2
		12	pendidikan dua arah 'two-way education'	1
		13	Pendidikan Kejuruan dan Pelatihan 'vocational education and training'	1
		14	pemerataan pendidikan 'educational equity'	1
		15	pendidikan yang layak 'adequate education'	1

Table 2. Transitivity.

Participant (actor)	No.	Process	Participant (recipient)	Circumstances
Dana desa 'Fund'	1	material: memperbaiki 'improves'	pelayanan pendidikan 'educational services'	–
–	2	material: dialokasikan 'allocated to'	bidang pendidikan 'education sector'	10,4 triliun '10.4 trillion'
Kita 'We'	3	material: menyaksikan 'witnessed'	program kerjasama 'cooperation program'	di bidang pendidikan 'in education sector'
Kami 'We'	4	material: memiliki 'have'	tenaga kerja yang berpendidikan 'educated workforce'	–
Saya 'I'	5	material: memperioritaskan 'prioritize'	kerjasama 'cooperation'	di bidang sosial dan budaya, terutama pendidikan 'in the social and cultural aspects, particularly education'
Saya 'I'	6	material: berkonsentrasi 'concentrate on'	Pendidikan 'education'	–
semua rakyat 'all citizens'	7	material: mengakses 'access'	pelayanan pendidikan 'educational services'	–
Rakyat 'citizens'	8	material: mengakses 'access'	pelayanan pendidikan 'educational services'	–
Menteri Pendidikan dan kebudayaan 'The Minister of Education and Culture'	9	material: menguatkan 'strengthen'	pendidikan karakter 'character building'	–
Kebijakan 'Policies'	10	material: meningkatkan 'increase'	akses pemerataan pendidikan 'access to educational equity'	–
–	11	material: ditanamkan 'imbedded'	kesadaran bela negara 'awareness to defend the country'	melalui pendidikan kewarganegaraan 'through civic education'
PMI 'The Red Cross'	12	material: memperkuat 'strengthen'	pendidikan tentang nilai-nilai kemanusiaan 'education on human values'	–
seluruh rakyat Indonesia 'all citizens of Indonesia'	13	material: berpendidikan 'are educated'	–	–
Kita 'We'	14	material: tanggulangi 'tackle'	kemiskinan dan pengangguran 'poverty and unemployment'	melalui pembangunan ekonomi, pembangunan pendidikan 'through development of economy and education'

(Continued)

Table 2. (*Continued*).

Participant (actor)	No.	Process	Participant (recipient)	Circumstances
–	15	material: diperlukan 'need'	investasi kita di sistem pendidikan dan bidang kesehatan 'our investment in educational and healthcare system'	supaya sumber daya manusia kita menjadi lebih baik. 'for better human resources'
semua warga bangsa 'all citizens'	16	material: memperoleh 'acquire'	pendidikan yang layak 'adequate education'	–

5. strengthening character building (no. 9)
6. imbedding awareness to defend the country (no. 11)
7. strengthening human values (no. 12)
8. tackling poverty through education (no. 14)
9. role of education in producing human resources (no. 15)

In the context of delivery, the speeches by President Joko Widodo were addressed to all the citizens of Indonesia. In this perspective, education in President Joko Widodo's political communications focused on persuasive actions which were intended to influence people of Indonesia to support his political policies regarding education.

Joko Widodo's administration began with the restructuring of the 2013 Curriculum which focused more on students' competences. The ideas of education in the 2013 Curriculum heavily emphasized aspects of neo-liberalism and puts less emphasis on aspects of nationalism. Meeting the needs of the markets and industries had become the basis for the development of the national education system in Indonesia.

Widodo and Kalla's agenda on education was explicitly summarized in their vision and missions called Nawa Cita. In Nawa Cita the policies related to education are set out in points 5, 6, and 8:

Point 5: improving the quality of life of Indonesian citizens by improving the quality of education and training under the program of Indonesia Pintar (Smart Indonesia) with free 12-year compulsory education.
Point 6: increasing productivity and competitiveness of Indonesian citizens at the international level by developing a number of Science and Techno Parks in several areas and polytechnic and vocational schools with advanced technological infrastructure.
Point 8: revolutionizing national character through the development of a national education curriculum.

The notion of education by President Joko Widodo was directed toward three basic aspects in accordance to the vision and missions under his administration (retrieved from: http://kpu.go. id/koleksigambar/VISI_MISI_Jokowi-JK.pdf), which are:

1. Sovereignty in (the aspect of) economy;
2. Independence in (the aspect of) economy; and
3. Character in (the aspect of) culture.

In the missions explained above, education has a very important role. Education in President Joko Widodo's perspective was an answer to the alarming state of Indonesia. In the area of education, the notion of national education was directed toward efforts to develop national capability of being economically independent. This challenge focused on the preparation of qualified human resources who would reliably develop the nation specifically in the aspect of education. Implicitly, the quality of human resources was reflected through several aspects, such as academic level, personal character and attitude, productivity, and welfare.

Generating qualified human resources based on Joko Widodo's perspective could be achieved by providing access to education for all. A number of education programs that emphasize the improvement of access to education in Indonesia has been introduced, such as the Indonesia Pintar Program (Smart Indonesia) and the compulsory 12-year education 'until secondary school', free tuition fees, and scholarships for diploma, bachelor, master, and doctoral degrees. This affirms that education is a priority of the government in developing Indonesia.

Widodo and Kalla (2014) implemented the discourse of character revolution through education by emphasizing the importance of revolutionizing Indonesia's character and morale. It was done through civic studies including course subjects such as Indonesian history, patriotism and the love for one's country, the spirit of nationalism, and ethics. The explanation above also affirms that there were values of liberalism in President Joko Widodo's concept of education.

These values of liberalism were ideally proportioned with the values of nationalism in the national curriculum under President Widodo's administration. The national curriculum is designed with 70% (for the elementary level) or 60% (for the practical and polytechnic education level) of the content on ethics and character building. This balance of values is in line with Widodo's vision of education as "a cultural character".

President Joko Widodo positioned education as an answer to the characterless social conditions in Indonesia. The pressure from foreign cultures caused the Indonesian culture to experience fundamental changes. Education according to President Joko Widodo was a specific medium for character building. There were ten priorities of President Joko Widodo's administration (source: http://kpu.go.id/koleksigambar/VISI_MISI_Jokowi-JK.pdf retrieved on 22nd August 2016). They were:

1. emphasis on national curriculum that prioritizes civic education;
2. affordable tuition fee;
3. abolition of uniformity model;
4. structuring both locally and nationally balanced curriculum;
5. improvement of educational facilities;
6. provision of qualified teaching resources;
7. provision of life assurance for teachers;
8. equity of education facilities;
9. implementation of compulsory 12-year education; and
10. enhancing science and technology.

In their vision and mission, President Joko Widodo and Vice-President Jusuf Kalla (2014) emphasized the spirit of restoration, reformation, and revolution in Indonesia. These notions stood out in their political rhetoric which dominantly utilized expressions related to revolution, and those expressions were repeatedly conveyed in President Joko Widodo's speeches. Under his administration, the policies in education were directed at saving the national education, which was previously considered unsuitable with the socio-cultural context of Indonesia. Globally, it is apparent that the education system introduced by President Joko Widodo actually rebuffed the education system by the former President Susilo Bambang Yudhoyono, as represented by the 2013 Curriculum.

The values of neo-liberalism, which had been the focus of the previous administration under President Susilo Bambang Yudhoyono, mostly applied the competence-based curriculum approach. Then under President Joko Widodo's administration, there were at least three competence-based curriculums that were implemented as a continuation of the former president's policy i.e. Kurikulum Berbasis Kompetensi (KBK), Kurikulum Tingkat Satuan Pendidikan (KTSP), and Kurikulum 2013. These curricula were based on the notions of neo-liberalism because they focused on the efforts to meet the needs of markets and industries for qualified and capable human resources.

However, the change of the competence-based curriculum by President Joko Widodo was not a rejection of the entire concept of neo-liberalism in education. President Joko Widodo continued to include subjects on skills and competences in the curriculum composition.

However, he also emphasized course subjects that were related to character building and nurturing the love for one's country. Considering the issues, his administration also issued the Regulation of the Minister of Education and Cultures 23/2015 which facilitated the values of nationalism that were formerly absent in the previous policies. In its implementation, President Joko Widodo planned to develop a curriculum that prioritizes character and citizenship building, followed by other studies that also focus on skill development.

5 CONCLUSION

Through a linguistic analysis, it is apparent that President Joko Widodo emphasized educational concept that focused on character building, human values, and citizenship by underlining the development of human resources' competences and quality. This model of education was implemented through an interactive model of education which involved a two-way education. The development of education must be simultaneous with the development and improvement of infrastructure to ensure accessibility of education for every citizen of Indonesia.

Widodo and Kalla's agenda on national education was explicitly summarized in points 5, 6, and 8 of Nawa Cita. As reflected in Nawa Cita, Indonesia's education is aimed to (a) improve the quality of life of Indonesian citizens, (b) increase productivity and competitiveness, and (c) revolutionize national character. These objectives could be attained by emphasizing the curriculum that prioritizes civic education, provides affordable tuition fees, abolishes uniformity in education, develops a locally and nationally balanced curriculum, increases educational facilities, provides qualified teaching resources, provides life insurance for teachers, provides equal education facilities, implements free 12-year compulsory education, and enhances science knowledge and technology.

In a broader perspective, the educational system in Indonesia may be viewed as a form of rejection against neo-liberalism in education. However, the education system under President Widodo's administration was a compromise of liberalist-nationalist. The values of nationalism were emphasized through the implementation of subjects that were based on character building and citizenship, followed by subjects that were based on the development of knowledge and competences. The values of liberalism as reflected in several subject courses were ideally-balanced with other subjects that promoted the values of nationalism since President Widodo's administration gave more emphasis on building students' character and spirit of nationalism than other academic competences.

REFERENCES

Fairclough, N. (2013) *Critical Discourse Analysis: The Critical Study of Language*. London, Routledge.

Halliday, M.A.K. & Matthiessen, C.M.I.M. (2014) *Halliday's Introduction to Functional Grammar*. 4th edition. London, Routledge.

Sukardjo & Komarudin. (2009) *Landasan Pendidikan: Konsep dan Aplikasinya*. (Education Platform: Concepts and Application) Jakarta, Rajawali Press.

Widodo, J. & Kalla, J. (2014) *Jalan Perubahan untuk Indonesia yang Berdaulat, Mandiri, dan Berkepribadian* (The Road to Change for a Sovereign, Independent and Strong Character of Indonesia) [Online] Available from: http://kpu.go. id/koleksigambar/VISI_MISI_Jokowi-JK.pdf [Accessed 26th September 2016].

Widodo, J. (2014) *Sambutan Presiden Republik Indonesia pada Menerima Peserta Program Pendidikan Reguler Angkatan (PPRA)*. (Speech of the President of the Republic of Indonesia in welcoming the participants of the Regular Education Program/PPRA) [Online] Available from: http://www.setneg. go.id/index.php?option=com_content&task=view&id=8376&Itemid=26.

Widodo, J. (2014) *Keterangan Pers Presiden Republik Indonesia mengenai Pengalihan Subsidi BBM*. (Press Release of the President of the Republic of Indonesia on the Subsidy Transfer of Fuel) [Online] Available from: http://www.setneg.go.id/index.php?option=com_content&task=view&id=8 332&Itemid=26.

Widodo, J. (2014) *Sambutan Presiden Republik Indonesia pada Pembukaan Kompas 100 CEO Forum*. (Speech of the President of the Republic of Indonesia at the Opening of KOMPAS 100 CEO Forum)

[Online] Available from: http://www.setneg.go.id/index.php?option=com_content&task=view&id=83 40&Itemid=26.

Widodo, J. (2015) *Sambutan Presiden Republik Indonesia Peringatan Hari Bela Negara*. (Speech of the President of the Republic of Indonesia in Commemorating the "Defending of the Country" Day) [Online] Available from: http://jatengprov.go.id/id/siaran-pers/sambutan-presiden-ri-selaku-inspektur-upacara-pada-acara-peringatan-hari-bela-negara.

Widodo, J. (2015) *Sambutan Presiden Republik Indonesia Penganugerahan Tanda Kehormatan Satya lencana Kebaktian Sosial Bagi Pendonor Darah Sukarela 100 Kali*. (Speech of the President of the Republic of Indonesia in Awarding Social Outreach Ribbons to Voluntary Blood Donors of more than 100 times Commemorating the "Defending of the Country" Day) [Online] Available from: http://setkab.go.id/pidato-presiden-joko-widodo-pada-penganugerahan-tanda-kehormatan-satyalencana-kebaktian-sosial-bagi-pendonor-darah-sukarela-100-kali-di-istana-bogor-jabar-18-sesember-2015/.

Widodo, J. (2015) *Pembukaan Musyawarah Nasional IV Asosiasi DPRD Kabupaten Seluruh Indonesia (Adkasi)*. (Opening of the 4th National Deliberation of the Association of Indonesian Regional Parliament Members) [Online] Available from: http://setkab.go.id/sambutan-presiden-joko-widodo-pada-pembukaan-musyawarah-nasional-iv-asosiasi-dprd-kabupaten-seluruh-indonesia-di-hotel-pullman-jakarta-17-desember-2015/.

Widodo, J. (2015) *Pembukaan RAPIM Komisi Penyiaran Indonesia (KPI)*. (Opening of the Management Meeting of the Indonesian Broadcast Commission) [Online] Available from: http://www.setneg.go.id/index.php?option=com_content&task=view&id=10658&Itemid=26.

Widodo, J. (2015) *Sambutan Presiden Republik Indonesia pada Pelantikan Perwira Remaja Tni-Polri, Di Akademi Kepolisian Semarang*. (Speech of the President of the Republic of Indonesia at the Inauguration of Youth Officers of the Army and Police) [Online] Available from: http://setkab.go.id/sambutan-presiden-ri-pada-pelantikan-perwira-remaja-tni-polri-di-akademi-kepolisian-semarang-30-juli-2015/.

Widodo, J. (2015) *Sambutan Presiden Republik Indonesia pada Acara Peringatan Nuzulul Qur'an Nasional*. (Speech of the President of the Republic of Indonesia in Commemorating Nuzulul Quran) [Online] Available from: http://www.setneg.go.id/index.php?option=com_content&task=view&id=9534.

Widodo, J. (2015) *Paparan Presiden Republik Indonesia pada Silaturahim dengan Dunia Usaha Bertema "Presiden Menjawab Tantangan Ekonomi"*. (Presentation of the President of the Republic of Indonesia at the Social Meeting with the Business Community with the theme: The President Addresses the Economic Challenges) [Online] Available from: http://setkab.go.id/paparan-presiden-joko-widodo-pada-silaturahim-dengan-dunia-usaha-presiden-men jawab-tantangan-ekonomi-di-jakarta-convention-center-jakarta-9-juli-2015/.

Widodo, Joko. (2015) *Keterangan Pers Presiden Republik Indonesia Mengenai Pembentukan Panitia Seleksi Calon Komisioner KPK*. (Press Release of the President of the Republic of Indonesia at the Inauguration of the Selection Committee for Selecting the KPK Commissioners) [Online] Available from: http://www.setneg.go.id/index.php?option=com_content&task=view&id=9431&Itemid=26.

Widodo, J. (2015) *Sambutan Presiden Republik Indonesia pada Peringatan Isra Mi'raj Nabi Muhammad Saw Tahun 1436 Hijriah*. (Speech of the President of the Republic of Indonesia in Commemorating Isra Mi'raj of the Prophet Muhammad 1436 Hijriah) [Online] Available from: http://www.setneg.go.id/index .php?option=com_content&task=view&id=9365.

Widodo, J. (2015) *Pengarahan Presiden Republik Indonesia pada Musyawarah Rencana Pembangunan Nasional*. (Guidelines from the President of the Republic of Indonesia regarding the National Development Plan) [Online] Available from: http://www.setneg.go.id/index.php?option=com_content&tas k=view&id=9286&Itemid=26.

Widodo, J. (2015) *Sambutan Presiden Republik Indonesia pada Silaturahim Pers Nasional*. (Speech of President of the Republic of Indonesia at the Meeting with the National Press) [Online] Available from: http://www.setneg.go.id/index.php?option=com_content&task=view&id=9178&Itemid=26.

Widodo, J. (2015) *Sambutan Presiden Republik Indonesia pada Pembukaan Kongres Partai Keadilan dan Persatuan Indonesia*. (Speech of President of the Republic of Indonesia at the Congress Opening of the Justice and Unity Party of Indonesia [Online] Available from: http://setkab.go.id/sambutan-presiden-ri-pada-pembukaan-kongres-iv-partai-keadilan-dan-persatuan-indonesia-di-medan-sumatera-utara-18-april-2015/.

Widodo, J. (2015) *Sambutan Presiden Republik Indonesia pada Peringatan Harlah Ke-55 Pergerakan Mahasiswa Islam Indonesia (PMII)*. (Speech of President of the Republic of Indonesia in Commemorating the 55th Anniversary of the Movement of PMII) [Online] Available from: http://www.setneg.go.id/index.php?option=com_content&task=view&id=9281&Itemid=26.

Widodo, J. (2015) *Sambutan Presiden Republik Indonesia pada Peresmian Kampus Institut Pemerintahan Dalam Negeri (IPDN)*. (Speech of President of the Republic of Indonesia at the Official Opening of IPDN) [Online] Available from: http://setkab.go.id/sambutan-presiden-joko-widodo-pada-pere-smian-gedung-ipdn-kampus-ntb-praya-kab-lombok-tengah-20-april-2015/.

Widodo, J. (2015) *Sambutan Presiden Republik Indonesia pada Penandatangan Nota Kesepakatan Rencana Aksi Bersama tentang "Gerakan Nasional Penyelamatan Sumber Daya Alam Indonesia.* (Speech of President of the Republic of Indonesia at the Signing of the MOU for the Action Plan of the National Movement to Save Indonesia's Natural Resources) [Online] Available from: http://www.setneg.go.id/index.php?option=com content&task=view&id=8968.

Widodo, J. (2015) *Keterangan Pers Presiden Republik Indonesia pada Joint Press Conference dengan Presiden Republik Filipina.* (Press Release of the President of the Republic of Indonesia in a Joint Press Conference with the President of the Philippines) [Online] Available from: http://www.setneg.go.id/index.php?option=com_content&task=view&id=8788&Itemid=26.

Widodo, J. (2015) *Keynote Speech Presiden Republik Indonesia Indonesia Outlook 2015, Jalan Perubahan Untuk Indonesia Yang Berdaulat, Mandiri, dan Berkepribadian.* (Keynote Speech of the President of the Republic of Indonesia in presenting Indonesia's Outlook for 2015, the road to Change for Indonesian Sovereignty, Independence and Strong Character) Online] Available from: http://www.setneg.go.id/index.php?option=com_content&task=view&id=8695 &Ite mid=26.

Widodo, J. (2015) *Sambutan Presiden Republik Indonesia pada Pembukaan Musyawarah Nasional Himpunan Pengusaha Muda Indonesia (Munas HIPMI) XV Tahun 2015.* (Speech of the President of the Republic of Indonesia at the Opening of the 15th National Deliberation of the Association of Indonesian Young Entrepreneurs in 2015) [Online] Available from: http://setkab.go.id/sambutan-presiden-joko-widodo-pada-pada-pembukaanmusyawa rah-nasional-himpunan-pengusaha-muda-indonesia-munas-hipmi-xv-tahun-2015-di-bandung-jawa-barat-12-januari-2015/.

Widodo, J. (2016) *Sambutan Presiden Republik Indonesia dalam Sosialisasi Program Pengampunan Pajak (tax amnesty) Surabaya, Jawa Timur.* (Speech of President of the Republic of Indonesia in Campaigning for Tax Amnesty Program in Surabaya East Java) [Online] Available from: http://www.setneg.go.id/index.php?option=com_content&task=view&id=11908&Itemid=26.

Widodo, J. (2016) *Sambutan Presiden Republik Indonesia dalam Rapat Terbatas tentang Dana Alokasi Khusus Kantor Presiden.* (Speech of the President of the Republic of Indonesia at the Limited Meeting on the Special Allocated Funds from the President's Office) [Online] Available from: http://www.setneg.go.id/index.php?option=com_content&task=view&id=11665&Itemid=26.

Widodo, J. (2016) *Pernyataan Pers Bersama Presiden Republik Indonesia dan Presiden Serbia Istana Merdeka.* (Joint Press Conference of the President of the Republic of Indonesia and the President of Serbia) [Online] Available from: http://www.setneg.go.id/index.php?option=com_content&task=vie w&id=11484&Itemid=26.

Widodo, J. (2016) *Sambutan Presiden Republik Indonesia dalam Pertemuan Bilateral Presiden Republik Indonesia dengan Presiden Republik Serbia Istana Merdeka.* (Speech of the President of the Republic of Indonesia at the Bilateral Meeting with the President of Serbia at Merdeka Palace)[[Online] Available from: http://www.setneg.go.id/index.php?option=com_content&task=view&id=11656&It emid=26.

Widodo, J. (2016) *Pengantar Presiden Republik Indonesia Rapat Terbatas tentang Keuangan Inklusif Kantor Presiden.* (Foreword from the President of the Republic of Indonesia at the Ministerial Meeting of the Cabinet on Inclusive Finance of the President's Office) [Online] Available from: http://setkab.go.id/pengantar-presiden-pada-rapat-terbatas-mengenai-keuangan-inklusif-di-kantor-presiden-jakarta-26-april-2016/.

Widodo, J. (2016) *Pengantar Presiden Republik Indonesia Rapat Terbatas Kabinet Kerja tentang Tindak Lanjut Hasil Kunjungan Kerja Presiden Ke Eropa Kantor Presiden.* (Foreword from the President of the Republic of Indonesia at the Limited Ministerial Meeting of the Cabinet as a follow up of the President's Visit to Europe- President's Office) [Online] Available from: http://setkab.go.id/pengantar-presiden-joko-widodo-pada-rapat-terbatas-membahas-tindak-lanjut-hasil-kunjungan-pke-eropa-di-kantor-presiden-jakarta-25-april-2016/.

Widodo, J. (2016) *Keterangan Pers Presiden Republik Indonesia Pascakunjungan Kerja Ke Jerman, Inggris, Belgia, Belanda Bandara Halim Perdanakusuma.* (Press Release from the President of the Republic of Indonesia after the Official Visit to Germany, England, Belgium, Holland- Halim Perdana Kusuma Airport) [Online] Available from: http://www.setneg.go.id/index.php?option=com_content &task=view&id=11387&Itemid=295.

Widodo, J. (2016) *Pengarahan Presiden Republik Indonesia Rapat Kerja Pemerintah dengan Gubernur-Wakil Gubernur, Bupati-Wakil Bupati, dan Wali Kota-Wakil Wali Kota Seluruh Indonesia Hasil Pilkada Serentak Istana Negara.* (Guidelines from the President of the Republic of Indonesia at the Meeting with the Elected Governors-Vice Governors, Regents-Vice Regent, and Mayors and Deputy

Mayors throughout Indonesia - State Palace). [Online] Available from: http://www.pikiran-rakyat. com/nasional/2016/04/08/hari-ini-jokowi-beri-pengarahan-kepala-daerah-terpilih–366154.

Widodo, J. (2016) *Kuliah Umum Presiden Republik Indonesia Dies Natalis Ke-47 dan Lustrum Ke-8 Univesitas Negeri Sebelas Maret Solo, Jawa Tengah*. (General Lecture of the President of the Republic of Indonesia at the 47th Inauguration and the 8th Lustrum of Universitas Sebelas Maret, Solo, Centra Java) [Online] Available from: http://setkab.go.id/transkripsi-kuliah-umum-presiden-joko-widodo-pada-tanggal-11-maret-2016-di-universitas-negeri-solo-jawa-tengah/.

Widodo, J. (2016) *Pengantar Presiden Republik Indonesia Rapat Terbatas Kabinet Kerja tentang Pemberantasan Narkoba dan Rehabilitasi Korban Kantor Presiden*, (Foreword from the President of the Republic of Indonesia at the Limited Ministerial Meeting of the Cabinet on Eradicating drugs and Victim Rehabilitation) [Online] Available from: http://setkab.go.id/pidato-pengantar-presiden-joko-widodo-pada-rapat-terbatas-tentang-pemberantasan-narkoba-dan-rehabilitasi-korban-penyalahgu-naan-narkoba-di-kantor-presiden-24-februari-2016-pukul-14–00-wib/.

Widodo, J. (2016) *Amanat Presiden Republik Indonesia Pelantikan Gubernur-Wakil Gubernur Hasil Pilkada*. (Address of the President of the Republic of Indonesia in Inaugurating Governors and Deputy Governors elected from The General Elections) [Online] Available from: http://www.setneg. go.id/index.php?option=com_ content&task=view&id=11047&Itemid=26.

Widodo, J. (2016) *Pengantar Presiden Republik Indonesia Rapat Terbatas Kabinet Kerja tentang Pencegahan Virus Zika*. (Foreword from the President of the Republic of Indonesia at the Ministerial Meeting of the Cabinet in Preventing the Zika Virus) [Online] Available from: http://setkab.go.id/ pidato-pengantar-presiden-joko-widodo-pada-rapat-terbatas-pencegahan-penyebaran-virus-zika-di-kantor-presiden-3-februari-2016/.

Widodo, J. (2016) *Pengantar Presiden Republik Indonesia Rapat Terbatas Kabinet Kerja tentang Pencegahan Tindak Kekerasan dan Penindasan Terhadap Anak*. (Foreword from the President of the Republic of Indonesia at the Ministerial Meeting on Prevention of Violence and Oppression toward Children) [Online] Available from: http://www.setneg.go.id/index.php?option=com_content&task=view&id=1 1054&Itemid=26.

Widodo, J. (2016) *Sambutan Presiden Republik Indonesia Rapat Terbatas Kabinet Kerja tentang Kebijakan Status PTS Menjadi PTN*. (Speech of the President of the Republic of Indonesia at the Ministerial Meeting on the Policy to Change the Status of Private Universities to State Universities) [Online] Available from: http://www.setneg.go.id/index.php?option=com_content&task=view&id=10719&It emid=26.

Widodo, J. (2016) *Pengantar Presiden Republik Indonesia Sidang Paripurna Kabinet Kerja Kantor Presiden*. (Speech of the President of the Republic of Indonesia at the Plenary Meeting of the State Cabinet at the President's Office) [Online] Available from: http://www.setneg.go.id/index. php?option=com_content&task=view&id=12201&Itemid=26.

Cultural Dynamics in a Globalized World – Budianta et al. (Eds)
© *2018 Taylor & Francis Group, London, ISBN 978-1-138-62664-5*

Nationalism and neoliberalism in education: A critical discourse analysis of presidential speeches in Indonesia's reform era

R.M.P. Silalahi, N.J. Malik & U. Yuwono
Department of Linguistics, Faculty of Humanities, Universitas Indonesia, Depok, Indonesia

ABSTRACT: Fundamental economic, social, and cultural changes have taken place in Indonesia since 1998. The decline of the 32-year-long rule of Soeharto paved way for the establishment of a new regime called the Reform era. This change of regime could not resolve all the problems that arose from the transition of power. Education was one of the most severely affected sectors under the new regime. Indonesia's education system, which was based on nationalism, is integrated with the values of neoliberalism. The poor education system, low competitiveness of human resources, and globalization are the reasons why Indonesia's education system began to adopt the values of neoliberalism. Neoliberalism resulted in Indonesia's education system being based on meeting the needs of the industries. Because nationalism and neoliberalism are incorporated in Indonesia's education system, this study aims to reconstruct and compare the practice of nationalism and neoliberalism during the Reform era in Indonesia. In order to achieve this objective, a descriptive qualitative analysis was conducted on 18 Indonesian presidential speeches from 1998 to 2016. We used Fairclough's three-dimensional frameworks incorporating linguistics.

Keywords: Description, discourse interpretation, and socio-cultural explanation

1 INTRODUCTION

The resignation of Soeharto as the president of the Republic of Indonesia in May 1998 marked the end of the authoritarian New Order era and the birth of the Reform era. This transition of power, however, did not necessarily address various social issues and challenges at that time. Furthermore, the people of Indonesia had to face difficult challenges in several aspects of life, including the economic, social, political, and cultural aspects.

One of the sectors that experienced significant change during this transition of power was the education sector. Together with the decline of Indonesia's economy, its education sector declined. The survey conducted by the Political and Economic Risk Consultancy (PERC) showed that Indonesia's education system was one of the worst in Asia (Kompas, 2001). The poor management of the education system is evident from the issuance of various instant government policies, which were not necessarily in accordance with the needs of the society (Rohman, 2002). Rather than addressing different social issues in the society, education became a new problem for the people of Indonesia.

The Law of the Republic of Indonesia Number 12 of 2012 (regarding education) (Ristekdikti, 2012), the Presidential Regulation Number 8 of 2012 (pertaining to the National Qualifications Framework (NQF) of Indonesia), and the Regulation of the Minister of Education and Culture Number 73 of 2013 (on the implementation of NQF in Higher Education) (Ristekdikti, 2013) are a few examples that reflect the fact that education in Indonesia was directed to meet the needs of the industry. Education, which is supposed to be directed toward the development of knowledge and character of students, had undergone changes, and at present, it is more oriented toward preparing students to work in the industries.

As stated in the Constitution of the Republic of Indonesia Number 20 of 2003 on the National Educational System, education should be directed toward not only the development of the competences of students but also building character and patriotism (Ristekdikti, 2003). On the basis of these two premises, there are two dimensions of education in Indonesia, namely nationalism and neoliberalism, both of which are incorporated in Indonesia's education system. The values of nationalism are in line with the efforts made to meet the needs of stakeholders (the industry) and are apparently reflected in several government policies directed to produce graduates with certain competences as required by the industry. As a result, to some extent, neoliberalism has influenced the education sector in Indonesia. Because of this dualism of ideology and the bad condition of the education sector in Indonesia, the researchers are interested in reconstructing the ideas of nationalism and neoliberalism in the context of education in the Reform era.

2 THEORETICAL FRAMEWORK

Fairclough (2013) stated that the discourse analysis is a three-dimensional analysis that includes the description of a text, interpretation of a discourse, and the explanation from the sociocultural context. The description of a text is related to the description of the language elements, which reflect certain social actions. The analysis in this level uses the elements of the systemic functional linguistics (SFL) developed by Halliday (1978), and it reflects the three functions of language, namely ideational, interpersonal, and contextual.

In this study, we emphasized and focused more on the ideational function, which is reflected in lexicalization and transitivity. Lexicalization is related to the usage of words or other lexical elements (O'Halloran, 2008). In the critical perspective, the repeated usage of words or other lexical configuration represents the ideology that the writer or the speaker of the message wishes to communicate.

Transitivity is related to the syntactical construction in a clause. Transitivity represents reality through elements of lexicogrammar. Halliday (2008) classified transitivity into several categories according to the verbs, namely (a) material process (describing the physical actions of the participants), (b) mental process (related to the actions involving the five senses), (c) verbal process (related to the actions using articulators through language elements), (d) relational process (related to the verbs describing the act of becoming, describing, identifying, and symbolizing), (e) behavioral process (related to verbs relating participants to certain acts), and (f) existential process (related to the verbs describing the existence of some entities).

The second dimension of CDA (critical discourse analysis) is the interpretation of a discourse. This dimension is related to the production and consumption of a text, and it aims at exposing the power communicated through a text. Explanation on sociocultural context (the third dimension of CDA) is related to the social phenomena that influence the formation of a text. This includes social contexts, such as the institutions, media, society, culture, or sociopolitical conditions.

3 METHODOLOGY

We used CDA to analyze the independence-day speeches delivered by the following presidents of Indonesia:

a. Bahrudin Jusuf Habibie (1998–1999)
b. Abdurahman Wahid (1999–2001)
c. Megawati Soekarno Putri (2001–2004)
d. Susilo Bambang Yudhoyono (2004–2014)
e. Joko Widodo (2014–present).

The speeches were delivered in front of the members of the House of Representatives. They contained information on the presidents' accountability, achievements, and development plan for the following year.

4 ANALYSIS

The analysis was conducted in two stages to identify the ideologies that were planned to study, namely nationalism and neoliberalism. In the first stage of the analysis, we sought for constituents representing nationalism in the speeches. In the second stage, we sought for components representing neoliberalism.

The presidential speeches in the Reform era did not emphasize the values of nationalism. Out of the 18 data items for study, we did not find any constituent that particularly reflects the values of nationalism that are commonly manifested in expressions such as *cinta tanah air* (love for the country), *nilai* Pancasila (the values of Pancasila as the core values of Indonesia), *nasionalisme* (nationalism), and *kebangsaan* (nationality). This affirms that the concept of nationalism in the discourse of education was less emphasized in the political communications of the presidents.

At the beginning of the Reform era (B.J. Habibie and A. Wahid), national development was directed toward stopping the decline of the national economy, and less emphasis was put on the development of national education. There was no clear implementation of educational policies. During that time, the schools still used the 1994 goal-based curriculum, which was used in the New Order era. In this curriculum, the development of morality, religion, skills, love for country, and other competences was emphasized. This curriculum was still applied under the presidentship of Susilo Bambang Yudhoyono.

Because the values of nationalism were considered inadequate in the education curriculum in the Reform era, President Joko Widodo instructed to emphasize more on the subject of history in the curriculum. The campaign on enriching history was aimed to promote the values of nationalism. By introducing the idea of character revolution, the Indonesian government plans to change the national curriculum by stressing various course subjects that are related to the history of Indonesia, nationality, and patriotism in a perfect balance with other competences (Dahana, 2014).

Similarly, neoliberalism in the Reform era also acquired more emphasis under the presidents' political persuasion. The values of neoliberalism were reflected in the constituents that collocate with the words *kerja* (work) and *sumber daya manusia (SDM)* (human resources). Both words were the manifestation of neoliberalism in the discourse of national education, because the education sector at that time was aimed at producing human resources capable of competing in the national market. Table 1 shows the total number of times the words "work" and "human resources" together with their collocations appear in the data.

From the data presented in Tables 1 and 2, it can be observed that there are clauses with relational processes and other constituents that reflect neoliberalism as a participant in the syntactical construction. This affirms that the educational ideology of the presidents during the Reform era emphasizes the role of education particularly in producing human resources capable of competing in the employment and market context. It is apparent that the idea of neoliberalism in Indonesian education includes several issues related to the following notions:

Table 1. Lexicalization.

No.	Lexicalization	Total	Configuration	Total
1	*Kerja* (work)	10	*Lapangan kerja* (employment)	4
			Kesempatan kerja (job opportunity)	4
			Akses (terhadap) pekerjaan (access to jobs)	2
2	*SDM* (human resources)	5	*Kualitas SDM* (quality of human resources)	3
			Perbaikan SDM (improvement of human resources) resources	1
			Pengembangan SDM (development of human resource) resources	1

Neoliberalism in the discourse of education was conveyed through the construction of relational processes.

Table 2. Transitivity.

No.	Lexicalization	Participant	Process (relational)	Participant
1	*Kerja* (work)	*Upaya itu* (That effort)	*mencakup* (includes)	*penyediaan lapangan kerja* (providing job opportunities)
		Satu tantangan utama lapangan kerja kita (One of the main challenges in our employment sector)	*adalah* (is)	*pekerja kita masih berpendidikan SD* (that our workforce only has primary school education)
2	*Sumber daya manusia* (human resources)	*lemahnya kualitas SDM* (poor quality of human resources)	*akibat* (is due to)	*Kemiskinan dan tidak adanya akses terhadap pendidikan* (poverty and lack of access to education)
		perbaikan mutu SDM (improving the quality of human resources)	*melalui* (through)	*pendidikan* (education)
		pengembangan SDM (developing human resources)	*melalui* (through)	*pendidikan* (education)
		Program ini bertujuan untuk mengurangi kemiskinan dengan cara meningkatkan kualitas SDM (This program aims at reducing poverty by enhancing the quality of human resources)	*melalui* (through)	*pendidikan* (education)
		lemahnya kualitas SDM (poor quality of human resources)	*sebagai akibat* (as the result of)	*tidak adanya akses terhadap pendidikan* (lack of access to education)

1. Education is seen as the medium for providing employment
2. Education counters the low quality of the Indonesian workforce
3. The poor quality of education is in line with the lack of access to education in Indonesia
4. The quality of human resources is developed and enhanced through education.

Considering these key points, the presidential speeches were delivered so that the development plans for the education sector would be approved and then implemented under the various forms of Regulations and Laws. Furthermore, these presidential speeches were delivered in front of two groups of audiences (members of the House of Representatives and the people of Indonesia) so that these education initiatives could acquire legitimacy from the Indonesian legislature in particular and gain support from the people.

Specifically, together with the ideology of neoliberalism, the presidents attempted to gain support from the Indonesian legislatures to sustain the concept of education in the country. The aim of this concept was to satisfy the needs of the industry by providing qualified workforce and human resources. This approach essentially was manifested through various governmental policies, such as the Law of the Republic of Indonesia Number 12 of 2012 (on Higher Education), the Presidential Regulation Number 8 of 2012 (on NQF of Indonesia), and the Regulation of the Minister of Education and Culture Number 73 of 2013 (on the implementation of NQF in Higher Education) (Ristekdikti, 2013). These policies could be considered neoliberalistic as they focused on producing graduates with certain competences to meet the needs of the industry. Yuwono (2016), in his research on the Website of NQF, found that the Qualification Framework of Indonesia uses lexicalization that reflects the values of neoliberalism, especially those that collocate with the words *kerja* (work), such as *pasar kerja* (job market), *pengalaman kerja* (work experience), and *pelatihan kerja* (job training). The NQF of Indonesia presented a list of qualifications that graduates of higher education should possess. From the beginning of its establishment, the NQF involved practitioners from the industry; thus, the parameters provided by the industry were used as the benchmark in determining the quality of graduates in Indonesia. Finally, this reflects the changes in the purpose of national education that is more oriented toward the job market satisfying the needs of the stakeholders.

The values of neoliberalism in national education were also reflected in the curriculum, which had undergone changes several times in the Reform era. Neoliberalism was reflected in the implementation of the competence-based curriculum. This curriculum was implemented in order to produce graduates with certain competences to meet the needs of the industry. The education sector in the Reform era had undergone the following three changes in the competence-based curriculum:

1. *Kurikulum Berbasis Kompetensi* (KBK), which is a concept of curriculum emphasizing the competencies of students to complete task-based assignments on particular performance standards.
2. *Kurikulum Tingkat Satuan Pendidikan* (KTSP), which is a competence-based curriculum compiled and implemented by each education unit in Indonesia.
3. *Kurikulum* 2013 (K-13), which is a curriculum based on four aspects, namely knowledge, skill, attitude, and behavior. The curriculum is implemented to produce graduates with competences equivalent to international standards.

In the autonomous regions, the education system in Indonesia was geared toward attempts to produce competent human resources in remote areas that are able to fulfill the needs of the industries. The difference lies in the role of the educational institutions. KBK could not give authority to educational institutions to develop their own curriculum, different from the situations in KTSP and *Kurikulum* 2013.

Under the administration of Joko Widodo, K-13 was no longer in effect. The Indonesian Minister of Education and Culture asked education institutions in Indonesia to use KTSP as K-13 could not be implemented. At that time, Indonesia did not have sufficient infrastructure to run the curriculum. Meanwhile, the quality of human resources is determined by the education level, personality, attitude, productivity, and social welfare. These indicators are achieved through social programs named "Indonesia Pintar" or "Smart Indonesia" that offer "free 12-year compulsory education". Widodo and Kalla (2014) also emphasized the importance of revolutionizing the nation's character through the program of civic study, which proportionally incorporates educational aspects such as the history of Indonesia, patriotism, spirit to defend the country, and ethics in the Indonesian curriculum. In the elementary education level, 70% of the content of the curriculum should consist of education of students' ethics and character building as part of the Character Revolution (*Revolusi Mental*).

Widodo and Kalla (2014) emphasized the following 10 aspects of education:

1. Emphasizing nationalism in the national curriculum by teaching national history and patriotism, nurturing the spirit to defend the country, and promoting the development of science and technology
2. Providing affordable education to all citizens
3. Terminating the uniformity in the national education system
4. Forming a curriculum that maintains the balance of local and national content
5. Improving education facilities to support the transfer of knowledge
6. Recruiting and distributing teachers evenly
7. Providing adequate life assurance for teachers
8. Providing education facilities equally and evenly
9. Proposing a law that specifies free 12-year compulsory education
10. Implementing education based on science and technology.

5 CONCLUSION

On the basis of the analysis of the presidential speeches in the Reform era, the education curriculum has put less emphasis on the values of nationalism although these values were clearly implemented in the education. The values of nationalism focusing on nation building are considered contradictory to the values of neoliberalism, which focus on meeting the needs of the market and the industry. In the Reform era, the values of neoliberalism in the education sector were regarded to satisfy the needs of qualified human resources that are competitive in the global market.

The need for such competitive human resources in the global market has paved the way for the development of the education system that focuses on building competences in line with the market parameters. The industries representing neoliberalism were given the authority and power to determine the direction and the policy of education. Neoliberalism in Indonesia's education sector was directed toward the following four initiatives: (a) education is the medium for providing employment; (b) education encounters the low quality of the Indonesian workforce; (c) the low quality of education is parallel to the low access to education in Indonesia; and (d) development and enhancement of the quality of human resources are achieved through education.

The national education system oriented towards the efforts made to develop one's potentials and character as well as the love for one's country has been changed to satisfy the needs of the industry and meet the demands for qualified workforce capable of competing in the national and global markets. Therefore, it can be concluded that the national education in Indonesia has deviated from its original purpose and intention.

Widodo's administration emphasizes the importance of improving the students' ethics and national character building as part of the character revolution. Nationalism was emphasized in the national curriculum through several actions, such as teaching the national history of Indonesia, values of patriotism, and the spirit to defend the country. Education should be focused on the efforts to not only produce competent human resources for the interests of the industry but also strengthen national character and morale.

REFERENCES

Dahana, T. (2014) Tantangan Visi Sejarah Jokowi Selengkapnya. (The Challenges of Jokowi's Vision of History) [Online] Available from: http://www.kompasiana.com/tr.dahana/tantangan-visi-sejarah-jokowi [Accessed 17th August 2016].
Fairclough, N. (2013) Critical Discourse Analysis: The critical Study of Language. London, Routledge.
Habibie, B.J. (1998) Pidato Kenegaraan Presiden Republik Indonesia di depan Sidang DPR 15 Agustus 1998. (State Address of the President of Republic Indonesia before the Assembly of the House of Representatives 15th August 1998) [Online] Available from: http://kepustakaan-presiden.perpusnas.go.id/
Habibie, B.J. (1999) Pidato Kenegaraan Presiden Republik Indonesia di Depan Sidang DPR 16 Agustus 1999. (State Address of the President of Republic Indonesia before the Assembly of the House of Representatives 16th August 1999) Jakarta, Directorate of Publications, DG Press and Graphics-Ministry of Information Republic of Indonesia.
Halliday, M.A.K. (1978) Language as Social Semiotic, London: Edward Arnold.
Halliday, M.A.K. (2008) An introduction to Functional Grammar. 3rd edition. Beijing, Foreign Language Teaching and Research Press.
Kompas (2001). Mendiknas: Sistem Pendidikan di Indonesia Memang Terburuk di Asia. (Ministry of National Education: Education System in Indonesia is the Worst in Asia) Jakarta, Kompas.
Megawati. (2001) Pidato Kenegaraan Presiden Republik Indonesia di depan Sidang DPR 16 Agustus 2001. (State Address of the President of Republic Indonesia before the Assembly of the House of Representatives 16th August 2001) Jakarta, BAPPENAS.
Megawati. (2002) Pidato Kenegaraan Presiden R.I. dan Keterangan Pemerintah atas RUU tentang RAPBN 2003 Serta Nota Keuangannya di Depan Sidang DPR RI Jakarta, 16 Agustus 2002. (State Address of the President of Republic Indonesia and the Government's Explanation on the Draft Law on the Draft National State Budget 2003 and its Financial Notes before the House of Representatives on 16th August 1998) Jakarta, National Library of the Republic of Indonesia.

Megawati. (2003) *Pidato Kenegaraan Presiden R.I. dan Keterangan Pemerintah atas RUU tentang RAPBN 2004 Serta Nota Keuangannya di Depan Sidang DPR RI, Jakarta, 15 Agustus 2003*. (State Address of the President of Republic Indonesia and the Government's Explanation on the Draft Law on the Draft National State Budget 2004 and its Financial Notes before the House of Representatives on 16th August 2003) [Online] Available from: http://onesearch.id/Record/IOS2789-oai:dev2.pnri.go.id:kepustakaan-presiden-2633.

Megawati. (2004) *Pidato Kenegaraan Presiden Republik Indonesia dan Keterangan Pemerintah atas Rancangan Undang-Undang tentang Anggaran Pendapatan dan Belanja Negara Tahun Anggaran 2005 Serta Nota Keuangannya di Depan Sidang Dewan Perwakilan Rakyat 16 Agustus 2004*. (State Address of the President of Republic Indonesia and the Government's Explanation on the Draft Law on the Draft National State Budget 2005 and its Financial Notes before the House of Representatives on 16th August 2004) [Online] Available from: http://www.kemlu.go.id/id/pidato/presiden/Pages/PIDA-TO-KENEGARAAN-PRESIDEN-REPUBLIK-INDONESIA-DAN-KETERANGAN-PEMER-INTAH-ATAS-RANCANGAN-UNDANG-UNDANG.aspx.

O'Halloran. (2008) Systemic functional-multimodal discourse analysis (SF-MDA): Constructing ideational meaning. *Visual Communication*. 7 (4), 443–476.

Ristekdikti (2003) *Undang-Undang no. 20 Tahun 2003 tentang Sistem Pendidikan*. (Law no 20 of 2003 on Education System) [Online] Available from: http://sumberdaya.ristekdikti.go.id/wp-content/uploads/2016/02/uu-nomor-20-tahun-2003-tentang-Sisdiknas.pdf. [Accessed 17th August 2016].

Ristekdikti (2012) *Undang-Undang Republik Indonesia nomor 12 Tahun 2012 tentang Pendidikan Tinggi*. [Online] (the Law of Republic Indonesia number 12 of 2012 on Higher Education) Available from: http://risbang.ristekdikti.go.id/regulasi/uu-12-2012.pdf/ [Accessed 17th August 2016].

Ristekdikti (2013) *Peraturan Menteri Pendidikan dan Kebudayaan no 73 tahun 2013 tentang Penerapan Kerangka Kualifikasi Nasional Pendidikan Tinggi*. (Regulation of the Ministry of Education and Culture No. 73 of 2013 on the Implementation of National Qualification Framework in Higher Education) [Online] Available from: http://sindikker.ristekdikti.go.id/dok/permendikbud/Permendikbud73-2013JuklakKKNI.pdf. [Accessed 17th August 2016].

Rohman, A. (2002) *Akar Ideologis Problem Kebijakan Pendidikan di Indonesia Fondasi Keputusan Presiden Nomor 8 tahun 2012 tentang KKNI*. (The Ideological root of the education policy problem in Indonesia based on the Presidential Decree Number 8 of 2012 on QF) [Online] Available from: http://www.polsri.ac.id/ [Accessed 17th August 2016].

Wahid, A. (2000) *Pidato Kenegaraan Presiden Republik Indonesia di depan Sidang DPR 16 Agustus 2000. (State Address of the President of Republic Indonesia before the Assembly of the House of Representatives on 16th August 2000)* [Online] Available from: http://kepustakaan-presiden.perpusnas.go.id.

Widodo, J. & Kalla, J. (2014) *Jalan Perubahan untuk Indonesia yang Berdaulat, Mandiri, dan Berkepribadian*. (The Road to Change for Indonesia's Sovereignty, Independence and Character [Online] Available from: http://abbah.yolasite.com/resources/VISI%20DAN%20MISI%20JOKOWI%20JK.pdf/ [Accessed 17th August 2016].

Widodo, J. (2015) Pidato Kenegaraan Presiden Republik Indonesia dalam Rangka HUT Ke-70 Proklamasi Kemerdekaan Republik Indonesia, di Depan Sidang Bersama DPR RI Dan DPD RI, Jakarta, 14 Agustus 2015. *(State Address of the President of Republic Indonesia in commemorating the 70th Anniversary of Independence before the General Assembly and Senators, Jakarta 14th August 2015)* [Online] Available from: http://setkab.go.id.

Yudhoyono, S.B. (2005) *Pidato Kenegaraan Presiden R.I. dan Keterangan Pemerintah Atas RUU tentang RAPBN Tahun 2006 serta Nota Keuangannya di Depan Rapat Paripurna DPR RI, Jakarta, 16 Agustus 2005*. (State Address of the President of Republic Indonesia and the Government's Explanation on the Draft Law on the Draft National State Budget 2006 and Financial Notes before the House of Representatives on 16th August 2005) Jakarta, National Library.

Yudhoyono, S.B. (2006) *Pidato Kenegaraan Presiden Republik Indonesia serta Keterangan Pemerintah atas Rancangan Undang-undang tentang Anggaran Pendapatan dan Belanja Negara tahun Anggaran 2007 Beserta Nota Keuangannya*. (State Address of the President of the Republic of Indonesia and the Draft Bill on the State Budget of 2007 and its Financial Notes) [Online] Available from: http://www.kemlu.go.id/losangeles/id/arsip/pidato/Pages/Pidato-Kenegaraan-Presiden-Republik-Indonesia-serta-Keterangan-Pemerintah-atas-Rancangan-Undang-unda.aspx.

Yudhoyono, S.B. (2007) *Pidato Kenegaraan Presiden Republik Indonesia serta Keterangan Pemerintah atas Rancangan Undang-Undang tentang Anggaran Pendapatan dan Belanja Negara Tahun Anggaran 2008 Beserta Nota Keuangannya*. (State Address of the President of the Republic of Indonesia and the Draft Bill on the State Budget of 2008 and its Financial Notes) Jakarta, State Secretary Republic of Indonesia.

Yudhoyono, S.B. (2008) *Pidato Kenegaraan Presiden Republik Indonesiaserta Keterangan Pemerintah atas Rancangan Undang-Undang tentang Anggaran Pendapatan dan Belanja Negara Tahun*

Anggaran 2009 Beserta Nota Keuangannya di Depan Rapat Paripurna Dewan Perwakilan Rakyat Republik Indonesia Jakarta, 15 Agustus 2008. (State Address of the President of Republic Indonesia and the Government's Explanation on the Draft Law on the Draft National State Budget 2009 and Financial Notes before the General Assembly of the House of Representatives on 15th August 2008) [Online] Available from: https://rbaryans.wordpress.com/2008/08/16/pidato-kenegaraan-presiden-ri-atas-ruu-tentang-apbn-tahun-anggaran-2009/.

Yudhoyono, S.B. (2009) *Pidato Kenegaraan Presiden Republik Indonesia dalam Rangka Peringatan Hari Ulang Tahun Ke 64 Kemerdekaan Republik Indonesia di Depan Rapat Paripurna Dewan Perwakilan Rakyat Republik Indonesia Tahun 2009.* (State Address of the President of the Republic of Indonesia in commemorating the 64th Anniversary of Indonesia's Independence before the General Assembly of the House of Representatives of 2009) Jakarta, BAPPENAS.

Yudhoyono, S.B. (2010) *Pidato Kenegaraan Presiden Republik Indonesia dalam Rangka Hut Ke-65 Proklamasi Kemerdekaan Republik Indonesia di Depan Sidang Bersama Dewan Perwakilan Rakyat Republik Indonesia dan Dewan Perwakilan Daerah Republik Indonesia.* (State Address of the President of the Republic of Indonesia in commemorating the 65th Anniversary of Indonesia's Independence before the General Assembly of the House of Representatives of 2009) [Online] Available from: http://www.setneg.go.id/index.php?option=com_content&task=view&id=4716.

Yudhoyono, S.B. (2011) *Pidato Presiden Ri pada Penyampaian Keterangan Pemerintah atas RUU tentang APBN Tahun Anggaran 2012 Beserta Nota Keuangannya di Depan Rapat Paripurna DPR RI Jakarta, 16 Agustus 2011. (State Address of the President of Republic Indonesia and the Government's Explanation on the Draft Law on the Draft National State Budget 2012 and Financial Notes before the House of Representatives on 16th August 2011)* [Online] Available from: http://setneg.go.id/images/stories/kepmen/jurnalnegarawan/jn20/20_artikel2.pdf.

Yudhoyono, S.B. (2012) *Pidato Kenegaraan Presiden Republik Indonesia Dr.H Susilo Bambang Yudhoyono dalam rangka HUT ke-67 Proklamasi Kemerdekaan Republik Indonesia di depan sidang bersama DPR RI dan DPD RI. (State Address of the President of Republic Indonesia Dr. H. Susilo Bambang Yudhoyono before the General Assembly and Senators)* Jakarta: President's Special Staff for Publication and Documentation.

Yudhoyono, S.B. (2013) *Pidato Kenegaraan Presiden RI dalam Rangka HUT Ke-68 Proklamasi Kemerdekaan RI 16 Agustus 2013 Di Sidang Bersama DPD-DPR RI.* (State Address of the President of Republic Indonesia in Commemorating the 68th Anniversary of Indonesia's Independence 16th August 2013 before the General Assembly and Senators) [Online] Available from: http://beritasore.com/2013/08/16/pidato-kenegaraan-presiden-ri-dalam-rangka-hut-ke-68-proklamasi-kemerdekaan-ri-16-agustus-2013-di-sidang-bersama-dpd-dpr-ri/.

Yudhoyono, S.B. (2014) *Salinan Pidato Kenegaraan Presiden RI dalam Rangka HUT Ke-69 Kemerdekaan RI.* 9 copy of State Address of the President of RI in Commemorating the 69th Anniversary of Independence [Online] Available from: http://www.pikiran-rakyat.com/nasional/2014/08/15/293104/salinan-pidato-kenegaraan-presiden-ri-dalam-rangka-hut-ke-69-kemerdekaan.

Yuwono, U. (2016) *Apa yang Digarisbawahi dalam Kerangka Kualifikasi Nasional Indonesia? Sebuah Analisis Wacana Kritis.* What are the Highlights in the National Qualification Framework? A Critical Discourse Analysis [Lecture] The International Undergraduate Symposium on Humanities and Arts, Depok, 29th–31st August 2016.

Cultural Dynamics in a Globalized World – Budianta et al. (Eds)
© 2018 Taylor & Francis Group, London, ISBN 978-1-138-62664-5

Values of neoliberalism in education: A comparative study of Indonesian presidential speeches in the New Order and Reform Era

R.M.P. Silalahi, U. Yuwono & Y.J. Aminda
Department of Linguistics, Faculty of Humanities, Universitas Indonesia, Depok, Indonesia

ABSTRACT: The rise of neoliberalism represents the desire of humans to dominate the various aspects of life. On one hand, this dominance will bring individual and corporate welfare. On the other hand, it will trigger the destruction of the market system. In Indonesia, the existence of neoliberalism is clearly reflected in education. Neoliberal education has been implemented since the New Order era and has continued to develop in the Reform era. By analysing the speeches of the President of Indonesia, the writers found that in the New Order era, neoliberal values were more oriented toward developing the country to support the demands of the industry. Currently, the values are more directed at meeting the needs of industries. With regard to the concept of Neoliberalism incorporated in the educational system of Indonesia and considering the impact of neoliberalism, this research aims to analyse and compare the neoliberal ideologies in the New Order era and in the Reform era. A comparative descriptive analysis was performed on 48 Indonesian presidential speeches from the former President Soeharto to President Joko Widodo. Fairclough's three dimensional frameworks (the text description, discourse explanation, and sociocultural interpretation) were applied to reconstruct neoliberal ideologies. This research is expected to contribute to the studies of inter-temporal Critical Discourse Analysis and to provide ideas for the improvement of education in Indonesia.

1 INTRODUCTION

Venugopal (2015) as cited in Anderson (2000) states that neoliberalism is the most successful ideology in the world. The ideology of neoliberalism represents humans' desire to dominate various aspects in life, particularly economy. Along with the advancement of technology and globalization, this ideology continues to develop. Moreover, the development of this ideology does not only occur in the aspect of economy. In Indonesia, this ideology is also evident in the education sector. Neoliberalism in education is deeply explicated through policies that dominantly support market needs.

Since the independence of Indonesia, neoliberalism has developed rapidly. This can be considered as a reaction of Indonesian people to achieve economic welfare. Neoliberalism in education is evident from various government policies which focus on generating human resources that meet the market needs. Neoliberalism in Indonesia was initially introduced through Indonesia's presidential speeches. Since the New Order era until the Reform era, presidential speeches that represent this ideology in education were further enacted in the form of laws, regulations, government policies, and incorporated in the curriculum.

The explanation above highlights the role of the presidents in adopting the ideology of neoliberalism in education. In the perspective of communication, speeches delivered by the presidents of the Republic of Indonesia can be considered as a form of communication process between the presidents and its people. The position of the presidents as the head of the nation would significantly affect the urgency of the speeches because the speeches of the presidents can influence the policies and drive the purpose of a nation. Based on that understanding, the issue pursued by this research is how the values of neoliberalism in education in Indonesia are represented through the presidential speeches from the New Order era until the Reformation era. This research aims to analyse and compare the ideology of neoliberalism from the New Order era to the Reform era reflected in the presidential speeches from 1968, the era of President Soeharto, to the present time of President Joko Widodo.

2 THEORETICAL FRAMEWORK

Discourse analysis is divided into three dimensions by Fairclough (1992), which are text description, discourse explanation, and sociocultural interpretation. Text description is related to the analysis of linguistic elements which reflect certain social actions. Discourse explanation is related to two phases, i.e. institutional process, which is related to the institutions that also participate in the text production, and discourse process, which is related to the production and consumption of a text (Fairclough, 1995). Finally, sociocultural interpretation is related to the contexts outside of the texts (Fairclough, 1992). Context, in this case, includes situations outside of the analysed text, such as the context of media institution, society, cultures, and politics.

In this research, the analysis in the dimension of text description is focused on two linguistic elements namely lexicalization and transitivity. Lexicalization in a text reflects the ideational function, especially experiential function which is used particularly to describe experience through linguistic elements (O'Halloran, 2008). Aidinlou (2014, p. 264) states that there are several sub-categories of lexicalization, which are distorted lexical items, lexical variations, over-lexicalization, under-lexicalization, euphemistic expressions, and additions and omissions. The repeated usage of lexical configuration represents ideology communicated by the writer or the speaker.

Halliday and Matthiessen (2014) describe transitivity as a fundamental linguistic feature that enables people to construct mental images out of the reality and to give meanings to experiences and events. In functional grammar (Halliday and Matthiessen, 2014), representation of experience consists of (1) process, (2) participant, and (3) circumstance.

Functional grammar recognises six types of process which are used to describe an event or condition, which are (a) material process (related to physical actions that involve at least two participants); (b) mental process (related to the acts of experiencing and feeling by using senses and perceptions); (c) verbal process (related to the acts of stating, telling, communicating, reporting, and projecting); (d) relational process (related to the act of becoming, describing, identifying, and symbolizing); (e) behavioural process (related to the verbs that relate participants to certain acts); (f) existential process (related to the act of creating something that involves one participant).

3 METHODOLOGY

This research utilised CDA (modelled by Fairclough) on speeches conducted by presidents of Indonesia in commemorating the independence of Indonesia, starting from the New Order era to the Reform era.

4 ANALYSIS

Presidential speeches in the New Order era and the Reform era emphasized the role of education in the development of Indonesia. In the speech texts, the idea of neoliberalism is reflected through lexical configuration such as *kerja* 'work', *industri* 'industries', *sumber daya manusia* 'human resources', *kompetensi* 'competence', and *keterampilan* 'skills' below.

From the table below, it can be observed that in both eras, the idea of neoliberalism is emphasized. It is clearly reflected through the use of the word *kerja* 'work' and *sumber daya manusia* 'human resources'. Globally, the communicated idea of neoliberalism in the New Order Era and the Reform Era is related to:

1. The role of education in providing competent workforce and human resources.
2. The provision of competent human resources through education that meets the requirements of the industries.

The difference between the New Order and Reform era lies on the emphasis of the neoliberal values. The New Order emphasized the importance of improving skills to meet the needs of industries. This idea is clearly reflected through the repetition of words (such as industry and skills) in Soeharto's presidential speeches. Meanwhile, the Reform era emphasizes the

Table 1. Lexicalization.

| Word | Lexicalization configuration | | | |
	New order era	Total	Reformation era	Total
Kerja 'work'	Kerja 'work'	6	Kerja 'work'	4
	Lapangan kerja 'employment'	2	Lapangan kerja 'employment'	6
	Ketenagakerjaan 'workforce'	2	Kesempatan kerja 'job opportunity'	2
	Tenaga kerja 'labor'		Pekerja 'workers'	1
	Angkatan kerja 'workforce'	2		
	Pelatihan kerja 'job training'	1		
Industri 'industries'	Industri 'industries'	3		
	Pendidikan dan pelatihan (untuk) Industri 'education and training for industries'	2		
	Kebutuhan industri 'needs of industries'	2		
Sumber daya manusia 'human resources'	Sumber daya manusia 'human resources'	3	Sumber daya manusia 'human resources'	2
	Menciptakan sumber daya manusia 'creating human resources'	1	Kualitas sumber daya manusia 'quality of human resources'	8
	Melahirkan sumber daya manusia 'creating human resources'	3	Mutu sumber daya manusia 'quality of human resources'	2
	Penghasilan sumber daya manusia 'income of human resources'	1	Pengembangan sumber daya manusia 'development of human resources'	1
	Pengembangan sumber daya manusia 'development of human resources'	1		
	Sumber daya manusia berkualitas 'qualified human resources'	4		
'Kompetensi' (competence)	Kompetensi kerja dan industri 'work and industrial competence'	3	Kompetensi guru 'teacher competence'	2
			Kompetensi pendidikan 'educational competence'	2
			Kompetensi dan profesionalisme 'competence and professionalism'	1
			Meningkatkan kompetensi 'improving competence'	2
'Keterampilan' (skills)	Pendidikan dan pelatihan keterampilan 'education and training for skills'	2		

importance of building competences. Furthermore, the importance of competences in the Reform era is clearly reflected in the implementation of the curriculum in the Reform era.[1]

There are several linguistic expressions in the clausal level that represent the idea of neoliberalism in education. Those clauses are presented in the table below.

Table 2.   Transitivity.

New order era			
No. Participant	Process	Participant	Circumstance
1 –	Material: akan diperluas 'will be expanded'	Pendidikan keterampilan 'education on skills'	Terbina tenaga-tenaga kerja 'develop workforce'
2 Peningkatan produktivitas 'enhancement of productivity'	Material: menunjukkan 'shows'	Peningkatan taraf pendidikan pekerja 'increase in workers' education level	–
3 Salah satu upaya yang penting 'one of the important efforts'	Relational: adalah 'is'	meningkatkan pendidikan dan latihan 'improving education and training'	–
4 Pengembangan sumber daya manusia 'development of human resources'	Existential: terletak 'lies on'	pada usaha-usaha pendidikan dan pengembangan kebudayaan 'the efforts towards the development of education and culture'	–
5 Hasil pendidikan 'the outcome of education'	Material: menciptakan 'creates'	Sumber daya manusia 'human resources'	–
6 Pendidikan 'education'	Relational: merupakan 'is'	tujuan dan sekaligus wahana 'both the purpose and the vehicle'	dalam meningkatkan kualitas sumber daya manusia 'in improving the quality of human resources'
7 Upaya peningkatan keterampilan 'the effort to improve skills'	Relational: melalui 'through'	berbagai program pendidikan dan pelatihan 'various educational and training programs'	–
Reformation Era			
No Participant	Process	Participant	Circumstance
1 Perbaikan mutu sumber daya manusia 'improvement of human resources'	Relational: melalui 'through'	Pendidikan 'education'	–
2 pengembangan sumber daya manusia 'development of human resources'	Relational: melalui 'through'	Pendidikan 'education'	–
3 Program ini 'this program'	Material: bertujuan 'aims to'	mengurangi kemiskinan 'reduce poverty'	dengan cara meningkatkan kualitas sumber daya manusia 'by improving the quality of human resources'

---

1. Will be explained in sociocultural interpretation.

The use of material and relational processes emphasizes that political persuasion in education was meant to build awareness toward the importance of education for national development. As explained above, it can be observed that the idea of neoliberalism in both eras emphasizes more on generating skilful human resources that meet the demands of the industries. In the Reform era, emphasis is placed on the development of human resources to address national issues, such as poverty and economy downturn.

In the context of discourse, the presidential speeches were delivered in front of the legislatures of Indonesia at that time, namely the *Dewan Perwakilan Rakyat* or 'House of Representatives'. In the context of communication, the presidential speeches were directed to address various purposes so that the ideas that were conveyed were accepted and enacted in various forms such as laws, regulations, or policies since these legislatures have the authority as lawmakers. The laws are thus oriented towards neoliberalism, specifically in the field of education.

Through sociocultural interpretation, the notion of education in both eras apparently emphasizes the role of education in generating skilful human resources that would meet the needs of the markets and industries. Education institutions are the place to produce competent individuals that are able to compete in the working world. This notion is a reaction toward the socio-economic condition of Indonesia, which has not turned for the better. To answer this socio-economic problem, knowledge and technology are considered as the solution and the notion of education was viewed as the medium to alleviate poverty by enhancing the quality of human resources (Yudhoyono, 2014).

The difference between the notion of neoliberalism in the New Order and the Reform era lies on the implementation. The New Order era emphasizes education and skills. Skills in this case do not only refer to neoliberal purposes but also to nationalistic purposes. The former President Soeharto (1976) emphasized that education on skills should also be directed toward national development.

The notion of nationalistic education can also be seen from the dissemination of school subjects during the New Order era, which continued to maintain school subjects in its structure of the 1975, 1984, and 1994 curriculum (Wirianto, 2014). The subjects emphasizing Pancasila (Indonesia's core values), character building, and Nationalism were in an ideal proportion with the subjects focusing on improving skills and basic academic capabilities.

Pancasila as the core value of Indonesia should be implemented holistically in all aspects of the nation's life. Therefore, Soeharto's administration provided the basic guidelines to improve the political awareness of Indonesian citizens through a program named Penataran Pedoman Pengayatan dan Pengamalan Pancasila (the guidelines to comprehending and implementing Pancasila) in all educational institutions in Indonesia (Soeharto, 1979).

Furthermore, under the New Order era, industries had the opportunity to assist educational institutions in implementing education by these corporations and by financing the education.

> *Pendidikan itu dapat dilaksanakan sendiri oleh perusahaan yang bersangkutan ataupun dengan menyediakan biaya untuk pendidikan tenaga-tenaga Indonesia guna mengikuti sekolah, latihan-latihan dan sebagainya.*
>
> 'Education can be conducted by the corporations, or the corporations may provide financial support for education of the Indonesian workforce to participate in schools, trainings, and so on.' (Soeharto, 1974)

In the Reformation era, apparently nationalism did not gain a position in national education. In this era, neoliberalism in education was more evident. Students were directed to master certain competences in accordance to the needs of the markets and industries. In fact, during the administration of the former President Yudhoyono (2004–2014), industries were given ample portion and role in shaping the curriculum. Referring to the vision of competing at global and national levels, the workforce as the products of educational institutions was generated based on the competences set by the industries. These competences were outlined in the *Kerangka Kualifikasi Nasional Indonesia* ('Indonesian National Qualification Framework').

In his research on the distributed media of KKNI, Yuwono (2016) found that the used lexical elements reflected the values of neoliberalism. It is clearly shown by the repetition of lexical items that collocate with the words 'work'. The values of neoliberalism are also indicated by the list of qualifications required from the graduates of higher education that is in line with the needs of the industries. Accordingly, the setting up of *KKNI* should involve practitioners from the industries. This marked the end of the education institutions' supremacy in producing graduates.

However, under President Joko Widodo's administration, a significant change occurred. The 2013 Curriculum (created during the former President Yudhoyono's term of office) was replaced with a new curriculum since the application of the 2013 Curriculum garnered a lot of problems (retrieved from: *https://m.tempo.co/read/news/2014/11/24/079623945/alasan-kenapa-kurikulum-2013-bermasalah*). Education under President Joko Widodo's administration must start with character building and nurturing the love for one's country, and then to be followed by developing skills and competences.

5 CONCLUSION

Education on one hand is aimed to promote one's intelligence, and to build morale and characters, on the other hand. In the history of Indonesia, education was oriented to be in line with the values of neoliberalism. Education was viewed as the solution to meet the needs of industries for workforce. Neoliberalism in education emphasizes the role of education in providing competent labour and human resources and in generating competent human resources in education which meets the demands of the industries. However, significant differences can be observed by comparing the education curriculum under the New Order era and the Reform era.

The New Order era recognizes the importance of other values besides the efforts to meet the needs of the markets. The values of nationalism and liberalism were both incorporated in the context of national education. Soeharto understood the importance of improving national character and the values of Pancasila through education. Therefore, his administration provided the guidelines in improving national political awareness through an education program in all education institutions. On the other hand, in the Reform era, education in Indonesia put more emphasis on the values of neoliberalism and less on the values of nationalism. In the Reform Era, the industries were given ample portion to shape the curriculum to fulfil their needs.

However, significant change is apparent under President Joko Widodo's administration. President Widodo prioritises national character building before developing specific competences that are required by the industries. Widodo's governance has changed the concepts of education that focus on neoliberal values with the new concept of education that focuses on developing character.

REFERENCES

Aidinlou, N.A., Dehghan, H.N. & Khorsand, M. (2014) Ideology, change & power in literature and society: A critical discourse analysis of literary translations. *International Journal of Applied Linguistics & English Literature.* 3 (6), 260–271.

Fairclough, N. (1992) *Discourse and Social Change*. Cambridge, Polity Press.

Fairclough, N. (1995) Critical Discourse Analysis: The Critical Study of Language. Harlow, Pearson Education Limited.

Habibie, B.J. (1998) *State Address of the President of RI before the Parliamentary General Assembly 15th August 1998.* [Online] Available from: http://kepustakaan-presiden.perpusnas.go.id/uploaded_files/pdf/speech/normal/Pidato_1998 0815.pdf.

Habibie, B.J. (1999) *State Address of the President of RI before the Parliamentary General Assembly 16 August 1999.* Jakarta, Directorate of Publications DG Press and Graphics –Ministry of Information RI.

Halliday, M.A.K. & Matthiessen, C.M.I.M. (2014) *Halliday's Introduction to Functional Grammar. (4th ed)*. London, Routledge.

Janks, H. (2006) *Critical Discourse Analysis as a Research Tool*. [Online] Available from: http://www.uv.es/gimenez/Recursos/criticaldiscourse.pdf.

Megawati. (2001) *State Address of the President of RI before the Parliamentary General Assembly 16 August 2001*. Jakarta, BAPPENAS.

Megawati. (2002) *State Address of the President of RI and the Government's Explanation on the 2003 State Budget and Financial Notes I before the Parliamentary General Assembly RI Jakarta, 16th August 2002*. Jakarta, National Library RI.

Megawati. (2003) *State Address of the President of RI and the Government's Explanation on the 2004 State Budget of and Financial Notes before of the Parliamentary General Assembly, Jakarta, 15th August 2003*. [Online] Available from: http://onesearch.id/Record/IOS2789-oai:dev2.pnri.go.id:ke pustakaan-presiden-2633.

Megawati. (2004) *State Address of the President of RI and the Government's Explanation on the 2005 State Budget of and Financial Notes before the Parliamentary General Assembly of RI 16th August 2004*. [Online] Available from: http://www.kemlu.go.id/id/pidato/presiden/Pages/PIDATO-KENEGARAAN-PRESIDEN-REPUBLIK-INDONESIA-DAN-KETERANGAN-PEMERIN-TAH-ATAS-RANCANGAN-UNDANG-UNDANG.aspx.

O'Halloran, K.L. (2008) Systemic functional-multimodal discourse analysis (SF-MDA): constructing ideational meaning. *Visual Communication*. 7 (4), 443–476.

Soeharto. (1968) *State Address of the President of RI General Soeharto at the Parliamentary General Assembly 16th August 1968*. Jakarta, Ministry of Information RI.

Soeharto. (1969) *Sate Address of President Soeharto in front of the Parliament DPR-GR RI 16th August 1969*. Jakarta, Ministry of Information RI.

Soeharto. (1970) *State Address of the President of RI General Soeharto in front of the Parliament DPR-GR 16th August 1970*. Jakarta, Ministry of Information RI.

Soeharto. (1971) *State Address of the President of RI General Soeharto in front of the Gotong Royong Parliament-DPRGR 16th August 1971*. Jakarta, Ministry of Information RI.

Soeharto. (1972) *State Address of the President of RI General Soeharto in front of the Parliament 16th August 1972*. Jakarta, Ministry of Information RI.

Soeharto. (1973) *State Address of the President of RI General Soeharto in front of the Parliamentary DPR Assembly 16th August 1973*. Jakarta, Ministry of Information RI.

Soeharto. (1974) *State Address of the President of RI General Soeharto in front of the Parliamentary Assembly DPR 15th August 1974*. Jakarta, Ministry of Information RI.

Soeharto. (1975) *State Address of the President RI General Soeharto in front of the Assembly of the Parliament DPR 16th August 1975*. Jakarta, Ministry of Information RI.

Soeharto. (1976) *State Address of the President RI Soeharto in front of the Assembly of the Parliament DPR 16th August 1976*. Jakarta, Ministry of Information RI.

Soeharto. (1977) *State Address of the President RI Soeharto in front of the Assembly of the Parliament DPR 16th August 1977*. Jakarta, Ministry of Information RI.

Soeharto. (1978) *State Address of the President RI Soeharto in front of the Assembly of the Parliament DPR 16th August 1978*. Jakarta, Ministry of Information RI.

Soeharto. (1979) *State Address of the President RI Soeharto in front of the Assembly of the Parliament DPR 16th August 1979*. Jakarta, Ministry of Information RI.

Soeharto. (1980) *State Address of the President RI Soeharto in front of the Assembly of the Parliament DPR 16th August 1980*. Jakarta, Ministry of Information RI.

Soeharto. (1981). *State Address of the President RI Soeharto in front of the Assembly of the Parliament DPR 15th August 1981*. Jakarta, Ministry of Information RI.

Soeharto. (1982) *State Address of the President RI Soeharto in front of the Assembly of the Parliament DPR 16th August 1982*. Jakarta, Ministry of Information RI.

Soeharto. (1983). *State Address of the President RI Soeharto in front of the Assembly of the Parliament DPR 16th August 1983*. Jakarta, Ministry of Information RI.

Soeharto. (1984) *State Address of the President RI Soeharto in front of the Assembly of the Parliament DPR 16th August 1984*. Jakarta, Ministry of Information RI.

Soeharto. (1985) *State Address of the President RI Soeharto in front of the Assembly of the Parliament DPR 16th August 1985*. Jakarta, Ministry of Information RI.

Soeharto. (1986) *State Address of the President RI Soeharto in front of the Assembly of the Parliament DPR 15th August 1986*. Jakarta, Ministry of Information RI.

Soeharto. (1987) *State Address of the President RI Soeharto in front of the Assembly of the Parliament DPR 15th August 1987*. Jakarta, Ministry of Information RI.

Soeharto. (1988) *State Address of the President RI Soeharto in front of the Assembly of the Parliament DPR 16th August 1988*. Jakarta, Ministry of Information RI.

Soeharto. (1989) *State Address of the President RI Soeharto in front of the Assembly of the Parliament DPR 16th August 1989*. Jakarta, Ministry of Information RI.

Soeharto. (1990) *State Address of the President RI Soeharto in front of the Assembly of the Parliament DPR 16th August 1990*. Jakarta, Ministry of Information RI.

Soeharto. (1991) *State Address of the President RI Soeharto in front of the Assembly of the Parliament DPR 16th August 1991*. Jakarta, Ministry of Information RI.

Soeharto. (1992) *State Address of the President RI Soeharto in front of the Assembly of the Parliament DPR 15th August 1992*. Jakarta, Ministry of Information RI.

Soeharto. (1993) *State Address of the President RI Soeharto in front of the Assembly of the Parliament DPR 16th August 1993*. Jakarta, Ministry of Information RI.

Soeharto. (1994) *State Address of the President RI Soeharto in front of the Assembly of the Parliament DPR 16th August 1994*. Jakarta, Ministry of Information RI.

Soeharto. (1995) *State Address of the President RI Soeharto in front of the Assembly of the Parliament DPR 16th August 1995*. Jakarta, Ministry of Information RI.

Soeharto. (1996) *State Address of the President RI Soeharto in front of the Assembly of the Parliament DPR 16th August 1996*. Jakarta, Ministry of Information RI.

Soeharto. (1997) *State Address of the President RI Soeharto in front of the Assembly of the Parliament DPR 16th August 1997*. Jakarta, Ministry of Information RI.

Tempo.com. (2014) *Alasan Kenapa Kurikulum 2013 Bermasalah*. (Reason Why the 2013 Curriculum is Problematic) [Online] Available from: https://m.tempo.co/read/news /2014/11/24/079623945/alasan-kenapa-kurikulum-2013-bermasalah.

Venugopal, R. (2010) *Neoliberalism as a concept*. [Online] Available from: http://personal.lse.ac.uk/venugopr/venugopal2014augneoliberalism.pdf.

Wahid, A. (2000) *State Address of the President RI before the Assembly of the Parliament 16 August 2000*. Retrieved from: http://kepustakaan-presiden.perpusnas.go.id/uploaded_files/pdf/speech/normal/gusdur4.pdf

Widodo, J. (2015) *State Address of the President RI in commemorating 70th Anniversary of RI Independence, in front of the Assembly of the Parliament with DPR RI and DPD RI, Jakarta, 14th August 2015*. Retrieved from: http://setkab.go.id/pidato-kenegaraan-presiden-republik-indonesia-dalam-rangka-hut-ke-70-proklamasi-kemerdekaan-republik-indonesia-di-depan-sidang-bersama-dpr-ri-dan-dpd-ri-jakarta-14-agustus-2015/.

Wirianto, D. (2014) Perspektif historis transformasi kurikulum di Indonesia. (Historical Perspective on the Curriculum Transformation) BUDGET *Islamic Studies Journal*. 2 (1), 133–147.

Yudhoyono, S.B. (2005) *State Address of the President of RI and the Government's Explanation on the State Budget of 2006 with its Financial Notes in front of the Parliament DPR RI, Jakarta, 16th August 2005*. Jakarta, National Library.

Yudhoyono, S.B. (2006) *State Address of the President RI and the Government's explanation on the Draft Law on the State Budget of 2007 with its Financial Notes*. Retrieved from: http://www.kemlu.go.id/losangeles/id/arsip/pidato/Pages/Pidato-Kenegaraan-Presiden-Republik-Indonesia-serta-Keterangan-Pemerintah-atas-Rancangan-Undang-unda.aspx.

Yudhoyono, S.B. (2007) *State Address of the President RI and the Government's explanation on the Draft Law on the State Budget of 2008 with its Financial Notes*. Jakarta, Sekretariat Negara RI.

Yudhoyono, S.B. (2008) *State Address of the President RI and the Government's explanation on the Draft Law on the State Budget of TA 2009 with its Financial Notes in front of the Parliament Jakarta, 15th August 2008*. Retrieved from: https://rbaryans.wordpress.com/2008/08/16/pidato-kenegaraan-presiden-ri-atas-ruu-tentang-apbn-tahun-anggaran-2009/.

Yudhoyono, S.B. (2009) *State Address of the President RI in commemorating the 64th Anniversary of RI Independence in front of the Assembly of the Parliament in 2009*. Jakarta, BAPPENAS.

Yudhoyono, S.B. (2010) *State Address of the President RI in commemorating the 65th Anniversary of RI Independence in front of the Assembly of the Parliament and Senators*. Retrieved from: http://www.setneg.go.id/index.php?option = com_content&task = view&id = 4716.

Yudhoyono, S.B. (2011) President's Speech in Delivering the Government's Explanation *on the Draft Law of the State Budget for 2012 With its Financial Notes in front of the Assembly of the Parliament RI Jakarta, 16th August 2011*. Retrieved from: http://setneg.go.id/images/stories/kepmen/jur nalnegarawan/jn20/20_artikel2.pdf.

Yudhoyono, S.B. (2012) *State Address of the President RI Dr. H Susilo Bambang Yudhoyono in commemorating the 67th Anniversary of RI Independence in front of the Assembly of the Parliament*. Jakarta, Special Staff of the President in Publications and Documentation.

Yudhoyono, S.B. (2013) *State Address of the President RI in commemorating the 68th Anniversary of RI Independence 16th August 2013 di Sidang Bersama DPD-DPR RI*. Retrieved from: http://beritasore.com/2013/08/16/pidato-kenegaraan-presiden-ri-dalam-rangka-hut-ke-68-proklamasi-kemerdekaan-ri-16-agustus-2013-di-sidang-bersama-dpd-dpr-ri/on16 July 2015).

Yudhoyono, S.B. (2014) Copy of the *State Address of the President RI in commemorating the 69th Anniversary of RI Independence*. Retrieved from: http://www.pikiran-rakyat.com/nasional/2014/08/1 5/293104/ salinan-pidato-kenegaraan-presiden-ri-dalam-rangka-hut-ke-69-kemerdekaan.

Meeting the challenges of global competition: Preserving and revitalizing the cultural heritage of Bagansiapiapi

A.A. Harapan & L. Mariani
Department of Area Studies, Faculty of Humanities, Universitas Indonesia, Depok, Indonesia
Department of Anthropology, Faculty of Political and Social Sciences, Universitas Padjadjaran, Bandung, Indonesia

ABSTRACT: Bagansiapiapi is a small town located in Riau Province, Sumatra, which has been designated as the capital of Rokan Hilir District since 1999. Bagansiapiapi has a long history as one of the important cities in the Indonesian archipelago. Founded in the 19th century by ethnic Chinese from Fujian Province, the town flourished as the second most important fish port in the world after Norway. The colonial government of the Netherland Indies then developed the town and built many important buildings, for example, banks, hospitals, and churches, many of which are still intact and functional today. The inhabitants, mostly Chinese, have many traditions that are still valid, and the most important is the ritual of *bakar tongkang*, which has been regarded as one of the five icons of Riau tourism by the government of Riau Province. However, Bagansiapiapi has much richer cultural heritage than just the ritual of *bakar tongkang*. Due to globalization, the preservation and revitalization of these cultural heritages is challenging.

1 INTRODUCTION

Bagansiapiapi is a small town located in the eastern part of Riau Province, which has been designated as the capital of Rokan Hilir Regency since 1999. In its heyday until several decades ago, the town had been famous as the second largest fishing port in the world after Norway. Thus, it is unsurprising that this small and almost forgotten town has a long and impressive history.

The first known notes about the region could be traced back to the days when the Portuguese invaded Melaka (Malacca) in 1511. Because of its strategic location (facing Malacca Strait near Singapore and Penang, Malaysia), Bagansiapiapi became one of the many ports in the busy trade routes along Malacca Strait. The colonial government of the Netherland Indies began to build offices in the town in 1858, which later, in 1890, served as head offices in the region, until the ouster of the Netherland Indies in 1942 (Arfan Surya, 2016, p. 10–12). Many buildings had been constructed in the colonial era, such as the old harbor, hospitals, churches, water treatment plant, and the Bank Rakyat Indonesia (BRI) office—the second bank office after the first established in Central Java.

As a coastal town, Bagansiapiapi has a multiethnic society consisting of the Malay, Chinese, Javanese, Bataknese, Minangnese, Bugisnese, and Niasnese, each of which has a unique cultural heritage, thereby shaping the multicultural character of the town. This multicultural trait becomes increasingly apparent considering the location of the town close to Singapore and Malaysia, making the interactions between the inhabitants of these two neighboring countries common in Bagansiapiapi. In such setting, cultural identity is an important aspect of the behavior of people as it is a distinguishing feature of the individuals, which differentiates them from one community to another on the basis of decisions and choices made in everyday life.

According to Koentjaraningrat (1985), culture is a holistic system of ideas, actions, and works of humans as social beings gained through the process of learning. In this regard, cultural identity decides the actions and reactions of people when faced with external challenges in a competitive and globalized world. Parsudi Suparlan (1986) considered culture as knowledge of an operational nature, an overall knowledge gained by humans as social beings whose contents are the instruments of knowledge models that can selectively be used to understand and interpret the environment encountered and to encourage and take necessary actions.

In the case of Bagansiapiapi, the people have multicultural heritage, both tangible and intangible, which contributes to their cultural identity. Having a strong cultural identity helps the people of Bagansiapiapi to remain strong and true to themselves amid the great waves of globalization, a fact that we, as citizens of the world, consider undeniable.

1.1 Research purpose

In this study, we discuss both the tangible and intangible cultural heritage of Bagansiapiapi to understand the actual condition and function. The result, in turn, will be used to determine the correct means to effectively preserve and revitalize the cultural heritage in order to strengthen the cultural identity of the multiethnic society of Bagansiapiapi. Because much effort has already been made in preserving and revitalizing certain cultural heritage in effect, we will also discuss these efforts and verify whether they are sufficiently effective in maintaining the local cultural identity.

1.2 Theoretical framework

The basic working concept used in this study is Stuart Hall's concept of cultural identity, which treats identity as being and becoming and states cultural identity as an entity jointly owned by the members of the community or as an original form of an individual that is also internalized in all of the people who have the same history and ancestry. Stuart Hall (1990) stated that cultural identity is a reflection of the oneness of history in the form of cultural codes that made a group of people into "one identity" although they appear different from the "outside". From this perspective, besides the oneness of history and cultural codes that unite the people, physical traits are also an integral part of cultural identity.

Here, we try to describe how the cultural heritage of Bagansiapiapi, both the tangible and intangible aspects, could be maintained and developed to create the "oneness" that will assist the people in becoming deeply rooted in their own community and able to use their cultural identity to improve their competitiveness to meet the challenges of globalization.

1.3 Method and technical procedure

We use mostly data-collecting procedures conducted directly through a field study, which is based on a previously conducted library study. The library research does not reveal much about the sociocultural life of Bagansiapiapi. A study by Shanty Setyawati (2009) focuses on the decline of the fishing industry in the 1930s. Another study by Syahrul (2001) discusses the conflict between ethnic Chinese and Malay in 1998. Only Anastasya Yolanda (2008) examined the Malay and Chinese acculturation in the housing area of Bagansiapiapi. The most recent study by R. Didin Kusdian (2015) deals with efforts to revitalize the port of Bagansiapiapi. However, all of the studies mention the very important annual ritual of the people in Bagansiapiapi, *bakar tongkang*, which is organized by the ethnic Chinese community.

The findings of the library research determine the plan and design of the field research. The main purpose of the field study, determined after conducting a quick preliminary study, is to collect data about the existing cultural heritage of Bagansiapiapi. Considering the significance of the ritual *bakar tongkang* to Bagansiapiapi cultural life, the field study was conducted in the week the ritual was held. Overall, the data-collecting activities were conducted before, during, and after the ritual.

The field study was conducted using the following various methods and techniques:

a. Participatory observation
b. Collection of documentation (audios, visuals, photos, and videos)
c. Interviews (both structured and unstructured)
d. Data analysis (from collected documents).

Qualitative approach was the method selected for the whole study. Consequently, the research design is flexible and not rigid, following the findings of the field study. Even though there are some statistics involved, they only complement the basic and primary data in the form of pictures, photos, descriptions, and narrations.

The data sources are notably the tangible and intangible cultural heritage themselves and the prominent figures in each ethnic community in Bagansiapiapi. In this study, the individuals interviewed are from three ethnic communities, namely Chinese, Malay, and Javanese. The decision was made because they are the largest and most dominant ethnic communities in Bagansiapiapi and because of the time constraint of the field study (6 days).

In the discussion section, we describe the tangible and intangible cultural heritage of Bagansiapiapi, its function, its meaning for the people, and the possibilities of employing the cultural heritage to build and construct the local cultural identity of Bagansiapiapi. The qualitative approach is also used in the discussion in a manner that is flexible and open to every possible effort for searching and finding answers.

2 CULTURAL HERITAGE OF BAGANSIAPIAPI

The data concerning the cultural heritage are divided into two groups, namely tangible and intangible cultural heritage. The tangible cultural heritage consists of objects, such as buildings, monuments, temples, old houses, and other objects of historical importance. The intangible cultural heritage consists of cultural elements inherent in every ethnic group, which have historical backgrounds and potentials to contribute to the local cultural identity.

2.1 Tangible cultural heritage

A total of 31 tangible cultural heritage sites are present in Rokan Hilir Regency, which includes Bagansiapiapi. In 2012, the Balai Pelestarian Peninggalan Purbakala (the Office for Preservation of Cultural Heritage, which is responsible for all historical and archeological sites) in Batusangkar, West Sumatra, enlisted five sites in their inventory. This inventory was made by the Batusangkar office, because the region is included in its working area.

The study concentrates only on historical objects around Bagansiapiapi, which were built in different periods. There are remnants of temples built in the classical Hindu–Buddha period, sacred tombs built in the Islamic sultanate period, buildings constructed in the Dutch colonial period, and a pillbox built in the Japanese colonial period. However, our findings show that monuments and buildings with historical values are predominantly of Chinese origin. This is normal because the town of Bagansiapiapi was built and developed by the ethnic Chinese community, who came from Songkhla, Thailand, in the late 18th century after leaving their home in southern China (Surya Arfan, 2016).

The conditions of these cultural heritage objects differ from one another. The remnants of the Hindu–Buddhist temples in Tanah Putih District, the Sintong temple, and Sidinginan site are in bad shape. Both temples are practically nonexistent; there is only a pile of stones in Sintong, remnants from what was a *stupa* in the olden days.[1] The Sidinginan site is worse with only several scattered old stones, which the local people claimed as the remnants of an old temple.

---

1. Information obtained from a 90-year-old local woman who claims to have seen the temple and its stupa in her younger days.

The cultural heritage objects from the Islamic sultanate period are in much better shape. In fact, there are about seven sacred tombs in the area, but we only visited one as the others are far from Bagansiapiapi. People come to pay their respect, and the tomb is in a very good condition. According to the locals, the other tombs are also in good condition.

The colonial cultural heritage buildings are also in good shape mainly because they are still functional and used as government offices and the official residence of the regent. Only the piers of the dock at the old harbor are unrecognizable, and parts of them have become the foundation of a mosque. Sedimentation has shifted the harbor approximately 4 km away from the sea, and the old harbor is now located at the town center. The water treatment plant (not in working condition) and the military dormitory (still functional but not well maintained) are other buildings that are in less good shape.

The Chinese cultural heritage buildings are mostly in very good shape and in full working condition. There are seven small monuments bearing Chinese inscriptions in the town center, the oldest temple or *kelenteng* Ing Hok Kiong, and other 77 *kelenteng*s and the old house of *Kapitan*, the leader of the Chinese community. Only the old house of the *Kapitan* is in a very bad shape because the family that owns the house now cannot afford to renovate and maintain it.

Not much is known about the pillbox from the Japanese colonial period. It is located in Pulau Jemur, a small island near the port of Bagansiapiapi. There seems to be little interest in this historical monument, and the locals rarely speak about the Japanese colonization.

On the basis of the findings, it is clear that the states of cultural heritage and historical objects depend on people's perspective and whether they have a meaning and serve the people of Bagansiapiapi. The sacred tombs are equally important to the locals and the visitors; thus, they are well maintained. The *kelenteng* is very functional, an integral part of the daily life among the Chinese ethnic group, and very well maintained. The colonial buildings serve a pragmatic purpose and have always been part of government maintenance task.

Clear policies on the part of the government about these sites and buildings would have been beneficial. A detailed inventory would help to determine whether a monument/building/ tomb fulfills all the needs to be named as a cultural heritage object. The criteria should be based on the objective conditions, local perspectives, and its meaning and purpose. Eventually, by involving all ethnic groups in these cultural heritage objects policies, people of Bagansiapiapi would be increasingly aware of their cultural heritage and likely work voluntarily to preserve and revitalize their tangible cultural heritage objects.

2.2 *Intangible cultural heritage*

The intangible cultural heritage, as mentioned above, covers cultural elements inherent in every community, which have historical backgrounds and potentials to contribute to the local cultural identity. On the basis of this definition, the ritual of *bakar tongkang* is an intangible cultural heritage of the ethnic Chinese, which dominates the town of Bagansiapiapi, mostly because the ritual is the biggest event in the town that attracts people from all parts of the country and even from other countries. This is because the ritual of *bakar tongkang*, which literally means "burning the boat", is the most important ritual for the ethnic Chinese originating from Bagansiapiapi. The ritual is a commemoration of the ancestors of the ethnic Chinese community, who landed on Bagansiapiapi, built and developed the town, and pledged never to return to China. As proof of their pledge, they burned the *tongkang* or boat that brought them to Bagansiapiapi. Thus, the ritual is organized every year to pay respect to the ancestors and has become an integral part of the ethnic Chinese in Bagansiapiapi.

The ritual of *bakar tongkang* has been designated by the provincial government as one of the five tourism icons of Riau Province. Consequently, the ritual becomes the town's identity and one of the local cultural identities. Other ethnic groups accept the fact and the decision and participate in obtaining benefits from the festivities, which begin about a week before the ritual. With the invasion of several visitors and tourists of different ethnicities, the locals try to provide them daily necessities and thus have an additional source of income. Eventually, all ethnic groups benefit from the ritual although it is an exclusive Chinese ritual.

Other ethnic groups do not have any cultural trait comparable to the *bakar tongkang,* but they do have a sense of their own cultural identity. The two major ethnic groups, namely the Malay and the Javanese, have their own cultural organization to develop their arts and culture as well as educate their younger generation on their cultural identity.

The Malay ethnic group has their own rituals, which mostly serve religious and healing purposes, including the *Sei Mambang Deo-deo* ritual organized in December, the *Si Lancang* and *Buyung Kayuang* healing rituals, and the warding off misfortunes by reading certain prayers known as *Ratib,* such as *Ratib Biasa* and *Ratib Berjalan/Ratib Kelambai.* However, for all its originality, there is no ritual like the *bakar tongkang* among the rituals of the Malay ethnic group. It should be noted that the Malay really preserve their dances and make efforts to revitalize them by creating new dances based on old ones, such as several *zapin* dances. The Malay group is the official art envoy of the Rokan Hilir Regency and represents the region in many art festivals throughout the country and in international events.

The Javanese ethnic group focuses on artistic activities to preserve their cultural identity. They do not practice any ritual and only perform Javanese art such as *wayang* and *reog Ponorogo* and a modern version of Javanese music called *campursari.* They have an organization called the *Hangudi Utomo* with voluntarily membership with the purpose of preserving Javanese arts by performing and training the younger generation. They also perform on calls and receive a sum, which they use to pay the artists and add to the organization's fund made up from the members' contribution. It is clear that the Javanese is more concerned with performing arts, which they considered as their cultural identity.

The ethnic Chinese community also has their social organization, called *Yayasan Multi Marga,* which actively organizes the ritual of *bakar tongkang.* This *yayasan* or foundation is formed to forge cooperation between all families or *marga*s in organizing any social activity, most importantly the ritual *of bakar tongkang.* The foundation gathers its members and forms the organizing committee aimed at preparing and executing the ritual, including managing funds from donors. Almost all activities of the Chinese community are conducted within the foundation, meaning that they are all recognized by the foundation.

The findings show that the observed ethnic groups have strong cultural identity and make genuine efforts to maintain and even strengthen their cultural identity. In this regard, the government has only to encourage these ethnic groups to continue their efforts, to realize a strong cultural identity.

As for the town's local cultural identity, it seems that the local government relies mostly on the ritual of *bakar tongkang* as the local cultural identity of Bagansiapiapi. This is a good policy considering the fact that the ritual has attracted many visitors and provided the inhabitants with a significant additional source of income. The question arises, however, whether this policy could be accepted completely by the other ethnic groups. This is due to the fact that the ritual is exclusively Chinese and it would be difficult for other ethnic groups to add the ritual to their cultural identity and make it their own. It is no less difficult for the ethnic Chinese themselves who have to compromise with the fact that their sacred ritual has been designated as one of the five provincial tourism icons. This designation means that they have to adapt the ritual to the concept of tourist attraction. So far, there has been no protest and the government has asked other ethnic groups to participate in the procession of the boat from the *kelenteng* Ing Hok Kiong to the burning grounds. The Javanese complies and they are the only other ethnic group participating in the procession.

3 CONCLUSION

Using Stuart Hall's concept of identity as being and identity as becoming, it seems that identity as becoming could be applied to raise awareness on cultural identity among the people of Bagansiapiapi. It does not mean that they have no cultural identity. On the contrary, it is because most of them have a strong sense of cultural identity. It is precisely because of their sense of cultural identity as an ethnic group that they lack a local cultural identity as the inhabitants of Bagansiapiapi. Thus, the concept of identity of becoming is practicable by socializing and communicating the history of Bagansiapiapi to the people, so that they will feel that they "own" the history of their town, where they are born and raised.

Selecting the tangible cultural heritage objects is crucial because the people have to feel that the objects "belong" to them and are part of their cultural identity. Consequently, a study should be conducted to collect all data available on potential cultural heritage objects, so that all ethnic groups will be represented by the objects. It would be ideal if the objects could represent the local cultural identity of Bagansiapiapi as a whole.

As for intangible cultural heritage, efforts made to include them in the local cultural identity are much more challenging. Because intangible cultural heritage includes rituals and all activities concerning concepts and perspectives, it is much more difficult to build and construct a common local cultural identity for all ethnic groups. The efforts made by the local government to include other ethnic groups in the procession of *bakar tongkang* are appreciable; however, there may be other possibilities to develop. Some of the possibilities are to organize an arts and culture festival during the week of the ritual of *bakar tongkang* and provide each ethnic group with a stage where they could perform. Thus, there will be a type of cultural exchange between the ethnic groups in Bagansiapiapi and there will be an understanding among them.

Another effort is to conduct a serious study on the ritual of *bakar tongkang* and the history of Bagansiapiapi so that the historical facts become known to the people as a whole and there would be one common version of the history of the town, which includes all ethnic groups. Identity as becoming could be applied as the working concept for the people of Bagansiapiapi in an effort to raise their awareness of local cultural identity, preserve and revitalize their cultural heritage, and improve their competitiveness in this globalized world.

REFERENCES

Anastasya, Y. (2008) Percampuran Budaya Melayu dan Cina Studi Kasus: Pemukiman di kota Bagan-siapi-api dan Pulau Halang (Acculturation of Malay and Chinese Culture Case Study: Bagansiapi-api and Halang Island Housing Area). Depok, Department of Architecture, Faculty of Engineering, Universitas Indonesia (unpublished bachelor thesis). Translation: Acculturation of Malay and Chinese Culture Case Study: Bagansiapi-api and Halang Island Settlements. Depok, Department of Architecture, Faculty of Engineering, Universitas Indonesia (unpublished bachelor thesis).
Didin, K. (2015) Revitalization and port master plan case study: Port of Bagansiapiapi. Article presented at the 18th FTSPT International Symposium, Unila, Bandar Lampung on 28th August 2015.
Hall, S. (1990) Cultural identity and diaspora. In: Jonathan Rutherford (ed.). *Identity, community, culture, difference*. London, Lawrence & Wishart.
Koentjaraningrat. (1985) Pengantar Ilmu Antropologi. Jakarta, Aksara Baru. Translation: Introduction to Anthropology. Jakarta, Aksara Baru.
Parsudi, S. (1986) Kebudayaan dan Pembangunan. Media IKA, 14, p. 2–19. Translation: Culture and Development. Media IKA, 14, pp. 2–19.
Shanty, S. (2008) Pasang Surut Industri Perikanan Bagansiapiapi 1898–1936. Depok, History Department, Faculty of Humanities, Universitas Indonesia (unpublished bachelor thesis). Translation: Ups and Downs of Bagansiapiapi Fishery Industry 1898–1936. Depok, History Department, Faculty of Humanities, Universitas Indonesia (unpublished bachelor thesis).
Syahrul. (2001) Konflik Etnis Cina dan Melayu Ditinjau dari Sosio-Kriminologi (Studi Kasus Konflik di Bagansiapi-api Kecamatan Bangko Kabupaten Rokan Hilir Propinsi Riau Tahun 1998). Depok, Program Studi Sosiologi, Pascasarjana, Fakultas Ilmu Sosial dan Ilmu Politik, Universitas Indonesia (unpublished thesis). Translation: Conflict of Chinese and Malay Ethnics Reviewed from Socio-Criminology (Conflict Case Study in Bagansiapi-api, Bangko Sub-District, Rokan Hilir Regency of Riau Province in 1998). Depok, Sociology Study Programme, Master's Degree, Faculty of Social and Political Science, Universitas Indonesia (unpublished thesis).

Cultural Dynamics in a Globalized World – Budianta et al. (Eds)
© *2018 Taylor & Francis Group, London, ISBN 978-1-138-62664-5*

Preserving and maintaining traditional culture through interior design: A case study of Torajan elements in a local restaurant

P. Salim
Bina Nusantara University, Indonesia

ABSTRACT: Focusing on the application of Torajan ornaments in interior space, this paper aims to discuss the roles of traditional Torajan decorative styles that have almost been forgotten. Toraja has many decorative motifs that can be applied to interior, furniture and exterior designs. In a typical local restaurant, the connection between art and tradition is established through aesthetic elements that act functionally with their specific and unique characteristics. There is much opportunity to develop them into new variations of ornaments. The methods used in this study are content analysis, observation, interview and experimental design with a hermeneutic approach supported by design and enriched by social, cultural and psychological data. The results of this study show that it is important to raise awareness in the community regarding the richness of Torajan culture, which can be applied in public spaces.

Keywords: Torajan culture, Torajan ornament, interior design, decorative styles

1 INTRODUCTION

Indonesia is an archipelago with various tribes and ethnicities. It is immensely rich in art and culture, ranging from the west to the east of the archipelago. The value of its culture needs to be appreciated and preserved and should be passed on to the next generation. Because of the rapidly spreading Western culture in Indonesia, it is necessary to preserve its own unique culture. It has become common nowadays for a community to take pride in foreign culture and disregard its own. By contrast, other countries are far more interested in Indonesian culture than Indonesians are. There are even some neighbouring countries that claim some parts of the Indonesian culture to be their own.

The Indonesian culture, which is a combination of various local cultures from Sabang all the way to Merauke, is unique and diverse. It includes the art of traditional dances, ceremonies, traditional clothing and traditional food. Indonesia's slogan on diversity is *Bhinneka Tunggal Ika* (Unity in Diversity). Recently, tourists have become more interested in knowing more about Bali Island, Yogyakarta, Surabaya and Bandung rather than other tourist destinations. However, there are still several equally captivating tribal cultures, one of which is the Torajan culture. The Torajan is a tribe that resides in the northern mountains of South Sulawesi, Indonesia, with an estimated population of 1 million people, 500 of whom still live in Tana Toraja, North Toraja Regency and Mamasa.

The effort made by the government to preserve and develop the tradition and culture of a specified area is reflected in the Decree of People's Consultative Assembly of the Republic of Indonesia No. IV/MPR/1999 regarding the social issues of the Indonesian culture. The government plays an important role in preserving the culture; however, community support is also inevitable in the process. Another effort made by the government is the registration of batik in UNESCO as a cultural heritage belonging to Indonesia in order to prevent neighbouring countries claiming it to be theirs. Furthermore, performing traditional art, such as dances or exhibitions, in events conducted at the Indonesian embassies abroad can also promote the Indonesian culture globally.

As one of the cultural heritages that should be preserved, traditional decoration should be developed as an element of interior design, both constructively and decoratively. The application of traditional decoration as an interior element often experiences change and development in terms of shapes, motifs, materials, manufacturing methods and colours.

Restaurants, a public space where people pay to sit and enjoy food, are considered as one of the premises where traditional ornaments can be displayed. Restaurants are diverse in appearance and services, including a wide variety of cuisines and service models ranging from inexpensive fast food restaurants and cafeterias to mid-priced family restaurants and high-priced luxury restaurants. Decorating a café, which is a simple and small restaurant influenced by the Western culture, with traditional ornaments may be a challenge in the process of preserving local culture because of its small scope and cultural origin.

1.1 *Purpose*

In this study, we focus on the application of Tana Torajan ornaments in interior furniture such as tables, chairs and cabinets, aiming at improving the local culture and introducing it into society. Our main aim is to make foreigners attracted to the furniture. The purpose of this study is to highlight the importance of preserving the Torajan culture in order to adapt to modernisation and increase its popularity among people. Here, we will mainly focus on analysing the diversity of Torajan carvings and the meanings of Torajan ornaments, which will in turn allow the society to understand and eventually gain awareness of the existence of Torajan ornaments in their surroundings.

2 METHODOLOGY

The methods used in this study are content analysis, observation, interview with Torajan locals and experimental design with a hermeneutic approach supported by design, social cultural norm and psychological approaches. Literature review will also be conducted to serve as the reference for this case study.

The studies were carried out in both Tana Toraja and Jakarta. Because of time constraint, the object of this study was analysed gradually. The first phase of this case study was on the decorative ornaments of Tana Toraja currently displayed in Tator Café. We captured all the ornaments, interpreted their meaning and analysed how those ornaments would revive the value of Toraja. Data were collected through a field study via conducting interviews on groups with sufficient data. The aim of the observation in Tator Café was to document the decorations used in the restaurant. In addition to field study, literary data were analysed. Books explaining the culture of Toraja were necessary to gain an insight into the Torajan ornaments that would be applied to the interior furniture.

3 RESULTS AND DISCUSSION

Public space is a social space where people gather. It has also become something of a touchstone for critical theory with respect to social studies and urban design. This place is also often misunderstood as a gathering place, which is an element of the larger concept of social space.

Public space can be divided into several typologies (Carmona, et al., 2003, p.111). External public space is defined as the public space outside the room that can be accessed by everyone (public), such as city parks, squares and pedestrian paths. Internal public space, on the contrary, is a public facility run by the government that can be accessed by citizens freely without any specific restrictions, such as post offices, offices, hospitals and service centres. The last type of public space is external and internal 'quasi'-public space, which is defined as public facilities that are usually managed by the private sector with restrictions or rules that citizens must comply with, such as malls, discotheques and restaurants.

Zhang and Lawson (2009) used three categories of activities in public spaces. The first category is an activity process carried out as a transition of two or more major activities and typically moves from one place to another, for example, moving from home to the stall, where consumption takes place. The second category is a physical contact, which takes place in the form of interaction between two or more people who communicate directly. The last category is a transitional activity, which is usually carried out by a person without any specific goal, such as sitting and watching the scenery.

Culture can be defined in many ways. It can be observed through the reflection of a society's ideas, values, attitudes and normative or expected patterns of behaviour. Culture is not genetically inherited and cannot exist on its own; however, it is always shared by the members of a society (Hall 1976, p. 16). Hofstede (1980, pp. 21–23) defined culture as 'the collective programming of the mind which distinguishes the members of one group from another', which is passed on from one generation to the other. Culture is ever-changing as each generation adds something of its own before passing it further on. It is common for a culture to be both assumed and correct because it is the only one, or at least the first, to be learnt. Culture is made of many complex elements, such as the system of religion, politics, languages, clothing, building architecture and art.

The Indonesian society should understand and claim the rights of the diverse culture of the nation. Most activists are concerned and continue to strive for the maintenance of the growing culture of Indonesia, whereas there are also some who are completely ignorant of the issue. People should have the experience and knowledge of a culture in order to sense it. The government can construct cultural centres for people who need information about the culture. This study is an example of how to increase public awareness of an almost forgotten culture.

In this era of globalisation, Indonesia has adopted many foreign cultures without filtering them, and not all of them are positive. The current rapid growth of globalisation significantly affects the acculturation process of the Indonesian culture. With high-tech information systems, people can easily get information about other cultures. Sometimes, this brings major changes to the lifestyle of most Indonesians, as the attitudes of people may be influenced by those foreign cultures. The invasion of foreign countries into Indonesia before its independence in 1945 had also brought foreign cultures to Indonesia. These foreign countries not only have their spices and occupations, but also instilled their culture in Indonesia, which eventually influences the Indonesian culture. One example is the architecture of the Yogyakarta palace, which was influenced by the Japanese (Laksmi, 2012). At present, the influence of the Western culture is a result of modern technology and unlimited information flow.

Some neighbouring countries have even claimed the Indonesian culture as theirs, such as Reog Ponorogo dance from East Java, gamelan music from Central Java, Ulos cloth from North Sumatra and angklung music instrument from West Java. Indonesians should be proud of having such great and rich cultural heritage by preserving it so that it can remain Indonesian. The noble values of the culture and local identity need to be preserved in order to mitigate the potential problem of being forgotten by future generations. The society should have rights over the local culture by preserving and reviving its values, as well as introducing it to the public.

Tana Toraja is located in the northern part of the South Sulawesi Province and situated between Latimojong Mountain Range and Mount Reute Kambola. The name Toraja has a Bugis origin which denotes different groups of people of the mountainous regions of the northern part of the south peninsula, which have been discovered recently (Soertoto, 2003). 'Toraja' is derived from words of the coastal language: *to* means *people* and *riaja* means *uplands*. This word was first used as an expression by lowlanders to refer to highlanders. The Torajan indigenous belief system is polytheistic animism, called *Aluk*. In the Torajan society, the funeral ritual is the most elaborate and expensive event; the richer and more powerful is the individual, the more expensive the funeral is. In the *Aluk* religion, only nobles have the right to have an extensive funeral.

All Torajan houses have unique architecture. Torajan houses are boat-shaped with its two ends resembling a bow. Torajan houses are compound buildings consisting of traditional houses (*Tongkonan*) and rice storage buildings (*Lumbung*). The construction of Tongkonan is a laborious work and is usually carried out with the help of the extended family. The buildings are sculpted with ornaments of various shapes, which are painted with traditional colour, mostly black and red. All of them contribute to the aesthetic value of the building of Torajan house (Syahwandi, 1983). The *Tongkonan* is the centre of Torajan social life. The rituals associated with the *Tongkonan* are important expressions of Torajan spiritual life; therefore, all family members are impelled to participate, because symbolically, the *Tongkonan* represents links to their past, present and future.

Figure 1 shows the *Tongkonan*, a traditional house of the Torajan people. The roof is curved and boat-shaped. It used to be a bamboo structure, but the Torajan people prefer to use zinc nowadays. A row of water buffalo horns can also be found in front of the house. The house serves as a place not only to sleep and cook but also to keep corpses. The name *Tongkonan* is derived from the word *tongkon* (meaning to sit together). A *Tongkonan* is divided based on the level or role in society (the social stratification of Torajan society); the wealthier people are, the larger their Tongkonan will be when they die. In front of the *Tongkonan*, a granary made of the bark of coconut tree with parts of its land casted can be found. In addition to the granary, a variety of carvings, including a picture of chicken and sun and a symbol of resolving the matter, can be found (Sumalyo, 2001).

One form of the beautiful Indonesian cultures that has to be explored is the Torajan carving. At present, Torajan sculptures are mostly found in tourist markets. Tourism has a positive impact on the preservation of traditional culture in Toraja. Until the early 2000s, many Torajans made money by selling Torajan carvings or becoming an hotelier, a tour guide and so on. Thus, when the number of tourists visiting Tana Toraja decreased, many locals moved to other places, seeking for employment as miners or in the transportation industry (Adams, 2006). This will eventually affect the preservation of the traditional Torajan culture if the situation prevails.

Torajan wood carvings are known for their regularity and order; furthermore, they are abstract and geometric. Nature is often used as the basis of Torajan ornaments, because they are full of abstraction and regular geometrics. Torajan ornaments have several meanings, some of which are togetherness, brotherhood, wealth and position. In an effort to preserve these ornaments, the Torajan society has decided to implement the elements of the *Tongkongan*, regardless of whether interior or exterior, as constructive and decorative elements. The decoration used will become the basic form of a recurring pattern in a craft or art that completes the exterior of the *Tongkongan*. Each ornament used will have a hidden meaning in its motifs. In Torajan wood carvings, each panel symbolises a good will. The motifs have various symbols representing folklore, celestial objects, sacred animals and so forth.

All of Torajan ornaments have unique meanings, such as *Ne'Limbonga* (lake) representing luck, *Pa' Tedong* (buffalo head) representing prosperity, *Pa' Barre Alo* (sun) representing pride,

Figure 1.   *Tongkonan*, house of Toraja.

Pa'Ulu Karua representing hope, *Pa'Bamboo Uai* representing speed and *Pa'Manik Manik* meaning jewellery (Bobbyn, 1979). A unique feature of Torajan sculptures and ornaments is that they are dominated by various colours, mainly red, yellow, white and black. The colours are made from natural ingredients that are blended with palm wine vinegar. Palm wine serves as an adhesive for the colour to be attached and durable in the carving. The colours are not randomly selected, and each one of them has a philosophical meaning. Red symbolises blood, black symbolises death, white symbolises man of flesh and bone and yellow symbolises glory.

The pictures in Figure 2 show a round table and a four-seater dining table (90 cm × 90 cm) at Tator Café implementing the motif of Torajan ornaments. This type of decoration resembles *Neqlimbongan*, which originates from one of their ancestors 'Limbongan' who was estimated to have lived 3000 years ago, while the word *Neq* means lake. Therefore, this motif can be thought of as an infinite source of water and life. The motif is shaped like water in a whirlpool and interpreted as luck that shall come from all directions, just like springs converging in lakes and providing happiness. The top of this dining table is significantly decorated with small patterns of Torajan ornaments.

The pictures shown in Figure 3 are images of the cashier counter in Tator Café and its backdrop with Torajan motifs on it. Each motif has its own meaning. This café is bold enough to implement Torajan decorative elements, aiming to paint an image of Toraja itself although it is worth noting that not all motifs are suitable to be used in a restaurant. Some motifs such as the ones representing death and funerals would definitely be inappropriate for restaurant use. Although visitors may not necessarily understand the meaning and significance of these motifs, it is necessary to avoid such inappropriateness, especially since history and culture experts might notice these motifs.

Dining tables or furniture with motifs or local ornaments must be specially ordered and custom made. With the lack of interest for furniture with local content on it, the role of designers is now to better explore local cultures, in this case the Torajan culture, to be applied to furniture. One perfect example is Tator Café located in several malls in Jakarta. Attracting consumers, especially foreign markets, there is a possibility for a flow of trade.

Figure 2. Application of Torajan ornaments in Tator Café (1).

Figure 3. Application of Torajan ornaments in Tator Café (2).

Applying the ornaments as interior decoration in public space is an indirect way of reviving the culture of Toraja to the public. Being unique and rich in ornaments, the Torajan culture has the potential to be promoted to the global market, allowing it to gain more exposure. As one of the tribes in Indonesia, Toraja is rich in ornaments with unique meanings. They will be suitable and applicable as interior ornaments in public spaces. One of such spaces is Tator Café with its soothing shades of Torajan ornaments representing the Torajan culture. Indonesians should start appreciating, understanding and reviving the culture of Toraja before others can follow. By this way, it will be easier to promote the culture in other forms of art in order to gain global recognition.

4 CONCLUSION

Public space is a common space involving humans as users. In this case, their background is diverse in culture, economy, education and so on. Tator Café as a public space with the application of decorative Torajan ornaments is one of the research objects in this study. The interior elements such as floors, walls, ceilings and furniture, and other supporting elements, such as employee uniform and accessories, are important in making visitors familiar with the traditional Torajan culture.

Decorative Torajan ornaments are a cultural heritage of Indonesia, which should be preserved. The motifs and colours existing in such ornaments contain a symbolic meaning closely related to the Torajan religion. More attention should be devoted to the application of those ornaments. Therefore, this becomes the guideline for the application in interior elements (floors, walls, ceilings and furniture) in order to avoid any violations of the existing rules and meanings contained in each of those ornaments.

This study shows that modern society can be persuaded to be more aware of Indonesian tradition and culture, especially that of Toraja through interior ornaments. Through the application of some ornaments that hold philosophical interpretation and their presentation in an aesthetic manner, people can enhance their knowledge and understanding on the basis of their experience and sensibility.

The application of Torajan ornaments with their meanings in the café can revive the local atmosphere. This can become a significant part of the effort to revive the culture and introduce it to the visitors. By understanding the culture, people can enhance their aesthetic sensibility and make their experience as a method of revitalising the value of traditional culture. Learning culture through art and design can influence the formation of human characteristics, including their sense and sensibility. Therefore, this study is intended to support modern society in order to increase their sensitivity and ability to appreciate the beauty of the Torajan culture. It also can support modern society in understanding the almost-forgotten Torajan culture, so that this preserved heritage can be known and recognised by both local and global communities.

REFERENCES

Adams, K.M. (1995) Making-up the Toraja? The appropriation of tourism, anthropology, and museums for politics in upland Sulawesi, Indonesia. Ethnology Journal, Vol.34./2, 143–153.
Adams, K.M. (2008) More than an ethnic market: Toraja art as identity negotiator. *American Ethnologist*, Vol 25, No.3, Article 1, 327–351.
Carmona, et al. (2003) Public places—urban spaces, the dimension of urban design. New York: Architectural Press.
Dimas. (2011) Analisis upaya melestarikan budaya (Analysis of efforts to preserve culture). Retrieved from: http://dimaspratama11.wordpress.com/2011/11/19/analisis-upaya-melestarikan-budaya-bangsa/.
Hall, E.T. (1976) Beyond culture. New York: Anchor Books/Doubleday.
Hartanti, G. and Nediari, A. (2014) Pendokumentasian aplikasi ragam hias Toraja sebagai konservasi budaya bangsa pada perancangan interior (Documenting application of Torajan decorative ornaments as conservation of culture in interior design). *Journal Humaniora*, Vol 05/2, 32–40.

Hofstede. G. (1991) *Cultures and organizations: Software of the mind.* London: McGraw-Hill.

Kusuma, L. (2012) Pengaruh pandangan sosio-kultural Sultan Hamengkubuwana IX terhadap eksistensi keraton Yogyakarta (The influence of socio-cultural view of Sultan Hamengkubuwana IX of the existence of Yogyakarta Palace). *Jurnal Masyarakat dan Kebudayaan Politik*, Vol 25/1, 56–63.

Nanda, S. & Warms, R.L. (2013) *Cultural anthropology.* Belmont: Wadsworth Cengage Learning.

Nediari, A., Hartanti, G., Polniwati. (2014) Pengaplikasian ragam hias Toraja pada restoran di Jakarta sebagai wujud pelestarian budaya Indonesia (The application of Torajan decorative ornaments in restaurants in Jakarta as a preservation of Indonesian culture). *Proceedings of Seminar Nasional Seni Tradisi Universitas Trisakti*, 527–538, 2014.

Sande, J.S. (2001) *Toraja in carving.* Ujung Pandang: Balai Penelitian Bahasa.

Soeroto, M. (2003) *Toraja: Pustaka budaya dan arsitektur (Toraja: Art and Architecture).* Jakarta: Balai Pustaka.

Sumalyo, Y. (2001) Kosmologi dalam arsitektur Toraja (Cosmology in Torajan architecture). *Jurnal Dimensi Teknik Arsitektur*, Vol 29.

Syafwandi, Soimun H.P., & Panannangan, M. (1993) *Arsitektur tradisional Tana Toraja (The traditional architecture of Tana Toraja).* Jakarta: Depdikbud.

Tangdilintin, L.T., (1991) *Tongkonan (rumah adat Toraja): Arsitektur dan ragam hias Tongkonan (traditional house of Toraja): Architecture and Decorative Ornaments.* Ujung Pandang: Yayasan Lepongan Bulan Tana Toraja.

Tjong, M.L. (2014) Kajian kosmologi pada rumah tradisional suku Mamasa (The cosmological study of traditional house of Mamasa ethnic group). *Proceedings of Seminar Nasional Seni Tradisi Universitas Trisakti*, 192–207, 2014.

Toekio, S.M. (1990) *Mengenal ragam hias Indonesia (Getting to know decorative ornaments in Indonesia).* Jakarta: Angkasa.

Zhang & Lawson. (2009) Meeting and greeting: activities in public outdoor spaces outside high-density urban residential communities. *Urban Design International*, volume 14, 4, 207–214.

Cultural Dynamics in a Globalized World – Budianta et al. (Eds)
© *2018 Taylor & Francis Group, London, ISBN 978-1-138-62664-5*

The toponymy of the ancient port city of Gresik in the northern coastal area of Java

Z. Muhatta & N. Soesanti
Department of Archaeology, Faculty of Humanities, Universitas Indonesia, Depok, Indonesia

ABSTRACT: This study focuses on Gresik, an ancient port city on the northern coastal area of the island of Java, which used to be well-known as a trading city and served as the maritime axis of Nusantara (15th to 19th century AD). This research is conducted as an interdisciplinary study to explore the toponymy from the standpoint of linguistics, in collaboration with epigraphy (archaeology), philology (literature), maritime history, and geography. Based on literature and theoretical studies, the research tries to answer this question: "What is the relation between the names of places and the maritime culture of an ancient port city in the northern coastal area of Java?" The objects of the study are the names of places in the ancient port city of Gresik, including the archaeology and the maritime socio-cultural landscapes between Demak and Surabaya, which are collected from geographical maps, old inscriptions, and old manuscripts. The analysis focuses on history, lexicon, morphology, and semantic structures of names of places within synchronic and diachronic perspectives, which is concluded with an assessment of the track record on toponymy. By interpreting the maritime culture of the past that built the marine network system uniting the archipelago, Indonesia's national identity will be reinforced and strengthened as the future global maritime axis.

1 INTRODUCTION

Indonesia is a country whose majority of territory comprises of seas dotted with islands (Lapian, 2008). The islands of Indonesia stretch on the tropical waters from the Indian Ocean to the Pacific Ocean, and from Southeast Asia to Northern Australia. Indonesia has a coastal line of approximately 81,000 kilometres in length, making it one of the countries with the longest tropical coastlines in the world. The geographical location and the territorial distribution composing of more than 17,000 islands around the equator have made Indonesia an ideal harbour for traders from various countries, such as those from Europe, China, India, and Africa. In the ancient time, Indonesia used to be known as *Nusantara* and was renowned for its role as the maritime axis of the world (Vlekke, 2016).

The international trading centres along the northern coasts of Java were the ports of call for foreign traders on their journeys across the Spice Routes and the Silk Road. These routes made Nusantara the maritime axis because of its strategic location. Routes that traversed Nusantara, namely the Spice Routes, the Europe-Asia trading routes along the southern parts of Asia, were taken by traders of exotic commodities, such as pepper, clove, sandalwood, rice, and textile. Spice traders from the eastern part of Nusantara brought ships loaded with spices to harbours along the northern coasts of Java. Meanwhile, in the Silk Roads, the overland trading routes in the north connected Europe, Central Asia, and China, and the popular commodities were silk and chinaware, such as plates, bowls, glasses, vases, and textiles from India.

During the Islamic expansion to the East, other Asian countries, such as Arabia, Iran, Turkey, and the Moslem India also conducted trading activities in Java, resulting in the island to be renowned as a hub frequently visited by foreign traders that came in large trading ships (Reid, 2000: 29). These foreign traders' visits resulted in the sprouting harbours along the northern coasts, which grew rapidly to become large port cities that also expanded as places of settlement, kingdoms, religious activities, and other activities. The social pattern became increasingly complex, and the harbours developed various cultural landscapes to meet the growing needs of the people living in the harbour areas.

In the 15th century, Nusantara was widely known as an international trading centre. As centres of trading activities conducted by foreign traders across the Spice Routes and the Silk Road, cities along the northern coasts of Java were known as international harbours, as referred to in the Portuguese book, *Suma Oriental*, written by Tomé Pires (1485 AD), Chinese chronicles written by Ma Huan, and the Dutch state reports. This region became ideal for ships due to the moderate waves of the sea, which enabled ships to access (Cortesao, 1990). Based on Tomé Pires's notes (1515), rice crops became an attractive trading commodity in Java. Raffles also stated this fact in his book, *The History of Java* (2015).

This study falls under the Culture and Indigenous Studies. The approach of studying names of places is known as toponymy, a branch of onomastics. Onomastics consists of two branches: anthroponymy that focuses on human names, and toponymy that focuses on names of places. Toponymy is related to the long history of human settlements. Humans indicate memories of humans and places by names. Names of places constitute geographic identities by using the local languages. Language and space form the identity of their speakers. Language is a vehicle for cultural background and supports the identity of its speakers. As stated by Lauder and Lauder (2012), the function and the role of history in naming topographic features are of major significance in the understanding of the local culture. The research is studying the relatively new geographic names or the existing names that have existed for hundreds of years, and what the cultural and linguistic backgrounds are. As an interdisciplinary study, this study was conducted with an etymological approach in examining the toponymy in the northern coastal areas of Java from the linguistic point of view in collaboration with epigraphy (archaeology), philology (literature), history, and geography.

The purpose of the study is to answer the question: "What is the relation between the place names and the social culture of the ancient port city of Gresik in the northern coastal area of Java?" To answer this main question, the names of places are examined with toponymy. Doing a toponymy study in this area is important to uncover its social culture and identities. The limitation of the study comprises areas that have been mentioned as the "Javanese Coastline" by Denys Lombard, a leading French historian (1996: 37). This coastline area stretches along the northern beaches from the west part of Cirebon to the east part of Surabaya. Lombard uses the term "the northern coastal area of Java" following the limitation of the linguistically geographic area of the Sundanese, the Javanese and the Madurese languages as suggested by Tomé Pires. Pires divided these three language areas based on the cultural aspects; Java as a "space" which is divided in three major cultural areas to indicate the geographical space. The use of the term "the northern coast of Java" is in accordance with the margin of linguistically geographic areas of the Javanese language users.

The role of Gresik as a big seaport is inseparable from the relation with other port cities, such as Tuban, Sidhayu, Hujunggaluh, and Surabaya. However, Gresik apparently held a major role as the international seaport. In the research on the inscriptions from the Airlangga era by Susanti (2010), it is indicated that the increasing economic progress had led to an international development through the seas, as made evident by the foreigners who settled and were taxed, who were known as *warga kilalan* (tax payers). They were the trade representatives of India, Campa, Khmer, Mon (Remen), and Sri Lanka. Their activities were also mentioned in the Chinese chronicles (Groeneveldt, 1960: 47). According to Ma Huan (1433 AD), the Chinese sailor who made notes in the chronicles, Gresik was a new village located approximately half-day journey towards east from Tuban, recorded as *Ko-erh-shi* (Groeneveldt, 1964). Ma Huan described how Gresik started up in an abandoned plot of land on a sandy seashore area, which was built as a port city. In 1350–1400 AD, people from the central part of China (the Chinese sailors and traders) built new settlements in this area. After 1400 AD, Gresik enjoyed a rapid growth, and when Ma Huan came to Gresik in the same year, the city had become the best and most important port city (Djaenuderadjat, 2013: 207). In 1411 AD, the Chinese rulers in Gresik sent letters and presents to the Chinese

Emperor. They were not the Chinese people who ruled Gresik, but the people who managed the administrative area inhabited by the Chinese people (Rahardjo, 2002: 22).

This study uses the map of the current administration. Gresik Regency stretches in an area of 1,191.25 square kilometres at the latitude 112°-113°E and longitude 7°-8°S geographic coordinates. Its administrative regions are composed of 18 districts, 330 *desa* (villages), and 26 *kelurahan* (sub-districts). Districts (*kecamatan*) are divided into *desa* (villages) or *kelura-han* (urban communities); both *desa* and *kelurahan* are of similar levels, but *desa* has more power in local matters than *kelurahan*.

The object of this study comprises the names of places of the villages in the ancient city, including the archaeological, maritime, and socio-cultural landscapes of Gresik. From the ancient port cities, various kinds of information can be obtained about the lives of the communities pertaining to the sea that have contributed to the maritime culture. The term maritime in this research is limited to any human activities in the sea or water that support livelihood. The geographical location is as seen in the map in the next page.

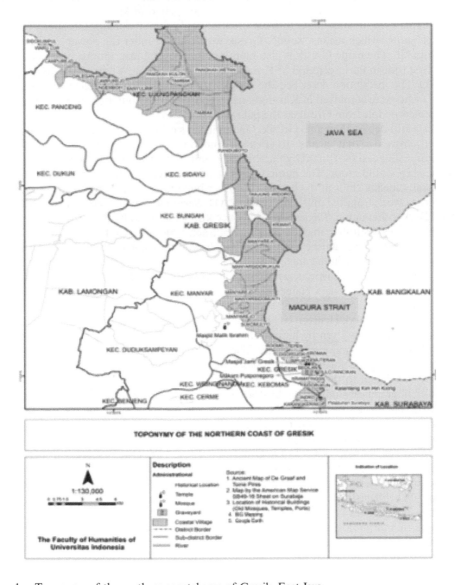

Figure 1.   Toponymy of the northern coastal area of Gresik, East Java.

In the early 11th century, Syech Maulana Malik Ibrahim brought and spread Islam in Gresik, at the same period as Fatimah Binti Maimun. Meanwhile, Nyai Ageng Pinatih, a harbour mistress, adopted a baby from Blambangan, who later grew up and was known as Sunan Giri, the authority of the Giri Kedaton kingdom (1487 AD). Sunan Giri had a position as a *sunan* or *wali* (a preacher of Islam), and also served as the Sultan or Prabu (a government ruler), whereas Syech Maulana Malik Ibrahim was the leader of kings and ministers (Poesponegoro, 1984).

In the 14th century, Gresik was also famous for its renowned Islam preacher, Sunan Maulana Malik Ibrahim. Gresik became a centre for Islam in the north coast of Java, attracting people from the east, such as Maluku and Ternate, to study Islam. Gresik-Jaratan was also well-known as the safest trading seaport in the Strait of Madura. The port of Gresik was adjacent to the delta of Bengawan Solo (Solo River), a very fertile area producing rice, which also facilitated trade and transport of these commodities to trading ships. Until the end of the 16th century AD, the port city of Grisse (Gresik-Jaratan) was famous as the important trading harbour of the Spice Routes (Meilink-Roelofs, 2016: 263).

The name Gresik is assumed to have been derived from a word in the ancient Javanese language, *gesik*, which means "sand". Gresik was initially under the power of the kingdom of Majapahit. The name Gresik is also mentioned in several inscriptions; it was mentioned for the first time in the inscription of Karangbogem or Trowulan VI, issued by Bhre Lasem in 1387 AD. The inscription described the people from Gresik (*"hana ta kawulaningong saking Gresik"*) who were employed in Karangbogem, and about a stipulation on *kawula*, or slaves, or redeemed people from Gresik at the palace.

Experts still debate the name Gresik. One opinion says that Gresik derived from the Arabic word *qarra syai* that means "to stick on something". This object is assumed to be the anchor of a ship cast overboard by the ship crew and sticks on the bottom of the sea as a sign that the ship would moor. The name Gresik is also related to the Javanese word *Giri-gisik* that literally means "hills at the beach". Another meaning of the word Gresik is mentioned in a literary work from the mid-19th century AD, *Serat Centini,* namely Giri-Gresik (Kusdi, 2014: 42). Another name, *Tandes*, was also once known, as used by the writer of the Chinese chronicles as a substitute name for Gresik (Rouffaer, 1906: 178–198).

The toponymy and linguistic names of places used in this study are related to the language used by the local people in the area. One media found to express human feelings as reflections of the cultural and social aspects, as well as the geographical areas where human beings have lived from the past is a village. Names of villages are footprints of progress toward human civilization and they indicate aspects of social, cultural, historical, and geographical information about human livelihood. Hence, it can be concluded that language as a tool of communications and e names of places given by humans are interrelated. Therefore, the linguistic analysis of names of places has a major significance. The data used in this study constitutes names of villages located along the northern coasts, based on which linguistic and archaeological analyses are made in relation to their locations in the geographical area (Lauder and Lauder, 2012). After a bibliographical study of the District of Gresik, data collection of the place names was conducted. The names used as the data for this study are those of the villages located along the northern coasts of the District of Gresik. These names and the archaeological landscape locations on maps (old mosques, old temples, and remains of ancient harbours) were collected based on an ancient map made by De Graaf and Tomé Pires (1615), American Map Service SB49–16 (Surabaya Sheet), BIG (Badan Informasi Geospatial), and Google Earth.

Two analysis was made in the data processing stage: the linguistic and archaeological analyses. The linguistic analysis consists of morphological analysis and the meaning of the names. The names of places are considered as lexicons, which have undergone through two kinds of morphological analysis, i.e. word formation and morphophonemic processes. Following Stageberg (2000) who stated that analysis of English words can be conducted by tracing the ancient Latin and Greek dictionaries, this study considers the lexicons as originated from Javanese. In finding the etymology, the meaning of the base morphemes is found by tracing the morphemes in the Old Javanese dictionaries. The names of places in East Java, particularly in Gresik, use regional or local language, namely the Javanese language and those inherited from their ancestors as the Old Javanese language users. The analysis is described in Table 1.

Table 1. Linguistic analysis of village names in the Northern Coastal areas of Java.

No.	Name of district	Name of village (Lexicon)	Morphological analysis		Dictionary meaning
			Word formation	Morphonemic	
1	District of Panceng	Sidokumpul	Sido (V) + kumpul (V)	–	*Sido* means "to become," *kumpul* means "to gather".
		Campurejo	Campur (N) + raharjo (N)	*rejo* from *raharjo*	*Campur* means "a combination" and *rejo* means "prosperous".
		Waru Lor	Waru (N) + lor (Adj)	–	*Waru* is a name of a plant, *lor* means "north".
		Dalegan	Daleg (V) + an (suffix)	–	*Daleg* means "sitting," or "a place to sit looking at the sea".
2	District of Ujung Pangkah	Ngemboh	ngemboh dari nga (affix) + imboh (V)	–	*Emboh* means "to add," from the word *imbuh*, with prefix *ng* that means "to add" (something).
		Banyuurip	Banyu (V)+ urip (V)	–	*Banyu* means "water," and *urip* means "life".
		Pangkah Kulon	Pangkah (N) + Kulon (Adj)	–	*Pangkah* means "mud," and *kulon* means "west".
		Pangkah Wetan	Pangkah (N) + Wetan (Adj)	–	*Pangkah* means "mud," and *wetan* means "east".
3	District of Sidayu	Randubroto	Randu (N) + broto (Adj)	–	*Randu* is another name of *kapok* (a name of plant); *broto* means "faithful, meditation".
4	District of Bungah	Bedanten	Bedah (V) + seganten (N)	Bedanten = Bedah + seganten	*Bedah* means "slice," *sedanten segoro* means "the sea".
		Tanjung Widoro	Tajung (N) + Widoro (N)	–	*Tajung* means "peninsula," *widoro* is a name of plant.
		Kramat	Kramat (N)	–	*Kramat* means "sacred, honourable, respectful, lofty".
5	District of Manyar	Manyarejo	Manyar (N) + rejo (Adj)	–	*Manyar* is a type of bird that lives near water, and *rejo* means "prosperous".
		Manyar sidorukun	Manyar (N)+ sido (V) + rukun (Adj)	–	*Manyar* is a type of bird that lives near water, *sido* means "to become," and *rukun* means "peaceful and cohesive".
		Manyar sidomukti	Manyar (N)+ sido (V) + mukti (Adj)	–	*Manyar* is a type of bird that lives near water, *sido* means "to become," and *mukti* means "lofty".
		Sukomulyo	Suko (V) + mulyo (Adj)	–	*Suko* means "to like," and *mulyo* means "lofty".
		Roomo	Roomo (N)	–	*Roomo* means "hair".

(*Continued*)

Table 1. (*Continued*).

No.	Name of district	Name of village (Lexicon)	Morphological analysis		Dictionary meaning
			Word formation	Morphonemic	
6	District of Gresik	Tepen	Tepen (N)	–	*Tepi* means "the edge," and *tepian* means "the edge part".
		Tlogopojok	Tlogo (N) + pojok	–	*Tlogo* means "lake," and *pojok* means "corner".
		Lumpur	Lumpur (N)	–	*Lumpur* means "soft and moist clay".
		Kroman	Ke (prefix) + room (N) + an (suffix)	–	*Karoman* means "hair," the synonym of *parambutan.*
		Kemuteran	Ke (prefix) + puter (N) + an (suffix)	–	*Puter* means "to rotate," *kemuteran* means "a place to rotate".
		Pekelingan	Pe + keling + an	–	*Keling* is a term for foreigners from the Indian peninsula.
		Bedilan	Bedil (N) + an (suffix)	–	*Bedil* means "fire arms," and *bedilan* means "shooting".
		Pulo Pancikan	Pulo (N) + Pancik (V) + an (suffix)	–	*Pulo* means "island," and *pancikan* is derived from the word *pancik* that means "to climb" or "to rise".
		Sidorukun	Sido (V) + Rukun (Suffix)	–	*Sido* means "to become," and *rukun* means "in agreement, peaceful".
		Kramatinggil	Kramat (N) + inggil (Adj)	–	*Kramat* means "sacred, honourable, respectful, lofty." *Inggil* means "high".
7	District of Kebomas	Indro	Indro (N or V)	–	*Indro* means "king or god," or "to go".
		Karangkering	Karang (N) + kering (V)	–	*Karnag* means "a coral stone" or "a place to wait," *kering* means "dry".

Based on the linguistic analysis, a tendency is indicated in the use names of places where there are lexicons constituting hydronym (names that contain words that denote an element of water and its derivatives), e.g. saltwater (*seganten*/N), freshwater (*banyu*/N, *tlogo*/N), and lexicons constituting non-hydronym (names that do not contain any words denoting an element of water and its derivatives), such as a lexicon of an original noun whose meaning refers to the sea and its family, e.g. *pangkah*/N, *lumpur*/N, *karang*/N, *pulo*/N, and lexicons that contain other elements, e.g. *roomo* (hair/N), *rejo*/Adj, *sido*/V.

3 TOPONYMY AND ARCHAEOLOGICAL LANDSCAPES

The archaeological method to study archaeological landscapes consists of observation, description, and interpretation (Deetz, 1967). In addition, we used interview methods at the sites in Gresik. The presence of an ancient temple built long ago indicates the scope of activities the local people carried out with foreigners in this area. The existing legacies of the cultural landscapes indicate foreign influence of the Chinese sailors who built temples to perform their worship according to the faith they brought from China, which also served as places for their community. The Kiem Hien Kiong temple remains an ancient temple that maintains its original use as a worship and social facility. Based on the information from the keeper, the temple was built in 1125 AD; unfortunately, no evidence was available to prove this.

Furthermore, Gresik and its surroundings are home to ancient mosques that indicate the influence of Islam and Arab culture. There are three ancient mosques in Gresik, namely the Pasucinan mosque in the village of Giri, the Jami Gresik mosque that is located near the public square, and the Maulana Malik Ibrahim mosque in the village of Gapuro (Sukolilo). These old mosques are still actively in use until today as places of worship and pilgrimage tour destinations - particularly because of the remaining ancient tombs, including the most famous one at the Maulana Malik Ibrahim mosque, namely the tomb of Regent Pusponegoro, the first regent originating from the District of Gresik (1617 AD). Based on interviews with the local inhabitants, this mosque is located about two kilometres from the seashore (formerly built on the seashore near the port), but the coastal line moved forward despite the sedimentation. In the olden days, there was no lighthouse to signal a land. Instead, the minaret of a mosque served as an aid to broadcast the call for prayers (*adzan*). This also indicates the life in the maritime glory in the eastern part of the northern coast of Java.

4 CONCLUSION

Based on the study, it can be concluded that toponymy indicates the closeness of the local people and the culture that have lived in the area. The proximity of the location to the sea is indicated by names of places that can be categorized as hydronyms and non-hydronyms that contain water as a marine element, showing the closeness of the inhabitants to the sea and its marine surroundings as their identity. On the other hand, the non-hydronyms also contain other elements of the sea, such as mud, island, and coral stones.

In addition to agriculture, the improvement of Indonesian national economy should also be based on marine resources. However, various political considerations have had an impact in the Indonesian maritime aspects. For quite a long time, Indonesia's maritime potentials lacked the attention of the government since more emphasis was placed on Indonesia as an agricultural country. The northern coastal areas of Java have held vital roles in the national economic activities. Various trades and other related activities conducted along the northern coasts of Java have proved the greatness of this area as the world's maritime axis back then. To support maritime-based development in the future, the ancient maritime culture needs to be studied by various other studies, since the understanding of the maritime culture of the past may provide us vast insights that may lead to new findings or conclusions based on the result of the studies. Based on this background, the directions of development in the future should be emphasised not only on agricultural aspects, but also on the optimisation of maritime potentials. By interpreting the maritime culture of the past that had built the marine network system which united the archipelago, Indonesia's national identity will be reinforced and strengthened as the global maritime axis in the future. The result of study could contribute to the revitalisation of seaport sites to strengthen the maritime potentials and support the government's attempts to develop tourism in Indonesia in the future.

REFERENCES

Cortesao, A. (1990) *The Suma Oriental of Tomé Pires, 1512–1515, 2 volumes set*. New York: Laurier Books.
de Graaf & Pigeaud, T.G.T. (1989) *Kerajaan-Kerajaan Islam Pertama di Jawa*. Jakarta: Pustaka Utama Grafitti. Translation: The First Islamic Kingdoms in Java Jakarta: Pustaka Utama Grafitti.

Deetz, J. (1967) *Invitation to Archaeology*. Garden City, NJ: Natural History Press [Doubleday] for the American Museum of Natural History.

Djaenuderadjat, E. (ed). (2013) *Atlas Pelabuhan-pelabuhan Bersejarah di Indonesia. Direktorat Sejarah dan Nilai Budaya*. Jakarta: Directorate General of Culture of the Ministry of Education and Culture. Translation: Maps of Historical Ports in Indonesia. Directorate of History and Cultural Value. Jakarta: Directorate General of Culture of the Ministry of Education and Culture.

Groeneveldt, W.P. (1960) *Historical Notes on Indonesia and Malaya Compiled from Chinese Sources*. Jakarta: Bhratara.

Kusdi. (2009) *Perlawanan Masyarakat Madura Atas Hegemoni Jawa. Yogyakarta: Penerbit Jendela*. Translation: Resistance of Maduranese Againts Java's Hegemony. Yogyakarta: Penerbit Jendela.

Lapian, A.B. (2008) *Pelayaran dan Perniagaan Nusantara Abad ke-16 dan 17*. Jakarta: Komunitas Bambu. Translation: *Shipping and Commerce in Nusantara in 16th and 17th* Centuries Jakarta: Komunitas Bambu.

Lauder, M.R.M.T. & Lauder, A.F. (2012) Writing Geographical Names in Indonesian. *Paper in 4th UNGEGN Names Training Course on Toponymy*. Yogyakarta, 17–21st September 2012. Indonesian Geospatial Information Agency (BIG).

Lombard, D. (1996) *Nusa Jawa: Silang Budaya. Kajian Sejarah Terpadu. Bag I: Batas-batas Pembaratan*. Jakarta: Gramedia Pustaka Utama. Translation: Java Nusantara: Cross Culture. Integrated Historical Study Part I: Borders of Westernisation. Jakarta: Gramedia Pustaka Utama.

Magetsari, N., et al. (1979) *Kamus Arkeologi 2. Jakarta: Proyek Penelitian Bahasa dan Sastra Indonesia dan Daerah*. Jakarta: Departemen Pendidikan dan Kebudayaan. Translation: Dictionary of Archaeology 2. Jakarta: Project of Indonesian and Local Language and Literature Research. Jakarta: (Department of Education and Culture).

Mardiwarsito, L. (1990) *Kamus Jawa Kuno Indonesia*. Ende: Nusa Indah. Translation: Dictionary of Indonesia's Ancient Java Ende: Nusa Indah.

Meilink-Roelofz, M.A.P. (1962) *Persaingan Eropa & Asia di Nusantara, Sejarah Perniagaan 1500–1680*. Translation: Competition of Europe & Asia in Nusantara, History of Commerce 1500–1680.

Raffles, T.S. (2015) *History of Java 1825*. Trans. Hamonangan, ed. Jakarta: Penerbit Narasi.

Rahardjo, S. (2002) *Peradaban Jawa*. Jakarta: Komunitas Bambu. Translation: Java Civilisation. Jakarta: Komunitas Bambu.

Rouffaer. (1906) *De Chineese NaamTs'e-Ts'un voor Gresik*. Gravenhage: Martinus Nijhoff.

Poesponegoro, M.D. (1984) *Sejarah Nasional Indonesia, jilid 4*. Jakarta: PN Balai Pustaka. Translation: National History of Indonesia, volume 4. Jakarta: PN Balai Pustaka.

Prawiroatmojo, S. (1981) *Bausastra Jawa – Indonesia*. Jakarta: Gunung Agung. Translation: Dictionary of Java – Indonesia. Jakarta: Gunung Agung.

Stageberg, N.C. & Dallin D.O. (2000) *An Introductory English Grammar*. Fifth edition. Fort Worth: Hartcourt College Publishers.

Susanti, N. (2010) *Airlangga, Raja Pembaru Jawa Abad 11*. Depok: Komunitas Bambu. Translation: Airlangga: King of Reformer of Java in 11th Century. Depok: Komunitas Bambu.

Vlekke, B.H.M. (2016) *Nusantara: Sejarah Indonesia*. Terj. Jakarta: Gramedia. Translation: Nusantara: History of Indonesia. Translated. Jakarta: Gramedia

Zoetmulder, P.J. & Robson, S.O. (2011) *Kamus Jawa Kuna Indonesia*. (Daru Suprapta and Sumarti Suprayitna, Translators) (6th print). Jakarta: Gramedia Pustaka Utama. Translation: Dictionary of Indonesia's Ancient Java (Daru Suprapta and Sumarti Suprayitna, Translators) (6th print). Jakarta: Gramedia Pustaka Utama.

Author index